2007 International Conference on Power Electronics and Drive Systems

Bangkok, Thailand
27-30 November 2007

Pages 480-948

IEEE Catalog Number:	CFP07PEL-PRT
ISBN 10:	1-4244-0644-7
ISBN 13:	978-1-4244-0644-9

Copyright © 2007 by The Institute of Electrical and Electronics Engineers, Inc.
All Rights Reserved

Copyright and Reprint Permissions: Abstracting is permitted with credit to the source. Libraries are permitted to photocopy beyond the limit of U.S. copyright law for private use of patrons those articles in this volume that carry a code at the bottom of the first page, provided the per-copy fee indicated in the code is paid through Copyright Clearance Center, 222 Rosewood Drive, Danvers, MA 01923.

For other copying, reprint or republications permission, write to IEEE Copyrights Manager, IEEE Operations Center, 445 Hoes Lane, Piscataway, New Jersey USA 08854. All rights reserved.

IEEE Catalog Number:	CFP07PEL-PRT
ISBN 10:	1-4244-0644-7
ISBN 13:	978-1-4244-0644-9
LOC:	2006933010

Additional Copies of This Publication Are Available from:

IEEE Service Center
445 Hoes Lane
Piscataway, NJ 08854

Phone:	(800) 678-IEEE
	(732) 981-1393
Fax:	(732) 981-9667
E-mail:	customer-service@ieee.org

Organizers/Committees

Organizers:

Chulalongkorn Univ.
Center of Excellence in Electrical Power Tech., Chulalongkorn Univ.
King Mongkut's Univ. of Tech. Thonburi
King Mongkut's Inst. of Tech. Ladkrabang
King Mongkut's Inst. of Tech. North Bangkok
IEEE Thailand Section
IEEE IAS/PELS Joint Chapter, Singapore Section

Technical Co-Sponsors:

IEEE Power Electronics Society
IEEE Industry Applications Society
IEEE Industrial Electronics Society
IEEE IAS/PELS/IES Joint Chapter, Thailand Section

Organizing Committees

General Chairman	Doncker, R. D.	RWTH-Aachen Univ.
Advisory Board	Lavansiri, D.	Chulalongkorn Univ.
	Jaovisidha, V.	Chulalongkorn Univ.
	Tandhavatana, S.	IEEE Thailand Section
	Pungprasert, V.	IEEE Thailand Section
	Leelarasmee, E.	Chulalongkorn Univ.
	Pichetchumroen, V.	King Mongkut's Inst. of Tech. Ladkrabang
	Yingwatana, A.	King Mongkut's Inst. of Tech. North Bangkok
	Liang, Y. C.	National Univ. of Singapore
	Panda, S. K.	National Univ. of Singapore
General Co-Chairmen	Karnasuta, K.	IEEE Thailand Section
	Vilathgamuwa, D. M.	Nanyang Tech. Univ.
Organizing Committee Chairman	Phoomvuthisarn, S.	Center of Excellence in Electrical Power Tech. Chulalongkorn Univ.
Co-Chairman	Kulvitit, Y.	Chulalongkorn Univ.
	Yungyuen, U.	King Mongkut's Univ. of Tech. Thonburi
	Khan-ngern, W.	King Mongkut's Inst. of Tech. Ladkrabang
	Chunkag, V.	King Mongkut's Inst. of Tech. North Bangkok
Technical Program Committee	Khan-ngern, W.	King Mongkut's Inst. of Tech. Ladkrabang
	King Jet, T.	Nanyang Tech. Univ.
	Sangwongwanich, S.	Chulalongkorn Univ.
	Chunkag, V.	King Mongkut's Inst. of Tech. North Bangkok
	Sirisukprasert, S.	Kasetsart Univ.
	Boonyaroonate, I.	King Mongkut's Univ. of Tech. Thonburi
Treasurers	Bunnagulrote, B.	Center of Excellence in Electrical Power Tech. Chulalongkorn Univ.
	Battul, D.	School of Electrical & Electronic Engineering Singapore Polytechnic
Publications	Tarateeraseth, V.	Srinakharinwirot Univ.
	Jangwanitlert, A.	King Mongkut's Inst. of Tech. Ladkrabang
Tutorials	Chunkag, V.	King Mongkut's Inst. of Tech. North Bangkok
	Liutanakul, P.	King Mongkut's Inst. of Tech. North Bangkok
Local Arrangements	Kinnares, V.	King Mongkut's Inst. of Tech. Ladkrabang
	Polmai, S.	King Mongkut's Inst. of Tech. Ladkrabang
	Yutthagowith, P.	King Mongkut's Inst. of Tech. Ladkrabang
	Kittiratsatcha, S.	King Mongkut's Inst. of Tech. Ladkrabang
	Fuengwarodsakul, N.	King Mongkut's Inst. of Tech. North Bangkok
Publicity	Suwankawin, S.	Chulalongkorn Univ
Exhibition	Jangwanitlert, A.	King Mongkut's Inst. of Tech. Ladkrabang
Secretariat	Suwankawin, S.	Chulalongkorn Univ.

iii

International Steering Committee

Acarnley, P.	Univ. of Newcastle	Kennel, R.	Univ. of Wuppertal
Akagi, H.	Tokyo Inst. of Tech.	Kolar, J. W.	Swiss Federal Inst. of Tech. (ETH) Zürich
Alex, Q. H.	North Carolina State Univ.	Lai, J. S.	Virginia Polytechnic Inst. and State Univ.
Amaratunga, G. A. J.	Univ. of Cambridge	Longya, X.	Ohio State Univ.
Bhat, A. K. S.	Univ. of Victoria	Lorenz, R. D.	Univ. of Wisconsin-Madison
Boroyevich, D.	Virginia Polytechnic Inst. and State Univ.	Matsuse, K.	Meiji Univ.
Bose, B. K.	Univ. of Tennessee	Mohan, N.	Univ. of Minnesota
Chan, C.C.	Univ. of Hong Kong	Nakaoka, M.	Yamaguchi Univ.
Clare, J. C.	Univ. of Nottingham	Ninomiya, T.	Kyushu Univ.
Dehong, X.	Zhejiang Univ.	Okuma, S.	Nagoya Univ.
Divan, D.	Georgia Inst. of Tech.	Qian, Z.	Zhejiang Univ.
Elbuluk, M. E.	Univ. of Akron	Rahman, M. A.	Memorial Univ. of Newfoundland
Enjeti, P.	Texas A&M Univ.	Schroeder, D.	Univ. of Munich
Ertan, B.	Middle East Technical Univ.	Sekiya, H.	Chiba Univ.
Forsyth, A.J.	Univ. of Manchester	Sen, P.C.	Queen's Univ.
Green, T. C.	Imperial College	Shoyama, M.	Kyushu Univ.
Guo, X. D.	Harbin Inst. of Tech.	Suetsugu, T.	Fukuoka Univ.
Holtz, J.	Univ. of Wuppertal	Teck, O. B.	McGill Univ.
Hui, R. S. Y.	City Univ. of Hong Kong	Tenti, P.	Univ. of Padova
Husain, I.	Univ. of Akron	Tsai, P. C.	National Tsing Hua Univ.
Jahns, T. M.	Univ. of Wisconsin–Madison	Undeland, T. M.	Norwegian Univ. of Science and Tech.
Jain, P.	Queen's Univ.	Wu, B.	Ryerson Univ.
Jezernik, K.	Univ. of Maribor	Wyk, J. D. V.	Virginia Polytechnic Inst. and State Univ.
Kazimierczuk, M. K.	Wright State Univ.	Zhengming, Z.	Tsing Hua Univ.

History of PEDS Conference

International Conference on Power Electronics and Drive Systems, PEDS Conference, originated in Singapore and the first PEDS conference was held in Singapore in 1995. The aim of the PEDS Conference is to provide a forum for participants from the industry and academia in the area of power electronics and drives to exchange ideas and have interactions. The conference is biennial and since 1995 the PEDS Central Committee, Singapore in collaboration with overseas organizing committees, have organized PEDS Conference series held in various Asia Pacific (IEEE Region 10) countries. All the PEDS conferences are being held in technical co-sponsorship with the IEEE Power Electronics Society and IEEE Industry Applications Society.

PEDS Conference	Venue
PEDS 1995	Singapore
PEDS 1997	Singapore
PEDS 1999	Hong Kong
PEDS 2001	Bali, Indonesia
PEDS 2003	Singapore
PEDS 2005	Kuala Lumpur, Malaysia
PEDS 2007	Bangkok, Thailand

List of Reviewers

Abbaszadeh, K.
Abe, S.
Acarnley, P. P.
Adnani, M. E.
Afjei, E.
Ahmad, G.
Ahmed, M.
Al-Haddad, K.
Amirifar, R.
Ang, S.
Apte, A. A.
Attaviriyanupap, P.
Awad, H.
Azli, N. A.
Baiju, M. R.
Bakan, A. F.
Beig, A. R.
Bhadra, S. N.
Bharanikumar, R.
Bhat, A. K. S.
Bina, M. T.
Biswas, S. K.
Boonyaroonate, I.
Bunlaksananusorn, C.
Chang, K.-T.
Chen, H.
Chen, J.-J.
Cheng, M.-Y.
Chengfeng, Y.
Cheung, N. C.
Chiba, A.
Chien, F. T.
Chiu, H.-J.
Choi, B.
Chou, J.-H.
Chunkag, V.
Clare, J. C.
Colli, V. D.
Corzine, K. A.
Covic, G. A.
Cruden, A.
Dahono, P. A.
Daming, Z.
Dianguo, X.
Doki, S.
Dong-Hee, L.
Duffy, M.
Dzung, P. Q.
Elbuluk, M. E.
Ertugrul, N.
Eskander, M.
Farhangi, S.
Filho, E. R.
Forsyth, A.

Fujiwara, O
Fukuda, S.
Garvey, S. D.
Grabner, H.
Griva, G.
Gueldner, H.
Guo, Y.
Hagh, M. T.
Hakimie, H.
Hamzah, M. K.
Hamzah, N.
Hanamoto, T.
Hava, A. M.
Hayashi, Y.
Hennen, M. D.
Higuchi, T.
Ho, S.-T.
Hofmann, W.
Hori, Y.
Howe, D.
Hsieh, G.-C.
Hua, S.
Huang, L.
Huang, S.-J.
Hung, J. Y.
Hur, J.
Hussien, Z. F.
Idris, N. R. N.
Jain, P. K.
Janakiraman, P. A.
Jangwanitlert, A.
Jerome, J.
Jianxin, S.
Khan, P. K. S.
Khan-Ngern, W.
Kim, I.-S.
Kim, Y.-H.
Kinnares, K.
Kobayashi, S.
Kolar, J. W.
Komurcugil, H.
Kubota, H.
Kulvitit, Y.
Kurokawa, F.
Lafoz, M.
Lai, C.-K.
Lai, J.-S.
Lecci, A.
Lee, D.-C.
Lee, E.-W.
Lee, S. C.
Lee, Y.-S.
Li, D. D.
Li, G.

Li, H.
Li, J.
Li, W.
Liang, Y. C.
Liaw, C.-M.
Lin, C.-H.
Lin, R.-L.
Liserre, C.
Lo, Y.-K.
Loh, A.
Lorenz, L.
Low, K.-S.
Manmek, T
Markadeh, G. A.
Marques, G. D.
Martins, J. F. A.
Matsui, M.
Matsuo, K.
Mekhilef, S.
Morales-Castorena, A.
Morimoto, M.
Morimoto, S.
Mukerjee, R.
Muni, B. P.
Murthy, S. S.
Mutoh, A.
Muyeen, S. M.
Nagaraju, J.
Narayanan, G.
Nho,N. V.
Noguchi, T.
Nussbaumer, T.
Okou, A. F.
Omar, A. M.
Pai, F.-S.
Palandurkar, M. V.
Pan, C.-T.
Panda, S. K.
Patel, H. K.
Phuong, L. M.
Pichetjamroen, V.
Ping, H. W.
Pires, A.
Pires, V.
Polmai, S.
Ponce, M.
Qian, Z.
Rafael, S.
Rahman, M. A.
Ramasamy, A. K.
Rashad, E. E. M.
Ratanapanachote, S.
Rizk, J.
Ruan, X.

Saied, B. M.
Sangwongwanich, S.
Saudemont, C.
Saxena, T. K.
See, K. Y.
Senthilkumar, R.
Shaojun, X.
Sharma, V. K.
Shieh, H.-J.
Shimizu, T.
Shin, G.-H.
Shing, C. S.
Shinnaka, S.
Shuhua, F.
Singh, B.
Sirisukprasert, S.
Soltani, J.
Sopavanit, C.
Staines, C. S.

Sumedha
Sun, K.
Suwankawin, S.
Tahami, F.
Takahashi, R.
Tanaka, T.
Tarnekar, S. G.
Tenti, P.
Thounthong, P.
Tomita, H.
Tseng, K.-J.
Tsui, M.
Vaclavek, P.
Vaez-Zadeh, S.
Veszpremi, K.
Vijayarajan, K.
Vilathgamuwa, D. M.
Villasenor, A. G.
Wang, C.-M.

Wang, H.-P.
Wang, L.
Weiming, M.
Wen, F.-L.
Wolbank, T. M.
Wu, L.
Wu, T.-F.
Xu, D. (David)
Xu, D. (Dehong)
Xu, L.
You, K.
Yousfi, D.
Zhang, X.
Zhengyu, L.
Zhong , Q.-C.
Zhu, J.
Zirn, O.
Zolghadri, M. R.

This page intentionally left blank.

Table of Contents

Fuel cell systems and applications 1
Bernard Davat

Recent Trends iin Power Qualliity Improvements Techniiques 58
Bhim Singh

Power Electronics for Future Utility Applications 213
Rik W. De Doncker, Christoph Meyer, Robert U. Lenke, Florian Mura

Digital Control Generations -- Digital Controls for Power Electronics through the Third Generation 221
Philip T. Krein

Power Electronics and Control of Renewable Energy Systems 226
F. Iov, M. Ciobotaru, D. Sera, R. Teodorescu, F. Blaabjerg

Design and Evaluation of a 60 000 rpm Permanent Magnet Bearingless High Speed Motor 249
T. Schneider, A. Binder

Performance Investigation of Two-, Three- and Four-Phase Bearingless Slice Motor Configurations 257
M.T. Bartholet, S. Silber, T. Nussbaumer, J.W. Kolar

Compensation of Pole Position Estimation Error for Sensor-less IPMSM Drives with DC Link Current Detection 265
Hisao Kubota, Yusuke Shibano, Takayuki Kobayashi

A Novel Dual-Stator Hybrid Excited Synchronous Wind Generator 270
Liu Xiping, Lin Heyun, Yang Chengfeng, Fang Shuhua, Guo Jian

Application of Multi-level Multi-domain Modeling in the Design and Analysis of a PM Transverse Flux Motor with SMC Core 275
Youguang Guo, Jianguo Zhu, Dikai Liu, Haiyan Lu, Shuhong Wang

New Approximate 2DOF Digital Controller for DC-DC Converter with Second-Order Differential Characteristics 280
Eiji Takegami, Kohji Higuchi, Kazushi Nakano, Satoshi Tomioka, Kazushi Watanabe, K.K. Densei-Lambda

High Accuracy CMOS Current Sensing Circuit for Current Mode Control Buck Converter 286
Yuang-Shung Lee, Chih-Jen Hsu

Small Signal Analysis of a dual-switch forward Converter with non-ideal transformer in Current-Programmed Control 291
Weiping Zhang, Yuzhou Lei, Xiaoqiang Zhang, Yuanchao Liu

High Frequency Transformer Designs for Improving Cross Regulation in Multiple-Output Flyback Converters 295
Kusumal Chalermyanont, Pairote Sangampai, Anuwat Prasertsit, Surapon Theinmontri

Operation of a wye Connected Three- Level Active Power Filter under Nonideal Conditions 299
H.B. Zhang, A.M. Massoud, S.J. Finney, B.W.Williams, T.C. Lim, H. Hotait

Application of GPRS Techniques for Wide-Area Power Quality Monitoring 305
Shun-Yu Chan, Jen-Hao Teng, David Chang, Li-Yuan Chin

Design and Development of Autotransformer Based 24-Pulse AC-DC Converter fed Induction Motor Drive 310
Bhim Singh, Vipin Garg, G.Bhuvaneswari

Power Quality Monitoring System Using Real-Time Operating System 318
Krisda Yingkayun, Suttichai Premrudeepreechacharn, Kosol Oranpiroj

Technology Performance Comparison of Triacs Subjected to Fast Transient Voltages 322
L. Gonthier, A. Passal

Table of Contents

On-line Junction Temperature Measurement of CoolMOS Devices...327
Andreas Koenig, Thomas Plum, Peter Fidler, Rik W. De Doncker

Analytical Design of High-Power MTO Thyristors...333
Thomas Plum, Rik W. De Doncker

A Novel Gate Driver with Output Voltage Having Double Source Voltage.................................338
K. I. Hwu, Y. T. Yau

Effects of Internal Feedback and Gate-Drive Signal on the Turn-off Loss of MOSFET ZVS..........342
Youthana Kulvitit, Puckapon Opanuruk, Tanvaa Tansatit

A Novel Bridge Type FCL Based on Single Controllable Switch...350
Wanmin Fei, Yanli Zhang, Qi Wang

A Novel Isolation Power Supply for Gating Multiple Devices in FACTS Equipment.....................354
Yanli Zhang, Wanmin Fei, Zhengyu Lu

Voltage and Frequency Controller for Parallel Operated Isolated Asynchronous Generators..........357
Bhim Singh, Gaurav Kumar Kasal

DSP controlled Semiconductor based High-Voltage Source..363
F. Martin, T. Leibfried, O. Kerz, K. Mossner

Open Switch Fault Diagnosis for a Doubly-Fed Induction Generator....................................368
W. Sae-Kok, D M Grant

Rapid Analysis & Design Methodologies of High- Frequency LCLC Resonant Inverter as Electrodeless Fluorescent Lamp Ballast...376
Yong-Ann Ang, David Stone, Chris Bingham, Martin Foster

Analysis and Control of Dual-Output LCLC Resonant Converters, and the Impact of Leakage Inductance.....382
Y. Ang, C. M. Bingham, M. P. Foster, D. A. Stone

A Novel QR ZCS Switched-Capacitor Bidirectional Converter...388
Yuang-Shung Lee, Yi-Pin Ko, Chien-An Chi

Analysis of a Half - Bridge Inverter for a Small- Size Induction Cooker Using Positive-Negative Phase- Shift Control under ZVS and NON-ZVS Operation...394
P. Achara, P. Viriya, K. Matsuse

Adaptive Phase Control Method for Load Variation of Resonant Converter with Piezoelectric Transformer...401
S. T. Yun, J. M. Sim, J. H. Park, S. J. Choi, B. H. Cho

Adaptation of Motor Parameters in Sensorless PMSM Drives...406
Antti Piippo, Marko Hinkkanen, Jorma Luomi

Development of 150000 r/min, 1.5 kW Permanent- Magnet Motor for Automotive Supercharger.........414
Toshihiko Noguchi, Masaru Kano

Analysis and Performance Evaluation of Radial Flux Air-Cored Permanent Magnet Machines with Concentrated Coils...420
P.J. Randewijk, M.J. Kamper, R-J. Wang

Analysis and Experimental Investigation for Field-Control Capability of a Novel Hybrid Excitation Claw- Pole Synchronous Machine..427
Yang Chengfeng, Lin Heyun, Liu Xiping, Fang Shuhua, Guo Jian

A single-Capacitor Turn-off Snubber for Interleaved Boost Converter with Coupled Inductor...........433
S.-Y. Tseng, J. Z. Shiang, Y.-H. Su

Buck-Boost Converter Associated with Active Clamp Forward Converter for PV Power System.........440
S. Y. Tseng, W. C. Chen, Y. J. Li, J. S. Kuo

Table of Contents

Comparison of Three-Phase DC-DC Converters vs. Single-Phase DC-DC Converters 448
Christian P. Dick, Andreas Konig, Rik W. De Doncker

Applying Modified One-Comparator Counter-Based PWM Control Strategy to Flyback Converter 456
K. I. Hwu,, Y. H. Chen

Analysis of Conducted EMI Reduction on a Boost Converter Using Progressive Inductor Winding Technique 460
Kritsada Saritsiri, Werachet Khan-Ngern

Practical Issues Concerned with Zero sequence component and Harmonic Compensation in Four-Wire systems 465
E. Pashajavid, K. Kanzi, M. Tavakoli Bina

Automated Design and Implementation of Resonant Controllers for Current Control of Shunt Active Filters 470
W. Lenwari, M. Sumner, P. Zanchetta

A Modular Structured Multilevel Inverter Active Power Filter with Unified Constant-Frequency Integration Control for Nonlinear AC Loads 475
P. Y. Lim, N. A. Azli

HCC PWM Control of the Single-Phase Bi- Directional Buck Converter giving IEEE 519 Compliance at any Power Factor 480
A. N. Arvindan, V. K. Sharma

Passive EMI Filter Performance Improvements with Common Mode Voltage Cancellation Technique for PWM Inverter 488
C. Khun, W. Khan-Ngern, M. Kando

Novel Auxiliary Diagnosis Method for State-of-Health of Lead-Acid Battery 493
Yu-Hua Sun, Hurng-Liahng Jou, Jinn-Chang Wu

Electromechanical Model of a Longitudinal Mode Piezoelectric Transformer 498
Shine-Tzong Ho

Latest Development of Transformer Parasitic Inductive Components and Lossless Inductive Snubber-Assisted Series Resonant High-Frequency ZCS-PFM DC-DC Converter for RF Generator 504
Hisayuki Sugimura, Manabu Ishitobi, Bishwajit Saha, Sang Pil Mun, Soon Kurl Kwon, Mutsuo Nakaoka

A General Method for Deciding the Input Filter Capacitance of Flyback Switching AC-DC Converter with Peak Current-Controlled Mode 510
Jiaxin Chen, Jianguo Zhu, Youguang Guo

Design of High Performance and Low Cost Line Impedance Stabilization Network for University Power Electronics and EMC Laboratories 515
D. Sakulhirirak, V. Tarateeraseth, W. Khan-Ngern, N. Yoothanom

A Robust Output Current Control Method with Disturbance Observer for Matrix Converter under Unbalanced Input Voltage 521
Kazuo Oka, Kouki Matsuse

FPGA Design of Single-phase Matrix Converter Operating as Cycloconverter 527
Z. Idris, M.K. Hamzah, A. Saparon, N.R.Hamzah, N.Y. Dahlan

Input and Output Ripple Analysis of AC Chopper 534
Arwindra Rizqiawan, Dessy Amirudin, Deni, Pekik Argo Dahono

A Three-level 4 × 3 Conventional Matrix Converter 541
Runjie Rong, Poh Chiang Loh, Peng Wang, Frede Blaabjerg

A novel primary-side controlled contactless battery charger 546
Yi-Hwa Liu, Shun-Chung Wang, Rong Ceng Leou

Table of Contents

Research on Digital Soft-switch Welding/Cutting Inverter Power Source 551
G.R. Zhu, Z. Liu, X. Li, B.Y. Liu, S.X. Duan, Y. Kang

Design of an Adjustable High Output Voltage Asymmetrically Switched Class D Converter 556
M. Rentzsch, H. Guldner, C. Ditmanson

New Direct High Frequency Soft-Switching Inverter-Fed AC-DC Converter with Voltage Doubler for Consumer Magnetron Drive 563
Hisayuki Sugimura, Bishwajit Saha, Hidekazu Muraoka, Sang Pil Mun, Tomokazu Mishima, Hideki Omori, Mutsuo Nakaoka

Complete loading Characteristics Modeling of an Axial Flux Permanent Magnet Synchronous Machine Using Ck Spline Functions 569
Z. Lakhdari, F. Amrane, L. Adélaide, Ph. Makany

The Bearingless 2-Level Motor 574
P. Karutz, T. Nussbaumer, W. Gruber , J.W. Kolar

Analysis and Design of a Sliding Mode Controller for Buck Converters Operating in DCM with Adaptive Hysteresis Band Control Scheme 581
Hung-Chih Lin, Tsin-Yuan Chang

Buck Converter Simulation Technique Based on the Fourier Transform 587
Acacio M. R. Amaral, A. J. Marques Cardoso

ANALYSIS OF HOPF BIFURCATION IN DC-DC LUO CONVERTER USING CONTINUOUS TIME MODEL 595
A.Kavitha, G.Uma

Analysis of a Mixed-Signal Control for DC-DC Converters based on Hysteresis Modulation And Estimated Inductor Current 600
D. Trevisan, S. Saggini, P. Mattavelli, L. Corradini, P. Tenti

Power Quality Study in Macao 607
Sio-Un Tai, Man-Chung Wong, Ming-Chui Dong, Ying-Duo Han

Some Findings on Harmonic Measurement in Macao 614
Sio-Un Tai, Man-Chung Wong, Ming-Chui Dong, Ying-Duo Han

Coordinated design of PSS and TCSC dynamics model for power system network oscillations 620
M. Tarafdar Haque, A. Roshan Milani, A. Lafzi

An Analytic Approach To Harmonic Analysis of 48-Pulse Voltage Source Inverter 626
B. Geethalakshmi, P. Dananjayan

Detailed losses Analysis of High-Frequency Planar Power Transformer 632
Yu Ma, Peipei Meng, Junming Zhang, Zhaoming Qian

Design of a Nuclear Magnetic Resonance Fast Field Cycling Air Cored Magnet 636
Duarte M. Sousa, Gil D. Marques, Pedro J. Sebastiao,, Antonio C. Ribeiro

Using DFT to Obtain the Equivalent Circuit of Aluminum Electrolytic Capacitors 643
Acácio M. R. Amaral, Gustavo M. Buatti, Hugo Ribeiro, A.J. Marques Cardoso

A Mathematical Analysis on Vector Inversion Generators 648
D. J. Thrimawithana, U. K. Madawala

Novel Multi-Level High Voltage Pulsed Power Generator 654
D. J. Thrimawithana, U. K. Madawala

Potential and Electric Field Distribution Analysis of Field Limiting Ring and Field Plate by Device Simulator 660
C.N. Liao, F.T. Chien, Y.T. Tsai

xii

Table of Contents

Wire and Wireless Linked Remote Control for the Group Lighting System Using Induction Lamps 665
Kyu Min Cho, Jae Eul Yeon, Ma Xian Chao, Hee Jun Kim

Induction Heating with Traveling Magnetic Field for Uniform Heating to Flat Metal 671
T. Sekine, H. Tomita, Y. Saito, S. Obata, S. Yoshimura

Three-Phase (LC)(L)-Type Series-Resonant Converter with Capacitive Output Filter 677
M. Almardy, A.K.S. Bhat

Analysis of a Full-Bridge Inverter for Induction Heating Using Asymmetrical Phase-Shift Control under ZVS and NON-ZVS Operation 685
N. Yongyuth, P. Viriya, K. Matsuse

FPGA-Based Phase-Shift ZVS Full-Bridge DC-DC Converter Using One-Comparator Counter-Based PWM Control Strategy 692
K. I. Hwu, Y. T. Yau

A Simplified Power Control Scheme for Resonant Inverter with Purely Resistive Load 697
Pramoch Dorkmai, Youthana Kulvitit, Tanvaa Tansatit

Voltage Injection Based Initial Rotor Position Estimation Method for Three-Phase Star- Connected Switched Reluctance Machines 703
P. Somsiri, P. Champa, P. Wipasuramonton, K. Tungpimonrut, P. Aree

Control Scheme for Switched Reluctance Drives with Minimized DC-Link Capacitance 710
Christoph R. Neuhaus, Rik W. De Doncker, Nisai H. Fuengwarodsakul

Multiphase Torque-Sharing Concepts of Predictive PWM-DITC for SRM 716
Helge J. Brauer, Martin D. Hennen, Rik W. De Doncker

A New Two Phase Configuration for Switched Reluctance Motor with High Starting Torque 722
E. Afjei, K. Navi, S. Ataei

Application of Power Electronics for Damping of Torsional Vibrations 726
T. Zoller, T. Leibfried, A. M. Miri

Application of Battery Energy Operated System to Isolated Power Distribution Systems 731
Bhim Singh, A. Adya, A.P. Mittal, J.R.P Gupta

Pulse Doubling in 18-Pulse AC-DC Converters 738
Bhim Singh, Sanjay Gairola

Magnetic Field Analysis and Control Strategy of Permanent Magnet Actuator for Low Voltage Vacuum Circuit Breaker 745
Fang Shuhua, Lin Heyun, Yang Chenfeng, Liu Xiping, Guo Jian

Analysis of Transformer Inrush Current under Harmonic Source 749
Chien-Lung Cheng, Jim-Chwen Yeh, Shyi-Ching Chern, Yi-Hung Lan

Voltage Sag Compensation Performance by DSTATCOM with Series Inductor and Energy Storage 755
Sumate Naetiladdanon

Cooperative Operation of Active Power Filters by Instantaneous Complex Power Control 760
Elisabetta Tedeschi, Paolo Tenti, Paolo Mattavelli

Impact of Adjustable Speed PWM drives on Operation and Harmonic Losses of Nonlinear Three Phase Transformers 767
M.A.S. Masoum, Paul S. Moses, Amir S. Masoum

Real-Time Implementation of Voltage Dip Mitigation using D-STATCOM with Fast Extraction of Instantaneous Symmetrical Components 773
Thip Manmek, Chathura P. Mudannayake

Table of Contents

Combined System of Static Synchronous Series Compensation and Passive Filter applied to Wind Energy Conversion System 781
A. Singer, W. Hofmann

Control of active injector for multi-pulse rectifiers operating on variable frequency supplies 788
Ismael Araujo-Vargas, Andrew J. Forsyth, F. Javier Chivite-Zabalza

36-pulse hybrid ripple injection for high performance aerospace rectifiers 796
F. Javier Chivite-Zabalza, Andrew J. Forsyth, Ismael Araujo-Vargas

A 48-pulse converter using dc-ripple injection 804
F. Javier Chivite-Zabalza, Andrew J. Forsyth

A Study of Different Possible Switched Mode Chopper Circuits for Multi-Magnet Based DC Electromagnetic Levitation System 812
Subrata Banerjee, Dinkar Prasad, Jayanta Pal

Power Supply with Potential Use in Magnetic Stimulation 817
Duarte M. Sousa, Antonio Ferraz

A Novel Maximum Power Point Tracking Method for the Photovoltaic System 824
Hurng-Liahng Jou, Wen-Jung Chiang, Jinn-Chang Wu

Maximum Power Point Algorithm in PV Generation: An Overview 829
Hardik P. Desai, H. K. Patel

A DC-Module-Based Power Configuration for Residential Photovoltaic Power Application 836
Bangyin Liu, Shanxu Duan, Yong Kang

Analysis and Improvement of Maximum Power Point Tracking Algorithm Based on Incremental Conductance Method for Photovoltaic Array 842
Bangyin Liu, Shanxu Duan, Fei Liu, Pengwei Xu

Application of Maximum Power Point Tracker with Self-organizing Fuzzy Logic Controller for Solar-powered Traffic Lights 847
Noppadol Khaehintung, Phaophak Sirisuk

Supply-side Current Harmonics Control of Three Phase PWM Boost Rectifiers Under Distorted and Unbalanced Supply Voltage Conditions 852
Xinhui Wu,, Sanjib K. Panda, Jianxin Xu

A Two-stage Converter with a Coupled-Inductor 858
Hirotaka Nakanishi, Yoshihiro Tomihisa, Terukazu Sato, Takashi Nabeshima, Kimihiro Nishijima, Tadao Nakano

Three-Phase AC to DC Converter with Minimized DC Bus Capacitor and Fast Dynamic Response 863
U. Kamnarn, Y. Kanthaphayao, V. Chunkag

A Simple Effective Duty Cycle Controller for High Power Factor Boost Rectifier 869
Hussain S. Athab, P. K. Shadhu Khan

A Cost Effective Method of Reducing Total Harmonic Distortion (THD) in Single-Phase Boost Rectifier 874
Hussain S. Athab, P. K. Shadhu Khan

Comparison of Different Methods to Detect Static Air Gap Asymmetry in Inverter Fed Induction Machines 880
T.M. Wolbank, P. Macheiner

Analysis of the Synchronous Torques in a Split Phase Induction Motor 886
P. Scavenius Andersen, D. G. Dorrell, N. C. Weihrauch, P. E. Hansen

On-Line Diagnosis of Three-Phase Closed Loop Induction Motor Drives Using an Eigenvalue aß-Vector Approach 894
J. F. Martins, V. Fernao Pires, A. J. Pires

xiv

Table of Contents

Design and Development of a 36-Pulse AC-DC Converter for Vector Controlled Induction Motor Drive..................899
Bhim Singh, Sanjay Gairola

Comparison of Outer- and Inner-Rotor Switched Reluctance Machines........................907
Martin D. Hennen, Rik W. De Doncker

Optimization of Predesign of Switched Reluctance Machines Cross Section Using Genetic Algorithms.................912
Satit Owatchaiphong, Christian Carstensen, Rik W. De Doncker

Shaft Position for an 8/6 Switched Reluctance Machine: Theoretical concept, FEM analysis and Experimental results.......................917
Silviano Rafael, P.J. Costa Branco, A.J. Pires

Sensorless Control of Brushless Doubly-Fed Reluctance Machines using an Angular Velocity Observer....................922
Milutin G Jovanovic, David G Dorrell

A Half-Bridge PV System with Bi-direction Power Flow Controlling and Power Quality Improvement....................930
C.L. Shen, S.T. Peng

Response of DSTATCOM under Voltage Flicker In Farm Wind........................937
K. Aodsup, P. N. Boonchiam, A. Sode-Yome, P. Kongsuk, N. Mithulananthan

A Comparative Study of Fixed Speed and Variable Speed Wind Energy Conversion Systems Feeding the Grid......................941
S.S. Murthy, Bhim Sing, P.K. Goel, S.K. Tiwari

Prediction of Wind Power Generation based on Chaotic Phase Space Reconstruction Models........................949
Dong Lei, Wang Lijie, Hu Shi, Gao Shuang, Liao Xiaozhong

Power Flow Control for Efficiency Improvement in a Forward-Flyback Mixed Converter.................954
Yoshito Kusuhara, Asahi Nakayama, Tamotsu Ninomiya, Shin Nakagawa

Hammerstein Model-Based Robust Control of DC/DC Converters........................959
F. Alonge, F. D'ippolito, T. Cangemi

A New Model Control DC-DC Converter to Improve Dynamic Characteristics........................968
F. Kurokawa, S. Sukita

Fuzzy Incremental Controller for the 3rd Order Buck Converter......................973
M. Veerachary, Deepen Sharma

Design of a Single-Stage Single-Switch Power- Factor-Corrected (S4-PFC) AC/DC Converter.................977
P. Kongthawornwattana, C. Bunlaksananusorn, S. Kittiratsatcha

A DSP-Based Unified Three Phase/Switch/Level Unity Power Factor Rectifier Using Feedback Linearization for DC-Bus Voltage Control......................983
Ali Moallem, Hesameddin Mirzaee Teshnizi, Mohammadreza Zolghadri

A Soft-Switched AC-DC Symmetrical Boost Converter with Power Factor Correction........................989
A. Jangwanitlert, J. Songboonkaew

Education Reforming for Power Electronics........................994
Weiping Zhang, Xiaohan Guan, Dongyan Zhang

A Novel Current Control System for PMSM Considering Effects from Inverter in Overmodulation Range.................999
Smith Lerdudomsak, Shinji Doki, Shigeru Okuma

Modelling of the Feeding Network of a Linear Synchronous Machine and Estimation of Model Parameters......................1006
J. Rost, H. Gueldner, R. Hellinger, A. Weller

Analysis of Losses in Inverter Fed Large Scale Synchronous Machines using 2D FEM Software..................1012
Samer Shisha, Chandur Sadarangani

Table of Contents

Position sensorless control of the Reluctance Synchronous Machine considering High Frequency inductances 1017
H.W. De Kock, M.J. Kamper, O.C. Ferreira, R.M. Kennel

Carrier PWM algorithm in overmodulation range for Multileg Multilevel Inverter 1027
Nguyen Van Nho, Hong Hee Lee

Carrier Based Single-state PWM Technique In multilevel Inverter 1033
Nguyen Van Nho, Quach Thanh Hai, Hong Hee Lee

Implementation of a Single-carrier Multilevel PWM Technique Using Field Programmable Gate Array (FPGA) 1041
N. A. Azli, L. Y. Teng, P. Y. Lim

SPACE VECTOR PWM FOR MULTILEVEL INVERTERS - A FRACTAL APPROACH 1047
Anish Gopinath, M.R. Baiju

Elimination of Harmonics in a Five-Level Diode-Clamped Multilevel Inverter Using Fundamental Modulation 1055
Sule Ozdemir, Engin Ozdemir, Leon M. Tolbert, Surin Khomfoi

Compensation of DC-Link Oscillations of Cascaded H-Bridge Converters 1060
M. Tavakoli Bina, B. Eskandari

Combined DC-Filter and optimized Modulation to Absorb DC-Link Oscillations of Cascaded H-Bridge Converters 1065
M. Tavakoli Bina, B. Eskandari

Control Strategies of a Hybrid Multilevel Converter for Expanding Adjustable Output Voltage Range 1070
Shoji Fukuda, Takatsugu Yoshida, Shigeta Ueda

High Efficiency Single Phase Multi-level Inverter by New Controlled Switch Signal 1078
Ruthapong Kumchaiyo, Itsda Boonyaroonate

FPGA Implementation of Quasi-BLDC Drive 1082
C.S. Soh, C. Bi, K.K. Teo

A Practical Method to Eliminate the Conduction Torque Ripple in BLDCM Using Cascade Topology 1088
Xiaofeng Zhang, Zhengyu Lu, Yu Ma, Zhaoming Qian

Program Architecture for Realizing Design Optimization of a BLDC Motor 1092
Dong-Hun Kim, Giwoo Jeung, Heung-Geun Kim, In Dong Kim

Stable Operation of the Brushless Doubly-Fed Machine (BDFM) 1096
Shiyi Shao, Ehsan Abdi, Richard Mcmahon

Sail Generator Feasibility Study 1102
Ha Pham Ngoc, Yasuaki Matsui, Pathom Attaviriyanupap, Osamu Iso

Braking Circuit of Small Wind Turbine Using NTC Thermistor under Natural Wind Condition 1109
Y. Matsui, A. Sugawara, S. Sato, T. Takeda, K.Ogura

Flywheel Energy Storage Drive for Wind Turbines 1115
K. Veszpremi, I. Schmidt

Theory, Simulation and Experimental Verification of a New Integral Cycle Robust Control Strategy for Self Excited Induction Generators 1123
S.S. Murthy, A.J.P. Pinto

Performance Comparison of DC Link Voltage Controllers in Vector Controlled Boost Type PWM Converter for Wind Turbine System 1129
W. Sudmee, B. Neammanee

Analysis and Design of Class DE Amplifier with Nonlinear Shunt Capacitance 1136
Hiroo Sekiya, Takayuki Watanabe, Tadashi Suetsugu, Marian K. Kazimierczuk

Table of Contents

A Novel Control Strategy of the Class-D Stereo Audio Amplifier..1142
Kyu Min Cho, Won Seok Oh, Hai Xu, Hee Jun Kim

Robust H_infinity Control Design for PFC Rectifiers..1147
F. Tahami, H. Molla Ahmadian, A. Moallem

Parallel Operation of Power Factor Corrected AC-DC Converter Modules With Two Power Stages................................1152
Aravind Pothana, Krishna Vasudevan

Noise Radiation of Switched Reluctance Drives..1160
K. A. Kasper, M. Bosing, R. W. De Doncker, S. Fingerhuth, M. Vorlander

Iron Losses in Electrical Machines Due to Non Sinusoidal Alternating Fluxes................................1167
J. A. Walker, D. G. Dorrell, E. Ritchie

Design Requirements for Doubly-Fed Reluctance Generators..1174
D. G. Dorrell

A Magnetic Gear Box for application with a Contra-rotating Tidal Turbine................................1182
Laxman Shah, A. Cruden, Barry W. Williams

Mechatronic . Advanced Computational Intelligence..1187
D. Schroder, H. Schuster, C. Westermaier

New Space Vector Control Approach for Four Switch Three Phase Inverter (FSTPI)................................1195
Phan Quoc Dzung, Le Minh Phuong, Pham Quang Vinh, Nguyen Minh Hoang, Tran Cong Binh

The Development of Artificial Neural Network Space Vector PWM for Four-Switch Three- Phase Inverter..1202
Phan Quoc Dzung, Le Minh Phuong, Pham Quang Vinh

Voltage Losses Compensation Using Artificial Neural Network for Estimation Nonlinear Characteristic of Switches..1208
N. Pothi, S. Premrudeepreechacharn, C. Rakpenthai

A Simple Carrier-Based PWM Method For Three-Phase Four-Leg Inverters Considering All Four Pole Voltages Simultaneously..1213
Nakharet Chudoung, Somboon Sangwongwanich

Inverted Sine Carrier Pulse Width Modulation for Fundamental Fortification in DC-AC Converters................................1221
R.Nandhakumar, S.Jeevananthan

Fault Detection and Reconfiguration Technique for Cascaded H-bridge 11-level Inverter Drives Operating under Faulty Condition..1228
Surin Khomfoi, Leon M. Tolbert

Investigation into Harmonic Losses in a PWM Multilevel cascaded H-Bridge Inverter Fed Induction Motor..1236
Prasopchok Hothongkham, Vijit Kinnares

Extend the Use of Auxiliary Circuit to Start up, Shut down, and Balance of the Modified Diode Clamped Multilevel Inverter..1242
Ahmed Ali Ashaibi, S.J. Finney, B.W. Williams, Ahmed Massoud

Five-Level Z-Source Neutral-Point-Clamped Inverter..1247
F. Gao, P. C. Loh, F. Blaabjerg, R. Teodorescu, D. M. Vilathgamuwa

Capacitor Voltage Balancing Using Redundant States for Five-Level Multilevel Inverter................................1255
Hadi A Hotait, Ahmed M Massoud, Steve J. Finney, Barry W. Williams

Sliding Mode Repetitive Control of PWM Voltage Source Inverter..1262
Sufen Chen, Y. M. Lai, Siew-Chong Tan, Chi K. Tse

Output Current Ripple Analysis of Five-Phase PWM Inverters..1267
Deni, E. G. Supriatna, P. A. Dahono

xvii

Table of Contents

An Improved 'DC-DC Type' High Frequency Transformer-Link Inverter by Employing Regenerative Snubber Circuit 1274
Z. Salam, S. M. Ayob, M. Z. Ramli, N. A. Azli

A Novel Dimming Technique for Cold Cathode Fluorescent Lamp 1278
K. I. Hwu, Y. H. Chen

Time Delay Compensation For A DSP-Based Current-Source Converter Using Observer-Predictor Controller 1284
Huu-Phuc To, Muhammed Fazlur Rahman, Colin Grantham

Implementation of Hysteresis Current Control for Single-Phase Grid Connected Inverter 1290
Krismadinata, Nasrudin Abd Rahim, Jeyraj Selvaraj

Use of Air-Cored Axial Flux Permanent Magnet Generator in Direct Battery Charging Wind Energy Systems 1295
F.G. Rossouw, M.J. Kamper

Transverse Flux Machines for Sustainable Development - Road Transportation and Power Generation 1301
D. Svechkarenko, A. Cosic, J. Soulard, C. Sadarangani

Low Voltage Ride-Through Capability for Wind Turbines based on Current Source Inverter Topologies.............1308
Pierluigi Tenca, Andrew A. Rockhill, Thomas A. Lipo

Optimal Control of Direct Driven Feed Axes with Flexible Structural Components 1316
Ekkehard Batzies, Tobias Scholler, Volkmar Welker, Oliver Zirn

Leakage Energy Recovered Narrow Pulsed Voltage Generator Associated with Ultrasound Generator for Liquid Food Sterilization 1321
S. Y. Tseng, Y. D. Chang, P. L. Huang, T. F Wu, Y. M. Chen

Energy Harvesting from Exercise Bicycle 1327
Suchart Janjornmanit, Samart Yachiangkam, Aswin Kaewsingha

Modeling and Analysis of Igniter for HID Lamps 1330
Weiping Zhang, Qiang Cheng

Design of a Single Bi-directional DC-DC Converter for Onboard Energy Improving of Zero Emission Electric Vehicles...1335
Werachet Khan-Ngern

Speed Sensorless Control with Neuron MRAS Estimator of an Induction Machine...1340
Dong Lei, Yang Dong, Liao Xiaozhong

Adaptive Flux model for commissioning of signal injection based zero speed sensorless flux control of induction machines...1346
T.M. Wolbank, M.A. Vogelsberger, R.H. Stumberger

Design and Performance of a Single Stator, Dual Rotor Induction Motor...1352
S. Sinha, N. K. Deb, N. Mondal, S. K. Biswas

Investigation of skew effect on the Performance of Self - Excited Induction Generators 1356
B. Sawetsakulanond, V. Kinnares

Analysis of Double Loops Discrete Single Input PI Fuzzy for Single phase Inverter 1363
S.M. Ayob, Z. Salam, N.A. Azli

A new three-phase varying-band hysteresis current controller for voltage-source inverters.........................1368
Vinciane Chereau, Francois Auger, Luc Loron

Diode-Assisted Buck-Boost Current Source Inverters 1376
F. Gao, C. Liang, P. C. Loh, F. Blaabjerg

Table of Contents

Single-Stage Fluorescent Lamps Electronic Ballast Using Class-DE Low dv/dt Rectifier for Power-Factor Correction...1383
Chainarin Ekkaravarodome, Adisak Nathakaranakule, Itsda Boonyaroonate

Output Impedance Design Consideration of Three Control Schemes for Bus Converter in On-Board Distributed Power System..1388
Seiya Abe, Masahiko Hirokawa, Tamotsu Ninomiya

Optimal Generation Rescheduling for Security Operation of Power Systems Using Optimal Control Theory...1394
J. Q. Sun, K. W. Chan, D. Z. Fang

Improvement of Transient Response of Thermal Power Plant Using VVVF Inverter......................1398
N. Matsui, F. Kurokawa

A Novel Circuit Topology for Three-Phase Four-Wire Distribution Electronic Power Transformer......................1404
H.Mirmousa, M.R.Zolghadri

A Half-Bridge DC/DC Converter for Plasma Cutting Machine...1412
N. Sanajit, A. Jangwanitlert

Ripple Estimation for Paralleled Converter System with Automatic Interleaving Function.............1417
Teruhiko Kohama, Ryota Tsunesada, Tamotsu Ninomiya

Design of a New Hysteretic PWM Controller for All Types of DC-to-DC Converters.....................1423
Min Lin, Takashi Nabeshima, Terukazu Sato, Kimihiro Nishijima

Implementation of Fuzzy Logic Controller with Bifurcation Control of a Current-mode Boost Converter.............1429
Noppadol Khaehintung, Phaophak Sirisuk, Anantawat Kunakorn

Phase Advance Approach to Expand the Speed Range of Brushless DC Motor...............................1434
Binhminh Nguyen, Minh C. Ta

Nonlinear Decoupled Control for a Six-Phase Series-Connected Two Induction Motor Drive Using the Sliding-Mode Technique..1442
J. Soltani, N. R. Abjadi, Gh. R. Arab Markadeh

The Decoupled Stator Flux and Torque Sliding-Mode Control of Induction Motor Drive Taking the Iron Losses into Account...1449
M.Hajian, J.Soltani, S.Hosein Nia, G.R.Arab

AN EFFICIENT DIRECT TORQUE CONTROL SCHEME FOR SPLIT PHASE INDUCTION MOTOR............1455
A. Khajeh, J. S. Moghani, M. Shahbazi

A Method of Speed Sensorless Vector Control Parallel -Connected Dual Induction Motors Fed by One Inverter in a Rotor Flux Feedback Control...1460
Jun Nishimura, Kazuo Oka, Kouki Matsuse

A Combined Model Flux Observer for Vector Control of Traction Asynchronous Motors...............1465
F. Tahami, S. Chini Foroosh

Torque Ripple Elimination for Doubly-Fed Induction Motors under Unbalanced Source Voltage.........1471
Hong-Geuk Park, Ahmed G. Abo-Khalil, Dong-Choon Lee, Kwang-Myoung Son

Online H8 Speed Control of Sensorless Induction Motors with Rotor Resistance Estimation............1477
Peda V Medagam, Farzad Pourboghrat

Analysis and Comparative Study on the Performance between Standard and High Efficiency Induction Machines operating as Self - Excited Induction Generators...1483
B. Sawetsakulanond, V. Kinnraes

A simple Approach to Capacitance Determination of Self - Exited Induction Generators for Terminal Voltage Regulation..1489
B. Sawetsakulanond, V. Kinnares

xix

Table of Contents

Symmetrical Components-Based Control Technique of Doubly Fed Induction Generators under Unbalanced Voltages for Reduction of Torque and Reactive Power Pulsations...1495
S. Wangsathitwong, S. Sirisumrannukul, S. Chatratana, W. Deleroi

A New Switching Technique for Direct Torque Control of Induction Motor using Four-Switch Three-Phase Inverter 1501
Phan Quoc Dzung, Le Minh Phuong, Pham Quang Vinh, Nguyen Minh Hoang, Nguyen Xuan Bac

Detection of Some Parameters of Induction Motors a Proposal and Its Verification...............................1507
H. Bulent Ertan, Volkan Sezgin, Baris Colak

Comparison of Basic Direct Torque Control Designs for Permanent Magnet Synchronous Motor............1514
M. N. Abdul Kadir, S. Mekhilef, W.P. Hew

Improved DSVM-DTC Based Current Sensorless Permanent Magnet Synchronous Motor Drive............1520
Bhim Singh, Devendra Goyal

A High Performance Direct Torque Control Scheme of Permanent Magnet Synchronous Motor............1527
Dong-Hee Lee, Young-Joo An, Eui-Chel Nho

Low Cost Position Sensor for Permanent Magnet Linear Drive 1533
Ralf Wegener, Florian Senicar, Christian Junge, Stefan Soter

Design of One Rotary-linear Permanent Magnet Motor with Two Independently Energized Three Phase Windings..1538
L. Chen, W. Hofmann

Position Estimation of Permanent Magnet Synchronous Motor Using Un-known Input Observer 1543
Masaru Hasegawa, Satoshi Yoshioka, Keiju Matsui

Switched Reluctance Motor Drive for Electric Motorcycle Using HFNN Controller 1549
Chih-Hong Lin

STATE - SPACE AVERAGING, SIMULATION, STABILITY STUDIES FOR STEP UP POSITIVE OUTPUT SWITCHED CAPACITOR DC-DC CONVERTER..1555
E. Jayashree, G. Uma, M. Vaigundamoorthi

Active Clamp Interleaved Boost Converter with Coupled Inductor for High Step-up Ratio Application 1560
S. Y. Tseng, J. Z. Shiang, W. S. Jwo, C. M. Yang

Active Clamp Interleaved Flyback Converter with Single-Capacitor Turn-off Snubber for Stunning Poultry Applications...1567
S. Y. Tseng, C. T. Hsieh, H. C. Lin

Novel Current Feedforward Average Current Mode Control Technique to Improve Output Dynamic Performance of DC-DC Converters 1575
P. Chrin, C. Bunlaksananusorn

Stability Analysis of Cascaded DC-DC Power Electronic System...1581
M. Veerachary, S. Bala Sudhakar

Averaged Switch Modeling of DC/DC Converters using New Switch Network.......................................1586
Chien-Min Lee, Yen-Shin Lai

Soft Transition Operation of UPS in High- Power-Factor Mode of Three-Phase Front- End Rectifier......1590
G. A. Dhomane, H. M. Suryawanshi

Specific Harmonic Power Suppression of Direct- Power-Controlled Current-Source PWM Rectifier......1595
Toshihiko Noguchi, Kohji Sano

Frequency-Controlled LCC Resonant Converter with Synchronous Rectifier 1601
Yu Ma, Xiaogao Xie, Zhaoming Qian

Selection of the Filter Capacitor for Power Supplies using 1-Phase Diode Rectifier...............................1605
N. Mondal, S. K. Biswas, S. Sinha, N. K. Deb

Table of Contents

High Performance Single-Phase Voltage Regulator with a Simple Circuit Topology 1610
Chien-Ming Wang, Ching-Hung Su, Chang-Hua Lin, Maw-Yang Liu, Kuo-Lun Fang

Small-Signal Modeling of Series Resonant Converter 1615
Weiping Zhang, Peng Mao, Yuanchao Liu

Modelling of Three phase Z-Source Boost Buck Rectifiers 1620
D M Vilathgamuwa, P C Loh, K Karunakar

A NEW SINGLE-PHASE CONTROLLED RECTIFIER USING SINGLE-PHASE MATRIX CONVERTER WITH REGENERATIVE CAPABILITIES 1626
R. Baharom, M.K. Hamzah, A. Saparon, S.Z. Mohammad Noor, N.R.Hamzah

Implementation of Space Vector Modulated 3. to 3 . Matrix Converter Fed Induction Motor 1632
S. Ganesh Kumar, S. Siva Sankar, S. Krishna Kumar, G. Uma

A Single-Phase High-Power-Factor Neutral-pointer Clamped Multilevel Rectifier 1636
Yun Xu, Yunping Zou, Chengzhi Wang, Wei Chen, Bangyin Liu

Two Phase Inverter Drive of Three Phase Motor 1641
Saksit Jangjaempradit, Masayuki Morimoto

Predictive Current Controller for Inverter Fed Medium Voltage Drives with LC Filter 1645
T. Laczynski, A. Mertens

Novel Control Strategy of Instantaneous Power Based CVCF Inverter 1651
Akira Sato, Toshihiko Noguchi

An Improved Parallel Processing UPS Using a Voltage-Controlled Voltage Source Inverter 1657
S.W. Lee, H. Dehbonei, S.H. Ko, S.R. Lee, B.H. Jang, Y.H. Moon, T.K. Ko

A PEMFC/Battery Hybrid UPS System for Backup and Emergency Power Applications 1662
Yuedong Zhan, Jianguo Zhu, Youguang Guo, Hua Wang

Design of the Two Parallel Inverter Modules by Circular Chain Control Technique 1667
K. Piboonwattanakit, W. Khan-Ngern

Investigation of Topologies of Low Voltage Multilevel Inverters 1672
Yanli Zhang, Wanmin Fei, Shoufang Wang

Solution for PWM converter switching for Voltage Source Inverter using Non- Traditional Method 1677
V. Jegathesan, Jovitha Jerome

Piecewise Linear Control Surface for Single Input Nonlinear PI-Fuzzy Controller 1682
S. M. Ayob, Z. Salam, N. A. Azli

Open-Loop Control of a Stepping Motor through IP Network 1686
K. Matsuo, T. Miura, T. Taniguchi

Fuzzy Logic Controller for Electric Vehicle Braking Strategy.....**Fig 4. adjusted due to text re-flow 1691
Xixi. Wang, K.W.Eric Cheng, Xiaozhong Liao, Norbert C. Cheung, Lei Dong

Skid Steering in 4-Wheel-Drive Electric Vehicle 1697
Gao Shuang, Norbert C. Cheung, K. W. Eric Cheng, Dong Lei, Liao Xiaozhong

A Flexible Multi-Pulse Control Strategy for Universal Nail Collator 1703
Chien-Lung Cheng, Shyi-Ching Chern, Jim-Chwen Yeh, Ming-Yi Wu

Cycloconverter Based Three Phase Induction Motor to Replace Flywheel of the Process Machine 1708
M.V. Palandurkar, M. A. Chaudhari, J. P. Modak, S. G. Tarnekar

A Novel Zero-Voltage-Switching Single-Stage High-Power-Factor Electronic Ballast 1712
Chien-Ming Wang, Ching-Hung Su, Chang-Hua Lin, Maw-Yang Liu, Kuo-Lun Fang

Opto-Mechatronic System Design of the LED Projector by Using Brushless DC Motor 1717
Jian-Long Kuo, Tzu-Hsuan Fang

Table of Contents

The Color Measurement System of PWM-Controlled LCD by Using Back-Propagation Neural Network.................1722
Jian-Long Kuo, Xian-Lin Liu

Gapped Air-cored Power Converter for Intelligent Clothing Power Transfer.........................1727
Y. Lu, K.W.E.Cheng, Y. L. Kwok, K. W. Kwok, K.W. Chan, N.C.Cheung

Simulation Program for Switching Converters Using Numerical Fourier Transform.....................1734
Yoshihiro Tomihisa, Hirotaka Nakanishi, Terukazu Sato, Takashi Nabeshima, Kimihiro Nishijima, Tadao Nakano

The Most Suitable Application of SiC Diode.................1740
Tomoaki Makino, Atsushi Hirota, Satoshi Nagai

Multi-Domain System Simulation and Rapid Prototyping of Digital Control Algorithms using VHDL-AMS.................1744
P.J. Randewijk

Reforming Power Electronics Laboratory.................1752
Xiaohan Guan, Weiping Zhang, Xusen Zhao, Yuanchao Liu

Online performance monitoring and testing of electrical equipment using Virtual Instrumentation.................1757
S.S. Murthy, Raghu K. Mittal, Avneesh Dwivedi, G. Pavitra, Sonika Choudhary

A Balancing Strategy and Implementation of Current Equalizer for High Power LED Backlighting.................1762
Chang-Hua Lin, Tsung-You Hung, Chien-Ming Wang, Kai-Jun Pai

Modeling of the Parasitical Capacitance Effect in LCD Panel and Corresponding Elimination Strategy.................1767
Chang-Hua Lin, Tsung-You Hung, Chien-Ming Wang, Kai-Jun Pai

On-line SOC Estimation of Battery for Wireless Tram Car.................1773
Hiroyuki Miyamoto, Masayuki Morimoto, Katsuaki Morita

Narrow- control-bandwidth Operation of Piezoelectric-transformer Converter.................1777
Weiping Zhang, Xiaoqiang Zhang, Yuzhou Lei, Yuanchao Liu

Modified Map of Variable Active Passive Reactance for Stability Evaluation with Consideration of Capacitor Mode.................1782
S. Mohammad Shariatmadar, Jalal Nazarzadeh

Design of the Longitudinal Mode Piezoelectric Transformer.................1788
Shine-Tzong Ho

The Comparison of Conducted EMI Emission and Electrical Performances of Lamps.................1794
C. Uyaisom, W. Khan-Ngern

Neural Identification of Average Model of STATCOM using DNN and MLP.................1799
M. Tavakoli Bina, S. Rahimzadeh

Hybrid Simulation of Power Systems with Dynamic Phasor SVC Transient Model.................1804
E. Zhijun, K. W. Chan, D. Z. Fang

CONTROL OF CURRENT- SOURCE ACTIVE POWER FILTER USING UNIT VECTOR TEMPLATE IN THREE PHASE FOUR WIRE UNBALNCED SYSTEM.................1810
K. Vadirajacharya, Pramod Agarwal, H.O. Gupta

Improved Control of Three Phase Active Filters Using Genetic Algorithms.................1816
Bhim Singh, Varun Singhal

A Fuzzy Adaptive Detecting Approach of Harmonic Currents for Active Power Filter.................1822
Yilong Qu, Weipu Tan, Yihan Yang

Comparative Evaluation of Harmonic Extraction Techniques for Three-Phase Three-Wire Active Power Filter.................1827
R. Chudamani, Krishna Vasudevan, C.S. Ramalingam

Table of Contents

Hybrid Passive Filter Design for Distribution Systems with Adjustable Speed Drives 1834
M.A.S. Masoum, A. Ulinuha, S. Islam, K. Tan

A Graphic User Interface-based Program for Voltage Sag Calculation 1840
T. Tayjasanant, K. Yossombut, P. Sawatpipat

Operational Characteristics of Fault Current Limiting Reactor Combined with Multi- Functional Inverter 1846
S. H. Ko, S. H. Lim, S. R. Lee, S. W. Lee, I. C. Kim, S. H. Ko, H. S. Kim

Low Cost AC Solid State Circuit Breaker 1851
W. Pusorn, W. Srisongkram, W. Subsingha, S. Deng-Em, P. N. Boonchiam

A Variable Gain Control Scheme of Digital Automatic Voltage Regulator for AC Generator 1857
Dong-Hee Lee, Jin-Woo Ahn, Tae-Won Chun

A Graphic User Interface-based Program for Harmonic Impedance Calculation 1862
T. Tayjasanant

The analysis and simulation of power circuits for AC high-voltage converters 1868
Y.Y. Skorokhod, S.I. Volskiy

A Single Stage Flyback PFC Converter for Testing Distance Relay Systems 1875
V. Fernao Pires, J. F. Martins, J. Fernando Silva

H-Infinity Control Theory Apply to New Type Arc-suppression Coil System 1880
Yilong Qu, Weipu Tan, Yihan Yang

Characteristics of a novel topology of a DC-AC Converter for Fuel Cells 1885
K. Fukushima, T. Ninomiya, I. Norigoe, Y. Harada, K. Tsukakoshi, Z. Dai

A Comparative Study of PWM Schemes for Grid Connected PV Cell 1891
Vineeta Agarwal, Alok Vishwakarma

This page intentionally left blank.

2007 International Conference on Power Electronics and Drive Systems

Pages 480-948

HCC PWM Control of the Single-Phase Bi-Directional Buck Converter giving IEEE 519 Compliance at any Power Factor

A. N. Arvindan*, V. K. Sharma**

* Department of Electrical and Electronics Engineering, S. S. N. College of Engineering, Kalavakkam – 603110, India
** Electrical Engineering Department, F/O Engineering & Technology, Jamia Millia Islamia, New Delhi – 110025, India

Abstract–This paper presents unity, lagging and leading power factor operations, with near sinusoidal line currents, of the bi-directional buck type improved power quality ac-dc converter (IPQC) that employs power MOSFET embedded two-quadrant switches (2QSWs). The topology of the IPQC itself is capable of providing variable bi-directional dc link voltage and non-reversible current i.e. two quadrant operation, however, application of the hysteresis-band current control (HCC) pulse width modulation (PWM) technique renders operation of the converter in either quadrant, with a harmonic profile at the utility interface conforming to the IEEE 519 stipulations, by appropriately modifying the switching pattern of the 2QSWs. The control strategy is evaluated for the IPQC by simulations implemented in the single-phase topology of the converter. Results confirm that the technique confers IEEE 519 compliance on the converter with regard to the Total Harmonic Distortion (THD), and, are presented for unity, and, leading and lagging power factor operations in the rectification and inversion modes respectively.

Index Terms--Hysteresis-Band Current Control (HCC), Improved Power Quality AC-DC converter (IPQC), Power Quality, Total Harmonic Distortion (THD).

I. INTRODUCTION

To mitigate some of the problems associated with conventional ac-dc converters and meet the contemporary stringent power quality standards [1]-[3] a new breed of ac-dc converters referred to as improved power quality ac-dc converters (IPQCs) are increasingly being used for various applications. These converters employ self-commutating devices such as MOSFETs, IGBTs, GTOs, etc. and have been classified [4] as unidirectional and bi-directional buck, boost, buck-boost, multilevel, and multipulse converters. IPQC technology has matured at a reasonable level for ac-dc conversion to feed loads ranging from fraction of kW to MW power ratings. T. Kataoka et al [5] are credited with the pioneering efforts associated with PWM control of the bi-directional buck type IPQC. The hysteresis-band current control (HCC) PWM technique though well established continues to evoke interest [6] for enhancement of power quality. The technique, referred to as delta modulation technique [7], [8] for buck converters, is effective in meeting the power quality standards and is far superior to the open loop control [9]. In this paper, direct sensing of the ac current

This work was supported by the S.S.N. Educational and Charitable Trust, Chennai – 600 004, India.

is used for simulating various operations of the converter.

Bi-directional buck converters provide a similar function as a conventional thyristor bridge converter but with improved power quality in terms of high input power factor and reduced harmonic currents at ac mains coupled with fast regulated bi-directional output voltage for reversible power flow. The buck converters provide an output dc voltage of an average value less than the amplitude of the ac line voltage in the rectification operation. In the inversion operation the average value of the ac output voltage is less than the dc link voltage.

This paper presents unity, leading and lagging power factor operations, with near sinusoidal line currents, of the single-phase bi-directional buck type improved power quality ac-dc converter that employs power MOSFET embedded two-quadrant switches (2QSWs). The topology of the IPQC is capable of providing unidirectional dc link current and variable bi-directional dc voltage i.e. two-quadrant operation, however, application of the hysteresis-band current control (HCC) based pulse width modulation (PWM) technique renders operation of the converter in either quadrant with an improved power quality at the ac interface by appropriately modifying the switching pattern of the 2QSWs. The control strategy is evaluated by simulations implemented in the single-phase topology of the converter including those pertaining to bilateral power transfer involving rectification and inversion operations. The simulations confirm that adoption of the control technique in the converter renders the harmonic profile at the ac utility interface compliant with the revised 1992, IEEE 519, stipulations with regard to the total harmonic distortion (THD).

II. TOPOLOGICAL CONSIDERATIONS

The power semiconductor switches, employed in bi-directional buck converters, are GTOs at higher power ratings and transistors at low power ratings with high switching frequency. The use of transistors viz. BJTs, MOSFETs and IGBTs necessitates a series diode with every transistor to provide the reverse voltage blocking capability. This two-device combination comprising series connection of a transistor and a diode constitutes a two-quadrant switch (2QSW) with controllable turn-on and turn-off of unidirectional current and bi-directional voltage blocking capability. The diodes have to be of the fast recovery type to ensure that the low turn-on and turn-

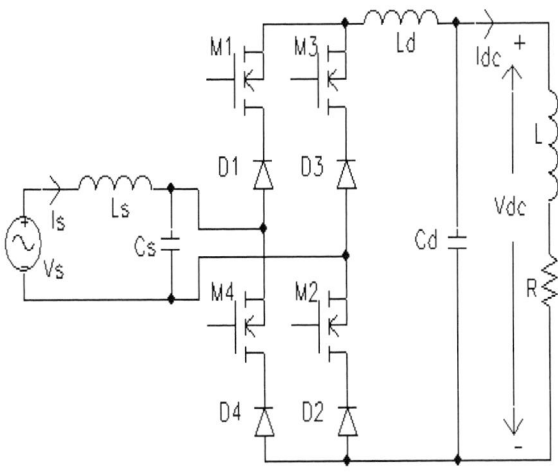

Fig. 1. Single-phase configuration of the bi-directional buck IPQC.

off times characteristic of the transistors, particularly of the MOSFETs, are not compromised. The series diodes apart from providing the bipolar blocking ability to the transistors also ensure that the relevant MOSFETs alone are effective in a particular half cycle.

The VSI bridge topology of the single-phase buck converter is shown in Fig. 1.

III. IMPLEMENTATION OF HCC PWM TECHNIQUE IN THE SINGLE-PHASE BI-DIRECTIONAL BUCK CONVERTER

Fig. 2 shows the control block diagram for the HCC PWM implementation in the single-phase bi-directional buck type IPQC. It is essentially a control technique that tracks a current reference, which is derived from a waveform template. From Fig. 2 it is evident that the current magnitude control I_m is considered to be an open loop controller. This is generally not the practice with

closed loop control associated with IPQCs because though primacy is given to power quality parameters at the ac interface of the converter the voltage regulation of the dc link is also of importance. Usually I_m is derived from the regulated voltage feedback wherein the error between the reference and dc link voltages is fed to a PID controller and then a current limiter. The feed-forward technique is another technique of I_m control. In this paper, the I_m value has been carefully selected after theoretical calculations and conducting several trials, therefore, the dc link voltage magnitude is not a problem. This also aids convergence of the simulations. The reference waveforms U_s is derived via the voltage transformer, T, and multiplied by I_m to obtain the template I_s^*. The switching patterns corresponding to operations pertaining to various power factors are generated by the hysteresis current controller based gate drive logic block that comprises logic gates supplemented by pulse generators apart from the comparators and relays. The block provides appropriate signals to the gate drive circuits (GD) of the two MOSFET sets- set1, comprising MOSFETs M1 and M2, and set2, comprising MOSFETs M3 and M4. The ON-OFF switching action of the MOSFETs in a set makes the line current I_s, which is measured by current sensor (CS), follow the template I_s^* within a hysteresis band of width h.

The HCC PWM technique implementation for the inversion operation in the single-phase bi-directional buck topology with a non-active ac load is shown in Fig. 3.

The conditions for switching the devices are:

Set1 MOSFETs M1 and M2 are on: $(I_s^* - i_s) > h/2$
Set2 MOSFETs M3 and M4 are on: $(I_s^* - i_s) < h/2$

The actual current is thus forced to track the sine reference wave within the hysteresis band by back- and-

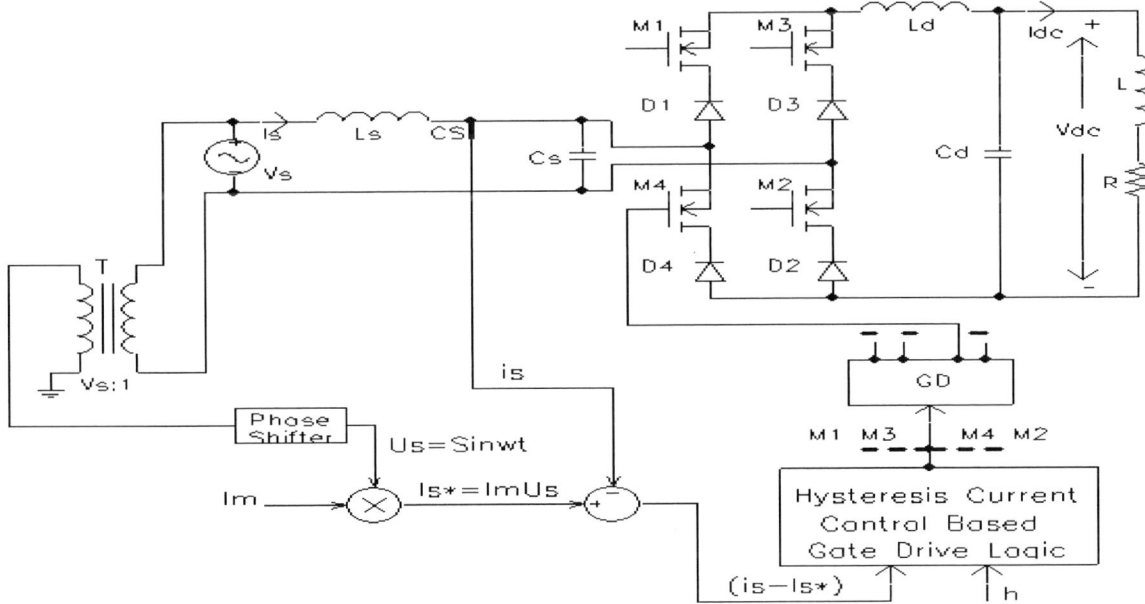

Fig. 2. Schematic of HCC PWM control in the single-phase bi-directional buck IPQC with power MOSFET embedded discrete 2QSWs.

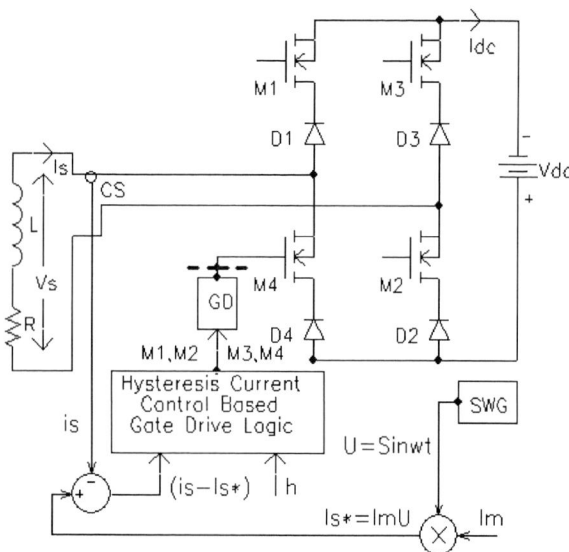

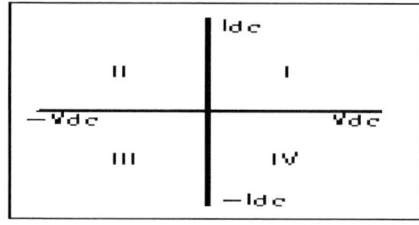

Fig. 3. Schematic of HCC PWM in bi-directional buck IPQC for inversion operation with a non-active ac load.

forth switching of the two pairs of MOSFETs. The inverter thus essentially becomes a current source with peak-to-peak current ripple, which is controlled within the hysteresis band irrespective of Vdc fluctuation. When the pair M1 and M2 is on, the current acquires a positive slope and when the other pair comprising M3 and M4 is on, it accquires a negative slope.

IV. CONVERTER SIMULATION

Simulations are conducted to obtain bilateral power flow i.e. rectification and inversion corresponding to quadrants I and II respectively, with sinusoidal line currents, at unity, and, leading and lagging power factor operations. The four quadrants with respect to the dc link current (I_{dc}) and voltage (V_{dc}) are indicated in Fig. 4.

Fig. 4. V_{dc}-I_{dc} plane with the quadrants demarcated.

The simulations are conducted with ideal a.c. voltage sources and batteries as dc voltage sources to facilitate simulation convergence. In the simulations, the line capacitors usually provided for filtering and improving the power factor, and dc link LC filters are not included as, primacy is given to the investigation of the efficacy of the control technique alone to provide a high power quality interface with the utility. A hysteresis band (h) of 0.25/0.5A width has been used in the simulations.
Peak value of the single-phase supply is 150V.
THD = Total harmonic distortion in the phase current(s).
R = Load resistance; L = Load inductance

V_{dc} = External dc voltage with reversed polarity (for II quadrant operation)
I_m = Peak value of reference current.
Q = Quadrant of operation.

The simulation data are given in Table I. The first column provides the OpCode that uses **ru** and **iu**, **d** and **g**, **90**, **30** and **45**, and, **ina**, to indicate rectification and inversion at unity power factor (UPF), leading and lagging power factors, 90°, 30° and 45° phase angles, and, inversion with non-active load, respectively.

TABLE I
SIMULATION DATA

Op Code	R Ohm	L mH	Vdc Volt	Im A	h A	Q	THD %
iu	8.75	0.1	-350	15.5	0.25	II	4.52
ru	8.75	0.1	-	15.5	0.5	I	2.62
ru	8.75	0.1	-	15.5	0.5	III	2.50
iu	8.75	0.1	350	15.5	0.25	IV	4.57
d90	8.75	0.1	-450	45.0	0.5	II	2.891
d30	8.75	0.1	-450	45.0	0.5	II	4.03
d45	8.75	0.1	450	45.0	0.5	IV	3.78
g90	8.75	0.1	-450	45.0	0.5	II	2.924
g30	8.75	0.1	-450	45.0	0.5	II	3.802
g45	8.75	0.1	450	45.0	0.5	IV	3.849
ina	12.0	0.025	-600	45.0	0.25	IV	4.359

V. RESULTS AND DISCUSSION

From Table I it is clear that the HCC PWM technique is able to limit the THD for operations in the two quadrants, at any power factor, to the permissible limit of 5% stipulated by [1].

A. Unity Power Factor Operation

The waveforms pertaining to the II-quadrant inversion operation of the converter, at unity power factor (UPF), are shown in Fig. 5. The THD of the ac side current, I$_s$, in

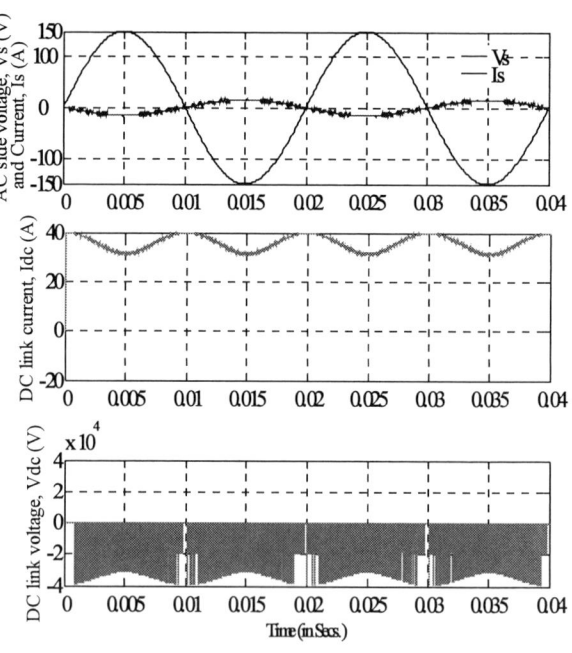

Fig. 5. Waveforms: II-quadrant inversion operation at UPF.

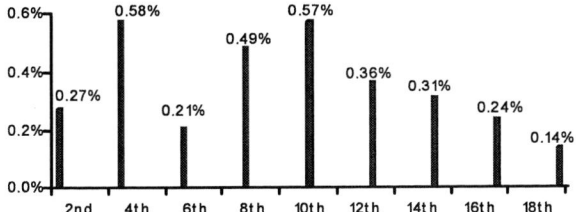

Fig. 6. Even harmonics by FFT analysis of ac side current I_s in Fig.5.

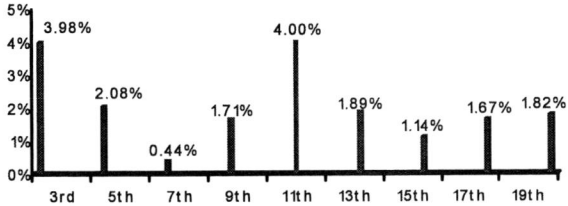

Fig.7. Odd harmonics by FFT analysis of ac side current I_s in Fig. 5.

Fig. 5 is determined by the fast fourier transform (FFT) analysis to be 4.52%. The magnitudes of the even and odd harmonics of I_s are shown in Figs. 6 and 7 respectively. It is clear that the even harmonics <17 and the 18th are much less than the limits, stipulated by IEEE 519, 1992, i.e. 1% and 0.375% respectively. The odd ones<17 are restricted to 4%, however, those of higher order i.e.17<order<23 exceed the 1.5% limit.

Fig. 8 shows the waveforms pertaining to I-quadrant operation that corresponds to rectification, at UPF. The THD of the ac side current, I_s, for the operation is 2.62%. It is clear that the ac side voltage, V_s, and current, I_s, waveforms are in phase, thus, in this operation, the converter presents itself as a resistor to the ac source. The dc link voltage, V_{dc}, and the dc link current, I_{dc}, are positive. Therefore, the dc link power is positive which is indicative of power flow from the ac side to the dc link. It

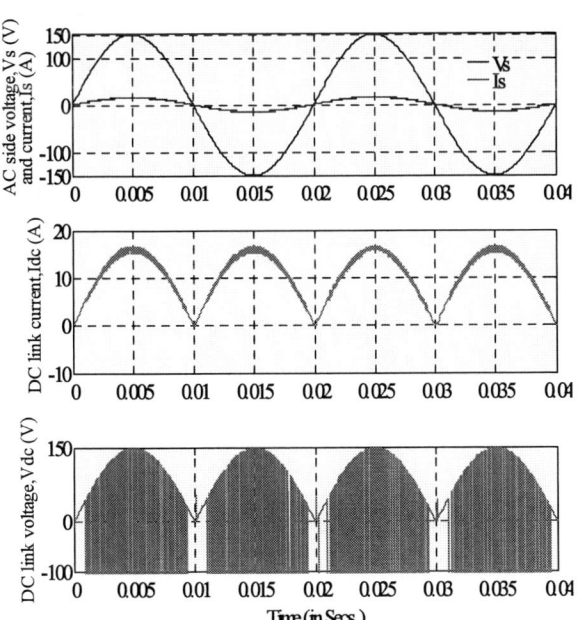

Fig. 8. Waveforms: I-quadrant rectification operation at UPF.

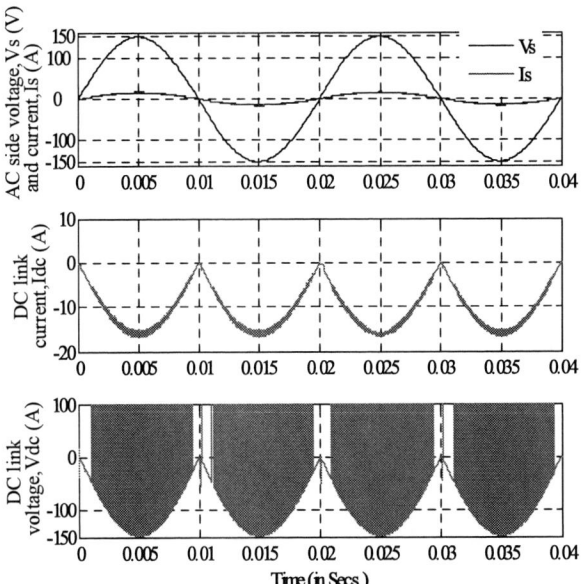

Fig. 9. Waveforms: III-quadrant rectification operation at UPF.

is significant that almost unity power factor with acceptable THD is achieved for bi-directional power flow between the ac source and the dc link i.e. inversion and rectification operations despite the presence of an inductor on the dc side. This is certainly not possible with the conventional phase-controlled converters.

In Fig. 1, if the devices i.e. the power MOSFETs and diodes are oriented in the opposite direction then the direction of the dc link current reverses for bi-directional dc link polarity. Thus, operations in the III and IV quadrants corresponding to rectification and inversion respectively become feasible. The waveforms for the rectification operation in the III quadrant at UPF are shown in Fig. 9. The positive dc link power (negative dc link voltage and negative dc link current) is indicative of power flow from the ac side to the dc link.

The waveforms for IV-quadrant inversion at UPF are shown in Fig. 10. The negative dc link power (positive dc link voltage and negative dc link current) is indicative of power flow from the dc link to the ac side.

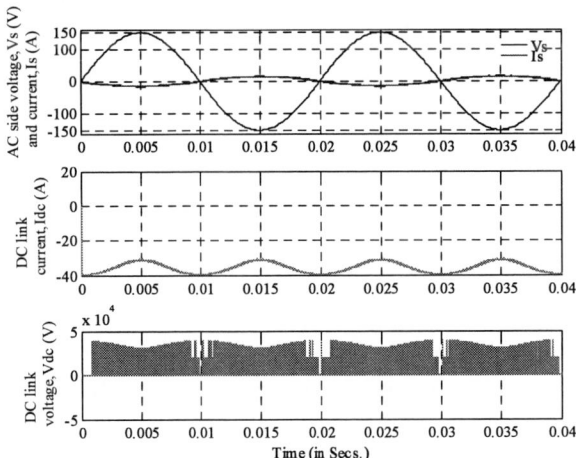

Fig. 10. Waveforms: IV-quadrant inversion operation at UPF.

B. Leading Power Factor Operation

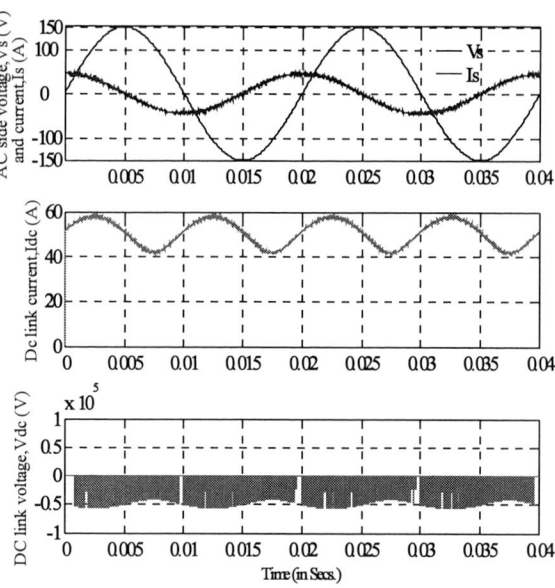

Fig. 11. Waveforms: 90° (lead) phase shift, II quadrant operation.

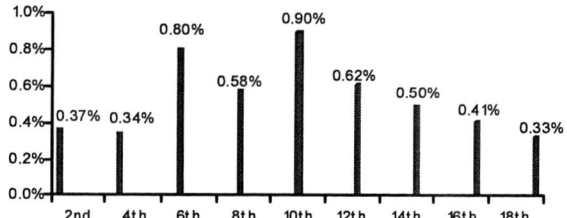

Fig. 13. Even harmonics by FFT analysis of ac side current I_s in Fig.12.

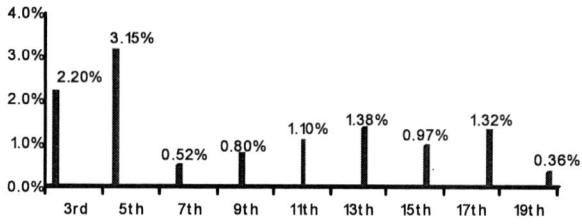

Fig.14. Odd harmonics by FFT analysis of ac side current I_s in Fig. 12.

Inversion operation at any power factor is possible in the II-quadrant; however, connection of an external dc source with reversed polarity across the dc link, is a prerequisite. The waveforms for 90° (lead) phase shift between the ac side voltage and current, corresponding to zero power factor (ZPF) (lead), for inversion operation are shown in Fig. 11. The THD of the ac side current, I_s, in this case is 2.891%. In the operation, from the ac side perspective the converter functions as a capacitor delivering lagging VARs i.e. a lagging VAR generator. The negative dc link power (negative dc link voltage and positive dc link current) is indicative of power flow from the dc link to the ac side.

The waveforms for 30° (lead) phase shift between the ac voltage and current, corresponding to power factor of 0.866 (lead), for inversion operation are shown in Fig. 12. The ac side current has a THD of 4.03% for the operation.

The magnitudes of the even harmonics of the current are shown in Fig. 13. It is clear that the even harmonics of order less than and equal to 17, and the 18th are limited to values less than the IEEE 519, 1992 limits of 1% and 0.375% respectively. The magnitudes of the odd harmonics of the current are shown in Fig. 14. From the figure it is evident that the odd harmonics till the 17th order are restricted to within the permissible 4% limit. In fact, the 19th harmonic is also well within the 1.5% limit.

As pointed out earlier, inversion operation in the IV-quadrant is possible if the power MOSFETs and diodes are oriented in the opposite direction. The external dc source with appropriate polarity is also required. The waveforms for inversion in the IV quadrant corresponding to a 45° (lead) phase shift between the ac side voltage and current, thus giving a power factor of 0.7071 (lead) are shown in Fig. 15.The THD of the ac side current for the operation is determined to be 3.78%. The dc link power is characteristically negative.

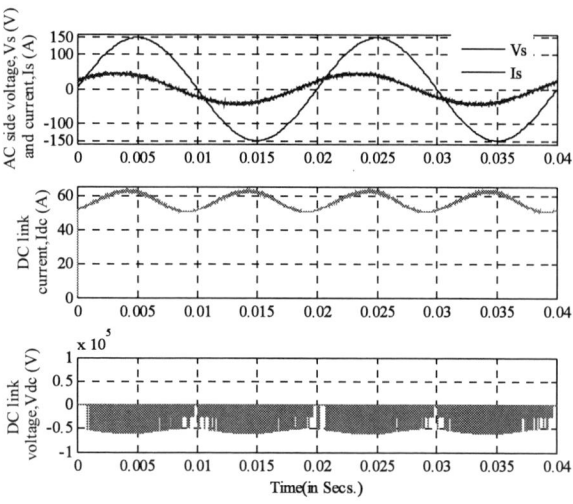

Fig. 12. Waveforms: 30° (lead) phase shift, II quadrant operation.

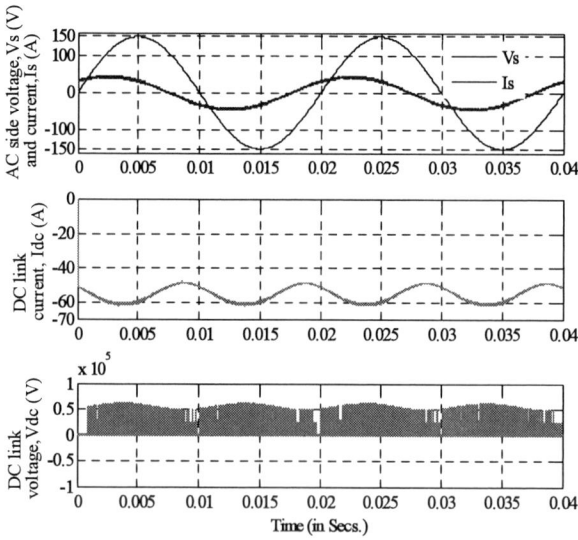

Fig. 15. Waveforms: 45° (lead) phase shift, IV quadrant operation.

C. Lagging Power Factor Operation

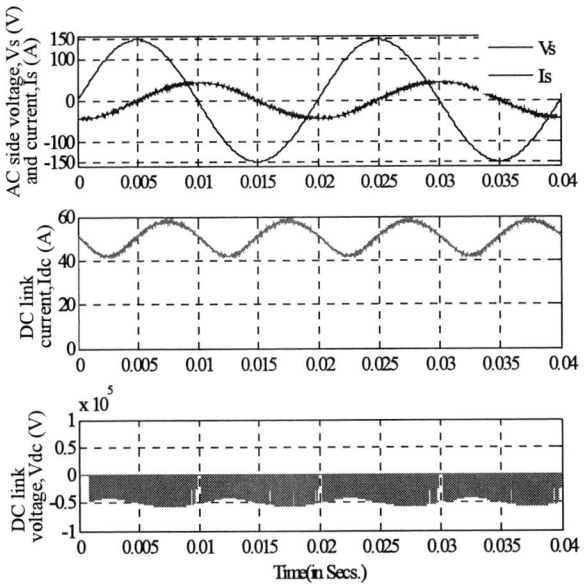

Fig. 16. Waveforms: 90° (lag) phase shift, II quadrant operation at ZPF.

The waveforms pertaining to 90° (lag) phase shift between the ac side voltage and current, corresponding to zero power factor (ZPF) (lag), for inversion operation are shown in Fig. 16. The negative dc link power (negative dc link voltage and positive dc link current) is indicative of power flow from the dc link to the ac side. The converter functions as a pure inductor from the ac side perspective. The THD of the ac side current for the operation is estimated to be 2.924%. The operations pertaining to ZPF (lag) and ZPF (lead) are important from the flexible alternating current transmission systems (FACTS) viewpoint as they could be used to control and regulate power transmission.

In Fig. 17 the waveforms relevant to II-quadrant operation with a 30° (lag) phase shift between the ac side

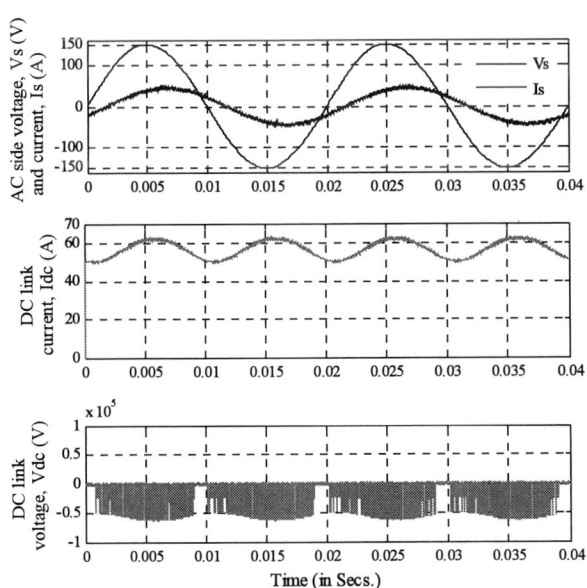

Fig. 17. Waveforms: 30° (lag) phase shift, II quadrant operation.

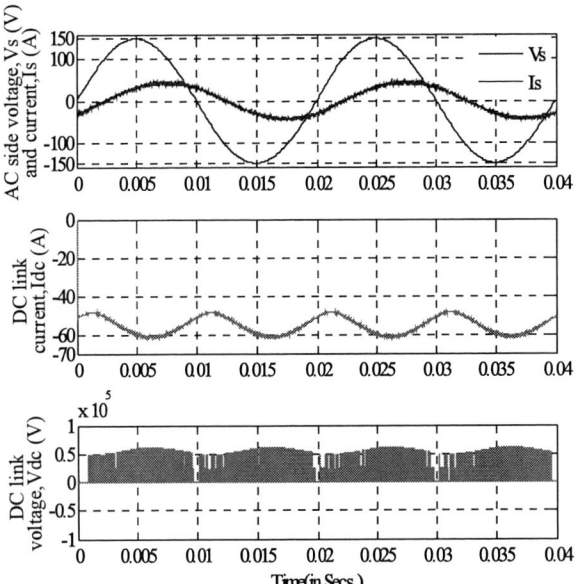

Fig. 18. Waveforms: 45° (lag) phase shift, IV quadrant operation.

voltage and current, corresponding to a power factor of 0.866 (lag), are shown. The THD of the ac side current for the operation is 3.802%.

The waveforms for inversion in the IV quadrant corresponding to a 45° (lag) phase shift between the ac side voltage and current, thus giving a power factor of 0.7071 (lag) are shown in Fig. 18. The ac side current has a THD of 3.849%.

The magnitudes of the even harmonics of the ac side current, I_s, in Fig. 18 are shown in Fig. 19. It is evident that the even harmonics are well within the limits stipulated by IEEE 519, 1992. The magnitudes of the odd harmonics of I_s are shown in Fig. 20. The odd harmonics of order less than 17 are restricted to levels well below the 4%, however, those of the 17th exceeds the 1.5% limit. The harmonic components of the ac side current till the 19th order have been considered in the result analyses.

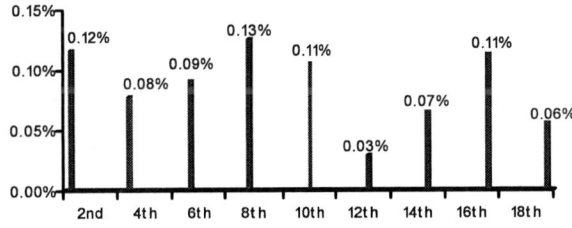

Fig. 19. Even harmonics by FFT analysis of ac side current I_s in Fig.18.

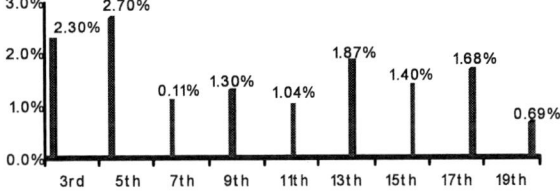

Fig. 20. Odd harmonics by FFT analysis of ac side current I_s in Fig. 18.

D. Inversion Operation With Non-Active AC Load

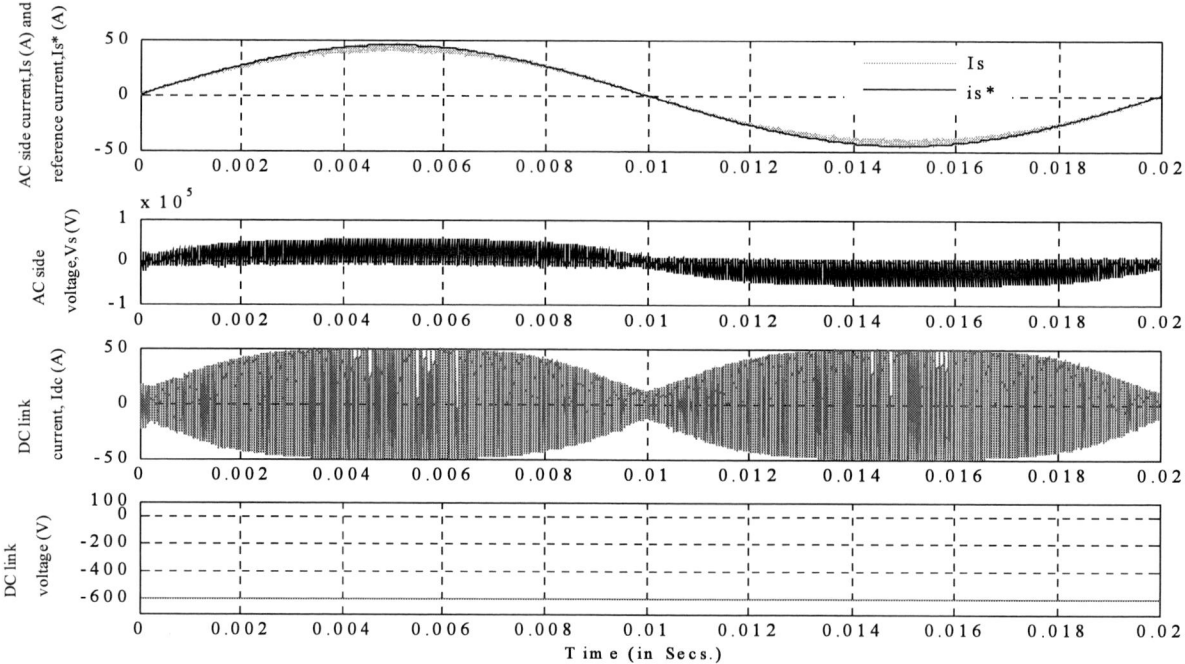

Fig. 21. Waveforms pertaining to inversion operation of the single-phase bi-directional buck IPQC with a non-active ac load.

Fig. 21 depicts the waveforms for inversion operation with a non-active ac load. The THD of the ac side current, I_s, in case is 4.359%. The magnitude of its fundamental component is 42.4A, and those of its odd and even harmonic constituents are shown in Figs. 22 and 23 respectively.

Fig. 24 shows the panned view of the currents in which the actual current is seen tracking its reference waveform. This illustrates the efficacy of the HCC PWM method with regard to bi-directional buck converters.

From Fig. 22 it is clear that the magnitudes of the odd harmonics are very small. The magnitudes of the even harmonics are much smaller. In fact their magnitudes are theoretically supposed to be zero because of the symmetrical nature of the switching in both half cycles of the current waveform. Practically the values are non-zero, however, the fast fourier transform (FFT) analysis of the

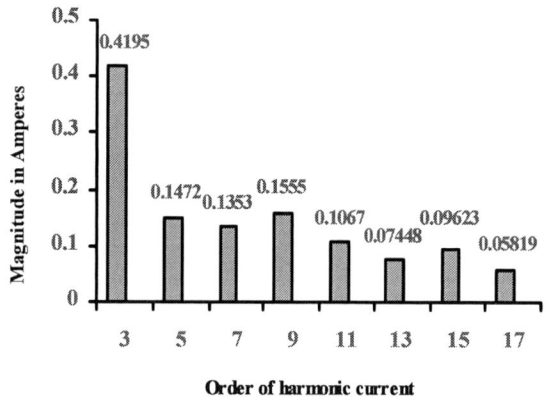

Fig. 22. Magnitudes of the odd harmonics of the current, I_s, in Fig. 21.

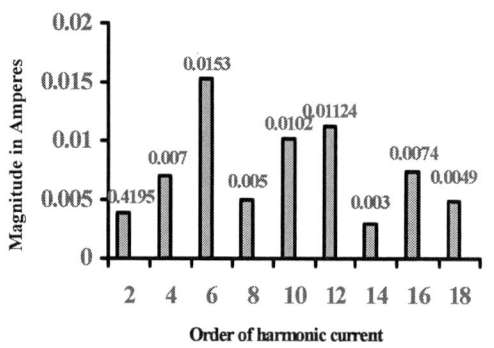

Fig. 23. Magnitudes of the even harmonics of the current, I_s, in Fig. 21.

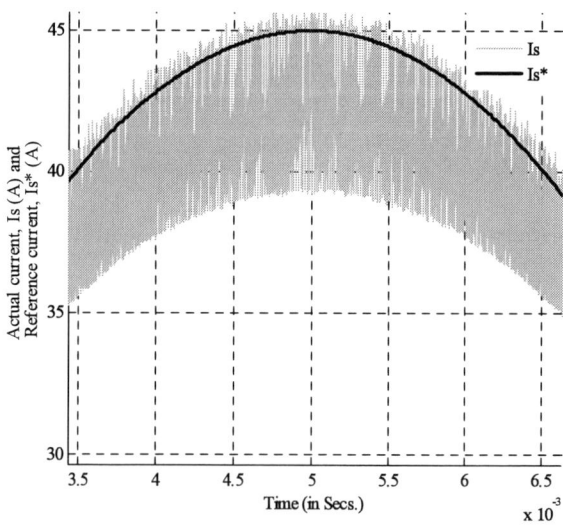

Fig. 24 Zoomed view of the actual and reference currents.

486

waveform reveals that their magnitudes are infinitesimal and, therefore, they can be considered inconsequential.

The magnitudes of the even harmonics are shown in Fig. 23. Their magnitudes are clearly very low relative to those of the odd harmonics. From Figs. 21, 22 and 23 and the above discussion it is clear that the actual current has almost unity displacement factor, low THD, and therefore, is near sinusoidal.

The output ac voltage is shown in Fig. 25. It is evident that the switching frequency of the devices is very high. This is because of the inherent characteristic of the HCC PWM strategy and the small value of h, the hysteresis band. A hysteresis band, h = 0.25A, which is only 0.55% of the peak value I_m of the reference sinusoidal current, is used.

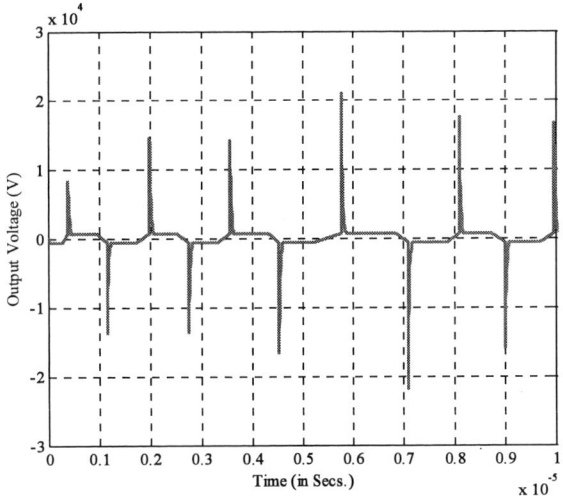

Fig. 26 Zoomed view of the output ac voltage waveform.

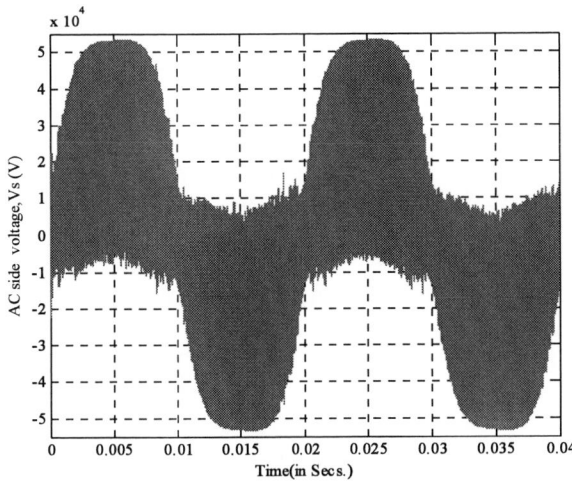

Fig. 25 Output voltage for inversion with non-active ac load.

The panned view of the output PWM ac voltage, obtained on a time base of 10μs, is shown in Fig. 26. This enables estimation of the switching frequency. From the figure the switching frequency is determined to be 0.5MHz approximately from the 0.2μs time interval.

VI. CONCLUSIONS

The HCC PWM technique is capable of conferring IEEE 519, 1992 compliant performance on the power MOSFET based single-phase bi-directional buck converter with regard to the THD for unity, leading and lagging power factor operations. Simulation based analyses reveal that it renders individual even harmonic currents within the stipulations of the standard; however, further refinement is required in it for limiting the higher order odd harmonics.

REFERENCES

[1] IEEE Recommended Practices and Requirements for Harmonics Control in Electric Power Systems, IEEE Std. 519, 1992.

[2] Electromagnetic Compatibility (EMC)—Part 3: Limits—Section 2: Limits for Harmonic Current Emissions (Equipment Input Current (16A per Phase), IEC1000-3-2, Dec., 1995.

[3] Draft-Revision of Publication IEC 555-2: Harmonics, Equipment for Connection to the Public Low Voltage Supply System, IEC SC 77A, 1990.

[4] Bhim Singh, B.N. Singh, A. Chandra, Kamal Al-Haddad, Ashish Pandey, and D.P. Kothari, "A Review of Single-Phase Improved Power Quality AC-DC Converters," *IEEE Trans. Ind. Electron.*, vol. 50, No. 5, pp. 962-981, October 2003.

[5] T. Kataoka, K. Mizumachi, and S. Miyairi, "A pulsewidth controlled AC-to-DC converter to improve power factor and waveform of AC line current," *IEEE Trans. Ind.Applicat.*, vol. IA-15, pp. 670-675, Nov./Dec.1979.

[6] Ali I. Maswood and F. Liu,"A unity power factor front-end rectifier with hysteresis current control," *IEEE Trans. Energy Conversion*, vol. 21, No.1, pp.69-76, Mar. 2006.

[7] R. Oruganti and M. Palaniapan,"Extension of inductor voltage control to three-phase buck-type ac-dc converter," *IEEE Trans. Power Electron.*, vol. 15, No.2, pp.295-302, Mar. 2000.

[8] B.M.M. Mwinyiwiwa, P.M. Birks, and B.T. Ooi, "Delta-modulated buck-type PWM converter," *IEEE Trans. Ind. Applicat.*, vol. 28, No.3, pp.552-557May/June 1992,.

[9] A.N. Arvindan and V.K. Sharma, "Modeling, Simulation and Performance Analysis of Improved Power Quality AC-DC Converter," *Proc. of IEE International Conference on Energy, Information Technology and Power Sector, (PEITSICON 2005)*, Kolkata (India), Jan. 2005, pp. 529-533.

Passive EMI Filter Performance Improvements with Common Mode Voltage Cancellation Technique for PWM Inverter

C. Khun*, W. Khan-ngern3* and M. Kando**

* Faculty of Engineering, ReCCIT, King Mongkut's Institute of Technology Ladkrabang, Faculty of Engineering, Bangkok, Thailand

** Electrical Department, Faculty of Engineering, Tokai University, Japan

Abstract– The common modulation strategy adopted for motor control in adjustable speed drives (ASDs) is pulse width modulation (PWM). The carrier frequency along with fast rise and fall time of the IGBTs employed results in non trivial common-mode (CM). The high dv/dt and di/dt causes shaft-voltage, which leads to bearing currents. This phenomenon has been well-known as one of the reason for premature bearing failure in PWM drive motors. To reduced EMI emission, this paper presents the experimental works that investigate the conducted emission from pulse width modulated (PWM) inverter feeding an induction motor using CM voltage cancellation technique to develop the ability of conventional passive EMI filter. The typical passive EMI filter also use in this work to compare and prove the effectiveness proposed method. The experimental results of two operation conditions: with load and without load are verified the effectiveness of proposed filters, respectively.

Index Terms— EMI Filter, CM Voltage Cancellation, PWM Inverter.

I. INTRODUCTION

The revolution of high-speed power electronic switching devices such as Transistors, MOSFETs, IGBTs are able to increase the carrier frequency of PWM inverter. It gives rise of ground current due to

♦ ground current escaping to earth through stray capacitors inside motors [1]

♦ conducted and radiated electromagnetic interference (EMI) [2].

♦ bearing current and shaft voltage.

The EMI emission generated by switch-mode power converters e.g. PWM inverter employed the most adjusted speed drives, can be transmitted to other electronic instruments or control and communication systems by means of conduction and radiation and may cause electronic systems to malfunction. It is well-known that conducted EMI emission from PWM inverter consists of the common-mode component and differential-mode (normal-mode) component. The CM noise current flows from line and neutral power wire through the stray capacitance inside the motor and other parasitic component of cables, inverter to the ground [2].

Whereas, the DM noise current flows through line and back to neutral of power main, as shown in Fig.1. Although, they should be separately consider, but the DM noise is not clearly discussed in this paper with

assumption that some appropriate DM components are installed for each design stage. So the CM noise becomes the main objective of this study.

There are many filtering techniques that have proven the effectiveness in suppressing EMI noises. However, this paper demonstrates a performance improvement of a conventional passive EMI filter using CM voltage cancellation techniques in motor drive system. So a typical passive EMI filter design is summarly discussed with CM transformer design. After that, other two CM voltage cancellation techniques are also considered and implemented within the same operated condition of motor drive system. The experiment is operated with two conditions: without load and with load. The conducted EMI noises are observed using a high frequency current probe to separate CM and DM noise emission. The analysis and experimental results are presented to verify and illustrate the effectiveness of each technique, respectively.

II. COMMON-MODE CURRENT/VOLTAGE CANCELLATION TECHNIQUES

A. Existing Techniques for CM voltage/current injection techniques

The size and performance of EMI filter components are important consideration in term of optimization. So there are many common-mode current cancellation techniques applied in PWM inverter motor drive. Yo-Chan Son and Seung-Ki Sul proposed an active CM EMI filter using two low voltage complementary transistors to inject CM current [4]. Moreover, in order to eliminate the dependence of the performance on the choice of the value of R and C, another emitter follower circuit was proposed by H. Akagi [1] Instead if passively connecting the CM potential to the auxiliary winding of the CM chock and grounding. On the other hand, I. Takahashi [3] proposed an active circuit for CM current injection of PWM inverter using an emitter follower circuit based on the principle of duality. But this suggested circuit is current injection method, opposite to the voltage cancellation of [1].

B. Proposed Circuit of CM Voltage Cancellation Techniques

978-1-4244-0644-9/07/$25.00 ©2007 IEEE

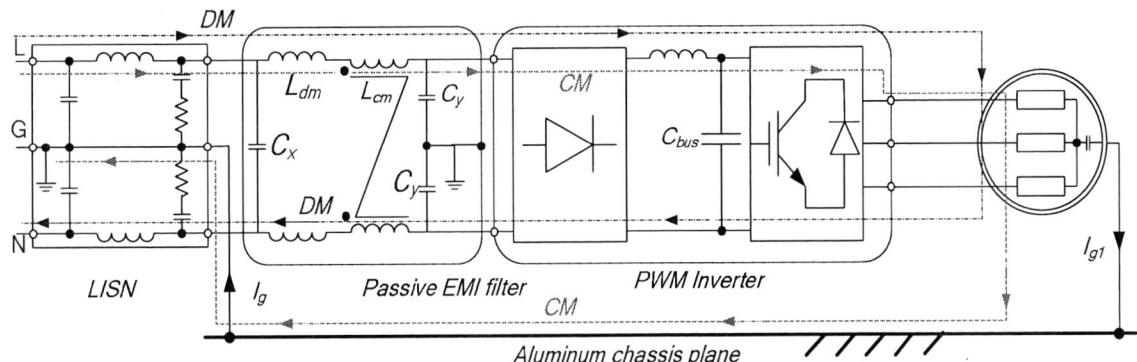

Fig. 1 Typical passive EMI filter with a PWM inverter system configuration

The proposed method in this paper is based on passive input EMI filter design using conventional CM chock (CM transformer) adding an auxiliary winding (secondary winding of CM chock) and other capacitors (C_x and C_y) and it is suitable use with different voltage levels. The proposed EMI filter circuit is shown in Fig. 2.

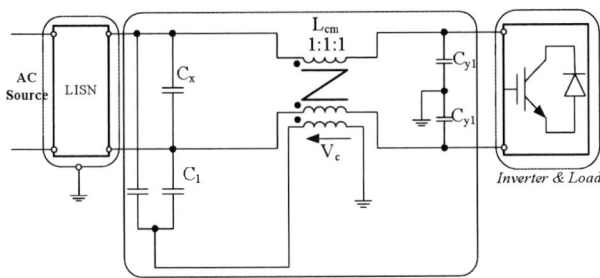

(a) Filter A: Proposed circuit.

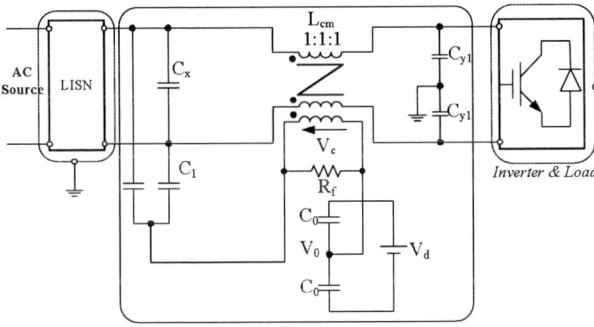

(b) Filter B: Proposed improvement circuit.

Fig.2. Proposed Filter circuits.

According to the proposed circuits as shown in Fig. 2, the design objectives of these filters have:

♦ Selection of C_1

The ceramic capacitors C_1 placed at the inverter input terminals of the inverter. Indeed, the resistors can be used to replace the capacitors C_1. However, there are not attractive because the resistors add more power losses to the system due to the flowing of the normal-mode currents.

The C_1 selection must be based on the maximum current that can be drawn from the main source to the inverter. If a large value of C_1 is chosen, the inverter power devices can be subjected to excessively high current pulse (capacitor charging current) [5]. Therefore, these capacitors should be selected as small as possible. The inverter for adjust speed drive is operated with the nominal current 6.4 A, and fed by a single-phase ac input system. Assuming that the inverter's switches are turned on within 500 ns, the maximum value of C_1 for 240 V is expressed:

$$6.4 = \frac{240}{500 \times 10^{-9}} \cdot C_{1,max} \qquad (1)$$

Hence, $C_{1,max} \approx 15$ nF

In the test setup, $C_1 = 10$ nF is selected to get high input impedance for CM voltage detection.

♦ Selection of C_y

CM capacitors C_y and C_1 are limited by safety considerations for ground leakage current that can calculate by equation (1).

$$C_y = \frac{I_{leakage}}{2\pi \times f \times 115 \times V} \qquad (2)$$

where: $I_{leakage}$ is the ground leakage current, V is ac line voltage, f is power line frequency that generally is equal to 50 or 60 Hz.

In this case, the capacitance should be restricted such that the ac leakage current safety requirements. With this reason, two 10 nF of C_{Y1} capacitors are employed to meet the 3.5 mA leakage current [2].

♦ Separated DC Power Supply

In the test setup, the separate 15 Vdc power supply is realized by a single-phase rectifier supplying the two capacitors C_0 connect in series. In order to remove the dc components, the capacitors C_0 are connected as illustrated

489

in Fig.3. The small capacitance of C_0 made the large variation of the neutral point potential V_0. Thus, C_0 is chosen as a value large enough to reduce the voltage variation. In the practical work, the capacitor of 3.3 µF is selected for C_0 [5].

♦ Common-mode transformer

In this case, the CM transformer is the same as a conventional CM choke of passive EMI filter, except for connecting a tightly coupled additional winding (auxiliary winding). The two primary windings of CM transformer with the same polarity are connected between LISN and inverter input plug. Then the polarity of the compensating voltage V_c is opposite to the CM voltage generated by the inverter. Because of the CM transformer played in role of CM choke of passive EMI filter, so the value of L_{cm} should be calculated using the system attenuation requirement. The attenuation requirement can be obtained from the CM noise measurement when the system operates with uninstalled any filter. The CM attenuation requirement can be determined using (3) [6]. CM noise spectrum and attenuation requirement is given in Fig. 3.

$$\left|V_{req,CM}\right|_{dB} = \left|V_{CM}\right|_{dB} - \left|V_{Limit}\right|_{dB} \qquad (3)$$

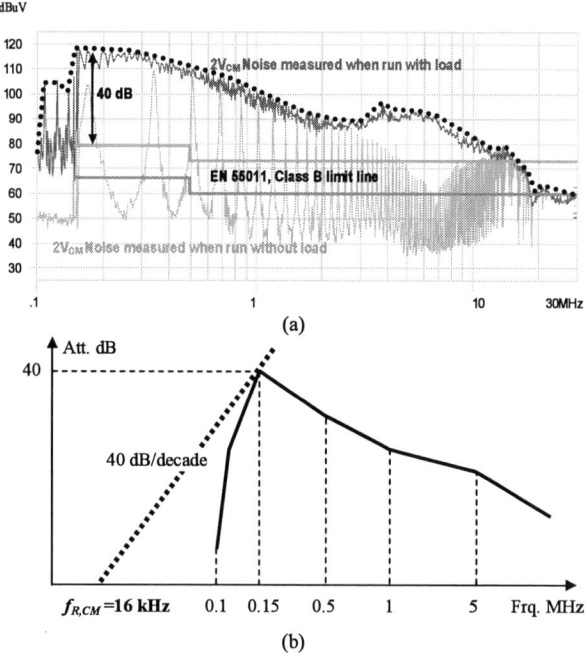

Fig. 3. (a) CM noise spectrum. (b) Attenuation Requirement.

As the combination of L_{cm} and $2C_{Y1}$ have a resonant frequency of $f_{R,CM}$ obtained in Fig. 3(b).

Therefore $\quad L_{cm} = \left(\dfrac{1}{2\pi \times f_{R,CM}} \right)^2 \cdot \dfrac{1}{2C_{Y1}}$

As $C_{Y1} = 10$ nF is selected, hence $L_{cm} \simeq 4.95$ mH. In the test setup, $L_{cm} = 4.5$ mH is selected within 1:1:1 winding ratio.

III. SYSTEM CONFIGURATION

In this work, a commercial PWM inverter motor drive feeding an induction motor is used as equipment under test (EUT). The measurement configuration is shown in Fig.4, including the following equipments:

♦ Induction motor: ½ hp, 220/380 V, 2.0/1.15 A, 50/60 Hz.

♦ A single phase commercial PWM inverter motor drive: 1.5 kW, input voltage 220 V, nominal current input of inverter is 6.4 A, switching frequency 16 kHz.

♦ Line Impedance Stabilization Network (LISN): single phase input voltage: 220 V, 10A.

♦ Grounding plane: aluminum sheet (1.2 m × 1.2 m).

♦ High bandwide current probe: 10 kHz to 250 MHz.

The LISN, PWM inverter interconnected power lines are placed on the ground plane in the configuration requirement and cases of other instruments that are used for waveform observation (spectrum analyzer, oscilloscope) are directly bolted to ground, respectively, as shown in Fig.5.

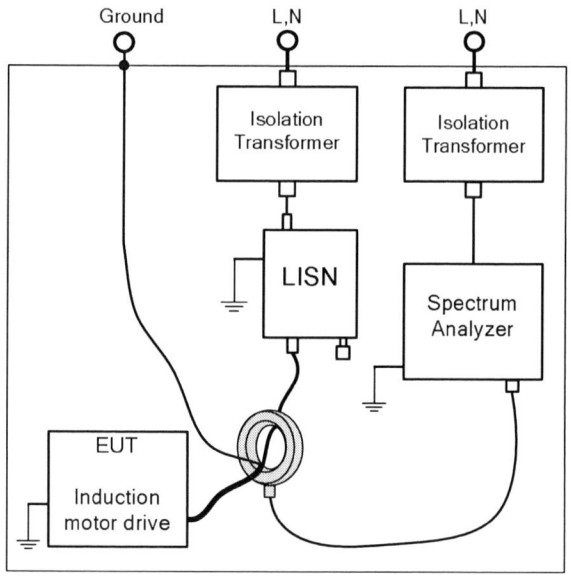

Fig. 4. Experimental Configuration Layout

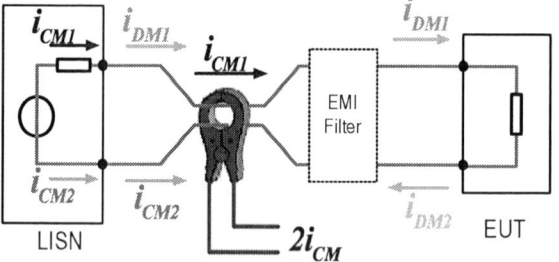

Fig.5. CM noise spectrum measurement.

490

A single-phase LISN 10 A is used to provide the stable source impedance at the high frequency. The high frequency current probe is also connected to EMC spectrum analyzer to observe CM noise spectrum as shown in Fig. 4. Otherwise, because of using current probe to measure CM noise, the received results from current probe are equal to $2I_{CM}$ [2]. The CM current/voltage attenuation can be calculated from equations (4) and (5).

$$\left| I_{CM} \right|_{dB} = \left| 2I_{CM} \right|_{dB} - 6\,\text{dB} \tag{3}$$

$$\left| V_{CM} \right|_{dB} = \left| 2V_{CM} \right|_{dB} - 6\,\text{dB} \tag{4}$$

where $\left| 2I_{CM} \right|_{dB}$, $\left| 2V_{CM} \right|_{dB}$ are CM current and voltage measured using high frequency current probe.

IV. Experimental Results and Discussion

The CM spectrum of the system without any EMI filter of both operations with load and without load, are shown in Fig. 3(a), and the conducted EMI spectrum of the both operations when the proposed circuit is installed in the system is demonstrated in Fig. 6. Furthermore, Fig. 8 present CM spectrum when the proposed improvement circuit installed.

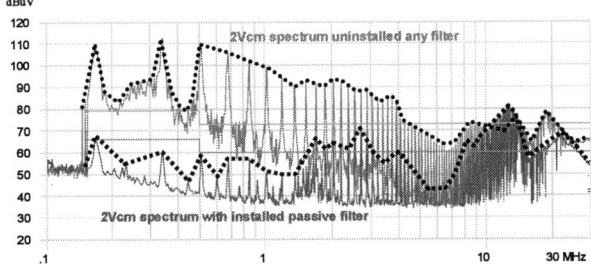

(a) CM noise spectrum when installed passive filter, run without load.

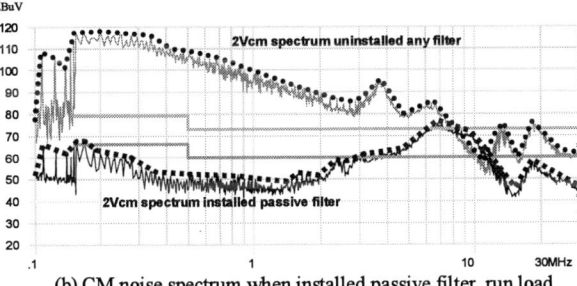

(b) CM noise spectrum when installed passive filter, run load.

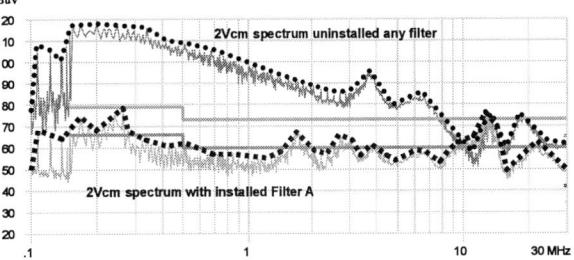

(c) CM noise spectrum when installed filter A, run without load.

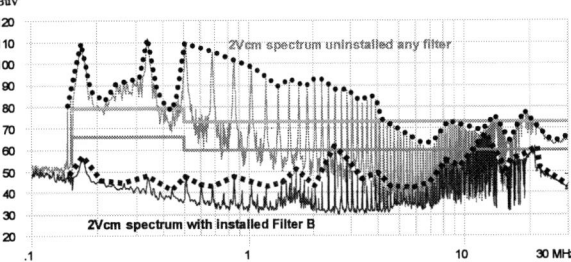

(d) CM noise spectrum when installed filter A, run with load.

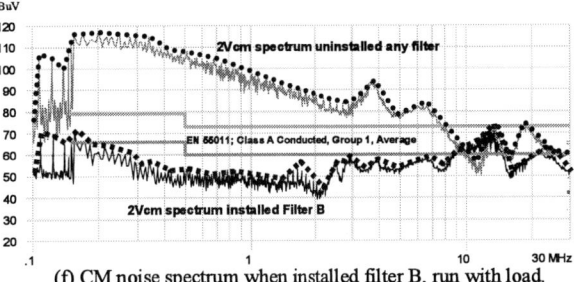

(e) CM noise spectrum when installed filter B, run without load.

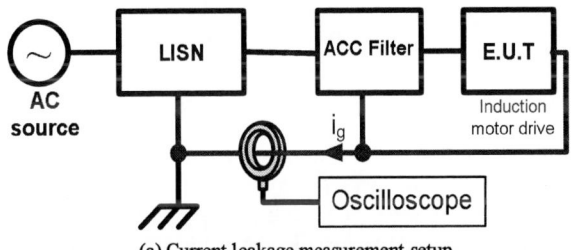

(f) CM noise spectrum when installed filter B, run with load.

Fig. 6. Experimental results of CM noise spectrums.

The leakage current, current flow through the ground cable is observed when installed filter and uninstalled filter. Fig. 7 shows the measurement test setup of the leakage current that flow through the ground cable.

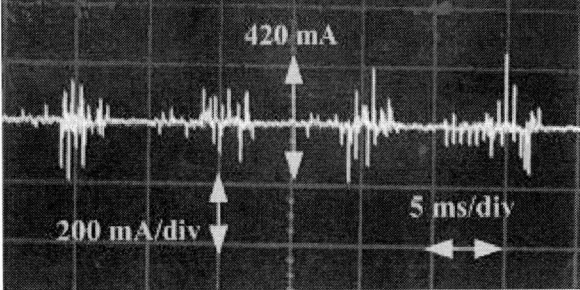

(a) Current leakage measurement setup

(b) Measured current leakage without filter, run with load.

491

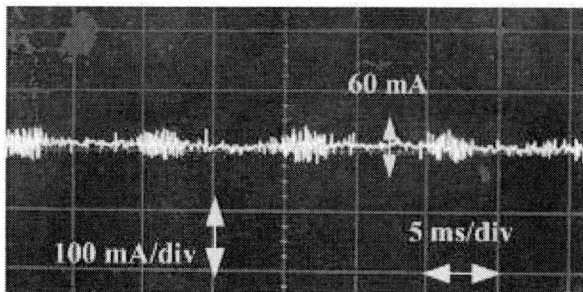

(c) Measured current leakage with proposed filter, run with load.

Fig. 7. Current leakage measurement.

The experiment is operated with three steps: the operation with normal filter, the operation with proposed circuit, Filter A and with proposed improvement circuit, Filter B. All of three condition operations are run with the same situation. According to the experimental results above, it can be analyzed:

1). First case of testing, when the passive filter is inserted into the system and operate without load, the result is given about 40 dB of the insertion loss (IL) at 150 kHz to 2 MHz, about 20 dB of IL at 2 MHz to 7 MHz and 12 MHz to 30 MHz of frequency range as shown in Fig. 6(a), respectively. But at high frequency range from 7 MHz to 12 MHz, the passive filter can attenuate very small insertion loss. Fig.6(b) shows the experimental results when the motor is run without torque, it also can results 40 dB at 150 kHz to 2 MHz and about 20 dB at 2 MHz to 7 MHz of frequency range as indicated in Fig. 6(b). The filter has no effect at high frequency range, cannot pass the limit line.

2). Second case of testing, when the proposed circuit, Filter A, is installed into the system, the IL can gain to 45 dB at 150 kHz to 4 MHz, 20 dB at 4 MHz to 10 MHz of frequency range, when the Filter A is installed into the system and operate without load as shown in Fig. 6(c). Fig. 6(d) demonstrates the experimental result when the motor is run; there is about 45 dB of IL at 150 kHz to 4 MHz, 25 dB at 4 MHz to 8 MHz and 15 MHz to 30 MHz. But the system performance still cannot pass around 12 MHz of high frequency range for the both operation, without load and with load as illustrated in Figs. 6(c) and (d).

3). Third case of testing, when the proposed improvement circuit, Filter B, is installed into the system and run without load, it gains about 50 dB of IL at 150 kHz to 4 MHz, 25 dB at 4 MHz to 10 MHz of frequency range. In case of operation with running the motor, the results are given about 60 dB of the insertion loss at 150 kHz to 2 MHz, 25 dB at 2 MHz to 9 MHz of frequency range. The system performance can pass the CISPR 11 standard limit line at whole conducted emission frequency range as presented in Figs. 6(e) and (f). Additionally, the leakage current to ground is also observed to express the filter performance. As the results that are shown in Fig. 7, the proposed improvement circuit, filter B, can reduce the leakage

current to ground or ground current from 420 mA to 60 mA of peak-to-peak values.

V. CONCLUSION

This paper presents a proposed circuit and proposed improvement circuit that are used to improve the typical passive EMI filter to mitigate CM noise in motor drive system and can comply with CISPR 11 standard. These circuits are based on the common-mode voltage cancellation technique. This proposed circuits are able suppress CM voltages generated by PWM inverter using voltage detecting and voltage compensation of secondary winding of common-mode choke. These techniques do not need any active circuits. The experimental results of two operated conditions, no load and full load of induction motor, are demonstrated the effectiveness of both proposed circuit using high frequency (HF) current probe over the frequency range 150 kHz to 30 MHz.

REFERENCES

[1] S. Ogasawara, H. Ayano, H. Akagi, "Measurement Reduction of Radiated by a PWM Inverter-Fed AC Motor Drive System," in *IEEE Transactions on IAS. July 1997,* Vol. 33, No. 4, pp. 860-867.

[2] Mark J. Nave, *Power Line Filter Design for switched-Mode Power Supplies.* Van Nostrand, 1991.

[3] Erkuuan Zhong, Thomas A. Lipo, "Improvement in EMC Performance of Inverter-Fed Motor Drives," in *IEEE Transactions on Industrial Applications,* vol. 31, No. 6, 2001, pp. 1247-1256.

[4] Yo-Chan son, Seung-Ki Sul, "Conducted EMI in PWM Inverter for Household Electric Appliance," in *IEEE Transaction on Industrial Applications, 2002,* vol. 38, pp. 1370-11379.

[5] I. Takahashi, Akoiro Ogata, "Active EMI Filter for Switching Noise of High Frequency Inverters," *Proc. PCC Nagaoko 1997,* pp. 331-334.

[6] Chhenggang Mei, Juan Carlos Balda, William P. Waite, "Cancellation of Common-Mode Voltages for Induction Motor Drives Using Active Method ," in *IEEE Transactions on Energy Conversion,* vol. 21, No. 2, 2006, pp. 380-386.

[7] Fu-Yuan Shit, Dan Y. Chen, "A Procedure for Designing EMI Filters for AC Line Applications," *IEEE Trans. Power Electronics.* vol. 11. no. 1. January 1996.

Novel Auxiliary Diagnosis Method for State-of-Health of Lead-Acid Battery

Yu-Hua Sun*, Hurng-Liahng Jou*, *Member, IEEE*, Jinn-Chang Wu**

* Department of Electrical Engineering
National Kaohsiung University of Applied Sciences
415 Jiangong Road, Kaohsiung 80778, Taiwan, Republic of China
** Department of Microelectronics Engineering
National Kaohsiung Marine University
142 Haijhuan Road, Nanzih District 81143, Taiwan, Republic of China

Abstract--A novel auxiliary diagnosis method for state-of-health (SOH) of a lead-acid battery unit is proposed in this paper. This method is based on the concept that the discharging curve for a health battery unit is smooth, however, the degradation of battery unit caused by the internal cell shorts, shorted cells open or the cell reversal will distort the discharging curve. Some experimental results are carried out to verify the proposed SOH diagnosis method. The experimental results show that the proposed method has the expected performance.

Index Terms-- lead-acid battery, state-of-health

I. INTRODUCTION

Lead-acid batteries have been widely used in electric vehicles [1], stand-alone renewable power generation, back-up power supplies for communication or computer related equipment [2] and uninterruptible power supply (UPS). State-of-health (SOH) of the battery is used to supply the degradation information of battery to the users. SOH indicates the ability of the battery to perform well in the charging, discharging power. If SOH can be estimated accurately, the users will know the state of the battery and whether the battery should be replaced or not. The degradation of battery may be caused by drying out, grid corrosion, thermal runaway, electrolyte leakage, internal cell shorts, shorted cells open, interconnected open or cell reversal [3]. There are many methods that have been developed to estimate the SOH of a battery. These methods can be categorized into three types:
(a) capacity
Capacity is the most popular method to indicate the SOH [4]. In the methods of this type, the life of battery is defined to be stopped when the capacity of battery is decreased to approximately 80% of the initial capacity

for general battery [5] and 60% of the initial capacity for small capacity battery [6]. SOH diagnosis by using the capacity is the ratio between the nominal capacity of the present time to the initial time, and it is represented as [7]:

$$SOH = \frac{(\text{nominal capacity at present time})}{(\text{nominal capacity at initial})} \quad (1)$$

(b) coup de fouet
The "coup de fouet" is a phenomenon particular to lead-acid batteries, which occurs at the beginning of the discharging.[8] This method involves measuring the trough voltage (low voltage point) in the "coup de fouet" region and using in determining the SOH of a battery [9].
(c) impedance techniques
Impedance techniques are based on the principle that the impedance and the resistance of a valve regulated lead-acid (VRLA) battery increase as the battery ages and loses capacity. Various impedance meters have been claimed to accurately measure the SOH of batteries [10]. The methods [11, 12] combined impedance measurements and fuzzy logic data analysis to estimate the SOH of the battery and have been published.

Besides the above methods, SOH monitoring based on battery voltage, temperature, capacity or internal impedance was developed [13,14] to obtain accurate results for estimating the SOH of the battery. However, there are still no economical methods which can accurately diagnose SOH of a battery.

In this paper, a novel auxiliary SOH diagnosis method to detect the degradation of lead-acid battery unit caused by the internal cell shorts, shorted cells open or the cell reversal is proposed. The salient feature of the proposed method is that the health status of the battery is estimated automatically at the end of each discharge cycle by measuring the battery voltage and current, and it does not require any complicated measurement. To verify the proposed auxiliary SOH diagnosis method, aging

---

The work was support by the ABLEREX Electronics Corporation, Ltd.

978-1-4244-0644-9/07/$25.00 ©2007 IEEE

experiments for lead-acid battery are developed. The experimental results show that the performance of the proposed method is as expected.

II. PROPOSED SOH DIAGNOSIS METHOD

In a healthy battery, the battery discharging voltage will be decreased smoothly. The degraded battery cells may result in internal short circuit [15, 16] or cell reversal [17], which will result in an unexpected voltage drop of a battery unit, and when internal short circuit open will result in the unexpected voltage rise of a battery unit. Hence, the degradation of battery caused by the internal cell shorts, shorted cells open or the cell reversal will distort the discharging curve of a battery unit. The proposed method is based on the concept that the above battery degradation phenomena will result in abrupt change of discharging voltage and discharging power. However, the battery current variation caused by the load change will also result in abrupt change of discharging voltage and discharging power. To avoid the un-corrected diagnosis caused by using the abrupt change of discharging voltage or discharging power, the development of a diagnosis method which can neglect the curves of discharging voltage and discharging power distorted by the change of load current (discharging current) is necessary. Because the proposed method cannot diagnose all battery degradation phenomena, such as: dry out or grid corrosion. Hence, the proposed method is an auxiliary diagnosis method.

The variation of discharging current and power can be expressed as (2) and (3).

$$\Delta I_{n+1} = I_{n+1} - I_n \tag{2}$$

$$\Delta P_{n+1} = P_{n+1} - P_n = V_{n+1}I_{n+1} - V_nI_n$$
$$= V_{n+1}(I_n + \Delta I_{n+1}) - V_nI_n$$
$$= (V_{n+1} - V_n)I_n + V_{n+1}\Delta I_{n+1} \tag{3}$$

where ΔI_{n+1} is the variation of discharge current, I_{n+1} is the discharging current value at $n+1^{th}$ sample, I_n is the discharging current value at n^{th} sample, ΔP_{n+1} is the variation of discharging power, P_{n+1} is the discharging power at $n+1^{th}$ sample and P_n is the discharging power at n^{th} sample. As seen in (3), the variation of discharging power not only depends on the variation value of discharging voltage but also the variation value of discharging current.

As seen in (3), the variation value of discharging power depends on the variation value of discharging current. The purpose of this paper is to develop a diagnosis method that can avoid the effect of discharging current variation, so that the SOH of battery can be diagnosed via only the variation value of the discharging power. The developed auxiliary diagnosis algorithm for

SOH (ADSOH) of battery at $n+1^{th}$ sample is shown as:

$$ADSOH_{n+1} = \Delta P_{n+1} - V_{n+1}\Delta I_{n+1} - (V_{n+1} - V_n)I_n * |\tanh(\Delta I_{n+1})|$$
$$= (V_{n+1} - V_n)I_n + V_{n+1}\Delta I_{n+1} - V_{n+1}\Delta I_{n+1} - (V_{n+1} - V_n)I_n * |\tanh(\Delta I_{n+1})|$$
$$= (V_{n+1} - V_n)I_n - (V_{n+1} - V_n)I_n * |\tanh(\Delta I_{n+1})| \tag{4}$$

where

$$\tanh(\Delta I_{n+1}) = \frac{\sinh(\Delta I_{n+1})}{\cosh(\Delta I_{n+1})} = \frac{e^{(\Delta I_{n+1})} - e^{-(\Delta I_{n+1})}}{e^{(\Delta I_{n+1})} + e^{-(\Delta I_{n+1})}}$$

Because the absolute value of the $\tanh(\Delta I_{n+1})$ function will approach unity when the variation value of discharging current (ΔI_{n+1}) is obvious. If $\tanh(\Delta I_{n+1})$ is approach to unity, then, $ADSOH_{n+1}$ will approach zero. This means that the ADSOH algorithm shown in (4) can avoid the effect of discharging current variation although the discharging voltage may be dropped due to the variation of discharging current caused by the load change. If the degradation of battery internal cells caused by the cell shorts, shorted cells open or the cell reversal appears, the battery voltage will vary immediately. At the same time, the ADSOH in (4) will be fluctuated evidently, so that the value of ADSOH can be converted into a SOH index.

Figure 1 shows the flow chart of the proposed ADSOH for a battery set. In discharging operation, measuring devices are employed to measure and retrieve the data of discharging voltages and currents of the batteries. The acceptable value of a predetermined sample period of time can be pre-set. The SOH of battery can be calculated via the equation ADSOH at the same time.

The data of discharging voltages, currents and powers of the batteries and ADSOH are stored in the database. The curves may be drawn by using the data of discharging voltages, currents and powers of the batteries and ADSOH. The compared data of the ADSOH was converted into a SOH index.

III. EXPERIMENTAL SYSTEM

An accelerated thermal aging experiment, developed to evaluate the charging and discharging characteristics of the lead-acid battery [18], is used to verify the proposed auxiliary SOH diagnosis method. Figure 2 shows the experimental system. This experimental system includes battery charger, thermostat, magnetic contact (MC), and battery measure system (BMS).

The conditions for battery aging experiments are defined as follows:

(1) Four independent strings of batteries are used in the experiment, and each string of batteries consisted of eight 7-Ah 12-V VRLA battery units connected in a series. Each battery unit consists of six cells internally connected in a series.

(2) Experiment environment temperature is 50 °C.

(3) Take a sample data every 10 sec when battery was discharged.

(4) Take a sample data every 5 min when battery was charged.

(5) The final discharging voltage of each battery set is 76.8V.

(6) The discharge load for each battery set is Halogen Lamp (120V 500W * 3).

(7) The experiment is stopped as the capacity of each sting of batteries is decreased to approximately 60% of the initial capacity.

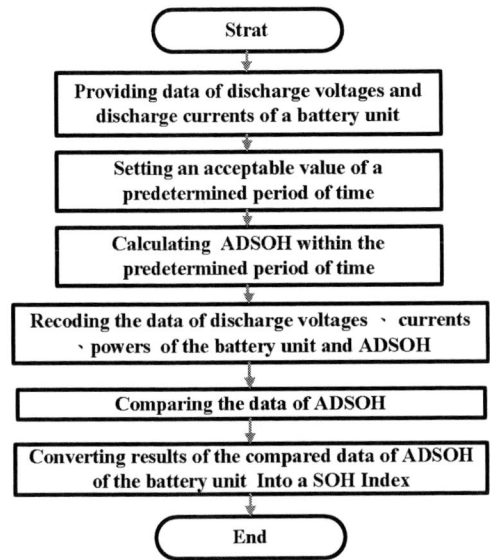

Fig. 1 The flow chart of the proposed SOH diagnosis method for a battery unit

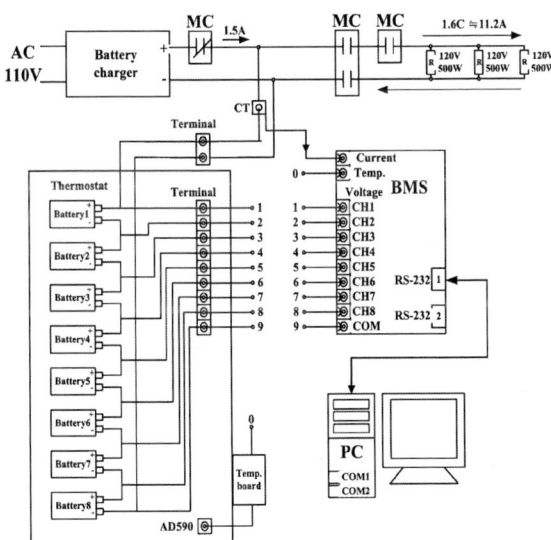

Fig. 2 Experimental system for a string of batteries

IV. EXPERIMENTAL RESULTS

The data of discharging current, voltage, power, and ADSOH of the battery (string 1, No.3) at 5th cycle are shown in Fig.3. The curve of discharging current

decreases gradually because the load used in the experiments is a constant resistance and the discharging current does not have obvious change. There exists an abrupt voltage drop of initial discharging at the instant when the battery is started to discharge. Instead of the abrupt voltage drop, the gradual decrease of discharging voltage of the battery appears within the predetermined sample period of time. Correspondingly, a curve in Figure 3(b) illustrates the discharging voltage of the battery in relation to a predetermined sample period of time, there exists an abrupt change of discharging voltage at the instant when the battery is started to discharge.

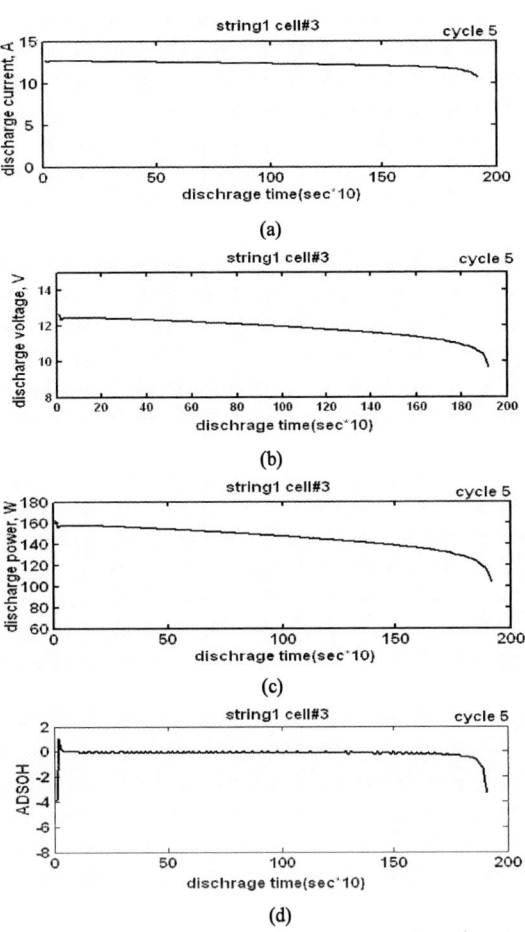

Fig. 3 Discharging curve of a battery (string 1, No.3) at 5th cycle（a）current,（b）voltage,（c）power,（d）ADSOH

Figure 3(c) illustrates the discharging power of the battery in relation to a predetermined sample period of time. Except the starting duration of discharging the battery, the change of the discharging power will be within a preferred range and the curve of discharging power is slight decrease. Figure 3(d) illustrates the ADSOH of the battery in relation to a predetermined sample period of time. Except the starting duration of to discharge the battery, the change of the ADSOH will be within a preferred range and the curve of discharging power is slight decrease. Because the abrupt change of discharging current, voltage and power at the instant of initial discharging is caused by the "coup de fouet". [8, 9] phenomenon, this is a normal phenomenon for the battery. Accordingly, the SOH of the battery can be determined "health" and no battery should be replaced.

Figure 4 shows the data of discharging current, voltage, power and ADSOH of the battery (string 1, No.3) at 159th cycle. Figure 4 illustrates the SOH of this battery is worsen. Figure 4(a) shows the curve of discharging current does not have obvious change. However, the curve of discharging voltage in Figure 4 (b) is distorted seriously. The abrupt drop of battery voltage near 400 and 500 sec can be explained as some cells in the battery are shorted, and the voltage rise near 1000 sec can be explained as a short cell opens. Figure 4(c) show the curve of discharging power. As seen in Figure 4 (c), the curve of discharging power in Figure 4 (c) is also distorted seriously. Figure 4(d) show the curve of the proposed ADSOH. The ADSOH curve showed in Fig. 4(d) is seriously fluctuated. This means that the health of battery is degraded..

may be distorted at the instant of load change. To verify the effect of load change (discharging current variation), the following experiment is carried out. In this experiment, only a battery unit is tested, electronic load is used as the variable load and the load current is changed in three steps during a discharge cycle..

Figure 5 shows the discharging curve of the battery under discharging current changed from 7A-14A-21A, the evident changes of discharging power and discharging voltage of the battery unit occur at the instant of load (discharging current) changes. As seen in Figure 5(b) and 5(c), the discharging curve of the battery voltage and battery power will be distorted at the instant of load change. Figure 5(d) shows the curve of ADSOH which will not be affected by the change of discharging current However, ADSOH curve will not be changed due to the changes of discharging current. Accordingly, the data of discharging voltage and discharging power cannot be used to determine the SOH of the battery due to the fact that it is affected by the change of the discharging current, but the ADSOH can avoid the effect caused by the variable discharging current.

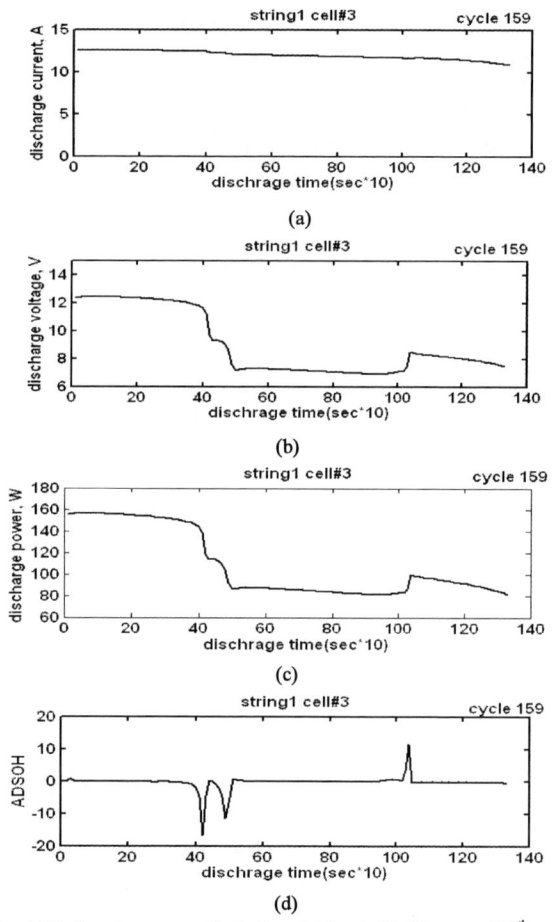

Fig. 4 Discharging curve of a battery (string 1, No.3) at 159th cycle, （a）current, （b）voltage, （c）power, （d）ADSOH.

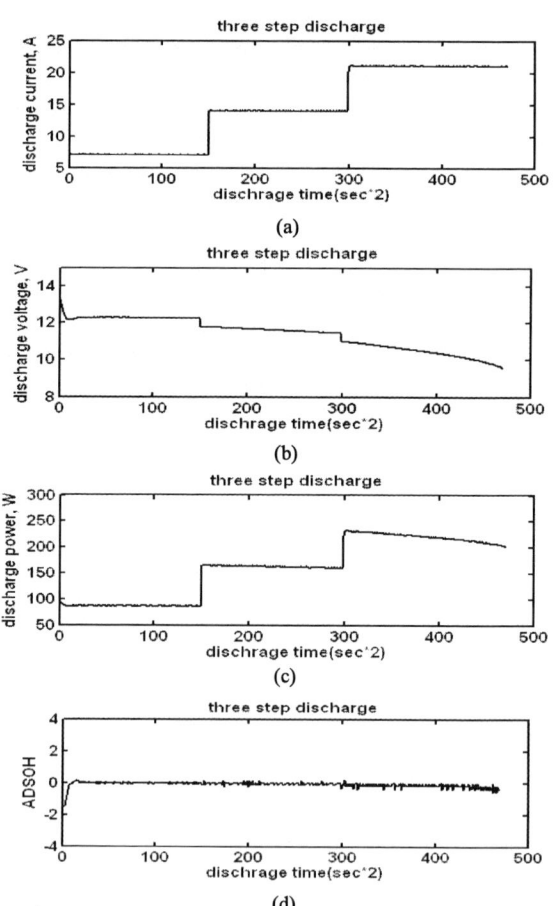

Fig.5 Discharging data of a battery at 7A、14A、21A three step discharge experiment （a）current, （b）voltage, （c）power, （d）ADSOH

The experimental results of above batteries show that they are discharged by a constant resistance. However, the discharging curve of battery voltage and battery power

From the above, it can be concluded that the data of

ADSOH can be used to determine the SOH of the battery and can avoid the un-corrected diagnosis caused by change of load.

V. CONCLUSIONS

This paper proposes a novel auxiliary diagnosis method for SOH of a lead-acid battery unit. In this method, the proposed ADSOH is used to diagnose the degradation of the lead-acid battery unit and indicate the degradation of the battery unit. The salient feature of the proposed method is that the SOH of a battery is estimated automatically at the end of each discharging cycle by measuring the battery voltage and current and it does not require any complicated measurement. The experimental results show that the proposed method can diagnose the degradation of battery unit caused by the internal cell shorts, shorted cells open and the cell reversal in the discharging process.

REFERENCES

[1] Caumont, P. Le Moigne; C. Rombaut; X. Muneret and P. Lenain, "Energy Gauge for Lead-Acid Batteries in Electric Vehicle," IEEE Trans. Energy Conversion, Vol. 15, No. 3, Sept. 2000, pp. 354-360.

[2] S. Manya, M. Tokunaga, N. Oda, T. Hatanaka and M. Tsubota, "Development of Long-Life Small-Capacity VRLA Battery Without Dry-Out Failure in Telecommunication Application Under High Temperature Environment," IEEE INTELEC, Sept. 2000, pp. 42-45.

[3] Stephen McCluer, "Battery Technology for Data Centers and Network Rooms: Battery Options," American Power Conversion, Rev 2005-10. pp. 1-8.

[4] P. E. Pascoe and A. H. Anbuky, "Standby VRLA battery reserve life estimation" IEEE INTELEC, Sept. 2004, pp. 516 – 523.

[5] IEEE Std. 1188–2005 IEEE Recommended Practices for Maintenance, Testing and Replacement of Valve Regulated Lead Acid (VRLA) Batteries in stationary applications.

[6] William E.; Rob E. and Barry R. "Testing of gel-electrolyte batteries for wheelchairs" Journal of Rehabilitation Research and Development, Vol . 25, No . 2, pp. 27-32.

[7] Tetsuro Okoshi , Keizo Yamadaa, Tokiyoshi Hirasawa and Akihiko Emori "Battery condition monitoring (BCM) technologies about lead–acid batteries" Journal of Power Sources, Vol. 158, No. 2, Aug. 2006, pp. 874-878.

[8] C. S. C. Bose, and F. C. Laman, "Battery state of health estimation through coup de fouet," IEEE INTELEC, Sept. 2000, pp. .597 – 601

[9] A. Delaille, M. Perrin, F. Huet and L. Hernout, "Study of the "coup de fouet" of lead-acid cells as a function of their state-of-charge and state-of-health," Journal of Power Sources, Vol. 158, No. 2, Aug. 2006, pp. 1019-1028.

[10] A. R. Waters, K. R. Bullock, C. S. C. Bose, "Monitoring the state of health of VRLA batteries through ohmic measurements," IEEE INTELEC, Oct. 1997, pp.:675 – 680.

[11] P. Singh, D. Reisner, "Fuzzy logic-based state-of-health determination of lead acid batteries" IEEE INTELEC, Oct.

2002, pp. 583 – 590.

[12] P. Singh, S. Kaneria, J. Broadhead, X. Wang, J. Burdick, "Fuzzy logic estimation of SOH of 125Ah VRLA batteries," IEEE INTELEC, Sept. 2004, pp. 524 – 531.

[13] A. H. Anbuky, P. E. Pascoe, P. M. Hunter, "Knowledge based VRLA battery monitoring and health assessment," IEEE INTELEC. Sept. 2000, pp. 687 – 694

[14] K. Takahashi and Y. Watakabe,." Development of SOH monitoring system for industrial VRLA battery string," IEEE INTELEC, Oct. 2003, pp. 664 – 670.

[15] E. Davis, D. Funk and W. Johnson, "Internal ohmic measurements and their relationship to battery capacity: EPRI's ongoing technology evaluation," Battcon, 2002, pp. 12-1 – 12-10.

[16] Robert E. Landwehrle, "In the final analysis: Post mortem tests and measurements on a VRLA battery," Battcon, 2005, pp. 19-1 – 19-6.

[17] J. Garche; A. Jossen; H. and Doring "The influence of different operating condition, especially over-discharge, on the lifetime and performance of lead acid batteries for photovoltaic systems," Journal of Power Sources Vol. 67, No. 1-2, 1997, pp. 201-212

[18] P. E. Pascoe and A. H. Anbuky, "Automated battery test system," Measurement, Vol. 34, No. 4, Dec. 2003, pp. 325-345.

Electromechanical Model of a Longitudinal Mode Piezoelectric Transformer

Shine-Tzong Ho*

* Mechanical Engineering Department, National Kaohsiung University of Applied Sciences,
415 Chien-Kung road, San-min district, Kaohsiung, Taiwan

Abstract– This paper presents an electromechanical model for a piezoelectric transformer (PT), which is driven at the longitudinal mode. The electromechanical model is established by the coupled equations of motion of the PT, which is derived by the Rayleigh-Ritz method and Hamilton's principle. The Rayleigh-Ritz method allows us to represent a continuous system by a discrete approximation. Thus, the variational expression of Hamilton's principle can be used to model the distributed piezoelectric structure for estimating the parameters of the equivalent circuit model of the PT. Furthermore, the equivalent circuit model is used to estimate the electrical characteristics of the PT. Such as input admittance, voltage step-up ratio, output power and efficiency are derived by the equivalent circuit model. Based on the model, the performance of the beam-type PTs can be predicted by the geometrical dimensions and material parameters in the design stage. The validity of the proposed model is experimentally confirmed.

Index Terms– Piezoelectric transformer, Longitudinal mode, Electromechanical model, Equivalent circuit.

I. INTRODUCTION

The idea of a piezoelectric transformer was first implemented by Rosen in 1956 [1], as shown in Fig.1(a). Due to the high energy density of piezoelectric materials, the high electromechanical coupling factors and high quality factor of the mechanical resonance, they tend to be lighter and more efficient than conventional electromagnetic transformers.

In the literature [1-10], many PTs have been proposed and a few of them found practical applications. Apart from switching power supply system, a Roson-type PT has been adopted in cold cathode fluorescent lamp inverters for liquid-crystal display [3,4]. Fuda investigated the relationship between heat generation and vibration velocity of PT by experiments in order to reduce the heat generation of the PT [3]. Kanayama proposed a PT which has an alternately poled structure in order to raise the output power of the PT [4,5]. The relation between the equivalent circuit parameters of the PT and the material property and size of the PT are discussed in Kanayama's paper, but the parameters still can not be accurately predicted. Fukunaga studied the relation between the characteristics and the dimension of the Rosen-type PT by an equivalent circuit in order to achieve a high efficiency and a large voltage step-up ratio

[6]. Sakurai proposed an improved equivalent circuit and used it to present the power transmission characteristics of a PT [7]. However, a high efficiency and a large voltage step-up ratio or the power transmission characteristics seem important to the design of the PT. And the conventional equivalent circuit method is a useful tool to analyze the characteristics of the PT, but the parameters of the equivalent circuit is necessary in the process of the analysis. Usually, the researchers get the parameters by measurement of an impedance analyzer. However, it is unavoidable to manufacture the prototype of the PT and measure it by several times of trial and error in order to improve the performance of the PT. A more powerful simulation method for predicting the performance of the PT is desired for this problem.

So far no electromechanical model of the PT is found to be capability of analyzing the distributed piezoelectric structure for estimating the parameters of the equivalent circuit model of the PT in literature. Therefore, the aim of this paper is to establish an electromechanical model for predicting the performance of the longitudinal mode PT by the geometrical dimensions and material parameters in the design stage. The electromechanical model will be established based on the coupled equations of motion of the PT, which is derived by the Rayleigh-Ritz method and Hamilton's principle. To verify the proposed model, the simulated results based on the model were compared with the experimental results and a good agreement is obtained.

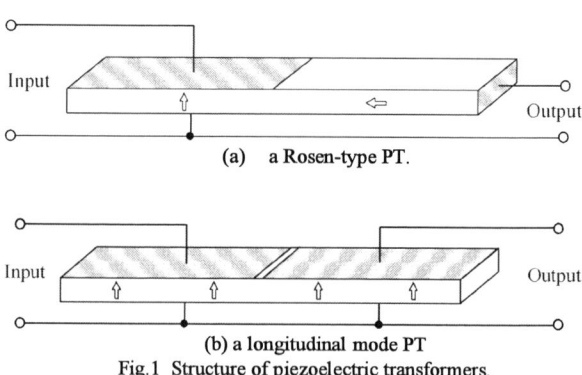

(a) a Rosen-type PT.

(b) a longitudinal mode PT

Fig.1 Structure of piezoelectric transformers.

II. THEORETICAL ANALYSIS

A. Longitudinal Vibration Modes

The longitudinal mode PT, as shown in Fig.1(a)(b), consists of two poled piezoceramic beams with equal cross sections rigidly boned together, or a single piezoceramic beam with both ends poled separately. A sinusoidal voltage is supplied to excite mechanical vibrations by the inverse piezoelectric effect via the input part. An output voltage can be induced in the output part due to the direct piezoelectric effect. A mode is excited if the supply frequency coincides with a particular vibration mode, and the frequency of these modes depends on the dimensions of the PT and its material properties. Considering a piezoelectric beam of length L with constant cross-sectional area A, its free vibration equation can be obtained as the following.

$$\frac{\partial^2 u(x,t)}{\partial x^2} = \beta^2 u(x,t) \tag{1}$$

where $\beta^2 = \rho\omega^2/E$. ρ, ω and E represent density, frequency and Young's modulus, respectively. The solution of Eq.(1) can be written as $u(x,t) = U(x)G(t)$ and the nth longitudinal mode $U_n(x)$ is determined by the free-free ends of boundary conditions, as the following.

$$U_n(x) = \cos\beta_n x \tag{2}$$

The resonant frequencies are given by

$$\sin\beta_n L = 0, \quad \beta_n = n\pi/L, \quad n=1,2,3... \tag{3}$$

where n represents the order of the mode.

B. Constitutive Equations

In Fig.1, either Fig.1(a) or Fig.(b) are operating in the longitudinal mode, but they have different design in the output part. Considering the poling and vibrating direction of the piezoceramic beam, the output part of Fig.1(a) is operating in the parallel longitudinal mode and the output part of Fig.1(b) is operating in the transverse length mode. Their performance will be different due to the difference of the coupling factors k_{31} and k_{33}. For convenience, we will analyze the PT of Fig.1(b) in the following.

Considering the transverse length mode, as shown in Fig.1(b), the magnitude of T_1 can be assumed to be mush larger than T_2 and T_3. Also, we can assume that only E_3, D_3 will be applied or detected. Therefore, the constitutive equations of a piezoelectric material can be simplified to one dimensional constitutive equations as in the following forms.

$$S_1 = s_{11}^E T_1 + d_{31}E_3, \tag{4}$$

$$D_3 = d_{31}T_1 + \varepsilon_{33}^T E_3,$$

where S_1 is the strain of x direction, T_1 is the stress of x direction, E_3 is the electric field of z direction, and D_3 is the electric displacement of z direction. d_{31} is the

piezoelectric constant, ε_{33}^T is the dielectric permittivity, and s_{33}^E is the elastic compliance coefficient.

III. ELECTROMECHANICAL MODEL OF THE PIEZOELECTRIC TRANSFORMER

A. Electromechanical Model of the PT

To model the dynamic behavior of the PT, equations of motion for the coupling system of the PT can be obtained by using Hamilton's principle. From Hagood's paper [11], we have a generalized form as the following.

$$\int_{t_1}^{t_2} \partial[T - U + W_e + W_i + W_o]dt = 0 \tag{5}$$

where T is the kinetic energy, U is the potential energy of the system, W_e is the electrical energy stored within the piezoceramic beam, W_i is the applied electric energy in the input section, and W_o is the applied electric energy in the output section. T, U, W_i, W_o can be written as

$$T = \frac{1}{2}\int_V \rho\dot{u}^2(x,t)dV \tag{6}$$

$$U = \frac{1}{2}\int_V S_1 T_1 dV \tag{7}$$

$$W_e = \frac{1}{2}\int_V E_3 D_3 dV \tag{8}$$

$$\partial W_i = -\partial\varphi_i \cdot q_i, \quad \partial W_o = -\partial\varphi_o \cdot q_o \tag{9}$$

where φ_i and q_i are the electric potential and the applied charge in the input section, respectively. φ_o and q_o are the electric potential and the applied charge in the output section. V means the whole volume of the PT. By substituting Eqs.(6)-(9) into Eq.(5), the equations of motion for the PT can be written in Laplace transform as

$$(m_n s^2 + d_n s + k_n)X_n + A_{on}V_o = A_{in}V_i \tag{10}$$

$$\sum_{n=1}^{m} A_{in}X_n + C_i V_i = \sum_{n=1}^{m} q_{in} \tag{11}$$

$$\sum_{n=1}^{m} A_{on}X_n = C_o V_o + \sum_{n=1}^{m} q_{on} \tag{12}$$

$$q_i = \sum_{n=1}^{m} q_{in}, \qquad q_o = \sum_{n=1}^{m} q_{on} \tag{13}$$

where V_i and V_o represent the input and output voltage. X_n is the generalized mechanical displacement. q_{in} and q_{on} represent the input and output charge due to the nth longitudinal mode vibration. The mass m_n, the stiffness k_n, the damping d_n, the input turn ratio A_{in}, the output turn ratio A_{on}, the input capacitance C_i, the output capacitance C_o for the equivalent circuit of PT can be obtained from the following.

499

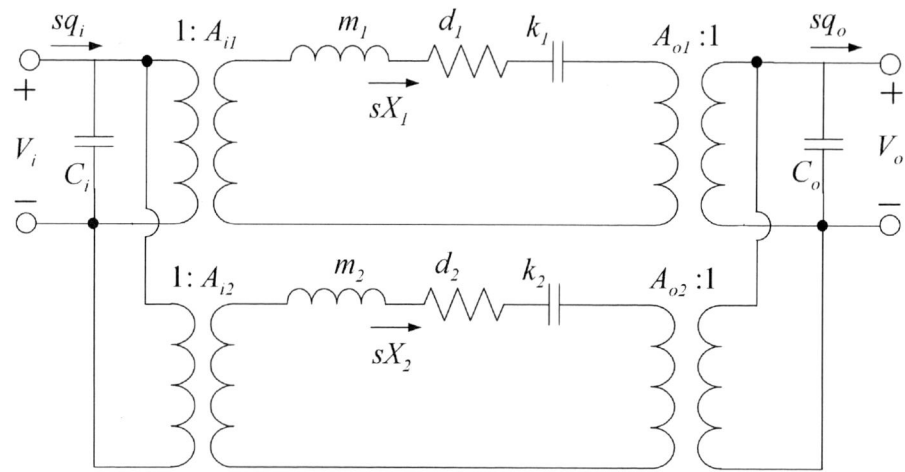

Fig. 2. Equivalent circuit of piezoelectric transformer.

$$m_n = \int_0^L \rho b h U_n^2 dx \qquad (14)$$

$$k_n = \int_0^L \frac{bh}{s_{11}^E}(\frac{\partial U_n}{\partial x})^2 dx \qquad (15)$$

$$A_{in} = \frac{1}{h}\int_{V_{TI}} \frac{d_{31}}{s_{11}^E}\frac{\partial U_n}{\partial x}dV_T \qquad (16)$$

$$A_{on} = \frac{1}{h}\int_{V_{TO}} \frac{d_{31}}{s_{11}^E}\frac{\partial U_n}{\partial x}dV_T \qquad (17)$$

$$C_i = \int_{V_{TI}} \frac{\varepsilon_{33}^T}{h^2}dV_{TI} \qquad (18)$$

$$C_o = \int_{V_{TO}} \frac{\varepsilon_{33}^T}{h^2}dV_{TO} \qquad (19)$$

$$d_n = \frac{k_n}{\omega_{nc}Q_m} \qquad (20)$$

where V_{TI} and V_{TO} represent the volume of the input section and the output section in the PT, respectively. The equivalent circuit model of the PT is shown in Fig.2, which is established by Eqs.(10)-(13). The model is considered only the first two modes here for convenience. From the equivalent circuit model, we can see that Eq.(10) satisfies Kirchhoff's voltage law equation, which shows that the input voltage $A_{in}V_i$ is the sum of the output voltage $A_{on}V_o$ and the voltage difference $(m_n s^2 + d_n s + k_n)X_n$. Eq.(11) satisfies Kirchhoff's current law equation in the input section, which shows that the input current I_i is the sum of the current flowing through $(m_n s + d_n + k_n/s)$ and the current flowing through C_i. Eq.(12) satisfies Kirchhoff's current law equation in the output section, which shows that the current flowing through $(m_n s + d_n + k_n/s)$ is the sum

of the current flowing through C_o and the output current I_o.

B. Admittance of the PT

After establishing the electromechanical equations of motion and the equivalent circuit model for the longitudinal mode PT, characteristics of the PT can be conducted more easily. Considering a load resistance R_L connected between the electrodes in the output part of the PT, Eq.(21) can be obtained by substituting $I_o=V_o/R_L$ into Eq.(12).

$$\sum_{n=1}^{m} sA_{on}X_n = sC_oV_o + V_o/R_L \qquad (21)$$

Substituting Eq.(21) into Eq.(10) to eliminate V_o, and eliminating $X_n(s)$ from Eqs.(10)(11). Then, the input admittance Y_i for the PT with a load resistance R_L in the output part can be obtained as the following.

$$Y_i = sC_i + \sum_{n=1}^{m} \frac{sA_{in}^2(sC_oR_L+1)}{(m_n s^2 + d_n s + k_n)(sC_oR_L+1) + sA_{on}^2 R_L} \qquad (22)$$

If the electrodes in the output part is short-circuited, the input admittance can be obtained as Eq.(23).

$$Y_i = sC_i + \sum_{n=1}^{m} \frac{sA_{in}^2}{m_n s^2 + d_n s + k_n} \qquad (23)$$

In case of the open-circuited output part, the input admittance can be obtained as Eq.(24).

$$Y_i = sC_i + \sum_{n=1}^{m} \frac{sA_{in}^2}{m_n s^2 + d_n s + k_n + A_{on}^2/C_o} \qquad (24)$$

The resonant frequency of the input impedance is changed from ω_{nc} to ω_{no} when the load resistance R_L varies from 0 to infinite. Here the resonant frequency $\omega_{nc} = \sqrt{k_n/m_n}$ and ω_{no} is as the following.

$$\omega_{no} = \sqrt{k_n/m_n + A_{on}^2/(C_o m_n)} \qquad (25)$$

$$\eta_{\max} = \frac{A_{on}^2}{2 d_n \omega_{nc} C_o + A_{on}^2} \cdot \qquad (33)$$

It is noted that the smaller the damping coefficient d_n, the higher the maximum efficiency. In other words, a high quality factor of the piezoelectric material is desired for the manufacture of a high efficiency of PT. On the other hand, a high efficiency of PT can be designed by raising its output turn ratio A_{on}. Depend on Eq.(17), we can conclude that it is necessary to choose a large value of d_{31}, and consider the relation between the output electrode and the vibration mode in order to obtain a large value of A_{on}.

IV. Simulation and Experimental Verification

A. Experimental Setup and the Impedance Measurements

To verify the electromechanical model, a beam-type PT with 10mm in width, 2mm in thickness and 35mm in length was used. The transformer structure was fabricated using the piezoelectric material APC855 by APC International, USA. The material is not a good choice for a good performance of PT because of its low quality factor. We used it herein to verify the proposed electromechanical model for convenience. The material properties provided by the supplier are listed in Table I. The PT has silver electrodes on two opposite surfaces and is poled along its thickness direction. According to Table.I, parameters of the equivalent circuit of the PT can be calculated by Eqs.(14)-(20) and shown in Table.II. In order to raise the output turn ratio A_{on} for the first two modes, the stress distributions of the mode shapes of the PT were considered, thus one of the electrodes of the PT is split into two regions on the half of length.

A HP 4194A Impedance Analyzer was used to measure the input impedance. The input impedance as a function of frequency at different load resistances are measured and shown in Fig.3. Based on Eqs.(22), the input impedance as a function of frequency at different load resistances are calculated and shown in Fig.4. However, the calculated values agreed well with the measurement values at the resonance frequencies no matter what the output electrode's electrical boundary condition is. On the other hand, it shows that the first resonant frequency is changed from 40310 Hz to 41309 Hz, and the anti-resonant frequency is changed from 41559 Hz to 42807 Hz in the input impedance of the PT with load resistance varied from short ($R_L=0$) to open ($R_L=\infty$).

In Fig.3 and 4, there exists an optimal load resistance $R_{L,opt}$, which shows the maximum damping ratio in the input impedance when compared with the other different load resistances. The optimal load resistance can calculated by Eq.(32). The optimal load resistance for the first mode is $R_{L,opt}=1.55$ kΩ, the optimal load resistance for the second mode is $R_{L,opt}=760$ Ω. The efficiency of the PT approaches to the maximum efficiency when the load resistance R_L approaches the optimal load resistance $R_{L,opt}$. They are in a good agreement between the theoretical analysis of Eq.(32) and the experiments.

C. Characteristics of the PT

Concerning the characteristics of the PT, the voltage step-up ratio, output power, and efficiency are all important to the PT. At the frequency band near the nth mode resonance frequency, the voltage step-up ratio for the PT with a load resistance R_L in the output part can be obtained by Eqs.(10) and (21) as the following.

$$\frac{V_o(s)}{V_i(s)} = \frac{s A_{in} A_{on} R_L}{(m_n s^2 + d_n s + k_n)(s C_o R_L + 1) + s A_{on}^2 R_L} \quad (26)$$

If the electrodes in the output part of the PT is short-circuited, the voltage step-up ratio for the PT can be obtained as zero by substituting $R_L=0$ into Eq.(26). In addition, Eq.(26) shows that the higher the load resistance R_L, the higher the voltage step-up ratio at the narrow band near its resonance frequency. The maximum voltage step-up ratio as a function of frequency can be obtained as Eq.(27) when the load resistance R_L approach infinite.

$$\frac{V_o(s)}{V_i(s)} = \frac{A_{in} A_{on}}{(m_n s^2 + d_n s + k_n) C_o + A_{on}^2} \quad (27)$$

On the other hand, if the natural frequency ω_{nc} is chosen as the operating frequency in the PT, then the voltage step-up ratio can be simplified as

$$\frac{V_o}{V_i} = \frac{A_{in} A_{on}}{j \omega_{nc} C_o d_n + d_n / R_L + A_{on}^2} \cdot \quad (28)$$

The output power of the PT can be calculated by the power consumption of the load resistance R_L as the following.

$$P_o = |V_o|^2 / R_L = \frac{A_{in}^2 A_{on}^2 V_i^2}{R_L[(\omega_{nc} C_o d_n)^2 + (d_n / R_L + A_{on})^2]} \quad (29)$$

According to the equivalent circuit of the PT, shown as in Fig.2, the input power of the PT can be calculated by the sum of the power consumption of the damping d_n and that of the load resistance R_L. Eq.(12) shows that the current flowing through d_n is $(s C_o V_o + I_o) / A_{on}$, thus the input power of the PT can be obtained as

$$P_i = V_o^2 [d_n (\omega_{nc} C_o / A_{on})^2 + d_n / (R_L^2 A_{on}^2) + 1 / R_L] \quad (30)$$

Therefore, the efficiency of the PT can be obtained by Eqs.(29) and (30).

$$\eta = \frac{P_o}{P_i} = \frac{A_{on}^2}{d_n R_L (\omega_{nc} C_o)^2 + d_n / R_L + A_{on}^2} \quad (31)$$

The maximum efficiency can be calculated by the differentiation of Eq.(31). Thus, the maximum efficiency can be obtained when the optimal load resistance $R_{opt,n}$ is

$$R_{opt,n} = 1 / (\omega_{nc} C_o). \quad (32)$$

Substituting Eq.(32) into Eq.(31) gives the maximum efficiency

TABLE I Properties of piezoelectric material

Piezoelectric coefficient d_{31}	$-270*10^{-12}$ C/N
Coupling factor k_{31}	0.38
Mechanical quality factor Q_m	65
Density ρ	7.6 kg/cm^3
Young's modulus Y_{11}^E	$5.9*10^{10}$ N/m^2
Dimensions ($h*b*L$)	2mm*10mm*35mm

TABLE II Parameters of the equivalent circuit

Input piezoelectric capacitance C_i	2.606 nF
Output piezoelectric capacitance C_o	2.606 nF
Input turn ratio A_{i1}	0.16
Output turn ratio A_{o1}	0.16
Effective mass m_1	$2.66*10^{-3}$ kg
Effective damping d_1	10.24 N-s/m
Effective stiffness k_1	$1.664*10^8$ N/m
Input turn ratio A_{i2}	0.3
Output turn ratio A_{o2}	0.3
Effective mass m_2	$2.66*10^{-3}$ kg
Effective damping d_2	20.47 N-s/m
Effective stiffness k_2	$6.655*10^8$ N/m

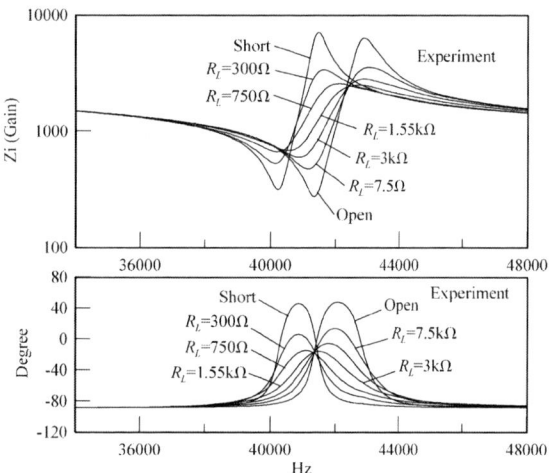

Fig.3 Measured input impedance at different load resistances.

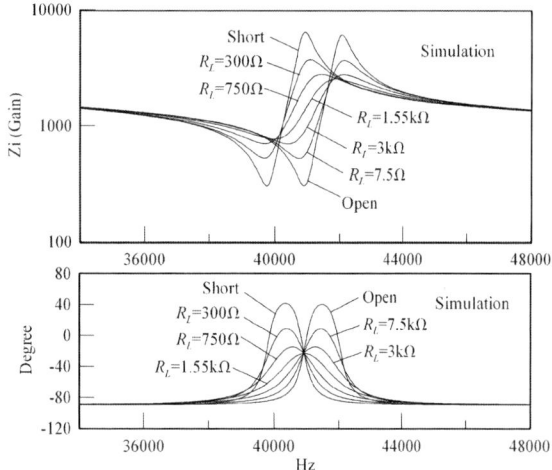

Fig.4 Calculated input impedance at different load resistances.

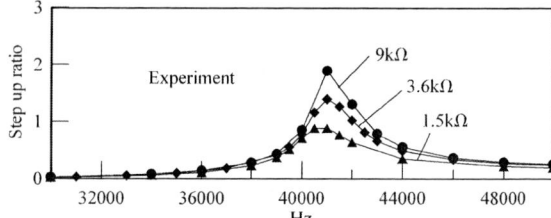

Fig.5(a) Measured voltage step-up ratio : 1st mode.

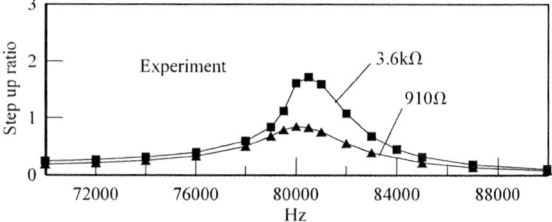

Fig.5(b) Measured voltage step-up ratio : 2nd mode.

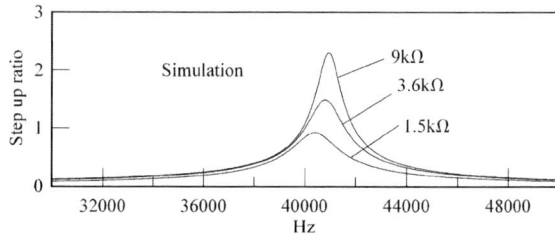

Fig.6(a) Calculated voltage step-up ratio : 1st mode.

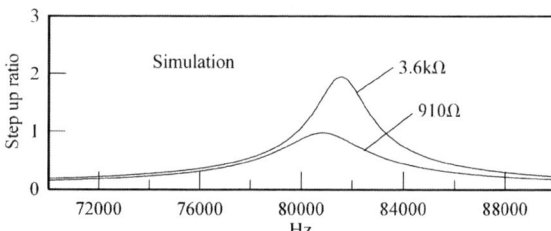

Fig.6(b) Calculated voltage step-up ratio : 2nd mode.

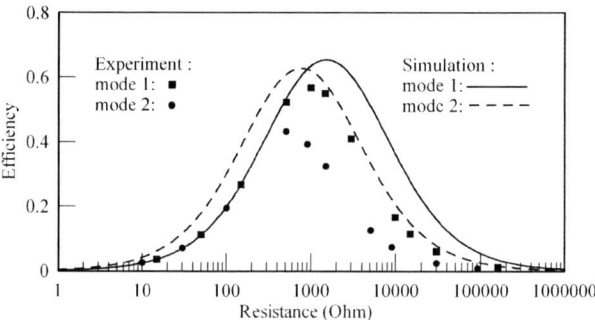

Fig.7 Efficiency as a function of load resistance

B. Voltage Step-up Ratio and Efficiency

The voltage step-up ratios as a function of frequency at different load resistances were measured and shown in Fig.5. The simulation of Fig.5 are calculated by Eq.(26) and shown in Fig.6. The measured peak value of the voltage step-up ratio curve is changed from 0.88 to 1.89 for the first mode when the load resistance varies from 1.5 kΩ to 9 kΩ. For the second mode, the measured peak value of the voltage step-up ratio curve is changed from

502

0.88 to 1.72 when the load resistance varies from 910 Ω to 3.6 kΩ. Depended on the simulated data, the peak frequency is slowly changed from 40400 Hz to 40920 Hz for the first mode, and the peak frequency is slowly changed from 80840 Hz to 81560 Hz for the second mode. The efficiencies as a function of the load resistance are measured and compared with the theoretical analysis of Eq.(31) and shown in Fig.7. There are some errors in the efficiency between the calculated values and the measured values. The errors may be caused by the simplification of Eq.(28). However, the optimal load resistances of experimental results are in a good agreement with that of the simulated results. It should be noted that the electromechanical model proposed in this paper can be used by a designer to predict the voltage step-up ratio by inserting the material parameters and geometrical dimensions into the equations. In this manner, the output power and the efficiency of the PT can also be predicted and the performance of a longitudinal mode PT can be improved more easily.

V. CONCLUSION

An electromechanical model of a longitudinal mode PT is presented. In contrast to the conventional lumped-equivalent circuit method, the proposed equivalent circuit model in this paper has the advantage of predicting the beam-type PT's performance as a function of design parameters. The parameters of the equivalent circuit model can be calculated from Eqs.(14)-(20). In addition, the voltage step-up ratio, input admittance, and output power of the PT are derived. The optimal load resistance and the maximum efficiency for the PT have been obtained. For verification of the proposed electromechanical model, some simulated results of the model are compared with the experimental results and a agreement was obtained.

REFERENCES

[1] C. A. Rosen, "Ceramic Transformers and Filters," *Proceedings of Electronic Comp. Symp.*, 1956, pp.205-211.

[2] Y. Hsu, C. Lee and W. Hsiao, "Optimizing Piezoelectric Transformer for Maximum Power Piezoelectric Transformer," *Smart Mater. Struct.*, vol.12, 2003, pp. 373-383.J.

[3] Y. Fuda, K. Kumasaka, M. Katsuno, H. Sato and Y. Ino, "Piezoelectric Transformer for Cold Cathode Fluorescent Lamp Inverter," *Jpn. J. Appl. Phys.*, vol.36, 1997, pp.3050-3052.

[4] K. Kanayama and N. Maruko, "Properties of Alternately Poled Piezoelectric Transformers ," *Jpn. J. Appl. Phys.*, vol.36, 1997, pp.3048-3049.

[5] K. Kanayama, N. Maruko and H. Saigoh, "Development of the Multilayer Alternately Poled Piezoelectric Transformers," *Jpn. J. Appl. Phys.*, vol.37, 1998, pp.2891-2895.

[6] H. Fukunaga, H. Kakehashi, H. Ogasawara and Y. Ohta, "Effect of Dimension on Characteristics of Rosen-type Piezoelectric Transformer," *IEEE Ann. Power Electron. Spec. Conf.*, 1998, pp. 1504-1510.

[7] K. Sakurai, K. Ohnishi and Y. Tomikawa, "Presentation of a New Equivalent Circuit of a Piezoelectric Transformer under High-power Operation," *Jpn. J. Appl. Phys.*, vol.38, 1999, pp.5592-5597.

[8] E. M. Syed, F. P. Dawson, E. S. Rogers, "Analysis and Modeling of a Rosen Type Piezoelectric Transformer," *IEEE Power Electronics Specialists Conference*, 2001, pp. 1761-1766.

[9] T. Asada and A. Takatsuka, "Piezoelectric Ceramic Transformer with the Suppression of the 2nd harmonic Vibration," *J. Acoust. Soc. Jpn*, vol.21, 2000, pp.271-273.

[10] S. Kawashima, O. Ohnishi, H. Hakamata, S. Tagami, A. Fukuoka, T. Inoue, and S. Hirose, "Third Order Longitudinal Mode Piezoelectric Ceramic Transformer and its Application to High-voltage Power Inverter," in *IEEE Ultrason. Symp.*, 1994, pp. 525-530.

[11] N. W. Hagood, W. H. Chung and A. V. Flotow, "Modeling of Piezoelectric Actuator Dynamics for Active Structural Control," *J. of Intell. Mater. Syst. And Struct.*, vol. 1, 1990, pp. 327-354.

Latest Development of Transformer Parasitic Inductive Components and Lossless Inductive Snubber-Assisted Series Resonant High-Frequency ZCS-PFM DC-DC Converter for RF Generator

Hisayuki Sugimura[1], Manabu Ishitobi[2], Bishwajit Saha[1],
Sang Pil Mun[1], Soon Kurl Kwon[1], Mutsuo Nakaoka[1, 3]

[1]Kyungnam University, Masan, Republic of Korea
[2]Nara National College of Technology, Japan
[3]Industrial College of Technology-University, Japan
E-mail: bsaha@ ieee.org

Abstract—The conventional series-resonant power frequency modulated DC-DC converter with a high-frequency (HF) transformer link designed for driving the magnetron of microwave ovens has the problem of the hard switching commutation at turn-on and turn-off the active power switching devices due to the influence of the magnetizing current of the high-frequency transformer. This paper presents a novel prototype of high-frequency transformer parasitic parameters with a lossless inductive snubber and a series resonant capacitor assisted series-resonant zero current switching pulse frequency modulated DC-DC power converter, which is designed for industrial use of high power magnetron for microwave ovens. In order to implement a complete and efficient soft switching commutation, the performance of the new converter topology is practically confirmed and evaluated in the prototype of power microwave generator for plasma application from practical point of view.

Index Terms—series resonant inverter, HF transformer link, voltage doubler rectifier, magnetron, ZCS-PFM

I. INTRODUCTION

With great advances of the power semiconductor switching devices such as MOSFETs, IGBTs, ESBTs and B-SITs as well as high-frequency circuit components, the latest research and development of the high-frequency inverter type switching mode DC-DC power conversion circuits and systems have attracted special interests for high voltage DC power applications [1]-[5]. From the downsizing point of view, the high-frequency resonant soft switching DC-DC converter using MOS gate power semiconductor devices are actively introduced in this new particular field of power electronic applications. On the other hand, IGBTs or MOS gate bipolar transistors are more suitable for zero current switching (ZCS) commutation rather than zero voltage switching (ZVS) due to their associated inherent tail currents at turn off switching commutations [6]-[12]. Thus, the series resonant inverters and converters operating under the principle of ZCS commutation are practically preferred from the switching losses point of view.

The conventional series-resonant circuit topologies of high frequency inverters can operate under a commutation principle of ZCS based on a discontinuous current mode (DCM) control. However, in the high frequency transformer link topologies it is difficult to implement the DCM in the conventional type circuit due to the influence of the magnetizing current that flows through the high-frequency transformer primary side. Especially, the influence of the magnetizing current is remarkable for the DC-DC converters with capacitor type output-smoothing filter and a constant high-voltage load as the power magnetron of the microwave appliances.

Under above technological background, this paper presents a high-frequency transformer parasitic parameters and a lossless inductive snubber assisted series-resonant DC-DC power converter with a voltage doubler rectifier circuit, which can operate under ZCS commutation based on a pulse frequency modulation (PFM) in order to improve the problem of the transformer magnetizing-inductive components-based hard switching commutation effectively. This new circuit topology can actively utilize the transformer parasitic circuit components as leakage and magnetizing inductance to achieve soft switching operation with the aid of a lossless inductive snubber and series resonance capacitor in the high voltage transformer primary side. The performance evaluations as the switching voltage and current waveforms of the power semiconductor switches, DC power regulation and power conversion efficiency characteristics are practically confirmed and evaluated as applied to the power magnetron of the microwave generator for plasma application.

II. CHARACTERISTICS AND ELECTRICAL EQUIVALENT CIRCUIT MODEL OF MAGNETRON

The heart of any microwave generator is the high voltage system. Its purpose is to generate microwave energy. The high-voltage components accomplish this by stepping up AC line voltage to high voltage, which is then changed to an even higher DC voltage. This DC power is then converted to the RF (microwaves) energy that cooks the food. Microwaves are very short waves of electromagnetic energy that travel at the speed of light. Microwaves used in microwave ovens are in the same family of frequencies as the signals used in radio and television broadcasting. Electromagnetic forms of energy, such as microwaves, radar waves, radio and TV waves, travel millions of miles through the emptiness of space without the need of any material medium through which to travel.

Figure 1 indicates the input voltage vs. input current characteristics of the power magnetron. As shown in this figure, the magnetron has non-linear input characteristics, what can be represented with the piecewise linear approximation which includes high resistance area in non-oscillating region and low resistance area in oscillating region for magnetron. When the voltage between anode and cathode exceeds about 7.4 kV (cut-off voltage), the magnetron anode current begins to flow from anode to cathode. On the other hand, when the voltage between anode and cathode is lower than the cut-off voltage, the anode current does not mostly flow. The cut-off voltage of magnetron a little fluctuates with the operating temperature which depends on the characteristics of the ferrite magnet in the magnetron, which are not considered here for the approximate v-i characteristics.

The electrical equivalent circuit model of the magnetron can simply represent by using pure resistances R_0 and R_1, an ideal diode D, an ideal battery VZ (cut-off

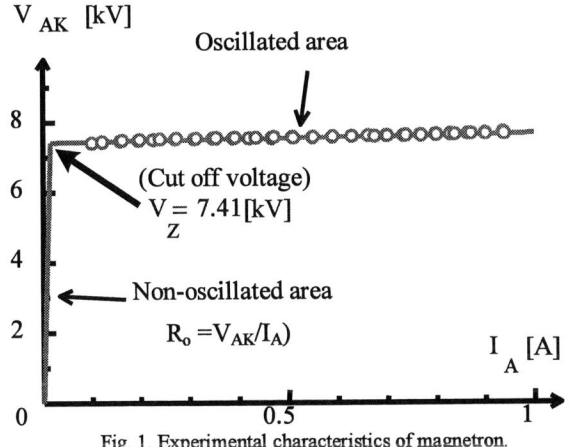

Fig. 1. Experimental characteristics of magnetron.

Fig. 2. Electrical equivalent circuit model of magnetron.

voltage) and an ideal switch as shown in Fig. 2. As illustrated in this figure, the position of the ideal switch in the equivalent circuit of the magnetron is selected

whether the voltage between anode and cathode is higher than the cut off voltage or not. In this paper, the stable microwave power is required for the semiconductor manufacturing production for industrial applications and the magnetron is used under the condition of continuous oscillation. Therefore, the only stable oscillation state is considered for the converter operation.

III. SERIES-RESONANT DC-DC CONVERTER WITH HIGH-FREQUENCY TRANSFORMER LINK

A. Operation in Discontinuous Current Mode (DCM)

Figure 3 shows the schematic circuit configuration of the conventional type DC-DC power converter with a high-frequency transformer link and voltage doubler circuit, which is designed for a high power magnetron drive in industrial power applications (this circuit is called circuit 1 in the following). This DC-DC converter circuit is composed of high frequency high voltage transformer, resonant high-frequency full bridge inverter with a series resonant capacitor C_r in the primary side of the high frequency transformer, full-wave voltage doubler type rectifier circuit, current smoothing inductor L_o, and a magnetron to generate microwave power in the high frequency transformer secondary side. The magnetron of Fig. 3 is represented by its electrical equivalent circuit shown in Fig. 2 under the condition of continuous oscillations. The circuit parameters and the design specifications of this circuit are indicated in table I.

TABLE I
DESIGN SPECIFICATIONS AND CIRCUIT PARAMETERS OF CIRCUIT 1.

Equivalent DC Voltage of Input Source	E	283 [V]
Leakage Inductance of HF-Transformer	L_e	2.3 [μH]
Primary Self-Inductance of HF-Transformer	L_1	26.4 [μH]
Turns-Ratio of HF-Transformer	N_2/N_1	20
Series Resonant Capacitor	C_r	467.3 [nF]
Output DC Filter Capacitor	C_1, C_2	11 [nF]
Output DC Filter Inductor	L_o	0.3 [H]
On Time of Gate Voltage Pulse	T_{on}	5.16 [μs]
Cut-off Voltage of Magnetron	V_Z	7.41 [kV]
Equivalent Resistance in Magnetron	R_1	266 [ ]

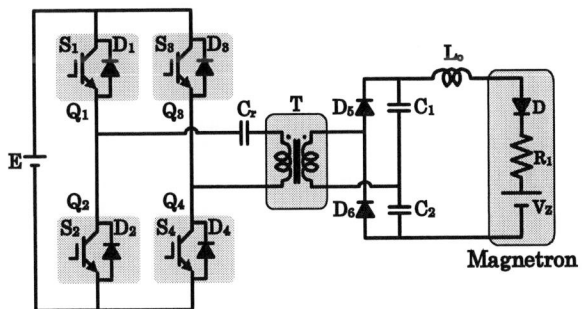

Fig. 3. High frequency link series resonant DC-DC converter (Ckt. 1).

In the oscillating state of the magnetron, its input side DC voltage can be c nearly considered constant since it is represented by a linear v-i characteristic with small slope as indicated in Fig. 1. Thus, the microwave power from the magnetron is proportional to its anode current.

505

Therefore, the output power of magnetron can be regulated by controlling the anode current with a closed feedback loop. Consequently, the runaway of the magnetron is prevented and the stable high microwave power can be effectively generated and controlled.

The steady state voltage and current waveforms in a DCM operation in the previously-developed series resonant DC-DC converter are illustrated in Fig. 4. In this case, the current through transformer primary winding is discontinuous. As shown in Fig. 4, the gate pulse signal sequences of IGBTs are designed to regulate the pulse frequency under constant on-time condition. The IGBTs are turned on with ZCS and turned off with hybrid ZVS&ZCS in all power regulation range when the switching frequency of this DC-DC converter is less than half of the resonant frequency decided by the resonant capacitor and the leakage inductance of high-frequency transformer. If the switching frequency is increased more than half of the resonant frequency, bridge current of this converter becomes continuous waveform and this converter operates in continuous current mode (CCM). Consequently, the operation becomes hard switching.

However, in case of using a high-frequency transformer with leakage and magnetizing parasitic circuit parameters, it is actually difficult to realize the DCM operation to achieve a zero current soft commutation. In order to implement this operating mode, it is necessary that the magnetizing current through the high-frequency transformer primary winding to be nearly zero. Nevertheless, the magnetizing current remarkably flows in the DC-DC converter with a capacitor input type output-smoothing circuit and a constant high voltage load as the magnetron is equal to zero. Because the output rectifier circuit is cut off when the secondary side voltage of the high-frequency transformer is less than the output-smoothing capacitor voltage, and the series resonant inverter is isolated separately from the voltage doubler rectifier with the capacitor input filter. In this case, only magnetizing current of the transformer circulates through the circuit of transformer primary side during the period of the cut off mode of the voltage doubler type rectifier. Consequently, the discontinuous current operating mode (see Fig. 5) of this converter can not appear and realize practically.

B. Hard Switching Operating Mode with Magnetizing Inductor

Figure 5 shows the measured operated waveforms of the current flowing through each IGBT switch and the voltage across it of the circuit 1 (see Fig. 3). In comparison with the simulation voltage and current waveforms in Fig. 4, the measured switch current waveforms observed in Fig. 5 are distorted. It is easily proven that this power converter operates under a hard switching mode due to the remarkably confirmed high voltage surges in the operating current and voltage waveforms.

At turn-on switching transition, the high-frequency

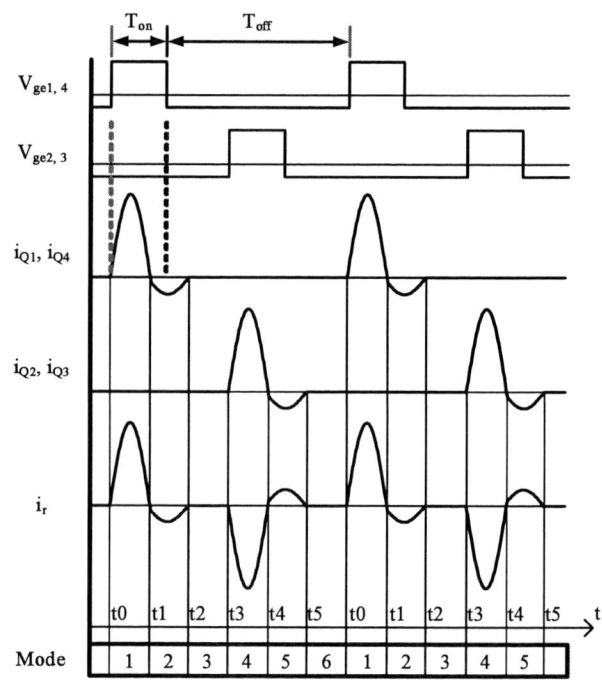

Fig. 4. Operating waveforms in discontinuous current mode.

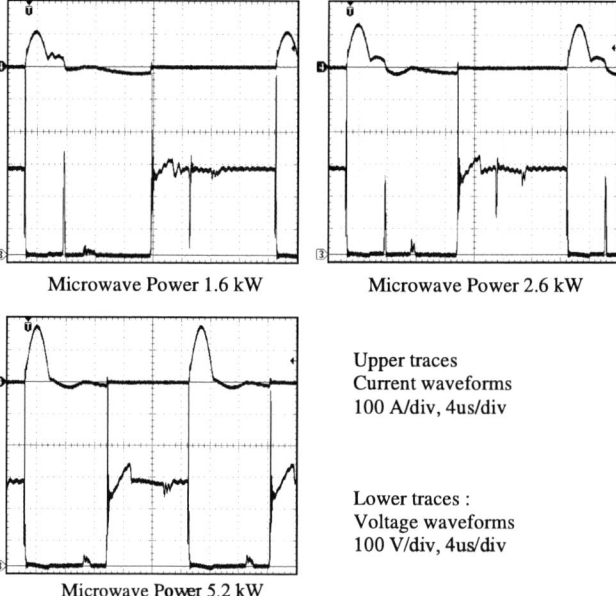

Microwave Power 1.6 kW   Microwave Power 2.6 kW

Microwave Power 5.2 kW

Upper traces
Current waveforms
100 A/div, 4us/div

Lower traces :
Voltage waveforms
100 V/div, 4us/div

Fig. 5. Experimental voltage and current waveforms of Q1 (Circuit 1).

transformer primary side current has an initial value due to the magnetizing current through the high-frequency high-voltage transformer.

Therefore, the current flowing through the switches jumps to this initial value, and IGBTs in the bridge arms of the series transformer resonant inverter has hard switching commutation at turn-on in all power regulation setting range. Extremely high voltage surges actually occur with this high di/dt stress. At turn-off switching transition, the case which occurs while the current flowing through the IGBT switches is forcibly cut off to zero before the zero current crossing point. Consequently, the IGBT switches are turned-off at hard switching commutations.

IV. IMPROVED SERIES TRANSFORMER RESONANT DC-DC POWER CONVERTER WITH LOSSLESS INDUCTIVE SNUBBER

In order to solve the significant problems mentioned above, a single inductive snubber assisted series-resonant ZCS-PFM DC-DC power converter with a high-frequency transformer link is proposed in Fig. 6 (this circuit is called circuit 2 in the following). This DC-DC power converter has a single inductive lossless snubber L_s in the input DC busline of the full bridge inverter. The current flowing through IGBT switches rises gradually at turn-on with the aid of this inductive snubber, and ZCS turn-on commutation can be achieved completely.

The gate pulse timing sequences of the IGBT switches of this DC-DC converter are illustrated in Fig. 7. The duty cycle of each gate pulse signal of IGBT is designed for a constant duty cycle of 50%. The pulse width of each gate signal varies with the switching frequency. As shown in Fig. 8, the IGBT switches can be always turned off while a current continues to flow through the antiparallel diodes and the primary winding of the transformer. Thereby, IGBT switches can be turned off with ZVS&ZCS hybrid commutation. Furthermore, even though the dead time between two gate pulse signals is set to zero, the input DC busline of this newly developed DC-DC converter can not be shorted due to the effect of the single lossless inductive snubber connected in the input side of the full bridge inverter. Therefore, this DC-DC converter circuit is possible to perform zero current soft switching commutation over wide power regulation range in PFM strategy.

The design specification of this DC-DC converter in this experiment is indicated in table II. The observed voltage and current operating waveforms of IGBT switches in the proposed converter are shown in Fig. 8. Observing Fig. 8, it is easy to prove that this proposed DC-DC converter can operate under a principle of ZCS operation at both turn-on and turn-off transitions.

The microwave power characteristic is shown in Fig. 9, while the measured efficiency characteristics for the output power of the circuit 1 and 2 are comparatively shown in Fig. 10. The power conversion efficiency of the proposed soft switching DC-DC converter is improved in comparison with that of the conventional DC-DC converter in the wide range of the high power region.

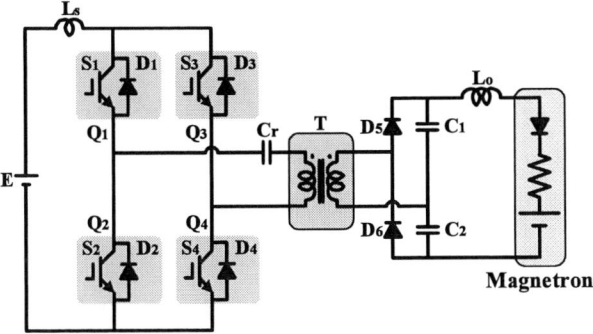

Fig. 6. Proposed soft switching DC-DC power converter (Circuit 2).

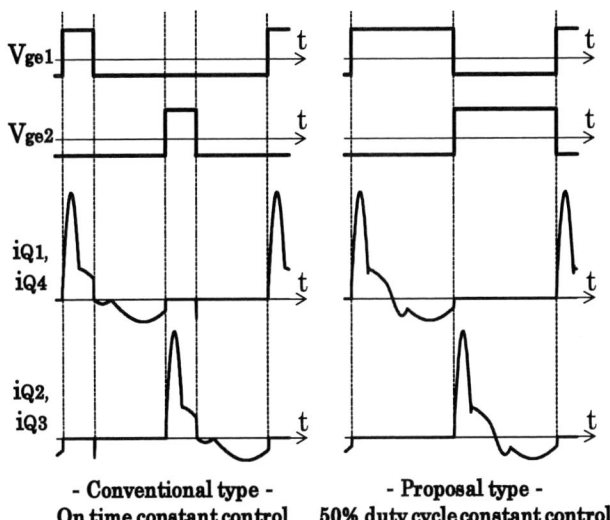

- Conventional type -
On time constant control

- Proposal type -
50% duty cycle constant control

Fig. 7. 50% Duty cycle constant PFM control scheme.

TABLE II
DESIGN SPECIFICATIONS AND CIRCUIT PARAMETERS OF CIRCUIT 2.

Equivalent DC Voltage of Input Source	E	283 [V]
Lossless Inductive Snubber	L_s	1 [nH]
Leakage Inductance of HF-Transformer	L_l	2.3 [nH]
Primary Self-Inductance of HF-Transformer	L_1	26.4 [nH]
Turns-Ratio of HF-Transformer	N_2/N_1	20
Series Resonant Capacitor	C_r	467.3 [nF]
Output DC Filter Capacitor	C_1, C_2	11 [nF]
Output DC Filter Inductor	L_o	0.3 [H]
Cut-off Voltage of Magnetron	V_Z	7.41 [kV]
Equivalent Resistance in Magnetron	R_1	266 [ ]

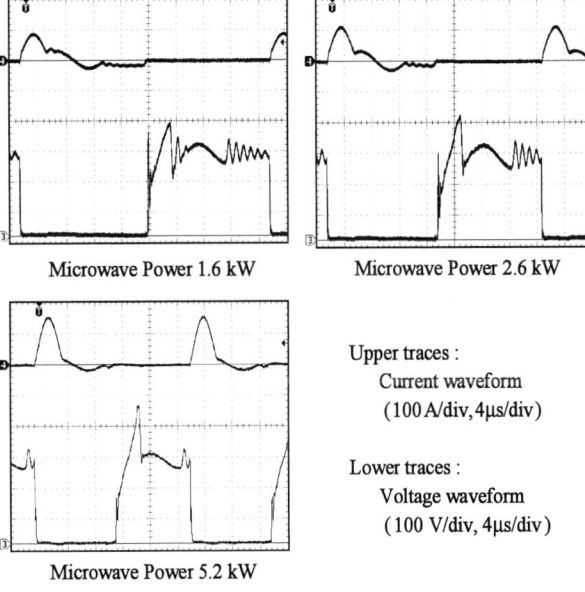

Microwave Power 1.6 kW Microwave Power 2.6 kW

Microwave Power 5.2 kW

Upper traces :
Current waveform
(100 A/div, 4μs/div)

Lower traces :
Voltage waveform
(100 V/div, 4μs/div)

Fig. 8. Experimental voltage and current waveforms of Q1 (Circuit 2).

V. REDUCING PEAK CURRENT STRESSES

As stated in section III, when the switching frequency is more than half the resonant frequency, the conventional DC-DC converter operates on hard-switching transition even in the good condition. However, the proposed DC-DC converter operates on soft-switching transition even

507

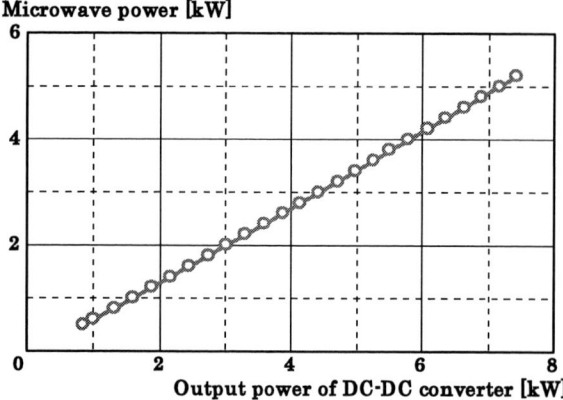

Fig. 9. Output power of DC-DC converter vs. microwave power.

in the good condition. However, the proposed DC-DC converter operates on soft-switching transition even if the switching frequency is more than half the resonant frequency, because its soft-switching operation bases on continuous current mode.

With a proper design of the circuit parameters, the peak value of the currents through the IGBT switches can be significantly reduced. Fig. 11 shows the observed voltage and current waveforms of the IGBT switches on the proposed converter with the proper circuit parameters (this circuit is called circuit 3). The design specification of this circuit is indicated in table III. These circuit parameters are designed to actively utilize the continuous current mode, in other words, the maximum of switching frequency on circuit 3 is set near the resonant frequency of the resonant circuit to increase the utilization factor of electric power. Observing Fig. 11, it is confirmed that the peak values of the current flowing through the IGBT switches of circuit 3 are much lower than those on the previous converter circuits (circuit 1 and circuit 2).

The measured peak current values for a wide output power control range of the conventional and proposed DC-DC converter are indicated in Fig. 12. It is proven from this figure that the peak current value of IGBT on circuit 3 is lower than the half of that on previous type

over all power regulation The measured peak current values for a wide output power control range of the conventional and proposed DC-DC converter are indicated in Fig. 12. It is proven from this figure that the peak current value of IGBT on circuit 3 are lower than the half of that on previous type over all power regulation range. The improvement of the efficiency can be expected by improving the characteristics of IGBT switches with reduced peak current values and rated values. Fig. 13 shows the power conversion

TABLE III

DESIGN SPECIFICATIONS AND CIRCUIT PARAMETERS OF CIRCUIT 3.

Items	Symbol	Value
Equivalent DC voltage of input source	E	283 V
Lossless inductive snubber	Ls	3.5 mH
Leakage inductance of HF transformer	L•	11.3 mH
Primary self-inductance of HF transformer	L_1	35.4 mH
Turn ratio of HF transformer	N_2 / N_1	20
Series resonant capacitor	Cr	274.8 nF
Capacitance of output filter capacitor	C_1, C_2	11 nF
Inductance of output filter inductor	Lo	0.3 H
Cut-off voltage of magnetron	Vz	7.41 kV
Equivalent resistance of magnetron	R_1	266 Ω

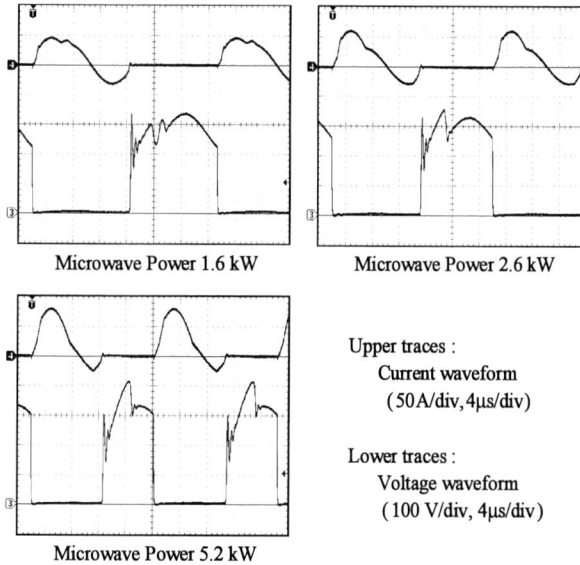

Upper traces :
Current waveform
(50A/div, 4µs/div)

Lower traces :
Voltage waveform
(100 V/div, 4µs/div)

Fig. 11. Experimental voltage and current waveforms of Q1 Circuit 3).

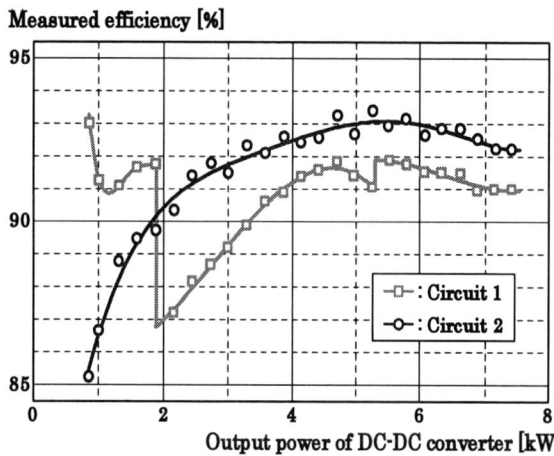

Fig. 10. Comparison of power conversion efficiency vs. output power characteristics for two converter topologies.

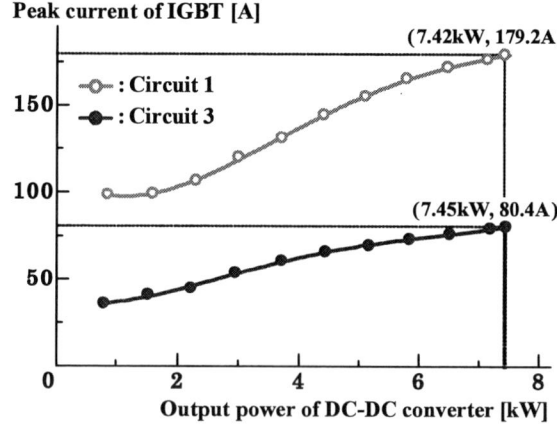

Fig. 12. Comparison of peak values of current flowing through IGBT.

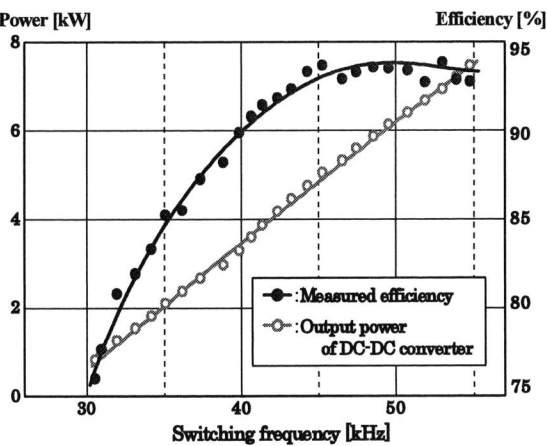

Fig. 13. Output power and measured efficiency vs. switching frequency characteristics of circuit 3.

efficiency and the output power characteristics of the experimental prototype of proposed DC-DC converter designed under the specification of table III. The maximum value of the efficiency of circuit 3 shown in Fig. 13 is about 94.3[%], and this value is higher than that of circuit 2.

VI. CONCLUSIONS

In this paper, a transformer parasitic parameter and a lossless inductive snubber assisted series-resonant ZCS-PFM DC-DC converter has been proposed in order to improve the significant problems of the hard switching commutation at turn-on and turn-off the active power switching devices in series-resonant PFM controlled DC-DC power converter with a high-frequency high-voltage transformer link. Based on the experimental results of the proposed pulse frequency modulated DC-DC power converter with a high-frequency high-voltage transformer, it was actually confirmed that all the active power switches could achieve ZCS commutation operation. The unique features of the proposed DC-DC power converter applied to a high power magnetron drive could be stated in the following:

ZCS commutation could be implemented in the all DC power regulation range and high power conversion efficiency could be performed in a wide power regulation range.

The transformer parasitic circuit components as leakage and magnetizing inductive components were effective to achieve soft switching (ZCS) operation with the aid of a single lossless inductive snubber.

The peak values of the current flowing through IGBT switches could be reduced more than the half of that on conventional DC-DC converter over all power regulation ranges by actively utilizing the continuous current mode operation with a proper design selection of circuit parameters.

REFERENCES

[1] A. Harada, "How to Use C.W. Magnetrons", *Proceedings of Microwave Effect and Application Symposium*, Japan, August, 2001.

[2] K. Harada, T. Ninomiya, "Fundamentals of Switched-Mode Converters", *Corona Book Publishing Co., Ltd.*, February, 1992,

[3] T. Matsushige, E. Miyata, M. Ishitobi, and M. Nakaoka, "Voltage-Clamped Soft-Switching Inverter-Fed DC-DC Converter for Microwave Oven and Utility AC Side Harmonic Current Evaluations", *Proceedings of IEEE-IAS International Appliance Technical Conference, IATC*, pp. 185-195, Kentucky, USA, May, 2000.

[4] T. Myoi, M. Ishitobi, L. Gamage, and M. Nakaoka, "Lossless Inductive Snubber-Assisted Series Resonant DC-DC Converter with ZCS-PFM Control Scheme", *Proceedings of International Power Electronics and Motion Control Conference (EPE-PEMC)*, Dubrovnik, Croatia, September, 2002.

[5] E. Hiraki, M. Nakaoka, "Practical Power Loss Analysys Simulator Development of Switching Mode Power Converter Using Measured Characteristic Values of Power Semiconductor Devices", *The Trans.of Institute of Electrical Engineering of Japan, IEEJ*, vol. 122-D, no. 12, December, 2002

[6] Yuzurihara, A. Takayanagi, and H. Fujikawa, "Microwave Generator System for Plasma Application (vol. 2)", *Kyosan Technical Circular,* vol. 53, no. 3, 2002.

[7] Shibata, "Industrial Microwave Power Engineering", Denkishoin Publishers Co., Ltd., December, 1986.

A General Method for Deciding the Input Filter Capacitance of Flyback Switching AC-DC Converter with Peak Current-Controlled Mode

Jiaxin Chen*,**, Jianguo Zhu**, and Youguang Guo**

* College of Electromechanical Engineering, Donghua University, Shanghai 200051, China
** Faculty of Engineering, University of Technology, Sydney (UTS), NSW 2007, Australia

Abstract–This paper presents a general method for deciding the input filter capacitance of flyback switching ac-dc converters with peak current-controlled mode. Firstly, a simulation model for flyback ac-dc converter is obtained by adding the rectifier and filter circuit to a flyback dc-dc converter model developed by the authors. The simulation results show that the processes of capacitor charging and discharging are independent, their boundary is near the maximal value of input voltage, and the part of flyback dc-dc converter can be seen as an approximately constant power load. Secondly, an analytic model for deciding the input power of flyback dc-dc converter with rated load and different input dc voltages is presented. Furthermore, the effect caused by the parasitical parameters in the electronic parts is studied, and the corresponding analytical method for deciding the input filter capacitance is given. The effect on the capacitance caused by the control delay is analyzed qualitatively.

Index Terms--Analytic model, flyback switching ac-dc converter, input filter capacitance, numerical simulation model.

I. INTRODUCTION

Flyback switching ac-dc converter with peak current-controlled model is commonly used as small power converter because of its simple structure. A comprehensive analysis for improving its performance is always desirable, so a great amount of work has been done on the operational principle, design methodology, modeling, and control of the flyback converter [1-5]. As the converter often works in wide input alternating voltage such as 85 - 264 VAC, the input capacitance has heavy impact on the output capability and other electrical specifications of converter. The flyback converter is a highly nonlinear and complicated system, and little literature introduced about the method for deciding its parameters. Although some papers [3-4] presented empirical formulae or principle to decide the capacitance, they are not accurate enough to meet the requirement of power converter design yet. In the actual process of designing a converter, engineers have to apply large numbers of trials and errors, which are costly.

With the development of the technique of computer aided design (CAD), many problems in nonlinear systems can be solved, and some new physical rules may be

obtained. In several current commercial CAD tools for control system design, Matlab/Simulink is one of the most widely used CAD systems for its convenience and strong performance. To analyze a nonlinear system, comparing with the analytic method, the numerical method based CAD often has higher accuracy but takes more time, and is harder to reveal the physical rules. As they are complementary, it is advantageous to obtain both of them for a system. For that, two methods for deciding the input filter capacitance of flyback switching ac-dc converters with peak current-controlled mode are introduced in this paper. Furthermore, the impact on the capacitance due to the parasitical resistance in the input circuit and the control delay is also analyzed.

II. NUMERICAL METHOD FOR DECIDING THE INPUT FILTER CAPACITANCE IN FLYBACK AC-DC CONVERTER

Fig. 1(a) shows the typical topology of a flyback ac-dc converter with peak current-controlled model, where the dashed line connects control block to the main circuit, C_0 is the input filter capacitance, and R_0 is the parasitical resistance in the input circuit.

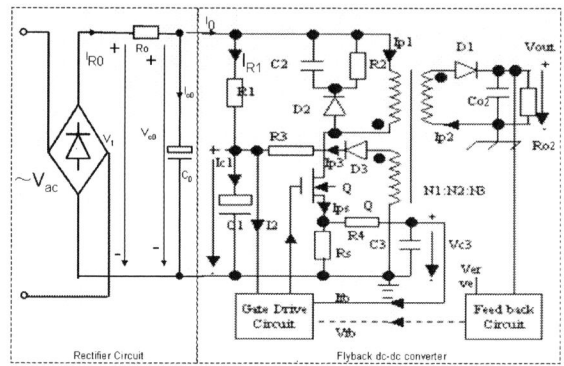

Fig. 1. Typical circuit of a flyback ac-dc converter

A. Modelling the Flyback Switching AC-DC Converter

Paper [1] introduced a simulation model of flyback dc-dc converter in Matlab/Simulink surrounding. The circuit of flyback ac-dc converter (Fig. 1) can be divided into two parts as shown in two dashed frames separately: one is the rectifier, and the other is the flyback dc-dc converter. The complete simulation model of flyback ac-dc converter is obtained by only adding the model of rectifier circuit, as shown in Fig. 2, to that in [1], which is in the subsystem of flyback dc-dc converter.

This work was partly supported by the Endeavour Australian Postgraduate and Postdoctoral Fellowships Program.

978-1-4244-0644-9/07/$25.00 ©2007 IEEE   510

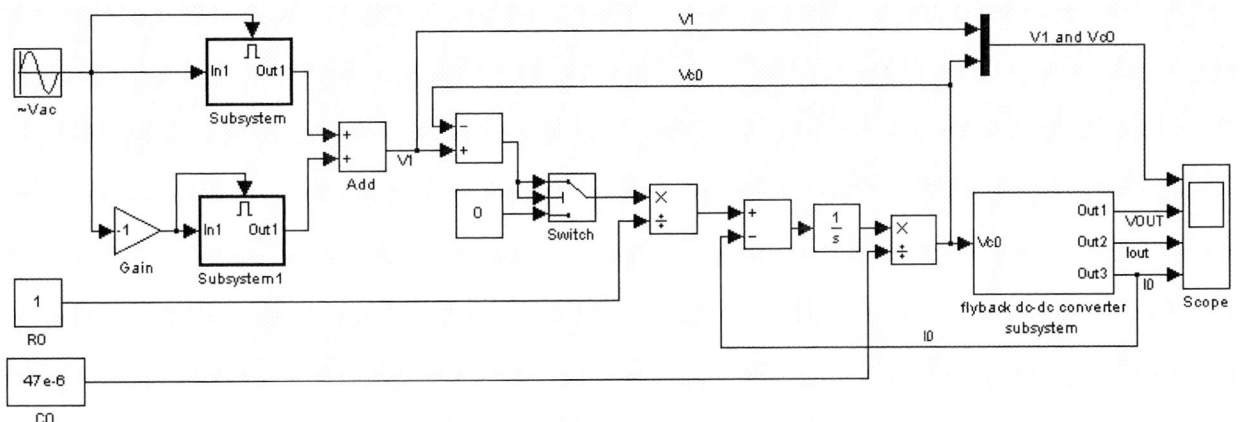

Fig. 2. Simulation model of the flyback ac-dc converter

B. Performance Analysis

The model shown in Fig. 2 is applied to the flyback switching ac-dc converter to analyze whether or not the value of the filter capacitance C_0 can meet the requirement of the output ability. The major data of the converter include [1]: Input voltage: V_{ac}=85-264 VAC/50 Hz; Nominal output voltage: V_{out}=5 VDC; Rated output current: I_{out}=3.6 A; Switching frequency: f=60 kHz; Peak value of the primary current: I_{p1max}=0.89 A; Input filter capacitance: C_0= 47 μF; Parasitical resistance in the input circuit: R_0=1Ω; Magnetic Core: EI25; Magnetic material: TDK PC40.

All the left parameters of electronic parts in the flyback dc-dc converters are the same as those in [1]. By putting all the parameters of electronic parts into the proposed simulation model and running the model in Matlab/Simulink surrounding, some important results are obtained and shown in Fig. 3(a)-(h).

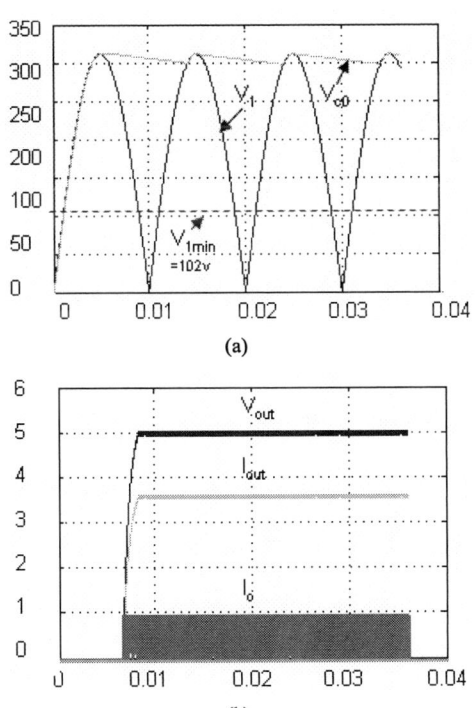

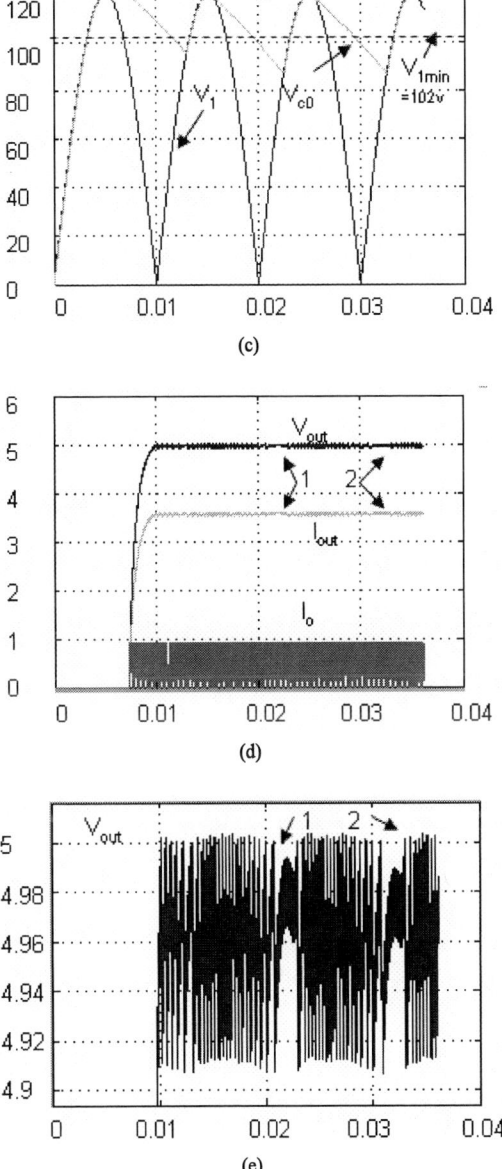

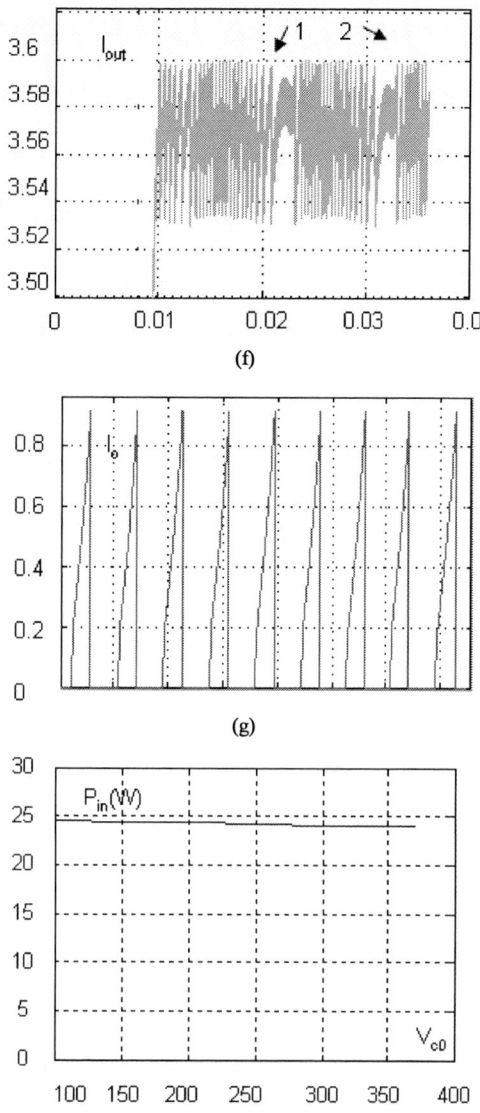

(f)

(g)

(h)

Fig. 3. Simulation results: (a) output voltage of rectifier V_1, and voltage across the capacitance C_0, V_{c0} when the input voltage V_{ac}=220 V; (b) output voltage V_{out}, output current I_{out}, and input current I_0 of flyback dc-dc converter when V_{ac}=220 V; (c) V_1 and V_{c0} when V_{ac}=85 V; (d) V_{out}, I_{out}, and I_o when V_{ac}=85 V; (e) detailed V_{out} when V_{ac}=85 V; (f) detailed I_{out} when V_{ac}=85 V; (g) detailed I_o when V_{ac}=85 V; and (h) input power of flyback dc-dc converter P_{in} versus input voltage V_{c0}.

According to the results, the following conclusions can be obtained. Firstly, as the parasitical resistance in the input circuit is very small and has little influence on the charging current to the filter capacitance, the processes of charging and discharging of the input filter capacitor are independent, and the boundary is near the maximal value of input voltage. Secondly, the input power of flyback dc-dc converter under different input alternating voltages can be seen as an approximately constant power load. Thirdly, according to the place pointed by arrows 1 and 2 in Fig. 3(d)-(f), it is found that when the voltage V_{c0} goes below the minimal voltage V_{1min}=102 V of flyback dc-dc converter, the output voltage of the converter with the rated load is smaller than the rated output voltage of 5 V. This also has strong impact on other performances of

converters. It indicates that the adopted C_0 is not large enough.

If ignoring the impact of the resistor R_0, according to the above three conclusions, the minimal value of the input filter capacitance C_{0min} can be approximately decided by the following equation:

$$C_{0min} = \frac{2P_{in}\left[\dfrac{T}{4} + \arcsin\left(\dfrac{V_{1min}}{\sqrt{2}V_{acmin}}\right)\right]}{2V_{acmin}^2 - V_{1min}^2} \quad (1)$$

where T is the period of the input alternating voltage (0.02 s here), and V_{acmin} is the minimal input alternating voltage of flyback ac-dc converter (V_{ac}=85 V here).

III. ANALYTIC METHOD FOR DECIDING THE INPUT FILTER CAPACITANCE IN FLYBACK AC-DC CONVERTER

In fact, the analytical method for deciding the input filter capacitance in flyback ac-dc converter has been given as (1). The only question left to be solved is the lack of theoretical verification that the input power of the flyback dc-dc converter can be seen as constant under different input dc voltages, even although it has been proved in a way by the simulation results shown in Fig. 3(h). For this, three parts of work are included in this section.

A. Theoretical Deduction under Ideal Assumptions

As an analytic method is often built based on the assumptions that all the electronic parts are ideal, the following assumptions and simplifications are made:

(1) The power MOSFET in the ON state is modeled by a zero resistance and in the OFF state by an infinite resistance, R_{INF}. The output capacitance and inductance of the leading wires are ignored.

(2) The diode in the ON state is modeled by a constant voltage source, V_F, and in the OFF state by an infinite resistance. The diode junction capacitance is negligible.

(3) All the leakage inductances and stray capacitances of the transformer are neglected.

(4) Passive components are linear, time invariant, and frequency independent.

Then the equivalent circuit of flyback dc-dc converter (a buck-boost dc-dc converter) is obtained and shown in Fig. 4.

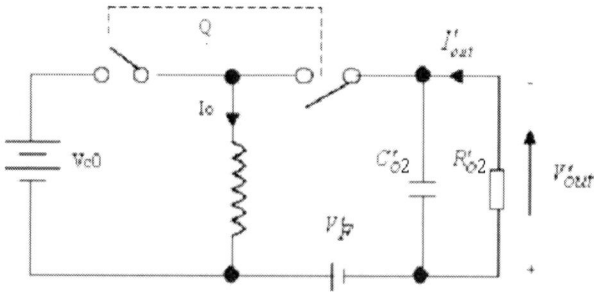

Fig. 4. Typical buck-boost converter circuit

The parameters of the secondary winding are referred to the primary side by

$$V'_F = \frac{N1}{N2}V_F \ , \ C'_{o2} = \left(\frac{N2}{N1}\right)^2 C_{o2} \ \text{and} \ V'_{out} = \frac{N1}{N2}V_{out}$$

where $N1$ and $N2$ are the numbers of turns of the primary and secondary windings of the transformer, respectively.

Because the operation state of dc-dc converter with the rated load and the maximal input dc voltage V_{1max} is often designed as the critical conduction mode, which can be processed as a special continuous conduction mode (CCM), then according to the analysis in [2], all the operation state with the rated load is CCM in the full input voltage range.

According to Fig. 4, the following equation can be obtained:

$$dI_o = \frac{V_{c0}}{L_1}DT = \frac{V'_{out} + V'_F}{L_1}(1-D)T \tag{2}$$

where I_o is the current in the primary winding of the transformer, V_{c0} is the input voltage, L_1 is the primary winding inductance, D is the duty ratio, and T is the time period of a duty cycle. From (2), the duty ratio can be calculated by

$$D = \frac{V'_{out} + V'_F}{V'_{out} + V'_F + V_{c0}} \tag{3}$$

The duty ratio of PWM only depends on the equivalent input voltage, equivalent voltage source of diode and output voltage; it is not affected by the load and the inductance.

According to the output equation during one period, i.e. $dV'_{out}=0$, the following equations are obtained:

$$\frac{i_o - i'_{out}}{C'_{o2}}(1-D)T = \frac{i'_{out}}{C'_{o2}}DT \tag{4}$$

$$i_o = i'_{out}/(1-D) \tag{5}$$

$$P_{DT} = V'_F i_o (1-D) = V'_F i'_{out} \tag{6}$$

$$P_{out} = V'_{out} i'_{out} \tag{7}$$

where P_{DT} is the steady-state power loss of diode D1, and P_{out} is the rated output power of the converter. As only two parts (R'_0 and V'_F) in Fig. 2 can consume power, the input power of the dc-dc converter can be calculated by

$$P_{in} = P_{DT} + P_{out} = (V'_F + V'_{out})i'_{out} \tag{8}$$

With the rated output load, all the variables of V'_F, V'_{out}, and i'_{out} are constant, so the input power of the converter P_{in} is also constant. This gives the theoretical verification that the input power of the flyback dc-dc converter can be seen as constant.

B. Impact on the Input Power by Parasitical Parameters in Actual Electronic Parts

Although there are many parasitical parameters in the converter, according to [3-4] and Fig. 1, the main parasitical parameters, which have heavy influence on the total power loss P_z, are constituted by five parts: the power loss in the transformer, $P_{cu\text{-}fe}$, the power loss in the snubber circuit of $R_2C_2D_2$, P_{rcd}, the switching power loss, P_{ms}, which also includes the power loss consumed by the current sensor resistor, R_s, the power loss of V_F in diode D1, P_{DT}, and the power loss of R_F in diode D1, P_{RT}. Before the further analysis, the following assumptions and simplifications are made:

(1) The transient process of switching ON and OFF is omitted, and then the switching power loss P_{ms} just takes the power loss consumed by the resistance R_{MON} of MOSFET in conducting state and R_s.

(2) Considering that all the leakage inductance and resistance in the transformer, the diode conducting resistor R_{DON} and the resistors R_{MON} and R_s have small influence on the duty cycle ratio D, so the duty cycle ratio can still be calculated by (3). Furthermore, all the energy in leakage inductance is approximately equal to that absorbed by the snubber circuit of $R_2C_2D_2$.

(3) Except for the power loss P_{DT}, as the sum of all the other power losses is very small, in order to simplify their calculation, all the currents here are considered by their average values.

Based on the above assumptions, the power losses can be obtained by the following equation group

$$\begin{cases} P_{cu\text{-}fe} = (\frac{i'_{out}}{1-D})^2 r_1 + P_{fe} \\[2mm] P_{RT} = i'^2_{out} R^2_{DON}(\frac{N_1}{N_2})^2 \\[2mm] P_{DT} = V'_F i'_{out} \\[2mm] P_{rcd} = \frac{L_s i'_{out} V_{c0}}{L_1 + L_s}\frac{D}{1-D} \\[2mm] P_{ms} = (\frac{i'_{out}}{1-D})^2 (R_{MON} + R_s)D \end{cases} \tag{9}$$

where r_1 is the resistance of the primary winding of the transformer, and P_{fe} is the core loss in the transformer with rated specifications, including eddy current loss, hysteresis loss and anomalous loss. According to the analysis in [5], the core loss can be seen as approximately constant.

As the duty cycle ratio is calculated by (3), all kinds of power loss in (9) can be calculated analytically. Along with (10) and (11), the total power loss P_z and the input power P_{in} can also be calculated analytically.

$$P_z = P_{cu\text{-}fe} + P_{RT} + P_{DT} + P_{rcd} + P_{ms} \tag{10}$$

$$P_{in} = P_{out} + P_Z = V'_{out}i'_{out} + P_Z \tag{11}$$

From (9) and (10), it is easy to find that when the input voltage V_{c0} decreases, the total power loss P_z will increase. This is consistent with that shown in Fig. 3(h). By replacing the input power P_{in} in (1) with (11), the minimal input filter capacitance with higher accuracy is obtained. It should be pointed out that according to (3), the power loss of diode D1 takes about 60% of total power loss, and so the input power is approximately constant with different input voltages.

C. Influence Caused by Control Delay

For the flyback dc-dc converter with peak current-controlled mode, the maximum current of I_o is often decided at the minimal input voltage V_{1min}. However, the circuit of RC filter such as R_4 and C_3, and transferring delay of switching control integrated chip (IC) will increase the influence on the maximum current of I_o. This ensures that the minimal value of C_o obtained from (1) can also satisfy the requirement of system output ability. As introduced in [1], for being limited by system reliability, the control delay caused by RC filter has been minimized, so its impact on the minimal value of C_o is very limited.

IV. Conclusions

This paper has presented a general method for deciding the input filter capacitance of flyback switching ac-dc converters with peak current-controlled model. Both numerical method and analytical method are given. Both the simulation results and theoretical deduction prove that the flyback dc-dc converter can be considered as a constant power load. The effects of parasitical parameters and control delay in the electronic parts are studied, and the corresponding analytical method for deciding the input filter capacitance is given. The impact on the capacitance caused by the control delay is analyzed qualitatively.

REFERENCES

[1] J. X. Chen, J. G. Zhu, Y. G. Guo, and J. X. Jin, "Modeling and simulation of flyback DC-DC converter under heavy load," in *Proc. IEEE Int. Conf. on Communications, Circuits and Systems*, Guilin, China, 25-28 June 2006, vol. 4, pp. 2757-2761.

[2] J. X. Chen, J. G. Zhu, and Y. G. Guo, "Calculation of power loss in output diode of a flyback switching DC-DC converter," in *Proc. ICES/IEEE Int. Conf. on Power Electronics and Motion Control*, Shanghai, China, 13-16 August 2006, vol. 1, pp. 1-5.

[3] Marty Brown, Power Supply Cookbook. 2nd Edition. England: Elsevier Science Ltd., 2001.

[4] Z. S. Zhang and X. S. Cai, Theory and Design of Switching Converters (in Chinese). Beijing: Electronics Industry Publishing Company, 1999.

[5] J. X. Chen, Y. G. Guo, and J. G. Zhu, "A generalized dynamic model for flyback switching converter based on nonlinear finite element analysis," in *Proc. IET Int. Conf. on Technology and Innovation*, Hangzhou, China, 6-8 November 2006, vol. 1, pp. 1-6.

Design of High Performance and Low Cost Line Impedance Stabilization Network for University Power Electronics and EMC Laboratories

D. Sakulhiririrak*, V. Tarateeraseth**, W. Khan-ngern* and N. Yoothanom***

*King Mongkut's Institute of Technology Ladkrabang, Faculty of Engineering, Bangkok, Thailand.
**Srinakharinwirot University, Faculty of Engineering, Ongkharak, Thailand.
***Sripatum University, Faculty of Engineering, Bangkok, Thailand.

Abstract– **This paper proposes how to analyze and design the Line Impedance Stabilization Network (LISN). The details are described component characteristics and how to design air coil inductors in single and multi layers based on a high self resonant frequency response concept. According to the CISPR 16-1 standard, the stabilized impedance 50 Ω of LISN at frequency range 150 kHz to 30 MHz is proved by simulation and experimental results. The proposed LISN is successful design the low cost by compare with commercial LISN. Finally, the performance of the proposed LISN is achieved comparing to CISPR standard and also with commercial LISN.**

Index Terms--**LISN, AMN, EMI, EMC**

I. INTRODUCTION

Line Impedance Stabilization Network: LISN or Artificial Mains Network: AMN is an important equipment for measuring the conducted Electromagnetic Interference (EMI) emissions following CISPR 16-1 standard. There are three main functions of LISN or AMN. Firstly to prevent incoming Radio Frequency (RF) disturbance from the mains supply. Secondly, to maintain the specified impedance at the equipment under test (EUT) terminal over the working frequency range and finally to couple RF voltage signal from the EUT to the measuring receiver.

There are two basic types of LISN, the v-network and the delta-network. The LISN v-network can be coupled the unsymmetric voltages: the amplitude of the vector voltage V_a or V_b where V_a is the vector voltage between one of the main terminals and earth and V_b is vector voltage between the other main terminals and earth. The LISN delta-network can be coupled the symmetric (Differential Mode: DM) and asymmetric (Common Mode: CM) voltages separately [1].

This paper proposes designing of the artificial mains 50 Ω /50 μH + 5 Ω V-network by improved self resonant frequency (SRF) of inductors as in [2] using air gap winding technique. To improve the previous researched work [2], the line/neutral selector switch as shown in Fig. 1 is removed because of two reasons: firstly, two output ports can provide voltage signal from line to ground and neutral to ground simultaneously as shown in equations (1-2), and lastly, this proposed LISN is designed

to apply with other conducted EMI emissions investigated devices such as DM/CM rejection network [3] or Paul-Hardin noise separation network [4] for more advanced research opportunities.

$$V_{Line-Ground} = 50 \cdot (I_{CM} + I_{DM}) \qquad (1)$$

$$V_{Neutral-Ground} = 50 \cdot (I_{CM} - I_{DM}) \qquad (2)$$

where I_{CM} is a common mode current

I_{DM} is a differential mode current

Fig. 1 shows a suitable circuit with the component values according to CISPR 16-1. These are the recommendable schematic and component values. It may be constructed for use with the current up to 100 A. Moreover, the recommended case dimension following the CISPR is 360x300x180 mm. [1]

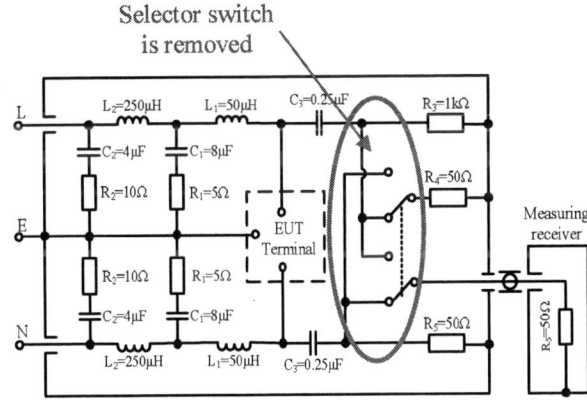

Fig. 1 Artificial mains 50 Ω / 50 μH + 5 Ω V-network.

II. THE PROPOSED SCHEMATIC AND COMPONENT VALUES

Fig. 2 shows proposed schematic of LISN by cut the selector switch off and adding some components such as R_7-R_{10}, C_7-C_8 and two BNC type N terminals (female). Table I shows the proposed LISN component values.

978-1-4244-0644-9/07/$25.00 ©2007 IEEE

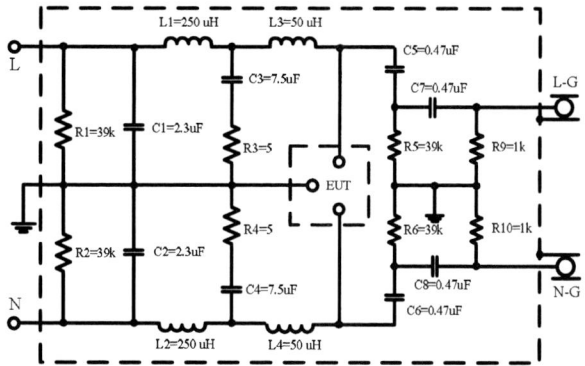

Fig. 2. The proposed schematic of LISN.

TABLE I

PROPOSED LISN COMPONENT VALUES

Component	Value
R_1, R_2	39 kΩ (1 W)
R_3, R_4	5 Ω (5 W)
R_5, R_6	39 kΩ (1 W)
R_7, R_8	1,000 Ω (1/2 W)
C_1, C_2	2.3 μF
C_3, C_4	7.5 μF
C_5, C_6, C_7, C_8	0.47 μF
L_1, L_2	250 μH
L_3, L_4	50 μH

L_3-L_4, C_3-C_4, R_3 and R_5-R_8 define the impedance; L_1-L_2, C_1-C_2 and R_1-R_2 provide the isolation to spurious mains signals and mains impedance variations, and C_6-C_6 decouples the measuring receiver from mains voltage.

III. THE COMPONENT DETAILS

A. Resisters

All resistors are the carbon type and the tolerance should as small as possible. In fact, the resistor, which made from carbon has small response with high frequency.

B. Capacitors

The capacitors, C_1-C_4, have high capacitance. They should have high rated voltage for the safety reason. Capacitors C_1-C_4 are the metallized polypropylene. Capacitors C_5-C_8 are metallized polyester. The SRF of capacitors are provided from manufacturer. The measured self resonant frequency (SRF) response, using impedance analyzer of C_1-C_2 are about 268 kHz, C_3-C_4 are about 169 kHz and C_5-C_8 are about 1 MHz, respectively.

C. Case of network

The case of network is mounted on a metal frame, which is then closed by metal lids. The bottom and side lids are perforated in order to improve the heat dissipation. The dimension of the case is about 300x280x170 mm. This case is smaller than the example case recommended by CISPR 16-1.

D. Inductor core and dimension

This is the key point to keep the proposed LISN to cheapest cost. Normally, the solenoidal winding of the inductor is wound on a coil former of an insulating material. Because of the product cost and temperature operation, in this paper, the unplasticized polyvinyl chloride (PVC) pipe is chosen to be as a coil former as shown in Fig. 3 [5].

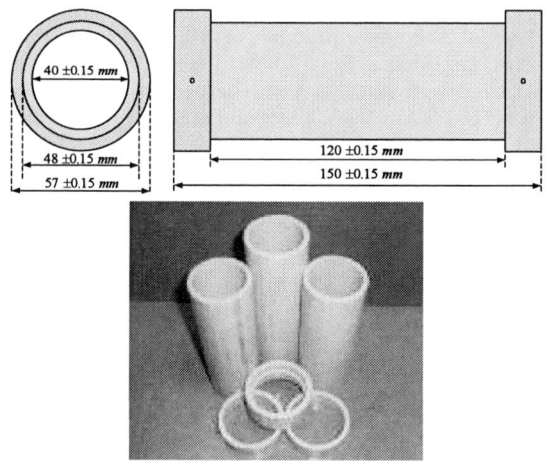

Fig. 3. The coil former of inductors.

IV. AIR CORE INDUCTOR DESIGN

A. Calculations

The inductors, used in proposed LISN, are an air core with solenoidal winding. The unplasticized polyvinyl chloride (PVC) pipe is chosen to be a coil former. The length, diameter and thickness are 120 mm, 48 ± 0.15 mm, and 1.5 ± 0.15 mm, respectively. The maximum operating temperature of PVC pipe is 60 °C [5]. The number of turn and wire size can evaluate by the equations as shown below [2]:

$$a_{wire} = \frac{I}{J} \tag{3}$$

where a_{wire} is a wire area (mm^2)

     I is a rated current (A)

     J is a current density (A/mm^2)

The wire size can be evaluated following equation (3). The rated current and current density are defined to be equal to 10 A and 3 A/mm^2, respectively. The wire diameter is 3.333 mm^2 and AWG 14 is chosen.

$$N = \sqrt{\frac{L l_m}{A_c \mu_0}} \qquad (4)$$

where N is a total numbers of turn (turns)

  L is an inductance (H)

  l_m is a mean magnetic length (m)

  A_C is a cross section of the core (m^2)

  μ_0 is a permeability of air (H/m)

Equation (4) is used specially for calculating the number of turn of L_3 and L_4. The calculated result is about 52 turns by define l_m equal to 120 mm.

L_1 and L_2 design in multi layer coil using the Wheeler's formula in equations (5-6) [6].

$$L(\mu H) = \frac{0.0315 \cdot N^2 \cdot \left(\frac{r_1 + r_2}{2}\right)^2}{6 \cdot \left(\frac{r_1 + r_2}{2}\right) + 9 l_m + 10(r_2 - r_1)} \qquad (5)$$

where r_1 is a core inner radius (mm)

  r_2 is an outer radius include the width of wire (mm)

$$N = (N_1 \cdot N_2) - (N_2 / 2) \qquad (6)$$

where N_1 is the number of turns per layer (turns)

  N_2 is the number of layers (layers)

Equation (5) is used for designing the number of turn for L_1 and L_2, The three layers were chosen. The result from equations (5-6) can be found that the number of turns per layer of L_1 and L_2 is about 39 turns per layer.

B. Winding Techniques

L_1 and L_2 are designed in multilayer. To limit inter-winding capacitance effect, air gap is added for reducing the turn-to-core and stray capacitance. The turn-to-core capacitance (C_{tc}) is equal to two times of total capacitance ($2C_{tt}$). The stray capacitance (C_s) is equal to $C_{tt}/n-1$ [7]. Figs. 4-5 show winding technique applied for L_1 and L_2. The SRF measurement result is about 960 kHz as shown in Fig. 6.

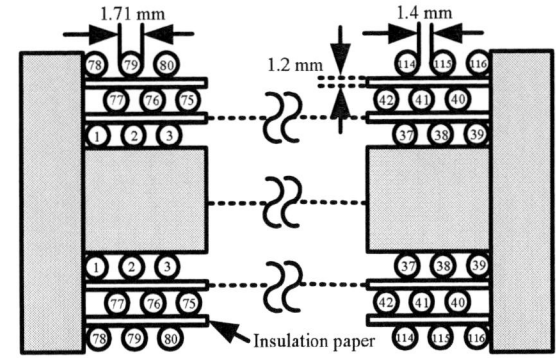

Fig. 4. Model multilayer winding technique of L_1 and L_2.

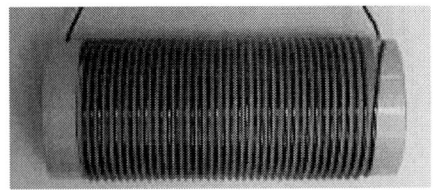

Fig. 5. Multilayer winding with air gap of L_1 and L_2.

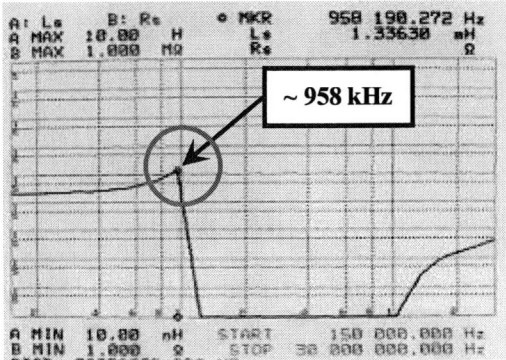

Fig. 6. Measured impedance characteristics of L_1 and L_2.

L_3 and L_4 are single layer with 0.61 mm gap as shown in Fig. 7-8 and the SRF measurement result is about 23 MHz as shown in Fig. 9.

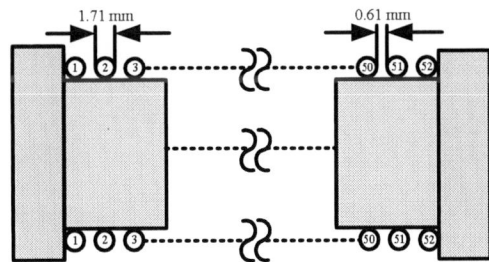

Fig. 7. Model winding technique of L_3 and L_4.

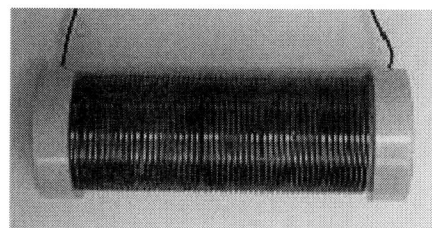

Fig. 8. Single layer winding with air gap of L_3 and L_4.

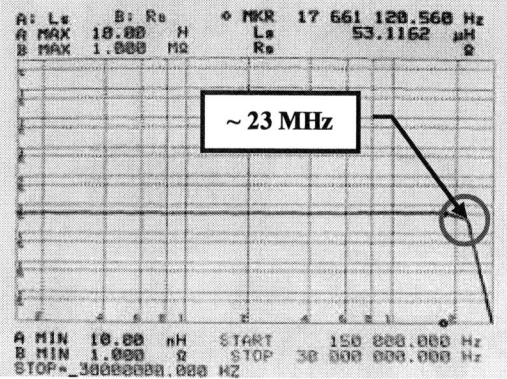

Fig. 9. Measured impedance characteristics of L_3 and L_4.

Actually, the SRF of L_3 and L_4 are good performance about 17 MHz even though they are not added the air gap. Then the improvement of the SRF also keep the size of core, the small air gap about 0.6 mm is added. But the air gap of L_1 and L_2, bigger than L_3 and L_4 is chosen to upgrade SRF from about 570 kHz to 960 kHz. Because of the previous research [2], it has some output impedance testing problems at high frequency which could not comply with the CISPR 16-1 requirement.

V. IMPEDANCE SIMULATION OF PROPOSED LISN

The simulated circuit of proposed LISN, neglected the effect of parasitic elements, is shown in Fig. 10. The AC sweep function of PSpice program is used to simulate the impedance at EUT port.

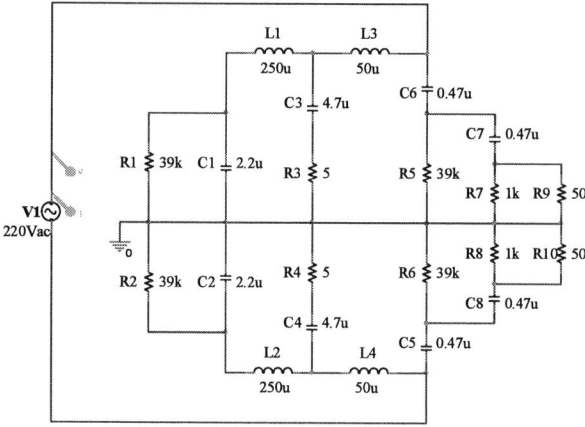

Fig. 10. Simulated circuit by PSpice program.

The simulated result is shown in Fig. 11. The curve increases rapidly in frequency range 30 kHz to 300 kHz and after that the curve provides the stabilized impedance about 47 Ω until 30 MHz. The simulated result of proposed LISN shows stabilized impedance following standard in frequency range of 150 kHz to 30 MHz.

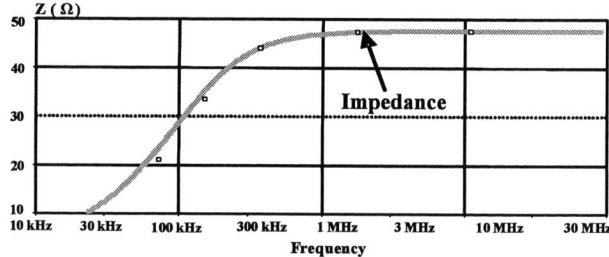

Fig. 11. The impedance at EUT connector.

VI. EXPERIMENTAL RESULTS

According to CISPR 16-1, the requirements for LISN or AMN testing are the output impedance and the insertion loss. But, due to limitation of equipment in laboratory, the insertion loss can not test. With these reasons, the experiment has divided to three subtopics. The first experiment is to confirm the CISPR 16-1 requirement by measured impedance at EUT terminals of proposed LISN. The second is measured the conducted EMI by using the proposed LISN comparing with commercial LISN. Finally, the temperature of air gap inductor is verified.

A. Impedance testing method

The impedance measurement, based on CISPR 16-1 methodology, is tested as shown in Fig. 12 [1]. Broadband load 50 Ω has to connect to BNC RF connector, normally connected to EMI receiver of proposed LISN. The impedance of proposed LISN is measured at EUT terminals between line to ground and neutral to ground respectively.

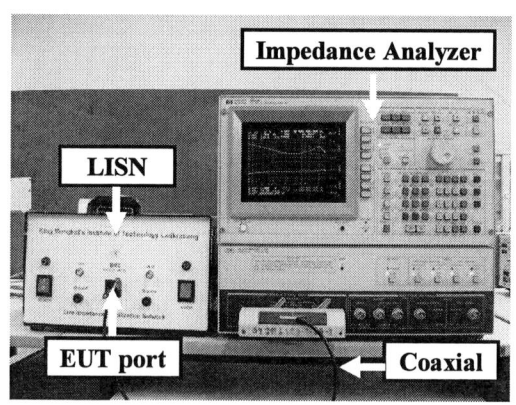

Fig. 12. Impedance testing setup.

518

The measured impedance from impedance/gain-phase analyzer, as shown in Figs. 13-14, guarantees that the proposed LISN can be provided the impedance about 50 Ω within ±20 % tolerance at frequency range 150 kHz–30 MHz for both line to ground and neutral to ground.

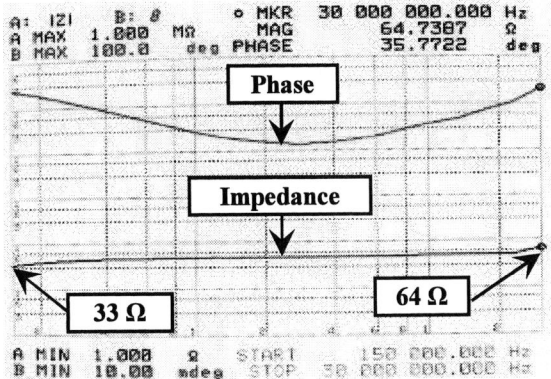

Fig. 13. The measured impedance of proposed LISN at EUT port Line to Ground side.

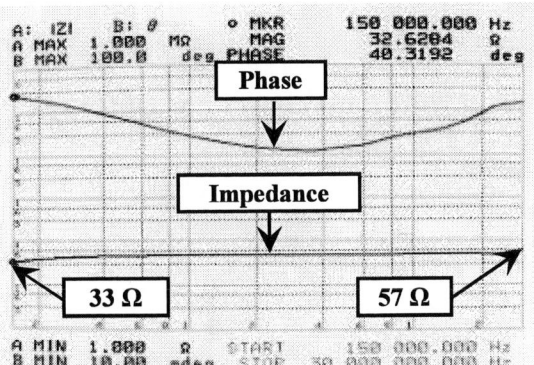

Fig. 14. The measured impedance of proposed LISN at EUT port Neutral to Ground side.

B. Conducted EMI testing method

In fact, LISN is a device used to transfer energy to load and to detect the conducted EMI to EMI receiver. From these reasons, the proposed LISN has to measure conducted EMI by comparing between the proposed LISN and the commercial LISN. The conditions are the frequency range 150 kHz–30 MHz following CISPR 22 standard and switching power supply is used as an EUT. Figs. 15-16 show line to ground and neutral to ground noise floor when measured before connecting the EUT. The conducted EMI measured by proposed LISN (blue line) is nearly close to commercial LISN (red line) as shown in Figs. 15-18.

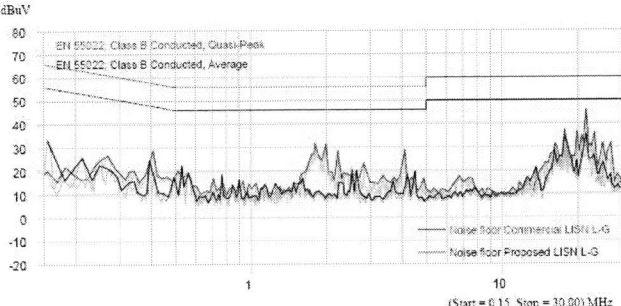

Fig. 15. The measured Line to Ground noise floor.

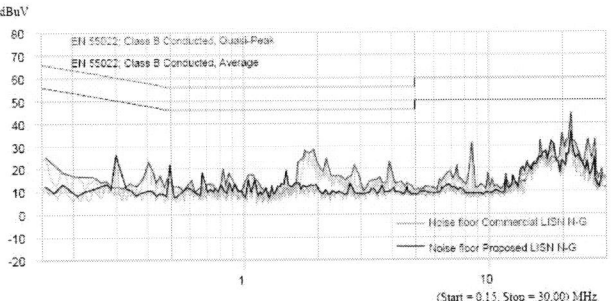

Fig. 16. The measured Neutral to Ground noise floor.

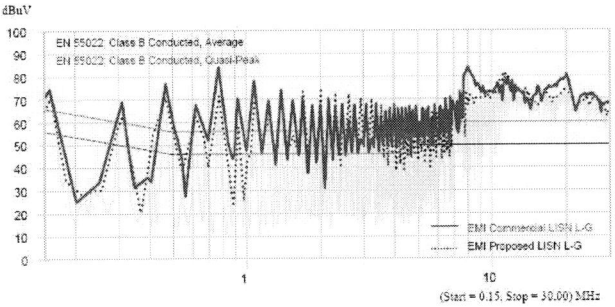

Fig. 17. The measured Line to Ground conducted EMI.

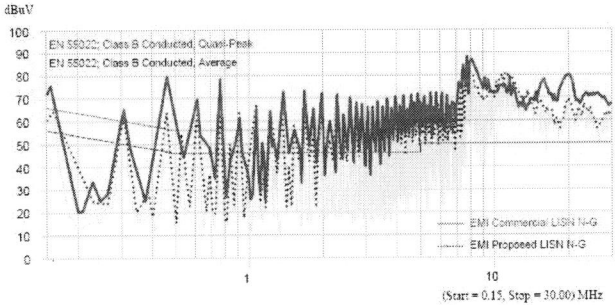

Fig. 18. The measured Neutral to Ground conducted EMI.

C. Operating temperature testing method

The operating temperature is measured at the middle of core winding of proposed LISN by using infrared thermometer. The testing time period is 60 minutes. The testing conditions are measured in every 5 minutes pass and connecting the continuous 8 A load (incandescent lamp) at EUT connector. From the experimental results as shown in Fig. 19, it can be seen that the operating temperature of proposed LISN is about 10 °C less than maximum temperature limit of PVC cores of proposed LISN.

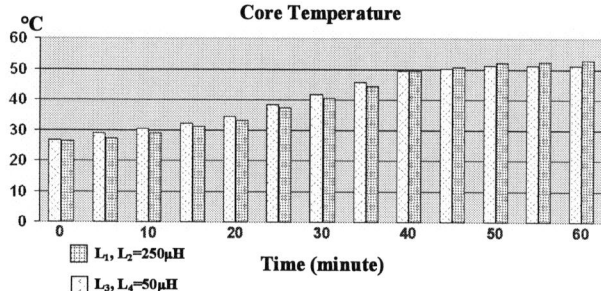

Fig. 19. Temperature measurement

VII. CONCLUSION

From all of the experimental results, it can be concluded that the proposed LISN can be provided the defined impedance 50 Ω ±20% and RF coupling signal as the functions of commercial LISN with a cheaper cost due to remove the selector switch and change type of inductor core material from bekalite to PVC. The total cost of the proposed LISN is about 100 US dollars. Finally, the proposed LISN has tested with continuous 8 A load at EUT connector for 60 minutes. The measured temperature at all devices is less than the temperature limit of devices.

Fig. 20. Top view of the proposed LISN.

REFERENCES

[1] C.I.S.P.R. Specification for Radio Disturbance and Immunity Measuring Apparatus and Methods, International Electrotechnical Commission, Geneva, Switzerland, 1999.

[2] D. Sakulhirirak et al. "An Affordable Line Impedance Stabilization Network Design for Academic Laboratory Institution," ECTI Conf., May 2006, pp 837-840.

[3] Marco Chiadó Caponet and Francesco Profumo "Devices for the Separation of the Common and Differential Mode Noise: Design and Realization", Applied Power Electronics Conference and Exposition (APEC) Seventeenth Annual IEEE, vol. 1, pp. 100-105, March 2002.

[4] Clayton R. Paul and Keith B. Hardin "Diagnosis and Reduction of Conducted Noise Emissions", IEEE Trans. on Electromagnetic Compatibility, vol. 30, no. 4, pp. 553-560, Nov 1988.

[5] Unplasticized polyvinyl chloride pipes for drinking water services, TIS 17-2532, 1989.

[6] The Magnetic Gun Club. (2002, June 16). Coil Parameter [Online]. Available: http://mgc314.home.comcast.net/coilparameters.htm

[7] A Massarini, M.K. Kazimierczuk and G. Granddi. "Lumped Parameter Models for Single-and Multiple-Layer Inductors," PESC'96, June 1996, vol. 1, pp. 295-301.

A Robust Output Current Control Method with Disturbance Observer for Matrix Converter under Unbalanced Input Voltage

Kazuo Oka*, Kouki Matsuse*

* Meiji University, 1-1-1 Higashimita Tama-Ku Kawasaki, JAPAN

Abstract– **This paper presents a robust output current control method with a disturbance observer for a matrix converter under an unbalanced input voltage. Adapting the disturbance observer to an output current control, the current control can be realized without an influence for an unbalanced input voltage and for a change of load parameters, a resistance or an inductance and so on. Finally, this current control is robustness for these conditions. The proposed method is effective for the applications of high performance AC motor drive employing a vector control because we assume a use of vector control. A validity of the proposed method is shown by a theoretical analysis using a Bode diagram and confirmed by the simulation results. The analysis with Bode diagram is first shown in this paper.**

Index Terms—**Matrix Converter, Unbalanced Input Voltage, Robust Current Control, AC Motor Drives**

I. INTRODUCTION

This paper presents a robust output current control method with a disturbance observer (DO) for a matrix converter (MC) under an unbalanced input voltage. The MC can directly convert a power from an input AC source to a load without any DC link. Since the MC has no reactive elements, the unbalance and distortion of the input voltage are immediately influenced to the load. Therefore, a high performance output current control of the load is impossible in this condition.

Many compensated methods for the unbalance and distortion of input voltage have been proposed. [1-3] But, they aren't considered about the robustness for a change of load parameters at all. For a realization of high performance AC motor drives, the change of load parameters must be considered too. Therefore, these methods are unsuitable for the high performance AC motor drive.

In this paper, we propose the robust output current control method for the MC under the unbalanced input voltage [5]. Adapting the DO to an output current control, the current control can be realized without an influence for the unbalanced input voltage and for the change of load parameters, a resistance or an inductance and so on. This current control is robustness for these conditions.

Financial support should be acknowledged here. Example: This work was supported by Japanese Ministry of Research.

The proposed method is effective for the application of high performance AC motor drive employing a vector control because we assume a use of vector control. A validity of the proposed method is shown by a theoretical analysis using a Bode diagram and confirmed by the simulation results. The analysis with Bode diagram is first shown in this paper.

II. STRUCTURE OF MATRIX CONVERTER

Fig.1 shows a structure of the three phases MC. The inductances L and capacitances C make an input AC filter for the reduction of a harmonic wave in an input current. The nine switches from Sua to Swc consist of the bi-directional switches. The load of the MC is one AC motor. In this paper, an induction motor (IM) is assumed as the load of MC.

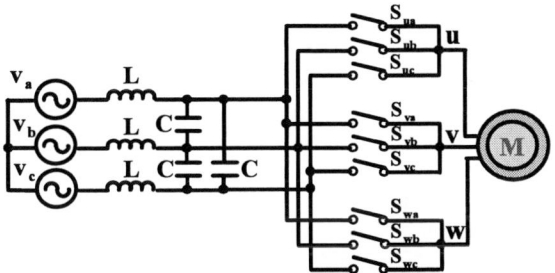

Fig. 1 Structure of Matrix Converter

III. MODULATION METHOD OF MATRIX CONVERTER

In this work, we assume an indirect modulation technique as a modulation method of MC. For this method, the MC is regarded as a cascade connection of two stages, a current source rectifier and a voltage source inverter. Finally, the space vector modulation method can be applied to the rectifier stage and inverter stage respectively [4].

Using this modulation method, $2f_i$ and $2f_i$-$2f_o$ frequency components are concluded in the output voltages of the synchronous reference frame under the unbalanced input voltage [2]. Where, f_i and f_o are a frequency of input voltage source and output voltage respectively.

IV. ROBUST OUTPUT CURRENT CONTROL WITH DISTURBANCE OBSERVER

A. Modeling of Induction Motor

An equation (1) expresses a circuit equation of IM for the synchronous reference frame in case of the unbalanced input voltage.

$$\mathbf{v}_{d(q)} + \mathbf{v}_{d(q)2f_i} + \mathbf{v}_{d(q)2f_i - 2f_o} = (R_s + P\sigma L_s)\mathbf{i}_{d(q)} + \omega\sigma L_s \mathbf{J}\mathbf{i}_{d(q)} + \mathbf{E_q} \quad (1)$$

$$\mathbf{v}_{d(q)} = \begin{bmatrix} v_d \\ v_q \end{bmatrix}, \mathbf{v}_{d(q)\_2f_i} = \begin{bmatrix} v_{d2f_i} \\ v_{q2f_i} \end{bmatrix}, \mathbf{v}_{d(q)\_2f_i} = \begin{bmatrix} v_{d2f_i-2f_o} \\ v_{q2f_i-2f_o} \end{bmatrix}, \mathbf{i}_{d(q)} = \begin{bmatrix} i_d \\ i_q \end{bmatrix}$$

$$\sigma = 1 - \frac{M^2}{L_s L_r}, \mathbf{J} = \begin{bmatrix} 0 & -1 \\ 1 & 0 \end{bmatrix}, \mathbf{E_q} = \begin{bmatrix} 0 \\ \omega\dfrac{M}{L_r}\Phi_{dr} \end{bmatrix}, P = \frac{d}{dt}$$

Where,

v_d, v_q : d- and q-axis components for stator voltage in the synchronous reference frame.

i_d, i_q : d- and q-axis components for stator current in the synchronous reference frame.

v_{d2fi}, v_{q2fi} : $2f_i$ components of d- and q-axis stator voltage in the synchronous reference frame.

$v_{d2fi-fo}, v_{q2fi-fo}$: $2f_i$-f_o components of d- and q-axis stator voltage in the synchronous reference frame.

$\mathbf{E_q}$: velocity electromotive force vector

These $2f_i$ and $2f_i$-f_o components only appear in case of the unbalanced input voltage.

R_s : stator resistance.
L_s : stator inductance.
L_r : rotor inductance.
M : mutual inductance.
ω: motor angular frequency expressed in electrical angle.
Φ_{dr}: d-axis component of rotor flux.
P : differential operator.

B. Conventional Output Current Control System

Fig.2 shows the block diagram of conventional output current control system. For the assumption in this work, the current control must be done for a flux current (d-axis current) and torque current (q-axis current). Since one PI (proportional-integral) controller is normally employed as the controller of current control system in the vector control, we regard the current control feedback system using the PI controller as a compared object for the proposed method. Therefore, we call this system "conventional method".

We express a transfer function of PI controller as follows.

$$C(s) = K_P + \frac{K_I}{s} \quad (2)$$

K_P : proportional gain.
K_I : integral gain.

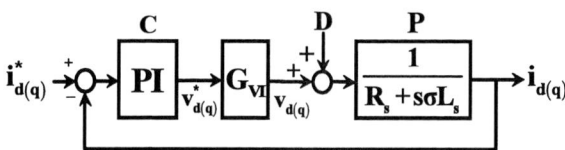

Where,

C : transfer function of controller.
G_{VI} : transfer function of inverter.
P : transfer function of actual plant.
D : transfer function of disturbance.
* : command.

Fig.2 Block Diagram of Conventional Output Current Control System

If K_P and K_I are set as an equation (3), a transfer function from d- and q- axis current command to actual those current will be the first-order lag system of time constant τ_i.

$$\begin{cases} K_P = \dfrac{\sigma L_s}{\tau_i} \\[3mm] K_I = \dfrac{R_s}{\tau_i} \end{cases} \quad (3)$$

From the equation (3),

● To obtain a desired response of d and q-axis current, the actual motor parameters must be use as the PI parameter.

● A current response characteristics and disturbance rejection characteristics can't design independently.

Finally, this system isn't robustness for the unbalanced input voltage and the change of motor parameters.

C. Proposed Robust Output Current Control System

a) Block Diagram of Proposed Robust Output Current Control System

Fig.3 shows the block diagram of proposed robust output current control system [5].

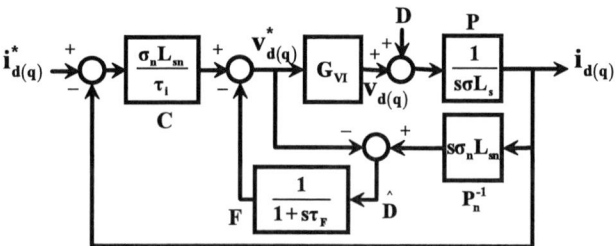

Where,

G_{VI} : transfer function of virtual inverter in MC.

P_n : transfer function of nominal plant.

F : transfer function of low pass filter.

τ_i : response time constant of d- and q-axis current.

τ_F : time constant for transfer function of low pass filter.

Subscript "n": nominal value.

^ : estimated value.

Fig.3 Block Diagram of Proposed Robust Output Current Control System

We call this system "proposed method".

b) Principle of Robust Output Current Controller with Disturbance Observer

From the equation (1), a following equation can be obtained.

$$P\sigma_n L_{sn}\mathbf{i}_{d(q)} = \mathbf{v}_{d(q)} - R_s\mathbf{i}_{d(q)} + P(\sigma_n L_{sn} - \sigma L_s)\mathbf{i}_{d(q)}$$
$$- \omega\sigma L_s\mathbf{J}\mathbf{i}_{d(q)} - \mathbf{E}_q + \mathbf{v}_{d(q)\_2f_i} + \mathbf{v}_{d(q)\_2f_i-2f_o} \cdots(4)$$

Now, we define the estimated disturbance as an equation (5).

$$\hat{\mathbf{D}} = -R_s\mathbf{i}_{d(q)} + P(\sigma_n L_{sn} - \sigma L_s)\mathbf{i}_{d(q)} - \omega\sigma L_s\mathbf{J}\mathbf{i}_{d(q)}$$
$$- \mathbf{E}_q + \mathbf{v}_{d(q)\_2f_i} + \mathbf{v}_{d(q)\_2f_i-2f_o} \cdots(5)$$

This equation shows that we regard a voltage drop with stator resistance, inductance error and velocity electromotive force as the estimated disturbance. In addition, regard $\mathbf{v}_{d(q)2fi}$ and $\mathbf{v}_{d(q)2fi-2fo}$ as that.

Then, the equation (4) is expressed as the follows.

$$P\sigma_n L_{sn}\mathbf{i}_{d(q)} = \mathbf{v}_{d(q)} + \hat{\mathbf{D}} \cdots(6)$$

Employing the equation (6), a block diagram of nominal plant model for the proposed method is drawn as Fig.4.

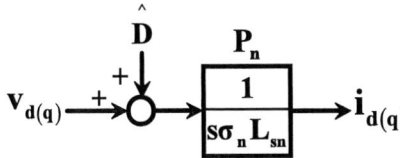

Fig.4 Block Diagram of Nominal Plant Model for Proposed Method

In Fig.3, the robust output current control is achieved with compensating an estimated disturbance calculated from the d (q)-axis voltage command and actual output current. Since an inverse model of nominal plant for proposed method has a differential characteristic, the estimated disturbance must be passed the low pass filter F to reject the high-order frequency components.

If a parameter of controller is chosen as shown in Fig.3, a transfer function from current command to actual value will be expressed by an equation (7).

$$\frac{\mathbf{i}_{d(q)}}{\mathbf{i}^*_{d(q)}} = \frac{1}{1+s\tau_i} \cdots(7)$$

This shows that the response of these current to be always a first-order lag system having a time constant τ_i, in case of the unbalanced input voltage and the change of motor parameter, resistance and inductance. Moreover, a transfer function from estimated disturbance to actual output current is as below.

$$\frac{\mathbf{i}_{d(q)}}{\hat{\mathbf{D}}} = \frac{\dfrac{1}{\sigma_n L_{sn}}s}{s^2 + \left(\dfrac{1}{\tau_F} + \dfrac{1}{\tau_i}\right)s + \dfrac{1}{\tau_F\tau_i}} \cdots(8)$$

Generally, mutual inductance is much bigger than leakage inductance. In this time, a transfer function from actual disturbance to estimated disturbance can be approximated to one. Then, we are able to obtain an equation (9).

$$\frac{\mathbf{i}_{d(q)}}{\mathbf{D}} = \frac{\dfrac{1}{\sigma_n L_{sn}}s}{s^2 + \left(\dfrac{1}{\tau_F} + \dfrac{1}{\tau_i}\right)s + \dfrac{1}{\tau_F\tau_i}} \cdots(9)$$

Otherwise, a transfer function from actual disturbance to actual output current for the conventional method is as an equation (10).

$$\frac{\mathbf{i}_{d(q)}}{\mathbf{D}} = \frac{\tau_i s}{1+\tau_i s}\frac{1}{R_s+\sigma L_s s} \cdots(10)$$

As the features of proposed system,

- A desired response of d and q-axis current can be obtained in case of the unbalanced input voltage.
- If the motor parameters are changed for a change of temperature and magnetic saturation, a desired response of these current can be obtained.
- We can independently design a disturbance rejection characteristics for these current responses with using τ_F. (2-degree of freedom control)

  Otherwise, since this characteristics for the conventional method is decided in τ_i and the actual circuit parameters, we can't independently design this characteristic for these current responses. (1-degree of freedom control)
- A disturbance rejection characteristics isn't always changed in these conditions.

Therefore, this system is robustness for these conditions.

V. DESIGN OF TIME CONSTANT FOR ESTIMATED DISTURBANCE FILTER

Since a disturbance frequency band that the system must be compensated is $2f_i$ in this work, we design the time constant τ_F for estimated disturbance filter as a transfer function gain from the actual disturbance to actual d-and q- axis current to be smaller than -20[dB] in this frequency band.

Using the equation (9), τ_F satisfied with this condition is given by a below inequality. But, the parameters shown in Table.1 and Table.2 are employed in calculating τ_F.

$$\tau_F \leq 0.53\left[ms\right]\cdots(11)$$

In this work, we set a value of τ_F to 0.3[ms]. Fig.5 shows a gain characteristic for the transfer function shown in equation (9) and (10). The parameters of Table.1 and Table.2 are employed for both characteristic. In this figure, a red line of horizontal axis expresses the disturbance frequency band that we must compensate.

As can be seen from Fig.5, a disturbance rejection characteristic of the proposed method is much superior compared with that of the conventional method. Therefore, the proposed output current system is robustness for the unbalanced input voltage and the change of motor parameters compared with the conventional method.

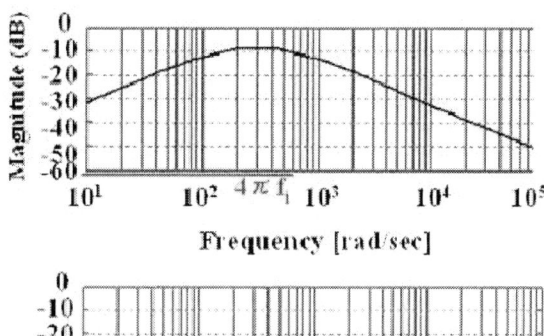

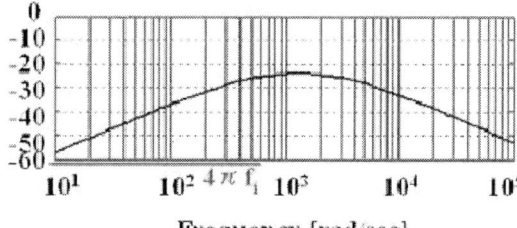

Fig.5 Gain Characteristic of Transfer Function From Actual Disturbance to d-and q-Axis Current

Left: Conventional Method
Right: Proposed Method

VI. SIMULATION RESULTS

We simulated to confirm the validity of proposed method. The PI controller is employed for the current controller of conventional method as shown Fig.2.

In this work, we assume that IM is driven with the vector control. Table.1 shows the circuit parameter and rating of IM used for the simulation. Moreover, Table.2 shows the simulation parameter.

TABLE.1 CIRCUIT PARAMETER AND RATING OF INDUCTION MOTOR USED FOR SIMULATION

Stator Resistance	0.7185[Ω]	Output	2.2[kW]
Rotor Resistance	0.5965[Ω]	Voltage	180[V]
Stator Inductance	63.38[mH]	Current	10[A]
Rotor Inductance	63.38[mH]	Rotor Speed	1750[rpm]
Mutual Inductance	61.28[mH]	Torque	12 [N.m]

TABLE.2 SIMULATION PARAMETERS

Amplitude of v_a	120[V]	τ_i	2[ms]
Amplitude of v_b	160[V]	τ_i	0.3[ms]
Amplitude of v_c	200[V]	K_p	2.065[1/s]
Input Voltage Frequency f_i	50[Hz]	K_I	359.25[Ω/s]

But, the circuit parameters expressed in Table.1 are nominal value.

Where, K_p and K_I are the proportional and integral gain of PI controller for the conventional method respectively. Though the actual circuit parameters are necessary in calculating the PI parameters as shown the equation (3), we calculated them with using those nominal values.

In addition, we simulated in case of stator resistance and mutual inductance to be changed 3 times, 0.8 times respectively compared with these nominal values. Where, we express as "nominal" in case of these motor parameters not to be changed and as "parameter change" in case of these to be changed.

Fig.6 shows the input current of MC, output current and output voltage in "parameter change" when a motor output is being kept to be constant. Where, a switching ripple of output voltage shown in Fig.5 is being rejected by a filter. Table.3 expresses the Fast Fourier Transformation (FFT) analyzed results of output current to confirm a rate of $2f_i\text{-}f_o$ and $2f_i\text{+}f_o$ components included in the output current for the unbalanced input voltage.

TABLE.3 RATE OF $2F_I\text{-}F_O$ AND $2F_I\text{+}F_O$ COMPONENTS INCLUDED IN OUTPUT CURRENT

$2f_i\text{-}f_o$ ($2f_i\text{+}f_o$) (%)	Conventional	Proposed
Nominal	15.6 (28.1)	3.12 (6.25)
Parameter Change	10.9 (4.69)	2.34 (2.34)

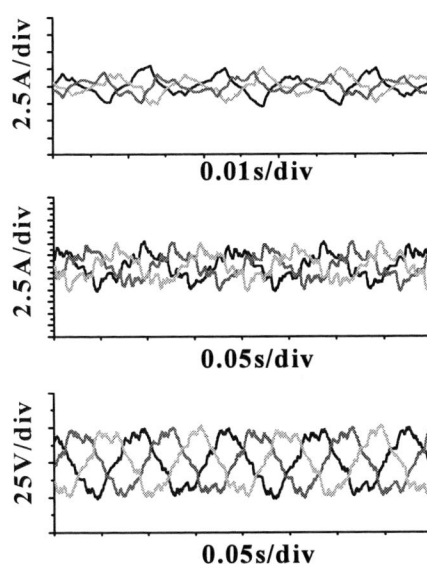

Fig.6 (a) Conventional Method

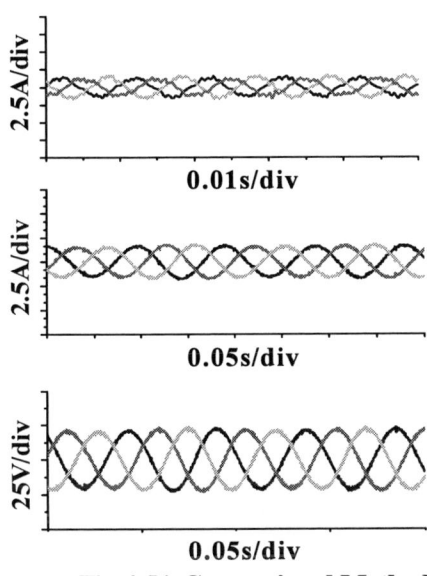

Fig.6 (b) Conventional Method

**Top: Input Current of MC
Middle: Output Current
Bottom: Output Voltage**

As can be seen from Fig.6 and Table.3,

○ The distortions of input current, output current and output voltage with the proposed method can be much decreased compared with the conventional method.

○ Using the proposed method, the rate of $2f_i\text{-}f_o$ and $2f_i\text{+}f_o$ components included in the output current, which are caused by the unbalanced input voltage can be much decreased compared with the conventional method in case of the motor parameter changes.

Fig.7 shows a step-response waveform of q-axis current.

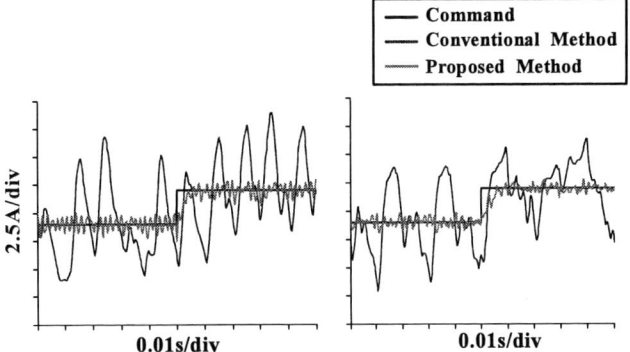

**Fig.7 Step-Response Waveform of q-Axis Current
(Left: Nominal Right : Parameter Change)**

As can be seen from Fig.7,

○ In the conventional method, many $2f_i$ and $2f_i\text{-}2f_o$ components, which causes by the unbalanced input voltage are included for the q-axis current in addition to the DC components.

○ In the proposed method, we can confirm that the q-axis current responses in the delay of τi for the current command in case of the "nominal" and "parameter change".

Finally, the proposed robust output current method is robustness for the unbalance input voltage and the change of motor parameters. Therefore, adapting the proposed method, a high performance AC motor drive can be realized.

VII. CONCLUSION

In this paper, we proposed the robust output current control method for the MC under the unbalanced input voltage. Adapting the disturbance observer to an output current control, the current control can be realized without an influence for the unbalanced input voltage and for the change of load parameters, a resistance or an inductance and so on. This current control is robustness for these conditions. The proposed method is effective for the application of high performance AC motor drive employing the vector control. A validity of the proposed method is shown by a theoretical analysis using a Bode diagram and confirmed by the simulation results.

REFERENCES

[1] J.K.Kang, H.Hara, E.Watanabe, A.M.Hava, T.J.Kume, "The matrix converter drive performance under abnormal input voltage", IEEE Trans. on Power Electronics, vol.5, no.4, pp234-240, 2002

[2] K.Sun, D.Zhou, L.Huang, K.Matsuse, "Compensation control of matrix converter fed induction motor drive under

abnormal input voltage conditions", IEEE IAS2004, vol.1 pp.623-630, 2004

[3] M.E.de.Oliveira Filho, E.R. Filho, K.E.B.Quindere, Jonas R. Gazoli, "A simple current control for matrix converter", IEEE IAS2006

[4] L.Huber, D.Borojevic, "Space vector modulated three-phase to three-phase matrix converter with input power factor correction", IEEE Trans. on Industry Appl., vol.31, NO.6, pp.1234-1246, 1995

[5] K.Oka, K.Matsuse: "A Robust Current Control Method of A Matrix Converter under An Unbalanced Input Voltage Source", THE 2007 ANNUAL MEETING RECORD I.E.E. JAPAN, 151-152 (2007)

FPGA Design of Single-phase Matrix Converter Operating as Cycloconverter

Z. Idris*, M.K. Hamzah*, A. Saparon*, N.R.Hamzah* & N.Y. Dahlan**

*Faculty of Electrical Engineering, Universiti Teknologi MARA, 40450 Shah Alam, Malaysia. mustafar@ieee.org
**Faculty of Electrical Engineering, Universiti Teknologi MARA Pulau Pinang,13500, Pulau Pinang, Malaysia

Abstract: - This paper is concerned on FPGA design for control implementations of the Single-Phase Matrix Converter (SPMC) operating as a Cycloconverter. The Sinusoidal Pulse Width Modulation (SPWM) technique is used to synthesize the output voltage. The power circuit uses the Insulated Gate Bipolar Transistor (IGBT) as switching device in the SPMC implementation. Safe-commutation strategy was incorporated to solve switching transients. A Xilinx Field Programmable Gate Array (FPGA) was used at the heart of the control electronics, implemented to verify operation with selected simulation and experimental results presented.

Keywords— *Sinusoidal Pulse Width Modulation (SPWM), Insulated Gate Bipolar Transistor (IGBT), Single-Phase Matrix Converter (SPMC), Cycloconverter, Field Programmable Gate Array (FPGA).*

1. INTRODUCTION

Matrix Converter (MC) is an advanced converter known to offer an "all silicon" solution for direct AC-AC conversion [1, 2], mainly with three-phase circuit topologies [3, 4, 5]. The single-phase version has subsequently been proposed called the Single-phase Matrix Converter (SPMC) for AC-AC conversion [6, 7, 8]. Limited publications had been found on SPMC, definitely negligible on switching strategies due to complex control requirements with its four bidirectional switch topology and the absence of simple safe commutation strategy. [9, 10]

Pulse width modulation (PWM) is a widely used technique for controlling the output of static power converters. They have different implementations, dynamics responses and PWM patterns [11]. Over the years, digital designs provide improvements over their analogue counterparts. They are immune to noise and less susceptible to voltage and temperature changes. Hence, a shift to digital implementation has been observed.

The design and development of a sinusoidal pulse-width modulation (PWM) generator suitable for Single-Phase Matrix Converter (SPMC) operating as a cycloconverter will be presented. It is based on the Xilinx chip XC4005XL Field Programmable Gate Array (FPGA) with IGBTs as the power switching device. The output voltage is synthesized using Sinusoidal Pulse Width Modulation (SPWM). The proposed design enables the modulation index and the switching frequency to be changed externally. Results are provided to demonstrate successful implementation of the design. Prior to hardware implementation, simulations were performed to predict the behaviour. A laboratory model test-rig of the SPMC was then developed to

experimentally verify the result. This paper introduces the steps and techniques for generating the SPWM patterns for SPMC operating as a cycloconverter system, which is placed in one chip without using external memory chips, where design re-uses is advantages in development.

2. CYCLOCONVERTER

The cycloconverter converts AC input power at one frequency to output power at different frequency (generally lower than input frequency) in one conversion, traditionally uses thyristors [12] as shown in Figure 1 with output of Fig 2. They are normally used in the high-power domain for variable frequency speed control of ac machines such as rolling steel mill [13], cement industry application [14] and ship propellers [15]. Various studies on cycloconverter has included, amongst others; improvements of harmonic spectrum in the output voltage with new control strategies [16], new topologies [17] and behaviours [18].

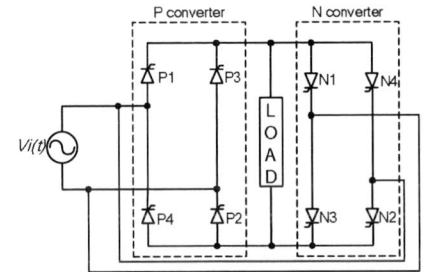

Fig. 1. Schematic of a classical single-phase cycloconverter

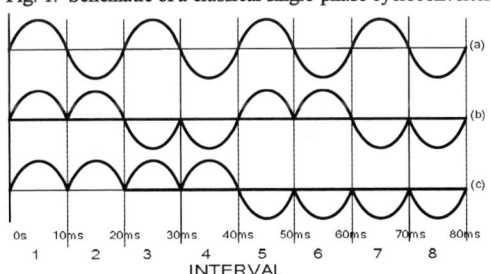

Fig. 2 : Cycloconverter Operation (a) input 50 Hz (b) output 25 Hz (c) output 12.5 Hz

3. SINGLE-PHASE MATRIX CONVERTER (SPMC)

The SPMC topology with its 4 bi-directional switches and its individual power switches; used in this work is as shown in Figs. 3 and 4 respectively; each capable of conducting current in both directions, blocking forward and reverse voltages [9]. To synthesize a variable voltage

978-1-4244-0644-9/07/$25.00 ©2007 IEEE

output the SPWM as illustrated in Fig. 5 is used. A safe commutation switching sequence as described in reference [10] is used as illustrated in Figs. 6 to 9. These avoid the occurrence of switching transients relating to use of reactive loads available during switch turn-off and commutation.

The switching sequence for operation of SPMC as the cycloconverter is tabulated in table 1 for an output of 25Hz and 12.5Hz with its associated illustrations as shown in Figs. 10 and 11 respectively.

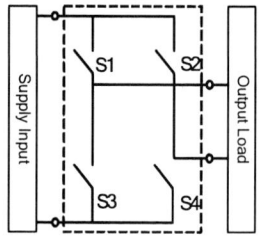

Fig. 3: SPMC circuit configuration

Fig.4: Bi-directional switch module

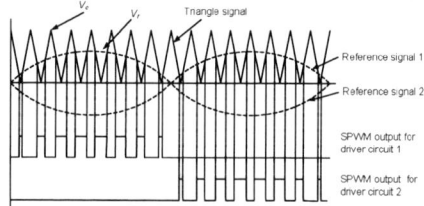

Fig. 5: Formation of SPWM

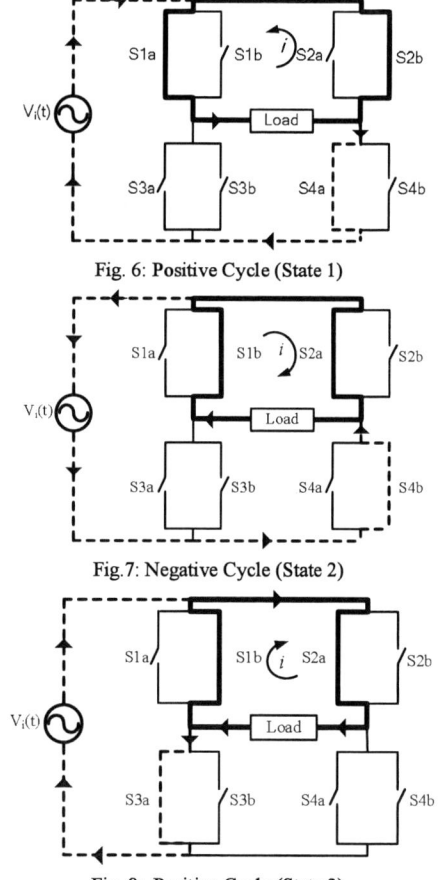

Fig. 6: Positive Cycle (State 1)

Fig.7: Negative Cycle (State 2)

Fig. 8: Positive Cycle (State 3)

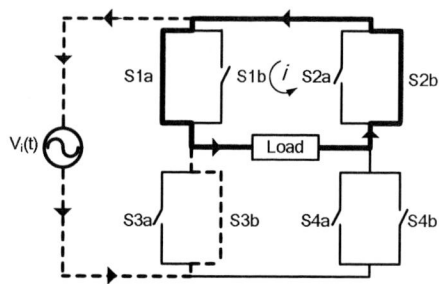

Fig. 9: Negative Cycle (State 4)

Table 1: Switching table (cycloconverter)

Input Frequency	Output Frequency	Time Interval	State	PWM Switch	Commutation Switch
50 Hz	25 Hz	1	1	S4a	S1a & S2b
		2	4	S3b	S1a & S2b
		3	3	S3a	S1b & S2a
		4	2	S4b	S1b & S2a
	12.5 Hz	1	1	S4a	S1a & S2b
		2	4	S3b	S1a & S2b
		3	1	S4a	S1a & S2b
		4	4	S3b	S1a & S2b
		5	3	S3a	S1b & S2a
		6	2	S4b	S1b & S2a
		7	3	S3a	S1b & S2a
		8	2	S4b	S1b & S2a

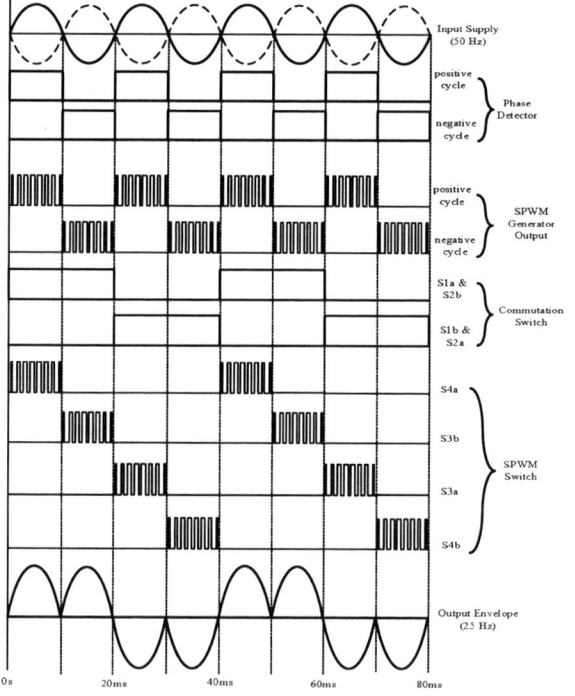

Fig. 10: Waveform of switching pattern for cycloconverter 25 Hz

4.   HARDWARE IMPLEMENTATION

A test-rig for the SPMC circuit was constructed for verifications of the work. They include; a single-phase supply, an input filter, a power circuit, a load and a controller (control electronics) as represented in Fig. 12. A 50 V_{rms} (50 Hz) single phase supply is used as a supply input with a low pass LC filter of 4 mH inductor and 10 µF capacitor filter. The Power Circuit is constructed using eight IGBTs of BUP 314D with common emitter configuration as previously described. The load used in this case is represented by resistive (R=50 Ω) and inductive (R=50 Ω and I=4mH).

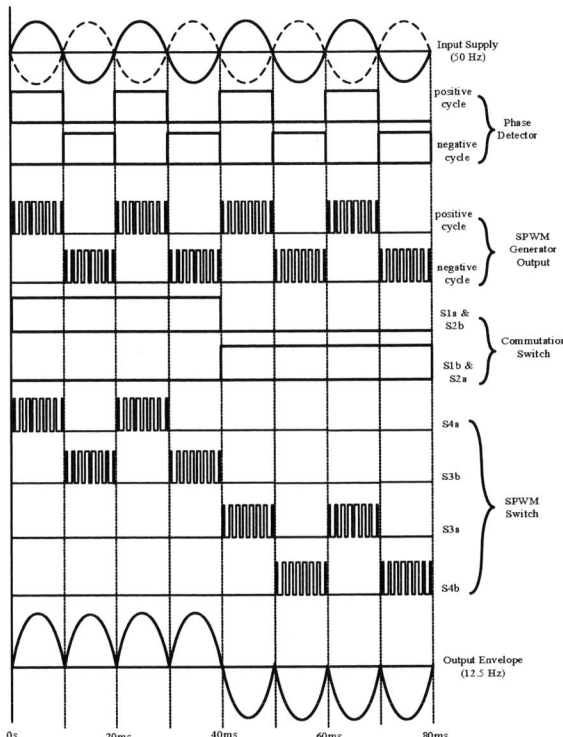

Fig. 11: Waveform of switching pattern for cycloconverter 12.5 Hz

5.   CONTROL ELECTRONICS

The Control Electronics comprise; a computer, a phase detector, isolated gate drives and a Xilinx FPGA at the heart of its digital control. This provides flexibility with the circuit design without hardwired modifications making it a favourable choice for Application Specific Integrated Circuits (ASIC) prototyping [19, 20]. A phase detector is used to effect the operation of both cycles by synchronising the supply waveform. The phase detector circuit diagram as shown in Fig. 13 is used to effect the operation of both cycles by synchronising the supply waveform.

6.   XILINX FPGA DESIGN

The functional block diagram of control algorithm developed for implementation of Xilinx to produce switching pattern for SPMC operating as cycloconverter is as shown in Fig. 14. Within the Xilinx is three major components; a) Desired Frequency, b) Switch Selector Unit (SSU) and c) Commutation Switch Selector (CSS).

The SPWM algorithm is converted into a digital form as illustrated in Figure 15. Due to their similarity nature of the sinewave between the negative and positive cycle; a half cycle (positive) could be used to optimise by repeating it in the negative cycle; performing required functions but at a reduced overheads.

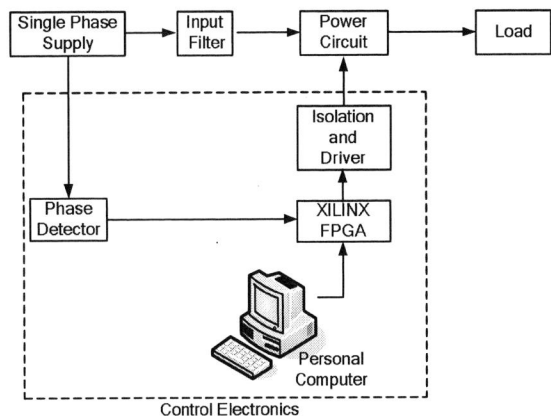

Fig. 12 : Block diagram of the SPMC (experimental set-up)

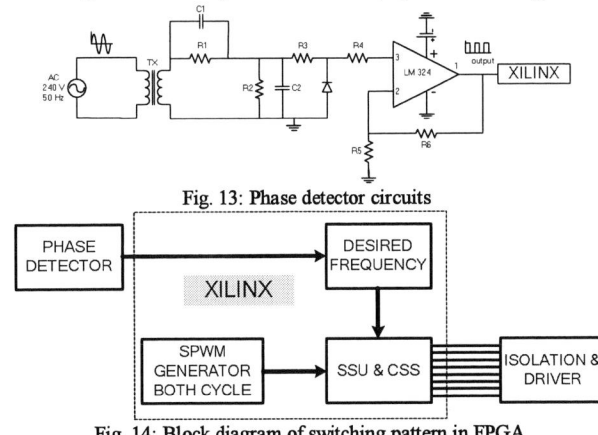

Fig. 13: Phase detector circuits

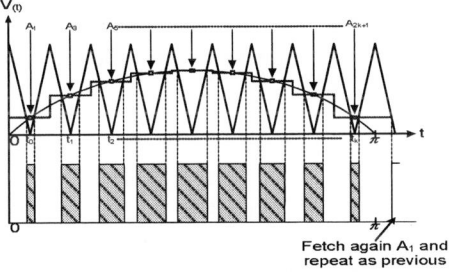

Fig. 14: Block diagram of switching pattern in FPGA

Fig. 15: Illustration of digital SPWM generation

A phase detector as previously described was used to observe the input supply waveform converting to a digital on-off signal that was subsequently used as part of the software initialisation routine for the operation of Xilinx FPGA; facilitating the output from Xilinx FPGA to be in synchronous with the frequency of single phase input voltage. It is also being used to determine the desired frequency pulse train being generated (e.g. 25 Hz or 12.5 Hz) by using FTC as illustrated in Fig. 16. Finally the SSU and CSS is used to implement the operation of the required switching sequence as in table 1; using the

529

various digital gate as shown in Fig. 17, thus generating the required gating pulse fed into the gate drive circuit.

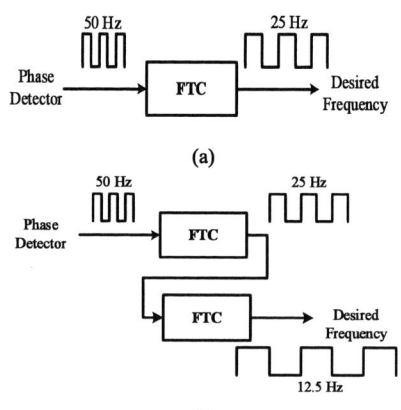

(a)

(b)

Fig. 16: Desired frequency unit (a) 25 Hz (b) 12.5 Hz

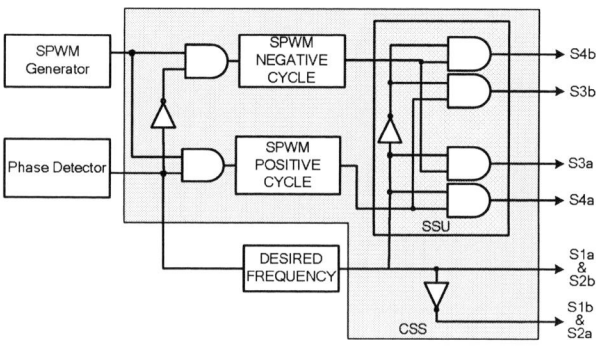

Fig. 17 : Schematic diagram of the switch selector unit (SSU) and Commutation Switch Selector (CSS)

The overall block diagram of SPWM generator is as shown in Fig. 18 with further details as in Figs. 19 to 21 with the top level schematic of the SPWM generator shown in Fig. 19 for the positive cycle; where the circuit consists of an up-down counter that generates an 'M' shape triangular waveform, produced by an eight bit counter i.e. CB8CLED. This digitally increments the counter value from 0 to 255 and then subsequently decrements it back to 0 again over a period of time. This 'M' shape carrier signal needs to be transformed to 'W' shape required using an inverter, which is then compared with the output of the multiplier; obtained from multiplication of the modulating signal from the look-up table (ROM) with an external modulation index input.

Good accuracy requires high bit number. For optimisation an eight bit counter with a clock speed of 2.55 MHz was determined based on equation (1). For a carrier of 5 kHz the use of sixteen bit counter is not possible; due to the very high clock frequency (655.35 MHz) requirements. XC4005XL on the other hand has a maximum clock speed 80 MHz. It has been further shown that an 8-bit counter provides acceptable accuracies [21, 22], resulting with a range of 0 to 255. To facilitate

variations in modulation index of 0 to 1 in steps of 0.1, the maximum value is divided by 10; i.e. 25.5.

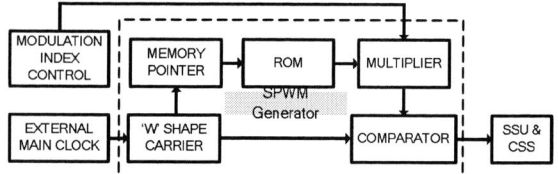

Fig.18: Block diagram of the proposed SPWM Generator

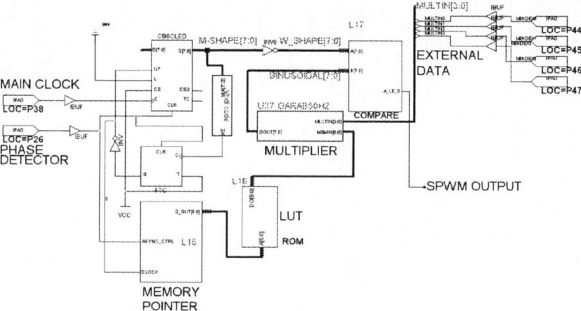

Fig. 19: The top level schematic diagram of SPWM generator in Xilinx FPGA

The Modulation Index control is needed to vary the modulation index of the SPWM. To achieve this, an externally 4-bit data input was used. This is then multiplied with the sample sinusoidal waveform that has been discretised and stored in the ROM to achieve variation in the amplitude of the reference signal.

An external main clock was used as the clocking signal for the FPGA due to frequency of carrier signal used in this work. An eight bit up-down counter is clocked at 2.55 MHz to produce a 5 kHz carrier frequency according to equation (1). The clock signal is locked and synchronized to the AC mains frequency of 50 Hz by using external phase detector circuit.

$$f_c = f_{clock}/[(2^n - 1) \times 2] \qquad (1)$$

The 'W' shape carrier waveform is designed using an eight-bit up-down counter representing a carrier signal ranging from 0 to 255 as shown in Fig. 20

The carrier unit provides information in terms of clock pulse to other units. The pulse train is generated at every zero point of 'W' shape carrier. The memory pointer unit relies on this information to fetch data from the ROM at an appropriate time as illustrated in Fig. 21 and this data is fed to the multiplier unit. Finally, the comparator unit is used to perform the required SPWM pulse train.

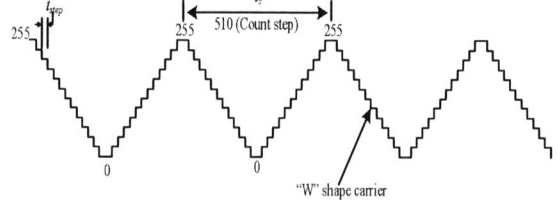

Fig. 20: Eight bit counter output

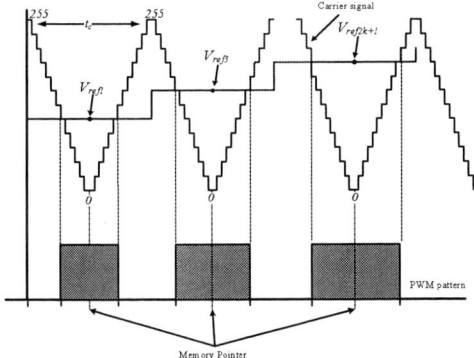

Fig. 21: Process of generating SPWM in digital control

For the sampled sinusoidal reference signal, data representing magnitude of the sinusoidal needs to be calculated first by using equations (2) and (3). These, are indicated by the small boxes and labelled A_1, A_3, A_5,…, A_{2k+1} as shown in Fig. 15, corresponding to dots of Fig. 15. The sampled data then store in lookup table (ROM). The lookup table are formed from the internal memory of the FPGA.

Process of generating the SPWM in digital control is shown in Fig. 21. Each data from the reference signal are sampled evenly between the half-cycle (0^0 to 180^0). This is done every time the carrier signal achieves zero. The time for each step of the counter t_{step} is given by equation (5).

$$Vref_{2k+1} = \left[round \frac{2^n - 1}{10} \left\{ sin\left(2\pi.f_r\left(2k+1.\frac{t_c}{2} \right) \right) \right\} \right] \quad (2)$$

$$t_c = \frac{1}{f_c} \quad (3)$$

$$t_r = \frac{1}{f_r} \quad (4)$$

$$t_{step} = \frac{t_c}{2(2^n - 1)} \quad (5)$$

Where 'k' is the carrier pulse position, '$2k+1$' is the phase angle, 'n' the bit size of up-down counter, 'f_c' the carrier frequency and 'f_r' the reference frequency .

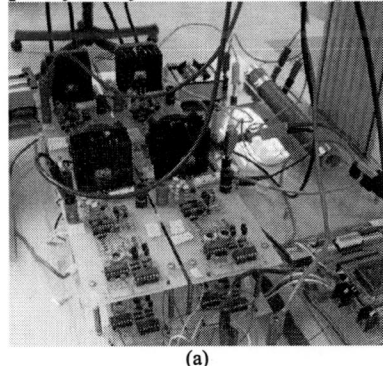

(a)

(b)

Fig. 22 : Experimental Test-Rig Setup (a) power circuit (b) phase detector circuit

8. RESULTS

To study the behaviour, a simulation model was developed in MATLAB/Simulink. A test rig was then constructed for experimental verification as shown in Fig. 22. Sample results from simulation and experimental work in the laboratory are as shown in Figs. 19-28.

A. R Load

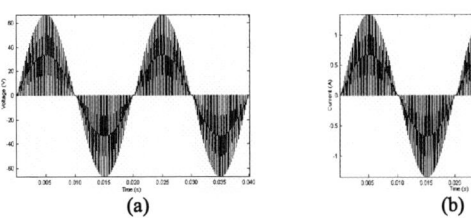

(a)　　　　　(b)

Fig. 19: Simulation result of SPMC at f_c = 50 Hz with m_i = 0.7 (a) output voltage (b) output current

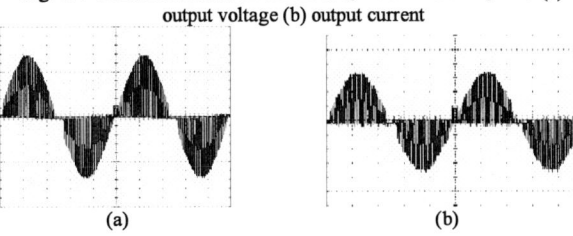

(a)　　　　　(b)

Fig. 20: Experimental result of SPMC at f_c = 50 Hz with m_i = 0.7 (a) output voltage, scale Y:50V/div, X: 4ms/div (b) output current, scale Y:500mA/div, X:4ms/div

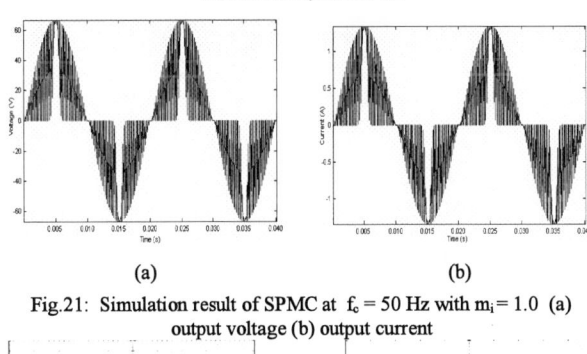

(a)　　　　　(b)

Fig.21: Simulation result of SPMC at f_c = 50 Hz with m_i = 1.0 (a) output voltage (b) output current

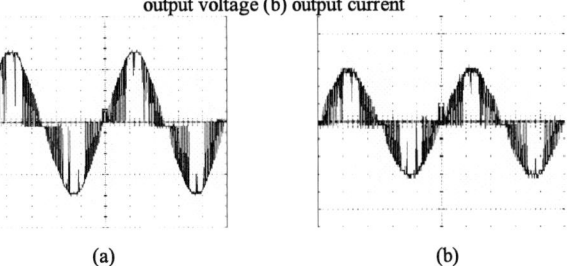

(a)　　　　　(b)

Fig. 22: Experimental result of SPMC at f_c = 50 Hz with m_i = 1.0 (a) output voltage, scale Y:50V/div, X: 4ms/div (b) output current, scale Y:500mA/div, X:4ms/div

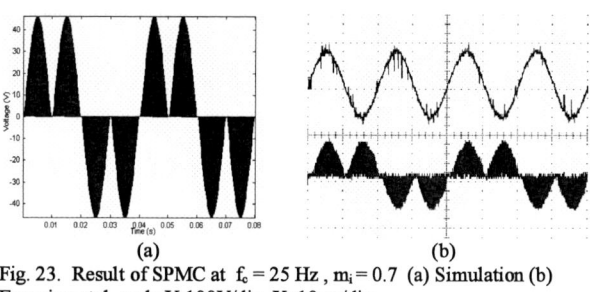

(a)   **(b)**

Fig. 23. Result of SPMC at $f_c = 25$ Hz , $m_i = 0.7$ (a) Simulation (b) Experimental, scale Y:100V/div, X: 10ms/div

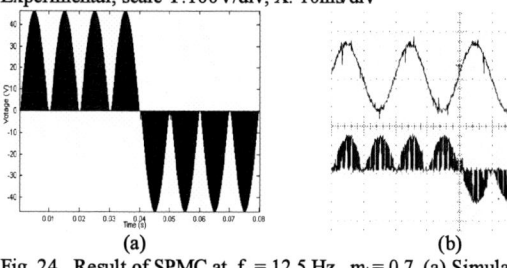

(a)   **(b)**

Fig. 24. Result of SPMC at $f_c = 12.5$ Hz , $m_i = 0.7$ (a) Simulation (b) Experimental, scale Y:50V/div, X: 10ms/div

B. R-L Load

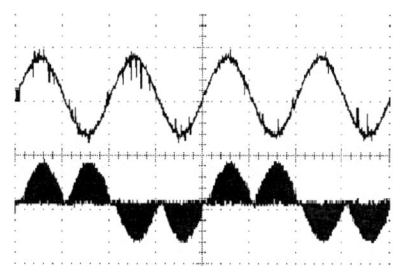

Fig. 26. Experimental result at 25 Hz with safe-commutation, scale Y:50V/div, X: 10ms/div

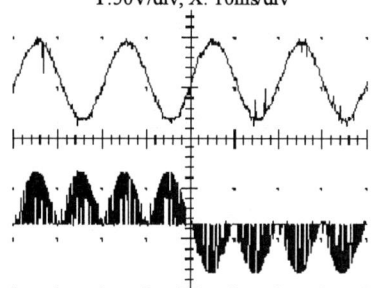

Fig. 28. Experimental result at 12.5 Hz with safe-commutation, scale Y:50V/div, X: 10ms/div

7. CONCLUDING REMARKS

Experience in designing the FPGA for implementation will outline. It will illustrate that the FPGA could effectively be used in SPMC with the four bidirectional switching arrangements. It will be discussed that the experimental results achieve good agreement with those predicted in simulations. SPWM controlling algorithm is placed on a single chip of XC4005XL FPGA and is capable of providing flexibility. The overall system is compact with no external memory system required. Tests have been carried out to show the effectiveness and flexibility of the proposed method.

8. ACKNOWLEDGEMENT

Financial support from Ministry of Science Technology and Innovation (MOSTI) Malaysia EScience Grant No : 03-01-01-SF0168 is gratefully acknowledged for implementation of this project.

REFERENCES

[1] M. Venturini, "A New Sine Wave in Sine Wave Out, Conversion Technique Which Eliminates Reactive Elements," Proceedings IEEE Powercon 7, 1980, pp.E3_1-E3_15.

[2] L. Gyugyi, and B.R. Pelly, "Static Power Chargers, Theory, Performance and Application," John Wiley & Son Inc, 1976

[3] P.W. Wheeler, J.C. Clare, L. Empringham, M. Bland, K.G. Kerris, "Matrix converters," IEEE Industry Applications Magazine, Vol. 10 (1), Jan-Feb2004, pp.59 – 65.

[4] T. Sobczyk, "Numerical Study of Control Strategies for Frequency Conversion with a Matrix Converter," Proceedings of Conference on Power Electronics and Motion Control, Warsaw, Poland, 1994, pp. 497-502.

[5] J.G. Cho and G.H. Cho, "Soft-switched Matrix Converter for High Frequency direct AC-to-AC Power Conversion," Int. J. Electron., 1992, 72, (4), pp. 669-680.

[6] A. Zuckerberger, D. Weinstock, A. Alexandrovitz, "Single-phase Matrix Converter," IEE Proc. Electric Power App, Vol.144 (4), Jul 1997 pp. 235-240.

[7] S.H. Hosseini, E. Babaei, "A new generalized direct matrix converter," IEEE International Symposium of Industrial Electronics, 2001. Proc. ISIE 2001. Vol(2), pp.1071-1076.

[8] A. Koei and S. Yuvarajan, "Single-Phase AC-AC Converter Using Power Mosfet's," IEEE Transaction on Industrial Electronics, Vol. 35, No.3, August 1988, pp.442-443.

[9] Z. Idris, M.K. Hamzah, M.F. Saidon, "Implementation of Single-Phase Matrix Converter as a Direct AC-AC Converter with Commutation Strategies"; 37th IEEE Power Electronics Specialists Conference, 2006. PESC06. 18-22 June 2006 Page(s):1 - 7

[10] Z.Idris, S. Z. M. Noor & M. K. Hamzah, "Safe Commutation Strategy in Single-phase Matrix Converter", IEEE Sixth International Conference PEDS 2005, Kuala Lumpur, Malaysia

[11] A. M. Omar "The Three-Phase Single Stage Flyback Converter," University of Malaya Ph.D Thesis 2002.

[12] A. Maamoun, "Development of cycloconverters," Canadian Conference on Electrical and Computer Engineering, 2003. IEEE CCECE 2003, Vol. 1, 4-7 May 2003, pp: 521 – 524.

[13] W. Timpe, "Cycloconverter drives for rolling mills," IEEE Trans. Industry Application, Vol. IA-18, no. 4, 1982, pp: 400-404, 1982.

[14] C.P. LeMone, M. Ehara, and L. Nehl, "AC adjustable speed application for the cement industry," in Proc. IEEE Cement Industry Technical Conf., Salt lake City, UT.,1986, pp: 335-362.

[15] W.A. Hill, G. Creelman, and L. Mischke, "Conrol strategy for an icebreaker propulsion system," IEEE Trans. Ind. Appl., Vol.28, no.4, pages: 887-892, Jul. /Aug. 1992.

[16] A. Karamat, T. Thomson, P. Mehta, "A novel strategy for control of cycloconverters," 22nd Annual IEEE on Power Electronics Specialists Conference, 1991. PESC '91, 24-27 June 1991, pp: 819 – 824.

[17] M.A. Choudhury, M.B. Uddin, A.R. Bhuyia, M.A. Rahman, "New topology and analysis of a single phase

delta modulated cycloconverter," Proceedings of the 1996 International Conference on Power Electronics, Drives and Energy Systems for Industrial Growth, 1996., Vol. 2, 8-11 Jan. 1996 pp: 819 – 825.

[18] W.A. Hill, E.Y.Y. Ho, I.J. Nuezil, "Dynamic behavior of cycloconverter system," IEEE Transactions on Industry Applications, Vol. 27, Issue 4, July-Aug. 1991, pp: 750 – 755.

[19] Y. -Y. Tzou, H. –J. Hsu, "FPGA based SVPWM Control IC for PWM Inverter," IEEE Trans. On Power Electronics, April 10, 1997, pp: 138-143.

[20] N.A Rahim, T.C. Green, B.W. Williams, "PWM ASIC Design for the Three-Phase Bidirectional Buck Converter," Int. J. Electronics, 1996, Vol. 81, No.5, pp: 603-615.

[21] S. Mekhilef, A.M. Omar, N.A. Rahim, "Modelling of three-phase uniform symmetrical sampling digital PWM for power converter,"Canadian Conference on Electrical and Computer Engineering, 2005. 1-4 May 2005, pp: 1505 – 1508.

[22] A.M. Omar, N.A. Rahim, "FPGA-based ASIC design of the three-phase synchronous PWM flyback converter," IEE Proceedings on Electric Power Applications, Vol. 150, Issue 3, May 2003 pp: 263 – 268.

Input and Output Ripple Analysis of AC Chopper

Arwindra Rizqiawan*, Dessy Amirudin, Deni, and **Pekik Argo Dahono**

*School of Electrical Engineering and Informatics, Institute of Technology Bandung
Tel. 62-22-2503315, Fax. 62-22-2508132
Jl. Ganesha 10, Bandung, INDONESIA
*email: windra@konversi.ee.itb.ac.id

Abstract – An input and output ripple analysis of ac chopper that is useful in filter design is presented in this paper. The analytical expressions of input and output ripples of single-phase and three-phase AC choppers are derived. The results are used to design the required input and output LC filter. As the unique values of inductance and capacitance of the filter cannot be determined based only on ripple specification, an additional criterion based on minimum total reactive power is proposed. Simulated and experimental results are included to verify the derived expressions.

Keywords: ripple analysis, ac chopper

I. INTRODUCTION

An ac voltage regulator is commonly used to control an output ac voltage for various usages. Covering from simple light dimmer until control of induction motor. Phase-angle control using thyristor was commonly used as ac regulator. This type offers such advantages as simplicity and the ability of controlling a large amount of power economically. But it also suffers from disadvantages such as high low order harmonic content in the both input and output voltages and currents, and discontinuity of power flow appears at both input and output side [1]-[2]. To overcome these drawbacks, PWM ac chopper can be used [3].

An AC chopper always has an input LC filter to suppress the generated input current ripple. If the AC chopper is used as an AC power supply, an output LC filter is also required. In recent years, various AC chopper topologies and its control methods were proposed in the literature [3]-[10]. Though a lot of works were published, only a few of them are concerned on ripple analysis of AC chopper. Ripple analysis is required to properly design the required LC filters. At present, most of ripple analysis was conducted by using Fourier series [2]. By using this method, an accurate result cannot be obtained without taking into account a large number of harmonics. Input and output ripple analysis of ac chopper is presented in this paper. Ripple analysis of single phase ac chopper begin with output side then continue with input side analysis. Since single phase and three phase system have special relation, the derived expressions for single phase ac chopper can be extended into three phase ac chopper using common relation. As the unique values of inductance and capacitance of the filter cannot be determined based only ripple specification, an additional criterion based on minimum total reactive power is proposed. Simulation and experimental results are included to verify the validity of the derived expressions.

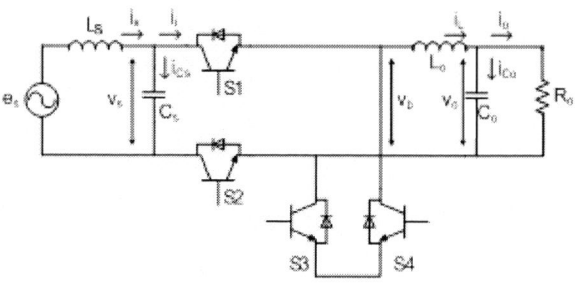

(a)

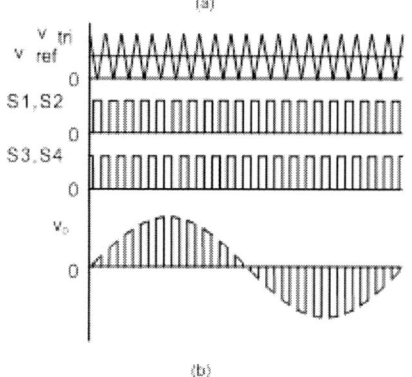

(b)

Fig, 1. (a) Scheme (b) Waveforms of single phase ac chopper

II. AC CHOPPER

Circuit topology for single phase buck ac chopper and its PWM scheme is shown in Fig. 1. In this topology, S1 and S2 will active during active mode, while S3 and S4 will active during the freewheeling mode. When switches S1 and S2 are turned ON (switches S3 and S3 are turned OFF), the voltage v_b is equal to the input voltage v_s. When switches S1 and S2 are turned OFF (switches S3 and S4 are turned ON) the voltage v_b is equal to zero.

In order to simplify the analysis, some assumption will be used in this paper. Those assumptions are such as, the power switches are assumed as ideal switches, the switching frequency is much higher than line frequency; and the ac voltage source is pure sinusoid with constant magnitude and frequency. The filter capacitor is assumed as pure capacitance without parasitic resistance.

III. OUTPUT RIPPLE ANALYSIS OF AC CHOPPER

Although buck ac chopper is used to explain the analysis, the same method can also applied to another type of ac chopper. The analysis is limited to continuous conduction mode only. For the purpose of output ripple analysis, the voltage ripple across the input filter

978-1-4244-0644-9/07/$25.00 ©2007 IEEE

capacitor is neglected. It is also assumed that voltage across input capacitor, C_s, is equal to the source voltage. Thus,

$$v_s = E_m \sin\theta \qquad (1)$$

From ac chopper circuit in Fig. 1 above, we can express the load side voltage as,

$$v_b = L_o \frac{di_L}{dt} + v_o \qquad (2)$$

If the load voltage and current are separated into the average and ripple terms, we get

$$v_b = \bar{v}_b + \tilde{v}_b \qquad (3)$$

$$i_L = \bar{i}_L + \tilde{i}_L \qquad (4)$$

Substituting (3)-(4) into (2) will result in an output voltage equation in terms of average and ripple components.

$$\bar{v}_b + \tilde{v}_b = L_o \frac{d\bar{i}_L}{dt} + \bar{v}_o + L_o \frac{d\tilde{i}_L}{dt} + \tilde{v}_o \quad (5)$$

Separating average and ripple components in the right and left hand side of (5), the followings are obtained

$$\bar{v}_b = L_o \frac{d\bar{i}_L}{dt} + \bar{v}_o \qquad (6)$$

$$\tilde{v}_b = L_o \frac{d\tilde{i}_L}{dt} + \tilde{v}_o \qquad (7)$$

The ripple voltage across the output capacitor is much smaller than the ripple voltage across the output filter inductor and, therefore, (7) can be simplified into

$$\tilde{v}_b = L_o \frac{d\tilde{i}_L}{dt} \qquad (8)$$

Based on (3) and (8), we can obtain the output ripple current as

$$\tilde{i}_L = \frac{1}{L_o} \int \left(v_b - \bar{v}_b \right) dt \qquad (9)$$

The load voltage is equal to source voltage during active mode (S1 and S2 ON), and equal to zero during freewheeling mode (S3 and S4 ON). The detailed output waveforms over one carrier period is shown in Fig. 2(a). Based on waveforms shown in Fig. 2(a), we can express the load voltage as follow

$$v_b = \begin{cases} v_s & \text{for } t_o \le t \le t_1 \\ 0 & \text{for } t_1 \le t \le t_2 \end{cases} \qquad (10)$$

Based on (9) and (10), output current ripple can be expressed as

$$\tilde{i}_L = \begin{cases} -\dfrac{\bar{v}_b}{L_o}\dfrac{T_{OFF}}{2} + \dfrac{v_s - \bar{v}_b}{L_o}\left(t - t_o\right) & \text{for } t_o \le t \le t_1 \\[4mm] \dfrac{\bar{v}_b}{L_o}\dfrac{T_{OFF}}{2} - \dfrac{\bar{v}_b}{L_o}\left(t - t_1\right) & \text{for } t_1 \le t \le t_2 \end{cases}$$

$$(11)$$

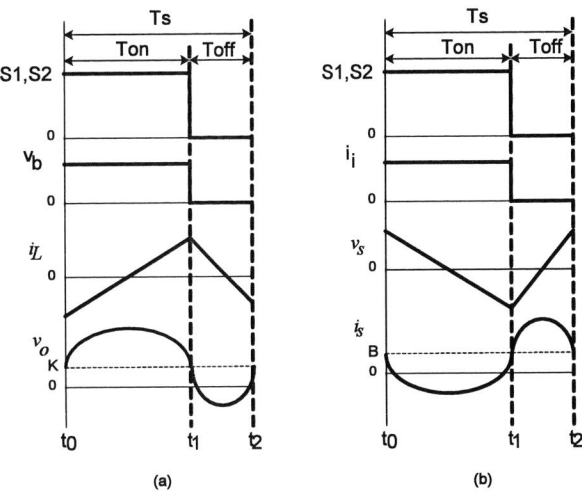

Fig. 2. Waveform details (a) output side (b) input side

The mean square value of output current ripple over one switching period can be obtained using following equations

$$\tilde{I}_L^2 = \frac{1}{T_s} \int_{t_o}^{t_o + T_s} \tilde{i}_L^2 \, dt \qquad (12)$$

$$\tilde{I}_L^2 = \frac{1}{12} \left(\frac{v_s \alpha (1-\alpha)}{f_s L_o} \right)^2 \qquad (13)$$

where $f_s = 1/T_s$ is switching frequency and $\alpha = T_{ON}/T_s$ is switch duty cycle. The root mean square (rms) value of output current ripple over one fundamental period can be determined using the following equation

$$\tilde{I}_{L,av} = \sqrt{\frac{1}{2\pi} \int_0^{2\pi} \tilde{I}_L^2 \, d\theta} \qquad (14)$$

Substituting (1) into (13) and the result is substituted into (14), the following is obtained after performing the integration

$$\tilde{I}_{L,av} = \frac{1}{2\sqrt{6}} \left(\frac{E_m \alpha (1-\alpha)}{f_s L_o} \right) \qquad (15)$$

From Fig. 1, the output filter capacitor current can be expressed as

$$i_L = i_{Co} + i_o \qquad (16)$$

If the above currents are separated into the ripple and average components, output side current equation of ac chopper becomes

$$\bar{i}_L + \tilde{i}_L = \bar{i}_{Co} + \tilde{i}_{Co} + \bar{i}_o + \tilde{i}_o \qquad (17)$$

Based on (17), the capacitor current ripple can be obtained as

$$\tilde{i}_{Co} = \tilde{i}_L - \tilde{i}_o \qquad (18)$$

The load current ripple is usually much smaller than the inductor current ripple and, therefore,

$$\tilde{i}_{Co} \approx \tilde{i}_L \qquad (19)$$

The capacitor voltage ripple can be calculated using

$$\tilde{v}_o = \frac{1}{C_o}\int \tilde{i}_{Co}\,dt = \frac{1}{C_o}\int \tilde{i}_L\,dt \qquad (20)$$

Substituting (11) into (20) and performing the integration, the following result are obtained

$$\tilde{v}_o = \frac{E_m \sin\theta}{L_o C_o}\begin{cases} K - \dfrac{\alpha T_{\mathit{off}}}{2}(t-t_0) + \dfrac{(1-\alpha)}{2}(t-t_0)^2 & \text{for } t_0 \le t \le t_1 \\[2mm] K + \dfrac{\alpha T_{\mathit{off}}}{2}(t-t_1) - \dfrac{\alpha}{2}(t-t_1)^2 & \text{for } t_1 \le t \le t_2 \end{cases}$$
$$(21)$$

where $K = \dfrac{\alpha^3(1-\alpha) - \alpha(1-\alpha)^3}{12}$. The mean square value of the capacitor voltage ripple over one carrier period can be obtained as

$$\tilde{V}_o^2 = \frac{1}{T_s}\int_{t_o}^{t_o+T_s} \tilde{v}_o^2\,dt \qquad (22)$$

Then, the rms output voltage ripple over one fundamental period can be obtained as

$$\tilde{V}_{o,av} = \sqrt{\frac{1}{2\pi}\int_0^{2\pi} \tilde{V}_o^2\,d\theta}$$
$$= \frac{E_m\sqrt{2}\alpha}{24 L_o C_o f_s^2}\left[\frac{1 - 5\alpha^2 + 6\alpha^3 - 2\alpha^4}{5}\right]^{1/2} \qquad (23)$$

The expression of output voltage ripple is useful in designing the required output LC filter, which will be presented in the later section.

IV. INPUT RIPPLE ANALYSIS OF AC CHOPPER

Now let us move to ripple analysis of input side of ac chopper. In this analysis it is assumed that output current is pure sinusoid and free of ripple, which is

$$i_L = I_o \sin(\theta - \phi) \qquad (24)$$

where ϕ is phase angle due to inductive dominant load. Thus the input current during one switching period can be written as

$$i_i = \begin{cases} I_o \sin(\theta - \phi) & \text{for } t_0 \le t \le t_1 \\ 0 & \text{for } t_1 \le t \le t_2 \end{cases} \qquad (25)$$

The detailed input current waveforms over one carrier period is shown in Fig. 2(b). Refer back to Fig. 1, we can write the current relation in the input side of ac chopper as

$$i_s = i_{Cs} + i_i \qquad (26)$$

Similar to output ripple analysis, we can also separate the currents into the ripple and average components as followings

$$i_s = \bar{i}_s + \tilde{i}_s \qquad (27)$$

$$i_{Cs} = \bar{i}_{Cs} + \tilde{i}_{Cs} \qquad (28)$$

$$i_i = \bar{i}_i + \tilde{i}_i \qquad (29)$$

Substituting (27)-(29) into (26) results in

$$\bar{i}_s + \tilde{i}_s = \bar{i}_{Cs} + \tilde{i}_{Cs} + \bar{i}_i + \tilde{i}_i \qquad (30)$$

Ripple and average components must be equal between the left and right hand sides of (30), thus

$$\bar{i}_s = \bar{i}_{Cs} + \bar{i}_i \qquad (31)$$

$$\tilde{i}_s = \tilde{i}_{Cs} + \tilde{i}_i \qquad (32)$$

Rewrite (32) in terms of capacitor current ripple, gives

$$\tilde{i}_{Cs} = \tilde{i}_s - \tilde{i}_i \qquad (33)$$

Due to LC filter in input side of ac chopper, source current ripple will be much smaller than capacitor current ripple. Neglecting source current ripple term in (33) results in

$$\tilde{i}_{Cs} \approx -\tilde{i}_i = \bar{i}_i - i_i \qquad (34)$$

From (34), we can see that capacitor current ripple is equal to negative of input current ripple. Average value of input current ripple can be obtained from (25) by using following equation

$$\bar{i}_i = \frac{1}{T_s}\int_{t_1}^{t_2} i_i\,dt \qquad (35)$$

$$\bar{i}_i = \alpha I_o \sin(\theta - \phi) \qquad (36)$$

The capacitor voltage ripple then

$$\tilde{v}_s = \frac{1}{C_s}\int \tilde{i}_{Cs}\,dt \qquad (37)$$

Substituting (25) and (36) for each time interval into (34) then calculate the capacitor voltage ripple using (37), will result in the followings

$$\tilde{v}_s = A\begin{cases} (\alpha-1)\left[-(t-t_0) + \dfrac{T_{ON}}{2}\right] & \text{for } t_0 \le t \le t_1 \\[2mm] -(\alpha-1)\dfrac{T_{ON}}{2} - \alpha(t-t_1) & \text{for } t_1 \le t \le t_2 \end{cases}$$
$$(38)$$

where $A = \dfrac{I_o \sin(\theta - \phi)}{C_s}$. Thus, we can obtain rms value of capacitor voltage ripple over one switching period as

$$\tilde{V}_s = \sqrt{\frac{1}{T_s}\int_{t_0}^{t_0+T_s} \tilde{v}_s^{\,2}\,dt} \qquad (39)$$

Substituting (38) into (39) results in

$$\tilde{V}_s = \frac{I_o \sin(\theta - \phi)(\alpha-1)\alpha}{C_s f_s\,2\sqrt{3}} \qquad (40)$$

The rms value of capacitor voltage ripple over one fundamental period can be calculated using following equation

$$\tilde{V}_{s,av} = \sqrt{\frac{1}{2\pi}\int_0^{2\pi} \tilde{V}_s^2\,d\theta} \qquad (41)$$

Substituting (40) into (41) will give us rms value of capacitor voltage ripple

$$\tilde{V}_{s,av} = \frac{1}{2\sqrt{6}} \frac{I_o \alpha (1-\alpha)}{C_s f_s} \qquad (42)$$

To analyze source current ripple, first let assume that voltage source is free of ripple. Source voltage equation in input side of ac chopper can be expressed as

$$e_s = L_s \frac{di_s}{dt} + v_s \qquad (43)$$

Source current can also be separated in ripple and average components, as stated in (27). Substituting (27) into (43) then separate between ripple and average terms, results in the followings equation

$$e_s = L_s \frac{d\overline{i_s}}{dt} + \overline{v}_s \qquad (44)$$

$$0 = L_s \frac{d\tilde{i_s}}{dt} + \tilde{v}_s \qquad (45)$$

Rearrange (45) will give us source current ripple equation, that is

$$\tilde{i_s} = -\frac{1}{L_s} \int \tilde{v}_s \, dt \qquad (46)$$

Substituting capacitor voltage ripple in (38) into (46), the integration result is

$$\tilde{i_s} = \frac{A}{2L_s} \begin{cases} \begin{aligned} &B + (1-\alpha)(t-t_0)^2 - \\ &(1-\alpha)(t-t_0)T_{on} \end{aligned} & \text{for } t_0 \le t \le t_1 \\ B + (1-\alpha)(t-t_1)T_{on} - \alpha(t-t_1)^2 & \text{for } t_1 \le t \le t_2 \end{cases}$$

$$(47)$$

where $B = (1-\alpha)\alpha \left(\dfrac{2\alpha-1}{6} \right) T_s^2$. Using the same

equation in (39) except replace capacitor voltage terms with source current ripple in (47) will result the rms value over one switching period, that is

$$\tilde{I_s} = \left(\frac{A\alpha(1-\alpha)}{2L_s C_s} \right) \sqrt{\frac{1+2\alpha-2\alpha^2}{180}} \qquad (48)$$

Now the rms value of source current ripple over one fundamental period can be obtained by integrating (48) over its fundamental period, results as

$$\tilde{I}_{s,av} = \frac{1}{12} \left(\frac{I_o \alpha (1-\alpha)}{f_s^2 C_s L_s} \right) \sqrt{\frac{1+2\alpha-2\alpha^2}{10}} \quad (49)$$

Then we have analytical expressions for input voltage and current ripple. These expressions will also useful to determine the input LC filter.

V. RIPPLE ANALYSIS OF THREE PHASE AC CHOPPER

Circuit topology for three phase ac chopper is shown in Fig. 3. This topology is extended version of single phase ac chopper topology shown in Fig. 1, since both topology have the same operation modes concept. In three phase ac chopper, during active mode S1, S2, and S3 will be in ON state while others in OFF state.

Reversely, during freewheeling mode S4, S5, S6 will have ON state and the others now turn to OFF state. The same assumption used in ripple analysis of single phase ac chopper above also used in this three phase ripple analysis. In order to simplify the analysis, the capacitor filter and load is assumed delta connected.

Output current ripple of three phase ac chopper can be obtained in a similar way for output current ripple in single phase ac chopper. Since we use phase to phase voltage as reference, thus output current ripple expression in the delta connection will equal with output current ripple in single phase ac chopper except now maximum voltage term in (15) will be replaced by maximum phase to phase voltage in three phase systems. Rewrite again (15) so it will suitable with three phase system

$$\tilde{I}_{AB,av} = \frac{1}{2\sqrt{6}} \left(\frac{E_{m,AB} \alpha (1-\alpha)}{f_s L_o} \right) \qquad (50)$$

Expression (50) also valid for the other phase. To obtain output line current ripple, we can simply use $\sqrt{3}$ relation between phase current and line current in three phase system, thus we get the expression for output line current ripple as

$$\tilde{I}_{L,av} = \frac{1}{2\sqrt{2}} \left(\frac{E_{m,AB} \alpha (1-\alpha)}{f_s L_o} \right) \qquad (51)$$

Similar extension analysis also can be used in input ripple analysis of three phase ac chopper. Since we assumed that voltage across input capacitor is equal to source voltage, in this case is phase to phase voltage, using similar derivation of single phase ac chopper we will get the result as

$$\tilde{V}_{ab,av} = \frac{1}{2\sqrt{6}} \frac{I_o \alpha (1-\alpha)}{C_s f_s} \qquad (52)$$

Similar voltage ripple expressions are used for other phases.

Derivation of source current ripple in single phase ac chopper in the previous section can be regarded as phase to phase configuration of three phase system. Thus to obtain source line current ripple in three phase ac chopper, we must convert phase to phase analysis into phase to neutral analysis. This can be done by using $\sqrt{3}$ relation between phases to neutral voltage to phase to phase voltage. Then we get the expression for input line current ripple for three phase ac chopper as

$$\tilde{I}_{a,av} = \frac{1}{12\sqrt{3}} \left(\frac{I_o \alpha (1-\alpha)}{f_s^2 C_s L_s} \right) \sqrt{\frac{1+2\alpha-2\alpha^2}{10}} \quad (53)$$

The same expression valid for other phase currents.

VI. DETERMINATION OF LC FILTER

In practice, only the output voltage ripple and source current ripple are specified. These ripples, however, are determined by the product of inductance and capacitance of the LC filter. Thus, unique values of inductance and capacitance of the filter cannot be determined. To solve this problem, an additional criterion based on minimum

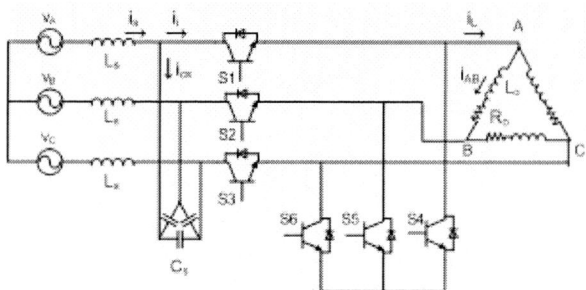

Fig. 3. Three phase ac chopper circuit topology

reactive power in the LC filter is used. Reactive power in the output LC filter can be expressed as

$$P_r = \omega_r L_o (\overline{I_L}^2 + \tilde{I}_{L,av}{}^2) + \omega_r C_o (\overline{V_o}^2 + \tilde{V}_{o,av}{}^2) \quad (54)$$

Since ripple components are much smaller than fundamental components, (54) is simplified to

$$P_r = \omega_r L_o \overline{I_L}^2 + \omega_r C_o \overline{V_o}^2 \quad (55)$$

Fundamental output current can be expressed as $\overline{I}_L^2 = \left(\mathsf{Re}(I_o)^2 + (\mathsf{Im}(I_o) - \overline{V}_o C_o \omega)^2 \right)$, then expressing (23) in term of output capacitor, substituting into (55) and then take the minimum value by

$$\frac{\partial P_r}{\partial L_o} = \frac{\partial \left[\omega L_o \left(\overline{I}_o^2 + \left(\mathsf{Im}(\overline{I}_o) - \omega \overline{V}_e \left(\frac{E_m f(\alpha)}{24 L_o f_s^2 \overline{V}_{e,av}} \right) \right)^2 \right) + \omega \overline{V}_e^2 \left(\frac{E_m f(\alpha)}{24 L_o f_s^2 \overline{V}_{e,av}} \right) \right]}{\partial L_o} = 0$$

(56)

Then, inductor value for LC filter in output side is obtained as

$$L_o = \frac{\overline{V}_o}{f_s \overline{I}_o} \sqrt{\left(\frac{E_m f(\alpha)}{24 \tilde{V}_o} \left(\frac{E_m f(\alpha)}{24 \tilde{V}_o f_s^2} - 1 \right) \right)} \quad (57)$$

Where $f(\alpha) = \alpha \sqrt{\dfrac{\alpha^2 - 5\alpha^4 + 6\alpha^5 - 2\alpha^6}{5}}$, then

value of C_o can be obtained by substituting (57) into (23) in terms of capacitor. Thus we have determine the output LC filter based on output ripple.

A similar method can be used to determine LC filter in input side of ac chopper. A similar method also can be used to derive the expressions of input and output ripples of three-phase ac-ac chopper.

VII. SIMULATION RESULTS

In order to verify the derived expression, simulation of single phase and three phase ac chopper were constructed. The source voltage for both ac chopper is set 110 sin ωt, while for three phase ac chopper this is phase to phase voltage. Inductor input filters of 1 mH were inserted series between source and 30 uF capacitor filter. In three phase ac chopper these L and C filter value also used, in form of series inductor and delta connected capacitor. In this simulation, an IGBT is used as switching device with frequency sampling of 5 kHz. The ac chopper is connected to a RL load contains of 5 mH inductance and 5 Ω resistor, for three phase ac chopper this load is connected in delta configuration.

Figs. 4, 5, and 6 show simulation and calculated results of output current ripple, output voltage ripple, and input current ripple of single phase ac copper as a function of duty cycle, respectively. Agreement between the calculated and simulation result is clearly shown.

While Figs. 7, 8, and 9 show simulation and calculated results of output phase current ripple, output voltage ripple, and source line current ripple of three phase ac copper as a function of duty cycle, respectively. Once again, agreement between the calculated and simulation result is clearly shown.

VIII. EXPERIMENTAL RESULTS

An experimental setup for single phase ac chopper as shown in Fig. 1 has been built to verify the derived expressions. This experimental setup uses transistor as switching device, operated in 1 kHz of switching frequency. 30 mH and 360 uF are used as input LC filter to assure good input source, while 10 mH and 250 uF are used as output LC filter. Resistor 5Ω is used as load of the single phase ac chopper. Input voltage is set on 35V, while output voltage is various as variable operating duty cycle.

Fig. 10 shows the resulted output current and voltage waveform. Figs. 11-14 show the experimental result of output current ripple, output voltage ripple, input voltage ripple, and input current ripple of single phase ac chopper, respectively. Agreement of the calculated and experimental result is clearly shown.

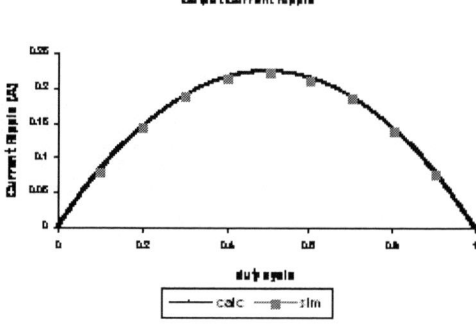

Fig. 4. Output current ripple of single phase ac chopper.

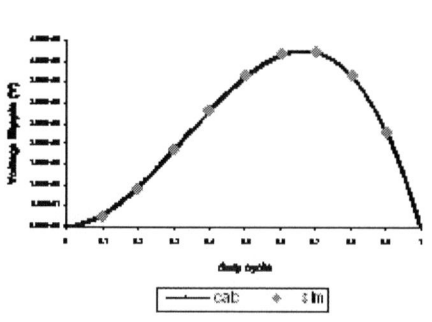

Fig. 5. Input voltage ripple of single phase ac chopper.

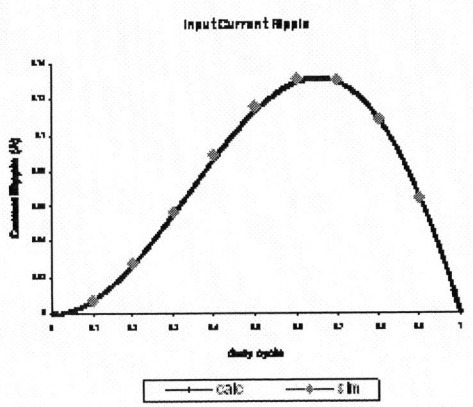

Fig. 6. Input current ripple of single phase ac chopper.

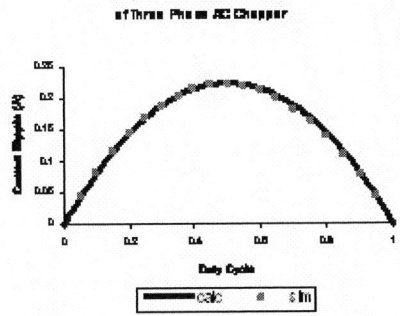

Fig. 7. Output phase current ripple of three phase ac chopper.

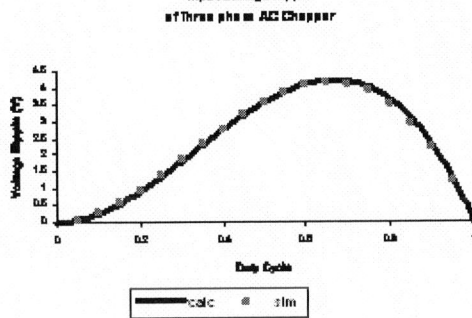

Fig. 8. Output voltage ripple of three phase ac chopper.

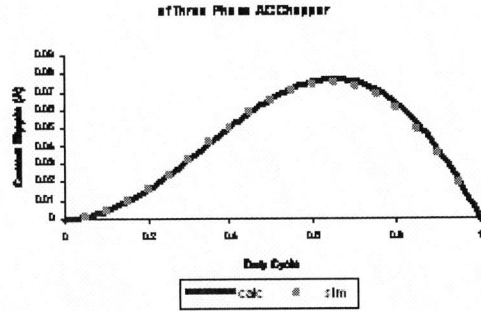

Fig. 9. Input phase current ripple of three phase ac chopper.

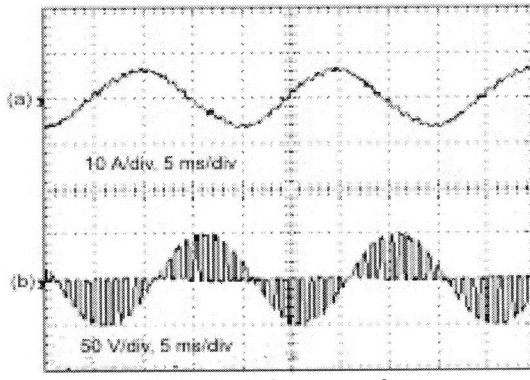

Fig.10. Resulted output waveforms
(a) current waveform, (b) voltage waveform

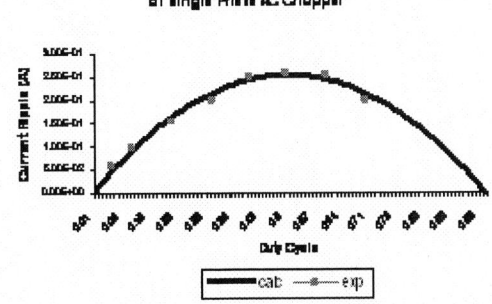

Fig. 11. Experimental result of output current ripple of single phase ac chopper

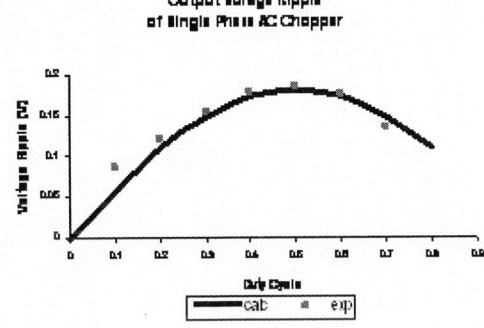

Fig. 12. Experimental result of output voltage ripple of single phase ac chopper

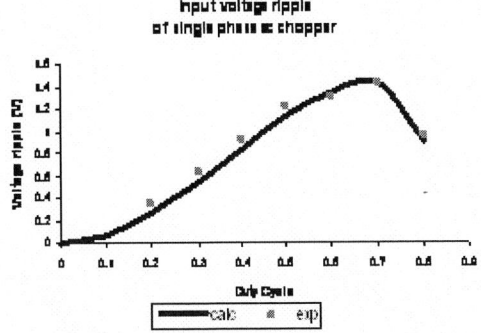

Fig.13. Experimental results of input voltage ripple of single phase ac chopper

539

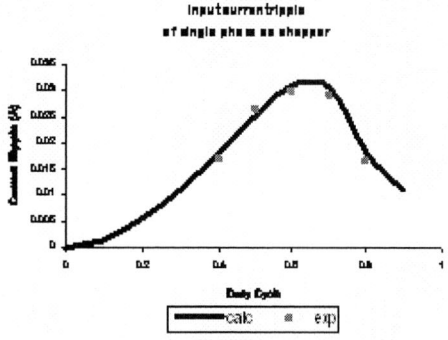

Fig.14. Experimental results of input current ripple of single phase ac chopper

IX. CONCLUSIONS

In this paper, analytical expressions for input and output side ripple in ac chopper has been derived. Ripple in three phase ac chopper can be obtained using the same approach used in single phase ac chopper. From the analysis we can see that there is no difference between ripple pattern as a function of duty cycle in single phase ac chopper and three phase ac chopper, since there is no difference of switching pattern between single phase and three phase ac chopper. The derived expression is useful to determine the required filter. The derived expression has been verified by simulation and experimental result.

REFERENCES

[1] N. A. Ahmed, K. Amei, and M. Sakui, "A New Configuration of Single-Phase Symmetrical PWM AC Chopper Voltage Controller", *IEEE Trans. Ind. Appl.* Vol. 46. No. 5, pp. 942-952, Oct. 1999

[2] K. E. Addoweesh, and A. L. Mohammadein, "Micropocessor Based Harmonic Elimination in Chopper Type AC Voltage Regulators", *IEEE Trans. Power Electron.*, Vol. 5, No. 2, pp. 191-200, April 1990

[3] B. H. Kwon, B. D. Min, and J. H. Kim, "Novel Topologies of AC Chopper", *IEE Proc. Eletr. Power. Appl.* Vol. 143, No. 4, July 1996

[4] D. H. Jang, and G. H. Choe, "Improvement of Input Power Factor in AC Choppers Using Asymmetrical PWM Technique", *IEEE Trans. Ind. Appl.*, Vol. 42, No. 2, pp. 179-185, Apr. 1995

[5] Z. Fedyczak, R. Strzelecki, and G. Benysek, "Single-phase PWM AC/AC Semiconductor Transformer Topologies and Applications", *Proc. Conf. PESC'02*, pp. 1048-1053, June 2002

[6] C. A. Petry, J. C. Fagundes, and I. Barbi, "New Direct Ac-Ac Converters Using Switching Modules Solving the Commutation Problem", *Proc. of IEEE ISIE 2006*, pp. 864-869, July 2006

[7] S. Srinivasan, and G. Venkataramanan, "Comparative Evaluation of PWM AC-AC Converters", *Proc. Conf. PESC'95*, pp. 529-535, June 1995

[8] S. Ben-Yaakov, Y. Hadad, and N. Diamantstein, "A Four Quadrant HF AC Chopper with no Deadtime", *Proc. Conf. APEC'06*, pp. 1461-1465, March 2006

[9] G. Venkataramanan, "A Family of PWM Converters for Three Phase AC Power Conditioning", *Proc. Conf. Power Electronic, Drives, and Energy System for Industrial Growth 1996*, pp. 572-577, June 1996

[10] E. Lefeuvre, T. Meynard, and P. Viarouge, "Robust Two-Level and Multilevel PWM AC Choppers", *Proc. Conf. EPE 2001*, pp. 1-8, 2001

[11] P. A. Dahono, S. Riyadi, A. Mudawari, and Y. Haroen, "Output Ripple Analysis of Multiphase dc-dc Converters", *Proc. Conf. PEDS'96*, pp. 626-631, July 1999

[12] P. A. Dahono, "Input Ripple Analysis of Multiphase DC-DC Converters", *Jurnal IECI*, Vol. 2, No. 2, pp. 57-62, 2000

A Three-level 4 × 3 Conventional Matrix Converter

Runjie Rong[1], Poh Chiang Loh[1], Peng Wang[1], Frede Blaabjerg[2]

[1]School of Electrical and Electronic Engineering
NANYANG Technological University
Singapore
rongrunjie@hotmail.com

[2]Institute of Energy Technology
Aalborg University
Aalborg East, Denmark
fbl@iet.aau.dk

Abstract. **This paper proposes a topology of a three-level 4 × 3 conventional matrix converter with 12 bi-directional switches. PWM control and modulation index compensation have been investigated. Operation theory has been verified by the simulation results using Matlab. The simulation results show that the switching output performance of the proposed matrix converter is more efficient than that of existing matrix converters.**

Keywords: matrix converter, rectifier stage, inverter stage.

INTROUCDUTION

In the past few years, matrix converter has been widely studied due to the following advantages:

1) Don't need any large energy storage component such as large DC capacitor or inductor.
2) By controlling the switching devices appropriately, four-quadrant operation is straightforward and both output voltage and input current are sinusoidal with only harmonics around or above switching frequency.

Two typical topologies of matrix converters, which have been developed, are AC-DC-AC indirect matrix converter (IMC) [1] and AC-AC conventional matrix converter (CMC) [2] shown in Fig.1 and Fig.2, respectively. For the indirect frequency conversion scheme, the CMC is divided into a voltage-fed rectifier input stage and an inverter output stage with impressed output currents which are directly connected on the DC side. The physical implementation of this basic idea results in the IMC topology. Both of them are capable of providing simultaneous amplitude and frequency transformation of a voltage system and allowing for bi-directional power flow and independent control of the displacement power factor without the use of the bulky and limited lifetime passive components such as large DC-link electrolytic capacitors or AC boost inductors [1], [3]. However, the two typical topologies have two-level "+" and "−" output voltages which leads to some low order harmonics in the output. To reduce the distortion of output current, a three-level matrix

converter which consists of 12 bi-directional switches as shown in Fig.3 is proposed in this paper. Operational principles of the PWM control for three-level converters are presented. The simulation results of two-level and three-level converter are compared.

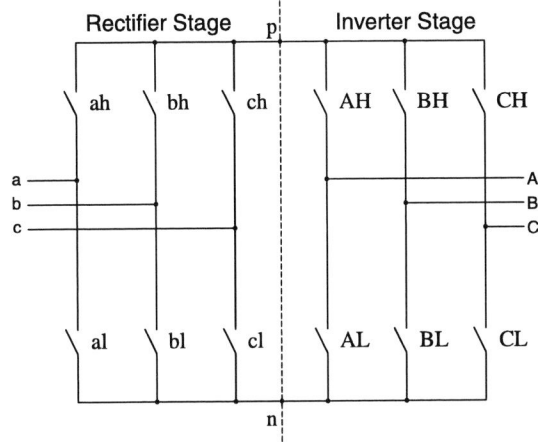

Fig.1 AC-DC-AC indirect matrix converter

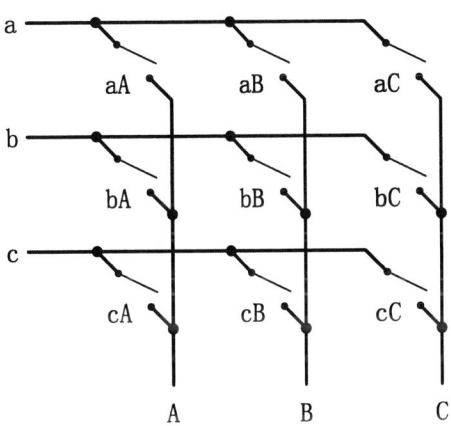

Fig.2 AC-AC conventional matrix converter

PWM CONTROL

Fig.3 shows the schematic diagram of a 4 × 3 switch matrix, which is developed from a 3 × 3 CMC. A neutral point is introduced in the switch matrix. The control scheme is divided into a rectifier input stage and an inverter output stage. In Fig.4 [3], we can see a bi-directional current source rectifier (CSR) is directly connected to a three-level voltage source inverter (VSI) without using any intermediate energy storage element. Fig.4 shows the topological arrangement of a three-level sparse matrix converter, where a bi-directional

978-1-4244-0644-9/07/$25.00 ©2007 IEEE   541

CSR is connected to a three-level neutral-point-clamped (NPC) inverter instead of the traditional two-level VSI as shown in Fig.1. Comparing the two figures, a second feature noted in Fig.4 is the clamping of the rear-end NPC neutral potential to the star-point of the input capacitive network used for filtering the three-phase chopping currents $\{I_{al}, I_{bl}, I_{cl}\}$ so as to produce a set of three-phase sinusoidal utility currents labeled as $\{I_a, I_b, I_c\}$. This second feature implies that only a star-connected filter capacitive network can be used since a delta-connected network does not provide the needed neutral star-point. Using the circuit shown in Fig.4 and noting that only an upper and a lower switch from different phases of the input CSR are turned on at any instant, the converter intermediate dc-link voltage V_i can assume any of the three input line voltages expressed as V_{XY}, where $\{X, Y\} = \{a, b, c\}$ and $X \neq Y$. For example, when upper switch of phase 'a' and lower switch of phase 'c' are commanded ON, line voltage $V_{ac} = V_{aN} - V_{cN}$ appearing across the 'a' and 'c' phase filter capacitors is imposed across the dc-link. However, the input split-dc source no longer has balanced upper and lower voltages. Instead, the upper and lower dc voltages are now respectively represented by the individual phase voltages V_{aN} and V_{cN} appearing across the 'a' and 'c' phase filter capacitors, which obviously are not equal (except at the six phase-intersection points per fundamental cycle). Because of this, modulation schemes previously reported for NPC inverter control cannot be applied directly, as explained in the next section before appropriate modulation and gain compensation schemes are designed for controlling the three-level sparse matrix converter with zero input rectification loss, and sinusoidal input currents and output voltages produced at all instances.

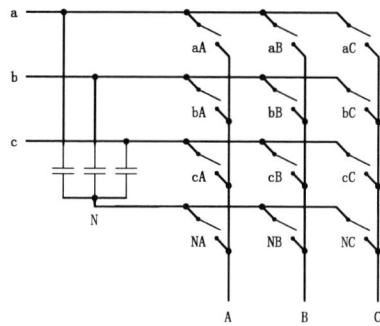

Fig.3 A 4 × 3 switches matrix

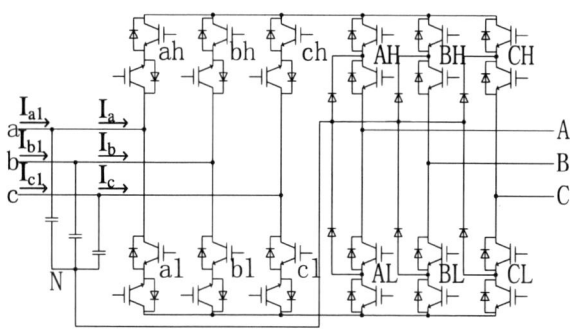

Fig.4 Three-Level Sparse Indirect Matrix Converter

Control of the rectifier input stage

The typical space vector modulation [1], [3] is used. The space vector diagram is shown in Fig.5. For the reference vector is in the sixth sextant, the nearest active vectors recommended for use are $R_{ch,bl}$ and $R_{ah,bl}$, which respectively impose line voltages V_{cb} and V_{ab} across the converter dc-link. Therefore, only V_{cb} and V_{ab} are used for a switching cycle, and only phase 'c' and phase 'a' are pulse-width modulated to the positive dc-rail by commutating phases 'c' and phase 'a'. On the contrary, phase 'b' is firmly clamped to the negative dc-rail. Given this scenario, duty ratios of the two modulated phases can be determined as:

$$V_a = V_m \cos \theta_a; V_b = V_m \cos \theta_b; V_c = V_m \cos \theta_c;$$

$$V_a + V_b + V_c = 0;$$

$$\Rightarrow -\frac{V_c}{V_b} - \frac{V_a}{V_b} = \delta_c + \delta_a = 1;$$

$$\delta_c = -\frac{V_c}{V_b} = -\frac{\cos \theta_c}{\cos \theta_b}$$
$$\delta_a = -\frac{V_a}{V_b} = -\frac{\cos \theta_a}{\cos \theta_b} \tag{1}$$

Where V_m is the voltage amplitude, θ_x (x=a, b, c) is the respective phase angular displacement, and δ_c and δ_a are the duty ratios of phases 'c' and 'a' respectively. The same duty ratio expressions given in (1) can equally be used for the third sextant during which phase 'b' is clamped to the positive dc rail, while phases 'c' and 'a' are commutated once per switching cycle. Further generalizing the expressions for other sextants, (1) can be rewritten as:

$$\delta_1 = \begin{cases} -\dfrac{V_{min}}{V_{max}}, & \text{positive dc rail clamping} \\[2ex] -\dfrac{V_{max}}{V_{min}}, & \text{negative dc rail clamping} \end{cases}$$

$$\delta_2 = \begin{cases} -\dfrac{V_{mid}}{V_{max}}, & \text{positive dc rail clamping} \\[2ex] -\dfrac{V_{mid}}{V_{min}}, & \text{negative dc rail clamping} \end{cases}$$

$$V_{max} = max(V_a, V_b, V_c);$$
$$V_{mid} = mid(V_a, V_b, V_c); \tag{2}$$
$$V_{min} = min(V_a, V_b, V_c).$$

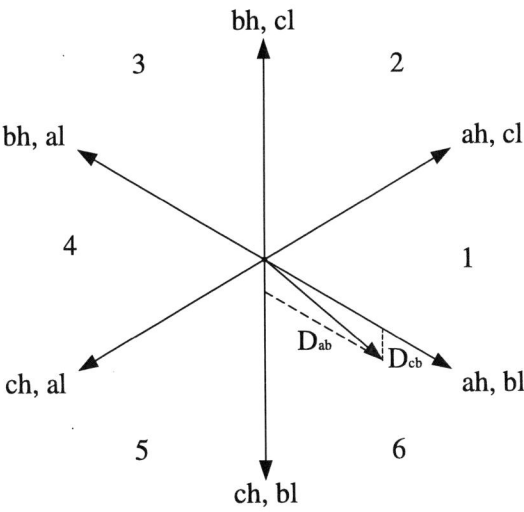

Fig.5 Space Vector diagram

Control of three-level inverter output stage

Because there will be three-level "+", "−", "N" in the output voltage, the inverter is modulated using carrier-based approach and a special carrier is introduced shown in Fig.6. The special carrier consists of two anti-triangular waves. The unique feature of the triangular wave is that the rising and falling triangular edges have different gradients so that it can force each triangular edge to span a complete rectifier stage switching state duration expressed as either $\delta_1 T$ or $\delta_2 T$, where δ_1 and δ_2 are the rectifier stage duty ratios. The three-phase reference signal: $Ref.V_a$, $Ref.V_b$ and $Ref.V_c$ are compared with the special triangular wave. During the time the reference wave greater than the upper triangular, the output is clamped to "+". When the reference wave is less than the lower triangular wave, the output is clamped to "−". When the reference wave is less than the upper triangular wave and greater than the lower one, the output is clamped to "N". Synchronizing in this way ensures that a complete inversion stage sequence comprising of "$N1\{+,N,N\} \leftrightarrow A1\{+,N,−\} \leftrightarrow A2\{N,N,−\} \leftrightarrow N2\{N,−,−\}$" as shown in Fig.6 is placed within each imposed rectifier stage switching state to always give the correct volt-sec average per half carrier cycle using the nearly constant applied dc-link voltage (assuming a reasonably high triangular frequency) corresponding to each rectifier stage state.

In addition to tracking accuracy, synchronizing the carriers with the CSR state sequence can conceptually eliminate the CSR switching loss by selectively placing either state $N1\{+,N,N\}$ or $N2\{N,−,−\}$ around each CSR commutation instant, depending on whether the upper or lower dc-rail is clamped. As a generic rule, for lower (upper) dc-rail clamping, state $N2\{N, −, −\}$ ($N1\{+, N, N\}$) should preferably be entered before commutating another CSR phase to the upper (lower) dc-rail since $N2$ ($N1$) effectively creates an interval with zero upper (lower) dc-link current

given that the NPC inversion stage is now connected across the neutral point 'N' and lower (upper) dc-rail only.

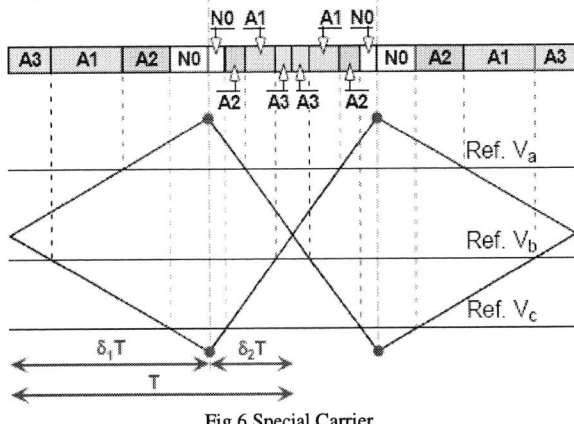

Fig.6 Special Carrier

Signals transmission

The PWM signals from the three-level IMC is transferred to the three-level CMC. It can be seen from Fig.4 that one phase of rectifier stage and one phase of inverter stage can be connected only if both of their two upper switches or lower switches are "on". So two "AND" gate and one "OR" are needed for the first nine switches to detect the signals from CMC. The signals from the IMC for the last three switches "NA", "NB", "NC", can be directly used without any process.

Commutation

A short circuit between two phase-sources will happen during every commutation period since semiconductor devices cannot be switched instantaneously due to propagation delays and finite switching times [2], [4], [5].

A reliable method to solve this problem is current direction based commutation. This method uses a four-step commutation strategy in which the direction of current flow through the commutation cells can be controlled. To implement this strategy the bi-directional switch must be designed in such a way as to allow the direction of the current flow in each switch to be controlled.

One of the structure, Common emitter anti-parallel IGBT, diode pair, is shown in Fig.7. This bi-directional switch arrangement consists of two diodes and two IGBTs connected in anti-parallel. The diodes are included to provide the reverse blocking capability. It should be noted that it is possible to independently control the direction of the current in the bi-directional switch. Each bidirectional switch requires an isolated power supply for the gate drives, but both devices can be driven with respect to the same voltage-the common emitter point.

543

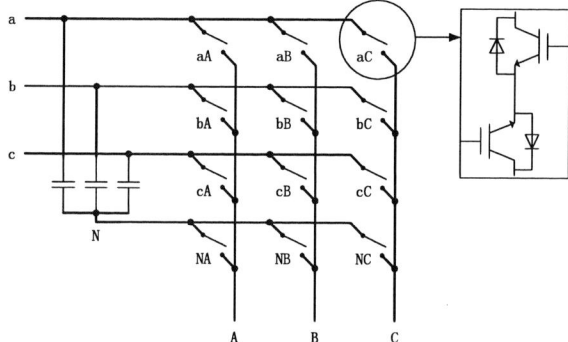

Fig.7 Common emitter anti-parallel IGBT, diode pair

Fig.8 shows a schematic of a two-phase to one-phase matrix converter. In steady-state, both of the devices in the active bi-directional switch are gated to allow both directions of current flow. The following explanation assumes that the load current is in the direction shown and that the upper bi-directional switch (S_1) is closed. When a commutation to S_2 is required, the current direction is used to determine which device in the active switch is not conducting. This device is then turned off. In this case, device S_{1b} is turned off. The device that will conduct the current in the incoming switch is then gated, S_{2a} in this example. The load current transfers to the incoming device either at this point or when the outgoing device (S_{1a}) is turned off. The remaining device in the incoming switch (S_{2b}) is turned on to allow current reversals. This process is shown as a timing diagram in Fig.9; the delay between each switching event is determined by the device characteristics.

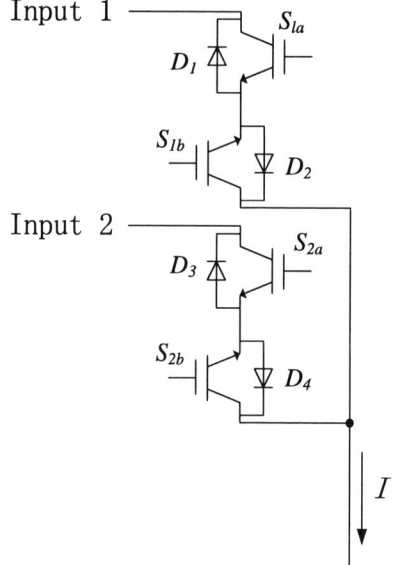

Fig.8 Two-phase to single-phase matrix converter

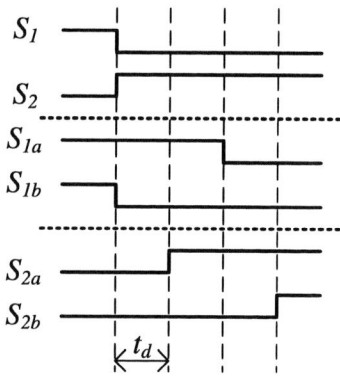

Fig.9 Timing diagram for four-step commutation between two bi-directional switches

SIMULATION RESULTS

The proposed three-level 4×3 matrix converter shown in Fig.3 and the AC-AC conventional matrix converter shown in Fig.2 has been implemented in Simulink. The parameters of the simulation are given as follow:

Supply: $V_{in-line} = 110V$; $f_{in} = 50Hz$;
Load: $R = 20\Omega$; $L = 6.3mH$;
Input filter: $L_f = 0.5mH$; $C_f = 15\mu F$;
$f_{out} = 120Hz$; $f_{carrier} = 5kHz$.

Fig.10 and Fig.11 show the results for two-level CMC and three-level 4×3 matrix converter, respectively. They are output line voltage and three phase output voltage. It is clear that the 3-level sparse IMC has a three-level output voltage generation capability as there are three distinct levels present in this waveform (positive and negative envelope of the rectified input voltage and zero) compared to only two for the two-stage CMC.

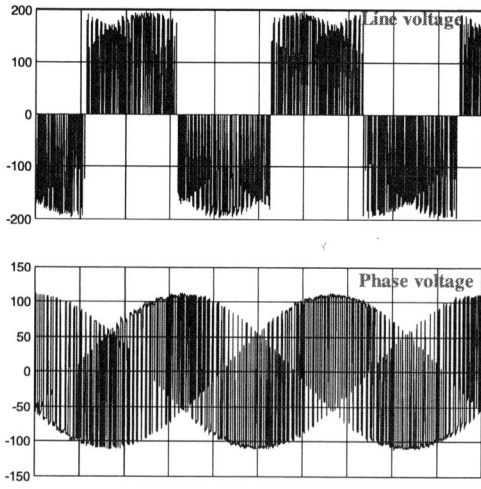

Fig.10 Two-level CMC

544

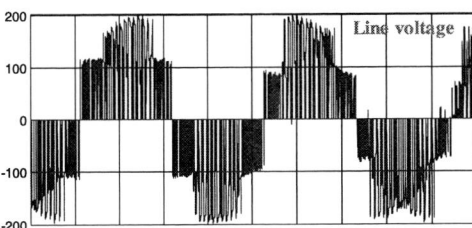

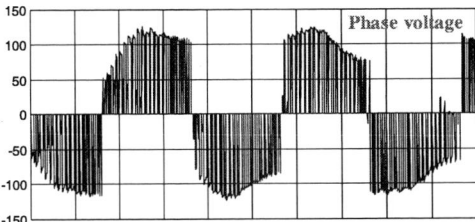

Fig.11 Three-level 4×3 matrix converter

The input side and output side performance of the three-level 4×3 matrix converter is illustrated in Fig.12 and Fig.13. The input currents are sinusoidal and balanced, which proves that the proposed modulation method provides a proper sine wave in/sine wave out operation.

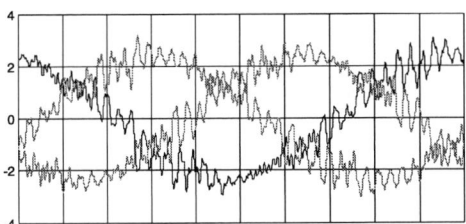

Fig.12 Three phase input currents

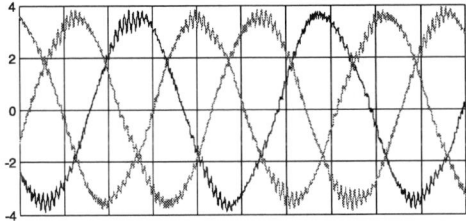

Fig.13 Three phase output currents

CONCLUTION

In this paper, a new three-level 4×3 matrix converter was proposed. The performance of the proposed converter and of a two-level CMC are compared using simulation models implemented in Simulink that clearly prove that the proposed topology has sinusoidal input currents and three-level output voltage generation capability, which could reduce significantly the size of output filter needed in some applications.

REFERENCES

[1]. J. W. Kolar, M. Baumann*, F. Schafmeister, H. Ertl* "Novel three-phase AC-DC-AC sparse matrix converter", Applied Power Electronics Conference and Exposition, 2002. APEC 2002. Seventeenth Annual IEEE;

[2]. Patrick W. Wheeler, Jon C. Clear, Lee Empringham, Michael Bland, Klaus G. Kerris "Matrix converters", IEEE Industry Applications Magazine, Jan/Feb 2004;

[3]. Christian Klumpner, Meng Lee, Patrick Wheeler "A new three-level sparse indirect matrix converter", IEEE Industrial Electronics, IECON 2006 - 32nd Annual Conference on Nov. 2006 Page(s):1902 – 1907;

[4]. Lixiang Wei, Thomas. A Lipo "A novel matrix converter topology with simple commutation", Industry Applications Conference, 2001. Thirty-Sixth IAS Annual Meeting. Conference Record of the 2001 IEEE Volume 3, 30 Sept.- 4 Oct. 2001 Page(s):1749 - 1754 vol.3;

[5]. Patrick Wheeler, Jon Clare, Lee Empringham, Maurice Apap, Michael Bland "Matrix converter", Power Engineering Journal, Dec 2002.

A novel primary-side controlled contactless battery charger

Yi-Hwa Liu*

*Department of Electrical Engineering, National Taiwan University of Science and Technology, Taipei, Taiwan.

Shun-Chung Wang**

** Department of Electrical Engineering, Lunghwa University of Science and Technology, Taoyuan, Taiwan

Rong Ceng Leou***

***Department of Electrical Engineering, Cheng Shiu University, Kaoshiug, Taiwan

Abstract- Contactless energy transmission system boasts the advantages such as safety, reliability, low maintenance, and long product life. However, with no connections between the input side and output side, the control of the power converters will be made difficult without the knowledge of the secondary voltage and current information. In this paper, a novel primary-side controlled contactless battery charger is proposed. Analysis of the presented system as well as the software algorithm will be provided in detail. The advantages of the proposed system include simplified design and reduced size of secondary side. Simulation and experimental results are then provided to validate the correctness of the proposed system.

I. INTRODUCTION

In many applications, contactless energy transmission system (CETS) has distinct advantages over the conventional energy transmission system which uses wires and connectors. The advantages of such systems are safety, reliability, convenience, low maintenance, no contact resistance, and long product life. CETS transmits electric power via magnetic coupling within a pair of coils. Its typical applications includes: robots, mining and underwater tools, linear movable systems and non-contact battery chargers for electric vehicles and consumer products [1-5].

Typically, the primary and secondary winding of the transformer used in CETS are wound on two magnetic pieces separated by an air gap. Therefore, the coupling coefficient is much lower than unity in these applications. This makes the leakage inductance have a similar magnitude to the magnetizing inductance, resulting in poor conversion efficiency. Resonant circuits are normally employed in the primary and/or secondary winding to increase the power transfer capability and minimize the required voltage and current ratings of the power supply [1]. Furthermore, with no connections between the input side and output side, the control and protection of the power converters is very much different than the control of the converters that employ a conventional feedback control which uses signal communication between the output and input. Several types of CETS have been proposed to deal with this problem [2-5]. In [2], a simple control circuit with an infrared LED is used to transmit the information of the secondary side. The output regulation is performed by a dual modulation with a locally regulated rectifier at the secondary side as presented in [3, 4]. In [5], a close-loop control is obtained using overlapped transmission of data on the power transmission, and an optimal charging is achieved. To summary, the methods presented in the literatures require additional hardware and/or complicated controller; therefore increase the complexity of the CETS.

In this paper, a novel primary-side controlled contactless battery charger is proposed. The proposed battery charger is able to properly and safely charge the lead-acid battery, which is still widely used in most commercially available electrical vehicles and movable systems [6]. On-line feedback of the battery status such as the battery voltage and the charging current is essential for charger design. In the proposed primary-side controller, this information is derived directly from the measured current and duty cycle command of the primary side. The advantages of the proposed system include simplified design and reduced size of secondary side. Detailed description of the proposed system and its experimental results will be provided in the following sections.

II. SYSTEM CONFIGURATION

Fig. 1 shows the system configuration of the proposed system. In Fig. 1, the power converter used in the primary side is a full-bridge converter, which is used to provide a square wave on the primary side of transformer. Information of the secondary side is obtained directly from measuring the primary side current, thus eliminates the feedback circuit from secondary side. The CETS uses series-resonant parallel-loaded (SRPL) topology to boost the power transfer capability and minimize the required voltage and current ratings of the power supply. The output stage contains only a bridge rectifier and the lead-acid battery. From Fig. 1, the whole system can be divided into four major parts: primary side controller, full-bridge power stage, CETS and output stage. Detailed descriptions about each part will be given in the following sections:

a. Primary-side controller: the primary-side controller samples the primary-side current signal, computes the required PWM command and then provides the gating signals of full-bridge converter. In the proposed system, the main controller is implemented using the TMS320LF2407 DSP device from Texas Instrument Corp. It should be noted that the controller obtains the gating signal only from primary-side current information; therefore no signal isolation circuits are required. Detailed software algorithm will be provided in section IV.

b. Full-bridge power converter: Fig. 2 shows the PWM controlled power stage. It is a conventional full-bridge MOSFET inverter. The PWM signal controls the H-bridge drivers, turning opposite pairs of MOSFETs off and on. In order to prevent the MOSFETs in the same inverter leg

978-1-4244-0644-9/07/$25.00 ©2007 IEEE

create a direct path from the supply line to ground, dead time should be provided. Because of the dead time, there are two separate signals for each MOSFET. The gate driver circuits used in this paper is the high and low side driver IR2010 by International Rectifier Corp.

c. Contactless energy transmission system: the CETS used in this system consists of a transformer with large air gap and resonant circuits. The core model and the equivalent magnetic circuit of the transformer are shown in Fig. 3. According to Fig. 3(b), the coupling coefficient k can be calculated using the following equations:

$$k = \frac{\Phi_2}{\Phi_1} = \frac{\Phi_2}{\Phi_2 + 2\Phi_{leakge}} \tag{1}$$

$$\Phi_2 = \frac{F}{2R_{Gap}} , \ \Phi_{leakge} = \frac{F}{R_{leakge}} \tag{2}$$

Substitute Eq. (2) into Eq. (1), one can obtain

$$k = \frac{\dfrac{F}{2R_{Gap}}}{\dfrac{F}{2R_{Gap}} + \dfrac{2F}{R_{leakge}}} = \frac{1}{1 + \dfrac{4R_{Gap}}{R_{leakge}}} \tag{3}$$

Where $R = \dfrac{l_e}{\mu A_e}$, l_e is the equivalent magnetic path length, A_e is the cross-sectional area and μ is the core permeability. Circuit model of the utilized transformer is shown in Fig. 3(c).

The proposed system uses the series resonant topology at the primary side and the parallel resonant topology at the secondary side. A series resonant primary side is selected to minimize the VA rating of the transformer; while the parallel resonant secondary side can improve the power transfer ability and supply a stable current [7]. Detailed analysis of the CETS will be provided in section III.

d. Output stage: The lead-acid battery requires regulated DC voltage and/or current to charge, while the resonant circuit provides an AC wave on the secondary side of the transformer. Therefore, a full-bridge diode rectifier is used to filter the AC waveform out and produce a DC waveform. Conventionally, the output stage of a contactless battery charger often contains a local battery charger which will regulate the rectified DC waveform. However, this will increase the complexity as well as the size of the secondary circuits. In this paper, the regulation of the charging voltage/current is attained by regulating the gating signals of the full-bridge power converter using the primary-side control algorithm; therefore, the only circuit needed at the secondary side is a typical full-bridge diode rectifier. The developed software will be presented in section IV.

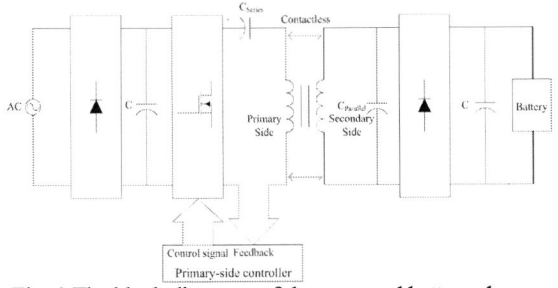

Fig. 1 The block diagrams of the proposed battery charger

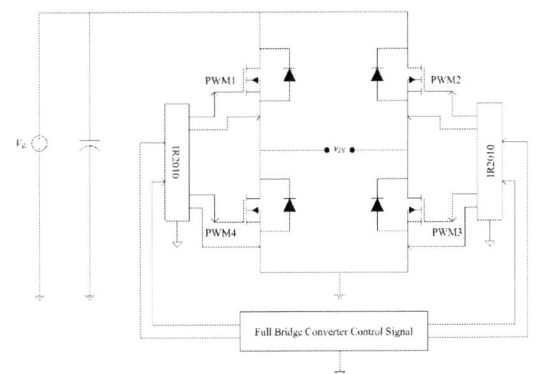

Fig. 2 Full-bridge power converter

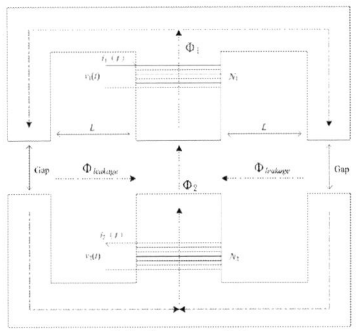

(a) Core model

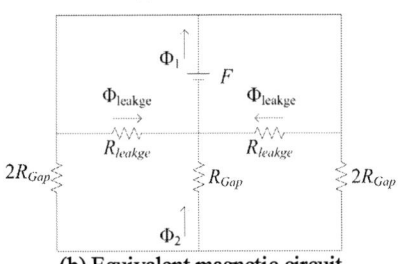

(b) Equivalent magnetic circuit

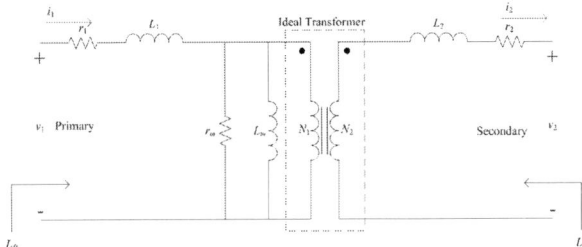

(c) Circuit model of the transformer
Fig. 3 Model of the transformer

III. ANALYSIS OF THE PROPOSED CETS

The circuit diagram of a series resonant CETS is shown in Fig. 4 (a). Using basic circuit analysis principles, the Thevenin's equivalent circuit of this system can be obtained by Eq. (4) and Eq. (5). The corresponding Thevenin equivalent circuit is shown in Fig. 4 (b) [8-9].

$$V_{th} = V_{IN}\left(\frac{n \cdot \omega^2 \cdot L_m (X + L_m)}{r_1^2 + \omega^2 (X + L_m)^2} + j\omega \frac{n \cdot L_m \cdot r_1}{r_1^2 + \omega^2 (X + L_m)^2}\right) \tag{4}$$

$$Z_{th} = \left[\frac{n^2 \cdot r_1 \cdot \omega^2 \cdot L_m^2}{r_1^2 + \omega^2 (X + L_m)^2} + r_2\right]$$
$$+ j\omega\left[\frac{n^2 \cdot L_m \cdot \left[r_1^2 + \omega^2 X (X + L_m)\right]}{r_1^2 + \omega^2 (X + L_m)^2} + L_2\right] \tag{5}$$

547

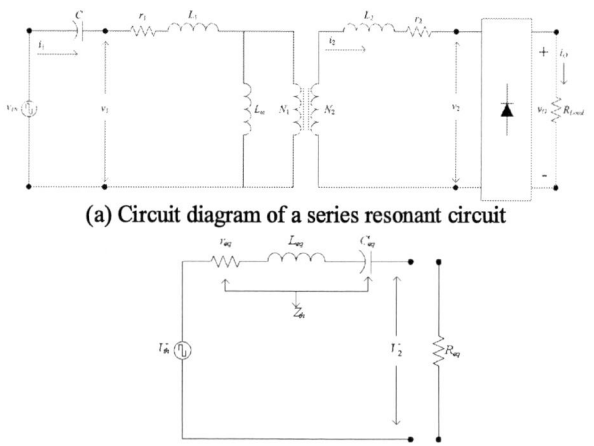

(a) Circuit diagram of a series resonant circuit

(b) Thevenin's equivalent circuit

Fig. 4 Circuit diagram and equivalent circuit of a series resonant CETS

Using the derived equivalent circuit, the voltage gain and current gain of the series resonant CETS can be derived as

$$\frac{V_2}{V_{IN}} = \frac{n \cdot L_m \cdot R_{eq}}{R_{eq}\left(L_m + L_1 - \frac{1}{\omega^2 C}\right) + j\omega\left[n^2 \cdot L_m \cdot \left(L_1 - \frac{1}{\omega^2 C}\right) + L_2\left(L_m + L_1 - \frac{1}{\omega^2 C}\right)\right]} \quad (6)$$

$$\frac{I_2}{I_1} = \frac{1}{n} \cdot k \cdot L_S \frac{\omega^2 L_S + j\omega\left(R_{eq} + r_2\right)}{\left(R_{eq} + r_2\right)^2 + \left(\omega L_S\right)^2} \quad (7)$$

From Eq. (6) and Eq. (7), the current gain of series resonant CETS is a nonlinear function of ω and R_{eq}. In other words, the current gain is a load-dependent function.

The resonant topology used in this paper is a SRPL resonant topology. Adding a parallel capacitor to the secondary makes the CETS behaves like a current source. Fig. 5 shows the circuit configuration of a SRPL resonant circuit. Using the same technique, the voltage gain and current gain of the series resonant CETS can be derived as

$$\frac{V_2}{V_{IN}} = \left(\frac{n \cdot \omega^2 \cdot L_m\left(X + L_m\right)}{r_1^2 + \omega^2\left(X + L_m\right)^2} + j\omega\frac{n \cdot L_m \cdot r_1}{r_1^2 + \omega^2\left(X + L_m\right)^2}\right) \cdot \frac{Z_{R\|C}}{Z_{th} + Z_{R\|C}} \quad (8)$$

$$\frac{I_2}{I_1} = \frac{k}{n} \cdot \omega \cdot L_S \frac{\omega L_S + j\left(Z_{R\|C} + r_2\right)}{\left(Z_{R\|C} + r_2\right)^2 + \left(\omega L_S\right)^2} \cdot \left(\frac{\frac{1}{j\omega C_P}}{R_{eq} + \frac{1}{j\omega C_P}}\right) \quad (9)$$

Where $Z_{R\|C} = \frac{1}{1 + \left(\omega C_P \cdot R_{eq}\right)^2}\left[\left(\omega C_P \cdot R_{eq}\right)^2 + j\omega C_P \cdot R_{eq}\right]$

Fig. 6 shows the bode plot of the current gain transfer function. From Fig. 6, it can be seen that the current gain is a constant when the operating frequency equals to the resonant frequency and is independent to the load. This is a desirable feature when designing a battery charger because most battery charger uses constant current charging. Detailed charging algorithm will be provided in the next section.

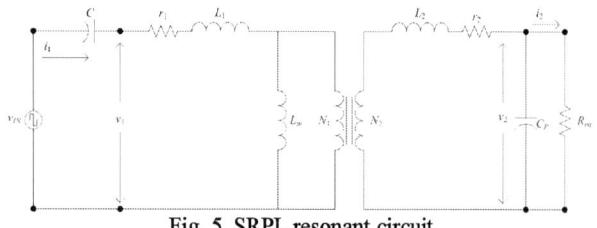

Fig. 5. SRPL resonant circuit

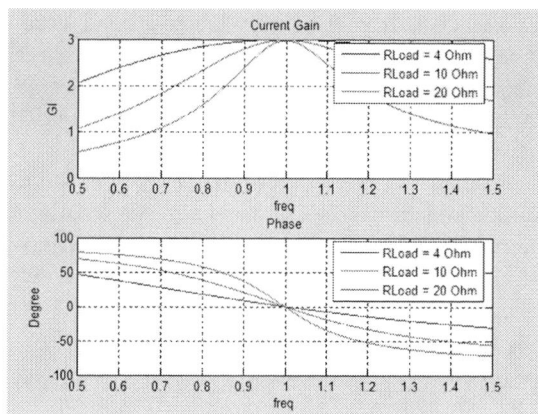

Fig. 6 Bode plot of the current gain

IV. SOFTWARE CONFIGURATION

Conventionally, lead-acid batteries' charging occurs in two steps, the battery is charged at a constant current until the battery voltage reaches the predefined upper voltage limit followed by a constant voltage charging until the current reaches a predetermined small value. This method is often called constant current-constant voltage (CC-CV) charging method. However, CC-CV is not suitable for rapid charging since the constant voltage charging seriously extends the charging time and also reduces the cycle life of the battery. Multi-stage constant current charging algorithm proposed in [10] can be utilized to shorten the charging time and prolong the cycle life. Fig. 7 illustrated the concept of the three-stage constant current charging algorithm used in this paper. From Fig. 7, the total charging period is divided into three stages. In each stage, the charging current is set to a pre-determined value. During charging, the voltage of battery will increase. When the voltage exceeds the preset limit voltage V_{limit}, the stage number will increase and a new charging current set value will be applied accordingly. This process will continue until stage number reaches 3.

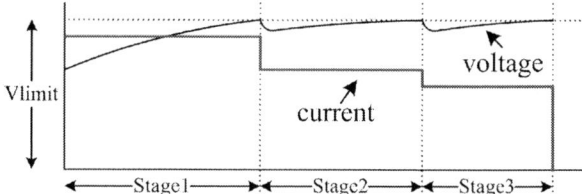

Fig. 7 The concept of multi-stage constant current charging

From section III one can see that the current gain can be kept constant if the operating frequency is fixed at the resonant frequency. According to this fact, if the information of the primary-side current can be obtained, the value of the secondary-side current can be calculated. Fig. 8 shows the software flowchart of the proposed system. From Fig. 8, the digital controller first reads into the measured primary-side current value i_1 using a 50-order digital FIR filter, then divides this value with a constant gain I_{gain}. The controller then determines the current command i_{com} according to the stage number. These two values are then processes by a digital PI controller to obtain the required gating signals of the full-bridge power converter. The same procedure will continue until the stage number exceeds 3, and the whole charging process will stop.

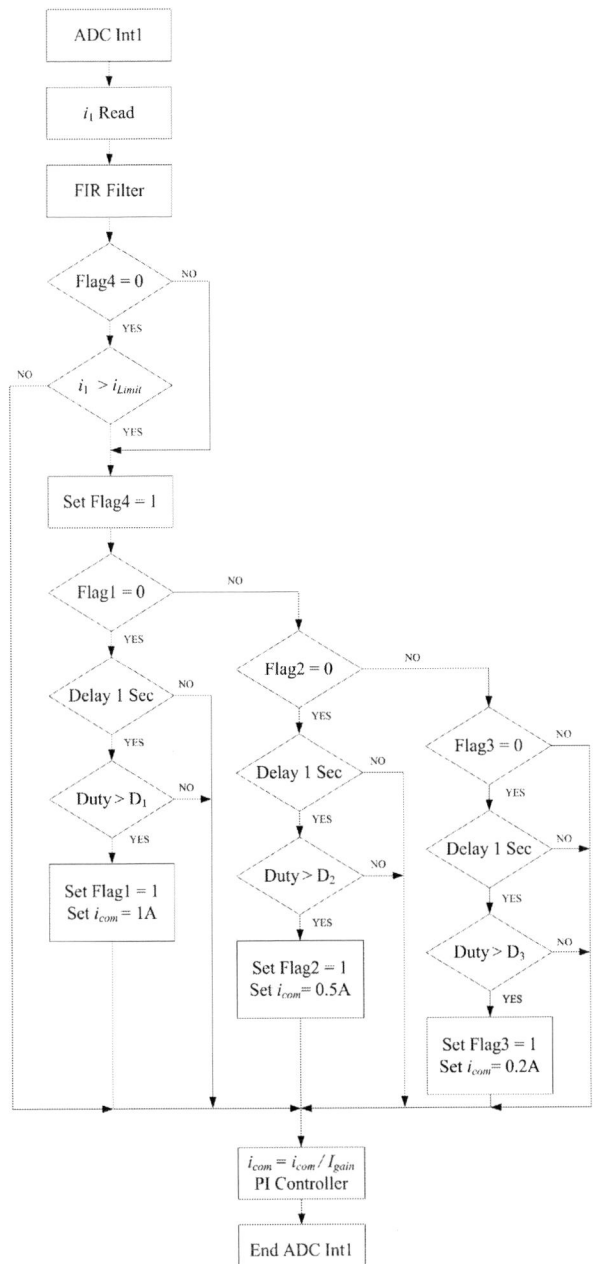

Fig. 8 Software flowchart of the proposed system

V. SIMULATION AND EXPERIMENTAL RESULTS

In order to validate the correctness of proposed system, a prototype system is built. The system has been designed according to the following specifications:

input voltage: 85 Vac ~ 150 Vac

operating frequency: 100 kHz

battery rating: lead-acid battery, 12 V nominal, 3 A

Fig. 9 shows the simulation circuit of the proposed system, the simulation is performed using the PSIM software. Fig. 10 shows the simulation result of the proposed charger. From Fig. 10, the output current can follow the current command correctly. It should be noted that the charging time of a lead acid battery can be several hours; therefore, a constant resistive load is used instead in this simulation. However, the current following capability of the proposed controller can still be well exhibited. Fig. 11 shows the

experimental results of the proposed system. From Fig. 11, one can observe that the charging current is kept constant during each stage, and the proposed controller can successfully perform the multi-stage charging without the knowledge of the battery voltage.

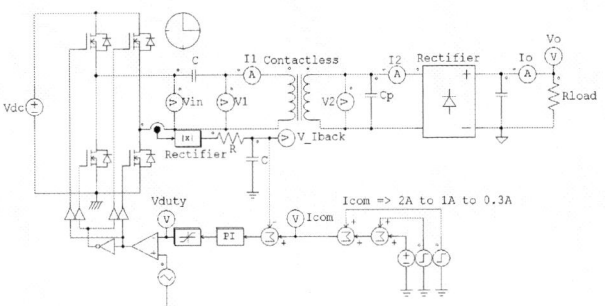

Fig. 9 Simulation circuit of the proposed system

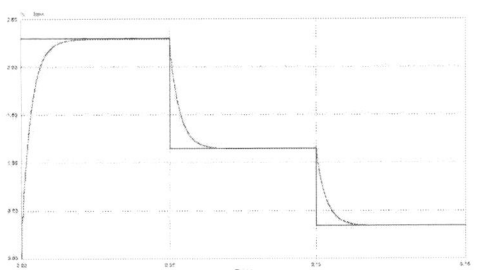

Fig. 10 Simulation result of the proposed system

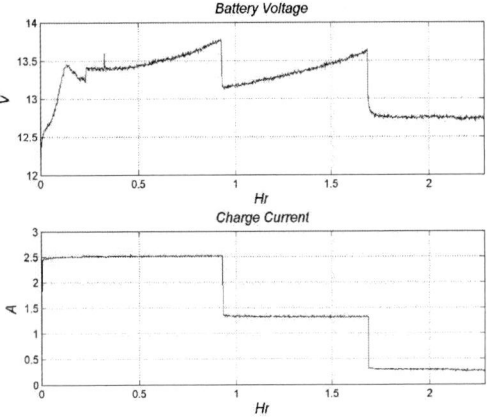

Fig. 11 Experimental results of the proposed system

VI. CONCLUSION

In this paper, a primary-side controlled contactless battery charger is proposed. Detailed analysis of the utilized CETS is provided and a multi-stage constant current charging algorithm is presented. The proposed controller obtains the gating signal only from the primary-side current information; therefore no signal isolation circuits are required. Using this technique, the only regulation circuit needed at the secondary side is a typical full-bridge diode rectifier. No additional power converter/regulator and controller is required. Therefore, the size of the secondary side can be minimized. The advantages of the proposed system include simplified design and reduced size of secondary side.

ACKNOWLEDGMENT

This work was sponsored by National Science Council, Taiwan, under research grant NSC 96-2623-7-011-002-ET.

REFERENCES

[1] C.S. Wang, O.H. Stielau, G.A. Covic, "Design considerations for a contactless electric vehicle battery charger," IEEE Transactions on Industrial Electronics, Vol. 52, No 5, Oct. 2005, pp. 1308-1314.

[2] C.G. Kim, D. H. Seo, J.S. You, J.H. Park, B.H. Cho, "Design of a contactless battery charger for cellular phone," IEEE Transactions on Industrial Electronics, Vol. 48, No 6, Dec. 2001, pp. 1238-1247.

[3] Y. Jang, M.M. Jovanovic, "A new soft-switched contactless battery charger with robust local controllers," Proceedings of the 25th International Telecommunications Energy Conference, 19-23 Oct. 2003, pp. 473-479.

[4] Y. Jang, M.M. Jovanovic, "A contactless electrical energy transmission system for portable-telephone battery chargers," IEEE Transactions on Industrial Electronics, Vol. 50, No 3, June 2003, pp. 520-527.

[5] J. Hirai, T.W. Kim, A. Kawamura, " Study on intelligent battery charging using inductive transmission of power and information," IEEE Transactions on Power Electronics, Vol. 15, No. 2, March 2000, pp. 335-345.

[6] P.M. Hunter, A.H. Anbuky, "VRLA battery rapid charging under stress management," IEEE Transactions on Industrial Electronics, Vol. 50, No 6, Dec. 2003, pp. 1229-1237.

[7] H. Abe, H. Sakamoto, K. Harada, "A noncontact charger using a resonant converter with parallel capacitor of the secondary coil," IEEE Transactions on Industry Applications, Vol. 36, No. 2, March-April 2000, pp. 444-451.

[8] C.S. Wang, G.A. Covic, O.H. Stielau, "Power transfer capability and bifurcation phenomena of loosely coupled inductive power transfer systems," IEEE Transactions on Industrial Electronics, Vol. 51, No 1, Feb. 2004, pp. 148-157.

[9] Y. Wu, L. Y, S. Xu, "Modeling and performance analysis of the new contactless power supply system," Proceedings of the Eighth International Conference on Electrical Machines and Systems, Vol. 3, 27-29 Sept. 2005, pp. 1983-1987.

[10] Y. H. Liu, J. H. Teng and Y. C. Lin, "Search for an optimal rapid charging pattern for lithium ion batteries using ACS algorithm," IEEE Transactions on Industrial Electronics, Vol. 52, No. 5, Oct. 2005 , pp. 1328 -1336.

Research on Digital Soft-switch Welding/Cutting Inverter Power Source

G.R.Zhu, Z.Liu, X.Li, B.Y.Liu, S.X.Duan and Y.Kang

College of Electrical and Electronics Engineering, Huazhong University of Science and Technology,
Wuhan, Hubei, China, 430074, zhgr_55@126.com

Abstract--According to the basic characteristics of arc welding and cutting, digital control system of soft-switch arc welding/cutting inverter power source is researched and developed. This paper analyzes the operation principle and presents hardware structure about digital control system of the soft-switch arc welding/cutting inverter power source. Then the software system is designed based on phase shift realization principle, and a detailed flowchart of the main program and the interrupt service routine are given. Through plentiful experiments and tests, it can be found that the soft-switch arc welding/cutting inverter power source based on digital control system operates well in various working situations. As a result, it is thought to be suitable for arc welding/cutting applications due to its effectiveness and robustness.

Index Terms--welding/ cutting, soft-switch, inverter, digital control system

I. INTRODUCTION

Generally, arc welding machines and cutting machines are widely used in industrial applications for joining or cutting materials. Inverter has become the focus in the field of welding and cutting power source because of its small volume, light weight, good controlling performance and aptness for real-time controlling of welding and cutting process. Compared with hard switch techniques, the soft switch is preferable in the high-power DC-DC converter application based on low power wastage and little EMI [1], so it becomes an ideal structure in the welding and cutting power supply. Now the manufacture of arc welding/cutting power source is mainly based on analog inverter [2], in which component parameter flutters and performance varies with the changing of the environment and time. Owing to the fact that digital control technology is flexible, exact and reliable, it is the up-to-date method used in soft-switch arc welding/cutting inverter power source. In recent years, some researches touch digital welding/cutting power sources, but most of them are still based on system simulation and lower power inverter [3]. In this laboratory, a large power soft-switch arc welding/cutting inverter power source has been researched and developed, and digital control system of soft-switch arc welding/cutting inverter power source is proposed in this paper.

II. THE CIRCUIT TOPOLOGY AND OPERATION PRINCIPLE

A. The circuit topology

In this soft-switch arc welding/cutting inverter power source, the Phase-Shift Full-Bridge ZVZCS (FB-ZVZCS-PWM) [4] inverter is employed. Although the volt-ampere characters and the ranges of voltage and current of arc welding machines and cutting machines are different, they both share the fundamental output characters of quickly declining voltage and invariable current [5]; therefore, a machine with the multi-functions of arc welding and cutting can be developed. The secondary side of high frequency transformer can either be shifted to output full-wave converter in arc welding, or be switched to output full-bridge converter in cutting. Changing output converter mode means changing the voltage ration of the high frequency transformer, which can meet the two work situations only by shifting a switch.

The topology of the soft-switch arc welding/cutting inverter is shown in Fig.1. Different from the conventional phase shifting full bridge inverter; there are a saturated inductor Lsat and a blocking capacitor Cb in the main circuit [2].

In Fig.1, Q1, C1, D1 and Q2, C2, D2 are leading leg switches, which is completely the same as the conventional phase-shifting full bridge inverter, and ZVS is also realized by paralleling capacitors to the switches; Q3, D3 and Q4, D4 are lagging leg switches. Since the lagging leg switches need realize ZCS, no capacitor that will store charges and add loss during switches turning on is paralleled. The topology shown in Fig.1 is mainly based on the following considerations:

(1)The blocking capacitor Cb, not only resets current as reverse blocking voltage source to help lagging leg switches realize ZCS, but also suppresses the bias of the power transformer;

(2)PWM (pulse width modulation) control strategy is adopted. Switching frequency is constant, so the designs of high frequency transformer and filtering links of input and output are easy.

In Fig.1, shifting the switch point to the C point

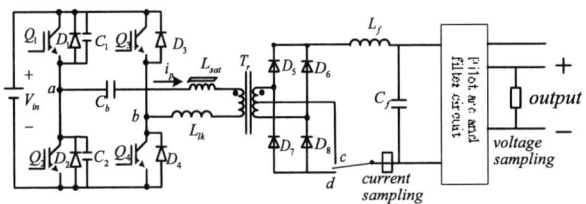

Fig.1. The topology of the soft-switch arc welding/cutting inverter

978-1-4244-0644-9/07/$25.00 ©2007 IEEE

means that the secondary side of high frequency transformer constructs output full-wave converter when working in arc welding. Shifting the switch point to the D point means that the secondary side of high frequency transformer constructs output full-bridge converter when working in cutting. A separate pilot arc circuit is in series to the main circuit in cutting, which can be started by the cutting gun.

B. Operation principle

The power switching component in the leading leg (the left leg) is turned on under zero voltage, the same as that in the FB-ZVS-PWM inverter; and the absorbing capacitor parallel to the switch, which helps to turn off the switch under low voltage. As for the lagging leg(the right leg), the block capacitor Cb makes the primary current rapidly decrease to zero and it maintains during the freewheeling interval; Therefore, the power components of the lagging leg is turned off under zero current. When the other switch is turned on shortly, the inductor has not been saturated, so an approximate Zero-Current-Switch is also realized.

Because a lot of documents deal with the operation principle of FB-PS-ZVZCS inverter [4], equivalent circuit and detailed analysis of various operating modes will not be presented here.

Preference [4] analyzes design procedures and carefully selects components, which provide the theoretic foundation for the manufacture of PS-FB-ZVZCS inverter power sources; and introduces the realization process of the output technique characteristics in arc welding/cutting.

III. DIGITAL CONTROL SYSTEM OF SOFT SWITCH ARC WELDING/CUTTING INVERTER

A. Digital Control System structure and phase shift realization principle

Digital control system of the soft-switch arc welding/cutting inverter power source is shown in Fig.2. In this study, Digital Signal Processor (DSP) TMS320LF2407 provided by Texas Instruments is selected for implementation because of its function and simple architecture [6]. The features of this DSP are: A/D converter (10-bit), two event managers to generate PWM signals, 4 timer/counter (16-bit). The core of the hardware system is DSP, around which the circuits, which includes sampling circuit, protection circuit, DSP external circuit and drive circuit, are designed in detail. The output voltage and current of the proposed inverter are sensed by sensors and converted by A/D of DSP as

feedback after being filtered by digital low pass filter.

As to full bridge phase shift circuit, the most important problem is how to create phase shift pulse in the digital control system. A direct phase shift pulse method based on the DSP symmetric PWM waveform generation with full compare units is applied. The method is shown in Fig.3.

In Fig.3, the direct phase shift pulse method with DSP

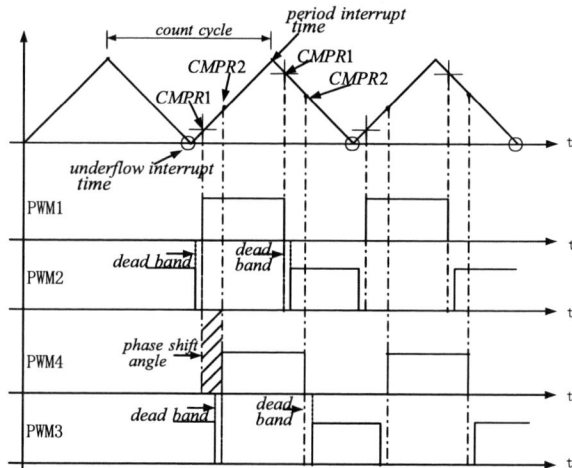

Fig.3. Direct phase shift pulse methods with DSP full compare units

full compare units is that the two full compare units of the DSP Event Manager A (EVA) directly produce four PWM pulses. The fundament theory of phase shift angle is that there is a periodic delay time from the leading leg drive to lagging leg drive. The two up/down switches drive pulses of the leading leg are produced by the full compare unit 1, and the two up/down switches drive pulses of the lagging leg are produced by the full compare unit 2. The up and down switching of each leg drive pulses are reverse and between them exists the dead band. If the given data of the leading leg register CMPR1 is fixed, the given data of phase shift angle register CMPR2 comes from full compare event, which can produce the lagging leg drive pulse. Therefore, this method can realize $0^0 \sim 180^0$ phase shift. The data of CMPR1&2, which is the compare register of the two full compare units, varies in the underflow interrupt and period interrupt with the demand of the system regulator. The falling edge compare data is given in the underflow interrupt, rising edge compare data is given in the period interrupt, and the counter data is the pulse period.

In the program, the control register is set by symmetric PWM waveform generation with full compare units, Timer 1 must be put in the continuous up/down counting mode, and dead band can be set directly through Dead-Band Timer Control Register (DBTCR).

In a word, the direct phase shift pulse method does not need more hardware to synthesize pulse [7], so it is very simple, flexible, convenient and reliable.

The phase shift PWM waves generated by the EVA module of the DSP and PI regulator are driven and amplified to control the power semiconductors IGBT of the high-frequency link inverter. Moreover, the system

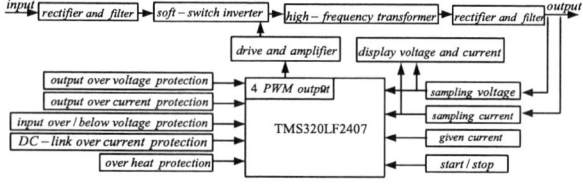

Fig.2. Digital control system of the soft-switch arc welding/cutting inverter

552

can control the arc welding/cutting voltage and current by zero switching [8].

B. Software flow of the digital control system

System soft mainly consists of main program and interrupt program. The main program accomplishes system initialization, start/stop detection and initialization, and circulation waiting for interrupts.

The interrupt programs consist of period interrupt, underflow interrupt and Power-Drive Protection Interrupt (PDPINTA). The period interrupt is made up of soft-start of unload voltage, voltage and current sampling, and arc welding and cutting operation condition shifting. In the period interrupt various regulation methods are applied in various operation conditions. Accordingly, the period interrupt finishes building unload voltage and keeping load current invariable. The underflow interrupt will update the data of the register CMPR1 and CMPR2. The frame of main program, T1 period interrupt and T1 underflow interrupt programs are shown respectively in Fig.4, Fig.5 and Fig.6.

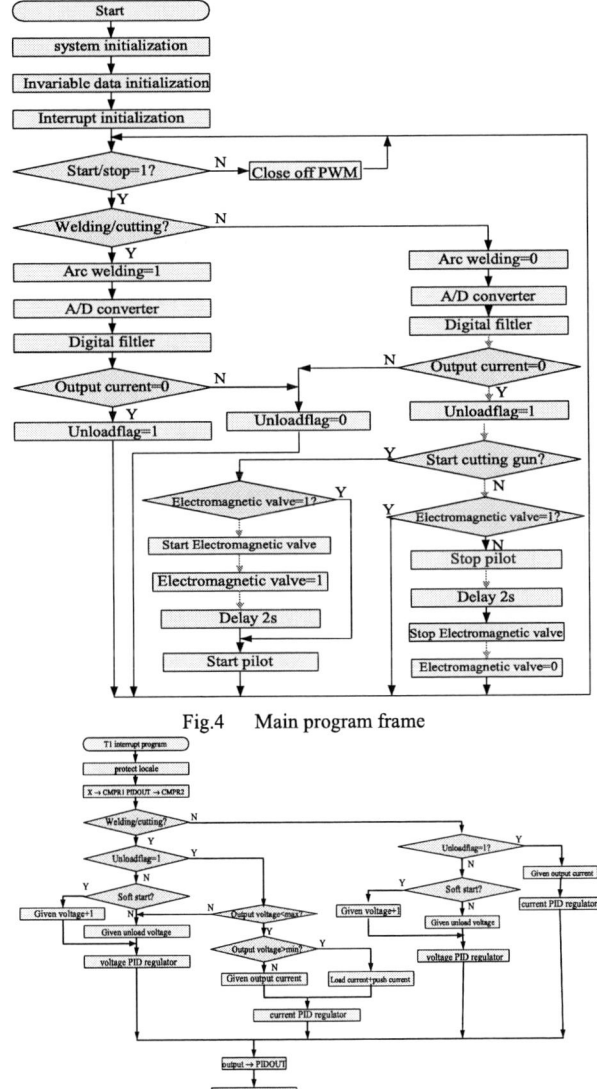

Fig.4   Main program frame

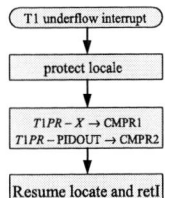

Fig.6 T1 underflow interrupt program frame

Achieve the purpose of possible fault protection in arc welding/cutting processing such as over current, over/below voltage, over heat, a corresponding signal can close off hardware drive amplifier circuit and produce PDPINTA software interrupt. In the PDPINTA software interrupt program, closing off pulse and signal alarm are produced.

In order to prevent the two switches of the same bridge leg from being switched on simultaneously, at the "starting" time, the PWM driving wave is opened by software first, and then by hardware; at the "end" time, the PWM driving wave is closed off by hardware first, and then by software. The drive system runs normally only after the software brings the PWM wave and hardware is ready, and drive system is forbidden either when the software doesn't bring the PWM wave or when hardware isn't ready.

IV. EXPERIMENT RESULTS AND ANALYSIS

According to the upper control process, an experimental digital soft-switch arc welding/cutting inverter power source system rated at 10kVA has been constructed to verify the proposed control strategy, and the specifications and designed components values are summarized in Tab.1.

TABLE I
SPECIFICATIONS AND COMPENTS USED IN EXPERIMENT

Vin (input voltage)	DC500V±20%
P_O (output power)	10kW
Unloaded voltage(arc welding)	70V
Output current(arc welding)	40-250A(adjustable)
Unloaded voltage(cutting)	160V
Output current(cutting)	40-100A(adjustable)
Switching frequency	20kHz
Controller	TMS320LF2407

The ZVS experiment waveforms of leading leg output current is 200A in arc welding process and 100A in cutting process shown in Fig.7 and Fig.8, in which

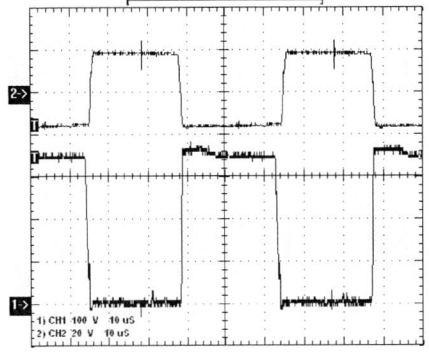

Fig.7. Q1 ZVS waveforms in arc welding operation with 200A

Fig.5 T1 period interrupt program frame

553

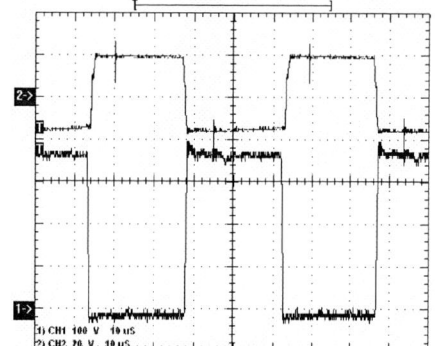

Fig.8. Q1 ZVS waveforms in cutting operation with 100A

Channels 1 denotes the Q1 voltage, and Channels 2 denotes the drive waveform of the corresponding Q1 switch.

In Fig.7 and Fig.8, when the leading leg IGBT is switched on, the Q1 voltage is zero, which means the realization of zero voltage switching on. When Q1 is switched off, the Q1 voltage increases slowly, this means the realization of soft switching off.

The experiment waveforms of lagging leg output current is 200A in arc welding process and 100A in cutting process shown in Fig.9 and Fig.10, in which Channels 1 denotes the primary side current waveform of the transformer, and Channels 2 denotes the drive waveform of Q3 switch.

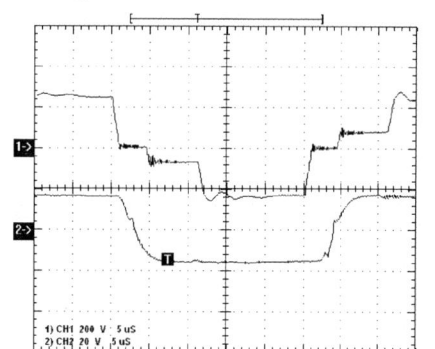

Fig.9. Q3 ZCS waveforms in arc welding operation with 200A

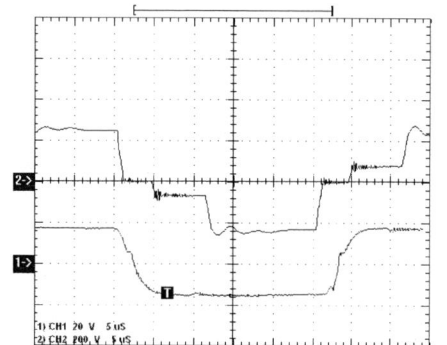

Fig.10. Q3 ZCS waveforms in arc welding operation with 100A

In Fig.9 and Fig.10, when Q3 is switched off, the primary side current already deceases to zero, which means the realization of zero current switching off. When Q3 is switched on, the primary side current will

increase after a period of time, which realizes soft switching on.

When the machine works in arc welding, where output current is 175A, and unloaded voltage is 70V, the process of the unloaded-short circuited-loaded welding voltage and welding current waveforms are shown in Fig.11, in which Channels 1 denotes the welding voltage and Channels 2 denotes the welding current.

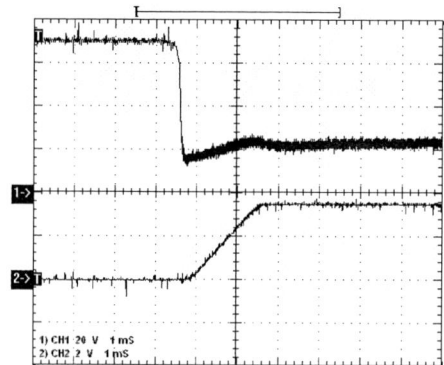

Fig.11. Output voltage and current waveforms of the cutting process

From Fig.11, the output current is zero, when the unloaded output voltage is 70V, and the time of from the short circuit to load is less than 2ms. After pilot arc succeeds, the machine works well as an invariable current power source. Therefore, the machine is suitable for the arc welding from these waveforms.

When the machine is used to cut 12mm thick mild steel work piece, in which output current is 65A, and unloaded voltage is 160V, the waveforms of the output voltage and output current are shown in Fig.12.

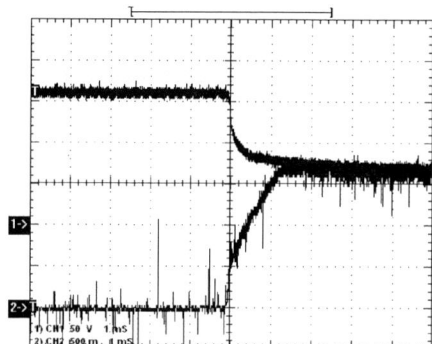

Fig.12. Output voltage and current waveforms of the welding process

It can be seen from Fig.12, (1) when unloaded voltage is 160V, output current is zero; (2) as soon as the plasma arc is piloted, the voltage quickly drops to about 80V, the current quickly increases to about 65A; (3) after pilot arc succeeds, the machine cuts well, the output voltage is about 70V, and the output current is invariable 65A. Therefore, the machine is suitable for cutting from these waveforms.

V.　CONCLUSIONS

The digital arc welding/cutting power source is designed according to the phase-shift full-bridge ZVZCS, and the output characteristics of invariable current can meet the technique need of arc welding/cutting. It operates well over a wide range of the unloaded-short circuited-loaded conditions. The digital control system is simple, flexible and reliable. The operation has been verified by experiments.

REFERENCES

[1] Chen Shujun. "Soft switching converter and harmonic elimination for arc welding power source" [Doctor thesis].Harbin:Harbin Institute of Technology,1999.

[2] Jung-Goo Cho, Sabate. J. A, Guichao Hua. "Zero-voltage and zero-current switching full bridge PWM converter for high-power applications". IEEE Transactions on Power Electronics, 1996, 11(4): 622~628

[3] Deshang Sha, Yunjie Bao, Bojin Qi. "Full Digitalized ZVZCS control technique for high-frequency link inverter" IEEE Power Electronics Specialists Conference, pp:1-6, June, 2006

[4] Cuadros.C, Lin.C.Y, Boyevich.D, Watson.R, Skutt.G, Lee.F.C, Ribardiere.P; "Design procedure and modeling

of high power, high performance, zero-voltage zero-current switched, full-bridge PWM converter" Applied Power Electronics Conference and Exposition, vol.2, Page(s):790 – 798, Feb. 1997.

[5] Zhu Guo-rong, Liu Zhao, Li Xun, Duan Shan-xu and Kang Yong. "The multi-functional arc welding cutting inverter based on PS-FB-ZVZCS 2nd IEEE Conference in Industrial Electronics and Applications". 2007.pp:1912-1916.

[6] Texas Instruments Incorporated. TMS320F/C240 DSP Controllers Reference Guide.1999: (6)2-95.

[7] Em-Soo.Kim, Tae.Jin.Kim, Young.Bok.Byun, Tae.Geun. Koo,Yoon.Ho.Kim. "High power full bridge DC/DC converter using digital-to-phase-shift PWM circuit" IEEE 32nd Power Electronics Specialists Conference. vol. 1, Page(s):221 – 225, June 2001

[8] Ben Hongqi, Yang Shiyan, Yuan Shubin, Zhao Junbao. "Application research on ZVZCS full-bridge converter technique". China Welding Transaction, 2005, Vol (14) 2:113-116.

Design of an Adjustable High Output Voltage Asymmetrically Switched Class D Converter

M. Rentzsch, H. Güldner, C. Ditmanson
Technische Universität Dresden,
Mommsenstrasse-13, 01069-Dresden, Germany

Abstract—**An asymmetrically switched class D inverter with a series-parallel resonant tank and a Walton Cockroft voltage multiplier for medium power, high voltage applications is presented. This converter operates in a self sustained oscillation mode above the resonant frequency and is controlled by varying the duty cycle of one switch. A straightforward procedure for designing the converter is given. Furthermore, this paper shows the benefits of digital implementation of the carrier signal.**

I. INTRODUCTION

Xray-equipment, cable testing devices, particle detector supplies and capacitor chargers need high voltage power supplies with adjustable output voltage and/or adjustable output current.

High voltage transformers need a larger isolation distance between each layer of windings. This causes a reduction of coupling between the primary and the secondary winding, which increases the leakage inductance. As a result of the high turns ratio of the high voltage transformer, the parasitic winding capacitances increase [1].

Resonant converters integrate the nonlinearities and parasitics of the high voltage transformer and enable the implementation of soft switching techniques. Thus, the switching frequency can be increased while the dynamic stresses on the power semiconductors remain the same. The resonant current is approximately sinusoidal, which reduces losses (magnetizing losses and dielectric losses) caused by higher order harmonics. To ensure zero voltage switching (ZVS), resonant converters have to operate above their resonant switching frequency [4].

For medium power applications with output power in the range 100W < P_{out} < 1000W, two-switch inverter topologies are preferred [2]. A power factor correction (PFC) input stage supplies the converter with a constant input voltage.

The resonant circuit consist of a series-parallel resonant tank. As the converter must handle open-circuit and short-circuit loads, the literature [3,6,7] indicates this type of resonant tank as the most suitable one.

The output voltage of a resonant converter can be regulated by adjusting either the switching frequency or the amplitude of the input voltage. This paper shows, that using an asymmetrically switched class D inverter with a series-parallel resonant tank offers the benefits of a wide-range adjustable output voltage and effective handling of open and short-circuit loads (the latter attained by regulation of the amplitude of the first harmonic of the inverter output voltage). The converter operates in a self-sustained mode above the resonant frequency in order to ensure ZVS.

In this paper, the converter is described in detail and a straightforward procedure for designing the components of the power stage—based on a first harmonics analysis—is given. The voltage multiplier is analysed and dimensioned with the help of the charge distribution method.

II. CIRCUIT DESCRIPTION

The topology of the 800W class D inverter with series-parallel (LCC) resonant tank and a Walton Cockroft voltage multiplier is shown in Fig. 1. Its output voltage is 20kV with a switching frequency of 200kHz at the rated operating point.

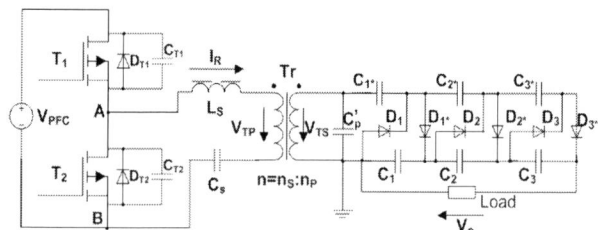

Fig. 1. Class D inverter with series-parallel resonant tank and a Walton Cockroft voltage multiplier

The parasitic capacitances C_{Tx} of the MOSFETs T_x are used for ZVS. The leakage inductance of high voltage transformer Tr and an auxiliary inductor form the series inductance L_s. The parallel capacitor C_p', referred to the secondary side of transformer Tr, is assembled by the parasitic winding capacitances of Tr and an additional discrete capacitor. The load, which will be considered in the analytical calculations as a pure resistor, can also be capacitive. A three stage voltage multiplier (Greinacher-Cascade or Walton Cockroft multiplier) is used to boost the output voltage of the resonant tank. This reduces the isolation demands and the turns ratio of transformer Tr.

This work was financially supported with resources of the European Fund for regional development 2000-2006 and with resources of Freistaat Sachsen, Germany.

978-1-4244-0644-9/07/$25.00 ©2007 IEEE

III. OPERATION OF THE ASYMMETRICALLY SWITCHED CLASS D INVERTER

As the quality factor Q of the resonant tank is high enough, the resonant network filters out the first harmonic of the rectangle voltage v_{AB}. Thus, in comparison to [8], the resonant current i_R and the voltage across C_s are assumed to be sinusoidal. The output voltage v_{AB} of the asymmetrically and the symmetrical switched class D inverter and their first harmonic are shown for comparison in Fig. 2a.

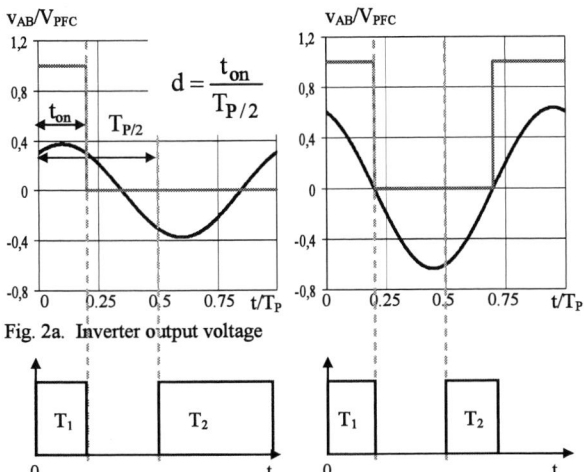

Fig. 2a. Inverter output voltage

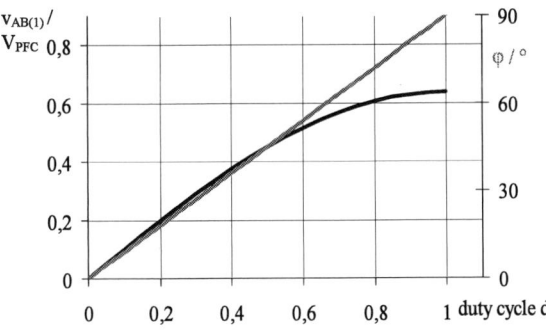

Fig. 2b. Switching signals of the power semiconductors

In Fig. 2b the corresponding switching signals of the power semiconductors to the inverter output voltage, Fig. 2a, are shown. The first harmonic of the inverter output voltage in the asymmetrical case and its dependency on the control variable d can be calculated as

$$V_{AB(1)} = \frac{V_{PFC}}{\pi} \cdot \sqrt{2(1-\cos(d \cdot \pi))} \cdot \cos\left(\vartheta - \frac{d \cdot \pi}{2}\right). \quad (1)$$

and is shown in Fig. 3.

$$\varphi = d \cdot \frac{\pi}{2} \quad (2)$$

indicates the phase angle between v_{AB} and $v_{AB(1)}$.

Fig. 3. $v_{AB(1)}/V_{PFC}$ and φ vs. duty cycle d respectively

IV. CHARACTERISTIC VOLTAGE AND CURRENT WAVEFORMS

Characteristic voltage and current waveforms are shown in Fig. 4. and were drawn with the following assumptions:

- As the switching behaviour is dominated by the resonant tank and not by the power semiconductors, T_1 and T_2 will be modelled with an ideal switch, a reverse Diode and a drain-source capacitor. All other elements of the converter are ideal.
- The resonant current i_R is sinusoidal.
- The sum of the capacitances of the voltage multiplier is much larger than C_P'.
- The converter operates in a continuous conduction mode

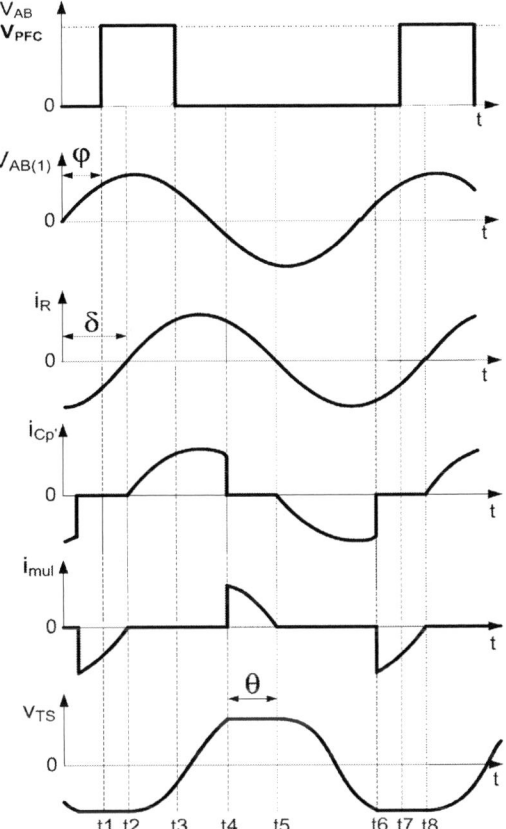

Fig. 4. Characteristic voltage and current waveforms

The behaviour of the converter during one switching period can be divided into five sections:

Section I ($t_2 < t < t_3$):

At t_2, i_R changes its sign and is now positive. To ensure ZVS, the power semiconductor T_1 was switched on slightly previous to time t_2. V_{AB} is positive and energy is fed into the resonant tank. Furthermore, on the voltage multiplier

$$v_{TS} + \sum_{k=1}^{3} V_{C_{k^*}} > \sum_{i=1}^{2} V_{C_i}, \quad (3)$$

557

and the voltage multiplier is in a blocking state. The current i_R will pass through C_p'.

Section II ($t_3 < t < t_4$):
At t_3, T_1 is switched off. After discharging C_{T2} and charging C_{T1}, i_R will flow through D_{T2}, and T_2 can be switched on with ZVS until time T_5.

Section III ($t_4 < t < t_5$) :
At t_4,

$$v_{TS} + \sum_{k=1}^{3} V_{C_{k^*}} > \sum_{i=1}^{3} V_{C_i} , \qquad (4)$$

and the voltage multiplier starts conducting.

Section IV ($t_5 < t < t_6$) :
At t_5, i_R changes its sign and is now negative.
The voltage multiplier enters its blocking state, as

$$v_{TS} + \sum_{k=1}^{3} V_{C_{k^*}} < \sum_{i=1}^{3} V_{C_i} , \qquad (5)$$

and i_R passes through C_p'.

Section V ($t_6 < t < t_7$) :
At t_6,

$$v_{TS} + \sum_{k=1}^{3} V_{C_{k^*}} < \sum_{i=1}^{2} V_{C_i} , \qquad (6)$$

and the voltage multiplier starts conducting again. Furthermore T_2 is switched of. After discharging C_{T1} and charging C_{T2}, i_R will flow through D_{T1} and T_1 can be switched on with ZVS until the time instance T_8.

V. DESIGN OF THE RESONANT TANK

The design of the series-parallel resonant tank is based on a first harmonic analysis. In [5], an analytical analysis and design of a half/full bridge inverter with LCC resonant tank and a capacitive filter has been proposed. Reference [6] extended this proposal for the full bridge inverter with a variable duty cycle. In this paper another extension of the analytical concept in [5] under consideration of the functionality of the asymmetrically switched class D inverter and the output voltage multiplier is derived. To this end, more simplifications are introduced:

- The parasitic capacitances of the power semiconductors are negligible.
- The turns ratio of the voltage multiplier is ideal, hence twice the number of multiplier stages.

An equivalent circuit of the LCC converter with a voltage multiplier is shown in Fig. 5.

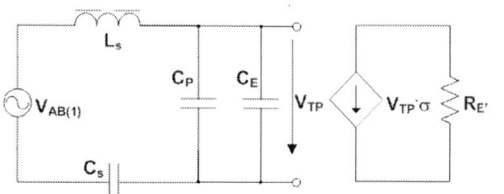

Fig. 5. Equivalent circuit of the LCC converter with voltage multiplier

Corresponding to [5], the elements of the circuit shown in Fig. 5 can be calculated as:

$$R_{E'} = 4n_t^2 n_m^2 \cdot R_E = R_o \frac{k_v^2}{2} \qquad (7)$$

where

$$k_v = 1 + 0.27 \sin\left(\frac{\theta}{2}\right), \qquad (8)$$

is a voltage waveform coefficient, defining the relationship between the output voltage referred to the primary side and the first harmonics of the transformer primary voltage. θ is the conduction angle of the voltage multiplier

$$\theta = 2 \tan^{-1} \sqrt{\frac{2\pi n_t^2 n_m^2}{\omega C_p R_o}} . \qquad (9)$$

σ is the product of the turns ratio of the high voltage transformer and the voltage multiplier

$$\sigma = 2 n_t n_m . \qquad (10)$$

As [5] proposed, the rectifier with a capacitive filter load can be approximately represented by an R-C equivalent circuit with

$$C_E = \frac{2\sigma^2}{\omega_s R_o k_v^2} \cdot \tan|\beta| , \qquad (11)$$

where β defines the angle between the first harmonics of the primary transformer voltage and current. The transfer function which relates the first harmonics of the transformer input voltage and the first harmonics of the inverter output voltage can be computed as

$$\frac{V_{TP(1)}}{V_{AB(1)}} = \qquad (12)$$

$$\frac{1}{\sqrt{\left[1 - \frac{C_P}{C_S}\left(\left(\frac{\omega}{\omega_s}\right)^2 - 1\right)\left(1 + \frac{\tan|\beta|}{\omega C_p R_E}\right)\right]^2 + \left[\frac{C_P}{C_S}\left(\left(\frac{\omega}{\omega_s}\right)^2 - 1\right) \cdot \frac{1}{\omega C_p R_E}\right]^2}}$$

where

$$\omega_s = \frac{1}{\sqrt{L_S C_S}} . \qquad (13)$$

The input phase angle δ of the first harmonics of the inverter output voltage and the resonant current

$$\delta = \frac{\pi}{2} - d \cdot \frac{\pi}{2} = \frac{\pi}{2} - \varphi , \qquad (14)$$

are further computed

$$\tan\delta = \frac{1}{\omega C_p R_E} \cdot \frac{C_P}{C_S}\left[\left(\frac{\omega}{\omega_s}\right)^2 \left(1 + \left(\omega C_p R_E + \tan|\beta|\right)^2\right) - 1\right] - $$

$$\left(\omega C_p R_E + \tan|\beta|\right)\left(1 + \frac{C_P}{C_S}\left(1 + \frac{\tan|\beta|}{\omega C_p R_E}\right)\right) . \qquad (15)$$

An important expression of [5] can be extended with (1), (10) and the ratio between the PFC output voltage and the converter output voltage is evaluated as

$$\frac{V_o}{V_{PFC}} = \frac{n_t n_m}{2} \cdot \sqrt{2(1 - \cos(d \cdot \pi))} \cdot \frac{\cos\left(\frac{\pi}{2} - d \cdot \frac{\pi}{2}\right)}{\sin^2\left(\frac{\theta}{2}\right)} . \quad (16)$$

The following constraints are necessary to achieve a reasonable design:

- The switching frequency should not exceed 400kHz in order to reduce gate drive problems, to ensure ZVS, and to limit core and dielectric losses
- The capacitance of the parallel capacitor referred to the secondary side should not exceed 50pF/n². This is the minimum achievable parasitic winding capacitance of the high voltage transformer [6]
- In order to reduce to isolation requirements of transformer Tr, its output voltage is limited to 4kV.
- The allowed AC voltage across the series and parallel capacitors decreases with increasing switching frequency, Fig. 6

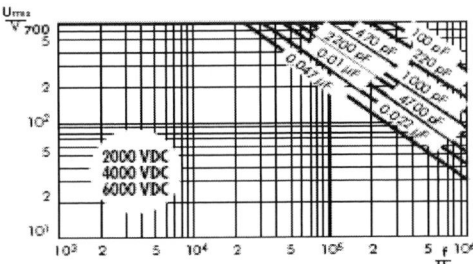

Fig. 6. Maximum AC-voltage of the capacitor as a function of the frequency (datasheet WIMA corporation, FKP1 capacitor)

With the rated operating frequency set at 200kHz, the switching frequency remains below 400kHz at open-circuit with low output voltage. A three stage voltage multiplier will boost the voltage v_{TS} to achieve the desired output voltage of 20kV under full load.

The design process is iterative and starts with initial values of d, θ and α=Cp/Cs. All following diagrams where obtained with the help of the equations (7...16). In order to retain a dynamic controller response, the duty cycle cannot be set too high. But a larger angle δ will result in an increase of the amplitude I_R, (Fig. 7), for the same output power, hence, the circulating energy which generates additional losses in all resistive elements will be larger and the converter efficiency smaller [7].

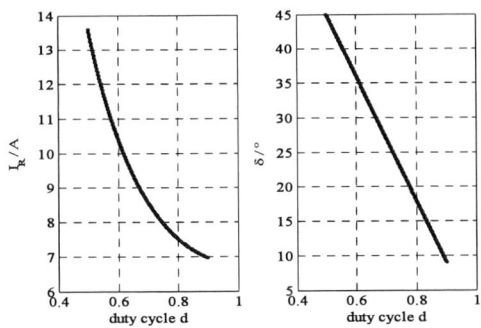

Fig. 7. Duty cycle d vs. I_R and δ respectively

An optimum design trade-off is to set the starting value of d to 0.6 at the rated operating point. The amplitude of i_R is independent off the voltage multiplier conduction angle and the parameter α.

The frequency range in which the converter operates under different load conditions depends strongly on α. For a larger value of α, the frequency range will be smaller but the AC-voltage stress of C_S increases. However, by reducing α the range of f_s will dramatically increase.

Fig. 8 clarifies the dependency of L_S, V_{CS}, C_S and n_t on θ and α respectively at the nominal operating point. A reasonable range for the design parameter θ can be evaluated with Fig. 8 and is between 80° and 100°.

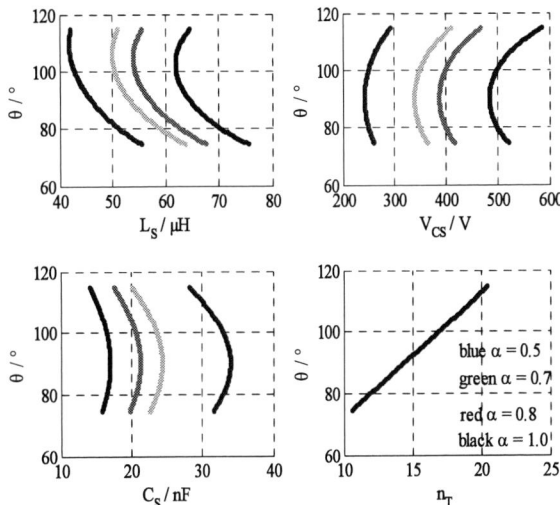

Fig. 8. L_S, V_{CS}, C_S and n_t as a function of θ and α respectively at nominal load, nominal output voltage and nominal duty cycle

The following values of the resonant tank elements are taken from Fig. 8:

$L_S = 55\mu H$, $C_S = 25nF$, $C_P = 18nF$, $n_t = 14$.

C_S must withstand AC-Voltage V_{CS} as given in Fig. 8. The required number of discrete capacitors is dependent on the AC stresses and can be determined based on Fig. 6. Furthermore, the series capacitor must withstand a DC-voltage of

$$V_{AB_0} = \frac{V_{PFC}}{2} \cdot d, \quad (17)$$

which is a result of the operation of the inverter.

Further optimization to fulfil certain design constraints, e.g. maximum allowable core temperature of the inductor, can be done in the next iteration step.

In Fig. 9, switching frequency is shown as a function of the output voltage and the load respectively.

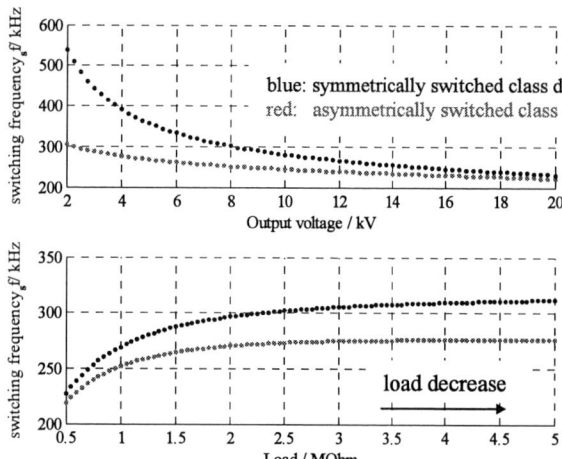

Fig. 9. Upper diagram: Switching frequency as a function of the output voltage at rated load
Lower diagram: Switching frequency as a function of the load at nominal output voltage

The advantage of using the asymmetrically switched class D inverter for a converter with adjustable output voltage is clarified.

TABLE I
RATIO OF SWITCHING FREQUENCIES FOR DIFFERENT OUTPUT VOLTAGES

	Asymmetrically switched inverter	Symmetrically switched inverter
f_S (2kV) / f_S (20kV) @ rated load	**1.36**	**2.39**

VI. DESIGN OF THE VOLTAGE MULTIPLIER

Analytical equations for the voltage multiplier to determine the output voltage and its output impedance are derived with the technique of charge reloading [9]. To supply the load with a constant output current, a charge Δq has to pass through the multiplier

$$\Delta q = \frac{I_o}{f_s}, \qquad (18)$$

where I_o is equal to the output voltage divided by the load resistance. This charge is taken out of every capacitor.

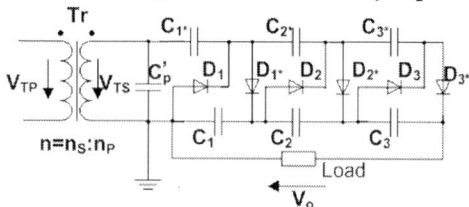

Fig. 10. Voltage multiplier with high voltage transformer and parallel capacitor

The output voltage ripple of the multiplier is the sum of all voltage drops of the capacitors C_1, C_2 and C_3 and can be computed as

$$\Delta v_o = \frac{I_o}{2 \cdot f_s} \left(\frac{3}{C_1} + \frac{2}{C_2} + \frac{1}{C_3} \right), \qquad (19)$$

where

$$2 \cdot \Delta v_o = V_{o\_max} - V_{o\_min}. \qquad (20)$$

The maximum achievable output voltage at a certain output current can be determined by adding the maximum voltages of the capacitors C_1, C_2 and C_3. While C_{1*} is loaded with twice the maximum transformer secondary voltage, C_1 will be charged with

$$V_{C1,max} = 2 \cdot V_{TS,max} - \frac{3 \cdot \Delta q}{C_{1*}}. \qquad (21)$$

The maximum voltage of the capacitor C_{2*}, that is loaded via C_1 is

$$V_{C2*,max} = 2 \cdot V_{TS,max} - \frac{3 \cdot \Delta q}{C_1 *} - \frac{3 \cdot \Delta q}{C_1} \qquad (22)$$

According to this scheme, it is possible to calculate the maximum voltage of each capacitor. Consequently, the maximum voltage of the multiplier can be calculated as

$$V_{o,max} = \sum_{i=1}^{n} V_{Cn,max} \qquad (23)$$

$$V_{o,max} = 6 \cdot V_{TS,max} - \frac{9 \cdot \Delta q}{C_{1*}} - \frac{6 \cdot \Delta q}{C_1} - \frac{4 \cdot \Delta q}{C_{2*}} - \frac{2 \cdot \Delta q}{C_2} - \frac{\Delta q}{C_{3*}} \qquad (24)$$

The average output voltage V_o is calculated by subtracting the voltage ripple from the maximum output voltage

$$V_o = 6 \cdot V_{TS,max} - \frac{I_o}{f_s} \cdot \left(\frac{9}{C_{1*}} + \frac{7.5}{C_1} + \frac{4}{C_{2*}} + \frac{3}{C_2} + \frac{1}{C_{3*}} + \frac{0.5}{C_3} \right) \qquad (25)$$

By using (25), an equivalent circuit of the voltage multiplier (Fig. 11) can be developed.

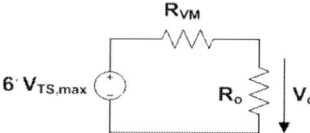

Fig. 11. Equivalent circuit of the voltage multiplier

$$R_{VM} = \frac{1}{f_s} \cdot \left(\frac{9}{C_{1*}} + \frac{7.5}{C_1} + \frac{4}{C_{2*}} + \frac{3}{C_2} + \frac{1}{C_{3*}} + \frac{0.5}{C_3} \right) \qquad (26)$$

The conversion ratio of the multiplier is

$$n_m = \frac{V_o}{V_{TS,max}} = \frac{6}{\left(1 + \frac{1}{R_o f_S} \cdot \left(\frac{9}{C_{1*}} + \frac{7.5}{C_1} + \frac{4}{C_{2*}} + \frac{3}{C_2} + \frac{1}{C_{3*}} + \frac{0.5}{C_3} \right) \right)} \qquad (27)$$

The output impedance R_{VM} of the multiplier is a function of the switching frequency, the capacitances, and the number of multiplier stages. For a given value of installed capacitance, the output impedance of the multiplier varies with the distribution of the capacitance [10], which is indicated by (29). To achieve a voltage drop of less than 1% using a constant capacitance distribution, each capacitor must have a capacitance 2.5nF. Reference (10) proposes that a graded capacitor distribution lowers R_{VM} and increases the bandwidth of the multiplier. With (19) the following obtainable values for the capacitors were chosen:
C_1 and C_{1*}=3.3nF; C_2 and C_{2*}=2.2nF; C_3 and C_{3*}=1.5nF.

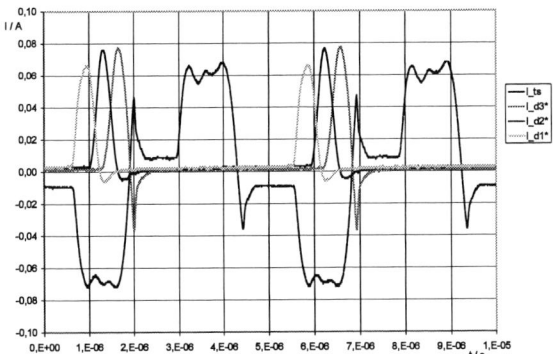

Fig. 12. Currents of the Diodes D_{1*}, D_{2*}, D_{3*} and the transformer

The voltage stress of a given diode is equal to the voltage stress of the corresponding capacitor, e.g. $V_{D3*,max}=V_{C3,max}$. The measurement results in Fig. 12 clarify the stresses on diodes D_{1*}, D_{2*}, and D_{3*} in light load operating point. D_{2*} and D_{3*} are turned off "hard" but need not block negative voltage immediately. The necessity to block voltage starts when the voltage multiplier enters its blocking state. This transition is dictated by the output voltage of the high voltage transformer. The electrical demands of fast switching can be accomplished with a series connection of diodes. However, due to their varying junction capacitances and blocking resistances, either an extra resistor and capacitor network for balancing must be installed or diodes with guaranteed avalanche energy absorption capabilities must be used.

VII. SAWTOOTH SIGNAL GENERATOR

To ensure converter operation above the resonant frequency under all load conditions, the switching frequency is self-adjusting. As a result, ZVS can be attained over the complete operating range. This operation is cited in the literature as the *self sustained oscillation mode above the resonant frequency* [4]. The measured inductor current is used to determine the actual switching frequency. References [4, 6] propose the usage of an analogous sawtooth signal generator, which adapts its slope to reach one given amplitude S_T with varying switching frequencies. A block diagram of such a generator is shown in Fig. 13.

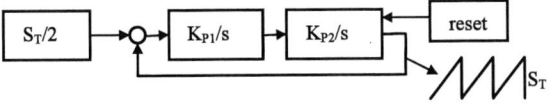

Fig. 13. Constant amplitude sawtooth generator

The second integrator of the generator is reset when the resonant current crosses zero. The sawtooth signal is then compared with a controller output signal, in this case the duty cycle, to generate switching signals for the power semiconductors. If the resonant frequency changes abruptly, (e.g. during a load change) the amplitude of the sawtooth signal goes through a transient state. The time of the transient condition depends on the parameters of the sawtooth signal generator and the load step of the converter and can be a couple of switching periods. During this time the switching signals are erroneous.

A digital implementation can minimize this erroneous effect. When the frequency changes, not the signal slope, but rather the controller output signal is adjusted. A VHDL algorithm for this control scheme suitable for FPGA implementation was developed. Simulation results are shown in Fig. 14. It can be seen that after only half a switching period the switching signals will be correct and the transient time of the digitally implemented sawtooth signal generator is dramatically reduced. The duty cycle is the turn off signal for the switch T_1. To ensure ZVS for both power semiconductors T_2 must also be turned off prior to the resonant current changing its sign.

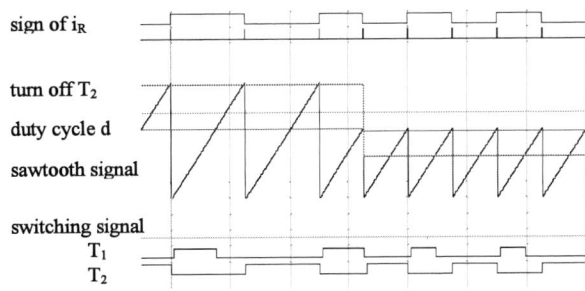

Fig. 14. Simulated VHDL code of the sawtooth signal generator

VIII. MEASUREMENT RESULTS

Fig. 16 shows a measurement result of the prototype, Fig.15, at nominal output voltage with an output resistance of 560kΩ.

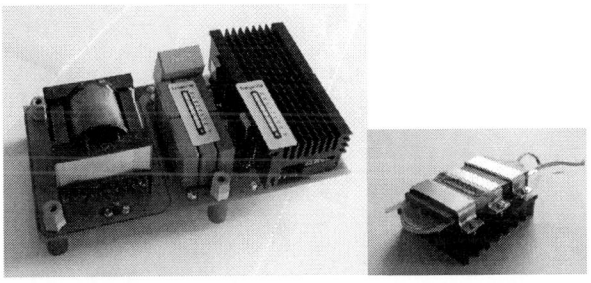

Fig. 15. Prototype without voltage multiplier

The DC-link voltage is 380V, the switching frequency 193kHz and the maximum of the resonant current 12.5A. The measured voltages of the MOSFETs show quasi no transients. The prototype was operated with a closed inner loop, self adjusting switching frequency, but an open outer loop. An FPGA created the switching signals.

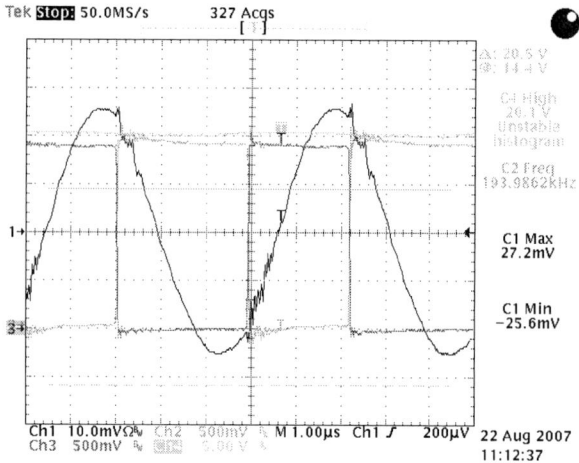

Fig. 16. Measurement result of the prototype at nominal output voltage

Fig. 17 shows another measurement result with a duty cycle of approximately 0.1 with the same load. The DC-link voltage is 380V, the output voltage 1.8kV. The self adjusted switching frequency is 324kHz.

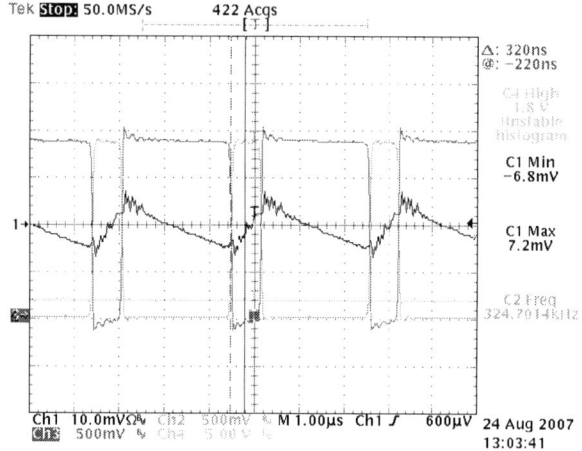

Fig. 17. Measurement result of the prototype with reduced duty cycle

IX. CONCLUSION

A straightforward design procedure for an asymmetrically switched class D inverter with an LCC resonant tank and a Walton Cockroft multiplier to generate adjustable high output voltages for medium power applications was developed.

The specialty of the asymmetrically switched inverter was clarified, and the comparison of this inverter with the symmetrical switched inverter showed the benefits of the asymmetrical switching of the power semiconductors. The digital implementation of a sawtooth signal generator improves the behaviour of the converter under transient conditions. Measurement results verify the analytical design process of the resonant tank components and their predicted electrical stress.

REFERENCES

[1] S.D. Johnson, A.F. Witulski, R.W.Erickson, "Comparison of Resonant Topologies in High-Voltage DC Applications", IEEE Transactions on Aerospace and Electronic Systems, VOL. 24, NO.3, May 1988

[2] J. M. Hancock, "Future Directions In Semiconductors For Power Electronics", APEC, Infineon, 2004.

[3] R.L. Steigerwald,"A Comparison of Half-Bridge Resonant Converter Topologies, IEEE Transactions on Power Electronics , Vol. 3, No. 2, April 1988, S. 174-183.

[4] H. Pinheirro, P.K. Jain, G. Joós,"Self-Sustained Oscillating Resonant Converters Operating Above the Resonant Frequency, IEEE Transaction on Power Electronics, Vol. 14, No 5, September 1999, S. 803-816.

[5] G. Ivensky, A. Kats, S. Ben-Yaakov,"A Novel RC Model of Capacitive-Loaded Parallel and Series-Parallel Resonant DC-DC Converters,"Power Electronics Specialists Conference, IEEE PESC, St. Louis, Missouri, USA, 22-27 June 1997 Page(s):958 - 964 vol.2

[6] F.S. Cavalcante, J.W. Kolar,"Design of a 5kW High Output Voltage Series-Parallel Resonant DC-DC Converter," in Proceedings of the 34th IEEE Power Electronics Specialists Conference, Acapulco, Mexico, 2003, vol.4, pp. 1807-1814.

[7] R. Gean, F. Canales, F.C. Lee, W.C. Tipton, "A High-Frequency High-Efficiency Three-Level LCC Converter for High-Voltage Charging Applications",Proc. IEEE PESC, 2004, pp. 4100-4106.

[8] W. Eberle, Y.F. Liu,"A Zero Voltage Switching Asymmetrically Half- Bridge DC/DC Converter with Unbalanced Secondary Windings For Improved Bandwidth" Power Electronics Specialists Conference, IEEE PESC, 2002, Volume 4, 23-27 June 2002 Page(s):1829 - 1834

[9] M. Beyer, W. Boeck, K. Moeller, W. Zangl, "Hochspannungstechnik", Springer Verlag Berlin, 1986

[10] F. Belloni, P. Maranesi, M. Riva, "Parameters Optimization for Improved Dynamics of Voltage Multipliers for Space", Power Electronics Specialists Conference, IEEE PESC, Aachen 2004, Volume 1, 20-25 June 2004 Page(s):439 - 443 Vol.1

New Direct High Frequency Soft-Switching Inverter-Fed AC-DC Converter with Voltage Doubler for Consumer Magnetron Drive

Hisayuki Sugimura[1], Bishwajit Saha[1], Hidekazu Muraoka[2], Sang Pil Mun[1], Tomokazu Mishima[3],
Hideki Omori[4], Mutsuo Nakaoka[1,5]

[1]Kyungnam University, Masan, KOREA
[2]Hiroshima National College of Maritime Technology, Hiroshima, JAPAN
[3]Kure National College of Technology, Hiroshima, JAPAN
[4]Matsushita Electric Industrial Co. Ltd., Osaka, JAPAN
[5]Industrial College of Technology-University, Hyogo, JAPAN
Email: bsaha@ieee.org

Abstract – The oscillation voltage (Ebm) is an important parameter of the characteristic constant of a high power microwave generator. When Ebm could be made high, it serves so as to improve the power conversion efficiency of a magnetron. Thus, the newly developed magnetron with 7% higher oscillation cut-off voltage than the present conventional one is manufactured by Matsushita Electric Industrial Co.Ltd, In addition to this, the utility AC power line current reduced harmonic type direct inverter AC-DC power converter suitable and acceptable for driving this magnetron is proposed in this paper. This direct high frequency inverter type soft switching PWM AC-DC converter has the voltage-boost function based on active clamp scheme with soft switching pulse modulation strategy topology. This paper describes a novel type active clamp AC-DC power converter circuit defined as direct high frequency inverter and the switching operation of this power converter is described and discussed on the basis of simulation and experimental results. The operating characteristics of this soft switching direct high frequency inverter type AC-DC power converter using IGBTs that incorporates the harmonic current improvement as well as power factor correction are evaluated and discussed from a practical point of view as compared with the conventional AC-DC power converter.

Index Terms—AC-DC power conversion, Consumer power electronics, Magnetrons, Direct high frequency inverters.

I. INTRODUCTION

The high frequency inverter type high voltage DC power supply to drive consumer magnetron is based upon the power electronics circuits technology for the microwave power generator, which handles relatively high power conversion process ranging from 1.0[kw] to 1.4[kw]. The high frequency inverter type AC-DC power converter for the microwave oven as well as high frequency inverter type AC-AC converter for induction heating type cooker and hot water producer has attracted special interest as one of consumer power appliances from global environmental conservation and reproduction viewpoints. In recent years, research and development have been actively carried out from a practical point of view for the purposes of realizing miniaturization, downsizing based on high efficiency, lowered electromagnetic noises, high performances related waveform quality and responsibility,

lower cost due to minimum circuit components, higher utilizations, and new multi-functions. Of these, the high power microwave output can be required for rapid cooking processing as one of the basic operating performances of a consumer microwave oven. Most of consumer power electronic appliances such as microwave oven, induction heating cooker and steamer, are connected to the circuit to 100V/200V utility AC power supply, which use the full-wave rectified DC output. As for the rectified smoothing output, making the most of non-smoothing operation can minimize the smoothing filter capacitance of a rectification smoothing circuit. In addition, because of non-smoothing filter design in this power converter, the harmonic current components in the utility AC power line grid side could be minimized so as to be sinewave shaping under a condition of unity power factor. Since the commercial power supply line current can be brought close to sinusoidal wave, the harmonics reduction can be performed by using the AC-DC converter design. But, because of non-linearity of i-v characteristic of the magnetron, high frequency inverter output voltage or the output voltage of voltage doubler could not raise up to magnetron oscillation operation voltage in the lower portion of the utility AC power supply instantaneous voltage. As a result, the magnetron device stops stable oscillation operation and the utility AC line current distortion causes because the utility AC input line current stops. Thus, in order to implement the high power microwave output, the new type magnetron with a high cut-off voltage design has just developed by Matsushita Electric Industrial Co. Ltd, Japan. We developed new high frequency switching AC-DC power conversion processing equipment that takes out the DC high voltage through a high frequency transformer and a voltage doubler rectifier in addition to the control implementation scheme for executing total power factor improvement and harmonic current reduction in the utility AC commercial frequency alternate current line side. The newly-developed direct soft switching high frequency inverter is directly changed into high frequency alternate current from a utility AC commercial alternate current power supply to drive this type of magnetron efficiently with low cost and downsizing. Finally, Proposed are hybrid modulation pattern selection method of the partially-modulated operating frequency in addition to its duty factor control strategy in

978-1-4244-0644-9/07/$25.00 ©2007 IEEE

order to improve harmonic current components and make total harmonic power factor unity under a practical condition of the peak limitation of the magnetron anode current.

II. LOAD CHARACTERISTICS OF THE MAGNETRON FOR MICROWAVE OVEN

A. Structure of Magnetron

Figure 1 shows the exterior appearance (see Fig.1 (a)) of magnetron structure in addition to its internal cross section (see Fig.1(b)). The internal construction includes cylindrical cathode and an anode on the concentric circle, and the permanent magnet with ferrite core gives the magnetic field toward axial direction as depicted Fig.2. The big difference between the new model magnetron and conventional magnetron indicates that the magnetic field intensity of the permanent magnet with 2512 [G] is larger as compared with 1862 [G] of the conventional one. The behaviour of the electron in action space is determined by this strong magnetic field intensity, and a magnetron comes to start an oscillation by the high cut-off voltage, so that magnetic field intensity becomes strong. In this case, the cut-off voltage designed for about 4.6 [kV] that reaches an oscillation mode is designed for higher value than that of the conventional magnetron designed for about 3.6 [kV]. Moreover, the conversion efficiency of the new type magnetron is also 78%. As a result, it is higher than conventional one; 75%.

B. i-v Characteristics and Electrical Equivalent Circuit Model of Magnetron

Figure 3 represents the terminal voltage vs. current characteristics of the new magnetron. Concerning to the magnetron v-i characteristics, the magnetron has non-linear v-i characteristics, what is called, piecewise linear characteristics

which include high resistance area in non-oscillating range for the magnetron and low resistance area in oscillating range for the magnetron. When the cut-off voltage between anode and cathode exceeds about 4.6[kV], the magnetron anode current begins to flow from anode toward cathode terminal. On the other hand, when the voltage between anode and cathode is lower than a cut-off voltage, the anode current of magnetron does not flow. The static model of the electrical equivalent circuit of the magnetron with these characteristics can simply represent by using two pure piecewise resistances, ideal diode to determine one quadrant operation and ideal battery voltage corresponding to the cut-off voltage Vz as depicted in Fig.4. As illustrated in Fig.4, the load type with two power switches; Load1 in non-oscillation mode and Load2 in oscillation mode is to be automatically selected whether the voltage between anode and cathode is higher than the cut-off voltage Vz 4.6[kV] or not.

III. HIGH FREQUENCY INVERTER WITH BOOST FUNCTION

A. System Consideration

The high efficient magnetron introduced newly has a specialized feature that the cut-off voltage of the magnetron is higher than the conventional one. Therefore, it is considered that the non-oscillating period of the magnetron increase near the zero valley point area of the utility AC input voltage. As a

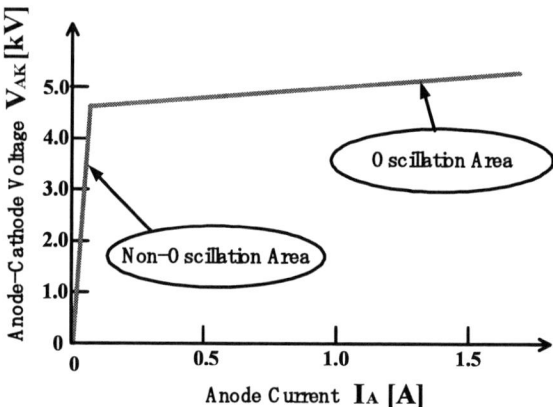

Fig.3. Current vs. voltage characteristics model of magnetron

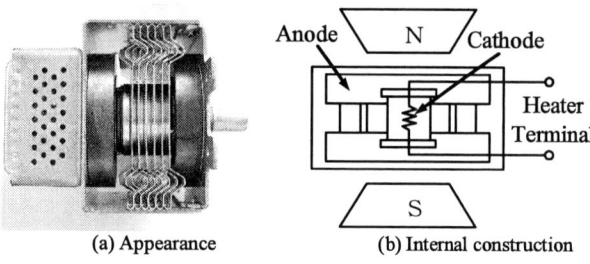

(a) Appearance   (b) Internal construction

Fig.1. Magnetron.

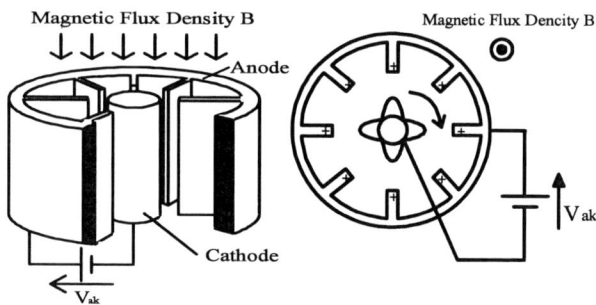

Fig.2 Operation principle of magnetron

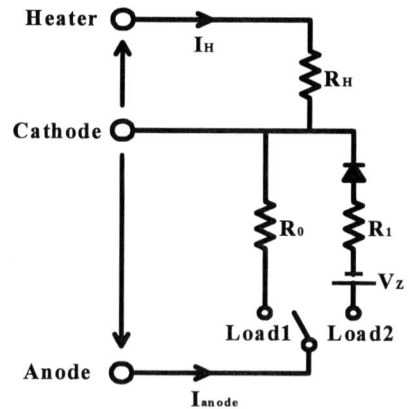

Fig.4 Electrical equivalent circuit modelling of magnetron

result, the power factor of the utility input current will be worse than conventional one. Especially, in the latest development of household microwave oven, because the utility input current is actually limited by the indoor wiring capacity, a deterioration of power factor is concerned with decrease of utility input power. As a result, the required microwave output could not provide. Then, a magnetron is needed to drive under a condition of high power factor by reducing the non-operating period of a magnetron and the distortion of the utility AC input line current waveform by maintaining the applied voltage across a high frequency transformer near zero voltage around valley of the utility AC input voltage by adding a particular voltage boost function. But, the power conversion efficiency actually decreases to use the some power conversion processing stages and the increase of circuit components results in the large-sizing volume. Thus, a direct high frequency inverter with the multi function of one stage power conversion that has three operating abilities of rectification, boost and frequency conversion is proposed by the authors. Consequently, this proposed direct inverter type AC-DC power converter has boost function without reducing the power conversion efficiency and increasing circuit components and power semiconductor switching devices.

B. Power Circuit and System Topology

The main power circuit and system topology of the direct high frequency inverter type AC-DC power converter for a microwave oven is shown in Fig.5 in which the utility AC power supply has three winding high frequency transformer, and the design specifications and circuit parameters are indicated in Table I. The multiple power conversion circuit stages consist of the utility power supply, non-smoothing filter, high frequency soft switching inverter, three winding high frequency boost transformer, full-wave voltage doubler rectifier circuit, magnetron circuit represented by electrical equivalent circuit with i-v characteristic as zener diode characteristics, and magnetron heater circuit. The commercial utility AC power source voltage supplies the electric power to a high frequency inverter through the low path filter as non-smoothing filter that consists of an inductor and a capacitor. This low path filter is inserted not to leak the noise by the high frequency inverter to a commercial utility power

source side. Moreover, since the capacitance is designed for the value that can maintain a certain constant voltage in the operating frequency area of high frequency inverter, the inductance of the filter inductor is set to 100[μH] and the capacitance of the filter capacitor is set to 6.0[μF]. Therefore, a commercial utility AC power source does not design so as to have smoothing ability for commercial AC frequency and can deliver to the output stage of direct high frequency inverter from 0[V] to 141[V] in non-smoothing filter operation. A high frequency transformer is connected to the primary winding at the middle point of the circuit of a filter capacitor in series with a power switching semiconductor device, and a reverse conducting diode is connected to a semiconductor switching device, respectively. The capacitor C2 is connected between collector-emitter of the semiconductor-switching device Q1, and works as lossless capacitive snubber at the time at a turn-off of the switching device Q1 and Q2. This capacitor performs a soft switching operation commutation. The capacitors C3 and C4 are connected in parallel with the diodes D1 and D2, respectively, and these capacitors have the clamp action which maintains the voltage applied for a high frequency transformer by charging the energy stored into the high frequency transformer through the reverse conducting diodes D3 and D4. Moreover, the secondary winding of a high frequency transformer is connected to the double voltage rectification circuit that consists of the capacitors C5 and C6 and the diodes D5 and D6.

C. High Frequency Three Winding Transformer

The appearance of a high frequency three winding transformer is shown in Fig.6. Although the independent power supply which is not influenced upon the inverter operation is actually adopted for the heater power supply in an industrial magnetron drive as a plasma generator, the magnetron heater power is to be supplied from installing the 3rd winding in a high frequency high-voltage transformer, from a viewpoint of small size, lightweight and low cost, for the household electric appliances.

D. Gate Pulse Control Implementation

Figure 7 gives the asymmetrical PWM signal pulse timing sequences. These gate pulse voltages are respectively supplied

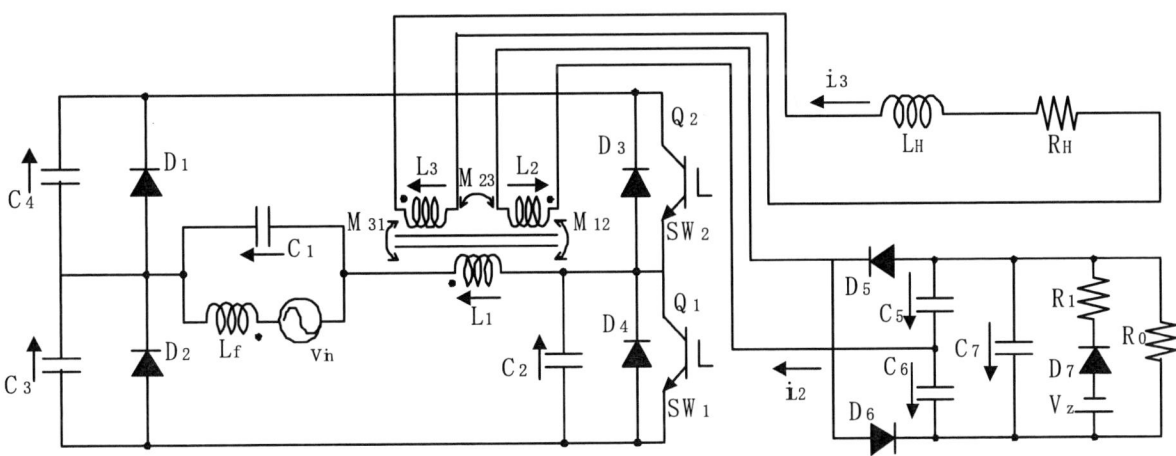

Fig.5 Circuit diagram of proposed converter

to the power semiconductor switching blocks; the main switching block Q1 (SW1/D4) and the subsidiary switching block Q2 (SW2/D3). In this Fig.8, the constant frequency duty factor (or duty cycle) control scheme is implemented for the continuous power regulation strategy of this high frequency quasi-resonant inverter using IGBTs. The duty factor is defined as the conduction time Ton1 including a dead time Td of the main active power switch SW1 during one period T. The control variable range of this duty factor is basically from 0 to 1. The output power of this high frequency inverter is controlled smoothly by varying this duty factor defined as the ON time of the gate voltage pulse signal for driving main active power switch SW1 as a control variable.

IV. EXPERIMENTAL RESULTS AND DISCUSSIONS

A. Steady State Operating Waveforms of the Switching Blocks; Q1 and Q2

TABLE I
DESIGN SPECIFICATIONS

Item		Symbol	Value
High frequency inverter		C1	0.6[μF]
		C2	0.1[F]
		C3	4.5[μF]
		C4	4.5[μF]
		Lf	100[μH]
HF-3 winding transformer	inductance	L1	60.0[μH]
		L2	24.0[mH]
		L3	0.60[μH]
	coupling coefficient	K12	0.81
		K23	0.66
		K31	0.55
Full-wave voltage doubler rectifier		C5	3600[pF]
		C6	3600[pF]
Penetration capacitor		C4	500[pF]
Magnetron	cut-off voltage	Vz	4600[V]
	non-oscillation Load	R0	500[kΩ]
	oscillation load	R1	300[Ω]

Fig.6. High frequency three winding transformer

Figure 8 (a) shows the appearance of newly developed high frequency inverter type AC-DC power converter, which is composed of the direct AC-HFAC power converter, high frequency transformer and full wave voltage doubler and the magnetron. Figure 8 (b) shows the IGBT (40[A]/600[V]) produced by Fairchild Semiconductor Co. Ltd for this high frequency inverter type AC-DC converter. Figure 9 (a) and (b) illustrate the measured voltage and current waveforms of the switching blocks; Q1 and Q2, and Figure 10 shows the simulating waveforms of these power switches. As understood clearly from these operating waveforms, it is noted that the entire switching block in this inverter type AC-DC power converter can achieve the soft switching. Moreover, the simulated operating waveforms have good agreement of the observed operating waveforms and it is clear that the validity of this simulation method is also more effective. It is confirmed that the soft switching commutation in all the operating area during one cycle period of the utility AC voltage can be achieved for this high frequency inverter type AC-DC power converter.

B. Partially Pulse Modulation

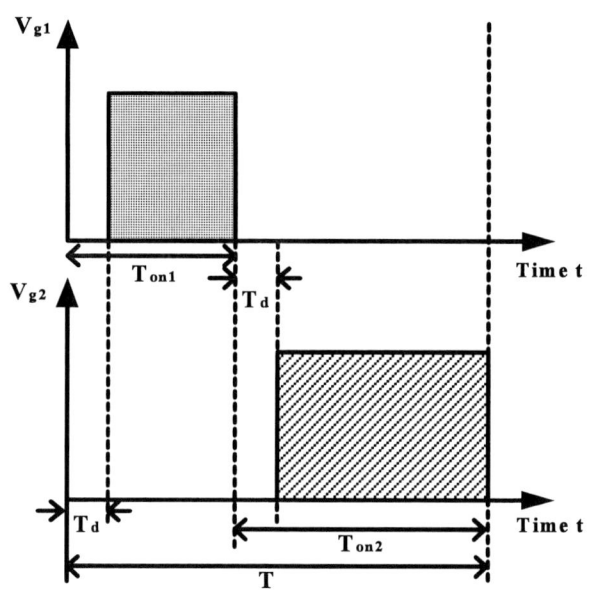

Fig.7. Gate voltage pulse signal timing sequence

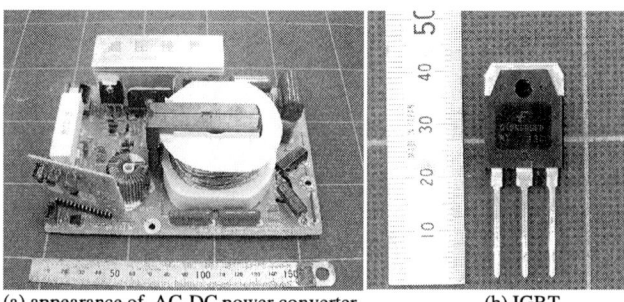

(a) appearance of AC-DC power converter   (b) IGBT

Fig.8 The appearance of newly developed high frequency inverter type AC-DC power converter and IGBT

Figure 11 demonstrates how to select modulation patterns in respect of the partially pulse modulated pattern scheme due to inverter operating frequency as well as its duty factor-based asymmetrical pulse width modulation in order to improve line current harmonic distortion and total power factor in the utility AC power grid side. This principle is based on the characteristics of this direct high frequency inverter; when its operation frequency is so as to be high, the output voltage decreases. On the other hand, when duty factor is so as to be high, the output voltage increases. This pulse modulation scheme is based on pulse modulation on the duty factor, which is changed so as to be a low value near peak value of utility input sinewave AC voltage. On the other hand, it is changed so as to be a high value near transient build-up and transient build-down of the utility input sinewave AC voltage. Moreover, this pulse modulation scheme is based on pulse modulation on the operation frequency, which is changed so as to be a high value near peak value of utility input sinewave AC voltage; on the other hand, it is changed so as to be a low value near transient build-up and transient build-down of

utility input sinewave AC voltage. A conduction angle of an input current makes wider for the utility sinewave AC input voltage by introducing this modulation strategy. The oscillation period of magnetron increases in accordance with the conduction angle of input line current for positive and negative utility AC voltage half cycle. In addition to this, the peak value of anode current is held with a low value near the peak absolute value of the utility AC input voltage. So this pulse modulation is more effective modulation method for excellent oscillation range of the magnetron and harmonic current reduction. Furthermore, the main control requirement relating to the operating life problem of the magnetron must hold down the peak value of anode current below to 1.2 [A] is fully met from an application point of view.

C. Evaluations of Input Line Current Waveforms and Comparative Studies

Figure 12 (a) shows the utility AC input line current waveform of the modulated high frequency inverter type AC-DC power converter during one cycle period of utility sinewave AC voltage, Figure 12 (b) shows utility AC input line current waveform of the conventional high frequency inverter. The utility AC input line current of newly developed high frequency inverter type AC-DC power converter shown in Fig.12 (a) becomes continuous near zero voltage of the power supply voltage, and these waveforms is good similar to a sinusoidal current wave. This high frequency inverter type AC-DC power converter can control voltage across transformer with boost operation function. Furthermore, since it can build up the oscillation voltage of the magnetron near zero area of utility AC power supply voltage to introduce partial pulse modulation pattern, the idling operating period of utility AC input line current could be minimized as small as possible. The results on FFT analysis of the utility AC input line current of this high frequency inverter type AC-DC power

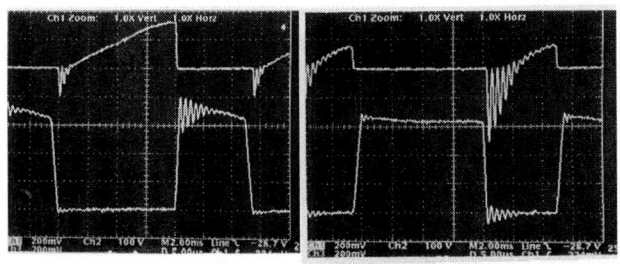

(a) Switching block Q1   (b) Switching block Q2
Current axis; 40A/DIV   Voltage axis; 100V/DIV   Time axis; 5μs/DIV

Fig.9. Experimental waveforms of current and voltage of Q1 and Q2

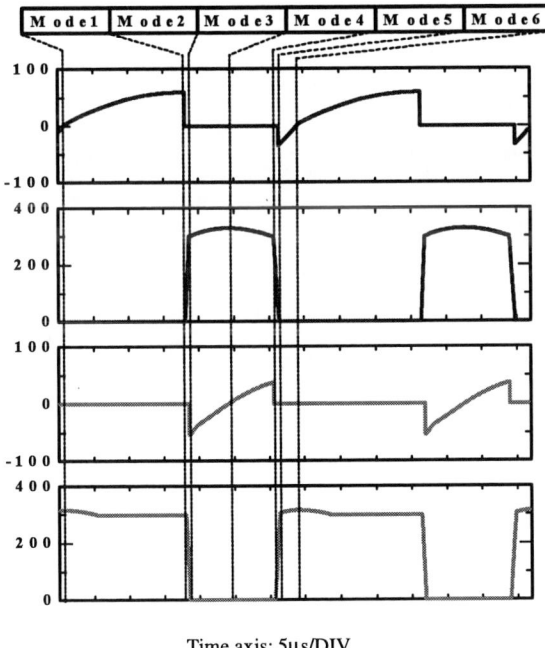

Time axis; 5μs/DIV

Fig.10. Simulation waveforms of current and voltage of Q1 and Q2

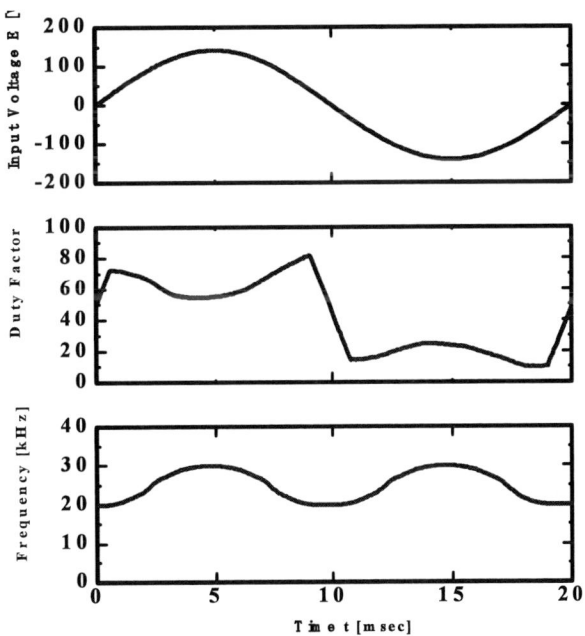

Fig.11. Pulse width and pulse frequency mixed pattern modulation

converter and conventional inverter type AC-DC power converter is shown in Fig.13. The harmonics distortion of the utility AC input line current can be suppressed under EN 61000-3-2 regulation specified as a guideline in Japan. The utility AC input line current of the developed inverter type AC-DC power converter to this standard value is settled within the standard value in all the current harmonic orders. It is noted that the good input line current waveform can be obtained for this AC-DC power converter. Furthermore, it is noted that harmonics current in utility AC line for newly inverter type AC-DC power converter is less than the conventional high frequency inverter type AC-DC power converter. This result is considered because the total power factor in the utility AC side of this high frequency inverter type AC-DC power converter can be improved as compared with that of conventional inverter type AC-DC power

converter. The total power factor and total system efficiency of this high frequency inverter and conventional high frequency inverter type AC-DC power converter are comparatively illustrated in Table II. It is proven that the high frequency inverter type AC-DC power converter becomes high power factor and high efficiency as compared with a conventional one.

V. CONCLUSIONS

The authors developed the new model high efficient high-power magnetron with high cut-off DC voltage so as to provide harmonic current reduction in the utility AC grid power side and total power factor improvement in utility AC power grid side in addition to the newly proposed high frequency inverter with boost and active voltage clamp function. Moreover, the AC-DC power converter including high frequency inverter ability, boost function, voltage clamp and a rectification function was integrated as one power converter. The high frequency inverter type AC-DC power converter treated here was more suitable for driving the new type magnetron. Finally, the harmonic components to the 40th order of the distorted harmonics input line current in the commercial sinusoidal wave AC side has been sufficiently suppressed below each regulation value specified by EN 61000-3-2 on harmonic current order. As a result, and AC-DC his power converter system was cost-effective for driving consumer high power magnetron for microwave oven.

TABLE II
COMPARATIVE EFFICIENCY AND POWER FACTOR OF CONVENTIONAL AND PROPOSED POWER CONVERTER SYSTEM

	This Developed Inverter	Conventional Inverter
Power Factor	0.99	0.96
Efficiency	92%	90%

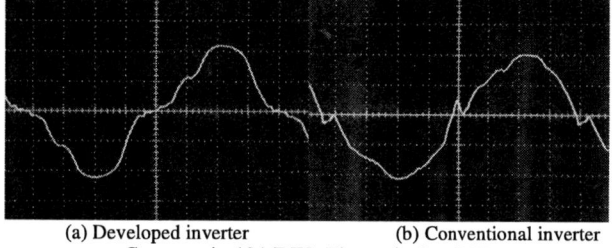

(a) Developed inverter　　　　(b) Conventional inverter
Current axis; 10A/DIV　Time axis; 2ms/DIV

Fig.12 Input side current waveforms

REFERENCES

[1] K. Yasui, D. Bessyo, H. Omori, H. Terai, M. Nakaoka" The Inverter Circuit Skills to Realize low-cost, compact-size Power Supply for Microwave Oven, and the Advantages of Improved Defrosting" International Appliance Technical Conference 2000

[2] K. Yasui, D. Bessyo, H. Omori, M. Nakaoka, "The Active-Clamp Quasi-resonant Inverter Power Supply for Magnetron Drive" International Power Electronics Conference 2000 Tokyo

[3] E. Miyata, M. Nakaoka, K. Yasui, D. Bessyo, H. Omori, "Performance Evaluations on an Active Voltage-Clampling Self-Excited Zero Voltage Soft Switching DC-DC Converters for Consumer Magnetron Drive." Energy Engineering in Electronics and Communications, Vol. 100 No.303, 304, pp. 33-40, 2000.

[4] T. Matsushige, M. Ishitobi, M. Nakaoka, D. Bessyo, H. Yamashita, H. Omori, H. Terai, "Pulse Width and Pulse Frequency Modulated Soft Commutation Inverter type AC-DC Power Converter with Lowered Utility 200V AC Grid Side Harmonic Current Components", Proceedings of 2001 International Conference on Power Electronics (ICPE), pp.484-488, October, 2001.

[5] T. Matsushige, S. Chandhaket, S. Moisseev, E. Hiraki, M. Nakaoka, H. Terai, D. Bessyo, K. Yasui, H. Omori, "Cost Effective PWM Soft Switching High Voltage Converter with Utility AC Side Harmonic Current Reduction Strategy for Microwave Oven", Proceedings of International Symposium on Industrial Electronics (ISIE), Vol.1, pp.118-123, December, 2000.

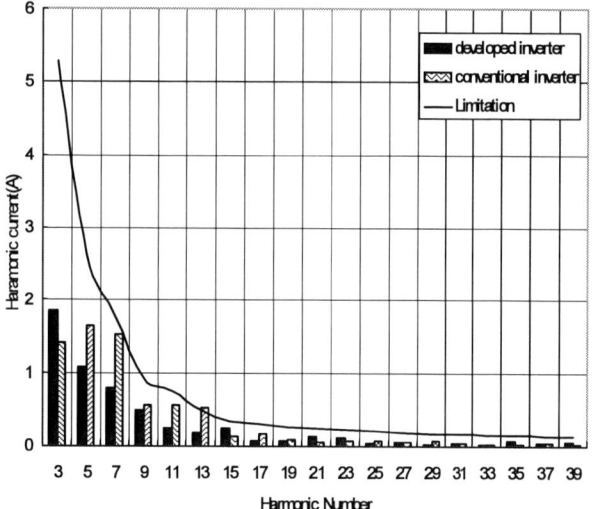

Fig.13. FFT spectrum of input line current in utility grid

Complete loading Characteristics Modeling of an Axial Flux Permanent Magnet Synchronous Machine Using Ck Spline Functions

Z. Lakhdari*, F. Amrane**, L. Adélaïde*** and Ph. Makany*

* Laboratoire Universitaire des Sciences Appliquées de Cherbourg, UCBN, Rue Aragon, BP 78, 50130 Cherbourg-Octeville, France
** EG-ELEC, av Paris 94300 VINCENNES, France
*** Laboratoire Central des Ponts et Chaussées (LCPC), 58, boulevard Lefèbvre, 75732 PARIS Cedex 15, France
Corresponding author, E-mail adress: zakaria.lakhdari@chbg.unicaen.fr

Abstract-- **This article presents for the first time the differential equations of the complete loading characteristics of three phase Axial Flux Permanent Magnet Synchronous Machine (AFPMSM) with broad air-gap. It shows that this characteristic leads to algebraic differential equations (ADE), which we know recently, the properties and integrated numerical means. These equations are governed by a multidistributed value problem (MDVP) which makes this problem a difficult numerical one, non accessible by traditional solvers. This article shows thanks to the properties of C^k spline functions that we can integrated such problems by judicious choices of the initial vectors and simulated annealing methods. The first results of simulations will be given.**

Index Terms-- **ADE and MDVP, AFPMSM, Ck spline functions, Modeling and simulation.**

I. INTRODUCTION

THE need of increasing the performances of electrical machines leads step by step to define precise analytical model of these machines, in order to carry out fast, precise and reliable controls. These constraints require for these kind of machines analytical knowledge so that fast and precise controllers can be elaborated. In this way, we present an analytical model of the AFPMSM with broad air-gap (Fig. 1.). We show that simple considerations of differential geometry coupled to a extended Lagrangian Formalism make it possible to obtain differential equations of the system with a good precision, as show our numerical simulations see Fig. 2. After some simples algebraic manipulations we show that our model leads to algebraic differential equations which properties and analytical methods to treat them are now known.

The complete loading characteristic in electrical machine like AFPMSM leads to MDVP (multi distributed value problems) which cannot be integrate by traditional solvers. For that we have developed spline functions known as Ck

This work was done under the financial participation of the European Fund for Research and Development (FEDER).

spline functions [1][2], which have, as main property, the fact that the coefficients of their functional development are the derivatives (total or partial) up to k, of the considered function itself at each point of discretization. This allows

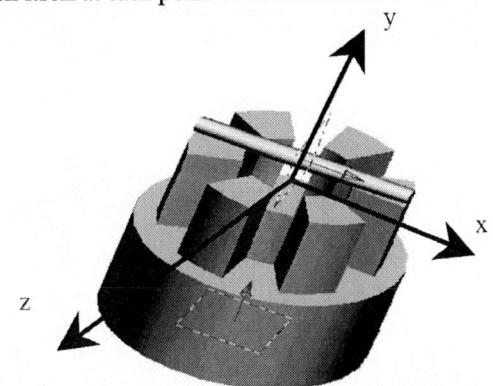

Fig. 1. Axial Flux Permanent Magnet Synchronous Machine (AFPMSM) geometry

introducing directly and with an arbitrary precision [3] the differential constraints of the systems as well as the intrinsic conditions with the geometry of the MDVP system. This approach coupled with a judicious choice of the initial functions obtain by geometrical considerations can be achieved by simple simulated annealing solvers [4]. This studies is motivated by the fact that in electrical engineering, the permanent magnet synchronous machines and variable reluctance machines gain ground, year after year, and that in automatic control, recent progress of the Coupled Algebraic Differential Equations (CADE) allow us to consider analytical modeling of these electrical machines, coupling the electrical engineering to automatic control. Our study will be undertaken on a AFPMSM with broad air-gap. In order to model this AFPMSM we will present an analytical method which uses electromagnetic Lagrangian. Classical variationnal calculus leads by applying enhance Lagrange equations to CADE which are analyzed and simulated by C_k

spline functions approach. Another difficulty of this approach is that the electromotive forces (EMF) of such MSAPFA are non sinusoidal. The machine was designed on request of local industry of Low-Normandy, France. In order to answer to the technological constraints the geometry and the dimensions of the magnetic bar (rotor), stay invariant, therefore all works are primarily related to the optimization of the stator and to the development of the power supply of this machine.

II. LARGE AIR GAP AFPMSM MODELING BY DIFFERENTIAL GEOMETRY

The machine is constituted of three identical poles. Our digital simulations show that most of the magnetic fields lines are concentrated on a very narrow line see Fig.2 . We have thus supposed magnetic field to be monodimensional manifold as shown by Fig. 3.

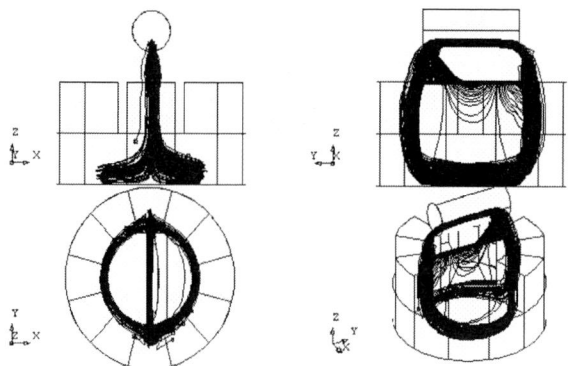

Fig. 2. Magnetic Fills Numerical Results Model

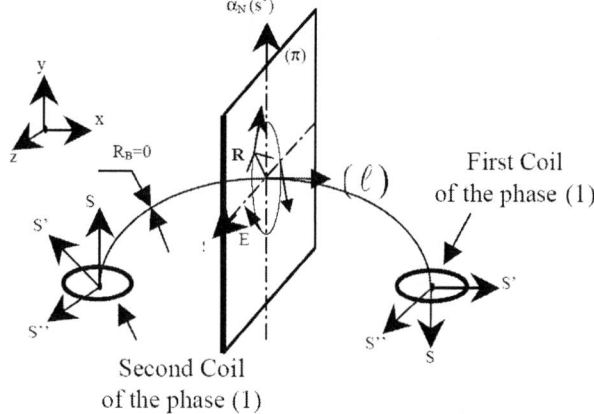

Fig. 3. The monodimensional Magnetic Fills Model

In this case the conservation of the magnetic flux and the relative magnetic permeability of air/cylinder head versus stator leads for one pole for the following equations:

$$H \oint_l d\ell = N.I$$
(1)

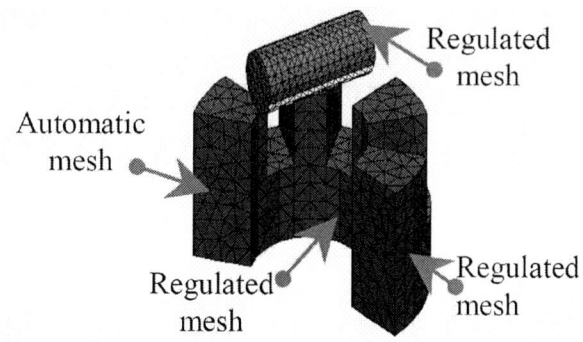

Fig. 4. AFPMSM mesh

N is the number of coil per pole and per phase
I is the stator phase current
ℓ is the curvilinear abscissa of the mono dimensional magnetic fields.

Then the value of the field H is given directly by the geometry of the line of the magnetic field which satisfies to the Lagrange equation. Geometrical considerations shown by Fig.1 related to the machine properties brought us to define this monodimensional manifold of field by the following functions,

$$y = g(f(\theta), \theta)$$
(2)

with :

$$x = \rho \cos(\theta)$$
$$z = \rho \sin(\theta)$$
$$\rho = f(\theta)$$
(3)

where equation (2) is the field equation for the consider pole of the machine and theta is the angular position of the rotor versus the symmetric axis of the consider pole. By definition we have

$$d\ell = \sqrt{\left[\frac{dx}{d\theta}\right]^2 + \left[\frac{dy}{d\theta}\right]^2 + \left[\frac{dz}{d\theta}\right]^2}$$

Referring to (1), (2), (3) and figure 1 and by assuming that $g(f(\theta), \theta) = g(\rho, \theta) = g(\theta)$,
so

$$dl = \sqrt{[f(\theta)]^2 + \left[\frac{\partial f(\theta)}{\partial \theta}\right]^2 + \left[\frac{\partial g(\rho, \theta)}{\partial \theta}\right]^2 \cdot \left[1 + \frac{\partial f(\theta)}{\partial \theta}\right]^2} \cdot d\theta$$
(4)

Hence equation (1) leads the following field equation for the consider pole,

$$\int_\theta H \cdot \sqrt{[f(\theta)]^2 + \left[\frac{\partial f(\theta)}{\partial \theta}\right]^2 + \left[\frac{\partial g(\rho, \theta)}{\partial \theta}\right]^2 \cdot \left[1 + \frac{\partial f(\theta)}{\partial \theta}\right]^2} \cdot d\theta = NI$$

because H is constant, we have,

$$H \oint_\theta \sqrt{[f(\theta)]^2 + \left[\frac{\partial f(\theta)}{\partial \theta}\right]^2 + \left[\frac{\partial g(\theta)}{\partial \theta}\right]^2 \left[1 + \frac{\partial f(\theta)}{\partial \theta}\right]^2} \, d\theta$$

(5)

then

$$H.\int_{\theta} \sqrt{f^2(\theta) + f'^2(\theta) + g'^2(\theta) \cdot (1 + f'^2(\theta))} \cdot d\theta$$

with

$$d\ell = \sqrt{f^2(\theta) + f'^2(\theta) + g'^2(\theta) \cdot (1 + f'^2(\theta))} \cdot d\theta$$

III. THE DYNAMICAL APPROACH OF THE FULL LOADING CHARACTERISTIC OF AFPMSM

The monodimensional field model of one pole leads easily to show that the sum of the three monodimensional fields created by the three poles of the AFPMSM machine leads to a spinning field pattern of the same frequency that the inverter fed currents [5]. This makes it possible to calculate the loading characteristic in the reference frame of the spinning field pattern and to look according to the load phase angle of the rotor compared to the spinning field pattern, the deformation of the spatial distribution of the magnetic field. This leads to consider in the reference frame of the spinning field pattern a calculation of deformation of static line of field presenting by the model given by Fig. 3.

Clearly the dynamics of the rotor in this spinning field pattern is given by the following electromagnetic Lagrangian density:

$$\pounds \left(\begin{array}{l} E^2(x, y, z, t) - c^2 B^2(x, y, z, t) + \\ j(x, y, z, t) A(x, y, z, t) - \rho(x, y, z, t) . V(x, y, z, t) \end{array} \right)$$

Where $\rho(x,y,z,t)$ is the spatial distribution of the static electric charges, $V(x,y,z,t)$ is the scalar potential resulting from the distribution of ρ, $j(x,y,z,t)$ is the current density of the machine, $A(x,y,z,t)$ is the resulting potential vector, $E(x,y,z,t)$ is the electrical field and $B(x,y,z,t)$ is the magnetic field.

Terms $\rho.V$ and $j.A$ are null because we neglect in the first approach the eddy current and magnetic losses in the rotoric and statoric bulk. Then $E = -gradV - \dfrac{\partial A}{\partial t} \equiv 0$, because $E = -gradV = 0$ (no charges) and we supposed no high frequencies transient phenomena, then $(-\dfrac{\partial A}{\partial t} \cong 0)$.

Hence the electromagnetic Lagrangian of the machine can be written as follows,

$$L = \int_{\theta} \int_{\rho} \mu^2 . B^2 d\rho \cdot d\theta = \int_{(\ell)} \mu^2 . B^2 d\ell$$

(7)

Because these kind of motors are controlled by fixed EMF, B is also fixed in our computation, then minimizing the magnetic field density $B(x,y,z,t)$ leads to minimize the magnetic field path $d\ell$.

With the associated Lagrangian equations

$$\begin{cases} \dfrac{\partial \ell}{\partial f(\theta)} - \dfrac{d}{d\theta} \left[\dfrac{\partial \ell}{\partial \dfrac{\partial f(\theta)}{\partial \theta}} \right] = 0 \\ - \dfrac{d}{d\theta} \left[\dfrac{\partial \ell}{\partial \dfrac{\partial g(\theta)}{\partial \theta}} \right] = 0 \end{cases}$$

(8)

Classical variationnal calculus and algebraic manipulations show that this above mentioned equations lead without loss of generality to the following coupled algebraic differential equations,

$$F_1(\theta) =$$

$$[f(\theta) - f''(\theta).[1 + g'^2(\theta)] + 2.f'(\theta).g'(\theta).g''(\theta)]$$
$$.\left[f^2(\theta) + f'^2(\theta) + [1 + f'^2(\theta)].g'^2(\theta) \right] +$$
$$f'(\theta).[1 + g'^2(\theta)].[f(\theta).f'(\theta) +$$
$$f'(\theta)f''(\theta).[1 + g'^2(\theta)] + [1 + f'^2(\theta)].g'(\theta).g''(\theta)] = 0$$

$$F_2(\theta) =$$

$$[-2f'(\theta).f''(\theta).g'(\theta) - [1 + f'^2(\theta)].g''(\theta)] +$$
$$\left[f^2(\theta) + f'^2(\theta) + [1 + f'^2(\theta)].g'^2(\theta) \right] +$$
$$g'(\theta).[1 + f'^2(\theta)].[f(\theta).f'(\theta) +$$
$$f'(\theta)f''(\theta).[1 + g'^2(\theta)] + [1 + f'^2(\theta)].g'(\theta).g''(\theta)] = 0$$

(9)

IV. THE MULTI-DISTRIBUTED VALUE PROBLEM (MDVP)

By using the assumptions given in the beginning of the section 3 and by considering Figure 5, the distribution of the magnetic field is Centrosymmetric from the axe y.

Then it is only necessary to compute the magnetic field in the following cylindrical coordinates, r, θ, y, (see Fig.1.), from $r = \dfrac{-L}{2}$ to $r = 0$ the other part of this magnetic field will be given by using the previous centrosymmetric properties.

Then in cartesian coordinates x, y, z given by Fig.1, we have: if $y(\rho, \theta)$ is the magnetic field function:

$$y(\frac{-L}{2}, 0) = 0 \tag{10}$$

$$y(\frac{-\ell}{2}, \theta_{max}) = e \tag{11}$$

$$\frac{dy(\frac{-L}{2}, 0)}{dx} = +\infty \tag{12}$$

$$\frac{dy(0, \theta_{max})}{dx} = 0 \tag{13}$$

571

see figure 5.

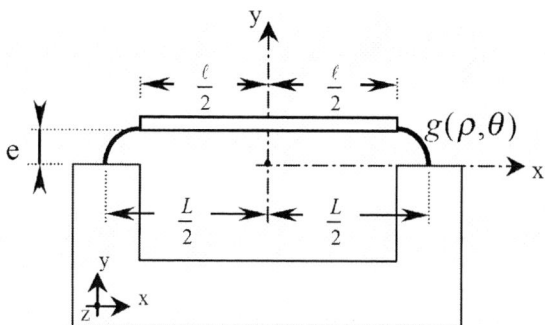

Fig. 5. XOY plane

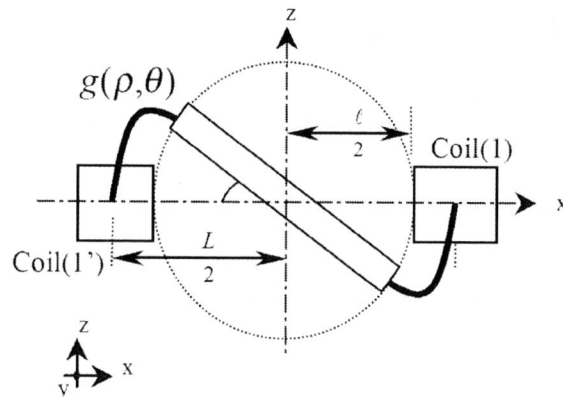

Fig. 6. XOZ plane

In the zox plane, see Fig. 6, we have,

$$z(\frac{-L}{2},0) = 0 \qquad (14)$$

$$z(\frac{-\ell}{2},\theta_{max}) = \frac{-\ell}{2}.\sin(\theta_{max}) \qquad (15)$$

then geometric considerations lead to the following multidistributed conditions in the algebraic space of representation:

Thus from Fig.5, we have, from (10),

$$g(\frac{-L}{2},0) = 0 \qquad (16)$$

from (11),

$$g(\frac{-\ell}{2},\theta_{max}) = e \qquad (17)$$

from (12),

$$\frac{dy}{dx} = \frac{dy/d\theta}{dx/d\theta} = \frac{\dfrac{\partial g(f(\theta),\theta)}{\partial\theta} + \dfrac{\partial g(f(\theta),\theta)}{\partial f(\theta)}.\dfrac{\partial f(\theta)}{\partial\theta}}{-f(\theta).\sin(\theta)} = +\infty$$

$$(18)$$

from (13),

$$\frac{dy(0,\theta_{max})}{dx} = 0 \qquad (19)$$

From Fig.6 by considering equation (3), condition (14) leads to a trivial condition, same for condition (15).

We then have only four MDVP given by (16), (17), (18) and (19), and we can remark that no relevant informations are given by Fig.6.

V. C^k SPLINE FUNCTIONS AND COUPLED ALGEBRAIC DIFFERENTIAL EQUATIONS (CADE)

C^k spline functional expansions have the remarkable property that for a considered function, the coefficients of its C^k spline functional expansion are only the set of the all derivatives (partial or total) up to k at each point of discretization on the open set Ω . This means that we can include in $H_k(\Omega)$, the Sobolev space on Ω , generated by C^k spline functions, the set of the all differential constraints, value conditions and boundary conditions as simple exact algebraic relations.

This implies that for a given process the algebraic inclusion of the various differential invariants leads to rebuilds the functional space $H_k(\Omega)$ in a well defined appropriate functional space (or manifold), specific to the considered process.

In this sens, Sobolev spaces generated by C^k spline functions can be compared for nonlinear differential equations, to Fourier space in which all linear differential equations are represented as algebraic relations of frequencies.

Moreover, algebraic properties of C^k spline functions allow us to introduce elegant and efficient algorithm formulations of the classical nonlinear differential problems.

Let us consider the classical linear state equation

$$\dot{q} = A.q + B.u \qquad (20)$$

where q is the state vector of dimension N , u is the control vector of dimension m , A is the dynamical real matrix of dimension $N \times N$, B is the control real matrix of dimension $N \times m$. The C^k spline development of equation (20) can be written as follows:

$$q(t) = \sum_{i=0}^{I}\sum_{v=0}^{k}\left[[A]^v.q_i^0 + \sum_{j=0}^{v-1}[A]^j.B.u_i^{(v-1-j)}\right] S_i^{k,v}(t)$$

$$(21)$$

where q_i^0 and $u_i^{(v-1-j)}$ are respectively 0th and $(v-1-j)^{th}$ time derivative of $q(t)$ and $u(t)$ at the discretization point i, $i \in \Omega \equiv [0,I]$.

The nonlinear explicit state equation can be defined as follows,

$$\dot{q} = F(q,u) \qquad (22)$$

where q and u are respectively the state vector and the control vector mentioned above and $F(q,u)$ is a nonlinear analytical function of dimension N.

By introducing $\tilde{F}$ the vector field associated to $F(q,u)$ and defined by,

$$\tilde{F} = \sum_{j=1}^{N} F^j(q,u).\frac{\partial}{\partial q_j} + \sum_{\ell=1}^{m}\frac{du}{dt}.\frac{\partial}{\partial u_\ell}$$

$$(23)$$

where $F^j(q,u)$, q_j, u_ℓ are respectively the jth and ℓ^{th} component of respectively the nonlinear analytical vector function $F(q,u)$, $q(t)$, $u(t)$.

The C^k spline development of equation (22) can be written as follows:

$$q(t) = \sum_{i=0}^{I} \sum_{v=0}^{k} \left. \widetilde{F}^{(v)}(q,u) \right|_i .S_i^{k,v}(t) \qquad (24)$$

where $\left. \tilde{F}^{(v)}(q,u) \right|_i$ is the vth iteration of the vector field defined by equation (23) at the point of discretization i.

At last the nonlinear implicit state equation can be defined as follows:

$$F(\dot{q},q,u) = 0 \qquad (25)$$

Then $q(t)$ can be written in its C^k spline functional expansion as:

$$q(t) = \sum_{i=0}^{I} \sum_{v=0}^{k} \left. \frac{d^v}{dt^v} F(\dot{q},q,u) \right|_i .S_i^{k,v}(t) \qquad (26)$$

where $\left. \dfrac{d^v}{dt^v} F(\dot{q},q,u) \right|_i$ is the vth time derivative of the implicit nonlinear state equation at the point of discretization i.

Clearly CADE's belong to this last case, and can be treated, using property (26) by classical minimization of the canonical functional distance associated to $H_k(\Omega)$.

VI. SIMULATION

We have then to solve $F_1(\theta) = 0$ and $F_2(\theta) = 0$ as defined by equations (9). The minimization of the canonical functional distance associated to $H_k(\Omega)$ (or it's associated reduced manyfold) leads to a large algebraic nonlinear functional of the successive derivatives of f and g versus θ at each point of discretization i, which can be solved under an appropriate initial vectorial function by simulated annealing methods [10][11] or Computer Algebra[6].

Annealing methods find numerically the global maximum of these kind of functional on $H_k(\Omega)$ but can failed for various reasons, including for example, inappropriate initial vector, and inappropriate tuning convergence parameters or simply by round off error.

Gröbner basis find by computer algebra, the exact set of solution of functional in $H_k(\Omega)$ (global and local). The global solution will then be find by introducing each local solutions on the functional of $H_k(\Omega)$. This last method is on implementation in our research group but requires large computation power.

This is why we have chosen to solve this problem by simulated annealing method with appropriate initial functions. These two functions are obtained as two polynomials of θ of the lowest order which minimize the energy and satisfy the MDVP (see paragraph 4).

For reasons of conciseness, results of these computations will be given in the conference.

VII. CONCLUSION

We have presented in this paper an analytical model of loading characteristic of MSAPFA motors with a large air gap. This model leads to a Coupled Algebraic Differential Equations (CADE) which can be handled by automatic control community in order to build appropriate non linear controllers. In simulations, we have shown that CADE can be simply written in the C^k spline functions formalism, in appropriate manifold resulting from the inclusions in $H_k(\Omega)$ of the various invariants, in order to give efficient and powerful algorithms.

In the first approach we have chosen to minimize the functional in $H_k(\Omega)$ simulated annealing methods which requires a good knowledge of the initial vector in $H_k(\Omega)$. The first simulations are in accordance with experimental results.

REFERENCES

[1] M. Rouff, The computation of C^k spline functions, *Computers Math. Applic. Vol. 23, N° 1, pp 103-110, 1992.*

[2] M. Rouff and W.-L. Zhang, C^k spline functions and linear operators, *Computers Math. Applic. Vol. 28, N° 4, pp 51-59, 1994.*

[3] M. Rouff, Y. Slamani , "The computation of C^k spline functions spectra: definitions and first properties", International Journal of Pure and Applied Mathematics. *Vol.8,N°3,pp.307-333,2003.*

[4] M. Rouff, and M. Alaoui, Computation of dynamical electromagnetic problems using Lagrangian formalism and multidimensional Ck spline functions", Zeitschrift für Angewandte Mathematik und Mechanik, Academie Verlag Berlin, *Vol. 76 No.1, 1996, pp. 513-14.*

[5] Fitzgerald, A.E., Kingsley, C., and Umans. S. D. Electrical Machinery. McGraw Hill, 4th Edition, 1983.

[6] J.C Faugère, P. Gianni, D. Lazard, and T. Morat, Efficient Computation of Zero-Dimensional Gröbner Basis by Change of Ordering. Journal of Symbolic Computation *Vol. 16, N° 4 (October 1993), 329-344.*

[7] M. Rouff , Z. Lakhdari and J.M Dequen, Computation of eddy current losses induces by magnetic domain walls motion with C^k spline functions, European Symposium on Numerical Methods in Electromagnetics , 6-8, March 2002 N°8211; Toulouse; France.

[8] J., Azzouzi, G. Barakat and B. Dakyo, B. Quasi-3-D analytical modeling of the magnetic field of an axial flux permanent-magnet synchronous machine, IEEE Transaction on Energy Conversion, Volume 20, Issue 4, Dec. 2005 Page(s): 746 – 752.

[9] Parviainen, A., Niemelä, M., Pyrhönen, J., "Modeling Axial-flux Permanent-Magnet Machines". IEEE Transaction on Industry Applications. Vol. 40, No. 5, 2004, pp. 1333-1340.

[10] M. Poloujadoff; J.C. Mipo and P. Siarry. "Designing p/2p windings by the simulated annealing method [induction machines], IEEE International Electric Machines and Drives Conference Record, 1997, Volume, Issue , 18-21 May 1997 Page(s):TA1/5.1 - TA1/5.3.

[11] S. Subramanian; R. Bhuvaneswari "Comparison of Modern Optimization Techniques with Applications to Single-Phase Induction Motor Design", Electric Power Components and Systems, Volume 34, Issue 5 May 2006 , pages 497 – 507

The Bearingless 2-Level Motor

P. Karutz*, T. Nussbaumer**, W. Gruber*** and J.W. Kolar*

* ETH Zurich, Power Electronic Systems Laboratory, 8092 Zurich, Switzerland, karutz@lem.ee.ethz.ch
** Levitronix GmbH, Technoparkstrasse 1, 8005 Zurich, Switzerland
*** Johannes Kepler University Linz, Altenbergerstr. 69, 4040 Linz, Austria

Abstract– Several processes in chemical, pharmaceutical, biotechnology and semiconductor industry require contactless levitation and rotation through a hermetically closed chamber wall. This paper presents a novel concept that combines crucial advantages such as high acceleration capability, large air gap and a compact motor setup. The basic idea is to separate a homopolar bearing unit axially from a multipolar drive unit on two different height levels. Hence, the proposed concept is denominated as "Bearingless 2-Level Motor". In this paper, the bearing and drive functionalities are explained in detail and design guidelines are given based on analytic equations and electromagnetic 3D simulations. Furthermore, the influence of non-idealities such as saturation and coupling effects are evaluated and included in the design. Finally, measurements on an experimental prototype exemplify the design considerations and prove the excellent performance of the new concept.

I. INTRODUCTION

In the past decades there have been a lot of research activities in the field of bearingless motor drives [1] - [3]. The implementation of bearingless motor technology includes key features such as contactless operation, online tuneable bearing parameters, almost unlimited life time, wearless and lubrication-free operation and therefore a high level of purity. In the pharmaceutical, chemical, biochemical and semiconductor industry several processes require the application of chemical substances on rotating objects under clean room conditions [4]. Here, bearingless motors are of high interest for these applications. The advantage of the bearingless

motor technology in these sensitive processes is its ability to spin a rotor in an encapsulated chamber, where the demand for high purity is satisfied and locally limited clean room space can be provided while avoiding failure susceptible seals.

Fig. 1 demonstrates schematically such a process, showing a levitated rotor carrying a process object. The process is enclosed by a chamber and the rotor is levitated and accelerated through the process chamber walls by the aid of electromagnetic bearings and drives, respectively. Basically, there are several requirements for these applications:

- A big air gap is required in order to ensure a minimum thickness and therefore mechanical robustness of the process chamber that is placed within the air gap.

- A compact motor setup is desirable due to the constantly increasing costs of clean room space.

- A high acceleration capability is needed in order to minimize the times between the process rotation speeds. This is directly influencing the efficiency and therefore the operating costs of the equipment.

- A maximum rotation speed required by the process has to be reached.

- A high temperature resistance is needed, including thermal expansion issues.

- A highly chemical resistant hardware setup avoids that the various strongly reactive chemicals degenerate the motor components.

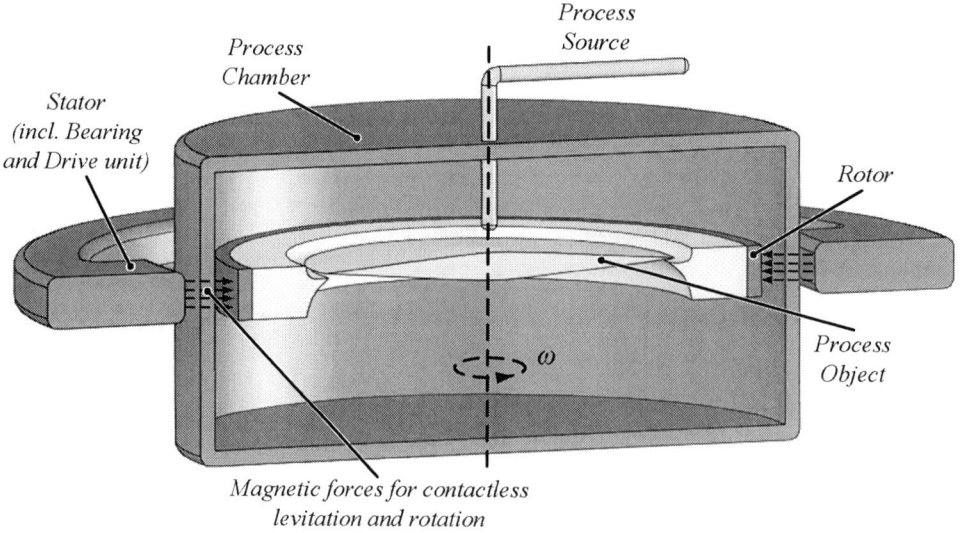

Fig. 1: Schematic cut view of an industry spinning process that is hermetically sealed within a process chamber, using magnetic bearing technology for the levitation of the rotor.

- A stable, vibration-free levitation and rotation has to be ensured within the whole operating range, i.e. potential axial, radial and tilting resonances must not effectuate significant axial and radial displacements of the rotor.

Obviously, not all requirements can be fulfilled simultaneously, since they are partially conflicting. However, in the past several concepts have been developed that showed good performance in one or more of the before-mentioned aspects.

In [5], a setup for a magnetically levitated pump system for use in semiconductor, chemical and pharmaceutical industry has been introduced. It incorporates a combined iron path for the drive and the bearing windings. Here, a high number of stator claws levitates and drives a single permanent magnet impeller, where its radial position is controlled actively and the axial position as well as the tilting around the radial axes is controlled passively. Due to the nature of the concept a very high number of stator claws would be necessary to levitate rotors with large diameters (i.e. number of pole pairs). Therefore, this concept has been adapted in pump applications with a pole pair number of one and impeller diameters smaller than 100 mm.

Another concept has been presented in [6]. The concept features the utilization of the permanent magnet field of the rotor on different height levels for the drive and the bearing. The rotor can be built in a very compact way; however, due to the operation principle this concept uses only the stray flux components for the driving of the rotor, which results in a relatively low motor torque and a poor acceleration performance.

The bearingless segment motor with a combined bearing and drive has been presented in [7] and shows very good acceleration behaviour in combination with a compact setup. However, due to the coupled windings for bearing and drive the control of the motor gets very complicated. Furthermore, the cogging torque is quite

critical for the segment motor. Additionally, the concept demands for a higher number of sensors and for increased power electronics effort.

In this paper, a new "Bearingless 2-Level Motor" is proposed, combining the advantages of the before-mentioned concepts. In the following, the concept will be referenced as B2M. The principle of the B2M is explained in more detail in section II. In section III, the functionality of the magnetic bearing is introduced and analytical descriptions of stability are given. This is followed by a design procedure of the permanent magnet synchronous drive presented in section IV. Finally, the outstanding performance of the B2M is proven by measurements on a laboratory prototype in section V.

II. PRINCIPLE OF THE 2-LEVEL MOTOR CONCEPT

The B2M concept introduced in this paper is based on the principle that the bearing and drive forces are applied on two different height levels (cf. Fig. 2). This enables a drive structure with significantly increased torque as compared to the concept presented in [6]. In comparison to the motors with integrated drive/ bearing functionality [7] the proposed 2-level concept shows a greatly reduced control effort and the advantage of separately optimised drive and bearing system.

A schematic cut view is depicted in Fig. 2. At the upper level, the magnetic bearing is located, consisting of rotor and stator permanent magnets (in order to provide a magnetic biasing) and the bearing windings around the four stator claws. The permanent magnet synchronous motor drive is positioned at an axially lower level. The rotor magnets are round-shaped and diametrically magnetized with alternately reversed polarisation direction. Additionally, the drive claws and windings are located between the bearing claws on the stator, wherefore a more compact setup can be achieved. Position and angular sensors are distributed around the stator for the detection of the radial position and the rotation speed.

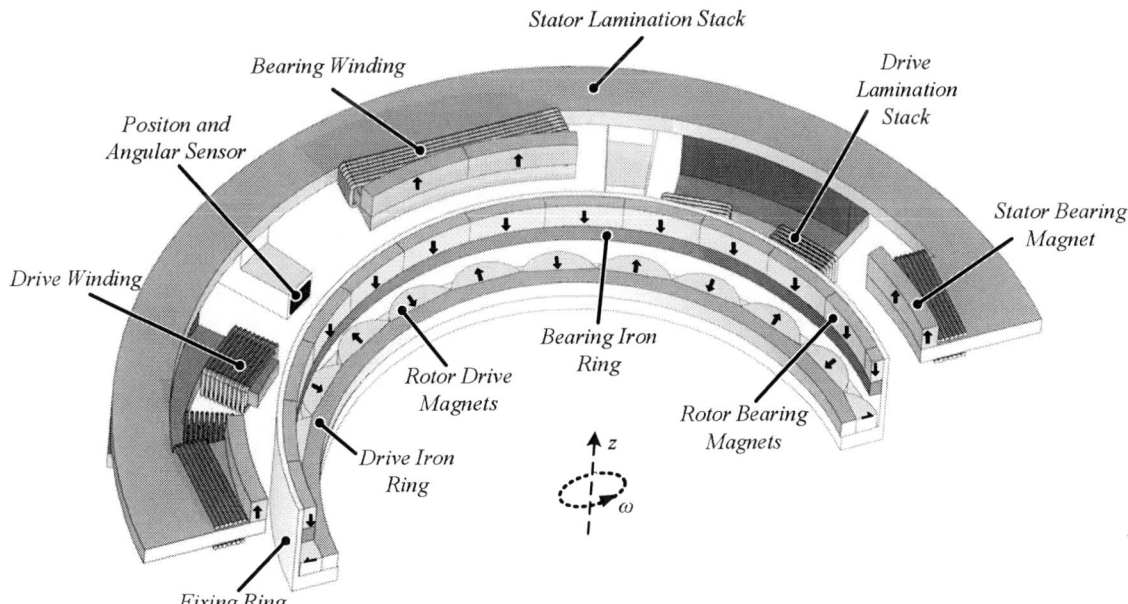

Fig. 2: Schematic cut view of the Bearingless 2-Level Motor including bearing and drive units on stator and rotor side.

III. HOMOPOLAR MAGNETIC BEARING

A. Design

In Fig. 3 the main components of the active radial bearing are depicted. The axial position and the tilting of the rotor is passively stabilized through reluctance forces in the air gap. The axial stiffness $k_{Z,B}$ [N/mm] describes the force [N] that is needed in order to move the rotor 1 mm out of its

stable position. The positive axial stabilization causes a negative destabilization in radial direction [8]-[10]. Hereby, the radial stiffness $k_{R,B}$ [N/mm] specifies the required force [N] needed to return the rotor back to its stable position after being displaced by 1 mm. Permanent magnets on both the stator and the rotor bearing iron ring are used for flux biasing and to define the flux path through the air gap. The flux density can be altered depending on the rotor position by supplying the bearing windings, thereby generating Maxwell-forces towards the target position [11]. The force-current factor $k_{I,B}$ [N/(A·turns)] describes the force that can be generated per ampere-turn of the bearing winding.

Generally, a high axial stiffness $k_{Z,B}$ is desired in order to counteract the weight force

$$k_{Z,B} \cdot \Delta z = m \cdot g \tag{1}$$

resulting in a minimum axial stiffness

$$k_{Z,B} > \frac{m \cdot g}{\Delta z_{\max}}, \tag{2}$$

where Δz_{max} is the maximum allowable displacement in the axial direction, m is the mass of the rotor and g is the gravitational constant.

However, as mentioned before, a high axial stiffness comes along with a destabilizing radial stiffness that has to be overcome by the stabilizing active magnetic force imposed by the bearing currents. Therefore, for allowing a maximum radial deflection Δr_{max} from the stable position, the force-current factor $k_{I,B}$ has to be larger than a minimum value given by

$$k_{I,B} > \frac{k_{R,B} \cdot \Delta r_{\max}}{N_B \cdot I_B}, \tag{3}$$

where N_B is the bearing coil winding number and I_B the bearing controller current. Here, it has to be considered that the force-displacement dependency is non-linear; therefore, evaluating (3) with a linear radial stiffness $k_{R,B}$ is only valid within a limited operating range.

In order to facilitate the fulfilment of (3), N_B has to be chosen as high as possible. However, a high number of bearing turns decreases the current rise capability in the bearing inductance L_B. The electrical time constant of the bearing is given by

$$\tau_E = \frac{L_B}{R_B}, \tag{4}$$

where R_B is the winding resistance. Since L_B scales with $N_B{}^2$ and R_B with N_B, the electrical time constant increases linearly with N_B. For achieving a stable system control the condition

$$\tau_E \ll \tau_M, \tag{5}$$

with

$$\tau_M = \sqrt{\frac{m}{k_{R,B}}} \tag{6}$$

has to be satisfied, i.e. a small number of bearing windings is desirable from this point of view. Therefore the selection of N_B will always be a trade-off between high dynamics (cf. Eqs. (4) and (5)) and the maximum force condition (cf. Eq. (3)).

B. Interference with the drive system

Due to the axial and circumferential separation of the drive and bearing system the mutual coupling effects can be assumed to be low, wherefore general design considerations can be carried out separately. However, interactions between the bearing and the drive unit can cause tilting problems, which have to be considered and are addressed here now shortly.

In addition to the bearing's radial stiffness $k_{R,B}$, the diametrically magnetized permanent magnets of the drive on the rotor cause a magnetic force towards the stator leading to an additional destabilizing radial stiffness $k_{R,D}$ that has to be considered additionally in the bearing design. Besides the linear deflection along the axial or radial axis the tilting tendency of the bearing has to be investigated. Hereby, the destabilizing radial force of the drive $F_{R,D}$ causes a torque $M_{R,D}$ around the mass balance point with the height h of the rotor being the lever. At the same time the rotor is stabilized through the stabilizing bearing axial force $F_{Z,B}$ (provided by two of the four bearing windings) acting with a lever with the length of the effective radius r, causing a torque $M_{Z,B}$. The tilting tendency k_{Tilt} can therefore be described as the ratio of the two torques by

$$k_{Tilt} = \frac{M_{R,D}}{M_{Z,B}} = \frac{F_{R,D} \cdot h}{2 \cdot F_{Z,B} \cdot r} = \frac{k_{R,D} \cdot h^2 \alpha}{2 \cdot \frac{k_{Z,B}}{4} \cdot r^2 \alpha} = \frac{2 \cdot k_{R,D} \cdot h^2}{k_{Z,B} \cdot r^2}. \tag{7}$$

In order to guarantee a stable operation the condition $k_{Tilt} \ll 1$ has to be ensured.

An additional destabilizing torque on the drive level may be caused by the superimposed electromagnetic forces resulting from the drive winding currents. However, as will be shown in following section, for the pre-

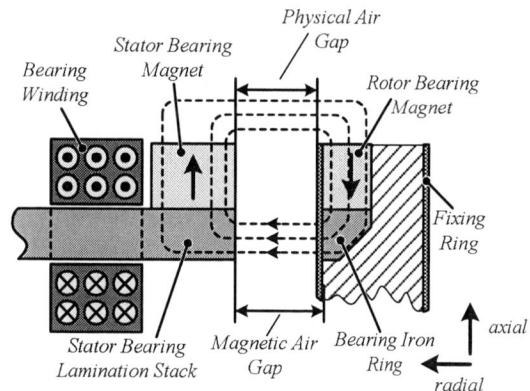

Fig. 3: Bearing principle of the Bearingless 2-Level Motor.

sented B2M drive the opening angle of the drive lamination stack φ_D is selected equal to the angle of 180° electrical. With this, there will always be the same amount of attracting and repellent radial forces caused by the drive ampere-turns for any rotor position. Therefore, no resulting radial force acting on the rotor is caused by the drive current, which is why it is not considered in (7).

IV. PERMANENT MAGNET SYNCHRONOUS DRIVE

A. Basics

The main components of the permanent magnet synchronous drive [13] are depicted schematically in Fig. 4. The flux path is defined by the stator pole shoes with the two contrarily wound and series connected drive windings, the air gap and the round-shaped and alternating, diametrically magnetized magnets located on the rotor drive magnet ring. A motor torque M_D is generated if the drive windings are supplied with a rotation speed synchronized sinusoidal current, resulting in a tangential force F_T. Since permanent magnets always attract iron independently of their magnetization direction, the drive has priority positions defined by the constructive design (φ_D, w_{Claw}, d_{Claw}) the stator drive claw with respect to the rotor magnet dimensions and strength as

well as the size of the air gap. This behaviour causes a cogging torque $M_{Cogging}$, which can lead to jerky rotation especially in the low rotational speed range and has to be prevented by an optimisation of the stator claw design (see section IV.B).

The main drive parameters introduced depend mainly on the flux density distribution in the air gap. Since this distribution is highly non-linear and it is inexpedient to describe them analytically the subsequent design considerations are carried out using the 3D finite element simulation tool Maxwell® 3D [12].

B. Design

The mayor degrees of freedom for the permanent magnet synchronous drive design are the shape of the stator claws, especially the drive stator claw width w_{Claw}, the number of turns N_D of the drive coils and the

thickness d_{IR} of the drive iron ring. By optimising these parameters the design aim of minimum cogging torque $M_{Cogging}$, acceptable radial stiffness $k_{R,D}$ and maximum motor torque M_D can be reached.

A first great reduction of the cogging torque can be achieved if the drive phases are circularly shifted by 90° electrically in order to avoid having two maximum field densities forcing the rotor into a preferred position. A further optimisation has to be carried out by simulations.

For simplifying the design simple U-shaped drive elements as depicted in Fig. 4 are considered here. As a detailed analysis shows, this shape does not lead to minimum cogging torque, however, to an optimal utilization of the available space for the drive. Since the cogging torque does not reach critical values due to the before-mentioned 90°-shifting of the phases, this shape is considered here. Furthermore, the opening angle φ_D is set to $\varphi_D = 180°$ el.. This maximizes the achievable torque and leaves w_{Claw} as the only dimensional optimization parameter.

Fig. 5 shows the simulation results of $M_{Cogging}$, M_D and $k_{R,D}$ for different stator claw widths w_{Claw}. As can be seen there, a minimum cogging torque is reached for $w_{Claw} = 20$ mm at a high motor torque. Furthermore, Fig. 5 shows the linear dependency of the negative radial stiffness on the stator claw widths, which justifies the selection of $w_{Claw} = 20$ mm rather than any wider stator claw.

Another design parameter is the thickness d_{IR} of the drive iron ring, which constitutes the feedback path for the drive flux Φ_D. If d_{IR} is selected very small, saturation effects in the drive iron ring will occur and will degrade the flux density in the air gap and consequently the induced voltage and the drive torque. The critical thickness for d_{IR} is given at the connection point between two rotor magnets, since the maximum drive flux has to pass through there (cf. Fig. 4). The resulting flux density defining the saturation in the iron ring is composed of two flux components, where the major component is the permanent flux density by the drive permanent magnets and the minor component is the flux imposed by the drive winding currents. The distance between two stator

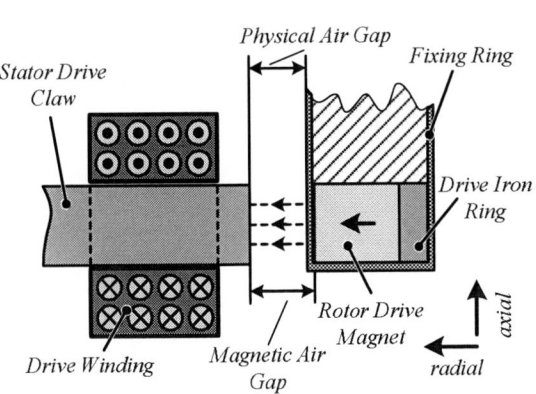
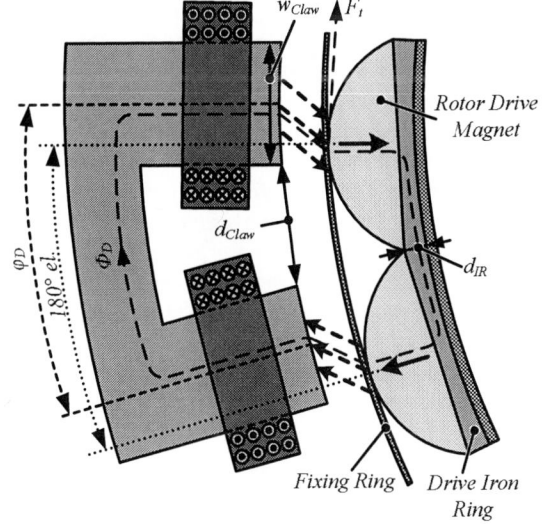

Fig. 4: Principle of the drive of the Bearingless 2-Level Motor with side view (left) and top view (right).

claws d_{Claw} is much smaller than the magnetic air gap δ_{mag} plus the length of the permanent magnets l_{PM} that add on to the flux path due to there air like permeability ($\mu_{mag} \approx 1$, cf. Fig. 6). Therefore, the current generated flux will mainly pass the through the space between the stator claws and hardly enter the rotor iron ring. Hence, the flux density in the iron ring will be clearly dominated by the permanent magnets. Thus, only this portion will be considered for the following design guidelines.

An integration of the approximately sinusoidal flux density distribution along the back side of half a drive magnet (cf. Fig.6) gives the total flux that passes from one magnet to the neighboured one through the iron ring

$$\Phi_D = \int B \cdot dA_{PM} = \int_0^{r\pi/2p} B_{PM} \cdot \sin\left(\frac{bp}{r}\right) \cdot h \cdot db = \frac{B_{PM} \cdot r \cdot h_{IR}}{p}, \quad (8)$$

In order not to saturate the iron material ($B < B_{Sat,Fe}$) at the position $\alpha = 0°$ the cross-section A_Φ has to fulfil the condition

$$A_\Phi \geq \frac{\Phi}{B_{Sat,Fe}}, \quad (9)$$

which gives with $A_\Phi = h_{IR} \cdot d_{IR}$ (cf. Fig. 6) the minimum thickness for the drive iron ring

$$d_{IR,min} = \frac{B_{PM} \cdot r}{B_{Sat} \cdot p}. \quad (10)$$

The exact value of the flux density B_{PM} can be ascertained only by electromagnetic simulations. However, analytical approximations can already give a rough guideline. The maximum value of B_{PM}, which represents the worst-case condition for the saturation in the iron, occurs, when the air gap between rotor and stator becomes minimal, i.e. when the rotor magnets lie exactly in front of the drive claws as shown in Fig. 6. In this position, the flux density can be estimated (with $\mu_R \to \infty$) by

$$B_{PM,max} \approx B_R \cdot \frac{l_{PM}}{l_{PM} + \delta_{mag}}, \quad (11)$$

where B_R is the remanence flux density of the permanent magnet, l_{PM} is the length of a permanent magnet, and δ_{mag} is the magnetic air gap (including the thickness of the sensor ring). Since in reality not all lines of the magnetic flux will follow the shortest way (and some not even enter the stator claw) and thus the average air gap will be larger than δ_{mag}, (11) represents a worst-case approximation. As a detailed analysis shows, at that considered maximum point the impressed force by the drive windings is zero, therefore the before-mentioned negligence of that influence has been correct.

Hence, (10) and (11) provide a guideline for the required iron thickness $d_{IR,min}$ in dependency of the pole pair number p and the radius r. However, selecting d_{IR} smaller than $d_{IR,min}$ relates to a weight reduction and can probably lead to an increased acceleration performance of the motor even though the air gap flux density is reduced.

Besides the discussed constructional parameters the winding number of the drive coils greatly influences the

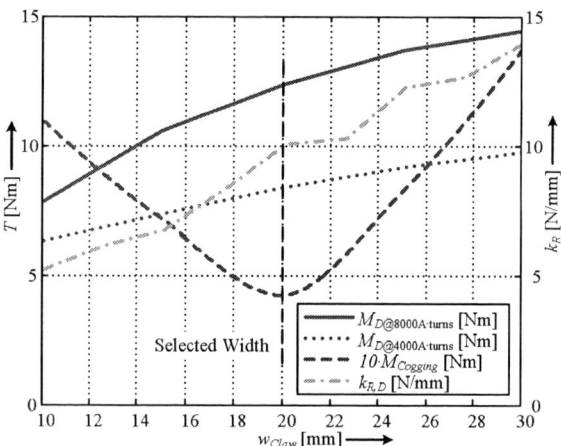

Fig. 5: Results of 3D finite element simulations for cogging torque $M_{Cogging}$, radial stiffness $k_{R,D}$ and motor drive torque M_D for two different ampere-turn ratios (per drive claw) in dependency on the stator claw width w_{Claw}.

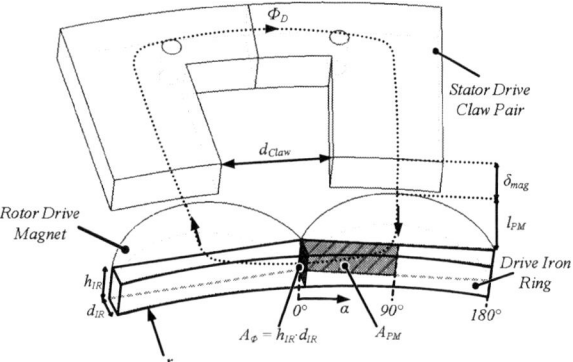

Fig. 6: Schematic view of a drive pole pair and the iron ring in front of a stator claw pair with flux cross section areas indicated.

acceleration behaviour of the B2M. The maximum applicable drive current is given by

$$I_D = \frac{-U_{ind} \cdot R_C \pm \sqrt{(R_C^2 + \omega^2 \cdot L_C^2) U_{DC}^2 - \omega^2 \cdot L_C^2 \cdot U_{ind}^2}}{R_C^2 + \omega^2 \cdot L_C^2}, \quad (12)$$

where U_{ind} is the rotation speed dependent induced voltage, R_C is the coil winding resistance, L_C the coil winding inductance, and $\omega = 2\pi n_R$ the electrical angular frequency.

As can be seen in Fig. 7 (a), for low rotational speeds the drive current is limited by the maximum current $I_{PE,max}$ provided by the power electronics, while for higher rotation speeds the current is decreasing due to the growing impedance $\omega \cdot L_C$ and due to the induced voltage which is increasing linearly with ω (cf. Fig. 7 (a)). Both $U_{ind} \sim N_D$ and $L_C \sim N_D^2$ are depending on the number of coil turns N_D, wherefore the available drive current is decreasing with increasing turns number (cf. Fig. 7 (a)). On the other hand, the drive power is given by the product of the induced voltage and the drive current

$$P_{Drive} = U_{ind}(N_D) \cdot I_{Drive}(N_D). \quad (13)$$

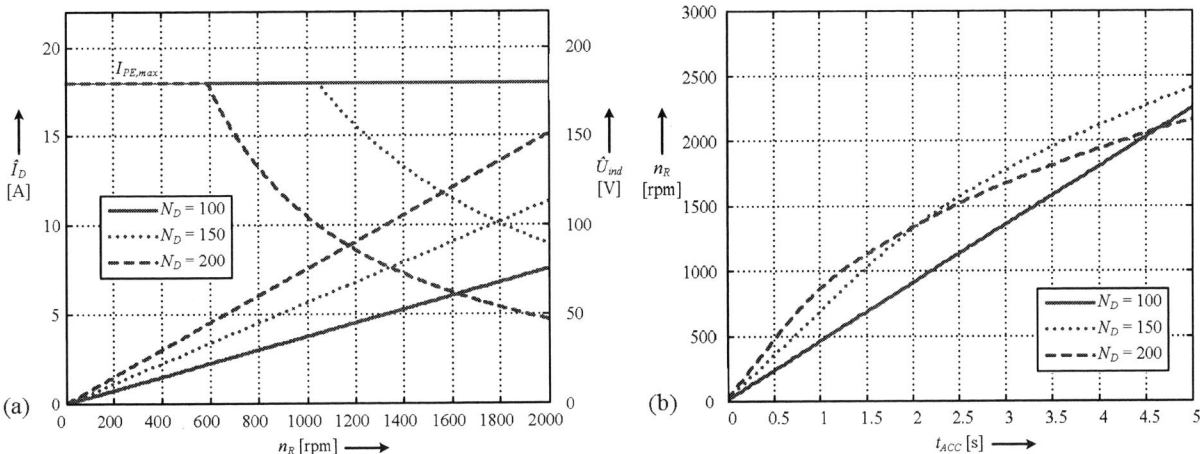

Fig. 7: (a) Achievable drive current $\hat{I}_D$ (for $I_{PE, max}$ = 18 A) and induced voltage $\hat{U}_{ind}$ in dependency of the rotation speed n_R for different coil winding numbers N_D and (b) estimated acceleration performance of the B2M.

Thus, an optimum number of turns can be identified for a certain rotation speed region. This is shown in Fig. 7 (b), where the acceleration times for different rotor speeds and winding numbers are plotted. It shows that a target rotational speed of 2000 rpm can be reached within 3.6 s for an optimum number of turns of N_D = 150. It has to be mentioned that this calculation is only correct for non-saturated material both in the stator claws and in the iron ring as discussed before.

V. EXPERIMENTAL PERFORMANCE

Based on the design guidelines that have been presented in the previous sections, a prototype has been built in order to verify the design considerations. In Table 1 the characteristic parameters of the chosen design are compiled. Fig. 8 shows the complete assembly including the rotor and the stator-sided bearing and drive system. In Fig. 9 a better insight is provided into the constructional details of the setup. One can see the shape of the stator lamination stack and the 24 round-shaped permanent magnets (p = 12) placed between the drive positioning ring and the iron ring. The measured values of the axial and radial stiffness and the force-current factor are compared in Fig. 10 (a)-(c) with the

simulated values and show generally a good agreement. For the axial stiffness (cf. Fig. 10(a)) the assumption of a linear factor $k_{Z,B}$ is correct in a wide area and the measured axial stiffness shows only a slight general deviation from the value predicted by the simulations. In contrast, the radial stiffness (cf. Fig. 10(b)), shows in reality a stronger nonlinear behaviour and therefore a bigger deviation from the linearized value in of the simulations. For the force-current factor, a perfect agreement between measurement and simulations can be seen in Fig. 10(c).

Finally, Fig. 11 shows the acceleration performance of the B2M drive from 0 rpm to 2000 rpm for $\hat{I}_{PE,max}$ = 18 A. For the run-up sequence, the final speed of 2000 rpm can be reached within 3.8 s, which is close to the value predicted by the simulations (cf. section IV.B), while the deceleration is accomplished within 2.8 s. This performance is very satisfactory considering the motor dimensions and the large air gap. With this, on the one hand the design procedure and correctness of the simulations could be verified, and on the other hand the excellent performance of the B2M concept could be proved.

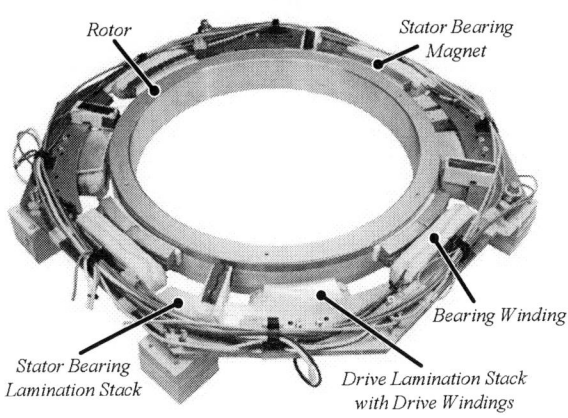

Fig. 8: Photography of completely assembled laboratory proto type with bearing and drive windings.

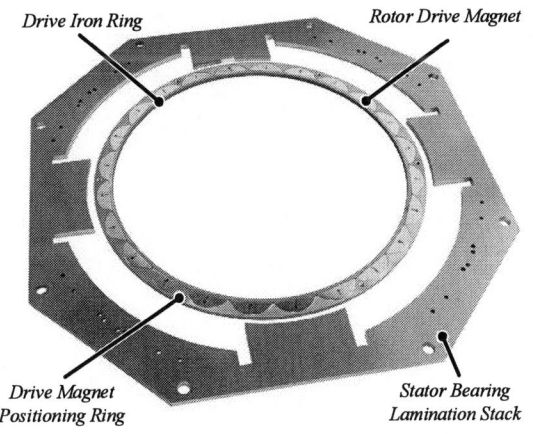

Fig. 9: Photography of rotor inside showing round-shaped drive magnets being placed between the drive iron ring and the drive magnet positioning ring.

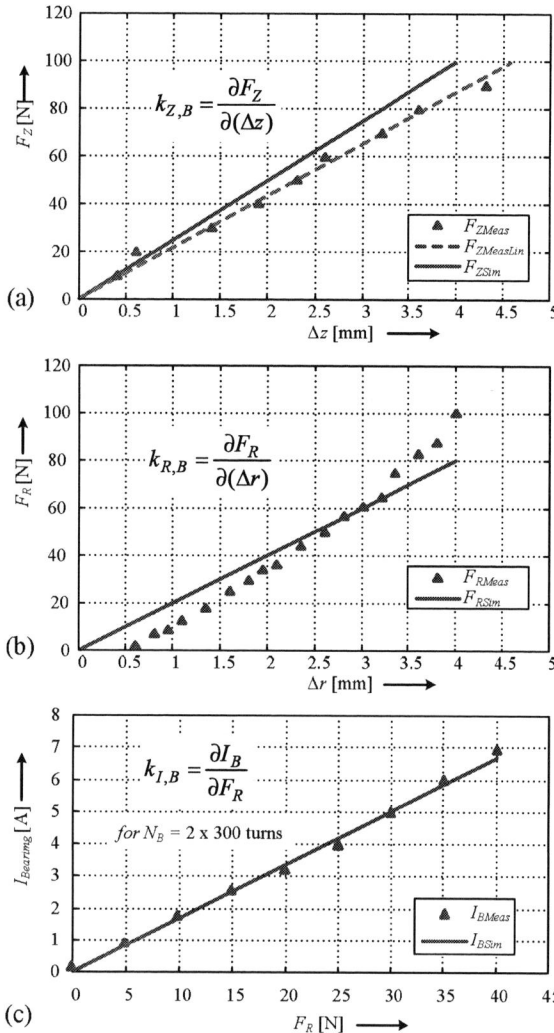

Fig. 10: (a) Measured and simulated axial stiffness; (b) measured and linearized simulated radial stiffness; and (c) measured and simulated force-current factor.

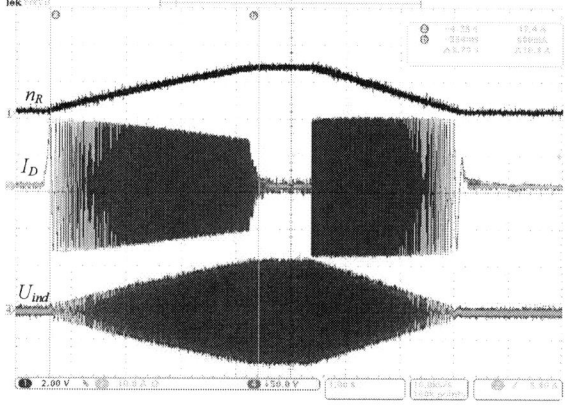

Fig. 11: Acceleration performance of B2M from 0 to 2000 rpm for $I_{PE,max} = 18A$ in 3.8s and deceleration in 2.8s (scales: 1600 rpm/div., 10 A/div., 50 V/div., 1 s/div.). For the measurement of the induced voltage a separate measurement coil with the winding number of a half phase (2 x 150 turns) has been used.

TABLE 1: DESIGN DATA OF THE EXPERIMENTAL SETUP

Outside rotor diameter	410 mm
Mechanical air gap δ_{Mech}	7 mm
Number of pole pairs p	12
Axial stiffness $k_{Z,B}$	25 N/mm
Radial stiffness $k_{R,B}$	-20 N/mm
Force-Current factor $k_{I,B}$	1 N/(100 A·turns)
Tilting stiffness $k_{\varphi,B}$	1 N/°
Motor Moment M_D for $I_D = 1A$	0.7 Nm
Cogging Torque $M_{Cogging}$	0.45 Nm
Bearing phase winding number N_B	2 x 300 turns
Drive phase winding number N_D	4 x 150 turns
Rotor mass m	5 kg

VI. CONCLUSIONS

The paper describes a new concept called "Bearingless 2-Level Motor" (B2M) that is of high interest for several industry branches, where contactless levitation and rotation in clean room environments is required. The new concept features high acceleration capability, a compact setup, low power electronics effort and a separate design and simple control of the bearing and drive units even for large air gaps. In this paper, the functionality of the B2M concept has been explained and guidelines for the design of the drive and bearing unit have been presented, also taking saturation and coupling effects between the drive and bearing system into account. Finally, the theoretical considerations have been verified on a prototype setup by measurements of design parameters and achievable acceleration times.

REFERENCES

[1] R. Schoeb, N. Barletta, "Principle and Application of a Bearingless Slice Motor," *JSME Int. Journal Series C*, pp. 593-598, 1997.

[2] A. Chiba, D.T. Power, M.A Rahman, "Chracteristics of a Bearingless Induction Motor," *IEEE Trans. Magnetics*, vol. 27, no. 6, Nov. 1991.

[3] S. Silber, W. Amrhein, P. Boesch, R. Schoeb, N. Barletta, "Design aspects of bearingless slice motors," *IEEE/ASME Trans. Mechatronics*, vol. 10, no.6, pp. 611-617, Dec.2005.

[4] Y. Chisti, M. Moo-Young, "Clean-in-place systems for industrial bioreactors: Design, validation and operation", *Journal of Industrial Microbiology and Biotechnology*, 1994.

[5] N. Barletta, R. Schöb, "Design of a Bearingless Blood Pump," *3rd Int. Symp. on Magnetic Suspsension Technology*, Tallahassee, 1995.

[6] T. Schneeberger, J. W. Kolar, "Novel Integrated Bearingless Hollow-Shaft Drive," *Proc. of the IEEE Ind. Applic. Conf. IAS*, Tampa (USA), 8 – 12 October 2006.

[7] W. Gruber, W. Amrhein, "Design of a Bearingless Segment Motor," *Proc. of the 10th Int. Symp. on Magnetic Bearings* Martigny, 2006.

[8] J. Delamare, E. Rulliere, J.P. Yonnet, "Classification and synthesis of permanent magnet bearing configurations," *IEEE Trans. Magnetics*, vol.31, no.6, pp 4190-4192, Nov. 1995.

[9] J.-P. Yonnet, "Permanent magnet bearings and couplings", *IEEE Trans. Magnetics*, vol.17, no.1, pp. 1169- 1173, Jan. 1981.

[10] S. Earnshaw, "On the nature of the molecular forces which regulate the constitution of the luminiferous ether," *Trans. Camb. Phil. SOC.*, vol. 7, no. 1, pp. 97-112, 1839.

[11] W. Amrhein, S. Silber, K. Nenninger, "Levitation forces in bearingless permanent magnet motors", *IEEE Trans. Magnetics*, vol.35, no. 5, pp. 4052-4054, Sep. 1999.

[12] Maxwell® 3D by Ansoft Corporation, http://www.ansoft.com.

[13] D.P.M. Cahill, B. Adkins, "The permanent magnet synchronous motor," *Proc. Inst. Elec. Eng.*, vol. 109, no. 48, pp. 483-491, Dec. 1962.

Analysis and Design of a Sliding Mode Controller for Buck Converters Operating in DCM with Adaptive Hysteresis Band Control Scheme

Hung-Chih Lin and Tsin-Yuan Chang
Department of Electrical Engineering
National Tsing Hua University, Hsin-Chu, 30013, TAIWAN
E-mail: hclin@larc.ee.nthu.edu.tw; tyc@ee.nthu.edu.tw

Abstract—This paper presents the analysis and design of a sliding mode controller for buck converters operating in discontinuous conduction mode. An adaptive hysteresis band control scheme is proposed to tightly regulate the output voltage under line and load variations. The proposed method can be employed for other dc-dc converters, and therefore the operation range of sliding mode controllers for dc-dc converters can be extended from continuous conduction mode to discontinuous conduction mode. Simulation results are provided to verify the proposed technique.

Index Terms—Adaptive hysteresis band control, buck converter, sliding mode control, discontinuous conduction mode.

I. INTRODUCTION

Thanks to the advance in power electronics technologies and the demand for high energy efficiency, the use of dc-dc converters has been growing rapidly in various fields. The dc-dc converters conventionally employ classical controllers (P, PI, or PID) based on linearized small signal models in order to stabilize the system and regulate the output voltage [1]. However, these controllers often fail to perform satisfactorily under parameter variations or load disturbances [2]. To maintain good performance under parameter and load variations, sliding mode (SM) control was adopted for dc-dc converters [3-5].

However, all the reported SM controlled converters so far are designed only to operate in continuous conduction mode (CCM) due to difficulty in constructing the sliding surface for the control design in discontinuous conduction mode (DCM) operation [6]. When operating in DCM, these SM controlled converters produce output voltage error. This restricts the operation range of the SM controlled converters. For high efficiency over a wide range, power converters have to be able to operate in CCM at heavy load condition and in DCM at light load condition. Moreover, DCM operation is frequently encountered, since converters are usually required to operate with their loads removed (at no load condition) [7]. Indeed, some converters are purposely designed to operate in DCM for all loads.

Hence, this paper proposes the analysis and design of a SM controller for buck converters operating in DCM with

This work was supported by National Science Council under project number 96-2220-E-007-027.

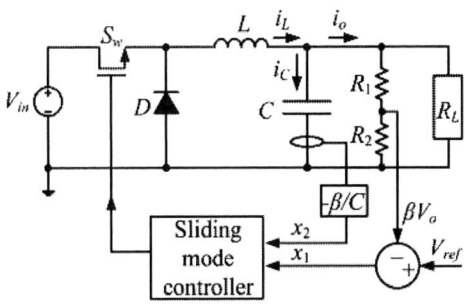

Fig. 1. A conventional SM controlled buck converter.

adaptive hysteresis band control scheme. The proposed controller can be employed for other dc-dc converters. Therefore, the operation range of SM controllers for dc-dc converters can be extended from CCM to DCM. This paper is organized as follows. In Section II, a brief review of conventional SM controllers for buck converters operating in CCM is introduced. The phase trajectories in phase plane are also analyzed in both CCM and DCM operation. Section III discusses the details of operation principle of the proposed SM controller. The mathematical model, stability concern, and adaptive hysteresis band control technique are presented. Simulation results are shown in Section IV. Finally, the conclusion is given in Section V.

II. REVIEW OF CONVENTIONAL SM CONTROLLERS FOR BUCK CONVERTERS OPERATING IN CCM

Fig. 1 shows a conventional SM controlled buck converter, where the control parameters are the output voltage error x_1 and the output voltage error dynamics x_2. The theoretical model and analysis are summarized below. More details can be found in [5].

A. Converter's Model

In CCM operation, x_1 and x_2 can be expressed as

$$x_1 = V_{ref} - \beta V_o \qquad (1)$$

$$x_2 = \dot{x}_1 = -\beta \frac{dV_o}{dt} = -\frac{\beta}{C} i_C = \frac{\beta}{C}\left(\frac{V_o}{R_L} - \int \frac{uV_{in} - V_o}{L} dt\right) \qquad (2)$$

where V_{ref}, βV_o, V_{in}, are the reference, sensed output and input voltage, respectively, C, L, R_L are the capacitance, inductance, load resistance, respectively, i_L, i_C, and i_o are the inductor, capacitor, and output current, respectively.

978-1-4244-0644-9/07/$25.00 ©2007 IEEE

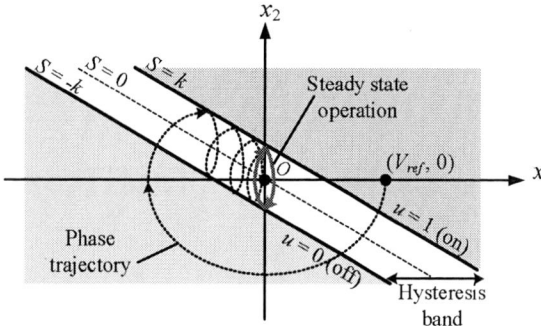

Fig. 2. Phase trajectory of the conventional SM controller with hysteresis band control in CCM operation.

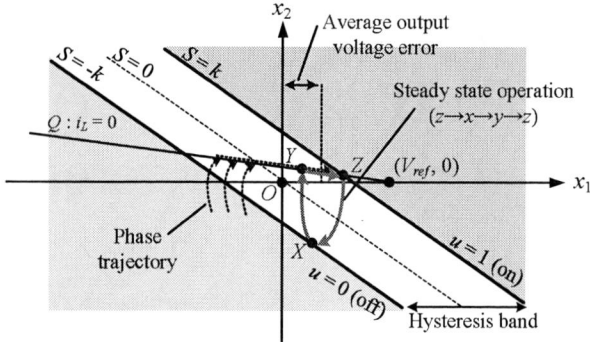

Fig. 3. Phase trajectory of the conventional SM controller with hysteresis band control in DCM operation.

$u = 1$ (on) or 0 (off) is the switching state of power switching device S_w. Then, the state space model under CCM can be obtained by differentiating (1) and (2) with respect to time

$$\begin{bmatrix} \dot{x}_1 \\ \dot{x}_2 \end{bmatrix} = \begin{bmatrix} 0 & 1 \\ -\frac{1}{LC} & -\frac{1}{R_LC} \end{bmatrix} \begin{bmatrix} x_1 \\ x_2 \end{bmatrix} + \begin{bmatrix} 0 \\ -\frac{\beta V_{in}}{LC} \end{bmatrix} u + \begin{bmatrix} 0 \\ \frac{V_{ref}}{LC} \end{bmatrix} \quad (3)$$

B. Controller Design

In SM control, the switching function is employed to decide the input states u to the system

$$S = \alpha x_1 + x_2 \quad (4)$$

where α is the sliding coefficient to be designed. For stability concern, the value of α should be greater than zero [5]. The basic control law is expressed as

$$u = \begin{cases} 1 & when & S > 0 \\ 0 & when & S < 0 \end{cases} \quad (5)$$

The basic control law in (5) determines the switching state u and directs the phase trajectory toward the sliding surface ($S = 0$) in phase plane. However, to maintain the phase trajectory on the sliding surface and force it toward the origin, the Lyapunov's second method must be obeyed [5]

$$\lim_{S \to 0} S \cdot \dot{S} < 0 \quad (6)$$

C. Hysteresis Band Control

The hysteresis band is commonly employed to limit the switching frequency to an acceptable range and alleviate the chattering effect when a converter enters the SM operation [5]. The control law with hysteresis band is

$$u = \begin{cases} 1 & when & S > k \\ 0 & when & S < -k \\ unchnaged & otherwise \end{cases} \quad (7)$$

where k is the hysteresis bandwidth. Fig. 2 details the phase trajectory in phase plane when operating in SM with hysteresis band control. The phase trajectory is directed by the hysteresis band $S = k$ and $S = -k$. The phase trajectory slides from starting position (V_{ref}, 0)

toward the origin O and settles around the origin O in steady state operation.

D. Problem Definition

To understand and analyze the operation of the conventional SM controlled converter in DCM, the phase trajectory in phase plane must be known firstly. Substituting (1) and (2) into (3) with $u = 0$ and $i_L = 0$, x_2 can be derived as

$$\dot{x}_2 = \frac{\beta}{R_LC} \cdot \frac{d}{dt}\left(\frac{V_{ref} - x_1}{\beta}\right) \quad (8)$$

Then integrating both sides of (8) leads to

$$x_2 = \frac{1}{R_LC}V_{ref} - \frac{1}{R_LC}x_1 \quad (9)$$

Being a straight line (line Q in Fig. 3) of $i_L = 0$ with a slope of $-1/(R_LC)$ in phase plane, equation (9) is very useful to understand the phase trajectories in DCM operation, since the inductor current goes to zero in every switching cycle.

As shown in Fig. 3, the phase trajectory is operated in SM control under DCM with hysteresis band $S = k$ and $S = -k$. Different to that in Fig. 2, the phase trajectory now follows line Q when the inductor current goes to zero. Note that any phase trajectory hitting line Q will continuously follow line Q until reaching the upper bound of hysteresis band $S = k$ (point Z in Fig. 3). Once the phase trajectory hits $S = k$ at point Z, the switch is turned on, and the phase trajectory moves down until reaching point X in the lower bound of hysteresis band $S = -k$. Then, the switch S_w is turned off, and the phase trajectory moves up until hitting line Q at point Y ($i_L = 0$). After that, the phase trajectory follows line Q until hitting $S = k$ at point Z. The above steps from points $Z \to X \to Y \to Z$ are repeatedly operated in steady state. Thus, the phase trajectory moves around the right side of the origin O in steady state operation, and the average value of control parameter x_1 (output voltage error) is nonzero.

From the above illustration, it can be concluded that the conventional SM controller with hysteresis band still stabilizes the system in DCM operation. But it can not ensure that the phase trajectory settles around the origin O, and therefore the output voltage error is produced.

582

III. PROPOSED SM CONTROLLER WITH ADAPTIVE HYSTERESIS BAND CONTROL SCHEME

To deal with the problem discussed in previous section, this paper proposes an SM controller for buck converters operating in DCM with adaptive hysteresis band control scheme. The theoretical model, analysis, and adaptive hysteresis band technique are discussed below.

A. Converter's Model in DCM

Fig. 4 shows the inductor current waveform during a switching period T_s in DCM operation. It has three characteristic periods: D_1T_s, D_2T_s, and D_3T_s which denote the switch S_w is active, non-active, and $i_L = 0$, respectively. Two switching state variables are therefore required at least to describe three switching states. By using two switching variables u and d, which are defined in Fig. 4, the SM control parameter x_2 in (2) is redefined as

$$x_2 = \dot{x}_1 = \frac{\beta}{C}\left[\frac{V_o}{R_L} - (1-d)\left(\int \frac{uV_{in} - V_o}{L}dt\right)\right] \quad (10)$$

Then, the state space model describing the buck converter in DCM can be obtained by differentiating (1) and (10) with respect to time

$$\begin{bmatrix} \dot{x}_1 \\ \dot{x}_2 \end{bmatrix} = \begin{bmatrix} 0 & 1 \\ -\frac{1}{LC} & -\frac{1}{R_LC} \end{bmatrix}\begin{bmatrix} x_1 \\ x_2 \end{bmatrix} + \begin{bmatrix} 0 & 0 \\ -\frac{\beta V_{in}}{LC} & -\frac{\beta V_o}{LC} \end{bmatrix}\begin{bmatrix} u \\ d \end{bmatrix} + \begin{bmatrix} 0 \\ \frac{V_{ref}}{LC} \end{bmatrix} \quad (11)$$

Although the above model shows up two switching variables u and d, there is only one independent control input u. Another input variable d is functions of u, the circuit elements, and the switching period T_s.

B. Design of an SM Controller in DCM

The basic control law of the proposed SM controller is identical to (5). To maintain the trajectory on the sliding surface in DCM operation, (6) still must be obeyed. Substituting (4), (5), and (11) into (6), the inequalities become

$$\lambda_1 = \left(\alpha - \frac{1}{R_LC}\right)x_2 - \frac{1}{LC}x_1 + \frac{V_{ref} - \beta V_i}{LC} < 0 \quad (12)$$

$$\lambda_2 = \left(\alpha - \frac{1}{R_LC}\right)x_2 - \frac{1}{LC}x_1 + \frac{V_{ref}}{LC} > 0 \quad (13)$$

where

$$\lambda_1 = \dot{S}_{u=1,d=0} \quad for \quad 0 < S < \xi \quad (14)$$

$$\lambda_2 = \dot{S}_{u=0,d=0} \quad for \quad -\xi < S < 0 \quad (15)$$

and ξ is an arbitrarily small positive value. When inductor current goes to zero, the inequality (13) becomes

$$\lambda_3 = \left(\alpha - \frac{1}{R_LC}\right)x_2 > 0 \quad (16)$$

where

$$\lambda_3 = \dot{S}_{u=0,d=1} \quad for \quad -\xi < S < 0 \quad (17)$$

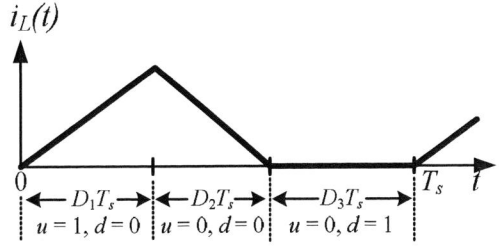

Fig. 4. The inductor current waveform during a switching period T_s in DCM operation.

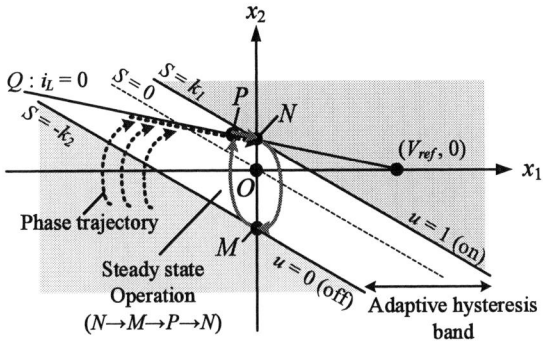

Fig. 5. Phase trajectory of the proposed SM controller with adaptive hysteresis band control in DCM operation.

The above inequalities represent the region of existence of the SM controller under DCM in phase plane. Note that the value of α must be greater than $1/(R_LC)$ in order to obey the inequality (16), since $x_2 = \beta V_o/(R_LC) > 0$ when inductor current goes to zero. If $\alpha < 1/(R_LC)$, there is no existence region in phase plane under DCM.

C. Proposed SM Controller with Adaptive Hysteresis Band Control Scheme in DCM

Fig. 5 shows the operation principle of the proposed SM controller with adaptive hysteresis band control scheme. The control law in (7) is redefined as

$$u = \begin{cases} 1 & when & S > k_1 \\ 0 & when & S < -k_2 \\ unchnaged & otherwise \end{cases} \quad (18)$$

where k_1 and k_2 are the proposed adaptive hysteresis bandwidth.

The phase trajectory now is directed by the adaptive hysteresis band $S = k_1$ and $S = -k_2$. To well control the phase trajectory for settling around the origin O in steady state operation and to regulate the output voltage, the upper bound of the adaptive hysteresis band $S = k_1$ is designed for passing point N which is the intersection of axis x_2 and line Q as shown in Fig. 5. Thus, when the inductor current goes to zero, any phase trajectory hitting line Q will continuously follow line Q until reaching the upper bound of the adaptive hysteresis band $S = k_1$ at point N as shown in Fig. 5.

Once the phase trajectory hits $S = k_1$ at point N, the switch is turned on, and the phase trajectory moves down until reaching point M in the lower bound of the adaptive hysteresis band $S = -k_2$. Then, S_w is turned off, and the

583

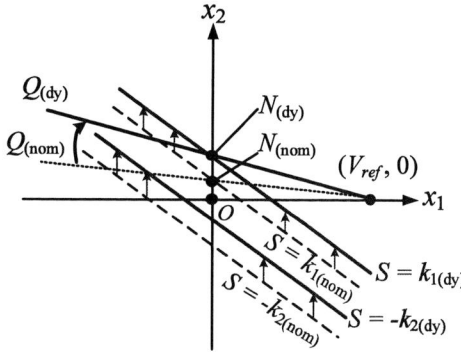

Fig. 6. Phase plane plots of the proposed adaptive hysteresis band control scheme for over-loaded operation.

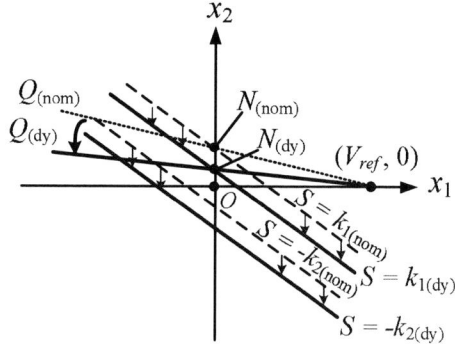

Fig. 7. Phase plane plots of the proposed adaptive hysteresis band control scheme for under-loaded operation.

Phase trajectory moves up until hitting line Q at point P ($i_L = 0$). After that, the phase trajectory follows line Q until hitting the upper limit $S = k_1$ at point N. The above steps from points $N \to M \to P \to N$ are repeatedly operated in steady state.

Hence, the output voltage error $x_1 (= V_{ref} - \beta V_o)$ is zero at point N in the end of every switching cycle. In this way, the output voltage is well regulated (due to $V_{ref} = \beta V_o$ at point N in every cycle), and any perturbation in V_o can be corrected within one cycle. The proposed controller therefore attains fast transient response and good load regulation. Furthermore, the proposed method still maintains the advantages of SM control.

Fig. 6 and Fig. 7 show the details about the proposed adaptive hysteresis band control scheme for over-loaded and under-loaded operations, respectively. $S = k_{1(nom)}$ ($S = k_{1(dy)}$) and $S = -k_{2(nom)}$ ($S = -k_{2(dy)}$) are the upper and lower bounds of the proposed adaptive hysteresis band for nominal (dynamic) load operation, respectively. $S = k_{1(dy)}$ and $S = -k_{2(dy)}$ are vertically shifted from $S = k_{1(nom)}$ and $S = -k_{2(nom)}$ according to load condition. For over-loaded condition in Fig. 6 (the dynamic load $R_{L(dy)}$ < the nominal load $R_{L(nom)}$), the slope of $Q_{(dy)}$ is steeper than that of $Q_{(nom)}$ and point $N_{(dy)}$ is above point $N_{(nom)}$. The hysteresis band $S = k_{1(dy)}$ and $S = -k_{2(dy)}$ are therefore vertically shifted up to ensure that $S = k_{1(dy)}$ passes point $N_{(dy)}$. For under-loaded condition in Fig. 7 ($R_{L(dy)} > R_{L(nom)}$), the slope of $Q_{(dy)}$ is smoother than that of $Q_{(nom)}$ and point $N_{(dy)}$ is under point $N_{(nom)}$. The hysteresis band $S = k_{1(dy)}$ and $S = -k_{2(dy)}$ are therefore vertically shifted down to ensure that $S = k_{1(dy)}$ passes point $N_{(dy)}$.

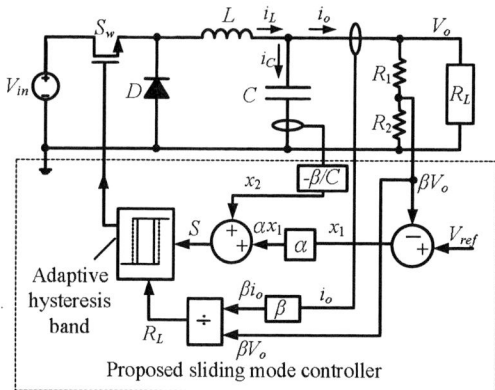

Fig. 8. Proposed SM controlled buck converter operating in DCM with adaptive hysteresis band control scheme.

D. System Implementation

Fig. 8 shows the overall schematic of the proposed SM controller. The proposed adaptive hysteresis band control scheme is adopted to operate the SM controlled in DCM. The proposed technique needs the information of instantaneous R_L that can be achieved by monitoring the output current i_o and output voltage V_o as shown in Fig. 8. Note that the value of α must be set greater than $1/(R_L C)$ as in previous discussions.

IV. SIMULATION RESULTS

As shown in Fig. 8, the proposed SM controller with adaptive hysteresis band control scheme is simulated in Matlab/Simulink and operated in DCM. The nominal system parameters are given as follows.

● Input voltage V_{in} = 24 V, desired output voltage V_{od} = 12 V.

● Capacitance C = 100 μF, inductance L = 30 μH, load resistance R_L = 36 Ω.

● Desired switching frequency f_d = 100 kHz.

Fig. 9 shows the output voltage and inductor current waveforms in steady state operation under nominal condition. In Fig. 9(a), the output voltage is well regulated. In Fig. 9(b), the output voltage is equal to the desired output voltage in the end of every switching cycle. At no load condition, the output voltage is still regulated tightly as shown in Fig. 10(a). In Fig. 10(b), the proposed SM controller shows fast transient response and good load regulation in a step load change from R_L = 72 Ω to R_L = 24 Ω. Fig. 11(a) and Fig. 11(b) show the average output voltage for different load resistance 22 Ω $\leq R_L \leq$ 80 Ω and different input voltage 13 V $\leq V_{in} \leq$ 48V, respectively. It can be concluded that the output voltage regulation is robust to load variation with only a 12 mV (0.1 %) deviation and line variation with only a 28 mV (0.23 %) deviation. Fig. 12(a) and Fig. 12(b) show the variations of the average switching frequency for different load resistance 22 Ω $\leq R_L \leq$ 80 Ω and input voltage 13 V $\leq V_{in} \leq$ 48 V, respectively.

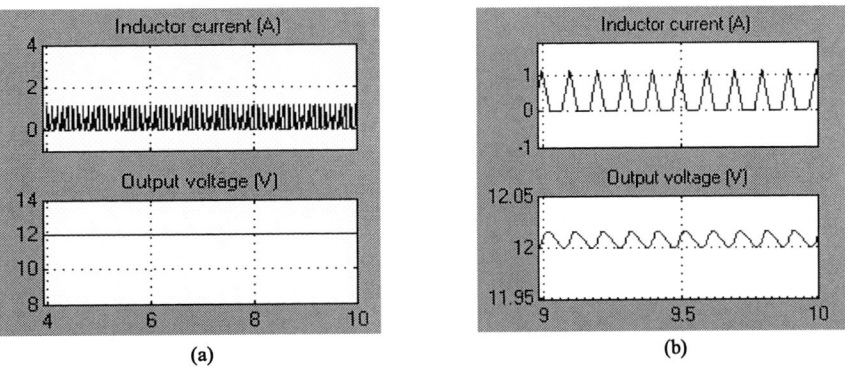

Fig. 9. (a) The output voltage and inductor current waveforms and (b) the output voltage and inductor current ripples in steady state operation.

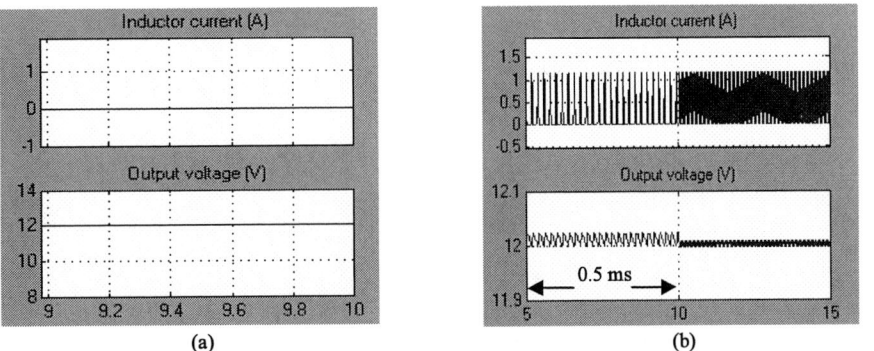

Fig.10. The output voltage and inductor current waveforms (a) at no load condition (b) at a step load change from R_L = 72 Ω to R_L = 24 Ω.

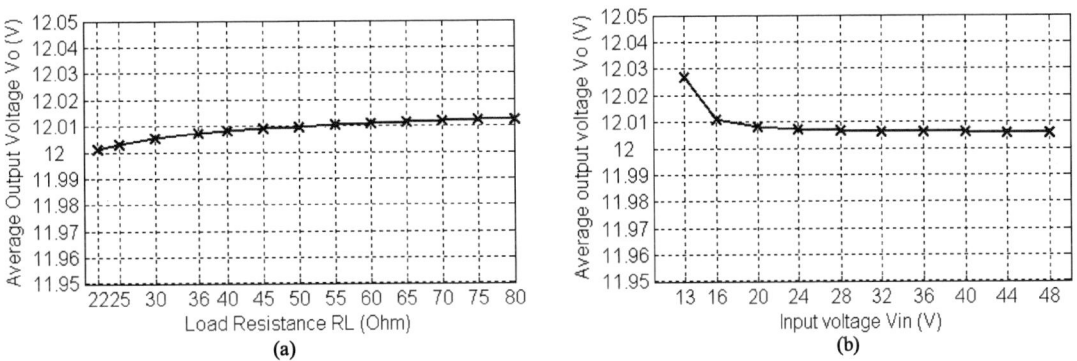

Fig. 11. The average output voltage for (a) different load resistance R_L (b) different input voltage V_{in}.

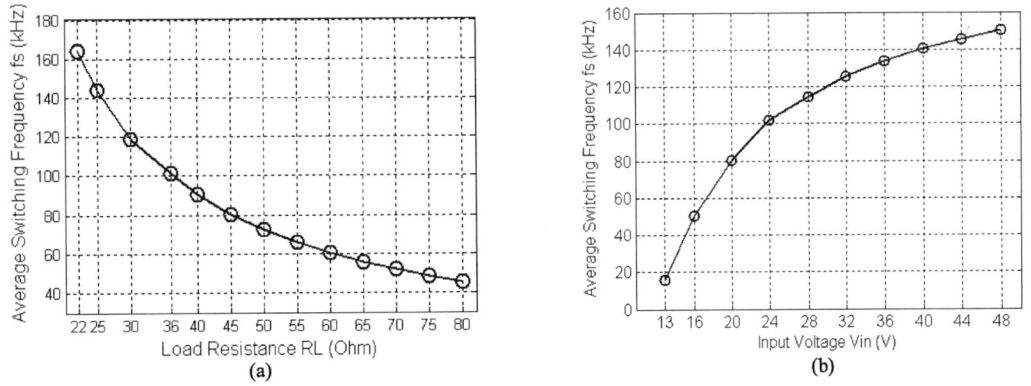

Fig. 12. The average switching frequency for (a) different load resistance R_L (b) different input voltage V_{in}.

585

V. Conclusions

A detailed analysis and design of an SM controller for buck converters operating in DCM is presented in this paper. Through the proposed SM controller with adaptive hysteresis band control technique, the converter provides fast dynamic response and good load regulation under load and line variations. The proposed controller provides a solution to extend the operation range of SM controllers for dc-dc converters from CCM to DCM.

References

[1] D. M. Mitchell, *DC-DC Switching Regulator Analysis*. NewYork: Mc-Graw-Hill, 1998.

[2] V. S. C. Raviraj and P. C. Sen, "Comparative study of proportional-integral, sliding mode, and fuzzy logic controllers for power converters," *IEEE Trans. Ind. Applicat.*, vol. 33, no. 2, Mar./Apr. 1997, pp. 518–524.

[3] R. Venkataramanan, A. Sabanoivc, and S. Cuk, "Sliding mode control of dc-to-dc converters," in *Proc. IEEE Conf. Ind. Electron. Control Instrumentations (IECON)*, 1985, pp. 251–258.

[4] V. M. Nguyen and C. Q. Lee, "Indirect implementations of sliding-mode control law in buck-type converters," in *Proc. IEEE Applied Power Electron. Conf. Expo (APEC)*, vol. 1, Mar. 1996, pp. 111–115.

[5] S. C. Tan, Y. M. Lai, M. K. H. Cheung, and C. K. Tse, "On the practical design of a sliding mode voltage controlled buck converter," *IEEE Trans. Power Electron.*, vol. 20, no. 2, Mar. 2005, pp. 425–437.

[6] Sreekumar C and V. Agarwal, "Hybrid Control of a Boost Converter Operating in Discontinuous Current Mode" in *Proc. IEEE Power Electron. Specialists Conference (PESC)*, Jun. 2006, pp. 1-6.

[7] W. Erickson and D. Maksimovic, *Fundamentals of Power Electronics*. Norwell, MA: Kluwer, 2001.

Buck Converter Simulation Technique Based on the Fourier Transform

Acácio M. R. Amaral (*) (**), A. J. Marques Cardoso (**)

* Polytechnic Institute of Coimbra, ISEC - DEIS, Rua Pedro Nunes – Quinta da Nora, P – 3030 199, Coimbra, Portugal
** University of Coimbra, FCTUC/IT, DEEC, Pólo II – Pinhal de Marrocos, P – 3030-290, Coimbra, Portugal

Abstract — **The aim of this paper is to present some tools that could simplify the design process of step down DC-DC converters. For that, a simple simulation technique based on the Fourier transform, as well as, an experimental technique that can be used to determine the equivalent circuit of the output filter, will be presented.**

To validate both simulation and experimental technique some experimental and simulated results will be presented.

Index Terms — **Modeling, simulation, design and step-down DC-DC converters.**

I. INTRODUCTION

In the design of switch mode power supplies, *smps*, the use of simulation tools as well as the knowledge of the equivalent circuit of the components used is essential to obtain the best design proposal.

Different simulation techniques based on the average behavior of *smps*, in low frequency range, have been proposed [1-5]. Their main objective was to study the control performance, and not the effect of parasitic elements of inductors and capacitors in high frequency response of the converter. Some integration techniques, based on the theoretical analysis of the converter [6], could be used to study the high frequency behavior of the converter. However, the influence of parasitic elements as function of frequency couldn't be analyzed. The use of Fourier Transform allows the computation of the high frequency response of *smps* without the use of switching models, and simultaneously, permits the study of the influence of parasitic elements with frequency.

In order to obtain the best simulation results, the equivalent circuit of the elements used in power stage design should be obtained at the operating conditions of the converter. For that, the technique proposed in [7] will be applied for capacitors, and a new technique based in [7] will be implemented for inductors.

II. THE MODEL OF STORAGE ELEMENTS

In order to reduce the size and height of *smps*, it was necessary to increase their operating frequencies, which introduce new problems related with the non-ideal behavior of the output filter at very high frequencies [8].

A. Inductors equivalent circuit

In Fig. 1 it is possible to observe the equivalent circuit of an inductor. The parasitic capacitances, C, are the result of the turn-to-turn capacitances, as well as of the turn-to-core capacitance. The parasitic resistances, R_T, are the sum of winding resistance, R_w, with the core resistance, R_C, which represents the core losses, and L represents the inductance value [9].

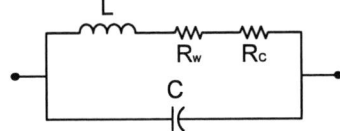

Fig. 1. Equivalent circuit of an inductor [9].

From the model presented in Fig. 1, it is possible to compute the equivalent impedance of the inductor as:

$$Z_{eq} = \frac{R_T + j\,w\,L\left(\left(1 - w^2\,L\,C\right) - \frac{R_T^2 C}{L}\right)}{\left(R_T\,w\,C\right)^2 + \left(1 - w^2\,L\,C\right)^2} \quad (1)$$

Since the resonance frequency of the inductors used in the design of DC-DC converters is much higher than its operating frequency, the parasitic capacitance can be neglected, and its equivalent circuit can be simplified as a resistor in series with an inductor.

B. Capacitors equivalent circuit

The behavior of a real capacitor can be very different according to its operating conditions. A capacitor is compounded by wires and an insulation material. Since wires have a resistance and an inductance, and insulators have a leakage resistance, the behavior of capacitors with frequency diverges from its ideal model. Fig. 2 shows the equivalent circuit of an aluminum electrolytic capacitor.

Fig. 2. Simplified equivalent circuit of an aluminum electrolytic capacitor [10].

The impedance of an aluminum electrolytic capacitor can be expressed as:

$$Z_C = ESR + jX_{cap} = ESR + j\,w\,ESL - \frac{j}{w\,C} \quad (2)$$

where C is the capacitance, ESR is the equivalent series resistance (which represents the wire resistance as well as leakage resistance) and ESL is the equivalent series inductance (wire inductance).

The majority of aluminum electrolytic capacitors manufacturers give the capacitance and the dissipation factor, DF, values. The DF indicates the quality of the capacitor, and represents the ratio of resistance to reactance.

$$DF = \frac{ESR}{X_C} \tag{3}$$

Using equation (3), it is possible to obtain the ESR intrinsic value of the capacitor. However, the DF value changes with frequency, as well as the capacitance and ESR values. Since the DF value is usually given at 120 Hz, it is only possible to compute the ESR intrinsic value using information given by manufactures at 120 Hz.

In this way, for simulation purpose of switch mode DC-DC converters, the equivalent model of inductors could be obtained using the information given by manufactures. Nevertheless, it is advised to compute the inductance value at the operating frequency of the converter, since most manufacturers give this value with a 10% of tolerance. For that, a new technique will be presented afterwards. In regard to capacitors, it is very important to determine the ESR value at the operating frequency of the converter, as well as its reactance value. Thus, the technique purposed in [7] was used.

III. EXPERIMENTAL LCR METER

The output filter of a switch mode DC-DC converter is compound by non-ideal elements like capacitors and inductors, which models change with frequency. Besides, the values given by the manufacturers present a tolerance. For this propose, two very simple experimental techniques were implemented in order to obtain the equivalent circuit of inductors and capacitors at the operating frequency of the converter.

The equivalent circuit of the output filter of a buck converter can be simplified as the one presented in Fig. 3, if medium and low capacitance aluminum electrolytic capacitors are considered.

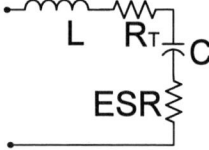

Fig. 3. Equivalent circuit of the output filter of a Buck converter.

The main portion of the impedance of those capacitors, at the operating frequency of the converter, is given by its ESR intrinsic value. To compute the ESR value the technique proposed [7] was used.

The technique proposed in [7] allows the determination of both gain and phase impedance values of the capacitor at the range of frequencies from 100 Hz to 100 kHz. To do so, it is necessary to submit a RC circuit (a resistor, R, in series with the capacitor to study) to a sinusoidal waveform with the desired frequency and power. The relation between the input voltage, V_{in}, and capacitor voltage, V_C, will give the gain and phase values of the capacitor at that frequency. The obtained waveform of V_C as function of V_{in} will look like the one presented in Fig. 4 – ellipse 1.

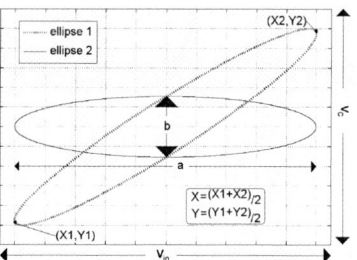

Fig. 4. Theoretical waveform of V_C as function of V_{in}.

To compute the ESR value, ellipse 1 is used, and it can be computed using equation (4)

$$ESR \cong m\,R \tag{4}$$

where m represents the slope of ellipse.

From ellipse 2, it is possible to reach the reactance value of the capacitor, which is given by equation (5).

$$X_{cap} = -\frac{b}{a}R \tag{5}$$

If b is approximately zero, the capacitor will be near its resonance frequency.

Despite the fact that the technique proposed in [7] allows the determination of the capacitor reactance, the computation of its capacitance is difficult at the converter operating frequency. If low and medium capacitance aluminum electrolytic capacitors are considered, the converter operating frequency could be very close to the resonance frequency of the capacitor or lower. If it is close, the reactance value is very small, and so, through a visual analysis of ellipse 2, it is very difficult to reach to a very accurate value. Besides, the ESL value can vary from 10nH to 30 nH [11], so its exact value is not known. In this way, a small a error in the computation of the reactance value will lead to a very large error in the computation of the capacitance. If the resonance frequency of the capacitor is much higher than the converter operating frequency, its ESR is much higher than its reactance, because the capacitor size is very small, so in both situations it is advised to use the capacitance value given by the manufacturer.

To compute the equivalent circuit of the inductor a new technique based in [7] will be used. Afterwards the formulas which lead to the determination of L and R_T will be obtained. In this way, using an RL circuit (a resistor, R, in series with the inductor to study), and considering the equivalent circuit of the inductor presented in Fig.3, it is possible to compute the relationship between the input voltage, V_{in}, and the one at the inductor terminals, V_L. From this relationship, it is possible to obtain the reactance value, X_L, and R_T of the inductor.

$$X_L \cong \frac{R - R\sqrt{1 - 4\left(\frac{b}{a}\right)^2}}{2\frac{b}{a}} \tag{6}$$

$$R_T \cong \frac{m\left(R^2 + X_L{}^2\right) - X_L{}^2}{R} \tag{7}$$

IV. FOURIER ANALYSIS AND SIMULATION TECHNIQUE

The Fourier series is a mathematical tool that could be used to represent an arbitrary periodic function through its decomposition in a sum of much simpler sinusoidal components functions, the harmonics. Equation (8) synthesizes that relationship:

$$f(t) = a_0 + \sum_{n=1}^{\infty} \left(a_n \cos\left(w_0\, n\, t \right) + b_n \sin\left(w_0\, n\, t \right) \right) \quad (8)$$

where, a_0, a_n and b_n are the Fourier coefficients, w_0 is the radian velocity, ($w_0 = 2\,\pi f_0$), and f_0 is the fundamental frequency of the function $f(t)$.

The Fourier coefficients can be computed for any periodic function as soon as Dirichlet conditions are satisfied, which occurs in most physical systems [12].

Fig. 5 shows the buck converter schematic, as well as, its equivalent circuits during conduction and non-conduction stage.

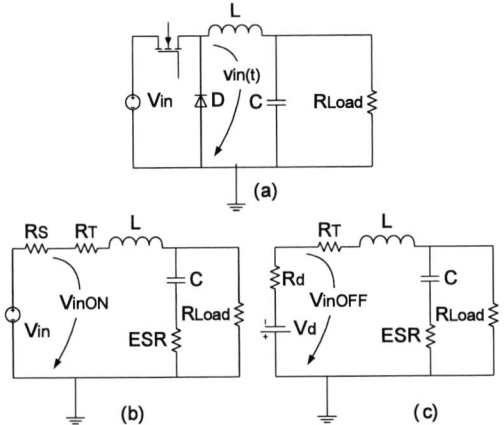

Fig. 5. Schematic of a Buck converter (a); equivalent circuit at conduction stage (b) and non-conduction stage (c).

where, R_S is the drain-source resistance, R_d diode resistance and the V_d the knee voltage.

From Fig. 5, it is possible to conclude that both semiconductors are settled in order to produce a square waveform at the input of the LC filter. Thus, the buck converter can be analyzed considering the linear system composed by the output filter and the square waveform formed by both semiconductors.

In this way, to implement the proposed simulation technique, firstly, it is necessary to compute the harmonics which represent the input voltage. After that, the gain and phase of the different electrical variables should be computed for each harmonic.

Since the LC filter represents a linear system, it is possible to use superposition theorem to determine the different electrical variables of the buck converter.

A. Input voltage

For simulation purpose the input voltage is considered a perfect square waveform. Through the analysis of Fig. 5, it is possible to determine the input voltage during conduction stage, V_{inON}, as well as the input voltage during non-conduction stage, V_{inOFF}.

$$V_{inON} = \frac{R_T + R_{Load}}{\left(R_T + R_{Load} + R_S \right)} V_{in} \quad (9)$$

$$V_{inOFF} = -\frac{R_T + R_{Load}}{\left(R_T + R_{Load} + R_d \right)} V_d \quad (10)$$

Using Fourier transform, it is possible to compute the input waveform as:

$$V_{in}(t) = \langle V_{in} \rangle + \sum_{n=1}^{\infty} \left(c(n) \cos\left(w_0\, n\, t - \varphi(n) \right) \right) \quad (11)$$

where,

$$\langle V_{in} \rangle = V_{inON}\, D + V_{inOFF}\, (1-D)$$

$$a(n) = \frac{V_{inON} - V_{inOFF}}{\pi\, n} \sin(2\,\pi\, n\, D)$$

$$b(n) = \frac{V_{inON} - V_{inOFF}}{\pi\, n} \left(1 - \cos(2\,\pi\, n\, D) \right)$$

$$c(n) = \sqrt{\left(a(n) \right)^2 + \left(b(n) \right)^2}$$

$$\begin{cases} if \left(a(n) < 0 \right) \Rightarrow \varphi(n) = \arctan\left(\dfrac{b(n)}{a(n)} \right) + \pi \\[2mm] if \left(a(n) \geq 0 \right) \Rightarrow \varphi(n) = \arctan\left(\dfrac{b(n)}{a(n)} \right) \end{cases}$$

Fig. 6a shows the simulated input waveform of $V_{in}(t)$ for converter C_1 computed using (11), which is the result of sum of the first 100 harmonics and considering a sampling frequency of 20 MHz. Fig. 6b shows the first 3 harmonics of $V_{in}(t)$.

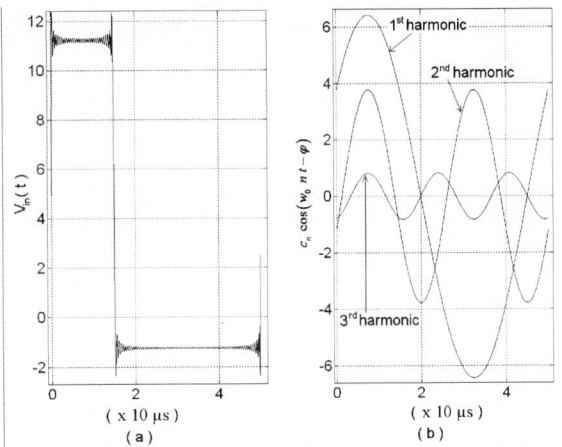

Fig. 6. Simulated input voltage of converter C_1 considering the first 100 harmonics (a) and the first three harmonics of $V_{in}(t)$ (b).

B. Sinusoidal analysis

At this moment, it is very important to choose which electrical variables wanted to be represented. In this study, the inductor current, i_L, the inductor voltage, v_L, and the output voltage ripple, Δv_O, were analyzed.

Afterwards, the relation between the n harmonic of i_L, v_L and Δv_O and the respectively n harmonic of $Vin(t)$ will be computed.

$$\frac{H_{\Delta v_O}}{H_{V_{in}}} = \frac{ESR\ R\ A - X_C\ R\ B}{A^2 + B^2} + \frac{-ESR\ R\ B - X_C\ R\ A}{A^2 + B^2}\ j \quad (12)$$

$$\left|\frac{H_{\Delta v_O}}{H_{V_{in}}}\right| = \sqrt{\left(\mathrm{Re}\left(\frac{H_{\Delta v_O}}{H_{V_{in}}}\right)\right)^2 + \left(\mathrm{Im}\left(\frac{H_{\Delta v_O}}{H_{V_{in}}}\right)\right)^2}$$

$$\begin{cases} if\left(H_{V_{in}} < 0\right) \Rightarrow phase\left(\frac{H_{\Delta v_O}}{H_{V_{in}}}\right) = \mathrm{arctang}\left(\frac{\mathrm{Im}\left(\frac{H_{\Delta v_O}}{H_{V_{in}}}\right)}{\mathrm{Re}\left(\frac{H_{\Delta v_O}}{H_{V_{in}}}\right)}\right) + \pi \\[2em] if\left(H_{V_{in}} \geq 0\right) \Rightarrow phase\left(\frac{H_{\Delta v_O}}{H_{V_{in}}}\right) = \mathrm{arctang}\left(\frac{\mathrm{Im}\left(\frac{H_{\Delta v_O}}{H_{V_{in}}}\right)}{\mathrm{Re}\left(\frac{H_{\Delta v_O}}{H_{V_{in}}}\right)}\right) \end{cases}$$

$$\frac{H_{i_L}}{H_{V_{in}}} = \frac{ESR\ A + R\ A - X_C\ B}{A^2 + B^2} - \frac{X_C\ A + ESR\ B + R\ B}{A^2 + B^2}\ j \quad (13)$$

$$\left|\frac{H_{i_L}}{H_{V_{in}}}\right| = \sqrt{\left(\mathrm{Re}\left(\frac{H_{i_L}}{H_{V_{in}}}\right)\right)^2 + \left(\mathrm{Im}\left(\frac{H_{i_L}}{H_{V_{in}}}\right)\right)^2}$$

$$\begin{cases} if\left(H_{V_{in}} < 0\right) \Rightarrow phase\left(\frac{H_{i_L}}{H_{V_{in}}}\right) = \mathrm{arctang}\left(\frac{\mathrm{Im}\left(\frac{H_{i_L}}{H_{V_{in}}}\right)}{\mathrm{Re}\left(\frac{H_{i_L}}{H_{V_{in}}}\right)}\right) + \pi \\[2em] if\left(H_{V_{in}} \geq 0\right) \Rightarrow phase\left(\frac{H_{i_L}}{H_{V_{in}}}\right) = \mathrm{arctang}\left(\frac{\mathrm{Im}\left(\frac{H_{i_L}}{H_{V_{in}}}\right)}{\mathrm{Re}\left(\frac{H_{i_L}}{H_{V_{in}}}\right)}\right) \end{cases}$$

$$\frac{H_{V_L}}{H_{V_{in}}} = \frac{E\ A + B\ F}{A^2 + B^2} + \frac{F\ A - B\ E}{A^2 + B^2}\ j \quad (14)$$

$$\left|\frac{H_{V_L}}{H_{V_{in}}}\right| = \sqrt{\left(\mathrm{Re}\left(\frac{H_{V_L}}{H_{V_{in}}}\right)\right)^2 + \left(\mathrm{Im}\left(\frac{H_{V_L}}{H_{V_{in}}}\right)\right)^2}$$

$$\begin{cases} if\left(H_{V_{in}} < 0\right) \Rightarrow phase\left(\frac{H_{V_L}}{H_{V_{in}}}\right) = \mathrm{arctang}\left(\frac{\mathrm{Im}\left(\frac{H_{V_L}}{H_{V_{in}}}\right)}{\mathrm{Re}\left(\frac{H_{V_L}}{H_{V_{in}}}\right)}\right) + \pi \\[2em] if\left(H_{V_{in}} \geq 0\right) \Rightarrow phase\left(\frac{H_{V_L}}{H_{V_{in}}}\right) = \mathrm{arctang}\left(\frac{\mathrm{Im}\left(\frac{H_{V_L}}{H_{V_{in}}}\right)}{\mathrm{Re}\left(\frac{H_{V_L}}{H_{V_{in}}}\right)}\right) \end{cases}$$

$$A = R_T\ ESR + R_T\ R + X_L\ X_C + ESR\ R$$
$$B = -R_T\ X_C + X_L\ ESR + R\ X_L - X_C\ R$$
$$E = R_T\ R + ESR\ R_T + X_L\ X_C$$
$$F = -X_C\ R_T + X_L\ R + X_L\ ESR$$

where $H_{\Delta vo}$, H_{vin}, H_{iL} and H_{VL} represent the n harmonic of the output voltage ripple, input voltage, inductor current and inductor voltage.

C. Simulation technique

Thus, computing the different sinusoidal components functions resulting from the input square wave, and summing these individual responses, it is possible to reach to $i_L(t)$, $v_{out}(t)$ and $v_L(t)$ of the buck converter in steady state regime as can be seen in (15), (16) and (17).

$$V_{OUT}(t) = \langle V_{out}\rangle + \sum_{n=1}^{\infty}\left(G_{Vout}(n)\cos\left(w_0\ n\ t + \delta_{Vout}(n)\right)\right) \quad (15)$$

where,

$$\langle V_{out}\rangle = \frac{\langle V_{in}\rangle R}{R + R_T} \quad \wedge \quad G_{Vout}(n) = c(n)\ G_Z(n)$$

$$G_Z(n) = \sqrt{\left(\mathrm{Re}\left(G_{out}(n)\right)\right)^2 + \left(\mathrm{Im}\left(G_{out}(n)\right)\right)^2}$$

$$\begin{cases} if\left(\mathrm{Re}\left(G_{out}(n)\right) < 0\right) \\[1em] \quad \delta_{Vout}(n) = \pi - \varphi(n) + \mathrm{arctang}\left(\frac{\mathrm{Im}\left(G_{out}(n)\right)}{\mathrm{Re}\left(G_{out}(n)\right)}\right) \\[1.5em] if\left(\mathrm{Re}\left(G_{out}(n)\right) \geq 0\right) \\[1em] \quad \delta_{Vout}(n) = -\varphi(n) + \mathrm{arctang}\left(\frac{\mathrm{Im}\left(G_{out}(n)\right)}{\mathrm{Re}\left(G_{out}(n)\right)}\right) \end{cases}$$

$$\mathrm{Re}\left(G_{out}(n)\right) = \frac{ESR\ R\ A(n) - X_C(n)\ R\ B(n)}{A(n)^2 + B(n)^2}$$

$$\mathrm{Im}\left(G_{out}(n)\right) = -\frac{ESR\ R\ B(n) + X_C(n)\ R\ A(n)}{A(n)^2 + B(n)^2}$$

$$i_L(t) = \langle i_L\rangle + \sum_{n=1}^{\infty}\left(G_{iL}(n)\cos\left(w_0\ n\ t + \delta_{iL}(n)\right)\right) \quad (16)$$

where,

$$\langle i_L\rangle = \frac{\langle V_{in}\rangle}{R + R_T}$$

$$G_{iL}(n) = c(n)\sqrt{\left(\mathrm{Re}\left(G_{iL}(n)\right)\right)^2 + \left(\mathrm{Im}\left(G_{iL}(n)\right)\right)^2}$$

$$\begin{cases} if\left(\mathrm{Re}\left(G_{iL}(n)\right) < 0\right) \\[1em] \quad \delta_{iL}(n) = \pi - \varphi(n) + \mathrm{arctang}\left(\frac{\mathrm{Im}\left(G_{iL}(n)\right)}{\mathrm{Re}\left(G_{iL}(n)\right)}\right) \\[1.5em] if\left(\mathrm{Re}\left(G_{iL}(n)\right) \geq 0\right) \\[1em] \quad \delta_{iL}(n) = -\varphi(n) + \mathrm{arctang}\left(\frac{\mathrm{Im}\left(G_{iL}(n)\right)}{\mathrm{Re}\left(G_{iL}(n)\right)}\right) \end{cases}$$

$$\text{Re}\left(G_{iL}(n)\right) = \frac{ESR\, A(n) + R\, A(n) + X_C(n)\, B(n)}{A(n)^2 + B(n)^2}$$

$$\text{Im}\left(G_{iL}(n)\right) = -\frac{ESR\, B(n) + X_C(n)\, A(n) + R\, B(n)}{A(n)^2 + B(n)^2}$$

$$v_L(t) \cong \sum_{n=1}^{\infty} \left(G_{VL}(n)\cos\left(w_0\, n\, t + \delta_{VL}(n)\right)\right) \qquad (17)$$

$$G_{VL}(n) = c(n)\sqrt{\left(\text{Re}\left(G_{vl}(n)\right)\right)^2 + \left(\text{Im}\left(G_{vl}(n)\right)\right)^2}$$

$$\begin{cases}
\textit{if } \left(\text{Re}\left(G_{vl}(n)\right) < 0\right) \\
\quad \delta_{VL}(n) = \pi - \varphi(n) + \arctan\left(\frac{\text{Im}\left(G_{vl}(n)\right)}{\text{Re}\left(G_{vl}(n)\right)}\right) \\
\textit{if } \left(\text{Re}\left(G_{vl}(n)\right) \geq 0\right) \\
\quad \delta_{VL}(n) = -\varphi(n) + \arctan\left(\frac{\text{Im}\left(G_{vl}(n)\right)}{\text{Re}\left(G_{vl}(n)\right)}\right)
\end{cases}$$

$$\text{Re}\left(G_{vl}(n)\right) = \frac{E(n)\, A(n) + B(n)\, F(n)}{A(n)^2 + B(n)^2}$$

$$\text{Im}\left(G_{vl}(n)\right) = \frac{F(n)\, A(n) - B(n)\, E(n)}{A(n)^2 + B(n)^2}$$

$$X_L(n) = 2\pi f\, n\, L \wedge X_C(n) = \left(2\pi f\, n\, C\right)^{-1}$$

$$A(n) = R_T\, ESR + R_T\, R + X_L(n)\, X_C(n) + ESR\, R$$

$$B(n) = -R_T\, X_C(n) + X_L(n)\, ESR + R\, X_L(n) - X_C(n)\, R$$

$$E(n) = R_T\, R + ESR\, R_T + X_L(n)\, X_C(n)$$

$$F(n) = -X_C(n)\, R_T + X_L(n)\, R + X_L(n)\, ESR$$

where n represents the harmonic number.

Usually, designers chose an aluminum electrolytic capacitor that operates very close to its resonance frequency. In this way, the ESL can be neglected for simulation purposes, and so the circuit of Fig. 3 can be used to model the output filter.

However, if that isn't the case, the ESL should be considered in order to get best simulation results. For that, equation (18) should be used.

$$X_C(n) = \frac{1}{\left(2\pi f\, n\, C\right)} - \left(2\pi f\, n\, ESL\right) \qquad (18)$$

V. Experimental and Simulated Results

In this section a comparison between experimental and simulated results will be presented. Thus, with respect to experimental results, it was necessary to design a buck converter. In Table I, it is possible to observe the characteristics, of the power stage elements of the converters used in this study.

The simulated results were obtained through the development of a small program in *Matlab* environment based in expressions (15), (16) and (17). For that, the data

presented in Table II and III, as well as, the first hundred harmonics were considered.

A. Equivalent circuit of the prototype

Table I shows the characteristics of power stage elements given by manufactures.

TABLE I.
CHARACTERISTICS OF POWER STAGE ELEMENTS, GIVEN BY MANUFACTURES.

Inductor	L (1kHz, ±10%)	22 μH
	R_{dc}	11 mΩ
Capacitor	C (120 Hz, 20 °C, ±20%)	1000 μF
	ESR (120 Hz, 20 °C, ±10%)	159 mΩ
MOSFET	$R_{drain\text{-}source}$	70 mΩ
Diode	$R_{conduction}$	10 mΩ
	V_d	0.8 V

However, as was discussed in section II, it is advised to compute the equivalent circuit of the output filter at the operating frequency of the converter studied (20 kHz).

In this way, with respect to the capacitor, the technique proposed in [7] was used, in order to obtain its equivalent circuit, at the operating frequency of the converter. Fig. 7 shows the relation between V_C and V_{in}.

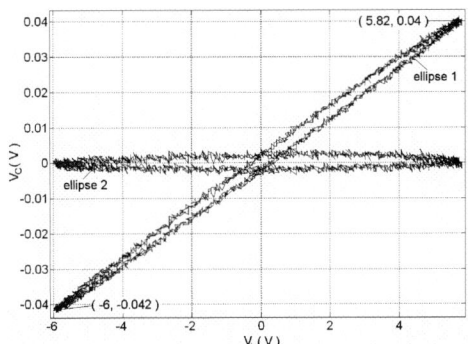

Fig. 7 Relation between V_{in} and V_C, at 20 kHz, for the capacitor presented in Table I.

For the inductor presented in Table I, the technique proposed in section III was used. Fig. 8 shows the relation between V_L and V_{in}.

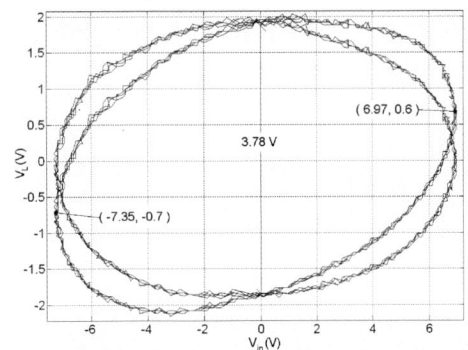

Fig. 8 Relation between V_{in} and V_L, at 20 kHz, for the inductor presented in Table I.

Afterwards, the ESR, L and R_T values are computed using the techniques presented in section III.

For that propose, a 10 Ω resistor was used, and the computed values were compared with the ones obtained using impedance Gain-phase analyzer HP 4294.

TABLE II.
COMPUTED VALUES OF ESR, L AND R_T.

Technique	ESR	L	R_T
Proposed	69.4 mΩ	22.72 μH	0.16 Ω
HP 4294	60 mΩ	22.34 μH	0.14 Ω

Table I, II and III shows the characteristics of the three converters developed for this study.

TABLE III.
CONVERTER CHARACTERISTICS.

Converter	V_{in}	T	T_{ON}	R
C_1	11.75 V		15 μs	
C_2	11.65 V	50 μs	25 μs	1 Ω
C_3	11.60 V		35 μs	

B. Experimental and simulated results

In Figs. 9-11, it is possible to compare the experimental and simulated waveforms of the inductor current, i_L, inductor voltage, v_L, and of the output voltage ripple, Δv_O, for the three converters.

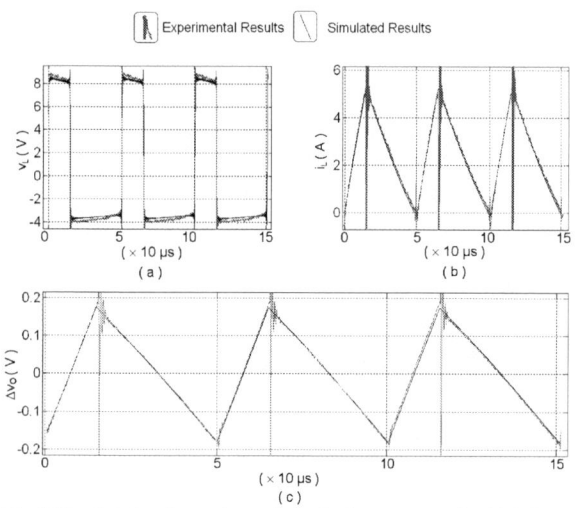

Fig. 9 Simulated and experimental results for converter C_1: (a) inductor voltage; (b) inductor current and (c) output voltage ripple.

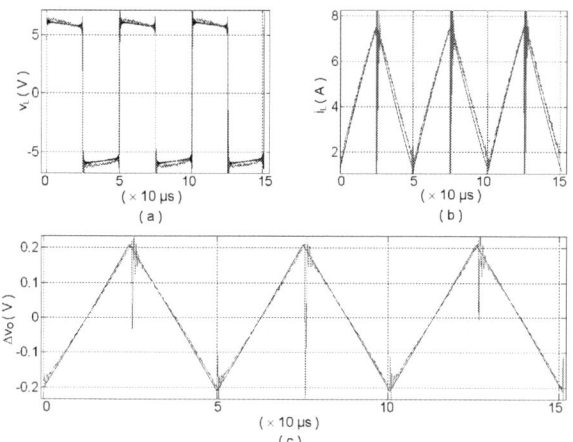

Fig. 10 Simulated and experimental results for converter C_2: (a) inductor voltage; (b) inductor current and (c) output voltage ripple.

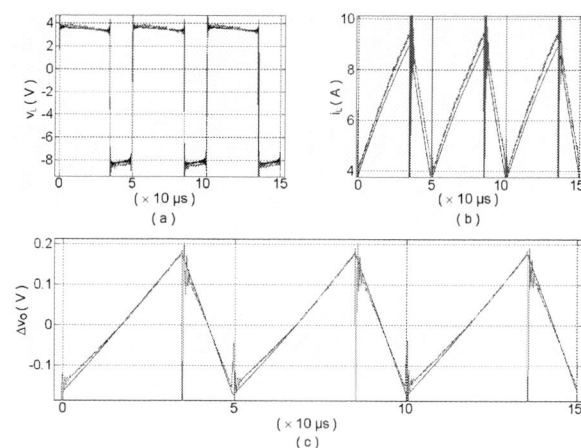

Fig. 11 Simulated and experimental results for converter C_3: (a) inductor voltage; (b) inductor current and (c) output voltage ripple.

From the similarity between experimental and simulated results, it is possible to conclude that the proposed technique can be used with much profit.

VI. FREQUENCY ANALYSIS

In this section, a study about the performance of the output filter is presented. For that, only simulation results related to converter C_1, using different output filters, are considered.

Thus, to show clearly the influence of each parameter in the output voltage ripple, different *ESL*, *ESR*, *C* and *L* are considered.

Table IV shows the characteristics of the different filters.

TABLE IV.
CHARACTERISTICS OF THE DIFFERENT FILTERS.

Filter	ESR	C	L	ESL
F_1	69.4 mΩ	1000 μF	23 μH	0
F_2	69.4 mΩ	1000 μF	46 μH	0
F_3	138.8 mΩ	1000 μF	23 μH	0
F_4	69.4 mΩ	1000 μF	23 μH	20 nH
F_5	20.0 mΩ	4700 μF	23 μH	0
F_6	20.0 mΩ	4700 μF	23 μH	20 nH
F_7	0.8 mΩ	180 μF	23 μH	18 nH

The *ESR* values presented in Table IV for F_1, F_2, F_4, F_5 and F_6 were obtained using an impedance gain-phase analyzer. The F_3 filter uses an old capacitor, while F_7 uses medium power film capacitor. The inductance and capacitance values were the ones given by the manufacturers. The typical *ESL* value for radial-leaded aluminum electrolytic capacitors varies from 10 nH to 30 nH [11].

Seeing that the presented study is based on a frequency analysis, first, it is necessary to determine the most significant harmonics of the input voltage and related then with its means value. Then, its possible to evaluate the output filter performance with frequency, for that $(G_Z(n))^{-1}$ should be computed.

Fig. 12 shows the first hundred harmonics of $V_{in}(t)$ and the relation between the first hundred harmonics of $V_{in}(t)$ and its mean value.

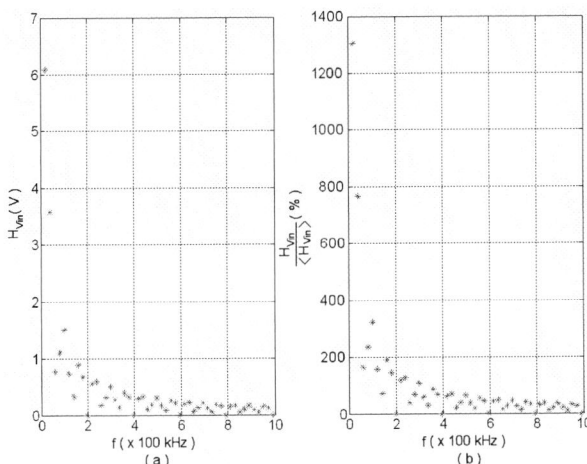

Fig. 12 $V_{in}(t)$ harmonics of C_1: (a) the first 100 harmonics of $V_{in}(t)$; (b) relation between the first 100 harmonics of $V_{in}(t)$ and its mean value.

From Fig.12b it is possible to evaluate the weight of each harmonic in the input voltage.

Fig. 13 shows a comparison between the performances of the different filters and F1, as function of frequency.

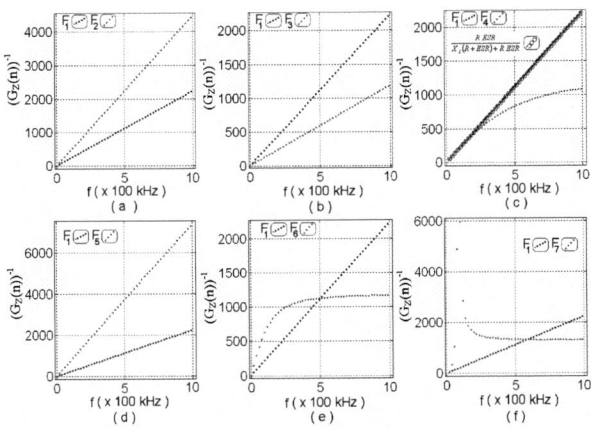

Fig. 13 Filter perfomance.

From Fig. 13, it is possible to conclude that filter models F3 and F4 show a worst performance than F1, which can be explained by the *ESR* and *ESL* influence, respectively.

To show clearly that the model presented in Fig. 3 is a very good approximation, for aluminum electrolytic capacitors, when low and medium capacitances are considered, equation (19) will be used to evaluate the influence of *ESL* in the F4 and F6 performance.

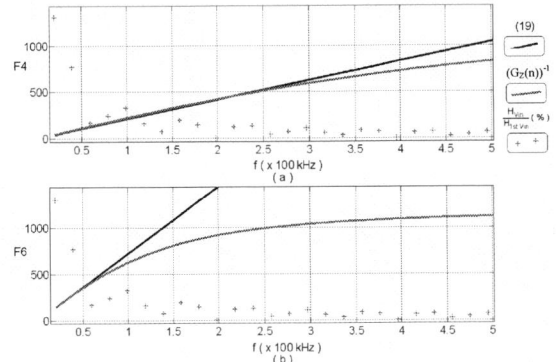

Fig. 14 Comparison of F4 and F6 filter performance with equation (19).

From Fig. 14a it is possible to conclude that Fig. 3 is a very good approximation for small and medium aluminum electrolytic capacitors.

$$\left(G_z\left(n\right)\right)^{-1} = \left(\frac{X_L\left(n\right)\left(R + ESR\right) + R\,ESR}{R\,ESR}\right) \tag{19}$$

However, when high capacitance capacitors are considered, the model should introduce the effect of *ESL*, as can be seen from Fig. 14b. This situation could be explained by its lower resonance frequency, as well as, its lower the *ESR*.

From Fig. 13a and 13b, it is possible to conclude that the decrease of L and the increase of the *ESR* are linear related with the increase of the output voltage ripple, for low and medium aluminum electrolytic capacitors.

From Fig. 13f, it is possible to conclude that the use of a medium power film capacitor, with much lower capacitance, reveals much more productive at lower frequencies. For this capacitor, the effect of *ESR* can be almost neglected and the *ESL* is not significant near the operating frequency of the converter. In this way, if the *ESR* and *ESL* values are neglected equation (20) could be used to compute the filter performance as function of frequency.

$$\left(G_z\left(n\right)\right)^{-1} = \left|\left(w(n)\right)^2 L\,C - 1\right| \tag{20}$$

However if the *ESL* value is considered equation (21) should be used.

$$\left(G_z\left(n\right)\right)^{-1} = \frac{\left|\left(w(n)\right)^2 C\left(ESL + L\right) - 1\right|}{\left|\left(w(n)\right)^2 C\,ESL - 1\right|} \tag{21}$$

Fig. 15 shows a comparison of both approximations, (20) and (21), and F7.

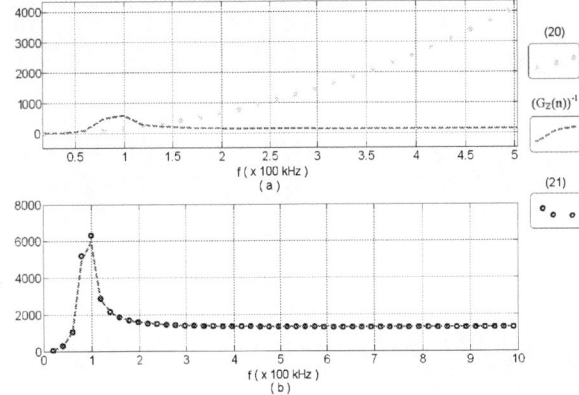

Fig. 15 Comparison of F7 with (20) and (21) approximations.

In this way, it is possible to conclude that equation (21) can be considered a good approximation when low capacitance and low *ESR* capacitors are considered. Fig.15 shows that the effect of *ESL* for frequencies lower than the resonance frequency of the capacitor is good, when low capacitances are considered, which could be explained its higher resonance frequency. However, for

593

frequencies higher than its resonance frequency the *ESL* effect becomes pernicious.

From the analysis of Fig. 13, it is possible to conclude, for small capacitors, that the effect of *ESR* limits the performance of the output filter of *smps*. However, when low *ESR* capacitors are used, the size and cost of the output filter increase significantly.

Fig. 16 shows the output voltage ripple of converter C_1 considering the F_2, F_3, F_4, F_5, F_6 and F_7 filters.

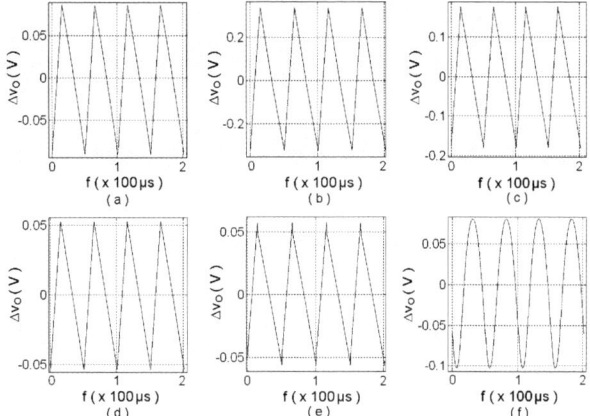

Fig. 16 Output voltage ripple of converter C_1, considering the output filters (a) F_2, (b) F_3, (c) F_4, (d) F_5, (e) F_6 and (f) F_7.

From Figs. 16a and 16b, it is possible to conclude that the increase of the inductor inductance to twice its initial value reduces the Δv_O to approximately half its initial value, and, on the opposite, the increase of *ESR* to twice its initial value increases to twice the Δv_O. From Fig. 16c it is possible to conclude that the effect of *ESL* is almost insignificant. However for high capacitance capacitor the same is not true, as can be seen from the comparison of Figs. 16d and 16e. Fig. 17 shows a zoom of both waveforms.

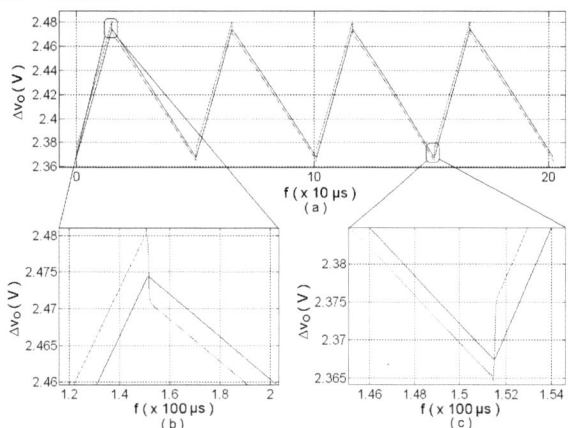

Fig. 17 Output voltage ripple of converter C_1, considering the output filter F_5 and F_6.

From the comparison of Figs. 9c and 16f, it is possible to conclude that the metal film capacitor, which has an eighth of the capacitance of the aluminum electrolytic capacitor used, produces less than half of the output voltage ripple of the aluminum electrolytic capacitor, which could be explained by its very small *ESR*. In fact, the effect of the *ESR* for this capacitor can be neglected, as can be seen in Fig. 15b.

VII. CONCLUSIONS

In this paper some tools that could simplify the design process of step down DC-DC converters were presented.

A simple simulation technique based on the Fourier transform was presented, which allows both time and frequency analysis of the converter in steady state regime.

An experimental technique, which could be used to determine the equivalent circuit of the output filter, was also presented. This technique reveled very accurate when the results were compared with the ones obtained with an impedance gain phase analyzer.

Both techniques, when used simultaneously, could be a very important tool in the design process of switch mode power supplies.

REFERENCES

[1] G. Wester and R. Middlebrook, "Low-frequency characterization of switched dc-to-dc converters", *in IEEE PESC Rec.*, pp. 9-20, 1972.

[2] R. Middlebrook and Cúk, "A general unified approach to modelling switched-converter power stage", *in IEEE PESC Rec.*, pp. 18-34, 1976.

[3] R. Tymerski, V. Vorperian, F. Lee, W. Bauman "Nonlinear modelling of the PWM switch", *in IEEE PESC Rec.*, pp. 968-979, 1988.

[4] Y. Lee, Y. Cheng, "Computer-aided analysis of electronic dc-dc transformers", *in IEEE Transactions on Ind. Electron.*, vol IE-35, n°.1, pp. 148-152, 1988.

[5] S. Ben-Yaakov, "SPICE simulation of PWM dc-dc converter systems: voltage feedback, continuous inductor conduction mode", *in IEE Electron. Lett.*, vol. 25, n° 16, pp. 1061-1063, Aug. 1989.

[6] A. Amaral, A. Cardoso, "Theoretical Analysis of the Behaviour of a Buck Converter in Steady State Regime", *Proceedings of The 20th International Congress & Exhibition on Condition Monitoring and Diagnostic Engineering Management*, Faro, Portugal, 13-15 June, 2007.

[7] A. Amaral, A. Cardoso, "An Experimental Technique for Estimating the ESR and Reactance Intrinsic Values of Aluminum Electrolytic Capacitors", *Proceedings of IEEE Instrumentation and Measurement Technology Conference*, Sorrento, Italy, 24-27 April, 2006.

[8] P. Krein, "Elements of Power Electronics", *Oxford University Press Inc*, 1997, Oxford.

[9] M. O'Hara, "Modelling Non-Ideal Inductors in Spice", *Newport Components*, United Kingdom, 8 November, 1993.

[10] S. Parter, "Improved Spice Model of Aluminum Electrolytic Capacitors for Inverter Applications", *IEEE Transactions on Industry Applications*, Vol. 39, N° 4, pp. 929-935, July 2003.

[11] Application Guide, *Aluminum Electrolytic Capacitors, Cornell Dubilier*, pp. 2.183-2.202.

[12] R. Thomas, A. Rosa, "The Analysis and Design of Linear Circuits", *Prentice-Hall*, New Jersey, 1998.

ANALYSIS OF HOPF BIFURCATION IN DC-DC LUO CONVERTER USING CONTINUOUS TIME MODEL

A.Kavitha*, G.Uma**

* Researcher and Lecturer, College of Engg., Guindy, Anna University, India.
** Assistant Professor/Power Electronics and drives division, College of Engg., Guindy, Anna University, India.

ABSTRACT-- **DC-DC Converters have been reported as exhibiting a wide range of bifurcations and chaos under certain conditions. This paper analyses the bifurcations in current controlled Luo topology operating in the continuous conduction mode by means of a continuous time model. The stability of the system is analyzed by studying the locus of the complex eigen values and the characteristic multipliers locate the onset of Hopf bifurcation. The 1-periodic orbit loses its stability via hopf bifurcation and the resulting attractor is a quasi -periodic orbit. This later bifurcates to chaos via border collision bifurcation. A computer simulation using MATLAB SIMULINK confirms the predicted bifurcations. It has also been inferred from the experimental results that the margin of system stability decreases as the load decreases.**

KEYWORD-: Border collision bifurcation, Continuous time model, Hopf Bifurcation, Luo Converter,

I. INTRODUCTION

Power electronics is a field rich in non-linear dynamics [1]. Chaos is an apparently disordered deterministic behavior, which is an universal phenomenon that is present in many systems in all areas of science. A rich variety of bifurcations and chaos are present, if the switching action is governed by feedback control as in regulated power supplies [2]. Many literature reports the presence of bifurcations in buck, boost, buck-boost and cuk converter topologies [4],[5]. Positive output Luo converters are a series of new step up dc-dc converters derived from buck-boost converters. It can step up and step down the voltage with high power density, high power efficiency and the topology of the converter is also very simple [7]. These converters are widely used in computer peripheral equipment and industrial applications, especially for high voltage projects [8]. In this paper, an attempt is made to study the bifurcation in a positive output elementary luo converter. The averaging approach is one of the most widely adopted modeling strategies for switching converters that yields a simple model [9]. Hence it is proposed to perform the analysis by considering the converter operating in a hysteretic current controlled mode.

II. CIRCUIT OPERATION OF LUO CONVERTER:

The circuit diagram of the positive output Luo converter is shown in figure 1.In the circuit S is the power switch and diode D is the freewheeling diode. The energy storage elements are inductors L_1, L2 and capacitors C_1, C_2.R is the load resistance.

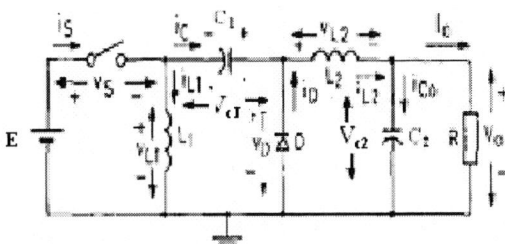

Fig 1.Circuit Diagram of Luo Converter

When the switch is on, the inductor L_1 is charged by the supply E. At the same time the inductor L_2 absorbs the energy from source and the capacitor C_1. The load is supplied by the capacitor C_2.The equivalent circuit of Luo converter in mode 1 operation is shown in figure 2.

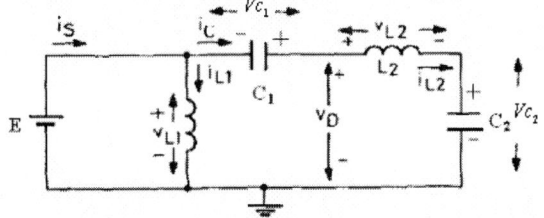

Fig 2. Equivalent Circuit of Luo Converter in Mode 1 Operation

During off condition, the current i_s drawn from the source becomes zero. Current i_{L1} flows through the freewheeling diode to charge the capacitor C_1. Current i_{L2} flows through C_2 –R circuit and free wheeling diode D to keep itself continuous.

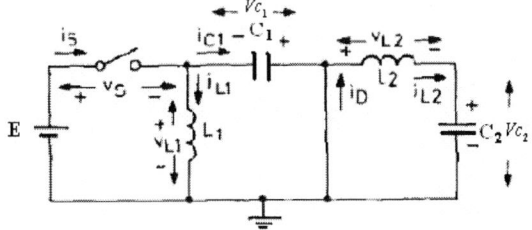

FIG 3. Equivalent Circuit of Luo Converter in Mode 2 Operation

II. HYSTERETIC CURRENT PROGRAMMED CONTROL

The general circuit diagram of hysteretic current-mode control is shown in figure 4. In actual implementation, the switch is turned on and off in a

978-1-4244-0644-9/07/$25.00 ©2007 IEEE   595

hysteretic fashion, when the sum of the inductor currents falls below or rises above a preset hysteretic band [10].

The output voltage is fed back to set the average value of the hysteretic band, forcing the control variable to be related by the following control equation

$$i_1 + i_2 = g(V_0) \qquad (1)$$

where i_1 and i_2 are the inductor currents
  V_0 is the output voltage
  g (.) is the control function

The control law is of the form

$$\Delta(i_1 + i_2) = -\mu \Delta V_o \qquad (2)$$

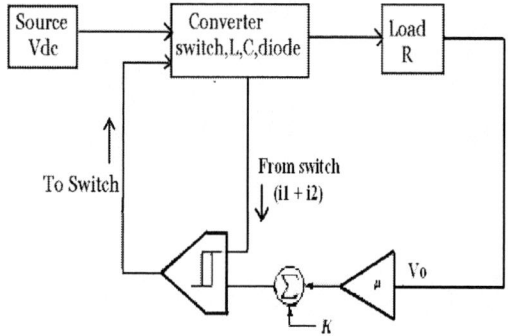

Fig 4: Hysteretic Current Programmed Control.

The inductor current $i_1 + i_2$ is related to V_0 by the following control equation

$$i_1 + i_2 = K - \mu V_o \qquad (3)$$

where K and µ are the control parameters.

III AVERAGED STATE SPACE EQUATIONS OF LUO CONVERTER

The system can be represented by the following state space equations where δ=1, When the switch is turned on and δ= 0 when the switch is turned off.

$$\frac{diL_1}{dt} = \frac{(1-\delta)}{L} Vc_1 + \frac{E}{L}$$

$$\frac{diL_2}{dt} = \frac{\delta Vc_1}{L} - \frac{Vc_2}{L}$$

$$\frac{dVc_1}{dt} = \frac{(1-\delta)}{C} i_1 - \frac{i_2}{C}\delta$$

$$\frac{dVc_2}{dt} = \frac{iL_2}{C_2} - \frac{Vc_2}{RC_2} \qquad (4)$$

Since i_1+i_2 can be related to v_2, given by equation (1), the fourth order system reduces to a third order system as follows,

$$\frac{diL_2}{dt} = \frac{\delta Vc_1}{L} + \frac{\delta E}{L} - \frac{Vc_2}{L}$$

$$\frac{dVc_1}{dt} = \frac{(1-\delta)}{C}(K - \mu Vc_2) - \frac{iL_2}{C}$$

$$\frac{dVc_2}{dt} = \frac{iL_2}{C_2} - \frac{Vc_2}{RC_2} \qquad (5)$$

Using the following dimensionless variables

$$X_1 = \frac{Ri_2}{E}; X_2 = \frac{V_1}{E}; X_3 = \frac{V_2}{E}; \tau = \frac{Rt}{2L}; \quad \xi = \frac{L/R}{CR};$$

$$K_1 = \mu R; K_0 = \frac{KR}{E}$$

the set of differential equations given in (5) can be rewritten as,

$$\frac{dX_1}{d\tau} = \frac{X_2[X_2 + X_3[1 + K_1\xi] - K_1 X_1\xi]}{1 + X_2} - 2X_3 +$$

$$\frac{[X_2 + X_3[1 + K_1\xi] - K_1 X_1\xi]}{1 + X_2}$$

$$\frac{dX_2}{d\tau} = -2X_1\xi + (2K_o\xi + 2X_3\xi K_1)\left[1 - \frac{[X_2 - K\xi X_1 + X_3[1 + K_1\xi]]}{2(1 + X_2)}\right]$$

$$\frac{dX_3}{d\tau} = 2\xi[X_1 - X_3] \qquad (6)$$

IV. DERIVATION OF JACOBIAN

The Jacobian matrix for the dimensionless system evaluated at the equilibrium point is derived by the following matrix.

$$J(X) = \begin{bmatrix} \dfrac{d\left[\left(\dfrac{dX_1}{d\tau}\right)\right]}{dX_1} & \dfrac{d\left[\left(\dfrac{dX1}{d\tau}\right)\right]}{dX_2} & \dfrac{d\left[\left(\dfrac{dX_1}{d\tau}\right)\right]}{dX_3} \\[2em] \dfrac{d\left[\left(\dfrac{dX_2}{d\tau}\right)\right]}{dX_1} & \dfrac{d\left[\left(\dfrac{dX_2}{d\tau}\right)\right]}{dX_2} & \dfrac{d\left[\left(\dfrac{dX_2}{d\tau}\right)\right]}{dX_3} \\[2em] \dfrac{d\left[\left(\dfrac{dX_3}{d\tau}\right)\right]}{dX_1} & \dfrac{d\left[\left(\dfrac{dX_3}{d\tau}\right)\right]}{dX_2} & \dfrac{d\left[\left(\dfrac{dX_3}{d\tau}\right)\right]}{dX_3} \end{bmatrix}$$

The Jacobian matrix is formed by differentiating the dimensionless autonomous equation as in the matrix mentioned above

$$J(X) = \begin{vmatrix} J_{11} & J_{12} & J_{13} \\ J_{21} & J_{22} & J_{23} \\ J_{31} & J_{32} & J_{33} \end{vmatrix}$$

Where,

$$J_{11} = X_2\left(\frac{-K_1\xi}{1 + X_2}\right) - \frac{K_1\xi}{1 + X_2}$$

$$J_{12} = \frac{X_2}{1+X_2} - \left[\frac{-K_1\xi X_1 + X_3(1+K_1\xi) + X_2}{(1+X_2)^2}\right]X_2 + \frac{1}{(1+X_2)}$$

$$+\left[\frac{-K_1\xi X_1 + X_3(1+K_1\xi) + X_2}{(1+X_2)}\right] - \left[\frac{-K_1\xi X_1 + X_3(1+K_1\xi) + X_2}{(1+X_2)^2}\right]$$

$$J_{13} = \frac{(1+K_1\xi)}{1+X_2}X_2 - 2 + \frac{(1+K_1\xi)}{1+X_2}$$

$$J_{21} = -2\xi + \frac{(2K_o\xi + 2X_3\xi K_1)}{1+X_2}\frac{K_1\xi}{2}$$

$$J_{22} = \frac{-1}{2(1+X_2)} + \frac{1}{2(-K_1\xi X_1 + X_3(1+K_1\xi) + X_2)(1+X_2)^2)}$$

$$J_{23} = 2K_1\xi\left[1 - \frac{1}{2(1+X_2)(-K_1\xi X_1 + X_3(1+K_1\xi) + X_2)}\right]$$

$$J_{31} = 2\xi$$

$$J_{32} = 0$$

$$J_{33} = -2\xi \qquad (7)$$

V. IDENTIFICATION OF HOPF BIFURCATION USING EIGEN VALUES

The stability of the system can be studied by deriving the eigen values of the system at the equilibrium point. The Luo converter designed with the values given in appendix is then analyzed for its stability. The system has one negative real eigen value and a pair of complex poles. The real part of the complex pole may be either positive or negative real part depending upon the values of K_0, K_1 and ξ.

The table 1 shows the typical scenario of the variation of the eigen values for various values of K_0. The same analysis is performed for the various values of ξ.
The following observations are made.

- For $\xi=0.0136$, the hopf bifurcation point is obtained at $K_0=9$.
- For $\xi=0.5$, the hopf bifurcation point is obtained at $K_0=5$.
- For $\xi=1$, the hopf bifurcation point is obtained at $K_0=4$.

TABLE 1. EIGEN VALUES FOR $\zeta = 0.0136$

Values of k_0	Eigen values for K=1	Remarks
$k_0 = 1$	-0.0137, $-0.0159 \pm 0.2318i$	stable
$k_0 = 3$	-0.0199, $-0.0097 \pm 0.2313i$	stable
$k_0 = 5$	-0.0221, $-0.0054 \pm 0.2310i$	stable
$k_0 = 7$	-0.0233, $-0.0020 \pm 0.2306i$	stable
$k_0 = 9$	-0.0240, $0.0008 \pm 0.2303i$	unstable
$k_0 = 11$	-0.0246, $0.0034 \pm 0.2301i$	unstable

It is inferred that the hopf bifurcation point [10] (i.e. the critical value of K_o) decreases as ξ value increases. For lower values of ξ or load resistance, the hopf bifurcation point is obtained at higher value of K_o

VI. MARGIN OF STABILITY CURVE

The critical value of K_o depends on the values of K_1 and ξ. The margin of stability curve is shown in the figure 5. It is inferred from this curve that as the values of K_o, K_1 and ξ crosses the critical value, the margin of stability decreases.

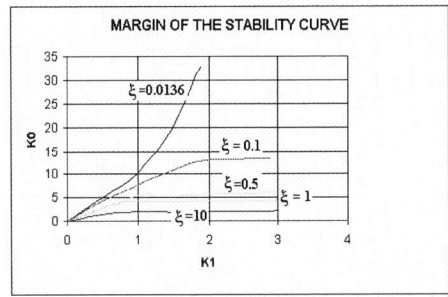

Fig 5: Margin of the Stability Curve for K_o, K_1 And ξ

The locus of the complex eigen pair for various values of K_o is shown in figure 6. The movement of the locus from left plane to the right plane shows that the system loses its stability when the load is decreased.

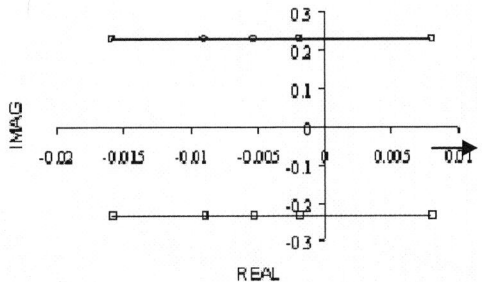

Fig 6 : Locus of the Complex Eigen Value Pair

597

V. HARDWARE IMPLEMENTATION OF POSITIVE OUTPUT LUO CONVERTER EXHIBITING STAGES LEADING TO CHAOS

To verify the analysis a prototype of Luo Converter is constructed and tested. The block diagram of the experimental setup and the hardware results are presented in the following section.

A BLOCK DIAGRAM OF THE HARDWARE CIRCUIT.

The block diagram of the experimental setup of the Luo converter is as shown in Figure 7. It consists of a power circuit and a control circuit.

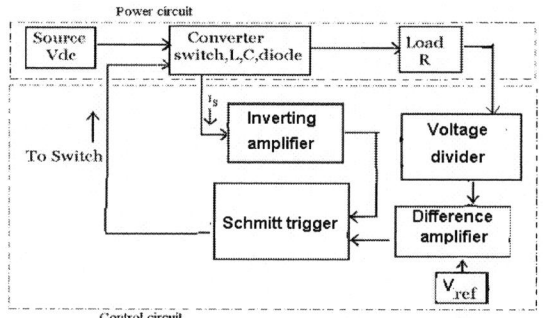

Fig 7 : Block Diagram of Experimental Circuit

The power circuit consists of two inductors L_1 and L_2, two capacitors C_1 and C_2, switch S, input voltage source E, diode D and Load resistance R. The converter is assumed to operate in continuous conduction mode.

The control circuit consists of the following blocks: Voltage Divider, V_{ref} Generation, Difference Amplifier , Inverting amplifier, and a Schmitt trigger.The output voltage is stepped down using a voltage divider circuit.

A reference voltage is generated and fed to non-inverting input of the difference amplifier. The voltage from the divider circuit is given to the inverting input of the difference amplifier LM358.

Schmitt trigger is implemented using µA741.The current in the switch is sensed by using an inverting amplifier and fed to inverting input of the Schmitt trigger. The output from the difference amplifier is given to non-inverting input of the Schmitt trigger. Schmitt trigger acts as a hysteresis controller and generates pulses based on the two inputs. Switch in luo converter is in series with supply hence optocoupler (MCT2E) is used to isolate the control circuit from the power circuit.

The Luo converter dynamics is studied by varying the Load, keeping the remaining parameters constant and the possible route to chaos is observed. The hardware results are also presented.

B. ROUTE TO CHAOS IN INDUCTOR CURRENT BY VARYING LOAD RESISTANCE

The various stages leading to chaos is studied both in simulation and in real time.

5.1 Fundamental operation

The Luo converter operating in continuous conduction mode is modeled and simulated using MATLAB/ SIMULINK software. The simulated fundamental inductor current obtained when R=20Ω is shown in figure 8.1.

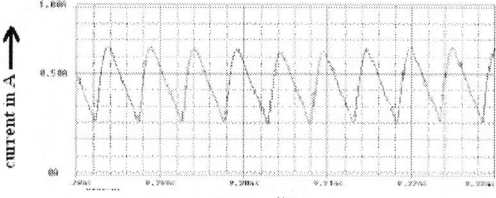

Fig 8.1 Simulated Fundamental Current Waveform

The fundamental and stable operation obtained in the experimental set up is as shown in figure 8.2.

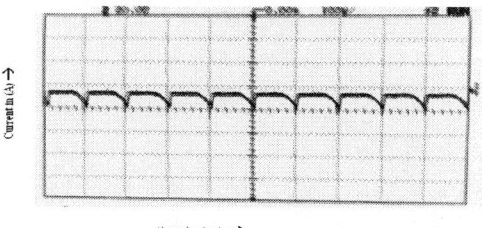

Fig 8.2 Fundamental Waveform of Inductor Current

5.2 Quasi-period operation

When load is decreased, the quasi-period operation of the inductor current is observed and the waveforms are shown in figure 8.3 and 8.4. Due to its non-periodic nature, quasi-periodic type of operation tends to induce noises, some of which fall in the audible range, and is thus rather undesirable for practical use. For this reason, this type of operation has never been allowed in the final product although it is frequently encountered during the development stage of a power supply.

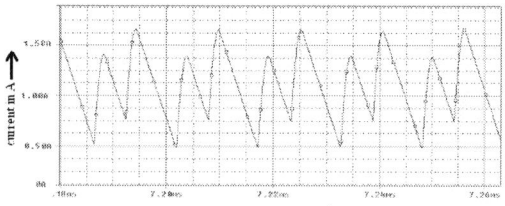

Fig 8.3 Simulated Quasi-Periodic Current Waveform

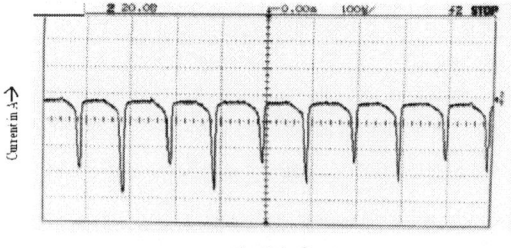

Fig.8.4 Experimental Quasi-Periodic Inductor Current

5.3 Chaotic operation

Chaotic attractor can be identified as a structure of long-term trajectories in a bounded region of phase space, which folds the bundle of trajectories back onto itself, resulting in mixing and divergence of nearby states [12]. Simulated waveform of the chaotic inductor current with further decrease in load is also shown in figure 8.5

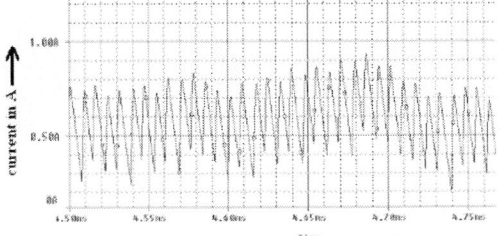

Fig 8.5 Simulated Chaotic Regime of Inductor Current

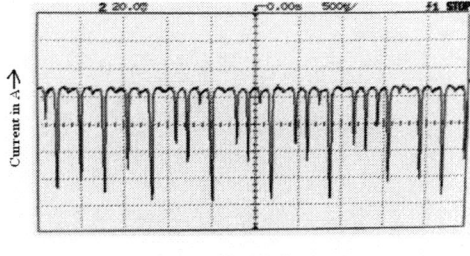

Fig.8.6 Experimental Chaotic Regime of Inductor Current

When load is decreased further, the inductor current of the proposed converter enters the chaotic regime, which is visualized in hardware as in fig 8.6.

VI. CONCLUSION

In this work, the hopf bifurcation analysis of a hysteretic current mode controlled Luo dc-dc converter is performed. It has been shown that as the control parameters are varied the nominal periodic orbit undergoes a hopf bifurcation, quasi periodicity and finally enters into chaotic regime. The simulated results obtained are to be verified with the experimental circuit. It has also been inferred from the experimental results that the margin of system stability decreases as the load resistance increases.

REFERENCES

[1] D.C.Hamill, "*Power electronics: A field rich in nonlinear dynamics,*" in proc. Int, spec. Workshop Non linear Dynamics of Electron. Syst, Dublin, Ireland, 1995, pp. 165-178.

[2] C.K.Tse, Mario DC Bernardo, "*Complex behavior in switching power converters*", IEEE proceedings, Vol.90, No.5, May 2002, 768-771.

[3] Ned Mohan, T.M.Undeland, W.P.Robbins, "*Power Electronics: converter, applications and design*", New York: Wiley, Second edition, 1995.

[4] Soumitro Banerjee and George C.Verghese, 2001, "*Nonlinear phenomena in Po wer Electronics*", IEEE Press

[5] David C. Hamill and David J. Jeffries, "*Sub harmonics and chaos in a controlled switched-mode power converter*", IEEE Transactions on Circuits and Systems, Vol.35, No.8, July 1988.

[6] Jonathan H.B. Deane and David C. Hamill, "Instability, sub harmonics and chaos in power electronic systems", IEEE Transactions on Power Electronics, Vol.5, No.3, July 1990.

[7] F.L.Luo,"*Advanced dc-dc converters*",

[8] F.L.Luo," *Positive output Luo converters: voltage lift technique*", IEEE Proceedings, power Applications.

[9] J.M.T Thompson and Stewart, " *Non Linear Dynamics and Chaos*", John Wiley.

[10] S.C.Wong and Y.S.Lee, "*SPICE modeling and simulation of the hysteretic current – controlled Cuk converter*", IEEE Trans. Power Electron., Vol.8, pp.580-587, Oct. 1993.

[11] C.K.Tse,"*Hopf Bifurcation and Chaos in a Free-Running Current-Controlled Cuk Switching Regulator*", IEEE Transactions on circuits and system, Vol 47, No 4, 2000

[12] C. K.Tse and William C.Y.Chan, 1995, "*Instability and chaos in current-mode controlled cuk converter*", IEEE.

[13] C.K.Tse, S.C.Fung and M.W.Kwan, "*Experimental confirmation of chaos in a current- programmed Cuk converter,*" IEEE Trans. Circuits Syst-Part I, Vol, 43, July 1996.

APPENDIX

Component values
The component values are chosen as
V_{in}=12V; $L_1 = L_2 = 0.01$H; $C_1 = C_2 = 20\mu$F; R = 40Ω; $f_s = 10$ KHz.

EXPERIMENTAL SET-UP

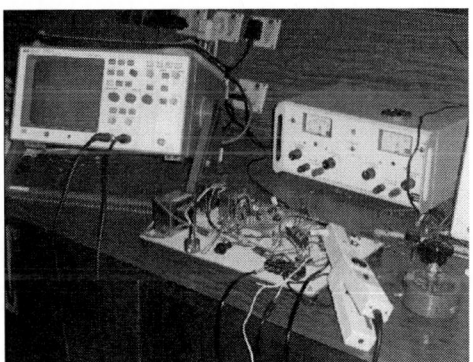

BIOGRAPHIES

A.Kavitha has been awarded the Master's degree from Indian Institute of Madras in the year 2002. She is a member of ISTE. Presently working as a Lecturer, EEE department, College of Engineering, Guindy, Anna University, Chennai, India. Her areas of interest are DC-DC Converters, Chaos and Bifurcations.

G.Uma born in Mayiladuthurai, Tamil Nadu, India on 31-07-1968.She completed Ph.D degree in DC-DC Converters from Anna University. Presently she is working as an Assistant Professor, EEE department, College of Engineering, Guindy, Anna University, Chennai, India. Her areas of interest are Power Quality, Resonant Converters, and Matrix Converters.

Analysis of a Mixed-Signal Control for DC-DC Converters based on Hysteresis Modulation And Estimated Inductor Current

D. Trevisan*, S. Saggini*, P. Mattavelli**, L. Corradini***, P. Tenti***

* DIEGM, University of Udine, Italy
** DTG, University of Padova, Italy
*** DEI, University of Padova, Italy

Abstract--This paper investigates a mixed signal voltage-mode controller for dc-dc converters based on hysteresis modulation. Both switch turn-on and switch turn-off instants are determined asynchronously through comparison of the converter output voltage with a voltage ramp driven by the digital controller and generated through Digital-To-Analog Conversion (DAC). With respect to a previously reported mixed-signal solution, the implementation proposed here generates different slope voltage ramps during the switch-on and switch-off phases and it fully approximates the dynamic capabilities of a multi-loop control with an internal hysteresis current control based on the estimated inductor current. Simulation and experimental results on a synchronous buck converter are also reported.

Index Terms—DC-DC Converters, Hysteresis Modulation, Mixed-Signal Controllers

I. INTRODUCTION

Integrated digital controllers for high-frequency Switch-Mode Power Supplies (SMPS) are gaining attention both from the scientific and industrial point of view. As discussed in previous works [1-8], they offer several potential advantages compared to their analog counterparts, such as immunity to component variations, ability to implement sophisticated control schemes, controller autotuning and system diagnostics features; some of these advantages have been also demonstrated by practical implementations.

On the other hand, most of the digital controllers developed up to now [1-10] are not always able to match the dynamic performance of their analog counterparts, at least in term of achievable closed-loop bandwidth, and several non-linear digital control algorithms are used in order to guarantee fast dynamic response under large-signal load step variations. For this reason, the development of alternative digital (or mixed-signal) control architectures, which potentially enable simpler control architecture and faster dynamic response, is an interesting trend in SMPS control field. Some examples in this direction are reported in [11,12], where switching instants are determined by a combinations of system clock and intersections of controlled state variables and digitally controlled voltage and current ramps. In [11] the analog peak current-mode modulator and the voltage-

mode modulator are controlled by two DACs, while in [12] a similar principle is applied to the voltage-mode control of a synchronous buck converter. In [13], the derivative term of the compensator is obtained in the analog domain using a signal which is the difference between the output voltage at the beginning of the switching period (obtained using sample-and-hold) and the instantaneous output voltage.

This paper proposes an extension of the mixed synchronous/asynchronous digital voltage-mode controller based on [12] – where the switch turn-off is determined by the digital clock using a digital PI (Proportional-Integral) control – and further developed in [15], where the use of a low-resolution DAC and a comparator were proposed for the emulation of a multi-loop control with the internal hysteresis current control, allowing both switch turn-on and turn-off to be determined asynchronously by the intersection of analog quantities. In this technique the additional current signal, needed for the internal current control, is internally estimated and generated by the DAC so as to emulate the inductor switching ripple without current sensing requirements. Moreover, the switching frequency is controlled by modulating the amplitude of the voltage ramp. In [15] a simplified implementation was proposed, based on the generation a voltage ramp that features slopes equal to their average value during both the turn-on and off phases. Such approximation led to a low-complexity frequency stabilization algorithm at the price of slower dynamics.

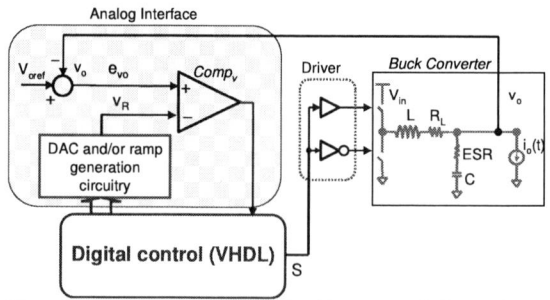

Figure 1 – Block diagram of control architecture applied to dc-dc converters.

978-1-4244-0644-9/07/$25.00 ©2007 IEEE

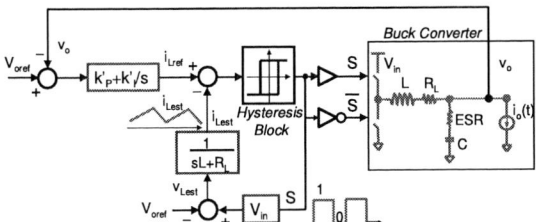

Figure 2 – Multi-loop control with hysteresis current control based on the inductor current estimation i_{Lest}.

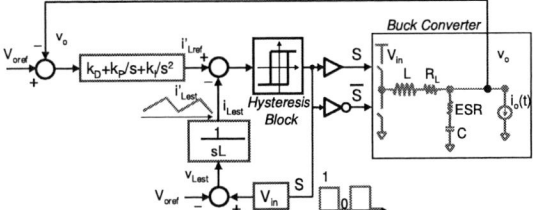

Figure 3 – Derivation of the control of Fig. 2 where the inductor estimation i'_{Lest} is performed using a pure integral action.

Instead, in this paper, because of the use of different slopes of the voltage-controlled ramps during the switch-on and switch-off intervals, the dynamic performances are fully comparable to those obtainable with pure hysteresis modulation. On the other hand, an additional switching frequency stabilization algorithm is needed compared with [15]. Simulation and experimental results have verified the effectiveness of the proposed control architecture.

II. BASIC PRINCIPLES OF OPERATION

Figure 1 shows the basic scheme of the proposed mixed-signal voltage-mode control applied to dc-dc converters. Switch turn-off and turn-on are determined by the intersection of the output voltage error $e_{vo}(t)$ and the reference ramp signal $v_R(t)$ generated by DAC or, more generally, by analog circuitry which implements the step-wise linear voltage ramp $v_R(t)$.

A. Basics of multi-loop control with hysteresis current control and estimation of inductor current [15]

The basic principle of operation is derived from a modified version of the hysteresis modulation in multi-loop control, as shown in Fig. 2, where an internal hysteresis current control is used and an external Proportional-Integral (PI) regulator on the output voltage error e_{vo} determines the inductor current reference i_{Lref}. Since only the output voltage sensing is used, the inductor current estimation is based only on the switch status S, the nominal input voltage V_{in} and the output voltage reference V_{oref}, as shown in Fig. 2. Using time-averaging and small-signal assumptions, the estimated inductor current i_{Lest} of Fig. 2 is given by:

$$i_{Lest}(s) = \frac{V_{in}}{sL+R_L}\delta(s) = \frac{V_{in}}{R_L(1+s\tau_I)}\delta(s), \quad (1)$$

where $\tau_I = L/R_L$ and the output voltage reference V_{oref} is assumed to be constant. Moreover, since the hysteresis block maintains the current error near zero, using time-averaging assumptions, $i_{Lref} = i_{Lest}$. Then, taking into account that $i_{Lref}(s) = (k'_P + k'_I/s)e_{vo}(s)$, the transfer function between the output voltage error e_{vo} and the duty cycle δ is given by:

$$\frac{\delta(s)}{e_{vo}(s)} = \frac{R_L}{V_{in}}(k'_P + \frac{k'_I}{s})(1+s\tau_I) \quad (2)$$

Thus, we have briefly recalled that the control shown in Fig. 2 represents, under small-signal assumptions, an equivalent PID (Proportional-Integral-Derivative) control on the output voltage error. The main advantages of the scheme of Fig. 2 are a fast dynamic response ensured by the hysteresis block with respect to a PWM based solution, and a fast large-signal response, as in any hysteresis modulation.

Under our point of view, the scheme of Fig. 2 and the equivalent small-signal behavior reported in (2) are relevant for understanding the dynamic characteristics of the system and for the design of the regulator coefficients, since the design criteria of PID controller parameters are well-known.

B. Equivalent control scheme when the inductor current is estimated using an integral action

From the implementation point of view, it is convenient to perform the estimation of the inductor current using a pure integral action. In fact, if the ramp signal $v_R(t)$ is obtained using a DAC, then an up-down counter can be used. Whenever an analog circuitry is used, then a capacitor can be charged/discharged by a constant current. In either implementation, a pure integral action of the current estimation is needed. When the inductor current estimation is performed using a pure integral action (i.e. substituting the block $1/(sL+R_L)$ with $1/(sL)$) and imposing the same small-signal dynamic behavior (2), the control block diagram of Fig. 3 is derived. The equivalence between the scheme of Fig. 2 and Fig. 3 is based on (2). In fact, from Fig. 3 we have:

$$i'_{Lref}(s) = (k_D + \frac{k_P}{s} + \frac{k_I}{s^2})e_{vo}(s) \quad (3a)$$

$$i'_{Lest}(s) = \frac{V_{in}}{sL}\delta(s) = \frac{(m_{on}+m_{off})}{s}\delta(s) = \frac{2m_v}{s}\delta(s) \quad (3b)$$

where m_{on} and $-m_{off}$ (being m_{off} positive) are the current slopes during the switch on and switch off period respectively, while m_v is the average slope between m_{on} and m_{off}. Since $i'_{Lref} = i'_{Lest}$, (2) becomes:

$$\frac{\delta(s)}{e_{vo}(s)} = \frac{1}{2m_v}(k_D s + k_P + \frac{k_I}{s}) \quad (4)$$

which again represents an equivalent PID control, as discussed for (2).

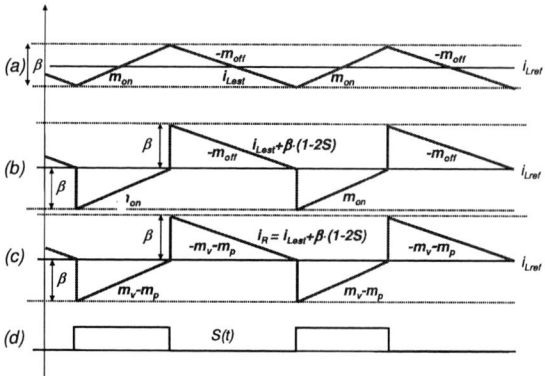

Figure 4 – Derivation of the ramp waveform i_R.

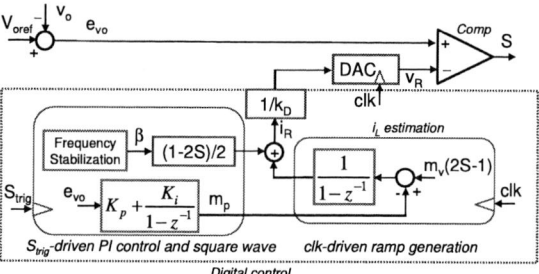

Figure 6 – Equivalent Block diagram of Fig. 5 where the digital implementation of the analog block is reported.

C. Replacement of the hysteresis block by a comparator

In Fig. 5, the hysteresis block has been replaced by a comparator, where on the negative terminal the signal $i_R(t)=i'_{Lest}+\beta/2(1\text{-}2S)$ is used, being β the hysteresis band and S the switch status ($S=1$, switch on, $S=0$, switch off). Indeed, as reported in Figs. 4a and 4b, the use of the hysteresis block or the use of the comparator with an added square-wave signal which depends on the switch status and whose peak-to-peak amplitude is β, are equivalent. Moreover, in Fig. 5 the PI part of the PID regulator has been moved to the negative input of the comparator.

Looking at Fig. 5 and at (4), it can be shown than the positive slopes m_{on} and m_{off} do not need to be strictly related to the effective current slopes. For example, both values can be set equal to a constant m_v (i.e. the average between the two) without affecting the equivalent small-signal behavior (4), as it is also evident looking at the last equality in (3b). Indeed, any slope difference from the theoretical ones is now compensated in scheme of Fig. 5 by the output m_p of the PI regulator, with no influence on the time-average dynamic behavior. This point is important from the implementation point of view, since the ramp slope during the on and off period can be set to m_v and $-m_v$ respectively, thus simplifying circuit implementation.

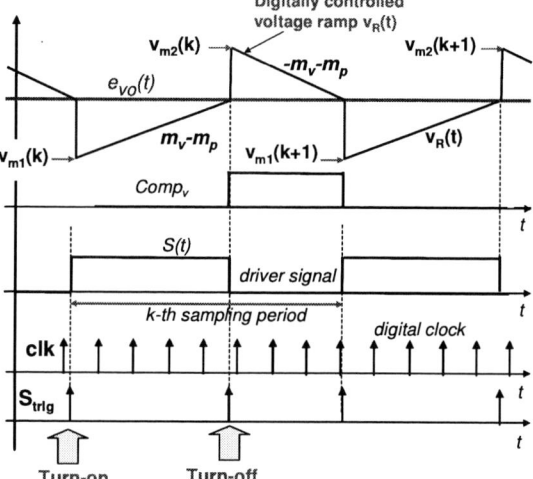

Figure 7 – Main waveforms under steady-state conditions

D. Digital implementation of the ramp generator using two different clock signals

The final step toward the proposed solution is the digital implementation of the integrator and the PI control of the scheme in Fig. 5. One example is reported in Fig. 6, where the integration is performed using two distinct clock signals: the first is based on a fast internal clock clk with period T_{clk}, and is used for the inductor current estimation or, equivalently, for the generation of the ramp with slopes equal to $-m_p+m_v$ and $-m_p-m_v$; the second, S_{trig}, is generated by the switching events (either turn on or turn off) and superimposes the square wave that defines the hysteresis window. The relevant steady-state waveforms of the proposed implementation are shown in Fig. 7, where the DAC generated ramp $v_R(t)$ is shown along with the switch status S and the two aforementioned clock signals clk and S_{trig}.

The choice of performing the integration of the PI control using signal S_{trig} is due to the fact that the PI contribution is not dominant at high-frequency and a larger integration step should not affect the dynamic performance, but this provision limits the digital controller complexity.

It is also worth to mention the possibility of generating the ramp $v_R(t)$ by means of analog circuitry (i.e. capacitor

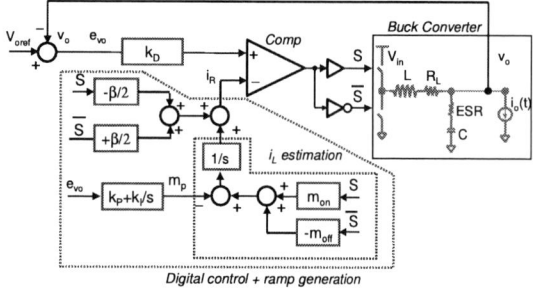

Figure 5 – Equivalent block diagram of Fig. 3 where a comparator has been substituted to the hysteresis block and part of the PID control has been moved to the negative part of the comparator.

with linear charge and discharge) and with a DAC that sets the initial value of the capacitor after each switch transition. This option, which may be effective also from the IC point of view, is interesting for high-frequency operation where the internal clock frequency is not much higher than the switching frequency.

In previously reported implementations of this technique [15], the generated voltage ramp had equal slopes $+m_v$ and $-m_v$ for the turn-on and turn-off switching phases respectively. This approximation led to a control law with an inherent frequency stabilization algorithm. However, this came at the price of a different control dynamics with respect to a pure hysteresis control. No such approximation has been undertaken for the present implementation, thus fully exploiting the dynamic capabilities of the discussed method. Moreover, having variable slopes in the voltage ramp generation also enhances the overall robustness against switching noise and ringing, which are disturbances especially critical when low or high duty cycle operation is considered. Indeed, as the turn-on or turn-off ramp slopes increase because of low or high duty cycle operation, the switching event sensitivity to disturbances is strongly reduced.

Provisions aimed to stabilize the switching frequency have to be included as indicated in Fig. 6. On this purpose, several techniques have been proposed in the literature that can be successfully employed. In order to avoid interactions between the frequency stabilization algorithm and the voltage control loop, the former has been realized with a simple up/down counter plus a saturated accumulator so as to provide slow dynamic response and limited impact on system complexity. The frequency control algorithm is presented in Fig. 8 where the gain k_{freq} contains the scaling factor $1/T_{clk}$ and furthermore can be used to tune the controller dynamic performance; the saturation block forces the definition of a minimal and maximal value for the hysteresis band thus effectively limiting the frequency modulation range.

In the scheme of Fig. 6, three major points should be highlighted:

1) the derivative gain K_D can be normalized to one, if all the terms in the digital control algorithm are divided by K_D. Thus, in the positive input of the comparator, the output voltage error is directly used, as shown in Fig.1;

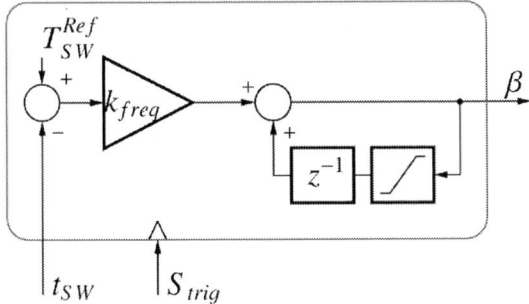

Figure 8- Implementation of the frequency stabilization algorithm

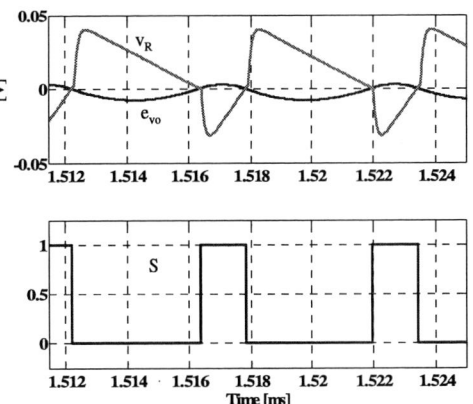

Figure 9 – Simulated steady-state waveforms: (top) output voltage error e_{vo} and ramp voltage v_R; (bottom) switching signal S

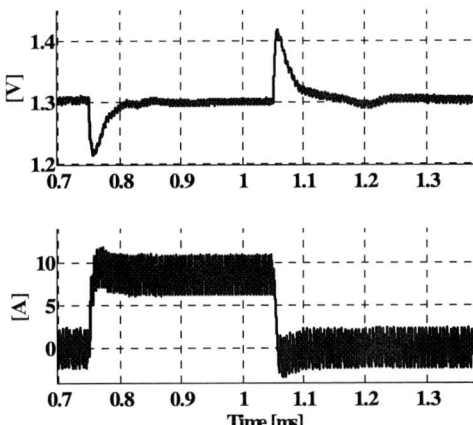

Figure 10– Simulated load step change ($I_o = 0$ A $\longleftrightarrow$ 8.5 A); output voltage (top) and inductor current (bottom)

2) the equivalent derivative action of the controller is not affected by sampling delays since it remains in the analog domain;

3) the sampling of the output voltage error $e_{vo}(t)$ is performed by the DAC and the comparator, since the DAC level is known at each commutation of the comparator; thus, the sampling of $e_{vo}(t)$ is performed twice per switching period.

It is worth to underline that simple analog components are required by the scheme of Fig. 1 and Fig. 6, i.e. a low-resolution DAC and a comparator. Thus the proposed technique inherits the advantages common to digital control techniques. Other advantages include the intrinsic programmability of the compensator and the possibility to implement controller tuning and diagnostic functions.

III. SIMULATION RESULTS

The proposed solution has been initially verified by simulation on a synchronous buck converter with the following parameters: V_{in}=5-12 V, V_o=1.3 V, L=1.2 µH, C=160 µF, f_{sw}=250 kHz. The capacitor value is quite small and not suitable to limit the overshoot and undershoot to the typical dynamic specifications for point-of-load applications. This investigation is however interesting to show the dynamic properties of the proposed solution. Controller parameters are designed based on (4) and imposing a bandwidth of around 60 kHz and two PID zeros at f_{z1}=10 kHz and f_{z2}=5 kHz.

Figure 9 reports the main steady-state waveforms while Fig. 10 shows the dynamic behavior during load transients (I_o = 0 A $\leftrightarrow$ 8.5 A). The converter response is fast and well damped after the load step change. Please note the slight filtering on the voltage ramp waveform $v_R(t)$ in Fig. 9, always present in real DACs and that reduce the impact of quantization effects on the system behavior.

In case of ceramic output capacitors (i.e. negligible *ESR*), the adopted modulation scheme presents the same behavior as the one presented in [12]. When the output voltage variation is limited to the region where switch on or switch off period variation cannot be measured, the controller action is purely proportional [12]; if the system results unstable in this situation, an oscillation between the boundaries of that region is expected.

Simulations also verified the effectiveness of the frequency stabilization algorithm proposed in II.D. Frequency stabilization during load transients could be restored in 8-10 switching cycles, confirming the validity of the discussed solution.

Finally, in order to verify that the small-signal model given by (4) is a good approximation, Fig. 11 compares the theoretical loop gain of the voltage loop with the results computed by numerical simulations (indicated with dots 'o'). Simulation results show a good modeling accuracy.

IV. EXPERIMENTAL RESULTS

The proposed solution has been tested using a FPGA by Xilinx (Spartan 3) and an experimental prototype has been realized with the following parameters: V_{in} = 5-12 V, V_o=1.3 V, L=1 µH, C= 160 µF (ceramic caps), f_{sw}~250 kHz, I_{onom}= 10 A, T_{clk}= 100 ns (f_{clk}=10 MHz).

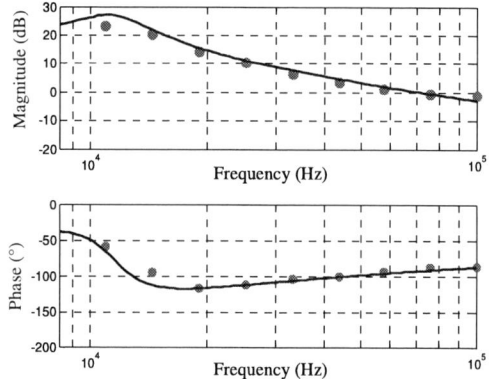

Figure 11 – Loop gain of the voltage loop: verification of (4) (dots 'o' indicate time-domain simulations)

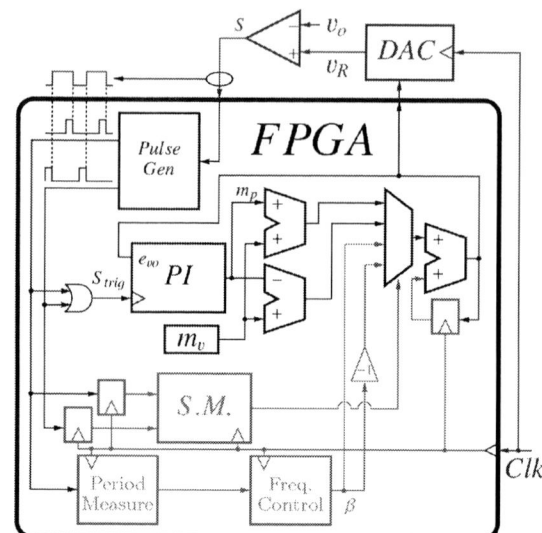

Figure 12 – FPGA internal architecture block diagram

The controller parameters are designed similarly to what reported in the simulation section, with the difference that f_{z1} has been placed at 2.3 kHz.

The control algorithm was developed in VHDL and it requires approximately 7,250 gates with no specific optimizations in word length and internal organization. As far as the latter is concerned, a fully parallel architecture for the PI controller has been selected in order to take full advantage of the derivative action inherently available in the modulation scheme. Figure 12 presents the block scheme of the FPGA architecture; the asynchronous part of the device is triggered by the output of the 'Pulse Gen' block. This one is necessary in order to provide sharp and noise-free pulses to the remainder part of the logic: this block requires 600 gates. A simple state machine and a multiplexer coordinate the input data to the accumulator that calculates the internal ramp representation; these blocks take up 4,400 gates. The switching frequency control requires 950 gates and 1,300 are necessary for the PI controller.

The load step responses to an 8 A load change are reported in Figs. 13 and 14 where it is shown that the initial transient response is very fast, well damped and with small limit cycle oscillations after load turn on.

Figure 15 presents the steady-state waveforms of the main quantities involved in the system characterization and control. Then, Fig. 16 and Fig. 17 present a detail of a positive and negative load step variation respectively. It is shown that the control action requires a sudden and wide f_{sw} and t_{on} variation. As previously mentioned, the frequency stabilization algorithm restores the target switching frequency in about ten switching periods after the load step.

The aforementioned robustness of the proposed solution against switching noise could be appreciated in the experimental tests, where a definite reduced sensitivity to disturbances was observed with respect to

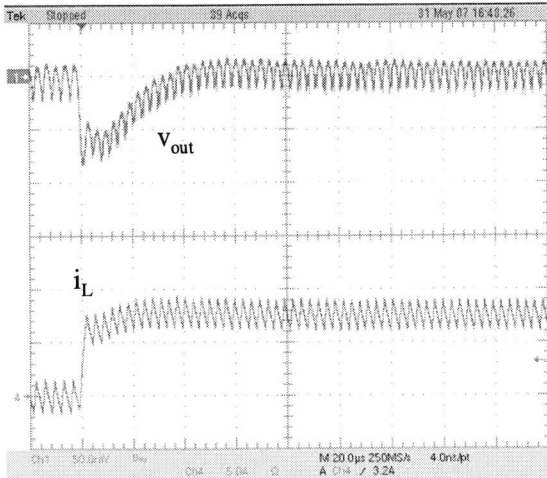

Figure 13– Experimental load step change (Io=0 A → 8 A) with the proposed control (v_O – 50mV/div, iL -5A/div, time 20μs/div)

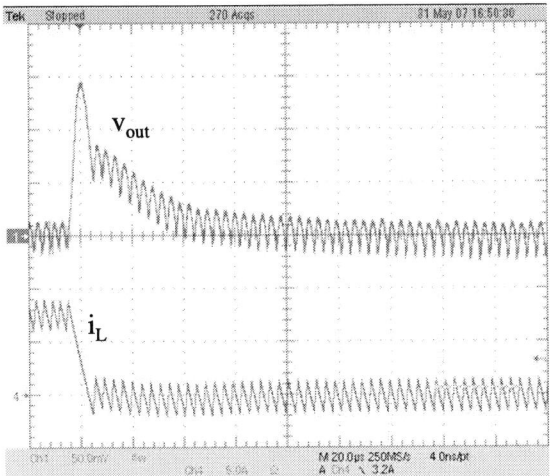

Figure 14– Experimental load step change (Io=8 A → 0 A) with the proposed control (v_O – 50mV/div, iL -5A/div, time 20μs/div)

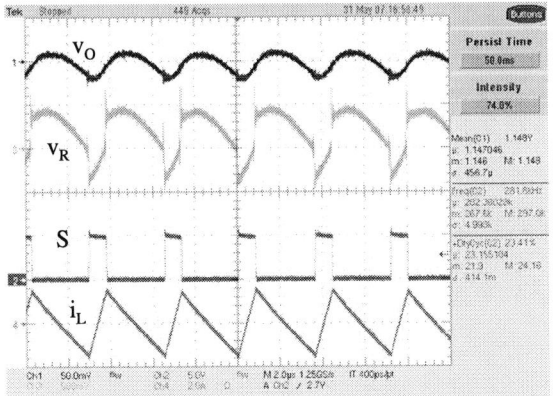

Figure 15– Experimental steady state waveforms (v_O – 50mV/div, v_R – 500mV/div, S – 5V/div, iL -2A/div, time 2μs/div)

the previously reported solution employing constant slopes for the voltage ramp generation.

The frequency modulation and duty-cycle jitter phenomena are presented with greater detail in Fig. 18 where the steady state waveforms have been acquired with infinite persistence. It can be easily noticed that there is a significant jitter in the steady-state waveforms, mainly due to the presence of the aforementioned limit-cycle oscillations, since the prototype has been equipped with only ceramic capacitors [12], and it is affected by the presence of some noise. As a consequence, the switching frequency is continuously controlled and regulated. It is worth to highlight that the variations in Fig. 18 are mainly due to switching frequency variations and to a limited extent due to duty-cycle adjustments; in fact statistical measurements on the duty-cycle in steady state conditions have shown that the average duty-cycle is 23.07% for the conditions reported in Fig. 18, with minimum and maximum values equal to 22.19% and 24.25%, respectively, and a standard deviation equal to 0.32% within a measurement window 10000 switching periods wide.

V. CONCLUSIONS

This paper has investigated a mixed signal voltage-mode controller for dc-dc converters that fully approximates the dynamic capabilities of classical analog hysteresis controllers. Both switch turn-on and switch turn-off are determined asynchronously by comparing the converter output voltage error with a digitally controlled voltage ramp. The resulting controller guarantees low-complexity, frequency modulation during transients and high dynamic performances, which approximately resemble those of a multi-loop control with an internal hysteresis current control based on the estimated inductor current. A frequency stabilization technique has been realized affecting the equivalent hysteresis band through a simple manipulation of the ramp amplitude. Simulation and experimental results on a digitally controlled synchronous buck converter have verified the properties of the proposed solution.

REFERNCES

[1] B.J. Patella, A. Prodic, A. Zirger, D. Maksimović, "High-frequency Digital Controller IC for dc/dc Converters", IEEE Trans. on Power Electronics, Vol. 18, No. 1, January 2003, pp. 438-446.

[2] J. Xiao, A.V. Peterchev, S.R. Sanders, "Architecture and IC implementation of a digital VRM controller", IEEE Trans. on Power Electronics, Vol. 18, No. 1, January 2003, pp. 356-364.

[3] A.V. Peterchev, S.R. Sanders, "Quantization resolution and limit cycling in digitally controlled PWM converters", IEEE Trans. on Power Electronics, Vol. 18, No. 1, January 2003, pp. 301-308.

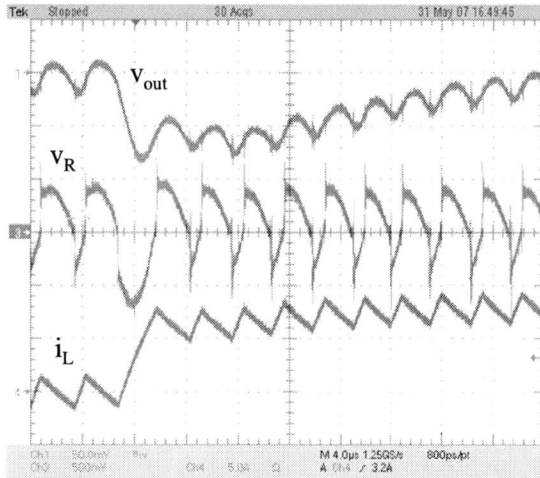

Figure 16– Experimental load step change (Io=0 A → 8 A) with the proposed control (v_O – 50mV/div, v_R – 500mV/div, iL -5A/div, time 4μs/div)

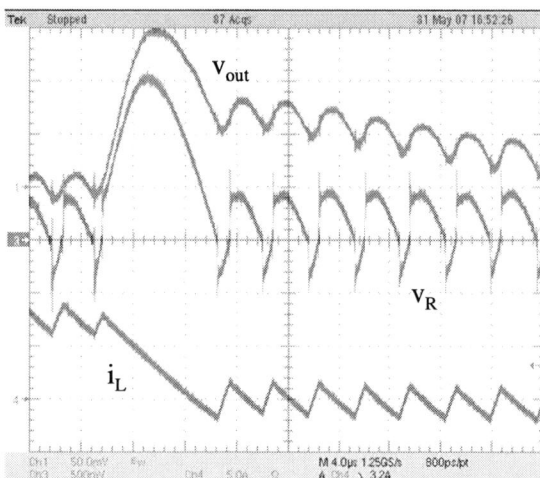

Figure 17– Experimental load step change (Io=8 A → 0 A) with the proposed control (v_O – 50mV/div, v_R – 500mV/div, iL -5A/div, time 4μs/div)

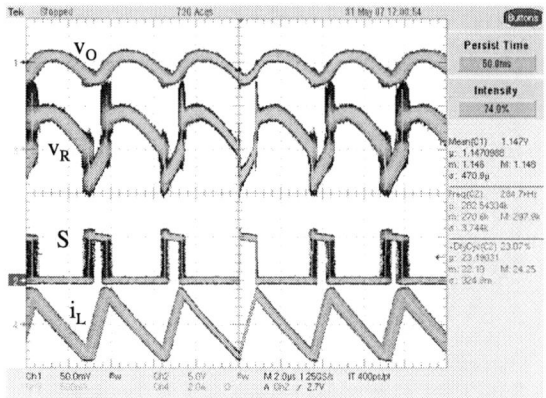

Figure 18– Experimental steady state waveforms acquired with infinite persistance (v_O – 50mV/div, v_R – 500mV/div, S – 5V/div, iL - 2A/div, time 2μs/div)

[4] Hao Peng, D Maksimovic, A. Prodic, E. Alarcon, "Modeling of quantization effects in digitally controlled DC-DC converters", Power Electronics Specialists Conference (PESC'04), June 20-25, 2004, pp. 4312 – 4318

[5] B. Miao, R. Zane, D. Maksimovic. "Practical On-Line Identification of Power Converter Dynamic Responses", IEEE Applied Power Electronics (APEC'05), Austin (TX), March 2005.

[6] A. Prodic, D. Maksimovic, R.W. Erickson "Design and implementation of a digital PWM controller for a high-frequency switching dc-dc power converter", IEEE IECON'01, 2001, pp. 893-898.

[7] F. Kurokawa, T. Sato, H. Matsuo, H. Eto, "Output characteristics of DC-DC converter with DSP control" 24th Annual International Telecommunications Energy Conference, 2002, (INTELEC'02), pp: 421 -426.

[8] C. Kranz, "Complete Digital Control Method for PWM dc-dc Boost Converter" IEEE Power Electronics Conference 2003 (PESC'03), Acapulco, Mexico, June 2003.

[9] J. Chen, A. Prodic , R. W. Erickson and D. Maksimovic, "Predictive digital current programmed control," IEEE Trans. on Power Electronics, Vol. 18, No. 1, pp. 441-419, January 2003.

[10] Huliehel, F.; Ben-Yaakov, S, "Low-frequency sampled-data models of switched mode DC-DC converters", IEEE Transactions on Power Electronics, Vol. 6, No. 1 , Jan. 1991, pp 55 -61.

[11] S. Saggini, M. Ghioni, A. Geraci, "An innovative digital control architecture for low-voltage, high-current dc-dc converters with tight voltage regulation" IEEE Trans. on Power Electronics, vol. 19, January 2004, pp. 210-218.

[12] D. Trevisan, P. Mattavelli, G. Garcea, M. Ghioni, S. Saggini "High-Performance Synchronous/Asynchronous Digital Voltage-Mode Control for dc-dc Converters" IEEE Applied Power Electronics Conference (APEC) '06, Dallas (TX), March 2006.

[13] S. Saggini, P. Mattavelli, M. Ghioni, "High-Performance Mixed-Signal Voltage-Mode Control for dc-dc Converters with inherent analog derivative action" IEEE Applied Power Electronics Conference (APEC'07), Anaheim, March 2007.

[14] Abu-Qahouq, J.A.; Hong Mao; Batarseh, I, "Novel control method for multiphase low-voltage high-current fast-transient VRMs" IEEE Power Electronics Conference 2002 (PESC'02), June 2002, pp. 1576-1581.

[15] D. Trevisan, S. Saggini, P. Mattavelli, "Hysteresis-Based Mixed-Signal Voltage-Mode Control For DC-DC Converters", IEEE Power Electronics Specialists Conference (PESC'07), Orlando, June 2007

Power Quality Study in Macao

Sio-Un Tai*, Man-Chung Wong*, Ming-Chui Dong*, Ying-Duo Han**

* Faculty of Science and Technology, University of Macau, Macau, SAR, P.R. China
** Department of Electrical Engineering, Tsinghua University, Beijing, P.R. China

Abstract—This paper reports power quality measurements in thirteen locations of Macao. The results are analyzed and compared with relevant standards for evaluating the quality of power in Macao. The data collected provides the initial core of documentation for future reference and follow-up of the trends at the end-users.

Index Terms—power quality, voltage RMS variation, voltage sag/swell, voltage imbalance, flicker, harmonics, neutral current, power factor, frequency.

I. INTRODUCTION

Power quality has become an increasing concern for utilities and their electrical customers, in recent years; there has been proliferation of modern electronics such as computer loads, variable speed drives and industrial logic controllers. While such devices are sensitive to the variation of the supply voltage, they are also the source for power quality disturbances. Due to their nonlinear nature, these loads inject harmonic current into the power system and cause voltage harmonic distortion. There is a need to understand how the disturbances will affect sensitive loads and develop appropriate specifications, or install appropriate power conditioning systems. Harmonics can result in equipment heating, communication interface and control malfunctions. Voltage sags of only few cycles can cause loss of computer data or errors. The increased concern for power quality has resulted in significant advances in monitoring equipment that can be used to characterize disturbances and power quality variations [1].

Macao is a part of China's territory. It is located on the Southeast coast of China to the west of the Pearl River Delta. Moreover, there are not any formal regulations for harmonics, imbalance and flicker in Macao. According to Macao power supply company (CEM) Statistics 2004 shows in Table 1, the energy consumption of hotel & commercial building is the most, the next is residential building, and then is public administrative building and industrial building. Therefore, a power quality study is recommended to perform at different types of buildings in Macao, they are: Industrial type: a special facility building (Fac), Public type: three public administrative buildings (Pub A, B & C), Residential type: five residential buildings (Res A, B, C, D & E), Other type: a commercial building (Com), a hotel (Hot), a middle school (Sch) and an indoor sport center (Cen).

The measurement terms include 15-minute averages voltage RMS variation, voltage sag/swell, 15-minute maximum voltage harmonic, 15-minute maximum voltage imbalance, voltage flicker Pst and Plt, 15-minute averages neutral current RMS, 15-minute maximum current harmonic, 15-minute averages power factor and 15-minute maximum and minimum frequency variation. On the other hand, the measurement period for each location is one week (seven days) for achieving the whole pattern, the measurement point is main power distribution panel of building and the measurement equipment is power quality analyzer ACE-4000.

This paper is divided into five sections, the first section is introduction, the second section is power quality standards, the third section is results of monitoring, the fourth section is estimation of power quality in Macao, and the final section is conclusions.

Table 1
Electricity consumption distribution in Macao in year 2004

Customer	No. of customer	Percentage	Sales energy	Percentage
Domestic	173,760	87.1%	588.9 GWH	31.3%
Wholesale & retail, hotels & recreation, commercial	21,119	10.6%	972.6 GWH	51.6%
Industrial	2,422	1.2%	141.9 GWH	7.5%
Public sector & street lighting	2,281	1.1%	180.7 GWH	9.6%

II. POWER QUALITY STANDARDS

In this power quality study, it uses the standard limits in Table 2 for evaluating the performance.

Table 2
Power quality standards

Item	Standard Limits	Standard From
Voltage RMS variation	230V phase to neutral +5% ~ -10% (207V ~241.5V)	Macao
	Neutral to ground <3V	[2]
Voltage sag/swell	207V (-10%) <V_{RMS}<253V (+10%)	IEEE 1159-1995
Voltage harmonic	THD_V% < 5%	IEEE 1159-1995
Voltage imbalance	<2%	Mainland China
Voltage flicker	Pst <1.0, Plt <0.8	Mainland China
Neutral current RMS	<20% of phase current RMS	[3]

Current harmonic	I_{SC}/I_L *
	    <20, TDD_I%<5%
	    20<50, TDD_I%<8%
	    50<100, TDD_I%<12%
	    100<1000, TDD_I%<15%
	    >1000, TDD_I%<20%
		IEEE 519-1992

Power factor	>0.85	Mainland China
Frequency variation	$\pm$2% (49 ~ 51Hz)	Macao

* I_{SC}=maximum short-circuit current at PCC
I_L=maximum demand load current at PCC

This work was supported by the Research Committee of University of Macau.

III. RESULTS OF MONITORING

Some of the major measurement results are summarized in this section. Measurement involved voltage RMS variation, voltage sag/swell, voltage harmonic, voltage imbalance, voltage flicker, neutral current, current harmonic, power factor and frequency variation.

A. Voltage RMS Variation

Table 3 shows the 15-minute averages phase voltage RMS variation measurement results. All results are within the limits. Moreover, in most cases the measured values are under 230V. Fig. 1 shows the measured figure of the public administrative building A which is always lower than 230 V and even lower than 220V during office hours.

Besides, one form of common-mode noise in three-phase power systems is the voltage difference between neutral and ground. The effect of this noise on the computer system is somewhat debatable, yet computer vendor specifications typically call for less than 0.5-3V [2]. In this study it finds that the high values in the public administrative building A, the residential building B, C & D and the hotel exceed 3V. Fig. 2 shows the measured 15-minute maximum neutral voltage RMS and maximum neutral current RMS figures of the public administrative building A, basically the voltage follows the current. Moreover, the neutral voltage values of this building are higher than the neutral voltage values of other measured buildings, most of the values exceed 3V during office hours. It is believed that it is caused by the impedance of the electric cable between transformer and main distribution panel of the building since system earthing is classified as TN-S in Macao. Fig. 3 shows the measured 15-minute maximum neutral voltage RMS and maximum neutral current RMS figures of the residential building D, the voltage values are generally between 1V and 3V except the one 8.378V. It is believed that the high value comes from the local power network since there is not obvious rising on the neutral current before, during and after the event. Moreover, similar status also happens in the residential building B& C and the hotel.

Table 3
Measured voltage RMS variation

Location	Phase A		Phase B		Phase B	
	Max.	Min.	Max.	Min.	Max.	Min.
Fac	229.0V	219.7V	228.8V	218.6V	229.3V	219.8V
Pub A	232.5V	213.5V	233.2V	213.9V	233.8V	216.2V
Pub B	232.9V	218.3V	232.6V	217.8V	233.6V	218.4V
Pub C	234.5V	220.8V	234.2V	220.6V	235.1V	221.8V
Res A	230.3V	220.0V	231.0V	222.3V	231.3V	222.6V
Res B	231.4V	222.5V	231.8V	223.1V	232.0V	222.7V
Res C	232.7V	223.4V	232.5V	222.5V	232.0V	222.6V
Res D	231.7V	221.6V	232.2V	222.6V	230.2V	220.9V
Res E	231.8V	221.9V	231.9V	221.7V	232.8V	222.4V
Com	233.4V	221.1V	233.4V	221.2V	234.2V	221.5V
Hot	232.6V	223.4V	232.8V	223.0V	233.6V	224.3V
Sch	232.5V	219.3V	233.2V	220.1V	233.3V	223.7V
Cen	234.8V	223.5V	235.0V	224.0V	235.5V	224.9V

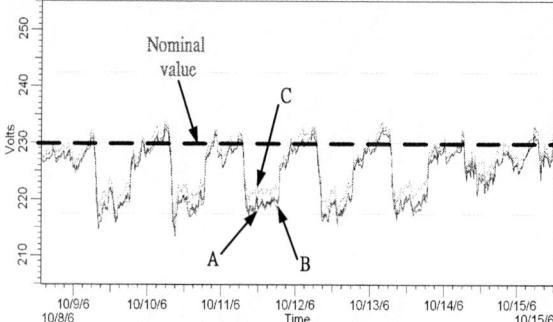

Fig. 1 Phase voltage RMS variation versus time in the public administrative building A

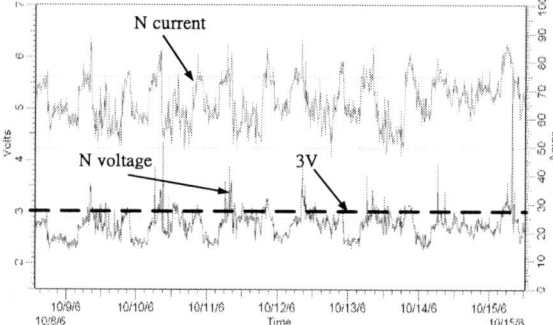

Fig. 2 Neutral voltage RMS and neutral current RMS versus time in the public administrative building A

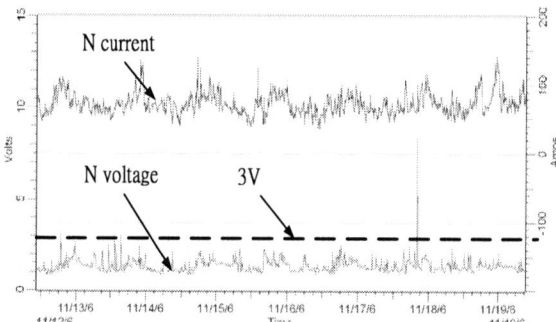

Fig. 3 Neutral voltage RMS and neutral current RMS versus time in the residential building D

B. Voltage Sag/Swell

Table 4 shows there are 79 and 8 voltage sag events have been recorded during the monitoring period in the special facility building and the public administrative building A respectively. All events are inside the CBEMA envelop. Therefore, it is believed they are not potentially hazardous for sensitive equipments. For special facility building, the sags are caused by the starting of the 320KW large capacity traditional water pump. For public administrative building A, the sags are mainly caused by the impedance of the electric cable between transformer and main distribution panel since the current RMS among phase lines have not large differences before, during and after the event, and no similar event happens in the other building which is using the same transformer with the public administrative building A. Fig. 4 shows one of the sags which is inside the CBEMA envelop at phase A of the public administrative building A.

Table 4
Measured voltage sags

Location	Phase	No. of sags	Min. voltage RMS sags & duration (cycles)
Fac	A	30	203V & 12 cycles
	B	25	201V & 13 cycles
	C	24	200.8V & 13 cycles
Pub A	A	8	206.6V & 1 cycle
	B	0	N/A
	C	0	N/A
Others	A	0	N/A
	B	0	N/A
	C	0	N/A

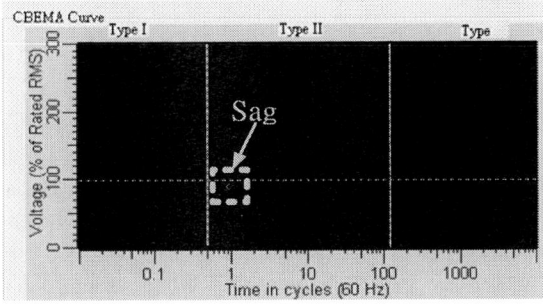

Fig. 4 CBEMA curve with recorded one voltage disturbance at phase A of the public administrative building A

Table 5
Measured voltage harmonic THD$_V$%

Location	Phase A (Max.)	Phase B (Max.)	Phase C (Max.)
Fac	3.205%	3.417%	3.545%
Pub A	2.322%	2.696%	2.851%
Pub B	6.823%	4.797%	3.766%
Pub C	1.873%	1.799%	2.073%
Res A	2.714%	2.210%	2.506%
Res B	3.026%	2.513%	2.801%
Res C	3.510%	3.127%	3.289%
Res D	4.198%	3.559%	3.884%
Res E	2.922%	2.549%	2.581%
Com	3.156%	2.649%	2.820%
Hot	3.293%	3.076%	3.101%
Sch	2.940%	2.788%	2.722%
Cen	2.677%	2.464%	2.516%

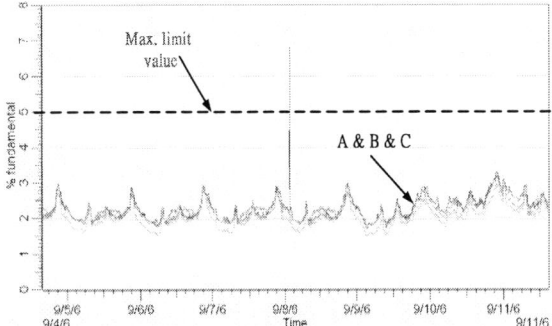

Fig. 5 Voltage harmonic THD$_V$% versus time in the public administrative building B

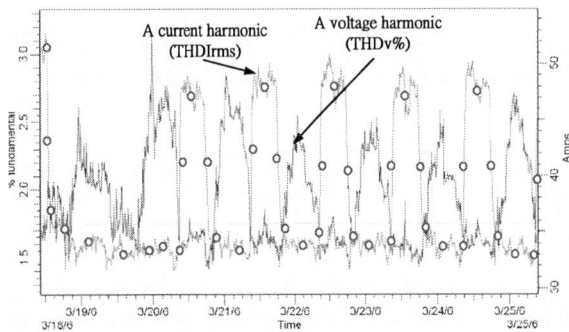

Fig. 6 Voltage harmonic THD$_V$% and current harmonic THD$_I$rms versus time in the commercial building

C. Voltage Harmonic

Table 5 shows the 15-minute maximum total voltage harmonic distortion THD$_V$% measurement results. Almost all results are within the limits except some values in the public administrative building B. Besides, order 3, 5 & 7 are main harmonic components of all the buildings. Fig. 5 shows the measured voltage THD$_V$% figure of the public administrative building B, the voltage distortion is generally between 2% and 3% except the one 6.823%, the periods of the event are 100 cycles. It is believed that the harmonic comes from the local power network since there is not obvious rising on the total current harmonic distortion THD$_I$rms before, during and after the harmonic event.

Moreover, there is a worth point to concern is the background voltage harmonic, it occurs in the commercial building, the hotel, the middle school and the indoor sport center. Fig. 6 shows the current harmonic has different patterns than the voltage ones in the commercial building. Current harmonic THD$_I$rms is high but voltage harmonic THD$_V$% is low during office hours, and it is opposite during non-office hours. This is the result of the fact that the voltage distortion at this location is not governed only by the harmonic current injected by the customer, but also is primarily determined by other loads on the same transformer feeder or the local power network [4]. Since the commercial building and the hotel are using their own transformers and there are no other loads from other users, and high background voltage harmonic occurs at night and it seems that it will not produce high current harmonic at that period, it is believed that it is caused by the local power network.

D. Voltage Imbalance

Table 6 shows the 15-minute maximum voltage imbalance measurement results. Almost all results are within the limits except some values in the public administrative building B, the residential building C and the middle school. Fig. 7 shows the measured figure of the residential building C, its imbalance is generally between 0.5% and 1% except the one 48.89%, the period of the high imbalance value is 1 cycle, it is believed that the imbalance comes from the local power network and the main imbalance is on phase angles since the voltage RMS and the current RMS among phase lines have not large differences before, during and after the event. Similar status also happens in the public administrative building B and the middle school.

Table 6
Measured voltage imbalance

Location	Max.	Min.
Fac	1.973%	0.344%
Pub A	1.336%	0.441%
Pub B	2.204%	0.267%
Pub C	0.836%	0.315%
Res A	0.893%	0.292%
Res B	0.760%	0.298%
Res C	48.890%	0.271%
Res D	0.640%	0.186%
Res E	0.769%	0.338%
Com	0.638%	0.326%
Hot	0.714%	0.362%
Sch	2.108%	0.402%
Cen	0.986%	0.314%

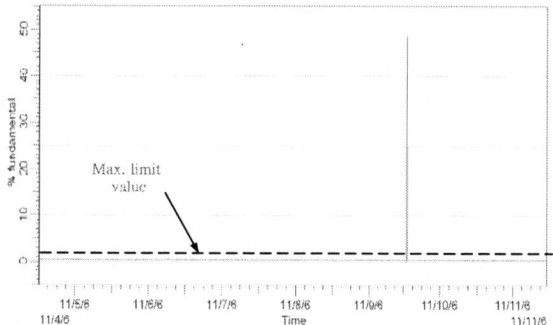

Fig. 7 Voltage imbalance versus time in the residential building C

E. Voltage Flicker

Table 7 shows the voltage flicker Pst and Plt measurement results. Most of the measured values are within the limits except some values in the special facility building and the public administrative building A. Fig. 8 shows the measured flicker Pst figure of the special facility building, the high values (around 0.9~1.5pu) are caused by 320KW large capacity water pump starting, in general its values are only around 0.2pu. On the other hand, it is believed that the high values of the public administrative building A are mainly caused by the impedance of the electric cable between transformer and main distribution panel.

Table 7
Measured voltage flicker Pst & Plt

Location	Phase A (Max.)		Phase B (Max.)		Phase C (Max.)	
	Pst	Plt	Pst	Plt	Pst	Plt
Fac	1.466	0.937	1.558	0.990	1.612	1.009
Pub A	1.532	1.286	1.223	0.889	1.109	0.519
Pub B	0.854	0.190	0.755	0.221	0.610	0.205
Pub C	0.701	0.244	0.724	0.252	0.734	0.254
Res A	0.246	0.122	0.247	0.136	0.251	0.128
Res B	0.285	0.142	0.278	0.154	0.277	0.148
Res C	0.225	0.122	0.250	0.145	0.236	0.126
Res D	0.390	0.136	0.460	0.155	0.408	0.143
Res E	0.247	0.108	0.268	0.123	0.253	0.122
Com	0.361	0.238	0.358	0.246	0.409	0.265
Hot	0.406	0.176	0.428	0.191	0.439	0.195
Sch	0.507	0.334	0.667	0.336	0.857	0.291
Cen	0.343	0.150	0.347	0.157	0.335	0.156

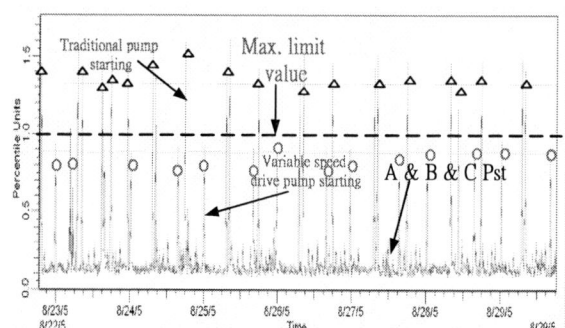

Fig. 8 Voltage flicker Pst versus time in the special facility building

F. Neutral Current RMS

In general the neutral current RMS should not exceed 20% of the phase current RMS [3]. Table 8 shows that the 15-minute averages neutral current RMS exceed 20% of the phase current in the residential building A, B, C, D & E, the hotel, the middle school and the indoor sport center, they are between 30.1% and 91.6% of phase current. Fig. 9 shows in the residential building A, the phase current RMS is around 160A during non-office hours and 70A during office hours, the neural current RMS is around 100A during non-office hours and 20A during office hours, the neutral current exceed 20% of the phase current.

Table 9 & 10 show the neutral current RMS including fundamental and harmonic measurement results during office and non-office hours of the public administrative building A & B & C, the residential building C & D & E, the commercial building and the indoor sport center. The tables show that current harmonic $THD_I\%$ is between around 60% and 244% during office hours and 5 of 8 locations exceed 100%. On the other hand, between around 47% and 172% during non-office hours and 6 of 7 locations exceed 100%. It is concluded that harmonic components on neutral current play a larger role than the linear load unbalances.

Table 8
Measured phase and neutral current RMS

Location	Phase A (Max.)	Phase B (Max.)	Phase C (Max.)	Neutral (Max.) (% of phase current)
Fac	1797A	1801A	1373A	121.3A (8.8%)
Pub A	305.1A	304.9A	308.5A	44.94A (14.7%)
Pub B	1.629KA	1.637KA	1.639KA	130.2A (7.9%)
Pub C	966.8A	985.6A	960.0A	101.2A (10.5%)
Res A	198.4A	154.5A	153.7A	106.5A (69.2%)
Res B	163.5A	120.8A	171.2A	94.23A (78.0%)
Res C	245.3A	241.9A	307.2A	112.7A (46.5%)
Res D	243.9A	263.5A	253.9A	99.04A (40.6%)
Res E	142.9A	147.8A	156.9A	67.1A (46.9%)
Com	454.0A	448.0A	419.7A	82.77A (19.7%)
Hot	310.6A	337.6A	261.9A	89.24A (34.2%)
Sch	227.6A	243.9A	254.5A	68.57A (30.1%)
Cen	52.06A	32.54A	31.69A	29.04A (91.6%)

610

Table 9
Measured neutral current RMS during office hours

Location	Neutral current RMS	Neutral current fundamental RMS	Neutral current harmonic RMS (THD$_I$%)
Pub A	37.2A	22.46A	29.49A (131.3%)
Pub B	111.2A	51.6A	98.53A (190.9%)
Pub C	84.42A	30.84A	75.39A (244%)
Res C	72.7A	59.75A	40.91A (68.46%)
Res D	30.46A	23.22A	18.47A (79.5%)
Res E	37.54A	20.31A	30.31A (149.2%)
Com	69.53A	47.9A	55.44A (115.7%)
Cen	24.43A	20.79A	12.85A (61.8%)

Table 10
Measured neutral current RMS during non-office hours

Location	Neutral current RMS	Neutral current fundamental RMS	Neutral current harmonic RMS (THD$_I$%)
Pub A	24.49A	12.18A	20.56A (168.8%)
Pub B	53.93A	36.62A	38.28A (104.5%)
Pub C	30.85A	14.83A	25.6A (172.6%)
Res C	85.67A	46.01A	71.27A (154.9%)
Res D	70.76A	41.48A	56.43A (136.0%)
Res E	43.55A	24.62A	35.5A (144.1%)
Com	34.23A	30.5A	14.47A (47.4%)

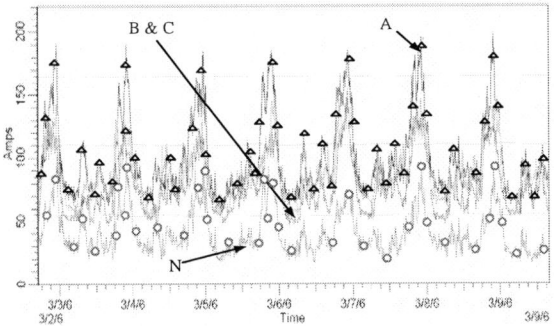

Fig. 9 Current RMS versus time in the residential building A

G. Current Harmonic

Table 11 shows the 15-minute maximum current harmonic THD$_I$rms measurement results. And according to the short circuit current and the demand load current of the locations, it gets the TDD$_I$% limits shows in Table 12. Table 12 also shows the calculated TDD$_I$%, almost all results are within the limits except some values in the special facility building, and the high values mainly are caused by the running of 320KW large capacity variable speed drive water pump and the starting of the 320KW large capacity traditional water pump. Fig. 10 shows the current harmonic THD$_I$rms in the special facility building is around 200A (TDD$_I$%=10%) during the running of variable speed drive water pump but some exceed 500A (TDD$_I$%=25%) or even 1000A (TDD$_I$%=50%) at traditional water pump starting. Besides, order 3, 5 & 7 are main harmonic components of all the buildings.

Table 11
Measured current harmonic THD$_I$rms

Location	Phase A (Max.)	Phase B (Max.)	Phase C (Max.)
Fac	1394A	1364A	254.5A
Pub A	17.89A	14.96A	23.33A
Pub B	72.92A	73.7A	108.5A
Pub C	47.35A	47.32A	58.14A
Res A	33.13A	34.85A	31.05A
Res B	28.43A	28.85A	30.18A
Res C	35.09A	39.86A	39.58A
Res D	34.68A	40.64A	32.84A
Res E	26.7A	26.49A	28.08A
Com	52.76A	52.99A	51.25A
Hot	77.24A	79.12A	61.79A
Sch	27.91A	27.12A	29.11A
Cen	6.434A	5.452A	10.59A

Table 12
Current harmonic TDD$_I$% limits and calculated current harmonic TDD$_I$% from measured current harmonic THD$_I$rms

Location	TDD$_I$% Limit	Phase A TDD$_I$%	Phase B TDD$_I$%	Phase C TDD$_I$%
Fac	5%	69.70%	68.20%	12.73%
Pub A	12%	3.58%	2.99%	4.67%
Pub B	8%	4.56%	4.60%	6.78%
Pub C	8%	4.74%	4.73%	5.81%
Res A	12%	5.52%	5.81%	5.18%
Res B	12%	4.74%	4.81%	5.03%
Res C	12%	5.85%	6.64%	6.60%
Res D	12%	5.78%	6.77%	5.47%
Res E	12%	4.45%	4.42%	4.68%
Com	5%	2.64%	2.65%	2.56%
Hot	8%	4.83%	4.95%	3.86%
Sch	15%	9.30%	9.04%	9.70%
Cen	15%	8.04%	6.81%	13.24%

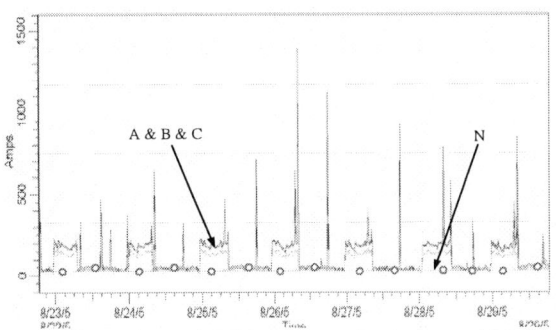

Fig. 10 Current harmonic THD$_I$rms versus time in the special facility building

H. Power Factor

Table 13 shows the 15-minute averages power factor measurement results. Some values exceed the limits except in the special facility building and the hotel. Fig. 11 shows the measured figure of the public administrative building B, its power factor is around 0.82 during office hours and 0.88 during non-office hours. In other words, it is high during non-office hours but low during office-hours. Since the measured power factor of two centralized air-conditioning systems in this research is around 0.82 during running and the running current of the systems exceeds 40% of the total running current of the buildings, it is believed that the low values during office hours is mainly caused by the centralized air-conditioning system.

611

Table 13
Measured power factor

Location	Phase A		Phase B		Phase C	
	Max.	Min.	Max.	Min.	Max.	Min.
Fac	0.939	0.879	0.944	0.876	0.926	0.862
Pub A	0.963	0.792	0.972	0.791	0.962	0.807
Pub B	0.885	0.735	0.879	0.727	0.922	0.758
Pub C	0.959	0.724	0.967	0.724	0.937	0.714
Res A	0.957	0.729	0.942	0.710	0.922	0.688
Res B	0.938	0.766	0.915	0.721	0.934	0.742
Res C	0.939	0.819	0.967	0.867	0.893	0.735
Res D	0.956	0.814	0.932	0.809	0.852	0.685
Res E	0.912	0.720	0.897	0.723	0.920	0.788
Com	0.953	0.808	0.950	0.825	0.915	0.758
Hot	0.997	0.910	0.996	0.911	0.993	0.887
Sch	0.953	0.761	0.996	0.750	0.969	0.627
Cen	0.741	0.634	0.807	0.400	0.898	0.330

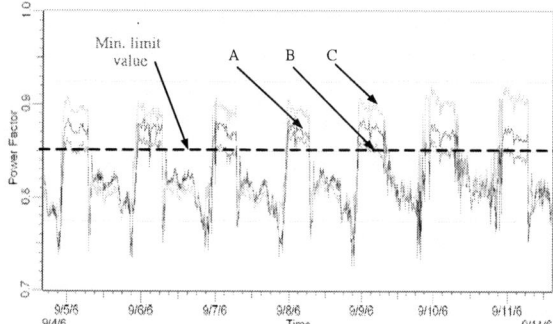

Fig. 11 Power factor versus time in the public administrative building B

I. Frequency Variation

The 15-minute maximum and minimum measured frequency is 50.2Hz and 49.8Hz respectively, it is ±0.4% of the nominal value and all results are within the limits. Fig. 12 shows the frequency variation in the residential building A.

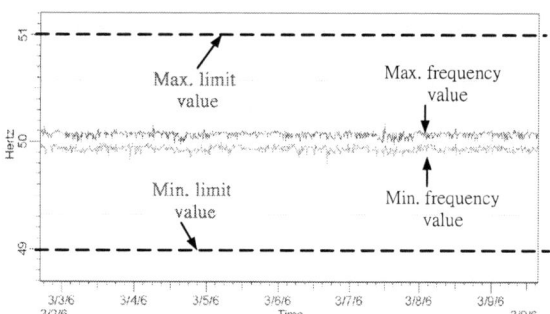

Fig. 12 Frequency variation versus time in the residential building A

IV. Estimation of Power Quality in Macao

Table 14 shows a summary of power quality in Macao which is based on the measured results in the above thirteen locations of Macao. Here, it neglects the findings on electric cable impedance which may cause high neutral voltage, voltage sag and flicker in the administrative building A since only one location has these problems and the status is not in usual in Macao. Moreover, for the power quality in the whole Macao, estimation is made as below:

- Voltage RMS variation: In general it is within the standard limits. Most voltage RMS values are under 230V. Lower voltage RMS values during office hours due to the running of large capacity equipments such as centralized air-conditioning system.

- Voltage sag/swell and flicker: In general they are mainly inside the industrial building. Besides, the sag and flicker will be reduced since industry will be reduced and increasing the use of variable speed drive (VSD) systems in Macao. On the other hand, the sags are within the CBEMA envelop. Moreover, there is not any swell event.

- Voltage harmonic and voltage imbalance: In general they are within the standard limits. But sometimes there are high values which exceed the limits and might come from the local power network. On the other hand, the voltage harmonic will be increased due to the use of VSD systems are increasing continuously in Macao.

- Neutral current RMS: Some values exceed 20% of phase current, and it is mainly in the residential building since in general the electrical appliance inside the building are single-phase equipment.

- Current harmonic: In general $TDD_I\%$ is within the limits. On the other hand, the current harmonic will be increased due to the use of VSD systems are increasing continuously in Macao.

- Power factor: In general most values exceed the standard limits. But it is believed that power factor will be improved since the end-users start to install capacitor banks in Macao for saving money.

- Frequency: All is within the standard limits.

Table 14
Summary of power quality in Macao based on measured results

Item	Type	Measurement Results
Voltage RMS Variation	Industrial, Public, Residential, Others	Phase voltage is within the limits, some neutral voltage values exceed 3V which may be caused by the local power network.
Voltage Sag/Swell	Industrial	Voltage sag events are mainly caused by large power capacity equipment. Moreover, all is within the CBEMA envelop.
	Public, Residential, Others	No sag/swell event.
Voltage Harmonic	Industrial, Public, Residential, Others	All is within the limits except some high values are caused by the local power network. Order 3, 5 & 7 are main voltage harmonic components. Somewhere background voltage harmonic are found.
Voltage Imbalance	Industrial, Public, Residential, Others	All is within the limits except some high values are caused by the local power network.
Voltage Flicker	Industrial	Some values exceed the limits and it is mainly caused by large power capacity equipments.
	Public, Residential, Others	All is within the limits.
Neutral Current RMS	Industrial, Public	All is within 20% of phase current and harmonic is main component.
	Residential, Others	Some values exceed 20% of phase current RMS and harmonic is main component.

Current Harmonic	Industrial	Some $TDD_I\%$ values exceed the limits and it is mainly caused by large power capacity equipments. Order 3, 5 & 7 are main current harmonic components.
	Public, Residential, Others	All is within the limits. Order 3, 5 & 7 are main current harmonic components.
Power Factor	Industrial, Public, Residential, Others	In most cases exceed the limits.
Frequency variation	Industrial, Public, Residential, Others	All is within the limits.

V. CONCLUSIONS

A power quality monitoring in the thirteen locations of Macao has been reported. In general the results shows that phase voltage and frequency are within the limits, neutral voltage, voltage harmonic $THD_V\%$ and voltage imbalance sometimes exceed the limits and may come from the local power network, voltage sag, flicker and current harmonic $TDD_I\%$ sometimes exceed the limits in industrial building and may come from large power equipment, somewhere neutral current exceeds 20% of phase current and harmonic is main component, most power factor exceeds the limits and somewhere background voltage harmonic are found, order 3, 5 & 7 are main voltage harmonic components, order 3, 5 & 7 are main current harmonic components. Moreover, an estimation of power quality for the whole Macao also be reported, it mainly points out that voltage sag and flicker will be reduced since the industry will be reduced and increasing the use of variable speed drive (VSD) systems in Macao, voltage harmonic and current harmonic will be increased due to the use of VSD systems and power factor will be improved since the end-users start to install capacitor banks in Macao.

ACKNOWLEDGEMENT

The authors would like to thank the Research Committee of University of Macao for their supports, and also thank the persons of the thirteen buildings for their support and helpful for this power quality monitoring.

REFERENCES

[1] Eloi Ngandui, and Cedric Meignant, "Power quality monitoring and analysis of a university distribution system," *IEEE CCECE,* vol. 2, pp. 863-867, May 2001.

[2] Gruzs, T.M., "A survey of neutral currents in three-phase computer power systems," *IEEE Trans. on industry applications,* vol. 26, pp. 719-725, July-August 1990.

[3] A. C. Liew, "Excessive neutral currents in three-phase fluorescent lighting circuits," *IEEE Trans. on industry applications,* vol. 25, no. 4, pp. 776-782, July/August 1989.

[4] Alexander E. Emanuel, John A. Orr, David Cyganski, Edward M. Gulachenski, "A survey of harmonic voltages and currents at the customer's bus," *IEEE Trans. on power delivery,* vol. 8, No. 1, pp. 411-421, January 1993.

Some Findings on Harmonic Measurement in Macao

Sio-Un Tai*, Man-Chung Wong*, Ming-Chui Dong*, Ying-Duo Han**

* Faculty of Science and Technology, University of Macau, Macau, SAR, P.R. China
** Department of Electrical Engineering, Tsinghua University, Beijing, P.R. China

Abstract—This paper reports some findings on harmonic measurement at several buildings of Macao. The findings are current harmonic, harmonic current is caused by variable refrigerant volume (VRV) air-conditioning system, harmonic effect on capacitor, even order current harmonic and neutral current harmonic, they are for reference for Macao to set up the harmonic standard limits in the future.

Index Terms—power quality, current harmonic.

I. INTRODUCTION

Power quality has become an increasing concern for utilities and their electrical customers, in recent years; there has been proliferation of modern electronics such as computer loads, variable speed drives and industrial logic controllers. While such devices are sensitive to the variation of the supply voltage, they are also the source for power quality disturbances. Due to their nonlinear nature, these loads inject harmonic current into the power system and cause voltage harmonic distortion. There is a need to understand how the disturbances will affect sensitive loads and develop appropriate specifications, or install appropriate power conditioning systems. Harmonics can result in equipment heating, communication interface and control malfunctions. Voltage sags of only few cycles can cause loss of computer data or errors. The increased concern for power quality has resulted in significant advances in monitoring equipment that can be used to characterize disturbances and power quality variations [1].

Macao is a part of China's territory. It is located on the Southeast coast of China to the west of the Pearl River Delta. Moreover, there are not any formal regulations for harmonics, imbalance and flicker in Macao. Therefore, a power quality study is recommended to perform at different types of buildings in Macao, they are: Industrial type: a special facility building (Fac), Public type: three public administrative buildings (Pub A, B & C), Residential type: five residential buildings (Res A, B, C, D & E), Other type: a commercial building (Com), a hotel (Hot), a middle school (Sch) and an indoor sport center (Cen). The measurement period for each location is one week (seven days) for achieving the whole pattern, the measurement point is main power distribution panel of building and the measurement equipment is power quality analyzer ACE-4000.

This paper introduces the results of harmonic monitoring and then reports the findings on current harmonic, harmonic current is caused by variable refrigerant volume (VRV) air-conditioning system, harmonic effect on capacitor, even order current harmonic and neutral current harmonic.

This paper is divided into four sections, the first section is introduction, the second section is results of harmonic monitoring, the third section is findings on harmonic measurement, and the final section is conclusions.

II. RESULTS OF HARMONIC MONITORING

This section reports an estimation of harmonic in Macao which is based on the monitoring results in the above several locations of Macao is as below [2]:

- High voltage harmonic (THD$_V$%) may be caused by the local power network;
- Order 3, 5 & 7 are main voltage harmonic components;
- Somewhere background voltage harmonic is found;
- High phase current harmonic (TDD$_I$%) may be caused by the large power capacity equipments inside buildings;
- Order 3, 5 & 7 are main phase current harmonic components;
- Some neutral current exceed 20% of phase current and harmonic is main component.

III. FINDINGS ON HARMONIC MEASUREMENT

This section reports some findings on harmonic measurement which are worth to concern and analysis, they are: current harmonic (THD$_I$%) vs. current harmonic (TDD$_I$%), more harmonic current is caused by VRV centralized air-conditioning system, harmonic effect on capacitor bank, even order current harmonic and neutral current harmonic. The findings may let the end-users to pay more attentions or more researches on these power quality terms in the near future.

A. Current Harmonic (THD$_I$%) vs. Current Harmonic (TDD$_I$%)

In this section it reports the finding on current harmonic (THD$_I$%) vs. current harmonic (TDD$_I$%) by using the measured current harmonic results in Macao.

Current harmonic distortion levels can be characterized by the complete harmonic spectrum with magnitudes and phase angles of each individual harmonic component [3]. It is also common to use a single quantity,

---

This work was supported by the Research Committee of University of Macau.

978-1-4244-0644-9/07/$25.00 ©2007 IEEE

the Total Current Harmonic Distortion (THD$_I$%) or Total Current Demand Distortion (TDD$_I$%) to evaluate the harmonic levels. IEEE 519-1992 recommends to use current harmonic (TDD$_I$%) but other areas such as Hong Kong recommends to use current harmonic (THD$_I$%) to evaluate the current harmonic levels. Table 1 shows the current harmonic standard limits from IEEE 519-1992 and Hong Kong (ESMD). The main differences of the two ways are: current harmonic (TDD$_I$%) value must be referred to a constant base (e.g. the rated load current or demand current), current harmonic (THD$_I$%) value must be referred to the fundamental current base which can vary over a wide range.

According to the power quality measurement, in general the buildings are equipped with centralized air-conditioning system (chiller unit), its current harmonic (THD$_I$%) is low during office hours but high during non-office hours since the injected harmonic current from the chiller unit generally is little, and the fundamental current of the chiller unit exceeds 40% of the total fundamental current of the building. Moreover, the increasing of harmonic current due to nonlinear load such as computer equipment is less than the increasing of fundamental current of chiller unit. As a result, the base current is increased too much but the harmonic current is increased little during office hours but it is opposite during non-office hours.

Fig. 1 shows the measured total current harmonic distortion (THD$_I$%) figure of the public administrative building C which is equipped with chiller unit, its harmonic (THD$_I$%) value is around 5% and 20% during office hours and non-office hours respectively, the difference is 15%. In other words, the values during office hours are within the limits (<5%) which mentioned in Table 2 but during non-office hours exceeds the limits. Moreover, the results seem is not reasonable and even it is wrong if only using the total current harmonic distortion (THD$_I$%) limits to evaluate the harmonic level of the building since in fact the injected harmonic current during non-office hours is lower than the injected harmonic current during office hours. On the other hand, the main concept of the current harmonic standards are to limit individual power consumers inject harmonic current for assuring that voltage distortion (THD$_V$%) at the PCC does not exceed 5%. Fig. 2 shows the measured total current harmonic distortion (THD$_I$rms) figure of the public administrative building C, its harmonic (THD$_I$rms) value is around 35A and 15A during office hours and non-office hours respectively, the difference is 20A. Moreover, when comparison between the trend of harmonic distortion (THD$_I$%) and the trend of harmonic distortion (THD$_I$rms), it finds that generally current harmonic distortion (THD$_I$rms) is opposite to current harmonic distortion (THD$_I$%). The above assumes the chiller unit is switched-on during office hours and is switched-off during non-office hours.

On the other hand, for other types of buildings such as residential building, the difference of current harmonic (THD$_I$%) between office hours and non-office hours is not as large as the difference in the building which is equipped with chiller unit. Fig. 3 shows the measured current harmonic (THD$_I$%) figure of the residential building B which harmonic (THD$_I$%) is around between 15% and 25%, the difference is 10%.

As a conclusion, for all types of buildings using current harmonic (THD$_I$%) to evaluate the current harmonic level seems is not reasonable. Current harmonic (THD$_I$%) seems is more suitable to use for evaluating the harmonic level of an equipment since the base fundamental current of an equipment would not vary over a wide range.

From the above, it seems that using total current demand distortion (TDD$_I$%) is a reasonable way to estimate current harmonic level since it could reflect the actual harmonic current level, but the calculation of current harmonic (TDD$_I$%) needs to get I$_{SC}$ and I$_L$ first, different selection of I$_{SC}$ and I$_L$ will affect the results. In general the end-users may not easy to get a correct I$_{SC}$ and I$_L$. Therefore, the selected I$_{SC}$ and I$_L$ may not be corrected and would evaluate the harmonic level incorrectly, or choosing different values under different persons. One of the good harmonic indices is characterized by the following: Harmonic indices should be simple and practical so that they can be widely used with ease. Thus it is recommended the local power supply company could acknowledge the end-users about their I$_{SC}$ and I$_L$ at interval.

Finally, Table 2 & 3 show the measured current harmonic (THD$_I$%) and calculated current harmonic (TDD$_I$%) from the power quality measurement in several buildings of Macao, it shows that generally harmonic (THD$_I$%) exceeds the limits but harmonic (TDD$_I$%) is within the limits.

Table 1
Current harmonic standard limits

Standard Limits		From
I$_{SC}$/I$_L$ *		IEEE 519-1992
<20,	TDD$_I$%<5%	
20<50,	TDD$_I$%<8%	
50<100,	TDD$_I$%<12%	
100<1000,	TDD$_I$%<15%	
>1000,	TDD$_I$%<20%	
I<40A,	THD$_I$%<20% **	Hong Kong (ESMD)
40A≦I<400A,	THD$_I$%<15%	
400A≦I<800A,	THD$_I$%<12%	
800A≦I<2000A,	THD$_I$%<8%	
I≧2000A,	THD$_I$%<5%	

* I$_{SC}$=maximum short-circuit current at PCC
   I$_L$=maximum demand load current at PCC
** I=rated current of building

Table 2
Current harmonic (THD$_I$%) limits and measured current harmonic (THD$_I$%)

Location	THD$_I$% Limit	Phase A THD$_I$% (Max.)	Phase B THD$_I$% (Max.)	Phase C THD$_I$% (Max.)
Fac	5%	55.30%	42.57%	29.21%
Pub A	12%	28.09%	50.75%	47.65%
Pub B	5%	19.27%	20.06%	27.12%
Pub C	5%	25.61%	28.84%	27.01%
Res A	12%	27.74%	34.43%	29%
Res B	12%	28.44%	34.64%	29.01%
Res C	12%	21.4%	26.91%	18.82%
Res D	12%	21.79%	17.89%	14.64%
Res E	12%	26.48%	32.3%	29.26%
Com	5%	37.04%	39.96%	45.41%
Hot	5%	33.21%	30.8%	28.66%
Sch	12%	46.83%	64.13%	63.73%
Cen	15%	120.1%	16.17K%	16.14K%

Table 3
Current harmonic (TDD$_I$%) limits and calculated current harmonic (TDD$_I$%) from measured current harmonic (THD$_I$rms)

Location	TDD$_I$% Limit	Phase A TDD$_I$%	Phase B TDD$_I$%	Phase C TDD$_I$%
Fac	5%	69.70%	68.20%	12.73%
Pub A	12%	3.58%	2.99%	4.67%
Pub B	8%	4.56%	4.60%	6.78%
Pub C	8%	4.74%	4.73%	5.81%
Res A	12%	5.52%	5.81%	5.18%
Res B	12%	4.74%	4.81%	5.03%
Res C	12%	5.85%	6.64%	6.60%
Res D	12%	5.78%	6.77%	5.47%
Res E	12%	4.45%	4.42%	4.68%
Com	5%	2.64%	2.65%	2.56%
Hot	8%	4.83%	4.95%	3.86%
Sch	15%	9.30%	9.04%	9.70%
Cen	15%	8.04%	6.81%	13.24%

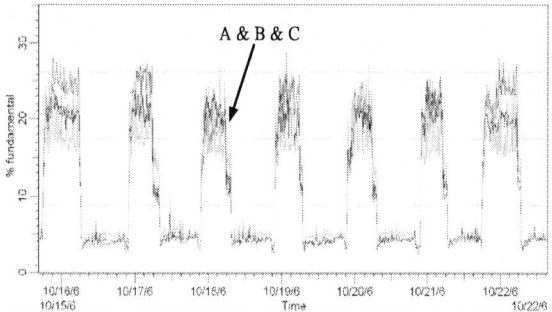

Fig. 1 Current harmonic (THD$_I$%) versus time in the public administrative building C

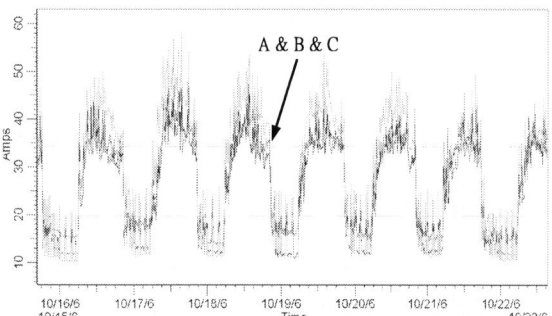

Fig. 2 Current harmonic (THD$_I$rms) versus time in the public administrative building C

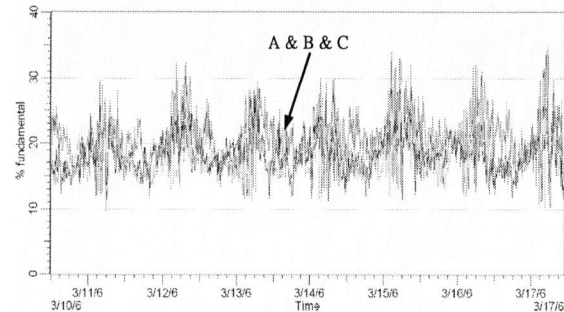

Fig. 3 Current harmonic (THD$_I$%) versus time in the residential building B

B. More Harmonic Current is caused by Variable Refrigerant Volume (VRV) Centralized Air-conditioning System

In this research, besides taking the measurement in the main power distribution panels of several buildings of Macao, it has also measured three types of popular centralized air-conditioning systems in Macao since most industrial, public and commercial type buildings in Macao are using the air-conditioning systems, and the operation current of the air-conditioning systems generally exceed 40% of the total operation current of the building, the power quality level of air-conditioning system for building is very important. The three types of systems are: Variable Refrigerant Volume (VRV) Air-Conditioning System, Air-Cooled Chiller Unit and Water-Cooled Chiller Unit. VRV system is a new and advanced air-conditioning system, it mainly uses variable speed drive technology, and the other two types of chiller units are traditional centralized air-conditioning systems. In this section it reports the measured current harmonic results in Macao from three types of centralized air-conditioning systems.

Table 4 shows a comparison table of the harmonic level among VRV system, air-cooled chiller unit and water-cooled chiller unit which are based on the measurement results. The results show that the current harmonic distortion (THD$_I$%) of VRV system is 38.5%, it exceeds the standard limits, but the harmonic distortion of air-cooled chiller unit and water-cooled chiller unit are 7.2% and 3.1% respectively, they are less than VRV system 31.3% and 35.4% respectively, they are within the standard limits. In other words, more harmonic current is caused by the VRV air-conditioning system.

Fig. 4 ~ 6 show the measured voltage waveform and current waveform of VRV system, air-cooled chiller unit and water-cooled chiller unit respectively. The figures show that the current harmonic distortion of VRV system is much larger than the current harmonic distortion of the two chiller units.

In the future, for the purpose of energy saving and more flexible use, more variable speed drive technology will be applied to different equipments such as lifts, air-conditioning systems and motors in Macao and other areas. At that time more harmonic current would be produced and injected into the power system to affect the operation of equipments. Therefore, it is recommended that the current harmonic of all variable speed drive products should be limited.

Table 4
Comparison of current harmonic among VRV system, air-cooled chiller unit and water-cooled chiller unit

Air-conditioning system	Operation current RMS	Harmonic current RMS	THD$_I$%
VRV system	16.3A	6.2A	38.5%
Air-cooled chiller unit	205A	14.7A	7.2%
Water-cooled chiller unit	143A	4.45A	3.1%

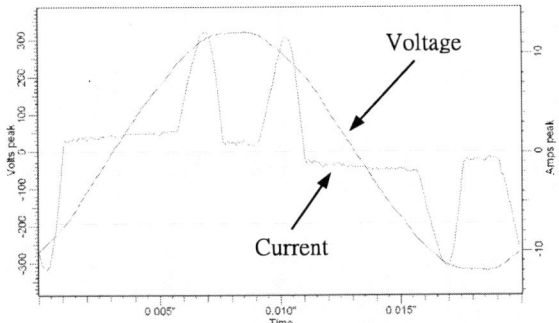

Fig. 4 Voltage and current waveform of VRV air-conditioning system

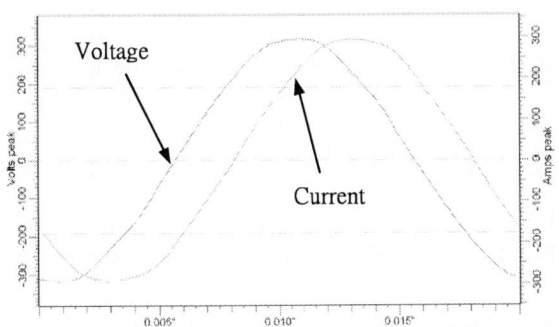

Fig. 5 Voltage and current waveform of air-cooled chiller unit

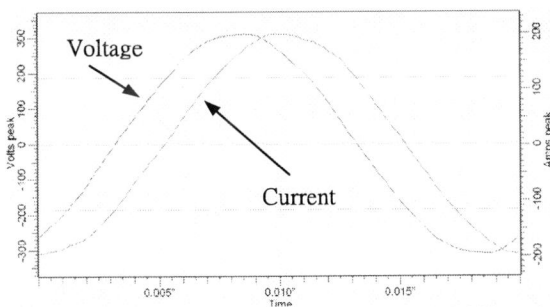

Fig. 6 Voltage and current waveform of water-cooled chiller unit

C. Harmonic Effect on Capacitor Bank

The use of capacitor is to improve power factor and voltage also has a significant influence on harmonic levels. Capacitor does not generate harmonics, but provide network loops for possible resonant conditions. If the addition of capacitor tunes the system to resonate near a harmonic frequency present in the load current or system voltage, large currents or voltage at that frequency will be produced.

According to the power quality measurement in several buildings of Macao, it seems that there is not any harmonic is amplified, one of the reasons is only 1 of 13 measured locations has installed capacitors due to nowadays there are not any formal regulations and standard limits for controlling the harmonics and instructing the installation of capacitor in Macao. Based on the measurement in Macao, it knows that order 3, 5 & 7 are main current harmonic components in Macao. Therefore, the end-users should note the harmonics before installing any capacitor. Moreover, the understanding of harmonic levels in Macao is not much until to the present, so it is also recommended the end-users should measure their individual order harmonic level once they need to install capacitor banks to increase the power factor or reduce the energy losses in the future.

D. Even Order Current Harmonic

In general the even order current harmonic level is less than the odd order current harmonic level. Therefore, many power quality compensators focus on how to reduce the odd order current harmonic but not focus to reduce the even order harmonic. In this research it reports the measured odd and even order current harmonic results, and uses the limits shows in Table 5 from IEEE 519-1992 to evaluate individual order harmonic status.

Table 6 and 7 show the maximum odd order current harmonic and even order current harmonic results in percent of I_L respectively from the power quality measurement in several buildings of Macao. It finds that some even order current harmonic results in the special facility building, the public administrative building C and the middle school exceed the limits. Therefore, it should start to care the even order current harmonic. Besides, the maximum even order current harmonic is near the maximum odd order current harmonic in the special facility building, the public administrative building A, the public administrative building C and the middle school, it shows that the even order harmonic also is a main component of harmonic.

Table 5
Individual order harmonic standard limits

Maximum Harmonic Current Distortion in Percent of I_L					
Individual Harmonic Order (Odd Harmonics)					
I_{SC}/I_L *	<11	11≤h<17	17≤h<23	23≤h<35	35≤h
<20	4.0	2.0	1.5	0.6	0.3
20<50	7.0	3.5	2.5	1.0	0.5
50<100	10.0	4.5	4.0	1.5	0.7
100<1000	12.0	5.5	5.0	2.0	1.0
>1000	15.0	7.0	6.0	2.5	1.4
Even harmonics are limited to 25% of the odd harmonic limits above.					

* I_{SC}=maximum short-circuit current at PCC
  I_L=maximum demand load current at PCC

Table 6
Measured maximum odd order current harmonic in percent of I_L

Location	Order Limit	Order	Phase A (Max.)	Phase B (Max.)	Phase C (Max.)
Fac	4%	3	27.96%	24.72%	7.47%
Pub A	10%	3	2.58%	2.13%	3.35%
Pub B	7%	3	4.15%	4.26%	6.43%
Pub C	7%	5	3.71%	3.41%	4.30%
Res A	10%	5	4.05%	4.18%	3.86%
Res B	10%	5	4.01%	4.14%	3.88%
Res C	10%	3	4.47%	5.01%	5.41%
Res D	10%	3	4.61%	5.01%	3.74%
Res E	10%	3	3.21%	3.01%	3.76%
Com	4%	5	2.27%	2.23%	2.20%
Hot	7%	5	3.97%	4.11%	3.06%
Sch	12%	3	7.18%	7.00%	7.25%
Cen	12%	3	7.37%	5.81%	9.37%

Table 7
Measured maximum even order current harmonic in percent of I_L

Location	Order Limit	Order	Phase A (Max.)	Phase B (Max.)	Phase C (Max.)
Fac	1%	2	28.62%	22.42%	4.75%
Pub A	2.5%	2	2.31%	2.26%	1.92%
Pub B	1.75%	2	1.20%	1.08%	1.06%
Pub C	1.75%	2	3.03%	2.86%	3.11%
Res A	2.5%	2	0.91%	1.49%	1.97%
Res B	2.5%	2	1.63%	0.56%	1.32%
Res C	2.5%	2	1.29%	1.30%	0.98%
Res D	2.5%	2	0.81%	0.90%	1.07%
Res E	2.5%	2	1.76%	1.60%	0.73%
Com	1%	2	0.54%	0.53%	0.63%
Hot	1.75%	2	0.73%	0.69%	0.66%
Sch	3%	2	5.67%	5.56%	6.14%
Cen	3%	10	0.46%	0.46%	1.53%

E. Neutral Current Harmonic

High neutral current can cause overload power feeder, distribution transformer and voltage distortion [4]. Table 8 shows the measured neutral current RMS is between 7.9% and 91.6% of the phase current RMS, most exceed the limits (< 20% of phase current) [5]. On the other hand, the harmonic components on neutral line play a larger role than the linear load unbalances since generally the neutral current harmonic ($THD_I\%$) is more than 100%.

Nowadays, usually it is focus on how to limit the phase current harmonic but not focus on limit the neutral current harmonic. Table 9 shows the neutral current harmonic ($THD_I rms$), neutral current harmonic ($THD_I\%$), neutral current harmonic ($TDD_I\%$) and phase A current harmonic ($THD_I rms$) results from the power quality measurement in several buildings of Macao. The table shows some neutral harmonic current ($THD_I rms$) is higher than phase harmonic current ($THD_I rms$), and some neutral current harmonic ($TDD_I\%$) values exceed the limits in the public administrative building C, the residential building C, the middle school and the indoor sport center even their phase current harmonic ($TDD_I\%$) are within the limits. The neutral harmonic current is mainly due to the triplen harmonic currents which add in the neutral conductor. Therefore, it should start to limit the neutral current harmonic. Besides, all neutral current harmonic ($THD_I\%$) exceeds the limits too much.

Fig. 7 shows the measured current harmonic ($THD_I rms$) of the public administrative building C, its neutral current harmonic is higher than the phase current harmonic.

Finally, due to the possibility of large triplen harmonic currents existing in the neutral conductor for building loads with a large proportion of non-linear equipment, it is not recommended to use neutral conductors with a cross-sectional area less than that of phase conductors in the main circuit.

Table 8
Measured phase and neutral current RMS

Location	Phase A (Max.)	Phase B (Max.)	Phase C (Max.)	Neutral (Max.) (% of phase current)
Fac	1797A	1801A	1373A	121.3A (8.8%)
Pub A	305.1A	304.9A	308.5A	44.9A (14.7%)
Pub B	1.629KA	1.637KA	1.639KA	130.2A (7.9%)
Pub C	966.8A	985.6A	960.0A	101.2A (10.5%)
Res A	198.4A	154.5A	153.7A	106.5A (69.2%)
Res B	163.5A	120.8A	171.2A	94.2A (78.0%)
Res C	245.3A	241.9A	307.2A	112.7A (46.5%)
Res D	243.9A	263.5A	253.9A	99.0A (40.6%)
Res E	142.9A	147.8A	156.9A	67.1A (46.9%)
Com	454.0A	448.0A	419.7A	82.7A (19.7%)
Hot	310.6A	337.6A	261.9A	89.2A (34.2%)
Sch	227.6A	243.9A	254.5A	68.5A (30.1%)
Cen	52.0A	32.5A	31.6A	29.0A (91.6%)

Table 9
Measured current harmonic

Location	$TDD_I\%$ Limit	Phase A $THD_I rms$ (Max.)	Neutral $THD_I rms$ (Max)	Neutral $THD_I\%$ (Max)	Neutral $TDD_I\%$ (Max)
Fac	5%	1.394KA	45.88A	18K%	2.29%
Pub A	12%	17.89A	36.81A	92.71K%	7.36%
Pub B	8%	72.92A	126.9A	485.9K%	7.93%
Pub C	8%	47.35A	91.02A	165.4K%	9.1%
Res A	12%	33.13A	58.95A	106.2K%	9.83%
Res B	12%	28.43A	57.09A	162.2K%	9.52%
Res C	12%	35.09A	82.82A	541.6K%	13.8%
Res D	12%	34.68A	66.06A	665.1K%	11.0%
Res E	12%	26.7A	53.58A	181.9K%	8.9%
Com	5%	52.76A	60.47A	712.9%	3.0%
Hot	8%	77.24A	46.88A	76.21K%	2.9%
Sch	15%	27.91A	62.55A	14.65K%	20.8%
Cen	15%	6.434A	18.55A	2.964K%	23.1%

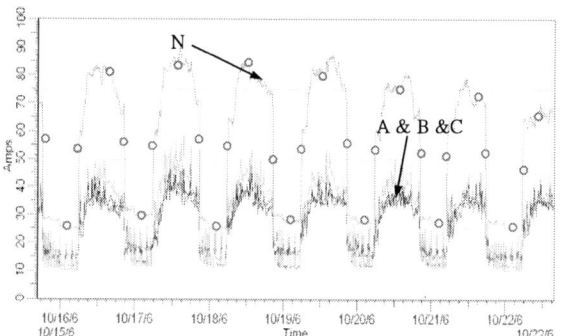

Fig. 7 Current harmonic ($THD_I rms$) versus time in the public administrative building C

IV. CONCLUSIONS

Five point's findings on harmonic measurement in Macao have been reported and it could be concluded as below:

- It should note the differences between current harmonic $THD_I\%$ and $TDD_I\%$, $TDD_I\%$ is more suitable to estimate the current harmonic level than $THD_I\%$ for building.

- More harmonic current is caused by variable refrigerant volume (VRV) air-conditioning system than traditional air-cooled chiller unit and water-cooled chiller unit.

- The end-users should measure their individual order harmonic level once they need to install capacitor banks to improve the power factor.

- Some even order current harmonics exceed the limits, thus it should start to care the even order current harmonic.

- Harmonic is a main factor on neutral current and some neutral current harmonic (THD_Irms) is more than phase current harmonic (THD_Irms).

ACKNOWLEDGEMENT

The authors would like to thank the Research Committee of University of Macau for their supports, and also thank the persons of the buildings for their support and helpful for this power quality monitoring.

REFERENCES

[1] Eloi Ngandui, and Cedric Meignant, "Power quality monitoring and analysis of a university distribution system," *IEEE CCECE,* vol. 2, pp. 863-867, May 2001.

[2] Sio-Un Tai, Man-Chung Wong, Ming-chui Dong, Ying-Duo Han, "Power quality study in Macao," *Proc. of 7th Int. Conf. on Power Electronics and Drive Systems (PEDS 2007),* Bangkok (Thailand), Nov. 2007. (Accepted)

[3] Vannoy, D.B., McGranaghan, M.F., Halpin, S.M., Moncrief, W.A., Sabin, D.D. "Roadmap for power quality standards development," *Petroleum and Chemical Industry Conference, Industry Applications Society 52nd Annual,* pp. 267-276, Sept. 2005.

[4] T.M. Gruzs., "A survey of the neutral current in three-phase computer power systems," *IEEE Trans. on industry applications,* vol. 26, pp. 719-725, July-August 1990.

[5] A. C. Liew, "Excessive neutral currents in three-phase fluorescent lighting circuits," *IEEE Trans. on industry applications,* vol. 25, no. 4, pp. 776-782, July/August 1989.

Coordinated design of PSS and TCSC dynamics model for power system network oscillations

M. Tarafdar Haque*, A. Roshan Milani**, A. Lafzi***

Islamic Azad University, Ahar Branch, Iran*
Azerbaijan Regional Electric Company, Tabriz, Iran**
University of Tabriz, Tabriz, Iran***

tarafdar@tabrizu.ac.ir

Abstract

This paper presents coordinated design of PSS and TCSC dynamics model for electric power systems oscillations damping. A simplified fundamental frequency model of TCSC is proposed and the model results are verified, initially. This paper presents an analytical dynamic model for TCSC to demonstrate how it consistently behaves in the large power system networks and efficiently stabilizes its oscillations in the case of a disturbance. The model for possible undesirable interactions between the PSS-type damping controllers and the TCSC damping controller is implemented by using the multi-machine model in the design stage. The proposed technique ensures that the designed controllers fulfill various practical requirements of the oscillation damping problem in power systems.

I. Introduction

TCSC is a series FACTS device which allows rapid and continuous changes of the transmission line impedance. It has great application potential in accurately regulating the power flow on a transmission line, damping inter-area power oscillations, mitigating sub synchronous resonance (SSR) and improving transient

stability. A typical TCSC module consists of a fixed series capacitor (FC) in parallel with a thyristor controlled reactor (TCR). The TCR is formed by a reactor in series with a bi-directional thyristor valve that is fired with a phase angle ranging between 90 and 180 with respect to the capacitor voltage. The overall scheme is shown in Fig. 1.

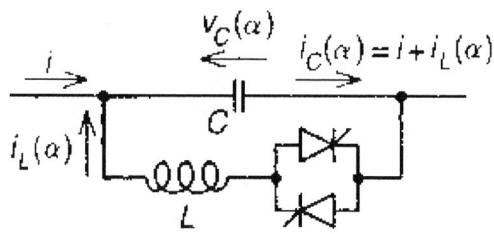

Fig. 1. Basic structure of TCSC

Electromechanical oscillations damping is recognized as an important issue in electric power system operation. Application of power system stabilizers (PSSs) is one of the first measures to enhance the damping of power swings. As increasing transmission line loading over long distances, the use of conventional power system stabilizers might not provide sufficient damping for inter-area power swings, in some cases. In these cases, other effective solutions are needed to be studied. In dynamic applications of TCSC's, various control techniques and designs have been proposed for damping power oscillations to improve system dynamic response. The power electronics development has allowed the application of new devices to improve power system performance. Some of these devices may be used to damp electromechanical oscillation. The thyristor Controlled Series Capacitor (TCSC) is a kind of FACTS device that has been successfully used to enhance damping in power systems [1,2]. The PSSs are effective in the oscillation damping. However, there may be cases where the system PSSs is not able to suitably damp inter-area oscillation modes. In such cases, the simultaneous use of both controller types (PSS and FACTS damping controller) are required to guarantee a good closed loop system performance. The operational limits such as environment, financial and market, associated with the deregulation and competitiveness of nowadays' power systems; require the efficient control strategies to provide operational reliability and financial profitability. Reliability and financial profitability of such operation may be achieved, in some cases, with the use of both devices together. However, separated design of PSS and FACTS damping controller may cause dynamic interactions between them. A coordination procedure may be required to avoid such possible dynamic interactions between PSS and FACTS damping controller. Nowadays' large power systems are usually constituted by many interconnected control areas. Such interconnections give flexibility to the system operation. On the other hand, the controller coordination in such kind of power systems may be rather complex due to system dimensions. However, such complexity may be overcome with the use of a multi-machine model in the simultaneous design of PSS and FACTS damping controller.

The paper is structured as follows: Section II TCSC modeling and basic control scheme. In Section III, the test system model is described, with a description of the analysis and simulation tools used in this paper. The results of applying proposed TCSC controller design for stability enhancement of the test system are also discussed in this section. Finally, Section IV summarizes the main contributions of this paper.

II. TCSC Modeling and Basic Control Scheme

A typical TCSC has two mode operation :
1) switches off
2) switches on

The corresponding state space equation $\dot{X} = AX + BU$ for the two topologies are as equations 1 and 2.

$$S\,V_c = 1/C\,(I_s - I_L) \qquad (1)$$

$$S\,I_L = 1/LC\,(K_1 * I_s - K_2 * I_L) \qquad (2)$$

Where $k_1 = k_1(\alpha)$ and $k_2 = k_2(\alpha)$ are the unknown model parameters dependent on the firing angle and it is presumed that the line current is constant over one fundamental cycle in according to[4]. The model is controlled by varying the phase delay of the thyristor firing pulses synchronized through a PLL to the line current waveform. The controller is of a PI type with a feedback filter and a series compensator.

The controller model consists of a second order feedback filter, PI controller, PLL, series compensator and a transport delay model, as shown in Fig. 2 [3].

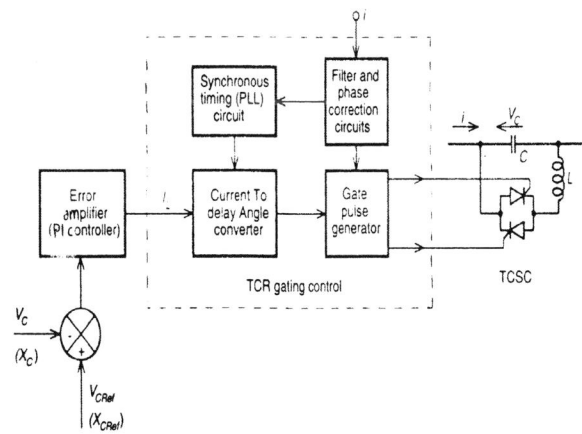

Fig. 2. Schematic of TCSC control model

The PLL synchronizes thyristor firings with the line current phase angle. For simplicity reasons, the TCSC voltage feedback control is used where v_{ref} can be a function of other parameters at higher control levels. Because of the thyristor firings at discrete time instants the system is actually a sampled data system with the sampling frequency $f_s = 360$ Hz [4,5]. The continuous model, therefore, includes a first-order delay, given by time constant T_{d1} to accommodate the phase lag introduced by sampling the firing angle signal as it is discussed in [4].

III. Test System Model

Building TCSC and its control model and applying Matlab software, a test network was built as shown in Fig. 3.

The sample system is a three phase type system, 60Hz, 400 kV, transmitting power from a power plant consisting of 3×350 MVA generators to another power pool represented by an equivalent network model through 450 km transmission line. There is one TCSC compensator considered in the transmission lines. Line is compensated around 40% of its series reactance by series variable capacitors. There are two 25 MVAR shunt reactor compensators in system connected to line ends. At Bus 2 there is a 150 MW local load.

We apply a three phase fault in transmission line on the sample system. The fault is applied at fault time (t_f=18/60s) and then is eliminated at clearing time (t_c= 24/60s). Two cases are studied and examined; first transmission power network without series compensation, i.e. when TCSC is bypassed by protective circuit breakers and

second when series compensator is in circuit and operating in control mode helping the power system operating condition. For simulation, at first stage TCSC is not implemented in the power system circuit. Figs 4 and 5 depict power system load angle and generator terminal voltage simulation results for this case. Figs 6 to 8 show simulation results for power system load angle, current and TCSC voltage and current wave forms when TCSC is applied in the power system. The tests to verify the performance of the proposed controllers were carried out in a well-known power system model, and the results obtained by means of the modal analyses and non-linear simulations are presented in this section.

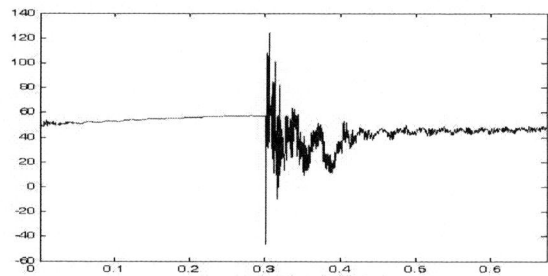

Fig. 4. Load angel of system (generator 1) without TCSC

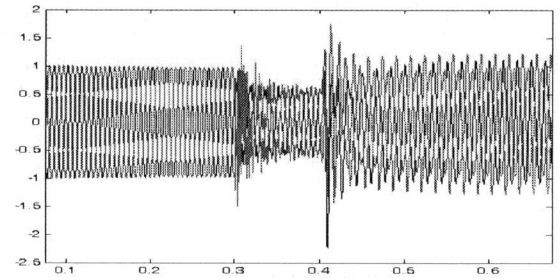

Fig. 5. Bus 1 voltage without TCSC

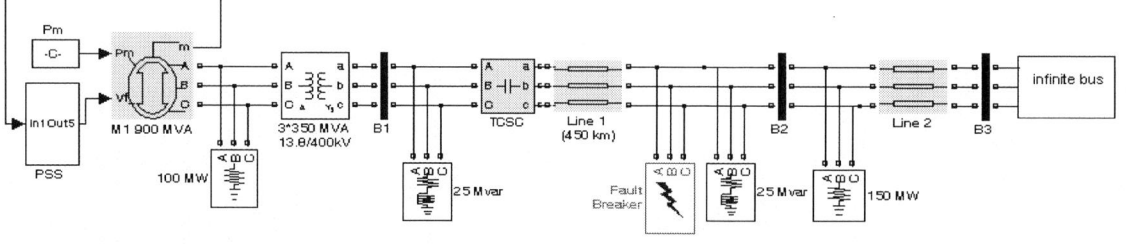

Fig 3. Diagram of sample system

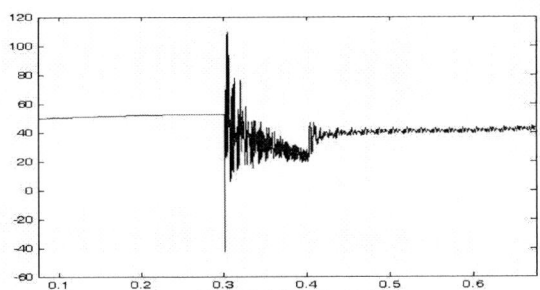

Fig. 6. Load angel of system (generator 1) with TCSC

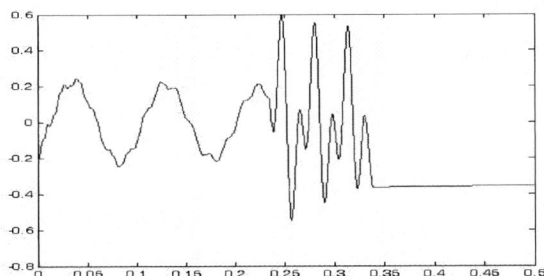

Fig. 7. Capacitor voltage of TCSC

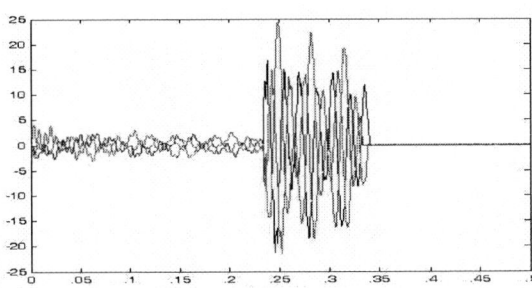

Fig. 8. Fault current throw TCSC

The adopted system is rather used in small signal stability studies, and it is constituted by two areas interconnected by a tie-line (as shown in the diagram of Fig(9). The complete data of this system can be obtained in [6].

The TCSC is included in the tie-line (bus 1-2), since such branch is a weak connection which limits the power transfer between the two areas. Besides, the tie-line has a significant influence in the inter-area mode that has to be damped. The system analyses and controller design may be carried out by means of linear models to improve small signal stability margin, since linear models are usually able to acceptably represent the

dynamic behavior regarding to low-frequency electromechanical oscillation.

Controller design for power systems are usually based on output feedback, since not all the model state variables are available for direct measurement in the real system. For this reason, the proposed damping controllers (PSS-type and TCSC damping controller) are based on the dynamic output feedback structure. Such control structure can be represented by a linear equation set, in state space form, given by

$$\dot{X}_c(t) = A\, X(t) + BY(t) \qquad (3)$$
$$U(t) = CX(t) \qquad (4)$$

The operating point of the electric power system plays an important role in the electromechanical oscillation dynamics. However, the system operating point is usually an uncertainty in design methodologies based on linear classical control techniques, since the power system is, in general, modeled as a Linear Time Invariant (LTI) system. The control problem consists basically in calculating dynamic controllers, represented by the matrices A, B, C which guarantee the specified performance index and robustness for the closed loop system.

The tests to verify the performance of the proposed controllers were carried out in a well-known power system model, and the results obtained by means of the modal analyses and non-linear simulations are presented.

The system loads in the operating conditions of the base case are P=350 MW, Q=50 MVAr. The parameters of the voltage regulators used in the respective system generators are Ke=300 and Te=0.01 s. Generator 4 was used as an infinite-bus, supplying an angular reference to the model. The perturbation used to stimulate the oscillation modes in the non-linear simulations is a short-circuit at bus 2 in t=18/60 (sec). In t=24/60 (sec) the short-circuit is eliminated and the system pre-fault operating condition is restored (the line involved with the fault was not turned off to avoid system islanding).

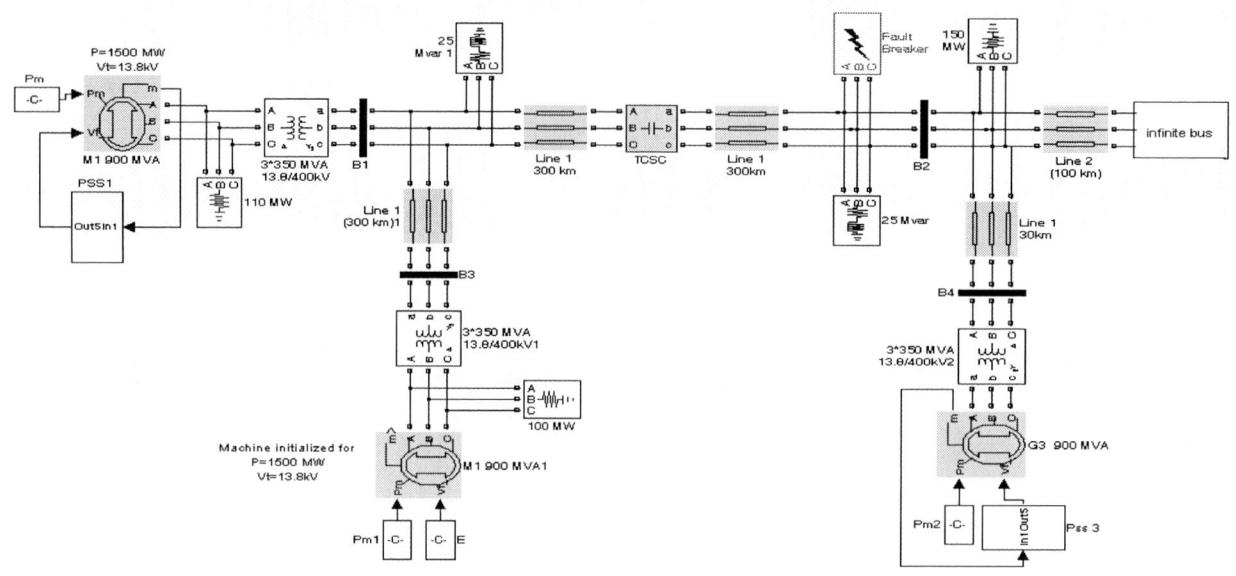

Fig. 9. Diagram of test system in Matlab

The test system in open loop (without damping controllers) is unstable in the base case operating condition. Therefore, the system requires controllers to operate in a stable way and with good performance. The eigen-values related to the local and inter-area modes of the open loop system, in the base case operating conditions, are shown in Fig. 10.

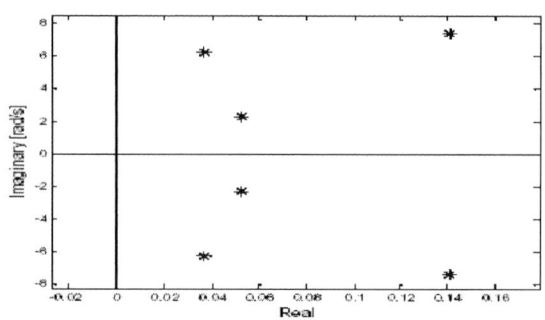

Fig. 10. Poles of test system in open loop

Two PSS type damping controller and one TCSC damping controller were simultaneously designed. The TCSC damping controller was design to the TCSC inserted in system tie-line and the PSS type damping controller were design for generator 1 and generator 3. The derivative of the generator speed deviation is used as input signal for the PSS type damping controller to avoid the action of such controllers in steady state conditions [5]. In order to access stable condition, the TCSC controller model with PSS and test operation of the coordinated two controllers with each other is explained. Figs.11 to 14 show simulation results for power system load angle, Bus 1 voltage, capacitor voltage and TCSC current when TCSC is applied in the power system.

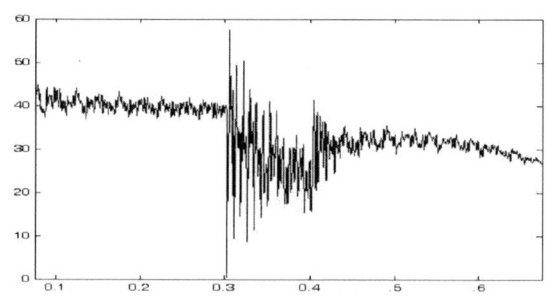

Fig. 11. Load angel of system(generator 1)

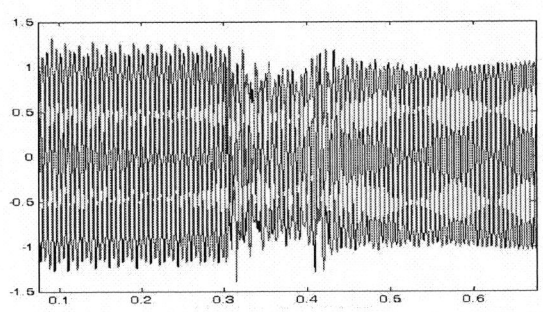

Fig. 12. Bus 1 voltage

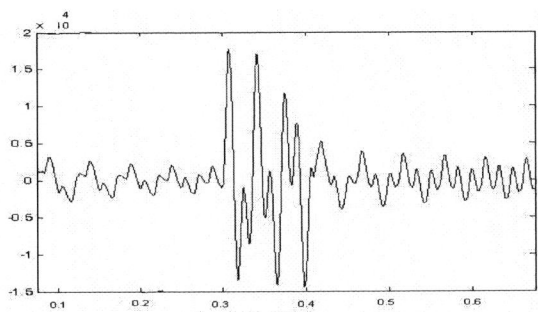

Fig. 13. Capacitor voltage of TCSC

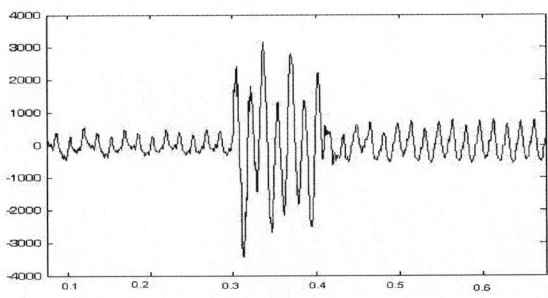

Fig. 14. TCSC current

IV. Conclusion

In this paper TCSC damping controller model is used by a second order feedback filter, PI controller and a PLL with PSS controller in power system. A simultaneous coordinated design of PSS-type and TCSC damping controllers for electric power systems is used. Such kind of design is important in the cases where only the PSSs are not able to adequately damp the system oscillations (mainly in those cases where inter-area modes are involved). It is evident by obtained results that the dynamic of system operating is improved, considerably.

References

[1] Yong Hua Song, Allan T Johns, FACTS IEE, IEE Publication, 1999.

[2] Ricardo Tenorio, "Improvements for Power System Performance Modelling, Analysis and Benefits of TCSC," IEEE 2001.

[3] Narain G.Hingorani, Laszio Gyugyi, Understanding FACTS, IEEE Press, 2000.

[4] Dragan Jovcic, and G. N. Pillai," Analytical Modeling of TCSC Dynamics," IEEE Transaction on Power Delivery, Vol. 20, no. 2, April 2005.

[5] Roman Kuiava, Ricardo V. de Oliveira, Rodrigo A. Ramos, and Newton G. Bretas, "Simultaneous Coordinated Design of PSS and TCSC Damping Controller for Power Systems,"IEEE 2006.

[6] P. Kundur, Power System Stability and Control. EPRI Editors, McGraw- Hill, New York, 1994.

An Analytic Approach To Harmonic Analysis of 48-Pulse Voltage Source Inverter

B. Geethalakshmi[*] and P. Dananjayan[**]

[*] Research Scholar, Dept. of ECE, Pondicherry Engg. College, Pondicherry, India
[**] Professor, Dept. of ECE, Pondicherry Engg. College, Pondicherry, India

Abstract– **Multi-pulse voltage source inverter topology having a number of 6-pulse VSIs as elementary units is widely used in high power applications. This paper is aimed to present a detailed analysis of the harmonic components present in individual VSIs of a 48-pulse voltage source inverter and mitigation of these harmonics so as to produce a sinusoidal wave output. The analytical expressions for the 48-pulse inverter output voltages using Fourier analysis have been obtained. Vector diagrams are drawn to clearly illustrate the mechanism of harmonic cancellation in multi-pulse inverters. The complete digital simulation of 48-pulse VSI is performed using MATLAB/Simulink and the simulation results closely agreed with the analytical results.**

Index Terms—**Multi-pulse inverter, Phase shifting transformer, Total Harmonic Distortion, Voltage source inverter.**

I. INTRODUCTION

A very simple 6-pulse inverter produces a square wave output [1] as it switches the direct voltage source on and off. However the basic objective of a voltage source inverter (VSI) is to produce a sinusoidal AC voltage with minimal harmonic distortion from a DC source. This is achieved in multi-pulse inverter configuration where harmonics are reduced significantly by increasing the number of 6-pulse units and phase shifting the quasi-square wave output of each unit with the help of phase shifting transformers.

In the multi-pulse inverter configuration the fundamental components of individual units are kept in phase and a proper phase shift is introduced between the harmonics of individual units, which are to be eliminated at the output of the electromagnetic interface. These multi-pulse inverters are suitable in high voltage and high power applications due to their ability to synthesize waveforms with better harmonic spectrum and higher voltages with a limited maximum device rating.

In general star and delta-connected windings have a relative phase shift of 30° and 6-pulse inverter bridges connected to each of these Y and Δ transformers will give an overall 12-pulse operation eliminating 5th and 7th harmonics. This principle can be extended for 24-pulse and 48-pulse operation.

So far 12-pulse and 24-pulse inverters are analyzed in the literature [2–5]. This paper attempts to analyze the 48-pulse operation in detail. A complete description of the phase shift of the gate pulse patterns as well as the configuration of the phase shifting transformers for each VSI is presented. Analytical expressions are obtained and vector diagrams are drawn to clearly analyze the neutralization of harmonic components in the 48-pulse voltage source inverter.

II. 48-PULSE INVERTER

48-Pulse inverter can be used without AC filters due to its high performance and low harmonic rate on the ac side. They are obtained by combining eight 6-Pulse VSIs with an adequate phase shifts between them. Each of the VSI needs a coupling transformer of which four of them require a Y-Y transformer with a turns ratio of 1:1 and the remaining four require a Δ-Y with a turns ratio of 1:√3. The output of the phase shifting transformers is connected in series to cancel out the lower order harmonics. The schematic diagram of 48-pulse inverter is shown in Fig. 1.

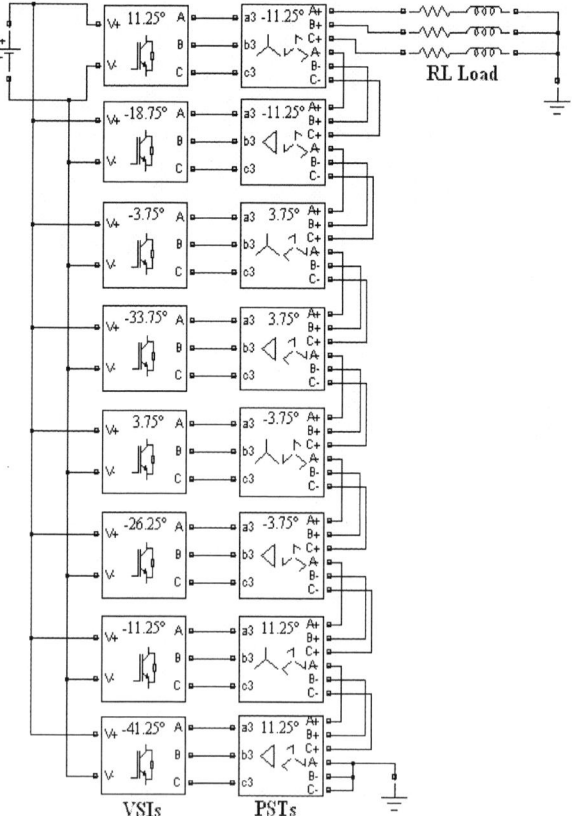

Fig. 1. 48-pulse inverter.

978-1-4244-0644-9/07/$25.00 ©2007 IEEE

To create a 48-pulse waveform with a harmonic content in the order of m = 48r±1, (where r = 0, 1, 2,...) the eight 6- pulse converter voltages need to be phase shifted. This is implemented by introducing appropriate phase shift in the phase shifting transformer and the gate pulse pattern of individual VSI. Table 1 shows the phase displacements applied to the gate pulse pattern of each VSI and the corresponding phase shifting transformer.

TABLE I
PHASE DISPLACEMENT FOR A 48-PULSE VSI

Coupling transformer	Gate pulse pattern	Phase shifting transformer
Y-Y	+11.25°	-11.25°
Δ-Y	-18.75°	-11.25°
Y-Y	-3.75 °	+3.75°
Δ-Y	-33.75°	+3.75°
Y-Y	+3.75°	-3.75°
Δ-Y	-26.25°	-3.75°
Y-Y	-11.25°	+11.25°
Δ-Y	-41.25°	+11.25°

III. HARMONIC ANALYSIS

Carrying out the Fourier analysis to the 6-pulse inverter output voltage, the instantaneous values of the phase-to-phase voltage and the phase to neutral voltage are given by:

$$V_{ab}(t) = \sum_{m=1}^{\infty} V_{ab_m} \sin\left(m\omega t + m\frac{\pi}{6}\right) \qquad (1)$$

$$V_{an}(t) = \sum_{m=1}^{\infty} V_{an_m} \sin m\omega t \qquad (2)$$

where $V_{ab_m} = \frac{4}{m\pi} V_{DC} \cos m\frac{\pi}{6}$

$$V_{an_m} = \frac{4}{3\pi m} V_{DC}\left(\cos m\frac{\pi}{3} + 1\right)$$

$$\forall m = 6r \pm 1, r = 0, 1, 2, ...$$

The voltages $v_{bc}(t)$ and $v_{ca}(t)$ exhibit a similar pattern except phase shifted by 120° and 240° respectively. Similarly the phase voltages $v_{bn}(t)$ and $v_{cn}(t)$ are also phase shifted by 120° and 240° respectively.

The 12-Pulse voltage source inverter gives better harmonic performance [3] and is obtained by combining two 6-Pulse inverters. The fundamental and harmonic components of the phase-to-phase voltages and phase-to-neutral voltages are phase shifted by 30° from each other [6]. If this phase shift is corrected, then the phase-to-neutral voltage harmonics, other than those of 12r±1, would be out of phase to those of the phase-to-phase voltage and with $1/\sqrt{3}$ times the amplitude. Hence if the phase-to-phase voltages of a second converter are connected to a transformer with delta-connected secondary and $\sqrt{3}$ times the turns compared to the Y-connected secondary, and the pulse train of one converter is shifted by 30° with respect to the other, the combined output voltage will have a 12-pulse waveform,

with harmonics of the order of 12r±1. Thus the 12-Pulse inverter will have 11th, 13th, 23rd, 25th....., harmonics with amplitudes of 1/11th, 1/13th, 1/23rd, 1/25th ......, respectively of the fundamental ac voltage.

The relationship between the phase-to-phase voltage and the phase-to-neutral voltage is expressed as:

$$V_{ab_m} = (-1)^r \sqrt{3} V_{an_m} \qquad (3)$$

If the VSI₁ output is connected to a Y-Y transformer with a 1:1 turn ratio, the line to neutral voltage using (3) can be expressed as:

$$V_{an}(t)_1 = \frac{1}{\sqrt{3}} \sum_{m=1}^{\infty} \frac{V_{ab_m}}{(-1)^r} \sin m\omega t \qquad (4)$$

$$\forall m = 6r \pm 1, r = 0, 1, 2, ...$$

Suppose if the VSI₂ produces phase-to-phase voltages lagged by 30° with respect to VSI₁ and with the same magnitude, it is given by

$$V_{ab}(t)_2 = \sum_{m=1}^{\infty} V_{ab_m} \sin m\omega t \qquad (5)$$

If this inverter output is connected to a Δ-Y transformer with a 1:1/√3 turn ratio, the line-to-neutral voltage in the Y-connected secondary would be

$$V_{anY}(t)_2 = \sum_{m=1}^{\infty} V_{an_m} \sin m\omega t \qquad (6)$$

Therefore line-to-line voltage in the secondary side is

$$V_{abY}(t)_2 = \sum_{m=1}^{\infty} \sqrt{3} V_{an_m} \sin\left(m\omega t + \frac{m\pi}{6}\right) \qquad (7)$$

The 12-Pulse inverter output is obtained by adding the equations (1) and (7).

$$V_{ab}(t)_{12} = V_{ab}(t) + V_{abY}(t)_2 \qquad (8)$$

$$V_{ab}(t)_{12} = \sum_{m=1}^{\infty} V_{ab_{12m}} \sin\left(m\omega t + \frac{m\pi}{6}\right) \qquad (9)$$

$$\forall m = 12r \pm 1, r = 0, 1, 2, ...$$

since $V_{ab_{12m}} = V_{ab_m} + \sqrt{3} V_{an_m}$

$$= 2 V_{ab_m}$$

$$\therefore V_{ab}(t)_{12} = 2 \sum_{m=1}^{\infty} V_{ab_m} \sin\left(m\omega t + \frac{m\pi}{6}\right) \qquad (10)$$

Two 12-Pulse inverters, phase shifted by 15° from each other, can provide a 24-Pulse inverter, with much lower harmonics on ac and dc side. Its ac output voltage will have 24r±1 order harmonics, i.e., 23rd, 25th, 47th, 49th....., harmonics, with magnitudes of 1/23rd, 1/25th, 1/47th, 1/49th...., respectively, of the fundamental ac voltage. Thus the output voltage of 24-Pulse inverter is obtained as:

$$V_{ab_{24}}(t) = 4 \sum_{m=1}^{\infty} V_{ab_m} \sin\left(m\omega t + 22.5^°m + 7.5^°x\right) \qquad (11)$$

where x = 1 for positive sequence harmonics
　　 x = -1 for negative sequence harmonics
$\forall m = 24r \pm 1, r = 0, 1, 2, ...$

Similarly, the 48-Pulse inverter is derived from two 24-Pulse inverters having a phase shift of 7.5° from each other. A phase shift of +3.75° is applied to the phase-shift windings of the first four transformers (24-Pulse inverter₁) and –3.75° to the remaining four transformers (24-Pulse inverter₂). It is also necessary to shift the firing pulses of first 24-Pulse inverter by 7.5° with respect to the other. Thus the 48-Pulse inverter output is obtained by giving proper phase shift as described in Table I to the PWM modulator and the phase shifting transformer of the eight individual voltage source inverters which co-ordinate to produce a sinusoidal output.

The instantaneous values of the phase-to-phase voltage $V_{ab}(t)$ and the phase-to-neutral voltage $V_{an}(t)$ of the 48-pulse inverter output voltage is obtained as:

$$V_{ab_{48}}\left(t\right) = 8 \sum_{m=1}^{\infty} V_{ab_m} \sin\left(m\omega t + 18.75^{\circ}m + 11.25^{\circ}i\right) \quad (12)$$

$$V_{an_{48}}\left(t\right) = \frac{8}{\sqrt{3}} \sum_{m=1}^{\infty} V_{ab_m} \sin\left(m\omega t + 18.75^{\circ}m - 18.75^{\circ}i\right) \quad (13)$$

$\forall m = 48r \pm 1, \; r = 0, 1, 2, \ldots\ldots$

i = 1, for positive sequence harmonics

i = -1, for negative sequence harmonics

IV. HARMONIC NEUTRALIZATION

Vector diagrams are drawn to illustrate the harmonic cancellation mechanism in the 48-pulse inverter configuration. The magnitude and phase angle of the harmonic components present at the outputs of the 6-Pulse voltage source inverters VSI₁ to VSI₈ are given in Figs. 2 - 9 respectively. Since the harmonic components 5, 7, 17, 19, 29, 31, 41, 43… present in adjacent 6-pulse voltage source inverters (VSI₁ and VSI₂, VSI₃ and VSI₄, VSI₅ and VSI₆ and VSI₇ and VSI₈) are out of phase and are having same magnitude, cancel each other. The in phase components are added resulting to a 12-pulse inverter configuration. Thus a 12-Pulse inverter which is obtained by cascading two 6-Pulse inverters are having harmonics in the order of 12r ±1.

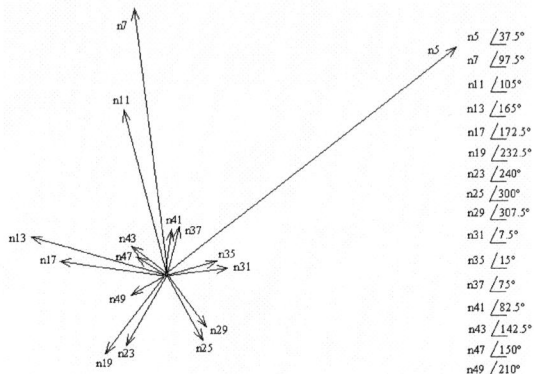

Fig. 3. 6-Pulse VSI₂ harmonics

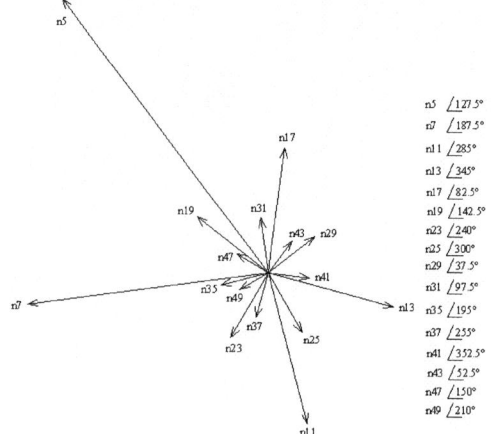

Fig. 4. 6-Pulse VSI₃ harmonics

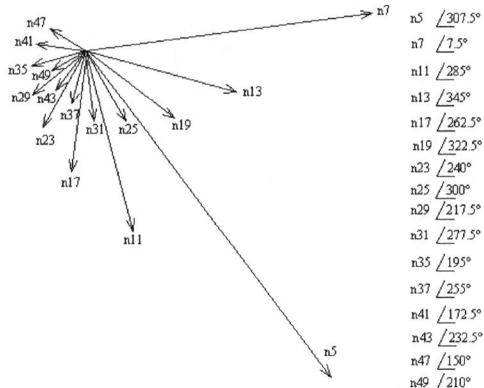

Fig. 5. 6-Pulse VSI₄ harmonics

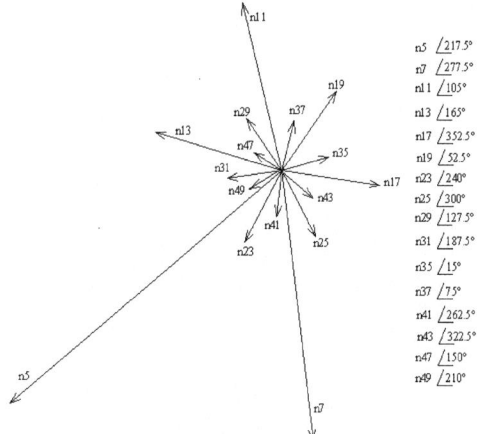

Fig. 2. 6-Pulse VSI₁ harmonics

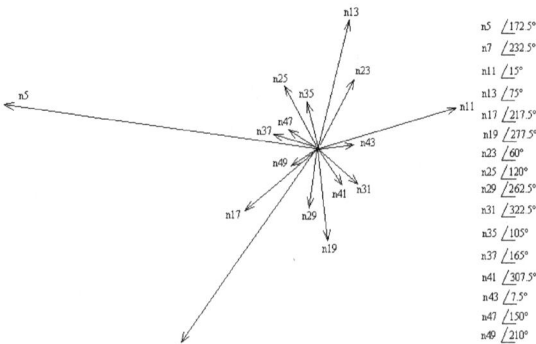

Fig. 6. 6-Pulse VSI₅ harmonics

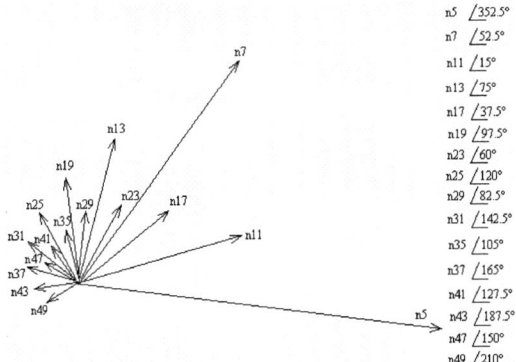

Fig. 7. 6-Pulse VSI$_6$ harmonics

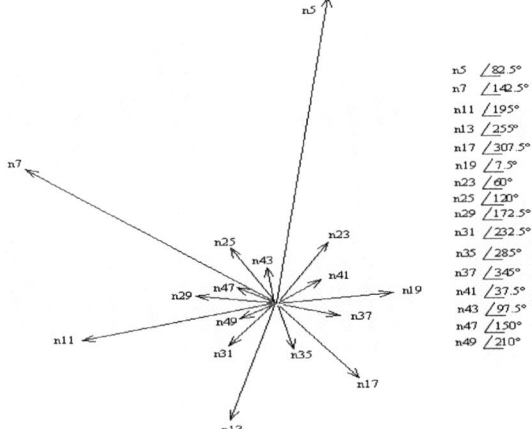

Fig. 8. 6-Pulse VSI$_7$ harmonics

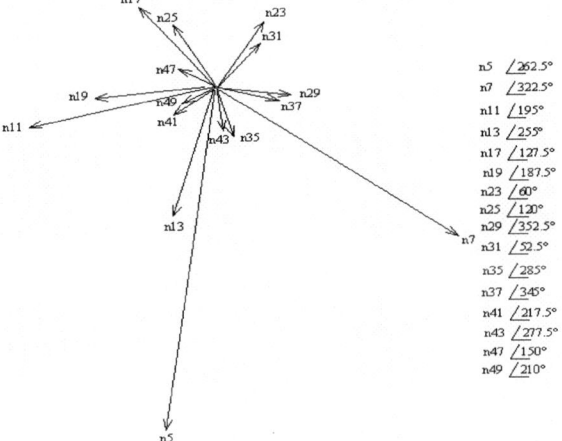

Fig. 9. 6-Pulse VSI$_8$ harmonics

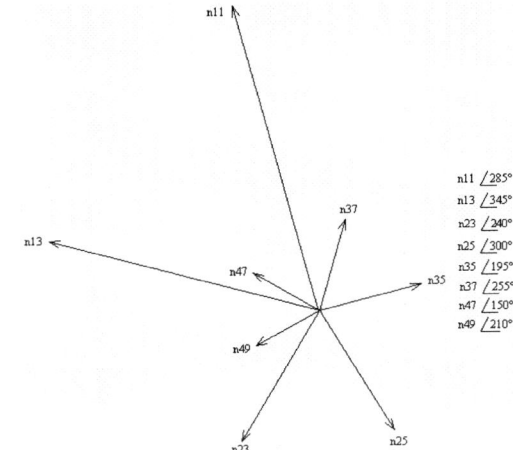

Fig. 10. 12-Pulse VSI$_1$ harmonics

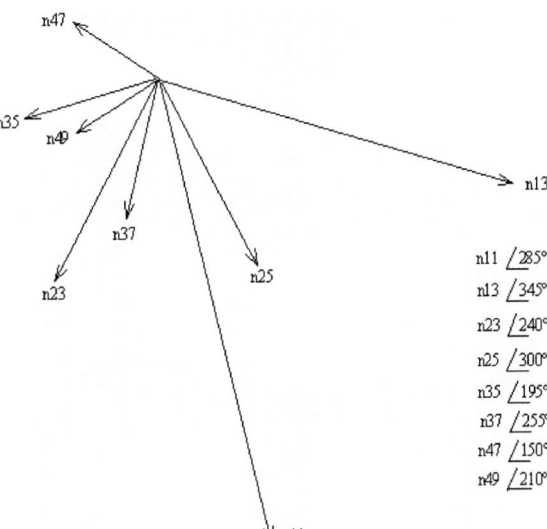

Fig. 11. 12-Pulse VSI$_2$ harmonics

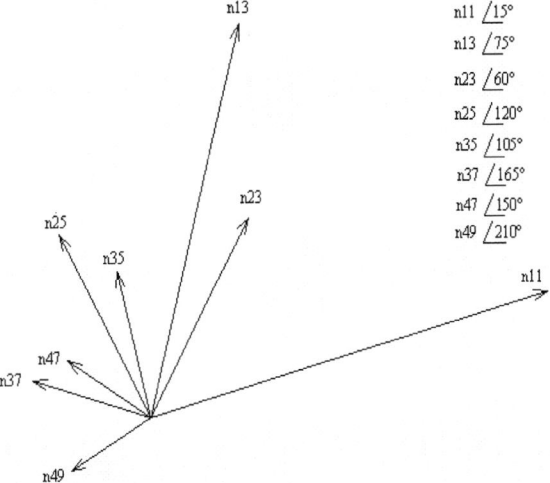

Fig. 12. 12-Pulse VSI$_3$ harmonics

The harmonics of 12-pulse VSIs are clearly described in Figs. 10 -13. Since the harmonics components 11th, 13th, 35th, 37th….., of adjacent 12-Pulse inverters are in opposite phase to each other they are cancelled out and the harmonics 23rd, 25th, 47th, 49th,…. are in phase are added. This harmonic mitigation results to a 24-Pulse inverter with the harmonic components in the order of 24r±1. Figs. 14 and 15. display the harmonic components of the 24-Pulse inverters.

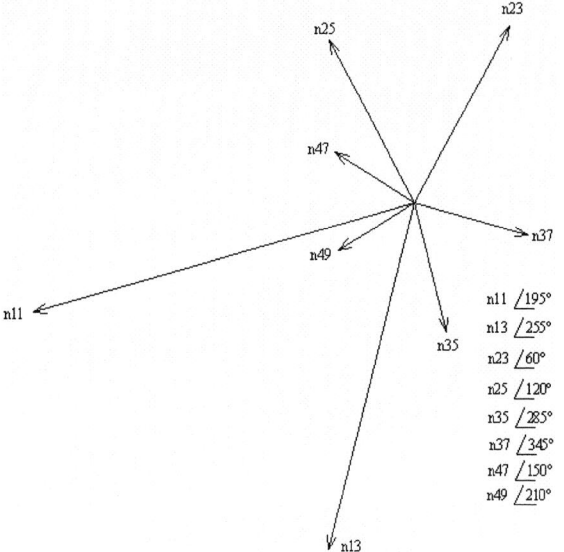

$$n11 \angle 195°$$
$$n13 \angle 255°$$
$$n23 \angle 60°$$
$$n25 \angle 120°$$
$$n35 \angle 285°$$
$$n37 \angle 345°$$
$$n47 \angle 150°$$
$$n49 \angle 210°$$

Fig. 13. 12-Pulse VSI$_4$ harmonics

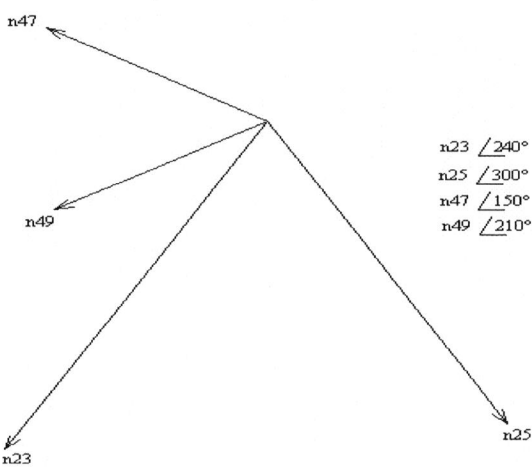

$$n23 \angle 240°$$
$$n25 \angle 300°$$
$$n47 \angle 150°$$
$$n49 \angle 210°$$

Fig. 14. 24-Pulse VSI$_1$ harmonics

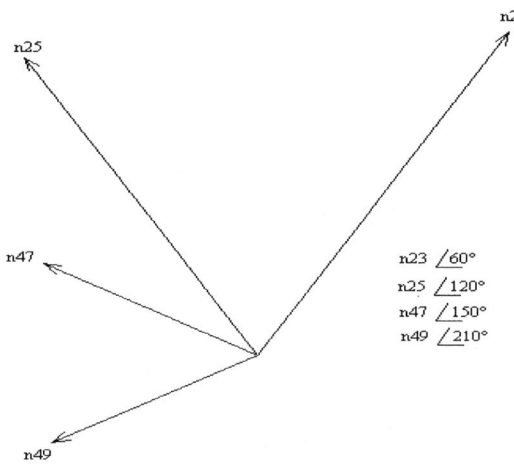

$$n23 \angle 60°$$
$$n25 \angle 120°$$
$$n47 \angle 150°$$
$$n49 \angle 210°$$

Fig. 15. 24-Pulse VSI$_2$ harmonics

Since the 23rd, 25th ,...., harmonics of the 24-Pulse inverter 1 and 2 are in opposite phase, they cancel each other resulting to a 48-Pulse inverter with the 47th, 49th,.... harmonics as shown in Fig. 16.

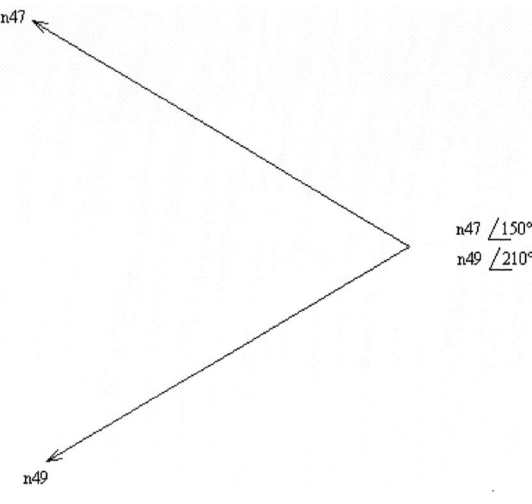

$$n47 \angle 150°$$
$$n49 \angle 210°$$

Fig. 16. 48-Pulse VSI harmonics

V. SIMULATION RESULTS AND DISCUSSION

The 48-pulse inverter is simulated using MATLAB/Simulink to analyze the harmonics in its output voltage. A DC source of 1000 volts is used at the input side. The load is a star connected RL load of 10ohm resistance and 0.1H inductance connected in series. The THD plot of the 48-Pulse inverter is shown in Fig. 17 and the THD is found to be 0.52%. This negligibly small value of THD results to the sinusoidal output voltage and current as depicted in Figs. 18 and 19.

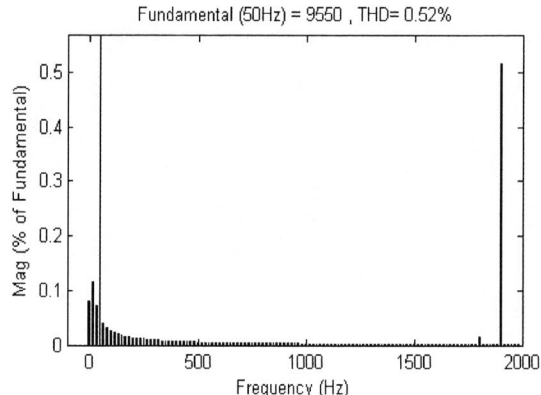

Fig. 17. 48-Pulse VSI THD

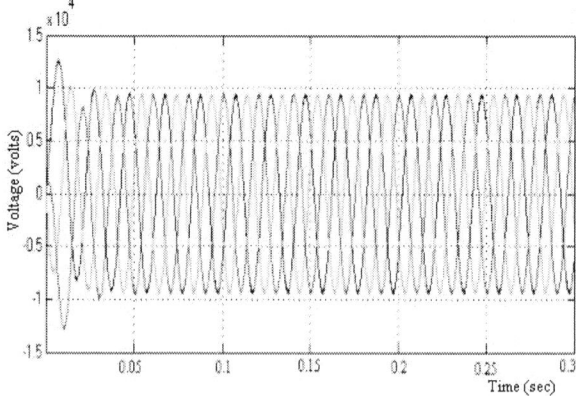

Fig. 18. 48-Pulse VSI output voltage

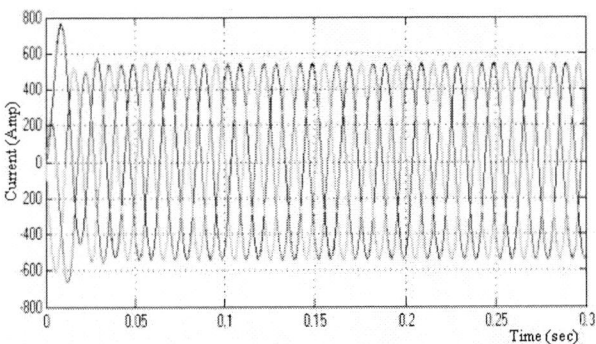

Fig. 19. 48-Pulse VSI output current

VI. CONCLUSIONS

For high power applications the best option is the 48-pulse inverter, although a 24-Pulse inverter with a filter tuned to the 23rd - 25th harmonics could be adequate. This paper described the various stages of development of 48-pulse voltage source inverter. The pulse pattern and transformer arrangement of individual VSIs for harmonic neutralization have also been discussed in detail. The analytic expressions for 48-Pulse inverter output voltage are derived using Fourier series. Vector diagrams are drawn to explain the harmonic mitigation in various stages. It is found that theoretical results are closely agreed with the simulated results.

REFERENCES

[1] N. Mohan, T.M. Undeland, and W.P. Robbins, "Power Electronics: Converters, Applications and Designing", John Wiley and Sons, 1995

[2] A. R. Bakhshai, G. Joos and P. Jain, "A Novel Single Pulse And PWM VAR Compensator For High Power Applications", Proceedings of IEEE Conference 0-780304943-1/98.

[3] Ricardo Davalos M., Juan M. Ramirez and Ruben Tapia O., "Three-phase multipulse converter STATCOM analysis", Electrical Power and Energy Systems 27 (2005) 39-51.

[4] Bhim Singh, G. Bhuvaneswari and Vipin Garg, "Harmonic mitigation using 12-pulse AC-DC converter in Vector Controlled Induction Motor Drives", IEEE Transaction on Power Delivery, vol.21, no. 3, July 2006.

[5] C. J. Hatziadoniu and F.E.Chalkidakis, "A 12-Pulse Static Synchronus Compensator for the distribution system employing the 3-level GTO inverter", IEEE Trans. on Power Delivery, vol. 12, no.4, pp. 1830-1835, October 1997.

[6] Narain G. Hingorani and Laszlo Gyugyi, "Understanding FACTS: Concepts and Technology of Flexible AC Transmission Systems. IEEE press, New York, 1999.

BIOGRAPHIES

B.Geethalakshmi received Bachelor of Engineering in 1996 and Master of Engineering in 1999 from Bharathidasan University. She is currently pursuing her Ph.D work in power electronics application in power systems. She published paper in international journal and presented research papers in various international conferences. Her areas of interest include power converters such as ac-dc-ac converters, matrix converter and power factor correction techniques.

P.Dananjayan received Bachelor of Science from University of Madras in 1978, Bachelor of Technology in 1982 and Master of Engineering in 1984 from the Madras Institute of Technology, Chennai and Ph.D degree from Anna University, Chennai in 1998. He is working as a Professor and Head of the Department of Electronics and Communication Engineering, Pondicherry Engineering College, Pondicherry, India. He has more than 42 publications in National and International Journals. He has presented more than 130 papers in National and International conferences. He has produced 3 Ph.D candidates and is currently guiding nine Ph.D students. His areas of interest include power electronics application in power system, ATM Networks, Wireless Communication and Spread spectrum Techniques.

Detailed losses Analysis of High-Frequency Planar Power Transformer

Yu Ma, Peipei Meng, Junming Zhang and Zhaoming Qian

College of Electrical Engineering, Zhejiang University, Hangzhou, China

Abstract– **As the operating frequency has been increased toward the megahertz range to improve the power density, the accurate losses estimation of power transformer becomes very important in the power converter design stage. The detailed losses of the high frequency planar transformer will be obtained taking into account both the high frequency effected conductor losses and the non-sinusoidal excitation core losses. The loss analysis of the planar transformer is verified by the Finite Element Method(FEM) Simulations and could be utilized to optimize the transformer design procedure.**

Index Terms—**Loss Analysis, Planar Transformer, High Frequency, FEM simulation**

I. INTRODUCTION

As the operating frequency has been increased toward the megahertz range to improve the power density, the accurate loss estimation of a power transformer becomes very important in the power converter design stage. Traditionally, transformer design is based on the power transformers with sinusoidal excitation. When the transformer is operated at a high frequency with the practical non-sinusoidal excitation, some loss mechanisms have to be taken into account: skin and proximity effects in windings and the increased eddy current and hysteresis losses in cores.

Eddy current losses in transformer windings have been previously reported in the literatures[1-5]. Many of them is based on the paper submitted by Dowell[1]. However, some articles[2,3] only deal with transformers operating with sinusoidal current and voltages. In most switch-mode power converters, waveforms have a broad spectrum and the analysis based on sinusoidal excitation will not be valid. Then the actual waveform can be decomposed into its Fourier components and the winding losses at each frequency will be summed up[4,5]. The similar problem could be also met when calculating the core losses. The excitation voltage waveforms could be anything but sinusoidal in the application of power electronics. If sinusoidal excitation and uniform flux distribution are assumed within the magnetic material, the core losses can easily be calculated from the well-known Steinmetz equation[6]. To overcome the disadvantages in estimating the non-sinusoidal excitation core losses by this equation, the modified Steinmetz equation(MSE) has been proposed[7,8]. This modification to Steinmetz equation has been shown accurate for some arbitrary waveforms, although it still has some problems when

dealing with extremely special waveforms as described in [9,10].

This paper presents a detailed methodology for analyzing the losses in high frequency planar transformer. The loss estimation takes account of switching-type waveforms encountered in power supplies, inclusive of high frequency skin and proximity effects in windings and non-sinusoidal excitation core losses. The loss analysis procedure has been verified by the Finite Element Method(FEM) simulation results.

II. LOSSES ANALYSIS OF HIGH FREQUENCY PLANAR TRANSFORMER

A. Winding Losses

When power converters operate at high frequency, the design difficulty for the transformer becomes much higher. The current density redistribution inside the winding wires (skin and proximity effects) strongly increases the copper losses. Therefore, more detailed winding loss estimation is necessary in order to optimize the transformer design.

Fig.1 shows the winding structure of a high frequency planar transformer. MMF diagrams[1] are used to find the H field at the surface of the conductors. As shown in the Fig.1, the magnetomotive forces on the bottom and top sides of the bottom layer are denoted F(0) and F(h), respectively.

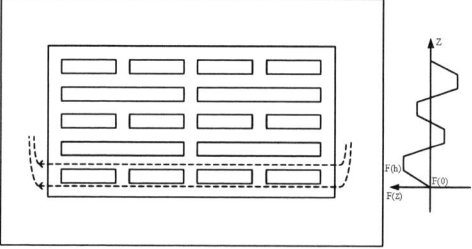

Fig. 1. The structure of the Planar Transformer

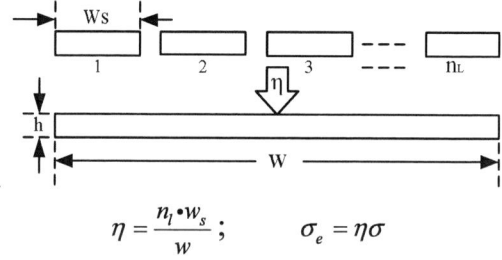

$$\eta = \frac{n_l \bullet w_s}{w}; \qquad \sigma_e = \eta\sigma$$

Fig.2 Porosity factor for a foil layer

Firstly, the foil windings which do not extend the full

This work was supported by China National Science Fund (No.50237030).

978-1-4244-0644-9/07/$25.00 ©2007 IEEE

winding window, may be treated as foils with equivalent conductivity $\sigma_e = \eta\sigma$, as shown in Fig.2. The winding porosity factor η is defined as the ratio of the actual layer copper area to the area of the effective foil conductor. For a sinusoidal excitation with frequency f, the skin depth is given by

$$\delta = \sqrt{\frac{1}{\pi\mu\sigma f}} \qquad (1)$$

where σ is the conductor conductivity and the permeability μ is equal to μ_0 for copper conductor.

The porosity effectively reduces the conductivity of conductor, and thereby increases the effective skin depth:

$$\delta_e = \frac{\delta}{\sqrt{\eta}} \qquad (2)$$

With sinusoidal current excitation, the copper loss of the bottom layer[11] in Fig.1 is given by

$$P = R_{dc}\frac{\varphi}{n_l^2}[(F^2(h)+F^2(0))\cdot G_1(\varphi)-4F(h)F(0)\cdot G_2(\varphi)] \qquad (3)$$

where n_l is the number of turns in the layer, R_{dc} is the dc resistance of the layer, $\varphi = \dfrac{h}{\delta_e}$.

If the winding carries current of rms magnitude I, the parameter m can be defined as the MMF $F(h)$ to the layer ampere-turns $n_l \cdot I$:

$$m = \frac{F(h)}{n_l I} \qquad (4)$$

Then the copper loss in the layer can be derived as

$$P = I^2 R_{dc}\varphi\{(2m^2-2m+1)G_1(\varphi)-4m(m-1)G_2(\varphi)\} \qquad (5)$$

$$P = I^2 R_{dc}\varphi\cdot Q(\varphi,m) \qquad (6)$$

where $\quad G_1(\varphi) = \dfrac{\sinh(2\varphi)+\sin(2\varphi)}{\cosh(2\varphi)-\cos(2\varphi)}\quad$, and

$G_2(\varphi) = \dfrac{\sinh(\varphi)\cos(\varphi)+\cosh(\varphi)\sin(\varphi)}{\cosh(2\varphi)-\cos(2\varphi)}$.

Assume that the transformer has M layer primary windings and N layer secondary windings. The total winding losses P_w can be found by summation over all layers

$$P_w = I_p^2\sum_{j=1}^{M}[R_{p,j}\cdot\varphi_j\cdot Q(\varphi_j,m_j)] + I_s^2\sum_{k=1}^{N}[R_{s,k}\cdot\varphi_k\cdot Q(\varphi_k,m_k)] \qquad (7)$$

where I_p is the rms value of the primary current, I_s is the rms value of the secondary current, $R_{p,j}$ is the dc resistance of j th primary layer and $R_{s,k}$ is the dc resistance of k th secondary layer. The subscripts j and k indicate the specific parameters for j th primary winding layer and k th secondary winding layer, respectively.

As can be seen from the MMF diagram shown in the Fig.1 and the layer loss equation (5), interleaving

windings can significantly reduce the proximity losses when the primary and secondary currents are in phase. Therefore, it is reasonable to minimize the winding losses by optimizing the winding design strategy. The above procedure is used to estimate the high frequency copper losses under the sinusoidal current excitation. Actually, the current waveforms in power converters contain significant harmonics, which will lead to increased winding losses. The effect of harmonics on the winding losses can be determined with the field harmonic analysis.

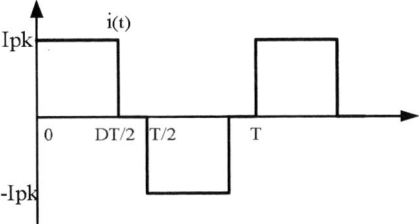

Fig.3 PWM winding current waveform

The copper losses of each individual harmonic can be calculated and summed to find the total winding losses. An arbitrary periodic current waveform, as shown in Fig.3, may be represented by its Fourier series:

$$i_{(t)} = I_{dc} + \sum_{i=1}^{\infty}\sqrt{2}I_i\cos(iwt) = \sum_{i=1,odd}^{\infty}\frac{4I_{pk}}{i\pi}\cdot\sin\frac{i\pi D}{2}\cdot\cos(iwt) \qquad (8)$$

where $I_{dc} = 0$, $w = \dfrac{2\pi}{T}$, and $I_i = \dfrac{2\sqrt{2}I_{pk}}{i\pi}\cdot\sin\dfrac{i\pi D}{2}$ is the rms magnitude of the i th harmonic.

Since the skin depth δ is smaller for high frequency harmonics than for the fundamental, the effective value of φ_i for i th harmonic is

$$\varphi_i = \sqrt{i}\cdot\varphi_1 \qquad (9)$$

The winding losses P_i due to harmonic i is given by

$$P_i = I_{p,i}^2\sum_{j=1}^{M}[R_{p,j}\cdot\sqrt{i}\varphi_j\cdot Q(\sqrt{i}\varphi_j,m_j)] + I_{s,i}^2\sum_{k=1}^{N}[R_{s,k}\cdot\sqrt{i}\varphi_k\cdot Q(\sqrt{i}\varphi_k,m_k)] \qquad (10)$$

The total copper losses due to all the harmonics is

$$P_{cu} = \sum_{i=1,odd}^{\infty}\left\{I_{p,i}^2\sum_{j=1}^{M}[R_{p,j}\cdot\sqrt{i}\varphi_j\cdot Q(\sqrt{i}\varphi_j,m_j)] + I_{s,i}^2\sum_{k=1}^{N}[R_{s,k}\cdot\sqrt{i}\varphi_k\cdot Q(\sqrt{i}\varphi_k,m_k)]\right\} \qquad (11)$$

B. Core Losses

Core losses in magnetic materials consist of three important components: a) classical eddy current loss, which results from eddy currents produced by the changing magnetic field inside the core; b) Hysteresis losses, which depend only on the peak flux density and linearly on the frequency; c) Excess eddy current losses. The origin of the excess eddy current loss mechanism is the magnetic domain wall motion that exist inside the magnetic material. The study of hysteresis model has shown that the overall core losses are directly related to the magnetization velocity. Therefore, the time derivative of the magnetic flux is averaged over one cycle to derive the equivalent frequency $f_{sin,eq}$ [7].

If $\frac{dB_w}{dt}$ denotes the weighted time derivative of flux B, the weighted time derivative of flux $\frac{dB_w}{dt}$ for the sinusoidal excitation is given by

$$\frac{dB_{w,sin}}{dt} = \frac{(B_{max}-B_{min})\pi^2}{2} \cdot f_{sin} \qquad (12)$$

In an arbitrary magnetizing current, the flux B can be described with piecewise linear function. Then the weighted time derivative of flux B can be derived as

$$\frac{dB_w}{dt} = \sum_{k=2}^{K} \frac{dB_k}{dt} \cdot \frac{(B_k-B_{k-1})}{B_{max}-B_{min}} \qquad (13)$$

where $\frac{B_k-B_{k-1}}{B_{max}-B_{min}}$ is the weighting factor and $\frac{dB_k}{dt} = \frac{B_k-B_{k-1}}{t_k-t_{k-1}}$ is the time derivative of magnetic flux. Compared with the sinusoidal weighted time derivative of flux $\frac{dB_{w,sin}}{dt}$ (12), the equivalent frequency for an arbitrary magnetizing current can be given by

$$f_{sin,eq} = \frac{2}{\pi^2} \sum_{k=2}^{K} [\frac{B_k-B_{k-1}}{B_{max}-B_{min}}]^2 \cdot \frac{1}{t_k-t_{k-1}} \qquad (14)$$

By inserting the $f_{sin,eq}$ in the modified Steinmetz equation, the total core losses could be obtained:

$$P_{core} = C_m \cdot f_{sin,eq}^{\alpha-1} \cdot B^\beta \cdot f(ct2 \cdot \tau^2 - ct1 \cdot \tau + ct) \qquad (15)$$

where f is the operating frequency, B is the amplitude of the flux density, τ is the operating core temperature. The other parameters could be easily obtained from the manufacturers data sheets. The modification to the conventional Steinmetz equation has been shown to be accurate for some arbitrary waveforms. It is also sufficient for the typical waveforms encountered in the normal power converters.

III. SIMULATION RESULTS

To validate the loss estimation mentioned above, the detailed losses in a high frequency planar transformer are studied using the Finite Element Method(FEM) in detail.

The parameters of the transformer studied are shown in Fig.4. For this planar transformer, the red plates in the model represent the eight primary turns and the yellow plates represent two secondary turns. The primary windings and secondary windings are interleaved to minimize the winding losses.

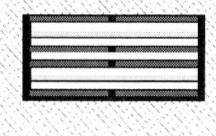

Core size: EI 18/4/10
Material: 3F4
Primary turns: 8
Secondary turns 2
Conductors Foils,140um
Insulator thickness 200um
Operating frequency. 1MHz

Fig.4 The parameters of Planar Transformer

This planar transformer is used in a half bridge converter and its associated voltage and current waveforms are shown in Fig.5. The voltage across the primary winding is a square wave with peak values of −40v and 40v. The current through the primary winding is also a square wave with peak values of −2.5A and 2.5A.

In order to obtain the accurate transformer losses using 2D FEA solvers, the simulation is based on the "Double 2D" approach presented in [12]. The approach is based on the division of the windings of magnetic component in two parts. Each part produces field distribution in different planes of the space. Then the simulated losses for each part could be summed up to obtain the total losses of the power transformer.

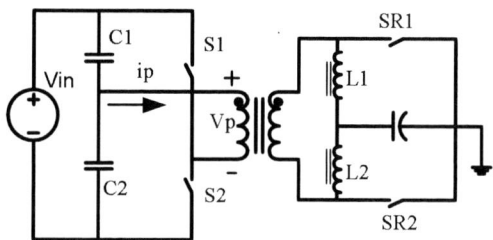

(a) Fixed duty cycle Half Bridge Converter

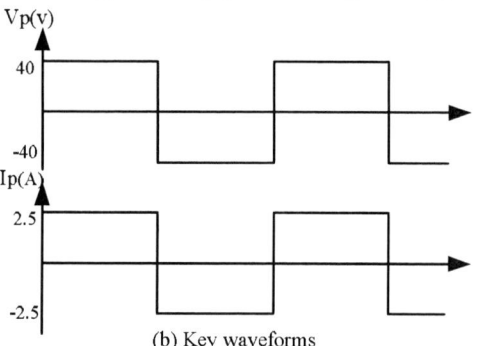

(b) Key waveforms

Fig.5 The Half Bridge Converter

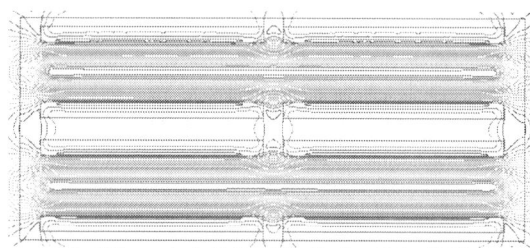

Fig.6 Magnetic field flux Distribution @ fs=1MHz

TABLE I. THE LOSSES COMPARISON

Losses type	Calculated	Simulated
Core loss	0.116w	0.125w
Winding loss	0.47w	0.524w

Fig.6 shows the magnetic flux distribution in frontal model of the planar transformer. The plot shows a small non-tangential field component at the outer layers. This can be explained by the spacing of the primary winding turns as shown in the Fig.6.

Table.1 gives the calculated and simulation results for the studied planar transformer. The operating frequency of the transformer is 1MHz. As can be seen, the theoretical Loss calculations are in agreement with the results obtained with the Finite Element simulations. As the two-dimensional finite element analysis allows the estimating of the fringing effect at the end of the conductors, the simulated losses may exceed the calculated transformer losses a little.

IV. CONCLUSIONS

This paper presents detailed loss analysis for a high frequency planar transformer. The loss estimation takes account of switching-type waveforms encountered in power supplies, inclusive of high frequency skin and proximity effects in windings and the non-sinusoidal excitation core losses. Accurate approximations have been provided to facilitate the calculations. The loss analysis has been verified by the Finite Element Method (FEM) simulation results.

ACKNOWLEDGEMENT

The author would like to thank the help of Astec Hong Kong Corporation.

REFERENCES

[1] P.L.Dowell, "Effects of eddy currents in transformer windings", *Proc Inst. Elect. Eng,* vol..113, No.8, Aug,1966.

[2] "Transformer and winding design", Part 3 of High Frequency Ferrite Power Transformer and choke Design, Philips Tech .Pub.207, Philips, The Netherlands,1986

[3] R.Petkov, " Optimum Design of a High-Power High-Frequency Transformer", *IEEE Transactions on Power Electronics*, vol.11, No.1, January 1996.

[4] P.S.Venkatraman, " Winding eddy current losses in switch mode Power transformers due to rectangular wave currents," *In Proc. Powercon 11*, 1984, Power Concepts Ins. Ventura,CA

[5] Johan Tjeerd Strydom, and Jacobus.D, " Electromagnetic Modeling for Design and loss Estimation of Resonant Integrated Spiral planar Power Passives(ISP3)," *IEEE Transactions on Power Electronics*, vol.19, No.3, pp.603-617, May 2004.

[6] C.P. Steinmetz, " On the law of hysteresis," *Proc. IEEE*, vol.72, pp.196-221, Feb 1984.

[7] M.Albach, A.Brockmeyer, " Calculating Core losses in transformers for arbitrary magnetizing currents: A comparison of different approaches," *In Proc. Power ESC96*, pp.1463-1468, June 1996.

[8] Jurgen Reinet, Ansgar Brockmeyer, "Calculation of losses in Ferro- and Ferrimagnetic Materials Based on the Modified Steinmetz Equation, " *IEEE Trans on Industry Applications*, vol.37, No.4, pp.1055-1061, Aug 2001.

[9] L.JieLi, T.Abdallah, and C.R.Sullivan, "Improved Calculation of core loss with nonsinusoidal waveforms, " in *Proc, Industry Applications Society Conference*, vol.4, pp.2203-2210, Oct 2001.

[10] Kapil. Venka, Charles.R.S, "Accurate Prediction of Ferrite core loss with Nonsinusoidal waveforms using only Steinmetz Parameters, " *IEEE workshop on Computers in Power Electronics*, 2002

[11] Robert W.Erickson, Dragan Maksimovic, " Fundamentals of Power Electronics- Part III Magnetics. " University of Colorado, Boulder, Colorado.

[12] R.Prieto, J.A.Cobos, O.Garcia, P.Alou, "Model of Integrated Magnetics by means of Double 2D Finite Element Analysis Techniques, " *Power Electronics Specialist Conference*(PESC) 1994.

Design of a Nuclear Magnetic Resonance Fast Field Cycling Air Cored Magnet

Duarte M. Sousa[1], Gil D. Marques[1], Pedro J. Sebastião[2], and António C. Ribeiro[2]

[1] Instituto Superior Técnico, DEEC AC-Energia, TULisbon – Av. Rovisco Pais – 1049-001 Lisboa – Portugal
[2] Instituto Superior Técnico, DF, TU Lisbon – Av. Rovisco Pais – 1049-001 Lisboa – Portugal and Centro de Física da Matéria Condensada, Av. Prof. Gama Pinto 2, 1649-003 Lisboa, Portugal

Abstract-- **A method for the design of a nuclear magnetic resonance fast field cycling air-cored magnet is presented in this paper.**

General analytic expressions relating the geometric parameters of the magnet with the flux density, the current density and the power losses were obtained. An optimization algorithm based on the thin circular loop coil model to determine the positions of the windings for a target flux density minimizing the power losses is described.

The results obtained in this paper show that, in order to obtain a magnet with a reasonable practical size, the highest possible current density should be used. A prototype that can generate a magnetic flux density of 1.6 T was designed and built. This magnet is cooled using a coolant fluid circulating on longitudinal ducts.

The magnet was tested in typical Fast Field Cycling Nuclear Magnetic Resonance experiments and presents a good performance.

***Index Terms*-- Air Cored Magnet Design, Fast Field Cycling NMR, Flux Density.**

I. INTRODUCTION

Nuclear Magnetic Resonance (NMR) Fast Field Cycling (FFC) magnets should generate the highest possible flux density (B), with the highest possible relative homogeneity ($B/\Delta B$) (10^6 ideally) [1-3].

To fulfill the FFC NMR technique requirements, the electrical current that drives the magnet, must cycle between at least two levels as exemplified in figure 1. Typically a FFC experiment has two major requirements: in the transient states the current must change at required rates. In the steady states the current must be constant, i.e., it should have a ripple as low as possible.

Since the seventies, some coils specially optimized for FFC NMR were designed to produce similar magnetic fields but present different geometries and dimensions.

FFC NMR magnets are typically made of copper and have air cores [4-8]. They should produce the required magnetic flux density B, should have low self inductance (Lm) and minimum power losses (P).

The minimization of the self inductance is necessary to allow fast commutations of the magnetic field (transient state) without the need of very high voltages. In fact, what should be minimized is the magnetic energy stored in the magnet.

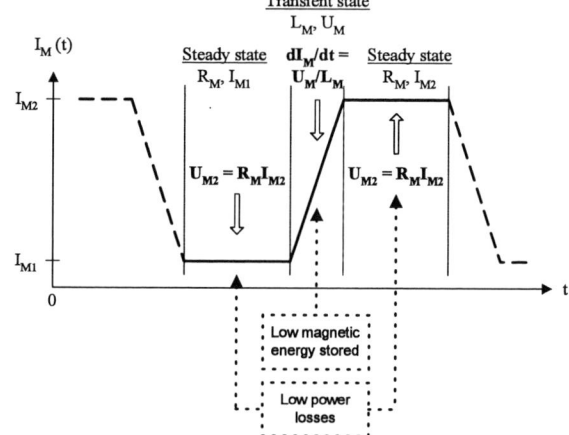

Fig. 1. Magnet current states.

The above conditions are mainly dependent on the number of windings and on the current. The minimization of Lm and P can be obtained using different techniques and algorithms [4-8]. In some of the reported studies the maximization of the relative homogeneity ($B/\Delta B$) was considered the most important requirement leading to algorithms of considerable complexity. On the other hand, the design and optimization of magnets has been widely studied taking into account the specificities and requirements of the magnetic resonance techniques. Different approaches and algorithms have been developed and are well described in the literature [9 -20].

In this work, a different approach is presented. The algorithm is implemented in two sequential tasks. In the first task, the design a coil that generates the target magnetic flux density with the minimum losses is considered. The position of each winding is determined using a given constant current and for a given cross section of the conductors considered. In the second task some windings are added or removed at defined positions to increase the relative homogeneity of the flux density. Surprisingly, it was verified that this approach is easily achieved with a non-symmetrical magnet.

As it is well known, in order to obtain an optimal magnet, a high current density should be used. However, high current density increases the difficulty of cooling the magnet. To solve this problem a magnet cooling system with longitudinal hydraulic ducts that allow the circulation of the cooling fluid was designed. A pump was used to impose forced convection in the hydraulic ducts.

The above procedure was used to obtain an air-cored magnet that generates a flux density up to 1.6 T with a minimum inner diameter of 3.0 cm. The inner diameter was set accordingly to the size of the probe head used.

II. OPTIMIZATION ALGORITHM

A. Preliminary Approach

In general, FFC NMR magnets are basically cylindrical in shape (Fig. 2), with an inner air-core with diameter D_{in}, an external diameter D_{ex}, and length L. The space between D_{in} and D_{ex} is filled with interlaced conducting coil layers of thickness d_c, and hydraulic layers of width d_l.

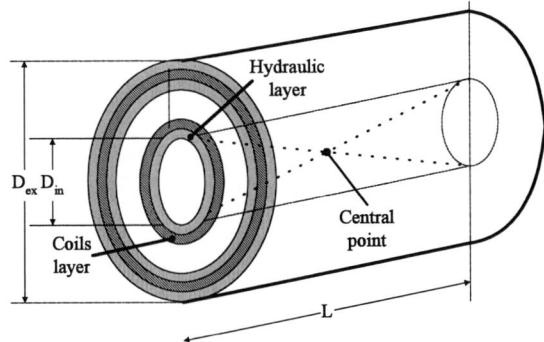

Fig. 2. Basic magnet structure.

From the Biot-Savart law it is possible to write the following expression for the flux density as a function of the current density J at the central point of the inner core of the magnet [21].

$$B = \left(\frac{d_c}{d_c + d_l} \right) \frac{\mu_0 L J}{2} \log \frac{D_{ex} + \sqrt{D_{ex}^2 + L^2}}{D_{in} + \sqrt{D_{in}^2 + L^2}} \quad (1)$$

The power losses can also be estimated by taking into account the dimensions of the magnet and the current density:

$$P = \left(\frac{d_c}{d_c + d_l} \right) \frac{\pi \rho J^2 L}{4} (D_{ex}^2 - D_{in}^2) \quad (2)$$

where ρ is the resistivity of the conducting material.

For a given target field B_{max}, the power losses given by equation (2) is minimized using the "MatLab/Optimization Toolbox" with the function CONSTR that finds the constrained minimum of a function of several variables. The function P = f (L, D_{ex}, J, d_l), equation (2), is minimized subject to the equality constraint given by equation (1) and the following inequality constraints:

1) $L \le L_{max}$
2) $D_{in} < D_{ex} \le D_{max}$
3) $J \le J_{max}$
4) $d_{l\_min} \le d_l \le d_{l\_max}$

$$(3)$$

The type of conductor (copper) and the thickness of the conducting coil layers d_c were previously defined taking into account the standard copper wires available.

The limits L_{max}, D_{max}, D_{in}, J_{max}, d_{l_min} and d_{l_max} were chosen according to the following practical reasons:
- inside the inner core of the magnet it is mandatory to place a r.f. detection coil and a sample thermally insulated. This set (generally assigned as probe head) is in general cylindrical and requires D_{in} from 2 cm up to 3 cm;
- minimum and maximum values to d_l are defined to allow continuous flowing of the cooling fluid;
- the limits L_{max} and D_{max} are needed by the CONSTR function but did not influence the final result;
- the current density cannot achieve unreasonable values.

It was observed that this optimization program always selected the minimum value of d_l and the maximum value of J, for any set of parameters (3) and any target field B_{max}.

The major result of the optimization process is that a ratio $1 < \frac{D_{ex}}{L} < 2$ was always verified independently of D_{in}, d_l or d_c. Therefore, the flux density at the central point of the magnet, given by equation (1), assuming that $D_{in} << L$ can be approximated by:

$$B \approx k_{B0} \mu_0 \left(\frac{d_c}{d_c + d_l} \right) L J , \quad (4)$$

with $0.2 < k_{B0} < 0.3$.

In addition, the magnet's volume, which is basically the volume of a cylinder of length L and external diameter D_{ex}, as function of B and J, is given by,

$$V \approx k_{V0} \frac{\pi}{\mu_0^3} \left(1 + \frac{d_l}{d_c} \right)^3 \left(\frac{B}{J} \right)^3 , \quad (5)$$

with $25 < k_{V0} < 87$.

The power losses equation (2) may also be simplified:

637

$$P \approx k_{V0} \frac{\pi}{\mu_0{}^3} \rho \left(1 + \frac{d_l}{d_c}\right)^2 \frac{B^3}{J} \qquad (6)$$

From the analysis of expressions (4), (5) and (6) it can be verified that the coil's volume and power losses, increase with d_l, decrease with J and increase with B^3 for the optimization achieved.

Therefore, a small magnet will be able to produce a required flux density inside the magnet air-core with low power losses provided that the current density is high enough. Generically, the optimization of the magnet parameters must be compatible with the lowest possible magnet's volume.

The ratio $\frac{P}{V} \approx \rho J^2 \frac{d_c}{d_c + d_l}$ is an important factor, which characterizes the relation between the magnet and the associated cooling system, since a high value of P/V cannot be obtained with arbitrary small values of the hydraulic layers' width d_l.

Using the above preliminary approach, several solutions can be obtained changing the target magnetic field, the current density or the width of the hydraulic layers. In figure 3 the magnet's volume is plotted as function of B for a constant current density $J = 100$ A/mm^2 and different values of d_l. As expected, the volume of the magnet increases considerably with decreasing J. The power losses and the magnetic energy increase.

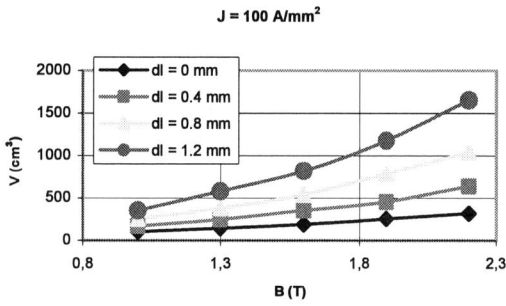

Fig. 3. Volume V = f (B, d$_l$) for J = 100 A/mm^2.

In this preliminary approach the relative homogeneity of the field was not considered. In fact, it may be shown that high values of J and small values of the volume lead to low homogeneities [21]. Nevertheless, the relative homogeneity $B/\Delta B$ must be included in the optimization process.

B. Final Approach

The optimization algorithm is based on the thin circular loop (TCL) coil model. As for most of the magnets referred in the literature, a symmetrical structure will be considered (Fig. 4).

In this case, on the longitudinal axis, the flux density amplitude using the TCL model, is [5]-[22]:

$$B_{jk} = A_{jk} I_j = \frac{\mu}{2} \left(\frac{r_j{}^2}{\left[r_j{}^2 + (a_k - z)^2\right]^{3/2}} + \frac{r_j{}^2}{\left[r_j{}^2 + (a_k + z)^2\right]^{3/2}} \right) I_j \qquad (7)$$

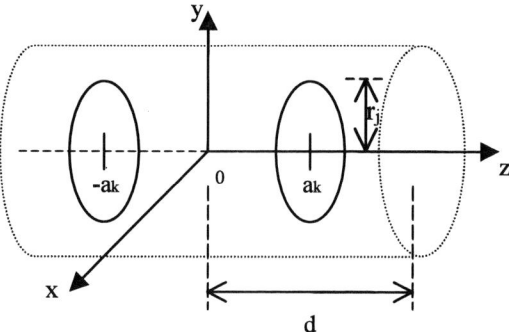

Fig. 4. Symmetrical magnet using the TCL coil model.

Where r_j, a_k, and z are indicated in figure 4 and μ is the magnetic permeability of the air.

The function $A_{jk}(a_k, z)$ is plotted in figure 5 for a constant radius r_j and as a function of a_k and z.

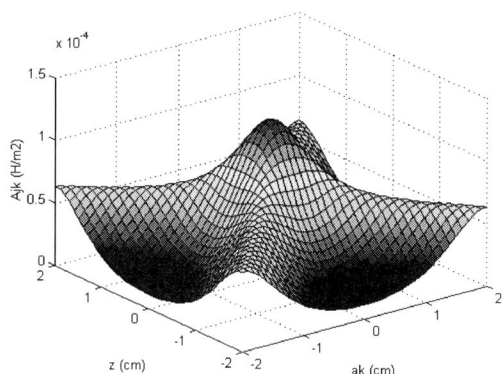

Fig. 5. A$_{jk}$ = f(a$_k$, z).

Therefore, according to the function A_{jk} properties, the maximum flux density amplitude is obtained when all the windings should be placed as near as possible to the center of the magnet.

The algorithm to optimize the desired field homogeneity at the central volume of the magnet (along 1 cm) is based on the properties of the function A_{jk}, which is used to compute the flux density.

During the magnet design, several solutions were tested. The magnets computed were mainly dependent on the following parameters: the current density J; the cross section of the conductors d_c; the hydraulic layer width d_l.

The current density, which is important to set the magnet's volume, depends on the current and cross section of the conductors. If J has a low value, a large and, therefore, undesirable magnet is obtained.

The hydraulic layers width d_l is important for the efficiency of the magnet cooling system. Its final value is a compromise between the magnet's volume and the cooling fluid characteristics.

The gap between successive windings should also be taken in account, because the small value of this parameter multiplied by the number of windings in each layer causes an important increase of the magnet volume.

1) Task one of the algorithm

The task 1 of the algorithm can be summarized in the following steps:

1 → Define: maximum flux density B_{max}, internal diameter D_{in}, conductors section d_c and current density J;

2 → Starting from a single symmetrical winding near the geometrical center of the magnet new windings (see figure 6.a) or new layers (see figure 6.b) are added to the magnet until the maximum target field is reached.

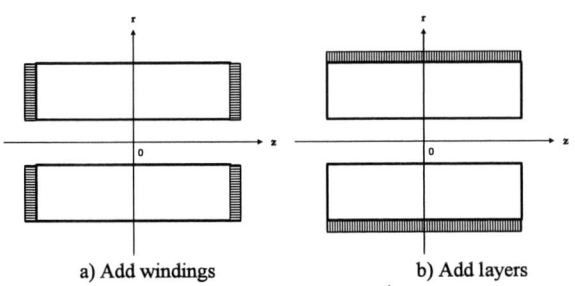

a) Add windings   b) Add layers

Fig. 6. Example of one step of the algorithm 1ˢᵗ task.

2) Task two of the algorithm

The desired field homogeneity is obtained in a second optimization task by adding or eliminating coils to the magnet obtained in the preliminary optimization stage.

This second optimization task can be summarized in the following steps and described by figure 7:

1) Starting from the preliminary magnet's distribution of the coils the magnet flux density is optimized inside a central cylindrical volume of 1cm×2.5cm inside the magnet by adding or removing single coils either the axial or radial boundaries;

2) In successive steps, every new field distribution inside the target volume was compared with the previous one and accepted if the field homogeneity converged to the target value maintaining B_{max}.

The procedure, adding or removing a single coil, gives rise to unsymmetrical magnets. To obtain symmetrical magnets, symmetrical pairs of coils should be used. The symmetry of the magnet was not considered decisive for the application in view.

III. DESCRIPTION OF THE PROTOTYPE

Based on the described algorithm, a magnet with the cross-longitudinal section of 131 cm² was obtained and is schematically represented in figure 8. According to the developed algorithm the magnet is slightly asymmetric. It was verified that this choice leads to a magnet with less copper and lower power losses than a symmetric magnet.

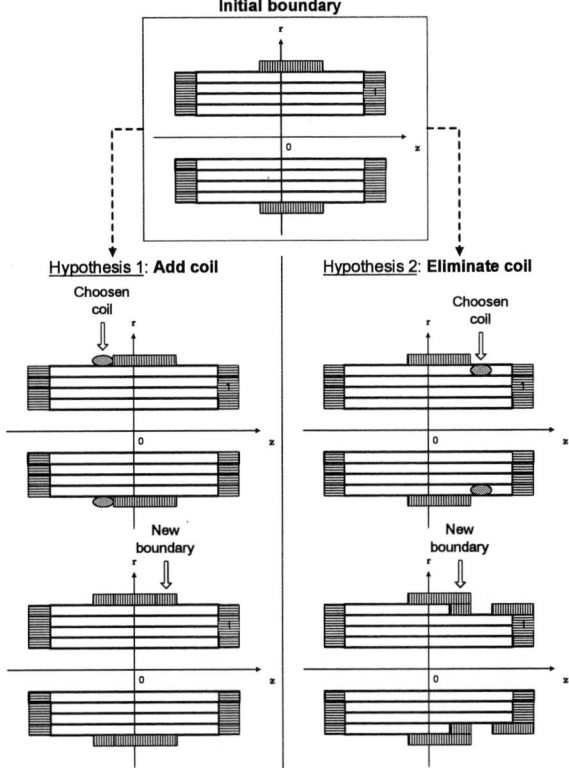

Fig. 7. Example of one step of the algorithm 2ⁿᵈ task.

Two photos of the prototype are presented in figure 9. The main parameters of the magnet are presented in table I.

TABLE I
MAGNET PARAMETERS

R_m (Ω)	1.96 ~2
L_m (mH)	18
B_{max} (T)	1.6
I_{max} (A)	200
P (kW)	78.4
W_m (J)	360
Inner radius (cm)	1.5
Length (cm)	10.4
Diameter (cm)	12.6
Conductive layers thickness (mm)	1
Hydraulic Layers width (mm)	1
Number of layers	23

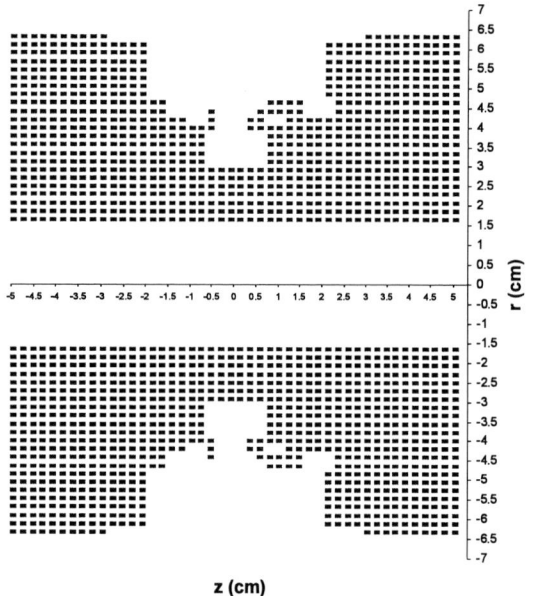

Fig. 8. Cross-longitudinal view of the magnet.

Fig. 9. Photos of the magnet.

A. Flux density at the longitudinal axis

The magnitude of the magnetic flux density calculated along the magnet longitudinal axis is plotted in figure 10.a and in figure 10.b for a few axial central points.

From figure 10 it can be observed that the longitudinal axis flux density is slightly asymmetric due to the asymmetrical magnet coils distribution. The flux density

deviation ΔB at longitudinal axis central points is $\approx 2\times10^{-4}$ (for 1 cm around the magnet central point). It was verified experimentally that in all FFC NMR studies performed this value was acceptable.

B. Cooling System

Temperature distribution along the magnet was obtained, taking into account the number of windings N_j, the ohmic value of each magnet's layer, assuming that the hydraulic layer width d_l is chosen so that it contributes to minimize the size of the magnet and, allow the proper cooling of the system. The distribution presented in figure 11 was calculated with the following expressions, derived from basic physical equations [23]-[24]:

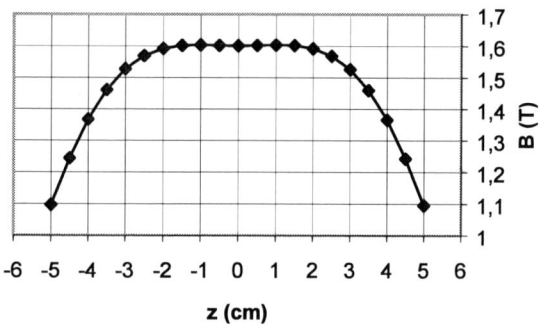

a) Along the longitudinal axis

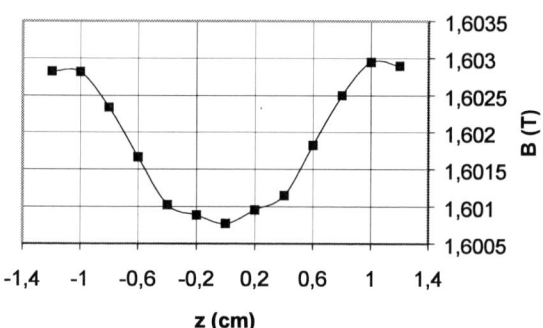

b) Central points

Fig. 10. Flux density along the magnet longitudinal axis.

$$\overline{T}_j^f(z) = \overline{T}_1^{fi} + (j-1)\frac{P\,j}{V\,j}\frac{d_c}{2h} + \frac{P\,j}{V\,j}\frac{1}{c_p}\frac{d_c}{\overline{v}\,\rho_f d_l}z \quad (10)$$

$$\overline{T}_j^c(z) = \overline{T}_1^{fi} + j\frac{P\,j}{V\,j}\frac{d_c}{2h} + \frac{P\,j}{V\,j}\frac{1}{c_p}\frac{d_c}{\overline{v}\,\rho_f d_l}z \quad (11)$$

Where: $\overline{T}_j^f(z)$ is the fluid temperature distribution for the layer j; $\overline{T}_j^c(z)$ is the coil temperature distribution for the layer j; $\overline{T_1^{fi}}$ is the input temperature of the fluid; h is the heat transfer coefficient; P_j are the power losses of the layer j; V_j is the volume of the layer j; c_p is the specific heat at constant pressure; $\overline{v}$ is the average flow velocity of the cooling fluid; ρ_f is the fluid density; and ρ is the copper resistivity.

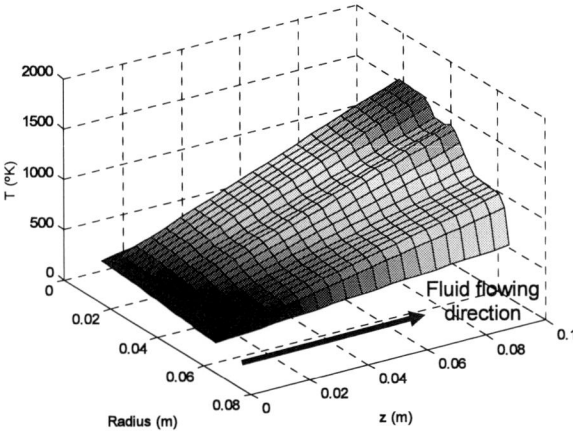

Fig. 11. Cooling fluid temperature distribution.

Since:

a) the number of coils N_j decreases from the inner to the outer of the magnet;

b) the power losses P_j change from layer to layer and are higher in the inner layer of the magnet;

c) the ratio P_j/V_j presents the highest value in the inner layer and the lowest in the outer layer;

d) the mass flow rate is greater in the external hydraulic layers because the average diameter D_j continuously increase from the internal hydraulic layers to the external ones.

It is possible to observe that:

• the cooling fluid temperature (and also the copper temperature) increases obviously with z;

• the radial temperature decreases from the internal layers to the external ones, more or less in steps which depend on the number of coils in each layer.

This temperature distribution was partially confirmed by experimental measurements done with forced air-cooling and lower current values [21]. A complete detailed experimental study of the temperature distribution inside the magnet at this point of the prototype development was severely limited by the use of a liquid fluid mandatory to achieve the nominal magnetic flux density.

IV. CONCLUSIONS

This paper presents the design procedure of a FFC NMR magnet using different optimization techniques. The magnet obtained presents a cylindrical structure and was made with copper bar wires with 1×2 mm^2.

The design of the magnet was made in two stages. First a preliminary study of the general basic relations between the physical and electrical parameters showed that the magnet volume should be minimized in order to have a low induction coefficient and low power losses. Therefore, the required current density to generate the target magnetic field could be obtained with lower values of current. This simple approach showed also that the ratio between the magnet external diameter and the magnet length converged to a value between 1 and 2 for all the starting conditions considered.

The optimization of the magnet's homogeneity was made in a second stage of the magnet's design process. A simple and yet effective algorithm was implemented which added or removed the necessary coils to improve the average homogeneity to about 10^{-4} inside a defined cylindrical volume of 1×2.5 cm^3 at the center of the magnet.

As final remark about the algorithm, it is important to point out that the implemented solution allows designing other magnets, starting from a simple symmetrical winding and defining only the maximum flux density, the internal diameter of the magnet, the conductors section and the current density.

The P/V ratio of the implemented magnet (13.5 W/cm^3) is also an important factor, which is strongly limited by the capacity of the cooling system, in particular, if the FFC spectrometer has to operate continuously.

The thermal model developed to characterize the magnet temperature distribution showed that the hydraulic layers width should have a minimum value, to assure both a good performance of the cooling fluid in a magnet with reasonable volume.

The temperature in each layer (both hydraulic and conductive) linearly increases along any direction parallel to the magnet axis and decreases with the increasing diameter of the layers. The slope of the axial temperature variation depends on physical and electrical dimensions of the conductive layers and on the characteristics of the cooling fluid.

Both the magnet and cooling system were tested in typical FFC NMR experiments and presented a performance that can be compared with other fast field cycling systems [1]-[3]-[4]-[8]. Comparisons between FFC magnets evolves the analysis of several factors like the volume, the ohmic value, the induction coefficient, the power losses the relative homogeneity and the power/volume ratio. Based on a global analysis of these factors and on the results collected from the experiments, the designed magnet fulfills the technique requirements and presents a good global performance [21]. FFC NMR experimental results obtained with this magnet were already published [25].

641

REFERENCES

[1] F. Noack, "NMR Field-Cycling Spectroscopy: Principles and Applications", *Progress in NMR Spectroscopy*, Vol. 18, pp. 171-276, 1986.

[2] R. Kimmich, "NMR – Tomography, Diffusometry, Relaxometry", Springer, 1997.

[3] E. Anoardo, G. Galli, G. Ferrante, "Fast- Field-Cycling NMR: Applications and Instrumentation", *Appl. Magn. Reson.*, 20, pp. 365-404, 2001.

[4] K.H. Schweikert, R. Krieg, and F. Noack, "A High-Field Air-Cored Magnet Coil Design for Fast-Field-Cycling NMR", *Journal of Magnetic Resonance*, 78, pp. 77-96, 1988.

[5] K.H. Schweikert, Thesis: "Aufbau und Erprobung eines Feldzyklus NMR Spektrometers für Deuteronen-Relaxionsuntersuchungen an Flüssigkristallen", Physikalishes Institut der Universität Stuttgart, Stuttgart, FGR, 1990.

[6] C. Job, J. Zajicek, M. F. Brown, "Fast field-cycling nuclear magnetic resonance spectrometer", *Review of Scientific Instruments.*, 67 (6), pp. 2113-2122, 1996.

[7] R.-O. Seitter, R. Kimmich, "Magnetic Resonace: Relaxometers", *Encyclopedia of Spectroscopy and Spectrometry*, pp. 2000-2008, London Academic Press, 1999.

[8] O. Lips, A. F. Privalov, S. V. Dvinskikh, F. Fujara, "Magnet design with high B0 homogeneity for a fast-field-cycling NMR Applications", *Journal of Magnetic Resonance*, 149, pp. 22-28, 2001.

[9] P. Alotto, P. Girdinio, P.;Molfino, M. Nervi: *"Mesh Adaption and Optimization Techniques in Magnet Design"*, IEEE Transactions on Magnetics, Vol. 32, No. 4, pp. 2954-2957, 1996.

[10] Y. Chu: *"Numerical Calculation for the Magnetic Field in Current-Carrying Circular Arc Filament"*, IEEE Transactions on Magnetics, Vol. 34, No. 2, pp. 502-504, 1998.

[11] S. Crozier, S. Dodd, D. M. Doddrell: *"A Novel Design Methodology for Nth Order, Shielded Longitudinal Coils for NMR "*, Meas. Sci. Technol. 7, pp. 36-41, 1996.

[12] L. K. Forbes, S. Crozier, D. M. Doddrell: *"Rapid Computation of Static Fields Produced by Thick Circular Solenoids"*, IEEE Transactions on Magnetics, Vol. 33, No. 5, pp. 4405-4410, 1997.

[13] M. W. Garrett: *"Thick Cylindrical Coil Systems for Strong Magnetic Fields with Field or Gradient Homogeneities of the 6th to 20th Order"*, Journal of Applied Physics, Vol. 38, No. 6, pp. 2563-2586, 1967.

[14] T. Ishikawa, M. Matsunami: *"An Optimization Method Based on Radial Basis Function"*, IEEE Transactions on Magnetics, Vol. 33, No. 2, pp. 1868-1871, 1997.

[15] D. B. Montgomery: *"Solenoid Magnet Design: The Magneticaspects and Superconducting Systems"*, Wiley, 1969.

[16] I. Munteanu, F. M. G. Tomescu: *"Optimisation of a Magnetic Device Based on Symbolic Analysis of the Field"*, IEEE Transactions on Magnetics, Vol. 33, No. 2, pp. 1840-1843, 1997

[17] L. S. Petropoulos, M. A. Morich: *"Novel gradient Coil Set with an Interstitial Gap for Interventional Nuclear Magnetic Resonance Applications"*, IEEE Transactions on Magnetics, Vol. 33, No. 5, pp. 4107- 4109, 1997.

[18] F. Shi, R. Ludwig: *"Magnetic Resonance Imaging Gradient Coil Design by Combining Optimization Techniques with the Finite Element Method"*, IEEE Transactions on Magnetics, Vol. 34, No. 3, 1998.

[19] H. Xu S. Conolly, G. Scott, A. Macovski: *"Homogeneous magnet design using linear programming"*, IEEE Transactions on Magnetics, Vol. 36, No. 2, pp. 476-483, 2000.

[20] J. H. Jensen: *"Minimum-volume coil arrangements for generation of uniform magnetic fields"*, IEEE Transactions on Magnetics, Vol. 38, No. 6, pp. 3579-3588, 2002.

[21] D. M. Sousa, PhD Thesis, Universidade Técnica de Lisboa, 2003.

[22] M. N. Wilson, "Superconducting Magnets", Oxford Science Publications, 1997.

[23] F. Incropera, D. P. Dewitt, "Fundamentals of Heat and Mass Transfer", 5th Ed., John Wiley & Sons, 2001.

[24] F. Kreith, M. S. Bohn, "Principles of Heat Transfer", 6th Ed., Brooks/Cole, 2001.

[25] D. M. Sousa, P. A. L. Fernandes, G. D. Marques, A. C. Ribeiro, and P. J. Sebastião, *"Novel pulsed switched power supply for a Fast Field Cycling NMR spectrometer"*, Solid State NMR", 25, pp. 160-166, 2004.

Using DFT to Obtain the Equivalent Circuit of Aluminum Electrolytic Capacitors

Acácio M. R. Amaral (*)(****), Gustavo M. Buatti (**), Hugo Ribeiro (***) and A.J. Marques Cardoso (****)

* Polytechnic Institute of Coimbra, ISEC - DEIS, Rua Pedro Nunes – Quinta da Nora, P – 3030 - 199, Coimbra, Portugal
** Alstom Transport, Rue Docteur Guinier, B.P. 4, 65600, Semeac, France
*** Polytechnic Institute of Tomar, ESTT/IT, DEE, Estrada da Serra, P – 2300 - 313, Tomar, Portugal
**** University of Coimbra, FCTUC/IT, DEEC, Pólo II – Pinhal de Marrocos, P – 3030 - 290, Coimbra, Portugal

Abstract- **This paper presents an economic and automatic experimental technique that allows the determination of the equivalent circuit of aluminum electrolytic capacitors. To implement the proposed technique it is necessary to arrange experimentally the capacitor under test in series with a resistor and connected to a sinusoidal voltage. The relationship between gain and phase of the sinusoidal voltage waveform applied to both capacitor and resistor, and the gain and phase of capacitor voltage give enough information to compute the capacitors equivalent circuit. To obtain both gain and phase of both waveforms with high accuracy, the discrete Fourier transform algorithm is used.**

Index Terms- **Aluminum electrolytic capacitors, DFT, equivalent circuit and measurement.**

I. INTRODUCTION

The aim of this paper is to develop a very simple and economic measurement tool that in an automatic way can be used to determine the equivalent circuit of aluminum electrolytic capacitors for a wide range of frequencies. The great advantage of such tool is that it can be used instead of an impedance gain-phase analyzer in a first analysis, since the latter is a very expensive equipment and not always available.

Aluminum electrolytic capacitors are very popular in power electronics due to their energy density and low cost per microfarad [1]. However, it is a component with very short lifetime [2], and which became a subject of great interest in the last years by researchers in many different fields. In order to give a picture of what was just said, in applications involving DC/DC converters more than 50% of the failures are related to the failure of the output filter electrolytic capacitor [3]. Thus, from the knowledge of some parameters of the component it is possible to monitor the condition of the capacitor, allowing predictive maintenance and avoiding more serious drawbacks such as the complete shutdown of the equipment.

From the point of view of the diagnosis of electrolytic capacitors, many approaches have been already proposed, which can be implemented online or offline.

Online techniques have the advantage that the capacitor does not need to be demounted from the board for measuring its parameters. Most of the proposed approaches deal only with the Equivalent Series Resistance (herein called *ESR*) as an indicator of a possible parametric failure, because this parameter has a stronger variation during its lifetime than the capacitance

itself. It is observed that the *ESR* increases a lot with time, sometimes becoming 2 or 3 times higher than its initial value (this is the criterion given by some suppliers to indicate its end of life). On the other hand, the capacitance shows a decrease normally in the order of 20 to 30 per cent.

In [2], for the first time the authors called the attention for the *ESR* estimation as an important parameter for diagnosis purposes, and has demonstrated how to evaluate the *ESR* in forward and buck-boost converters. Later, in [4] the same was done for the boost converter, and in all these cases the estimation could be done online with the addition of current and/or voltage sensors. In [5], for frequencies in the order of the switching frequency of the converter, usually the impedance is almost only due to the *ESR*. Thus, after filtering the voltage and current through the same, through the knowledge of the fundamental components it is possible to evaluate the *ESR* value. In that approach, the authors needed to characterize the *ESR* with temperature in order to evaluate the left lifetime before the failure of the capacitor, and so, offline techniques capable to extract the parameters of electrolytic capacitors start to appear as very useful techniques as well, when more sophisticated equipments are not available.

In [6], the authors proposed a real time method for estimating the *ESR* in Adjustable Speed Drives (*ASDs*) and Uninterruptible Power Supplies (*UPSs*), which can be implemented in any other power electronics converter if the current and voltage signals through the capacitor are available. The starting point of the work is that at steady state condition the power losses in the capacitor are only due to the *ESR*.

Recently, not only *ESR* has been regarded anymore as the unique parameter to be monitored, but also the capacitance since it can reinforce the conclusions about the condition of the capacitor. Hybrid models were proposed in [7] for extracting the parameters of power electronics converters, and between them, *ESR* and capacitance. With the same scope, in [8,9] simpler approaches than hybrid models are presented and discussed for parameter estimation covering most of non-isolated DC/DC converters, and *ESR* and capacitance are also extracted for diagnosis purposes.

Regarding offline techniques they are not only important for diagnosis purposes [10], but also for designing purposes. That is to say that normally, suppliers do not give enough information about the

978-1-4244-0644-9/07/$25.00 ©2007 IEEE

parameters in a wide range of frequencies, but they give information about only one specific frequency. In this way, if the designer can have enough information about the component that he is dealing with, it is possible to obtain more accurate information regarding the behavior of the designed circuit, mainly at simulation stage.

In the next section, a more detailed description of offline techniques is found. Then the approach proposed in this paper is described, followed by its validation through experimental results. Finally, conclusions are presented and discussed.

II. REVIEW OF OFFLINE TECHNIQUES

An aluminum electrolytic capacitor is made of wires and an insulation material. The wires have a resistance and an inductance, and insulators have a leakage resistance. Thus, the behavior of capacitors with frequency diverges from their ideal model. Fig. 1 shows the equivalent circuit of an aluminum electrolytic capacitor.

Fig. 1 - Simplified equivalent circuit of an aluminum electrolytic capacitor.

The impedance of an aluminum electrolytic capacitor can be expressed as:

$$Z_C \cong ESR + j\, X_{CAP} \wedge X_{CAP} = \left(w\, ESL - \frac{1}{w\, C} \right) \qquad (1)$$

where C is the capacitance, ESR is the equivalent series resistance (which represents the wire resistance as well as leakage resistance) and ESL is the equivalent series inductance (wire inductance).

The typical ESL value for radial electrolytic capacitors and screw-terminals capacitors is about 1-2 nH/mm for terminal spacing and dos not change significantly with temperature and frequency [11].

In [12] a more complex model for electrolytic capacitors is presented. However, the model adopted in this paper is simplified and leads to satisfactory results, as it will be proven during the validation stage of the proposed technique in this paper.

In [13] an experimental technique that allows the determination of the ESR intrinsic value of aluminum electrolytic capacitors near their resonance frequency was reported. For doing that, the used circuit was composed by an inductor, a resistor, the capacitor to be monitored, and a sinusoidal voltage for feeding the circuit. Using equation (2) as well as the input voltage and the capacitor voltage, it is possible to determine the ESR intrinsic value.

$$ESR = \frac{\sqrt{\left(\dfrac{V_C\, L\, C\, w}{V_{in}} \right)^2 - \dfrac{1}{w^2}}}{C} \qquad (2)$$

where V_{in} is the sinusoidal voltage applied to the LCR circuit, V_C is the capacitor voltage, L is the inductor value, C is the capacitance value and w is the radian velocity of the sinusoidal waveform.

However the previous technique presented some drawbacks, such as the fact that both capacitance and inductance have a tolerance, which in some cases could reach 30% of the nominal value. Not only, the capacitance is frequency dependent, and its value is reduced for increasing frequencies [11].

In order to overcome this problem, a new technique was presented in [14], which allows the computation of both ESR and reactance values of aluminium electrolytic capacitors without the previous drawbacks. For that purpose, a very simple RC circuit connected to a signal generator followed by a power amplifier is needed (Fig. 2). The scope is to create a sinusoidal waveform with enough power and the desired frequency to feed the RC circuit, in order to extract the intrinsic ESR value and the reactance value of the aluminum electrolytic capacitor.

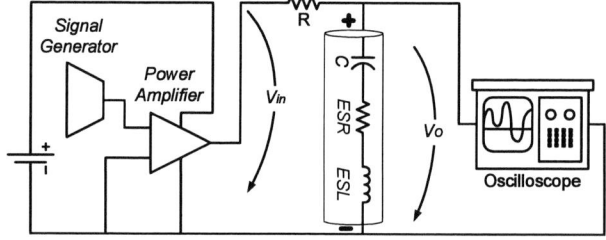

Fig. 2 - Schematic of the experimental technique proposed in [14].

Thus, the relationship between the output voltage, V_O, and the input voltage, V_{in}, gives all the required information to compute both ESR and the reactance values of the component. The resulting curve of V_O as function of V_{in} is going to be a rotated ellipse, (Fig. 3 – ellipse 1).

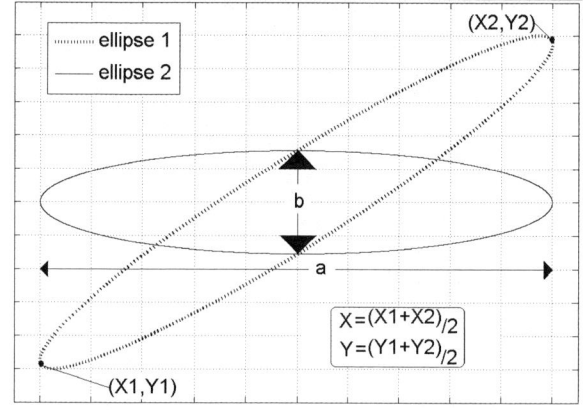

Fig. 3 - Resulting curve of V_O as function of V_{in}.

Using the slope of this ellipse, m, and equation (3), it is possible to obtain the intrinsic ESR value, if the conditions (4) and (5) below are respected.

$$ESR \cong m\, R = \frac{Y}{X} R \qquad (3)$$

644

$$(R + ESR) \gg |X_{cap}| = \left| \left(ESL\, w_o - \frac{1}{w_O\, C} \right) j \right| \tag{4}$$

$$ESR\, w^2 \gg \left(1 - ESL\, C\, w^2 \right)^2 \tag{5}$$

To obtain the reactance of the capacitor, first it is necessary to rotate the ellipse to a horizontal position (Fig. 3 – ellipse 2). The reactance value could be computed through the relation between both axes, and using equation (6):

$$X_{CAP} = -\frac{b}{a} R \tag{6}$$

where b and a represent Y and X axis, respectively, of ellipse 2.

However, to implement the just described technique it is necessary to compute the R value as a function of the operation frequency and of the capacitance of the capacitor used, in order to respect equations (4) and (5). To avoid using the previous calculations, capacitor current and capacitor voltage could be used directly; however a current sensor will be needed, which increases the cost and complexity of the technique.

For low frequencies, and considering capacitors with low and medium capacitance, it is possible to conclude that the computed R values should be higher than the ones used for high capacitance capacitors. In such case, since the capacitance is smaller, the reactance value is much higher. Thus, the ratio between the axis of ellipse 2 is much higher than the slope of ellipse 1. If it is so, from the graphical analysis it will be very easy to introduce errors in the ESR estimation.

The technique proposed in this paper is not based in a graphical analysis, but in the Discrete Fourier Transform algorithm, herein called DFT. The new proposed approach not only can be used for any capacitance and frequency range, but also eliminates the probability of introducing human errors when analyzing the graphics, because both ESR and X_{CAP} are computed directly from the data waveforms.

Considering that the studied waveforms are sinusoidal waveforms with high frequency noise, it is only necessary to compute the first harmonic to obtain both gain and phase displacement between the input and capacitor voltages. In this way, the computation effort is very small and direct.

In a first moment it is needed to obtain simultaneously the input and capacitor voltages during a period. Following this, it is used the DFT algorithm to evaluate both gain and phase displacement of the input and capacitor voltages for the first harmonic, F_1. For doing that, (7) must be used:

$$F_1 = \frac{1}{N} \sum_{n=0}^{N-1} f_n\, e^{-j\, w_o n} \;,\; w_o = \frac{2\,\pi}{T} \tag{7}$$

where N represents the number of equidistant subintervals in a period T, of the signal under analysis represented by f

and which can be the input voltage or the capacitor voltage.

Afterwards equations (8) and (9) could be used for evaluating both ESR and X_{CAP}.

$$ESR \cong R\, G_{in\_out} \cos\left(\phi_o - \phi_{in} \right) \tag{8}$$

$$X_{CAP} \cong R\, G_{in\_out} \sin\left(\phi_o - \phi_{in} \right) \tag{9}$$

where, G_{in_out} represents the gain (ratio between the modulus of the 1st harmonic of V_O and V_{in} and (ϕ_o-ϕ_{in}) the phase displacement of the 1st harmonic between the input and capacitor voltage.

III. Experimental Results

In this section a comparison between the new proposed approach and the one presented in [14] is done, in order to show the advantages of the first one.

Table I presents the characteristics given by the manufacturer of the capacitor used in this study, C_1.

TABLE I
CHARACTERISTICS OF THE CAPACITOR, C_1, AT 120Hz AND 20 ºC.

Capacitor	Capacitance	ESR_{max}	DF_{max}
C_1	100 µF	1.061 Ω	0.08

In order to implement the technique proposed in [14], first of all it is necessary to compute the R value that respects (4). For doing that, the operating frequency as well as the capacitance C_1 (Table I) should be taken into account. It was considered that one of the values is much greater than the other one, when the first one is at least 50 times higher than the second one.

Both techniques where implemented for the range of frequencies of 100 Hz to 10 kHz.

In Table II it is possible to observe the minimum R values that respect (4), as well as the standard resistor chosen for the experimentations.

TABLE II
CALCULATED R VALUES AS FUNCTION OF THE FREQUENCY USED IN THE EXPERIMENTAL MEASUREMENTS.

Frequency	Rectance Value, X_C	Minimum R Value	Standard R Value
100 Hz	15.92 Ω	795 Ω	1 kΩ
120 Hz	13.26 Ω	663 Ω	1 kΩ
250 Hz	6.36 Ω	318 Ω	500 Ω
500 Hz	3.20 Ω	160 Ω	220 Ω
750 Hz	2.12 Ω	106 Ω	220 Ω
1 kHz	1.6 Ω	80 Ω	100 Ω
2.5 kHz	637 mΩ	31.8 Ω	47 Ω
5 kHz	318 mΩ	15.8 Ω	20 Ω
7.5 kHz	212 mΩ	10.6 Ω	20 Ω
10 kHz	159 mΩ	7.7 Ω	10 Ω

Fig. 4 shows the experimental curves of V_O as function of V_{in}, for frequencies of 120 Hz, 1kHz and 5 kHz.

645

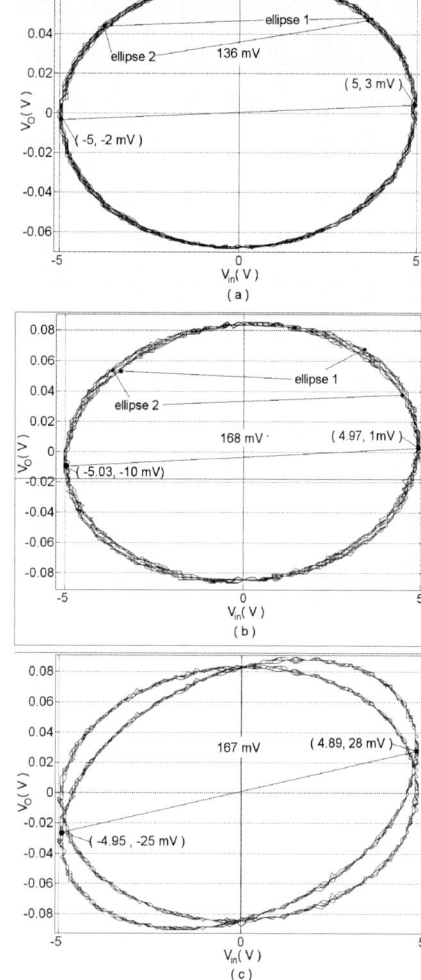

Fig. 4 – Experimental results of V_O as function of V_{in}, for the frequencies of: (a) 120Hz, (b) 1 kHz and (c) 5 kHz.

Fig. 5 shows a comparison between the computed *ESR* value using the technique proposed in [14], the new proposed approach and an impedance gain-phase analyzer HP 4294.

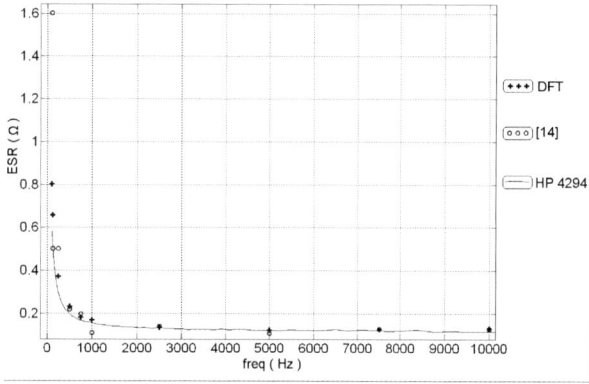

Fig. 5 – Computed *ESR* value using the new proposed approach, the technique proposed in [14] and the impedance gain-phase analyser.

From Fig. 4, it is possible to conclude that ellipse 1 and ellipse 2 are approximately equal when low frequencies are considered, due to the very high values of X_{CAP} and R.

In such case, the obtained *m*, *a* and *b* values from the visual analysis could have a significant error, that is even increased by the high value of the resistor.

In Figures 5, 6 and 7, in their legends "DFT" represents the *ESR* values computed using the new proposed approach, "[14]" the results obtained from the technique proposed in [14] and "HP 4294" the *ESR* values acquired from the impedance gain-phase analyzer.

From Fig. 5 it is possible to conclude that for low frequencies range, the *ESR* values computed using [14] are far from the ones computed using the impedance gain phase analyzer. On the other hand, the ones computed with the new proposed approach matches very closely to the results obtained with the gain-phase analyzer.

Fig. 6 shows a comparison between the computed X_{CAP} values using the technique proposed in [14], the new proposed approach and an impedance gain-phase analyzer HP 4294.

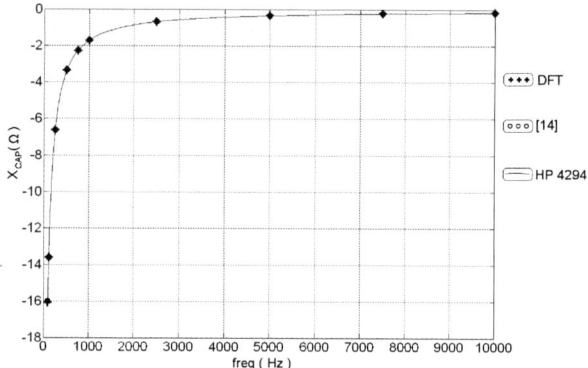

Fig. 6 – Computed X_{CAP} value using the new proposed approach, the technique proposed in [14] and the impedance gain-phase analyser.

Fig. 6 shows that both techniques exhibit a very good accuracy, when the computed X_{CAP} is compared with the one obtained from the impedance gain phase analyzer.

Fig. 7 shows the errors between the computed *ESR* and X_{CAP} values using both techniques and the ones obtained using the impedance gain-phase analyzer. From Fig. 7, it is possible to conclude that the new proposed approach is much more accurate at low frequencies than the one presented in [14], for both *ESR* an X_{CAP} evaluations.

The new proposed approach can also be used as a fault diagnosis technique, through the comparison of the obtained values of *ESR* and *C* with the ones given by manufacturers at 120Hz.

IV. CONCLUSIONS

In this paper, an experimental off-line technique for estimating the equivalent circuit of aluminum electrolytic capacitors was presented. The proposed technique shows to be accurate for low capacitance capacitors.

To implement the proposed technique, it is necessary to use a power amplifier together with a signal generator and a RC circuit. From the resulting curves and using DFT algorithm, it is possible to obtain the equivalent circuit of the capacitor at the operating frequency of the experience.

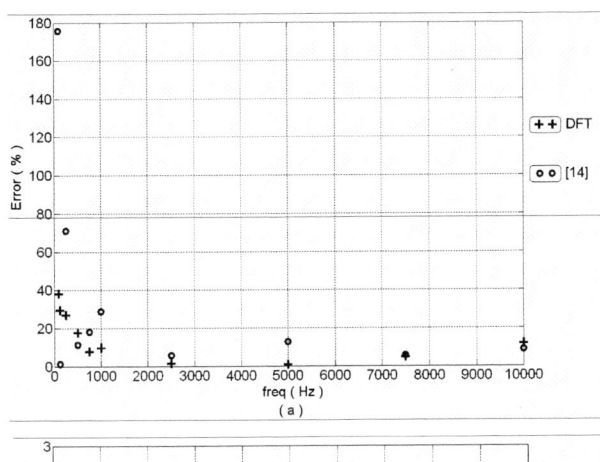

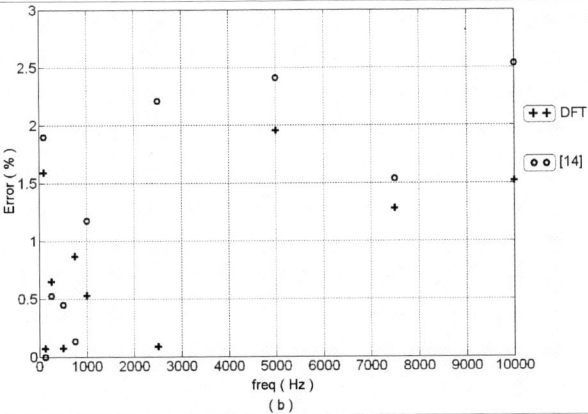

Fig. 7 – Computed errors: (a) *ESR* and (b) X_{CAP}.

REFERENCES

[1] W. J. Sarjeant, J. Zimheld and F. W. Macdougall "Capacitors", *on IEEE Trans. on Plasma Science*, Vol. 26, n° 5, October 1998, pp. 1368-1392.

[2] K. Harada, A. Katsuki and M. Fujiwara, "Use of ESR for Deterioration Diagnosis of Electrolytic Capacitor", *on IEEE Trans. on Power Electronics*, Vol. 8, n° 4, October 1993, pp. 355-361.

[3] A. Lahyani, P. Venet, G. Grellet, P. Viverge, "Failure prediction of electrolytic capacitors during operation of a switch mode power supply", *on IEEE Transaction on Power Electronics*, Vol. 13, n° 6, November 1998, pp. 1199 – 1207.

[4] A. M. R. Amaral and A. J. M. Cardoso, "Use of ESR to Predict Failure of Output Filtering Capacitors in Boost Converters", *in Proceedings of IEEE International Symposium on Industrial Electronics 2004*, Ajaccio – France, 4-7 May, 2004, Vol. 2, pp. 1309-1314.

[5] P. Venet, F. Perisse, M. Husseini, G. Rojat, "Realization of smart electrolytic capacitor circuit", *IEEE of Industry Applications Magazine*, n° 1, January/February 2002, pp. 16-20.

[6] E. Aeloiza, J. H. Kim, P. Ruminot and P. N. Enjeti, "A real time Method to Estimate Electrolytic Capacitor Condition in PWM Adjustable Speed Drives and Uninterruptible Power Supplies", *in Proceedings of 2005 IEEE Power Electronics Specialist Conference*, pp. 2867-2872.

[7] H. Ma. X. Mao, N. Zhang and D. Xu, "Parameter Identification of Power Electronics Circuits Based on Hybrid Models", *in Proceedings of 2005 IEEE Power Electronics Specialist Conference*, pp. 2855-2860.

[8] G. Buiatti, A. M. R. Amaral and A. J. M. Cardoso, "An Online Technique for Estimating the Parameters of Passive Components in Non-Isolated DC/DC Converters", *in 2007 IEEE International Symposium on Industrial Electronics*, 4-7 June 2007, Vigo, Spain.

[9] G. Buiatti, A. M. R. Amaral and A. J. M. Cardoso, "Parameter Estimation of a DC/DC Buck Converter using a continuous time model", *in Proceedings of 12th European Power Electronics and Drives Conference*, 2-5 September 2007, Aalborg, Danemark.

[10] A. M. R. Amaral and A. J. M. Cardoso, "Using Newton-Raphson Method to Estimate the Condition of Aluminum Electrolytic Capacitors", *in 2007 IEEE International Symposium on Industrial Electronics*, 4-7 June 2007, Vigo, Spain.

[11] S. G. Parler, "Improved Spice Models of Aluminum Electrolytic Capacitors for Inverter Applications", *on IEEE Trans. on Industry Applications*, Vol. 39, n° 4, July/August 2003, pp. 929-935.

[12] M. L. Gaspari, "Life Prediction Modeling of Bus Capacitors in AC Variable-Frequency Drives", *on IEEE Trans. on Industry Applications*, Vol. 41, n° 6, November/December 2005, pp. 1430-1435.

[13] A. M. R. Amaral and A. J. M. Cardoso, "An ESR Meter for High Frequencies", *in Proceedings of the 6th IEEE Conference on Power Electronics and Drive Systems*, Kuala Lumpur – Malaysia, 24 November - 4 December, 2005.

[14] A. M. R. Amaral and A. J. M. Cardoso, "An Experimental Technique for Estimating the ESR and Reactance Instrinsic Values of Aluminum Electrolytic Capacitors", *in Proceedings of IEEE Instrumentation and Measurement Technology Conference*, Sorrento – Italy, 24-27 April, 2006, pp. 1820-1825.

A Mathematical Analysis on Vector Inversion Generators

D. J. Thrimawithana*, and U. K. Madawala*

*Department of Electrical & Computer Engineering, The University of Auckland, Auckland, New Zealand

Abstract– This paper presents a mathematical model for a vector inversion generator, through which explicit analytical expressions for component stresses and output characteristics can be obtained. The derivation of a simplified equivalent circuit for transient analysis of the converter is presented in detail. The equivalent circuit model is used to analyze the performance of a 4-stage vector inversion generator. The results are presented and compared with simulated results to validate the model. Methods that could be used to improve some of the limitations imposed by this model are also presented and verified.

Index Terms– pulsed power, high voltage, vector inversion generator

I. INTRODUCTION

High voltage pulsed power systems are increasingly used in a broad spectrum of disciplines. Industrial applications such as food processing and electric fencing systems; medical instruments such as X-ray machines and defibrillators; military appliances such as radar systems and active denial technology (ADT) and also scientific applications such as particle accelerators and ion implantation systems are just a few application that are based on pulsed power systems [1-5].

A system that is capable of generating short duration, high voltage and high power pulses is termed a high voltage pulsed power supply (HVPPS). These systems operate by storing energy in an energy compression element and discharging the stored energy through the target device as a short high power pulse. The energy compression is based on either reactive elements such as inductors and capacitors or pulse forming transmission lines [1-7].

The vector inversion technique is one of the few high voltage pulse generation technologies that are used in current pulsed power applications. Common high voltage pulse generation technologies include direct discharge type, pulse transformer type and the Marx generator type generators. All these technologies utilise reactive energy compression elements to generate high peak power pulses repetitively. In comparison to other HVPPS technologies, the vector inversion technique requires only a single power switch (inversion switch) to generate an output. Hence vector inversion generators are the preferred type of HVPPS in a number of high power applications, even though these generators tend to be bulky in comparison to the counterparts [8-12].

Hence an analytical model that could predict the output characteristics of a vector inversion generator is an invaluable tool for engineers designing HVPPS systems. This could help designers to avoid time consuming computer simulations, and also to optimize the design parameters to improve efficiency. For example such an analytical model could be used for resonant frequency compensation of the converter to achieve efficiencies over 90% [8-12].

To the best knowledge of the authors, a paper that presents a comprehensive analytical model for an N-stage transformer coupled vector inversion converter has not been published to date. This paper intends to present a simplified circuit model for an N-stage converter, from which explicit analytical solutions for the component stresses and output characteristics can be derived.

II. PRINCIPLE OF OPERATION

A schematic diagram of a 2N-stage transformer coupled vector inversion generator is shown in Fig. 1(a). When switch S_1 is closed while S_2 is open, the charger system charges the energy storage capacitors C_1-C_{2N}, to the required voltage through the magnetizing inductance of the transformers. During the charging phase, the magnetizing inductance of the transformers appears as a short circuit, resulting in the charging arrangement shown by Fig. 1(b). This arrangement ensures that the odd numbered capacitors are charged with a positive polarity where as the even numbered capacitors are charged with a negative polarity, as indicated in Fig. 1(b). Hence the net voltage appearing across the output load is zero [8-12].

A pulse can now be generated at the output by closing the inversion switch S_2, while S_1 is left open. When S_2 is closed, it causes the even numbered capacitors to resonate with the leakage inductances of the transformers, inverting the capacitor voltages. In comparison to the fast voltage swing of even numbered capacitors, the voltages across odd numbered capacitors change slowly due to large magnetizing inductance that is in parallel with them. The result is an amplified ringing output voltage, which would ideally be equal to 2N times the initially charged voltage. Practically the peak load voltage is less than 2N due to losses and secondary resonances. The ringing is mainly determined by the load impedance and could not be controlled without extra switches. The efficiency of the converter can be improved by increasing the transformer coupling and using resonant frequency compensation [8-12].

This work was supported by Technology for Industry Fellowship (TIF), New Zealand.

978-1-4244-0644-9/07/$25.00 ©2007 IEEE

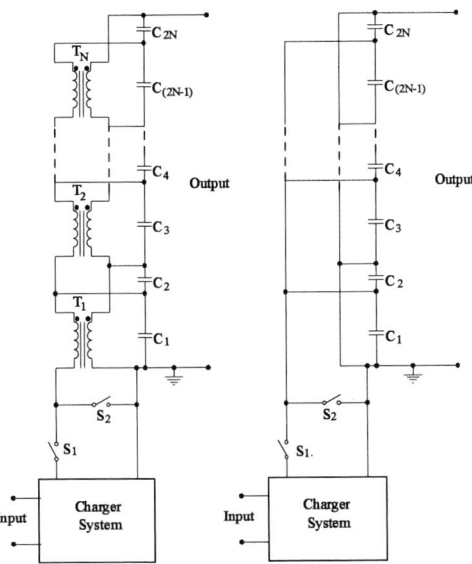

Fig. 1 A vector inversion generator (a) Schematic representation
(b) Charging phase equivalent circuit

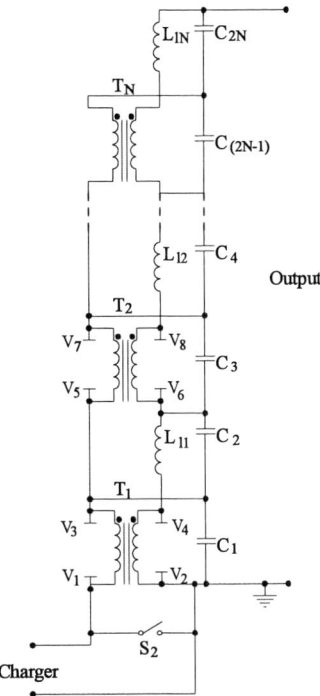

Fig. 2 A vector inversion generator with transformer leakage
inductance indicated externally

III. A Simplified Model

The operation of the converter is primarily governed by the leakage inductance of the transformers. It is assumed that the transformers are identical and have a 1:1 turn's ratio, which would be the case for a typical design. By transforming the primary leakage inductances to the secondary of the transformers, the equivalent circuit illustrated in Fig. 2 could be derived to analyze the operation of a vector inversion generator. But the presence of the transformers in Fig. 2 makes it rather difficult to obtain a mathematical model for the converter. However, this difficulty can be overcome by further simplifying the circuit presented in Fig. 2, as described below.

Consider the moment immediately after the inversion switch S_2 is closed. The nodes V_1 and V_2 are shorted together and C_1 is connected across the primary of transformer T_1. The transformer will induce a secondary voltage V_{C1} across V_2-V_4, which is the voltage across the capacitor, C_1. Since V_1 and V_2 are held at the earth potential by the switch, voltage that appears at V_3 and V_4 relative to the earth is V_{C1} during the pulsing state, even though V_{C1} might change as C_1 discharges through both the load and magnetizing inductance of the transformer. Hence the components looking into V_3-V_4 will see it as a short circuit, and for the simplified model shown in Fig 3, T_1 and C_1 are replaced by a short circuit across V_3-V_4. To simplify the analysis, it assumed that the voltage across the odd numbered capacitors is constant during pulsing, which would be the case for light-medium loads.

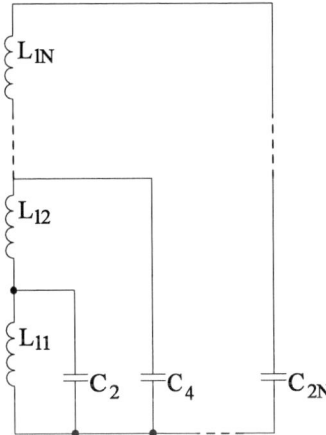

Fig. 3 A simplified equivalent circuit for a vector inversion
generator

A similar argument applies to the subsequent stages, resulting in a simplified model shown in Fig. 3 that can be used to derive a mathematical mode for the converter. For example, consider T_2 and C_3 of the system. As the transformer maintains the same voltage across both the primary and secondary;

$$(V_7 - V_5) = (V_8 - V_6) \tag{1}$$

The voltage at V_7 and V_8, relative to V_5 can be given by (2) and (3) respectively, where V_{C3} is the voltage across C_3. By combining (1) to (3), $V_{8,5}$ is simplified to obtain (4).

649

$$V_{7,5} = V_7 - V_5 = (V_6 - V_5) + V_{C3} \tag{2}$$

$$V_{8,5} = (V_6 - V_5) + (V_8 - V_6) \tag{3}$$

$$V_{8,5} = 2(V_6 - V_5) + V_{C3} \tag{4}$$

Equations (2) and (4) reveal that the voltage across V_8-V_7 is equal to the voltage across V_6-V_5 during the pulsing period. The nodes V_7-V_5 and V_8-V_6 are separated from each other by two dependent, 1:1 voltage sources, generated by the transformer T_2. Since these two voltage sources are in opposite polarity to each other, they cancel out each other during the transient analysis of the even numbered capacitors. Thus, T_2 and C_3 can be removed from the circuit by connecting nodes V_7-V_5 and V_8-V_6 together as shown by Fig. 3, to model the transient behavior of the circuit. This approach simplifies the complicated vector inversion topology, shown in Fig. 2, to a transient time equivalent model given in Fig. 3, facilitating a mathematical solution.

IV. THE DERIVATION OF SOLUTIONS

The validity of the simplified model in Fig. 3 is verified by analyzing a 4-stage vector inversion generator shown in Fig. 4. It is assumed that the converter is lightly loaded and hence the voltages across C_1 and C_3 are approximately constant at V_C, over the pulsing period. A complete list of circuit parameters are given in the appendix.

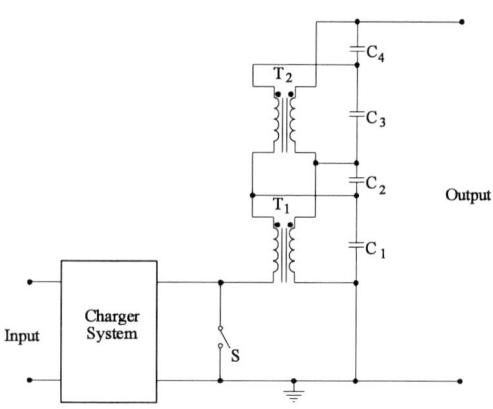

Fig. 4 A 4-stage vector inversion generator

The simplified transient model that is used to obtain an analytical solution is shown in Fig. 5. During the analysis it is assumed that the leakage inductances L_{l1} and L_{l2} are equal and have a value of L.

By considering the voltages across the components and currents flowing through them, (5)-(8) can be derived.

$$v_{C2} = L \frac{di_3}{dt} \tag{5}$$

$$v_{C2} = -\frac{1}{C} \int i_1 dt \tag{6}$$

$$v_{C4} - v_{C2} = L \frac{di_2}{dt} \tag{7}$$

$$v_{C4} - v_{C2} = -\frac{1}{C} \int i_2 dt \tag{8}$$

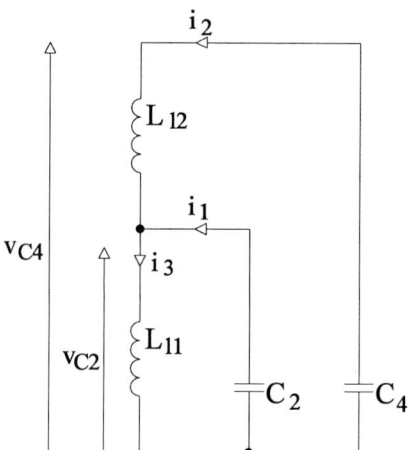

Fig. 5 A simplified equivalent circuit

By substituting (5) in (7) and noting that $i_3 = i_1 + i_2$,

$$v_{C4} = L \frac{d(i_1 + 2i_2)}{dt} \tag{9}$$

By manipulating (5) to (9),

$$2L \frac{d^2 i_2}{dt^2} + \frac{i_2}{C} = -L \frac{d^2 i_1}{dt^2} \tag{10}$$

$$L \frac{d^2 i_1}{dt^2} + \frac{i_1}{C} = -L \frac{d^2 i_2}{dt^2} \tag{11}$$

Solving (10) and (11) in time-domain can be tedious, and therefore they are solved in S-domain.

The S-domain equivalents of (10) and (11) are given by,

$$2L \left\{ s^2 I_{2(s)} - \frac{di_{2(0)}}{dt} - si_{2(0)} \right\} + \frac{I_{2(s)}}{C} \\ = L \left\{ s^2 I_{1(s)} - \frac{di_{1(0)}}{dt} - si_{1(0)} \right\} \tag{12}$$

$$L \left\{ s^2 I_{1(s)} - \frac{di_{1(0)}}{dt} - si_{1(0)} \right\} + \frac{I_{1(s)}}{C} \\ = L \left\{ s^2 I_{2(s)} - \frac{di_{2(0)}}{dt} - si_{2(0)} \right\} \tag{13}$$

Since the initial currents flowing through the inductors are 0, and the voltages across the capacitors are initially V_C, the initial conditions for the system are,

$$i_{1(0)} = i_{2(0)} = 0 \tag{14}$$

$$L \frac{di_{2(0)}}{dt} = v_{C4(0)} - v_{C2(0)} = 0 \tag{15}$$

650

$$L\frac{di_{1(0)}}{dt} = \frac{v_{C2(0)}}{L} - \frac{di_{2(0)}}{dt} = \frac{V_C}{L} \qquad (16)$$

By applying the initial conditions (14)-(16), (12) and (13) can be simplified and solved to obtain,

$$I_{1(s)} = \frac{V_C s^2/L + V_C/L^2 C}{s^4 + 3s^2/LC + 1/(LC)^2} \qquad (17)$$

$$I_{2(s)} = \frac{V_C/L^2 C}{s^4 + 3s^2/LC + 1/(LC)^2} \qquad (18)$$

Equations (17) and (18) can now be converted to time domain to obtain the time-domain analytical solutions as given by.

$$i_{1(t)} = A_1 \sin(\omega_1 t) - A_2 \sin(\omega_2 t) \qquad (19)$$

$$i_{2(t)} = B_1 \sin(\omega_1 t) - B_2 \sin(\omega_2 t) \qquad (20)$$

where,

$$A_1 = \frac{V_C}{\sqrt{5}L\omega_1}(1-k_1), \quad A_2 = \frac{V_C}{\sqrt{5}L\omega_2}(1-k_2)$$

$$B_1 = \frac{V_C}{\sqrt{5}L\omega_1}, \quad B_2 = \frac{V_C}{\sqrt{5}L\omega_2}$$

$$\omega_1 = \frac{1}{\sqrt{LC}}k_1, \qquad \omega_2 = \frac{1}{\sqrt{LC}}k_2$$

$$k_1 = \frac{3-\sqrt{5}}{2}, \qquad k_2 = \frac{3+\sqrt{5}}{2}$$

By applying (20) in (6),

$$v_{C4} = \frac{V_C}{\sqrt{5}k_1}\cos(\omega_1 t) - \frac{V_C}{\sqrt{5}k_2}\cos(\omega_2 t) \qquad (21)$$

From (7), (20) and (21),

$$v_{C2} = v_{C2} - \frac{V_C}{\sqrt{5}}\cos(\omega_1 t) + \frac{V_C}{\sqrt{5}}\cos(\omega_2 t) \qquad (22)$$

V. OUTPUT CHARACTERISTICS

The output voltage characteristics of the converter for light loads can be determined with the aid of (21) and (22). As indicated in Fig. 2, the output load voltage is equal to the addition of voltages across the capacitors. Hence,

$$v_{out} = 2V_C + v_{C2} + v_{C4} \qquad (23)$$

Equations (21) and (22) are substituted in (23) to obtain,

$$v_{out} = 2V_C + \Phi\cos(\omega_1 t) + \Psi\cos(\omega_2 t) \qquad (24)$$

where,

$$\Phi = V_C\frac{1+\sqrt{5}}{\sqrt{5}(3-\sqrt{5})}, \quad \Psi = V_C\frac{\sqrt{5}-1}{\sqrt{5}(3+\sqrt{5})}$$

The switch current can be found by adding (19) and (20) and is given by,

$$i_{switch(t)} = C_1 \sin(\omega_1 t) - C_2 \sin(\omega_2 t) \qquad (25)$$

where,

$$C_1 = \frac{V_C}{\sqrt{5}L\omega_1}(2-k_1), \quad C_2 = \frac{V_C}{\sqrt{5}L\omega_2}(2-k_2)$$

VI. VERIFICATION OF THE MODEL

The 4-stage vector inversion generator, shown in Fig. 4 was simulated using PSPICE to analyze the operation of the converter and to verify the analytical solutions given by (24) and (25). The voltages V_{out}, V_{C2}, and V_{C4} obtained from the simulator are compared with the theoretical results as shown in Fig. 6. For lightly loaded conditions, the theoretical results are in very good agreement with the simulations, validating the analysis presented in sections IV and V.

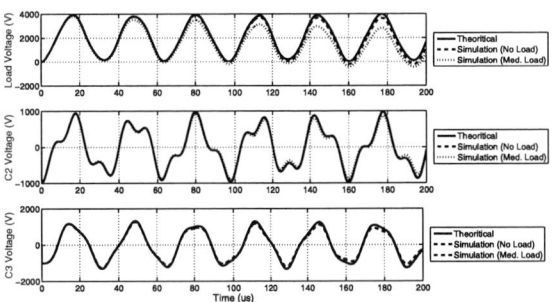

Fig. 6 Voltages across the output and capacitors

However, the simulated results for medium loads deviate from the theoretical results, as the simulation time is increased. This deviation is expected, as the model does not account for the load current. It could also be seen from the results that the load voltage is mainly governed by the leakage inductance of the transformers and it exhibits vigorous ringing.

The simulated and theoretical switch current waveforms are shown in Fig. 7. Similar to the voltages, the theoretical solution closely follows the simulation for a lightly loaded converter, but deviates slightly for medium loads. The deviation of theoretical results from the simulations under medium loads increases with increasing time. It is also noted that the switch current stress, which is determined by the leakage inductance, is very high.

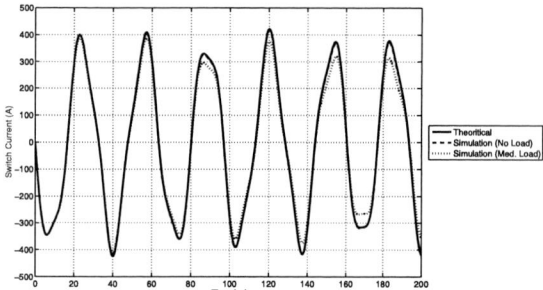

Fig. 7 Current through the inversion switch

VII. An Improved Model For High Loads

The model introduced in section III, and validated in section VI, could only be used to model lightly loaded vector inversion generators. As both the load and pulse period increase, the theoretical results become less accurate. However, the equivalent circuit shown in Fig. 3 can be modified to incorporate the loading effect on the resonance. This is achieved by introducing current sources at each LC node as indicated in Fig. 8. The value of the current sourced by each source is given as,

$$I_L = NV_C + \sum_{n=1}^{N} \frac{V_{C(2n)}}{R_L} \qquad (26)$$

The introduction of the voltage dependent current sources complicates the analysis and the derivation of a complete mathematical solution to obtain output characteristics of a vector inversion generator is extremely difficult and beyond the scope of this paper. Hence the validity of the proposed model, shown in Fig. 8 is validated by comparing the results of a 6-stage vector inversion circuit with simulations.

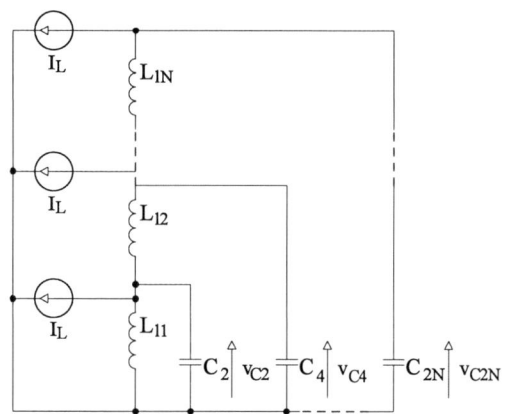

Fig. 8 Modified equivalent circuit

The comparisons of simulated and theoretical voltages of V_{out}, V_{C2}, V_{C4}, and V_{C6} are shown in Fig. 9 and Fig. 10. Because of high load the ringing waves are significantly attenuated within 200 us time period. As evident from the

comparison, both theoretical and simulated results are in very good agreement and, therefore, verify the accuracy and the validity of the proposed theoretical model over the entire load range.

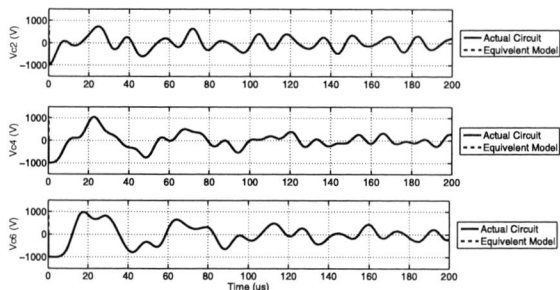

Fig. 9 Capacitor voltages

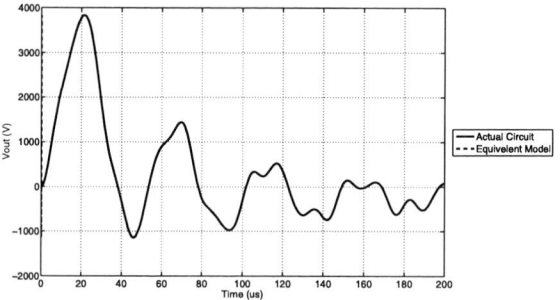

Fig. 10 Load voltage

VIII. Conclusions

A simplified equivalent model for a vector inversion generator has been introduced, and a theoretical solution for the output characteristics of a lightly loaded 4-stage converter was derived. The validity of the solutions was verified through PSPICE simulations. An improved equivalent model that is suitable for analysis under heavy loads has also been presented and verified with simulations.

Appendix

TABLE I
SIMULATION PARAMETERS

Property	Value	Unit
Capacitance per stage	1	μF
Transformer magnetizing inductance	10	mH
Transformer coupling coefficient	0.999	
V_C	1000	V
R_L (heavy)	100	Ω
R_L (medium)	1000	Ω
R_L (light)	5000	Ω

ACKNOWLEDGEMENT

The authors acknowledge the contributions of R.C.B Woodhead, P.C. Lunenburg, and financial support by TIF NZ.

REFERENCES

[1] D. J. Thrimawithana, U. K. Madawala, and R. C. B. Woodhead, "Pulsed power generation techniques," *in 32nd annual IEEE IECON'06 Conf*, 2006, pp. 2014-2019.

[2] D. Shmilovitz, and S. Singer, "Pulsed power generation by means of transmission lines," *IEEE Trans. Power Electronics*, vol. 18, pp. 221-230, Jan. 2003.

[3] S. T. Pai, and Q. Zhang, "Introduction to high power pulse technology," World Scientific Publishing Co. Pte. Ltd., Singapore, 1995.

[4] A Kempkes, J. A. Casey, M. P. J. Gaudreau, T. A. Hawkey, and I. S. Roth, "Solid-state modulators for commercial pulsed power systems," *in 25th Int. IEEE Conf.*, 2002, pp. 689-693.

[5] M. Kanter, S. Singer, R. Cerny, and Z. Kaplan, "Multikilojoule inductive modulator with solid-state opening switches," *IEEE Trans. Power Electronics*, vol. 7, pp. 420-424, April 1992.

[6] M. V. Fazio, and H. C. Kirbie, "Ultracompact pulsed power," *Proc. of the IEEE*, vol. 92, pp. 1197-1204, July 2004.

[7] M. Buttram, "Some future directions for repetitive pulsed power," *IEEE Trans. Plasma Science*, vol. 30, pp. 262-266, Feb. 2002.

[8] B. Meyer, A. Watson, T. G. Engel, and M. Kristiansen, "A single gap transformer coupled L-C generator with resonant frequency compensation," *in Proc. 7th Pulsed Power Conf.*, 1989, pp. 749-752.

[9] T. G. Engel and M. Kristiansen, "A compact high voltage vector inversion generator," *in 10th IEEE Int. Pulsed Power Conf.*, Vol. 2, pp. 1389-1393, July 1995.

[10] S. A. Marryman, M. F. Rose, and Z. Shotts, "Characterization and application of vector inversion generators," *in 14th IEEE Int. Pulsed Power Conf.* Vol. 1, pp. 249-252, June 2003.

[11] T. G. Engel, C. Kaplicki, and W. C. Nunnally, "High-Voltage Pulse production using transformer-coupled LC vector inversion generators," *IEEE Trans. Plasma Science*, Vol. 28, pp. 1377-1381, October 2000.

[12] D. J. Thrimawithana, and U. K. Madawala, "A Mathematical Analysis on Vector Inversion Generators," Tru-Test Ltd, Auckland, New Zealand, Tech. Rep., Jan, 2006

A Novel Multi-Level High Voltage Pulsed Power Generator

D. J. Thrimawithana*, and U. K. Madawala*

*Department of Electrical & Computer Engineering, The University of Auckland, New Zealand

Abstract-- This paper presents a novel solid-state and high power pulse generation technique that is suitable for a wide range of pulsed power applications. The technique, termed as multi-level pulsed power converter, can be considered as a hybrid of the direct discharge type and the Marx generator but with considerably less complexity in both control and circuitry. It has the ability for generating pulses with flexible amplitude and duration similar to that of a Marx generator, and unlike the direct discharge type requires no voltage balancing and snubbing circuitry. As voltage balancing is inherent to the technique, the timing between switching events is not critical to balance the voltage stresses and as such the driving circuitry of the converter is relatively simple. The validity of the proposed technique is verified by presenting both theoretical analysis and simulation results of a 3-stage, 3 kV converter.

Index Terms--pulsed power, high voltage, multi-level converter

I. INTRODUCTION

A system that is capable of generating short duration, high voltage and high power pulses is termed a high voltage pulsed power supply (HVPPS). These systems operate by storing energy into an energy compression element and discharging the stored energy through the target device as a short high power pulse. A variety of HVPPS technologies could be found in literature, which utilise different switching and energy compression schemes to generate high power pulses. These technologies can be divided into two broad categories based on the type of energy compression elements used. The first category uses reactive elements such as inductors and capacitors for energy compression, whereas the second category is based on pulse forming transmission lines to produce the same results. The difference between the two categories is found in the pulse width. Whilst typical reactive element based topologies are capable of producing pulses that last for hundreds of microseconds, the practical pulse width of transmission line based topologies is limited to a few microseconds [1-7].

Applications of pulsed power systems are found in a broad spectrum of disciplines. Industrial applications such as food processing and electric fencing systems; medical instruments such as X-ray machines and defibrillators; military appliances such as radar systems and active denial technology (ADT) and also scientific applications such as particle accelerators and ion implantation systems are just a few examples [1-8].

The focus of this paper is to introduce a new high voltage pulse generation technique, based on capacitive

energy storage, which could be used in electric fencing energizers. According to literature and to the best knowledge of the authors, such a technique has not been reported to date. The new technology could be used to design a new breed of solid state electric fence energizers that would overcome some of the deficiencies found in the current technology. This technique could also be used in many other pulsed power applications that require reliable, compact and efficient solid state HVPPS.

II. CURRENT TECHNOLOGY

The direct discharge type, pulse transformer type, Marx generator type, and vector inversion type are the four basic capacitive energy storage HVPPS topologies that are used in current designs [1-7]. A comprehensive comparison of these four topologies is presented in [1].

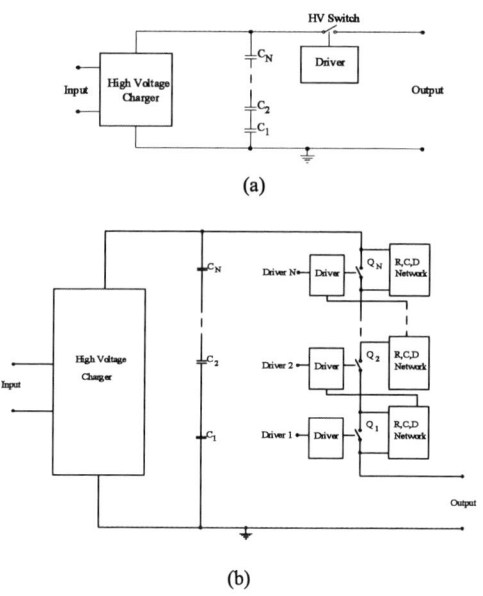

Fig. 1 A direct discharge type HVPPS (a) Conceptual circuit (b) A solid-state implementation

As illustrated in Fig. 1(a), the direct discharge type concept is simple, but the implementation of the converter is somewhat complicated. The design of a solid state high voltage switch that is required to generate the output pulse is complicated as current semiconductor technology does not offer switches with adequate ratings that suit the design. Hence modern HVPPSs based on this technology often utilize a chain of series connected switches with proper balancing and driver circuitry to meet the design specifications, and such an implementation of the converter is shown in Fig. 1(b). This method is unreliable

---

This work was supported by Technology for Industry Fellowship (TIF), New Zealand.

978-1-4244-0644-9/07/$25.00 ©2007 IEEE

as failure of a single component in the switching circuit could lead to a catastrophic system failure. Hoever, the ability to generate pulses with variable pulse widths is an advantage of this topology, in comparison to other HVPPS topologies [1, 10-12].

In contrast, the implementation of a pulse transformer type HVPPS shown in Fig. 2 is simple and reliable, and as a result it is one of the most widely used technologies. The topology is similar to the direct discharge type, but includes an output step-up pulse transformer to reduce the voltage stress on the pulse forming switch at the expense of current stress. In comparison to direct discharge and Marx types, this technology is very inefficient and bulky due to the pulse transformer [1,3]. Also it is not practical to generate pulses with variable pulse parameters using this technique.

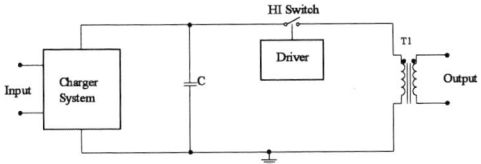

Fig. 2 A pulse transformer type HVPPS

The Marx generator concept, originated by E. Marx in 1924, overcomes some of the deficiencies presented by direct discharge and pulse transformer types by employing complex circuitry. Marx generator operates by charging a number of energy compression capacitors in parallel and then arranging them in series to generate the desired high voltage pulse at the output. This scheme reduces both the voltage and current stresses on the components. Another advantage of this topology is the ability to generate pulses with variable amplitude and duration. A conceptual circuit of an N-stage Marx generator is shown in Fig. 3 [13-15].

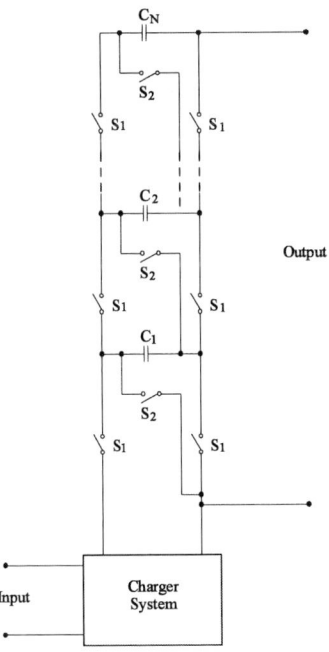

Fig. 3 A Marx generator type HVPPS

A schematic diagram of a 2N-stage vector inversion type pulse generator is shown in Fig. 4. This technique utilizes a number of tightly coupled transformers and a switch (S_2) to generate a high voltage output. The requirement for a number of large transformers, less controllability over the output and the high current stress on the switch has made this type less attractive, in comparison to other methods [16-17].

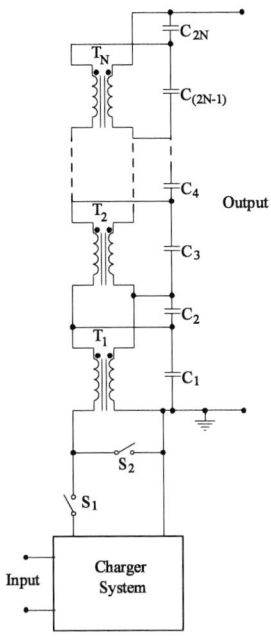

Fig. 4 A vector inversion type HVPPS

III. THE PROPOSED TOPOLOGY[1]

The proposed high voltage pulse generation topology as shown schematically in Fig. 5 (a), is derived from a multi-level converter topology and hence it is called the multi-level HVPPS. Although this technique appears to be similar to the direct discharge type, it has the unique ability of generating pulses with flexible amplitude and duration similar to that of the Marx type, but with less circuit complexity [9].

The multi-level HVPPS begins operation by charging a series connected storage capacitor bank (C_1-C_N) to the desired voltage through a high voltage charging circuit. It is preferred to use a multi-output charger to charge the stages separately and improve the efficiency of the design with better charge balancing between the capacitors. The number of stages required by the converter is determined by the component voltage ratings and the specified peak output voltage.

After the capacitors were charged to the required voltage, the switches are turned-on sequentially from Q_1 to Q_N and turned-off sequentially from Q_N to Q_1 to generate the desired output pulse. A delay between the switching of subsequent stages could easily be introduced to generate an output pulse with a variable amplitude and

---

[1] This technology is protected by New Zealand patent application 535719 and foreign equivalents

pulse width. An example of such an output pulse shape that could be generated across a resistive load is shown in Fig. 5(b).

The diode links between the energy storage capacitors and corresponding switches ensure proper voltage sharing across the switches, eliminating the requirement for a complicated balancing and driver circuitry. For example assume all the capacitors are charged to an equal voltage of V_c and the two bottom switches (Q_1 and Q_2) are turned-on to generate an output pulse with a peak voltage of $2V_c$. Under these conditions capacitors C_1 and C_2 are discharging through the diode chain $D2_1$-$D2_{N-2}$, the switches Q_1-Q_2 and the load. Switches Q_3-Q_N are in off-state disconnecting C_3-C_N from the output and the voltage seen by these switches is $(N-2)V_c$ as the diode chain $D2_1$-$D2_{N-2}$ clamps the emitter of Q_3 to C_3. Also the diode chains restrict the maximum voltage stress across each switch to V_c during all states of circuit operation. It is assumed that the voltage rating of the diodes is as same as the switches, hence multiple diodes are used for each connection to meet the required voltage rating.

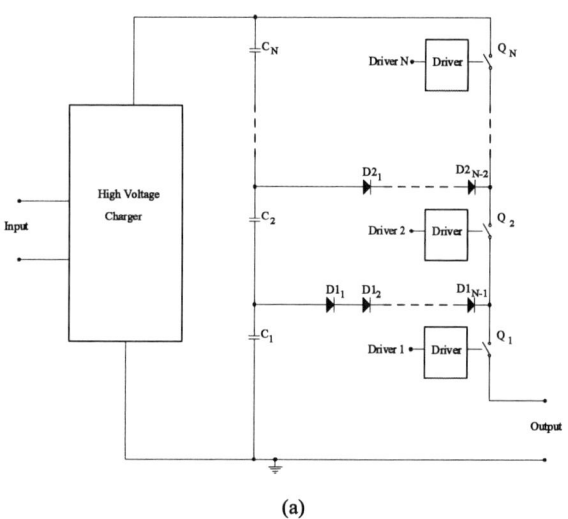

(a)

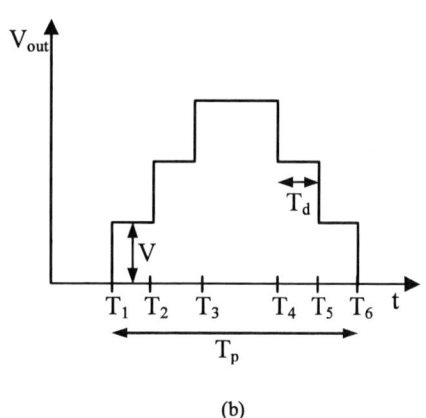

(b)

Fig. 5 (a) A multi-level HVPPS (b) An output pulse generated by a 3 stage converter

In comparison to all existing pulse generation technologies, this technique has many advantages. Mainly, it allows for an implementation of a reliable solid-state HVPPS by utilizing the multi-level concept with a simple switching scheme as the component voltage and current stresses are low. Moreover this topology also eliminates the need for expensive snubbing circuits, balancing circuits, and driver circuits. The converter could also generate waveforms with variable amplitude and duration, similar to the Marx generator, but with a less circuit complexity and cost [9].

IV. CIRCUIT ANALYSIS

This section analyses a multi-level converter, deriving the requirements for the high voltage charger, voltage and current stresses on devices, and the output characteristics.

Consider an N-stage converter that has a total capacitance of C_{eq} per stage, with all stages are charged to an equal voltage of V_c. The converter is designed to generate output pulses (V_{out}) as indicated in Fig. 5(b), at a repetition rate of f_{rp}, and with a T_p pulse width.

The total amount of energy stored in the converter, which is be delivered to the load, is given by (1). The output characteristics of the converter are determined by the RLC network formed by the load and the storage capacitors. For a pure resistive load of R, the output voltage can be expressed by (2), where U(t) denotes the unit step function.

$$E = \frac{1}{2}NC_{eq}V_c^2 \tag{1}$$

$$
\begin{aligned}
V_{out} &= V_c \sum_1^N X_n e^{-\alpha_n(t-T_n)}\left[U(t-T_n)-U(t-T_{n+1})\right] \\
&+ V_c \sum_1^{N-1} Z_n e^{-\alpha_n(t-T_{2N-n})}\left[U(t-T_{2N-n})-U(t-T_{2N})\right]
\end{aligned}
\tag{2}
$$

where
$$X_1 = 1$$
$$X_2 = \left(1 + e^{-\alpha_1(T_2-T_1)}\right)$$
$$X_3 = 1 + \left(1 + e^{-\alpha_1(T_2-T_1)}\right)e^{-\alpha_2(T_3-T_2)}$$
$$\vdots$$
$$X_n = 1 + \left\{1 + \left[\frac{1+\left(1+e^{-\alpha_1(T_2-T_1)}\right)\times}{e^{-\alpha_2(T_3-T_2)}}\right] \times e^{-\alpha_3(T_4-T_3)}\dots \right\}e^{-\alpha_{n-1}(T_n-T_{n-1})}$$

$$Z_n = Z_{n+1} - C_{n+1}$$
$$Z_N = X_{N+1} - 1$$
$$D'_n = \left[X_n - (X_{n+1}-1)\right]/n$$
$$D''_n = Z_n\left[1-e^{-\alpha_n(T_{2N}-T_{2N-n})}\right]/n$$
$$C_m = 1 - \sum_{n=1}^{(N+1-m)} D'_{m+1-n} - \sum_{n=1}^{(N-m)} D''_{m-n}$$
$$\alpha_n = n/RC_{eq}$$

The load current and the energy dissipated in the load under this condition are given by (3) and (4). The minimum power throughput of the charger that is required to charge the storage capacitors is a function of the stored energy as given by (5).

$$I_{out} = \frac{V_{out}}{R} \tag{3}$$

$$E_{out} = \int_0^t \frac{V_{out}^2}{R} \, dt \tag{4}$$

$$P_{Charger} = \frac{Ef_{rp}}{(1 - T_p f_{rp})} \tag{5}$$

Since all the switches are in series, the load current flows through all of them. In addition to the load current, during the turn-on of a switch, the reverse recovery current that is required to commutate off the diode chain prior to the switch, flows through the switch. Hence the maximum current stress experienced by the switches during the turn-on of the device is given by,

$$I_{sw(n)} = I_{out}\big|_{t=T_n} + I_{rr} \tag{6}$$

V. SIMULATION RESULTS

A 3-stage HVPPS, shown in Fig. 6 and based on the proposed multi-level concept, was designed and simulated to validate the proposed technique. The driver circuits for the switches were energized through a diode chain (D_1-D_3) to minimize circuit complexity. This method ensures that the turning on of a switch takes place only after the switches below it have been turned on, and hence provides protection against false triggering. Opto-isolators were used to control the switch drivers through a microcontroller that is powered with reference to the bottom switch. The opto-isolators reduce the parasitic coupling between stages that could cause false triggering of switches. A complete list of circuit parameters are given in the appendix.

The load voltage and current, obtained from PSPICE simulations and the theoretical model, for a heavy load and a light load are shown in Fig. 7 and Fig. 8, respectively. The waveforms obtained from the theoretical derivations follow the simulations closely, validating the theoretical analysis. It is noted that the voltages across the capacitors in lower stages can reverse in polarity under heavy loads or when the switching delay is comparatively large, as depicted in Fig. 7. This could be avoided by having diodes across the storage capacitors to clamp the negative voltage.

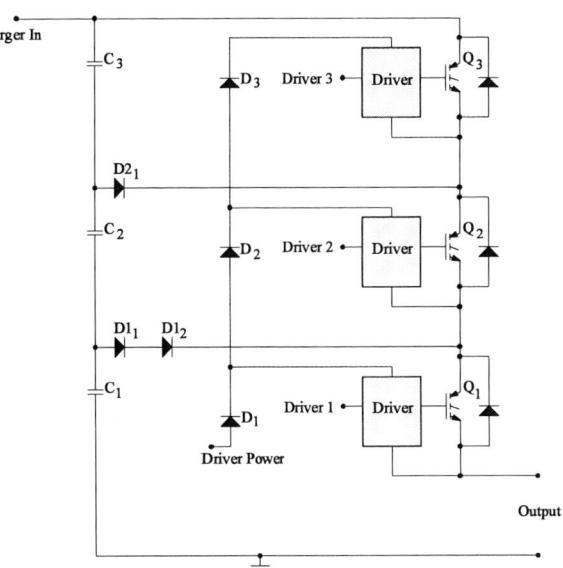

Fig. 6 A 3-stage pulse generator

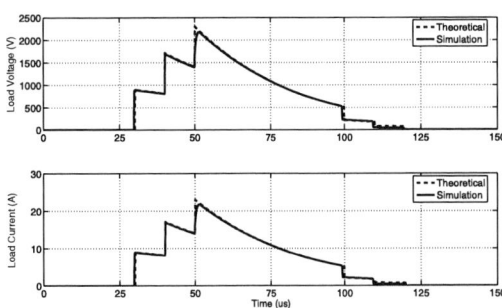

Fig. 7 Load voltage for a heavy load

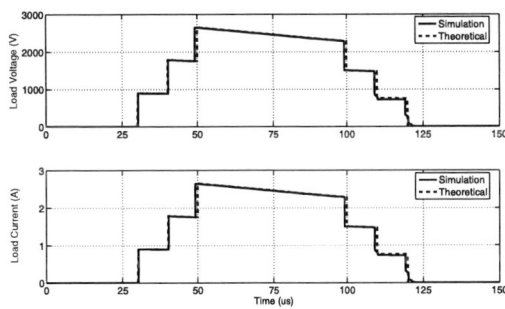

Fig. 8 Load voltage for a light load

Fig. 9 illustrates the voltage stress across the three switches when the converter is operated with a high load. The voltage stresses across the two diode chains under these conditions are shown in Fig. 10. As expected, the maximum voltage stress across a switch is equal to the charged voltage of the corresponding storage capacitor (V_c). Thus, it confirms the ability of the proposed concept to share the voltage stress equally across the switches without requiring an additional voltage balancing circuitry. The voltage stress on any given diode chain is equal to the sum of voltages across the capacitors above this diode chain that are contributing to the output. Hence the results suggest that there is a need for multiple diodes in lower stages to withstand the high voltage stress and also to reduce the reverse recovery current.

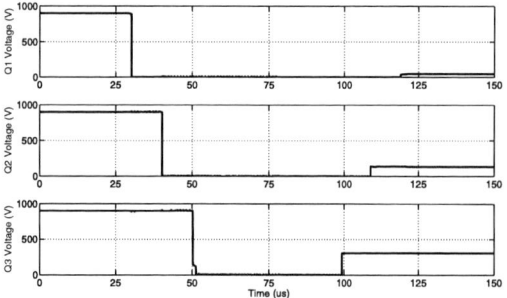

Fig. 9 Switch voltage stress

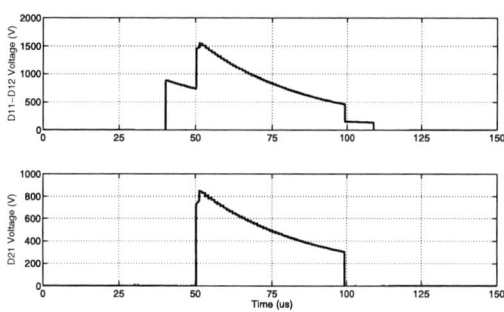

Fig. 10 Diode voltage stress

The current stresses experienced by the switches are shown in Fig. 11. Initially there is a large current flowing through the switches, which could be attributed to the reverse recovery current of the diode chain prior to switches. After the diode is commutated off the load current dominates the switch current and decreases as the capacitors discharge through the load causing the output voltage to decrease.

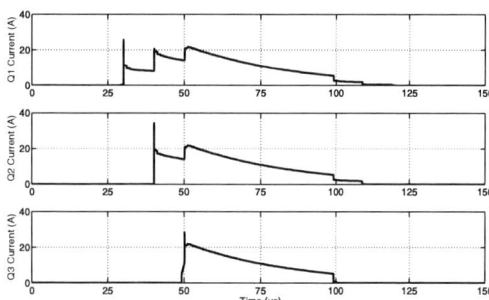

Fig. 11 Switch current stress

VI. CONCLUSIONS

A novel solid state pulsed power supply, termed multi-level pulsed power converter, has been introduced. The operation of the multi-level converter has been described in detail and theoretical solutions for the output characteristics and component stresses have been derived. The ability of the proposed concept to generate output pulses with flexible amplitude and duration has been demonstrated through PSPICE simulations and theoretical analysis. The inherent protection and component voltage stress balancing properties of the converter, which would be useful to design simple yet more reliable high power pulse generators, have also been demonstrated through simulations.

APPENDIX

TABLE I
CIRCUIT PARAMETERS

Property	Value	Unit
C_{eq}	1	µF
V_C	900	V
f_{rp}	0.66	Hz
T_p	90	µs
R_L (heavy)	100	Ω
R_L (light)	1000	Ω
IGBT	IXGH10N100	
Diode	BYT12P-1000	

ACKNOWLEDGEMENT

The authors acknowledge the contributions of R.C.B Woodhead, P.C. Lunenburg, and financial support by TIF NZ.

REFERENCES

[1] D. J. Thrimawithana, U. K. Madawala, and R. C. B. Woodhead, "Pulsed power generation techniques," *in 32nd annual IEEE IECON'06 Conf*, 2006, pp. 2014-2019.

[2] D. Shmilovitz, and S. Singer, "Pulsed power generation by means of transmission lines," *IEEE Trans. Power Electronics*, vol. 18, pp. 221-230, Jan. 2003.

[3] S. T. Pai, and Q. Zhang, "Introduction to high power pulse technology," World Scientific Publishing Co. Pte. Ltd., Singapore, 1995.

[4] A Kempkes, J. A. Casey, M. P. J. Gaudreau, T. A. Hawkey, and I. S. Roth, "Solid-state modulators for commercial pulsed power systems," *in 25th Int. IEEE Conf.*, 2002, pp. 689-693.

[5] M. Kanter, S. Singer, R. Cerny, and Z. Kaplan, "Multikilojoule inductive modulator with solid-state opening switches," *IEEE Trans. Power Electronics*, vol. 7, pp. 420-424, April 1992.

[6] M. V. Fazio, and H. C. Kirbie, "Ultracompact pulsed power," *Proc. of the IEEE*, vol. 92, pp. 1197-1204, July 2004.

[7] M. Buttram, "Some future directions for repetitive pulsed power," *IEEE Trans. Plasma Science*, vol. 30, pp. 262-266, Feb. 2002.

[8] J. C. McCutchan, *"Electric Fence Design Principles,"* Department of Electrical Engineering, University of Australia, Australia, 1980.

[9] New Zealand patent application 535719 and foreign equivalents.

[10] J. H. Kim, M. H. Ryu, S. Shenderey, J. S. Kim, and G. H. Rim, "High voltage-pulse power supply using IGBT stacks," *in 30th Annual Conf. of IEEE, Industrial Electronics Society*, vol. 3, pp. 2843-2847, Nov. 2004.

[11] M. P. Gaudreau, J. A. Casey, I. Roth, T. Hawkey, M. Mulvaney, and M. A. Kempkes, "Solid-state pulsed power systems for the Next Linear Collider," in *IEEE Conf. Pulsed Power Plasma Science*, vol. 1, pp. 298-301, June 2001.

[12] R. Richardson, R. J. Rush, S. M. Iskander, and P. Gooch, "Compact 12.5 MW, 55 kV solid state modulator," *in 13th Int. Pulsed Power Conf.,* 2001, pp. 636-639.

[13] G. E. Lehy, "Prototype development progress toward a 500 kV solid state Marx modulator," *in 35th Annual IEEE Conf.*, vol. 1, pp.831-834, June 2004.

[14] R. J. Richter-Sand, R. J. Adler, R. Finch, and B. Ashcraft, "Marx-stacked IGBT modulators for high voltage, high power applications," *in 25th Int. IEEE Conf.,* 2002, pp. 390-393.

[15] A. Krasnykh, R. Akre, S. Gold, and R. Koontz, "A solid state Marx type modulator for driving a TWT power modulator symposium," *in 24th Int. Conf.,* 2000, pp. 209 – 211.

[16] B. Meyer, A. Watson, T. G. Engel, and M. Kristiansen, "A single gap transformer coupled L-C generator with resonant frequency compensation," in *Proc. 7th Pulsed Power Conf.,* 1989, pp. 749-752.

[17] T. G. Engel and M. Kristiansen, "A compact high voltage vector inversion generator," *in 10th IEEE Int. Pulsed Power Conf.*, Vol. 2, pp. 1389-1393, July 1995.

Potential and Electric Field Distribution Analysis of Field Limiting Ring and Field Plate by Device Simulator

C.N. Liao*, F.T. Chien**, and Y.T. Tsai*

*Dep. of Electrical Engineering, National Central University
No. 300, Jhongda Rd., Jhongli City, Taoyuan County, Taiwan 32001, R.O.C.
** Dep. of Electronic Engineering, Feng Chia University
No. 100, Wenhwa Rd., Seatwen, Taichung, Taiwan 40724, R.O.C

Abstract-- **Potential and strength of surface electric field distribution have strongly influence on breakdown voltage and reliability of power semiconductor devices. Potential distribution can be determined by different field-limiting ring and field plate design which can be described by solving Poisson's equation in one dimension briefly. In this paper, the influence of design factors such as spacing between main junction and ring, ring width, and field plate width on potential and strength of surface electric field distribution are analyzed. From the simulation results, the relationship between those factors and potential and strength of surface electric field distribution can be found. Understanding the effect of design factors upon the junction termination edge, multi field-limiting rings and field plates of high breakdown power devices can be designed.**

Index Terms-- **termination, field- limiting ring, field plate, surface potential distribution, surface electric field.**

I. INTRODUCTION

For high-voltage power semiconductor devices, the breakdown voltages were limited severely by junction curvature effect which results in electric field crowing [1]-[2]. Therefore, in order to overcome the electric field crowing effect at the periphery of a power device, the junction termination edge structure technologies are developed to improve the blocking voltage of power devices. There structures are field-limiting ring (FLR) [3]-[4], field plate (FP) [1]-[5], semi-insulating poly-crystalline silicon (SIPOS) [6], junction termination extension (JTE) [7]-[8], reduced surface field (RESURF) [9] and variation lateral doping (VLD) [10] to use as junction termination edge. Among them, FLR combined with FP structures are used widely because the processes are simpler and can be integrated with the active area cells without more masks.

The blocking voltage of a FLR-FP structure is influenced by several factors which include ring junction depth, the spacing between each ring, ring width, surface charge density, field oxide thickness, field plate width and epitaxy concentration. Therefore, to design a FLR structure, the device simulator plays an important role in

This work was supported by National Science Council.

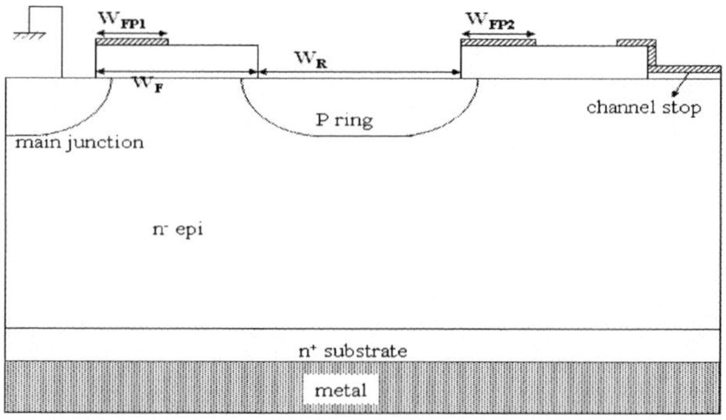

Fig.1 Cross-section of field-limiting ring and field plate structure.

analyzing the electrical characteristics inside a device. Further more, surface electric field and hot spot location of a device is hard to be observed by the measurement equipments. Using a simulator also helps to save the cost for designing a high breakdown device.

A lot of literatures focused on the factors we mentioned above and investigated how they affect their breakdown voltage. However, for a good FLR-FP structure design, not only the required breakdown voltage should be reached, but also potential and surface electric field distribution should be verified to achieve a reliable device. A designer must let the potential distribution and surface electric field averagely over each ring by modifying these factors when designing. Moreover, it also has to avoid high electric field crowding and hot spot to occur at the surface due to the boundary of FP structure. In this paper, we will simulate and analyze the potential distribution and the surface electric field from four factors: (1) spacing between main junction and ring (W_F), (2) ring width (W_R), and (3-4) two field plate widths $(W_{FP1}$ and $W_{FP2})$, which are shown in Fig. 1. Here, we use 200V n-type devices for this analysis. In next section, we will describe potential and surface electric field distribution briefly and present the simulated results which affected by each factor in the third section.

II. POTENTIAL AND SURFACE ELECTRIC FIELD DERIVATION BY ONE DIMENSION NUMERICAL ANALYSIS

In this study, we use 200 V junction termination edge structure which includes one FLR and two FP components. In order to achieve breakdown voltage over

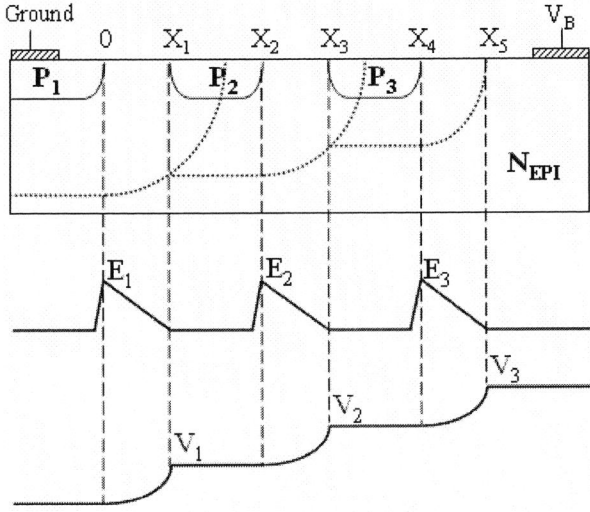

Fig.2 (a) Cross-section of field-limiting ring, (b) One dimension electric field, (c) One dimension potential.

200 V, thickness of 15 μm and resistance of 6 Ω -cm epitaxy (EPI) is used. The termination dimension for simulation is 60μm. The boron with the concentration of 5×10^{13} cm^{-2} is implanted and driven to form the main junction and the p rings.

The potential and electric field can be derived briefly from the basic p-n junction principle [11], [12] as shown in Fig. 2. Here, we ignore other non-ideal factors and consider the junction termination edge is an ideal p-n structure.

Fig.2 shows a multi-FLR structure which involves 2 rings (P$_2$ and P$_3$) and the main junction (P$_1$). The electric field and potential can be integrated readily from Poisson's equation as shown in (1) and (2)

$$\rho(x) = q(p - n + N_d^+ - N_a^-) \tag{1}$$

$$\frac{\rho(x)}{\varepsilon} = \frac{dE(x)}{dx} = -\frac{d^2V(x)}{dx^2} \tag{2}$$

The electric field and potential at P$_1$ junction can be derived respectively and shown in (3) and (4).

$$E_1(x) = \frac{qN_{EPI}}{\varepsilon}(x_1 - x) \tag{3}$$

$$V_1 = \frac{qN_{EPI}}{2\varepsilon}x_1^{\,2} \tag{4}$$

where X$_1$ is the distance from P$_1$ to P$_2$, N$_{EPI}$ is the EPI concentration.

Similarly, electric field and potential of P$_2$ and P$_3$ can be derived by the same approach and shown as followed.

$$E_2(x) = \frac{qN_{EPI}}{\varepsilon}(x_3 - x_2) \tag{5}$$

$$V_2 = \frac{qN_{EPI}}{2\varepsilon}(x_3 - x_2)^2 + V_1 \tag{6}$$

$$E_3(x) = \frac{qN_{EPI}}{\varepsilon}(x_5 - x_4) \tag{7}$$

$$V_3 = \frac{qN_{EPI}}{2\varepsilon}(x_5 - x_4)^2 + V_2 = V_{BR} \tag{8}$$

From the above equations, we can observe that epitaxy concentration affects the intensity of electric field and potential directly. The electric field and potential increase or decrease with EPI concentration. With the breakdown voltage increases, the EPI concentration will decrease which results in the electric field and potential of a ring decreases. Hence, the number of ring must be increased to meet breakdown voltage.

The intensity of surface electric field of junction termination edge structure must to be designed as low as possible for achieving reliable devices. To achieve a reliable high breakdown voltage device, we also required that the surface electric field must be equalized. Furthermore hot spot has to occur at the bottom of first ring, which can avoid the reliability issue from process vibration and make sure the avalanche breakdown to occur inside the termination. In addition, the breakdown voltage must be supported averagely by each ring, in other words,

$$V_1 = V_2 - V_1 = V_3 - V_2 \tag{9}$$

In next section, we will show how these factors affect the potential distribution and surface electric field.

III. RESULTS

Based on the cross section shown in Fig. 1 in Section I and the one dimension analysis presented in Section II, we can derive the surface potential and electric field distribution briefly. Now, we show the influence of these four factors on the surface electric field and surface potential, respectively.

A. Spacing Between Main Junction and Ring, W_F

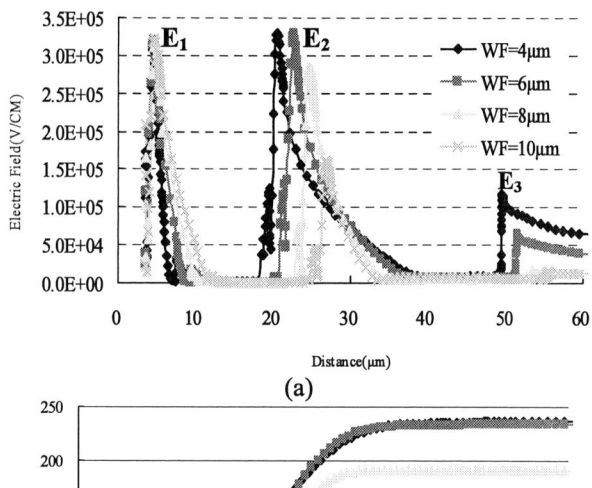

(a)

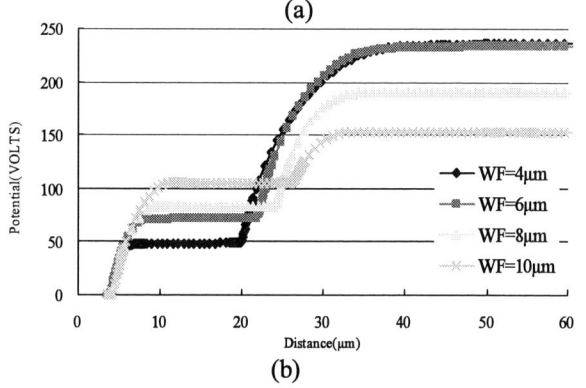

(b)

Fig.3 Electric field and potential of field-limiting ring with different W_F (a) surface electric field (b) surface potential distribution.

The surface electric field and surface potential with different W_F are shown in Fig. 3. For surface electric field, the strength of E_1 does not change obviously but E_2 and E_3 change a lot. The peak of E_2 not only decreases from 3.2×10^5 V/cm to 1.6×10^5 V/cm but also moves outside when W_F increases. For surface potential distribution, potential across p ring changes from 48 V to 106 V when W_F from 4 μm to 10 μm. But the breakdown voltage decreases from 233 V to 153 V when W_F increases. This might due to a limited termination length can't extend the electric field sufficiently.

B. Ring Width, W_R

The surface electric field and surface potential with different W_R are shown in Fig. 4. For surface electric field, the strength of E_1 and E_2 do not change. The peak moves outside with increased W_R. For surface potential distribution, the potential across p ring and breakdown voltage do not change (48 V and 235 V). Therefore, different W_R will not affect surface electric field and surface potential distribution. One conclusion can be obtained that the ring width can be designed as small as possible when designing for a shorter termination length.

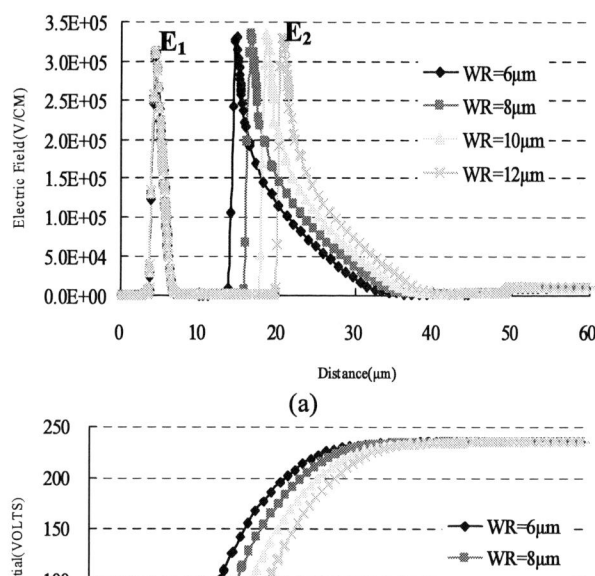

(a)

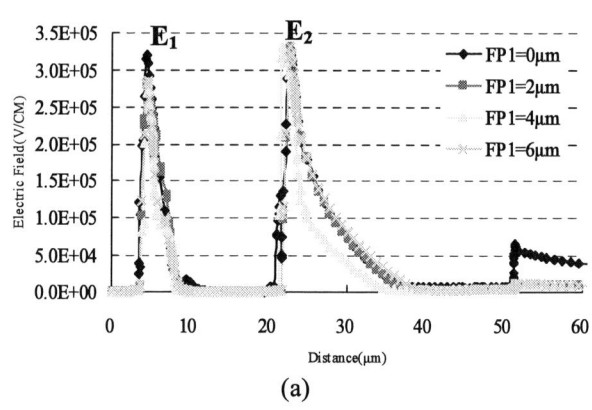

(b)

Fig.4 Electric field and potential of field-limiting ring with different W_R (a) surface electric field (b) surface potential distribution.

C. First Field Plate Width, W_{FP1}

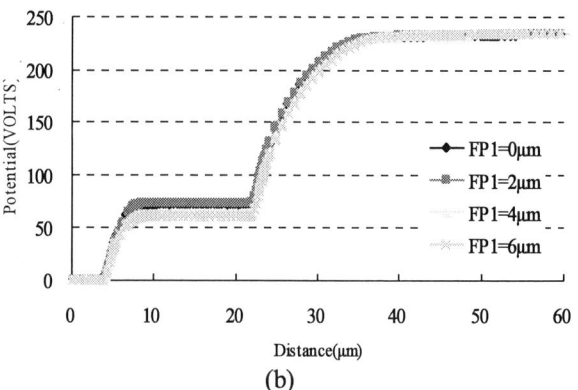

(b)

Fig.5 Electric field and potential of field-limiting ring with different W_F (a) surface electric field (b) surface potential distribution.

662

The surface electric field and surface potential with different W_{FP1} are shown in Fig. 5. For surface electric field, E_1 and E_3 change obviously but E_2 not. The peak of E_1 reduced from 3.2×10^5 V/cm to 2.8×10^5 V/cm due to FP structure. For surface potential distribution, the potential across p ring reduced from 73 V to 61.5 V when W_{FP1} increases. The breakdown voltage does not change despite the change of W_{FP1}.

D. Second Field Plate Width, W_{FP2}

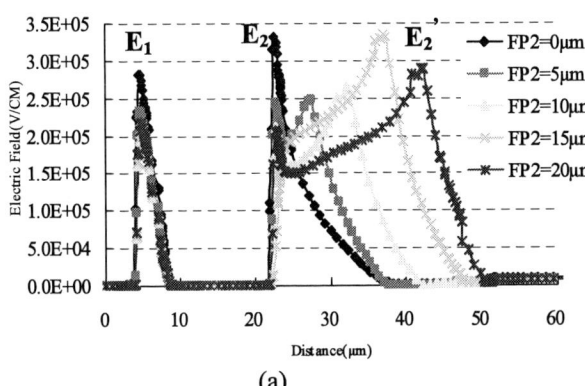

(a)

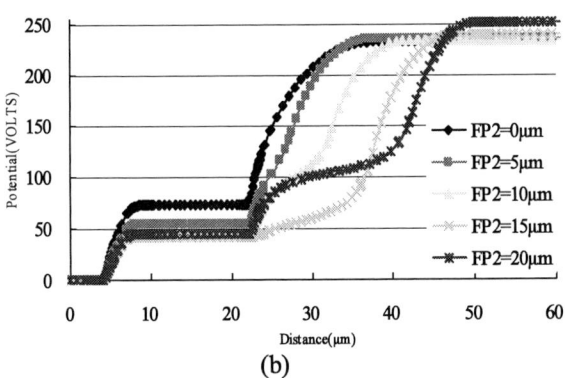

(b)

Fig.6 Electric field and potential of field-limiting ring with different W_F (a) surface electric field (b) surface potential distribution.

The surface electric field and surface potential with different W_{FP2} are shown in Fig. 6. Fixed FP1 length and different FP2 length were chosen and compared with no FP structure. For surface electric field, E_1 decreases from 2.8×10^5 V/cm to 2.1×10^5 V/cm, and E_2 decreases from 3.3×10^5 V/cm to 2.1×10^5 V/cm in the meanwhile FP2 add and further drop when W_{FP2} increases. Nevertheless, another peak, E_2' which results from the boundary of FP2, occurs and reaches maximum when W_{FP2} is 15 μm. Then it decreases again due to the channel stop as show in Fig. 1. For surface potential distribution, the potential across p ring is diminished from 73 V to 42.5 V. However, the breakdown voltage almost equals but reaches maximum in this design when W_{FP2} is 20 μm. A conclusion can be obtained that the distance between FP2 and channel stop should be as small as possible to keep outer electric field low. The result is the same as that shown in [5].

IV. MUITI FLRs-FPs DESIGN

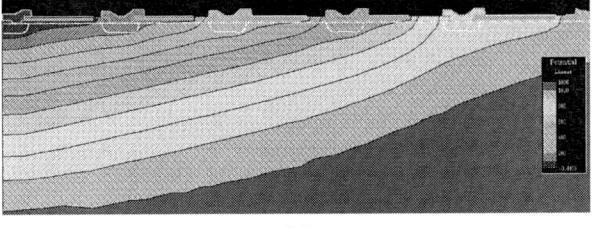

(a)

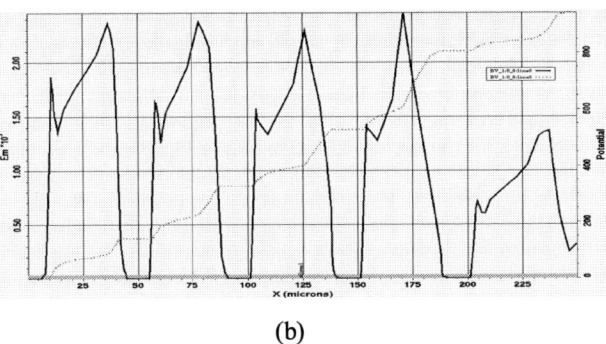

(b)

Fig.7 A multi FLR-FP structure (a) cross-section of potential (b) surface potential and surface electric field distribution.

Fig.7 (a) shows a cross section of multi FLRs-FPs (4FLRs-5FPs) junction termination edge design for 900V power MOSFET. In fig. 7 (b), the x-axis is the distance, and the left y-axis is electric field and potential for right. The surface potential and electric field spread averagely on each ring. Also, surface electric field is lower than 2.5×10^5 V/cm. It is good for reliability of device.

V. CONCLUSIONS

Surface potential and electric field have strong influence on breakdown voltage and reliability of devices. Different length of the spacing and the FP design lead to different surface potential and electric field. The potential have to be shared averagely, and electric field peak also need to be equalized and as low as possible. In this paper, we focus on field plate width, ring width and spacing between main junction and ring. The spacing between rings must be long as possible to spread potential on each ring averagely but not too long to decrease surface electric field. Besides, FP structure lowers surface electric field efficiently. The distance between outer FP and channel stop should be shorted to increase breakdown voltage and lower electric field.

ACKNOWLEDGEMENT

The authors would like to thank Yen-Chih Huang, who provided a helpful simulation model for this work.

REFERENCES

[1] V.A.K. Temple and M.S. Alder, "Calculation of diffusion curvature related avalanche breakdown in high-voltage planar p-n junctions", *IEEE Trans. Electron Devices*, vol. ED-22, pp.910-916, 1975.

[2] V. Anantharam and K.N. Bhat, "Analytical solutions for the breakdown voltage of punched through diodes having curved junction boundaries at the edges", *IEEE Trans. Electron Devices*, vol. ED-27, pp.939-945, 1980.

[3] M. S. Alder, V.A.K. Temple, A.P. Ferro and R.C. Rustay, "Theory and breakdown voltage for planar devices with single field limiting ring", *IEEE Trans. Electron Devices*, vol ED-24, pp.107-113, 1977.

[4] V.C. Kao and E.D. Wolley, "High voltage planar p-n junctions", *Proc. IEEE*, vol.55, pp.1409-1414, 1967.

[5] C. Basavana Goud and K.N. Bhat, "Two-dimensional analysis and design considerstions of high-voltage planar junctions equipped with field plate and guard ring", *IEEE Trans. Electron Devices*, vol.38, pp.1497-1504, 1999.

[6] T. Matsushita, T. Aoki, T. Ohtsu, H. Yamoto, H. Hayashi, M. Okayama and Y. Kawana, "Highly reliable high-voltage transistors by use of the SIPOS process", *IEEE Trans. Electron Devices*, vol. ED-23, pp.826-830, 1976.

[7] V.A.K. Temple, "Junction termination extension (JTE), a new technique for increasing avalanche breakdown voltage and controlling surface electric fields in p-n junctions", in *IEEE Int. Electron Device Meet (IEDM)*, pp.423-426, 1977.

[8] V.A.K. Temple and W. Tantraporn, "Junction termination extension for near-ideal breakdown voltage in p-n junction", *IEEE Trans. Electron Devices*, vol. ED-33, pp.1601-1608, 1986.

[9] J.A. Appels and H.M.J. Vaes, "High-voltage thin layer devices (Resurf devices)", in *IEEE Int. Electron Device Meet (IEDM)*, pp.238-241, 1977.

[10] R.Stengl, U. Gosele, C. Fellinger, M. Beyer and S. Walesch, "Variation of lateral doping as a field terminator for high-voltage power devices", *IEEE Trans. Electron Devices*, vol. ED-33, pp.426-428, 1986.

[11] W.C. Lin, K. Petrosky and D. Lampe, "Estimate of increase of planar junction breakdown voltage with field limiting ring", IEEE, pp.674-677, 1988.

Wire and Wireless Linked Remote Control for the Group Lighting System Using Induction Lamps

Kyu Min Cho*, Jae Eul Yeon**, Ma Xian Chao***, and Hee Jun Kim***

* Dept. of Information and Communications, Yuhan College, Korea
** Visual System Team, Fairchild Korea Semiconductor, Korea
*** Div. of Electrical and Computer Engineering, Hanyang University, Korea

Abstract--This Paper presents a wire and wireless linked remote control system for the group lighting system using induction lamps. Ethernet based network communication is used for long distance management and 2.4GHz RF network is adopted for the local area communication between the main network and the ballast. For the effective remote control and management including dimming using wire and wireless linked digital communication networks, the control circuit of the ballast is implemented with fully digital circuit using MCU and EPLD. In this paper, the applicable system configuration is proposed for the group lighting-control system and the detailed system configurations including fully digital controlled electronic ballast for the induction lamp are described.

Index Terms— wire and wireless linked, remote control, Ethernet, 2.4GHz RF, dimming, induction lamp

I. INTRODUCTION

Since the induction lamp has no electrodes, it has very long lifetime more than 60,000hours. Therefore, the induction lamp system is usually used in the case where the system maintenance is difficult such as high ceiling lighting system. Induction lamp is driven by using a high-frequency resonant inverter and its driving frequency is over several hundreds of kilo hertz. Therefore, the remote control has to be needed for the dimming of the induction lamp. For the effective management of the group lighting system including the dimming and monitoring, central management system using a personal computer is proper.

This paper presents a wire and wireless linked remote control system for the group lighting using induction lamps. Ethernet based network communication is used for the long distance management using personal computer and 2.4GHz digital radio frequency network is adopted for the local area communication between the main network and the ballast. For the effective remote control and management including dimming using wire and wireless linked digital communication networks, control circuit of the ballast is implemented with fully digital circuit using MCU and EPLD.

II. PROPOSED LIGHTING CONTROL SYSTEM

A. Electronic Ballast for Induction Lamps

Fig. 1 shows the configuration of the induction lamp and its lighting principle. High frequency magnetic flux induces electromotive force in the vessel, and then the discharging is occurred. Normally, the induction lamp is driven by using a high-frequency resonant inverter and its

frequency is over several hundreds of kilo hertz and up to several Giga hertz. Fig. 2 shows the equivalent circuit of the induction lamp. Since the equivalent resistance is almost infinite before lighting, high voltage must be applied to the primary coil for the ignition. However, the equivalent resistance is rapidly decreased after ignition.

Fig. 3 shows a half bridge resonant inverter including all sensing circuit for the control of the ballast. V_{S1} is used for fuse status monitoring, and over/under voltage protection. V_{S2} and I_S are used to calculate the dc-link power in the case of the closed-loop power control mode. v_L is used for protection of switching devices in the ignition stage and also used for detecting the lamp fault.

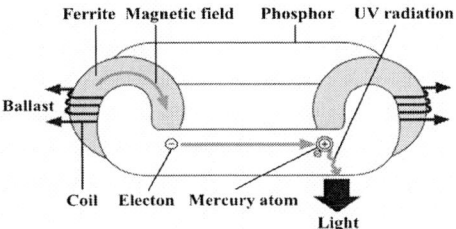

Fig. 1. Configuration of the induction lamp and its lighting principle.

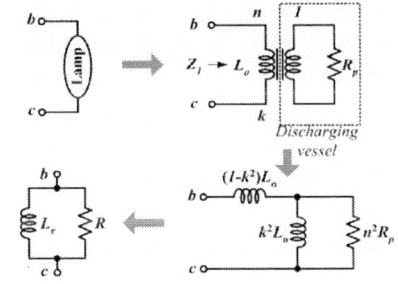

Fig. 2. Equivalent circuits of the induction lamp.

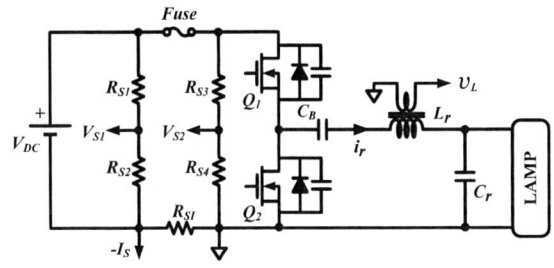

Fig. 3. Main circuit diagrams of the electronic ballast.

978-1-4244-0644-9/07/$25.00 ©2007 IEEE

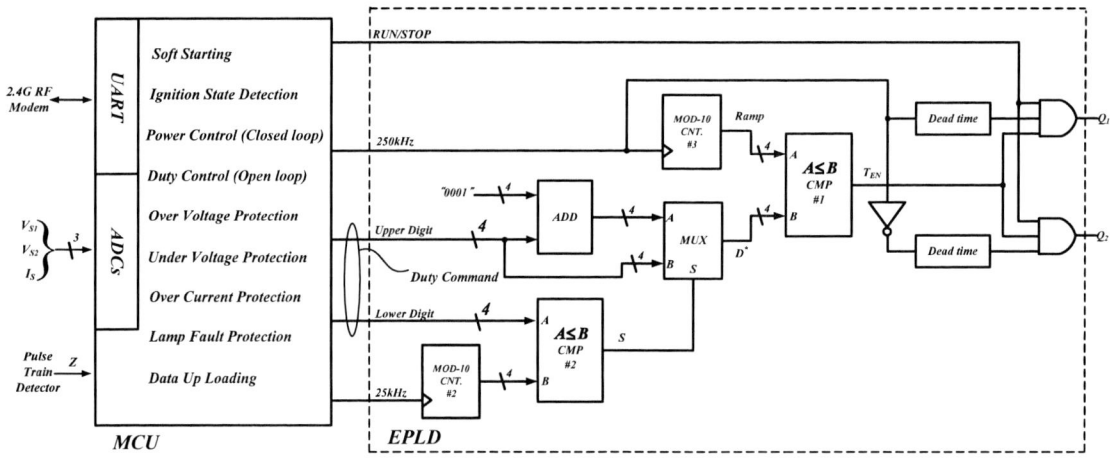

Fig. 4. Block diagrams of the proposed digital controller.

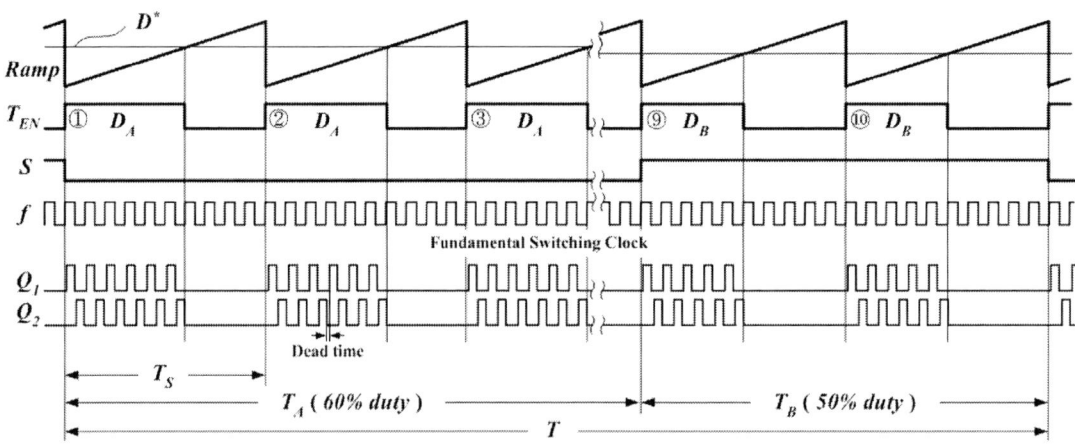

Fig. 5. Specific time chart of the proposed average burst PWM method (in the case of 58% duty).

B. Digital Controller for the Induction Lamp Ballast

Fig. 4 presents block diagrams of the proposed digital controller for the dimming of the induction lamp. Low price 8-bit one-chip MCU can be used for the main controller. Voltage and current of the dc link of the ballast are gathered by using built-in A/D converters of the MCU and digitized data are transferred to the host PC through the wire and wireless communication network. 115.2-kps UART interface is used for all information exchanges between the remote RF modem and the MCU. For the dimming of induction lamp, a novel average burst duty control method is implemented by using an EPLD. A novel average burst duty control method is used for the dimming. The proposed control method can produce variable PWM duty with 1-% of steps. All functions are implemented by using a low cost small size EPLD.

Fig. 5 shows the specific time chart of the proposed average burst PWM method and it is in the case of 58-% duty. Since 25-kHz of burst PWM frequency is adopted, the proposed averaging burst PWM method does not produce sound noise caused by PWM.

In the ignition mode, the switching frequency is gradually decreased from 500-kHz to 250-kHz during predetermined duration. So the lamp is ignited softly.

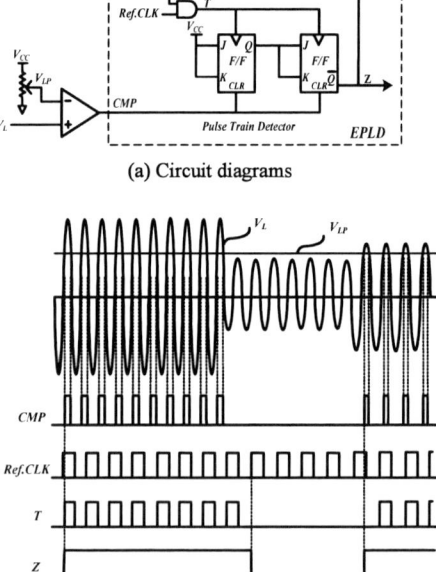

(a) Circuit diagrams

(b) Timing chart

Fig. 6. Circuit diagrams and timing chart of the pulse train detector for the ignition status decision and lamp fault decision.

666

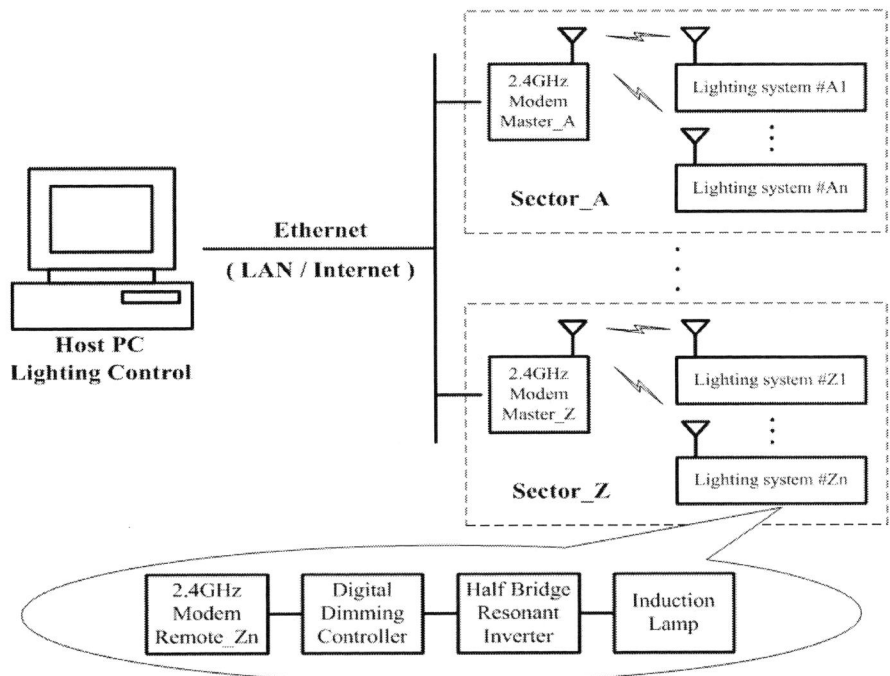

Fig. 7. Configurations of the proposed wire and wireless linked remote control system using induction lamp.

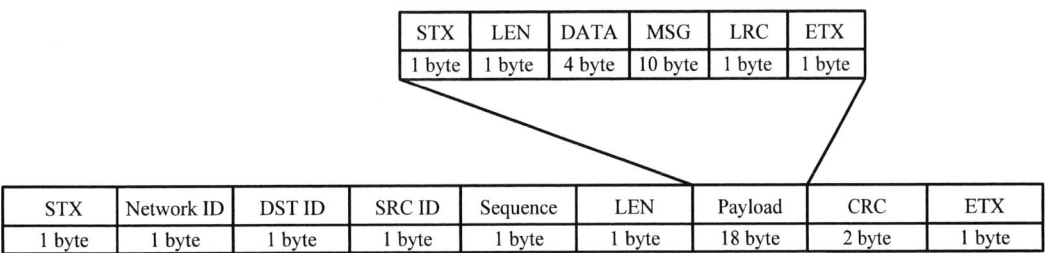

Fig. 8. Data frame structure in the RF communications.

Fig. 6(a) shows circuit diagrams of the pulse train detector, which can decide ignition status and lamp fault status. In the case of the successful ignition, the output signal will be low state after ignition, because the voltage of the inductor shown in Fig. 3 is rapidly decreased and then there are no signals at the output of the comparator. Otherwise, the high state of the output of the pulse train detector means ignition fault, no lamp or lamp fault. In that case the driving signal is masked and the switching devise will be protected from high resonant current. Since this lamp status decision method uses the output signal of auxiliary winding, it is useful compare with the other lamp fault detection circuit such as a method using the voltage of the lamp directly.

C. Wire and Wireless Linked Remote Control System

Since the induction lamp has long life time, it is very useful, especially in the fields of high ceiling and tunnel lighting system. Sometimes those lighting system needs controlling the illumination. For those cases, remote control system must be needed because induction lamps are driven at very high frequency. Therefore the wire and wireless linked remote control system for the group lighting system using induction lamps is proposed.

Fig. 7 shows the overall system configurations of the proposed remote lighting control system. In order to use Internet network or local area network for the main long distance communication network, the main network is designed with Ethernet based network. Therefore the system management can be easily achieved by using personal computer to handle the proposed system. And the commercial mobile communication network can be also combined with the proposed system. In that case, the system manager can handle the lighting system with mobile devices such as cellular phone and/or PDA which loaded application soft-ware to control and monitor the system.

For the local area communication between the main network and the each ballast, wireless network is more proper than wired network. 2.4GHz RF network is built in the proposed system. We need, therefore, two kinds of RF modem. Ethernet interfaced master modem and

667

UART interfaced local modem linking the wireless main network and ballasts. Therefore ballast includes the local modem. In the future, to work with another management system such as IBS, standard protocol like a ZigBee is more proper for the proposed system. In this stage, however, simple user protocol is adopted in the proposed system. Fig. 8 represents the data frame structure, which is adopted in the proposed RF modem.

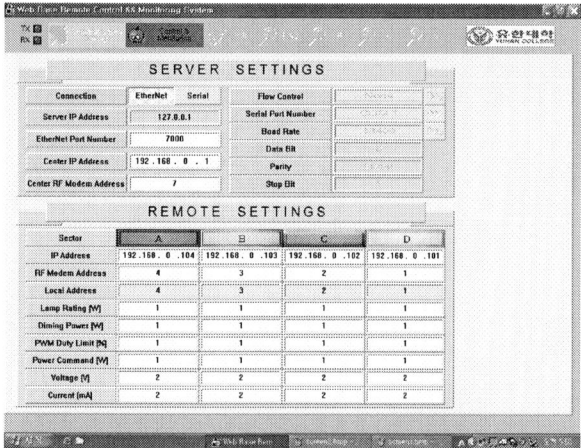

(a) Screen for server and remote modem setting.

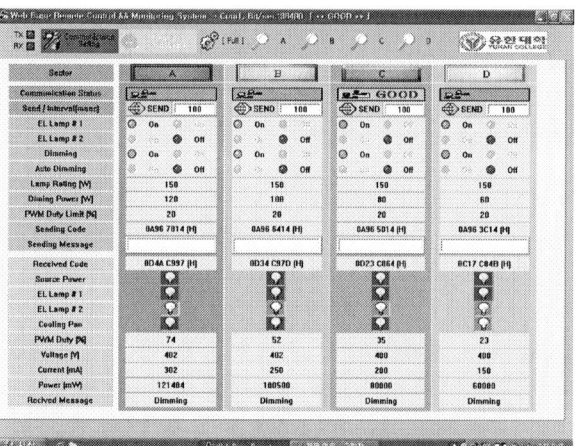

(b) Screen for the controlling and system monitoring.

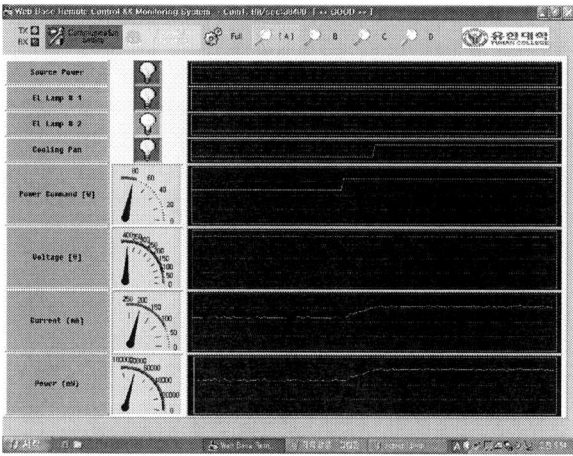

(c) Screen for the monitoring of each ballast status

Fig. 9. Images of the prototype management program screen

III. EXPERIMENTAL RESULTS

For the experimentation, proto lighting system is constructed, which is composed of four sectors. Each sector is composed with four lamps. In actual application system, the number of sectors is not limited and the number of lamps is dependent on addressable capability of the RF modem. In the experimental setup, 150-W and 100-W induction lamps manufactured by OSRAM(they usually call it electrode-less fluorescent lamp).

The principal parameters of the experimental setup are as follows:

- DC link voltage, V_{DC} : 400 [V]
- Switching frequency, f_s : 250 [kHz]
- Burst PWM frequency, f_b : 25 [kHz]
- Blocking capacitance, C_B : 2.2 [nF]
- Resonant inductance, L_r : 232 [uH]
- Resonant capacitance, C_r : 2.2 [nF]
- Switching devices : IRF840
- Starting frequency : 500 [kHz]
- Successful ignition determining time : 1 [Sec]
- Over voltage protection level : 450 [V]
- Under voltage protection level : 200 [V]
- Over current protection level : 500 [mA]

Fig. 9 shows some images of the prototype server management program screen. Fig. 9(a) is a screen for the parameter settings such as IP address of the center modem, local address of the remote modem, and some gains of the control/monitoring factors.

Fig. 9(b) is a screen for overall system control and monitoring screen. Settings constructed in this screen are as follows:

- Communication interval
- Lamp on/off
- Control mode
- Lamp power ratings
- Reference power level
- Minimum PWM duty
- Message to the local RF modem

Monitoring factors in this screen are as follows:

- Source power stauts
- Lamp on/off status
- Cooling pan staus
- PWM duty
- DC link voltage
- DC link current
- DC link power
- Message from the local RF modem

Fig. 9(c) is a screen for detailed graphical monitoring factors from the each ballast. Since all data can be captured into a file in the PC, system manager can check the past status, especially, when an accident is occurred in the ballast. Moreover, error message is automatically sent to system manager by using SMS(short message service).

Fig. 10 shows some images of master RF modem with Ethernet interface and local RF modem with UART interface used in the prototype system. The RF modem uses 2.4GHz ISM band radio frequency and the RF power is 10-mW. The data transfer rate of the RF modem is 1-Mbps. The RF modem is worked by using the direct sequence spread spectrum modulation method. Since CSMA/CA, automatic re-transmission, and CCITT- 16bit CRC algorithms are implemented in the MAC protocol level, reliable communication can be achieved. The data transfer rate of UART interface is up to 115.4-kbps.

(a) Master modem with Ethernet interface

(b) Remote modem with UART interface

Fig. 10. Images of 2.4GHz radio frequency data modem.

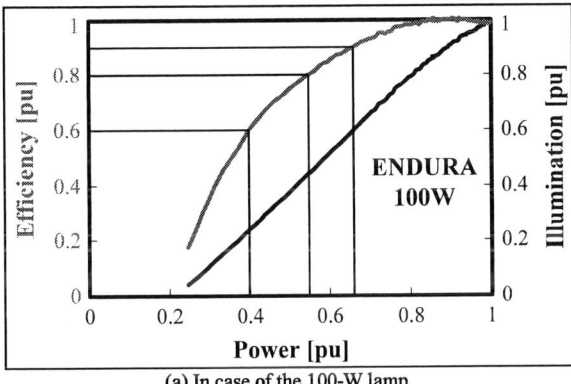

(a) In case of the 100-W lamp.

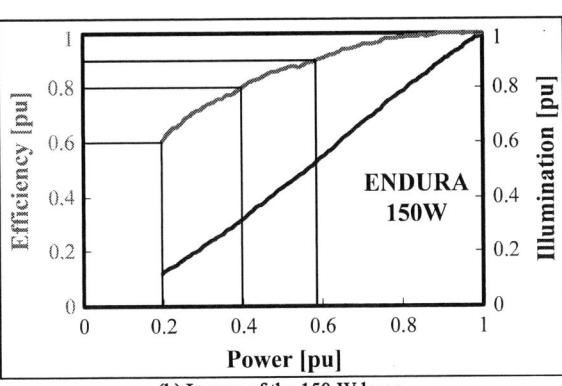

(b) In case of the 150-W lamp.

Fig. 11. Illumination and Efficiency according to the power.

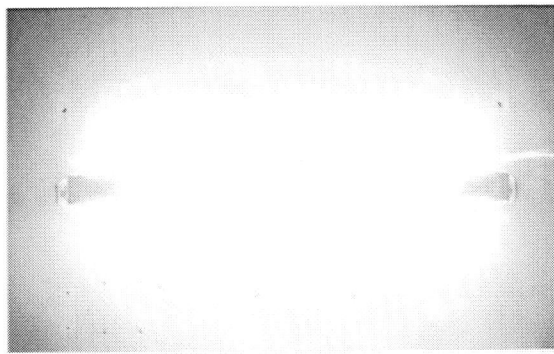

(a) In case of 100[%] power.

(b) In case of 60[%] power.

(c) In case of 30[%] power.

Fig. 12. Images of the lamp lighting according to the power.

Fig. 11 represents variations of illumination and efficiency according to the various power. Fig. 11(a) and Fig. 11(b) are in the case of 100-W lamp and 150-W lamp, respectively. The trends of illumination and efficiency according to the power percentage are different from each other. The reason for that is minimum power dissipation for exciting the light is nearly same even though the power ratings are different from each other. It can be seen that the case of 150-W lamp is more effective in dimming than the case of 100-W lamp.

In actual, the dc link power is not same with the lamp power because the dc link power includes not only the lamp power but also some losses in the resonant inverter. Since the lamp is driven at high frequency, it is nearly impossible to calculate the exact lamp power using by MCU. Therefore, the lamp power is indirectly controlled by controlling the dc link power in the power control mode. If the illumination is directly controlled by using a

669

light sensor, the maximum dc power should be controlled to protect the ballast. Normally, the indirect lamp power control is sufficient in the lamp dimming system.

Finally, images of the lighting lamp at various powers are shown in Fig. 12. Since the ratings of lamp power is 150-W, 60-% power and 30-% mean 90-W and 45-W, respectively.

IV. CONCLUSIONS

This paper proposes a wire and wireless linked remote controlled group lighting system using induction lamps, which is driven by fully digital controlled electronic ballast. And a novel average burst PWM duty control algorithm is proposed for dimming. Since the proposed PWM method can produce the output with the 1-% of duty resolution and with the 25-kHz burst frequency, the lamp power is stably controlled without sound noise.

For the remote controlling and monitoring the group lighting system, Ethernet based communication and 2.4GHz RF communication are constructed.

With the result of experimentation using prototype setup, the usefulness of the proposed remote control system and effectiveness of the proposed dimming method for the induction lamp are confirmed and verified.

REFERENCES

[1] R. Louis, "A Novel Ballast for Electrodeless Fluorescent Lamps," in *Conf. Rec. of IEEE Industry Applications*, vol. 5, pp. 3330-3337, 2000.

[2] H. Kido, "A study of electronic ballast for electrodeless fluorescent lamps with dimming capabilities," in *Conf. Rec. of IEEE Industry Applications*, vol. 2, pp. 889-894, 2001.

[3] H. Y. Wang, A. V. Stankovic, D. Kachmaric, L. Berone, "A novel discrete dimming ballast for linear fluorescent lamps," in *Conf. Rec. of IEEE PESC*, pp. 815-820, 2004.

[4] M. K. Kazimierczuk, W. Szaraniec, "Electronic Ballast for Fluorescent Lamps," *IEEE Trans. on Power Electronics*, pp. 386-395, 1993.

[5] D. Wharmby, "Electrodeless lamps for lighting: a review," in *Conf. Rec. IEEE APEC*, pp. 948-954, 2002.

[6] B. Cook, "New developments and future trends in high efficiency lighting," *J. of Eng. Science and Educ.*, 207-217, 2002.

[7] S. Ben-Yaakov, M. Shvartsas, "A Behavioral SPICE Compatible Model of and Electrodeless Fluorescent Lamp," in *Conf. Rec. of IEEE APEC*, pp.948-954, 2002.

Induction Heating with Traveling Magnetic Field for Uniform Heating to Flat Metal

T. Sekine*, H. Tomita*, Y. Saito*, S. Obata*, S. Yoshimura*
* Tokyo Denki University, Ishisaka Hatoyama, Saitama, 350-0394 Japan

Abstract--As one of the effective dismantlable adhesion method, "the ALLOVER method" for interior construction using induction heating (IH) and thermoplastic adhesive has been proposed. The study of suitable IH coil is most important for complete adhesion and dismantlement of materials, because IH area in metal load becomes mostly adhesion area. In generally a conventional spiral IH coil is applied, but the uniform heating to long steel stud or wide steel plate is inadequately. For these adhesions, we propose the effective coils with three-phase traveling magnetic field. This paper describes for the experiments and analysis of the new IH coils and applications.

Index Terms--Induction Heating, Traveling magnetic field, Three-phase high frequency inverter, Dismantlable adhesion, Allover method.

I. INTRODUCTION

Authors are investigating a downsize and portable induction heating (IH) device for application to interior construction[1]-[5]. About ten years ago, we have proposed a novel dismantlable adhesion method which was named as the ALLOVER method for interior construction of wainscot or floor panel, tile etc[6]-[11]. The method is applied a portable IH device and a thermoplastic adhesive (that is normally named as a hot-melt adhesive) which is melted by IH.

When a Joule-heat is caused by IH on metal plate pasted a hot-melt adhesive, the adhesive is also melted by thermal energy. And to get cold again, adhesion will be completed. The method is able to easy assemble in adhesion process and to dismantle again by same process as shown Fig. 1. This is a strong point of the method.

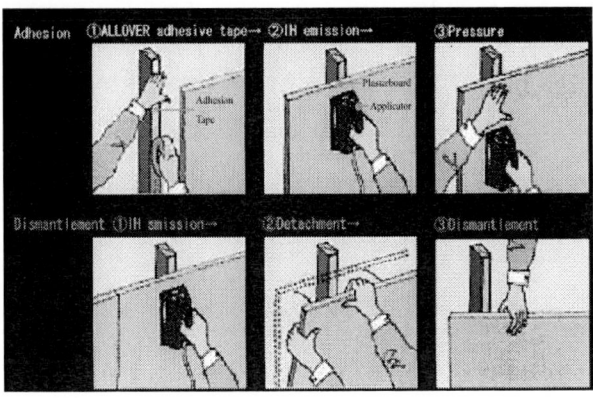

Fig. 1. Process of adhesion and dismantlement.

Everyone can easily use this method, not only in building or house constructions but in other many industry applications. Recently, IH cookers as home appliance with 10~100kHz inverters are rapidly progressed, because compact and powerful power devices are developed. As the ALLOVER method's device, we have developed a hand-held IH device using the latest compact IPM (intelligent power module) and conventional spiral coil as shown in Fig. 2[7]-[11].

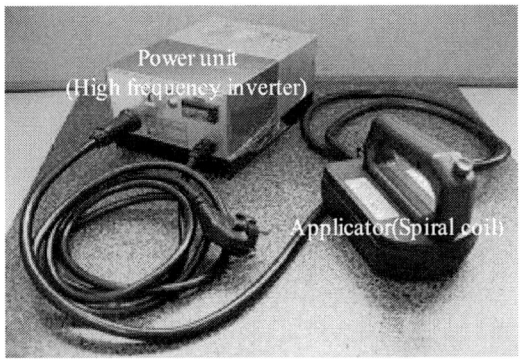

Fig. 2. IH device for interior construction.

Moreover we have proposed a compact and powerful IH system with traveling magnetic field. For example, there is an adhesion of interior finishing material to long steel stud as shown in Fig. 3. To generate a traveling magnetic field along the stud, an application of three phase linear induction coil driven by high frequency inverter is effective for uniform heating in metal surface, at compare with conventional IH device using spiral coil[12]-[14]. Therefore, simple and high quality uniform heating/adhesion is possible to long metal. In the recent paper[15], new IH device consists of the 3-phase, 200V, 20kHz full-bridge inverter and linear induction coil with short pitch winding was introduced and experimented.

In this paper, the usefulness and electric characteristics of the proposed 3-phase system are proved by some experiments and analysis with a finite element method. And the uniform heating not only straight metal but also uniform heating for large flat metal is reported.

II. IH WITH TRAVELING MAGNETIC FIELD

A poly-phase induction motor induces a rotor current depend upon sinusoidal rotating magnetic field. The motor generates a rotating torque between the magnetic field and the rotor current. If the rotor is locked, the rotor will be overheated by an excessive induced current in the rotor. This fundamental principle of rotor heating is available for our IH system, and the rotating magnetic field is changeable to the straight-line magnetic field with linear induction coil. Thus a magnetic field moves linearly with uniform velocity, and a locked secondary metal is also uniformly heated. Therefore this method is applicable to uniform heating and adhesion in long steel stud as shown in Fig. 3.

In addition, as in Fig. 16, it is also possible to heat extended area like a large steel plate with extending direction-d of the coil width as shown in Fig. 14. Therefore a uniform heating in wide area of flat steel could be enabled. It is usefulness for speedy and uniform adhesion of tile as shown in Fig. 4 and cork-board, for examples.

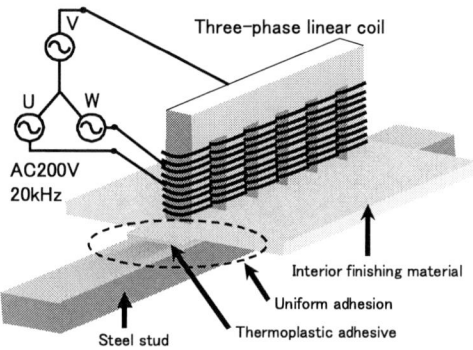

Fig. 3. IH by 3-phase linear induction coil.

A. Heating target and the conditions

In the experiments and analysis of IH characteristics, a thin galvanizing steel plate of width: 212mm, length: 300mm and thickness: 0.115mm[15] is tested by proposed coils and inverter. This steel is realized to flooring or tile etc. as shown in Fig. 4. By considering the combination of tile and steel plate in Fig. 4, the steel plate is applied adhesives on the both sides. One side's adhesive of the steel has a strong contact property as epoxy resin, and adheres strongly to reverse side of tile. And another side's adhesive of the steel is applied a hot melt resin which is melted by IH. There are various interior finishing materials which have different heat conduction properties. But these heat conduction properties are disregarded in the paper, the steel plate is used with only air space instead of material's thickness in the experiments and analysis.

Fig. 4. Product tile pasted steel plate with adhesives.
Tile size: 98 × 98[mm],
Tile thickness: 4mm, Steel thickness: 0.115mm.

B. Simulation of thermal distribution by IH

In the electromagnetic-heat transition analysis with computer, we applied the finite element method by the "ANSYS ver10.0" [16]-[18]. The parameters of the steel and ferrite core[19] are listed in Table I.

TABLE I
PARAMETERS OF RELATIVE PERMEABILITY

Relative permeability of core	2500
Relative permeability of steel plate	150

III. LINE HEATING BY TRAVELING MAGNETIC FIELD

A. Suitable coil and inverter for line heating

Heating power in the steel plate is calculated by (1). If magnetic flux is sinusoidal waveform, the induced voltage in the steel is shown by the faraday's law as (3) from (2). So the rms value of the voltage is shown by (4).

$$W = I^2 R = \frac{V^2}{R} \tag{1}$$

$$\phi = BS \sin 2\pi f t \tag{2}$$

$$V = N \frac{d\phi}{dt} = 2\pi f BSN \cos 2\pi f t \tag{3}$$

$$V = 2\pi f BSN / \sqrt{2} \tag{4}$$

where W: Heating energy, I: Eddy current, R: Resistance, V: Induced voltage, ϕ: Magnetic flux, B: Magnetic flux density, S: cross section of ferrite core, f: Frequency, N: Number of coil turn.

The frequency of the moving flux and the inverter frequency are same. From (1)-(4), a higher inverter frequency and a lot of turns of coil winding which is able to get high-MMF (magneto-motive force) are necessary for getting induced voltage for effective IH. But the current which is necessary for high-MMF is decreased in constant output voltage of inverter, because the coil has higher impedance to depend on the big-turns and high frequency. So, to get the both benefits of low impedance and high-MMF, each coil windings are connected with parallel. And due to portable size of coil, the two-poles short-pitch winding is applied as shown in Fig. 5 and 6[15]. The weight of the applicator coil is about 1kg, and the number of coil turns in single pole and impedance per phase are 33turns and 7.03 Ω at 20 kHz.

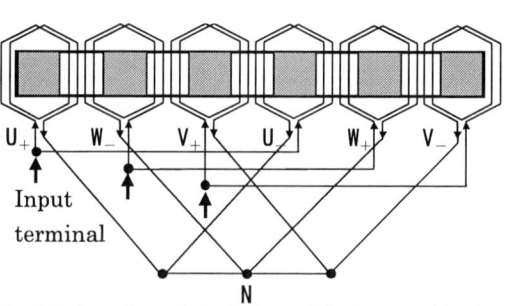

Fig. 5. 3-phase short-pitch winding coil for long metal heating.

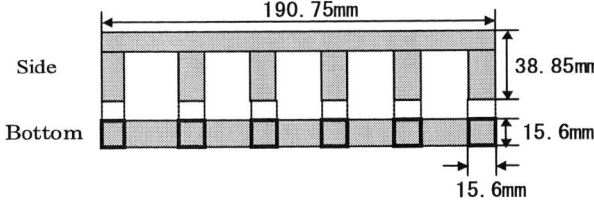

(a) Measure of ferrite core.

(b) Configuration of assembled coil.

Fig. 6. Experimented coil for long metal heating.

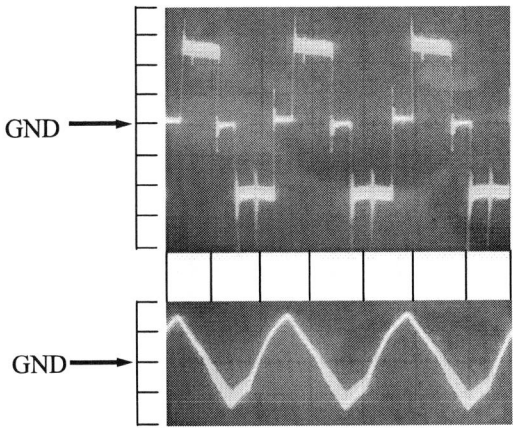

Fig. 8. Output waveforms of inverter.
Upper: W-U voltage (100V/div) , Lower: Current (20A/div),
Time: 20 μ s/div.

IV. LINE HEATING CHARACTERISTICS

A. Thermal conditions in experiment and analysis

The thickness of finishing materials is important parameter for heating power in IH. That is meaning the distance between the heating coil and steel plate. We experiment on the distances that are from 2mm to 21mm with air spacing. The experiments and computer simulations are performed by the parameter of the space which is called "Gap" in this paper.

The examples of thermal conditions on the steel plate in the analysis and experiment are shown in Fig. 9. From the Fig.9, the thermal conditions are observed in a long heating area on the steel plate, and these depend on the our IH coil.

We also experimented the IH of conventional steel stud applied hot-melt adhesive. Wide melting area of adhesive on steel stud was observed as shown in Fig. 10.

B. Design of the inverter and frequency

The 90% of heating energy by eddy currents in a metal plate is spent inside depth- δ by skin effect. The peculiar rule of the effect on IH is shown (5). In our working method of tile or floor panel, the reverse side of the steel plate must be heated by IH until melting condition of hot melt adhesive. Therefore the δ is equal to the thickness-γ of steel plate in (5), the suitable frequency-f of inverter is deduced.

$$\gamma = \delta = \frac{\sqrt{\sigma \times 10^7}}{2\pi\sqrt{\mu \cdot f}} \qquad (5)$$

where σ : resistivity, μ : relative permeability, f : frequency.

From the parameters of steel plate ($\gamma = \delta$ =0.115m, μ =150, σ =1.7 $\times$ 10^{-7} Ω m), the frequency is deduced 21.7 kHz. But IGBT's maximum switching frequency of supplied IPM is 20 kHz. Therefore in the experiments and analysis the inverter frequency is also decided three-phase 20 kHz, although an error rate is 7.8%. For the system protection from over current and short-circuit, the dead time in switching IGBT is set to 8 μ sec. The duty ratio becomes 67.7%.

The circuit diagram is shown in Fig. 7. The part of inverter is used IPM for simple circuit and protection. The output waveforms of inverter are shown in Fig. 8. The current seems as sinusoidal waveform.

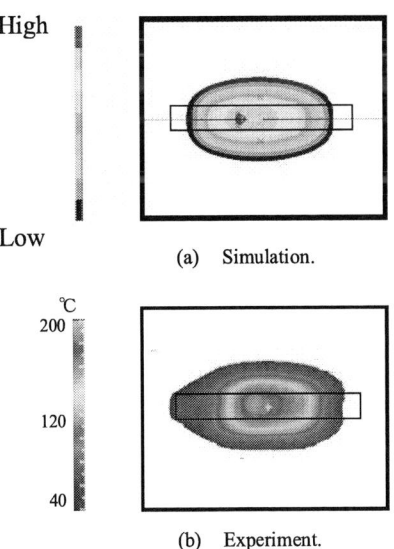

(a)   Simulation.

(b)   Experiment.

Fig. 9. Thermal distribution on steel plate.
Gap: 15mm,   Steel thickness:0.115mm, Heating time:10sec,
Outside frame is steel plate size. Inside frame is coil size.

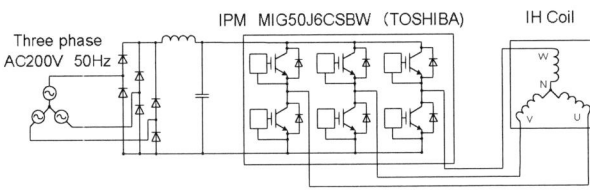

Fig. 7. Inverter circuit

673

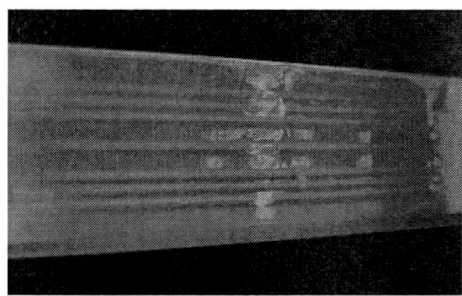

Fig. 10. Melted adhesive on a steel stud by IH.
Heating time: 8sec, Gap: 12mm,
The stud size (Width: 45mm, Length: 900mm, Thickness: 0.2mm).

B. Heating characteristics of steel plate

By the experiment, the transient heating characteristics of the steel plate with gap parameter are shown in Fig. 11. The heating time is decided as 16sec for considering working process.

When the gap is expanded, the permeance of magnetic circuit is also decreasing. Therefore, induced electromotive force ($d\phi/dt$) in steel plate becomes small and eddy current in steel plate decreases. Thus at wide gap, the temperature of the steel plate is going down.

Fig. 12 shows the heating time characteristics with the parameter of rising temperatures (100℃, 130℃, 180℃, 200℃, these are examples of melting points of adhesives). These characteristics will be given as suitable information for bonding to working operators.

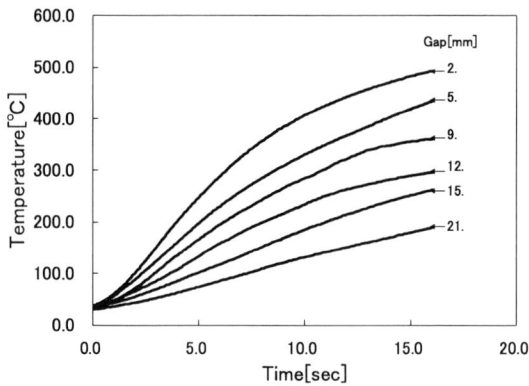

Fig. 11. Transient heating characteristic of steel plate.

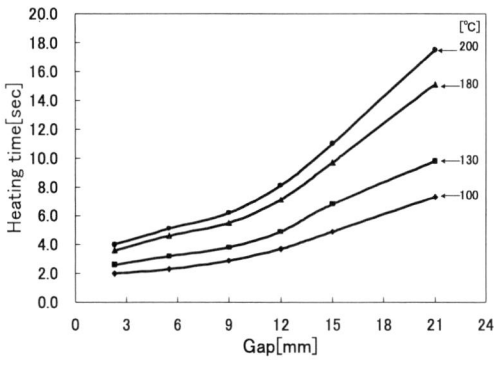

Fig. 12. Heating time characteristics at rose temperatures.

C. Analysis method of heating energy in steel plat at IH

Conditions of measurement and calculation: The electric characteristics of the inverter output power are measured by high frequency power-meter. Although the voltage waveform is square-wave as shown in Fig. 8, current waveform is like sine-wave in same figure. So, in calculations of this chapter, we suppose as pseudo-sine waveform. The measured values of voltage or current in each phase are almost same. Therefore, the output power per each phase of inverter is considered as balancing. The output power of inverter are measured in primary heating coil, under the no-load (coil only) and under the load of steel plate (IH condition). The power loss of the steel plate is important and calculated by subtracting the primary coil loss (core loss and copper loss) from the inverter output power.

Core iron loss of the primary coil: The each coil in the core tooth are independent as shown in Fig. 5. So, magnetic flux density-B in one-teeth is calculated by (6) using input voltage V_1 (output line voltage of inverter). From the calculated flux-B, the core loss-W_B is solved by the [19]. As the result, the W_B becomes about 9W. Output voltage-V_1 of inverter is usually constant, then.

$$B = \frac{V_1}{\sqrt{3} \cdot 2\pi f N S} \qquad (6)$$

where V_1: line to line voltage, f: frequency, N: number of coil turn, S: cross section of core.

Solves coil resistance: The no-load testing is experimented. The inverter excites only the coil without steel plate. The input coil energy-W is lost by the core loss-W_B and the copper loss-W_1 then. W_1 is shown in $W_1 = 3I_0^2 R$, where I_0 is measured current of no-load in a phase. So, the coil resistance-R is calculated by (7).

$$W_1 = W - W_B$$
$$R = \frac{W - W_B}{3 \cdot 17.1^2} \qquad (7)$$

I_0 is 17.1A from measurement. Therefore, R becomes 0.3Ω.

Solves the steel plate loss: From the obtained resistance and core loss of primary coil, secondary loss-W_2 which means of steel plate loss is solved by (8), where I_1 is inverter output current at IH.

$$W_2 = W - (W_B + 3 \cdot I_1^2 R) \qquad (8)$$

D. Electric characteristics by IH

From measurement of inverter output power at experiment of without steel plate (that is no-load), the power was spent by the coil resistance of Litz-wire and core loss [19]. Next, we experimented the load conditions of a steel plate with parameter of air-gap, the heating energy on a steel plate at IH were calculated as shown in Fig. 13.

When the gap is widely expanded, the load heating energy will be so small, because most of the magnetic flux becomes to leakage flux. And the widest gap is mean of no-load condition. At the time, few increasing input current is taking in, because of decreasing of the permeance of magnetic circuit. Also the copper loss of the winding coil is increase. As a result, the power loss of the coil is larger than the steel load loss, and efficiency is decreasing at wide gap.

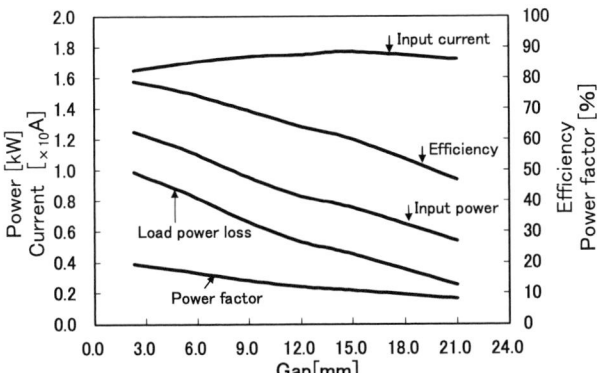

Fig. 13. Electric characteristics of primary coil and steel load.

V. WIDE METAL HEATING BY SUITABLE COIL

A. Adhesion of tile to wainscot by IH

For adhesion of wide flat plate as tile shown in Fig. 4 and cork-board etc., an extension of heating area is essentially. We propose the new coil as shown in Fig. 14 which is improved by same method in chapter II with traveling magnetic filed.

(a) Picture

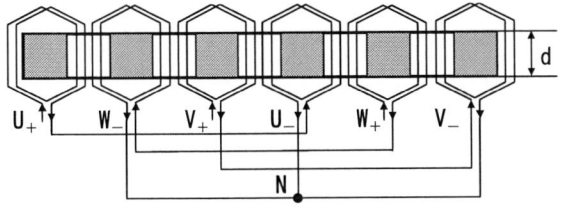

(b)Winding structure

Fig. 14. Wide linear coil for flat metal heating.
Size: 140(Length)×105(d)×40(Height)[mm].

In Fig. 14, the structure of winding is also a short-pitch winding and two poles. And the width-d of coil's length is extended, so that cross section of generated magnetic flux in core is also extended. Therefore extended uniform heating area will be accomplished. This principle is usefulness for wide metal heating.

The number of turn per one coil and the impedance per phase are 9-turn and 7.09 Ω at 20kHz. We called the coil as wide linear coil and made the prototype fitted to one's tile of Fig. 4. The size is shown in Fig. 14. When the coil surface meets to the tile at heating, the coil receives a high heating energy from the tile, because of the high heat-conductivity of tile. Therefore as the picture of Fig. 14(a), the coil has been separated by the wood boards among the tile and coil. The air-gap between coil and tile became to 6mm in the experiment. The total gap between coil and steel plate was 11mm, because of tile thickness: 5mm and air-gap: 6mm. A heating target is the steel plate of thickness 0.115mm as chapter II. In this chapter, experiment is performed at gap of 11mm only.

B. Heating characteristics by wide linear coil

We have set a heating time: 4sec. And the simulation is performed by the same conditions of table I.

In the experiment and simulation, the thermal distributions of the steel plate are shown in Fig. 15.

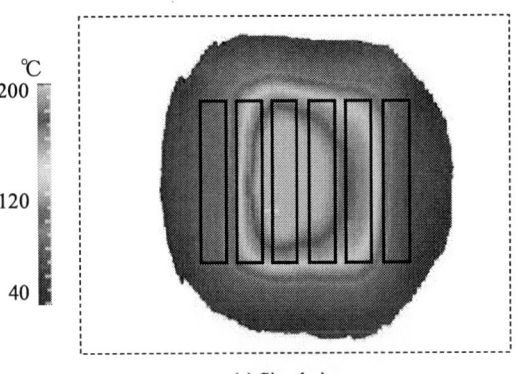

(a) Simulation.

(b) Experiment.

Fig. 15. Thermal distribution on steel plate by wide liner coil
Gap:11mm, Steel thickness:0.115mm, Heating time:4sec.

It is understood that the thermal distribution is extending gradually from center of the steel plate to over

675

the coil area. Although a little bias is observed by the stress from traveling magnetic field, the distribution is almost same in experiment and simulation. In an actual practice, bonding of a 5mm thickness tile was experimented. The heating time for melting of the hot-melt adhesive was 10sec, because the thermal discharge from steel to tile was large. The melting condition of hot-melt adhesive on the steel was observed in Fig. 16.

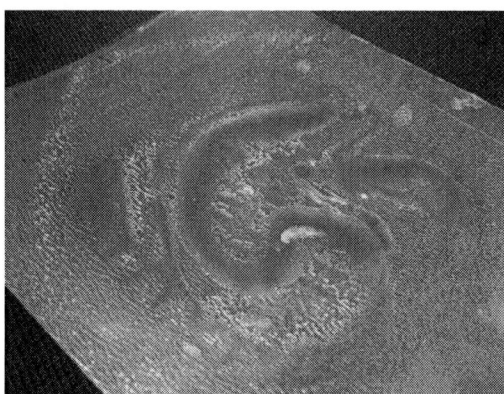

Fig. 16. Melting condition of hot-melt adhesive on steel plate at adhesion of tile.

C. Electric characteristics of the wide linear coil

By the same analytical process as the line heating, heating energy of the steel plate is obtained as shown in Table II. The heating energy to steel plate was calculated as 75% of coil input power. And coil loss was almost copper loss.

TABLE II

ELECTRIC CHARACTERISTIC IN WIDE LINEAR COIL.
(measuring output power from inverter)

coil input power at no-load	138W
coil input power at heating steel plate	689W
heating energy on steel plate	542W

VI. CONCLUSION

An uniform heating of long or wide area on steel plate for adhesion in interior construction was realized by the new coils with traveling magnetic field using short-pitch winding and 20kHz three-phase inverter. And from measurement of the output power of inverter, the heating energy to steel plate was calculated simply. Consequently, the expected characteristics on the uniform heating of steel plate were realized.

In future, a precise analysis and experiment of the IH in various loads will be performed by obtaining circumstantial various parameters (Relative permeability, resistivity and heat conduction coefficient at particular temperature etc.). And various IH characteristics will be organized systematically.

Many applications of the proposed method may be appeared in various assembly process or dismantlement, not only constructions.

ACKNOWLEDGMENT

For accomplishing the investigation, Mr. K. Suzuki of president the SAIHIT Corporation Japan, leading member of The Consortium of Dismantlable Adhesion Japan and study group of The ALLOVER method Japan cooperated with us kindly. We owe them a lot.

REFERENCES

[1] T. Kobari, S. Motegi, S. Yoshimura, H. Tomita, "The study on a exciting coil for adhesion system with induction heating", *Proc. of IEEJ- JIASC2002*, Kagoshima (Japan), Aug. 2002, p.1056.

[2] Y. Arakawa, S.Motegi, S. Yoshimura, H. Tomita, "The interior finishing material construction method by induction heating", *Proc. of IEEJ-JIASC2002*, Kagoshima (Japan), Aug. 2002, p.1057.

[3] S. Obata, H. Tomita, "Numerical calculations of metal foil induction heating and analysis of edge bake phenomena", *IEEJ Trans. FM*, Vol.123, No.10, 2003, pp.1002-1009.

[4] S.Obata, H.Tomita, M.Tanimitsu, "Theoretical analysis of rectangular metal foil heating applied a long rectangular induction coil", *Proc. of IPEC-Niigata 2005*, S38-1, Niigata (Japan), 2005, pp.1327-1332.

[5] S.Obata, T.Sekine, H.Tomita, "Hot-melt adhesive method using metal foil induction heating and performance of long E-type core head", *11th European Conference on Power Electronics and Applications (EPE2005)*, Dresden (Germany), Sep. 2005, No.283.

[6] H.Tomita, "The advanced adhesion and joint technologies, No.5 Future adhesives", Book, N.G.T. Publishing Co., 2000, pp.73-79.

[7] Hirata, "The ALLOVER method and the ALLOVER tape", *Japan Energy & Technology Intelligence*, Vol.50, No.11, 2002, pp.125-127.

[8] M. Tanimitsu, Qi Wu-Bo, S. Yoshimura, H. Tomita, "Metal bonding method for dismantlable adhesion using electromagnetic induction heating", *Proc. of 6th Japan-France congress on mechatronics & 4th Asia-Europe congress on mechatronics*, Japan, 2003, pp.576-579.

[9] H. Tomita, "A dismantlable adhesion method using thermoplastic or thermosetting glue", *Jounal of the adhesion society of Japan*, Vol.39, No.7, 2003, pp.271-278.

[10] M. Tanimitsu, H. Tomita, S. Yoshimura, S. Obata, Qi Wu-Bo, "A recycle technique with dismantlable adhesion using electromagnetic induction heating", *Proc. of the First Asia International Symposium on Mechatronics Theory*, Method and Application, China, 2004, pp.247-252.

[11] H.tomita, S.Obata, M.Tanimitsu, "Recyclable adhesion method using induction heating with hot-melt adhesive tape", *proc. of IPEC-Niigata 2005*, S38-3, Niigata (Japan), 2005, pp.1341-1345.

[12] T. Sekine, H. Tomita, S. Obata, S. Yoshimura, "Study on induction heating adhesion device using traveling magnetic field", *Journal of the Institute of Power Electronics*, Vol.31, 2005, pp.29-35.

[13] T. Sekine, H. Tomita, S. Yoshimura, S. Obata, "Dismantlable adhesion using induction heating with traveling magnetic field", *Proc. of IEEJ- JIASC2005*, Vol.1, No.21, Fukui (Japan), Aug. 2005, pp.119-122.

[14] H. Tomita, T. Sekine, S. Obata, "Induction heating using traveling magnetic field and three-phase high-frequency inverter", *11th European Conference on Power Electronics and Applications (EPE2005)*, Dresden (Germany), Sep. 2005, No. 371.

[15] T. Sekine, H. Tomita, S. Obata, Y. Saito, "An induction heating method with traveling magnetic field for long structure metal", *Proc. of IEEJ- JIASC2006*, Nagoya (Japan), Aug. 2006, 1-38.

[16] Oszka'r Bi'ro', Kurt Preis, "On the use of the magnetic vector potential in the finite element analysis of three-dimensional eddy current", *IEEE Trans. Magn.*, Vol.25, No.4, pp.3145-3159, 1989.

[17] P. Robert, M. Ito, T. Takahashi, "Numerical solution of three dimensional transient eddy current problems by the A-ϕ method", *IEEE Trans. Magn.*, Vol.28, pp.1166-1169, 1992.

[18] K. Ishibashi, "A least residual approach for 3D eddy current analysis by BEM", *IEEE Trans. Magn.*, Vol.29, pp.1512-1515, 1993.

[19] Product Catalogue JFCS-5-G, *Ferrite cores for power supply EMI prevention / EMC & Pulse transformer*, JFE FERRITE CORP, 2005.

Three-Phase (LC)(L)-Type Series-Resonant Converter with Capacitive Output Filter

M. Almardy, and A.K.S. Bhat

Department of Electrical and Computer Engineering, University of Victoria,
Victoria, BC, V8W 3P6, Canada

Abstract–This paper presents a three-phase (LC)(L)-type DC-DC series-resonant converter with capacitive output filter. Operation of the converter has been presented using the operating waveforms and equivalent circuit diagrams during different intervals. An approximate analysis is used, and design procedure is presented with a design example. SPICE simulation results for the designed converter are shown for input voltage and load variations. Major advantages of this converter are the leakage and magnetizing inductances of the high-frequency transformer are used as part of resonant circuit and the output rectifier voltage is clamped to the output voltage. Also, the converter operates in soft-switching for the inverter switches with a narrow frequency control range and the tank current decreases with the load current.

Index Terms– Approximate analysis, LCL-type, three-phase, zero-voltage switching.

I. INTRODUCTION

Single-phase high-frequency (HF) transformer isolated dc-to-dc resonant converters have been proposed and analyzed by several authors [1,2]. However, for medium to high power levels, resonant converters using a single-phase HF transformer isolation face severe component stresses. Therefore, three-phase HF transformer isolated dc-to-dc converters have been proposed [3-10] and they have numerous advantages over single-phase HF transformer isolated converters.

Fixed-frequency operation [3-7] can not maintain zero-voltage switching (ZVS) for wide variation in load and supply voltage changes. Therefore, several three-phase dc-dc resonant converters with three-phase HF transformer isolation using variable frequency control are reported in [5-10]. Three-phase LCC-type resonant converters with inductive output filters have been proposed in [6-8]. Three-phase LCC-type resonant DC-DC converter with capacitive output filter was proposed in [5,9]. A multiphase topology of the dc-to-dc series-resonant converter was introduced in [10]. In this topology when the operating points are close to or lower than a half of resonant frequency, the sub converters are less sensitive to the switching frequency.

An ac-to-dc converter employing three-phase modified series-parallel resonant inverter (MSPRC) operating in high input line power factor is proposed in [8]. The proposed converter achieves ZVS for full-load to half-load. However, the variation in switching frequency to regulate the output is high.

Single-phase HF transformer isolated (LC)(L)-type SRC with capacitive output has been proposed and analyzed in the literature [11,12]. However, analysis and design of 3-phase HF transformer isolated (LC)(L)-type SRC is not available in the literature. In this paper, a three-phase (LC)(L)-type series resonant converter (SRC) with capacitive output filter (Fig. 1) is proposed. A variable frequency control with 180° wide gating pulses is adopted for regulating the output. This converter has all the advantages of a 3-phase HF transformer isolated LCC-type resonant converter: (1) Input and output ripple is six times the switching frequency reducing the input/output filter requirements. (2) Uses a single 3-phase HF transformer for isolation requiring smaller size and weight. (3) Leakage inductances of the HF transformer are used as part of resonant inductances. (4) Operates in ZVS (i.e., above resonance or lagging pf mode) for entire load range and for supply variations.

In addition, this converter has the following advantages: (1) Magnetizing inductance of the HF transformer can be profitably used as part of the resonant circuit. (2) Capacitive output filter limits the output rectifier voltage ratings to output voltage (unlike phase-shifted ZVS PWM converters). (3) The required variation in switching frequency is narrow.

The outline of this paper as follows: Section II presents the operation of the proposed converter. This is followed by the modeling and analysis of the converter in Section III. Section IV presents the converter design that is illustrated by a design example. Detailed SPICE simulation results for the designed converter are given in Section V and are compared with the theory.

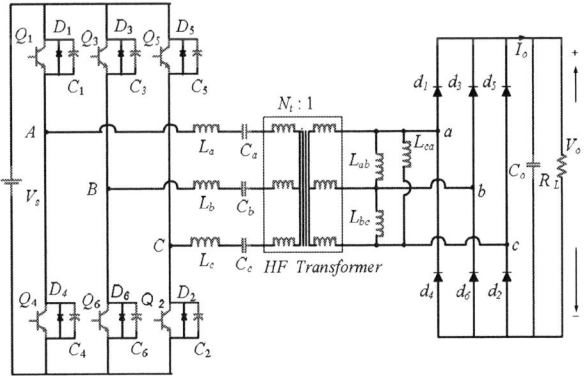

Fig. 1 Three-Phase dc-to-dc (LC)(L)-type series resonant converter with capacitive output filter ($L_a=L_b=L_c=L_{eq}$, $C_a=C_b=C_c=C_s$, $L_{ab}=L_{bc}=L_{ca}$).

This work was partly supported by a research grant from NSERC, Canada.

978-1-4244-0644-9/07/$25.00 ©2007 IEEE

II. CONVERTER OPERATION

The full-load operating waveforms of the proposed 3-ϕ (LC)(L)-type SRC (Fig. 1) with 180^0 wide gating pulse scheme are shown in Fig. 2. Different devices conducting during different intervals of operation are also marked in Fig. 2. The converter is operating in above resonance with all the switches turned-on with ZVS. In a full cycle (i.e., one switching period), there are 12 intervals of operation and the devices conducting during the first half-period and second half-period are symmetrical. Therefore, operation during one half-period is considered and the equivalent circuit models for one half-period are shown in Fig. 3. The operation of the converter during different intervals can be understood by referring to the waveforms shown in Fig. 2 and the equivalent circuits of Fig. 3.

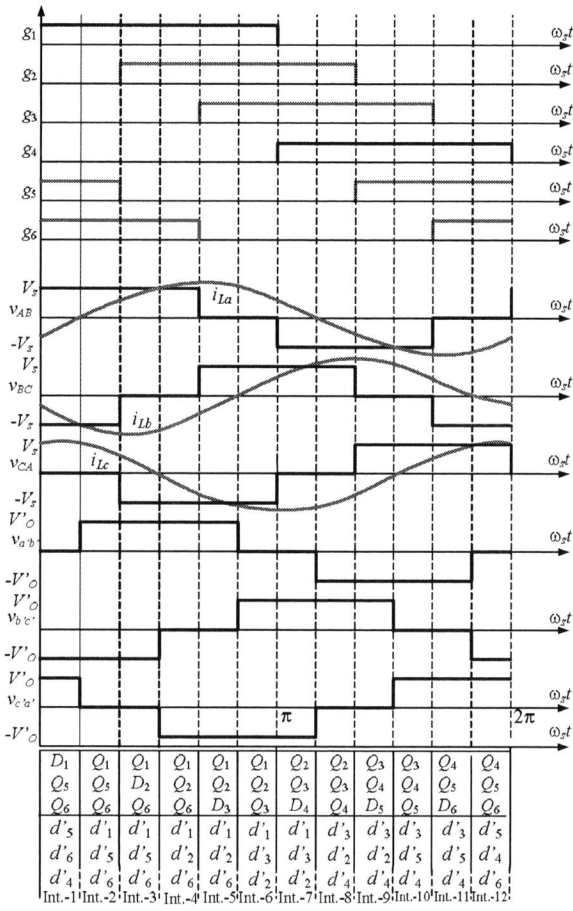

Fig. 2 Operating waveforms of three-phase (LC)(L)-type series-resonant dc-dc converter (Fig. 1) using 180° wide gating pulses.

III. MODELING AND ANALYSIS OF THE CONVERTER

A. Assumptions

In the modeling and analysis, the following assumptions are made:
1. The switches, diodes, inductors, and capacitors used are ideal.
2. Only fundamental components of the waveforms are used in the analysis, effect of higher order harmonics is neglected.

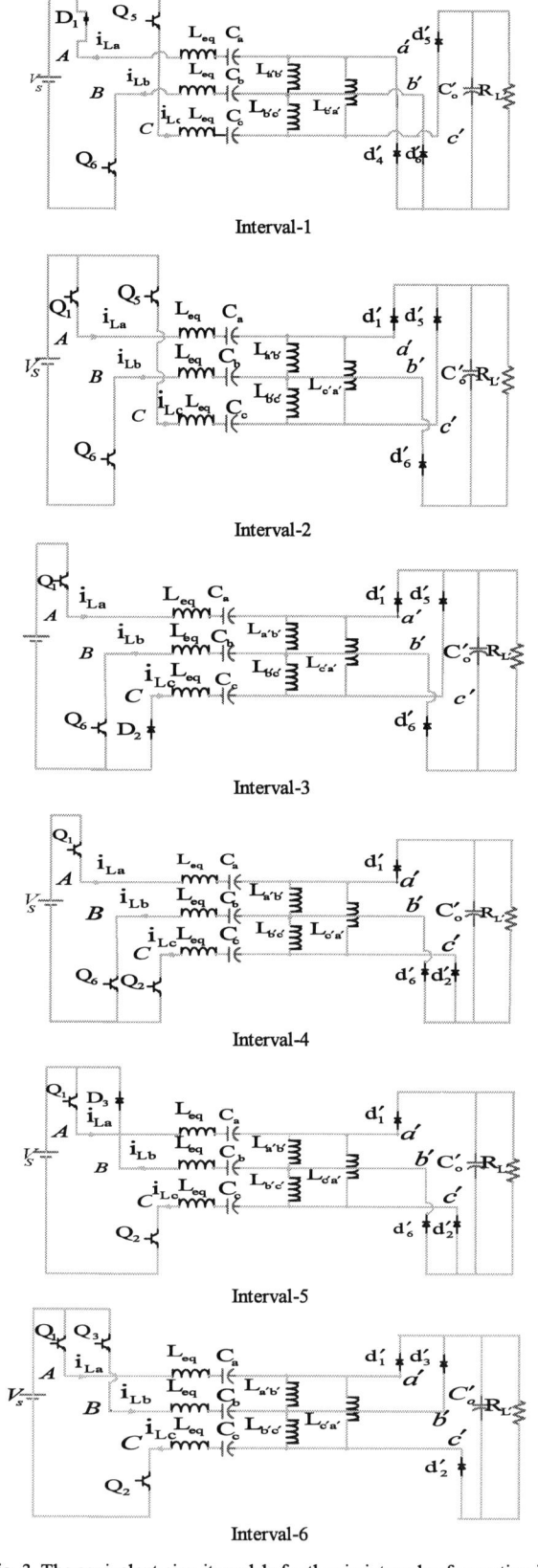

Fig. 3. The equivalent circuit models for the six intervals of operation in one HF half-period with 180° gating pulse control for the waveforms shown in Fig. 2.

3. The input and output voltages are assumed to be constant without any ripple.

4. All the three phases are identical and the following relations are valid: $L_a = L_b = L_c = L_{eq}$, $C_a = C_b = C_c = C_s$, $L_{ab} = L_{bc} = L_{ca}$, $L_p = L_{a'N} = L_{a'b'}/3$.

5. The magnetizing inductor is considered as part of the parallel resonant inductor.

B. Modeling

The proposed converter shown in Fig. 1 is analyzed using the approximate complex ac circuit analysis [1,7]. Based on the waveforms of Fig. 2 and since the converter is operated with 180^0 wide gating pulse control scheme, typical waveforms of the converter for one phase can be drawn as shown in Fig. 4. Since a capacitive output filter is used, $v_{a'b'}$ can be considered as a quasi-square-wave of amplitude V'_o ($= N_t V_o$). The three-phase full wave output bridge rectifier stage in Fig. 1 can be equivalently considered as a combination of two three-phase half wave rectifiers. Delta-to-Wye transformation is used on the secondary side. All the components are reflected to the primary side to give the per-phase equivalent circuit as shown in Fig. 5(a). Once the ac resistance, R_{ac} seen by the inductor $L_{a'N}$ ($= L_p$) is derived, the per-phase phasor equivalent circuit model is shown in Fig. 5(b) can be drawn. In the per-phase equivalent circuit model shown in Fig. 5(b), the resonant circuit input voltage is represented by the fundamental component of the line-to-neutral square-wave voltage across AN, that is converted from line-to-line voltage. Assuming that the input current to the rectifier is a sinusoidal current and the rectifier input voltage is represented by the fundamental component of the quasi-square-wave input voltage across the rectifier bridge, the derivation [5] of the ac resistance R_{ac} is given in Appendix 1 and is given by

$$R_{ac} = V_{a'N1}/I_{a',rms} = (6/\pi^2)R'_L \qquad (1)$$

C. Base values and Normalization

All the equations presented in the analysis are normalized using the following base values:
Base voltage, $V_B = V_{s,min}$; base impedance, $Z_B = R'_L$; base current, $I_B = V_{s,min}/R'_L$.

All the normalized quantities are denoted by an extra subscript "pu". The normalized reactances are:

$$X_{Leq,pu} = QF, \quad X_{Cs,pu} = Q/F, \quad X_{Lp,pu} = QF(L_p/L_{eq}) \quad \text{p.u.} \qquad (2)$$

where,

$$Q = \omega_{rs}L_{eq}/R_L = \sqrt{L_{eq}/C_s}/R'_L \qquad (3)$$

The ratio of switching frequency f_s to series resonance frequency f_{rs} is given by:

$$F = \omega_s/\omega_{rs}, \quad \omega_{rs} = 1/\sqrt{L_{eq}C_s}, \quad \omega_s = 2\pi f_s \qquad (4)$$

The converter gain,

$$M = V'_{opu} = V'_o/V_B, \qquad V'_o = N_t V_o \qquad (5)$$

where N_t : 1 is the transformer turns ratio.

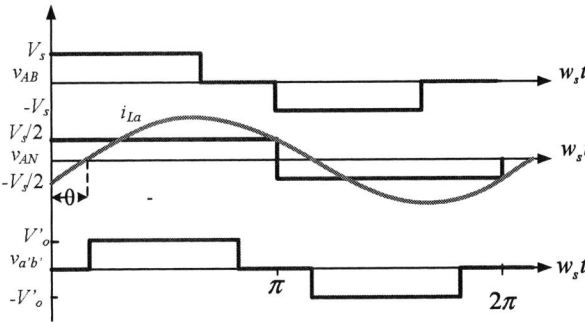

Fig. 4 Typical operating waveforms for one phase of the three phases at the output of the converter.

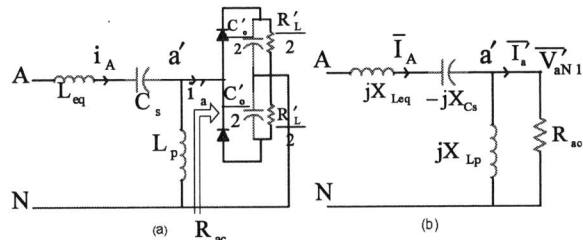

Fig. 5: (a) Equivalent circuit for one of the three phases at the output of the converter. (b) The per-phase (line–to–neutral) phasor equivalent circuit of the three-phase converter (Fig. 1).

D. Converter Gain and Component Stresses

The fundamental components of the inverter line to line voltages can be expressed as [5]

$$v_{AB} = (2\sqrt{3}/\pi)(V_s)\sin(\omega_s t + \pi/6) \qquad (6)$$

$$v_{BC} = (2\sqrt{3}/\pi)V_s \sin(\omega_s t - \pi/2) \qquad (7)$$

$$v_{CA} = (2\sqrt{3}/\pi)V_s \sin(w_s t - 7\pi/6) \qquad (8)$$

Therefore, the RMS values of the fundamental components of the inverter line-to-line, and line-to-neutral voltages V_{AB} and V_{AN} can be expressed as

$$V_{AB1} = (\sqrt{6}/\pi)V_s \qquad (9)$$

$$V_{AN1} = (\sqrt{2}/\pi)V_s \qquad (10)$$

Therefore, the output voltage in per unit referred to the primary side using equations (A2) and (10):

$$V'_{opu} = (V'_{a'N1}/V_{AN1}) \qquad \text{p.u.} \qquad (11)$$

But from Fig. 5(b),

$$\frac{\overline{V}_{a'N1}}{\overline{V}_{AN1}} = \frac{1}{1 + \dfrac{X_{Leq}}{X_{Lp}} - \dfrac{X_{Cs}}{X_{Lp}} + j\left(\dfrac{X_{Leq}}{R_{ac}} - \dfrac{X_{Cs}}{R_{ac}}\right)}$$

$$\frac{\overline{V}_{a'N1}}{\overline{V}_{AN1}} = \frac{6/\pi^2}{D_1 + jD_2} \qquad \text{p.u.} \qquad (12)$$

Therefore, the converter gain is

$$V'_{o,pu} = \frac{6/\pi^2}{(D_1^2 + D_2^2)^{1/2}} \qquad \text{p.u.} \qquad (13)$$

where,

$$D_1 = (\frac{6}{\pi^2})[1 + \frac{L_{eq}}{L_p}(1 - \frac{1}{F^2})] \quad , \quad D_2 = Q(F - \frac{1}{F}) \qquad (14)$$

The impedance looking into terminals A and N is determined as

$$Z_{AN} = j(X_{Leq} - X_{Cs}) + \frac{R_{ac}(jX_{Lp})}{R_{ac} + jX_{Lp}} \qquad (15)$$

Therefore, the impedance in per unit is

$$Z_{AN,pu} = \frac{B_1 + jB_2}{B_3} \qquad p.u. \qquad (16)$$

where,

$$B_1 = (\frac{6}{\pi^2})[QF(\frac{L_p}{L_{eq}})]^2, B_2 = QF(\frac{L_p}{L_{eq}}) + Q(F - \frac{1}{F})B_3,$$

$$B_3 = 1 + [(\frac{\pi^2}{6})QF(\frac{L_p}{L_{eq}})]^2 \qquad (17)$$

The inverter output peak current through the inverter switches [5] is

$$I_{Appu} = I_{Leqp,pu} = \frac{2}{\pi |Z_{AN,pu}|} \qquad \text{p.u.} \qquad (18)$$

The initial inverter current is given by

$$I_{Leq0,pu} = I_{Leqp,pu} \sin(-\theta) \qquad \text{p.u.} \qquad (19)$$

where,

$$\theta = \tan^{-1}(\frac{B_2}{B_1}) \qquad \text{rads.} \qquad (20)$$

For operation in lagging pf mode, the initial current should be of negative value. The peak voltage across the inductors ($L_{a'b'}$ & L_{eq}) and across the capacitor C_s are

$$V_{La'b'p,pu} = V'_{opu} \qquad \text{p.u.} \qquad (21)$$
$$V_{Leqp,pu} = I_{Leqp,pu}X_{Leq,pu} \qquad \text{p.u.} \qquad (22)$$
$$V_{Csp,pu} = I_{Leqp,pu}X_{Cs,pu} \qquad \text{p.u.} \qquad (23)$$

The peak current through the inductor $L_{a'b'}$ can be expressed as

$$I_{La'b'ppu} = \frac{V_{La'b'p,pu}}{3X_{Lp,pu}} \qquad \text{p.u.} \qquad (24)$$

IV. CONVERTER DESIGN

A simple design procedure is illustrated by a design example. Design curves (Fig. 6) obtained from the analysis is used to find the optimum point of operation of the designed converter with the following specifications:

Minimum input voltage, $V_{s,min}$ = 110 V ; maximum input voltage, $V_{s,max}$ = 130 V ; nominal input voltage $V_{s,nom}$ = 120 V. Output power P_o = 1000 W; output load voltage V_o = 48 V. Inverter switching frequency f_s = 100 KHz.

All the design curves obtained for the optimum value of inductor ratio L_{eq}/L_p = 0.1 are plotted with variation in switching frequency ratio F, for varying values of Q. Converter must operate in lagging power factor (above resonance) mode in order to reduce switching losses and to take this advantage, the initial current at the output of the inverter must be negative. Frequency modulation is used to regulate the output voltage (i.e., constant gain) for load change from full-load to 10% load. The kVA rating of the resonant tank circuit decreases as Q_s decreases for a given F. It was also observed that as the value of Q of the full-load is increased, the decrease in the peak inverter output current with load current is large. However, this decrease is small for $Q > 4$. The following compromised values are chosen: L_{eq}/L_p = 0.1, F = 1.05, Q = 4 (Q at full load). The component values and their ratings can be calculated from the analysis based on these design values. The load voltage reflected to the primary side is V'_0 = 91.947 V. Therefore, the transformer ratio that is required to obtain the required output voltage of 48 V is $N_t : 1$ = 1.916. The calculated values for the designed converter are: L_{eq} = 56.5μH, C_s = 49.42 nF. Since L_{eq}/L_p = 0.1, L_p = 565 μH. The RMS and average current through the switches are 5.583 A and 3.328 A, respectively; and the average current through anti-parallel diode of the switch is 0.298 A. The average current for rectifier diodes is 6.944 A. The switches used are MOSFETs IRFB4103s ($R_{DSon}@25^0$ = 139 mΩ and I_D = 17 A, fall time t_f = 5.4 ns), and the semiconductors used in the output rectifier bridge are schottky diodes 15TQ060 ($I_{F(av)}$ = 15 A, V_R = 60 V, and V_d = 0.56 V).

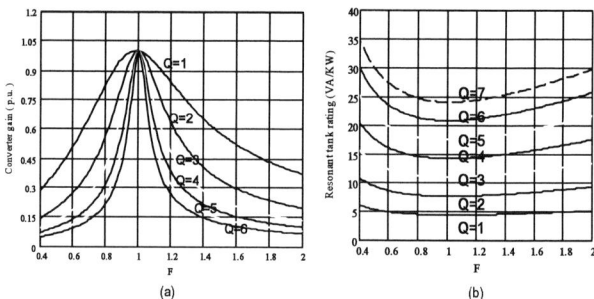

Fig. 6 Design curves obtained for L_{eq}/L_p = 0.1. (a) Converter gain versus normalized switching frequency F. (b) Total kVA/kW rating of tank circuit versus F.

V. SPICE SIMULATION RESULTS

The component values obtained from the design are used for SPICE simulation of the three-phase (LC)(L)-type series-resonant dc-dc converter. The behavior of the converter for variation in load and input voltage has been evaluated from the analysis and simulation. The simulation waveforms obtained for the converter with minimum input voltage ($V_{s,min}$ = 110 V) and for three load conditions (full-load, half-load and 10%-load) are shown in Figs. 7 to 9.

680

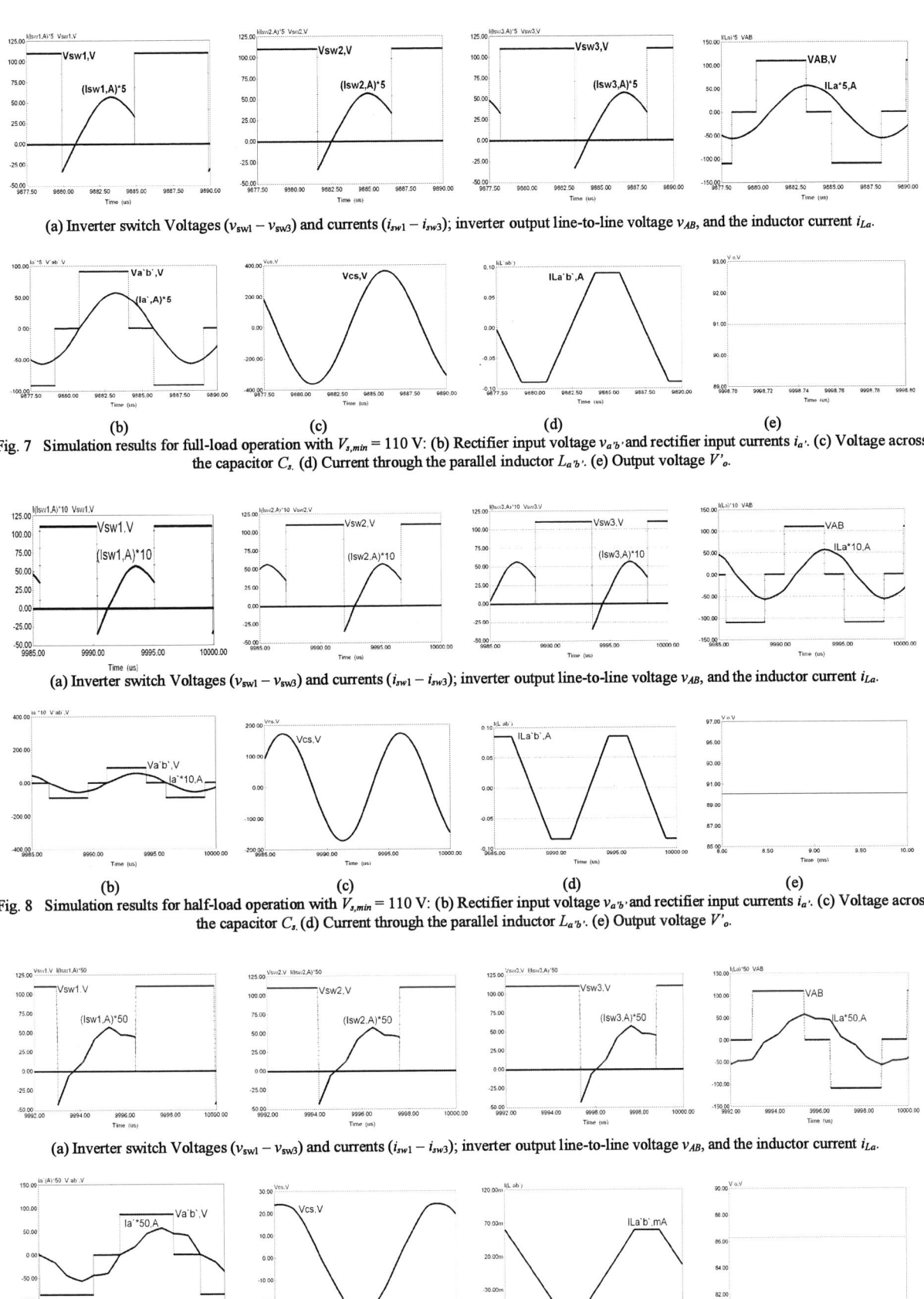

(a) Inverter switch Voltages ($v_{sw1} - v_{sw3}$) and currents ($i_{sw1} - i_{sw3}$); inverter output line-to-line voltage v_{AB}, and the inductor current i_{La}.

(b) (c) (d) (e)

Fig. 7 Simulation results for full-load operation with $V_{s,min} = 110$ V: (b) Rectifier input voltage $v_{a'b'}$ and rectifier input currents $i_{a'}$. (c) Voltage across the capacitor C_s. (d) Current through the parallel inductor $L_{a'b'}$. (e) Output voltage V'_o.

(a) Inverter switch Voltages ($v_{sw1} - v_{sw3}$) and currents ($i_{sw1} - i_{sw3}$); inverter output line-to-line voltage v_{AB}, and the inductor current i_{La}.

(b) (c) (d) (e)

Fig. 8 Simulation results for half-load operation with $V_{s,min} = 110$ V: (b) Rectifier input voltage $v_{a'b'}$ and rectifier input currents $i_{a'}$. (c) Voltage across the capacitor C_s. (d) Current through the parallel inductor $L_{a'b'}$. (e) Output voltage V'_o.

(a) Inverter switch Voltages ($v_{sw1} - v_{sw3}$) and currents ($i_{sw1} - i_{sw3}$); inverter output line-to-line voltage v_{AB}, and the inductor current i_{La}.

(b) (c) (d) (e)

Fig. 9 Simulation results for 10% load operation with $V_{s,min} = 110$ V: (b) Rectifier input voltage $v_{a'b'}$ and rectifier input currents $i_{a'}$. (c) Voltage across the capacitor C_s. (d) Current through the parallel inductor $L_{a'b'}$. (e) Output voltage V'_o.

681

The simulation waveforms obtained of the converter at maximum input voltage $V_{s,max}$ = 130 V and three load conditions (full-load, half-load and 10%-load) are given in Figs. 10 to 12. In all these waveforms, switching frequency (f_s) was varied to keep the load voltage approximately the same as the full-load value. These results verify that the proposed converter operates with ZVS turn-on for all the switches from full load to light load condition. The variation of the frequency required for load regulation is narrow.

Tables I and Table II compare the results obtained from the simulation and theory with $V_{s,min}$ = 110 V and $V_{s,max}$ = 130 V, respectively; for three different load conditions. The results in these tables show that the theoretical results are reasonably close to the simulation results.

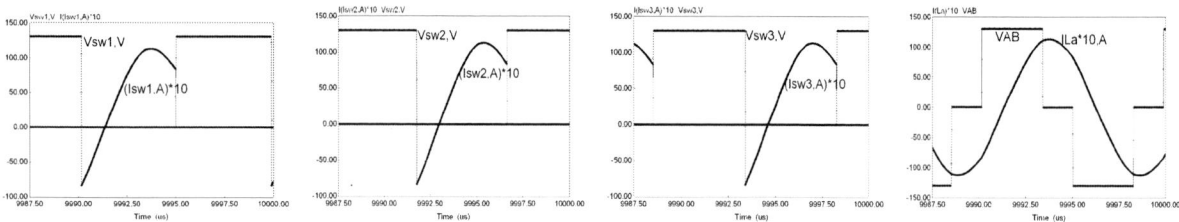

(a) Inverter switch Voltages ($v_{sw1} - v_{sw3}$) and currents ($i_{sw1} - i_{sw3}$); inverter output line-to-line voltage v_{AB}, and the inductor current i_{La}.

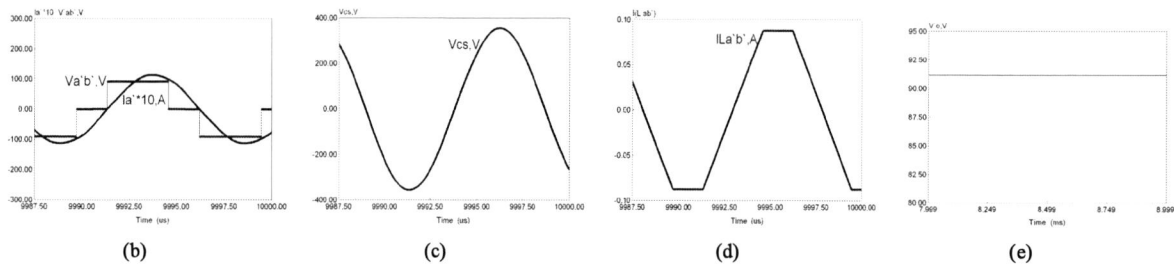

(b)   (c)   (d)   (e)

Fig. 10 Simulation results for full-load operation with $V_{s,max}$ = 130 V: (b) Rectifier input voltage $v_{a'b'}$ and rectifier input currents $i_{a'}$. (c) Voltage across the capacitor C_s. (d) Current through the parallel inductor $L_{a'b'}$. (e) Output voltage V'_o.

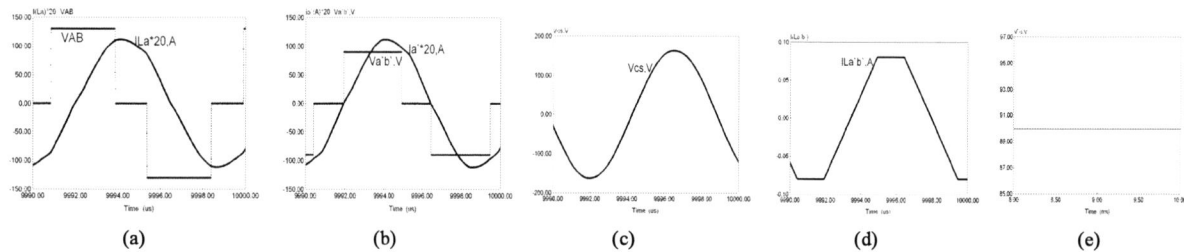

(a)   (b)   (c)   (d)   (e)

Fig. 11 Simulation results for half-load operation with $V_{s,max}$ = 130 V. (a) Inverter output line-to-line voltage v_{AB} & inductor current i_{La}. (b) Rectifier input voltage $v_{a'b'}$ and rectifier input current $i_{a'}$ (c) Voltage across the capacitor C_s. (d) Current through parallel inductor $L_{a'b'}$. (e) Output voltage V'_o.

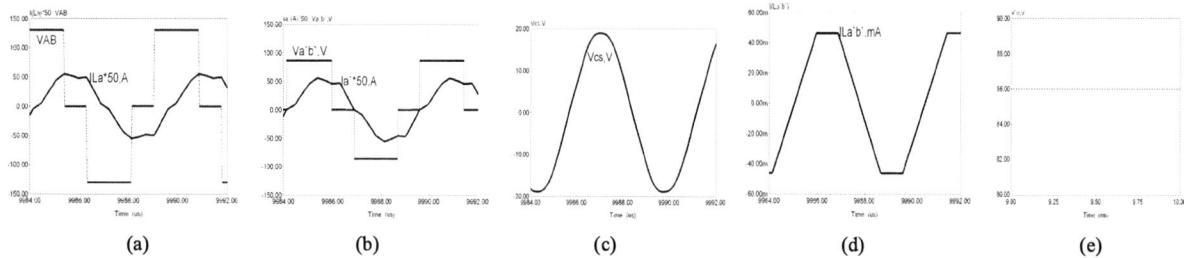

(a)   (b)   (c)   (d)   (e)

Fig. 12 Simulation results for 10% load operation with $V_{s,max}$ = 130 V. (a) Inverter output line-to-line voltage v_{AB} & inductor current i_{La}. (b) Rectifier input voltage $v_{a'b'}$ and rectifier input current $i_{a'}$. (c) Voltage across the capacitor C_s. (d) Current through parallel inductor $L_{a'b'}$. (e) Output voltage V'_o.

VI. CONCLUSIONS

A three-phase LCL-type series resonant converter with capacitive output filter has been proposed. The converter has been analyzed using the complex ac circuit method.

Based on the analysis, design curves have been obtained and a design example was given. The designed converter was simulated using SPICE simulation program. Various waveforms obtained for changes in load and input voltage have been presented. The simulation results show ZVS turn-on for all the switches for the complete load and supply range. The proposed converter has shown narrow frequency range (compared to LCC-type) to regulate the output voltage. The tank current decreases with the load current. An experimental converter is under construction and results will be presented in the future.

TABLE I COMPARISON OF THE ANALYSIS AND SPICE SIMULATION RESULTS WITH VARIABLE FREQUENCY CONTROL FOR THE 1 kW, 3-ϕ LCL-TYPE SRC WITH CAPACITIVE OUTPUT FILTER, FOR DIFFERENT LOAD CONDITIONS WITH $V_{s,min}$ = 110 V.

Parameters	Full-Load		Half-Load		10% Load	
	Analysis	Simulation	Analysis	Simulation	Analysis	Simulation
Switching frequency, f_s (kHz)	100	100	104.8	104.8	144.2	144.2
Load, R'_L (Ω)	8.45	8.45	16.91	16.91	84.54	84.54
Output voltage, V'_o (V)	91.947	90.954	91.96	90.05	91.96	86.31
Peak current, I_{Leqp} (A)	11.39	11.338	5.698	5.65	1.145	1.138
RMS current, I_{Leqrms} (A)	8.05	7.944	4.03	3.877	0.809	0.776
Peak voltage, V_{Csp} (V)	366.84	362.82	175.13	171.49	25.58	24.08
RMS voltage, V_{Csrms} (V)	259.398	256.706	123.837	120.67	18.086	17.377
Peak voltage across parallel inductor, $V_{La'b'p}$(V)	91.947	90.954	91.96	90.05	91.96	86.31
Peak current through parallel inductor, $I_{La'b'p}$ (A)	0.086	0.089	0.082	0.084	0.06	0.058
RMS current through parallel inductor, $I_{La'b'rms}$(A)	0.061	0.066	0.058	0.062	0.042	0.043

TABLE II COMPARISON OF THE ANALYSIS AND SPICE SIMULATION RESULTS WITH VARIABLE FREQUENCY CONTROL FOR THE 1 kW, 3-ϕ LCL-TYPE SRC WITH CAPACITIVE OUTPUT FILTER, FOR DIFFERENT LOAD CONDITIONS WITH $V_{s,max}$ = 130 V.

Parameters	Full-Load		Half-Load		10% Load	
	Analysis	Simulation	Analysis	Simulation	Analysis	Simulation
Switching frequency, f_s (kHz)	102.6	102.6	110.4	110.4	183	183
Load, R'_L (Ω)	8.454	8.454	16.908	16.908	84.54	84.54
Output voltage, V'_o (V)	91.947	91.139	91.947	89.941	91.947	86.032
Peak current, I_{Leqp} (A)	11.391	11.265	5.694	5.55	1.142	1.095
RMS current, I_{Leqrms} (A)	8.055	7.944	4.026	3.914	0.808	0.779
Peak voltage, V_{Csp} (V)	357.422	354.165	166.154	162.735	20.101	18.90
RMS voltage, V_{Csrms} (V)	252.736	249.464	117.489	113.181	14.213	13.55
Peak voltage across parallel inductor, $V_{La'b'p}$(V)	91.947	91.139	91.947	89.941	91.947	86.032
Peak current through parallel inductor, $I_{La'b'p}$ (A)	0.084	0.087	0.078	0.08	0.047	0.046
RMS current through parallel inductor, $I_{La'b'rms}$ (A)	0.059	0.064	0.055	0.06	0.033	0.035

APPENDIX 1
DERIVATION OF AC RESISTANCE R_{ac} [5]

The rectifier input voltage $v_{a'N}$ is a square wave of amplitude voltage $V'_o/2$ (Fig. 5). The peak and RMS values of the fundamental component are given by

$$v_{a'N1p} = 2V'_o / \pi \qquad (A1)$$

$$V_{a'N1} = \sqrt{2}V'_o / \pi \qquad (A2)$$

Assuming approximate sinusoidal input currents ($i_{a'}$, $i_{b'}$ & $i_{c'}$) to the rectifier, the average output load current referred to the primary is [5]

$$I'_o = (6/2\pi) \int_{\pi/3}^{2\pi/3} \sqrt{2}I_{a',rms} \sin(\omega_s t) d(\omega_s t)$$
$$= (3\sqrt{2}/\pi)I_{a',rms} \qquad (A3)$$

Therefore, RMS value of the rectifier input current is

$$I_{a',rms} = \left[\pi/(3\sqrt{2})\right]I'_o \qquad (A4)$$

Therefore, the ac resistance seen by the inductor L_p is

$$R_{ac} = V_{a'N1}/I_{a',rms} = (6/\pi^2)R'_L \qquad (A5)$$

REFERENCES

[1] R. L. Steigerwald, "A comparison of half-bridge resonant converter topologies," *IEEE Trans. Power Electronics*, vol. 3, no.2, pp. 174-182, April 1988.

[2] A. K. S. Bhat, "Analysis and design of a series-parallel resonant converter," *IEEE Trans. On Power Electronics*, vol. 8, no.1, pp. 1-11, January 1993.

[3] R. W. A. A. De Doncker, D. M. Divan, and M. H. Kheraluwala, "A three-phase soft-switched high-power-density dc/dc converter for high-power applications," *IEEE Trans. Industry Applications*, vol. 27, no. 1, pp. 63-73, January/February 1991.

[4] A. R. Prasad, P. D. Ziogas, and S. Manias, "A three-phase resonant PWM DC-DC converter," in *Proceeding of the IEEE Power Electronics Specialists conf.*, 1991, pp. 463-473.

[5] A. K. S. Bhat, "A three-phase series-parallel resonant converter with capacitive output filter," *IEEE Second International Power Electronics and Motion Control Conference*, Hangzhou, China, pp. 626-631, Nov. 1997.

[6] A. K. S. Bhat and R. L. Zheng, "A three-phase series-parallel resonant converter — Analysis, Design, Simulation, and Experimental Results," *IEEE Trans. Industry Applications*, vol. 32, no. 4, pp. 951-960, July/August 1996.

[7] A. K. S. Bhat and R. L. Zheng, "Analysis and design of a three-phase LCC-type resonant converter," *IEEE Trans. On Aerospace and Electronic Systems*, vol. 34, no. 2, pp. 508-519, April 1998.

[8] S. S. Tanavade, M. A. Chaudhari, H. M. Suryawanshi, and K. L. Thakre, "A Three-Phase Modified Series-Parallel Resonant Converter-Analysis, Design and Simulation," *National Power Electronics Conference*, India, 2005, pp. 693-696.

[9] A. Sunil, G. E. Michael, and J. W. Michael, "Analysis and design of a new three-phase LCC-type resonant DC-DC converter with capacitor output filter," vol. 2. *IEEE PESC*, 2000, pp. 721-728.

[10] V. Nguyen, J. Dhyanchand, and P. Thollot, "A multiphase topology of series-resonant DC-DC converter."

Proceedings of power conversion International, 1985, pp. 45-53.

[11] A.K.S. Bhat, "Analysis and design of a fixed-frequency LCL-type series-resonant converter with capacitive output filter", *IEE Proceedings, Circuits, Devices and Systems*, vol. 144, no. 2, pp. 97-103, April 1997.

[12] A.K.S. Bhat, "Analysis and design of a fixed frequency LCL-type series resonant converter", *IEEE Trans. on Aerospace and Electronic Systems*, vol. 31, no. 1, 125-137, Jan. 1995.

Analysis of a Full-Bridge Inverter for Induction Heating Using Asymmetrical Phase-Shift Control under ZVS and NON-ZVS Operation

N. Yongyuth* , P. Viriya* and K. Matsuse**

* Dept. of Electrical Engineering, Faculty of Engineering, King Mongkut's Institute of Technology Ladkrabang, Bangkok, 10520, Thailand, Tel. 662-7373000 EXT. 3515, 3516 Fax. 662-3264550, E-Mail : kpviriya@kmitl.ac.th

** School of Science and Technology, Meiji University, 1-1-1 Higashimita, Tama-ku, Kawasaki-shi 214, Japan, Tel. +81-44-934-7293, Fax. 03(3296)4339, E-Mail : matsuse@ics.meiji.ac.jp

Abstract—**This paper presents a detailed analysis of circuit operation under ZVS and NON-ZVS switching conditions in a high-frequency full-bridge inverter for induction heating, using the principle of asymmetrical phase-shift control over a wide control range both in positive and negative directions. A variety of modes of circuit operation with the voltage and current equations during phase-shift power control under the operating conditions of ZVS and NON-ZVS are analyzed as a first step and the output voltage and current waveforms are obtained by MATLAB program. These waveforms will be analyzed by Fourier analysis which can lead further to the calculation of ac output power P_O , dc input power P_d, and hence the conversion efficiency η of the full-bridge inverter. The analysis results shows that the control ranges of ac output power P_o and dc input power P_d are limited by the occurrence of NON-ZVS operating condition, which changes according to the switching frequency f_s.**

Index Terms — **full-bridge, induction heating, asymmetrical phase-shift, inverter, ZVS, NON-ZVS**

I. INTRODUCTION

The concept of this paper is achieved by considering further as a continuous idea from the research work [1-6] starting from a single phase full-bridge phase-shift power control under the switching conditions of ZVS and NON-ZVS for induction heating. The circuit operation and its corresponding output voltage and current waveforms for phase-shift power control at phase-shift $\phi = 40°$ are illustrated in the upper-part of Fig. 1 with two repeated periods of powering (P), free-wheeling (F), and regenerating (R). Then, reducing the phase-shift ϕ from $\phi = 40°$ to zero phase-shift $\phi = 0°$, the output waveform with only two repeated periods of powering (P), and regenerating (R) is obtained with the operating circuit of two free-wheeling modes (F) eliminated [3]. The output waveform with zero phase-shift ($\phi = 0°$) will be used as a mid-point between the positive and negative directions of phase-shift control as shown by the output waveforms of asymmetrical phase-shift $\phi = 40°$ and phase-shift $\phi = -40°$ in the lower part of Fig. 1. So, in this paper, the full-bridge inverter with the load of induction heating of Fig. 2 which operates under the switching conditions of ZVS and NON-ZVS during asymmetrical phase-shift

control both in positive and negative directions will be analyzed in details for various quantities of circuit parameters, since the phase-shift control range in both directions will be limited when ZVS circuit operation becomes NON-ZVS operation. These quantities are, for examples, ac output voltage v_o , ac output power P_o, dc input power P_d and inverter efficiency η , etc.

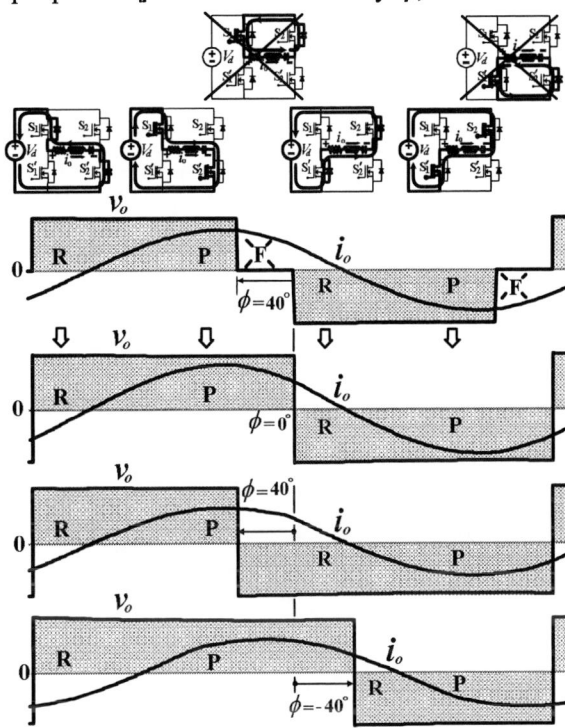

Fig. 1 Circuit operation of asymmetrical phase-shift related to that of symmetrical phase-shift full-bridge inverter

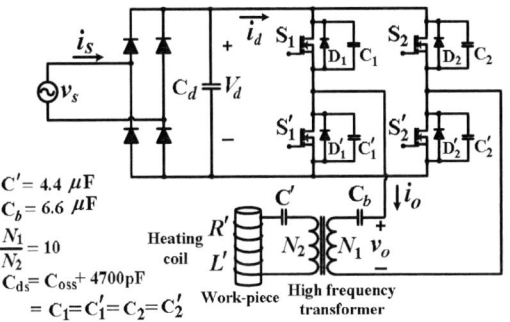

Fig. 2 Full-Bridge Inverter fed induction heating

II. ANALYSIS OF CIRCUIT OPERATION

First, we show a variety of modes of circuit operation for the main power circuit which are illustrated in Fig. 3 for the case of phase-shift control ($\phi = 0 \sim 180°$ and $\phi = 0 \sim -180°$). The circuit operation in one cycle of output voltage and current waveforms v_o, i_o is shown in 6 modes under the case of ZVS operation (Modes ①②③①'②'③'). Fig. 4 shows the linear variation of the phase difference between the fundamental output voltage $v_{o,1}$ and the front edge of square wave output voltage v_o during phase-shift control both in positive and negative directions. It can be seen that with positive phase-shift from 0° to +180°, the phase angle of the fundamental output voltage $v_{o,1}$ will move away from the font edge of output voltage v_o with an increasing leading angle. This makes the zero-crossing of output current i_o move toward the font edge of output voltage v_o, where NON-ZVS operation may occur, but in case of negative phase-shift control, the NON-ZVS operation may occur at the tailing edge of positive half-cycle of square wave voltage v_o, and in case of zero phase-shift control, NON-ZVS operation may occur both at the front and tailing edges, especially when the operating frequency is not high enough. In Fig. 5, we also show the circuit operation in one cycle of output voltage and current waveforms v_o, i_o in another 6 modes under the cases of the following NON-ZVS operation : (1) Modes

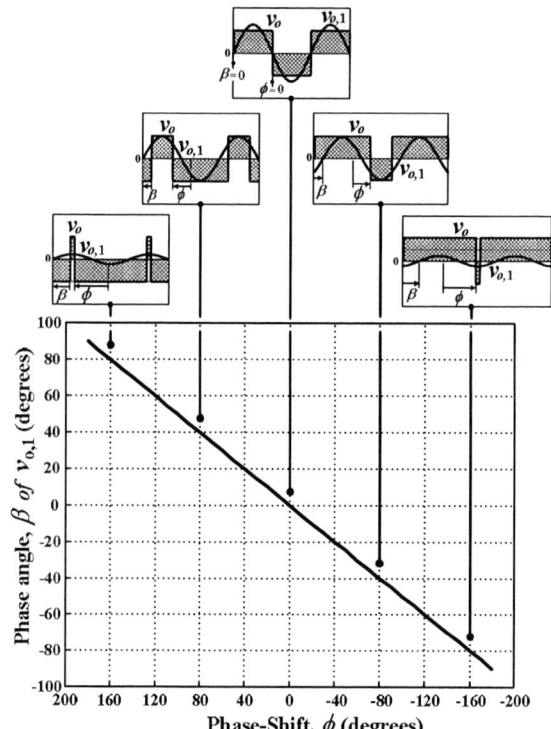

Fig. 4 Linear variation of phase angle β of $v_{o,1}$ vs. phase-shift angle ϕ

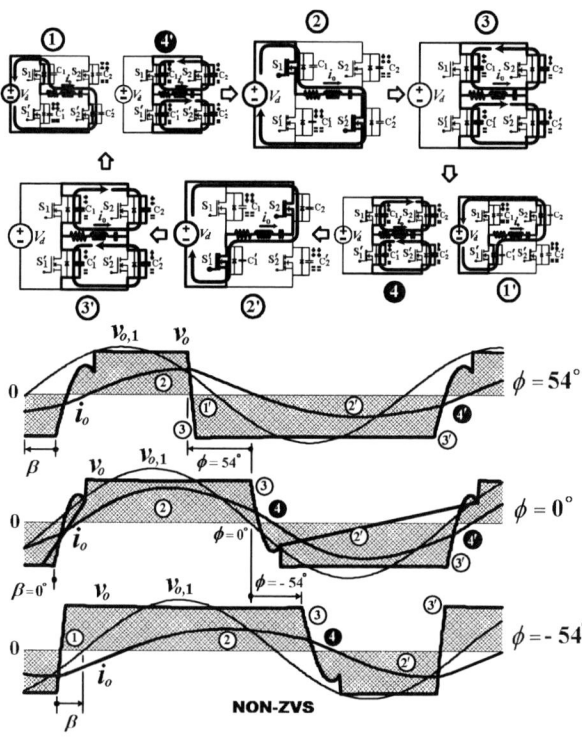

Fig. 5 Circuit operation for the case of **NON-ZVS** at phase-shift $\phi = 54°$, 0° and $-54°$

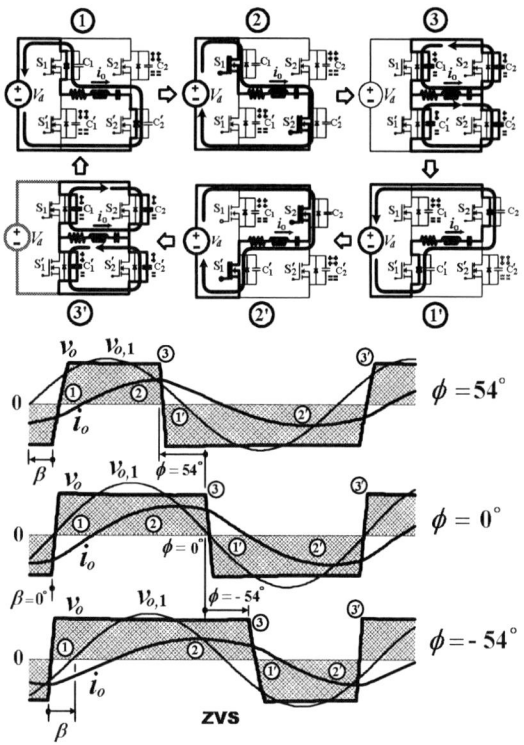

Fig. 3 Circuit operation for the case of **ZVS** at phase-shift $\phi = 54°$, 0° and $-54°$

②③①'②'③'❹ for phase-shift $\phi = 54°$ can cause a breakdown to the switch pair S_1-S_2', by considering Mode ❹. (2) Modes ②③①❷②'③'❹ for phase-shift

$\phi = 0°$ can cause a breakdown to the switch pair S_1-S_2', by considering Mode ❹ and the switch pair S_2-S_1', by considering Mode ❹. (3) Modes ①②③❹②'③' for phase-shift $\phi = -54°$ can cause a breakdown to the switch pair S_2-S_1', by considering Mode ❹. So, there are three possibilities for the full-bridge inverter switches to become breakdown (the switch pair S_1-S_2', the switch pair S_2-S_1', or both).

III. VOLTAGE AND CURRENT EQUATIONS IN EACH MODE OF CIRCUIT OPERATION

From these modes of circuit operation, various equations of output voltage v_o and output current i_o can be also calculated and obtained in the following equations :

Case (1) Equations ① ② ③ ①' ②' ③' for ZVS operation
Case (2) Equations ② ③ ①' ②' ③' ❹ for $\phi > 0°$
Case (3) Equations ② ③ ❹ ②' ③' ❹ for $\phi = 0°$
Case (4) Equations ① ② ③ ❹ ②' ③' for $\phi < 0°$

$$v_o = V_d$$
$$i_o = e^{-\alpha t}\left[\left(\frac{V_d - V - \alpha L I}{\omega_1 L}\right)\sin\omega_1 t + I\cos\omega_1 t\right] \quad \text{①②}$$

$$v_o = \left(\frac{C_1+C_2}{C_1 C_2}\right)\left(\frac{1}{\alpha^2+\omega_2^2}\right)\left[e^{-\alpha t}(A_1\sin\omega_2 t + B_1\cos\omega_2 t)+D_1\right] + (V_2-V_1)$$
$$i_o = \frac{e^{-\alpha t}}{\omega_2}\left[\left(\frac{-2V-V_1+V_2+V_1'-V_2'}{2L}-\alpha I\right)\sin\omega_2 t + \omega_2 I\cos\omega_2 t\right] \quad \text{③❹}$$

$$v_o = -V_d$$
$$i_o = e^{-\alpha t}\left[\left(\frac{-V_d - V - \alpha L I}{\omega_1 L}\right)\sin\omega_1 t + I\cos\omega_1 t\right] \quad \text{①'②'}$$

$$v_o = \left(\frac{C_1'+C_2'}{C_1' C_2'}\right)\left(\frac{1}{\alpha^2+\omega_2^2}\right)\left[e^{-\alpha t}(A_2\sin\omega_2 t + B_2\cos\omega_2 t)+D_2\right] + (V_1'-V_2')$$
$$i_o = \frac{e^{-\alpha t}}{\omega_2}\left[\left(\frac{-2V-V_1+V_2+V_1'-V_2'}{2L}-\alpha I\right)\sin\omega_2 t + \omega_2 I\cos\omega_2 t\right] \quad \text{③'❹'}$$

Where ;

V_1 : the initial value of voltage V_{C1}
V_1' : the initial value of voltage V_{Ci}
V_2 : the initial value of voltage V_{C2}
V_2' : the initial value of voltage V_{Ci}
I : the initial value of load current i_o in each mode of circuit operation
V : the initial value of load capacitor voltage in each mode of circuit operation

$$\alpha = \frac{R}{2L}$$
$$\omega_1 = \sqrt{\frac{1}{LC}-\left(\frac{R}{2L}\right)^2}$$
$$\omega_2 = \sqrt{\left(\frac{1}{LC}+\frac{1}{LC_{ds}}\right)-\left(\frac{R}{2L}\right)^2}$$

$$A_1 = \left\{RC_{ds}(V_1-V_2+V_1'-V_2')(\alpha^2+\omega_2^2)-2LI(\alpha^2+\omega_2^2) -2\alpha(V+V_1-V_2)-\alpha(C_{ds}/C)(V_1-V_2+V_1'-V_2') -\alpha C_{ds}L(V_1-V_2+V_1'-V_2')(\alpha^2+\omega_2^2)\right\}(1/4\omega_2 L)$$

$$B_1 = \left\{(-C_{ds}/C)(V_1-V_2+V_1'-V_2')-2(V+V_1-V_2) +C_{ds}L(V_1-V_2+V_1'-V_2')(\alpha^2+\omega_2^2)\right\}(1/4L)$$

$$D_1 = \left\{(C_{ds}/C)(V_1-V_2+V_1'+V_2')+2(V-V_1-V_2)\right\}(1/4L)$$

$$A_2 = \left\{RC_{ds}(V_2'-V_1'+V_2-V_1)(\alpha^2+\omega_2^2)-2LI(\alpha^2+\omega_2^2) -2\alpha(V+V_2'-V_1')-\alpha(C_{ds}/C)(V_2'-V_1'+V_2-V_1) -\alpha C_{ds}L(V_2'-V_1'+V_2-V_1)(\alpha^2+\omega_2^2)\right\}(1/4\omega_2 L)$$

$$B_2 = \left\{(-C_{ds}/C)(V_2'-V_1'+V_2-V_1)-2(V+V_2'-V_1') +C_{ds}L(V_2'-V_1'+V_2-V_1)(\alpha^2+\omega_2^2)\right\}(1/4L)$$

$$D_2 = \left\{(C_{ds}/C)(V_2'-V_1'+V_2+V_1)+2(V-V_2'-V_1')\right\}(1/4L)$$

The above equations in cases (2), (3) and (4) are those of NON-ZVS. Fig. 6 shows the calculated and experimental results of output voltage and current waveforms under ZVS operation, using the equations in case (1) with the use of MATLAB program. The calculated waveforms are obtained with phase-shift $\phi = 54°, 0°, -54°$ at switching frequency 68 kHz. It can be observed that the peak value of output current waveforms becomes the highest at phase-shift $\phi = 0°$ and then become decreasing with the increase or decrease of phase-shift from $0°$ to $54°$ or $0°$ to $-54°$, respectively. Moreover, at only the mid-point of phase-shift $\phi = 0°$, the output current i_o can be obtained with almost a sinusoidal waveform. These calculated waveforms with the principle of circuit operation are also verified by comparison with the

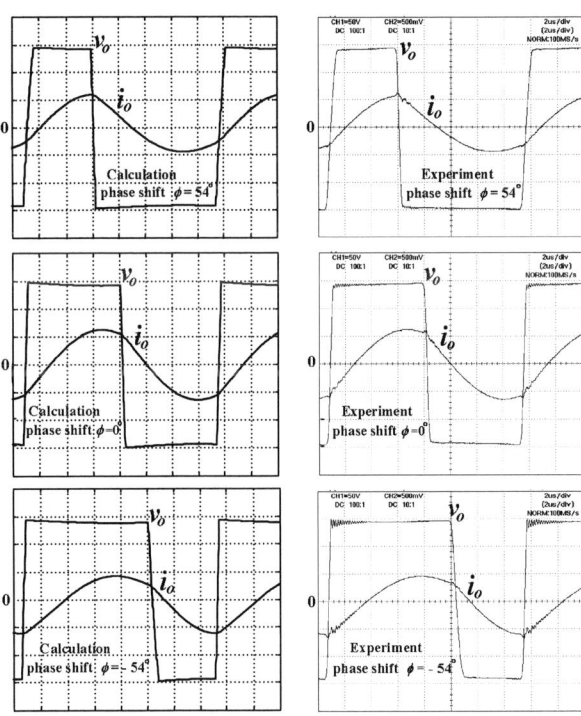

Fig. 6 Calculated and experimental output voltage and current waveforms under ZVS operating condition at $f_s = 68$ kHz

50 V/div , 5 A/div , 2 μs/div

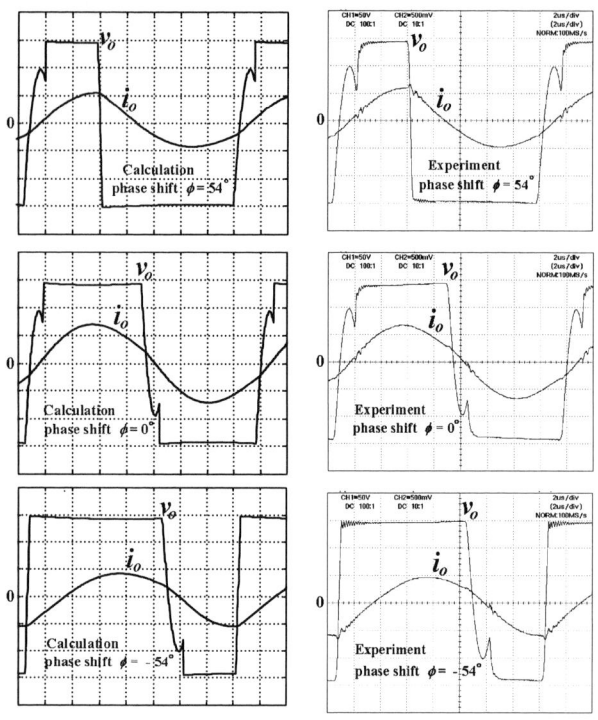

Fig. 7 Calculated and experimental output voltage and current waveforms under **NON-ZVS** operating condition
50 V/div , 5 A/div , 2 µs/div

experimental ones. Fig. 7 shows the calculated and experimental results of output voltage and current waveforms under NON-ZVS operation, using the equations in cases (2), (3) and (4) with the use of MATLAB program. The calculated waveforms are obtained with phase-shift $\phi = 54°$, $0°$, $-54°$ at switching frequency 64 kHz, 58 kHz , and 64 kHz, respectively. These calculated waveforms with the principle of circuit operation are also verified by comparison with the experimental ones. Also, it is observed that at different phase-shift control, NON-ZVS operation occurs at different position of the output voltage (front edge at $\phi = 54°$, tailing edge at $\phi = -54°$, and both at $\phi = 0°$).

IV. ANALYSIS OF OUTPUT POWER P_o, DC INPUT POWER P_d AND EFFICIENCY

The calculated output voltage v_o of Fig. 6 can be analyzed into various component waveforms as shown in (5), by Fourier analysis. Then, applying these component waveforms as the input voltage to the RLC load equivalent circuit, the current equation i_o can be obtained as shown in (6). Again, the voltage v_o in (5) can be used to find out the rms values of output voltage in terms of ac-dc components ($V_{o,rms(ac,dc)}$), dc component ($V_{o,rms(dc)}$), ac component ($V_{o,rms(ac)}$), and fundamental component ($V_{o,rms(1)}$) as shown in Fig. 8.

$$v_o = -\frac{\phi}{\pi}V_d + \frac{V_d}{\pi}\Big[\{2\sin(\pi-\phi)\}\cos\omega_s t$$
$$+\{2-2\cos(\pi-\phi)\}\sin\omega_s t\Big]$$
$$+\frac{V_d}{2\pi}\Big[\{2\sin 2(\pi-\phi)\}\cos 2\omega_s t$$
$$+\{2-2\cos 2(\pi-\phi)\}\sin 2\omega_s t\Big] \qquad (5)$$
$$+\frac{V_d}{3\pi}\Big[\{2\sin 3(\pi-\phi)\}\cos 3\omega_s t$$
$$+\{2-2\cos 3(\pi-\phi)\}\sin 3\omega_s t\Big]+\cdots$$

$$i_o = \frac{V_d}{\pi Z_1}\Big[\{2\sin(\pi-\phi)\}\cos(\omega_s t-\theta_1)$$
$$+\{2-2\cos(\pi-\phi)\}\sin(\omega_s t-\theta_1)\Big]$$
$$+\frac{V_d}{2\pi Z_2}\Big[\{2\sin 2(\pi-\phi)\}\cos(2\omega_s t-\theta_2)$$
$$+\{2-2\cos 2(\pi-\phi)\}\sin(2\omega_s t-\theta_2)\Big] \qquad (6)$$
$$+\frac{V_d}{3\pi Z_3}\Big[\{2\sin 3(\pi-\phi)\}\cos(3\omega_s t-\theta_3)$$
$$+\{2-2\cos 3(\pi-\phi)\}\sin(3\omega_s t-\theta_3)\Big]+\cdots$$

$$\cos\theta_n = \cos\left(\tan^{-1}\left(\frac{\omega_s L-(1/\omega_s C)}{R}\right)\right), \cos\left(\tan^{-1}\left(\frac{2\omega_s L-(1/2\omega_s C)}{R}\right)\right),$$
$$\cos\left(\tan^{-1}\left(\frac{3\omega_s L-(1/3\omega_s C)}{R}\right)\right),\cdots \qquad (7)$$

$$P_o = V_{o1}I_{o1}\cos\theta_1 + V_{o2}I_{o2}\cos\theta_2 + V_{o3}I_{o3}\cos\theta_3 +\cdots \qquad (8)$$

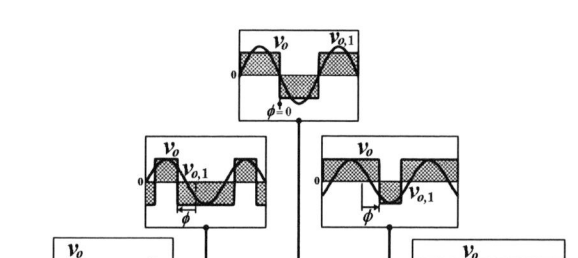

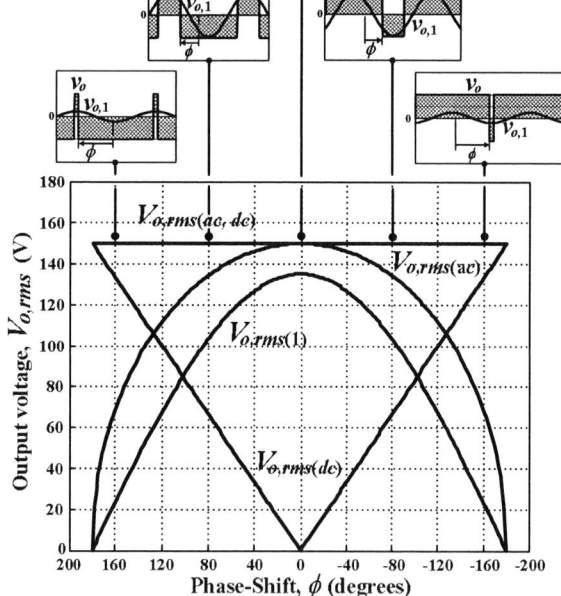

Fig. 8 Variation of various output voltages $V_{o,rms(ac,dc)}$, $V_{o,rms(dc)}$, $V_{o,rms(ac)}$ and $V_{o,rms(1)}$ vs phase-shift ϕ

These equations can also lead to the calculation of ac output power P_o, using the definition of P_o in (8). Fig. 9 shows the output voltage and current waveforms v_o, i_o with the phase difference angle of each harmonic. With phase-shift $\phi = +54°$ and $\phi = -54°$ there will be harmonic orders 1, 2, 3, 4, 5, . . . and with phase-shift $\phi = 0$ there will be harmonic orders 1, 3, 5, ... From this harmonic content, it can be seen that from the second harmonic order upward, each pair of output voltage and current waveforms will have the phase difference angle almost equal to $90°$, since the RLC equivalent circuit of the load now becomes equivalent to almost a pure inductive reactance and consequently can not generate

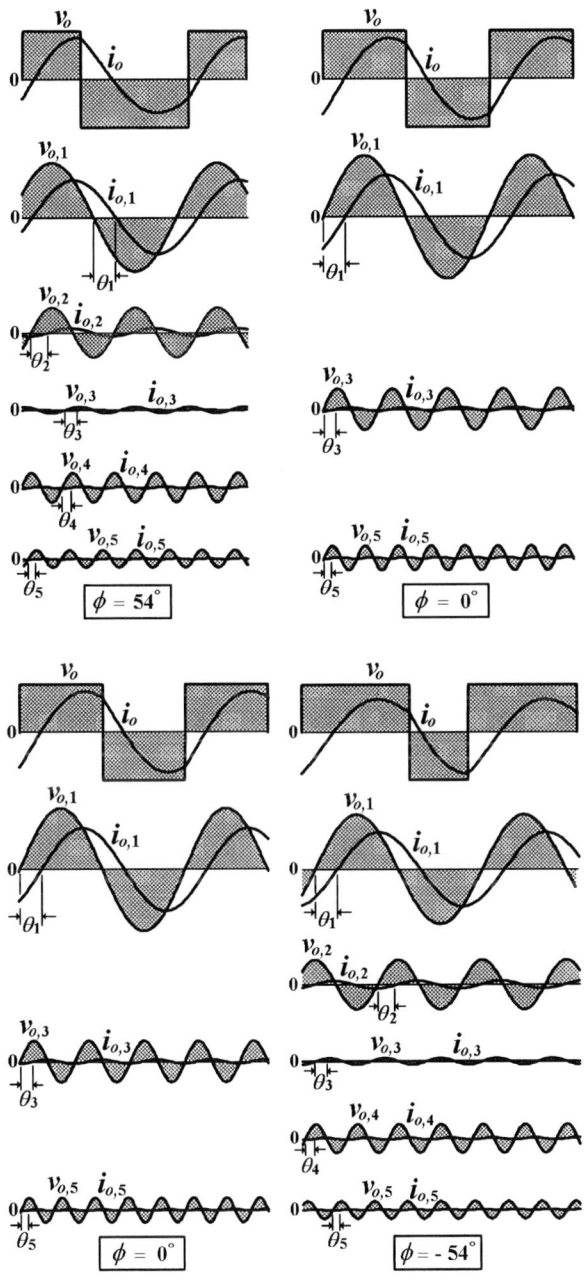

Fig. 9 Output voltage and current waveform v_o, i_o with the phase difference angle of each harmonic

any output power. So, it is quite reasonable to calculate approximately the output power from the fundamental pair of output voltage and current waveforms without considering other harmonic orders. The calculated result and the experimental one are shown for comparison in the same graph of Fig. 10 with each switching frequency held constant at 66 kHz, 67 kHz, 68 kHz, 69 kHz and 70 kHz, while changing the phase-shift ϕ starting from $0°$ to both the positive and negative directions. From the starting point of phase-shift ($\phi = 0°$) to both the positive and negative directions, it is observed that there is a certain limitation for the control range of phase-shift for each characteristic curve of output power P_o under a constant operating frequency. This can be understood by considering first the waveforms v_o, i_o at zero phase-shift ($\phi = 0°$) and switching frequency $f_s = 66$ kHz and also the waveforms v_o, i_o at the same zero phase-shift but at switching frequency $f_s = 70$ kHz. It can be seen that the phase difference between the output voltage and current waveforms v_o, i_o for these two cases are quite different. Higher switching frequency of 70 kHz can results in a larger phase difference due to lower level of output power. So, higher switching frequency can result in a wider control range of phase-shift under ZVS operation and when NON-ZVS operation is encountered the control range begin to be terminated.

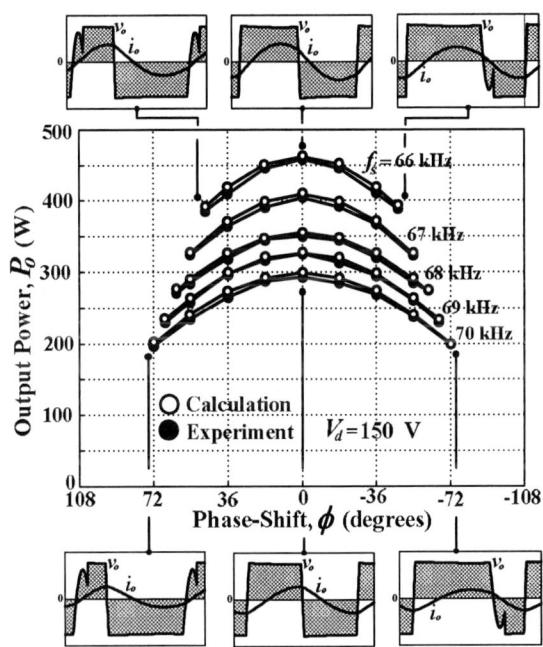

Fig. 10 AC output power P_o vs phase-shift ϕ at switching frequency $f_s = 66$ kHz, 67 kHz, 68 kHz, 69 kHz and 70 kHz

For the calculation of dc input current I_d, the output current i_o can be used again to calculate the dc input current I_d, since the current flow on the dc input side for a certain time duration is the same as that on the ac output side. The calculated result of this dc input current I_d is obtained as shown by an equation in (9) and is also obtained as shown in a graph of Fig. 11.

$$I_d = \frac{V_d}{2\pi^2 Z_1}\Big[\{2\sin(\pi - \phi)\}$$
$$\times\{2\sin(\pi - \phi - \theta_1) - \sin(-\theta_1) - \sin(2\pi - \theta_1)\}\Big]$$
$$+\frac{V_d}{2\pi^2 Z_1}\Big[\{2 - 2\cos(\pi - \phi)\}$$
$$\times\{-2\cos(\pi - \phi - \theta_1) + \cos(-\theta_1) + \cos(2\pi - \theta_1)\}\Big]$$
$$+\frac{V_d}{2\times 2^2\pi^2 Z_2}\Big[\{2\sin 2(\pi - \phi)\}$$
$$\times\{2\sin(2\pi - 2\phi - \theta_2) - \sin(-\theta_2) - \sin(2\times 2\pi - \theta_2)\}\Big]$$
$$+\frac{V_d}{2\times 2^2\pi^2 Z_2}\Big[\{2 - 2\cos 2(\pi - \phi)\}$$
$$\times\{-2\cos(2\pi - 2\phi - \theta_2) + \cos(-\theta_2) + \cos(2\times 2\pi - \theta_2)\}\Big]$$
$$+\frac{V_d}{2\times 3^2\pi^2 Z_3}\Big[\{2\sin 3(\pi - \phi)\}$$
$$\times\{2\sin(3\pi - 3\phi - \theta_3) - \sin(-\theta_3) - \sin(3\times 2\pi - \theta_3)\}\Big]$$
$$+\frac{V_d}{2\times 3^2\pi^2 Z_3}\Big[\{2 - 2\cos 3(\pi - \phi)\}$$
$$\times\{-2\cos(3\pi - 3\phi - \theta_3) + \cos(-\theta_3) + \cos(3\times 2\pi - \theta_3)\}\Big] + \cdots$$

(9)

$$P_d = V_d I_d \qquad (10)$$

Where $\quad Z_n = \sqrt{R^2 + \left(n\omega_s L - \dfrac{1}{n\omega_s C}\right)^2}$

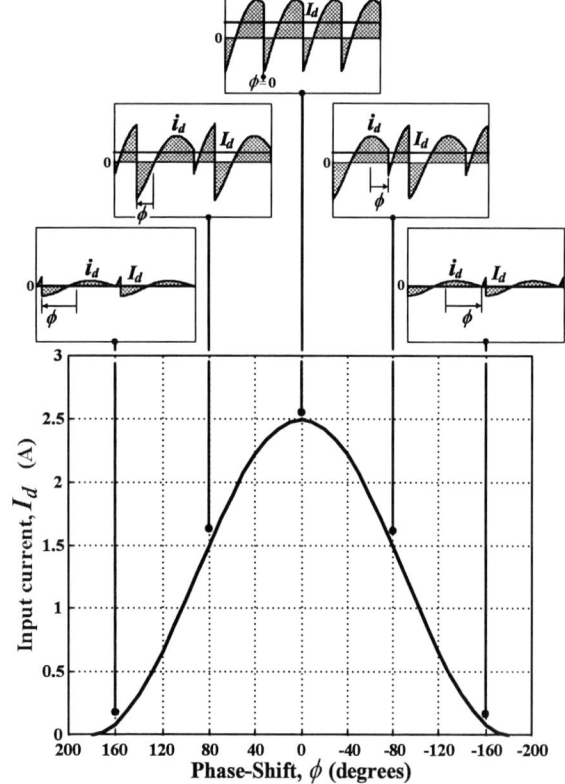

Fig. 11 Variation of dc input current I_d vs phase-shift ϕ
at switching frequency $f_s = 68$ kHz

The dc input current I_d can be also used to calculate the dc input power P_d. The calculated result with the experimental one are also shown in the graph of Fig. 12 with each switching frequency held constant at 66 kHz, 67 kHz, 68 kHz, 69 kHz and 70 kHz, during change of phase-shift ϕ from 0° to both the positive and negative directions. From the starting point of phase-shift ($\phi = 0°$) to both the positive and negative directions, it is observed that there is a certain limitation for the control range of phase-shift for each characteristic curve of dc input power P_d under each constant operating frequency.

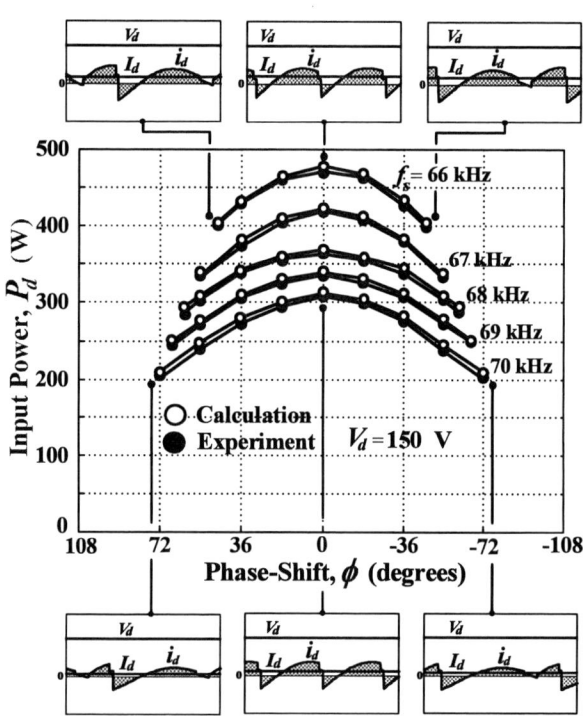

Fig. 12 DC input power P_d vs phase-shift ϕ at switching frequency f_s = 66 kHz, 67 kHz, 68 kHz, 69 kHz, 70 kHz

Then, the ratio of ac output power P_o and dc input power P_d makes possible the calculation of the full-bridge inverter efficiency η vs. phase-shift ϕ which is plotted in various curves, each of which is held at constant switching frequency of 66 kHz, 67 kHz, 68 kHz, 69 kHz and 70 kHz. The calculated results are shown as some examples in Figs. 13 and 14 for 66 kHz and 70 kHz respectively and they are verified by comparing with the experimental ones in the same figure. The result shows that the efficiency is almost constant at 96 % over the whole control range of phase-shift ϕ which is not constant but the control range will change according to the switching frequency ; that is, a wider control range of phase-shift ϕ will be obtained with a higher operating frequency.

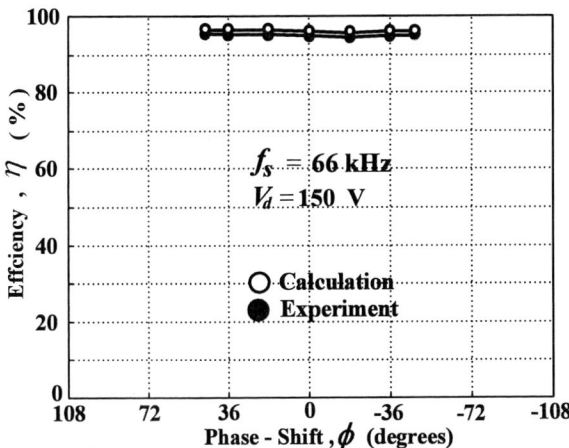

Fig. 13 Inverter efficiency vs phase-shift ϕ at switching frequency $f_S = 66$ kHz

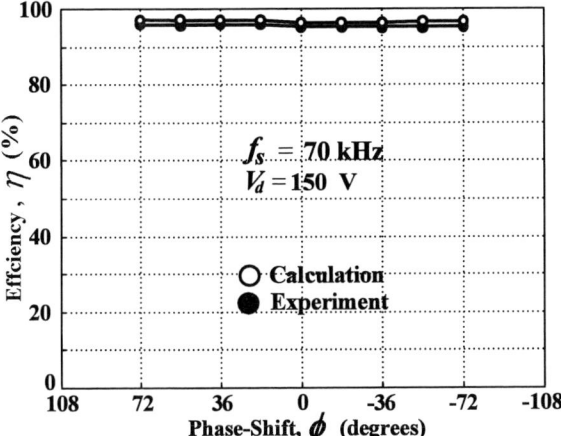

Fig. 14 Inverter efficiency vs phase-shift ϕ at switching frequency $f_S = 70$ kHz

V. CONCLUSION

The detailed analysis of circuit operation under ZVS and NON-ZVS switching conditions in a high-frequency full-bridge inverter for induction heating using asymmetrical phase-shift control has been already presented both theoretically and experimentally. There are four main important points to be concluded here as follows :

1. For the positive phase-shift control of output voltage v_o, the phase angle of fundamental output voltage $v_{o,1}$ always lead the front edge of square wave voltage v_o, which makes the zero-crossing of output current i_o move toward the front edge, where NON-ZVS operation may occur. For the negative phase-shift control, the fundamental output voltage $v_{o,1}$ always moves to the leading direction, which makes the zero-crossing of output current i_o move toward the tailing edge of square wave voltage v_o, where NON-ZVS operation may occur. For zero phase-shift control, NON-ZVS operation may

occur both at the front and tailing edges, especially when the operating frequency is not high enough. All these three cases of the occurrence of NON-ZVS operation can be avoided by increasing the operating frequency in order to move the zero-crossing of current i_o away from the front and tailing edges.

2. In phase-shift control, the maximum ac output voltage will be obtained at phase-shift $\phi = 0$ and when phase-shift angle ϕ is increased away from zero degree, the ac output voltage, current and power $V_{o,rms(ac)}$, $I_{o,rms}$, P_o and consequently the dc input current and power I_d, P_d will become decreasing symmetrically both in positive and negative phase-shift control.

3. In power control by phase-shift, the control range of phase-shift is limited by the occurrence of NON-ZVS operating condition for each operating frequency. At this point, if further increase of phase-shift is required, this is also possible by increasing the switching frequency to a higher value and when NON-ZVS is encountered the same process can be repeated again.

4. Higher switching frequency can results in a larger phase difference due to lower level of output power. So, higher switching frequency can result in a wider control range of phase-shift under ZVS operation and when NON-ZVS operation is encountered the control range begins to be terminated.

REFERENCES

[1] P. Viriya , N. Yongyuth , I. Miki and K. Matsuse "Analysis of Circuit Operation under ZVS and NON-ZVS Conditions in Phase-Shift Inverter for Induction Heating," *IEEJ Trans. IA.*, vol. 126, no. 5, pp. 560-567, May 2006.

[2] L. Grajales, J. A. Sabate, K. R. Wang, W. A. Tabisz, and F. C. Lee, "Design of a 10 kW, 500 kHz Phase-Shift Controlled Series-Resonant Inverter for Induction Heating," *Proc. IEEE Industry Applications Soc. Annu. Meeting, Toronto, Canada,* 1993, pp. 843-849.

[3] P. Viriya , N. Yongyuth and K. Matsuse "Analysis of Transition Mode from Phase Shift to Zero-Phase Shift Under ZVS and NON-ZVS Operation for Induction Heating Inverter," *Proc. Power Conversion Conf. (PCC), Nagoya, Japan,* April 2007, pp. 1512-1519.

[4] L. Grajales and F. C. Lee, "Control System Design and Small Signal Analysis of a Phase-Shift-Controlled Series-Resonant Inverter for Induction Heating," *Proc. IEEE Power Electronics Specialist Conf. (PESC),* 1995, pp. 450-456.

[5] P. Viriya, and T. Thomas, "Power Transfer Characteristics of a Phase-shift Controlled ZVS Inverter for the Application of Induction Heating," *Proc. Int. Power Electron. Conf. (IPEC),* 2000, pp.423-428.

[6] J. M. Burdio, L. A. Barragan, F. Monterde, D. Navarro, and J. Acero, "Asymmetrical Voltage-Cancellation Control for Full-Bridge Series Resonant Inverter," *IEEE Trans. Power Electron.,* vol. 19, no. 2, pp. 461-469, Mar. 2004.

[7] J. A. Sabate, R. W. Farrington, M. M. Jovanovic, and F. C. Lee, "Effect of Switch Capacitance on Zero-Voltage Switching of Resonant Converters," *Proc. Applied Power Electron. Conf.,* 1992, pp. 213-220.

FPGA-Based Phase-Shift ZVS Full-Bridge DC-DC Converter Using One-Comparator Counter-Based PWM Control Strategy

K. I. Hwu[1], Y. T. Yau[2]

[1]Center for Power Electronics Technology, National Taipei University of Technology, Taiwan
[2]Industrial Technology Research Institute, Taiwan
eaglehwu@ntut.edu.tw

Abstract-**This paper applies the one-comparator counter-based PWM (Pulse-Width-Modulated) control strategy to controlling the FPGA (Field Programmable Gate Arrays)-based phase-shift ZVS full-bridge DC-DC converter based on two output inductors with direct coupling. To verify the performance of the proposed control schemes, some transient load variations are imposed on the constructed converter.**

I. INTRODUCTION

As generally acknowledged, the controller for the switching power supply consists conventionally of analog components such that its performance is significantly influenced by electromagnetic interference, variations of temperature, component aging etc. Therefore, the digital controller has been presented to overcome such problems. Up to now, the digital signal processor (DSP) [1-6] or the microcontroller [7-11] or the field programmable gate arrays (FPGA) [12-15], together with the analog-to-digital converter (ADC) to sample the desired signal, are widely used. However, a high-speed and high-resolution ADC is expensive and is not easy to be combined into an integrated circuit. Therefore, one-comparator counter-based pulse-width-modulated (PWM) control without any ADC has been presented [15]. Such a sampling method extracts the information from the output voltage, which is send to one comparator compared with the output voltage reference, and the resulting digital signal from the comparator is sent to the FPGA to create a suitable driving signal to control the switch, so as to get the desired output voltage.

Consequently, in this paper, the sampling method used in [15] is applied to phase-shift ZVS full-bridge DC-DC converter using one-comparator counter-based feedback sampling, based on two output inductors with direct coupling [16]. As generally recognized, inverse coupling of two inductors is widely used to double the frequency of the current flowing through the output capacitor and hence to reduce the output voltage ripple. But in actually, the output voltage ripple by this way is not only reduced but distorted, which is not beneficial in one-comparator counter-based feedback sampling. That is why direct coupling of two inductors is used herein. There are five sections to be discussed. Section II presents the main power stage, section III illustrates the proposed control topology, section IV describes design of key

parameters, section V gives some experimental results and section VI makes a final conclusion.

II. MAIN POWER STAGE

Fig. 1 shows that the main power stage takes the phase-shift current-doubler ZVS full-bridge Dc-Dc converter with synchronous rectification (SR) and coupling inductors. On the one hand, the primary side contains the four switches S-A, S-B, S-C and S-D with their corresponding anti-diodes D-A, D-B, D-C and D-D, the parasitic capacitances C-A, C-B, C-C and C-D, the resonant inductance L_r composed of leakage inductance L_{lk} of the main transformer MT and an additional inductance L_a. On the other hand, the secondary side contains the two switches S-AB and S-CD used as synchronous rectifiers with the corresponding anti-diodes D-AB and D-CD, the output inductors L_1 and L_2 coupled together to reduce the required size.

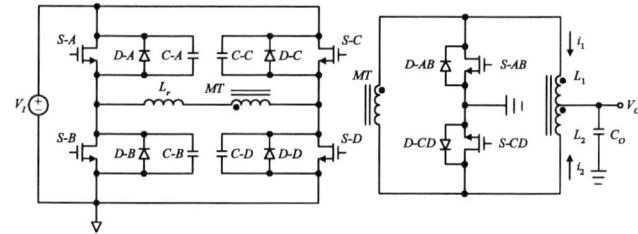

Fig. 1. Main power stage.

III. PROPOSED CONTROL TOPOLOGY

In Fig. 2, the proposed FPGA-based digital control strategy is presented. The system clock is set to 20MHz from OSC, which is changed after PLL to 100MHz for the main controller and the PID controller and to 200MHz for the 9-bit digital PWM (DPWM) signal generator. Besides, the output voltage V_O is sampled via a voltage sampling circuit referred to [15] and the resulting signal VFB is generated and sent to FPGA. The output voltage information is obtained via the voltage sampling circuit through one-comparator counter-based feedback sampling without any ADC used. The signal VFB along with the PID controller creates appropriate DPWM signals Q-A, Q-B, Q-C and Q-D to drive the switches S-A, S-B,

978-1-4244-0644-9/07/$25.00 ©2007 IEEE

S-C and S-D respectively. Since the current-doubler SR and the two coupled inductors are utilized, the direct coupling shown in Fig. 3 is employed herein instead of the inverse coupling shown in Fig. 4, so as to apply the mentioned sampling method to this constructed converter. This is because the AC components of the currents flowing through i_1 and i_2 have the same sign of the current slopes for any time under direct coupling and hence the resultant current i_{Co} and the output voltage ripple are larger than that under inverse coupling. Based on the mention above, the sampling method to be used is very suitable for this constructed converter with output inductors directly coupled.

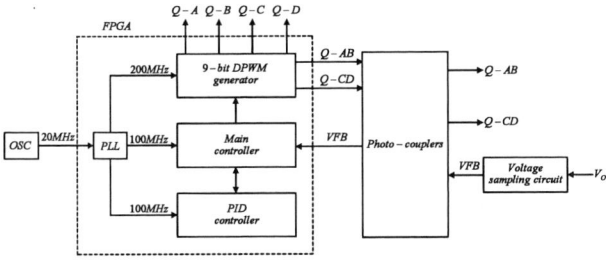

Fig. 2. Proposed FPGA-based digital control.

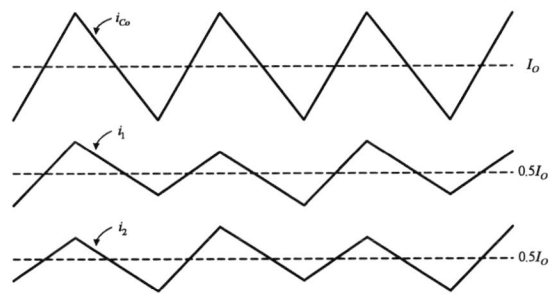

Fig. 3. Direct coupling

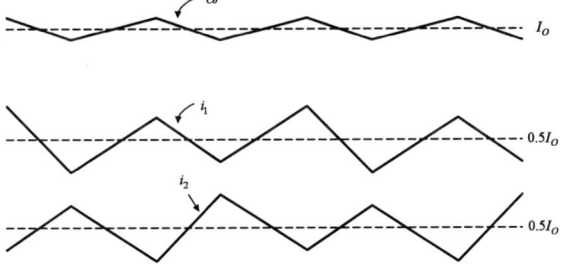

Fig. 4. Inverse coupling

IV. DESIGN OF KEY PARAMETERS

Before entering into designing the key parameters, the corresponding specifications are described as follows: (i) rated input voltage: $V_I = 24\text{V}$; (ii) rated output voltage: $V_O = 5\text{V}$; (iii) turn ratio of MT: $n = N_s/N_p = 5/6$; (iv) rate output current: $I_{O-rated} = 10\text{A}$; (v) minimum output current: $I_{O-\min} = 0.5\text{A}$; (vi) switching frequency: $f_s = 195\text{kHz}$; (vii) output current

slew rate from no load to rated load: $SR = 10\text{A/}\mu\text{s}$; (viii) output voltage ripple: $\Delta V_O \leq 50\text{mV}$; and (ix) parameters of controller: $k_p = 1.5$, $k_i = 0.625$ and $k_d = 1$.

A. Calculation of Effective Duty Cycles for L_1 or L_2

In the phase-shift topology, since this converter always operates in CCM due to synchronous rectification, the relationship between V_O and V_I in the steady state can be expressed as

$$V_O/V_I = nD/2 \qquad (1)$$

Hence, the ideal effective duty cycle D for L_1 and L_2 is kept constant at 0.5 after calculation.

B. Calculating Resonant Capacitance and Inductance

Since IRFZ44NS MOSFETs are used as switches, the value of C_{oss} of each switch is about 360pF with respect to the corresponding datasheet. Also, the parasitic capacitance C_{XFMR} and the leakage inductance L_{lk} for the main transformer MT are about 10pF and $0.05\mu\text{H}$ respectively based on INTERSIL datasheet estimation. According to the application note of TI U-136A, the resonant capacitance C_r can be represented as

$$C_r = \frac{8}{3}C_{oss} + C_{XFMR} \qquad (2)$$

And hence, C_r is set to 970pF.

Also, the resonant inductance L_r is the sum of the leakage inductance L_{lk} and the additional inductance L_a. And hence, the desired inequality to satisfy conditions of ZVS is

$$L_r I_{PRI}^2 \geq C_r V_I^2 \qquad (3)$$

where I_{PRI} is the minimum initial current flowing through L_r at resonance.

Furthermore, the converter is intended to operate with ZVS under the condition of above 25% of $I_{O-rated}$, so I_{PRI} can be expressed as

$$I_{PRI} = 0.125nI_{O-rated} \qquad (4)$$

And hence, I_{PRI} is set approximately to 1A.

Based on the mention above, the minimum value of L_r is calculated to be $0.558\mu\text{H}$. And hence, L_a is $0.508\mu\text{H}$. Therefore, three 180nH inductors, having product name of IHLP-2525 made by VISHAY Ltd., are chosen and connected in series. And hence, L_r is calculated to be $0.6\mu\text{H}$.

C. Calculation of Dead Time Between Switches

According to the obtained values of L_r and C_r, the dead time t_d between two MOSFET switches of the same lag is

$$t_d = 0.5\pi\sqrt{L_r C_r} \qquad (5)$$

And hence, t_d is calculated to be 37.7ns.

D. Calculation of Output Coupling Inductors L_1 and L_2

In this converter, coupling inductors are used at the output terminal. L_1 is magnetized and the corresponding current flows through L_2 based on the transformer principle. The following equations are under the condition that L_1 and L_2 are equal and set to L. M is the mutual inductance between L_1 and L_2, and di_1/dt and di_2/dt are set to $4A/\mu s$ and $3A/\mu s$.

$$\begin{cases} L\dfrac{di_1}{dt} - M\dfrac{di_2}{dt} = nV_I - V_O \\[2mm] L\dfrac{di_2}{dt} - M\dfrac{di_1}{dt} = -V_O \end{cases} \quad (6)$$

And hence, L and M are obtained to be $10.7\mu H$ and $9.3\mu H$, respectively. That is to say, the coupling coefficient between L_1 and L_2 is about 0.87.

E. Calculation of Output Capacitor

The output capacitor C_O is determined by the output current slew rate from no load to rated load, SR. As shown in Fig. 5, the slew rate of the ideal load current $I_{O\text{-}ideal}$ is far from larger than that of the actual load current $I_{O\text{-}actual}$. In Fig. 5, ΔQ is defined as the total electric charge required to offer the load as $I_{O\text{-}actual}$ rises from 0% to 90% of the variation in the load current from no load to rated load, ΔI_O, with the elapsed time Δt, and hence

$$\Delta Q = \frac{1}{2}\cdot\Delta I_O \cdot \Delta t = \frac{1}{2}\cdot(\frac{9}{10}\cdot\Delta I_O)\cdot\frac{(\frac{9}{10}\cdot\Delta I_O)}{SR} = \frac{81\cdot\Delta I_O^{\,2}}{200\cdot SR} \quad (7)$$

Also,

$$\Delta Q \cong C_O \cdot \Delta V_{CO} \quad (8)$$

Therefore,

$$C_O \cong \frac{81\cdot\Delta I_O^{\,2}}{200\cdot SR \cdot \Delta V_{CO}} \quad (9)$$

where ΔV_{CO} denotes the peak-to-peak voltage created from C_O due to ΔI_O.

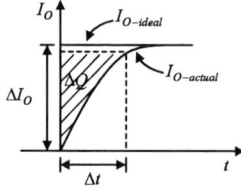

Fig. 5. Slew rate of the actual load current.

And hence, C_O is determined according to (9) and under the condition that $\Delta V_{CO} \leq 10mV$, the value of C_O is larger than $365\mu F$. Therefore, C_O is chosen to be three $330\mu F$ OSCON capacitors paralleled with two $22\mu F$ MLCC capacitors, so as to reduce the equivalent series resistance (ESR) and hence to meet the output voltage requirements.

V. EXPERIMENTAL RESULTS

The following experimental results contain measured waveforms pertaining to ZVS for the switch *S-A*, primary-side current, currents flowing through the coupled inductors and the output capacitor, and transient load responses. Figs. 6 and 7 depict this converter operating with ZVS at half and rated loads respectively. Besides, Figs. 8 and 9 show the primary-side currents at half and rated loads respectively. On the other hand, Fig. 10 show measured AC currents flowing through the output capacitor and coupled inductors at rated load. Figs. 11 and 12 show transient load responses due to load change from 0% to 100% and 100% to 0% of the rated load, which is used to tune the parameters of the PID controller. Moreover, the transient load responses shown in Figs. 13 to 18 are utilized under the variation of 50% of the rated load at the different initial loads, to further demonstrate the proposed control scheme.

VI. CONCLUSIONS

In this paper, the one-comparator counter-based PWM control scheme is used to control the FPGA-based phase-shift ZVS full-bridge DC-DC converter based on two output inductors with direct coupling. By means of some transient load responses, the effectiveness of the proposed control scheme is demonstrated.

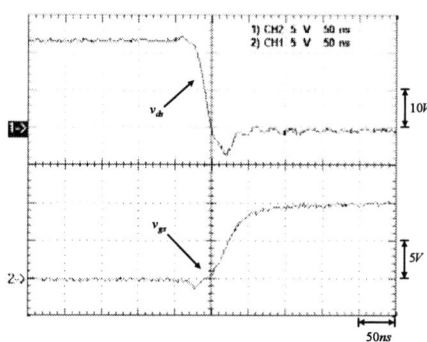

Fig. 6. Measured ZVS waveforms of *S-A* at turn-on under half load:
(1) voltage across *S-A*; (2) gate driving signal for *S-A*.

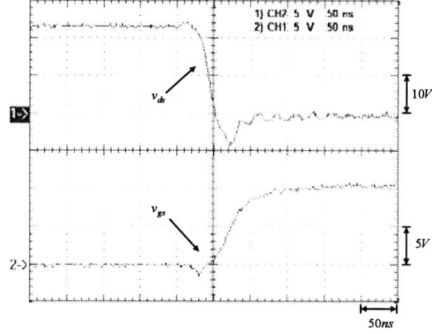

Fig. 7. Measured ZVS waveforms of *S-A* at turn-on under rated load:
(1) voltage across *S-A*; (2) gate driving signal for *S-A*.

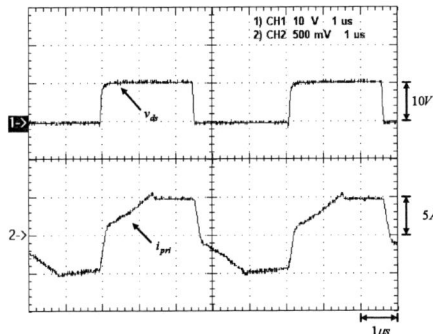

Fig. 8. Under half load: (1) gate driving signal for *S-A*;
(2) primary-side current.

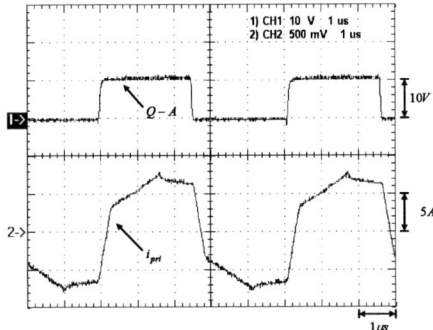

Fig. 9. Under rated load: (1) gate driving signal for *S-A*;
(2) primary-side current.

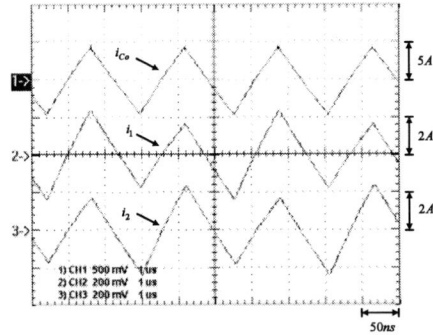

Fig. 10. Measured AC currents under rated load:
(1) flowing through C_O; (2) flowing through L_1; (3) flowing through L_2.

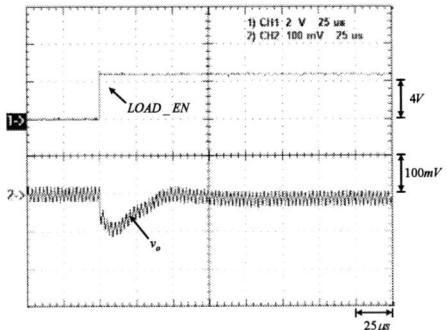

Fig. 11. Transient load responses due to load change from 0% to 100% of the
rated load.

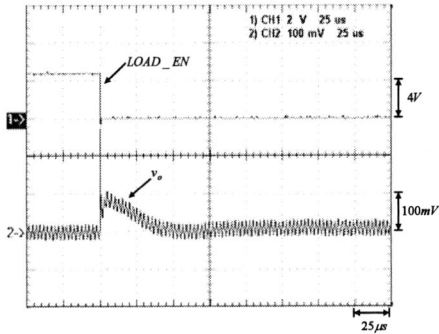

Fig. 12. Transient load responses due to load change from 100% to 0% of the
rated load.

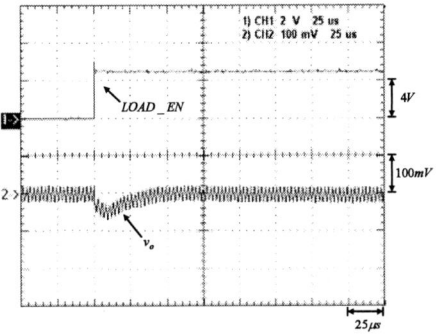

Fig. 13. Transient load responses due to load change from 0% to 50% of the
rated load.

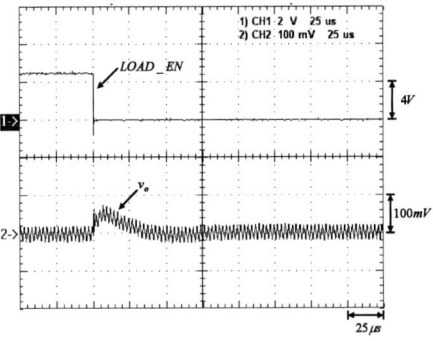

Fig. 14. Transient load responses due to load change from 50% to 0% of the
rated load.

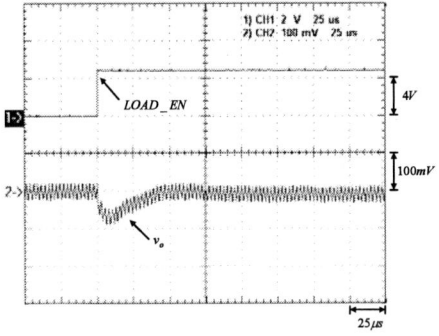

Fig. 15. Transient load responses due to load change from 25% to 75% of the
rated load.

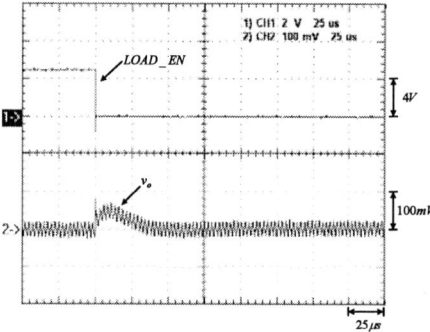

Fig. 16. Transient load responses due to load change from 75% to 25% of the rated load.

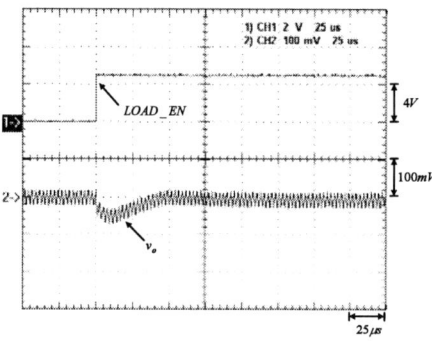

Fig. 17. Transient load responses due to load change from 50% to 100% of the rated load.

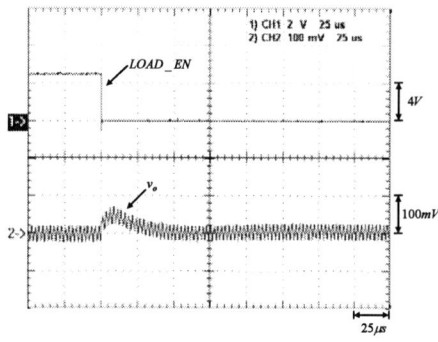

Fig. 18. Transient load responses due to load change from 100% to 50% of the rated load.

REFERENCES

[1] P. F. Kocybik and K. N. Bateson, "Digital control of a ZVS full-bridge DC-DC converter," *IEEE PESC'95*, vol. 2, pp. 687-693, 1995.

[2] C. H. Chan and M. H. Pong, "DSP controlled power converter," *IEEE Proc. Power Electronics and Drive Systems*, vol. 1, pp. 364-369, 1995.

[3] P. T. Tang and C. K. Tse, "Design of DSP-based controller for switching power converters," *IEEE TENCON'96*, vol. 2, pp. 889-894, 1996.

[4] A. Prodic, D. Maksimovic and R. W. Erickson, "Design and implementation of a digital PWM controller for a high-frequency switching DC-DC power converter," *IEEE IECON'01*, vol. 2, pp. 893-898, 2001.

[5] Guo Liping, J. Y. Hung and R. M. Nelms, "PID controller modifications to improve steady-state performance of digital controllers for buck and boost converters," *IEEE APEC'02*, vol. 1, pp. 381-388, 2002.

[6] D. Liu and H. Li, "Design and implementation of a DSP based digital controller for a dual half bridge isolated bi-directional DC-DC converter," *IEEE APEC'06*, pp. 695-699, 2006.

[7] A. Prodic and D. Maksimovic, "Design of a digital PID regulator based on look-up tables for control of high-frequency DC-DC converters," *IEEE COMPEL'02*, pp. 18-22, 2002.

[8] D. He and R. M. Nelms, "Peak current-mode control for a boost converter using an 8-bit microcontroller," *IEEE PESC'03*, vol. 2, pp. 938-943, 2003.

[9] D. He and R. M. Nelms, "Current-mode control of a DC-DC converter using a microcontroller: implementation issues," *IEEE IPEMC'04*, vol. 2, pp. 538-543, 2004.

[10] D. He, W. Dilliard and R. M. Nelms, "Microcontroller Implementation of current-mode control for a discontinuous mode boost converter," *IEEE IECEC'04*, pp. 255-260, 2004.

[11] K. Leung and D. Alfano, "Design and implementation of a practical digital PWM controller," *IEEE APEC'06*, pp. 1437-1442, 2006.

[12] S. Saggini, G. Garcea, M. Ghioni and P. Mattavelli, "Analysis of high-performance synchronous-asynchronous digital control for dc-dc boost converters," *IEEE APEC'05*, vol. 2, pp. 892-898, 2005.

[13] H. Peng and D. Maksimovic, "Digital current-mode controller for DC-DC converters," *IEEE APEC'05*, vol. 3, pp. 899-905, 2005.

[14] K. Wang, N. Rahman, Z. Lukic and A. Prodic, "All-digital DPWM/DPFM controller for low-power DC-DC converters," *IEEE APEC'06*, pp. 719-723, 2006.

[15] K. I. Hwu and Y. T. Yau, "Applying a counter-based PWM control scheme to an FPGA-based SR forward converter," *IEEE APEC'06*, vol. 3, pp. 1396-1400, 2006.

[16] Pit-Leong Wang, Peng Xu, Bo Yang and Fred C. Lee, "Performance improvements of interleaving VRMs with coupling inductors," *IEEE Trans. Power Electron.*, vol. 16, no. 4, pp. 499-507, 2001.

A Simplified Power Control Scheme for Resonant Inverter with Purely Resistive Load

Pramoch Dorkmai*, Youthana Kulvitit*, and Tanvaa Tansatit**

* Dept. of Electrical Eng., Chulalongkorn University. 254 Phyathai Rd., Bangkok 10330 Thailand.
e-mail: youthana.k@chula.ac.th
** Dept. of Medicine., Chulalongkorn University 1873 Rama IV Rd., Bangkok 10330 Thailand.
e-mail: tansatit@yahoo.com

Abstract--**This paper presents a simplified power control scheme for resonant inverter with purely resistive load and low-harmonic output waveforms. The control scheme was implemented in the power control of radio-frequency generator--the power source of an electrosurgical unit. As the proposed control scheme does not conform to standard feedback control scheme, the standard controller design technique can not directly be applied. By applying small-signal perturbation and linearization technique to the simplified control scheme, its small-signal model does reveal power feedback, and the standard controller design procedure can be used. The proposed control scheme features feed forward of power control signal, and only voltage controller design is required. Steady-state errors and transient responses of the proposed control scheme were investigated. The theoretical calculations were verified by computer simulations and hardware realization.**

Index Terms--**Modeling, power control, purely resistive load, resonant inverter, sinusoidal, static error.**

I. INTRODUCTION

RF generator finds applications in different variety fields: industrial, communication, medical, etc. Electrosurgical unit (ESU) using RF generator as its power source requires output power control. Because the equivalent impedance of human and animal tissues may be considered as purely resistive, load voltage and current waveforms are similar, and elapsed time between the two signals is negligible. This paper proposes a simplified power control scheme (SPCS) for a resonant inverter with purely resistive load. Fundamental frequency approximation and modified phasor transformation[2],[3] will be used to convert frequency-modulated signals into time-varying phasors. The well-known perturbation and linearization technique will be applied to the system state equations to derive small-signal model of the frequency-controlled inverter. A closed-form solution of the frequency-to-output voltage transfer function will be calculated. The controller of the SPCS will be designed. Basic performance criteria: system stability, transient response, and steady-state error will be investigated.

II. POWER CONTROL SCHEME

A. Standard Power Control Scheme

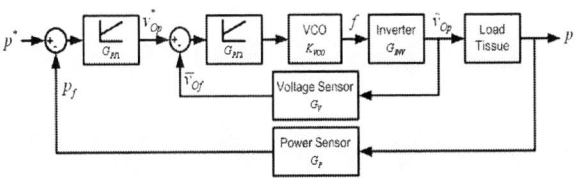

Fig. 1. Standard power control scheme.

In a standard feedback control scheme, output-controlled variables are measured and fed back to compare with reference signals. Errors are fed through controllers in order to regulate the controlled variables. Fig. 1 is a standard power control scheme. Instantaneous load voltage and current waveforms are fed to a signal multiplier. Output signal of the multiplier, representing instantaneous output power, is fed through a lowpass filter in order to filter out switching frequency harmonics--representing reactive power. Low frequency signal--representing active power--, is fed back to compare with power reference signal. Power error signal is input to power controller in order to generate voltage reference signal, if the output power control is implemented via load voltage control. Output power control via load current control is also possible.

B. Simplified Power Control Scheme

Standard power control scheme can be implemented for any kinds of load and signal's waveforms. Nevertheless, as the operating frequency of the inverter gets higher, signal multiplication becomes harder, phase shift may be introduced into voltage and current signals during the measurement process. Uneven phase shift of the voltage and current signals will degrade the accuracy of the output power signal. Electrosurgical unit (ESU) uses RF power source to avoid nerve stimulation[1]. The RF generators are frequently resonant inverter with operating frequency around 500 kHz. Output voltage waveforms of ESU are almost sinusoidal with negligible phase shift. It can be shown that, output power $p_O(t)$ of an inverter with sinusoidal output voltage and purely resistive load is proportional to product of peak value of the output voltage $V_{Op}(t)$ and current $I_{Op}(t)$ as in (1).

978-1-4244-0644-9/07/$25.00 ©2007 IEEE

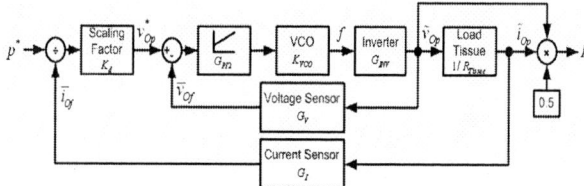

Fig. 2. SPCS for inverter with purely resistive load.

$$p_O(t) = \frac{1}{2} \cdot \left(V_{Op}(t) \right) \cdot \left(I_{Op}(t) \right), \tag{1}$$

$$V_{Op}(t) = 2 \cdot \frac{P_O(t)}{I_{Op}(t)} \tag{2}$$

$$I_{Op}(t) = 2 \cdot \frac{P_O(t)}{V_{Op}(t)} \tag{3}$$

Base on equations (1), (3) three SPCSs can be implemented: Firstly, a simplified version of standard power feedback control scheme of Fig. 1, where power feedback signal is product of low-frequency envelope of voltage and current signals; secondly, an alternative power control scheme where output power is controlled through voltage reference signal generated from power reference and output current signal as shown in Fig. 2; and finally, an alternative power control scheme similar to that of Fig. 2, but voltage and current signals are interchanged. Because the first simplified control scheme conforms to standard feedback control scheme, standard analytical techniques and design procedures can be applied directly. As standard analytical techniques and design procedure are well-known, the first SPCS will not be investigated. The alternative power control scheme presented in Fig. 2 will be explored. In the SPCS of Fig. 2, power-command reference signal $p^*(t)$ is divided by load current signal $i_{Of}(t)$, the output signal of the divider is voltage reference signal $v^*_{Op}(t)$.

$$v^*_{Op}(t) = K_d \cdot \frac{p^*(t)}{i_{Of}(t)} \tag{4}$$

where K_d is divider scaling factor

Load current signal is obtained from load current waveform by rectifying and filtering. If the corner frequency of the lowpass filter is high compares to load current signal frequency, load current signal is proportional to peak-value of load current.

$$i_{Of}(t) = K_I \cdot I_{Op}(t) \tag{5}$$

where K_I is low-frequency gain of load current sensing and processing network
Substitute (5) into (4) one obtains

$$v^*_{Op}(t) = \frac{K_d}{K_I} \cdot \frac{p^*(t)}{I_{Op}(t)} \tag{6}$$

As the SPCS is nonlinear and does not conform to standard feedback control scheme, standard analytical tools and design procedures are not applicable. To simplify the analysis, small-signal perturbation and linearization technique is applied. Steady-state and small-signal variation of voltage references are obtained as in (7) and (8):

$$V^*_{Op} = \frac{K_d}{K_I} \cdot \frac{P^*}{I_{OP}} \tag{7}$$

$$\hat{v}^*_{op}(t) = \frac{K_d \cdot P^*}{K_I \cdot I_{Op}} \cdot \left[\left\{ \frac{\hat{p}^*(t)}{P^*} \right\} - \left\{ \frac{\hat{i}_{op}(t)}{I_{Op}} \right\} \right] \tag{8}$$

Lets K_{VLoop} be steady-state closed-loop gain of voltage control loop:

$$K_{VLoop} = \frac{V_{OP}}{V^*_{OP}} = \frac{K_{PI} \cdot K_{VCO} \cdot K_{INV}}{\left[1 + K_V \cdot K_{PI} \cdot K_{VCO} \cdot K_{INV} \right]} \tag{9}$$

Then:

$$\frac{V_{OP}}{P^*} = \frac{K_d}{K_I \cdot I_{OP}} K_{VLoop} \tag{10}$$

$$\frac{P}{P^*} = \frac{K_d}{2 \cdot K_I} K_{VLoop} \tag{11}$$

By perturbing and linearizing (1), small-signal variation of output power for a given operating point is found to be a linear function of load voltage and load current signals as in (13)

$$\hat{p}(t) = \frac{1}{2} \cdot \left[I_{Op} \cdot \hat{v}_{op}(t) + V_{Op} \cdot \hat{i}_{op}(t) \right] \tag{12}$$

$$\hat{p}(t) = I_{Op} \cdot \hat{v}_{op}(t) = V_{Op} \cdot \hat{i}_{op}(t) \tag{13}$$

$$\hat{i}_{op}(t) = \frac{\hat{p}(t)}{V_{Op}} \tag{14}$$

$$\hat{v}_{op}(t) = \frac{\hat{p}(t)}{I_{Op}} \tag{15}$$

By using (8), (11) and (14) it can be shown that

$$\hat{v}^*_{op}(t) = \frac{K_d}{K_I \cdot I_{Op}} \cdot \left[\hat{p}^*(t) - \hat{p}_f(t) \right] \tag{16}$$

where

$$\hat{p}_f(t) = \left\{ \frac{K_I}{K_d \cdot K_{VLoop}} \right\} \hat{p}(t) \tag{17}$$

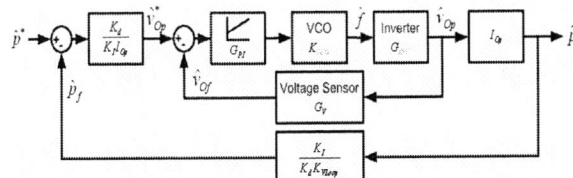

Fig. 3. Block diagram for signal variation of the SPCS.

Equation (16) shows that, voltage-command signal for load voltage control loop is proportional to the different between power-command signal $\hat{p}^*(t)$ and output power feedback signal $\hat{p}_f(t)$. Block diagram for signal variation of the SPCS based on (16) is shown in Fig. 3.

The SPCS in Fig. 2 seems to process no power feedback, but its small-signal model in Fig. 3 does reveal power feedback.

Block diagram for signal variations of the SPCS in Fig. 3 shows that, the proposed control scheme features feed forward of power control reference signal and load variation. Transient responses of the proposed control scheme are faster than the standard power control scheme with power controller.

C. Static Accuracy or Steady-State Error

In order to assess power controllability of the proposed control scheme, its static accuracy, and determine parameters governing power control static accuracy, steady-state error (SSE) of the SPCS should be calculated. Standard formulae of SSE can not be used, because neither the power feedback signal nor error signal exist in the SPCS. Basic definition will be used to calculate SSE.

$$SSE \quad \frac{P^* - \bar{P}}{P^*} = \frac{P^* - K_P \cdot P}{P^*} = 1 - K_P \cdot \frac{P}{P^*} \quad (18)$$

Because there is no power signal $\bar{P}$ which is proportional to output power P in the simplified control scheme, power transducer gain K_P must be determined indirectly. The product $(K_P \cdot P)$ may be considered as apparent power signal $\bar{P} = (K_P \cdot P)$.

In order to calculate SSE, closed-loop gain of power control loop P/P^* and power sensor's gain K_P must be known. As P/P^* is known as in(11), only K_P will be determined.

Power sensor's gain K_P is the ratio of power-command to output power for zero SSE condition. In the SPCS, SSE can take place only in voltage control loop. All transducer in the system must be considered as error free. For zero SSE condition, voltage reference signal $V_{OP\infty}^*$ is equal to the product of output voltage $V_{OP\infty}$ and voltage sensor's gain K_V as in (19)

$$V_{OP\infty}^* = \frac{K_d \cdot P^*}{K_I \cdot I_{OP\infty}} = K_V \cdot V_{OP\infty} \quad (19)$$

$$P_\infty = \frac{V_{OP\infty} \cdot I_{OP\infty}}{2} = \frac{K_d \cdot P^*}{2 \cdot K_I \cdot K_V} \quad (20)$$

$$K_P = \frac{P^*}{P_\infty} = \frac{2 \cdot K_I \cdot K_V}{K_d} \quad (21)$$

Substitute (11) and (21) in (18)

$$SSE = 1 - K_P \cdot \frac{P}{P^*} = 1 - \frac{2 \cdot K_I \cdot K_V}{K_d} \cdot \frac{K_d}{2 \cdot K_I} \cdot K_{VLoop} \quad (22)$$

$$SSE = 1 - K_V \cdot K_{VLoop} \quad (23)$$

If the open-loop gain of the voltage control loop approaches infinity, large-signal closed-loop gain K_{VLoop} is independent of operating point and approaches the reciprocal of voltage sensor's gain $1/K_V$. The product $K_V \cdot K_{VLoop}$ will approach unity. SSE is reduced to zero.

III. MODELING RESONANT INVERTER

A. Inverter Circuit

Circuit diagram of class D series resonant inverter[4] for an ESU is shown in Fig. 4(a). For a parallel loaded series resonant circuit with high Q factor, the fundamental frequency approximation can be applied as long as the inverter operating frequency is higher but not far from resonance. Fundamental frequency approximation removes the switching frequency's harmonics from the fundamental frequency sinusoidal waveforms, and the switching network can be replaced by a sinusoidal voltage source as shown in Fig. 4(b).

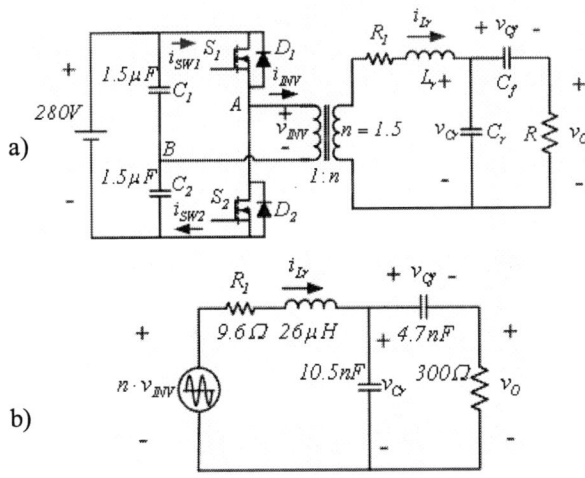

Fig. 4. Circuit diagram of class-D series resonant inverter and its equivalent circuit.

B. Inverter Model and Transfer Function

From the circuit diagram in Fig. 4(b), inverter's state equations can be obtained.

$$i_{Lr} = -\frac{R_l}{L_r}i_{Lr} - \frac{1}{L_r}v_{Cr} + \frac{1}{L_r}n \cdot v_{INV} \tag{24}$$

$$\dot{v}_{cr} = \frac{1}{C_r}i_{Lr} - \frac{1}{C_r R}\left(v_{Cr} - v_{Cf}\right) \tag{25}$$

$$\dot{v}_{cf} = \frac{1}{C_f R}\left(v_{Cr} - v_{Cf}\right) \qquad v_O = \left(v_{Cr} - v_{Cf}\right) \tag{26}$$

When modified phasor transformation is applied [2],[3], all state variables in the state equations are replaced by their corresponding time-varying phasors, and decomposed the state equations into direct (d) and quadrature (q) axis components:

$$i_{Lr1} = -\frac{R_l}{L_r}i_{Lr1} - \omega(t)\cdot i_{Lr2} - \frac{1}{L_r}v_{Cr1} + \frac{n}{L_r}\cdot\frac{2}{\pi}\cdot v_{DC} \tag{27}d$$

$$i_{Lr2} = \omega(t)\cdot i_{Lr1} - \frac{R_l}{L_r}i_{Lr2} - \frac{1}{L_r}v_{Cr2} \tag{28}q$$

$$\dot{v}_{Cr1} = \frac{1}{C_r}i_{Lr1} - \frac{1}{C_r R}v_{Cr1} - \omega(t)\cdot v_{Cr2} + \frac{1}{C_r R}v_{Cf1} \tag{29}d$$

$$\dot{v}_{Cr2} = \frac{1}{C_r}i_{Lr2} + \omega(t)\cdot v_{Cr1} - \frac{1}{C_r R}v_{Cr2} + \frac{1}{C_r R}v_{Cf2} \tag{30}q$$

$$\dot{v}_{Cf1} = \frac{1}{C_f R}v_{Cr1} - \frac{1}{C_f R}v_{Cf1} - \omega(t)\cdot v_{Cf2} \tag{31}d$$

$$\dot{v}_{Cf2} = \frac{1}{C_f R}v_{Cr2} + \omega(t)\cdot v_{Cf1} - \frac{1}{C_f R}v_{Cf2} \tag{32}q$$

$$v_{O1} = \left(v_{Cr1} - v_{Cf1}\right) \qquad v_{O2} = \left(v_{Cr2} - v_{Cf2}\right) \tag{33}$$

Applying small-signal perturbation to the system equations by replacing all the state variables with their corresponding combinations of steady-state and small-signal variations. The equations (27) to (33) are then decomposed into steady-state and small-signal transient equations. Only signals are considered.
Transient equations:

$$\frac{d\hat{x}(t)}{dt} = A_t\hat{x}(t) + B_t\hat{u}(t) \tag{34}$$

$$\hat{y}(t) = C_t\hat{x}(t) + D_t\hat{u}(t) \tag{35}$$

where

$$\hat{x} = \begin{bmatrix} \hat{i}_{lr1} & \hat{i}_{lr2} & \hat{v}_{cr1} & \hat{v}_{cr2} & \hat{v}_{cf1} & \hat{v}_{cf2} \end{bmatrix}^T \tag{36}$$

$$A_t = \begin{bmatrix} -\dfrac{R}{L_r} & -\Omega & -\dfrac{1}{L_r} & 0 & 0 & 0 \\[2mm] +\Omega & -\dfrac{R}{L_r} & 0 & -\dfrac{1}{L_r} & 0 & 0 \\[2mm] \dfrac{1}{C_r} & 0 & -\dfrac{1}{C_r R} & -\Omega & \dfrac{1}{C_r R} & 0 \\[2mm] 0 & \dfrac{1}{C_r} & \Omega & -\dfrac{1}{C_r R} & 0 & \dfrac{1}{C_r R} \\[2mm] 0 & 0 & \dfrac{1}{C_f R} & 0 & -\dfrac{1}{C_f R} & -\Omega \\[2mm] 0 & 0 & 0 & \dfrac{1}{C_f R} & +\Omega & -\dfrac{1}{C_f R} \end{bmatrix} \tag{37}$$

$$\hat{u} = \begin{bmatrix} \hat{v}_{dc} & \hat{f} \end{bmatrix}^T \qquad \hat{y}(t) = \hat{v}_{op} \tag{38}$$

$$B_t = \begin{bmatrix} \dfrac{n}{L_r}\dfrac{2}{\pi} & 0 & 0 & 0 & 0 & 0 \\[2mm] -2\pi I_{Lr2} & 2\pi I_{Lr1} & -2\pi V_{Cr2} & 2\pi V_{Cr1} & -2\pi V_{Cf2} & 2\pi V_{Cf1} \end{bmatrix}^T \tag{39}$$

$$C_t = \begin{bmatrix} 0 & 0 & \dfrac{V_{O1}}{V_{OP}} & \dfrac{V_{O2}}{V_{OP}} & \dfrac{V_{O1}}{V_{OP}} & -\dfrac{V_{O1}}{V_{OP}} \end{bmatrix} \tag{40}$$

$$D_t = \begin{bmatrix} 0 & 0 \end{bmatrix} \tag{41}$$

$$\hat{x}(s) = \begin{bmatrix} sI - A_t \end{bmatrix}^{-1} \cdot B_t \cdot \hat{u}(s) \tag{42}$$

$$\hat{y}(s) = \begin{bmatrix} C_t \cdot \left\{ \left(sI - A_t\right)^{-1} \cdot B_t \right\} + D_t \end{bmatrix} \cdot \hat{u}(s) \tag{43}$$

$$\hat{v}_{op} = \begin{bmatrix} \dfrac{V_{O1}}{V_{OP}} \end{bmatrix}\hat{v}_{Cr1} + \begin{bmatrix} \dfrac{V_{O2}}{V_{OP}} \end{bmatrix}\hat{v}_{Cr2} - \begin{bmatrix} \dfrac{V_{O1}}{V_{OP}} \end{bmatrix}\hat{v}_{Cf1} - \begin{bmatrix} \dfrac{V_{O2}}{V_{OP}} \end{bmatrix}\hat{v}_{Cf2} \tag{44}$$

Bode plot of inverter's control-to-output transfer function $v_o(s)/f(s)$ is essential for voltage controller design. Inverter's dynamic equations (35) to (44) were used to calculate control-to-output transfer function of the inverter. The parameters of inverter circuit components are shown in the circuit diagram of Fig 4.

IV. POWER CONTROL SYSTEM DESIGN

The SPCS was implemented in the power control of an ESU. In the power control loop, only proportional gain of the signal divider circuit is required. On the other hand, inverter output voltage controller should be properly designed. Bode plot of inverter's control-to-output transfer function is used for voltage controller design. In order to minimize SSE, standard PI controller is the choice. The transfer function of the controller is

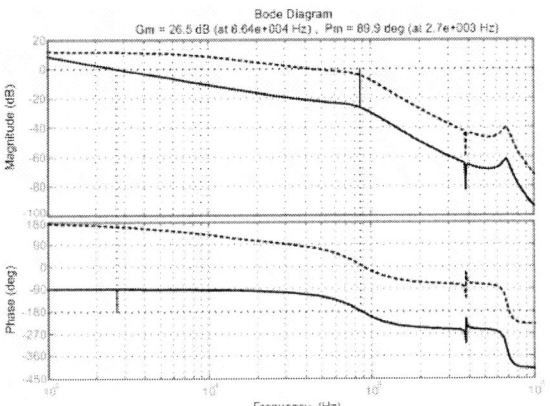

Fig. 5. Open-loop gain of voltage control loop with (lower) and without (upper) controller.

$$G_{vc}(s) = \frac{K_0(\frac{s}{\omega_z}+1)}{s} = \frac{K_0}{\omega_z} + \frac{K_0}{s} \qquad (45)$$

The voltage controller is design for the least favorable condition of load resistance-light load condition. The parameters of the controller are

$$K_0 = -4.24 \times 10^3 \quad , \qquad \omega_z = 57.6 \times 10^3 \quad rad/\sec \quad (46)$$

Fig. 5 shows Bode plot of open-loop gain of voltage control loop with (lower) and without (upper) voltage controller. Gain margin and phase margin of voltage control loop is 26.5 dB and 90 degrees respectively.

Fig. 6 shows Bode plot of open-loop gain of power control loop. Gain margin and phase margin of power control loop is 29.7 dB and 180 degrees respectively. Theoretical calculation of transient response for 50 W step command (250W to 300W) is shown in Fig. 7(a). The transient response of the system seems to dominate by single dominant pole. The rise time is approximately 0.11 ms which is corresponding to rise time of a first order system with a corner frequency of 3.18 kHz.

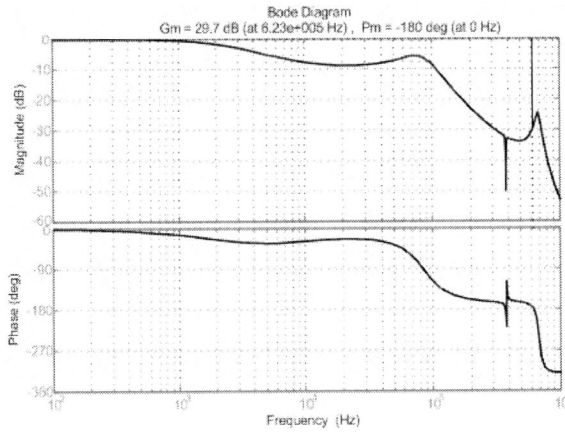

Fig. 6. Open-loop gain of power control loop.

V. SIMULATION AND EXPERIMENTAL RESULTS

Computer simulation and hardware realization are used to verify the theoretical calculation. Envelopes of the original time-domain waveforms are presented Fig. 7(b) and 7(c) show transient response for 50W step command (250W to 300W, load resistance R = 210 Ω). Fig. 8 show transient response for step load (R is decreased from 240 Ω to 210 Ω). Good agreement between theoretical analysis, computer simulation, and hardware realization ratifies the simplified power control scheme.

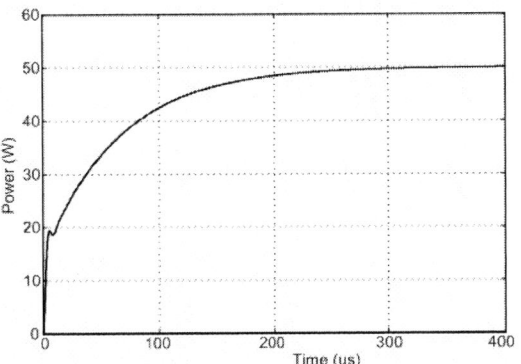

(a) Theoretical calculation.

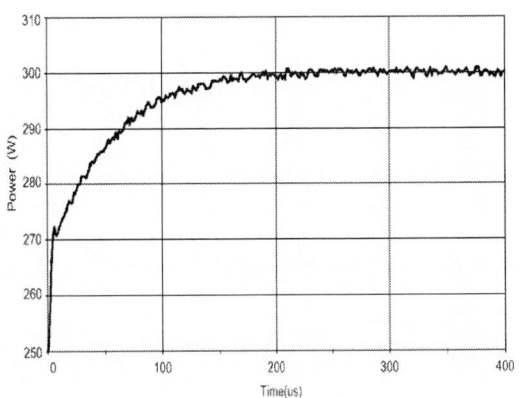

(b) Simulation result.

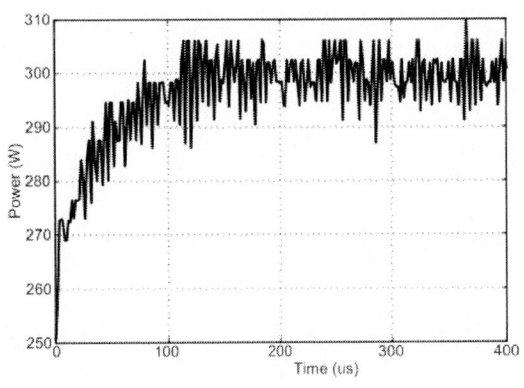

(c) Experimental result

Fig. 7. Transient response for 50 W step command.

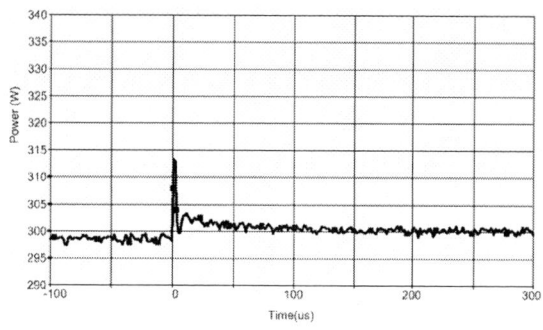

(a) Simulation result.

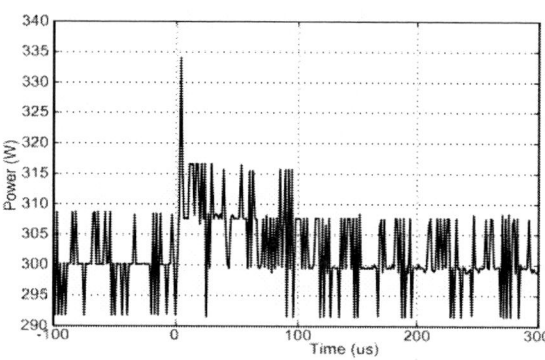

(b) Experimental result.

Fig. 8. Transient response for step load.

As SSE is very sensitive to circuit parameters, only simulation results will be used to compare with the theoretical calculations. SSEs of a nonzero SSE power control system for different values of power-reference signal and load resistance are tabulated in TABLE I. Agreement between theoretical calculations and computer simulations is satisfactory.

TABLE I SEE OF NONZERO SSE SYSTEM

SSE of voltage				
P*signal	P*out/set	R	simulation	calculation
1.35V	300W	300Ω	4.24	4.24
1.35V	300W	200Ω	4.91	4.91
0.675V	150W	300Ω	4.51	4.51
0.675V	150W	200Ω	4.77	4.77
SSE of calculated power				
P*signal	P*out/set	R	simulation	calculation
1.35V	300W	300Ω	4.24	4.24
1.35V	300W	200Ω	3.14	4.91
0.675V	150W	300Ω	5.99	4.51
0.675V	150W	200Ω	4.52	4.52
Parameters				
K_V=0.015	K_I=1.5		K_d=10	K_P=0.0045

Transient response of voltage error and calculated power error signal for a nonzero SSE condition (finite open-loop gain of voltage control loop) is presented in Fig. 9. SSE of both voltage and power control loop are approximately 5 percent.

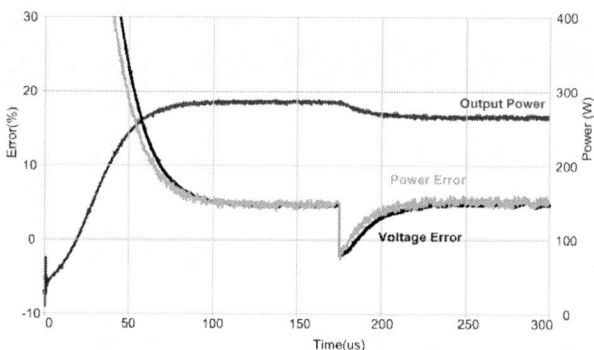

Fig. 9. Transient response of voltage error and calculated power error signal of nonzero SSE system

VI. CONCLUSION

SPCS of resonant inverter relies on a simple relationship—output power is proportional to product of the peak of load voltage and current. The simple relationship is valid for an inverter supplying purely resistive load with sinusoidal voltage. There is no explicit power feedback in the SPCS. Output power control takes effect through voltage feedback loop, as the voltage reference signal is a result of power reference signal divided by output current signal. If there is power reference signal and/or load variations, the variations is fed forward to voltage control loop in order to regulate output power. Static accuracy analysis shows that, SSE of the power control scheme depends on static accuracy of voltage loop control. As open-loop gain of voltage loop control approach infinity, SSE of both voltage loop and power loop control approach zero. Zero static accuracy of power control loop confirms that, the SPCS features power feedback. Computer simulation and experimental results ratify controllability of the SPCS.

REFERENCES

[1] J.R. LaCOURSE, M.C. Vogt, T. Miller, and S.M. Selikowitz, "Spectral Analysis Interpretation of Electrosurgical Generator Nerve and Muscle Stimulation," IEEE Trans. Biomed. Eng., vol. 35, no. 7, July 1988.

[2] Chun T. Rim and Gyu H. Cho, "Phasor Transformation and its Application to the DC/AC Analysis of Frequency Phase-Controlled Series Resonant Converters," IEEE Trans. Power Electron., vol. 5, no. 2, April 1990.

[3] Yan Yin, Regan Zane, John Glaser, and Robert W.Erickson, "Small-signal Analysis of Frequency-Controlled Electronic Ballasts," IEEE Trans. circuits and system, vol. 50, no. 8, August 2003.

[4] Sayed-Amr El-Hamamsy, "Design of High-Efficiency RF Class-D Power Amplifier," IEEE Trans. Power Electron., vol. 9, no. 3, May 1994.

Voltage Injection Based Initial Rotor Position Estimation Method for Three-Phase Star-Connected Switched Reluctance Machines

P. Somsiri[1,2*], P. Champa[1,2*], P. Wipasuramonton[1], K. Tungpimonrut[1], P. Aree[2]

[1]Industrial Control and Automation Lab, National Electronics and Computer Technology Center, Thailand
[2]Department of Electrical Engineering, Thammasat University, Thailand
E-mail: pakasit.somsiri@nectec.or.th

Abstract—This paper presents a simple method to determine the initial rotor position of a three-phase star-connected switched reluctance machine (SRM) at standstill without position and current sensors. The principle of the initial rotor position estimation is based on the stator inductance variation related to the actual rotor position. In the proposed method, only two narrow voltage pulse injections are applied in sequence. In this method the machine parameters and phase current detection are not required. The method is suitable for low-cost SRM-drive applications. The experimental results were carried out on a 6/4 poles three-phase SRM and show the satisfactory accuracy of the proposed initial rotor position estimation.

Index Terms—Switched reluctance motor, rotor position estimation, standstill

I. INTRODUCTION

Demand for high efficiency and low cost variable speed motor drives is increasing particularly in household and consumer applications. A switched reluctance machine (SRM) drive is considered as a competitive candidate to the other conventional variable speed drives due to several advantages [1]-[2]. SRM offer various attractive features such as, simple stator and rotor constructions, simple stator windings, no permanent magnet inside, simple cooling system, low cost, high reliability, and good performance over a wide speed range. However, the converter circuit most commonly used for SRM drives is the asymmetric half bridge converter, as shown in Fig. 1. This topology offers a unidirectional current; the voltage and current supplied to one phase are completely independent of the remaining phases by sequential excitation of each one of the SRM phases. Although this topology is simple, it is somewhat expensive due to the lack of a standardized power electronic converter for SRM drives, which would be available on the market as a single module. In recent years, due to the cost reduction and performance improvement, the three-phase full-bridge converter, which has long been used for ac drives, has been applied for SRM drives [3]-[5]. A three-phase SRM, with the star-connected stator windings, can be driven by a standard three-phase full-bridge converter, as illustrated in Fig. 2. Compared to the asymmetric half-bridge converter, the three-phase full-bridge converter has several advantages, including cost reduction due to the use of a standardized converter and compactness. In the

case of 6/4 poles three-phase SRM, the number of power cables between converter and SRM is reduced to the half.

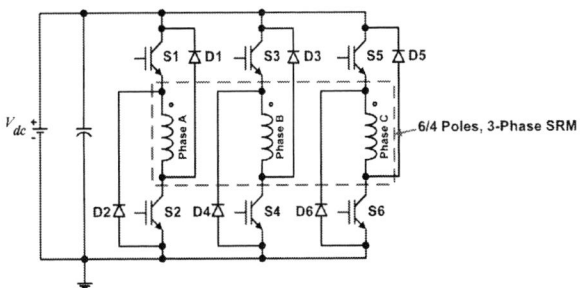

Fig.1 Three-phase SRM drives using an asymmetric half-bridge converter

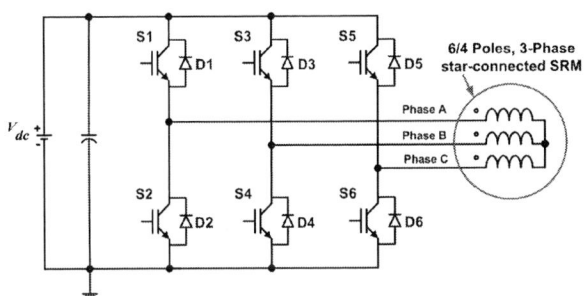

Fig.2 Three-phase star-connected SRM drives using a standard three-phase full-bridge converter.

The rotor position plays a key role in an efficient SRM operation. It needs to be synchronized with the phase excitations of the SRM. In general, it can be measured by using a position sensor. However, existence of a position sensor such as an optical encoder or a resolver not only adds cost and increases system size, but it also degrades the overall SRM drive system reliability. In the case of SRM, the starting from an unknown rotor position may be oscillation and the rotor kicking-back may cause a starting failure. Thus, smoothly and reliably starting the SRM from standstill without using a position sensor remains a research problem. Various methods for rotor position estimation at standstill to solve this problem were proposed [6-8]. The first method, each SRM phase was excited with a narrow voltage pulse. Then, the amplitude of each phase current measured was used to detect the initial rotor position [7]. This method required three current sensors. In either case, due to its limited

978-1-4244-0644-9/07/$25.00 ©2007 IEEE

resolution, the test current with very small amplitude may be detected incorrectly. In the second method, the initial rotor position was based on the flux linkage-current-position characteristic of the SRM [8]. However it required a large test signals and a comprehensive set of magnetic data to detect the rotor position at standstill. Additionally, this method was based on complex calculations. Furthermore, machine parameters and information of the phase currents are required.

In this paper, a simple method to determine the initial rotor position of a three-phase star-connected SRM at standstill is presented. This method is so simple that only two narrow voltage pulse injections are applied in sequence. In addition, no machine parameter and information of phase currents, as well as complex calculations, are required. Consequently, position and current sensors are not necessary in the proposed initial rotor position estimation. Hence SRM drives with the proposed method can be employed for low cost applications.

The next section of this paper describes the basic principle of the SRM. Section III presents the proposed initial rotor position estimation method. Section IV presents experimental results to validate and to demonstrate the effectiveness of the proposed method, followed by the conclusion in Section V.

II. BASIC PRINCIPLE OF SRM

The structure of the 6/4 three-phase SRM at aligned position of phase-A ($\theta = 0^\circ$) is shown in Fig. 3, where θ is the rotor position angle. The stator and rotor have different salient poles. While the rotor rotates, the phase inductance of the stator winding is varied. The relationship between the rotor position and the ideal phase inductance can be shown in Fig. 4. As shown in the figure, the sector numbers are defined for each 15° interval. Therefore, it can be expressed as:

Sector no.1: $1^\circ \leq \theta \leq 15^\circ$;   ($L_B < L_C < L_A$)

Sector no.2: $16^\circ \leq \theta \leq 30^\circ$;   ($L_B < L_A < L_C$)

Sector no.3: $31^\circ \leq \theta \leq 45^\circ$;   ($L_A < L_B < L_C$)

Sector no.4: $46^\circ \leq \theta \leq 60^\circ$;   ($L_A < L_C < L_B$)

Sector no.5: $61^\circ \leq \theta \leq 75^\circ$;   ($L_C < L_A < L_B$)

Sector no.6: $76^\circ \leq \theta \leq 90^\circ$;   ($L_C < L_B < L_A$)

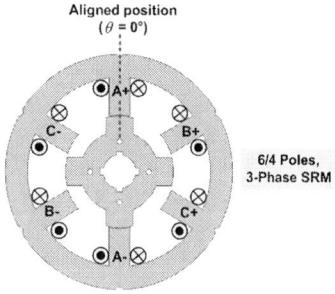

Fig. 3 Simplified structure of 6/4 three-phase SRM.

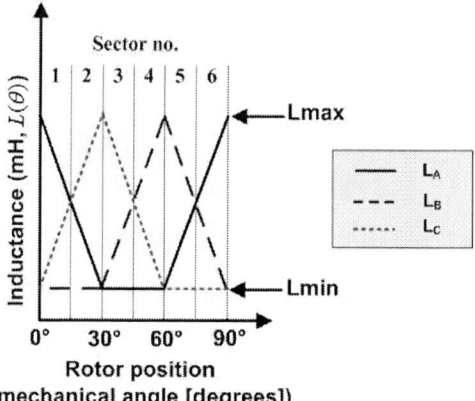

Fig.4 Idealized unsaturated inductance versus rotor position of 6/4 poles three-phase SRM.

In SRM drives, electromagnetic torque (T) is primarily controlled by tuning the commutation instants and the profile of phase current. It must be noted that the nonlinear effect of magnetic saturation are neglected in this study as expressed in (1). The torque sign depends on the position of current pulse with respect to the inductance profile $L(\theta)$, and it does not depend on the polarity of current. Where i is the phase current of the stator winding.

$$T \approx \frac{1}{2} i^2 \frac{\partial L(\theta, i)}{\partial \theta} \qquad (1)$$

III. PROPOSED INITIAL ROTOR POSITION ESTIMATION METHOD

The principle of the proposed method for detecting the rotor position at standstill is to measure the stator voltage and then compare them in each phase. The measured voltage varies due to the influence of the variation of the stator inductance related to the position of the rotor. A sequence of two voltage pulses is injected to a pair of selected phase windings. The operations of this process can be explained as follows.

A) First Voltage Pulse Injection

The voltage pulse injection consists of two intervals: the pulse injecting interval and the free-wheeling interval.

1) The Pulse Injecting Interval

A narrow voltage pulse is injected through the phase-A and B windings by connecting phase-A to the positive bus (V_{DC+}) and phase-B to the negative bus (V_{DC-}) (switches S1 and S4 "on", Fig. 5(a)) for a short interval time, while the phase-C is floating without any current flow. Therefore, the neutral voltage (V_N) or voltage across the phase-B winding can be measured through the terminal of a phase-C. Then the sample voltage is recorded, ($v_{NB(on)}^{(1)}$). The simplified circuit is shown in Fig. 5(b). Since the influence of winding resistance for a short duration of voltage injection is negligibly small, the voltage equation during the pulse injecting interval can be expressed as:

$$V_{DC} = V_{AN(on)}^{(1)} + V_{NB(on)}^{(1)}$$

$$\approx \left[L_A(\theta_0) + L_B(\theta_0)\right]\frac{di(on)}{dt}$$

$$\approx L_A(\theta_0)\frac{\Delta i(on)}{T_{S(on)}} + L_B(\theta_0)\frac{\Delta i(on)}{T_{S(on)}} \qquad (2)$$

where the superscript (1) is the first voltage pulse injection, The subscript (on) is the pulse injecting interval, V_{DC} is the DC-bus voltage, $\Delta i(on)$ is the current change during this pulse injecting interval, $T_{S(on)}$ is the voltage pulse injecting interval time, θ_0 is the initial rotor position, L_A and L_B are the phase-A and B winding inductances respectively, and finally $V_{AN(on)}^{(1)}$ and $V_{NB(on)}^{(1)}$ are the voltage across the phase-A and B winding during the pulse injecting interval respectively. From (2), the phase winding voltages, $V_{AN(on)}^{(1)}$ and $V_{NB(on)}^{(1)}$ can be rewritten in terms of V_{DC} as:

$$V_{AN(on)}^{(1)} = \left(\frac{L_A}{L_A + L_B}\right)V_{DC} \qquad (3)$$

and

$$V_{NB(on)}^{(1)} = \left(\frac{L_B}{L_A + L_B}\right)V_{DC} \qquad (4)$$

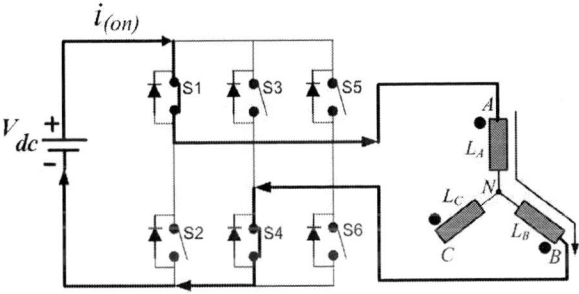

(a)   Current paths of switches S1 and S4 "on"

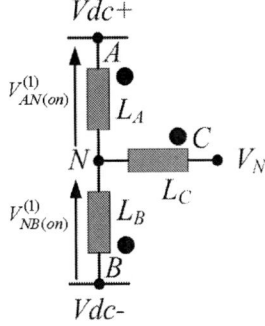

(b)   Simplified circuits

Fig. 5 Current paths and simplified circuits during pulse injecting interval of the first voltage pulse injection

2) The Free-Wheeling Interval

After the first pulse injecting interval, the switches S1 and S4 are then turned off, and its equivalent circuit is shown in Fig. 6(a). Consequently, the corresponding free-wheeling interval occurs as illustrated in Fig. 6(b). Then the sample voltage is recorded, ($V_{NA(off)}^{(1)}$). The voltage equation in the free-wheeling interval can be derived as:

$$V_{DC} = V_{NA(off)}^{(1)} + V_{BN(off)}^{(1)}$$

$$\approx -L_A(\theta_0)\frac{\Delta i(off)}{T_{S(off)}} - L_B(\theta_0)\frac{\Delta i(off)}{T_{S(off)}} \qquad (5)$$

where the subscript (off) is the free-wheeling interval, $\Delta i(off)$ is the current change during this free-wheeling interval, $T_{S(off)}$ is the free-wheeling interval time, and finally $V_{NA(off)}^{(1)}$ and $V_{BN(off)}^{(1)}$ are the voltage across the phase-A and B winding during this free-wheeling interval, respectively.

Fig. 5-6 shows the current paths and the simplified equivalent circuits arisen from the first injected voltage pulse, during both the injected pulse and the free-wheeling intervals. It should be noted that the total voltage drop across both the phase-A and B windings during the free-wheeling interval is opposite to that of the pulse-injecting interval, thus the immediate change of the DC-bus voltage also reflects to the phase-A and B windings in the same manner. Therefore, it can be concluded from (2) and (5) that:

$$V_{AN(on)}^{(1)} = V_{NA(off)}^{(1)}$$

and

$$L_A(\theta_0)\frac{\Delta i(on)}{T_{S(on)}} = -L_A(\theta_0)\frac{\Delta i(off)}{T_{S(off)}} \qquad (6)$$

$$V_{NB(on)}^{(1)} = V_{BN(off)}^{(1)}$$

and

$$L_B(\theta_0)\frac{\Delta i(on)}{T_{S(on)}} = -L_B(\theta_0)\frac{\Delta i(off)}{T_{S(off)}} \qquad (7)$$

Also, the voltages across the phase-A and B windings during the free-wheeling interval, Fig. 6(b), are

$$V_{BN(off)}^{(1)} = \left(\frac{L_B}{L_A + L_B}\right)V_{DC} \qquad (8)$$

and

$$V_{NA(off)}^{(1)} = \left(\frac{L_A}{L_A + L_B}\right)V_{DC} \qquad (9)$$

Finally, from (4) and (9), L_A and L_B, can be compared to each other according to the following equation:

$$V_{NA(off)}^{(1)} - V_{NB(on)}^{(1)} = \left(L_A - L_B\right)\frac{V_{DC}}{\left(L_A + L_B\right)} \qquad (10)$$

From (10), it is obviously seen that the values of L_A and L_B can be compared to each other by voltage comparison between $v_{NA(off)}^{(1)}$ and $v_{NB(on)}^{(1)}$ which can be done simply since both of them are reference to V_{DC-}.

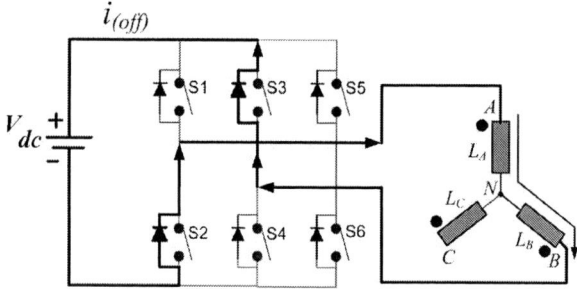

(a) Current paths of switches S1 and S4 "off"

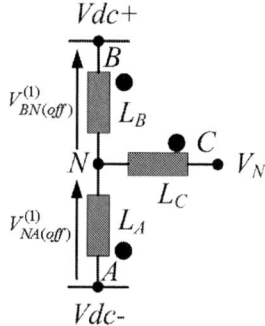

(b) Simplified circuits

Fig. 6 Current paths and simplified circuits during free-wheeling interval of the first voltage pulse injection

B) Second Voltage Pulse Injection

After disappearance of the first free-wheeling interval, with the same principle, the second voltage pulse with the same interval is injected to the phase-A and C windings. The phase-A connected to the positive bus and phase-C to the negative bus (switches S1 and S6 "on"), while phase-B is floating without any current flow. Then the sample voltage is recorded, ($v_{NC(on)}^{(2)}$), similar to those of the first injection, where the superscript (2) is the second voltage pulse injection, $v_{AN(on)}^{(2)}$ and $v_{NC(on)}^{(2)}$ are the voltage across the phase-A and C winding during the pulse injecting interval respectively. Then, the switches S1 and S6 are turned off and, consequently, the corresponding free-wheeling interval occurs. The sample voltage is recorded, ($v_{NA(off)}^{(2)}$), similar to those of the first free-wheeling interval. Therefore, the detected voltages across phase-A and C windings, and hence L_A and L_C, can be compared to each other according to the following equation:

$$V_{NA(off)}^{(2)} - V_{NC(on)}^{(2)} = \left(L_A - L_C\right)\frac{V_{DC}}{\left(L_A + L_C\right)} \quad (11)$$

where $v_{NC(on)}^{(2)}$ is the voltage across the phase-C winding during the second pulse-injecting interval and $v_{NA(off)}^{(2)}$ is the voltage across the phase-A winding during the second free-wheeling interval. These voltages can be simply measured through the phase-B terminal with reference to V_{DC-}.

Additionally, the different values of L_B and L_C ($L_B - L_C$) can be determined by the voltage difference between $v_{NB(on)}^{(1)}$ and $v_{NC(on)}^{(2)}$ according to the following equation:

$$V_{NB(on)}^{(1)} - V_{NC(on)}^{(2)}$$
$$= \left(L_B - L_C\right)\left(\frac{L_A}{\left(L_A + L_B\right)\left(L_A + L_C\right)}\right)V_{DC} \quad (12)$$

Finally, it can be summarized that the winding inductance values L_A, L_B and L_C, can be compared to each other by using (10) - (12). The measured voltage and inductance profile comparisons are related to the initial rotor position using the proposed method can be summarized in Table I.

TABLE I
Determination of Initial Rotor Position Using the Proposed Method

Phase voltage comparison			Inductance comparison	Initial rotor position	Chosen phase for excitation (CW)
$v_{NB(on)}^{(1)} < v_{NA(off)}^{(1)}$	$v_{NC(on)}^{(2)} < v_{NA(off)}^{(2)}$	$v_{NB(on)}^{(1)} < v_{NC(on)}^{(2)}$	$L_B < L_C < L_A$	$1°\leq\theta<15°$	B and C
$v_{NB(on)}^{(1)} < v_{NA(off)}^{(1)}$	$v_{NC(on)}^{(2)} = v_{NA(off)}^{(2)}$	$v_{NB(on)}^{(1)} < v_{NC(on)}^{(2)}$	$L_B < L_C = L_A$	$\theta = 15°$	B and C
$v_{NB(on)}^{(1)} < v_{NA(off)}^{(1)}$	$v_{NA(off)}^{(2)} < v_{NC(on)}^{(2)}$	$v_{NB(on)}^{(1)} < v_{NC(on)}^{(2)}$	$L_B < L_A < L_C$	$16°\leq\theta<30°$	B and C
$v_{NB(on)}^{(1)} = v_{NA(off)}^{(1)}$	$v_{NA(off)}^{(2)} < v_{NC(on)}^{(2)}$	$v_{NB(on)}^{(1)} < v_{NC(on)}^{(2)}$	$L_B = L_A < L_C$	$\theta = 30°$	B and C
$v_{NA(off)}^{(1)} < v_{NB(on)}^{(1)}$	$v_{NA(off)}^{(2)} < v_{NC(on)}^{(2)}$	$v_{NB(on)}^{(1)} < v_{NC(on)}^{(2)}$	$L_A < L_B < L_C$	$31°\leq\theta<45°$	A and B
$v_{NA(off)}^{(1)} < v_{NB(on)}^{(1)}$	$v_{NA(off)}^{(2)} < v_{NC(on)}^{(2)}$	$v_{NB(on)}^{(1)} = v_{NC(on)}^{(2)}$	$L_A < L_B = L_C$	$\theta = 45°$	A and B
$v_{NA(off)}^{(1)} < v_{NB(on)}^{(1)}$	$v_{NA(off)}^{(2)} < v_{NC(on)}^{(2)}$	$v_{NC(on)}^{(2)} < v_{NB(on)}^{(1)}$	$L_A < L_C < L_B$	$46°\leq\theta<60°$	A and B
$v_{NA(off)}^{(1)} < v_{NB(on)}^{(1)}$	$v_{NA(off)}^{(2)} = v_{NC(on)}^{(2)}$	$v_{NC(on)}^{(2)} < v_{NB(on)}^{(1)}$	$L_A < L_C < L_B$	$\theta = 60°$	A and B
$v_{NA(off)}^{(1)} < v_{NB(on)}^{(1)}$	$v_{NC(on)}^{(2)} < v_{NA(off)}^{(2)}$	$v_{NC(on)}^{(2)} < v_{NB(on)}^{(1)}$	$L_C < L_A < L_B$	$61°\leq\theta<75°$	A and B
$v_{NA(off)}^{(1)} = v_{NB(on)}^{(1)}$	$v_{NC(on)}^{(2)} < v_{NA(off)}^{(2)}$	$v_{NC(on)}^{(2)} < v_{NB(on)}^{(1)}$	$L_C < L_A = L_B$	$\theta = 75°$	A and C
$v_{NB(on)}^{(1)} < v_{NA(off)}^{(1)}$	$v_{NC(on)}^{(2)} < v_{NA(off)}^{(2)}$	$v_{NC(on)}^{(2)} < v_{NB(on)}^{(1)}$	$L_C < L_B < L_A$	$76°\leq\theta<90°$	A and C
$v_{NB(on)}^{(1)} < v_{NA(off)}^{(1)}$	$v_{NC(on)}^{(2)} < v_{NA(off)}^{(2)}$	$v_{NC(on)}^{(2)} = v_{NB(on)}^{(1)}$	$L_C = L_B < L_A$	$\theta = 90°$ or $\theta = 0°$	A and C

IV. EXPERIMENTAL RESULTS

The experimental system which was set up to validate the proposed method is shown in Fig. 7. It can be seen that the only four additional resistors, without any expensive current sensor, are integrated in the system. The duration of the injected voltage pulses applied in the experiment is $150\,\mu s$ by using dsPIC30F4011, with 50 volts of the DC-bus voltage. The SRM is star-connected, 6/4 poles type. The machine parameters are summarized in Table II.

Table II
Parameters of SRM

3 phase stator/rotor poles	6/4
Rated power	1,500W
Rated voltage	48V
Rated speed	1,500r/min
Stator pole-arc	22.4°
Rotor pole-arc	30.8°

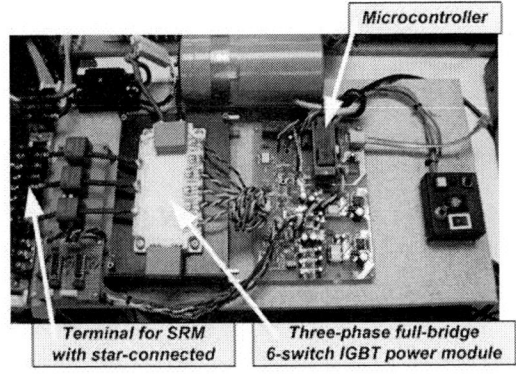

(a)   Hardware of three-phase full-bridge converter for SRM drive

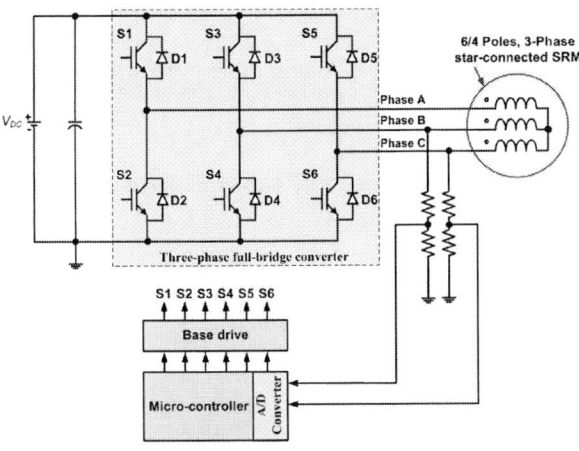

(b)   Configuration of SRM drive system

Fig. 7 Experimental SRM drive system

To avoid adverse affects during transient state due to parasitic parameters and non-idealities in the test system, the neutral voltages, with reference to the negative DC-bus, are sampled at the middle of the injected voltage pulse interval at 75 μs after the beginning of the voltage injection and around the middle of free-wheeling interval at 75 μs after the end of the voltage injection as shown in Fig. 8.

Fig. 8 shows the measured neutral voltages resulting of the injections at 10° of the actual position, and from the first (switches S1 and S4 "on", Fig. 8a, then the two sample voltages are recorded ($v_{NB(on)}^{(1)}$ and $v_{NA(off)}^{(1)}$)), and the second (switches S1 and S6 "on", Fig. 8b, then the two sample voltages are recorded ($v_{NC(on)}^{(2)}$ and $v_{NA(off)}^{(2)}$)). From the figure, it can be seen that the condition of phase

voltage comparison is $v_{NB(on)}^{(1)} < v_{NA(off)}^{(1)}$, $v_{NC(on)}^{(2)} < v_{NA(off)}^{(2)}$, $v_{NB(on)}^{(1)} < v_{NC(on)}^{(2)}$ and therefore, according to Table I, the estimate rotor position is 1°≤θ<15°.

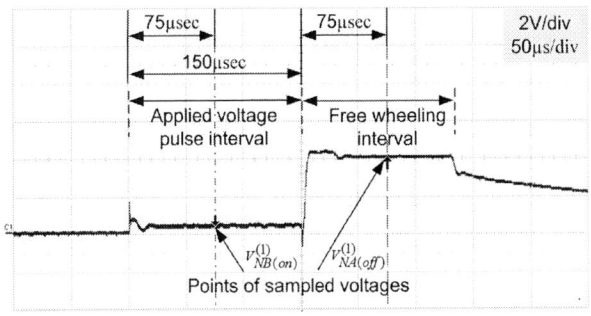

(a) Neutral voltage measured through phase-C terminal when the voltage pulse is injected across phase-A and B terminals (phase voltage comparison is $v_{NB(on)}^{(1)} < v_{NA(off)}^{(1)}$).

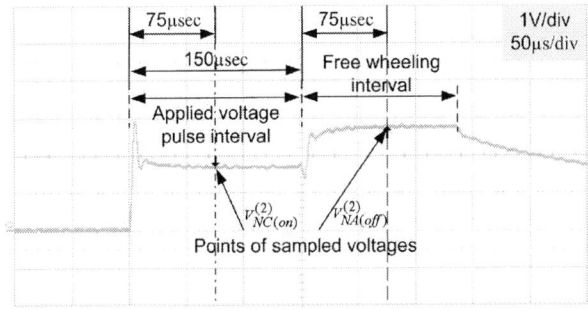

(b) Neutral voltage measured through phase-B terminal when the voltage pulse is injected across phase-A and C terminals (phase voltage comparison is $v_{NC(on)}^{(2)} < v_{NA(off)}^{(2)}$).

Fig. 8 Neutral voltages with reference to the negative DC-bus when $\theta_0 = 10°$ (inductance comparison is $L_B < L_C < L_A$)

In order to verify the accuracy of the proposed estimation method which have about 1 degree of resolution, the experiment is investigated at the 29°, 30° and 31°. Figs. 9-11 show the waveform of neutral voltages resulting from the first and the second injections in the case that the actual rotor position is located at 29°, 30° and 31°, respectively.

- Fig. 9 shows the measured neutral voltage resulting of the injections at 29°, it can be seen that the condition of phase voltage comparison is $v_{NB(on)}^{(1)} < v_{NA(off)}^{(1)}$, $v_{NA(off)}^{(2)} < v_{NC(on)}^{(2)}$, $v_{NB(on)}^{(1)} < v_{NC(on)}^{(2)}$ and therefore, according to Table I, the estimate rotor position is 16°≤θ<30°.

- Fig. 10 shows the measured neutral voltage resulting of the injections at 30°, it can be seen that the condition of phase voltage comparison is $v_{NB(on)}^{(1)} = v_{NA(off)}^{(1)}$, $v_{NA(off)}^{(2)} < v_{NC(on)}^{(2)}$, $v_{NB(on)}^{(1)} < v_{NC(on)}^{(2)}$ and therefore, according to Table I, the estimate rotor position is θ =30°.

- Fig. 11 shows the measured neutral voltage resulting of the injections at $31°$, it can be seen that the condition of phase voltage comparison is $v_{NA(off)}^{(1)} < v_{NB(on)}^{(1)}$, $v_{NA(off)}^{(2)} < v_{NC(on)}^{(2)}$, $v_{NB(on)}^{(1)} < v_{NC(on)}^{(2)}$ and therefore, according to Table I, the estimate rotor position is $31°\leq\theta<45°$.

The estimate rotor position is simply clarified. Therefore, it can be indicated the proposed method gives the satisfactory accuracy of the estimate rotor position. This is useful for choosing the phases to be excited, according to Table 1, in the machine starting up.

Fig.11 shows the relationship from experiments between the estimated values against the actual position. The middle value of each rotor position range (Table I) is use to represent its range. The result indicates the ranges of the rotor angle that enough for motor starting.

Fig.12 shows the resulting of position error, where the maximum of position error is $7.5°$.

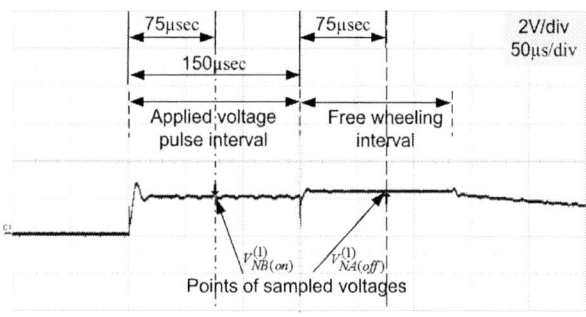

(a) Neutral voltage measured through phase-C terminal when the voltage pulse is injected across phase-A and B terminals (phase voltage comparison is $v_{NB(on)}^{(1)} < v_{NA(off)}^{(1)}$).

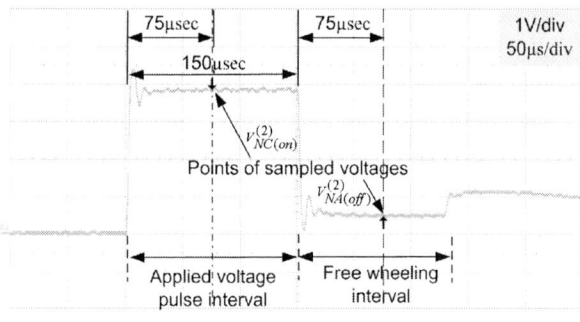

(b) Neutral voltage measured through phase-B terminal when the voltage pulse is injected across phase-A and C terminals (phase voltage comparison is $v_{NA(off)}^{(2)} < v_{NC(on)}^{(2)}$).

Fig. 9 Neutral voltages with reference to the negative DC-bus when $\theta_0 = 29°$ (inductance comparison is $L_B < L_A < L_C$)

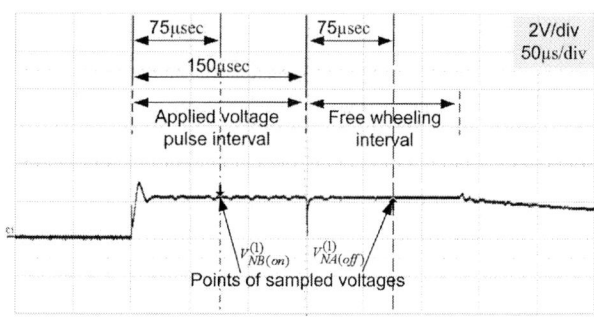

(a) Neutral voltage measured through phase-C terminal when the voltage pulse is injected across phase-A and B terminals (phase voltage comparison is $v_{NB(on)}^{(1)} = v_{NA(off)}^{(1)}$).

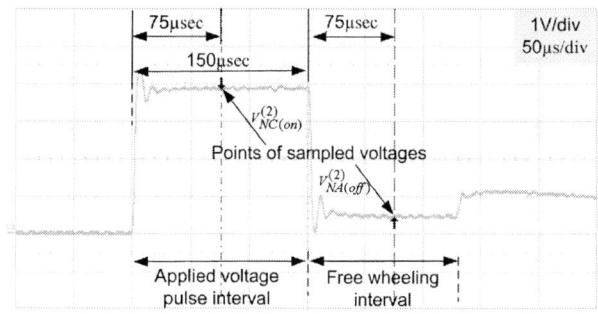

(b) Neutral voltage measured through phase-B terminal when the voltage pulse is injected across phase-A and C terminals (phase voltage comparison is $v_{NA(off)}^{(2)} < v_{NC(on)}^{(2)}$).

Fig. 10 Neutral voltages with reference to the negative DC-bus when $\theta_0 = 30°$ (inductance comparison is $L_B = L_A < L_C$)

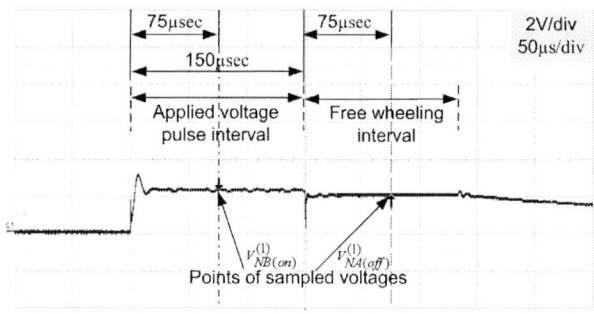

(a) Neutral voltage measured through phase-C terminal when the voltage pulse is injected across phase-A and B terminals (phase voltage comparison is $v_{NA(off)}^{(1)} < v_{NB(on)}^{(1)}$).

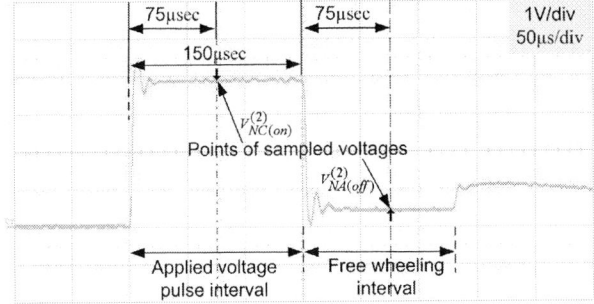

(b) Neutral voltage measured through phase-B terminal when the voltage pulse is injected across phase-A and C terminals (phase voltage comparison is $V^{(2)}_{NA(off)} < V^{(2)}_{NC(on)}$).

Fig. 11 Neutral voltages with reference to the negative DC-bus when $\theta_0 = 31°$ (inductance comparison is $L_A < L_B < L_C$)

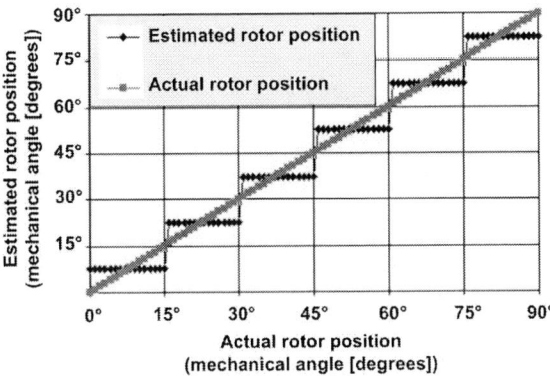

Fig. 12 Estimated versus actual position.

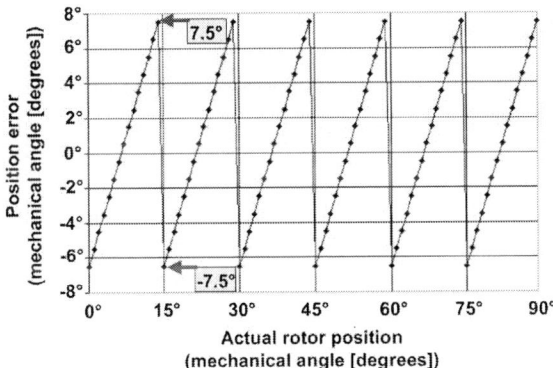

Fig. 13 Position estimation error

V. CONCLUSIONS

A simple initial rotor position estimation method at standstill is introduced in this paper. It is based on the stator inductance variation according to the rotor position. In the proposed method, only two narrow voltage pulses are applied in sequence to the phase windings to determine the rotor position. It is proved from the experiments that the satisfactory accuracy of the estimate

rotor position can be achieved. It is applicable for three-phase star-connected switched reluctance machines, in which low cost is the major concern. Moreover, the machine parameters and phase current detection are not required. As a result, no expensive position and current sensors are required in the proposed method.

REFERENCES

[1] T. J. E. Miller, *Switched Reluctance Motors and their Control*, Oxford: Science Publications, 1993.

[2] R. Krishnan, *Switched Reluctance Motor Drives*, CRC Press, 2001.

[3] A. C. Clothier and B. C. Mecrow, "The use of three phase bridge inverters with switched reluctance drives," Proc. Conf. Rec. of the International Conference on Electrical Machines and Drives, pp. 351-355, Sept., 1997.

[4] J. W. Ahn and S. G. Oh, "A three-phase switched reluctance motor with two-phase excitation," IEEE Trans. Ind. Applicat., pp.1067- 1075, Sept. /Oct., 1999.

[5] Y. -C. Kim, Y. -H. Yoon, B. -K. Lee, J. Hur, C. -Y. Won, "A new cost effective SRM drive using commercial 6-switch IGBT modules," IEEE Power Electronics Specialists Conference, pp.1-7, Jun., 2006.

[6] R. Visinka, "3-phase switched reluctance (SR) sensorless motor control using a 56F80x, 56F8100," Freescale semiconductor application note, 2005

[7] H. Gao, F. R. Salmasi., and M. Ehsani, "Sensorless control of SRM at standstill", IEEE Applied Power Electronics Conference and Exposition, pp.850 - 856, Mar., 2001.

[8] J. Bu, and L. Xu, "Eliminating starting hesitation for reliable sensorless control of switched reluctance motors," IEEE Trans. Ind. Applicat., Jan./Feb., 2001.

Control Scheme for Switched Reluctance Drives with Minimized DC-Link Capacitance

Christoph R. Neuhaus, Rik W. De Doncker
Institute for Power Electronics and Electrical Drives
RWTH Aachen University
Jaegerstrasse 17-19, 52066 Aachen, Germany
Phone: +49 241 8097154
Fax: +49 241 8092203
E-mail: ne@isea.rwth-aachen.de

Nisai H. Fuengwarodsakul
The Sirindhorn International Thai-German Graduate
School of Engineering
King Mongkut's Institute of Technology North Bangkok
1518 Pibulsongkram Road, Bangsue, 10800 Thailand
Phone: +66 2 9132500
Fax: +66 2 9135805
E-mail: nisaif@kmitnb.ac.th

Abstract—This paper presents a control scheme which allows minimization of dc-link capacitance in Switched Reluctance Drives. In SRMs the demagnetization energy of the outgoing phase causes a voltage increase in the dc-link. Therefore, large dc-link capacitors are required in conventional drives to absorb this energy to avoid overvoltages. The proposed control scheme balances the power flow between phases in such a way that the demagnetization energy of the outgoing phase does not flow back into the dc-link but directly into the incoming phase. Using this scheme, electrolytic capacitors can be replaced by film or ceramic capacitors leading to longer life-time and higher reliability.

Keywords: **Switched Reluctance Drives, Capacitance, Control**

I. INTRODUCTION

Switched reluctance drive (SRD) converters – as it is typical for adjustable speed drives – need a dc-link capacitor to smooth the rectified voltage. Normally in SRDs the size of the dc-link capacitor is comparatively large due to the large amount of magnetic energy that oscillates between the dc-link and the machine. This constraint is often the reason for the utilization of electrolytic capacitors which feature large capacitances at high nominal voltages. However, these devices are disadvantageous concerning their large size and also problematic concerning reliability [1], [2].

Hence, a minimization of the dc-link capacitance offers an opportunity of using alternative types of capacitors such as ceramic capacitors which overcome the mentioned reliability drawback. Many efforts have been spent on the dc-link minimization in 3-phase AC/DC/AC-Converters with special focus on input performance optimization by reduction of grid harmonics [3]–[5]. The proposed control schemes use a coupled control approach for both the inverter and the converter stage to balance the dc-link power at every sampling instance. However, these methods only work for converters with active rectifiers that enable reverse power flow into the grid.

However, for SRM converters with diode rectifiers which allow only motoring operation few publications focus on the minimization of the dc-link capacitance [6], [7]. One approach is the use of a special converter topology with reduced capacitances [8]. Though, it still has a comparatively large voltage overshoot in the dc-link.

This paper addresses the capacitance minimization problem from the control side by an active power balancing scheme during phase commutation.

II. DIMENSIONING OF DC-LINK CAPACITORS

DC-Link capacitors are utilized to provide a stable and relatively smooth voltage source for a converter. Since the input voltage to a dc-link normally fluctuates, the capacitor has to act as an energy buffer keeping its voltage up in periods where the input voltage is too low. The worst case is the operation on a single phase ac line where the rectified input voltage varies between zero and the absolute amplitude of the grid voltage. In such cases the dc-link capacitor has to be designed large enough to bridge the time between two half-waves of the grid voltage while providing the average power of the load. The lower the required voltage ripple is the larger the capacitor has to be.

In a motor drive application a dc-link voltage sag leads to a reduced capability of building up current in the machine windings. This in turn means diminished torque production. However, in many cases, especially for switched reluctance drives the specifications on the allowed dc-link voltage ripple are not so restrictive. Due to the large moment of inertia of the rotating masses, voltage sags down to zero can be bridged with little but tolerable speed oscillations.

More critical than voltage drops in the dc-link are overvoltages, because they lead to hardware damages of power electronic switches if they exceed the device's rated voltage. These overvoltages result from energy being fed back into the dc-link every time a phase is demagnetized. Usually, during the commutation, the demagnetization power of the outgoing phase is larger than the magnetization power of the incoming phase. In drives that are only used for motoring applications (pumps, fans, etc.) the dc-link is connected to the grid via a passive diode rectifier and cannot deliver any energy back into the grid. Therefore, the dc-link capacitor has to be designed to absorb all the demagnetizing magnetic energy while keeping the voltage below critical levels.

978-1-4244-0644-9/07/$25.00 ©2007 IEEE

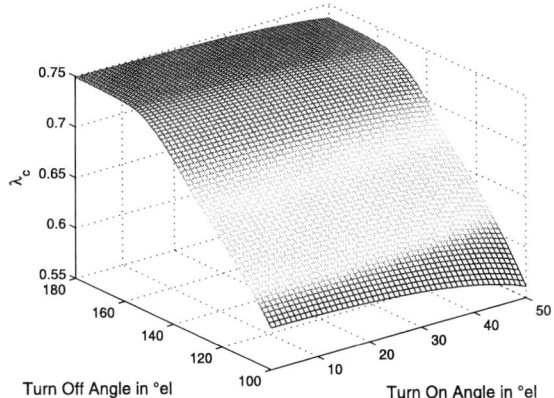

Fig. 1. Energy conversion factor λ_c at $n = 100\,\text{rpm}$ and $I = 40\,\text{A}$ as a function of switching angles

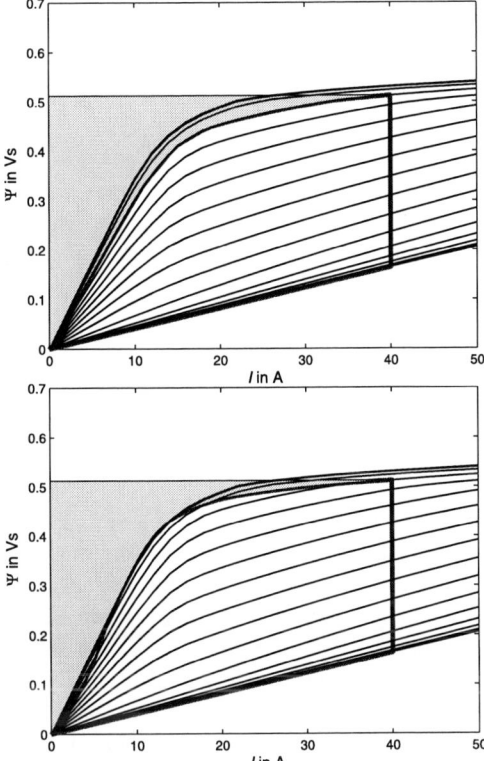

Fig. 2. Magnetic energy at low (100 rpm) and high (3500 rpm) speed at 40 A turn-off current and 140 °el turn-off angle

The key to the design of a well-sized dc-link capacitor for SRM converters is the knowledge of the magnetic energy W_{mag} that flows back out of the phase during demagnetization, This can be roughly estimated using the energy conversion factor of the machine, as expressed in (1).

$$\lambda_c = \frac{W_{\text{mech}}}{W_{\text{mech}} + W_{\text{mag}}} \tag{1}$$

λ_c represents the ratio between the converted mechanical energy W_{mech} and the total energy that has been fed into one phase. It also depends on the control parameters and

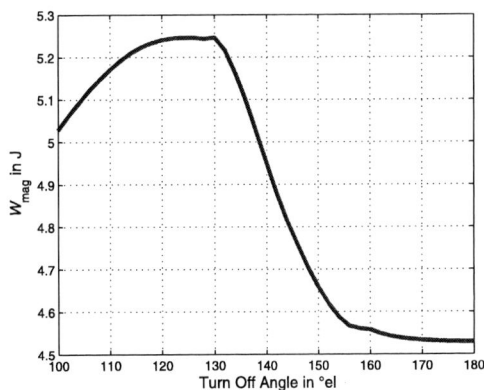

Fig. 3. Magnetic energy as a function of turn-off angle at $n = 100\,\text{rpm}$ and $I = 40\,\text{A}$

the operating point, i.e. turn-on and turn-off angle, current and speed, as illustrated in Fig. 1. For well designed SRMs with regular air gap lengths the average conversion factor lies above 0.7 for appropriate switching angles. Using this average conversion factor, the average demagnetized magnetic energy can be estimated as follows.

$$W_{\text{mag}} = \bar{T}_{\text{rated}} \frac{1 - \lambda_c}{\lambda_c} \frac{2\pi}{N_r N_{\text{ph}}} \tag{2}$$

$\bar{T}_{\text{rated}}$ is the average rated torque of the machine, N_r the number of rotor poles and N_{ph} the number of phases. Using (2) an estimation of the expected average magnetic energy is possible. However, this estimation does not cover the worst case with the nominal torque at low speed. A straight-forward method to determine the maximum expected demagnetization energy can be done using the flux-linkage characteristic of the machine as follows.

Knowing the flux-linkage characteristic of the machine the amount of magnetic energy can be ascribed to three quantities that are *turn-off current*, *turn-off angle* and *speed* [9], [10]. Figure 2 shows the coenergy curves of one machine for the same turn-off parameters at low speed (above) and higher speed (below). The highlighted area represents the magnetic energy. In case of low speed the machine phase is instantaneously demagnetized which means that all the stored magnetic energy in the phase flows back into the dc-link. In case of higher speed the demagnetization remains over a wider range of rotor position. Thereby, a noticeable amount of the stored magnetic energy is converted into torque production.

The critical operating region for the dc-link capacitor is therefore an operation at nominal torque in the base speed region. The magnetic energy can be calculated using the characteristic $i(\Psi, \vartheta)$ of the machine.

$$dW_{\text{mag}} = i(\Psi, \vartheta)\,d\Psi \tag{3}$$

$$W_{\text{mag}}\big|_{\vartheta = \vartheta_{\text{off}}} = \int\limits_{\Psi(\vartheta_{\text{off}})}^{0} i(\Psi, \vartheta_{\text{off}})\,d\Psi \tag{4}$$

Figure 3 shows the magnetic energy in one phase as a function of turn-off angle at a speed of $100\,\text{rpm}$ and a turn-off

711

current of 40 A. It illustrates that for earlier turn-off angles the magnetic energy gets larger but still stays in a narrow margin.

Once the magnetic energy is calculated, the minimum dc-link capacitance can be derived. Together with the peak rectified voltage $\hat{u}_{\text{rect}}$ that is applied to the dc-link and the maximum tolerable dc-link voltage $\hat{u}_{\text{dc}}$ the dc-link capacitance can be calculated using Equation (5).

$$C_{\min} = \frac{2\,|W_{\text{mag}}|}{\hat{u}_{\text{dc}}^2 - \hat{u}_{\text{rect}}^2} \tag{5}$$

III. POWER BALANCING PHASE COMMUTATION

As stated above the magnetic energy in the demagnetizing phase is responsible for the voltage increase in the dc-link capacitor. Conventionally the capacitor has to be designed large enough to take this energy while its voltage does not exceed tolerable limits. In order to allow a design with reduced dc-link capacitance the feed-back of the whole magnetic energy into the dc-link has to be avoided. Therefore, a control scheme is developed that allows the estimation and the control of the power flow between phases during phase commutation in order to minimize the fed back demagnetization power.

To estimate the power which the dc-link has to deliver to the machine it is necessary to know the differential change of energy in the phase that depends on the switching states. It is assumed that the machine is operated using pulse width modulation (PWM). The general dependency between the differential change of magnetic energy and the terminal quantities of the machine is given in (3). In order to balance the power that the dc-link has to deliver the absolute value of the rate of change of magnetic energy over time has to be equal for the outgoing (Index n) and for the incoming (Index $n+1$) phase.

$$-\frac{\mathrm{d}W_{\text{mag}}^{\text{n}}}{\mathrm{d}t} = \frac{\mathrm{d}W_{\text{mag}}^{\text{n}+1}}{\mathrm{d}t} \tag{6}$$

$$-i^{\text{n}}\frac{\mathrm{d}\Psi^{\text{n}}}{\mathrm{d}t} = i^{\text{n}+1}\frac{\mathrm{d}\Psi^{\text{n}+1}}{\mathrm{d}t} \tag{7}$$

The absolute value of the product of phase current and change of flux linkage over time has to be equal for both the incoming and outgoing phase. The change of flux linkage in a phase can be estimated easily when ohmic losses are neglected using the general equation of an inductance.

$$\frac{\mathrm{d}\Psi}{\mathrm{d}t} = V_{\text{ph}}$$

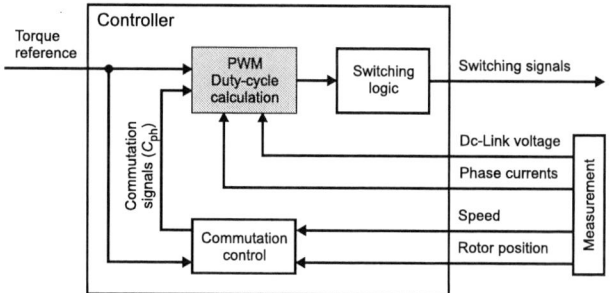

Fig. 4.   Block diagram of the proposed control

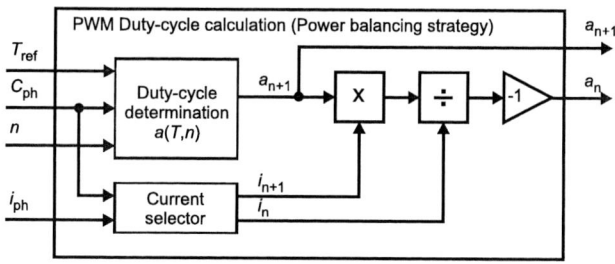

Fig. 5.   Power balancing control strategy

Using a PWM control method, as it is assumed here, the change of flux linkage over time in one sampling period can be calculated as follows.

$$\frac{\mathrm{d}\Psi}{\mathrm{d}t} = aV_{\text{dc}} \tag{8}$$

In the above equation a is the actual duty cycle of the PWM used to magnetize or demagnetize the phase. V_{dc} represents the dc-link voltage. Therefore, (7) can be simplified to

$$-i^{\text{n}}a^{\text{n}}V_{\text{dc}} = i^{\text{n}+1}a^{\text{n}+1}V_{\text{dc}}\,. \tag{9}$$

The incoming phase has to raise its current without interruption in order to generate torque quickly. Therefore, the duty cycle of the incoming phase is mainly determined by the required torque and the actual speed and should not be manipulated externally. However, the outgoing phase provides some freedom of delaying the reduction of current. Thus, (10) can be utilized as a control law for the duty cycle of the outgoing phase in order to slow down its demagnetization.

$$a^{\text{n}} = -a^{\text{n}+1}\frac{i^{\text{n}+1}}{i^{\text{n}}} \tag{10}$$

While the duty cycle of the incoming phase is kept constant during magnetization, the duty cycle of the outgoing phase has to be calculated at every sampling instant, since the fraction of the phase currents increases when the outgoing phase reduces its current. The sign of the duty cycle for the outgoing phase is always negative. That means that the only two possible switching states are *freewheeling* and *turn-off*. The duty cycle is then equal to the ratio between turn-off time and one period of the PWM. Strictly speaking (10) is only exact when the phase current does not change too fast during one sampling period. Therefore, the switching frequency of the PWM has to be high enough that current change during one sampling period can be neglected.

Figure 4 shows the block diagram of the proposed control. The duty cycle of the incoming phase is adjusted corresponding to the speed and reference torque. The commutation signals C_{ph} determine which phases must be active to produce torque. Fig. 5 illustrates the strategy of the PWM duty cycle calculation corresponding to (10).

IV. SIMULATION RESULTS

Computer simulations were carried out to investigate the behavior of the proposed control scheme. The general functionality is verified by a comparison between a conventional

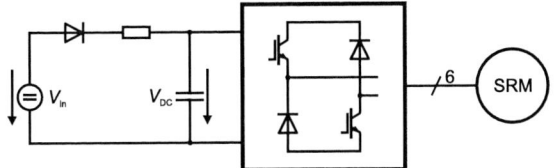

Fig. 6. Simulated drive configuration using a simplified dc-link model

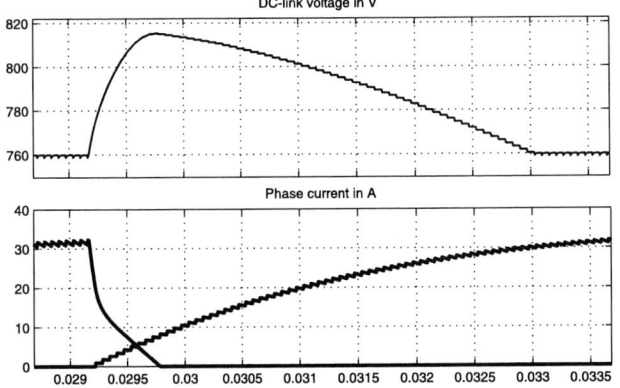

Fig. 7. Simulated phase currents and dc-link voltage at 100 rpm with conventional commutation

commutation and a commutation procedure using the proposed control scheme. This comparison is conducted for both low and high speed operation. The proposed control scheme was simulated with the MATLAB/Simulink environment. In the simulations the use of an 8 kW 3-phase SRM was assumed. The machine was driven by a 3-phase asymmetrical half-bridge converter. The frequency of the PWM switching signals was set to 16 kHz.

Fig. 6 illustrates the simplified circuit diagram of the drive system. As aforementioned, the main focus is the load-sided influence on the dc-link capacitor. Therefore, the dc-link capacitor was charged via a rectifier diode by an ideal voltage source to exclude the influence of source-sided voltage fluctuations. The supply voltage for the simulations was set to $V_{in} = 760$ V and the capacitance of the dc-link was chosen to

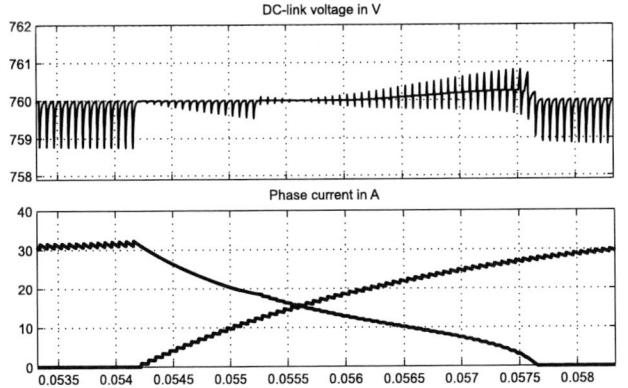

Fig. 8. Simulated phase currents and dc-link voltage at 100 rpm using the proposed control scheme

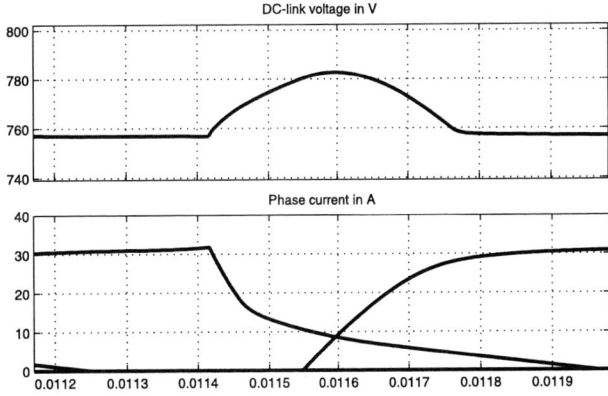

Fig. 9. Simulated phase currents and dc-link voltage at 3500 rpm with conventional commutation

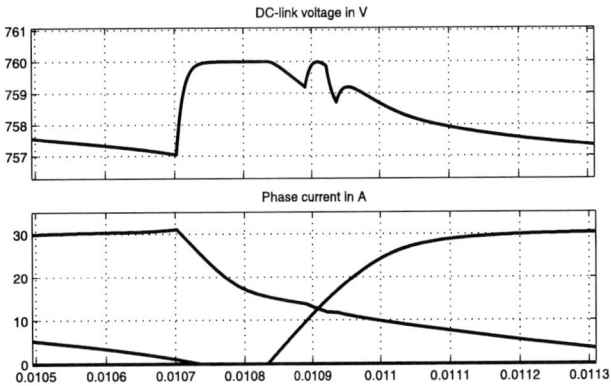

Fig. 10. Simulated phase currents and dc-link voltage at 3500 rpm using the proposed control scheme

be 100 μF.

Figures 7 and 8 show the simulated waveforms of phase currents and dc-link voltage during commutation at a speed of 100 rpm. The first of the two figures shows the waveforms for the conventional commutation, where the outgoing phase is turned off without delay. The second figure shows a commutation process using the proposed control scheme. In both cases the switching angles were $\vartheta_{on} = 20\,°$el and $\vartheta_{off} = 140\,°$el.

In Fig. 7 the outgoing phase is demagnetized immediately with full duty cycle. However, the incoming phase is magnetized very slowly with a small duty cycle of $a = 0.08$ to avoid too large currents in the incoming phase due to the low back-emf voltage. Therefore, the power consumption of the magnetizing phase is much lower than the delivered power of the outgoing phase. As a result, the voltage of the dc-link rises very fast up to a value of 815 V, which is equal to a relative increase of 7 %.

In Fig. 8 the duty cycle of the outgoing phase is calculated according to (10) after turn-off time. As long as the current of the incoming phase is low the current ratio in (10) is small which leads to a small duty cycle. If this duty cycle is lower than 5 % the phase remains in freewheeling in order to avoid too short turn-off or turn-on times for the power switches. Hence, at the beginning of commutation the outgoing phase

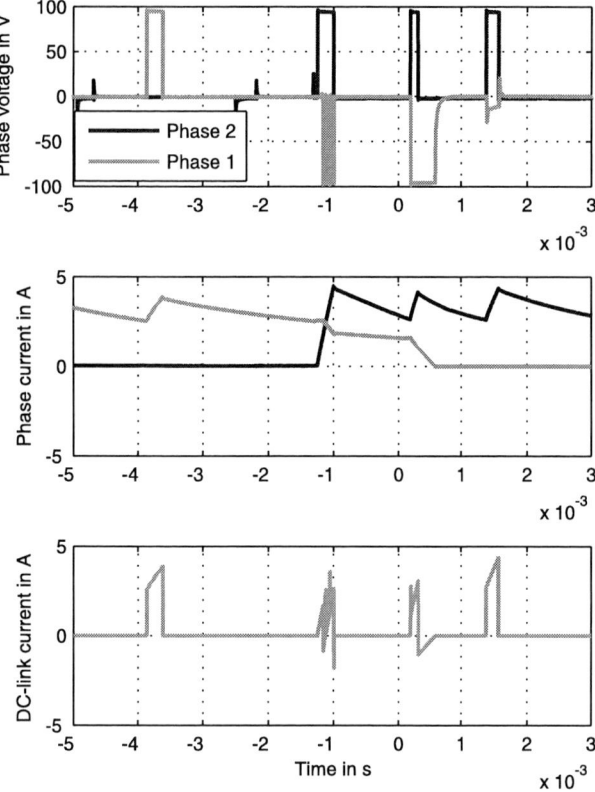

Fig. 11. Measured waveforms of phase voltages, currents and dc-link current in pulsed mode at 300 rpm using the proposed control scheme

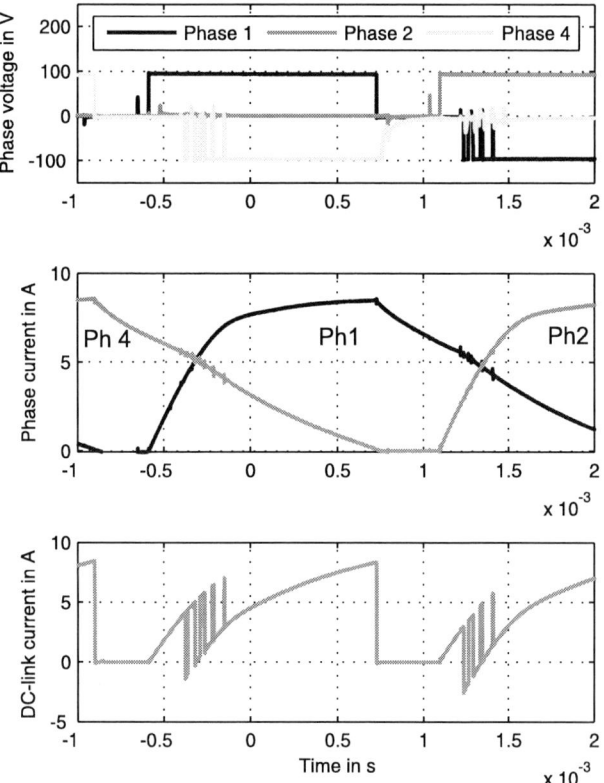

Fig. 12. Measured waveforms of phase voltages, currents and dc-link current in single-pulse mode at 1500 rpm using the proposed control scheme

does not feed any energy back into the dc-link. When the incoming current rises further the calculated duty cycle for the outgoing phase becomes larger and exceeds the protective limit which corresponds to the start of demagnetization. As the upper waveform shows, the dc-link voltage stays very close to the nominal value of 760 V.

Figures 9 and 10 show the dc-link voltage and phase current waveforms for a speed of 3500 rpm. At that speed the machine is already operated in single pulse mode which has a duty cycle of $a = 1$ during magnetization. The switching angles for both simulations were set to $\vartheta_{\mathrm{on}} = 20\,^{\circ}\mathrm{el}$ and $\vartheta_{\mathrm{off}} = 118\,^{\circ}\mathrm{el}$.

In Fig. 9 the outgoing phase is turned off immediately with a duty cycle of $a = -1$, corresponding to the conventional commutation technique. Since the turn-on angle of the next phase lies behind the turn-off angle of the outgoing phase, the dc-link is being fed with the complete actual current of the outgoing phase. This leads to an increase of the voltage up to 780 V.

In contrast to this, the proposed commutation scheme turns the outgoing phase into freewheeling when it excesses the turn-off angle as shown if Fig. 10. This is due to the small ratio between the two phase currents. Hence, the resulting dc-link power stays zero once the voltage has reached its nominal value. Only when the two phase currents have nearly the same value the outgoing phase starts to demagnetize actively. During the whole commutation, the dc-link voltage stays below or equal to the nominal value.

V. EXPERIMENTAL VERIFICATION

An experimental test setup was used to verify the proposed control scheme. It consisted of two motors, one induction machine used as the load drive and one SRM which was used as the test machine for the control. The power rating of this test setup was approximately 1 kW. The SRM was a two-pole four phase motor with a configuration of 8 stator teeth and 6 rotor teeth. The converter that was used to feed the SRM was an asymmetrical half-bridge connected to a dc-link with a nominal voltage of 90 V.

A DSP-based rapid control prototyping system was utilized to implement the proposed control strategy. It was configured to measure the most relevant machine status data such as phase current and rotor position at a sampling frequency of 16 kHz. At this frequency also the switching signals for the SRM converter were calculated and refreshed.

In contrast to the simulations in this paper, not PWM but current hysteresis control was implemented as the basic control mode for the SRM. During one sampling period of 62.5 µs a phase could either be turned on, off or into free-wheeling for the whole period which corresponds to equivalent duty cycles of $+1$, -1 or 0. As a consequence, (10) was simplified to $a^{\mathrm{n}} = -\frac{i^{n+1}}{i^{\mathrm{n}}}$ while phase n+1 is being turned on and $a^{\mathrm{n}} = 0$ while phase n+1 is being turned into free-wheeling.

The described test setup allowed to adjust and control speed of the load machine in a range from 300 rpm to 1500 rpm.

Due to the special type of converter that was used to supply the SRM it was not possible to measure the dc-link current directly. Instead, all phase currents and the corresponding phase voltages were measured externally. In a post-processing step, the dc-link component of each phase current was calculated by multiplying the phase current by the corresponding switching states. Subsequent summation of all dc-link current components yields the total dc-link current.

Figure 11 shows the measured waveforms of the voltages and currents of two neighboring phases during commutation. The bottom curve shows the total dc-link current. Due to the low speed the machine is operated in pulsed current mode. The current plot in the middle highlights the pulsed demagnetization of the outgoing phase when the incoming phase builds up current. According to (10) the duty cycle of the outgoing phase is determined in order to keep the charge balance in the dc-link positive during the next PWM period. When the active phase is free-wheeling due to hysteresis control, the outgoing phase also stays in free-wheeling mode in order to keep the power balance in the dc-link zero.

Figure 12 shows the measured waveforms of the phase voltages and currents and the dc-link current during commutation at 1500 rpm. The outgoing phase is not immediately turned off, but stays in free-wheeling mode until the incoming phase has built up a certain current. This is the result of a protective measure for the power switches. Too little duty cycles, as they would occur to the outgoing phase at the beginning of a commutation, are prohibited to avoid too large stress on the power switches. Therefore, the outgoing phase does not start the pulsed demagnetization immediately.

It is obvious that this method of current commutation takes more time than the immediate turn-off of a phase. Thus, the turn-off angle of a phase has to be chosen carefully in order to provide enough time for demagnetization. At low speed in current hysteresis control mode the turn-off angle must lie before the turn-on angle of the next phase, since an incoming phase draws the maximum charge out of the dc-link at its initial turn-on. This initial charge can be delivered completely by the outgoing phase. Otherwise, the demagnetization will take significantly more time.

VI. CONCLUSION

During commutation of two neighboring phases, magnetic energy of up to 30 % of the delivered mechanical energy is normally fed back into the dc-link capacitor which leads to a significant increase of the dc-link voltage. Therefore the dc-link capacitance has to be designed to take this energy without voltage overshoot.

By estimating and balancing the total dc-link power during commutation, the magnetic energy can be kept inside the machine and used to magnetize the incoming phase without flowing through the dc-link capacitor. It is shown by simulations that using the proposed control scheme the dc-link voltage could be kept in a very narrow band around the nominal value, which provides the opportunity of using dc-link capacitors of significantly reduced size.

Implementation and testing of the proposed control scheme on a test SRM proved that the dc-link current could be kept positive almost during the entire commutation process.

REFERENCES

[1] L. Malesani, L. Rossetto, P. Tenti, and P. Tomasin, "AC/DC/AC PWM converter with reduced energy storage in the DC link," *IEEE Transactions on Industry Applications*, vol. 31, no. 2, pp. 287–292, March 1995.

[2] R. Kros, "Capacitors in power electronic applications," in *Industry Applications Conference, 1997. Thirty-Second IAS Annual Meeting, IAS '97., Conference Record of the 1997 IEEE*, vol. 2, October 1997, pp. 1101–1108.

[3] B.-G. Gu and K. Nam, "A DC-link capacitor minimization method through direct capacitor current control," *IEEE Transactions on Industry Applications*, vol. 42, no. 2, pp. 573–581, March 2006.

[4] N. Hur, J. Jung, and K. Nam, "A fast dynamic DC-link power-balancing scheme for a PWM converter-inverter system," *IEEE Transactions on Industrial Electronics*, vol. 48, no. 4, pp. 794–803, August 2001.

[5] P. Liutanakul, S. Pierfederici, and F. Meibody-Tabar, "Load power compensations for stabilized DC-link voltage of the cascade controlled rectifier/inverter-motor drive system," in *Industrial Electronics Society, 2005. IECON 2005. 32nd Annual Conference of IEEE*, November 2005.

[6] W. S. Heglund and S. R. Jones, "Performance of a new commutation approach for switched reluctance generators," in *Energy Conversion Engineering Conference, 1997. IECEC-97. Proceedings of the 32nd Intersociety*, vol. 1, Honolulu, HI, July/Aug. 1997, pp. 574–579.

[7] M. Barnes and C. Pollock, "Power electronic converters for switched reluctance drives," *IEEE Transactions on Power Electronics*, vol. 13, no. 6, pp. 1100–1111, Nov. 1998.

[8] W. Thong and C. Pollock, "Two phase switched reluctance drive with voltage doubler and low dc link capacitance," *2005. Fourtieth IAS Annual Meeting. Conference Record of the 2005 Industry Applications Conference*, vol. 3, pp. 2155–2159, October 2005.

[9] R. B. Inderka and R. W. A. A. De Doncker, "High-dynamic direct average torque control for switched reluctance drives," *IEEE Transactions on Industry Applications*, vol. 39, no. 4, pp. 1040–1045, July/Aug. 2003.

[10] D. A. Staton, W. L. Soong, and T. J. E. Miller, "Unified theory of torque production in switched reluctance and synchronous reluctance motors," *IEEE Transactions on Industry Applications*, vol. 31, no. 2, pp. 329–337, Mar./Apr. 1995.

Multiphase Torque-Sharing Concepts of Predictive PWM-DITC for SRM

Helge J. Brauer, Martin D. Hennen and Rik W. De Doncker
Institute for Power Electronics and Electrical Drives
RWTH Aachen University
Jaegerstrasse 17-19, 52066 Aachen, Germany
Phone: +49 241 80-971 57
Fax: +49 241 80-922 03
E-mail: be@isea.rwth-aachen.de

Abstract—Instantaneous torque control for Switched-Reluctance-Machines (SRMs) with an arbitrary number of phases is introduced in this paper. Direct portability of this control to machines with any number of phases is achieved by the developed *Multiphase-Torque-Sharing* concept. This concept is also necessary for SRMs where maximum torque can only be achieved by phase overlap of more than two phases.

Furthermore, a new torque-sharing strategy for *Predictive PWM-DITC* is presented that minimizes losses in the machine without use of pre-computed current or torque profiles. This *Low-Loss-Commutation* strategy also works without fixed switching angles.

I. INTRODUCTION

Switched reluctance machines (SRMs) offer a robust structure and economic assembling. However, the relation of output and terminal quantities is non-linear which makes the analytical description of the machine rather complex and hard to implement on drive systems. To control the instantaneous torque in SRMs, basic achievements were made in 1985 by Byrne *et al.* [1], who developed a drive design to produce a smooth total torque [2]. Two years later Ilic'-Spong *et al.* [3] introduced a concept to share the total torque between the phases of the SRM during phase commutation. This idea was seized in various concepts afterwards. Some of these torque control concepts are based on off- or on-line calculated phase current or torque profiles [4], [5]. However, the implementation of these controls is data intensive and offers only limited flexibility during operation. Another method, called *Direct-Instantaneous-Torque-Control* DITC, was presented in [6]. It estimates the instantaneous torque and controls it with a simple hysteresis control. As the switching frequency is limited by the converter and the controller, the torque-ripple increases if phase inductance is low.

The developments presented in this paper are an enhancement of the *Predictive PWM-DITC* concept [7] that works with a fixed switching frequency in contrast to original variable frequency DITC. *Predictive PWM-DITC* estimates the flux-linkage and rotor-position to determine the necessary switching signals for each phase. However, this strategy was designed for a three-phase SRM as the concept controls the torque in just two phases at a time. In SRMs exceeding three phases this strategy will limit the ability to control the total torque and

lead to a distinctive torque-ripple during commutation. Some machine designs require the active control of more than two phases to achieve their maximum torque due to the designed phase overlap.

In this study, this control method is transferred to SRMs with an arbitrary number of phases. To guarantee the direct portability of the control to machines with different number of phases a newly-developed *Multiphase-Torque-Sharing* concept is introduced (Fig. 1). Its task is to optimize the total torque generation of the SRM and avoid a torque-ripple that might also result in displeasing acoustics.

In addition, a new torque-sharing strategy for *Predictive PWM-DITC* is presented that minimizes losses in the machine without time-consuming off-line calculations or simulations. For this *Low-Loss-Commutation* strategy no fixed switching angles are needed. The introduced algorithm can determine the optimum commutation path by estimating the efficiency in future sample-steps.

II. DYNAMIC TORQUE-SHARING CONTROL FOR AN ABITRARY NUMBER OF PHASES

To control the torque instantaneously, detailed information on the SRM characteristics is necessary. One of them is the inductance characteristic of a SRM, which defines whether a phase can contribute to torque at its actual rotor position. The contribution is limited to a specific rotor-angle region

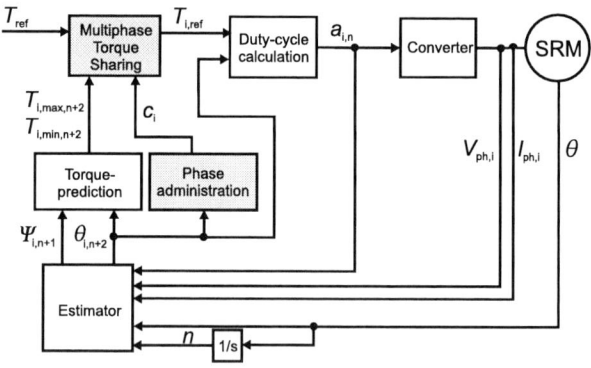

Fig. 1. Block diagram of enhanced Predicitve PWM-DITC

978-1-4244-0644-9/07/$25.00 ©2007 IEEE   716

that varies with the operating point. If the rotor-position lies within this active-region the phase is a so-called active phase. Therefore, the torque controller can adjust its reference torque within physical boundaries. In contrast to this, the reference torque of an inactive phase is set to zero as its torque contribution is not useful for the total torque generation. The active-region of the phase is marked by the switching angles that are a basic element of several control concepts (e.g. [6], [8]).

In SRMs with three or more phases the active-regions of at least two phases overlap and offer the possibility to share the total torque between the active phases. During this overlap the generation of reference torque will commutate from a previous - demagnetising - phase to a newly activated - magnetising - phase.

The controller requires a phase sequencer to determine the active phases (Fig. 1). The previous implementation [9] used a selector-table to select the demagnetising in dependency on the magnetising phase and the operation point. The dimension of the table rises on increase of the number of phases and therefore the table is hard to create for more than three phases. However, it is also only possible to select two phases in every PWM-step due to limitations of the former torque-sharing algorithm.

In SRMs exceeding three phases the phase-overlap increases and the torque-sharing algorithm needs to share the torque between more than two phases. Therefore, a new phase sequencer is introduced that is based on a sorted dynamic first-in-first-out (FIFO) list, called phase-list, which contains the active phases (see example in Fig. 2). The sorting of the phase-list indicates the phase's priority (top of the list = high priority; bottom of the list = low priority). With additional knowledge about the phase switching sequence based on the rotating direction the phase sequencer just needs to check the state (active/inactive) of one phase to update the phase-list.

III. TORQUE-SHARING ALGORITHM FOR AN UNSPECIFIED NUMBER OF ACTIVE PHASES

The phase-list created by the phase sequencer can be used by the *Multiphase-Torque-Sharing* algorithm. Its task is to allocate the reference torque to the active phases in dependence of the phase priority. This priority is given by the phase's position in the list. Priority decreases if moving to lower positions.

The former developed Torque-Sharing algorithm [9] allocates the torque between only two phases with every PWM-step. Thereby it starts with the newly activated phase - magnetising - and sets the phase torque to the required value. Afterwards the remaining reference torque is assigned to the previously activated - demagnetising - phase. Sometimes it is not possible for the previously activated phase to reduce its torque as demanded by algorithm due to physical restrictions. In this case the torque-sharing algorithm has to correct the reference torque that was set previously for the other phase. This previously used sharing algorithm cannot be used if more than two phases are active. For SRMs with three or more phases the new *Multiphase-Torque-Sharing* algorithm is introduced as displayed in Fig. 3.

First of all, the minimal torque $T_{\min}$ that will be reached in the sample-period, is determined by summing up all minimal torque amounts of each phase $T_{i,\min}$ (equation (1); N_{ph}=number of phases).

$$T_{\min} = \sum_{i=1..N_{\mathrm{ph}}} T_{i,\min} \qquad (1)$$

The difference of $T_{\min}$ and reference torque T_{ref} results in the remaining torque share ΔT_1, which has to be delivered by the active phases. This way also torque of inactive phases is taken in account. For further sharing between active phases the created phase-list is used. The Multiphase-Torque-Sharing algorithm starts with the first phase in the phase-list and compares the remaining torque share ΔT_1 with the maximum additional torque capacity $\Delta T_{q=1,\max}$ (equation(2)), whereas q is the index of the phase-list position.

$$\Delta T_{q,\max} = T_{q,\max} - T_{q,\min} \qquad (2)$$

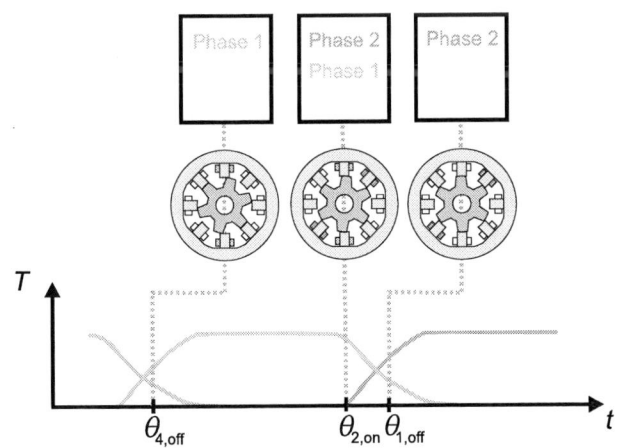

Fig. 2. Phase sequencer of active phases (four-phase SRM)

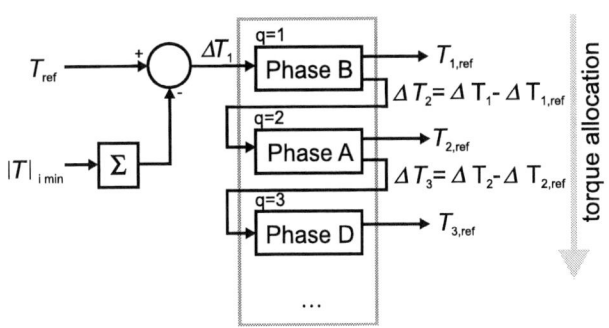

Fig. 3. Multiphase-Torque-Sharing algorithm

717

If the phase is able to deliver at least this difference ΔT_q, the reference phase torque $T_{q,ref}$ will be raised (equation (3)).

$$T_{q,ref} = T_{q,min} + \Delta T_{q,ref} \tag{3}$$

If the phase cannot deliver the difference ΔT_q it produces the maximum torque possible $T_{q,max}$. A detailed overview of the different torque allocation cases can be seen in equation (4).

$$\Delta T_{q,ref} = \begin{cases} 0 & : \Theta_1(\Delta T_q) \neq \Theta_1(T_{ref}) \\ \Delta T_{q,max} & : |\Delta T_q| > |\Delta T_{q,max}| \\ & \quad \wedge \Theta_1(\Delta T_q) = \Theta_1(T_{ref}) \\ \Delta T_q & : 0 \leq |\Delta T_q| < |\Delta T_{q,max}| \\ & \quad \wedge \Theta_1(\Delta T_q) = \Theta_1(T_{ref}) \end{cases} \tag{4}$$

Whereas Θ_1 is the Heaviside function (equation (5)).

$$\Theta_c(x) = \begin{cases} 0 : & x < 0 \\ c : & x = 0 \\ 1 : & x > 0 \end{cases} \tag{5}$$

The difference of ΔT_q and $\Delta T_{q,ref}$ (equation (6)) will be passed to the next active phase in the list and the torque assignment starts again.

$$\Delta T_{q+1} = \Delta T_q - \Delta T_{q,ref} \tag{6}$$

Due to the implemented summing of the minimal phase torque at the start of the torque-sharing algorithm, the reference torque assignment sequence starts with the high prioritized phase at the top of the list and ends with the low prioritized phase at the bottom of the list. This way no belated changes of already allocated reference torque values can occur.

Especially at high speed in generating mode the turn-on angle is pulled forward to the area of a positiv $\frac{dL}{d\theta}$ due to on-line calculations. More time to build the phase's flux-linkage becomes available. However, phases will contribute torque to the opposite direction during the first sample steps. Hence, the *Multiphase-Torque-Sharing* algorithm allows the phase with highest priority to increase its phase current also its torque contribution is to the opposite direction (equation (7)).

$$T_{1,ref} = T_{1,max} \quad : \quad \Theta_1(T_1) \neq \Theta_1(T_{ref}) \tag{7}$$

Figure 4 presents simulation results using *Predictive-PWM-DITC* extended by the introduced *Multiphase-Torque-Sharing* algorithm by means of a five-phase machine. It can be seen that the torque-ripple in the tested operating point $T_{ref} = 1800\,\text{Nm}$ and $n = 350\,\text{rpm}$ is less than one percent of the reference-torque. Furthermore, the phase torque curves indicate that the algorithm is able to handle the control of more than two phases at a time.

IV. EXTENDED TORQUE-SHARING ALGORITHM FOR LOW LOSS COMMUTATION

Techniques to raise the efficiency of a switched reluctance drive by optimizing control parameters are described in [10], [11] and [12]. These optimizations are based on modifications in the torque-sharing process that result in specific current-profiles. In [10] these modifications are derived by an iterative

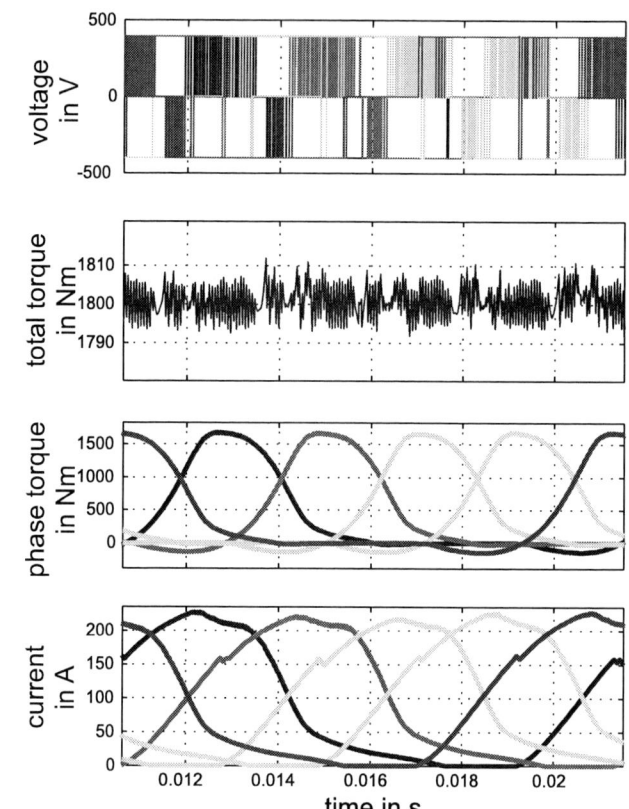

Fig. 4. Simulation of operation ($T_{ref} = 1800\,\text{Nm}; n = 350\,\text{rpm}$) using *Predictive PWM-DITC* extended by the *Multiphase-Torque-Sharing* algorithm

calculation that can be processed on-line. In contrast [11] and [12] present methods that are based on off-line determinations of the switching angles. These calculated angles can be modified on-line based on the reference torque and actual speed to increase the efficiency. However these control parameters are optimized for certain operating range and have limited performance at others.

The aim of the new *Low-Loss-Commutation* Torque Sharing algorithm is to achieve an efficient commutation in every operating point without previous off-line calculations. Converter losses, eddy currents, and hysteresis losses remain unaccounted for this concept. The rms phase current is used to rate the improvement of this *Low-Loss-Commutation* algorithm in reducing copper-losses in the machine windings. The optimization is achieved within the commutation area where total torque can be shared between two or more phases. Especially at low speed the available commutation time is often not used.

The theory of the commutation algorithm is introduced by means of a four-phase SRM. The calculated phase overlap for a four-phase SRM is $90\,°\text{elec}$. With respect to the dimensions of rotor and stator pole the possible commutation angle area is much smaller (see marked area in Fig. 5). An efficient operation can only be achieved if the possible torque contribution of each phase over the whole commutation area is taken into account based on the resulting losses in each sample

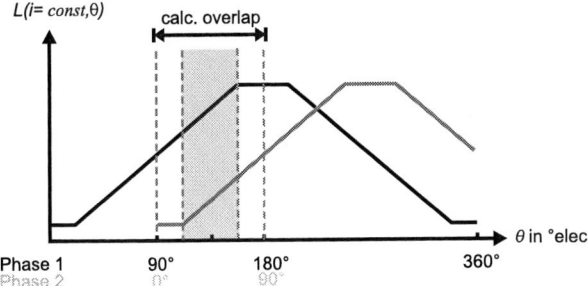

Fig. 5. Overlap of ideal phase inductances (four-phase SRM)

step. The total copper losses in a four-phase SRM during one commutation step of two phases are calculated by equation (9)

$$\theta_{ph} = \frac{360°\,elec}{N_{ph}} \qquad (8)$$

$$P_{total} = P(\sigma T_{ref}, \theta) + P([1-\sigma]T_{ref}, \theta + \theta_{ph}). \qquad (9)$$

with $\theta_{ph} = 90°\,elec$ (phase-position offset), θ as the rotorposition and σ as the share-ratio of the torque between magnetizing and demagnetizing phase. The calculations are based on measured characteristics of the SRM and the measured phase resistance, which is assumed to be constant. If the resulting losses are determined for each share-ratio σ and each rotor-position θ with a constant reference torque, an optimal commutation path can be detected (Fig. 6). The control can follow this path until a certain speed level. This speed limit is given by the ability to reduce the flux linkage in the demagnetizing phase which is linked to the maximum possible flux linkage change $\frac{\partial \Psi}{\partial \theta}$ in each rotor position. In case energy can be recuperated into the DC-link without considerations of rotor position the maximum possible change in the flux linkage can be calculated using equation (10) with the DC-link voltage V_{DC} and the angular velocity ω_{el}. The result should be less or equal to the maximum fluxlinkage change $max\left(\frac{\partial \Psi_{opt}}{\partial \theta}\right)$ that is needed to follow the optimal commutation path (Fig. 6). The

curve of the flux linkage Ψ_{opt} can be determined inserting the commutation path into the machine characteristic $\Psi(T, \theta)$.

$$max\left(\frac{\partial \Psi_{opt}}{\partial \theta}\right) \leq \frac{\partial \Psi}{\partial \theta} = \frac{V_{DC}}{\omega_{el}} \qquad (10)$$

$$\omega_{el,max} = \frac{V_{DC}}{max\left(\frac{\partial \Psi_{opt}}{\partial \theta}\right)} \qquad (11)$$

Therefore, the maximum speed that still offers a possibility for a low loss commutation follows in equation (11). If the speed limit is exceeded the machine losses will increase and the control strategy may not be able to govern a constant output torque with the same efficiency as regular *Predictive PWM-DITC*. This efficiency decrease can be explained by a later start of demagnetization of the phase compared to regular *Predictive PWM-DITC* where the demagnetization is forced by the turn-off angle θ_{off}. At higher speeds this efficiency decrease is noticeable by means of unwanted torque contributions in the opposite direction which even may lead to torque dips if the speed is further increased. However, if the speed limit is restrained the *Low-Loss-Commutation* control offers an optimized operation over the whole allowed speed range.

The implementation of the *Low-Loss-Commutation* control strategy extends the *Predictive PWM-DITC* structure. This way the control is able to react on environmental and input parameter changes with the same latency as *Predictive PWM-DITC*. A block diagram of the implementation is shown in figure 7. In order to cover a wide speed range the implemented control is able to switch to the regular *Predictive PWM-DITC* strategy if the speed limit for the *Low-Loss-Commutation* strategy is exceeded. As on-line calculations of the machine losses which are presented in equation (9) need to be conducted for various torque sharing-ratios σ and at least two phases, processing will get rather time consuming. Therefore, a figure of merit is introduced that is able to determine phases which contribute to the total torque with best efficiency. This is achieved by the so called efficiency factor (equation (12)) expressed by the ratio of the maximum additional torque capacity (equation (2)) and the difference of the squared currents corresponding to the maximum and minimum torque $I_{i,n+2,max}$, $I_{i,n+2,min}$ at the after next PWM-sample. As the squared current can be used

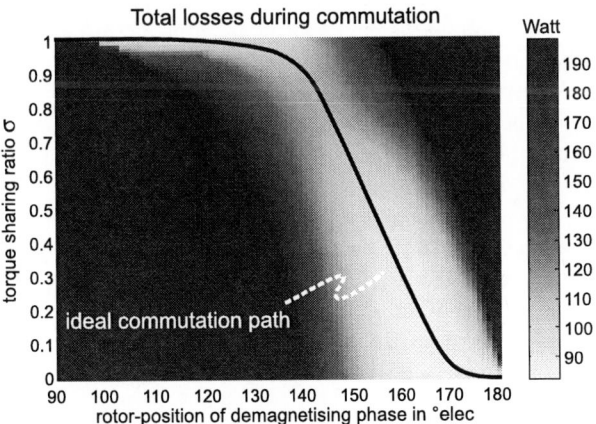

Fig. 6. Total losses during commutation (reference: demagnetising phase) $T_{ref}=4\,\mathrm{Nm}$; $R_{ph}=3,1\,\Omega$

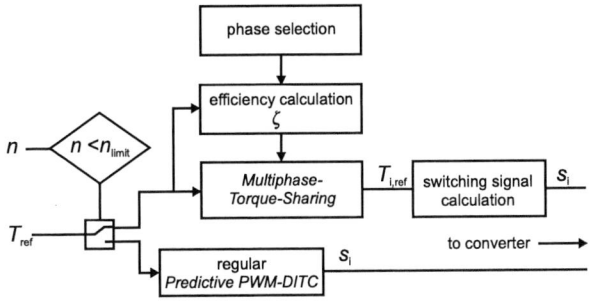

Fig. 7. Block diagram of the extended *Predictive PWM-DITC* with *Low-Loss-Commutation* strategy

to analyze the instantaneous losses, the calculated efficiency factor will rate the additional torque capacity for each phase based on the expected additional losses.

$$\zeta = \frac{T_{i,n+2,\max} - T_{i,n+2,\min}}{I_{i,n+2,\max}^2 - I_{i,n+2,\min}^2} \qquad (12)$$

Based on the results, the phase-list can be resorted as the position in the list is equal to the phase's priority. Changing the sorting of the phase-list puts the phase with the best efficiency on top of the list. Afterwards, this optimized list is processed by the *Multiphase-Torque-Sharing* algorithm and will result in the commutation, which is shown in the experimental results. As phase priority might change with every sample step an increase of iron losses due to a magnetization of the demagnetizing phase might occur. By limiting the maximum torque contribution of the demagnetizing phase to its contribution of the previous sample step additional iron losses should be avoided. Simulations indicated that the efficiency improvement of the *Low-Loss-Commutation* algorithm is not constant when compared with the regular *Multiphase-Torque-Sharing* concept. The comparison of the simulated results showed that the rms phase current, which is used to rate the efficiency, is up to 9% smaller than the regular sharing concept if the *Low-Loss-Commutation* algorithm is used. Figure 8 shows the relative difference between rms-phase-currents in the tested four-phase SRM with and without the *Low-Loss-Commutation* strategy for different operating points. It can be seen from the figure that the reduction varies in dependence of the operating point due to the fact that the switching angles of *Predictive PWM-DITC* without *Low-Loss-Commutation* are only optimized for certain operating points. In this simulation, the switching angels were optimized for a total torque around 3 Nm.

V. EXPERIMENTAL RESULTS

The described torque sharing methods and the control algorithm have been verified on a four-phase 8/6 SRM with a DC-link voltage of 300 V. In this case, the implemented control algorithms used previously measured machine characteristics

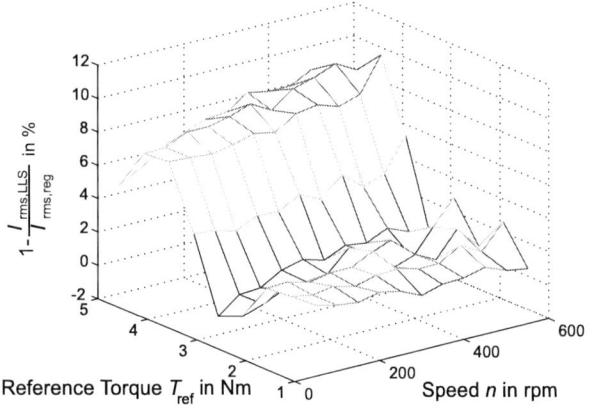

Fig. 8.   Reduction of rms-current by using *Low-Loss-Commutation*

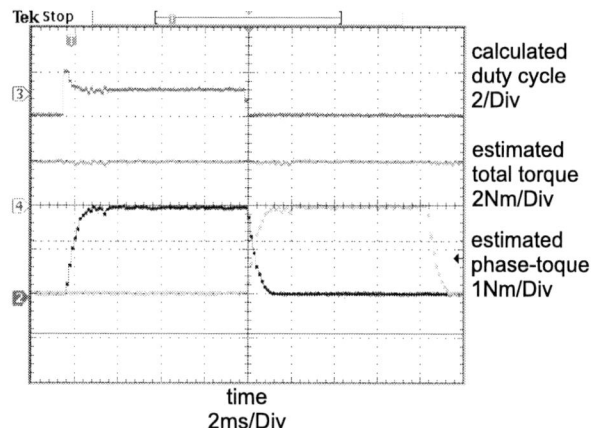

Fig. 9.   Measurement results with *Multiphase-Torque-Sharing* algorithm ($T_{\mathrm{ref}} = 2\,\mathrm{Nm}$; n=300 rpm)

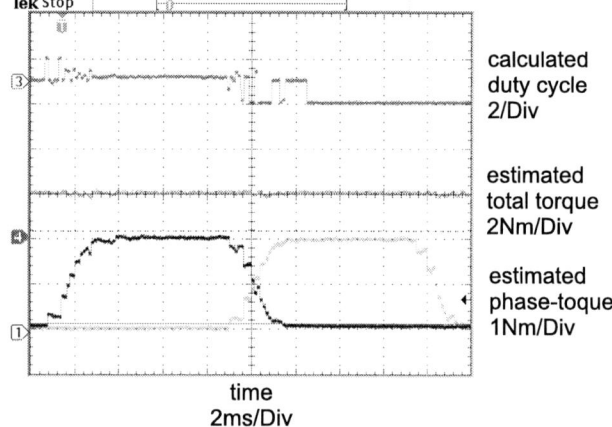

Fig. 10.   Measurement results with *Low-Loss-Commutation* algorithm ($T_{\mathrm{ref}} = 2\,\mathrm{Nm}$; n=300 rpm)

in order to enhance the accuracy of the torque predictions. Tests were conducted for PWM switching frequencies between 4 and 16 kHz. The estimated instantaneous torque was determined by current measurement in combination with the use of the SRM $T(I,\theta)$-characteristic. Due to the low-pass characteristic of torque sensors a measurement of the instantaneous torque is not possible. By using a $T(I,\theta)$-characteristic figure 9 presents a measurement with the developed *Multiphase-Torque-Sharing* algorithm and the new phase sequencer. The reference torque was set to 2 Nm at a speed of 300 rpm. The total instantaneous torque is fairly constant. With use of the *Low-Loss-Commutation* algorithm the total torque is also constant but differences can be noticed in the estimated phase-torque (Fig. 10). The commutation-time is longer and the curve shape of the sharing ratio varies. The measurements agree with previous calculations. However, the phase-torque shows a superposed slight zigzag curve shape that results from the predictive nature of the algorithm over just two PWM sample steps. A comparison between measured current profiles of the two torque-sharing algorithms confirmed the efficiency improvement that can be achieved with the *Low-Loss-*

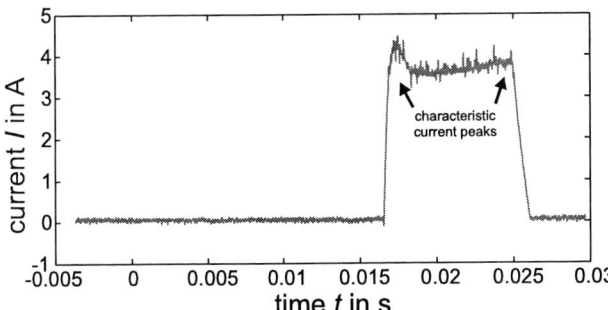

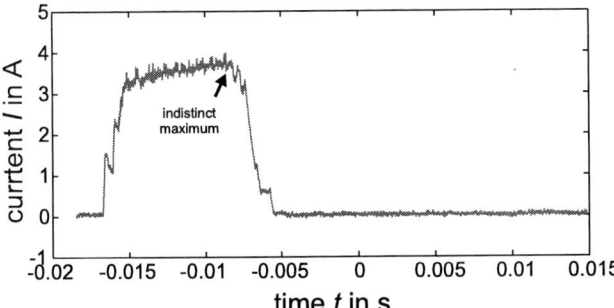

Fig. 11.
top: phase current without use of *Low-Loss-Commutation* strategy
($n = 300\,\text{rpm}; T_{\text{ref}} = 2\,\text{Nm}; I_{\text{rms}} = 1.89\,\text{A}$)
bottom: phase current with use of *Low-Loss-Commutation* strategy
($n = 300\,\text{rpm}; T_{\text{ref}} = 2Nm; I_{\text{rms}} = 1.80\,\text{A}$)

Commutation algorithm. Figure 11 shows the measured current profiles with and without use of the *Low-Loss-Commutation* strategy. It can be seen that the phase current has two distinct current peaks if the regular *Predictive-PWM-DITC* algorithm is used. This results from the used switching angles, that were derived by calculation methods described in [11]. However, using the *Low-Loss-Commutation* algorithm only a indistinct current maximum can be recognized. Also the current slope is smaller and longer compared to the regular algorithm. These facts lead to a smaller rms phase current I_{rms} and therefore indicate the loss reduction inside the machine.

VI. CONCULSION

With the newly-developed control algorithms the advantages of *Predicitve PWM-DITC* were utilized for machines exceeding three phases. The introduced *Multiphase-Torque-Sharing* algorithm uses a dynamic list to govern the phases, thereby the flexibility of the control has been increased.

Furthermore, it was discovered that efficient operation of SRMs is not only dependent on the switching angles that are used as control variables in, for example, the regular *Predicitve PWM-DITC* control. Moreover, the torque commutation path is important to achieve an efficient operation especially at low speed.

Therefore, the *Low-Loss-Commutation* algorithm was introduced. It offers a good control ability concurrent with maximum efficiency. The algorithm allows an optimized phase commutation in a wider operating area. Switching angles are not needed in this novel torque-sharing concept and therefore time-consuming off-line calculations, e.g. current-profile determination, can be avoided. However, the algorithm requires precise information about the machine characteristics to make reasonable efficiency predictions.

REFERENCES

[1] J. V. Byrne, F. McMullin, F. Devitt, and J. O'Dwyer, "Variable speed variable reluctance electrical machines," U.S. Patent 4670696, Oktober 1985.
[2] T. J. E. Miller, *Electronic Control of Switched Reluctance Machines.* Newnes, 2001.
[3] M. Ilic'-Spong, R. Marino, S. Peresada, and D. Taylor, "Feedback linearizing control of switched reluctance motors," *Automatic Control, IEEE Transactions on,* vol. 32, no. 5, pp. 371–379, May 1987.
[4] H. Ishikawa, Y. Kamada, and H. Naitoh, "Instantaneous torque regulation for switched reluctance motors for the use in EVs," in *Advanced Motion Control, 2004. AMC '04. The 8th IEEE International Workshop on,* 25-28 March 2004, pp. 65–69.
[5] I. Husain and M. Ehsani, "Torque ripple minimization in switched reluctance motor drives by PWM current control," *Power Electronics, IEEE Transactions on,* vol. 11, no. 1, pp. 83–88, Jan. 1996.
[6] R. Inderka and R. De Doncker, "DITC-direct instantaneous torque control of switched reluctance drives," *Industry Applications, IEEE Transactions on,* vol. 39, no. 4, pp. 1046–1051, July-Aug. 2003.
[7] C. Neuhaus, N. Fuengwarodsakul, and R. de Doncker, "Predictive PWM-based Direct Instantaneous Torque Control of Switched Reluctance Drives," in *Power Electronics Specialists Conference, 2006. PESC '06. 37th IEEE,* 18-22 June 2006, pp. 1–7.
[8] R. Inderka and R. De Doncker, "High-dynamic direct average torque control for switched reluctance drives," *Industry Applications, IEEE Transactions on,* vol. 39, no. 4, pp. 1040–1045, July-Aug. 2003.
[9] N. H. Fuengwarodsakul, "Predictive PWM-based Direct Instantaneous Torque Control for Switched Reluctance Machines," Ph.D. dissertation, ISEA, RWTH-Aachen, 2007.
[10] P. Kjaer, P. Nielsen, L. Andersen, and F. Blaabjerg, "A new energy optimizing control strategy for switched reluctance motors," *Industry Applications, IEEE Transactions on,* vol. 31, no. 5, pp. 1088–1095, Sept.-Oct. 1995.
[11] R. Inderka, "Direkte Drehmomentregelung Geschalteter Reluktanzantriebe," Ph.D. dissertation, ISEA, RWTH-Aachen, 2002.
[12] M. AbdulKadir and A. Yatim, "Maximum efficiency operation of switched reluctance motor by controlling switching angles," in *Power Electronics and Drive Systems, 1997. Proceedings., 1997 International Conference on,* vol. 1, 26-29 May 1997, pp. 199–204vol.1.

A New Two Phase Configuration for Switched Reluctance Motor with High Starting Torque

E. Afjei *, K. Navi**, and S. Ataei*

* Shahid Beheshti university, Iran
** Saadat Research Institute, Iran

Abstract—The Switched reluctance (SR) motor is a simple and robust machine, which has found application over a wide power and speed ranges in different shapes and geometries.

This paper briefly reviews the different types of SR motors with different geometries and then presents a new configuration for a two phase SR motor with shaped and skewed rotor poles. This motor has the ability to start and run in a specified direction without any difficulties. In another words the motor will always have starting torque no matter where the rotor position is. It also presents a centrifugal switch mounted on the motor shaft for a sudden advancement of current-pulses relative to rotor position after reaching a preset motor speed in order to develop higher torque at starting. To evaluate the motor performance, two types of analysis, namely numerical technique and experimental study have been utilized. In the numerical analysis, due to highly non-linear nature of the motor, a three dimensional finite element analysis is employed, whereas in the experimental study, a proto-type motor has been built and tested.

Index Terms-- Switched reluctance motor, reluctance motor, high starting torque.

I. INTRODUCTION

The Development of semiconductor rectifier and power switching technology in early 1960s let to its rapid and successful application to variable speed drives and simulated an interest in possible alternative and simpler motor/control configurations, of which the switched reluctance motor was one of them. A reluctance motor consists of a rotor, which has no windings of any kind, and is free to rotate between the pole pieces of a stationary singly or multiply excited magnetic structure known as stator. Torque is produced by the tendency of the rotor to align itself with the stator magnetic field [1].

This type of motors offers a number of advantages including simplicity in construction, cooling, geometric versatility, durability, and higher permissible rotor temperature [2].In general, there are four distinct types of switched reluctance motors: namely, regular doubly salient cylindrical [3-4], disc-type [5-6], multi-layer [7], and linear motors [8]. This classification stems from the general shape of the motor. A 6 by 4 regular cylindrical type motor and a 6 by 8 external rotor type motor have salient poles on both stator and rotor and the windings are wrapped around the stator poles. Currents in the stator

circuits are switched on and off in accordance with the rotor position. In order to increase the torque especially in small diameter fractional horsepower motor, the rotor is placed outside of the stator. Direct current motors with disc rotors are widely used and is has been proposed for SR motors as well. The need for disc type arises in applications where the spacing is of the primary concern and the production of torque requires a sufficiently thick rotor relative to the air gaps. An isolated multilayer SR motor consists of three magnetically independent layers or phases. Each layer comprises of a stationary part and a rotating piece known as stator and rotor, respectively. The term "isolated phase" or multilayer is derived from the fact that the motor itself has been composed of three isolated sections. In a three phase motor, each rotor section has a 15^0 angular shift in position from the next phase, whereas in the stator phase there is no shift. This motor has eight stator poles as well as eight rotor poles, which will be engaged in the torque production mechanism. A simple version of a linear SR motor consist of a stator section made up of three u shape structures with coils wrapped around them. The rotor has teeth with proper distance determined by the stator poles. Each stator coil is turned on in proper order to produce force to move the rotor forward.

This paper presents a new two phase configuration for switched reluctance motor with high starting torque as well as experimental results obtained for the new motor with and without using a centrifugal switch for the fast advancement in firing time.

II. THE NEW CONFIGURATION

The new motor has four poles on the stator like regular two phase SR motor and two shaped and skewed poles on the rotor. A three dimensional view of the motor is shown.

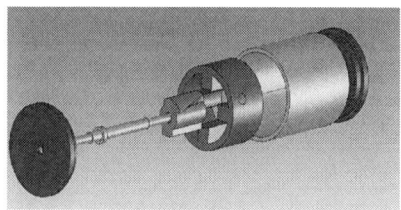

Fig. 1. A three dimensional view of the motor.

This work was supported by Shahid Beheshti university and Saadat Research Institute.

978-1-4244-0644-9/07/$25.00 ©2007 IEEE

The stator geometry is the same as a regular two phase SR motor but the rotor is shaped in such a way to produce starting torque. The shape of rotor is shown in Fig. 2

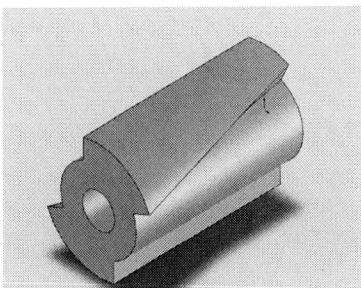

Fig. 2. The rotor shape.

The arc of rotor pole is the same as stator pole in one side and twice as big in the other side.

The design of SRM becomes complicated due to complex geometry and material saturation. The reluctance variation of the motor has an important role on the performance; hence an accurate knowledge of the flux distribution inside the motor for different excitation currents and rotor positions is essential for the prediction of motor performance. The motor is highly saturated under normal operating conditions. To evaluate properly the SRM design and performance a reliable model is required. The finite-element technique can be conveniently used to obtain the magnetic vector potential values throughout the motor in the presence of complex magnetic circuit geometry and nonlinear properties of the magnetic materials. These vector potential values can be processed to obtain the field distribution, torque, and flux leakage

The motor specifications considered for the study is:

stator core outer diameter	= 30	mm
stator core inner diameter	= 25	mm
stator arc	= 45	deg.
air gap	= 0.25	mm
rotor core outer diameter	= 13.5	mm
rotor shaft diameter	= 4.0	mm
rotor larger arc	= 90	deg.
Rotor smaller arc	= 45	deg.
number of turns per pole	= 50	

A fast acting mechanical governor mounted on the motor shaft is introduced to change the dwell angle just enough under different speeds, so that the phase commutation begins sooner and ends sooner. This method is very effective in creating sufficient time for the phase current to rise to the desired value therefore, the motoring torque will not fall off, and also, the current will be out of the winding before the rotor reaches the negative torque region.

III. NUMERICAL ANALYSIS

The field analysis has been performed using a Magnet CAD package which is based on the variational energy minimization technique to solve for the magnetic vector potential. The partial differential equation for the magnetic vector potential is given by;

$$-\frac{\partial}{\partial x}\left(\gamma\frac{\partial A}{\partial x}\right)-\frac{\partial}{\partial y}\left(\gamma\frac{\partial A}{\partial y}\right)-\frac{\partial}{\partial z}\left(\gamma\frac{\partial A}{\partial z}\right)=J \qquad (1)$$

where, A is the magnetic vector potential.

In the variational method (Ritz) the solution to (1) obtained by minimizing the following functional

$$F(A)=\frac{1}{2}\iiint_{\Omega}[\gamma(\frac{\partial A}{\partial x})^2+\gamma(\frac{\partial A}{\partial y})^2+\gamma(\frac{\partial A}{\partial z})^2]d\Omega-\iiint_{\Omega}JAd\Omega \qquad (2)$$

where Ω is the problem region of integration.

The motor configuration used for the numerical simulation is shown in Fig. 3.

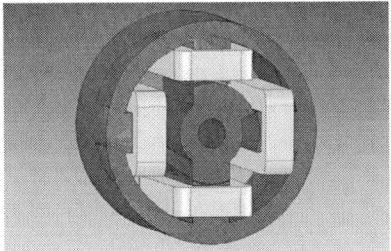

Fig. 3. The motor configuration.

The 3-D field analysis has been performed using a commercial finite element package [14], which is based on the variational energy minimization technique to solve for the magnetic vector potential.

The plots of magnetic flux density and its direction for the unaligned, half aligned, and fully aligned cases for a current magnitude of 1.5 A are shown in Figs 4,5, and 6, respectively

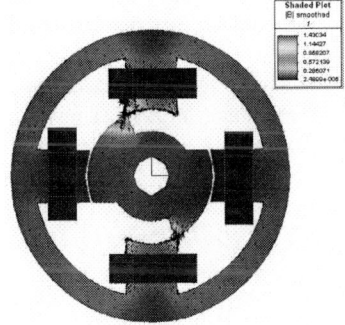

Fig. 4. magnetic flux density for the unaligned case

Fig. 4 shows the aligned corner of rotor and stator poles are in saturation and maximum magnetic field density in there is about 1.43 Tesla. The direction of magnetic flux inside the motor is also shown in the same Figure which starts from a stator pole and then to the rotor poles and passes through the yoke to the corresponding poles.

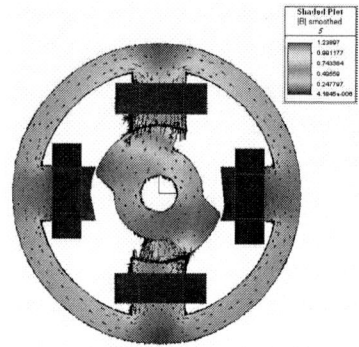

Fig. 5. magnetic flux density for half-aligned case.

As shown in Figure 5, the stator and rotor poles are in saturation with magnetic flux densities of 1.1 and 1.0 Tesla, respectively. Maximum magnetic flux density is about 1.25 Tesla at small local points on the very tip of the stator pole.

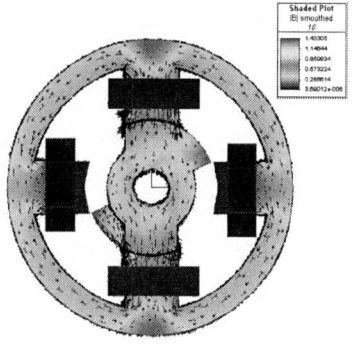

Fig. 6. magnetic flux density for the fully aligned case.

Finally, Figure 6 shows the stator and rotor poles are in saturation with magnetic flux densities of 1.43 and 1.1 Tesla, respectively.

The plot of static torque versus rotor positions developed by the reluctance motor is shown in Fig.7.

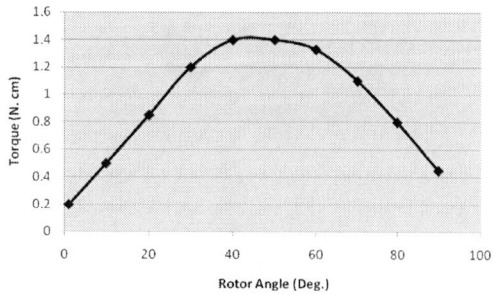

Fig. 7. Static torque versus rotor angle.

In the Fig. 7 zero degree is considered as unaligned case.

IV. EXPERIMENTAL RESULTS

The motor has been fabricated and tested for

Fig. 8. The fabricated motor.

performance and functionality in the laboratory.

Fig. 8 illustrates the novel two-phase SRM fabricated in the laboratory.

The static torque of the isolated phase was obtained by blocking the motor at different angle. The maximum static torque for a rated current of 1.5 A was measured to be about 1.N.cm. It was observed that the static torque shows lower value than computed which is expected, since, the silicon sheet steel material used to build the motor is not quite what is used for the numerical analysis.

Switched reluctance motor (SRM) drive has remarkable characteristics that make it attractive for high-speed applications. As the motor's speed increases the shape of the current waveform changes in such way that limits the production of motoring torque. At high speeds, it is possible for the phase current never reaches the desired value due to the self e.m.f. of the motor, therefore, the torque falls off. In order to remedy this problem, the phase turn on angle is advanced in such way that the phase commutation begins sooner. Advancing the commutation angle offers the advantages of getting the current into the phase winding while the inductance of the phase is low, and also of having a little more time to get the current out of the phase winding before the rotor reaches the negative torque region.

A fast acting governor is mounted on the motor shaft in order to act as a switch. This governor opens up fast at a pre-set value of about 100 rpm, which will cause a position advancement of the rotor poles with respect to the stator poles of about 8 degrees. It is possible to use other means of changing firing time such as utilization of different set of opto-couplers or employment of some kind of microcontroller. Using a torque meter, the dynamic torque for both motors versus speed have been measured by loading the motors. The torque speed characteristics of the motors without employing the fast centrifugal switching action are shown in Figs. 9 and 10, respectively

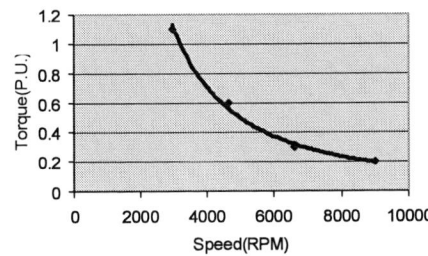

Fig. 9. Torque-speed characteristics without governor.

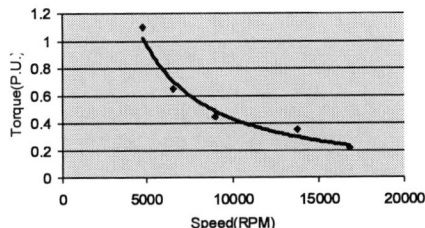

Fig.10. Torque-speed characteristics with governor.

As seen from the Fig 8, the speed torque curve is much higher than the normal curve

V. CONCLUSION

In this paper a novel two phase SR motor was fabricated in the laboratory. The motor parameters experimentally measured and tested. The two main objectives of this paper namely, introduction of a new two phase SR motor configuration with high starting torque, and use of centrifugal switch for fast turn-on time advancement were achieved. The experimental analysis shows the functionality of the motor in its new configuration, meaning, it has the ability and the potential of becoming a motor comparable with other types of switched reluctance motor in the industry. The fast acting switch will cause the motor to have larger over lap area between rotor and stator poles at the beginning, hence the stating torque will be higher than a regular SR motor.

REFERENCES

[1] B. Fahimi, A. Emadi, R. Sepe, "A switched reluctance machine based starter/alternator for more electric cars" Energy Conversion, IEEE Transactions on Energy Conversion, Volume: 19 , Issue: 1 , March 2004, pp.116 – 124.

[2] T.J.E. Miller, Switched Reluctance Motor Drive, Ventura, CA, Intertec Communications Inc, 1988.

[3] R. Krishnan, Switched Reluctance Motor Drive: Modeling, simulation, Analysis, Design and application, Magna physics publishing, 2001.

[4] F. Liang, Y.liao, and T. A. Lipo, "A new Variable Reluctance motor Utilizing an Auxilary Commutation

Winding",IEEE Tran. Ind. pp., Vol. 30, No. 2, pp. 423-432, March/April 1994.

[5] J.P. Bastos, R. Goyet, J. Lucidarme, C. Quichaud and F. Rioux-Damidau "Performance of a Multi- Disc Variable Reluctance Machine," International Conference on Electrical Machines, Budapest, 1982, pp. 254-257.

[6] E.Afjei, Yousefi Azad, "A Novel Disc Type Reluctance Motor", International journal of engineering, vol.10, Feb. 1997, pp. 11-17].

[7] E. Afjei & H. Toliyat, "A Novel Multilayer Switched Reluctance Motor", IEEE Transaction on Energy Conversion, Vol. 17, No. 2, June 2002,pp 1-5.

[8] Bae, H. K. Bae, P Vijaraghavan, and R Krishnan, "Design of a linear Switched Reluctance Machine, IEEE Industry Appl. Conf. (IAS '99'), vol. 1 Oct. 3-7, 1999, Phoenix Az., pp. 2267-2274.

[9] Magnet CAD Package: User Manual, Infolytica Corporation Ltd., Montreal, Canada, 2006.

Application of Power Electronics for Damping of Torsional Vibrations

T. Zöller*, T. Leibfried* and A. M. Miri*

*University of Karlsruhe, Institute of Electric Energy Systems and High-Voltage Technology (IEH)
Kaiserstr. 12, 76128 Karlsruhe, Germany

Abstract--Torsional vibrations occur in mechanical shafts whenever they couple rotating masses. Especially shaft assemblies with large inertias and long shafts, e.g. in turbo-generating sets of power plants or in drives, these oscillations can be a major problem due to the weak natural damping and the sharp resonance points of these systems [1]. It is impossible to obtain a desirable damping of these torsional vibrations by means of a mechanical method. However, by using a power electronics converter and an inductive [2] or capacitive energy storage, an electrical method can be used to implement an active damping. The following paper illustrates the needed configuration and the necessary controls of the required converters for damping torsional vibrations.

Index Terms-- active damping, mechanical shafts, power electronics, torsional vibrations

I. INTRODUCTION

Extended shaft assemblies own many mechanical eigenfrequencies, but there are only a few which are dominant. In most cases, it is possible to represent such shaft assemblies with a mass-spring system of less order. A typical turbo-generating set of a power plant normally has three to six dominant eigenfrequencies. These frequencies are located both subsynchronous and supersynchronous to the line frequency.

An excitation of these eigenfrequencies of the shaft system can be caused by mechanical or electrical disturbances. Predominantly, in the domain of power plants, failures in the electric supply network cause an excitation of torsional oscillations. They often come along with a strong change of the generator's delivering active power. According to

$$T_{el}(t) = \frac{p_{GEN}(t)}{\omega_{GEN}(t)} \quad (1)$$

the failure produces a shock excitation of the mechanical frequency response system via the electrical torque T_{el} [3] [4]. Subsequently, the triggered torsional vibration decreases according to the natural damping of the system. The degradation of the shaft and the couplings depends on the oscillation amplitude and the time to decay the vibration. Furthermore, so-called subsynchronous oscillations and subsynchronous resonances can result in significant shaft damages. In this case oscillating active power whose frequency spectrum includes one or more

eigenfrequencies of the mechanical shaft assembly drives to a steadily rising torsional vibration.

This vibration can reach critical material limits within seconds. Therefore the shaft diameters of turbo-generating sets have to be designed for much larger values than the steady state torque. But even then some kinds of excitations can be only handled by tripping the generator [5] [6].

The oscillations by drive systems can be excited by cascade controlled converters. But quite often, the oscillations appears as a mechanical problem, because they are mechanical initiated or excited. For example torsional oscillations in paper machine greatly affect the performance [7].

For damping subsynchronous oscillations in two island power systems of the Max-Planck-Institute for Plasma Physics (IPP) in Garching in Germany a novel damping circuit was developed. Presently, this system is applied successfully in several flywheel generators of the IPP. In this case the damping system suppresses oscillation excitations activated by a feedback of a plasma experiment. In the meantime, further simulations proved the suitability of this system for other configurations [8]. The up-to-now investigated and used Parallel Connected Damper Circuit (PCDC) is based on inductive energy storage. However, this system possesses some structural disadvantages compared to a PCDC with a capacitive storage. Thus new converter structures and control methods were investigated to build a PCDC with capacitive energy storage.

The paper is organized as follows; to begin with, a description of the basic approach of the active damping system (Chapter 2). In Chapter 3, the existing and presently used system, which is based on inductive energy storage, will be presented. A short summary about the disadvantages of this system will lead over to the next chapter describing the new PCDC with capacitive energy storage and showing the general adequacy of such a system. The paper ends with a conclusion of the investigations and the results.

II. THE PARALLEL CONNECTED DAMPER CIRCUIT - PCDC

The fundamental idea of the PCDC is the intention to arise the natural damping of the mechanical shaft assembly by an action at the electrical part of the shaft-generator-system. Torsional oscillation in a shaft system

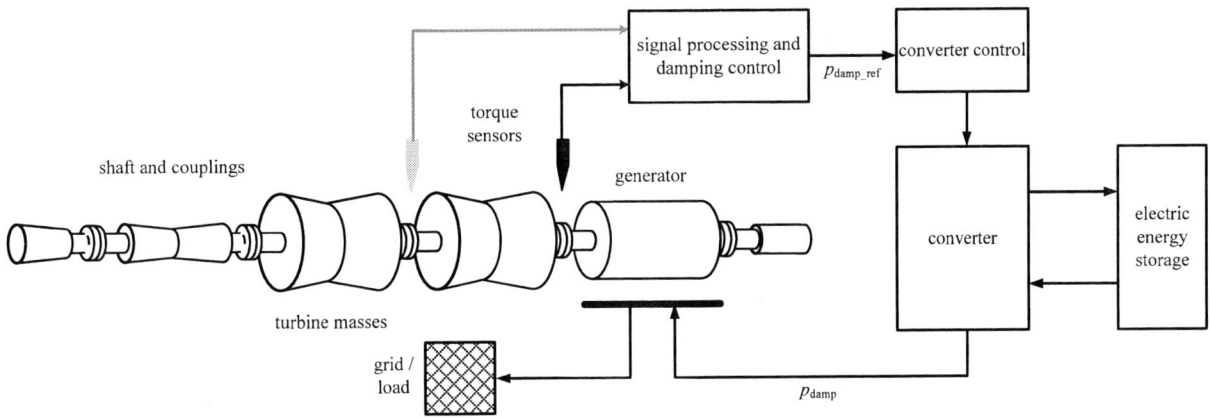

Fig 1. Schematic overview of the PCDC

can be described by the following n-dimensional differential equation system:

$$J\ddot{\phi} + D\dot{\phi} + K\phi = Bu \qquad (2)$$

$\phi(t)$ torsion angles of the shaft
$u(t)$ externally applied torques
J matrix of moments of inertias
D damping matrix
K stiffness matrix
B input matrix for the external torques

The damping of the torsional oscillations occurs by feeding an appropriate external torque Bu, that amplifies the existent natural damping $D\dot{\phi}$. This torque is injected electromagnetically via the stator windings of the generator and the motor respectively. In order to generate a counter torque, it is necessary to measure the torsional vibration which one wants to be damped, with sufficient accuracy. This can be one or more torsional modes at different locations of the shaft assembly. Incidentally, it is possible to calculate torsional torques of different locations from one measurement with a state observer. A prerequisite for this is a detailed modelling of the shaft assembly. The measurement of the required torque is not as simple. However, in the meantime, there are commercially available, accurate and reliable contact-less sensors for diverse shaft types. Figure 1 shows the configuration of a PCDC in principle. In order to get an electrical counter torque the motor and the generator, respectively, has to be loaded with a corresponding active power accordingly (1). This damping power $p_{\text{damp\_ref}}$ is calculated by the device "signal processing and damping control" of Fig. 1. The result is fed to the converter control. The signal $p_{\text{damp\_ref}}$ is a single or multi mode sinusoidal waveform dependent on the torsional vibration that oscillates symmetrically around zero. The requirement of the power electronic converter is to feed the generator with the preset active power p_{damp}. One needs an electric energy storage because the converter has to deliver and to absorb power.

Due to the fact that the damper circuit produces damping power, because of the measurement of the actual tor-

sional torques and according to this a counter torque with a resonance frequency of the mechanical shaft assembly, a high efficiency of the damping system is achieved. Thus, the nominal power of the damping device can be up to 1000 times lower than the nominal power of the generator and motor, respectively. In addition, the damper circuit works independent of the load, the excitation and the revolution speed of the machine. Once all control parameter are properly set, such a PCDC can be continuously used on demand to damp torsional vibrations.

III. PCDC WITH INDUCTIVE ENERGY STORAGE

The electric energy storage can be build either inductive or capacitive with an inductor and a capacitance, correspondingly. According to this, the type of converter has to be chosen. In combination with an inductive energy storage it is advisable to choose a B6C thyristor bridge. This rectifier is connected in parallel to the stator windings of a synchronous or asynchronous machine (Fig 2).

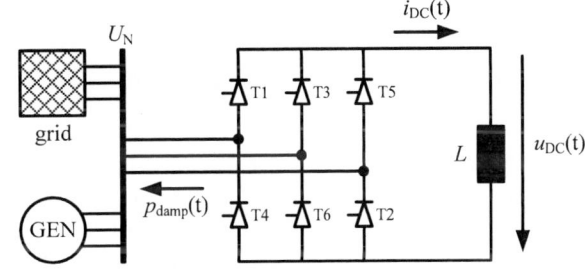

Fig 2. PCDC with inductive energy storage

The B6C thyristor bridge is a line commutated 6-pulse rectifier and is based on non sheddable valves. The output direct voltage $U_{\text{DC}}(t)$ can be tuned via the firing angle α according to

$$u_{\text{DC}}(t) = \frac{\sqrt{18}}{\pi} \cdot U_{\text{N}} \cdot \cos(\alpha(t)). \qquad (3)$$

The delivered and absorbed power of the bridge is calculated by

$$p_{\text{damp}}(t) = i_{\text{DC}}(t) \cdot u_{\text{DC}}(t). \qquad (4)$$

Dependent on the size of the inductor L and the direct current i_{DC} one supposes that the current is constant. Thus the power equation results in

$$p_{damp}(t) = \hat{p}_{damp} \cdot \cos(\alpha(t)) . \tag{5}$$

Here $\hat{p}_{damp}$ is a simply function of the nominal voltage U_N as well as the constant current I_{DC}. With an appropriate control of the firing angle α, the B6C thyristor birdge delivers the desired damping power p_{damp_ref}. However, the damping power is restricted to a particular frequency range. This is because of the line commutation of the B6C thyristor bridge. For the output voltage u_{DC} the line commutation allows only to commutate to a line-to-line voltage of higher potential. Figure 3 clarifies this interrelationship. There the output voltage u_{DC} at a preset frequency f_{damp} of the damping power is shown at 15 Hz and 50 Hz. The line frequency is 50 Hz.

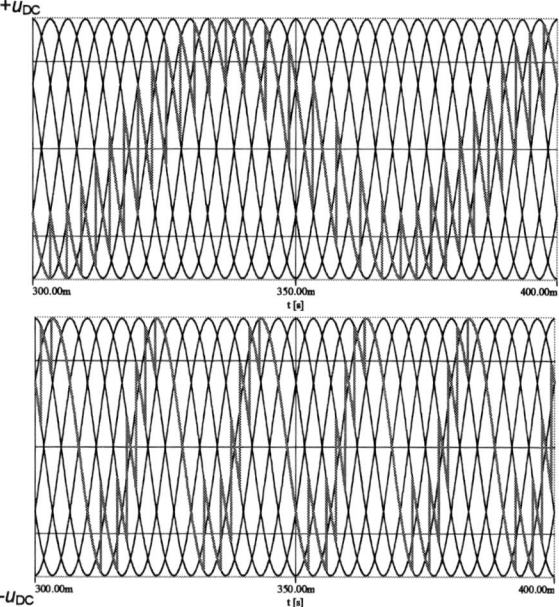

Fig 3. Curve of voltage $u_{DC}(t)$ (red) as well as the line-to-line voltages (gray) at a damping frequency of 15 Hz (top) and 50 Hz (bottom)

As can be seen in Fig 3, the maximum damping frequency f_{damp} is limited by the line frequency, because the quickest possible decrease of the voltage $u_{DC}(t)$ from $+u_{DC}$ to $-u_{DC}$ is determined by the run of the line-to-line voltage. Also, the ratio between the desired damping power p_{damp_ref} and the delivered actual damping power p_{damp} of the rectifier gets worse with a rising frequency. With this setup, it is safe to say that the damper circuit is able to generate damping power of good quality in a frequency range up to the line frequency.

The frequency range could be extended by the use of a higher pulse rectifier. However, a 12-pulse rectifier could double the maximum damping frequency f_{damp}, but this would complicate the system and furthermore would increase the price significantly.

Besides the frequency limits there are some other disadvantages of this system. An application of the PCDC at machines with a nominal voltage in the high voltage range would be relatively cost-intensive. In order to be able to buffer sufficient energy in the inductive storage one has to install either a converter transformer to reduce the maximum output voltage of the rectifier or to implement a high voltage coil with a big inductance. Another drawback is the reactive power demand of this system. To assure a high dynamic of the damper circuit at anytime, it is necessary to keep the coil charged i.e. the coil and the rectifier are current-carrying (I_{DC}). At this standby-mode the bridge would permanently absorb the reactive power according to

$$Q_{standby} = \frac{\sqrt{18}}{\pi} \cdot U_N \cdot I_{DC} . \tag{6}$$

Besides, the direct current I_{DC} would cause significant active losses in the rectifier and the coil.

In addition the line commutation always generates undesired harmonics in the line currents.

IV. PCDC WITH CAPACITIVE ENERGY STORAGE

The application of the described B6C thyristor bridge is not possible in combination with a capacitive energy storage because it only delivers a positive direct current i_{DC}. However, to discharge the capacitance and thus to output power, negative currents are required. A line-commutated four quadrant converter could do that but with this device the maximum damping frequency f_{damp} would be still restricted to the line frequency. Therefore, it is necessary to use a self-commutated converter with sheddable valves, e.g. Mosfet, IGBT or IGCT.

Figure 4 shows the basic configuration of a self-commutated three phase inverter with a capacitive electric energy storage.

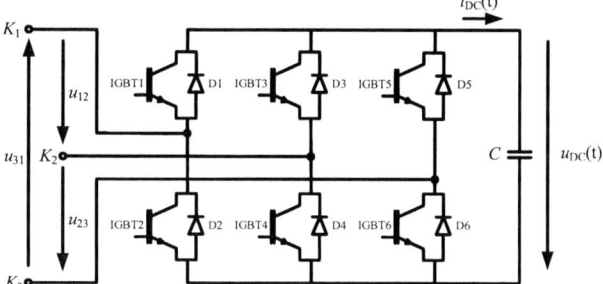

Fig 4. Three phase self-commutated inverter with IGBTs and capacitive energy storage

Let us assume a constant voltage u_{DC} of the capacitance C. By a suitable switching of the valves a three phase voltage of variable amplitude, phase and frequency can be generated at the clamps K_1, K_2 and K_3. Thereby the line-to-line voltages u_{12}, u_{23} and u_{31} are pulsed square wave signals with the voltage levels $-u_{DC}$, $+u_{DC}$ and 0 V. The average values of this pulse pattern result in sinusoidal voltages. If the inverter is coupled to a synchronous machine two three phase systems are interlinked via a reactance X. A simplified equivalent network is shown in Fig 5. There, the output three phase voltages u_G and u_{WR} of the generator and the inverter are represented as ideal monophase voltage sources.

One possibility in order to exchange power between the two systems is a modulation of the angle δ. There is the following analytical interrelationship [9]:

$$p_{\text{damp}}(t) = \frac{u_G \cdot u_{WR}}{X} \cdot \sin(\delta(t)) \tag{7}$$

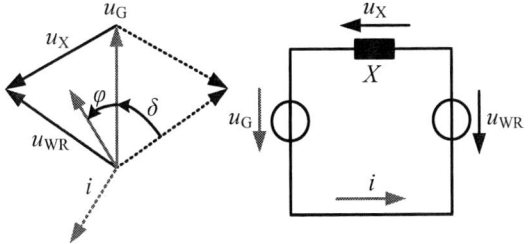

Fig 5. monophase equivalent network of the system with vector diagram

In power delivery the current i and the voltage u_G have the equal sign and the angle δ is negative (solid arrows in Fig 5). By contrast the angle δ is positive and the sign of i and u_G are opposed for power absorption by the damper circuit (dashed arrows in Fig 5). Due to this fact there is a great demand for reactive power q_{damp} according to (8)

$$q_{damp}(t) = \frac{u_G \cdot (u_G - u_{WR} \cdot \cos(\delta(t)))}{X} \tag{8}$$

This results again in losses and a higher nominal power of the damper circuit.

An optimal energy balance could be realized by maintaining the current i and voltage u_G always in-phase so that there is only an interchange of active power. Therefore it is advisable to consider the current i as the control parameter of the inverter. In a three phase grid the active power is calculated form the actual values of the phase voltages and the phase currents according to

$$p(t) = u_1(t) \cdot i_1(t) + u_2(t) \cdot i_2(t) + u_3(t) \cdot i_3(t). \tag{9}$$

A modulation of the current as shown in (10)

$$i_1(t) = \hat{u} \cdot \sin(\omega t) \cdot c \cdot p_{\text{damp\_ref}}(t)$$
$$i_2(t) = \hat{u} \cdot \sin\left(\omega t + \frac{2\pi}{3}\right) \cdot c \cdot p_{\text{dam\_ref}}(t)$$
$$i_3(t) = \hat{u} \cdot \sin\left(\omega t - \frac{2\pi}{3}\right) \cdot c \cdot p_{\text{dam\_ref}}(t) \tag{10}$$

leads to the damping power p_{damp}:

$$p_{\text{damp}}(t) = \frac{3}{2} \hat{u}^2 \cdot c \cdot p_{\text{damp\_ref}}(t) \tag{11}$$

This means that the required phase currents are obtained by the modulation of the respective phase voltage with the desired value of the damping power $p_{\text{damp\_ref}}$. Prerequisite of this, is the right choice of the constant c. In the vector diagram of Fig 5 now the voltages u_G and u_X are orthogonal and the angle φ is constant zero. Therefore the vector voltage u_{WR} has to be ever greater than the voltage u_G. This means that the voltage u_{DC} of the capacitance C has to be always greater than the amplitude of the line-to-line voltage of the three phase voltage system u_G. At a first glance, this is not possible because the maximum capacitance voltage u_{DC} is limited to

$$u_{\text{DC\_max}} = u_G \cdot \sqrt{2} \cdot \sqrt{3} \tag{12}$$

if the uncontrolled inverter is coupled to the voltage system u_G. Nevertheless, the inverter is able to work as a boost converter with an appropriate control. Thus much higher voltages u_{DC} can be achieved.

For the drive of a PCDC with capacitive energy storage, it is recommended to implement a closed-loop control for the phase currents. In the case a symmetric assembly, it is solely necessary to control two of the three phase currents. The third phase current results from the sum of the other two currents according to

$$0 = i_1(t) + i_2(t) + i_3(t). \tag{13}$$

An overview of the complete system is shown in Fig 6. To control the voltage u_{DC} a superposed closed-loop control is applied. For this purpose a setpoint / actual-value comparison is implemented. The result weighted by the transfer function $G(s)$ affects accordingly the desired value of the damping power $p_{\text{damp\_ref}}$. The functional block „current control" generates eleven binary signals, which were assigned via a logic to the six IGBT switching commands. An additional interlock unit offers special tools to prevent switching failures and provides control of various switching times.

Simulation results of the presented PCDC with capacitive energy storage are shown in Fig 7. Therefore, the PCDC with a rated power of 150 kW was connected exemplarily to a three phase voltage source with a nominal voltage of 400 V. This rated damping power should be capable to damp torsional oscillations in a generator set with a nominal power up to 10 MVA. The capacitor C has a value of 30 mF and the inductor of 400 µH. The

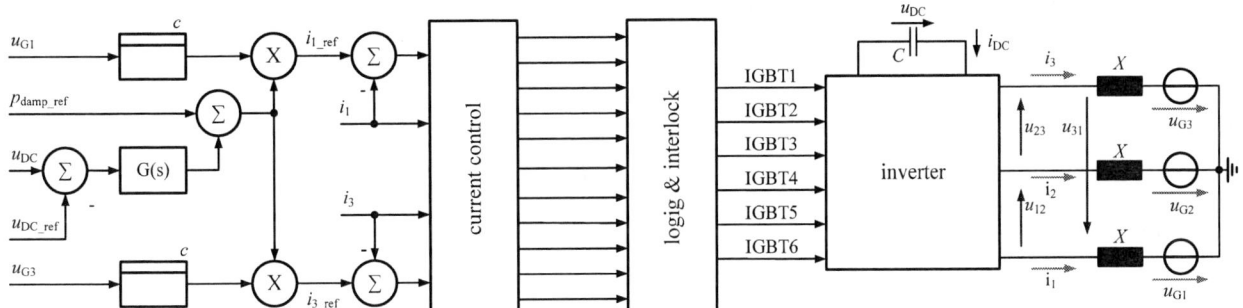

Fig 6. Schematic overview of the converter control

frequency range of the damping power p_{damp} is located between 10 Hz and 500 Hz at rated power.

In this simulation scenarios (Fig 7 and Fig 8) the intentions was to produce a damping power of various frequencies with a rising amplitude up to 100 kW.

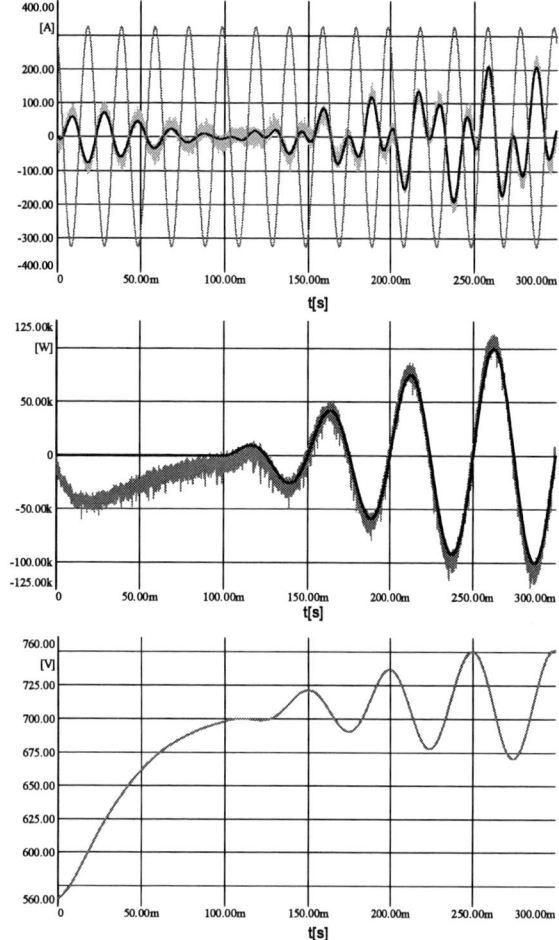

Fig 7. Simulation PCDC with f_{damp} = 20 Hz; *top*: i_{1_ref} (black), i_1 (green) and u_1 (red, dashed); *middle*: p_{damp_ref} (black) and p_{damp} (red); *bottom*: u_{DC}

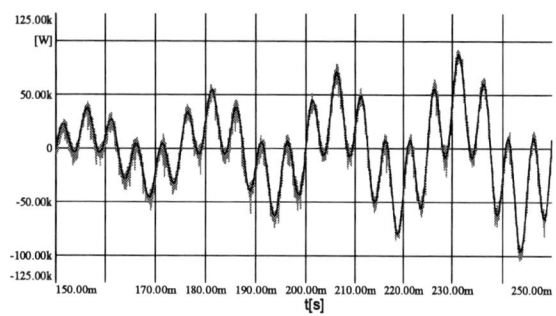

Fig 8. Simulation PCDC with f_{damp} = 200 Hz and 40 Hz; curves of the power p_{damp} (red) und p_{damp_ref} (black)

At first, in order to have enough energy and voltage reserves for the later damping mode, the voltage of the capacitor is raised to a level of about 700 Volts. The real damping mode starts at the time t = 100 ms with a frequency f_{damp} = 20 Hz. The amplitude of the damping power rises within 250 ms from zero to 100 kilowatts (Fig 7). Of course there is also a delivery of multimode

damping power possible. An appropriate simulation shows Fig 8. In this example the damping power has a 40 Hz as well as a 200 Hz mode.

V. CONCLUSIONS

The damping of torsional vibrations in shaft assemblies of power plants or drives is possible via a parallel connected damper circuit. The so far investigated and used method with a line-commutated B6C rectifier and an inductive energy storage possesses some disadvantages. The main points are the restricted frequency range of the damping power and the quite high demand for reactive power. Furthermore the required converter transformer increases the costs significantly. Simulative investigations showed the possibility to construct a damping system based on a PCDC with a capacitive energy storage. Therefore a self-commutated inverter with sheddable valves is required. The drive of the damping power was realized with a closed-loop control of the phase currents.

With the capacitive PCDC it is now possible to generate damping frequencies above the line frequency. A frequency range of several kilohertz is conceivable. The efficiency of this PCDC is essentially improved because there is only a small demand for reactive power. In addition, the standby can be handled smoothly. The recharge current of the capacitor is also quite small in real applications. The costs of this PCDC with capacitor in comparison to a PCDC with inductor can be estimated to be significantly lower.

For further experimental verifications a downscaled turbo-generator set is build up at the moment. This experimental assembly will afford a test of the PCDC under almost real conditions in the future.

REFERENCES

[1] IEEE State-of-the-art-Symposium, "Turbine Generator-Shaft Trosionals," IEEE /ASME / ASCE Joint Power Generation Conference, 1978.

[2] C. Sihler, A. M. Miri, A. Harada and ASDEX Upgrade Team, "Damping of Torsional Resonances in Generator Shafts Using a Feedback Controlled Buffer Storage of Magnetic Energy", in *Proc. 5th IPST, New Orleans, LA 2003, Paper 6b-3.*

[3] I.M. Canay, H.J. Roher. K.E. Schnirel, "Effect of Electrical Disturbances, Grid Recovery Voltage and Generator Inertia on Maximization of Mechanical Torques in Large Turbogenerator Sets", IEEE Trans., vol. PAS-99, No. 4, pp. 1357-1370, July/Aug. 1980.

[4] M. Humer, A. Wirsen, "Online Monitoring von Torsionsschwingungen in Wellensträngen von Kraftwerksturbosätzen", Schwingungssymposium 2006.

[5] F. Joswig, S. Kulig, "Preseptions about new kinds of subsynchronous resonances," in *Proc. 4th IPSPT, Rio de Janeiro, Brazil*, pp. 228-233, 2001.

[6] P. M. Anerson, B. L. Agrawal, J. E. Van Ness, "Subsyncronous Resonance in Power Systems," New York: IEEE Press, 1990, pp. 1-269.

[7] M. A. Valenzuela, J. M. Bentley, R. D. Lorenz, "Evaluation of Torsional Oscillations in Paper Machine Sections", IEEE Trans., vol. 41, No 4, pp. 493-501, March/April 2005.

[8] A. M. Miri, C. Sihler, T. Zöller, "Suppression of Subsynchronous Resonance by a Parallel Connected Damper Circuit", in *Proc. 6th IPST,Montreal, LA 2005, Paper IPST05-22*

[9] P. Kundur, "Power System Stability and Control" New York: McGraw-Hill, 1993.

Application of Battery Energy Operated System to Isolated Power Distribution Systems

Bhim Singh

Deptt. of Electrical Engineering
Indian Institute of Technology
Hauz Khas, New Delhi-110016,India
bsingh@ee.iitd.in

A. Adya

Deptt. of Electrical & Electronics
Maharaja Agrasen Institute of Technology
Rohini Sector 22, New Delhi, India
alkaadya@gmail.com

A.P. Mittal and J.R.P Gupta

Deptt. of Instrumentation and Control
Netaji Subhas Institute of Technology
Sector 16, Dwarka, New Delhi, India
mittalap@yahoo.com, jrpg83@yahoo.com

Abstract—**This paper deals with model of battery energy operated system for a 42.5kVA DG set using Simulink and Power System Block-set in MATLAB environment. Battery Energy Storage System (BESS) is employed for compensation along with a small synchronous generator of 42.5kVA capacity coupled to a diesel engine as a prime mover. The DG set feeds a wide variety of loads. The performance of the system is simulated for linear, non-linear balanced and unbalanced loads. Simulation results justify enhanced power quality of the system with BESS application.**

Keywords-BESS; load balancing; voltage regulation; power quality.

I. INTRODUCTION

There is an increased emphasis on deregulation and dispersed generation worldwide to meet growing energy requirement especially in developing nations. Recently, small distributed generation using conventional as well as unconventional energy sources has attracted attention even in developing countries. This is especially due to the fact that the centralized energy generation involves large gestation periods and huge costs. Isolated generating systems though set up quickly at a fraction of cost, face power quality problems such as poor voltage profile and load unbalancing. Poor power quality results in primarily due to harmonic pollution from increasing use of power electronic based loads. All this leads to distribution system problems such as harmonic distortions in voltages and currents and load unbalancing.

Power engineers consider improving power quality and providing reliable power at the lowest cost a major challenge. Possible solutions to power distribution problems have been suggested in the form of a number of power electronic based devices for improved power quality. Distribution Static Compensator (DSTATCOM), Distribution Voltage Regulator (DVR), Unified Power Quality Compensator (UPQC), BESS, HVDC Light are some of the prominent custom power devices used at distribution level [1-4].

The requirement for BESS arises mainly due to load fluctuations at the consumer level. BESS holds a very promising future to act as alternative power source to counteract the uncertainty in power supply. It has the ability to operate in all four quadrants. Singh et al [5,6]

have developed a model of BESS considering the battery is being charged if load is less than stipulated 80% of full load whereas BESS feeds extra power to loads when the load exceeds nominal 100% rating. Control techniques [5-15] for BESS include adaline based controller[5], SRF, P-Q, indirect current control theory [13-15] etc. Tsang et al [8] have used ANN based back-propagation technique for controlling BESS for enhancing damping. In this paper, BESS has been controlled to absorb or generate active and reactive powers in diesel generator based isolated system. BESS provides load leveling and improved power system performance even during adverse fluctuations in consumer loads requirement due to faults etc.

The application of BESS is new and still in developing stage; however a substantial amount of literature on configuration, progress, working, control and practical applications is available [10-12]. Till date, 0.5–40 MW rating of BESS has been commercially installed [8-12]. Miller etal [11,12] have demonstrated 5MVA, 2.5MWh BESS that allows large induction motor loads to function up-to one hour in the event of normal power failure.

Application of BESS to SEIG for power quality improvement is reported in [13]. The benefits of BESS are immense such as curtailing peak energy use, reduction of stress on the entire system and also it being emission free source, make it all the more attractive. Industrial and commercial users are going for this improved technology for greater backup generation and also to reduce the production downtime to minimum.

Research work on BESS integrates the technology developments in the areas of battery and power electronics and real time computer control for load management in power distribution system. An attempt is made in this paper to model an isolated system comprising a small alternator driven by diesel engine with BESS. It focuses on the modeling, simulation and control aspects of BESS for improvement of power quality features viz. voltage regulation, load balancing, harmonic reduction in case of non-linear loads.

II. SYSTEM CONFIGURATION

The basic structure of BESS is similar to that of DSTATCOM which is a popular shunt compensator. The major difference between the DSTATCOM and BESS

978-1-4244-0644-9/07/$25.00 ©2007 IEEE

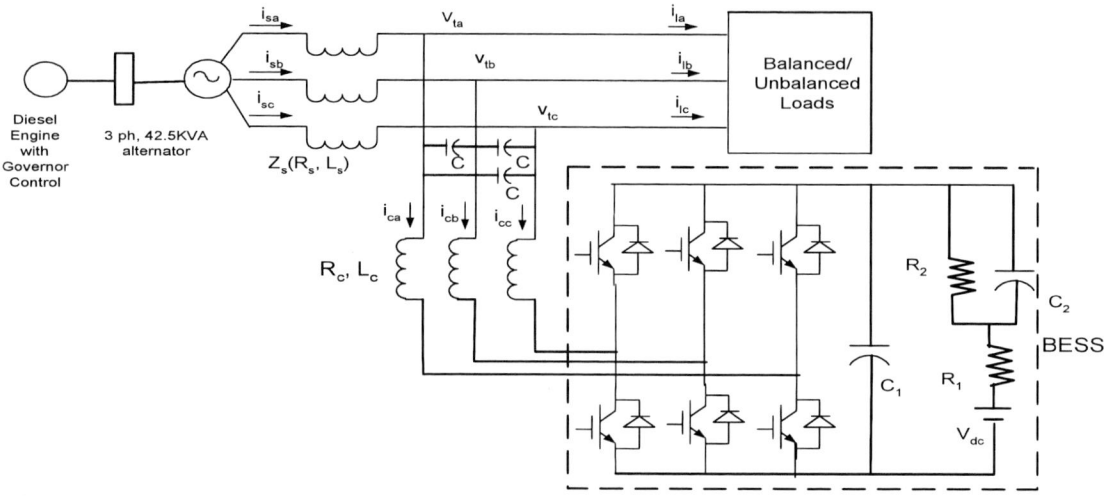

Fig. 1. Schematic diagram of system with BESS connected in shunt configuration

configurations is at the DC side. In DSTATCOM, a voltage source inverter is used and there is a DC link capacitor which is replaced by a battery in BESS.

Fig.1 shows an isolated distribution system with BESS installed on it. A small diesel engine acts as a prime mover for the alternator (42.5kVA). The BESS in shunt configuration acts as a source of leading or lagging vars and is applied to regulate the voltage at the point of common coupling (PCC). When the consumer load is less than the fixed power generated by the synchronous machine, BESS absorbs the surplus power whereas when the consumer load requirement exceeds the synchronous machine capacity, BESS also acts as a source of power. BESS provides the necessary reactive component of the supply current thereby maintaining voltage at PCC terminals as well as harmonic compensation. The choice of suitable controllers on BESS can be used to provide load leveling.

III. CONTROL SCHEME

The configuration of BESS as shown in Fig.1 consists of a battery at the DC link; hence the DC link voltage remains practically constant. A DC battery (700V) has a associated small series resistance (R_1) connected in with a parallel combination of a resistor (R_2) and a large capacitor (C_2). A DC link capacitor (C_1) is also connected as shown. The values of the parameters for DG set, battery and controller are mentioned in Appendix. The block diagram for the control scheme of BESS is shown in Fig.2. It utilizes one proportional-integral (PI) controller for regulating the ac terminal voltage. The in-phase components of the BESS reference currents are required for charging the dc capacitor to the level of reference dc bus voltage and to meet its losses. The amplitude of in-phase component of the reference supply currents (I_{spdr}) is kept constant at a particular value depending on real power requirement of the load.

The instantaneous values for in-phase components of supply reference currents are obtained by multiplying I_{spdr} with the in-phase unit current vectors (u_a, u_b, u_c) derived from three phase sensed terminal voltages.

One PI controller is applied over the sensed and reference ac mains voltage. Its output is considered as the amplitude of quadrature component of the supply reference currents (I_{spqr}). The instantaneous values are obtained by multiplying the output of this PI controller with the quadrature unit current vectors (w_a, w_b, w_c) derived from unit in-phase current vectors (u_a, u_b, u_c) which are calculated from three-phase sensed terminal voltages. The total reference supply currents are obtained by adding respective in-phase and quadrature components.

PWM based hysteresis current controller is employed over instantaneous reference supply currents and sensed supply currents. If $i_{sa} < (i_{sar}-h_b)$, the upper switch is turned 'OFF' and lower switch is turned 'ON'. If $i_{sa} > (i_{sar}+h_b)$, the upper switch is turned 'ON' and lower switch is turned 'OFF'. In this manner, the switching logic for other two phases is obtained and the controller is able to regulate the currents in a band around the desired reference value.

IV. MATHEMATICAL MODELING OF BESS

Three-phase reference supply currents are computed using three-phase supply voltages. These reference supply currents consist of two components, one in-phase and another in quadrature with the supply voltages.

A. Computation of In-Phase Components of Reference Supply Currents

The amplitude of in-phase component of reference supply currents (I_{spdr}) is kept fixed at a particular value so that BESS supplies fixed real power. Three-phase in-phase components of the reference supply currents are computed using the in-phase unit current vectors (u_a, u_b,

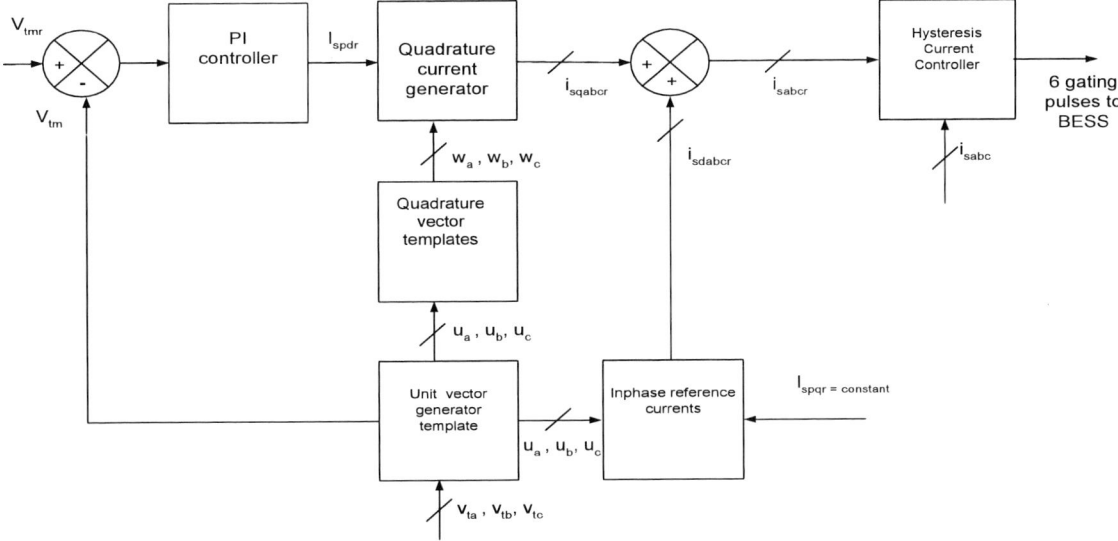

Fig.2. Control scheme for BESS

u_c) derived from three-phase terminal voltages (v_{ta}, v_{tb}, v_{tc}) using the following equations.

$$u_a = v_{ta}/v_{tmn}$$
$$u_b = v_{tb}/v_{tmn}$$
$$u_c = v_{tc}/v_{tmn} \qquad (1)$$

The amplitude of supply voltage (v_{tmn}) is computed as:

$$v_{tmn} = \sqrt{2/3}(v_{ta}^2 + v_{tb}^2 + v_{tc}^2) \qquad (2)$$

The amplitude of in-phase component of reference supply currents is computed as:

$$i_{sadr} = I_{spdr} * u_a$$
$$i_{sbdr} = I_{spdr} * u_b$$
$$i_{scdr} = I_{spdr} * u_c \qquad (3)$$

B. Computation of Quadrature Components of Reference Supply Currents

The amplitude of quadrature component of reference supply currents is computed using a PI controller over the average value of amplitude of supply voltage (v_{tm}) and its reference counterpart (v_{tmr}).

$$I_{spqr(n)} = I_{spqr(n-1)} + K_{pq}\{v_{ae(n)} - v_{ae(n-1)}\} + K_{iq} v_{ae(n)} \qquad (4)$$

where $v_{ae(n)} = v_{tmr} - v_{tm(n)}$ denotes the error in v_{tm} calculated over reference v_{tm} and average value of v_{tm}. K_{pq} and K_{iq} are the proportional and integral gains of the PI controller.

The quadrature unit current vectors are derived from in-phase unit current vectors as:

$$w_a = (-u_b + u_c)/\sqrt{3}$$
$$w_b = (u_a\sqrt{3} + u_b - u_c)/2\sqrt{3}$$
$$w_c = (-u_a\sqrt{3} + u_b - u_c)/2\sqrt{3} \qquad (5)$$

Three-phase quadrature components of the reference supply currents (i_{saqr}, i_{sbqr}, i_{scqr}) are computed using their amplitude and quadrature unit current vectors as:

$$i_{saqr} = I_{spqr} * w_a$$
$$i_{sbqr} = I_{spqr} * w_b$$
$$i_{scqr} = I_{spqr} * w_c \qquad (6)$$

C. Computation of Total Reference Supply Currents

Three phase instantaneous reference supply currents are computed by adding in-phase and quadrature components expressed as:

$$i_{sar} = i_{sadr} + i_{saqr}$$
$$i_{sbr} = i_{sbdr} + i_{sbqr}$$
$$i_{scr} = i_{scdr} + i_{scqr} \qquad (7)$$

A carrier-less PWM hysteresis current controller is employed over the reference and sensed supply currents to generate gating pulses of IGBT's of the BESS. This gives appropriate gating signals for all the three legs of VSI.

V.   MATLAB BASED MODELING OF SYSTEM

This section illustrates the model of BESS along with the isolated system. Small isolated power system consisting of an alternator feeding variety of loads is shown in Fig.1. The alternator is driven by diesel engine with governor control. A PI controller is tuned to regulate the ac terminal voltage at the PCC. The power as well as control circuit are modeled in Matlab / Simulink and power system block-set. Fig.3 shows the Simulink diagram representing BESS and the load on the distribution system. A small capacitor filter is connected at the PCC.

The BESS configuration has a voltage source inverter modeled using universal bridge from PSB toolbox library. It uses six IGBTs each shunted by a reverse parallel connected fast switching free wheeling diode. The output of BESS is coupled in parallel to the power system network through inductances of the coupling transformer. The alternator system feeds a variety of balanced and unbalanced loads. The linear load on the system is represented by three-phase resistive-inductive (R-L load) for lagging power factor. Switches are appropriately connected for making the load either balanced or unbalanced. The non-linear load is represented in the form of resistive load connected across a three-phase diode rectifier.

The PWM current controller on BESS is used to obtain appropriate pulses for the IGBTs. BESS controller block basically involves several subsystems like measurement system, reference current generation,

733

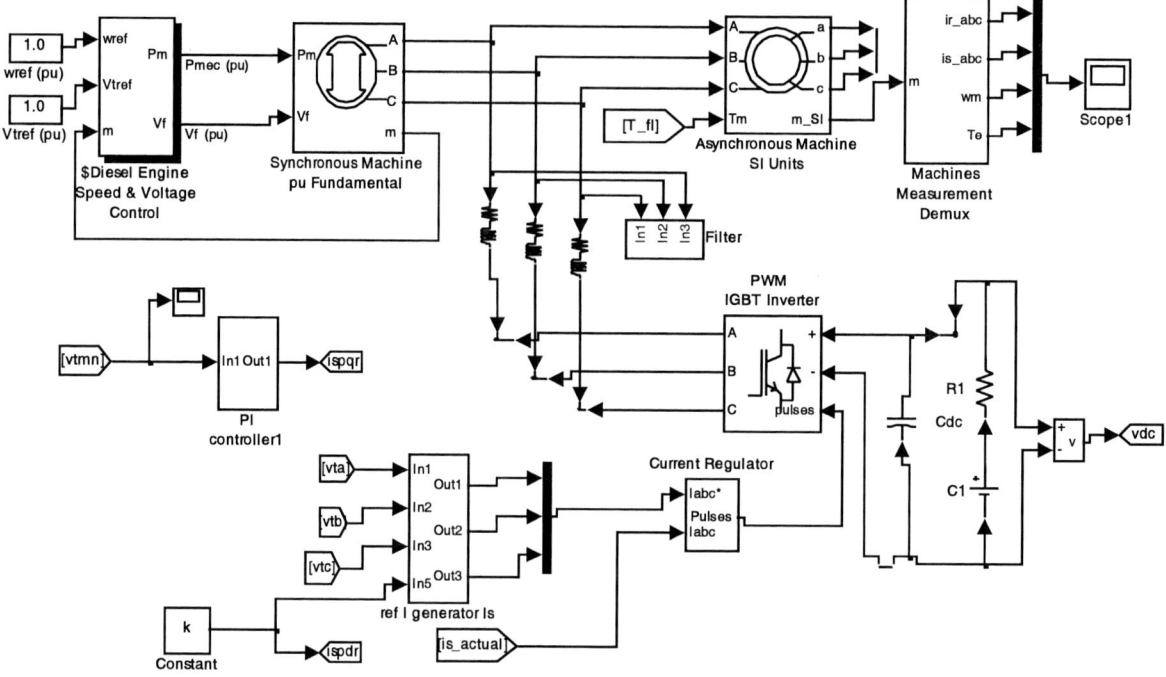

Fig. 3. MATLAB based model of battery energy storage system

ac voltage regulation loop and hysteresis current controller.

VI. PERFORMANCE OF BESS

Performance characteristics for the BESS system are given in Fig. 4-5 to illustrate its steady state and transient behavior. The necessary parameters of the system are given in Appendix.

A. *Performance of BESS with Linear Loads*

Performance of BESS connected to an isolated system feeding linear loads is shown in Fig.4 for voltage regulation and load balancing. Fig.4 shows variation of various quantities viz. three–phase supply voltages (v_s), terminal voltages (v_t), supply currents (i_s), load currents (i_l), BESS currents (i_c) and voltage at the point of common coupling (v_{tm}) along with reference value and BESS current (i_{bb}) for a variety of load changes. At t=0.2sec, a three-phase load of 30.4 kW is changed to two-phase load of 20.27kW and then to single-phase load 10.14kW at t=0.26sec.

The load is changed again from single-phase to two-phase and back to three-phase at t=0.32sec and t=0.38sec respectively. The amplitude of the in-phase component of currents (I_{spdr}) is kept at constant value of 30A. The PI controller on the terminal voltage regulates the voltage at PCC to reference value of 328V. The sub-plot for supply current (i_s) shows that the BESS system is able to maintain nearly constant three-phase balanced supply

currents under varying load conditions. At t=0.45sec, load on the system is doubled to 60.84kW. It is observed that supply currents (i_s) remain unchanged whereas the load currents and BESS currents are correspondingly increased. The load leveling capability of BESS is evident from Fig.4.

B. *Performance of BESS with Non-Llinear loads*

Fig.5 shows the response of the BESS for ac voltage regulation at PCC and harmonic reduction with non-linear load. Fig.5 shows variation of various quantities viz. three–phase supply voltages (v_s), terminal voltages (v_t), supply currents (i_s), load currents (i_l), BESS currents (i_c) and voltage at the point of common coupling (v_{tm}) along with reference value for a variety of load changes. The BESS system regulates v_{tmn} at its reference value. Fig.6 and Fig.7 show the harmonic spectra and total harmonic distortion of supply current as well as load currents. It is observed that the THD in the supply current has been reduced to 1.77% from THD in load current of 23.08%. The amplitude of in-phase component of current (I_{spdr}) is kept at a constant value of 30A to meet the active load component. The supply currents are sinusoidal, balanced and slightly leading with respect to supply voltages which is necessary to compensate for the line impedance drop.

At t=0.16sec, the non-linear load on the system is increased to 45kW (R=8Ω) and at t=0.22sec, the load is brought back to 30kW (R=12Ω). It is observed from the figure that small voltage changes are observed at

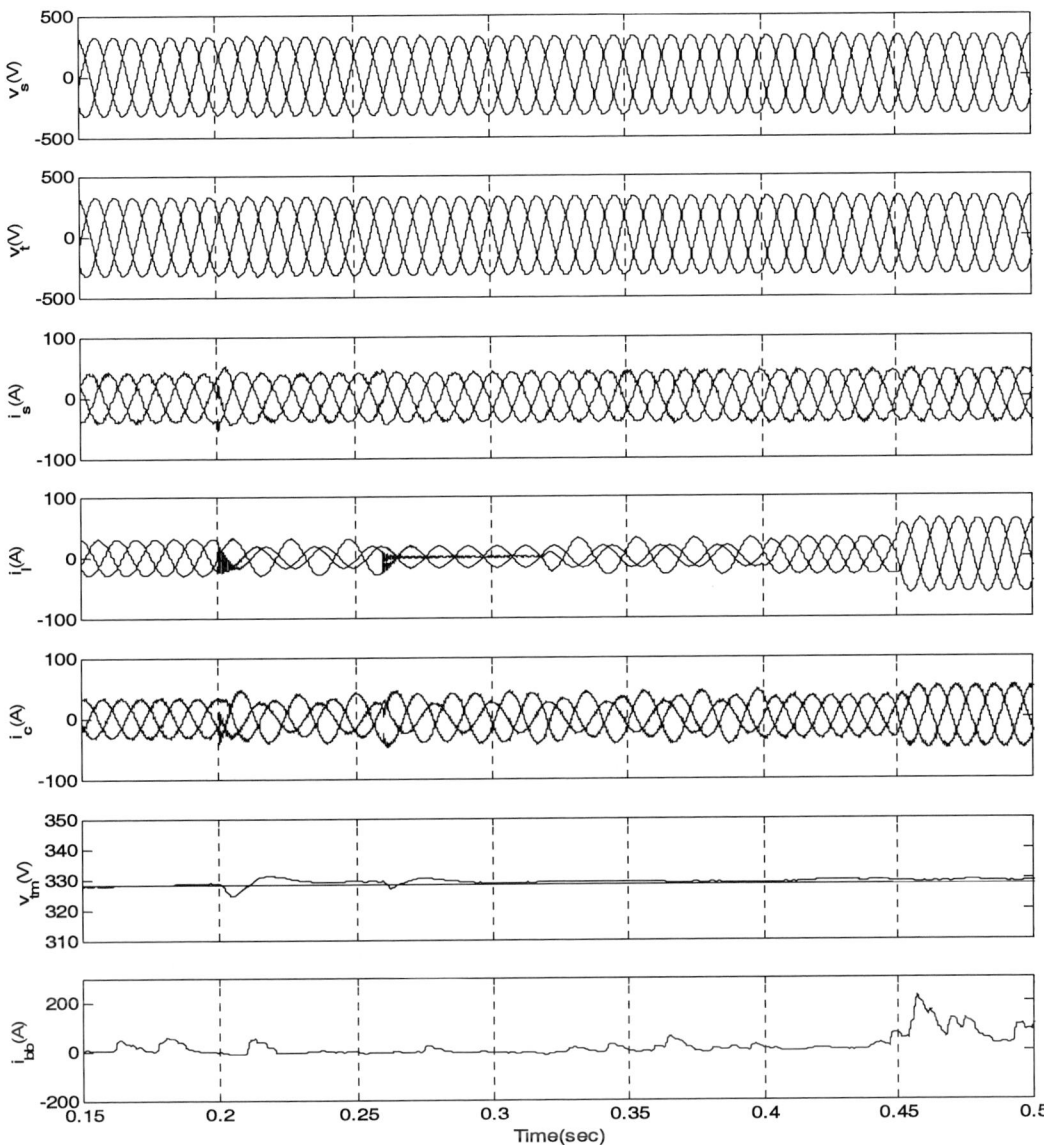

Fig. 4. Performance of system with BESS with linear loads

t=0.16sec and t=0.22sec; however, the voltage at the PCC is regulated back to reference value of 328V. The controller maintains voltage even under transient load conditions of sudden load change. Moreover, the supply currents are nearly sinusoidal even though load currents have high THD values. The BESS controller is able to regulate the THD of the system to meet IEEE 519 standards.

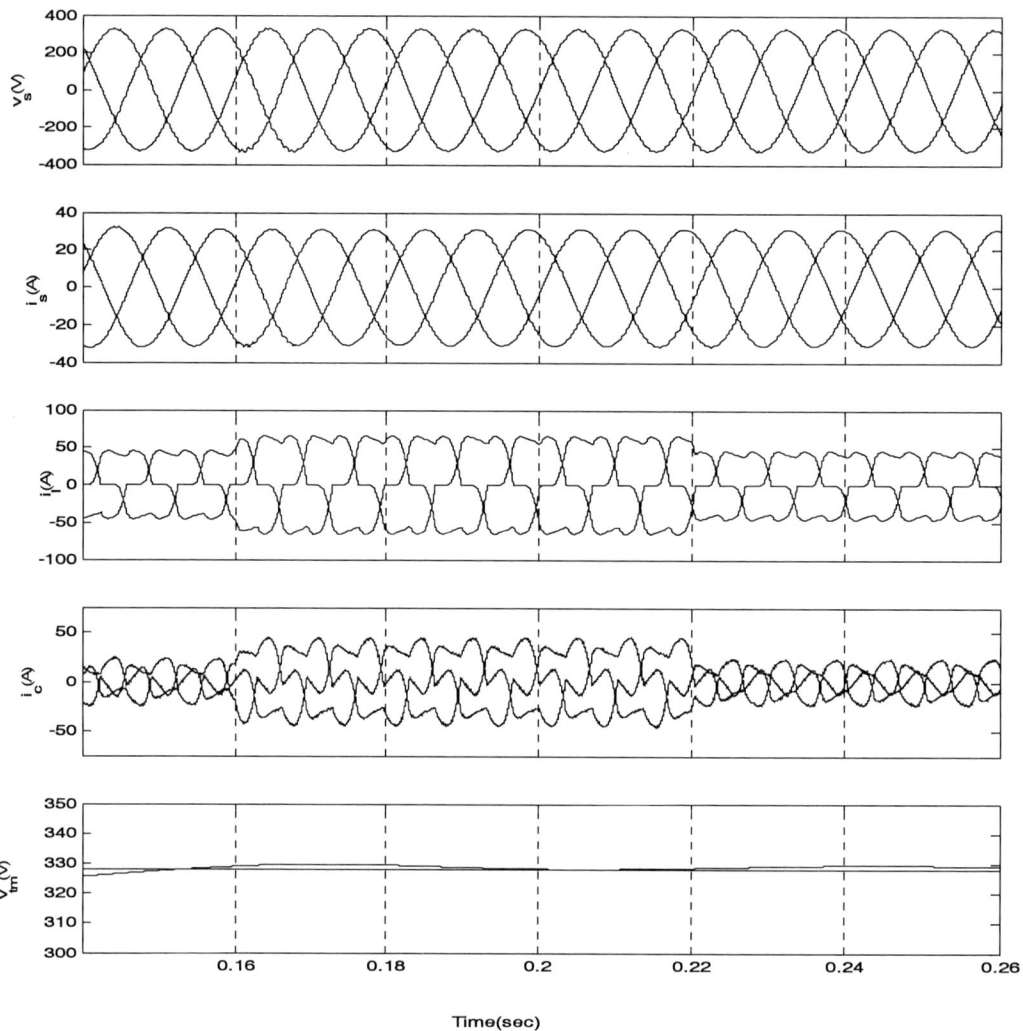

Time(sec)

Fig. 5. Performance of system with BESS with non-linear loads

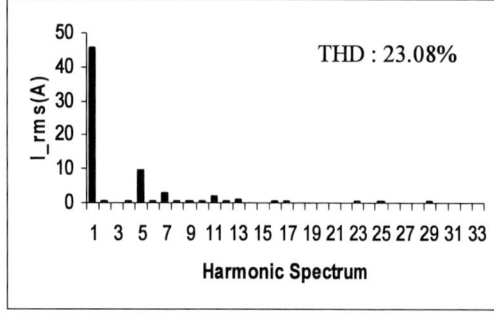

Fig. 6. Harmonic Spectra for load currents

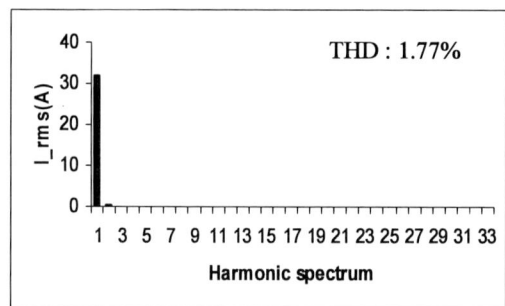

Fig. 7. Harmonic Spectra for supply currents

VII. CONCLUSIONS

A model of battery energy storage system has been developed in MATLAB environment using Power System Block-set. The performance of the developed model is tested under a wide variety of loading conditions. It is found that BESS is capable of maintaining the ac terminal voltage along with load balancing. It is able to regulate the voltage at PCC even under sudden load disturbances. The proposed control algorithm of the BESS is found suitable to provide load balancing and reduce harmonics in the supply currents and terminal voltages. Indirect current control technique has been applied over the sensed and reference supply currents for BESS and it has been found to be a simple technique. Only one PI controller is required to regulate terminal voltage and thus reduces computation effort. The control algorithm of the BESS is flexible and has been tested for power quality improvement for linear as well as nonlinear loads. BESS is able to reduce harmonics in voltage at PCC and supply currents to less than 5% IEEE 519 standards. BESS reduces harmonics in supply current to a large extent and provides quality power. It is found that BESS is able to provide more benefits in terms of improved voltage, energy management and protection from interruptions. It is hoped that BESS application will help in improving the power quality of isolated generating systems.

APPENDIX

3 phase, salient pole machine, 42.5kVA, 2 pole pairs
Stator: R_s=0.04808pu, L_l=0.08pu, L_{md}=2.11pu, L_{mq}=0.93pu.
Field winding: R_f=0.02662pu, L_{lfd}=0.1582pu
Dampers: R_{kd}=0.0754pu, L_{lkd}=0.1098pu,
R_{kd1}=0.07311pu, L_{lkq1}=0.06414pu ; H=0.1157sec,
Friction factor =0.01916
Battery and Controller parameters
V_{dc} =700V, R_1=0.1Ω, R_2=10000Ω, C_2=270000F, C_1=5000 F
L_c=5mH, R_c=0.1 Ω, h_b=0.5A

REFERENCES

[1] N.G. Hingorani and L. Gyugyi, (2001) *Understanding FACTS*, Delhi: Standard Publishers.

[2] A. Ghosh, and G. Ledwich, *Power Quality Enhancement Using Custom Power Devices*, London: Kluwer Academic Publishers, 2002.

[3] T.J.E Miller, *Reactive Power Control in Electric Systems*, Toronto, Ontario, Canada: Wiley Publications, 1982.

[4] R.M. Mathur, Static Compensators for Reactive Power Control, Winnipeg, Canada, Contexts Publications, 1984.

[5] B. Singh, J. Solanki, A. Chandra, "Adaline based control of battery energy storage system for diesel generator set," *Proc. of IEEE Power India Conference*, April 2006, pp.5.

[6] B. Singh, J. Solanki, A. Chandra, K. Al-Haddad, "A solid state compensator with energy storage for Isolated Diesel generator set," *Proc. of IEEE International Symposium on Industrial Electronics*, Vol.3, July 2006, pp. 1774-1778.

[7] O. Lara and E. Acha, "Modeling and analysis of custom power systems by PSCAD/ EMTDC," *IEEE Transactions on Power Delivery*, January 2002, Vol.17, No.1, pp 266-272.

[8] M.W. Tsang, D.Sutanto, "ANN controlled battery energy storage for enhancing power system stability," *Proc. of IEEE*

International Conference on Advances in Power System Control, Operation and Management, Oct/ Nov 2000, pp. 327-331.

[9] K.K.Leung, D. Sutanto, "Using battery energy storage system in a deregulated environment to improve power system performance," *Proc. of IEEE International Conference on Electric Utility Deregulation and Restructuring and Power Technologies, 2000*, April 2000, pp. 322-326.

[10] C.E. Lin, Y.S Shiao, C. L. Huang, P.S. Sung, "A real and reactive power control approach for battery energy storage system," *IEEE Transactions on Power systems*, Volume 7, Issue 3, Aug. 1992, pp. 1132 – 1140.

[11] N.W. Miller, R.S. Zrebiec, G. Hunt, R.W. Deimerico, "Design and commissioning of a 5 MVA, 2.5 MWh battery energy storage system," *Proc. of IEEE Conference on Transmission and Distribution*, Sept. 1996, pp. 339 – 345.

[12] N.W. Miller, R.S. Zrebiec, G. Hunt, R.W. Deimerico, "Battery energy storage systems for electric utility, industrial and commercial applications," *Proc. of 11th IEEE Battery Conference on Applications and Advances*, Jan 1996, pp. 235-240.

[13] D.K. Jain, S.P. Jain, R.S. Bhatia and B. Singh, "Battery energy storage system for improved   performance of self excited induction generators," *Proc. of IEEE PCI India Conference*, Nov 2004, pp.9-16.

[14] B.N.Singh, K.Al-Haddad, A.Chandra, "DSP based indirect-current controlled STATCOM –I multifunctional capabilities" *IEE Proceedings*, Vol. 147, No. 2, March 2000, p107-112.

[15] B.N.Singh, K.Al-Haddad, A.Chandra, "DSP based indirect-current controlled STATCOM –I multifunctional capabilities" *IEE Proceedings*, Vol. 147, No. 2, March 2000, p113-118.

Biography

Bhim Singh graduated from University of Roorkee in 1977 with BE (Electrical), MTech in Power Apparatus and Systems from IIT Delhi in 1979 and Ph.D. from IIT Delhi in 1983. He is currently working as a Professor in IIT Delhi. He is a Fellow of Institution of Engineers (India), Fellow of INAE and IETE, a Life Member of ISTE, SSI and NIQR and Senior Member IEEE. His research interests include power electronics, electrical machines, induction generator, active filters, static VAR compensator, analysis and digital control of electrical machines, FACTS, electric drives.

Alka Adya graduated from Delhi College of Engineering in 1996 with BE (Electrical), MTech in Power Systems from IIT Delhi in 2001 and Ph.D. from Delhi University in 2006. She is presently working as an Assistant Professor in Maharaja Agasen Institute of Technology, IP University. Her research interests include FACTS, power systems and power quality.

A.P.Mittal received his B.E. in 1978 from M.M.M Engg. College, Gorakhpur, M.E. in 1980 from University of Roorkee and PhD in 1991 from IIT Delhi. He has teaching experience of more than twenty years. He is presently Professor and Head of Instrumentation and Control Engineering Department in Netaji Subhas Institute of Technology. He is a Fellow of Institution of Engineers (India). His research interests include active filters, FACTS, electric machines and drives.

J.R.P Gupta graduated from Muzaffarpur Institute of Technology (M.I.T) and received his B.Sc. degree in 1972 and completed his Ph.D. degree from University of Bihar in 1983.He has been in Netaji Subhas Institute of Technology for the last ten years and is presently holding the position of Professor and Head of Instrumentation and Control Engineering Department in Delhi University. His research interests include power electronics, active filters, power quality, electric drives.

Pulse Doubling in 18-Pulse AC-DC Converters

Bhim Singh, *Senior Member, IEEE*, and Sanjay Gairola

Abstract— **In this paper, a novel 36-pulse AC-DC converter is designed, modeled and simulated using pulse multiplication in an 18-pulse AC-DC converter using DC ripple re-injection technique. The proposed technique is suitable for large current rating rectifiers such as electrowinning, electrochemical processes, induction heating, plasma torches, etc., where isolation is required mainly for stepping down the supply voltage. It consists of a parallel 18-pulse AC-DC converter configuration involving three phase shifted uncontrolled diode bridges and interphase reactors tapped with diodes. A set of interphase reactors is used which is capable of effectively doubling the pulse number in the converter. A prototype of 36-pulse converter is developed to validate the design and its model. It improves the power quality to meet IEEE-519 standard at varying loads.**

Index Terms—**18-pulse, 36-pulse, clean power, delta/triple delta-polygon transformer.**

I. INTRODUCTION

LARGE current rectifiers (kilo-Amperes) in 12-pulse and 18-pulse configurations are commonly used in several important applications like AC drives, electro-chemical processes, DC arc furnaces, plasma torches, etc [1-5]. Thyristor rectifiers have been used for controlled operation and the technology is well established [4]. A combination of rectifier-chopper (where diode rectifiers are used at front end and chopper for output control) is also becoming popular for large controlled DC currents due to reduced input current harmonics and high power factor (PF). A dual 18-pulse rectifier for high power multilevel inverters is also reported by Cheng et. al.[6]. It is common practice to use multiple 12-pulse or 18-pulse converter units fed from phase–staggered transformers to meet IEEE-519 standard [7] requirements as the total harmonic distortion (THD) of input line current of single unit are still high and may not qualify as clean power at high loads. An 18-pulse rectifier can be fed from Delta/Delta/Double polygon transformer [8] as shown in Fig. 1. Delta/ Polygon transformer based 18-pulse and 38-pulse are also proposed by Hammond et. al. [9]. Appropriate fork connections could also be employed to provide 18–pulse characteristics. However, polygons are easy to use for low voltage outputs because its design gives higher number of

Bhim Singh is with Department of Electrical Engineering, Indian Institute of Technology, Delhi, New Delhi-110016, India (e-mail: bhimsinghr@gmail.com).
Sanjay Gairola is with Department of Electrical and Electronics Engineering, Krishna Institute of Engineering and Technology, Ghaziabad (U.P.)-201106, India (e-mail: sanjaygairola@gmail.com).

turns that facilitates selection of the turn ratio. As the power to these loads is transferred at lower voltage levels, the use of parallel bridge configuration is justified.

The harmonics in input and output of conventional 12-pulse controlled rectifiers fed from delta/star-delta transformers can be reduced with thyristor-tapped interphase reactors [10]. A 12-pulse controlled AC-DC converter can be converted to 36-pulse AC-DC converter by using an inter-phase reactor tapped with three thyristors [10]. But this is possible only with controlled AC-DC converter and for firing angle above 5°.

Some applications (Navy needs THDi < 3% for its special application) have stringent power quality specifications and it is inevitable to go beyond 24-pulse AC-DC converter system configuration. Therefore, it is suggested that higher pulse configuration must be used so that AC-DC conversion meets IEEE-519 standard requirements. Although 36-pulse AC-DC converters are also popular for higher rating (that have six number of three-phase diode bridges, i.e., 36 diodes) but large number of diodes are required. With this view, an 18-pulse AC-DC converter is designed and it is modified to a 36-pulse AC-DC converter by using a novel pulse doubler that uses only three extra diodes.

In this paper, to increase the pulse number of an 18-pulse rectifier, the input transformer is modified and small rating specially tapped inter-phase transformers are introduced with three additional diodes connected as shown in Fig. 2. The taps on the inter-phase transformers are chosen such that a 36-pulse characteristic appears in the input line currents. The input transformer secondary windings are made exactly similar so that the reactance in each of the 9-phases remains same.

The proposed technique is based on ripple re-injection where the power of the circulating ripple frequency is fed back to the DC system via the interphase reactor which acts as an autotransformer. The DC voltage ripple acts as the frequency source for the derivation of appropriate voltage and current waveforms capable of modifying AC current and DC voltage to eliminate 18–pulse related harmonics. Detailed design of

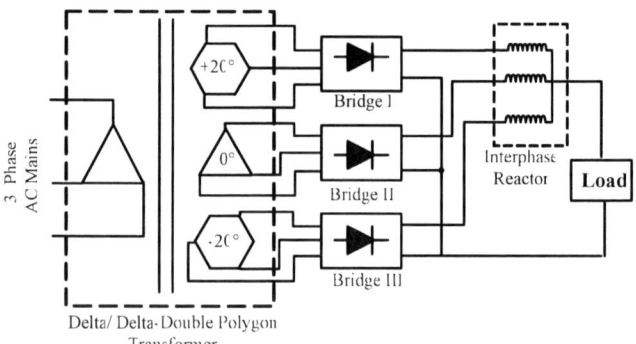

Fig. 1 A delta/delta-double polygon transformer configuration for 18-pulse

the tapped inter-phase transformer and resulting 36-pulse diode rectifier system is carried out to study the behavior of the AC-DC converter. The designed system is modeled and simulated in MATLAB to demonstrate its power quality improvement at AC mains. The design and model of proposed system is validated by test results obtained from prototype developed in the laboratory.

II. PROPOSED 36-PULSE AC-DC CONVERSION APPROACH

Fig. 2 shows the proposed 36-pulse AC-DC converter system, which is identical to the conventional 18-pulse AC-DC converter system, with the exception of a modified transformer configuration and three diodes are connected to a specially tapped inter-phase reactors. The secondary windings of input transformer are configured in delta-polygon to generate three balanced set of three-phase voltages with 20° phase shift for the diode rectifiers. The delta-polygon secondary windings provide equal leakage reactance to six pulse diode bridge converters I, II and III.

A. Delta/Triple Delta-Polygon Transformer Scheme

The commonly used input transformers for 18-pulse converters have non-identical secondary windings and this leads to different leakage reactance to the 20° phase shifted diode-bridge converters. To avoid this difference Delta/Triple Delta-Polygon transformer arrangement as shown in Fig. 2 is designed.

The stepped-down output voltage (it may be stepped-up also) is attained with symmetric delta-polygon secondary as depicted in Fig.3. Three diode bridge rectifiers are connected to three secondary windings at A_{-20}, B_{-20}, C_{-20}; A_0, B_0, C_0; A_{+20},

B_{+20}, C_{+20} for -20°, 0° and +20° phase shift respectively, where the subscripts denote phase angle with respective phase voltage. The number of turns for every winding is determined as a function of the required secondary voltage, V_s.

The winding voltages V_{K1}, V_{K2} and V_{K3} are defined by equations (1-3) as:

$$V_{K1} = \frac{V_s \sin(20°)\sin(10°)}{\sqrt{3}\sin(80°)\sin(120°)} = \frac{0.0402 V_p}{a} \tag{1}$$

$$V_{K2} = \frac{V_s \sin(20°)\sin(50°)}{\sqrt{3}\sin(80°)\sin(120°)} = \frac{0.17736 V_p}{a} \tag{2}$$

$$V_{K3} = \frac{V_s \sin(80°)}{\sqrt{3}\sin(50°)} = \frac{0.74227 V_p}{a} \tag{3}$$

where, a=transformation ratio of the transformer $=V_p/V_s$.

V_p, V_s =primary and secondary line voltages.

The winding voltages and currents of the transformer used for one of the three phases in prototype is shown in Fig. 4.

B. Operation of the Tapped Inter-Phase Reactor (IPR)

A necessary condition to achieve pulse doubling is to ensure that the average output voltages of the three converters are the same and they are displaced by an angle of 20° (or 40°). It is already known that an inter-phase reactor with two diode- taps [11] can effectively double the pulses in 12-pulse converters where the two bridges are fed from 30° phase shifted voltages.

A newly designed tapped inter-phase reactor set for pulse doubling in an 18-pulse AC-DC converter is shown in Fig. 5. From waveform of Figs. 6a and 6b it can be seen that the pulses can be effectively doubled if the output voltage V_{dc} is

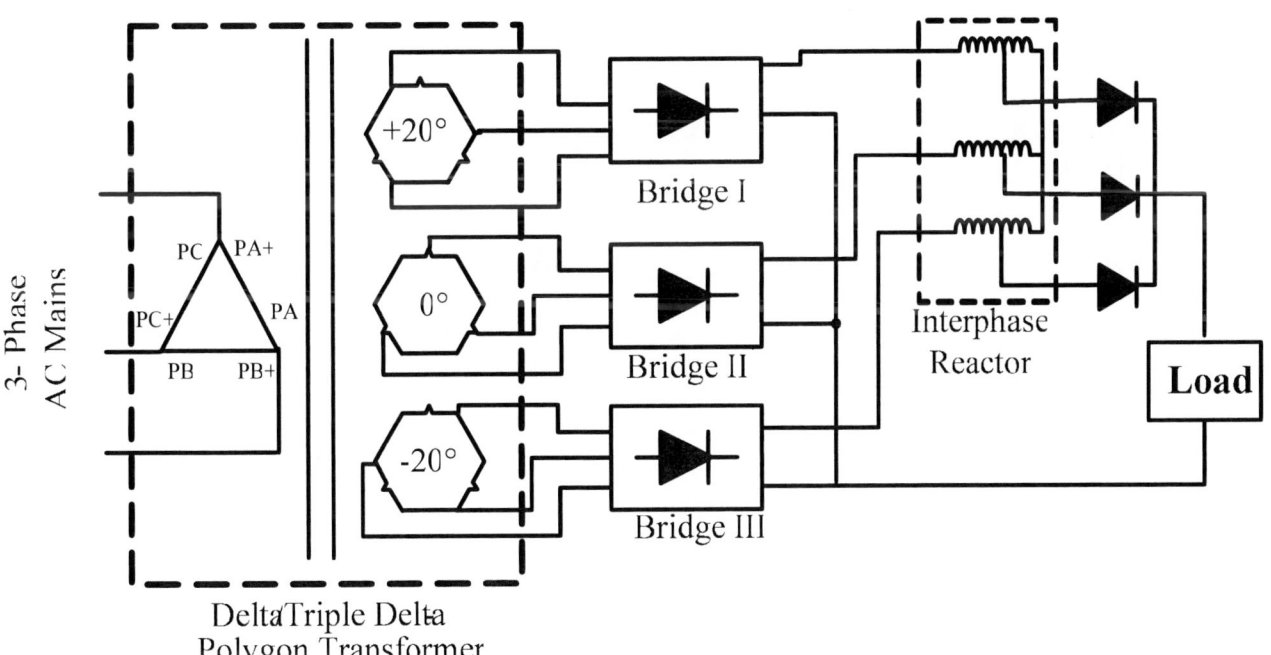

Fig. 2 Proposed topology for pulse doubling in 18-pulse configuration using delta/triple-delta-polygon input transformer and inter-phase reactor with three diode taps (Topology D).

made up of average of maximum and minimum of three rectifier output voltages V_{dc1}, V_{dc2} and V_{dc3} such that transition occurs at regular interval of 10°. For this the tapping must be suitably selected so that the diodes naturally conduct to produce a sequence of sine wave portions for 36-pulse output

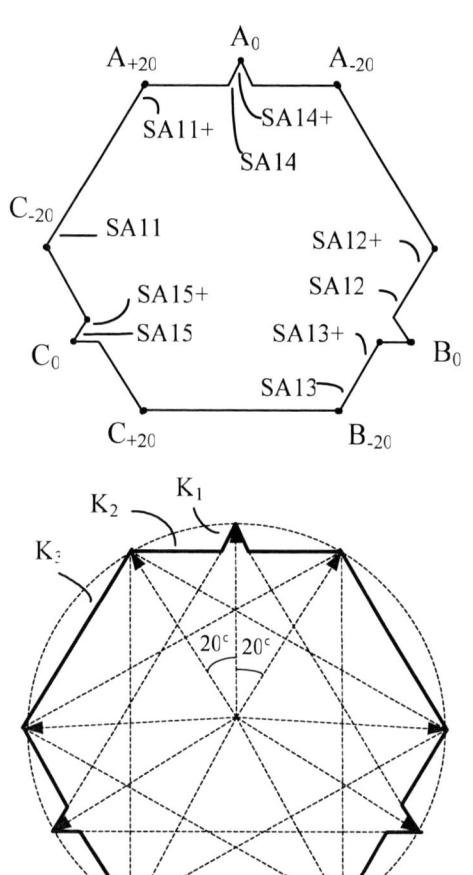

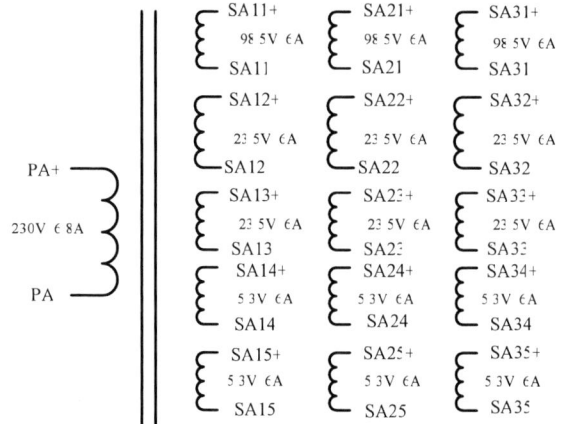

Fig. 3 Single Delta-polygon secondary arrangement of 18-pulse input transformer and its graphical representation.

Fig. 4 Winding voltage and current ratings of a phase used for delta/triple polygon used for proposed 18 and 36-pulse AC-DC converter system. [The letters in naotation stands as follows- P for Primary winding, S for Secondary winding, A for Phase A, + for polarity marking of windings, two digit number (say 23) for delta-polygon secondary(connecting bridge II) and winding number (winding 3), respectively].

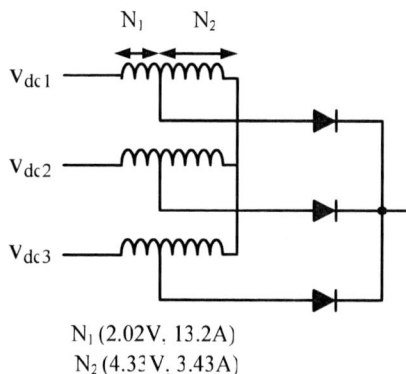

N_1 (2.02V, 13.2A)
N_2 (4.33V, 3.43A)

Fig. 5 Tapped inter-phase reactor arrangement.

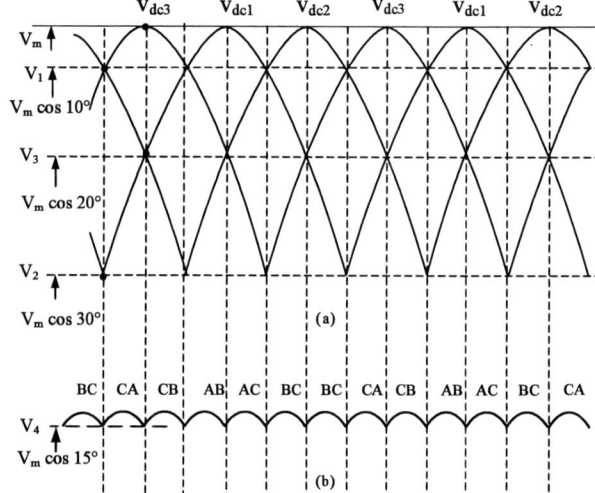

Fig. 6 Waveforms for derivation of tap positions. (a) Three 20° phase shifted bridge output voltages vdc1, vdc2, vdc3. (b) Desired 36-pulse waveform possible if taps are turned on in the marked sequence.

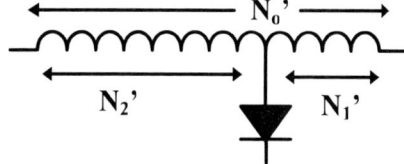

Fig. 7 Tapped inter-phase reactor corresponding to a single pulse of 10° duration.

as shown in Fig. 6b. The tap position can be found from following relations.

The DC output voltages of converters I, II and III are segments of phase-displaced sine waves and can be expressed for 60° interval as:

$$v_{dc1} = V_m \cos(\theta - 10°) \qquad (4)$$

$$v_{dc2} = V_m \cos(\theta - 30°) \qquad (5)$$

$$v_{dc3} = V_m \cos(\theta + 10°) \qquad (6)$$

The value of tap position from waveform of Fig. 6a to achieve desired output waveform of Fig. 6b gives turn ratio N_1' and N_2' of a single tapped reactor shown in Fig. 7 as:

$$N_1' = \frac{(V_1 - V_4)}{V_1 - V_2} = 0.1589 \qquad (7)$$

$$N_2' = 1 - N_1' = 0.8411 \qquad (8)$$

This tap position shall require six tapping diodes and longer interphase reactors. The value of taps N_1' and N_2' must now be modified, as if delta to star conversion is done, so that the continuous 10° pulses are obtained and the required number of diodes are reduced to half i.e., three. This leads to value of N_1 and N_2 in Fig. 5 as

$$N_1 = \frac{N_1'\{2\cos(30°)\}}{(\sqrt{3}/2)} = 0.13178 \tag{9}$$

$$N_2 = 1 - N_1 = 0.6822 \tag{10}$$

Mean rectified voltage is as:

$$V_{dc} = \frac{36}{2\pi} \int_{-\pi/36}^{\pi/36} V_m \frac{\cos(15°)}{\cos(7.5°)} \cos(\omega t) d(\omega t) = 0.9739 V_m = 1.377 V_s \tag{11}$$

III. MATLAB BASED SIMULATION

The proposed 36–pulse AC-DC converter is modeled and simulated in MATLAB environment along with Simulink and Power System Blockset (PSB) toolboxes. The AC-DC converter system having R-L load ($150V_{dc}$, 5kW) is fed from 230V, 50Hz AC supply. A R-L-V_c load can also be used as model for electrolytic plant where R stands for the resistance of the electric circuit, L for the inductance produced by conducting bars and V_c for the chemical reaction voltage of the series connected cells.

The proposed 36-pulse AC-DC converter shown in Fig. 2 is modeled in MATLAB. The 18-pulse AC-DC converter system using delta/triple delta-polygon transformer and three parallel connected 6-pulse diode bridges without diode tapped IPR is also simulated for comparison. The results obtained from the simulations are shown in Table I. The simulated waveforms at full-load for the 18 and 36-pulse AC-DC converters are shown in Figs. 8(i) and 9(i) respectively. The harmonic spectra of AC mains current waveforms along with its THD are shown in Figs. 8(ii) and 9(ii).

IV. EXPERIMENTAL PERFORMANCE OF PROPOSED 36-PULSE CONVERTER

A prototype of the proposed 36-pulse AC-DC converter configuration is developed in the laboratory for 5kW load for three-phase supply voltage of 230V, 50Hz. The transformation ratio is chosen as 1: 0.47. Various tests have been carried out to validate the model and the results have been recorded using 'Fluke 43B' power quality analyzer. The main transformer rating for the prototype is 6.5kVA while the total interphase reactor rating is negligible (62VA).

The recorded waveforms for the 18-pulse converter configuration are shown in Figs. 8(iii)-(vi). The power measurements along with AC mains voltage and current waveforms can be seen in Fig. 8(iii). Fig. 8(iv) shows the

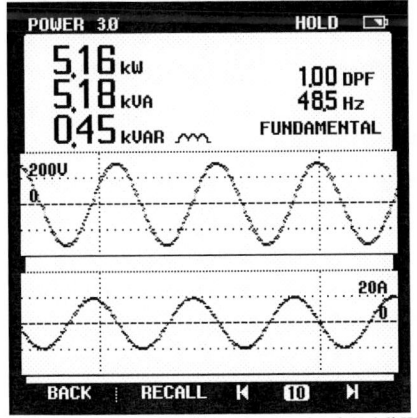

(i). Simulated AC mains and DC, voltage and current waveform.

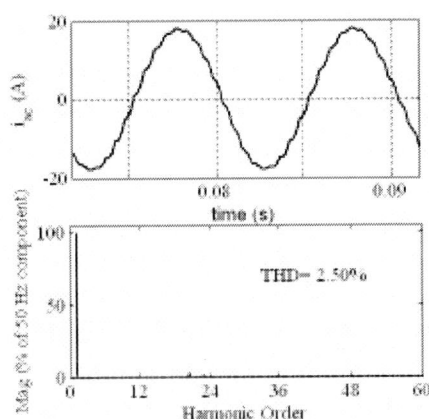

(ii). AC mains current spectrum at full-load obtained from simulation.

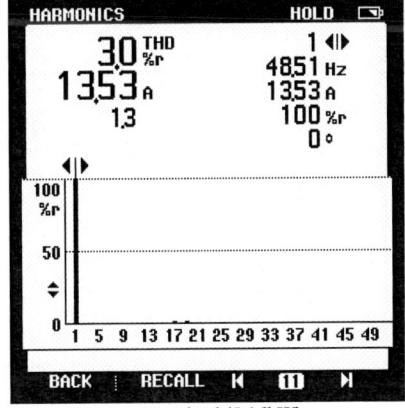

(iii). AC mains voltage and current waveform at load (5.16kW).

(iv). AC mains current spectrum at load (5.16kW).

Fig. 8(i-iv) Simulated and experimental results observed from the isolated 18-pulse AC-DC converter system at full-load.

741

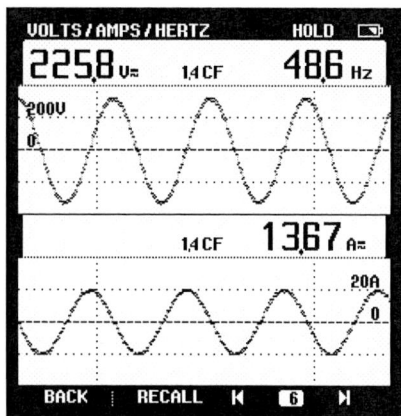

(v). AC mains voltage and current waveform at load (5.16kW).

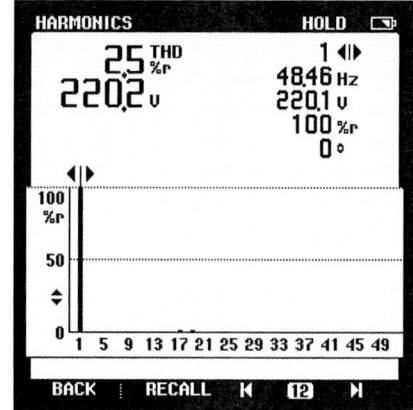

(vi). AC mains voltage spectrum at load (5.16kW).

Fig. 8(v-vi) Experimental results observed from the isolated 18-pulse AC-DC converter system at full-load.

harmonic spectrum of AC mains current. Fig 8(v) show the voltage and current waveforms with the crest factor. The input voltage spectrum from the implementation at full-load is shown in Fig. 8(vi). The similar waveforms for the 36-pulse converter configuration are shown in Figs. 9(iii)-(vi). The power quality indices obtained from test results with varying loads in two configurations are tabulated in Table II.

V. RESULTS AND DISCUSSION

The power quality indices obtained from simulations of proposed 18-pulse and 36-pulse AC-DC converters are given

in Table I. The corresponding experimental results are shown in Table II. The full-load simulated and test result waveforms of the 18-pulse converter topology shown in Fig. 8 can be compared with that of 36-pulse converter as shown in Fig. 9. The AC mains current waveform has improved in 36-pulse AC-DC converter. The simulation results show that THD_i (total harmonic distortion of AC mains current) has improved from 2.5% to 1.36% from 20% to 100% load. The test results confirm these improvements in power quality. The simulated and hardware results clearly show that the dominant 17th and 19th harmonics seen in the harmonic spectra (Figs. 8 (ii) and

TABLE I
COMPARISON OF POWER QUALITY PARAMETERS OF 18 AND 36-PULSE AC-DC CONVERTERS OBTAINED FROM MATLAB SIMULATIONS.

Topology	Load (% of Load Power)	THD V_{AC} (%)	AC Mains Current I_{AC} (A)	THD of I_{AC} (%)	Distortion Factor, DF	Displacement Factor, DPF	Power Factor PF	DC Voltage (V)	Load Current I_{dc} (A)
18-pulse	25	1.097	3.216	4.30	0.9990	.9944	.9934	151.2	8.359
	50	1.527	6.546	2.91	0.9795	.9995	.9790	149.8	16.65
	60	1.703	7.793	2.69	0.9995	.9789	.9784	149.3	19.9
	80	2.068	10.25	2.53	0.9994	.9767	.9761	148.2	26.35
	100	2.496	12.68	2.50	0.9994	.9744	.9738	147.2	32.71
36-pulse	25	1.126	3.387	4.21	0.9989	.9841	.9830	151.7	8.424
	50	1.386	6.326	2.016	0.9997	.9901	.9898	149.0	16.55
	60	1.357	7.552	1.357	0.9998	.9888	.9886	148.4	19.79
	80	1.365	9.986	1.055	0.9999	.9876	.9875	147.5	26.23
	100	1.360	12.39	1.368	0.9998	.9859	.9857	146.6	32.59

TABLE II
COMPARISON OF POWER QUALITY PARAMETERS OF HARDWARE IMPLEMENTATION RESULTS OBTAINED FOR 18-PULSE AND 36-PULSE CONVERTERS.

Sr. No.	Topology	Load, (kW)	THD Vs (%)	AC Mains Current I_s (A)	THD of I_s (%)	Crest Factor, CF	Displacement Factor, DPF	Power Factor, PF	DC Voltage (V)	Load Current I_{dc} (A)
1	18-pulse	1.26	1.2	3.15	3.9	1.4	1.0	0.9960	149.6	8.71
		2.36	1.6	6.02	3.7	1.4	1.0	0.9980	145.1	15.24
		3.13	1.8	8.11	3.6	1.4	1.0	0.9982	141.2	20.38
		4.00	2.2	10.39	3.5	1.4	1.0	0.9981	136.8	26.28
		5.16	2.5	13.53	3.0	1.4	1.0	0.9962	130.4	34.52
2	36-pulse	1.26	1.2	3.18	2.7	1.4	1.0	0.9997	151.1	8.10
		2.36	1.3	6.06	2.5	1.4	1.0	0.9997	144.0	15.13
		3.13	1.4	8.08	2.4	1.4	1.0	0.9989	140.2	20.22
		4.04	1.5	10.55	2.3	1.4	1.0	0.9988	135.4	26.14
		5.17	1.8	13.58	2.2	1.4	1.0	0.9981	129.2	33.91

742

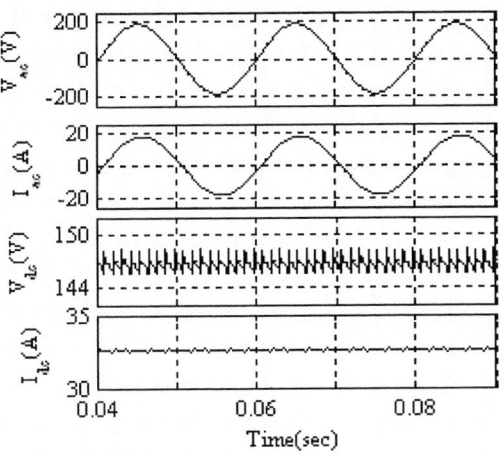

(i). Simulated AC mains and DC, voltage and current waveform.

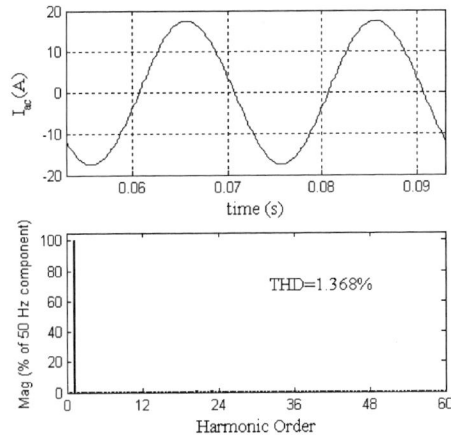

(ii). AC mains current spectrum at full-load obtained from simulation.

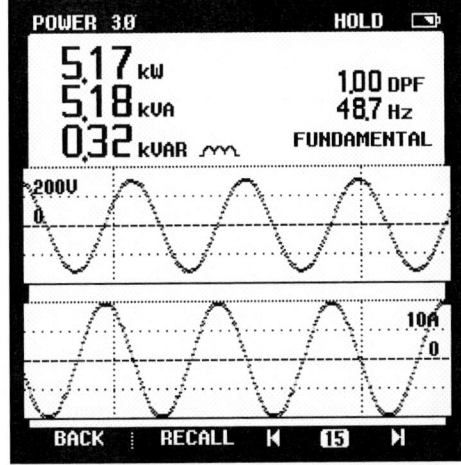

(iii). AC mains voltage and current waveform at load (5.17kW).

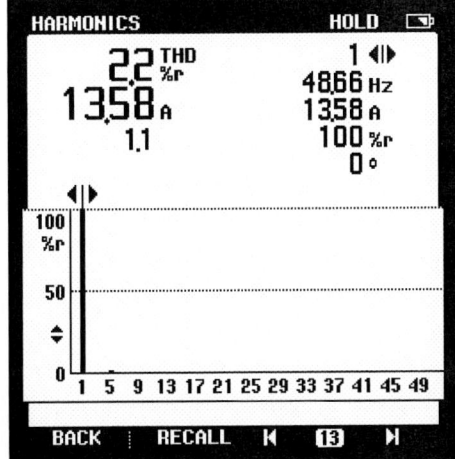

(iv). AC mains current spectrum at load (5.17kW).

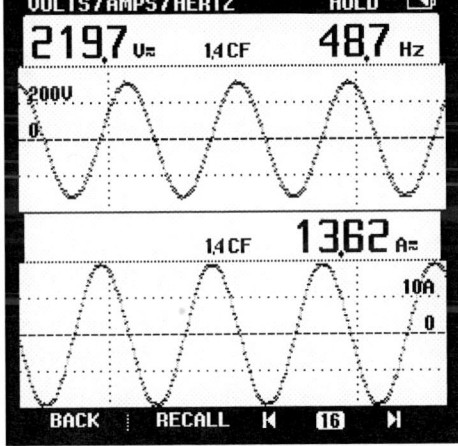

(v). AC mains voltage and current waveform at load (5.17kW).

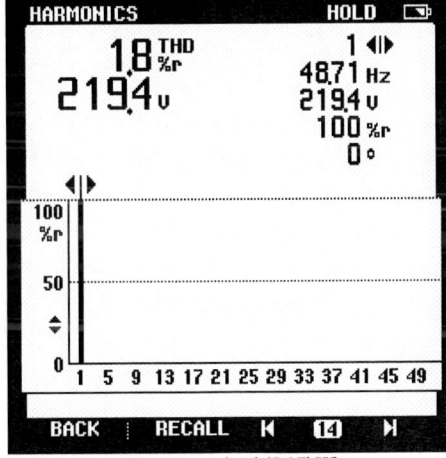

(vi). AC mains voltage spectrum at load (5.17kW).

Fig. 9 Simulated and experimental results observed from the isolated 36-pulse AC-DC converter system at full-load.

(iv)) are suppressed by the use of proposed diode-tapped interphase reactors. The THD of input voltage waveform has also improved from 2.49 to 1.36. The improvement in other power quality indices in 36-pulse converter can be seen at full load in Tables I and II in simulation and implementation respectively.

The power quality indices (THD$_i$, THD$_v$, Distortion Factor (DF), Displacement Factor (DPF) and Power Factor (PF)) have improved at most of loads in 36-pulse AC-DC converter in comparison to the 18-pulse converter configuration. This can be observed from Table I and Table II that show simulation and test results respectively.

743

TABLE III
COMPARISON OF POWER QUALITY PARAMETERS OF FOUR AC-DC CONVERTERS.

Sr. No.	Topo-logy	% THD of V_{ac}	AC Mains Current I_{ac} (A)		% THD of I_{ac} at		Distortion Factor, DF		Displacement Power Factor, DPF		Power Factor, PF		DC Voltage (V)	
			Light Load	Full Load	Light Load	Full Load	Light Load	Full Load	Light Load	Full Load	Light Load	Full Load	Light Load	Full Load
1	A	7.925	3.269	12.32	26.83	24.95	.9655	.9682	.9922	.9830	0.9580	.9517	146.8	143.5
2	B	2.445	3.389	12.66	4.491	2.55	.9988	.9994	.9833	.9740	.9822	.9734	151.5	147.0
3	C	2.496	3.216	12.68	4.30	2.50	0.9990	0.9994	.9944	.9744	.9934	.9738	151.2	147.2
4	D	1.360	3.387	12.39	4.21	1.368	0.9989	0.9998	.9841	.9859	.9830	.9857	151.7	146.6

(where, A is 6-pulse AC-DC converter fed from delta/delta transformer; B is 18-pulse AC-DC converter fed from delta/delta-double polygon Transformer; C is 18-pulse AC-DC converter fed from delta/triple delta-polygon Transformer; D is 36-pulse AC-DC converter).

The comparison of power quality indices is also made with 6-pulse and 18-pulse AC-DC converters (using delta/polygon-delta-polygon transformer of Fig.1) in Table III.

It can be seen that the performance of proposed converter configuration is much superior to the six-pulse and the 18-pulse converter configurations. The proposed pulse-doubler is observed to be quite effective at load higher than 50% of full-load.

VI. CONCLUSIONS

Simulated results have shown that by employing the proposed tapped inter-phase reactors with three additional diodes in an 18-pulse AC-DC converter can be extended to 36-pulse AC-DC converter operation. The resulting system has exhibited high level of performance with clean power characteristics not seen in diode based front end rectifiers. The 18-pulse converter related harmonics are suppressed and the input line current has 35th and 37th harmonics as the dominant harmonics in 36-pulse AC-DC converter. The hardware results have shown that the total harmonic distortion of input current is less than 2.7% at varying loads and it meets the stringent power quality requirements.

REFERENCES

[1] J. Schaeffer, *Rectifier Circuits: Theory and Design*. New York: Wiley-Interscience, 1965.

[2] G. Seguier, *Power Electronic Converters: AC/DC Conversion*, New York: McGraw Hill, New York, 1986.

[3] R. W. Lye (Editor), *Power Converter Hand Book-Theory, Design, Applications*, Power Delivery Department, GE Canada, Ontario, March 1990.

[4] Bin Wu, *High-Power Converters and AC Drives, IEEE Press*, Wiley-Interscience, 2006.

[5] *IEEE Standard Practices and Requirements for Semiconductor Power Rectifier Transformers*, IEEE Standards C57.18.10-1998.

[6] Z. Cheng and Bin Wu, "Dual 18-pulse Rectifier for High Power Multilevel Inverters," in *Proc. of IEEE Conf. IECON 2005*, 6-10 Nov. 2005, pp. 525-530.

[7] *IEEE Recommended Practices and Requirements for Harmonic Control in Electric Power Systems*, IEEE Standard-519, 1992.

[8] D. A. Paice, *Power Electronic Converter Harmonics: Multipulse Methods for Clean Power*, IEEE Press, New York, 1996.

[9] R. Hammond, L. Johnson, H. Shimp and D. Harder, "Magnetic Solution to Line Current Harmonic Reduction," in *Proc. Power Conversion*, Sep. 1994, pp. 354-364.

[10] M. Villablanca, J. del Valle, J. Rajar, J. Aborca and W. Rojas, "A modified back to back HVDC system for 36 pulse operation," *IEEE Transactions on Power Delivery*, vol. 15, no. 2, pp. 641-645, 2000.

[11] B. Singh, G. Bhuvaneswari, V. Garg, "Power-Quality Improvements in Vector-Controlled Induction Motor Drive Employing Pulse Multiplication in AC-DC Converters," *IEEE Trans. on Power Delivery*, vol. 21, no. 3, pp. 1578 – 1586, July 2006.

Bhim Singh (SM'99) was born in Rahamapur, U. P., India in 1956. He received B. E. degree in Electrical engineering from University of Roorkee, India in 1977 and M. Tech. and Ph. D. degrees from Indian Institute of Technology (IIT), New Delhi, in 1979 and 1983, respectively. In 1983, he joined as a Lecturer and in 1988 became a Reader in the Department of Electrical Engineering, University of Roorkee. In December 1990, he joined as an Assistant Professor, became an Associate Professor in 1994 and Professor in 1997 at the Department of Electrical Engineering, IIT Delhi. His field of interest includes power electronics, electrical machines and drives, active filters, static VAR compensator, analysis and digital control of electrical machines.

Prof. Singh is a Fellow of Indian National Academy of Engineering (INAE), Institution of Engineers (India) (IE (I)) and Institution of Electronics and Telecommunication Engineers (IETE), a Life Member of Indian Society for Technical Education (ISTE), System Society of India (SSI) and National Institution of Quality and Reliability (NIQR) and Senior Member of IEEE (Institute of Electrical and Electronics Engineers).

Sanjay Gairola was born in Chandigarh, India in 1968. He received B.E. degree in Electrical engineering from M.N. Regional Engineering College, Allahabad in 1991 and M.Tech. degree from Indian Institute of Technology (IIT), New Delhi, in 2001. In 1997, he joined as a Lecturer in the Department of Electrical Engineering, Krishna Institute of Engineering and Technology (KIET), Ghaziabad, U.P., India. In January 2004, he became Assistant Professor. He is a Life Member of Indian Society for Technical Education (ISTE). Presently he is also a research scholar in the Department of Electrical Engineering, IIT Delhi, pursuing for his Ph.D. degree. His field of interest includes power electronics, electric machines and drives.

Magnetic Field Analysis and Control Strategy of Permanent Magnet Actuator for Low Voltage Vacuum Circuit Breaker

Fang Shuhua, Lin Heyun, Yang Chenfeng, Liu Xiping and Guo Jian

School of Electrical Engineering, Southeast University, Nanjing 210018, P. R. China

Abstract—Permanent magnet actuator has such advantages as free-maintenance and high reliability. In this paper, a new type of permanent magnet actuator used for lower voltage circuit breaker is analyzed based on 3-D finite element method. Typical flux density distributions are computed with different currents. Two control strategies, open-loop control and closed-loop control, for the permanent magnet actuator to make or break breaker is presented in detail. The experimental results of two control strategies show that closed-loop control can achieve good performance than open-loop control.

Index Terms—permanent magnet actuator, finite element method, vacuum circuit breaker, open-loop control, closed-loop control

I. INTRODUCTION

Vacuum circuit breakers have been widely used in power systems and other control fields. However, most conventional actuators of the breakers have many mechanical components and their reliability is very low. In recent years, a novel actuator, called permanent magnet actuator (PMA) which only has several components, high reliability and free-maintenance with newly operation principle, has been developed [1]-[2]. The PMA can be classified into mono-stable and bi-stable styles which strongly couple magnet, circuit and mechanic equations [3]-[9]. In order to optimize the dynamic performance, it needs to analyze the magnetic field and propose effective control strategy. In the paper, a novel PMA for vacuum breaker is presented and analyzed using 3-D finite element method (FEM). The flux density distributions are computed and resultant electro-magnetic forces are achieved when the coil is supplied with different currents at different positions. Accordingly, open-loop and closed-loop control strategies are proposed and experimental results are achieved for the two control strategies.

II. STRUCTURE AND MAGNETIC FIELD ANALYSIS

A. Structure

Fig.1 (a) shows half model of the developed PMA. The actuator has several components, such as static iron, moving iron, coil and PM. Fig.1 (b) shows the assembly of the low voltage PM vacuum circuit breaker. The

breaker mainly consists of PMA, opening spring, over-travel spring and vacuum tube.

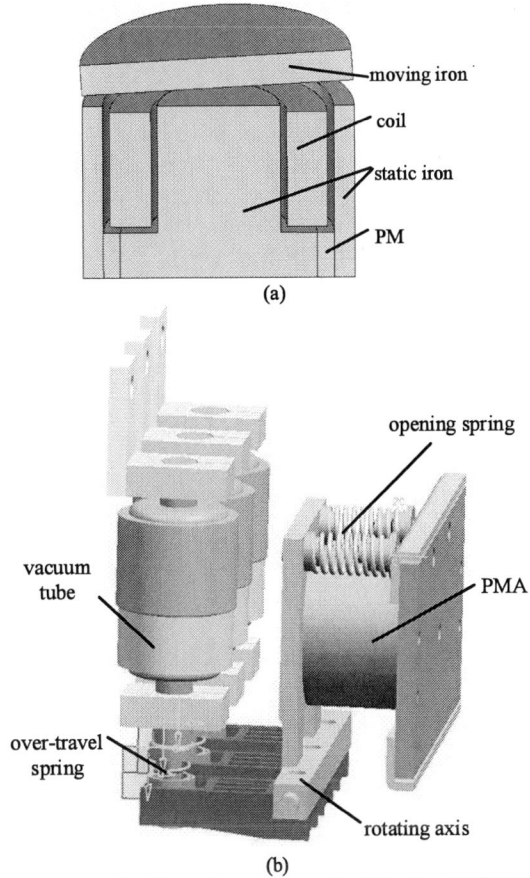

(a)

(b)

Fig.1. The sketch of low voltage vacuum circuit breaker (a) Half of the PMA (b) Assembly

B. Magnetic Field Analysis

When the breaker fulfills making or breaking courses, the moving iron acted by consultant force of spring force and electro-magnetic force rotates along the rotating axis. In dealing with the electro-magnetic problem, the PMA model can be simplified as half volumes to analyze. Before computation, some hypothesizes of ignoring eddy current and hysterisis are made. 3-D FEM is applied to analyze the magnetic field. Fig.2 shows the meshes of the PM actuator at the open position. There are 95646 elements and 131548 nodes, respectively. The PM is magnetized in radial direction in cylindrical coordinate system.

This work was supported by the National High Research and Development of China under Project 2006AA05Z224, and the foundation for Excellent Doctoral Dissertation of Southeast University, China.

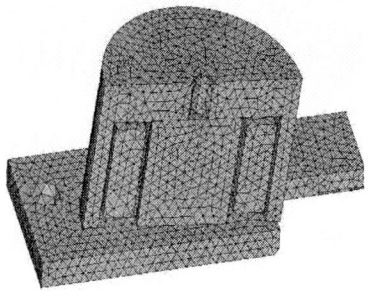

Fig.2. Meshes of PMA at the open position

Fig.3 shows the typical flux density distributions with the moving iron. Fig.3 (a) is the result of PM working only. Fig.3 (b) and Fig.3 (c) are the results of -2A at the close position and +4A at the open position, respectively. When the coil is supplied with positive current, magnetic field is strengthened so that moving iron can be attracted to the close position. Conversely, at the close position, supplied with negative current, magnetic field is weakened so that the opening spring can drive the moving iron to open the contactors of vacuum tubes.

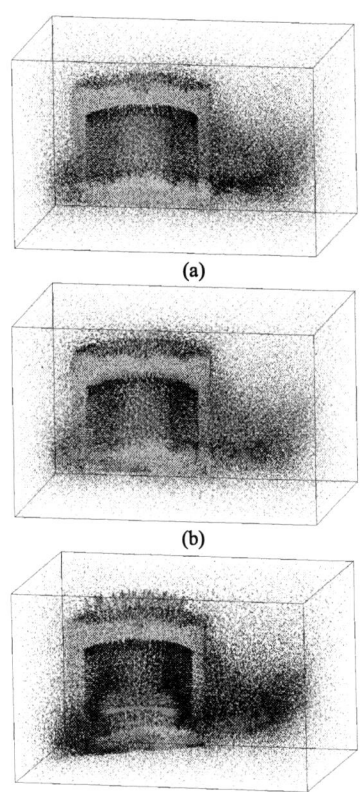

(a)

(b)

(c)

Fig.3. Flux density distributions. (a) PM only at close position (b) -2A at close position (c) +4A at open position

Fig.4 shows the computed electro-magnetic force at different positions when the coil is supplied with positive or negative currents. In Fig.4 (a), the electro-magnetic force decreases with negative current increasing, and the resultant force is close to zero when coil current is near to -3A. Then with current increasing, the resultant force begins to increase. However, in Fig.4 (b), the electro-magnetic direction is same with that of

permanent magnet, hence, with the coil current increasing, the resultant force is always strengthened enough to overcome mechanical anti-force and friction force to close the contactors of vacuum tubes.

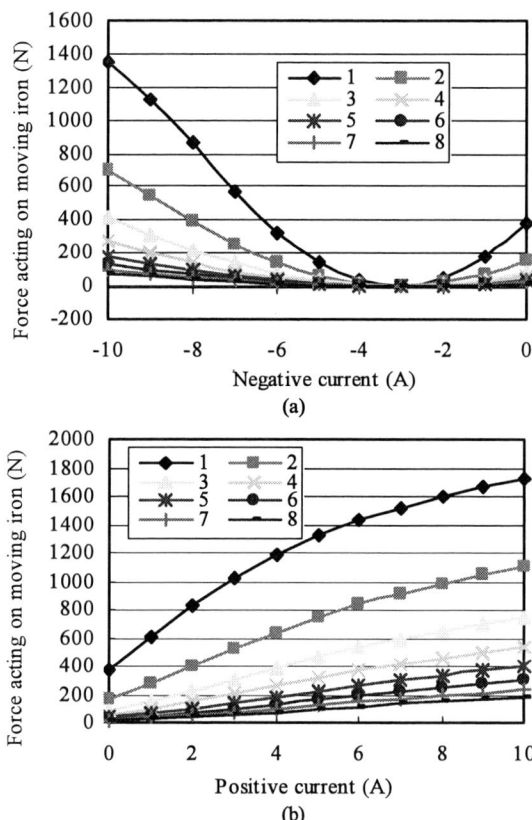

Fig.4. Force acting on moving iron at different angles (a) Supplied negative current (b) Supplied positive current

III. CONTROL STRATEGY

A. Control Strategy

Fig.5 shows the topology of driving circuit for making and breaking courses according to the developed PMA. In the topology, S1 and S2 are used to switch on the making course, while S3 and S4 are used to switch off the breaking course. The state of S1 and S2 is contrary to that of S3 and S4. Their control states are listed in table I.

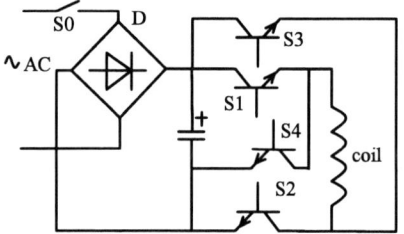

Fig.5. The topology of the driving circuit

TABLE I   CONTROL STATES

Item	State	
S1,S2	ON	OFF
S3,S4	OFF	ON
Breaker state	ON	OFF

746

B. Experimental Results

The experiments are carried out for making and breaking courses at different voltages. Fig.6 and Fig.7 show the results under capacitor voltage Uc=90V and Uc=160V. Three channels of the Tek oscillogragh are used to measure displacement of movable contactor of the vacuum tube, capacitor voltage and coil current together. From Fig.6, if the PMA approaches to close or open position, the energy stored in the capacitor does not consume completely. Subsequently, the capacitor will continue to discharge through coil.

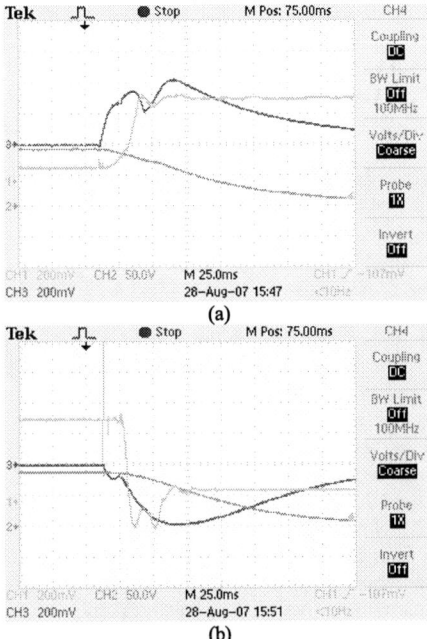

(b)

Fig.6. Measured waveforms at Uc=90V (a) Making course (b) Breaking course

CH1: displacement; CH2: capacitor voltage; CH3: coil current; below figures are the same.

However, when the capacitor voltage rises, control failure may happen in breaking course. This is because negative current flowing through the coil generates negative magnetic field and weakens permanent magnetic field severely enough to attract the moving iron to close position again. Fig.7 (b) is one of measured waveform of the failure control, the displacement waveform shows that the breaker is firstly opened, and then closed again under the work of continuing current through the coil.

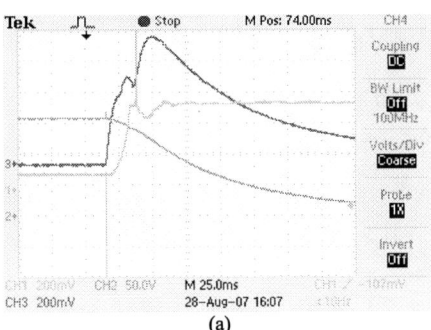

(a)

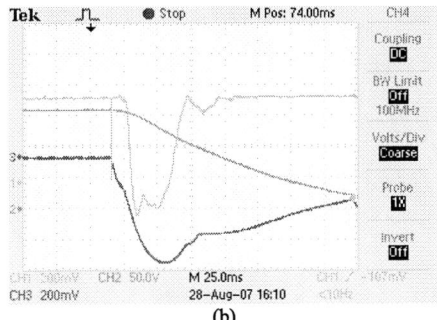

(b)

Fig.7. Measured waveforms at Uc=160V (a) Making course (b) Breaking course

C. Improved Control Strategy

Since the open-loop control strategy has disadvantage, closed-loop control is adopted to switch off the coil current when the breaker finishes making or breaking course. It can be seen that capacitor voltage drops from 210V to 200V and the making course only consume little energy in Fig.8 (a). Fig.8 (b) shows the breaking course, since moving iron is repulsed by opening spring together, consumed energy is much less than that of the making course. In fact, the waveform of the capacitor voltage in Fig.8 (b) clearly shows that the breaking course needs little energy.

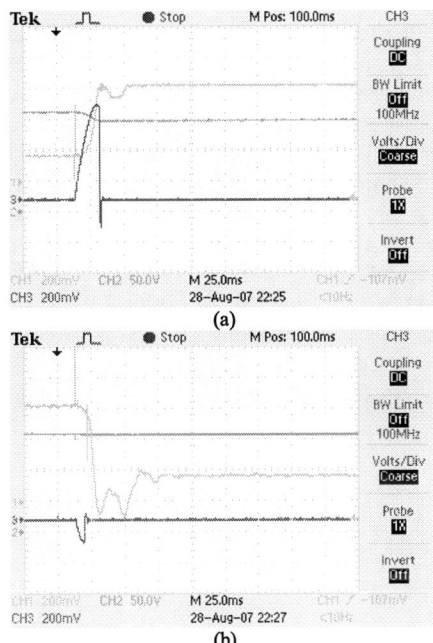

Fig.8. Measured waveforms at Uc=210V (a) Making course (b) Breaking course

In order to make sure whether the capacitor voltage rising has effect on the breaking course, experiments on several higher voltages are carried out as well. Fig.9 is one of the experimental results of making and breaking courses. Both in making course and in breaking course, the breaker has good performance for the closed-loop control.

747

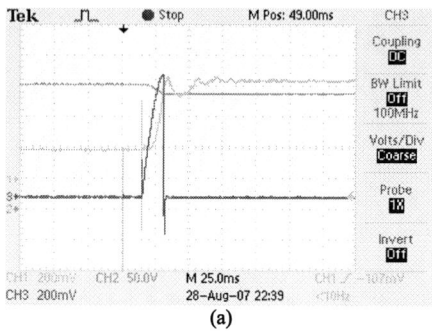

(a)

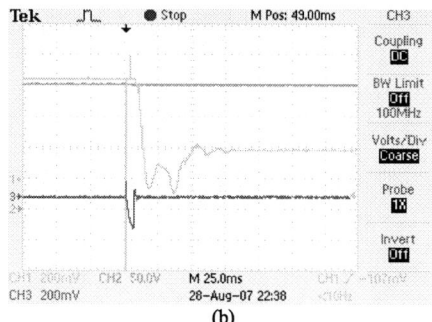

(b)

Fig.9. Measured waveforms at Uc=200V (a) Making course (b) Breaking course

IV. CONCLUSION

The paper develops a new type of PMA for low voltage vacuum circuit breaker. 3-D FEM is successfully used to analyze the magnetic field and electro-magnetic force is computed with coil supplied by different currents at different positions. Control strategies, including open-loop control and closed-loop control, are presented to satisfy the making and breaking processes. The experimental results show that in range of working voltage, open-loop control sometimes may fail to open the breaker, but closed-loop control has no such phenomena. So the proposed strategy with closed-loop control is an excellent solution to improve control reliability for the PMA of low voltage vacuum circuit breaker.

REFERENCES

[1] E. Dullni, "A vacuum circuit-breaker with permanent magnetic actuator for frequent operations," *Proc. ISDEIV. XVIIIth Int. Symp. on Discharges and Electrical Insulation in Vacuum,* Eindhoven (Netherlands), Aug. 1998, pp.688-691.

[2] B A R Mckean and C Reuber, "Magnets & vacuum - the perfect match," *Proc. of Fifth Int. Conf. on Trends in Distribution Switchgear: 400V-145kV for Utilities and Private Networks,* London. Nov. 1998, pp.73-79.

[3] X. Lin, H. J. Gao and Z. Y. Cai, "Magnetic field calculation and dynamic behavior analyses of the permanent magnetic actuator," *Proc. ISDEIV. XIXth Int. Symp. on Discharges and Electrical Insulation in Vacuum,* Xi'an (China), Sept. 2000, pp.532-535.

[4] I. W. Kyung and I. K. Byung, "Characteristic analysis and modification of PM-type magnetic circuit breaker," *IEEE Trans. on Magnetics,* vol. 40, no. 2, pp. 691-694, Mar 2004.

[5] S. A. Ruhland, "Vacuum circuit breaker with asymmetrical actuator," *Transmission and Distribution Conference and Exhibition 2002: Asia Pacific. IEEE/PES,* Yokohama(Japan), Oct. 2002, pp. 909-913.

[6] J. Sato, O. Sakaguchi, N. Kubota, et al. "New technology for medium voltage solid insulated switchgear," *Transmission and Distribution Conference and Exhibition 2002: Asia Pacific. IEEE/PES,* Yokohama (Japan), Oct. 2002, pp. 1791-1796.

[7] J. H. Kang, C. Y. Bae, and H. K. Jung, "Dynamic behavior analysis of permanent magnetic actuator in vacuum circuit breaker," *Proc. of 6th Int. Conf. on Electrical Machines and Systems (ICEMS 2003),* Beijing (China), Nov. 2003, pp.100-103.

[8] Y. D. Cao, C. G. Hou, X. M. Liu and E. Z. Wang, "Design on permanent magnet actuator and simulation of dynamic characteristic for vacuum circuit breaker," *Proc. ISDEIV. XXIst Int. Symp. on Discharges and Electrical Insulation in Vacuum,* Yalta (Ukraine), Sep.27-Oct.1, 2004, pp. 652-655.

[9] C. G. Hou, J. Sun, Y. D. Cao, et al, "Design and analyses on permanent magnet actuator for mining vacuum circuit breaker," *Proc. ISDEIV. XXIInd Int. Symp. on Discharges and Electrical Insulation in Vacuum (ISDEIV 2006),* Matsue (Japan), Sept. 2006, pp. 512-515.

Analysis of Transformer Inrush Current under Harmonic Source

Chien-Lung Cheng, *Member, IEEE,* Jim-Chwen Yeh*, Shyi-Ching Chern, Yi-Hung Lan

Department of Electrical Engineering
* Power Mechanical Engineering
National Formosa University
Huwei, Yunlin, 632, Taiwan
clcheng@nfu.edu.tw

Abstract

Inrush current is an important issue for three-phase transformers security and stability. It has close relationship with flux variation of three-phase transformers. Harmonic effects in power systems have been significantly increasing in the last decade. It results in more complicated for investigating inrush current. In this paper, $SIN^m(\omega t)$ waveforms represent non-sinusoidal waveforms because studies on $SIN^m(\omega t)$ waveforms appear to be simpler for more predictable results. A flux analysis of three-phase transformer for inrush current is proposed in this paper. The results are very helpful to estimate harmonics effect for inrush current.

Keyword: Transient analysis; Transformers; Harmonics

1. Introduction

Three-phase transformers are key components in power systems and power plants. They are also widely applied to uninterruptible power supply. Security and stability of transformers are both important and necessary to system operation. The large transient current of transformers due to flux saturation in the core often causes the mal-function of the protective relaying system, costing time and money as the engineers have to examine closely the transformer and the protective system, to check for faults. The large transient current also causes serious electromagnetic stress impact and shortens the life of transformers. The overvoltage [1]-[2] resulting from the inrush current could happen and cause serious damage to power apparatus. So it is very important to solve the effect of inrush current.

In recent years, various protective systems for transformers, based on the differential relaying system, were developed. Various techniques based on complex circuits or microcomputers [3]-[6] are proposed to distinguish inrush current from fault current. However, the transformer still must bear with large electromagnetic stress impact caused by the inrush current.

Non-sinusoidal voltage and current have significantly increased in power system due to broad application of solid-state controlled devices and fluorescent lamps.

Transformer is the most sensitive component in response to power system harmonics. As non-sinusoidal harmonics have been generated from many sources, harmonics flow through many transformers and cause a compound effect to the power system.

Focusing on the drawbacks of the above protection, an active suppression method [7]-[8] is developed by our group. The suppression method is very simple and effective. The flux of transformers is controlled to non-saturation by developed method. Non-sinusoidal excitation strongly influences the flux variation of three-phase transformers. Therefore, it is necessary to investigate the effect of non-sinusoidal excitation to flux variation for inrush current. In this paper, a flux analysis of three-phase transformers for inrush current under non-sinusoidal excitation is proposed. The maximum flux value of three-phase transformers is calculated and discussed in detail. It is very helpful to estimate the effect of active suppression method for suppressing inrush current.

2. The Proposed Method

2.1 Flux Analysis under Sinusoidal Excitation

Suppose that the connection circuit between three-phase transformer ($\Delta-Y$) and power source is shown in Fig. 1. The source line voltages in Fig. 1 are

$$V_{RS}=V \sin\omega t \qquad (1)$$

$$V_{ST}=V \sin(\omega t-2\pi/3) \qquad (2)$$

$$V_{TR}=V \sin(\omega t-4\pi/3) \qquad (3)$$

Where the symbol V represents the peak value under sinusoidal excitation. It is used as compared base value in

978-1-4244-0644-9/07/$25.00 ©2007 IEEE

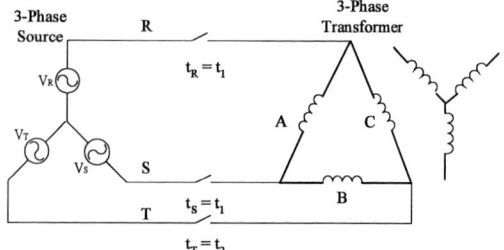

Fig. 1. The connection circuit between three-phase transformer and power source.

following discussion. The source has RST phase sequence. The switches on lines R, S, T are closed at t_1, t_1 and t_2 respectively. It is used as compared base value in following discussion. The flux linkages of A, B, C windings of the three-phase transformer, are computed as

$$\lambda_A = \int_1 V_{RS} dt = \frac{V}{\omega}(-\cos\omega t + \cos\omega t_1) \qquad (4)$$

$$\lambda_B = \int_1^2 (-\tfrac{1}{2}V_{RS})dt + \int_2 V_{ST} dt$$

$$= \frac{V}{\omega}\{[\frac{1}{2}(\cos\omega t_2 - \cos\omega t_1) + \cos(\omega t_2 - 2\pi/3)] - \cos(\omega t - 2\pi/3)\}$$

$$= \frac{V}{\omega}\{[\frac{\sqrt{3}}{2}\cos(\omega t_2 - \pi/2) - \frac{1}{2}\cos\omega t_1] - \cos(\omega t - 2\pi/3)\} \quad (5)$$

$$\lambda_C = \int_1^2 (-\tfrac{1}{2}V_{RS})dt + \int_2 V_{TR} dt$$

$$= \frac{V}{\omega}\{[\frac{1}{2}(\cos\omega t_2 - \cos\omega t_1) + \cos(\omega t_2 - 4\pi/3)] - \cos(\omega t - 4\pi/3)\}$$

$$= \frac{V}{\omega}\{[\frac{\sqrt{3}}{2}\cos(\omega t_2 + \pi/2) - \frac{1}{2}\cos\omega t_1] - \cos(\omega t - 4\pi/3)\} \quad (6)$$

Suppose that the switches on lines R, S, T are closed at the same time, that is $t_2 = t_1$, the flux linkages of A, B, C windings of the three-phase transformer become

$$\lambda_A = \int_1 V_{RS} dt = \frac{V}{\omega}(-\cos\omega t + \cos\omega t_1) \qquad (7)$$

$$\lambda_B = \frac{V}{\omega}[\cos(\omega t_1 - 2\pi/3)] - \cos(\omega t - 2\pi/3)] \qquad (8)$$

$$\lambda_C = \frac{V}{\omega}\{[\frac{\sqrt{3}}{2}\cos(\omega t_1 + \pi/2) - \frac{1}{2}\cos\omega t_1] - \cos(\omega t - 4\pi/3)\}$$

$$= \frac{V}{\omega}[\cos(\omega t_1 - 4\pi/3)] - \cos(\omega t - 4\pi/3)] \qquad (9)$$

Observing Eqs. (7)-(9), the reachable maximum value of flux linkages is different depending on t_1. Suppose that switches on lines R, S and T are closed at $\omega t_1 = \pi/2$, simultaneously. The reachable maximum absolute values of flux linkages on A, B, C windings are computed as

$$|\lambda_A| \le 1\frac{V}{\omega}$$

$$|\lambda_B| \le 1.866\frac{V}{\omega}$$

$$|\lambda_C| \le 1.866\frac{V}{\omega}$$

The reachable maximum value flux linkage on winding A is $\frac{V}{\omega}$. It is the least value, the same as steady-state condition. However, the reachable maximum value flux linkage on B and C windings is $1.866\frac{V}{\omega}$, much more than normal value $\frac{V}{\omega}$. The relation between line voltage and flux linkage at $\omega t_1 = \omega t_2 = \pi/2$ is shown in Fig. 2. The flux saturation results in serious inrush current. Besides, the reachable maximum value of flux linkage on A, B, C windings cannot increase or decrease simultaneously. This means that the reachable maximum value of flux linkage cannot be suppressed by powering on switches A, B, C, simultaneously.

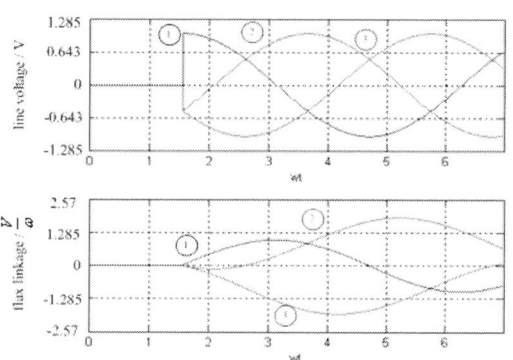

Fig. 2. The relation between line voltage and flux linkage at $\omega t_1 = \omega t_2 = \pi/2$, m= 1. (line voltage ①:$V_{RS}$ /V, ②:V_{ST} /V, ③:V_{TR} /V, flux linkage ①: $|\lambda_A / \frac{V}{\omega}| \le 1$, ②:$|\lambda_B / \frac{V}{\omega}| \le 1.866$, ③:$|\lambda_C / \frac{V}{\omega}| \le 1.866$)

After analyzing flux linkage variation, selecting $\omega t_1 = \pi/2$, $\omega t_2 = \pi$, the following results can be obtained from Eqs. (4)-(6).

$$|\lambda_A| \le \frac{V}{\omega}$$

$$|\lambda_B| \le \frac{V}{\omega}$$

$$|\lambda_C| \le \frac{V}{\omega}$$

The reachable maximum absolute values of flux linkages on A, B, C windings are the least value $\frac{V}{\omega}$, the same as steady-state condition. The relation between line voltage and flux linkage at $\omega t_1 = \pi/2$, $\omega t_2 = \pi$, is shown in Fig. 3. This cannot cause flux saturation. Hence the inrush current can be effectively eliminated.

750

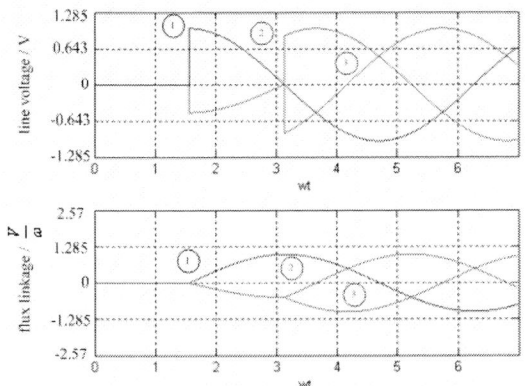

Fig. 3. The relation between line voltage and flux linkage at $\omega t_1=\pi/2$, $\omega t_2=\pi$, m= 1. (line voltage ①:V_{RS} /V, ②:V_{ST} /V, ③:V_{TR} /V,

flux linkage ①: $|\lambda_A / \dfrac{V}{\omega}| \leq 1$, ②:$|\lambda_B / \dfrac{V}{\omega}| \leq 1$, ③:$|\lambda_C / \dfrac{V}{\omega}|$ ≤ 1)

2.2 Flux Analysis under Non-sinusoidal Excitation

The distorted sinusoidal waveforms are described by m-order of sine functions in a convenient and generalized manner as:

$$V_{RS}=V_{max} SIN^m(\omega t) \qquad (10)$$

$$V_{ST}=V_{max} SIN^m(\omega t-2\pi/3) \qquad (11)$$

$$V_{TR}=V_{max} SIN^m(\omega t-4\pi/3) \qquad (12)$$

Where V_{max} is the peak value of voltage, and m is the order of sinusoidal waveform. Graphs of (10) are presented in Fig. 4. This function covers a wide spectrum of distortions, which exist in power system. The advantage of this approach consists of the fact that only one parameter of the exponent m can describe the distortion clearly. Strictly speaking, (10)-(12) are not typical for all distorted waveforms. However, it is more generalized and predictable for analyses and simulations. In this paper, only the generalized functions of (10)-(12) are used.

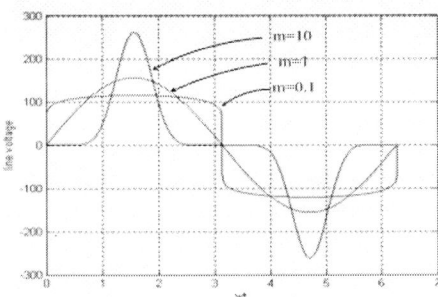

Fig. 4. The distorted excitation function SIN^m(wt) with same root-mean-square value V_{rms} =110 V.

Observing Fig.4, distortion parameter m=1 represents pure sinusoidal excitation. The discussion of sinusoidal excitation is presented in above section. Under non-sinusoidal excitation, distortion parameter m=10 and m=0.1 are discussed as following respectively. Suppose that three excitations is measured at same effective value V_{rms} =110 v. The peak value V_{max} of line voltage can be

calculated with respect to distortion parameter m=1,10,0.1 under same root-mean-square value V_{rms} =110 V. The calculated results are shown in Table 1. Table 1 shows the peak value V_{max} = 155.7 volt at m=1, V_{max} = 262.1 volt at m=10 and V_{max} = 121.2 volt at m=0.1.

Table 1. The peak value V_{max} of line voltage with respect to distortion parameter m, under same effective voltage V_{rms} =110 V.

distortion parameter m	1	5	10	0.2	0.1
V_{rms} (volt)	110	110	110	110	110
V_{max}(volt)	155.7	221.7	262.1	143.9	121.2

2.2.1 Distortion parameter m=10

$$V_{RS}=V_{max}sin^{10}\omega t \qquad (13)$$

$$V_{ST}=V_{max}sin^{10}(\omega t-2\pi/3) \qquad (14)$$

$$V_{TR}=V_{max}sin^{10}(\omega t-4\pi/3) \qquad (15)$$

The source has RST phase sequence. The switches on lines R, S, T are closed at t_1, t_1 and t_2 respectively. The flux linkages of A, B, C windings of the three-phase transformer, are computed as

$$\lambda_A = \int_1^t V_{RS}dt$$
$$= \frac{V_{max}}{\omega}\frac{1}{10240}(-2\sin10\omega t + 25\sin8\omega t -150\sin6\omega t$$
$$+600\sin4\omega t -2100\sin2\omega t +2520\omega t) \Big|_{t_1}^{t} \qquad (16)$$

$$\lambda_B = \int_1^2 (-\tfrac{1}{2}V_{RS})dt + \int_2^t V_{ST}dt$$
$$= -\frac{V_{max}}{\omega}\frac{1}{20480}(-2\sin10\omega t + 25\sin8\omega t -150\sin6\omega t + 600$$
$$\sin4\omega t -2100\sin2\omega t$$
$$+2520\omega t) \Big|_{t_1}^{t_2} + \frac{V_{max}}{\omega}\frac{1}{10240}[-2\sin10(\omega t-2\pi/3) +25\sin8(\omega t-2$$
$$\pi/3) -150\sin6(\omega t-2\pi/3) +600\sin4(\omega t-2\pi/3) -$$
$$2100\sin2(\omega t-2\pi/3) +2520(\omega t-2\pi/3)] \Big|_{t_2}^{t} \qquad (17)$$

$$\lambda_C = \int_1^2 (-\tfrac{1}{2}V_{RS})dt + \int_2^t V_{TR}dt$$
$$= -\frac{V_{max}}{\omega}\frac{1}{20480}(-2\sin10\omega t + 25\sin8\omega t -150\sin6\omega t +600$$
$$\sin4\omega t -2100\sin2\omega t$$
$$+2520\omega t) \Big|_{t_1}^{t_2} + \frac{V_{max}}{\omega}\frac{1}{10240}[-2\sin10(\omega t-4\pi/3) +$$
$$25\sin8(\omega t-4\pi/3) -150\sin6(\omega t-4\pi/3) +600\sin4(\omega t-4\pi/3) -$$
$$2100\sin2(\omega t-4\pi/3) +2520(\omega t-4\pi/3)] \Big|_{t_2}^{t} \qquad (18)$$

Observing Eqs. (16)-(18), the reachable maximum value of flux linkages is different depending on t_1. Suppose that switches on lines R, S and T are closed at $\omega t_1=\omega t_2=\pi/2$, simultaneously. The reachable maximum absolute values of flux linkages on A, B, C windings are computed as

$$|\lambda_A| \le 0.651\frac{V}{\omega}$$

$$|\lambda_B| \le 1.289\frac{V}{\omega}$$

$$|\lambda_C| \le 1.302\frac{V}{\omega}$$

The reachable maximum value flux linkage on winding A is $0.651\frac{V}{\omega}$. It is a small value, less than steady-state condition. This cannot cause flux saturation. However, the reachable maximum value flux linkage on B and C windings is $1.289\frac{V}{\omega}$ and $1.302\frac{V}{\omega}$, more than normal value $\frac{V}{\omega}$. The relation between line voltage and flux linkage at $\omega t_1 = \omega t_2 = \pi/2$ is shown in Fig. 5. The flux saturation results in medium inrush current.

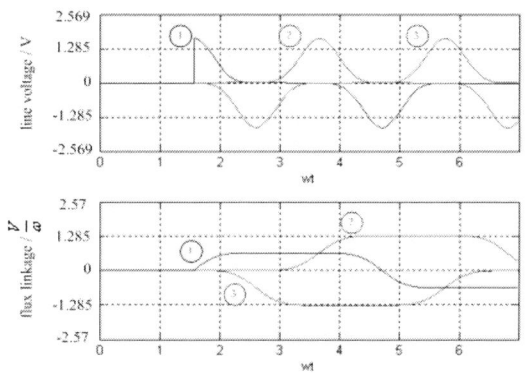

Fig. 5. The relation between line voltage and flux linkage at $\omega t_1 = \omega t_2 = \pi/2$, m= 10. (line voltage ①:V_{RS} /V, ②:V_{ST} /V, ③:V_{TR} /V, flux linkage ①: $|\lambda_A / \frac{V}{\omega}| \le 0.651$, ②:$|\lambda_B / \frac{V}{\omega}| \le 1.289$, ③:$|\lambda_C / \frac{V}{\omega}| \le 1.302$)

Selecting $\omega t_1 = \pi/2$, $\omega t_2 = \pi$, the following results can be obtained from Eqs. (16)-(18).

$$|\lambda_A| \le 0.651\frac{V}{\omega}$$

$$|\lambda_B| \le 0.911\frac{V}{\omega}$$

$$|\lambda_C| \le 0.910\frac{V}{\omega}$$

The reachable maximum absolute values of flux linkages on A, B, C windings are less than $\frac{V}{\omega}$. The relation between line voltage and flux linkage at $\omega t_1 = \pi/2$, $\omega t_2 = \pi$ is shown in Fig. 6. This cannot cause flux saturation. Hence the inrush current can be effectively eliminated.

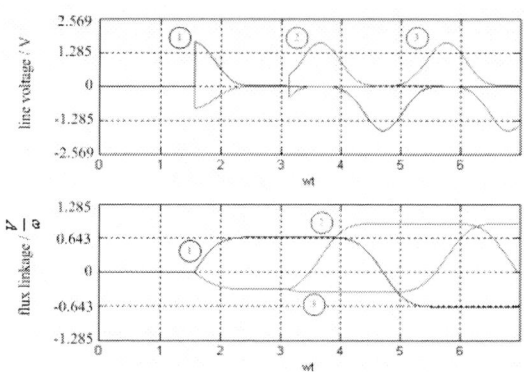

Fig. 6. The relation between line voltage and flux linkage at $\omega t_1 = \pi/2$, $\omega t_2 = \pi$, m= 10. (line voltage ①:V_{RS} /V, ②:V_{ST} /V, ③:V_{TR} /V, flux linkage ①: $|\lambda_A / \frac{V}{\omega}| \le 0.651$, ②:$|\lambda_B / \frac{V}{\omega}| \le 0.911$, ③:$|\lambda_C / \frac{V}{\omega}| \le 0.910$)

2.2.2 Distortion parameter m =0.1

$$V_{RS} = V_{max}\sin^{0.1}\omega t \tag{19}$$

$$V_{ST} = V_{max}\sin^{0.1}(\omega t - 2\pi/3) \tag{20}$$

$$V_{TR} = V_{max}\sin^{0.1}(\omega t - 4\pi/3) \tag{21}$$

The source has RST phase sequence. The switches on lines R, S, T are closed at t_1, t_1 and t_2, respectively. The flux linkages of A, B, C windings of the three-phase transformer, are computed as

$$\lambda_A = \int_1^t V_{max}\sin^{0.1}\omega t\, dt \tag{22}$$

$$\lambda_B = \int_1^2 (-\tfrac{1}{2}V_{max}\sin^{0.1}\omega t)dt$$
$$+ \int_2 (V_{max}\sin^{0.1}(\omega t - 2\pi/3))dt \tag{23}$$

$$\lambda_C = \int_1^2 (-\tfrac{1}{2}V_{max}\sin^{0.1}\omega t)dt$$
$$+ \int_2 (V_{max}\sin^{0.1}(\omega t - 4\pi/3))dt \tag{24}$$

Based on MATLAB computation from (22)-(24), the reachable maximum value of flux linkage is different depending on t_1. Suppose that switches on lines R, S and T are closed at $\omega t_1 = \omega t_2 = \pi/2$, simultaneously. The reachable maximum absolute values of flux linkages on A, B, C windings are computed as

$$|\lambda_A| \le 1.143\frac{V}{\omega}$$

$$|\lambda_B| \le 1.939\frac{V}{\omega}$$

$$|\lambda_C| \le 1.939\frac{V}{\omega}$$

The reachable maximum value flux linkage on B, C winding is $1.939 \frac{V}{\omega}$, much more than normal value $\frac{V}{\omega}$. The relation between line voltage and flux linkage at $\omega t_1 = \omega t_2 = \pi/2$ is shown in Fig. 7. The flux saturation results in serious inrush current.

Selecting $\omega t_1 = \pi/2$, $\omega t_2 = \pi$, the following results can be obtained from Eqs. (22)-(24).

$$|\lambda_A| \le 1.143 \frac{V}{\omega}$$

$$|\lambda_B| \le 1.309 \frac{V}{\omega}$$

$$|\lambda_C| \le 1.309 \frac{V}{\omega}$$

The reachable maximum value of flux linkage on B, C winding is $1.309 \frac{V}{\omega}$, more than normal value $\frac{V}{\omega}$. The relation between line voltage and flux linkage at $\omega t_1 = \pi/2$, $\omega t_2 = \pi$ is shown in Fig. 8. This causes minor flux saturation. The minor flux saturation results in smaller inrush current.

3. Results and Discussion

A 330VA, 110V/220V single-phase transformer and a 1KVA, 110V/220V three-phase transformer are used for on-site measurement of inrush current in our laboratory.

3.1 Results under Sinusoidal Excitation
3.1.1 Improper Switching-on
For three-phase transformers, Fig. 9. shows the experimental result of inrush current (Ip=16A), under $\omega t_1 = \omega t_2 = \pi/2$, three-phase transformer banks. The peak value of the inrush current of winding A, in Fig.9, is effectively suppressed due to no flux saturation. However, the inrush current of phases B and C cannot effectively be suppressed simultaneously. It can be observed that the peak value of the inrush current of winding C in Fig. 8, Ip= 16A, is more than 5 times the rated phase current.

3.1.2 Proper Switching-on
According to analysis results of flux linkage variation, selecting $t_1 = 4.2ms(\omega t_1 \approx \pi/2)$, $t_2 = 8.8ms(\omega t_2 \approx \pi)$, the experimental result of the inrush current (Ip=2A) is shown in Fig. 10. The inrush current is effectively suppressed.

3.2 Influence of Non-sinusoidal Excitation
In this paper, the influences of non-sinusoidal excitation to flux linkage are investigated in detail. The results are helpful for suppressing inrush current. Table 2 shows the comparison of reachable maximum absolute values of flux linkages on A, B, C windings under various conditions. Observing Table 2, non-sinusoidal excitation strongly

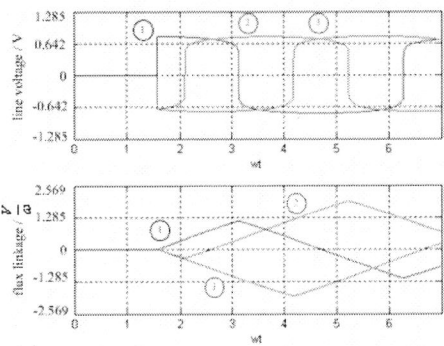

Fig. 7. The relation between line voltage and flux linkage at $\omega t_1 = \omega t_2 = \pi/2$, m= 0.1. (line voltage ①:V_{RS} /V, ②:V_{ST} /V, ③:V_{TR} /V, flux linkage ①: $|\lambda_A / \frac{V}{\omega}| \le 1.143$, ②:$|\lambda_B / \frac{V}{\omega}| \le 1.939$, ③:$|\lambda_C / \frac{V}{\omega}| \le 1.939$)

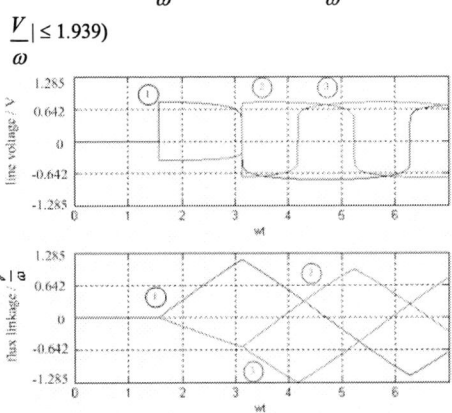

Fig. 8. The relation between line voltage and flux linkage at $\omega t_1 = \pi/2$, $\omega t_2 = \pi$, m= 0.1. (line voltage ①:V_{RS} /V, ②:V_{ST} /V, ③:V_{TR} /V, flux linkage ①: $|\lambda_A / \frac{V}{\omega}| \le 1.143$, ②:$|\lambda_B / \frac{V}{\omega}| \le 1.309$, ③:$|\lambda_C / \frac{V}{\omega}| \le 1.309$)

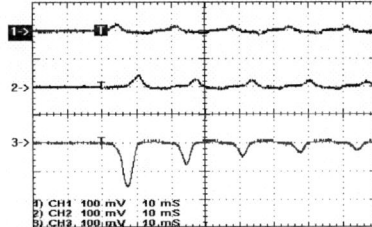

Fig. 9. Experimental result of inrush current (Ip=16A), under $\omega t_1 = \omega t_2 = \pi/2$, three-phase transformer banks. (1: phase A, 2: phase B, 3: phase C)

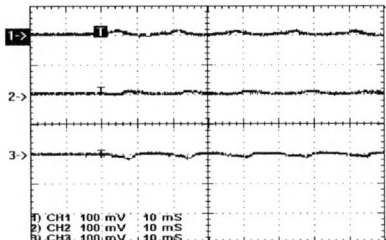

Fig. 10. Experimental result of inrush current (Ip=2A), under t_1 = 4.2ms, t_2 =8.8ms, three-limb three-phase transformers. (1: phase A, 2: phase B, 3:phase C)

influences flux saturation of core in three-phase transformers. Under $\omega t_1 = \pi/2$, $\omega t_2 = \pi/2$, the maximum flux linkage on A, B, C windings is $1.866\frac{V}{\omega}$ at m=1, $1.302\frac{V}{\omega}$ at m=10, $1.939\frac{V}{\omega}$ at m=0.1. The flux saturation is more serious at m=0.1 and much less serious at m=10. The similar results exist on other switching-on conditions. In conclusion, the non-sinusoidal excitation, more close to m=0.1, causes more serious flux saturation. On the contrary, the non-sinusoidal excitation, more lose to m=10, cause less flux saturation. Under $\omega t_1 = \pi/2$, $\omega t_2 = \pi$, the maximum flux linkage on A, B, C windings is $1\frac{V}{\omega}$ at m=1, $0.911\frac{V}{\omega}$ at m=10, $1.309\frac{V}{\omega}$ at m=0.1. Flux linkage $1\frac{V}{\omega}$ is the smallest value among all sinusoidal excitations m=1. Consequently, switching-on angle of $\omega t_1 = \pi/2$, $\omega t_2 = \pi$, can be applied to suppress inrush current very well for three-phase transformers. However, for non-sinusoidal excitation m=0.1, the flux linkage $1.309\frac{V}{\omega}$ at $\omega t_1 = \pi/2$, $\omega t_2 = \pi$, is not the least value. Comparing the flux linkages under non-sinusoidal excitation m=0.1, flux linkage $1.309\frac{V}{\omega}$ at $\omega t_1 = \pi/2$, $\omega t_2 = \pi$, is more than $1.275\frac{V}{\omega}$ at $\omega t_1 = 95$, $\omega t_2 = 180$. The switching-on angle of $\omega t_1 = \pi/2$, $\omega t_2 = \pi$, is no more the best selecting for suppressing inrush current. By changing switching-on angles, the maximum flux linkage can be decreased from $1.309\frac{V}{\omega}$ to $1.275\frac{V}{\omega}$ at $\omega t_1 = 95$, $\omega t_2 = 180$. In conclusion, non-sinusoidal excitation strongly influences flux saturation and flux saturation can be improved by proper switching-on.

4. Conclusion

This paper proposed an effective flux analysis of three-phase transformers in detail for estimating inrush current. The effect of non-sinusoidal excitation to inrush current is investigated and discussed in detail. In this paper, m-order sine functions bring some representative discussions for this topic. The results can be applied to improve the protective relaying system of three-phase transformers [9]. Furthermore, the experimental examination of non-sinusoidal excitation to inrush current for three-phase transformers is being investigated in our lab.

5. Acknowledgment

Table 2. The comparison of reachable maximum absolute values of flux linkages (V / ω) on A, B, C windings under various conditions.

	m=1			m=10			m=0.1		
	A	B	C	A	B	C	A	B	C
$\omega t_1 = \omega t_2 = 90°$	1.	1.866	1.866	0.651	1.289	1.302	1.143	1.939	1.939
$\omega t_1 = 90°$ $\omega t_2 = 180°$	1.	1	1	0.651	0.911	0.910	1.143	1.309	1.309
$\omega t_1 = 90°$ $\omega t_2 = 170°$	1	1.152	1.150	0.651	0.952	0.810	1.143	1.160	1.391
$\omega t_1 = 90°$ $\omega t_2 = 190°$	1	1.151	1.151	0.651	0.800	0.965	1.143	1.391	1.162
$\omega t_1 = 100°$ $\omega t_2 = 180°$	1.174	1.087	1.087	0.923	1.060	1.060	1.279	1.241	1.241
$\omega t_1 = 85°$ $\omega t_2 = 180°$	1.087	1.044	1.044	0.796	0.850	0.850	1.210	1.342	1.342
$\omega t_1 = 95°$ $\omega t_2 = 180°$	1.088	1.044	1.044	0.797	0.996	0.996	1.211	1.275	1.275

This work is supported by the National Science Council under research project: NSC93-2213-E150-025.

References

[1] G. Sybille, M. M. Gavrilovic and J. Belanger, Transformer Saturation Effect on EHV System Overvoltages, *IEEE Trans. PAS, Vol. PAS-104, No. 3*, March 1985, 671-680.

[2] R. Yacamini and A. Abu Nasser, Transformer Inrush Current and Their Associated Overvoltage in HVDC Schemes, *IEE Proc. Vol. 133, Pt. C, No. 6*, September 1986, 353-358.

[3] P. Lin, O. P. Malik, D. Chen, G. S. Hope, Y. Guo, Improved Operation of Differential Protection of Power Transformers for Internal Faults, *IEEE PES winter Meeting*, New York, January 26-30, Paper Number 92 WM 206-3 PWRD.

[4] T. S. Sidhu and M. S. Sachdev, 1992, On-Line Identification of Magnetizing Inrush and Internal Faults in Three-Phase Transformers, *IEEE PES winter Meeting*, New York, January 26-30, Paper Number 92 WM 205-5 PWRD.

[5] Guzman, A., Zocholl, Z., Benmouyal, G., Altuve, H.J., A current-based solution for transformer differential protection. I. Problem statement, *IEEE Trans. on PWRD. Vol. 16, No. 4*, 2001, 485 -491.

[6] Guzman, A., Zocholl, Z., Benmouyal, G., Altuve, H.J., A current-based solution for transformer differential protection. II. Relay description and evaluation, *IEEE Trans. on PWRD. Vol. 16, No. 4*, 2002, 886 -893.

[7] C. L. Cheng, J. C. Yeh, S.C. Chern, 2005, "A Simple Suppressing Method for Transformer Inrush Current", Proceedings of the IASTED International Conference on POWER AND ENERGY SYSTEMS, April 18-20, 2005, Krabi, Thailand, pp.14-18.

[8] C. L. Cheng, C. E. Lin, K. C. Hou, Y. F. Hsia, 2002, "Effective Suppressing Method of Three-Phase Transformers Inrush Current", IEEE Region 10 Technical Conference on Computers, Communications, Control and Power Engineering Proceedings, 2002, Beijing, China, pp. 2030-2033.

[9] C. L. Cheng, C. E. Lin, 2002, "A New Scheme of Protective System for Three-Phase Transformers", IEEE/PES Transmission and Distribution Conference and Exhibition 2002: Asia Pacific, Conference Proceedings, 2002, Yokohama, Japan, pp. 1808-1813.

Voltage Sag Compensation Performance by DSTATCOM with Series Inductor and Energy Storage

Sumate Naetiladdanon*

* Department of Electrical Engineering, King Mongkut's University of Technology Thonburi
126 Pracha Utit Rd. Bangmod, Toongkru, Bangkok 10140 Thailand

Abstract– **In this paper, the voltage sag compensation performance by distribution static synchronous compensator (DSTATCOM) is presented. The DSTATCOM injects the current into the system so that the load voltage can be instantaneously compensated. The three types of current injection schemes are described. The voltage sag compensation capability by DSTATCOM with different configurations is explained and analyzed. Simulation results confirmed that DSTATCOM system with additional series inductor and energy storage can improve voltage sag compensation performance and reduce the compensator rating as well.**

Index Terms– **voltage sag compensation, DSTATCOM, power quality.**

I. INTRODUCTION

Voltage sags are considered as the most important power quality problems faced by utilities and industrial consumers, especially in the term of high costs from lost productivity and downtime. It contributes more than 80% of power quality (PQ) problems that exist in power systems [1]. It has been reported that, high intensity discharge lamps used for industrial illumination get extinguished at voltage sags of 20% and industrial equipments like PLC and ASD are about 10% [2]. The most effective way to mitigate for voltage sag at the customer level is to apply a voltage sag mitigation equipment at the customer site. A dynamic voltage restorer (DVR) is recognized as an effective voltage sags mitigating device for its fast response, simple control, and fewer transients [3]. However, since voltage sags generally only occur a few times each year at any particular location, a DVR system will generally spend most of its time in standby mode waiting for a sag to occur [4]. It would be advantageous if the mitigating device has the multiple flexible functions that can be used in various conditions. The load compensation by DVR is not preferable because it will increase the loss for keeping the dc bus voltage constant.

Recently, the fast voltage sag compensation by DSTATCOM without using the static switch has been presented [5]-[6]. The voltage sag compensation is based on injecting the active and/or reactive current from DSTATCOM. It has the advantage of optimized energy which DVR does not have since DSTATCOM can be controlled to inject only the reactive power to restore the load voltage. However, a DVR system requires much smaller compensator rating for the same voltage sag compensation [7]. The configuration of DSTATCOM might be applied with additional series inductor and energy storage. The additional series inductor can be used to reduce the compensator rating and the energy storage can enhance the compensation performance as well as reduce the compensator rating.

This paper deals with the performance of DSTATCOM for voltage sag compensation. Three different current injection schemes of voltage sag compensation are described. The voltage sag compensation capability for these schemes is determined for different DSTATCOM configuration system. Steady state performances with the effect of additional series inductor and energy storage are shown based on simulation results.

II. VOLTAGE SAG COMPENSATION PRINCIPLE

The DSTATCOM system consists of a voltage source converter (VSC) with an additional energy storage device in parallel with the load and the additional inductor in series with the line reactance and the supply voltage, as shown in Fig. 1. A small capacitor is present on the dc side of the converter where the capacitor voltage is kept constant, by the energy from the energy storage reservoir or the incoming line power supply. The converter acts as a controllable current source connected in parallel with the load. The DSTATCOM can instantaneously compensate voltage sags by regulating the load voltage using the injected current from the converter and the voltage across equivalent inductance (additional series inductance and line inductance) bridges the sag. The

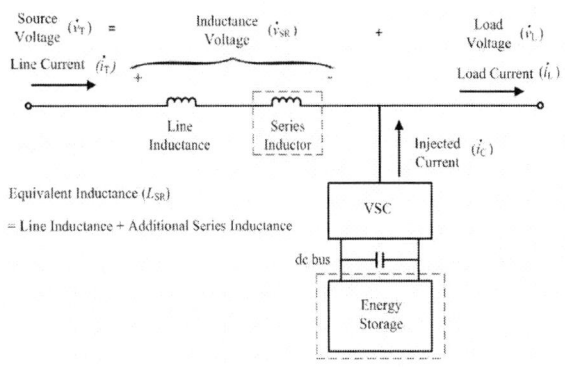

Fig. 1. DSTATCOM with series inductor and energy storage.

magnitude of the converter's output ac voltage (i.e. load voltage) is equal to the magnitude of the pre-sag load voltage. The relation of load, line current and injected current can be expressed in (1).

$$\dot{i}_L = \dot{i}_T + \dot{i}_C \tag{1}$$

And the relation of supply, load and inductance voltage can be expressed in (2)

$$\dot{v}_T = \dot{v}_L + \dot{v}_{SR} \tag{2}$$

There are three current injection schemes for voltage sag compensation as follows.

A. Phase-invariant current injection scheme

The purpose of this scheme is to protect load voltage from both magnitude and phase disturbances of the supply. The output voltage from the VSC, which is the voltage across the load, will be the same as the pre-sag load voltage ($\dot{V}_{L-pre}$). The injected current ($\dot{i}_C$) equals to the difference of the load current ($\dot{i}_L$), which equals the pre-sag load current ($\dot{i}_{L-pre}$), and the during-sag supply current ($\dot{i}_T$) and it can be derived as

$$\left. \begin{aligned} \dot{I}_L &= \dot{I}_{L-pre} \\ \dot{I}_C &= \dot{I}_L - \dot{I}_T = \dot{I}_L - \frac{\dot{V}_T - \dot{V}_L}{j\omega L_{SR}} \end{aligned} \right\} \tag{3}$$

or,

$$I_C e^{j\alpha} = I_L e^{j\varphi} - I_T e^{j\beta} = I_L e^{j\varphi} - \frac{V_T e^{j\theta} - V_L e^{j0}}{\omega L_{SR} e^{j\pi/2}} \tag{4}$$

where φ is the load power factor angle, β is the line current phase angle, α is the injected current phase angle, and θ is the source voltage phase angle by using the pre-sag load voltage as a reference ($\dot{V}_L = \dot{V}_{L-pre} = V_L e^{j0}$). Fig. 2 illustrates the phasor diagram of the phase-invariant current injection scheme.

B. Minimum energy current injection scheme

The purpose of this scheme is to minimize active power supplied to the load by the compensator. From phase-invariant injection scheme, the injected active power can be expressed by

$$P_C = 3V_L\left(I_L \cos\varphi - I_T \cos(\beta - \gamma)\right) = 3V_L\left(I_L \cos\varphi - \frac{V_T}{\omega L_{SR}}\sin(\theta - \gamma)\right) \tag{5}$$

where γ is the phase angle of the load voltage command, which equals zero in phase-invariant scheme. Note that the sign of γ is negative, if γ is measured by clockwise direction as shown in Fig. 3. The minimum injected active power can be achieved if the difference of θ and γ equals $\pi/2$, which equals zero in phase-invariant scheme. However, this could result as the active power absorption from the system in the case of $I_L\cos\varphi \doteq V_T/(\omega L_{SR})$. Thus, the condition of minimum active power injection can be accomplished by setting γ to γ_1 as

$$\gamma_1 = \begin{cases} \theta - \dfrac{\pi}{2} & \text{, when } I_L \cos\varphi > \dfrac{V_T}{\omega L_{SR}} \\[3mm] \theta - \sin^{-1}\left(\dfrac{I_L \cos(\varphi) \times \omega L_{SR}}{V_T}\right) & \text{, when } I_L \cos\varphi \le \dfrac{V_T}{\omega L_{SR}} \end{cases} \tag{6}$$

Fig. 3 illustrates the phasor diagram of the minimum energy current injection scheme. In a specific case where $I_L\cos\varphi \le V_T/(\omega L_{SR})$, the injected active power can be reduced to zero by setting the injected current phasor to be perpendicular with the load voltage phasor. Instant change of injected current from this scheme might cause the sudden phase angle shift or discontinuity of the load current and voltage during the start-up of injection.

C. Minimum apparent power current injection scheme

The purpose of this scheme is to minimize apparent power supplied to the load by the compensator (i.e. to minimize the injected current magnitude). The injected current magnitude can be solved by (4), and the condition of minimum apparent power injection is shown in (7).

$$\frac{\partial I_C}{\partial \gamma} = 0 \tag{7}$$

Solving (7) yields the angle γ as γ_2 which equals

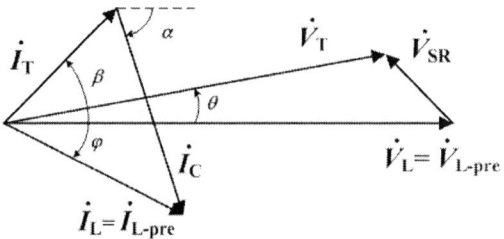

Fig. 2. Phasor diagram of phase-invariant injection scheme.

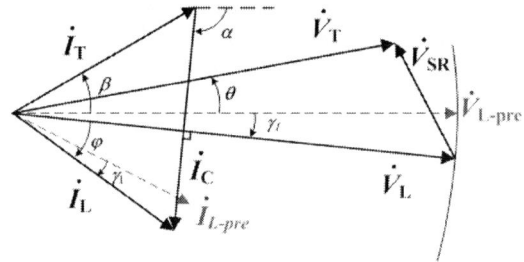

Fig. 3. Phasor diagram of minimum energy injection scheme.

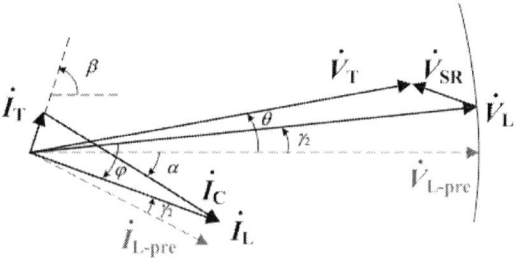

Fig. 4. Phasor diagram of minimum apparent power injection scheme.

$$\gamma_2 = \tan^{-1}\left(\frac{Z_L \sin\theta + \omega L_{SR}(\sin\theta \sin\varphi - \cos\theta \cos\varphi)}{Z_L \cos\theta + \omega L_{SR}(\cos\theta \sin\varphi + \sin\theta \cos\varphi)}\right) \quad (8)$$

where Z_L is the magnitude of the load impedance. Fig. 4 illustrates the phasor diagram of the minimum apparent power current injection scheme. Instant change of injected current from this scheme might also cause the sudden phase angle shift or discontinuity of the load current and voltage during the start-up of injection.

III. VOLTAGE SAG COMPENSATION CAPABILITY

During the voltage sag compensation, the voltage across equivalent inductance ($\dot{V}_{SR}$) bridges the sag voltage and it can be expressed by

$$\dot{V}_{SR} = j\omega L_{SR} \times \left(\frac{P_T - jQ_T}{\dot{V}_L}\right) = \frac{\omega L_{SR} Q_T}{\dot{V}_L} + j\frac{\omega L_{SR} P_T}{\dot{V}_L} \quad (9)$$

where P_T and Q_T are the incoming line active power and reactive power. If $\dot{V}_L = V+j0$ is taken as the reference phasor, the magnitude of the supply voltage during voltage sag compensation can be expressed as

$$V_T = \sqrt{\left[V + \frac{\omega L_{SR} Q_T}{V}\right]^2 + \left[\frac{\omega L_{SR} P_T}{V}\right]^2}$$

$$= \sqrt{\left[V + \frac{\omega L_{SR}}{Z_L} \times \frac{Q_T}{S_L} \times V\right]^2 + \left[\frac{\omega L_{SR}}{Z_L} \times \frac{P_T}{S_L} \times V\right]^2} \quad (10)$$

where S_L is the load apparent power. It can be expressed as the ratio of supply voltage with the load voltage which yields

$$v = \frac{V_T}{V} = \sqrt{\left[1 + x_{SR} q_T\right]^2 + \left[x_{SR} p_T\right]^2} \quad (11)$$

where x_{SR} is the ratio of equivalent reactance with the load impedance, q_T is the ratio of incoming line reactive power with the load apparent power and p_T is the ratio of incoming line active power with the load apparent power. Thus, the ratio of compensated voltage with the load voltage, as the function of equivalent reactance, load power factor angle and injected power, can be expressed as

$$v_{COMP} = 1 - v = 1 - \sqrt{\left[1 + x_{SR} q_T\right]^2 + \left[x_{SR} p_T\right]^2}$$

$$= 1 - \sqrt{\left[1 + x_{SR}(\sin\varphi - q_C)\right]^2 + \left[x_{SR}(\cos\varphi - p_C)\right]^2} \quad (12)$$

where q_C is the ratio of injected reactive power with the load apparent power and p_C is the ratio of injected active power with the load apparent power. The voltage sag compensation capability of DSTATCOM due to the presence of additional series inductor and energy storage can be described as follows.

A. DSTATCOM without Energy Storage

The minimum energy current injection scheme can only be used because only the reactive power can be exchanged; that is the incoming line active power equals the load active power ($p_C = 0$). And the injected reactive power mostly flows to the supply side (i.e. to supply the equivalent inductance). The voltage sag compensation capability in (12) can be rewritten as (13).

$$v_{COMP} = 1 - \sqrt{\left[1 + x_{SR}(\sin\varphi - q_C)\right]^2 + \left[x_{SR} \cos\varphi\right]^2} \quad (13)$$

Effect of equivalent reactance: For the voltage sag compensation purpose, the injected reactive power is much higher than the load reactive power (that is $\sin\varphi$ is less than q_C). As x_{SR} increases, the first term in the square root of (13) will decrease implying that the voltage sag compensation capability will increase. While the second term in the square root will increase but the voltage sag compensation capability will decrease. Because connecting high equivalent reactance in series with the incoming line will decrease the power transmission capability. This reduction of power availability at load is also applied during the normal operation of the load. The poor power factor and voltage drop problem will also occur.

Effect of load power factor angle: For the inductive load, the first term in the square root will increase implying that the voltage sag compensation capability will decrease. While the second term in the square root will cause the voltage sag compensation capability slightly increase. Because DSTATCOM is needed to supply the reactive power to the equivalent reactance, the inductive load will share the injected reactive power where the capacitive load will supply the reactive power to the equivalent reactance. As a result, the inductive load will decrease the voltage sag compensation capability while the capacitive load will increase the voltage sag compensation capability.

Effect of DSTATCOM rating: From the term of injected power in (12) implies how to determine the rating of DSTATCOM. In this case, the reactive power is only injected and it is considered as the rating of DSTATCOM. From the first term in the square root of (13), the voltage sag compensation capability will increase as the rating of DSTATCOM increase. Notice that the voltage sag compensation capability will certainly increase as long as the active power from the incoming can support the load at 100%.

B. DSTATCOM with Energy Storage

The phase-invariant and minimum apparent power current injection scheme can be applied because both active and reactive power can be exchanged. The phase-angle jump effect can be mitigated by using the phase-invariant current injection scheme, thus the transient compensation performance can be improved. However, the ratio of injected power term in (12) will be varied as the variation of the source voltage phase angle θ. Thus, this injection scheme is not considered in this paper. The voltage sag compensation capability by minimum apparent power current injection scheme can be expressed as (14) and the required injected reactive power is shown in (15) where s_C is the ratio of compensator rating with the load capacity.

$$v_{\text{comp}} = 1 - \sqrt{[1 - x_{SR}(\sin\varphi - q_C)]^2 + \left[x_{SR}\left(\cos\varphi - \sqrt{s_C^2 - q_C^2}\right)\right]^2} \tag{14}$$

$$q_C = s_C\left(1 + \left(\frac{x_{SR}\cos\varphi}{1 - x_{SR}\sin\varphi}\right)^2\right)^{-\frac{1}{2}} \tag{15}$$

Effect of equivalent reactance: As x_{SR} increases, the first term in the square root of (14) will cause the voltage sag compensation capability to increase and the second term will cause the voltage sag compensation capability to decrease.

Effect of load power factor angle: As DSTATCOM is needed to supply the reactive power to the equivalent reactance, the inductive load will decrease the voltage sag compensation capability while the capacitive load will increase the voltage sag compensation capability.

Effect of DSTATCOM rating: In this case, the active and reactive power can be injected and the injected apparent power S_C is considered as the rating of DSTATCOM. The increase of the DSTATCOM rating will decrease the second term in the square root of (14), as the active power is injected. Thus, the voltage sag compensation capability will be increased.

IV. SIMULATION RESULTS

The system in Fig. 1 is used to demonstrate the voltage sag compensation performance in steady state of DSTATCOM. It is considered that the pre-sag load voltage magnitude is 1.0 pu and the system load is 1.0 pu. The voltage sag compensation capability due to the presence of additional series inductor and energy storage can be described as follows.

A. DSTATCOM without Energy Storage

The minimum energy current injection scheme is considered. For fixed equivalent reactance case, it is considered to be the same as the load capacity ($S_C = 1.0$ pu). For fixed DSTATCOM rating case, it is considered to be half of the load impedance ($x_{SR} = 0.5$). The voltage sag compensation capability as the function of equivalent reactance and DSTATCOM rating with various load power factor are shown in Fig. 5a and, 5b respectively. The compensated voltage can be increased as the equivalent reactance becomes higher where the supply side absorbs the excess reactive power during the equivalent reactance is between 0 and 0.5 pu. The maximum compensated voltage is reached at the equivalent reactance = 0.5 pu, then it decreases as the supply side starts supplying the reactive power to the equivalent reactance. For fixed equivalent reactance, the compensated voltage can be increased as the DSTATCOM rating is increased. The negative compensated voltage occurred at the small compensator capacity because the size of series inductor is too big for the compensation. As DSTATCOM is needed to supply the reactive power to the equivalent reactance, the inductive load will decrease the voltage sag compensation capability while the capacitive load will increase the voltage sag compensation capability as shown in both cases.

B. DSTATCOM with Energy Storage

The minimum apparent power current injection scheme is considered. Using the same parameters for fixed equivalent reactance case and fixed DSTATCOM rating case, the voltage sag compensation capability as the function of equivalent reactance and compensator rating with various load power factor are shown in Fig. 6a and, 6b respectively. The injected active and reactive power with various load power factor in both cases are shown in Fig. 7a and, 7b respectively. The compensated voltage can be increased as the series inductor size. As well as the DSTATCOM rating increases, the compensated voltage can be increased due to the more injected power can support the equivalent reactance and the load. As the results, the DSTATCOM with energy storage give a better compensation performance. The voltage sag compensation capability will increase with the capacitive load. However in case of fixed DSTATCOM rating, the compensated voltage decreases the equivalent reactance exceeds 0.83 pu because the supply side starts supplying the reactive power to the equivalent reactance.

V. CONCLUSIONS

The steady state performance for voltage sag compensation by DSTATCOM with series inductor and energy storage has been studied in this paper. Three current injection schemes for voltage sag compensation with different purposes are discussed. Simulation results confirmed that DSTATCOM system with additional series inductor and energy storage can improve voltage sag compensation performance. Thus, DSTATCOM will be a prominent feature in power system in mitigating power quality related problems in the near future. The appropriate design of DSTATCOM with series inductor and energy storage applied with minimized energy current injection algorithm will be the future work.

REFERENCES

[1] R.C. Dugan, M. F. McGranaghan, H. W. Beaty, *Electrical Power Systems Quality*, McGraw Hill Companies, Inc., 1996.

[2] M. F. McGranaghan, D. R. Mueller, and M. J. Samotyj, "Voltage sags in industrial systems," *IEEE Trans. on Industry Applications*, vol. 29, no.2, Mar./Apr. 1993.

[3] S.S. Choi et al., "Dynamic Voltage Restoration with Minimum Energy Injection," *IEEE Trans. on Power System*, Vol. 15, pp.51– 57. Jan. 2000.

[4] M.J. Newman, D.G. Holmes, J.G. Nielsen, F. Blaabjerg, "A dynamic voltage restorer (DVR) with selective harmonic compensation at medium voltage level," *IEEE Trans. on Industry Applications*, Vol. 41, Issue 6, pp.1744 – 1753, Nov.-Dec. 2005.

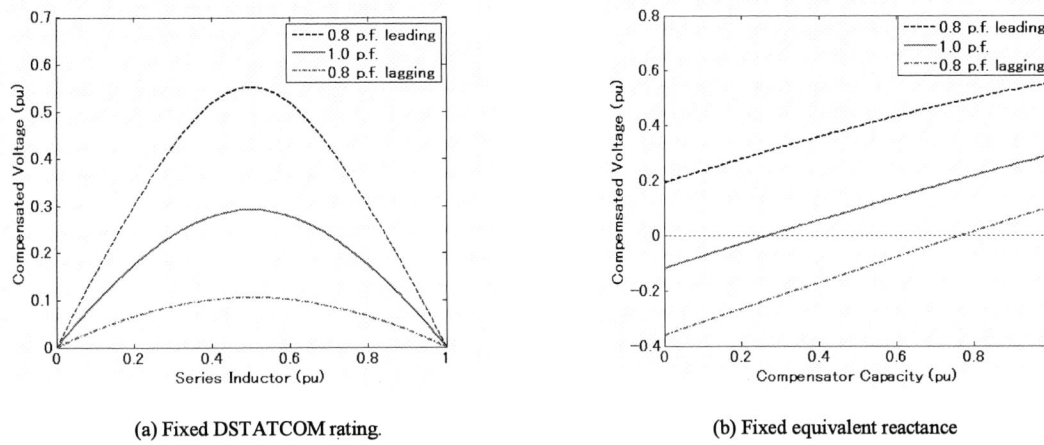

(a) Fixed DSTATCOM rating.   (b) Fixed equivalent reactance

Fig. 5. Voltage sag compensation capability of DSTATCOM without energy storage.

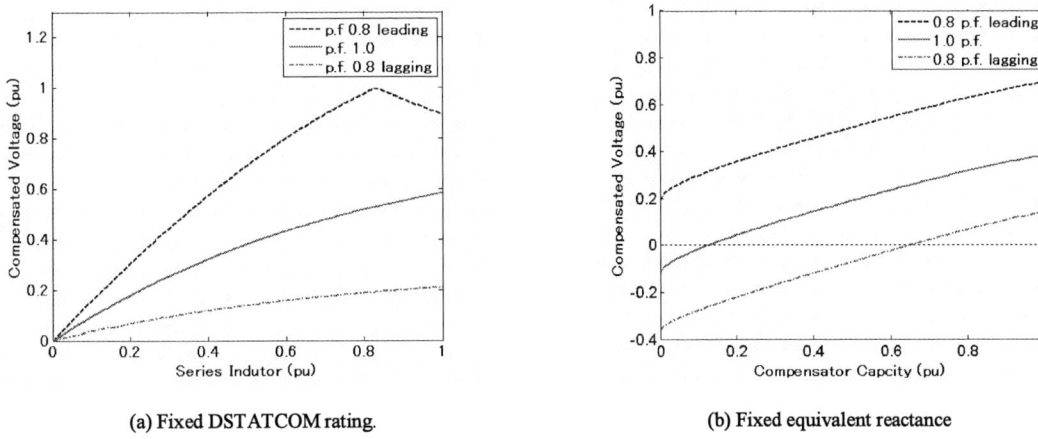

(a) Fixed DSTATCOM rating.   (b) Fixed equivalent reactance

Fig. 6. Voltage sag compensation capability of DSTATCOM with energy storage.

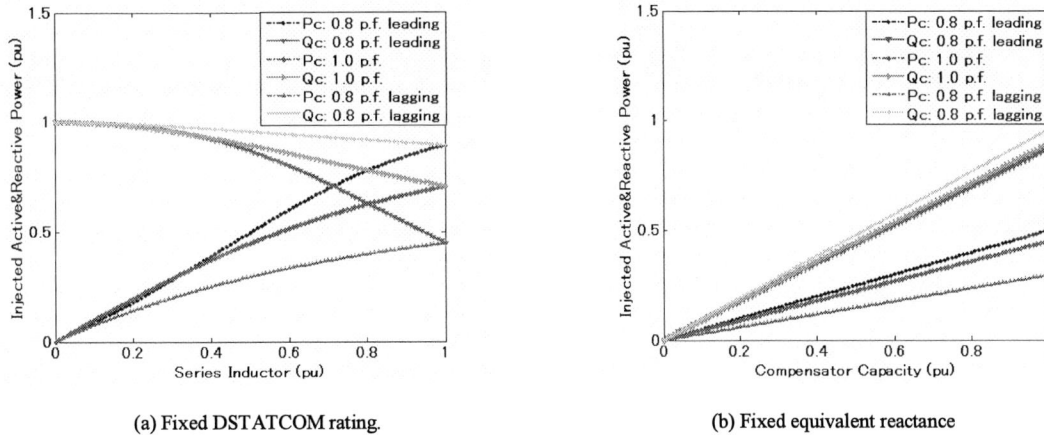

(a) Fixed DSTATCOM rating.   (b) Fixed equivalent reactance

Fig. 7. Injected active and reactive power of DSTATCOM with energy storage.

[5] A. Arulampalam et al., "Application study of a STATCOM with energy storage," *IEE Proceedings on Generation, Transmission and Distribution*, Vol. 150, Issue: 3, pp. 373 – 384, 2003.

[6] J. Xiaohua et al., "A 0.3 MJ SMES magnet of a voltage sag compensation system," *IEEE Transactions on Applied Superconductivity*, Vol. 14, Issue 2, pp. 717 – 720, 2004.

[7] M.H. Haque, "Compensation of distribution system voltage sag by DVR and D-STATCOM", *Proceeding of Power Tech Conference*, Volume 1, 2001.

Cooperative Operation of Active Power Filters by Instantaneous Complex Power Control

Elisabetta Tedeschi*, Paolo Tenti**, and Paolo Mattavelli*

*Department of Technologies and Management of Industrial Systems, University of Padova, Italy
**Department of Information Engineering, University of Padova, Italy

***Abstract.* The paper describes an approach to cooperative control of multiple active compensators acting in the same network. The goal is to take full advantage of the installed harmonic and reactive compensation capability to optimize the operation at the point of common coupling, even in presence of non-sinusoidal voltage supply and distorting loads. The cooperative operation is based on the conservation of instantaneous complex power, a quantity which envelopes power and energy information of the related electrical network. Based on the properties of instantaneous complex power, a synergistic control technique of thyristor-controlled reactive compensators and active power filters is derived, which sets the basis for cooperative operation.**

***Index Terms.* Harmonic and reactive compensation, distributed compensators, active power filters control, instantaneous complex power.**

I. INTRODUCTION

The problem of understanding the physical meaning of reactive and distortion power under non-sinusoidal conditions dates back to the first decades of 20th century [1]; in the same years the problem to compensate for reactive loads to improve the power factor was also approached for the first time [2].

Since then, several theories have been developed for the definition of reactive power under periodic non-sinusoidal conditions [3-5], each one with advantages and limitations. A fundamental decomposition of current and power terms was presented in [6].

Relevant efforts were also dedicated to the definition of instantaneous power terms, in view of dynamic compensation of reactive and distorting currents [7-12], and to the identification of solutions to improve the power factor by compensating unbalance, reactive and harmonic currents [13-16].

The use of active power filters as a general tool to compensate for unwanted current components was deeply analyzed in literature [17-19], and their control to help damping of harmonic oscillations was also investigated [20-21].

All the above subjects merge together when facing the problem of selecting and designing a cost-effective solution for harmonic and reactive compensation, based on a combination of passive filters, thyristor-controlled reactive compensators and active power filters.

In fact, cost-effective solutions normally privilege use of the cheapest compensation elements, i.e., passive resonant filters for selective harmonic elimination and low-pass filters for high-frequency harmonics limitation (*stationary compensators*). Thyristor-controlled reactors (*TCR*) and thyristor-switched capacitors (*TSC*) are also widely used, because they feature high power density, low cost and reactive power control capability (*quasi-stationary compensators*). At last, active power filters are also considered (*dynamic compensators*), which provide fast and accurate compensation capability, but are considerably more expensive than the other solutions.

It must also be considered that, usually, each compensator is designed to perform locally, either at the load terminals (*load compensation*) or at the point of common coupling (*system compensation*). If more compensators are connected to different ports of the same network (*distributed compensation*), it's generally difficult to fully exploit their compensation capability, also because detrimental interactions between grid, loads and the compensators themselves can happen, resulting in parasitic oscillations and unexpected voltage and current stresses.

Coordinated selection and cooperative operation of the various units is generally needed to implement an effective distributed compensation system. This is a challenging issue, analyzed in the literature in recent years [22], but not solved yet.

This paper presents an approach to cooperative operation of a system of distributed compensators, which results in optimum performance at the point of common coupling (*PCC*).

In particular, a dynamic control technique of active power filters is devised, which is based on complex power tracking principle and allows full and synergistic utilization of their compensation capability within a system of distributed compensators.

II. BASIC OPERATORS AND THEIR PROPERTIES

This paper makes use of a theory presented in [23] and [24], which refers to the case of non-sinusoidal periodic regime of period T and angular frequency ω .

In [23] the concept of *instantaneous complex power* was introduced and its main properties were investigated. In particular, it was shown that it is a conservative

978-1-4244-0644-9/07/$25.00 ©2007 IEEE

quantity in every electrical network, and represents an extension of the usual complex power, defined for sinusoidal operation, in the domain of non-sinusoidal variables. In addition, a voltage and current decomposition was introduced, where every term is related to a specific physical phenomenon (power absorption, energy storage, voltage and current distortion).

In [24], a generalized definition of reactive power under distorted conditions was discussed and applied to quasi-stationary harmonic and reactive compensation.

In this section, the main results of [23-24] will be summarized.

A. Homogeneous operators

For any given periodic quantity $x(t)$ of period T we define the *homogeneous integral operator* as:

$$\hat{x} = \hat{x}(t) = \omega\left(x_\int - \overline{x_\int}\right) \tag{1}$$

where: $x_\int = x_\int(t) = \int_0^t x(\tau)\,d\tau$ is time integral of x,

and $\overline{x_\int}$ is the average value of $x_\int(t)$ over period T.

Similarly, the *homogeneous differential operator* is:

$$\check{x} = \check{x}(t) = \frac{1}{\omega}\frac{dx(t)}{dt} \tag{2}$$

Note that $\hat{x}$ and $\check{x}$ are dimensionally homogeneous to basic quantity x.

B. Properties of homogeneous operators

Recall that the *internal product* of two generic periodic variables $x(t)$ and $y(t)$ is defined by:

$$\langle x, y \rangle = \frac{1}{T}\int_0^T x(t)\,y(t)\,dt \tag{3.a}$$

and correspondingly the *norm* is:

$$\|x\| = \sqrt{\langle x, x\rangle} = \sqrt{\frac{1}{T}\int_0^T x^2\,dt} = X \tag{3.b}$$

where X is the rms value of variable $x(t)$.

It is easy to verify that the homogeneous operators defined above have the following properties:

$$\langle x, \hat{x}\rangle = \langle x, \check{x}\rangle = 0 \tag{4.a}$$

$$\langle \hat{x}, \check{x}\rangle = -\|x\|^2 \tag{4.b}$$

$$x = \check{y} \quad \Leftrightarrow \quad \hat{x} = y \tag{4.c}$$

$$\langle \hat{x}, y\rangle = -\langle x, \hat{y}\rangle \tag{4.d}$$

$$\langle \check{x}, y\rangle = -\langle x, \check{y}\rangle \tag{4.e}$$

$$\langle \hat{x}, \check{y}\rangle = \langle \check{x}, \hat{y}\rangle = -\langle x, y\rangle \tag{4.f}$$

Moreover, if x and y are *sinusoidal quantities* with rms values respectively equal to X and Y and phase difference equal to φ, the following properties hold:

$$\|x\| = \|\hat{x}\| = \|\check{x}\| = X \tag{5.a}$$

$$\hat{x} + \check{x} = 0 \tag{5.b}$$

$$x^2 + \hat{x}^2 = x^2 + \check{x}^2 = x^2 - \hat{x}\check{x} = 2\|x\|^2 = 2X^2 \tag{5.c}$$

$$x\,y - \hat{x}\,\hat{y} = x\,y - \check{x}\,\check{y} = 2\,X\,Y\cos\varphi \tag{5.d}$$

$$\hat{x}\,y - x\,\hat{y} = x\,\check{y} - \check{x}\,y = 2\,X\,Y\sin\varphi \tag{5.e}$$

III. POWER TERMS UNDER DISTORTED CONDITIONS

A. Instantaneous complex power

Given voltage u and current i at a generic network port, we define the *instantaneous complex power* as:

$$\dot{s} = s_r + j\,s_i = \frac{u\,i - \hat{u}\,\check{i}}{2} + j\frac{u\,\check{i} - \check{u}\,i}{2} \tag{6}$$

where s_r and s_i are the *real* and *imaginary* power terms.

It is noticeable that *power $\dot{s}$ is conservative in every real network*. In fact, voltage terms $u, \hat{u}, \check{u}$ comply with *KLV* (Kirchhoff's Law for Voltages) and current terms $i, \hat{i}, \check{i}$ comply with *KLC* (Kirchhoff's Law for Currents), thus the conservation (Tellegen's) theorem applies.

Moreover, under *sinusoidal conditions*, owing to properties (5.d) and (5.e), $\dot{s}$ coincides, at each instant of time, with usual complex power, i.e.:

$$\dot{s}(t) = P + j\,Q \quad \forall t \in [0, T] \tag{7}$$

P and Q being the *active* and *reactive power*.

B. Active and reactive power terms under distorted conditions

Owing to property (4.f), the average value of real power s_r coincides with *active power P* even for distorted voltage and current waveforms. In fact:

$$\bar{s}_r = \frac{\langle u, i\rangle - \langle \hat{u}, \check{i}\rangle}{2} = \langle u, i\rangle = \frac{1}{T}\int_0^T u\,i\,dt = P \tag{8.a}$$

The average value of imaginary power is:

$$\bar{s}_i = \frac{\langle u, \check{i}\rangle - \langle \check{u}, i\rangle}{2} = \langle u, \check{i}\rangle = -\langle \check{u}, i\rangle \tag{8.b}$$

This latter quantity does not coincide with the *generalized reactive power Q* defined in [24]:

$$Q = \langle \hat{u}, i\rangle = -\langle u, \hat{i}\rangle \tag{9}$$

The difference is called *distortion reactive power* and is given by either one of the following expressions:

$$Q_d = \bar{s}_i - Q = \begin{cases} -\langle \check{u}, i\rangle - \langle \hat{u}, i\rangle = -\langle \hat{u} + \check{u}, i\rangle \\ \langle u, \check{i}\rangle + \langle u, \hat{i}\rangle = \langle u, \hat{i} + \check{i}\rangle \end{cases} \tag{10.a}$$

In [23, 24] it was demonstrated that generalized reactive power Q is related to the total average energy stored in the network. Instead, distortion reactive power Q_d exists only in presence of voltage distortion and current distortion. In fact, for sinusoidal voltage or current, Q_d vanishes owing to property (5.b). In conclusion we have:

$$\bar{s} = P + j(Q + Q_d) \tag{10.b}$$

C. Relations between instantaneous power terms and current at a given port

It is easily demonstrated that:

i) *If real power term s_r absorbed at a given network port is zero at any time, then the current entering that port is purely reactive*, i.e.:

$$s_r(t) \equiv 0 \quad \forall t \quad \Leftrightarrow \quad i = G\,\hat{u} \tag{11.a}$$

761

ii) *If imaginary power term s_i absorbed at a given network port is zero at any time, then the current entering that port is purely active, i.e.:*

$$s_i(t) \equiv 0 \quad \forall t \quad \Leftrightarrow \quad i = G u \qquad (11.b)$$

These relations are also called *basic theorems of compensation*. They clearly show that a purely active or reactive current absorption can be achieved, at a given port, by controlling the instantaneous power terms at that port. Since the power is conservative, the same result can also be obtained by controlling the power absorption of a compensating equipment installed everywhere in the network connected to the port.

D. Case of ohmic-inductive current

A purely ohmic-inductive current is expressed by:

$$i(t) = G u(t) + B \hat{u}(t) \qquad (12.a)$$

where, considering property (4.a):

$$\langle u, i \rangle = G \|u\|^2 \quad \Rightarrow \quad G = \frac{P}{\|u\|^2} \qquad (12.b)$$

$$\langle \hat{u}, i \rangle = B \|\hat{u}\|^2 \quad \Rightarrow \quad B = \frac{Q}{\|\hat{u}\|^2} \qquad (12.c)$$

Substituting (12.a) in (6) we find:

$$\dot{s} = (G + jB) \frac{u^2 - \hat{u}\,\breve{u}}{2} \qquad (13.a)$$

Considering property (4.f), the average value of $\dot{s}$ becomes:

$$\bar{\dot{s}} = (G + jB) \|u\|^2 \qquad (13.b)$$

and, substituting (12.b) and (12.c):

$$\left| \bar{\dot{s}} \right| = \sqrt{G^2 + B^2} \, \|u\|^2 = \sqrt{P^2 + Q^2 \frac{\|u\|^4}{\|\hat{u}\|^4}} \qquad (13.c)$$

Considering again property (4.a) we also find:

$$\|i\|^2 = G^2 \|u\|^2 + B^2 \|\hat{u}\|^2 = \frac{P^2}{\|u\|^2} + \frac{Q^2}{\|\hat{u}\|^2} \qquad (14.a)$$

The apparent power A is therefore given by:

$$A = \|u\| \|i\| = \sqrt{P^2 + Q^2 \frac{\|u\|^2}{\|\hat{u}\|^2}} \qquad (14.b)$$

Assuming now a *limited voltage distortion* (e.g., less than 5%), we have $\|u\| \approx \|\hat{u}\|$, thus (13.c) and (14.b) give:

$$\left| \bar{\dot{s}} \right| \approx \sqrt{P^2 + Q^2} \approx A \qquad (15)$$

This equation shows that the magnitude of the average complex power can be taken as an approximation of the apparent power. Accordingly, the *power factor* can be estimated as:

$$\lambda = \frac{P}{A} \approx \frac{P}{\sqrt{P^2 + Q^2}} \qquad (16)$$

IV. DISTRIBUTED COMPENSATION

A. Distributed compensation principle

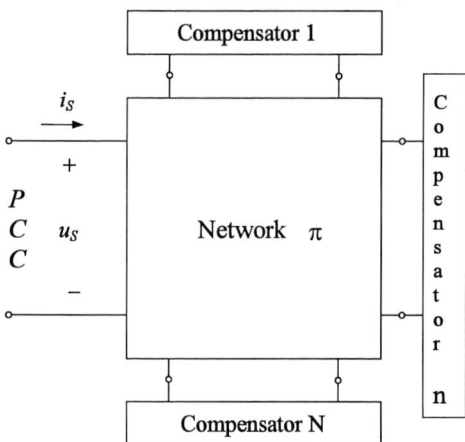

Fig. 1 – Representation of a distributed compensation system

The principle of distributed compensation is illustrated in Fig.1.

Given network π including linear, non-linear and time-varying loads and a set of distributed compensators C_1-C_N, the goal of the compensation is to keep the network behaviour at the supply port (Point of Common Coupling, *PCC*) within the given specifications in terms of power factor, total harmonic distortion, phase displacement at the fundamental frequency, etc. To this aim, the compensators must be designed and operated cooperatively.

According to the type of compensator, three compensating actions can be devised:

i) *Stationary (passive) compensation.* Is performed by passive filters and/or fixed reactors and capacitors, designed to provide a given amount of reactive and harmonic compensation under rated conditions. This solution is cheap and quite standardized, however it must be optimized for each specific load and location, taking into account that the behavior of passive elements is strongly affected by surrounding network parameters.

ii) *Quasi-stationary (reactive) compensation.* Is performed by thyristor-controlled reactors (*TCR*) and thyristor-switched capacitors (*TSC*), which can regulate the reactive power absorption by controlling their equivalent reactance. In [23, 24] it was demonstrated that reactive power Q_L absorbed by an inductor is a positive quantity proportional to its average energy:

$$Q_L = \frac{2\omega}{T} \int_0^T \frac{L i_L^2}{2} dt = \omega L \|i_L\|^2 \qquad (17.a)$$

Instead, reactive power Q_C absorbed by a capacitor is a negative quantity proportional to its average energy:

$$Q_C = -\frac{2\omega}{T} \int_0^T \frac{C u_C^2}{2} dt = -\omega C \|u_C\|^2 \qquad (17.b)$$

By properly gating the thyristors, *TCR* and *TSC* compensators can therefore provide positive or negative reactive power absorption from zero to a rated amount. Reactive power regulation is necessarily slow, since thyristors can be gated only once per line voltage cycle. As mentioned before, such compensation units are cheap; however, reactive power control is normally associated to a detrimental injection of current harmonics at the connecting point.

iii) Dynamic (active) compensation. Is performed by Active Power Filters (*APF's*), which provide fast control and can therefore compensate also for the reactive and harmonic currents absorbed by time-varying and nonlinear loads. Due to their high cost, however, such compensators are less diffused than the other ones.

B. Application problems

A successful distributed compensation approach must face two main problems:

- *Cost effectiveness*: this requires that the distributed compensation system makes use of cheap units whenever possible, while minimizing application of expensive active power filters; redundancy and overrating should be avoided and already installed compensating equipment should be fully exploited.

- *Cooperative operation*: coordinated design and synergistic control must be devised and implemented for the entire compensation system, allowing full exploitation of the installed compensation capability while avoiding detrimental interactions of the various compensating units (oscillations due to resonance, negative impedances generated by control, etc.).

Such problems are extremely complex and have not been fully analyzed yet. As a basic step towards the solution, a control approach is described hereafter, which gives the basis for cooperative operation of multiple compensators acting in the same network.

V. Cooperative Operation of Distributed Compensators

With reference to Fig.1, owing to the conservation theorem, we can write:

$$\dot{s}_S = \dot{s}_\pi + \sum_{n=1}^{N} \dot{s}_n \qquad (18)$$

where $\dot{s}_\pi$ is the instantaneous complex power absorbed by network π, $\dot{s}_n$ is that absorbed by generic compensator C_n and $\dot{s}_S$ is total power absorbed at the *PCC*.

The goal of the compensation is usually to make the network appearing, at the *PCC*, as a purely ohmic-inductive load with suitable power factor λ_S. This means that the instantaneous complex power $\dot{s}_S$ must comply with equation (13.a), where terms G and B are obtained according to (12.b) and (12.c). In (12.b) the active power term accounts for power P_π absorbed by network π (it is assumed that compensators absorb only a marginal amount of active power), while in (13.c) the reactive power term corresponds to desired power factor λ_S, i.e., according to (16):

$$P_S = P_\pi, \quad Q_S \approx P_\pi \sqrt{\frac{1}{\lambda_S^2} - 1} \qquad (19)$$

Of course, the various types of compensators perform differently:

- *Stationary compensators* have fixed parameters and provide a non-controllable amount of reactive and harmonic compensation.

- *Quasi-stationary compensators* control the absorption of reactive power in every cycle of the line voltage. They can be used to adjust reactive power Q_S absorbed at the *PCC*. The power sharing among the various reactive compensators is made according to their nature (TCR or TSC) and volt-ampere ratings.

- *Dynamic compensators* are driven to perform the remaining compensation duty, i.e., they compensate for the remaining unwanted components of instantaneous complex power absorbed at the *PCC*. The compensation duty is distributed among the various *APF* units according to their individual power rating, energy storage capability, distance from *PCC*, control bandwidth, etc.

Note that the effects of network dynamics and control delays should be carefully taken into account to avoid misoperation due to unwanted interactions and resonances between network and compensators. These aspects, although interesting, will not be treated here.

In the next section we will approach the problem of controlling an *APF* based on instantaneous complex power command.

VI. Dynamic Control of Active Power Filters Based on Instantaneous Complex Power Command

A. Definition of control problem

According to the approach described above, every *APF* receives a power reference command $\dot{s}^*$. The task of the controller is to determine the *APF* current reference i^* which provides an actual power absorption $\dot{s}$ as close as possible to $\dot{s}^*$. Recalling (2) and substituting $\dot{s}^*$ and i^* in (6), we find the relationship between power reference $\dot{s}^*$ and current reference i^*:

$$\dot{s}^* = s_r^* + j s_i^* \quad \begin{cases} s_r^* = \dfrac{1}{2}\left(u i^* - \hat{u}\,\dfrac{1}{\omega}\dfrac{di^*}{dt} \right) \\[2ex] s_i^* = \dfrac{1}{2}\left(\hat{u}\,\dfrac{1}{\omega}\dfrac{di^*}{dt} - \breve{u}\,i^* \right) \end{cases} \qquad (21)$$

This is a system of two differential equations with time-varying coefficients. It cannot be generally solved in analytical form to find current reference i^*. A solution can however be found in the discrete time domain.

B. Discrete time domain approach

Let's assume that the *APF* is pulse-width-modulated with switching frequency f_c. In every switching period T_c, we solve system (21) according to the time-averaging technique. We therefore limit our search to *average* value $\tilde{i}^*$ of current reference i^* in each switching period, i.e.:

$$\tilde{i}^* = \frac{1}{T_c} \int_0^{T_c} i^*(t)\, dt \qquad (22.a)$$

In fact, assuming that switching frequency f_c is much higher than line frequency f, term $\tilde{s}^*$ fully describes the operation of the *APF* as a compensator. The actual behavior of current i^* during a switching period is irrelevant, because every term added to $\tilde{i}^*$ only contributes to the high-frequency harmonic content of the current, which is then filtered out. Without loss of generality we can therefore assume, in each switching period, a linear behavior of reference current i^*, which makes easier determination of $\tilde{i}^*$. Let:

$$i^*(t) = i_0 + i_\Delta^* \frac{t}{T_c}, \quad t \in [0, T_c] \qquad (22.b)$$

where i_Δ^* is reference current variation along T_c and i_0 the actual (sampled) current at the beginning of the switching period, we also have:

$$\frac{di^*}{dt} = \frac{i_\Delta^*}{T_c}, \quad t \in [0, T_c] \qquad (22.c)$$

and:

$$\tilde{i}^* = i_0 + \frac{i_\Delta^*}{2} \qquad (22.d)$$

Discretization and averaging of equation (21), with the assumption that voltages $u, \hat{u}, \check{u}$ remain nearly constant in interval T_c, give:

$$\tilde{s}_r^* = \frac{1}{2\omega T_c}\left[\hat{u}_\Delta i_0 - \left(\hat{u} - \frac{\hat{u}_\Delta}{2}\right)i_\Delta^*\right] = \alpha_r i_0 + \beta_r i_\Delta^*$$

$$\tilde{s}_i^* = \frac{1}{2\omega T_c}\left[-u_\Delta i_0 - \left(u - \frac{u_\Delta}{2}\right)i_\Delta^*\right] = \alpha_i i_0 + \beta_i i_\Delta^* \qquad (23)$$

where u_Δ is the variation of supply voltage as compared to the previous switching period, and similarly for integral voltage variation $\hat{u}_\Delta$.

Coefficients α and β depend only on supply voltage behavior, while i_0 is a sampled value. The only unknown in equations (23) is therefore current variation i_Δ^*. Thus, system (23), which has two equations and one unknown, cannot be generally solved in exact form.

C. Approximate solution

An approximate solution can be found, in each switching period, by minimizing the distance between average complex power $\tilde{s}$ and reference $\tilde{s}^*$, i.e., by minimizing the function:

$$\varphi = a(\tilde{s}_r - \tilde{s}_r^*)^2 + b(\tilde{s}_i - \tilde{s}_i^*)^2 \qquad (24)$$

where a and b are suitable weighting factors. The value of i_Δ^* which minimizes φ is:

$$i_\Delta^* = \frac{a\beta_r(\tilde{s}_r^* - \alpha_r i_0) + b\beta_i(\tilde{s}_i^* - \alpha_i i_0)}{a\beta_r^2 + b\beta_i^2} \qquad (25)$$

D. Validation of solution

In order to check the validity of solution (25), which is computed for every switching cycle (*dynamic control*), a comparison has been made with an optimization technique in the frequency domain, which determines, in every line period, the amplitude and phase of the harmonic components of the *APF* current so as to minimize function φ (*optimum control*).

The comparison showed that dynamic control performs very similar to optimum control. In all validation tests, the errors introduced by dynamic control as compared to optimum control remained within a few percent. Such errors are mainly caused by the sampling delays introduced by dynamic control.

E. Control instability

The proposed dynamic control is affected by an instability which can occur during transients. In fact, the theory was developed under the assumption of periodic operation, but resulted in a control technique that looks applicable even under transient conditions. The discrepancy between periodic and transient operation is conveyed by term i_0, which does not generally coincide, under transient conditions, with stationary value i_0^*. Let $\varepsilon_0 = i_0 - i_0^*$ be such current error, an instability occurs if ε_0 propagates, from a switching cycle to the next one, with increasing amplitude. The reference value of current increase i_Δ^* is obtained by substituting i_0^* in place of i_0 in (25), which corresponds to periodic operation. Instead, under transient operation, equation (25) becomes:

$$i_\Delta^* + \varepsilon_\Delta = \frac{a\beta_r\left[\tilde{s}_r^* - \alpha_r(i_0^* + \varepsilon_0)\right] + b\beta_i\left[\tilde{s}_i^* - \alpha_i(i_0^* + \varepsilon_0)\right]}{a\beta_r^2 + b\beta_i^2}$$

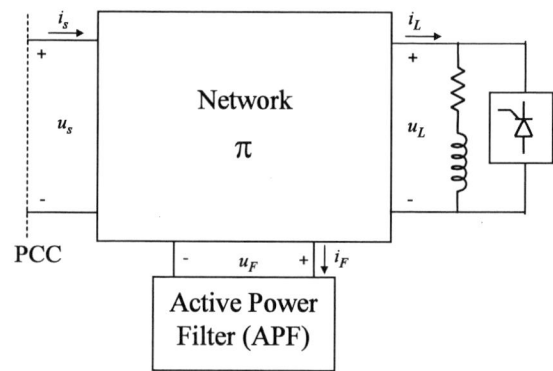

Fig. 2 – Application example

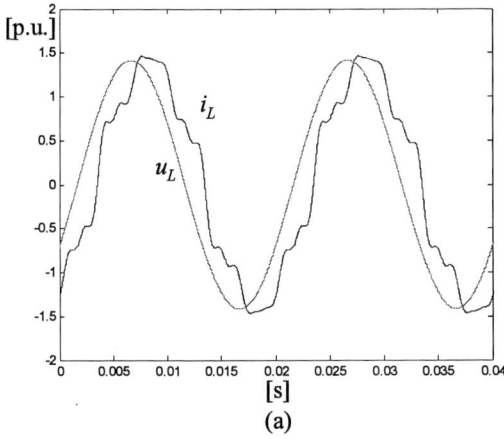
(a)

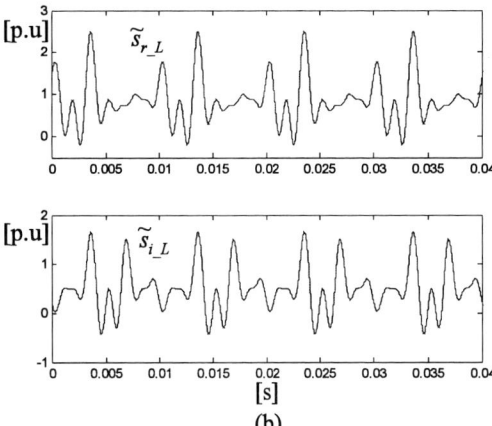
(b)

Fig. 3 – Load waveforms: voltage and current (a) and real and imaginary power terms (b)

Removing the stationary terms we get:

$$\varepsilon_\Delta = -\frac{a\,\alpha_r\,\beta_r + b\,\alpha_i\,\beta_i}{a\beta_r^2 + b\beta_i^2}\varepsilon_0 \qquad (26)$$

which expresses the increase of error term along the switching period. The condition for stability is that $(\varepsilon_0 + \varepsilon_\Delta)$ is less than ε_0, thus:

$$\left|\frac{\varepsilon_0 + \varepsilon_\Delta}{\varepsilon_0}\right| < 1 \quad \Rightarrow \quad 0 < \frac{a\,\alpha_r\,\beta_r + b\,\alpha_i\,\beta_i}{a\beta_r^2 + b\beta_i^2} < 2 \quad (27)$$

This inequality must be satisfied in each switching cycle. For this purpose, weights a and b can be dynamically selected to avoid instability; this provision, however, can cause some non-uniformity in real and imaginary power tracking along the line period.

VII. APPLICATION EXAMPLE

The network shown in Fig. 2 was considered, including a distorting load, made up of an ohmic-inductive load and a thyristor rectifier, an *APF* and a distribution network, which includes transformers causing a phase shift between the voltages appearing at the *PCC* (0°), the load terminals (-30°), and the *APF* terminals (30°). For the sake of simplicity, it is assumed that the distribution network absorbs a negligible amount of instantaneous power.

Load voltage and current waveforms are given in Fig.3.a, while 3.b shows the corresponding instantaneous power terms.

Fig.4 shows the power references of the *APF* (thin line) and the actual power absorption (thick line), assuming that unity power factor is required at the *PCC* and that weighting factors a and b are dynamically selected to ensure stability while privileging tracking of imaginary power reference s_i^*. This complies with the second basic theorem of compensation, which ensures that if the imaginary power is identical to zero at the *PCC*, then the power factor is unity. This choice is clearly reflected in Fig.4, which shows a tracking of the imaginary power term which is better than that of the real power term.

Finally, Fig.5 shows the total power terms delivered at

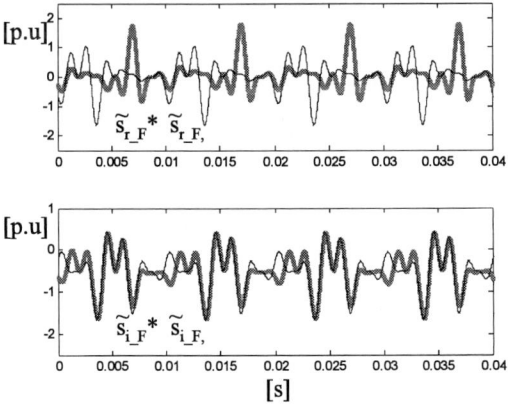

Fig. 4 – Real and imaginary power references and actual power absorption of the Active Power Filter

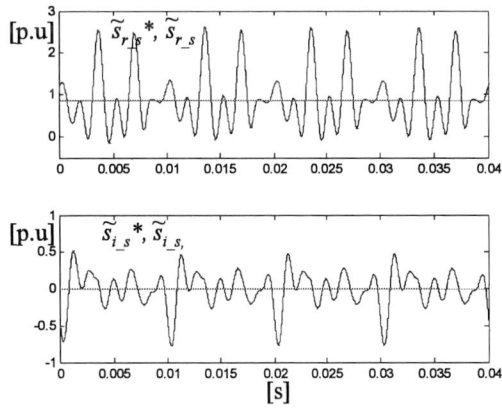

Fig. 4 – Real and imaginary power delivered at PCC compared with power references

the *PCC* together with the corresponding constant references.

While the load power factor is 0.858, the power factor at the *PCC* according to (16) results unity (0.99994). As expected, the tracking error is high for the real power term (rms value = 1.24 p.u.), and low for the imaginary power term (rms value = 0.36 p.u.).

It is apparent that the proposed dynamic control technique results in a good performance at the *PCC*.

VIII. CONCLUSIONS

A general approach to the problem to select, design and operate various kinds of compensators distributed in the same network was presented. The problem is complex, and the approach lies on a general theory related to the definition of power and current terms under non-sinusoidal conditions.

In particular, reference was made to the instantaneous complex power, a conservative quantity which extends the properties of the complex power to the non-sinusoidal domain. It provides the basis for cooperative operation of a system of quasi-stationary and dynamic compensators.

Control of active power filters based on instantaneous complex power tracking was then discussed, showing that the power absorption at the point of common coupling can be effectively controlled by acting on compensators connected elsewhere in the network.

Some application examples were developed, which demonstrate the capabilities of the proposed approach.

REFERENCES

[1] C.I.Budeanu: Puissances reactives et fictives, Institute Romain de l.Energie, Bucharest, 1927.

[2] S.Fryze: Wirk-, Blind-, Scheinleistung in Elektrische Stromkreisen mit nichtsinusformingen Verlauf von Strom und Spannung, ETZ, Bd. *53*, pp.596-599, 625-627, 700-702, 1932.

[3] W.Shepherd, P.Zakikhani: Suggested definition of reactive power for nonsinusoidal systems. Proc. Inst. Elec. Eng., vol. 119, pp.1361-1362, Sept. 1972.

[4] N.L.Kuster, W.J.M.Moore: On the Definition of Reactive Power under Non-Sinusoidal Condition. IEEE Trans. on Power Apparatus and Systems, PAS-99 (1980), pp.1845-1854.

[5] C.H.Page: Reactive Power in Non-Sinusoidal Situations. IEEE Trans. of Instrumentation and Measurement, Vol IM-29, No.4, Dec 1990, pp 420-423.

[6] L.S.Czarnecki.: Orthogonal Decomposition of the Currents in a 3-Phase Nonlinear Asymmetrical Circuit with a Nonsinusoidal Voltage Source. IEEE Trans. on Instrumentation and Measurements, vol.IM-37, n.1, pp.30-34, March 1988.

[7] H.Akagi, Y.Kanazawa, A.Nabae: Generalized theory of the instantaneous reactive power in three-phase circuits, Proc. of the Int. Power Electron. Conf., (JIEE IPEC) Tokyo/Japan, 1983, pp. 1375-1386.

[8] H.Akagi, Y.Kanazawa, A.Nabae: Instantaneous reactive power compensators comprising switching devices without energy storage components, IEEE Trans. Ind. Appl., IA-20, 1984, No. 3, pp. 625-630.

[9] H.Akagi, A.Nabae: The p-q theory in three-phase systems under non-sinusoidal conditions, European Trans. on Electric Power, ETEP, Vol. 3, No. 1, January/February 1993, pp. 27-31.

[10] J.L.Willems: Mathematical foundations of the instantaneous power concept: a geometrical approach, European Trans. on Electrical Power, ETEP, Vol. 6, No. 5, Sept./Oct.1996, pp. 299-304.

[11] L.Cristaldi, A.Ferrero: Mathematical foundations of the instantaneous power concept: an algebraic approach, European Trans. on Electrical Power, ETEP, Vol. 6, No. 5, Sept./Oct. 1996, pp. 305-309.

[12] L.Rossetto, P.Tenti: Evaluation of instantaneous power terms in multi-phase systems: techniques and application to power-conditioning equipment, European Trans. on Electrical Power, ETEP, Vol. 4, No. 6, Nov./Dec. 1994, pp. 469-475.

[13] L.S.Czarnecki: Reactive and unbalanced currents compensation in three-phase asymmetrical circuits under non-sinusoidal conditions. IEEE Trans. on Instrumentation and Measurements, vol.IM-38, 1989.

[14] L.S.Czarnecki, L.S.: Power factor improvement of three-phase unbalanced loads with nonsinusoidal supply voltages, European Trans. on Electrical Power Engineering (ETEP), Vol. 3, No. 1, Jan./Feb. 1993, pp. 67-72.

[15] J.L.Willems: Power factor correction for distorted bus voltages. Electr. Mach. a. Power Syst. 13 (1987), pp.207-218.

[16] S.J.Merhej, W.H.Nichols, "Harmonic Filtering for the Offshore Industry", *IEEE Trans. on Industry Applications*, Vol. 30, No. 3, May/June,1994, pp. 533-542.

[17] H.Akagi, A.Nabae, "Control Strategy of Active Power Filters Using Multiple Voltage Source PWM Converters", *IEEE Trans. on Ind. App.*, Vol.IA-22, no. 3, May/June 1986, pp.460-465.

[18] F.Z.Peng "Application Issues of Active Power Filters", *IEEE Industry Applications Magazine*, vol. 4, no. 5, September/October 1998, pp. 21-30.

[19] F.Z.Peng, H.Akagi, A.Nabae "A New Approach to harmonic Compensation in Power Systems - A Combined system of shunt passive and series active filters*", IEEE Trans on Industry Applications*, vol.26, no.6, 1990, pp. 983-990.

[20] H.Akagi, "Control strategy and site selection of a shunt active filter for damping of harmonic propagation in power distribution systems," *IEEE Trans. Power Delivery*, vol. 12, pp. 354–363, Jan. 1997.

[21] H.Akagi, H.Fujita, K.Wada, "A shunt active filter based on voltage detection for harmonic termination of a radial power distribution line", *IEEE Trans on Industry Applications.*, vol. 35, pp. 638–645, May/June 1999.

[22] P.Jintakosonwit, H.Fujita, H.Akagi, S.Ogasawara: "Implementation and Performance of Cooperative Control of Shunt Active Filters for Harmonic Damping Throughout a Power Distribution System" *IEEE Trans on Industry Applications.*, vol. 39,no.2, pp. 556–563, March/April 2003.

[23] P.Tenti, P.Mattavelli: "A Time-Domain Approach to Power Term Definitions under Non-Sinusoidal Conditions", *L'Energia Elettrica*, vol. 81, pp.75.84, 2004.

[24] P.Tenti, E.Tedeschi, P.Mattavelli, "Compensation Techniques based on Reactive Power Conservation", Seventh International Workshop on Power Definition and Measurements under Nonsinusoidal Conditions, Cagliari (Italy), July 2006.

Impact of Adjustable Speed PWM drives on Operation and Harmonic Losses of Nonlinear Three Phase Transformers

M.A.S. Masoum, *Senior Member, IEEE,* **Paul S. Moses**
Department of Electrical and Computer Engineering
Curtin University of Technology, WA, Australia

Amir S. Masoum, *Student Member, IEEE*
Department of Electrical Engineering
University of Western Australia, WA, Australia

Abstract— Impact of harmonics generated by adjustable speed PWM drive systems on the losses and power quality of three-phase nonlinear transformers is investigated. The combined effects of transformer magnetic core nonlinearity, non-sinusoidal input excitation, asymmetric operation, as well as harmonics injected by the PWM action of the drive are included and analyzed. A time domain nonlinear model for three-phase three-leg transformers is implemented and used to compute non-sinusoidal input and output waveforms in the presence of PWM drives. Additional power losses due to harmonics generated by the iron core, non-sinusoidal (input) excitation and nonlinearity of drive system are computed.

Index Terms— Adjustable speed drive, PWM, transformer, power quality, harmonics losses, asymmetric operation.

I. INTRODUCTION

PROPAGATION and generation of voltage and current harmonics in three phase transformers has received considerable attention in literature. Much is known about transformer additional losses due to harmonics and the fact that they are symptomatic of nonsinusoidal operation, leading to thermal damage in the insulation, iron core and windings [1-18]. However, past analyses have been mainly limited to balanced and symmetric operation of transformers with approximate modeling of nonlinear loads (e.g., ac drives).

The majority of transformer harmonic models to date assume perfectly symmetrical conditions; that is, the voltage and current harmonic magnitudes in each phase are identical. This is seldom true as asymmetry in the iron core and existing imbalances in the power system do permeate transformers. The complete extent of symptoms from unbalanced, asymmetric and nonsinusoidal operation is not so apparent in literature and limited documents are available to investigate such nonlinearities [8-14].

A significant step forward was the development of the nonlinear three phase magnetic circuit model. It was first derived by Fuchs et al. [9, 10] and later modified by Pedra et al. [15] to include the asymmetric nonlinear core reluctances. Clua et al. [8] investigated unbalanced harmonic power flow in three phase transformers by using admittance matrices and sequence component equivalent models. The impact of unbalanced nonlinear loads on voltage harmonics on system buses was simulated. However, this model neglected core nonlinearity, assumed core limbs' behavior was symmetrical and no harmonic interaction occurred between the network and nonlinear loads.

The impact of symmetrical and unsymmetrical voltage sags on three phase transformers has been carefully studied by

Pedra et al. [14]. The transformer model is similar to the one implemented in this paper, which includes asymmetric nonlinear core branches (and PWM switching of the ac drives). Preceding this work, Medina et al. [12, 13] derived a Norton harmonic representation of three phase transformers. This frequency domain technique pioneered modeling together with asymmetrical core magnetization, harmonic cross coupling in the limbs and harmonic power flow.

Drive systems are nonlinear devices with non-sinusoidal waveforms that inject low order harmonic currents into transformers and distribution system. Electric utilities are very concern about the fast growth of large electric drives in the industrial sectors of power system and their impacts on losses and life time of distribution transformers. Variable frequency and PWM drives are considered as one of the biggest contributors to power quality problems due to their high-power ratings [19-21]. The nonlinear v-i characteristics of drive systems may result in even and triplen harmonic currents, conductor losses at harmonic frequencies, transformer saturation/overheating, power-factor and shunt capacitor failures, unsatisfactory performance of protection devices, harmonic and sub-harmonic torques in electrical machines, etc. Impact of harmonics generated by ac drives on power system devices, transformers, loads and equipments deserves more attention and requires detailed nonlinear modelling to suppress or prevent their generation.

In this paper, a nonlinear transformer model [15] is modified to include nonsinusoidal excitation, (non)linear loads transformer core nonlinearity, core asymmetry and harmonic cross coupling effects in the legs. The developed model is used to investigate the effects of linear loads and nonlinear PWM-operated drive systems on the operation, losses and power quality of three-phase transformers under symmetric and asymmetric operating conditions.

II. NONLINEAR TRANSFORMER MODEL

The proposed transformer model for unsymmetrical, imbalanced and nonsinusoidal operation is derived from reference [15]. It is based on the simultaneous solution of electric and magnetic equivalent circuits of three phase three leg transformers. The nodal equations of the circuits are solved in time domain using iterative techniques such as Newton-Raphson. The electric circuit governs the electrical connections of the source, load and the transformer itself (e.g., star or delta configuration). For three phase transformers, the magnetic circuit is necessary to represent the multiple flux paths, reluctances and magnetomotive forces within multi-legged iron core constructions [18]. It also models the

978-1-4244-0644-9/07/$25.00 ©2007 IEEE

asymmetric magnetizing behavior of the core.

A. Electric Equivalent Circuit

The electric circuit of Figure 1 can be simulated by most software packages such as PSIM or PSPICE. Typically a nodal matrix of equations describing the circuit is formed and solved through iterative time domain numerical techniques such as the Newton-Raphson algorithm.

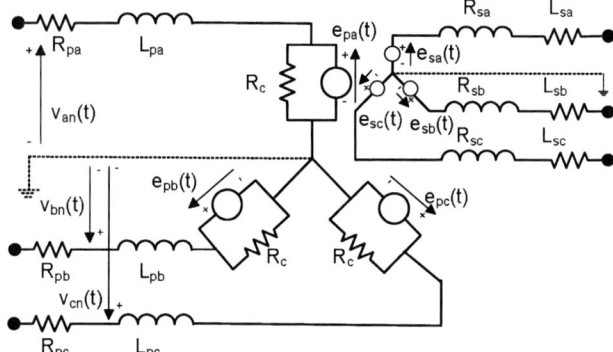

Figure 1: Electric equivalent circuit of three-phase transformers

The induced primary and secondary voltages are modeled as voltage sources controlled by the time derivative of the magnetic fluxes (Faraday's Law). This establishes a link between the magnetic and electric circuits.

In this paper, the electric circuit parameters, winding resistances, core loss resistances and leakage inductances are estimated from three phase open circuit and short circuit tests.

B. Magnetic Equivalent Circuit

The circuit in Figure 2 is an approximation of the equivalent circuit proposed by [9]. In this model, the seven reluctances of the core are reduced to three reluctance parameters which can be easily measured. This circuit can be programmed into PSPICE as an electrical circuit using the magnetic-electric duality principle.

The nonlinear reluctances in Figure 2 are implemented in PSPICE as flux sources dependent on their own MMF drops;

$$\Phi_k\{f_k\} = f_k \; \Re_k\{f_k\}^{-1} \qquad (1)$$

F_a, F_b and F_c are the MMFs developed in the limbs by electric circuit currents in the primary and secondary windings. This is another link between the two equivalent circuits.

The nonlinear reluctance functions can be fitted to each leg's magnetizing characteristic. The following function was proposed by Pedra et al. [15],

$$\Re\{f\}^{-1} = K_1 \Bigg/ \left[1 + \left(\frac{|f|^p}{f_0} \right)^{1/p} \right] + K_2 \qquad (2)$$

where the empirically determined parameters K_1, K_2, p and f_0 are constants that shape the function to any measured saturation curve $(\Phi - f)$. K_1 and K_2 are associated with the slope of the linear and nonlinear regions, respectively, p influences the smoothness of the knee region and f_0 defines where saturation starts.

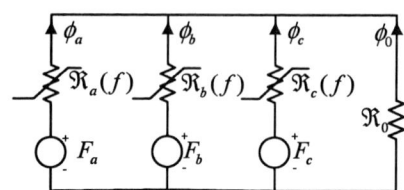

Figure 2: The approximate magnetic equivalent circuit of three-phase three leg transformers.

III. MODELING OF AC DRIVES

The literature has many documents on classification, modeling and analyses of nonlinear drive systems [19-21]. Modern electric drives include ac and dc types and utilize rectifier circuits and PWM switching technology to achieve better performance at lower cost and higher efficiency.

DC drives employ controlled rectifiers to realize variable dc voltages while ac drives usually have PWM inverters with variable voltage and variable frequency technology. In ac drives, a dc capacitor is normally applied between the rectifier and the PWM inverter to limit the low-ripple dc voltage. Many industrial loads use controlled ac drives with voltage-source inverter (VSI) that operate based on PWM switching. However, the dc capacitor magnifies line harmonics and may cause power quality problems. Harmonics distortions of ac drives become more severe under light load conditions.

A counterpart to the PWM-VSI electric drive is the current-source inverter (CSI)-based PWM-operated drive system; the dc bus includes a dc inductor (with no dc capacitor) and the load side is connected to a three-phase ac capacitor in parallel with the motor [21]. With this improved configuration, the capacitor and dc inductor act as a filter unit while the ac inductor serves as an energy storage element.

For analysis and modelling of large power systems with large number and high penetration of ac and dc drive systems, imprecise and approximate models are adequate. This is mainly due to memory storage and convergence problems associated with of harmonic power flow algorithms. However, for detailed harmonic analysis and power quality analysis, as performed in this paper, accurate modelling procedures and tools are required.

In this paper a voltage source inverter PWM drive system is used as the nonlinear load at the secondary terminals of a nonlinear three-phase transformer. The PSPICE source code of the nonlinear transformer model (devolved in the pervious section) is combined with AC-DC rectifier and DC-AC PWM Inverter. The PSPICE model for PWM inverter circuits [25] is slightly modified and used. The final result is a variable frequency VSI-PWM-based induction motor drive fed from the nonlinear transformer model (implemented in PSPICE source code), as shown in Fig.3.

768

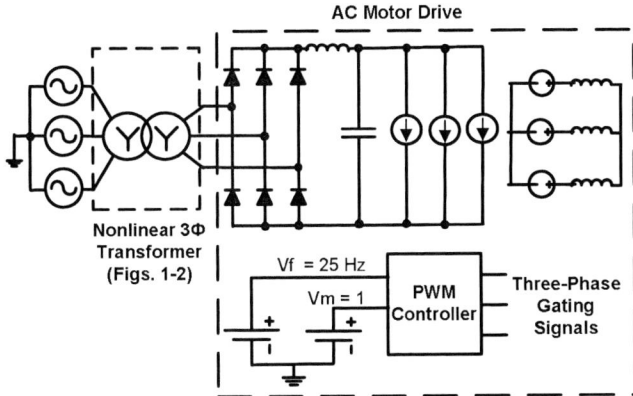

Figure 3: VSI-PWM-based induction motor drive fed from the three-phase nonlinear transformer (Figs. 1-2)

IV. SIMULATION RESULTS

Three sources of nonlinearity and harmonic generation in three-phase transformers will be considered; nonlinearity of the magnetic core (e.g., internally generated nonsinusoidal magnetizing current), nonlinear loading (e.g., harmonics currents injected by the VSI-PWM-based induction motor drive), and non-sinusoidal excitation (e.g., input harmonic voltages). To investigate the impacts of these sources on the performance and losses of three-phase transformers, nine cases are considered. Due to the page limitations, the waveforms are only presented for selected cases.

Cases 1-3: Sinusoidal Excitation with Linear Loads

With rated sinusoidal input voltages, four types of wye-connected loads (R, series RL and parallel RC) are simulated (Table 1). For these cases, the only sources of harmonics are the nonlinearities associated with transformer magnetic core as shown by the nonsinusoidal magnetizing current of Fig. 4 for rated resistive load with sinusoidal excitation.

Case 4: Sinusoidal Excitation with a Nonlinear Load

Most harmonics in power transformers are due to nonlinear loads at the output terminals. To demonstrate this phenomenon, a three phase diode bridge rectifier (with rated resistive load on the DC side) is placed at the output terminals of the transformer. The load is selected such that rated fundamental output kVA is delivered. Simulation results are shown in Fig. 5. The nonsinusoidal load current has deteriorated the power quality of the output current (THD$_i$=24.2%) and output voltage (THD$_v$=5.2%).

Cases 5-7: Nonsinusoidal Excitation with Linear Loads

Transformer is excited by nonsinusoidal voltages with rated fundamental component and different orders and magnitudes of harmonic components. Simulation were performed for no-load, linear (R, RL and RC) and nonlinear load conditions (Tables 1-2). The RC load combined with 20% of 5th harmonics results in current waveforms oscillating and peaking at great then 2 times the rated value. This is due to circuit resonance with the 5th harmonic causing the impedance at the frequency (250 Hz) to cancel and large currents to flow.

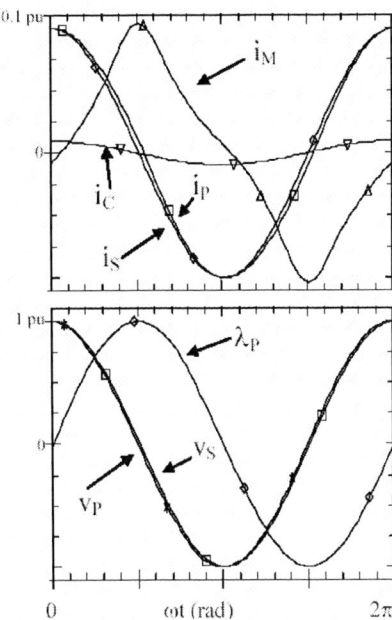

Figure 4: Case 1- transformer operation with rated sinusoidal excitation and rated wye connected resistive load. Full load magnetizing (i$_M$) and core-loss (i$_C$) currents, primary (i$_P$) and secondary (i$_S$) currents, primary flux linkages (λ$_P$), primary (V$_P$) and secondary (V$_s$) voltages are shown in per unit of rated values.

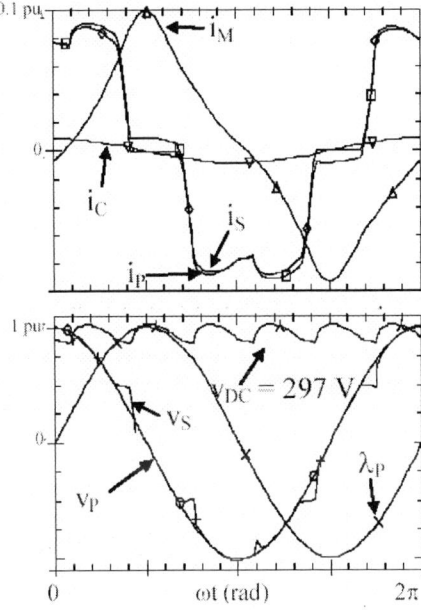

Figure 5: Case 4- transformer operation with rated sinusoidal excitation and rated kVA delivered to a three phase diode bridge rectifier with DC side resistor. Full load magnetizing (i$_M$) and core-loss (i$_C$) currents, primary (i$_P$) and secondary (i$_S$) currents, primary flux linkages (λ$_P$), primary (V$_P$) and secondary (V$_s$) voltages are shown in per unit of rated values.

TABLE 1: SUMMARY OF SIMULATION RESULTS FOR LOAD CASE STUDIES 1-8

Case	V_{IN}	Load	KVA^{LOAD} [%] pu	PF	Transformer Secondary Current					Transformer Secondary Voltage				
					harmonic order h					harmonic order h				
					$h=1$	$h=3$	$h=5$	$h=7$	THD_I %	$h=1$	$h=3$	$h=5$	$h=7$	THD_V %
1	Rated freq.	R	~100	1	99.4	0.09	0.03	0.01	0.10	99.4	0.09	0.03	0.01	0.10
2		R-L	~100	0.8 lag	100.0	0.05	0.01	0.00	0.05	100.0	0.10	0.03	0.02	0.10
3		R||C	~100	0.8 lead	100.1	0.21	0.23	0.13	0.34	100.0	0.11	0.07	0.03	0.13
4		NL*	~100	-	100.3	0.06	22.4	9.3	24.2	99.5	0.27	4.4	2.6	5.2
5	|V^5|=20%	R	~100	1	99.4	0.13	19.2	0.08	19.3	99.4	0.13	19.2	0.08	19.3
6		R-L	~100	0.8 lag	100.0	0.06	6.1	0.02	6.1	100.0	0.13	19.1	0.09	19.0
7		R||C	~100	0.8 lead	100.1	0.29	140.1	0.79	139.9	100.1	0.18	44.8	0.13	44.7
8		NL*	~100	-	100.1	0.53	21.1	8.7	22.9	99.0	0.25	23.5	2.5	23.8

*) NL = Nonlinear load consisting of a three phase diode bridge rectifier with rated resistive load, connected to load bus (transformer secondary).

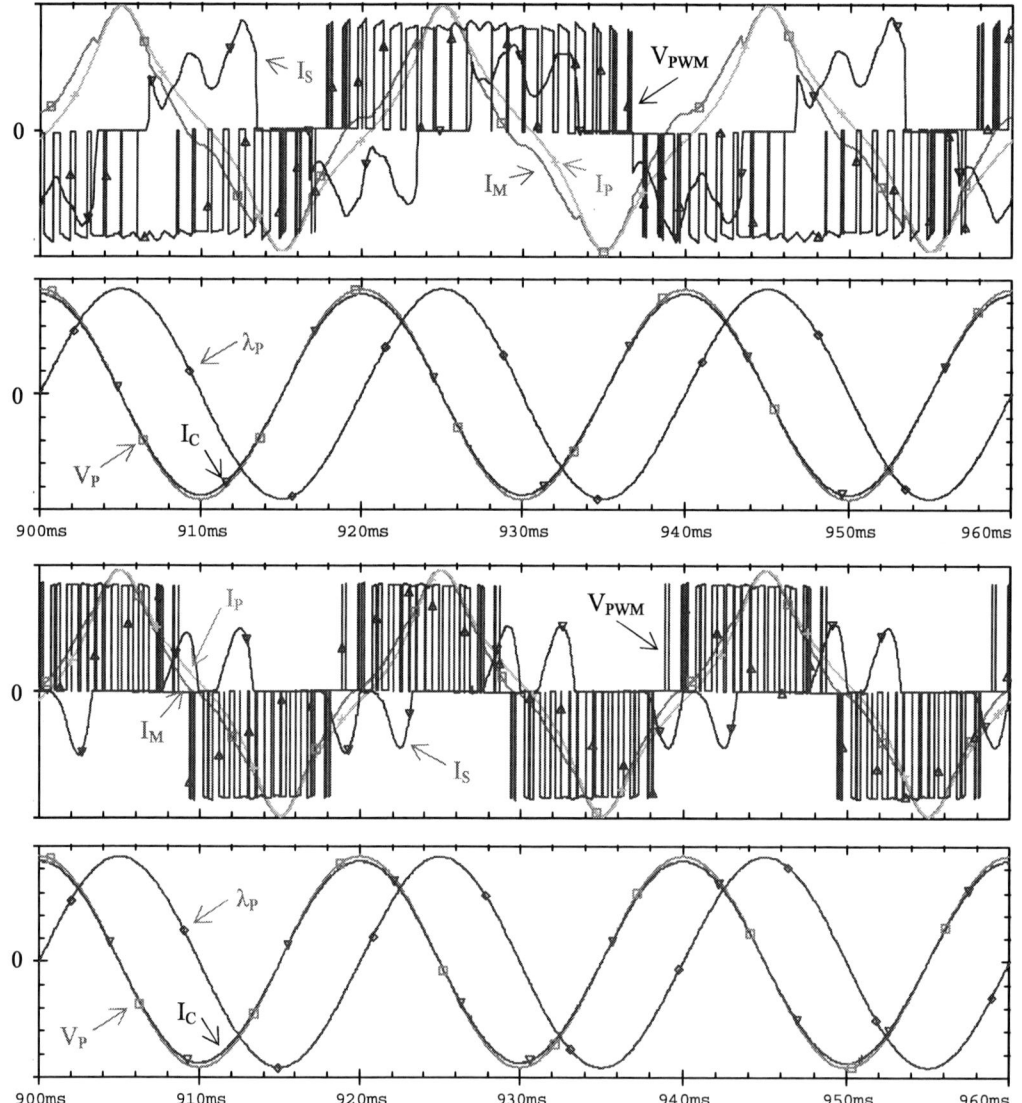

Figure 6: Case 9- Transformer operation with rated sinusoidal excitation and three phase VSI-PWM-based induction motor drive with PWM frequency of 25 Hz (top two waveforms) and 50 Hz (bottom two waveforms). Full load magnetizing (i_M) and core-loss (i_C) currents, primary (i_P) and secondary (i_S) currents, primary flux linkages (λ_P), primary (V_P) and secondary (V_s) voltages are shown in per unit of rated values.

Case 8: Nonsinusoidal Excitation with a Nonlinear Load

Transformer is connected to a three phase diode bridge rectifier (with rated resistive load on the DC side) and is excited by nonsinusoidal voltages with rated fundamental component and 20% of 5th harmonic at a phase shift of zero degree (tables 1-2).

Case 9: Sinusoidal Excitation with VSI-PWM-Based induction Motor Drive Load

Transformer is connected to a three phase VSI-PWM-based induction motor drive and is excited by rated sinusoidal voltages. Simulation results are show in Fig. 6 and Table 2.

TABLE 2: SUMMARY OF SIMULATION RESULTS FOR LOAD CASE

Case	Load	$P_{copper\ loss}$ [W]	$P_{core\ loss}$ [W]	$P_{total\ loss}$ [W]	ΔP_{LOSS}* [%]
1	R	180	518	698	0
2	R-L	189	536	724	3.8
3	R‖C	176	511	687	-1.5
5	R	187	537	724	3.8
6	R-L	190	556	746	6.9
7	R‖C	527	563	1090	56.3**
9	PWM	295	520	815	17.0

*) Increased in total transformer losses, compared with rated losses.
**) High losses due to resonance (between C_{load} and $L_{transformer}$).

V. CONCLUSION

Impact of harmonics generated by adjustable speed PWM drive systems on the losses and power quality of three-phase nonlinear transformers is investigated using a nonlinear transformer model that has been adapted to compute input and output voltage and current waveforms and total (fundamental and harmonic) losses under linear and nonlinear load conditions. The combined effects of transformer magnetic core nonlinearity, non-sinusoidal input excitation, asymmetric operation, as well as harmonics injected by the PWM action of the drive are included and analyzed. Additional power losses due to harmonics generated by the iron core, non-sinusoidal (input) excitation and nonlinearity of drive system are computed. Simulation results indicate that ac drive systems may cause considerable increase in transformer lasses due to high percentage of current harmonic injections.

VI. REFERENCES

[1] E.F. Fuchs, D. Lin and J. Martynaitis, "Measurement of three-phase transformer derating and reactive power demand under nonlinear loading conditions", *IEEE Transactions on Power Delivery*, vol. 21, pp. 665-672, 2006.
[2] D. Yildirim and E.F. Fuchs, "Measured transformer derating and comparison with harmonic loss factor approach", *IEEE Transactions on Power Delivery*, vol. 15, pp. 186-191, 2000.
[3] E.F. Fuchs, D. Yildirim and W.M. Grady, "Measurement of eddy-current loss coefficient PEC-R, derating of single-phase transformers, and comparison with K-factor approach", *IEEE Transactions on Power Delivery*, vol. 15, pp. 148-154, 2000.

[4] L.W. Pierce, "Transformer design and application considerations for nonsinusoidal load currents", *IEEE Transactions on Industry Applications*, vol. 32, pp. 633-645, 1996.
[5] M.A.S. Masoum, E.F. Fuchs and D. J. Roesler, "Impact of nonlinear loads on anisotropic transformers," *IEEE Transactions on Power Delivery*, vol. 6, pp. 1781-1788, 1991.
[6] M.A.S. Masoum, E.F. Fuchs and D. J. Roesler, "Large signal nonlinear model of anisotropic transformers for nonsinusoidal operation; part II: magnetizing and core-loss currents", *IEEE Transactions on Power Delivery*, vol. 6, pp. 1509-1516, 1991.
[7] E.F. Fuchs, M.A.S. Masoum and D. J. Roesler, "Large signal nonlinear model of anisotropic transformers for nonsinusoidal operation: part I: lambda-i characteristics", *IEEE Transactions on Power Delivery*, vol. 6, pp. 1874-1886, 1991.
[8] J. Clua, L. Sainz and F. Corcoles, "Three-phase transformer modeling for unbalanced harmonic power flow studies", Ninth International Conference on Harmonics and Quality of Power, 2000.
[9] E.F. Fuchs and Y. Yiming, "Measurement of lambda-i characteristics of asymmetric three-phase transformers and their applications", *IEEE Transactions on Power Delivery*, vol. 17, pp. 983-990, 2002.
[10] E.F. Fuchs, Y. You and D.J. Roesler, "Modeling and simulation, and their validation of three-phase transformers with three legs under DC bias", *IEEE Transactions on Power Delivery*, vol. 14, pp. 443-449, 1999.
[11] M.A.S. Masoum and E. F. Fuchs, "Transformer magnetizing current and iron-core losses in harmonic power flow", *IEEE Transactions on Power Delivery*, vol. 9, pp. 10-20, 1994.
[12] A. Medina and J. Arrillaga, "Generalised modelling of power transformers in the harmonic domain", *IEEE Transactions on Power Delivery*, vol. 7, pp. 1458-1465, 1992.
[13] A. Medina and J. Arrillaga, "Simulation of multi-limb power transformers in the harmonic domain", *IEE Proceedings on Generation, Transmission and Distribution, IEE Proceedings C*, vol. 139, pp. 269-276, 1992.
[14] J. Pedra, L. Sainz, F. Corcoles and L. Guasch, "Symmetrical and unsymmetrical voltage sag effects on three-phase transformers", *IEEE Transactions on Power Delivery*, vol. 20, pp. 1683-1691, 2005.
[15] J. Pedra, L. Sainz, F. Corcoles, R. Lopez and M. Salichs, "PSPICE computer model of a nonlinear three-phase three-legged transformer", *IEEE Transactions on Power Delivery*, vol. 19, pp. 200-207, 2004.
[16] "IEEE recommended practice for establishing transformer capability when supplying nonsinusoidal load currents", *ANSI/IEEE Std C57.110-1986*, 1988.
[17] "IEEE recommended practice for establishing transformer capability when supplying nonsinusoidal load currents", *IEEE Std C57.110-1998*, 1998.
[18] G. Chang, C. Hatziadoniu, W. Xu, P. Ribeiro, R. Burch, W. M. Grady, M. Halpin, Y. Liu, S. Ranade, D. Ruthman, N. Watson, T. Ortmeyer, J. Wikston, A. Medina, A. Testa, R. Gardinier, V. Dinavahi, F. Acram and P. Lehn, "Modeling devices with nonlinear voltage-current characteristics for harmonic studies", *IEEE Transactions on Power Delivery*, vol. 19, pp. 1802-1811, 2004.
[19] M. Villablanca, W. Flores, C. Cuevas and P. Armijo, "Harmonic reduction in adjustable-speed synchronous motors", *IEEE Transactions on Energy Conversion*, vol. 16, no. 3, pp. 239-245, 2001.
[20] J. Faiz, H. Barati and E. Akpinar, "Harmonic analysis and performance improvement of slip energy recovery induction motor drives", *IEEE Transactions on Power Electronics*, vol. 16, no. 3, pp. 410-417, 2001.
[21] Y. Yin and A.Y. Wu, "A low-harmonic electric drive system based on current-source inverter", *IEEE Transactions on Power Industry Applications*, vol. 34, no. 1, pp. 227-235, 1998.

[22] "IEEE Recommended Practice and Requirements for Harmonic Control in Electric Power Systems", IEEE Standard 519, 1992.

[23] N. Mohan, T. M. Undeland and W. P. Robbins, *Power Electronics: Converters, Applications, and Design*, New York: Wiley, 1989.

[24] M. H. Rashid, *Power Electronics, Circuits, Devices, and Applications*, 2nd edition, Englewood Cliffs, NJ: Prentice-Hall, 1993.

[25] L. Salazar and G. Joos, "PSPICE simulation of three-phase inverters by means of switching functions", *IEEE Transactions on Power Electronics*, vol. 9, no. 1, pp. 35-42, 1994.

Real-Time Implementation of Voltage Dip Mitigation using D-STATCOM with Fast Extraction of Instantaneous Symmetrical Components

Thip Manmek* and Chathura P. Mudannayake**

* Department of Control and Instrumentation Engineering, Mahanakorn University of Technology, Bangkok, THAILAND.
** School of Electrical Engineering and Telecommunication, University of New South Wales, Sydney, AUSTRALIA.

Abstract— This paper presents the application of the proposed efficient least squares algorithm in power supply voltage dip and unbalance detection for mitigation using a D-STATCOM. The proposed method is capable of identifying the instantaneous symmetrical components of the fundamental frequency accurately even though the point of common coupling voltage is strongly corrupted by the voltage harmonics. Also, it fulfils the specific requirements of the fast transient response, accuracy and robustness in order to ensure the satisfactory performance of the mitigation system. The proposed method extracts detecting the positive- and negative-sequence components and then using those sequence components for generating reference values of current that need to be injected into the point of connection D-STATCOM in order to compensate the voltage errors. Furthermore, the suitability of the proposed method in balanced/unbalance voltage dip compensation is verified by a experimental studies.

Index Terms—D-STATCOM, voltage unbalance and voltage dip detection method, instantaneous symmetrical component, least squares algorithm.

I. INTRODUCTION

Recently, the distribution static synchronous compensator (D-STATCOM) has been introduced to distribution networks to manage the system reactive power and regulate the voltage at the distribution buses. A D-STATCOM usually consists of a shunt connected voltage source converter (VSC) [1]. The benefits of using a VSC are sinusoidal currents, high current bandwidth, controllable reactive power to regulate bus-voltage level and to minimize the resonances between the grid and the converter. A system with these characteristics can be used to inject a controllable current into the grid. By injecting a current into the point of common coupling (PCC), a shunt-connected VSC can boost the voltages at that point during a voltage dip. Even though the theory, control and modeling of conventional static compensator (STATCOM) have been broadly discussed in the literature, more preference is given to the D-STATCOM due to its simple connection requirement [2],[3]. Furthermore, an unbalance correction can also be added to the functions of the D-STATCOM [4], [5].

The extraction and tracking technique of voltage dip is the core of the D-STATCOM mitigating control strategy. In order to obtain the required information to control the D-STATCOM further processing during voltage unbalance or dip is required and the processed information is required to be updated as fast as possible. Moreover, the choice of techniques for the voltage dip detection is highly dependent on the real-time implementation, the available computational hardware, and the amount of computational effort.

The typical standard information tracking or detection methods such as the Fourier transform [6] and the phase-locked-loop (PLL) [7] are generally used in D-STATCOM systems. The Fourier transform technique can return information regarding the state of system supply. The advantage of this method is that it can return magnitude and phase of the fundamental and harmonics component of the supply voltage. However, it takes at least one cycle of the fundamental when a dip has commenced before information regarding the magnitude and phase angle can be assumed accurately [8]. A voltage dip detection technique that utilizes the phase-locked-loop (PLL) to each supply phase independently has been introduced in [9], [10]. This technique can be combined with any other technique to detect the magnitude of the dip voltage. The PLL technique does not give good results if the voltage dip are associated with a phase angle jump such as in unbalance voltage dip [11].

The advantages of obtaining the dq–components via the symmetrical components instead of the direct dq–transformation to solved problem in the dq–current controllers of D-STATCOM. In the case of the grid voltage or the load voltages are unbalanced, a ripple of double the grid frequency will occur in the dq–reference frame. In the case of unbalanced three-phase voltages, breaking the voltage signals into positive- and negative-sequence components and then transforming into the dq– synchronous reference frames (SRF) results in dc-quantities.

In this paper the voltage dips detection and unbalances detection based on the proposed efficient least squares algorithm for using in D-STATCOM applications is introduced. The proposed voltage dip detection system extract the fundamental component and positive and negative-sequence component of the voltage to generate a current reference signal for the D-STATCOM to compensate voltage dips. The experimental results with practical parameter are provided to demonstrate that the proposed balanced and unbalanced voltage dip detection system is an excellent tool for extracting the required information for a D-STATCOM application.

978-1-4244-0644-9/07/$25.00 ©2007 IEEE

II. VOLTAGE DIP DETECTION BASED ON PROPOSED EFFICIENT LEAST SQUARES METHOD

A. Overview of the Proposed Efficient Least Squares Method

The instantaneous voltages waveforms in a three-phase power system can be generally expressed as:

$$v_{Sa}(t) = \sum_{i=1,5,7,\ldots}^{K} V_{Sai} \cos\left(\omega_i t + \beta_{ai}\right) \tag{1}$$

$$v_{Sb}(t) = \sum_{i=1,5,7,\ldots}^{K} V_{Sbi} \cos\left(\omega_i t + \beta_{bi}\right) \tag{2}$$

$$v_{Sc}(t) = \sum_{i=1,5,7,\ldots}^{K} V_{Sci} \cos\left(\omega_i t + \beta_{ci}\right) \tag{3}$$

In which V_{Sai}, V_{Sbi} and V_{Sci} are the unknown magnitude values of the three-phase voltages, and β_{ai}, β_{bi}, and β_{ci} are the unknown phase angles. Subscript $i = 1,5,7,\ldots K$ refers to the fundamental and harmonic components respectively. The symbol ω_i is known angular frequency of the ith harmonic.

The proposed method is used to extract the fundamental component of the three-phase voltages. The discrete-time version of three-phase voltage can be written in matrix notation as

$$y = \mathbf{A}x \tag{4}$$

$\mathbf{A}$ is a complex rotation matrix, x is a complex vector consisting of magnitude and phase angle of the input signal. The vector y is the input signal. The proposed method is based on the conventional least squares algorithm. The aim is to solve the system equations without inverting any matrix with real number elements. The proposed least squares algorithm can calculate harmonic components by simply multiplying each set of input signals by a constant matrix. Consequently, the proposed method is immune to transient distortions and unbalanced conditions.

The vector x which consists of a complex signal of fundamental and harmonic component can be found using the proposed method as follows:

$$x = \left(\left(\mathbf{R}\right)^{\mathsf{T}} \mathbf{R}\right)^{-1} \left(\mathbf{R}\right)^{\mathsf{T}} y = \mathbf{C}y \tag{5}$$

where $\mathbf{R} = \begin{bmatrix} H_1^0 & H_{-1}^0 \\ H_1^1 & H_{-1}^1 \\ H_1^2 & H_{-1}^2 \\ \vdots & \vdots \\ H_1^L & H_{-1}^L \end{bmatrix}$

H_1^L is the rotation matrix , ie. $\begin{bmatrix} \cos\omega_1 nT & -\sin\omega_1 nT \\ \sin\omega_1 nT & \cos\omega_1 nT \end{bmatrix}$

L is the number of measured samples,

$\mathbf{C}$ is a constant matrix, ie. $\left(\mathbf{R}^{\mathsf{T}}\mathbf{R}\right)^{-1}\mathbf{R}^{\mathsf{T}}$

The proposed method is computationally efficient, the size of the matrix is required is only $2 \times L$, since only the fundamental component needs to be identified.

B. Instantaneous positive/Negative Sequence voltage component Detection Method Based on Proposed Efficient Least Squares Method

The fundamental component in term of the complex signals can thus be obtained as expressed in (6). This method is capable of identifying the fundamental component accurately even though the point of common coupling voltage is strongly corrupted by voltage harmonics.

$$\begin{bmatrix} v_{Sa1}^{\cos} \\ v_{Sa1}^{\sin} \\ \vdots \\ v_{Sb1}^{\cos} \\ v_{Sb1}^{\sin} \\ \vdots \\ v_{Sc1}^{\cos} \\ v_{Sc1}^{\sin} \end{bmatrix} = \begin{bmatrix} V_{Sa1}\cos(\omega_1 t + \beta_{a1}) \\ V_{Sa1}\sin(\omega_1 t + \beta_{a1}) \\ \vdots \\ V_{Sb1}\cos(\omega_1 t + \beta_{b1}) \\ V_{Sb1}\sin(\omega_1 t + \beta_{b1}) \\ \vdots \\ V_{Sc1}\cos(\omega_1 t + \beta_{c1}) \\ V_{Sc1}\sin(\omega_1 t + \beta_{c1}) \end{bmatrix} = \mathbf{C} \times \begin{bmatrix} v_{Sa}(n) \\ v_{Sa}(n-L+1) \\ \vdots \\ v_{Sb}(n) \\ v_{Sb}(n-L+1) \\ \vdots \\ v_{Sc}(n) \\ v_{Sc}(n-L+1) \end{bmatrix} \tag{6}$$

The fundamental positive- and negative- sequence components are extracted by using equations (7) and (8) respectively. The D-STATCOM application that is discussed in the next section uses synchronously rotating reference frame controllers (i.e. dq – controllers). Therefore, transformation of the instantaneous positive- and negative-sequence voltage components into the dq – axes is required.

$$\begin{bmatrix} v_a^+ \\ v_b^+ \\ v_c^+ \end{bmatrix} = \frac{1}{3}\begin{bmatrix} 1 & 0 & -\frac{1}{2} & \frac{\sqrt{3}}{2} & -\frac{1}{2} & \frac{\sqrt{3}}{2} \\ -\frac{1}{2} & \frac{\sqrt{3}}{2} & 1 & 0 & -\frac{1}{2} & -\frac{\sqrt{3}}{2} \\ -\frac{1}{2} & -\frac{\sqrt{3}}{2} & -\frac{1}{2} & \frac{\sqrt{3}}{2} & 1 & 0 \end{bmatrix}\begin{bmatrix} v_{Sa1}^{\cos} \\ v_{Sa1}^{\sin} \\ v_{Sb1}^{\cos} \\ v_{Sb1}^{\sin} \\ v_{Sc1}^{\cos} \\ v_{Sc1}^{\sin} \end{bmatrix} \tag{7}$$

$$\begin{bmatrix} v_a^- \\ v_b^- \\ v_c^- \end{bmatrix} = \frac{1}{3}\begin{bmatrix} 1 & 0 & -\frac{1}{2} & \frac{\sqrt{3}}{2} & -\frac{1}{2} & -\frac{\sqrt{3}}{2} \\ -\frac{1}{2} & -\frac{\sqrt{3}}{2} & 1 & 0 & -\frac{1}{2} & \frac{\sqrt{3}}{2} \\ -\frac{1}{2} & \frac{\sqrt{3}}{2} & -\frac{1}{2} & -\frac{\sqrt{3}}{2} & 1 & 0 \end{bmatrix}\begin{bmatrix} v_{Sa1}^{\cos} \\ v_{Sa1}^{\sin} \\ v_{Sb1}^{\cos} \\ v_{Sb1}^{\sin} \\ v_{Sc1}^{\cos} \\ v_{Sc1}^{\sin} \end{bmatrix} \tag{8}$$

The transformation of the positive- and negative- sequence components into the dq – reference frame is given in (9) and (10) respectively.

$$\begin{bmatrix} v_d^+ \\ v_q^+ \end{bmatrix} = \frac{2}{3}\begin{bmatrix} \sin(\omega t) & \frac{1}{2}\left(-\sin(\omega t)-\sqrt{3}\cos(\omega t)\right) & \frac{1}{2}\left(-\sin(\omega t)+\sqrt{3}\cos(\omega t)\right) \\ \cos(\omega t) & \frac{1}{2}\left(-\cos(\omega t)+\sqrt{3}\sin(\omega t)\right) & \frac{1}{2}\left(-\cos(\omega t)-\sqrt{3}\sin(\omega t)\right) \end{bmatrix}\begin{bmatrix} v_{Sa}^+ \\ v_{Sb}^+ \\ v_{Sc}^+ \end{bmatrix} \tag{9}$$

$$\begin{bmatrix} v_d^- \\ v_q^- \end{bmatrix} = \frac{2}{3}\begin{bmatrix} -\sin(\omega t) & \frac{1}{2}\left(\sin(\omega t)-\sqrt{3}\cos(\omega t)\right) & \frac{1}{2}\left(\sin(\omega t)+\sqrt{3}\cos(\omega t)\right) \\ \cos(\omega t) & \frac{1}{2}\left(-\cos(\omega t)-\sqrt{3}\sin(\omega t)\right) & \frac{1}{2}\left(-\cos(\omega t)+\sqrt{3}\sin(\omega t)\right) \end{bmatrix}\begin{bmatrix} v_{Sa}^- \\ v_{Sb}^- \\ v_{Sc}^- \end{bmatrix} \tag{10}$$

where v_d^+ and v_q^+ are the dq – components of the positive-sequence voltage, and

v_d^- and v_q^- are the dq – components of the negative-sequence voltage.

The overview of the complete proposed voltage dip detection method is illustrated in Fig. 1.

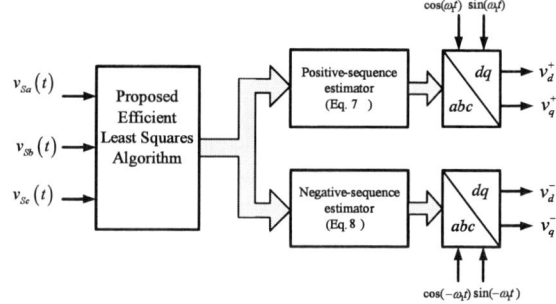

Fig. 1.   Block diagram of voltage dip detection method based on proposed efficient least squares algorithm.

III. APPLICATION OF PROPOSED INSTANTANEOUS POSITIVE/NEGATIVE SEQUENCE VOLTAGE COMPONENT DETECTION METHOD IN D-STATCOM

A. System Configuration

The configuration of the D-STATCOM with the proposed voltage dip detection method is illustrated in Fig.2. The D-STATCOM consists of a three-phase voltage source converter (VSC), a dc-side capacitor C_{dc} with its leakage resistance R_{dc}, and an inductance L_F on the *ac*-side of the converter. The resistance R_F represents the cable resistance of the *ac*-side of the converter. A shunt filter capacitor with capacitance C_F and an inductance L_{tr} that represents transformer leakage inductance are added to the ac- side of the voltage source converter that forms a LCL filter. This filter helps in effectively filtering out the switching ripple in the output voltage waveform. The grid is represented by using an ideal voltage source and impedance. This impedance consists of an inductance L_{line} and a resistance R_{line} which characterizes the transformer and power line respectively. The analysis and design of the D-STATCOM controller are conducted in the rotating reference frame which is synchronized to the voltage vector. The D-STATCOM mitigates the voltage dips by dynamically injecting a current of desired amplitude and phase angle into the grid line. The two inner dq – current regulators in Fig.2 force the converter currents i_{Fd} and i_{Fq} to follow the command currents i_{Fd}^* and i_{Fq}^* respectively. The command i_{Fd}^* to the d – axis current loop is obtained by summing the *dc*-link V_{dc}^2 -controller output and d – axis component of the reactive power controllers output i_{Cd}.

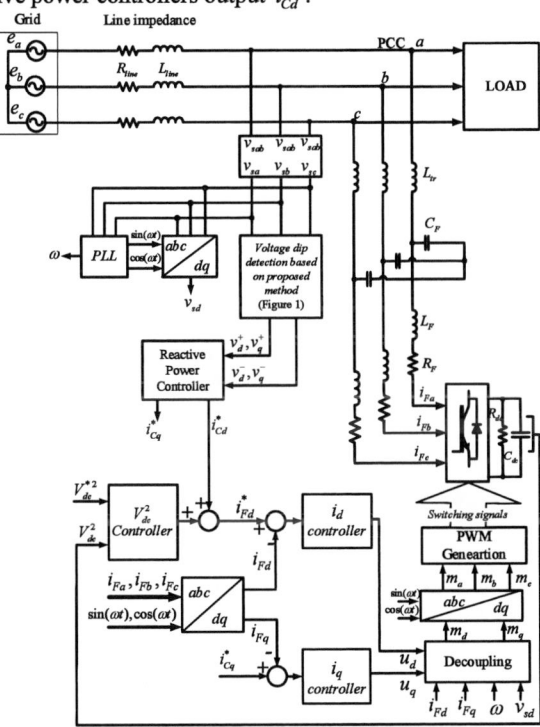

Fig. 2. Overall schematic diagram of D-STATCOM with proposed voltage dip detection method.

The command i_{Fq}^* is obtained from the q – axis component of the reactive power controller output i_{Cq}. The purpose of the outer loop V_{dc}^2 -controller is to regulate the *dc*-link voltage to a required level. The modulation signals m_a, m_b and m_c for the PWM generator are derived from the output of the current controllers. The decoupling terms are added to the output of the current controllers in order to remove the coupling and are

then transformed into the abc – stationary reference frame to obtain the modulation signal for the VSC. The $\sin(\omega_1 t)$ and $\cos(\omega_1 t)$ terms required for the transformation between the abc – and dq – reference fames are obtained via a phase locked loop (PLL) which is synchronized to the fundamental component of the voltage. All of the above are illustrated in Fig. 2.

B. Modeling of Three-Phase PWM Converter

The currents of the three-phase PWM converter in the synchronous reference frame can be obtained by applying Kirchoff's Voltage Law (KVL) and Kirchoff's Current Law (KCL) to the ac-side of the converter. These are represented in (11) and (12).

$$\frac{d}{dt}i_{Fd} = -\frac{R_F}{L_F}i_{Fd} + \omega i_{Fq} - \frac{V_{dc}}{2L_F}m_d + \frac{1}{L_F}V_{sd} \quad (11)$$

$$\frac{d}{dt}i_{Fq} = -\frac{R_F}{L_F}i_{Fq} - \omega i_d - \frac{V_{dc}}{2L_F}m_q \quad (12)$$

where

i_{Fd}, i_{Fq} are $d-q$ axis components of the converter current,
m_d, m_q are $d-q$ axis components of modulation signals,
V_{sd} is d – axis component of PCC voltage,
V_{dc} is dc-link voltage, and
ω is angular frequency in rad/s.

The power delivered from the ac-side of the converter must be balanced with the power received at the dc-side. By neglecting losses in the converter, the following equation for dc-link voltage can be written.

$$\frac{dV_{dc}}{dt} = -\frac{V_{dc}}{R_{dc}C_{dc}} + \frac{3V_{sd}i_{Fd}}{2C_{dc}V_{dc}} \quad (13)$$

C. Design of d-q Current Controllers

Equations (11) and (12), which described the dynamics of the d–q axes currents are coupled to each other (i.e. i_q depends on i_d and visa versa). It is necessary to decouple them for proper control design. Decoupling can be achieved by introducing new terms u_d and u_q to (11) and (12) respectively. The terms u_d and u_q are shown in (14) and (15).

$$u_d = -\omega i_{Fd} - \frac{V_{dc}}{2L_F}m_q \quad (14)$$

$$u_q = \omega i_{Fq} - \frac{V_{dc}}{2L_F}m_d + \frac{1}{L_F}V_{mp} \quad (15)$$

Fig. 3 shows the transfer function (G_T) for the decoupled converter in the synchronous reference frame together with the PI (proportional plus integral) controller (G_c) for both $d-q$ axes. The closed loop transfer function of the $d-q$ current feedback loops (inner loops) can be described by:

$$G_{cloop}(s) = \frac{G_c G_T}{1 + G_c G_T} = \frac{K_P s + K_I}{s^2 + \left(\frac{R_F}{L_F} + K_P\right)s + K_I} \quad (16)$$

By comparing the denominator of (16) with the optimum coefficients of ITAE (integral of time multiplied by absolute magnitude of the error) criterion for a ramp input for 2nd-order transfer function in [12], the parameters of the PI controller (G_c) can be selected as follows:

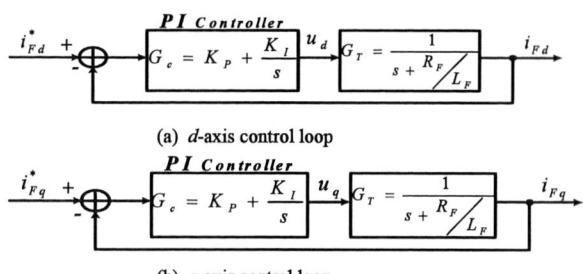

(a) d-axis control loop

(b) q-axis control loop

Fig. 3. dq −axis closed-loop current control diagrams.

$$K_P = 3.2\omega_{ni} - \frac{R_F}{L_F} \quad K_I = \omega_{ni}^2 \qquad (17)$$

where ω_{ni} is the natural frequency of the closed-loop response.

K_P, K_I are proportional gain and integral gain respectively. The dynamic response of the current controller depends on the natural frequency (ω_{ni}), and hence the value of ω_{ni} is chosen for the desired dynamic response.

D. Positive-Negative-Synchronous Reference Frame Controller (Reactive Power Controllers)

In the case of balanced three-phase voltage, the direct transformation of abc voltages into the $d - q$ reference frame will result in dc-quantities. Hence, the D-STATCOM can use a conventional PI-controller to control the injected reactive currents. However, if the grid voltage or the load voltages are unbalanced, a ripple of double the grid frequency will occur in the $d - q$ reference frame. In the case of unbalance three-phase voltages, breaking the voltage signals into positive- and negative- sequence components and then transforming into $d - q$ synchronous reference frame (SRF) results in dc-quantities and these $d - q$ feedback signals allow in control design reactive power control.

Transformation of three-phase balance voltages with unity magnitude into the positive- and negative- $d - q$ synchronous reference frame results in dc-quantities with following values.

$v_d^+ = 1\,\mathrm{pu}$, $v_q^+ = 0\,\mathrm{pu}$,

$v_d^- = 0\,\mathrm{pu}$, $v_q^- = 0\,\mathrm{pu}$

where v_d^+, v_q^+ - dq components of positive-sequence voltage, and v_d^-, v_q^- - dq components of negative-sequence voltage.

The voltage dips (both balanced and unbalanced) in power systems can be corrected by regulating positive- and negative-synchronous dq components to the values corresponding to the balanced case given above. Three controllers are utilized to regulate these dq components of the positive- and negative-sequence voltages as shown in Fig. 4. As shown in Fig. 4 v_d^+ is regulated to 1 pu via i_q^{+*} in the positive-sequence synchronous reference frame. v_d^- and v_q^- are regulated to zero via the i_q^{-*} and i_d^{-*} in the negative-sequence synchronous reference frame. These dq - current components in the positive- and negative- synchronous reference frames are converted into three axes components and added those together to obtain three- phase currents that need to be injected to the line in order to compensate for the voltage dips.

These abc current commands are then converted to the positive- sequence synchronous reference frame dq components to generate the current commands (i.e i_{Cd}^* and i_{Cq}^*) for the current controllers. All these are indicated in Fig. 4.

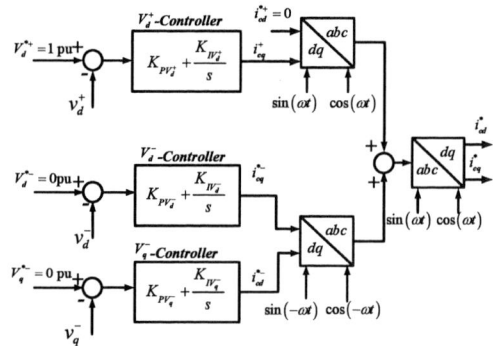

Fig. 4.   Block scheme of reactive power controller for D-STATCOM.

E. Design of Outer Loop DC Voltage Controller

The voltage control loop regulates the dc-link voltage at a required level. Equation (13) which describes the relationship between V_{dc} and i_{Fd}, is a non-linear equation. This equation is rearranged so that it can be treated with linear control theory. Rearranging (18),

$$\frac{d}{dt}(V_{dc})^2 + \frac{2}{R_{dc}C_{dc}}(V_{dc})^2 = p \qquad (18)$$

where $p = \dfrac{3V_{sd}}{C_{dc}}i_{Fd}$

Equation (18)can be written in Laplace domain as:

$$G_{TV} = \frac{V_{dc}^2(s)}{P(s)} = \frac{1}{s + 2/R_{dc}C_{dc}} \qquad (19)$$

Now, V_{dc}^2 instead of V_{dc} can be used for control design. This does not cause any technical problems since V_{dc} is unidirectional [13]. The d-axis inner loop is assumed to be very fast compared to the outer voltage loop so that the inner loop can be replaced with unity gain.

The closed-loop transfer function of the outer voltage loop (G_{dc_cloop}) is shown below in

$$G_{dc\_cloop}(s) = \frac{G_C G_T}{1 + G_C G_T} = \frac{K_{Pv}s + K_{Iv}}{s^2 + \left(\dfrac{2}{R_{dc}C_{dc}} + K_{Pv}\right)s + K_{Iv}} \qquad (20)$$

This closed-loop transfer function has 2nd order characteristics. The proportional and integral constants (i.e. K_{Pv} and K_{Iv}) of the PI controller can be obtained by comparing the denominator of this transfer function to the ITAE-criterion for a ramp input for 2nd-order transfer function [12]as follows:

$$K_{Pv} = 3.2\omega_{nv} - \frac{2}{R_{dc}C_{dc}}, K_{Iv} = \omega_{nv}^2 \qquad (21)$$

where ω_{nv} is the natural frequency of the closed voltage loop.

The natural frequency of the outer loop (ω_{nv}) is chosen to be considerably lower than the natural frequency of the inner current controller (ω_{ni}) to avoid any possibility of formation of higher order systems.

IV. EXPERIMENTAL RESULTS FOR THE D-STATCOM WITH THE PROPOSED INSTANTANEOUS POSITIVE/NEGATIVE SEQUENCE VOLTAGE COMPONENT DETECTION

The overview of prototype D-STATCOM system with the proposed efficient least squares algorithm based voltage dip detection is given in the block diagram shown in Fig. 5. The experimental setup consists of a three-phase IGBT converter and ac-side inductors, an additional resistive load bank (i.e. load bank-A) and a switch. The arrangement is used for generating voltage dips and unbalances that are required for the experimental studies. The load current drawn into the load bank-A causes voltage drop in the line inductance and thereby generates voltage dip in the PCC. The system parameters of the prototype D-STACOM are given in TABLE I. The proposed efficient least square algorithm based dip detection method and the control algorithm required for the D-STATCOM system have been implemented on the dSPACE 1104 R&D board. The experimental of the D-STATCOM have been carried out for three different cases: unbalanced grid voltage, balanced voltage dip and unbalanced voltage dip.

A. Experimental results for unbalanced grid voltage mitigation

The D-STATCOM system is tested for the mitigation of unbalanced grid voltages. The unbalanced grid voltage in the experimental setup is generated by applying an unbalanced load at the load bank-A. Fig. 6 shows the unbalanced phase voltages of phases-*a*, -*b* and -*c*. As may be seen, the voltages of phase -*a* and -*b* are reduced by about 15% and 7% respectively. Fig. 7 shows the steady-state voltages of the grid after the compensation using the D-STATCOM with the proposed method. The unbalances are not apparent in the compensated voltages.

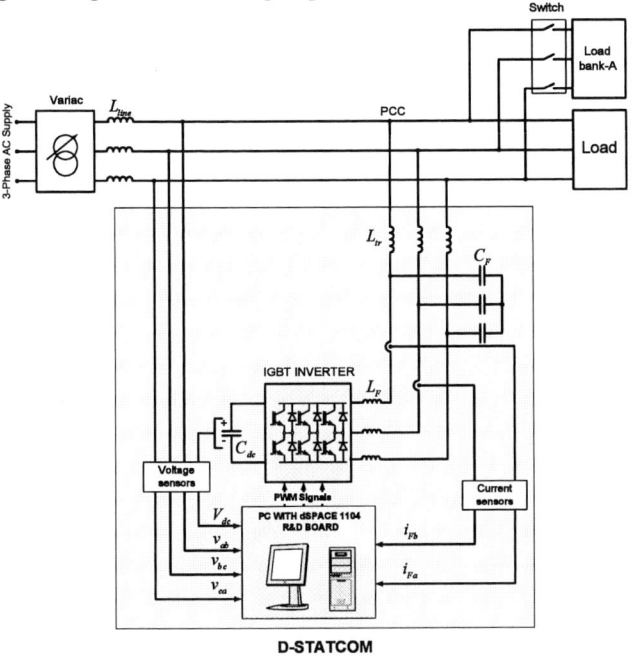

Fig. 5.   Overview of the experimental setup of the D-STATCOM system.

TABLE I.   DESIGN SPECIFICATIONS AND CIRCUIT PARAMETERS OF THE PROPOSED D-STATCOM.

Phase-voltage of three-phase supply: (V_s) (*rms*)	210V=1 pu, 50Hz
Sampling frequency of D-STATCOM control algorithm (f_s)	8000 Hz
Converter switching frequency (f_{sw})	8000 Hz
dc-link voltage (V_{dc})	700 V
Line inductance (L_{line})	11 mH
ac-side inductance (L_F)	6 mH
Filter capacitance (C_F)	8 μF
Transformer leakage inductance (L_{tr})	2 mH
Natural frequency of *dq*- current control loops (ω_{ni})	942.47 rad/s
Natural frequency of V_{dc}^2 -control loop (ω_{nv})	15.7 rad/s
Proportional gain of the positive-SRF d − voltage controller	0.3
Integral gain of the positive-SRF d − voltage controller	20
Proportional gain of the positive-SRF dq − voltage controllers	0.12
Integral gain of the positive-SRF dq − voltage controllers	10
Sampling frequency of the proposed efficient LS algorithm	2000 Hz
Number of samples of the proposed efficient LS algorithm (L)	20

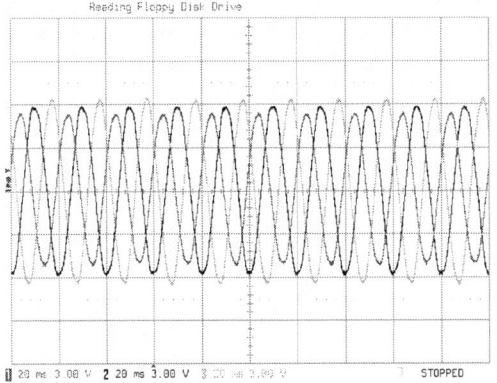

Fig. 6.   Unbalanced phase voltages of phases-*a*, -*b* and -*c* (100 V/div).

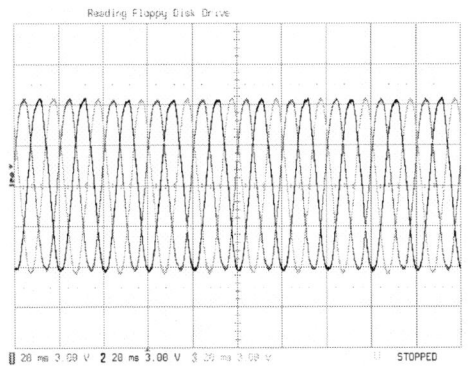

Fig. 7.   Compensated voltage of phases-*a*, -*b* and -*c* using D-STATCOM with proposed method (100 V/div).

Fig. 8 shows the dynamic response of the D-STATCOM during the grid unbalance mitigation. The unbalance mitigation is enabled at t = 0.06 sec. Fig. 9 shows the corresponding d − component of the positive-sequence voltage in the positive SRF (i.e. v_d^+) and the dq − components of the negative-sequence voltage in the negative SRF (i.e. v_d^- and v_q^-). During the unbalanced period, the v_d^+, v_d^- and v_q^- have values of about 0.9 p.u., 0.1p.u. and 0.03 p.u. respectively (1.0 p.u. = 210 V).

As may be seen, the values of v_d^+, v_d^- and v_q^- are regulated to 1.0 p.u., 0 p.u. and 0 p.u. by the reactive power controllers when the unbalance mitigation is enabled. The v_d^+ shows fast transient response. The v_d^- and v_q^- have slower transient response compared to that of the v_d^+.

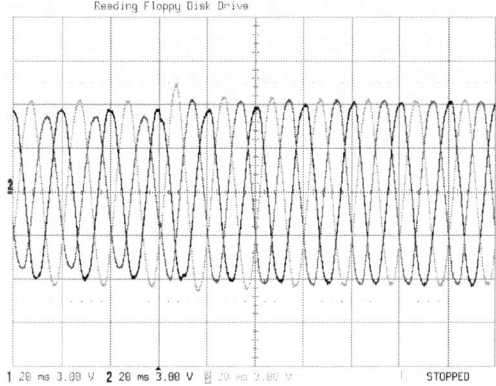

Fig. 8.   Dynamic response when unbalance mitigation started at t = 0.06 sec.: voltage of phases- *a*, -*b* and -*c* (100 V/div).

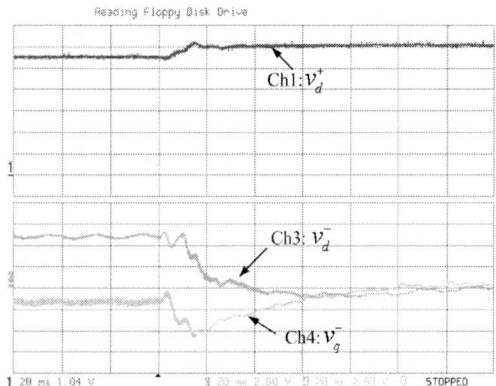

Fig. 9.   Dynamic response when unbalance mitigation started at t = 0.06 sec.: (Ch1) *d* – component of positive-sequence voltage in positive SRF (34.67 V/div); (Ch3) *d* – components of negative-sequence voltage in negative SRF (8.67 V/div); (Ch4) *q* – components of negative-sequence voltage in negative SRF (8.67 V/div).

B.   Experimental results for balanced voltage dip mitigation

A first set of experiments is carried out when the grid voltage is affected by a balanced voltage dip. The balanced voltage dip is generated by applying the same load to each phase of the resistive load bank-A shown in Fig. 5. The line inductance L_{line} causes the same voltage drop in each phase due to this balanced load. The resistive load bank-A is switched on at t = 0.06 sec. The generated balanced voltage dip is shown in Fig. 10. The magnitude of this voltage dip is about 17%.

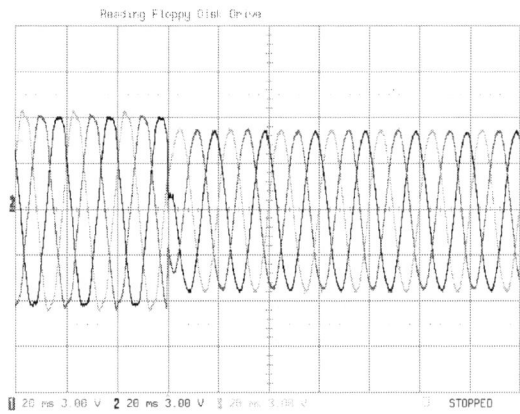

Fig. 10.   Balanced voltage dip of 17% occurred.

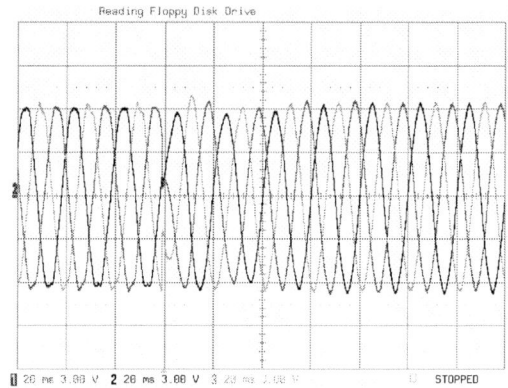

Fig. 11.   Compensated voltage waveforms of phases-*a*, -*b*, and -*c* at PCC in case of 17% balanced voltage dip at t = 0.06 sec. (100 V/div)

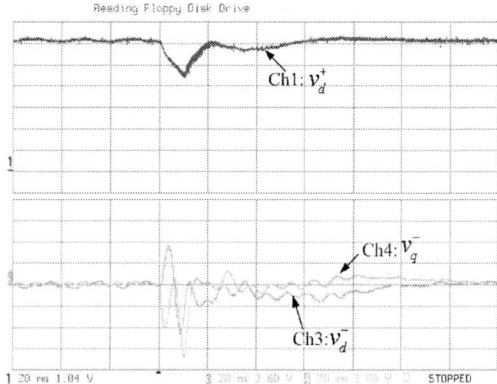

Fig. 12.   Dynamic response of balanced voltage dip compensation when dip started at t = 0.06 sec.: (Ch1) *d* – component of positive-sequence voltage in positive SRF (34.67 V/div); (Ch3) *d* – components of negative-sequence voltage in negative SRF (8.67 V/div); (Ch4) *q* – components of negative-sequence voltage in negative SRF (8.67 V/div).

Fig. 11 shows the compensated voltage waveforms of phases-*a*, -*b*, and -*c* using the D-STATCOM with the proposed voltage dip detection method. Fig. 12 shows the corresponding *d* – component of the positive-sequence voltage in the positive SRF (i.e. v_d^+) and the *dq* – components of the negative-sequence voltage in the negative SRF (i.e. v_d^- and v_q^-). The value of v_d^+ is regulated to 1.0 p.u. after a short transient to result in balanced three-phase voltages. The values of v_d^- and v_q^- are zero for any balanced three-phase voltages including balanced dips. As may be seen in Fig. 12, v_d^- and v_q^- have settled back to zero after the transient disturbance.

C.   Experimental results for unbalanced voltage dip mitigation

This section presents the results for the unbalanced dip mitigation using the D-STATCOM. The unbalanced voltage dip for this experiment is generated by switching on an unbalanced load in the resistive load bank-A shown in Fig. 5. The switch is turned on at t =0.06 sec. Fig. 13 shows the generated unbalanced voltage dip using the experimental setup. As may be seen, the voltages of the phases-*a*, -*b* and -*c* are reduced by 28%, 15% and 8% respectively. Fig. 15 shows the v_d^+, v_d^- and v_q^- components corresponding to the compensated waveform given in Fig. 14. Fig. 14 shows the compensated voltage waveforms of phases -*a*, -*b* and –*c* using the D-STATCOM with the proposed voltage dip detection method.

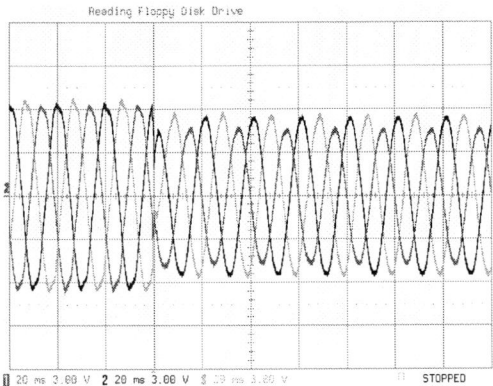

Fig. 13.   Voltages of phases-*a*, -*b* and -*c* for unbalanced dip occurring at t = 0.06 sec. (100 V/div)

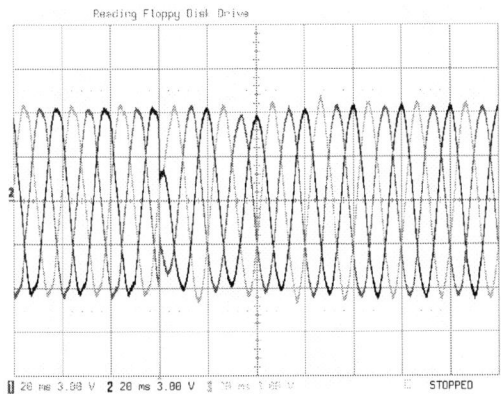

Fig. 14.   Compensated voltage waveforms of phases-*a*, -*b*, and -*c* at PCC in case of unbalanced voltage dip at t = 0.06 sec. (100 V/div)

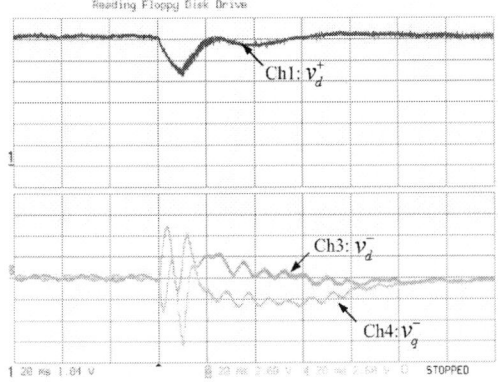

Fig. 15.   Dynamic response of unbalanced voltage dip compensation when dip started at t = 0.06 sec.: (Ch1) d – component of positive-sequence voltage in positive SRF (34.67 V/div); (Ch3) d – components of negative-sequence voltage in negative SRF (8.67 V/div); (Ch4) q – components of negative-sequence voltage in negative SRF (8.67 V/div).

V.   CONCLUSIONS

This paper has shown that the proposed efficient least squares algorithm can be used for voltage dip and unbalanced mitigation in the D-STATCOM. The discussed D-STATCOM consists of a voltage source converter which injects reactive current into the grid in order to mitigate the voltage dips and unbalances. The D-STATCOM consists of positive- and negative-synchronous reference frame controllers (i.e. reactive power controller) that regulate the dq – components of the positive and negative voltage to result in balanced three-phase voltages at the point of common coupling. The dc-link voltage is regulated to a set value via the d – axis current. The synchronous reference frame dq – axes current controllers are utilized to force the currents into the grid.

The proposed efficient least squares algorithm outputs the instantaneous cosine and sine terms of the fundamental component. In the voltage dip and unbalance detection methods, the size of the constant matrix (i.e. $\mathbf{C}$) required is only $2 \times L$, since only the fundamental component needs to be identified (i.e. $K = 1$). The instantaneous positive- and negative- sequence components of the voltages are obtained using the instantaneous cosine and sine terms of the fundamental components of phases -*a*, -*b* and -*c*. The identified positive- and negative-sequence components are then represented in the positive- and negative- synchronous reference frames respectively (i.e. v_d^+, v_q^+, v_d^- and v_q^-). These voltage components are regulated to set values that correspond to the three-phase balance voltages in order to mitigate the voltage dips and unbalances. The proposed voltage dip and unbalance detection method allows for extraction of these sequence component within a half fundamental cycle. In addition the proposed method is capable of identifying the voltage dips and unbalances accurately even though the point of common coupling is strongly corrupted by the voltage harmonics.

A laboratory experimental setup for the prototype D-STATCOM is built using an IGBT voltage source converter in order to verify the performance. The proposed voltage dip detection method and the D-STATCOM control algorithm are implemented on the dSPACE 1104 R&D board. The prototype D-STATCOM system is tested for the mitigation of balanced and unbalanced voltage dips. The experimental results verified the successful real-time application of the proposed voltage unbalance and dip detection method in extracting information for the D-STATCOM application.

VI.   REFERENCES

[1]   O. C. Montero-Hernandez and P. N. Enjeti, "A fast detection algorithm suitable for mitigation of numerous power quality disturbances," *Thirty-Sixth IAS Annual Meeting IEEE Industry Applications Conference, 2001*, vol. 4, pp. 2661 - 2666 2001.

[2]   P. G.-G. a. A. G-Cerrada, "Control system for a PWM-based STATCOM," *IEEE Transactions on Power Delivery*, vol. 15, pp. 1252-1257, October 2000.

[3]   P. S. Sensarman, K. R. Padiyar, and V. Ramanarayanan, "A STATCOM for composite power line conditioning," *Proceedings of IEEE International Conference on Industrial Technology 2000.* , vol. 1, pp. 542-547, 2000.

[4]   C. Su, G. Joos, and L. T. Moran, "Dynamic performance of PWM STATCOMs operating under unbalance and fault conditions in distribution systems," *IEEE Power Engineering Society Winter Meeting*, vol. 2, pp. 950-955, 2001.

[5]   C. Wei-Neng and Y. Kuan-Dih, "Design of D-STATCOM for fast load compensation of unbalanced distribution systems," *4th IEEE International Conference on Power Electronics and Drive Systems*, vol. 2, pp. 801-806, 2001.

[6]   A. Campos, G. Joos, P. D. Ziogas, and J. F. Lindsay, "A Dsp-based Real-time Digital Filter For Symmetrical Components," *Proceedings. Joint International Power Conference Athens Power Tech, 1993 (APT 93)*, vol. 1, pp. 75-79, 1993.

[7] M. R. K.-G. Iravani, M. , "Online estimation of steady state and instantaneous symmetrical components.," *IEE Proceedings-Generation, Transmission and Distribution,* , vol. 150, pp. 616-622, 2003.

[8] R. M. Gnativ and J. V. Milanovic, "Identification of voltage sag characteristics from the measured responses," *10th International Conference on Harmonics and Quality of Power, 2002,* vol. 2, pp. 535-540, 2002.

[9] V. B. B. a. P. N. Enjeti, "An active line conditioner to balance voltage in a three-phase system," *IEEE Transections on Industry Applications,* vol. 32, pp. 287-292, 1996.

[10] V. Kaura and V. Blasko, "Operation of a phase locked loop system under distorted utility conditions," *IEEE Transactions on Industry Applications,* vol. 33, pp. 58-63, 1997.

[11] P. Wang, N. Jenkins, and M. H. J. Bollen, "Experimental investigation of voltage sag mitigation by an advanced static VAr compensator," *IEEE Transactions on Power Delivery,* , vol. 13, pp. 1461-1467, 1998.

[12] M. K. Masten, "The control handbook [Book Review]," *Automatic Control, IEEE Transactions on,* vol. 45, pp. 1581, 2000.

[13] Y. Yang, M. Kazerani, and V. H. Quintana, "Modeling, control and implementation of three-phase PWM converters," *IEEE Transactions on Power Electronics,* , vol. 18, pp. 857-867, 2003.

Combined System of Static Synchronous Series Compensation and Passive Filter applied to Wind Energy Conversion System

A. Singer, W. Hofmann*
*Chair of Electrical Machines and Drives
Chemnitz University of Technology Germany
amr.singer@s2003.tu-chemnitz.de
wilfried.hofmann@e-technik.tu-chemnitz.de

Abstract- **Permanent magnet salient pole synchronous connected dc network produces less power than the rated power even at rated excitation. Static synchronous series compensation (SSSC) provides a solution to this problem. Since SSSC compensates the reactance voltages drop of the generator. The passive filters are used to eliminate the harmonics which are generated by nonlinear loads. This paper presents simulation and experimental results when applying static synchronous series compensation and passive filter to permanent magnet salient pole synchronous generator (PSG) in order to increase the output power, stabilize the output voltage and get rid of the harmonic respectively.**

I. INTRODUCTION

Due to the Kyoto protocol commitment to reduce the carbon dioxide emission, many countries have future plans to replace their thermal power stations with wind power plant. The growth rate of wind energy use worldwide increased sharply in 2005. With a newly installed power from wind turbines of 11,407 MW last year's installation figures were exceeded by 40 %. In Europe, 6,372 MW wind power were newly installed, corresponding to an increase as against 2004 of approx. 7.6 % [1] .The PSG offers a very attractive solution to offshore wind plants due to high power to weight ratio, maintenance free and high number pole pair so there is no need for gearbox [2], [3], [4]. The advantages of this solution are high efficiency and reliability, which compensate the disadvantage of its high investment cost. The main disadvantage of PSG is that the magnet flux is constant. The measurement of 25KVA electrical excited synchronous generator reveals that generator delivers less than its rated power at rated excitation. To overcome this problem static synchronous series compensation (SSSC) should be provided to PSG [5], [6], [7]. Since SSSC compensates the reactance voltage drops of the generator and stabilizes the terminal voltage of the generator. In order to get rid of the harmonic which are produced due to the presence of the rectifier, the passive filters are used [8]. The proposed system is depicted in fig.1.

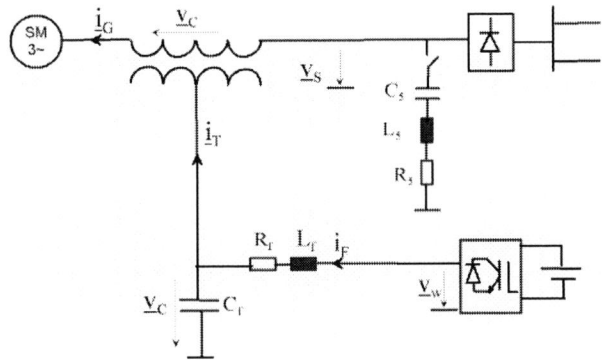

Fig.1: Scheme of SSSC with passive filter applied to PSG

II. STATIC SYNCHRONOUS SERIES COMPENSATION AND PASSIVE FILTER

A. Static synchronous series Compensation

The main function of SSSC is to change the impedance as if we are adding a capacitance or inductance by applying a compensation voltage which increases and decreases the voltage across the impedance respectively and thereby the current and power will increase or decrease respectively. It is discussed in details in [9], [10]. The compensation voltage V_C lags the current angle by 90 degree to act as a capacitor as shown in the phasor diagram in fig. 2. where θ current angle, δ power angle E internal induced voltage and Vs terminal voltage. The power equation is presented in eq. (1)

$$P = -\frac{3EV_S \sin\delta + 3V_S V_C \cos\theta}{X} \quad (1)$$

B. Passive Filter

Since the generator feeds a dc network through a rectifier bridge, as result the generator current contains harmonic components. The harmonic currents flowing through the generator reactance produce harmonic voltage distortion. The installation of passive filter for the fifth harmonic, since it is the dominant harmonics, reduce this distortion by diverting harmonic currents in low impedance paths.

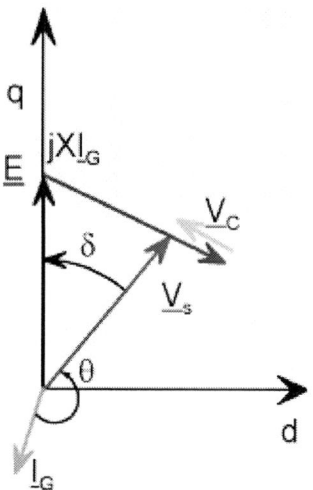

Fig.2: Phasor diagram of SSSC applied to PSG

As result the generator terminal voltage will be almost free from the harmonics. Furthermore passive filters improve the power factor of the generator in case of non sinusoidal current, the power factor is expressed in eq. (2). This is due reduction of the displacement reactive power D as the harmonic currents are reduced as in eq. (3) and passive filters are capacitive at the fundamental frequency and provide reactive power Q_c as in eq. (4). The generator apparent power S, which is presented by eq. (5), can be converted more to active power Also this will reduce the rating of the SSSC.

$$\cos\theta = -\frac{I_{Gf}}{I_G} \qquad (2)$$

$$D = V_S\sqrt{I_{G5}^2 + I_{G7}^2\ldots} \qquad (3)$$

$$Q_C = \frac{V_S^2}{X_C}\frac{n^2}{(n^2-1)} \qquad (4)$$

$$S^2 = P^2 + (Q_1 - Q_c)^2 + D^2 \qquad (5)$$

Where I_{Gf} the fundamental component of the generator current I_G , V_S the terminal voltage before the rectifier n harmonic order f_n/f_1, X_C capacitor reactance at f_1

C. Dimensioning the SSSC with and without passive filter

In order to dimension the required compensation voltage in the presence and absence of the passive filter, the generator current must be first analyzed, to facilitate analysis, the following assumptions are made:

1) Valves are treated as ideal switches
2) The dc current is not interrupted and free from ripple component.
3) The direct axis and quadrature axis impedance are assumed to equal

At the beginning the dc current I_d is calculated based on the equivalent circuit of fig 3. The dc current I_d is expressed in eq (6). where the E_{do} average dc voltage for the internal induced voltage and the compensation voltage at no load, R_x hypothetical resistance and V_{dc} DC network voltage [11],[12].

$$I_d = \frac{E_{do} - V_{dc}}{R_X} \qquad (6)$$

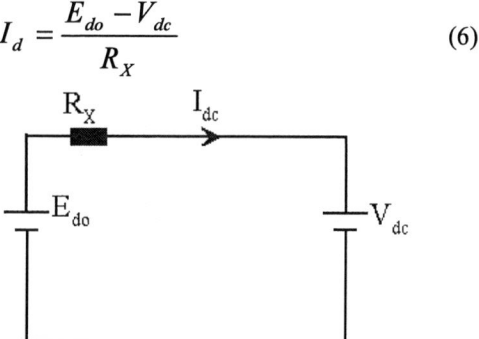

Fig.3: Dc equivalent circuit

The load current I_L produced by the soothed dc current I_d in the first half cycle is given as follows [13].The fourier expansion of i_L is expressed in eq.(7)

$$i_L = \sum_{k=1,5,7\ldots}^{\infty}(A_{Lk}\cos k\theta + B_{Lk}\sin k\theta) \qquad (7)$$

$$A_{Lk} = \frac{\sqrt{3}I_d(-1)^{l+1}}{\pi}[\frac{2\sin ku}{k} + \frac{1}{1-\cos u}\{\frac{-2\sin ku}{k}$$
$$+\frac{\sin(k+1)}{k+1} + \frac{\sin(k-1)u}{k-1}\}]$$

$$B_{Lk} = \frac{\sqrt{3}I_d(-1)^{l}}{\pi}[\frac{2\cos ku}{k} + \frac{1}{1-\cos u}\{\frac{2(1-\cos ku)}{k}$$
$$-\frac{1-\cos(k+1)}{k+1} - \frac{1-\cos(k-1)u}{k-1}\}]$$

$$k = 6l \pm 1(l = 0,1,2\ldots)$$

where u the overlap angle and calculated based on the assumption that the inductance is infinitely large and given by

$$u = \cos^{-1}(1 - \frac{2X_S I_d}{\sqrt{3}E})$$

Each harmonic component is calculated using eq. (8). Then the generator fundamental current is calculated using the equivalent circuit of the fundamental component which is presented in fig. 4. The harmonics of the generator current are calculated using the equivalent circuit of the harmonic which is presented in fig. 5 where the rectifier current source I_L. The generator harmonic current I_{Gk} and terminal harmonic voltage V_{Sk} are given as follows in eq (9) and eq (10) respectively. Since the impedance Z_F filter impedance is very small at the resonance frequency the harmonic current will flow in the filter and the terminal harmonic voltage V_{Sk} will be reduced.

$$I_{LK} = \sqrt{(A_{Lk}^2 + B_{Lk}^2)/2} \qquad (8)$$

$$I_{Gk} = \frac{Z_F}{Z_S + Z_F} I_{Lk} \qquad (9)$$

$$V_{Sk} = \frac{Z_F Z_S}{Z_S + Z_F} I_{Lk} \qquad (10)$$

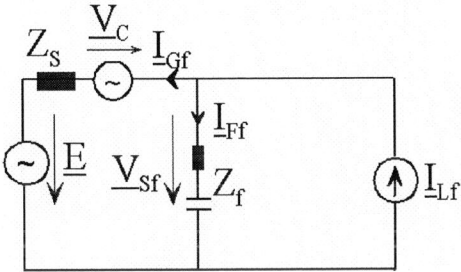

Fig.4: Scheme of SSSC with passive filter applied to PSG

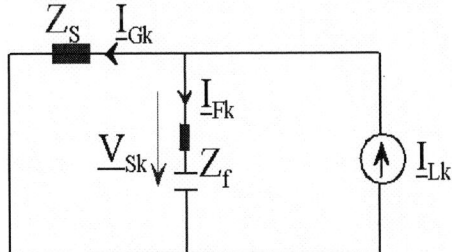

Fig.5 : Equivalent circuit of SSSC and passive filter to applied to PSG

The calculation is made for the machine with parameter in Table 1 for both cases with SSSC only and SSSC and passive filter. For both cases the internal induced voltage was 1.75 p.u and the machine was connected 500V dc network.

The required compensation voltage with and without passive is presented in fig. 6. It is clear that needed compensation voltage decreases with the installation of passive filter. This is due to the improvement in the power factor and the absence of the harmonic in the generator current and terminal voltage. The required compensation voltage is less when then the filter capacitance of the passive is 0.25pu than with capacitance 0.2 pu. The reason is that the passive filter will provide more reactive power as previously mentioned. The passive filters with bigger capacitance are able to provide more reactive power.

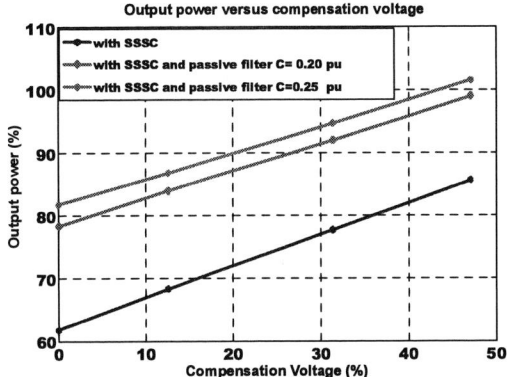

Fig.6 : Calculation of the required compensation voltage for SSSC alone and SSSC with different passive filter

III. SIMULATION RESULT

A. Simulation result with SSSC

The following simulation results show the effects of applying the SSSC to the PSG. The generator current increases compared to without compensation at fixed DC network voltage. As expected the generator current and the output power increase. The current of the generator has harmonic the fifth and seventh harmonic [14],[15].These harmonics are present in the terminal voltage also. The generator current and terminal voltage with compensation are presented in fig 7,8 respectively.

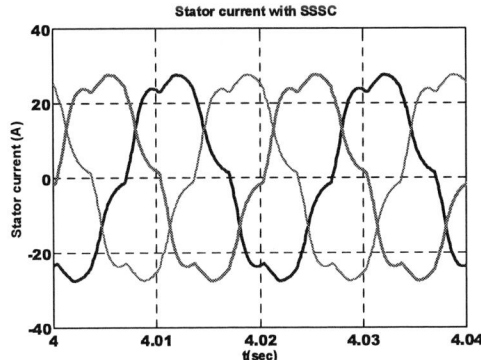

Fig.7: Simulation of stator current with SSSC

783

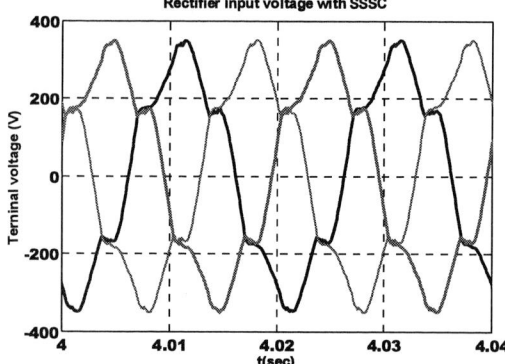

Fig.8 : Simulation of terminal voltage with SSSC

B. Simulation result with SSSC and passive filter

When applying the SSSC and passive filter for the fifth harmonic, the generator current increases. As result the output power increases compared to the case with only SSSC. The harmonics in the generator current decreases. This is due to that fact the passive filters eliminate the harmonics generated by nonlinear load by providing low impedance path. The amount of the harmonics in the current depends on the quality of the filter also . Filter with better quality can extract more harmonic current. The current and terminal voltage of the generator, with the same compensation voltage as without filter, are shown in fig 9,10 respectively. It is clear that the passive filter reduces the rating of the SSSC.

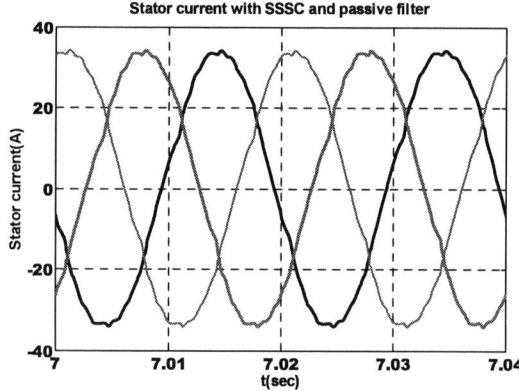

Fig. 9: Simulation of generator current with SSSC and passive filter

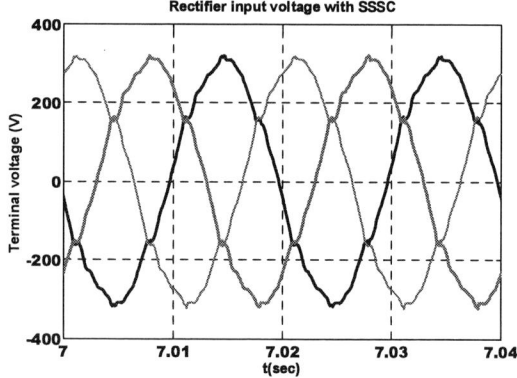

Fig. 10: Simulation of terminal voltage with SSSC and passive filter

IV. EXPERIMENTAL RESULTS

A. Experimental setup

Our small-scale experimental system of wind power generation consists of electrical excited synchronous generator connected to DC network as shown in fig. 11. A 58 KW dc machine fed by a 3phase thyristor controlled bridge with a separated field winding by a single-phase controlled rectifier were used to drive the generator shaft. The rotation speed of the dc motor is controlled by thyristor bridge to simulate the fluctuation of wind speed. Specification of synchronous generator used   for the experiment is shown in Table 1. The compensation voltage is realized by an inverter output, whose output voltage is fed to LC filter. Specification of inverter and filter is given in Table 2.

Rated Power S_n	25KVA
Rated voltage V_n	390V
Rated current I_n	37A
Rated power factor	0.8
Frequency f_n	50Hz
Direct axis reactance	11.5Ω
Quadrature axis reactance	6.5Ω

Table1 Specification of synchronous generator

Collector emitter voltage V_{CE}	1200V
Collector current I_c	25A
Filter inductance L_f	4m
Filter capacitance C_f	22μ F
Switching frequency	4KHz
Filter capacitance C_5	100 μ F
Filter inductance L_5	4mH
Q filter Quality	5

Table 2 Specification of Inverter and filter

Then output of the filter is connected in series with the generator through three single-phase transformers. The tapping changing of the transformers is set 1:1.5 turns' ratio. Hall effect sensors were used to measure the currents and the voltage. The voltage sensors have a ratio from 1000V/25mA, while the current transducers have a ratio 100A/25mA. As these sensors deliver a current signal, this latter must be converted to voltage. The processor AD converter is composed of 4 converters with 4 multiplexed inputs each with 16 bits resolution and 4μs sampling time. The rotor speed signal lead to an incremental encoder inputs together with AD converter, belong to DS1103 hardware board. This system enables one to work with models developed in matlab /simulink directly. The processor code is generated automatically. The inputs and outputs interfacing hardware and software are available as simulink blocks in real time interface.

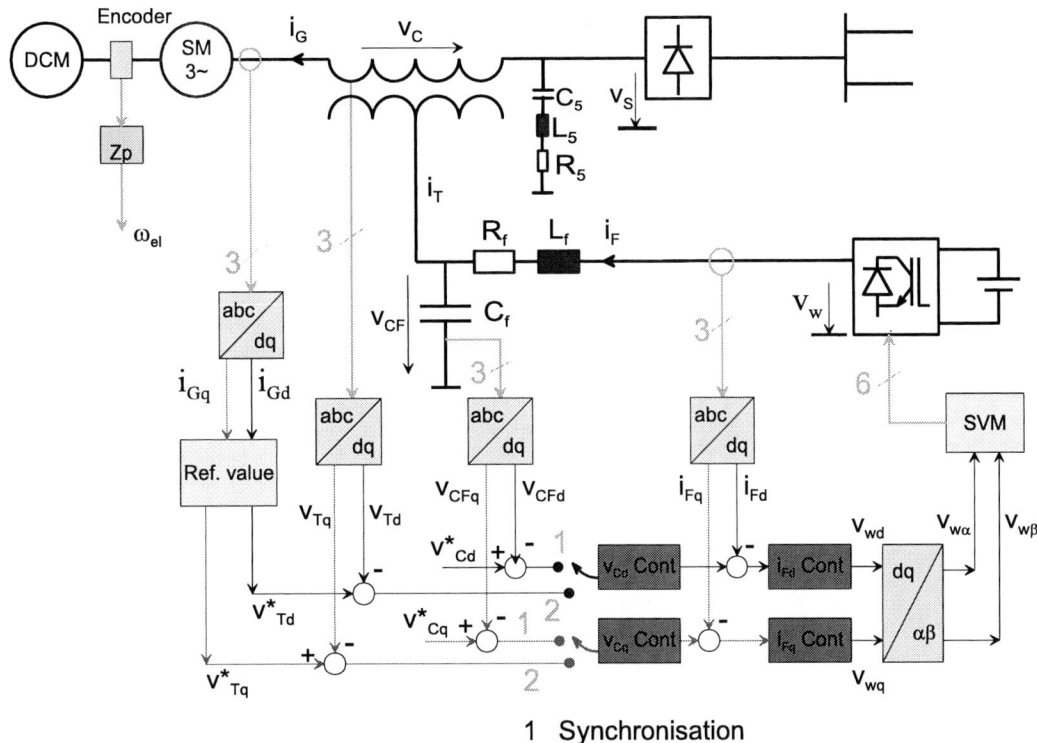

Fig. 11: Small scale experimental system of wind power plant

B. Measurements with SSSC

As shown in fig 12,13 which presents the measured stator of the generator current and terminal voltage with SSSC. The generator current and the terminal voltage in case of the SSSC have harmonic component, which agrees the simulation results.

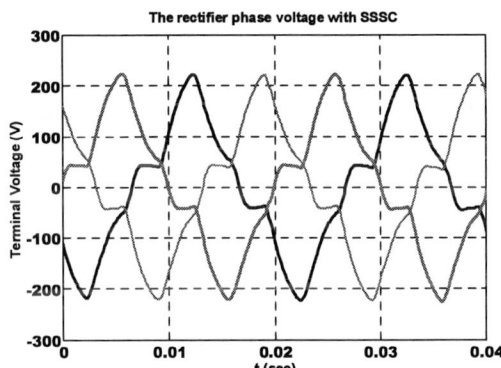

Fig. 13: Measured rectifier input voltage with SSSC

C. Measurements with SSSC and passive filter

When applying the SSSC and passive filter for the fifth harmonic, the generator current increases. As result the output power increases compared to the case with only SSSC. The measured stator of the generator current and measured terminal voltage with SSSC and passive filter is shown in fig 14,15. The harmonic component of the current decreases. Terminal voltage harmonics also decrease. This is due to the presence the passive filter.

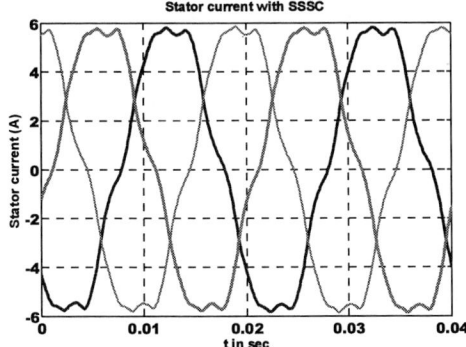

Fig. 12: Measured stator current with SSSC

785

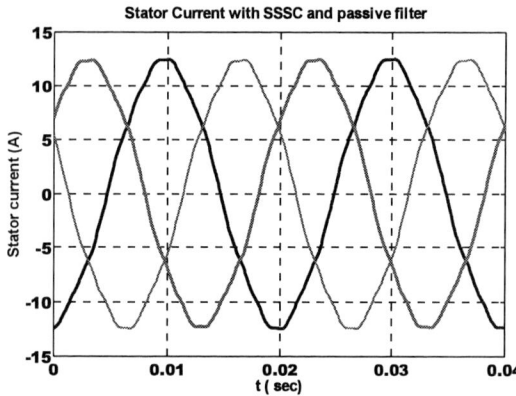

Fig. 14: Measured stator current with SSSC and passive filter

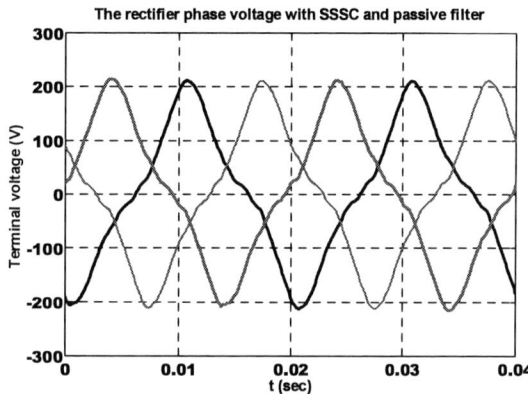

Fig. 15: Measured rectifier input voltage with SSSC and passive filter

The FFT of the terminal voltage harmonic with and without passive filter are presented in fig 15. The fifth harmonic of the terminal voltage decreases from 12% to 2.5% and the seventh from 7% to 1.1%. The THD of the terminal voltage decreases from 33.42% to 23.6%.

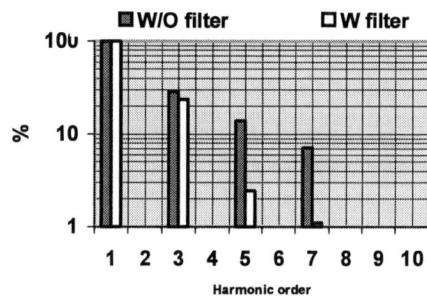

Fig. 15: Frequency spectrum the terminal voltage

The FFT of the generator current harmonic with and without passive filter are presented in fig 16. The fifth harmonic of the terminal voltage decreases from 10.3% to 2.7% and the seventh from 7% to 4 %. The THD of the generator current decreases from 12.5% to 5.16%

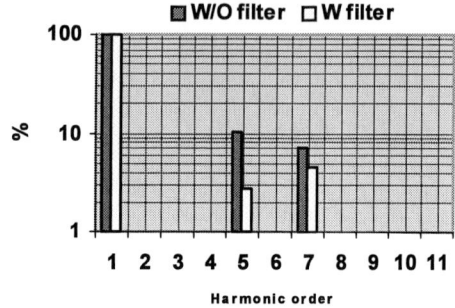

Fig. 16: Frequency spectrum the generator current

V. CONCLUSION

The measurements and simulation indicated that in order to obtain the rated power of the PSG with connection to DC network, we must exceed the rated excitation. SSSC is an attractive solution to this problem. SSSC increases the output power and stabilizes the output terminal voltage. Passive filter eliminates the harmonics and provides reactive power at the fundamental frequency. As result passive filters reduce the rating of SSSC. The reduction of the rating of SSSC depends the capacitance of the filter and the quality of the filter. This enables the design of actual wind turbine system with PSG more economical. The experimental results show that there is possibility to employ this to offshore wind power plants, so we will be able to build PSG and make use of its high power to weight ratio. Also its fixed flux problem is solved and got rid of the harmonics.

References

[1] C. Ender : Windenergienutzung in Deutschland. DEWI Magazin Nr. 28 Februar 2006 S. 10-21

[2] Spooner, E.; Williamson, A.C.: Direct Coupled, permanent magnet generators for wind turbine applications. IEE Proc. Electrical Power Applications Vol.143, No.1 1996 p.1-8

[3] Vilsboll,N.; Pinegin, A.; Bugge, J.: Zur Weiterentwicklung des Konzeptes von vielpoligen direktgetriebenen Permanent-Magnet-Generatoren für Windkraftanlagen mit variabler Drehzahl. Wind Kraft Journal 1996 H.6 S. 14-19

[4] Heier.S.; "Grid Integration of Wind Energy Conversion Systems ".John Wiley &Sons 1998.

[5] Gyugyi L.: Converter-Based FACTS Technology: Electric Power Transmission in 21st Century. IPEC 2000 Tokyo, pp. 15-26.

[6] Narian G. Hingorani, Laszlo Gyugyi "Understanding FACTS, Concepts and Technology of Flexible AC Transmission Systems " 1999 IEEE press".

[7] Yong Hua Song ; Allan T Johns, "Flexible ac transmission systems (FACTS)" 1999 IEE power and energy series 30.

[8] Y.S.Kim ,J.S Kim, Three phase three —wire series active power filter which compensates for harmonics and reactive power IEE Proc.-Electr. Power Appl. Vol. 151 No3 P276 -282

[9] Singer A. , Hofmann W. : Comparison between Different Compensation Methods Applied to Permanent Magnet synchronous Generator PCIM 2005 CD

[10] Singer A., Hofmann W.: Local Compensation of Wind Energy Conversion System. EPE 2005 Dresden CD

[11] Zhifang Sun, Masaaki Sakui : Calculation of Harmonic Currents in Series Connected Converter with AC Filters Under under unbalanced Power Supply , IEEE 1999 International Conference on Power Electronics and Drives Systems , PEDS 99 July 1999 , Hong Kong

[12] Masaaki Sakui ,Mitsuo Shioya : A Method for Calculating Harmonic Currents of a Three Phase Bridge Uncontrolled Rectifier with DC Filter , IEEE trans Ind Electr Vol 36 No3 pp. 436-440 August 1999.

[13] A. Ametani : Harmonic Reduction in Thyristor by Harmonic Current Injection , IEEE Trans on power app. Vol PAS –95, no. 2 pp. 441-450 , March/April 1976

[14] Singer A., Hofmann W.: Static Synchronous series Compensation applied to Permanent magnet Synchronous generator . PCIM 2007 CD

[15]Singer A., Hofmann W.: Static Synchronous series Compensation applied to Small Wind Energy Conversion System EPE 2007 Aalbog CD

Control of active injector for multi-pulse rectifiers operating on variable frequency supplies

Ismael Araujo-Vargas*, Andrew J. Forsyth**, and F. Javier Chivite-Zabalza**

*High School of Mechanical and Electrical Engineering, National Polytechnic Institute of Mexico, ESIME Culhuacan, Av. Santa Ana No. 1000, Col. San Francisco. Culhuacan, C.P. 04430, Mexico City, Mexico.
** School of Electrical and Electronic Engineering, The University of Manchester, PO Box 88, Sackville Street, Manchester M60 1QD, UK

Abstract - **A control strategy is described to drive an active injector for multi-pulse rectifiers operating on variable frequency aerospace supplies. The injection technique utilises a single low-rated active device that carries approximately 2.9% of the load current, and this may be driven with either low-frequency pulses or high-frequency, pulse-width modulation to produce 24-pulse operation or multi-level PWM operation respectively. The resultant input currents are almost sinusoidal, the line current THD being 2.36% for the 24-pulse operation and 1.06% for the PWM operation. The control principle, design and performance are presented in this paper. Experimental results of the controller with a 4 kW prototype are shown.**

I. Introduction

A 12-pulse rectifier with harmonic frequency injection is one solution to the increasingly strict power quality specifications being placed on rectifier equipment, for example in aerospace systems. The circuits are simple, robust and offer input currents with 24-pulse or better characteristics, [1] to [4]. The circuit in Fig. 1, [5], uses one low-rated transistor to provide the injection waveform, however, the transistor drive signal must remain synchronized with the 6th harmonic of the supply under all conditions; in aerospace applications the supply frequency can range between 400 to 800 Hz.

This paper presents the control strategy used to drive the transistor of the circuit shown in Fig. 1 operating on a variable frequency supply. The operation of the controller for the active injector is analysed for two operating modes, 24-pulse and multi-level PWM, showing how the active injector waveforms are used to synchronise the controller. The final part of the paper presents experimental results of the controller with a 4 kW laboratory prototype in both operating modes, in steady state and transient conditions.

II. Multi-Pulse Rectifier with Active Injection

A. Summary of the converter

The description of the circuit operation, Fig. 1, assumes lossless components, negligible transformer magnetizing current and output voltage ripple, sinusoidal source currents and equal capacitor voltages on C_{f1} and C_{f2}. The detailed operation of the active injector is explained with reference to the equivalent circuit in Fig. 2 and the idealised waveforms in Fig. 3. The current

sources i_1 and i_2, Fig. 3, and the parallel bypass diode paths represent the output of the main rectifiers. i_1 and i_2 are the full-wave-rectified input currents of the two rectifiers, which are phase shifted by 30° due to the transformer.

i_1 and i_2, and their instantaneous difference, $i_{dif} = i_1 - i_2$, are shown as the first and second waveforms in Fig. 3. The switching state of the active injector, v_{gs}, is the third waveform in Fig. 3, and the last two waveforms are the bi-directional switch voltage and current, v_{MG} and i_{inj} respectively.

When the active injector is in the off state with $i_1 > i_2$, i_{dif} flows in the top bypass diode clamping the voltage v_{O2} to zero and resulting in a voltage of V_O across the output of the bottom rectifier; whilst with $i_1 < i_2$, i_{dif} flows in the bottom bypass diode, v_{O1} is clamped to zero and v_{O2} equals the output voltage V_O. In contrast, when the active injector is in the on state, v_{O1} and v_{O2} are clamped to $V_O/2$ by the conducting switch and the split dc-link capacitors. The bypass diodes are reverse biased and i_{dif} flows through the bi-directional switch.

The circuit operation is modified by the bi-directional switch, or active injector, when this is driven using either low frequency pulses of 50% duty ratio at 12 times the AC supply frequency, Fig. 3(a), or high frequency pulses modulated in width, Fig. 3(b), producing the waveforms v_{MG} and i_{inj} shown in Figs. 3(a) and 3(b) respectively. In Fig. 3(a), it is seen that v_{MG} is converted to a quasi-squarewave, which causes 24-pulse operation of the rectifier; whereas in Fig. 3(b), v_{MG} is a three-level PWM waveform, which causes multi-level PWM operation of the rectifier.

v_{O1} and v_{O2} are given by $V_O/2 \pm v_{MG}$ and have complementary shapes, switching between zero, $V_O/2$ and V_O, the fundamental component of v_{O1} and v_{O2} being locked at six times the supply frequency.

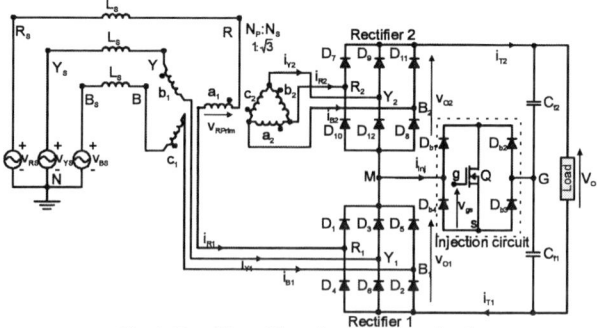

Fig. 1 Rectifier with active injection circuit.

This work was financially supported by Goodrich Corporation Electromagnetic Systems Technical Centre, Birmingham UK, the National Council of Science and Technology (CONACyT) and the National Polytechnic Institute (IPN) of Mexico.

978-1-4244-0644-9/07/$25.00 ©2007 IEEE

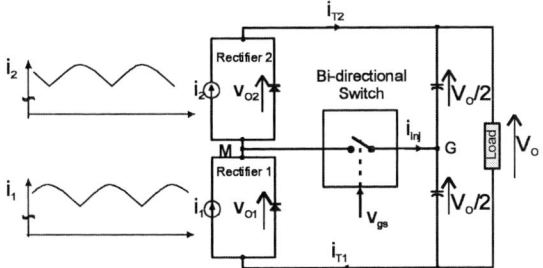

Fig. 2 Equivalent circuit of the multi-pulse rectifier with active injection.

B. Production of the 24-pulse and multi-level PWM waveforms

Fig. 4(a) shows how the modification of v_{MG} to a quasi-squarewave results in 24-pulse voltage waveforms at the right hand side of the input inductors with respect to the supply neutral, whilst Fig. 4(b) shows how PWM operation of the transistor in the active injector creates multi-level PWM waveforms. The sketches in Fig. 4(b) show the same waveforms as those presented in Fig. 4(a) for straightforward comparison. For simplicity Fig. 4 only shows the derivation of the v_{RN} waveform in both operating modes.

In Fig. 4, the first three waveforms show the lower rectifier input voltages referred to the mid-point of the dc link, v_{R1G}, v_{Y1G} and v_{B1G}, which are clamped to $-V_O/2$ when the respective line current is negative, and when the line current is positive the upper diode in the bridge leg is in conduction and the rectifier input voltage is equal to v_{MG}, shown in Fig. 3.

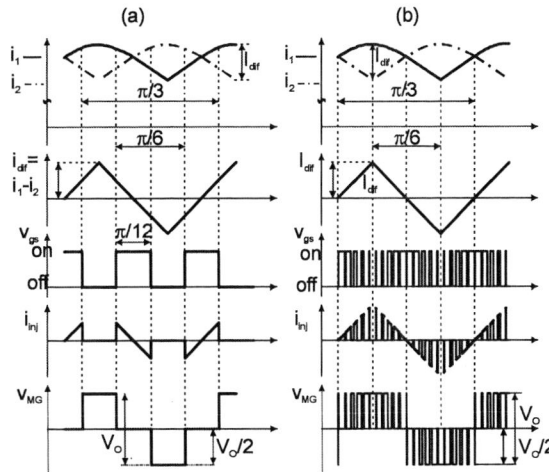

Fig. 3 Idealised waveforms of the active injection circuit in (a) 24-pulse operation, and (b) multi-level PWM operation.

The fourth waveform shows the $R1$ voltage referred to the supply neutral, v_{RIN}, which is obtained by subtracting the common mode voltage, $v_{RIN} = (2v_{R1G} - v_{Y1G} - v_{B1G})/3$. v_{R2G} and v_{B2G} are the fifth and sixth waveforms shown in Fig. 4, which are similar in form to v_{R1G}, v_{Y1G} and v_{B1G}, but are clamped to $-V_O/2$ when the respective input current is positive. The seventh waveform depicted in Fig. 4 is v_{RPrim}, the transformer primary voltage, which is calculated using $v_{RPrim} = (v_{R2G} - v_{B2G})/\sqrt{3}$, and lastly, v_{RN} is calculated as the sum of v_{RIN} and v_{RPrim}.

v_{RN} is seen to be a 24-pulse waveform in Fig. 4(a), whilst in Fig. 4(b) v_{RN} is seen to be a multi-level PWM with the same voltage levels as those obtained for the 24-

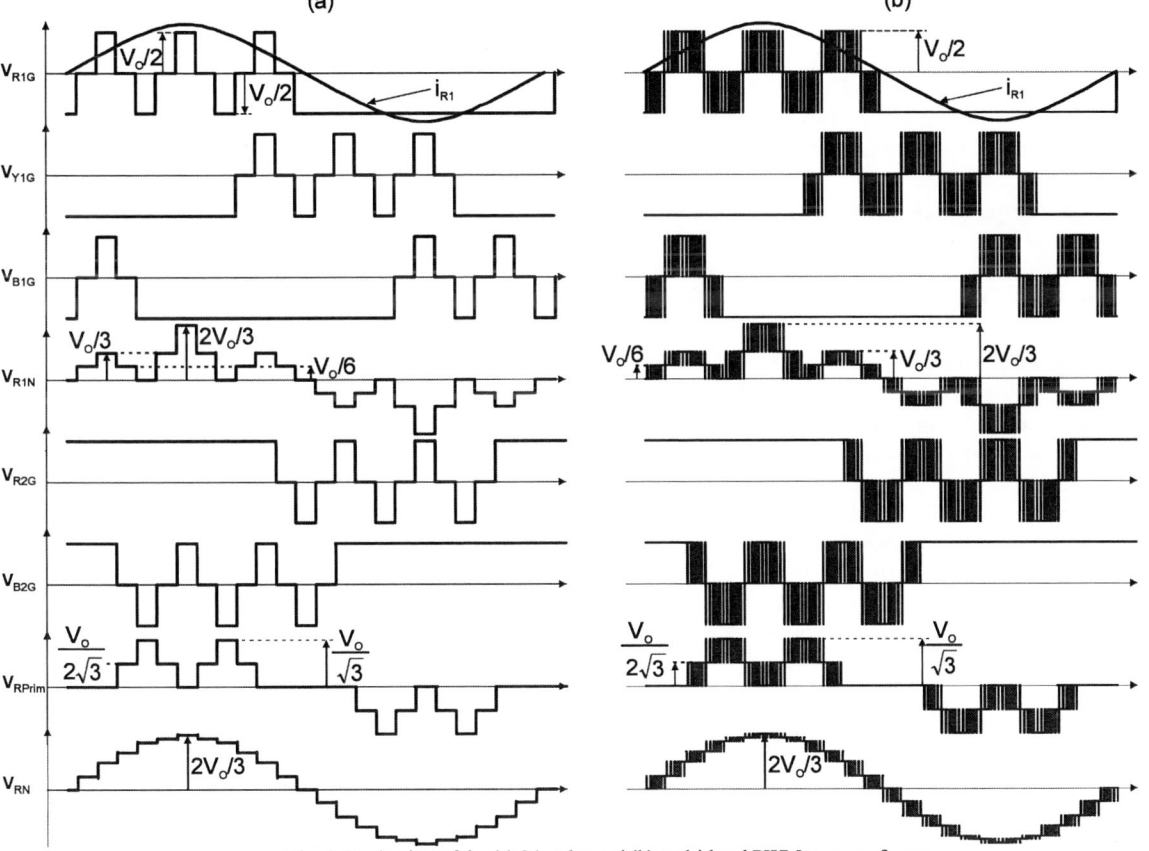

Fig. 4 Production of the (a) 24-pulse and (b) multi-level PWM v_{RN} waveforms.

pulse operation. The waveforms v_{YN} and v_{BN} have identical shapes but are phase shifted from v_{RN} by 120° and 240° respectively.

The optimum value of the angular pulse width, δ, to activate the bi-directional switch in the 24-pulse mode was determined by investigating the minimum total harmonic distortion (THD) of the v_{RN} waveform, and a minimum THD of 7.7% was found for $\delta = 15°$. In contrast, for the PWM mode, the optimum modulating waveform for v_{MG} was derived by attempting to minimize the error between the space vector of the set of multi-level voltage waveforms at the right hand side of the line inductors and an ideal space vector of balanced fundamental frequency sinewaves. This error, also defined as the harmonic space vector, was derived in terms of v_{MG}, and a minimum harmonic space vector was obtained when the modulating waveform for v_{MG} was seen to be approximately a triangular waveform of amplitude $\pm V_O/2$, [5].

III. CONTROL SYSTEM OF THE ACTIVE INJECTOR

A. Control strategy for the 24-pulse operation

The control system synchronises the active injector with the ripple-frequency of the rectifier over the 400 – 800 Hz frequency range and under supply transient conditions and changing load. Fig. 5 shows the control system for the 24-pulse circuit, which consists of a phase-locked loop (PLL) and a bipolar modulator to generate the pulses for the bi-directional switch, v_{gs}. The PLL consists of a phase detector, a loop filter, a voltage controlled oscillator (VCO) and a zero crossing detector (ZCD). The bipolar modulator compares the output triangular waveform of the VCO, v_{tri}, with the fixed levels $\pm V_{th}$ to generate v_{gs}. The reference signal for the PLL of the 24-pulse circuit is the current i_{inj}, which is sensed through a transducer of gain k_t placed between the bi-directional switch and the dc-link mid-point G. The phase detector comprises two double-pole/single-throw switches in series that connect the loop filter between $\pm k_t i_{inj}$ and zero depending on the state of the ZCD and the bipolar modulator output, v_{gs}. V_{DO} is the phase detector offset which is assumed zero for the 24-pulse circuit.

Suitable design of the PLL allows the VCO output frequency, ω_O, to be synchronised to the six-pulse ripple such that the activation pulses for the bi-directional switch occur at the correct positions. The operation of the system of Fig. 5 and the bi-directional switch is explained with reference to the idealised waveforms of Fig. 6, which correspond to the signals of the control diagram in Fig. 5, and it is assumed that the full-wave-rectified input currents, i_1 and i_2, for the two rectifiers are sinusoidal, such that i_{dif} is a triangular waveform at six-times the supply frequency with no dc offset.

In Fig. 6, the frequency of v_{tri}, ω_O, is slightly lower than that of i_{dif}, ω, and thereby, the PLL is out of lock. The ZCD output, v_{ZCD} in Fig. 6, determines the switching state of the phase detector. Together with v_{ZCD}, a ± 1 quasi-squarewave is shown, f_{PD}, which is equivalent to the switching function of the phase detector. v_{gs}, the

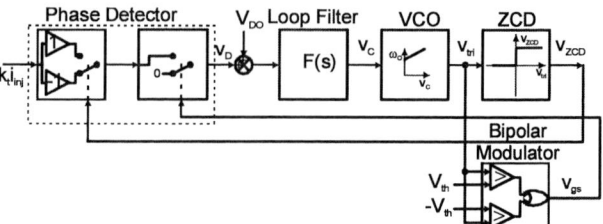

Fig. 5 Control system for the 24-pulse rectifier.

fourth waveform in Fig. 6, drives the active injector. It is seen that v_{MG} and i_{inj} are not steady waveforms due to the different frequencies of i_{dif} and v_{tri}. The shape of i_{inj} varies continuously throughout the period shown in Fig. 6; whereas the shape of v_{MG} is steady in a small period, which occurs when the current i_{dif} reverses whilst the bi-directional switch is on, otherwise the shape of v_{MG} varies continuously. This characteristic makes v_{MG} unsuitable to be the reference signal for the PLL. Since i_{inj} varies continuously with phase error between i_{dif} and v_{tri}, it was selected as the reference for the PLL.

The phase detector output, v_D, is the last waveform in Fig. 6, where $v_D = k_t i_{dif} f_{PD}$, which has a low-frequency component since ω_O is relatively close to ω. When these frequencies become equal, the low-frequency component, or local average v_D, becomes a DC offset, which is proportional to the phase error between $k_t i_{dif}$ and f_{PD}. The purpose of the loop filter in the closed loop diagram of Fig. 5 is to remove the high-frequency components of v_D, such that the VCO input signal, v_C, is steady and the VCO frequency, ω_O, is locked to the frequency of i_{dif}, ω. To place the pulses of v_{gs} in the correct phase, v_{ZCD} and f_{PD} should be 90° away from i_{dif}, which results in symmetrical positive and negative pulses of v_D, and, therefore, v_D must have a zero average, implying that $V_{DO} = 0$.

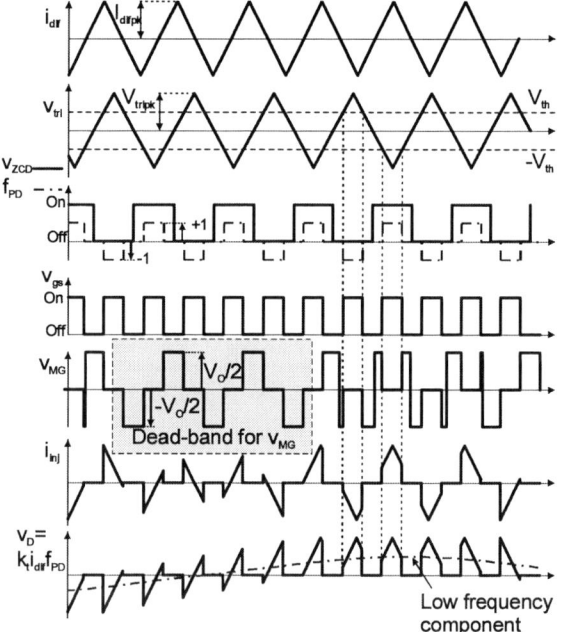

Fig. 6 Idealised waveforms of the control system for the 24-pulse circuit, with $\omega > \omega_O$.

790

B. Phase detector characteristic for the 24-pulse operation

The relationship between v_D and θ_e, the phase error between $k_t i_{dif}$ and f_{PD}, may be derived using eq. (1),

$$v_D = \frac{1}{2\pi}\int_{-\pi}^{\pi} k_t i_{dif}(\omega t) f_{PD}(\omega t) d(\omega t) \qquad (1)$$

which is calculated assuming that the waveforms $k_t i_{dif}$ and f_{PD} have equal frequencies, and that f_{PD} is delayed behind $k_t i_{dif}$ by an angle θ_e. By expressing one cycle of the waveforms $k_t i_{dif}$ and f_{PD} of Fig. 6 in a piece-wise linear manner, the integration of eq. (1) may be undertaken resulting in the phase detector characteristic shown in Fig. 7. v_D in Fig. 7 may be approximated using the Fourier series to describe the phase detector characteristic, and it was found that the amplitude of its fundamental is $0.365 k_t I_{difpk}$, the 3rd harmonic amplitude is 3.75% of the fundamental, and the rest of the harmonic magnitudes are much lower than 1%. Since the harmonics of v_D are very small, v_D may be described as $v_D \cong 0.365 k_t I_{difpk}\sin(\theta_e)$ and written as $v_D = k_D \theta_e$ if a linear approximation is applied for small values of θ_e. $k_D = 0.365 k_t I_{difpk}$ is the phase detector gain, such that v_D becomes proportional to θ_e.

C. Control strategy for the multi-level PWM operation

The control system for the multi-level PWM circuit, shown in Fig. 8, has a similar structure to that for the 24-pulse circuit. The system uses the bi-directional switch voltage, $k_{MG}v_{MG}$, as the reference signal for the PLL, where k_{MG} is the voltage transducer gain. v_{tri}, is synchronised to twelve-times the supply frequency, and then divided down to six-times the supply frequency, v_{sq6}, to drive the phase detector. The pulses for the bi-directional switch are produced by a fast comparator, which operates as a naturally-sampled, asynchronous pulse width modulator, comparing v_{tri}, the modulating signal, with a high-frequency triangular carrier waveform, $v_{carrier}$. The phase detector is driven by v_{ZCD} and v_{sq6}.

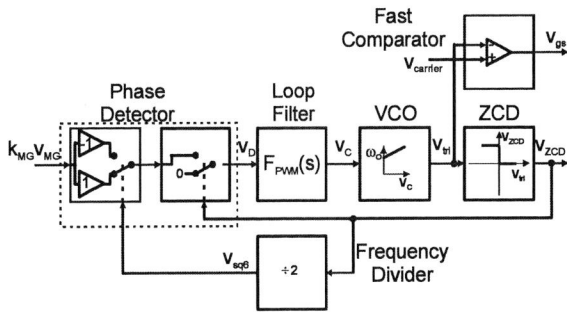

Fig. 8 Control system for the multi-level PWM rectifier.

The principle of operation of the PLL shown in Fig. 8 is the same as that described for the 24-pulse circuit. The waveform v_D represents the phase error between v_{MG} in the power circuit and the divided down VCO output, v_{sq6}. This is illustrated in Fig. 9, where the control loop waveforms of Fig. 8 are laid out assuming that ω_o, the frequency of v_{sq6}, is slightly lower than ω_i, the frequency of v_{MG}.

i_{dif} is the first waveform shown in Fig. 9. The second and third waveforms are v_{tri} and v_{ZCD}, which are at twice the frequency of v_{sq6}, ω_o, the fourth waveform shown in Fig. 9. The phase-detector switching function, f_{PD}, the fifth waveform shown in Fig. 9, is determined by the states of v_{ZCD} and v_{sq6}. v_{gs} is a naturally-sampled PWM waveform produced by v_{tri} and $v_{Carrier}$. The last two waveforms sketched in Fig. 9 are v_{MG} and v_D, the phase detector output.

Since ω_o is different to ω_i, v_{MG} does not have a stable shape and changes continuously with phase error in such a way that it is a suitable reference signal for the PLL. In Fig. 9, v_D is the instantaneous product of f_{PD} and $k_{MG}v_{MG}$, which is seen to have a low-frequency component since ω_o is relatively close to ω_i. When $\omega_o = \omega_i$, the low-frequency component becomes a DC offset, proportional

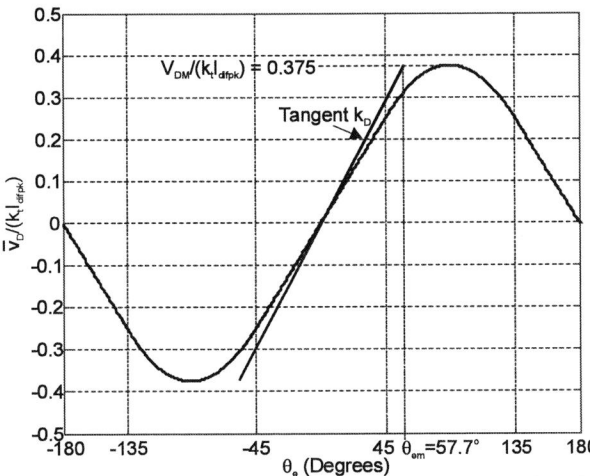

Fig. 7 Phase detector characteristic of the control system for the 24-pulse circuit.

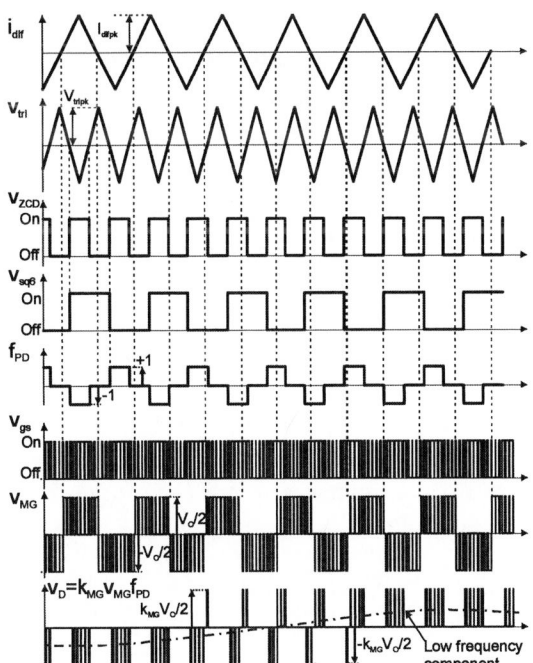

Fig. 9 Idealised waveforms of the control system of Fig. 8 for the multi-level PWM circuit, with $\omega_i > \omega_o$.

791

to the phase error between f_{PD} and $k_{MG}v_{MG}$. When v_D is zero, the waveforms f_{PD} and $k_{MG}v_{MG}$ are 90° out of phase, which is the relationship required for the injection method of the multi-level PWM operation. The loop filter removes the high frequency components of v_D, such that v_C is a steady signal and ω_O is locked to ω_i.

D. Phase detector characteristic for the multi-level PWM operation

To calculate the phase detector characteristic, the local average of the product $v_D = k_{MG}v_{MG}f_{PD}$, $\bar{v}_D$, may be derived using eq. (2):

$$\bar{v}_D = \frac{1}{2\pi}\int_{-\pi}^{\pi} k_{MG}v_{MG}(\omega t)f_{PD}(\omega t)d(\omega t) \qquad (2)$$

which is calculated assuming that the waveforms $k_{MG}v_{MG}$ and f_{PD} have equal frequencies, and that f_{PD} is delayed behind $k_{MG}v_{MG}$ by an angle θ_e. By expressing one cycle of the waveforms $k_{MG}v_{MG}$ and f_{PD} of Fig. 9 in a piece-wise linear manner, the integration of eq. (2) may be undertaken and approximated using the Fourier series to describe the phase detector characteristic. Again, like the phase detector characteristic for the 24-pulse circuit, it was found that the amplitude of the fundamental of $\bar{v}_D$ is $0.067k_{MG}V_O$. Therefore, the phase detector gain may be written as $k_D = 0.067k_{MG}V_O$ by using a linear approximation to the fundamental component of $\bar{v}_D$ for small values of θ_e.

E. Linear model of the control system

In the region of its steady-state operating point, the control system for both operating modes may be represented by the linear model shown in Fig. 10. The input and output frequencies are assumed to be equal on average, $\omega_O = \omega_i$, where v_C is the local average input voltage to the VCO from the filter, V_{CO} is the in lock value of v_C, k_V is the small-signal gain of the VCO, $k_V = \partial \omega_O / \partial \bar{v}_C$, and thereby, $\omega_O - \omega_i = \Delta \omega_O = d\theta_O/dt$, [6] and [7].

Fig. 10 also shows an input frequency deviation $\Delta \omega_i$, which is given by $\Delta \omega_O = d\theta_O/dt$. The small-signal model for the PLL is obtained by eliminating the dc offsets from Fig. 10, and therefore, the open loop transfer function for the PLL may be written as:

$$T(s) = \frac{\theta_O(s)}{\theta_e(s)} = \frac{k_D k_V F(s)}{s} \qquad (3)$$

Appropriate selection of the loop filter transfer function $F(s)$ determines the bandwidth, ω_m, of the PLL

and the phase margin of $T(j\omega)$, ϕ_m. ω_m defines the maximum frequency at which θ_i can vary and still be tracked satisfactorily by θ_O, and the maximum frequency step for $\Delta \omega_i$ that the PLL can tolerate to keep frequency lock.

IV. VERIFICATION OF THE CONTROL SYSTEM FOR THE ACTIVE INJECTOR

A. Circuit implementation

The operation and performance of the active injector controller were verified using a 4 kW laboratory prototype. The prototype was operated from a variable-frequency electronic power supply, a star-delta transformer with a 45:77 turns-ratio, the total source inductance per phase being 480 μH, or 0.13 pu at 400Hz, thereby ensuring continuous conduction operation of the rectifier. Each DC-link output filter consisted of a 470 μF electrolytic capacitor in parallel with polypropylene components. The main rectifier bridges were formed by two VUE-3506NO7 fast recovery modules and the bi-directional switch was formed by four ultra-fast diodes 60EPU04 and an IRFP22N50A MOSFET.

For simplicity, the controllers of Figs. 5 and 8 were implemented using analogue circuitry. Two double-pole/single-throw analog switches AD7512 formed the phase detector of the active injector control system. The current i_{inj} for the 24-pulse mode was sensed using a current transducer Honeywell CSNE151, which was externally arranged to provide a gain of $k_t = 5$ V/A; whereas the PLL for the PWM mode, the waveform v_{MG} was sensed with a voltage transducer of gain $k_{MG} = 1/19.2$. The gain k_D for each operating mode was calculated for a 4 kW load, giving $I_{difpk} = 2.2$ A and $V_O = 240$ V, and thereby $k_D = 4.125$ V/rad for the 24-pulse mode and $k_D = 0.75$ V/rad for the PWM mode.

The VCO was an ICL8038, and it was configured, for the 24-pulse mode, to operate at six times the 400 – 800 Hz supply frequency, that is from 2.4 kHz to 4.8 kHz; whilst for the PWM mode, the VCO was configured to operate at 4.8 kHz to 9.6 kHz, twelve times the 400 – 800 Hz supply frequency. The small-signal gain of the VCO, k_V, was 754 rad/Vs for the 24-pulse mode; whereas for the PWM mode k_V was 838 rad/Vs. The individual loop filter transfer function for the 24-pulse and PWM control systems were designed to be:

24-pulse circuit $\qquad F(s) = \dfrac{720(s+111)}{s(s+3.56k)} \qquad (4)$

PWM circuit $\qquad F(s) = \dfrac{1.56k(s+111)}{s(s+3.56k)} \qquad (5)$

giving a bandwidth of 100 Hz and a phase margin of 70° for both control systems. The bandwidth was chosen to be small to avoid jittering effects in the waveforms.

B. Experimental results

To allow frequency acquisition for the PLL, the variable-frequency supply was initially set to 600 Hz, since the average of the VCO control voltage, $\bar{v}_C$, at

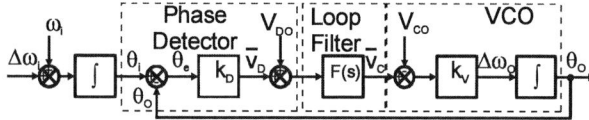

Fig. 10 Linear model of the PLL.

power up conditions is close to 0 V, such that ω_b is relatively near to 3.6 kHz. At this stage, the frequencies of the PLL, ω_b, and that of the six-pulse ripple, ω_i, are slightly different, but the PLL was able to acquire frequency lock.

After the PLL reached lock acquisition, the supply frequency was varied gradually over the 400 – 800 Hz range to confirm the system operation. Figs. 11(a) and 11(b) show the three-phase line current, v_{MG} and v_{RN} measured waveforms for the 24-pulse and multi-level PWM modes respectively in steady-state conditions, obtained with a 115 V, 400 Hz supply and a 4 kW load. A low-current 70 mH inductor was connected in parallel with the bi-directional switch to balance the voltage of the dc-link capacitors.

A slight current imbalance of less than 0.6 A is evident in Figs. 11(a) and 11(b) and was attributed to transformer asymmetry, however this is well within typical aerospace power quality limits, [8] and [9]. A power analyser measured the line current THD to be 2.36% and 1.06%, the angle of the line current was 7.06° and 6.75° with respect to the supply voltage, and the power factor was 0.992 and 0.993 for the 24-pulse and PWM modes respectively. The measured power loss was 193 W, for the 24-pulse, and 230 W, for the PWM mode. The slightly increased power loss in the PWM mode was due to the high-frequency PWM operation of the bi-directional switch, which was fixed at 100 kHz, increasing the circuit losses by approximately 15%. The additional losses occurred in the transformer, the main rectifiers and the bi-directional switch. The lower plots in Figs. 11(a) and 11(b) show the v_{RN} waveforms which have the 24-pulse and multi-level PWM shapes predicted in Figs. 4(a) and 4(b), although it is seen that there is a high-frequency oscillation on the rising edge of some of the voltage steps and width modulated pulses. These effects were attributed to the presence of parasitic inductance in the converter, but principally to interwinding capacitance of the three-phase transformer.

Fig. 12 shows a comparison of the normalised line current harmonics in the 12-pulse, 24-pulse and high-frequency PWM operating modes. The presence of very

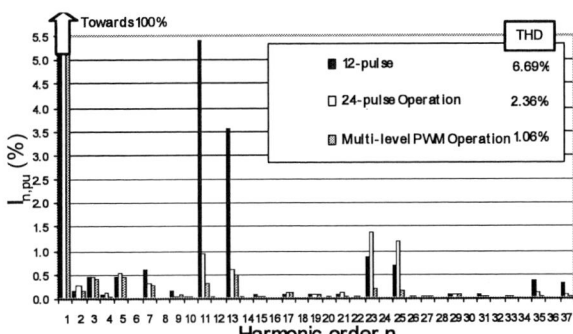

Fig. 12 Comparison of normalised line current harmonics for the 4 kW prototypes, 115 V, 400 Hz supply.

small third, fifth and seventh harmonics was attributed to imperfect harmonic cancellation, principally caused by the small current drawn for the transformer excitation and core losses. Fig. 12 shows that the 24-pulse mode virtually eliminates the 11th, 13th, 35th and 37th harmonics, but at the expense of an increase in the 23rd and 25th, whilst the PWM mode reduces all the harmonics below 0.5% of the fundamental.

A supply frequency step was applied to verify the control system performance under frequency deviations. Using eq. (3), the maximum input frequency step that the injector controller will support at 4 kW was calculated, and it was found that for the 24-pulse circuit the frequency step changes in i_{inj} were restricted to ±144 Hz, which translates to maximum step changes in the supply frequency of ±144 Hz/6 = ±24 Hz. Figs. 13(a) and 13(b) show the measured results of i_{R1}, v_{MG}, v_D and v_C for a 400 – 410 Hz frequency step at 0 ms, with a 115 V supply and a 4 kW load. i_{R1} and v_{MG} in Fig. 13(a) are seen to maintain the waveform quality throughout the transient; whereas v_D, in Fig. 13(b), transiently has a non-zero mean value to reduce v_C such that the VCO frequency and phase are adjusted. The mean value of v_D is restored to zero when the VCO has the correct frequency and phase.

Similarly, a supply frequency step was applied to verify the controller performance of the multi-level PWM circuit under frequency deviations. Figs. 13(c) and 13(d) show the measured results of i_{R1}, v_{MG}, v_D and v_C for a 400 - 410 Hz frequency step at 0 ms, with a 115 V supply and a 4 kW load. Again, the maximum input frequency step that the injector controller will support at 4 kW was calculated and it was found that step changes in v_{MG} were restricted to be within ±117 Hz, which translates to maximum step changes in the supply frequency of ±117 Hz/6 = ±19.5 Hz. In Fig. 13(c), the waveform quality of i_{R1} is seen to be maintained throughout the transient; whereas in Fig. 13(d), the local average of v_D is non-zero during the transient and this returns to zero after the transient decays away.

The controller performance relied on the amplitude of the current i_{inj} for the 24-pulse circuit, and, for the PWM circuit, on the amplitude of v_{MG}. These determine the phase detector gain k_D, the magnitude of the open-loop transfer function $T(s)$, the bandwidth ω_n, and thereby, the maximum step of input frequency, supply voltage or load that the injector controller supports for each operating mode.

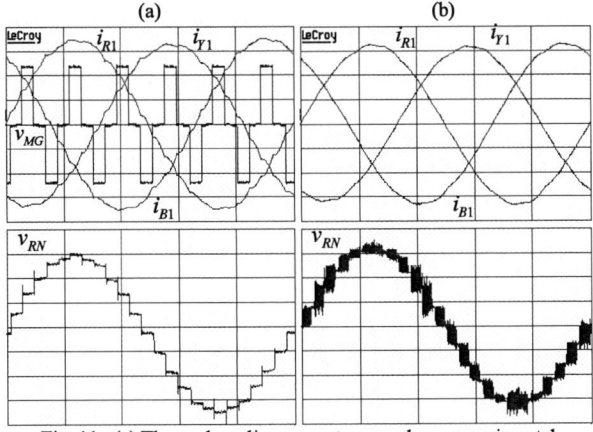

Fig. 11. (a) Three-phase line current, v_{RN} and v_{MG} experimental waveforms for the (a) 24-pulse circuit and (b) multi-level PWM circuit. 115 V, 400 Hz supply and a 4 kW load. 5 A/div, 50 V/div and 500 μs/div.

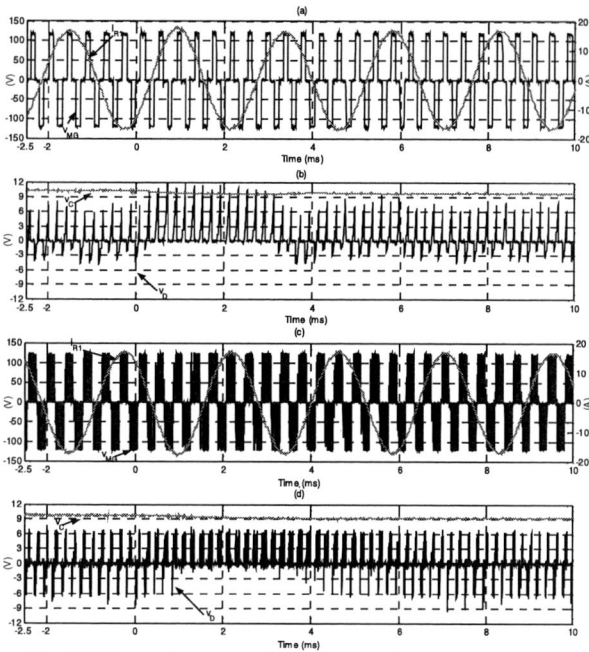

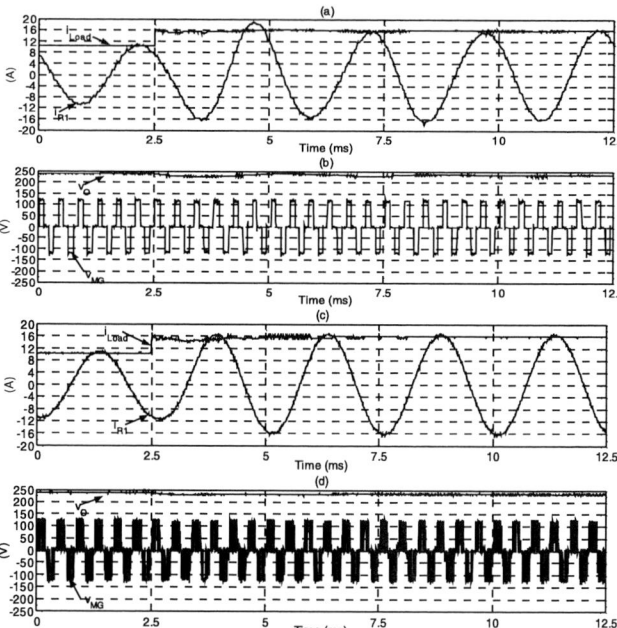

Fig. 13 Measured response of v_{MG}, i_{R1}, v_D and v_C to a 400 - 410 Hz frequency step at 0 ms for the 24-pulse circuit, (a) and (b), and for the multi-level PWM circuit, (c) and (d), with a 115 V supply and a 4 kW load.

Fig. 14 Response of v_{MG}, i_{R1}, v_D and v_C to a 2.5 - 4 kW load step at 2.5 ms for the 24-pulse circuit, (a) and (b), and for the multi-level PWM circuit, (c) and (d), with a 115 V supply.

To illustrate the performance of the active injector controller under transient load conditions, Figs. 14(a) and 14(b) show the response of i_{R1}, v_{MG}, the load current, i_{Load}, and the output voltage, V_O, to a load step for the 24-pulse circuit, from 2.5 kW to 4 kW at 2.5 ms with a 115 V, 400 Hz supply; whereas Figs. 14(c) and 14(d) show those for the multi-level PWM circuit under the same supply and load conditions. The measured results of Figs. 14(a) and 14(c) show that the operation of the active injector and the waveform quality of i_{R1} are maintained throughout the transient and that there is a small drop in the output voltage, which is due to the steady-state regulation characteristics of the converter.

V. CONCLUSION

The control systems for the 24-pulse and the multi-level PWM rectifiers of [5] have been presented, which consist of a PLL and a modulator. The principle of operation of the controller with the 24-pulse and PWM circuits was described, explaining how synchronisation of the active injector with the main rectifier is achieved, and then, the controller was modelled.

The controller operation was experimentally verified for the 24-pulse and PWM circuits. This resulted in steady-state, high-quality line currents with a THD of 2.36% and 1.06% for the 24-pulse and PWM operating modes respectively, under a 4 kW load and a 115V, 400 Hz supply. Moreover, it was demonstrated that the controller could maintain the bi-directional switch correctly synchronised and high current waveform quality over the supply frequency range of 400 – 800 Hz and under supply and load transient conditions. In contrast with typical control schemes for ac-dc three-phase converters, which need three reference signals, this

strategy uses a single reference signal to lock the active injector operation with the rectifier.

Future research on this control system could consider techniques for tracking rapid supply frequency variations, which would be particularly important in aircraft systems. In addition, other control techniques for robust synchronisation could be developed to accommodate larger supply frequency and load changes.

ACKNOWLEDGMENTS

The authors are grateful to Goodrich Corporation Electromagnetic Systems Technical Centre, Birmingham UK, the National Council of Science and Technology (CONACyT) and the National Polytechnic Institute (IPN) of Mexico, for their encouragement and the realisation of the prototype.

REFERENCES

[1] K. Mino, Y. Nishida, J. W. Kolar: "Novel Hybrid 12-Pulse Line Interphase Transformer Boost-Type Rectifier with Controlled Output Voltage and Sinusoidal Utility Currents". International Power Electronics Conference, Nigata, Japan, 2005, CD-ROM, ISBN: 4-88686-065-6.

[2] F. J. Chivite-Zabalza, A. J. Forsyth, D.R. Trainer, "A simple, passive 24-pulse ac-dc converter with inherent load balancing", IEEE Transactions on Power Electronics, Vol. 21-2, 2006, pp: 430 – 439.

[3] S. Choi, B. S. Lee, P. N. Enjeti, "New 24-pulse diode rectifier systems for utility interface of high-power ac motor drives", IEEE Transactions on Industry Applications, Vol. 32 - 2, 1997, pp: 531 – 541.

[4] B. Singh, G. Bhuvaneswari, V. Grag, and S. Gairola, "Pulse multiplication in AC-DC converters

for harmonic mitigation in vector-controlled induction motor drives," IEEE Transactions on Energy Conversion, Vol. 21 - 2, 2006, pp: 342 - 352.

[5] I. Araujo-Vargas, A. J. Forsyth, F. J. Chivite-Zabalza, "High performance multi-pulse rectifier with single transistor active injection", approved for publication in the IEEE Transactions on Power Electronics, TPEL-2006-12-0620, 2006.

[6] Wolaver, Dan H., "Phase-locked loop Circuit Design", First Edition, Prentice Hall, Englewood Cliffs, New Jersey, 1991, USA.

[7] A. Blanchard, "Phase locked loops: Applications to coherent receiver design", Wiley-Interscience, Translation from French, New York, U.S.A., 1976.

[8] E. Matheson, K. Karimi, "Power Quality Specification for More Electric Airplane Architectures", SAE Power Systems Conference, Florida, USA, 2002, 2002-01-3206.

[9] International Standarts, ISO1540:1984, Characteristics of aircraft electrical systems.

36-pulse hybrid ripple injection for high performance aerospace rectifiers

F. Javier Chivite-Zabalza   Andrew J. Forsyth   Ismael Araujo-Vargas

Javier.Chivite-Zabalza@manchester.ac.uk
The University of Manchester
School of Electrical and Electronic Engineering
Sackville Street, PO BOX 88 Manchester M60 1QD UK

Abstract – **This paper presents a 3-phase, voltage-sourced, 36-pulse converter that draws almost sinusoidal currents. The converter results from the combination of a series connected, 12-pulse, voltage-sourced rectifier with a passive voltage injection circuit and a bi-directional switch. The voltage injection circuit uses a single-phase rectifier bridge and a single-phase transformer. Both the voltage injection circuit and the bi-directional switch operate at six times the supply frequency and have a low rating. In this paper, the converter operation is explained and analysed. Subsequently, the converter is evaluated experimentally using a 4 kW, 400 Hz prototype, where the THD$_I$ of the line currents was measured to be below 1.2 %.**

I. INTRODUCTION

The amount of electrically powered equipment in airborne vehicles is increasing in the move towards the More Electric Aircraft (MEA), which is expected to yield lower fuel consumption, higher reliability, reduced maintenance costs and a reduction in size and weight [1]. This is having a considerable impact on the generators, the distribution networks and on the load equipment, and has created new technical challenges. The management of the power quality in variable frequency (VF) systems is one of those challenges. Consequently, the air-frame manufacturers are imposing strict power quality specifications upon the suppliers of equipment [2]. Aerospace electrical power systems are typically fed by a variable frequency (360Hz to 800Hz), making the use of passive filters an unsuitable solution. To comply with these regulations, the use of 18-pulse auto-transformer rectifier units (ATRU) has become a preferred solution by many manufacturers [3]. ATRUs are thought to be simple, reliable and, if no input-to-output isolation is required, as it is usually the case, they can be relatively small and light [4]. In practice nevertheless, they tend to be complex and expensive to design and manufacture, and their performance often fails to meet theoretical expectations, requiring even more complex winding arrangements or additional inductors to meet the specifications [5].

An interesting area of research in rectifiers is the use of low rating harmonic injection circuits to enhance performance. For instance, improving the performance of a 12-pulse circuit to achieve input currents that are characteristic of a 24 or 36-pulse converter [6-10].

This paper presents a new converter that is based on a combined active and passive harmonic injection circuit. When that circuit is added to a 12-pulse, series connected, voltage-sourced rectifier, the input currents experience a significant improvement becoming typical of those of a 36-pulse converter. The injection circuit combines previous work on a 24-pulse voltage injection technique [8] with a bi-directional switch. The resulting 36-pulse converter is relevant to ground and future aerospace systems that require input currents of a very high purity, using a simple and reliable circuit.

The converter proposed in this paper has a magnetic component VA rating of 50% of the power throughput, similar [5], or even higher [7] than that of other similar converters. However, since it offers a compact and simple solution, its overall weight and size are lower or comparable, offering a good performance at a competitive cost.

The operation of the 24-pulse voltage harmonic injection circuit shown in Figure 1 is reviewed first, followed by the operation of the 36-pulse converter. Subsequently, the main circuit waveforms are presented, and the Fourier expression of the input waveforms is obtained. That expression is used to determine the optimum turns-ratio of the injection transformer and the conduction angle of the bi-directional switch. Finally, experimental results are shown.

II. CONVERTER OPERATION

The operation of the converter is explained assuming idealised converter waveforms. That is, the conduction sequence of the diodes is based on sinusoidal currents, the output dc-link capacitors are big enough to filter the current ripple, the components in the circuit are ideal, the transformer turns-ratios have been perfectly achieved, the magnetising currents and core losses are negligible and steady-state operation has been achieved.

A. *Operation of the 24-pulse converter with harmonic voltage injection*

Figure 1 shows the 24-pulse converter that uses passive harmonic voltage injection [8]. The converter is based on a series connected, 12-pulse, voltage-sourced converter that is enhanced by the addition of a passive injection circuit. Due to the series inductance being located at the input terminals and to the dc-link filter capacitors being connected at the output of the two three-phase bridge rectifiers, the converter becomes a voltage-source topology. The input currents are defined by the voltage impressed across the input inductors, that is, by the difference between the sinusoidal supply voltage and the 24-pulse voltage synthesised at the right-hand-side of the input inductors. One of the rectifiers is connected to the delta windings of the main transformer, whereas the other is connected in series with the star-primary

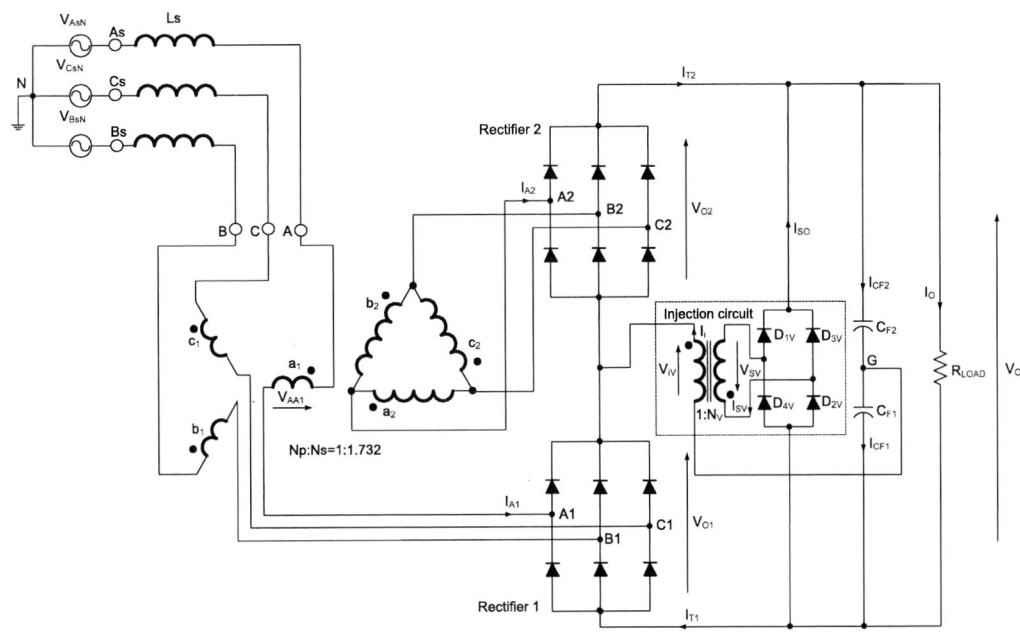

Figure 1 24-pulse voltage injection converter

windings. Due to the transformer connection and by choosing a primary to secondary turns ratio of $1:\sqrt{3}$, the rectifiers are fed with currents of equal magnitude, but phase-shifted by $30°$. The injection circuit consists of a single phase transformer that has a VA rating of 2% of the power throughput, and of a low current single-phase rectifier bridge. One of the transformer coils, the injection winding, is connected between the mid-point of the two main rectifiers and the mid-point of the dc-link capacitors. The other coil, the sensing winding, is connected to the ac terminals of the injection rectifier bridge, which has its dc output terminals directly connected to the dc output of the converter.

To reduce the weight and size of the converter, an important feature in aerospace applications, only one of the rectifiers is directly fed by the main transformer. Consequently, the VA-rating of the main transformer is approximately 50% of the power throughput. However, there is no galvanic isolation between the input and output of the converter, and the output voltage cannot be adjusted by the transformer turns-ratio. The injection techniques proposed in this paper are also valid for a fully isolated version of the transformer of Figure 1.

The operation of the converter is explained based on the equivalent circuit of Figure 2 that represents the topological configurations of the circuit, and with respect to Figure 3, that depicts the idealised waveforms. The first waveforms in Figure 3 are the output currents of the main rectifiers I_1 and I_2. These waveforms have the usual six-pulse ripple but are out of phase by $30°$ due to the star-delta connection of the main transformer. The second waveform in Figure 3 represents the injection current I_i that flows through the injection winding of the injection transformer, which is the difference between I_2 and I_1, it has six-times the supply frequency and can be approximated by a triangular shape. By Ampere's law, a current of $(I_2-I_1)/N_V$ will flow through the sensing winding, where N_V is the turns-ratio of the injection transformer. The third waveform in Figure 3 depicts the voltage across the injection winding of the voltage injection transformer. That voltage is obtained by dividing the voltage across the injection winding by the N_V turns-ratio. Finally, the forth waveform represents the output voltage across the bottom rectifier V_{O1}. That waveform is obtained as the addition of the injection voltage V_{iV} and $V_O/2$. The output voltage of the top rectifier V_{O2} has a complementary shape, so that $V_{O1}+V_{O2}$ equal the converter output voltage V_O.

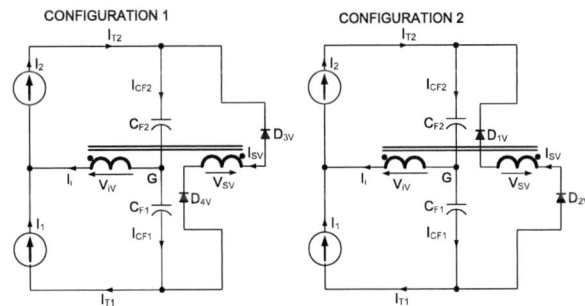

Figure 2 Equivalent circuits for the cyclic operation of the injection circuit

When the current I_1 is greater than I_2, the converter is in configuration 1, Figure 2, and both, the injection current I_i and the sensing current I_{SV} are negative. Consequently, diodes D_{3V} and D_{4V} are conducting and the voltages across the sensing and injection windings equal V_O and V_O/N_V respectively. Therefore, the output voltages across the main rectifiers become:

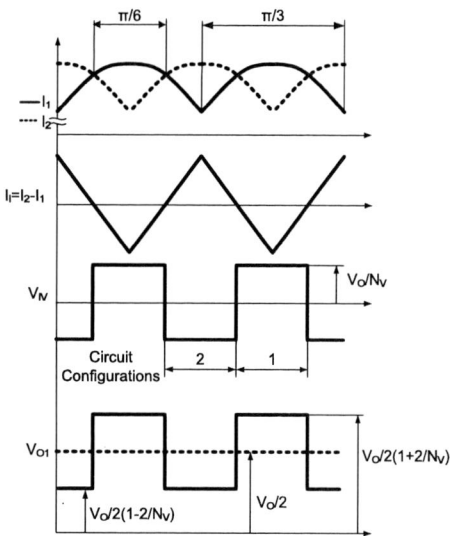

Figure 3   Idealised waveforms for the voltage injection circuit

$$V_{O1} = \frac{V_O}{2}\left(1 + \frac{2}{N_v}\right) \qquad (1)$$

$$V_{O2} = \frac{V_O}{2}\left(1 - \frac{2}{N_v}\right) \qquad (2)$$

When the current I_2 becomes greater than I_1, the circuit enters configuration 2, Figure 2, and the currents I_i and I_{sv} reverse direction. Consequently, the injection voltage becomes negative, V_{O1} equals (2) and V_{O2} (1). By appropriate choice of the injection turns-ratio Nv, the injected $\pm V_O/N_V$ voltage waveforms result in 24-pulse waveforms at the right-hand-side of the input inductors, having harmonics of $24k\pm1$, where k is a positive integer number. The turns-ratio N_V is shown in [8] to have an optimum value of 4.07.

To prove the converter operation, a 15 kW, 400 Hz prototype was built and tested using a solid-state voltage supply at 115 V rms line-to-neutral. The leakage inductance of the input transformer was solely used as the input inductance, and its value was measured to be 111μH, which represents 0.01 pu. for an output power of 15 kW at 400 Hz. The injection transformer was would on an EPCOS N87 ETD 59 ferrite core due to availability reasons, but a further reduction in weight and size could be obtained by choosing alternative magnetic materials.

The line currents and output voltage obtained in the experiment are shown in Figure 4. The line currents, which were well balanced, had a current total harmonic distortion (THD$_I$) of 2.57%, the power factor was measured to be 0.991 with the converter output voltage being 221 V. The rms current flowing through the sensing winding of the injection transformer had an rms value of 5.36 A and the voltage across the injection winding had a maximum value of 54 V. That resulted in a VA rating of the injection transformer of 2.2% of the power throughput. The shapes of those waveforms were similar to those of the predicted idealised waveforms.

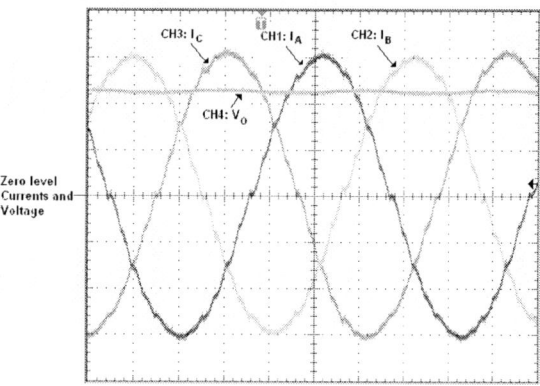

Figure 4   Input line currents and output voltage. V_{AsN}=115 V rms, F_S=400 Hz, P_O= 15 kW, Current scale: 20 A/div., voltage scale :100 V/div., time scale: 400 μs/div.

B. Operation of the 36-pulse converter with combined active and passive harmonic voltage injection

The 36-pulse hybrid converter shown in Figure 5 combines the 24-pulse converter depicted in Figure 1 with a bi-directional switch connected across the injection winding of the injection transformer. The switch operates at six times the supply frequency and was realized as seen in Figure 6. However, there could be other possible ways of implementing it. The converter of Figure 5 has a similar operation to the one presented in [10], where 4 active switches that form an H-bridge inverter drive the injection voltage. That converter has bi-directional characteristics, however, if only rectifier operation is required, the converter shown in Figure 5 offers a simpler and a potentially more reliable option.

The operation of the 36-pulse converter in Figure 5 is almost the same as that of the circuit of Figure 1. In this case, there is one additional topological configuration to those presented in Figure 2. That configuration, referred to as configuration 3, represents the instances where the injection switch is turned on, and has the equivalent circuit shown in Figure 7. The idealised waveforms that describe the operation of the 36-pulse converter are shown in Figure 8. The top waveform represents the full-wave rectified input currents. The second waveform represents the injection current I_i, which is the difference between I_2 and I_1 and has the approximate shape of a triangular waveform. The third waveform represents the voltage across the injection winding of the voltage injection transformer, the fourth waveform is the voltage across the output of the bottom rectifier, and finally, the bottom waveform represents the activation pulses for the injection switch.

When the injection switch is turned off, the operation of the circuit is exactly the same as that of the voltage injection circuit explained in Section II-A. However, when the injection switch is turned on, the circuit operates in configuration 3, Figure 7, and the common point of the rectifiers is directly connected to the mid-point of the dc-link capacitors, regardless of the direction of the injection currents. Consequently, the output voltages of the main

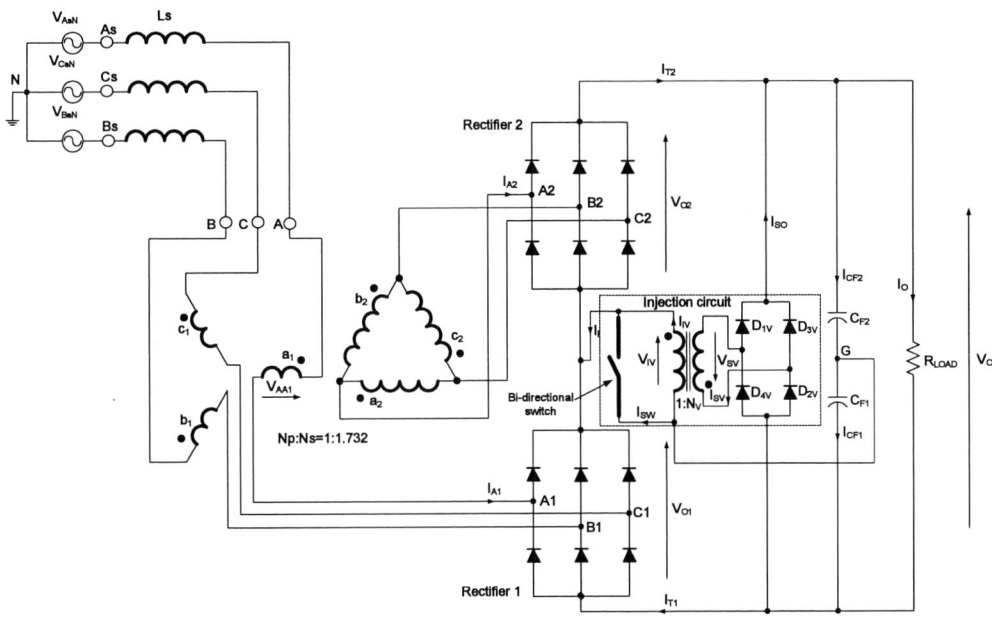

Figure 5   36-pulse voltage injection converter

rectifiers equal half the dc-link voltage, that is $V_{O1}=V_{O2}=V_O/2$. The angular duration of the activation of the switch is represented by an angle γ, which is centred on the zero crossing of the injection current I_i. The optimum values for the angle γ and the injection transformer turns-ratio are determined in Section III.

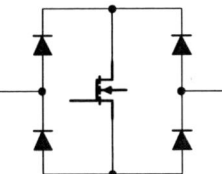

Figure 6   Implementation of the single switch using an uni-directional switch

CONFIGURATION 3

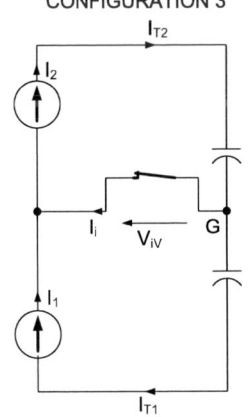

Figure 7   Additional equivalent circuit for the 36-pulse converter, corresponding to the conduction of the bi-directional switch

The main waveforms of the circuit are presented in Figure 9.

The first three waveforms show the input voltages of the lower rectifier with respect to the mid-point of the output dc-link, V_{A1G}, V_{B1G} and V_{C1G}. When the input current I_{A1} is negative, the bottom diode of the A1 leg conducts and the V_{A1G} voltage equals $-V_O/2$. When the input current becomes positive, the terminal A1 is connected to the positive rail of rectifier 1, and the voltage V_{A1G} equals the injection voltage V_{iV}, which switches between V_O/N_V, 0 and $-V_O/N_V$ at six times the supply frequency. The waveforms V_{B1G} and V_{C1G} are identical to V_{A1G} but are phase-shifted by 120° and 240° respectively.

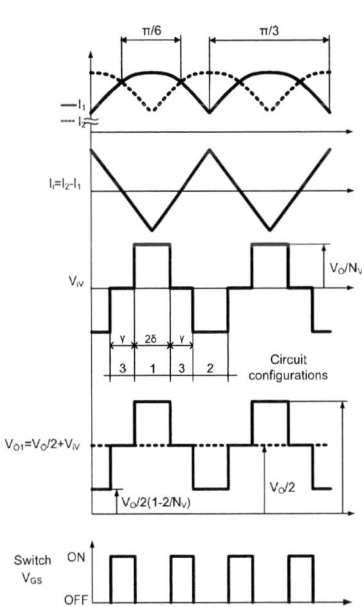

Figure 8   Idealised waveforms for the 36-pulse injection circuit

The fourth waveform presents the common mode voltage of the converter V_{NG} and is determined using equation (3).

$$V_{NG} = \frac{V_{A1G} + V_{B1G} + V_{C1G}}{3} \qquad (3)$$

V_{A1N}, shown as the fifth waveform, is obtained as $V_{A1N} = V_{A1G} - V_{NG}$. The voltage at the input terminal A2 of the top rectifier with respect to G, shown as the sixth waveform, is obtained in a similar manner. When the input current to terminal A2, I_{A2}, is positive, the top diode of that leg conducts and the V_{A2G} voltage equals $V_O/2$. When I_{A2} is negative, the bottom diode conducts and V_{A2G} equals $V_O/2 - V_{O2}$. V_{C2G} is the seventh waveform in Figure 9 and is obtained by phase-shifting V_{A2G} by 240°. The V_{AA1} voltage is derived using $V_{AA1} = (V_{A2G} - V_{C2G})/\sqrt{3}$ and, the V_{AN} voltage is

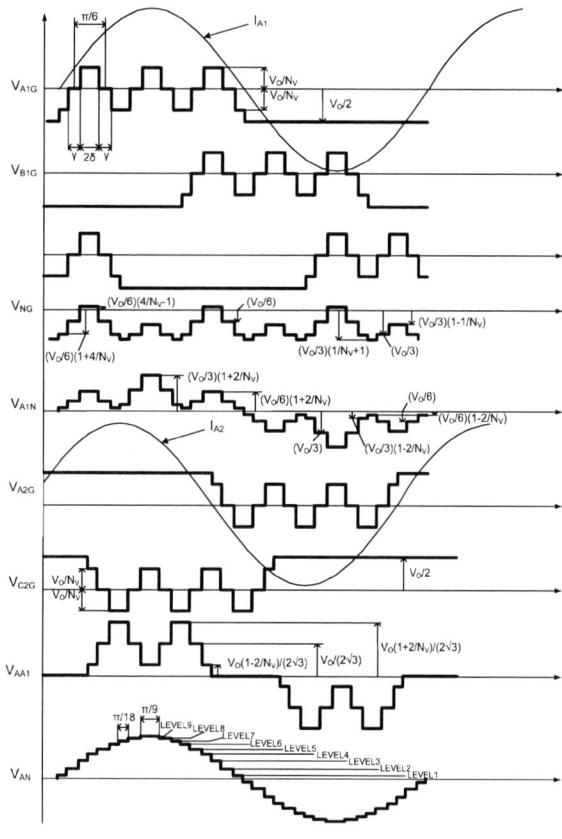

Figure 9   Production of the 36-pulse V_{AN} waveform.

$V_{AN-LEVEL1}$	$V_{AN-LEVEL2}$	$V_{AN-LEVEL3}$
$\dfrac{V_O}{3}\left(\dfrac{1}{2} - \dfrac{1}{N_V}\right)$	$\dfrac{V_O}{6}$	$\dfrac{V_O}{6}\left(\dfrac{2}{N_V} + 1\right)$
$V_{AN-LEVEL4}$	$V_{AN-LEVEL5}$	$V_{AN-LEVEL6}$
$\dfrac{V_O}{6}\left[\dfrac{2(1-\sqrt{3})}{N_V} + (1+\sqrt{3})\right]$	$\dfrac{V_O}{6}(1+\sqrt{3})$	$\dfrac{V_O}{6}\left[\dfrac{2(\sqrt{3}-1)}{N_V} + (1+\sqrt{3})\right]$
$V_{AN-LEVEL7}$	$V_{AN-LEVEL8}$	$V_{AN-LEVEL9}$
$\dfrac{V_O}{6}\left[(2+\sqrt{3}) + \dfrac{2\sqrt{3}-4}{N_V}\right]$	$\dfrac{V_O}{6}(2+\sqrt{3})$	$\dfrac{V_O}{6}\left[(2+\sqrt{3}) + \dfrac{2(2-\sqrt{3})}{N_V}\right]$

Table 1   Voltage levels for the 36-pulse V_{AN} waveform.

formed from the addition of V_{A1N} and V_{AA1}.

The voltage levels of the multi-level V_{AN} waveform follow directly from the waveforms in Figure 9 and are listed in Table 1. A similar set of waveforms may be drawn for the other supply phases, the only difference being the 120° and 240° phase shift.

III.   OPTIMUM INJECTION TRANSFORMER TURNS-RATIO AND SWITCH CONDUCTION ANGLE

The turns-ratio of the voltage injection transformer N_V and the conduction angle of the switch γ have been optimised to obtain line input currents with a minimum harmonic content. The line currents are determined by the voltages impressed across the input line inductors. For phase A, that voltage is the difference between the sinusoidal supply voltage V_{AsN}, and the voltage V_{AN}. Therefore, assuming a sinusoidal supply, a voltage V_{AN} that has 24-pulse characteristics will produce a current I_A that has also 24-pulse characteristics. However, the precise harmonic content of the line currents will be determined by the interaction of the V_{AN} harmonics with the impedance of the line inductors. If the series resistance component of the input inductors is neglected, the current harmonics become proportional to $V_{AN(n)}/n$, where n is the harmonic number. However, the results obtained by this approach become difficult to analyse as the harmonic number is increased, since they tend to have small magnitudes. Consequently, the purity of the voltage V_{AN}, or its THD_V, has been used for a general analysis of the circuit. It is expected that the most significant harmonics will be eliminated, or greatly minimised for the lowest THD_V. The purity of the $V_{AN(n)}/n$ waveform, related to the input currents, has also been subsequently verified to be optimum for the optimum values of N_V and γ.

The harmonic Fourier analysis of the line-to-neutral voltage V_{AN} at the right-hand side of the line inductor has been undertaken using the standard complex Fourier coefficient technique, the complex Fourier coefficients An_{AN} of the waveform being calculated using the formula in (4).

$$An_{AN} = \frac{1}{2\pi}\int_0^{2\pi} V_{AN}(\alpha)e^{-jn\alpha}d\alpha \qquad (4)$$

The analysis calculates the Fourier coefficients of V_{AN} from the analysis of the V_{A1G} and V_{A2G} waveforms, following the graphical procedure used to obtain the waveforms of Figure 9. Taking the beginning of the waveform I_A as the angular reference, they result in:

$$An_{AN} = \frac{-2}{3n\pi}j\left[K_A - K_B\right]\left\{\frac{3}{2}sin\left(\frac{n\pi}{2}\right) + \sqrt{3}\,sin\left(\frac{n\pi}{3}\right)\right\}$$
$$for\,(n \neq 2k)\,\&\,(n \neq 3k) \qquad (5)$$
$$An_{A1N} = 0\,for\,(n \neq 2k)\,or\,(n \neq 3k)\,k = 1,2,3...$$

where:

$$K_A = \frac{1}{2}sin\left(\frac{n\pi}{2}\right) \qquad (6)$$

$$K_B = -\frac{8}{N_V}sin\left[\frac{n}{4}\left(\frac{\pi}{6} - \gamma\right)\right]sin\left[\frac{n}{4}\left(\frac{\pi}{6} + \gamma\right)\right]sin\left(\frac{n\pi}{12}\right) \qquad (7)$$
$$for\,(n \neq 2k)\,\&\,(n \neq 3k)$$

To determine the optimum design point for the injection

circuit, the THD$_V$ of the V$_{AN}$ waveform has been calculated in MATLAB using equations (5-7) for the first 1000 harmonics for a wide range of values of N$_V$ and γ. The results of the calculation are presented in the three-dimensional plot of Figure 10. A V$_{AN}$ THD$_V$ of 5.037% is obtained for a turns-ratio of N$_V$= 3.057 and an angle γ=9.97°. For these optimum values, the converter input voltages V$_{AN}$, V$_{BN}$ and V$_{CN}$ exhibit 36-pulse characteristics, with harmonics of the order of 36 k±1. The line input current has a minimum THD$_I$ at slightly different values of N$_V$= 3.114 and γ=9.8°. For the purposes of the practical realisation of the circuit, this small difference can be neglected, and the γ angle can be approximated to 10°.

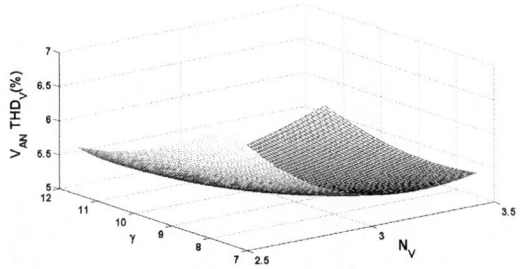

Figure 10 V$_{AN}$ THD against N$_V$ and γ

For clarity, Figure 11-(a) presents the variation of the V$_{AN}$ THD$_V$ with respect to the angle γ, by keeping the voltage injection ratio N$_V$ at its optimum value. Conversely, Figure 11-(b) presents the V$_{AN}$ THD$_V$ against N$_V$, with γ kept at its optimum value. The relatively flat-minimum in Figure 11 shows that the optimum design point is not especially sensitive to the values of N$_V$ and γ, indicating that the circuit performance will be fairly robust to small errors in the turns-ratio and in the duration of the switch conduction period.

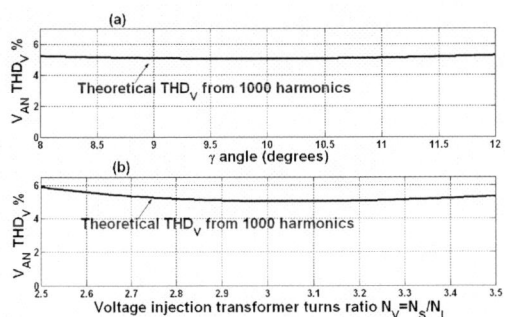

Figure 11 THD (%) of V$_{AN}$; (a)-versus switch ON angle γ with N$_V$=3.057, and (b)- versus voltage injection turns-ratio N$_V$ with γ=9.97°

The ideal no-load output voltage can be calculated from equations (5-7) using the equivalent fundamental, single-phase, equivalent circuit described in [8] for the optimum values of N$_V$ and γ. The value is 250.6 V for a 400 Hz, 115 V line-to neutral rms supply.

IV. EXPERIMENTAL RESULTS

To verify the converter operation, a prototype was built, consisting of a Y/Δ transformer with a primary-to-secondary turns-ratio of 1/√3. The primary-referred leakage inductance of the transformer was measured to be 35 μH per phase and the series resistance was 0.12 Ω per phase. Three line inductors of 330 μH were also connected, giving a total inductance value of 9.1% of the base impedance of the converter for a frequency of 400 Hz and an output power of 3.95 kW. The components used to build the converter are summarised in Table 2, and the details for the voltage injection transformer design are given in Table 3.

Part	Quantity	Reference
Main rectifier bridge	2	IXYS VUE-3506NO7
Bi-directional switch diodes	4	International rectifier 60EPU0, 60 A 400 V
Bi-directional switch MOSFET	1	International rectifier IRFP22N50A, 22 A 500 V
dc-link capacitors	2	470μF, in parallel with 10μF polypropylene
Voltage injection rectifier bridge	1	International rectifiers 36MT40, 35 A 400 V

Table 2   Prototype design details.

	Parameter	Value
Core	Size	Epcos ETD 59
	Material	Ferrite N87
Injection winding	Number of turns	14
	Wire size	3.142 mm²
	Inductance (measured)	1.04 mH
sensing winding	Number of turns	43
	Wire size	2.27 mm²
	Inductance (measured)	9.8 mH
Dimensions (L x H x W) (mm)		66.9 x 49.2 x 66.2
Weight		0.47 kg

Table 3   Injection transformer design details.

The control circuitry used to synchronise the switch operation is shown in the diagram block of Figure 12. It was implemented using analogue circuitry and was based on standard PLL techniques [11]. The injection current I$_i$ is used as the reference signal for the phase locked loop (PLL) circuit, which ultimately provides a triangular waveform V$_{TRI}$ that is locked to I$_i$, and has a phase difference of 90°. The triangular waveform is then compared with a reference voltage using a bi-polar modulator that generates the activation pulses V$_{GS}$ for the bi-directional switch shown in Figure 5. The angular duration of the switch conduction angle γ is adjusted by varying the level of the reference voltage V$_{REF}$.

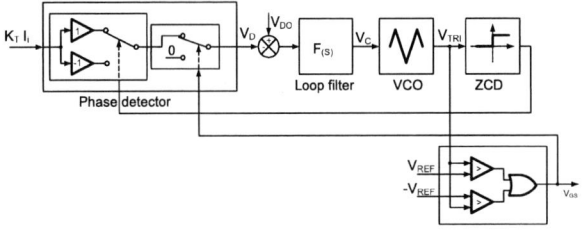

Figure 12 Control circuit for the 36-pulse rectifier

The PLL circuit to obtain V$_{TRI}$ consists of a phase detector, a loop filter, a voltage controlled oscillator (VCO) and a zero crossing detector (ZCD). The VCO generates a triangular waveform V_{TRI} of constant magnitude with a frequency proportional to the V_C input, and the ZCD generates a digital signal that is in phase with V_{TRI}. The phase detector generates a signal, V_D, which has an average value proportional to the phase difference between the V_{TRI} and I_i waveforms. V_D has a value of zero when I_i and V_{TRI} are 90° out of phase. This is accomplished by combining the reference waveform $K_T I_i$, where K_T is the current transducer gain, with the switch activation signal V_{GS} and V_{ZCD}. By careful design of the loop filter, the phase angle between I_i and V_{TRI} will equal the phase angle demand V_{DO} once steady state operation has been reached. Therefore, by setting $V_{DO}=0$, V_{TRI} and I_i will eventually be 90° out of phase as required.

Figure 13 shows the three input line currents and the injection current I_i at an output power level of 3.95 kW, the supply frequency being 400 Hz. A power analyser measured the current THD_I to be 1.18%.

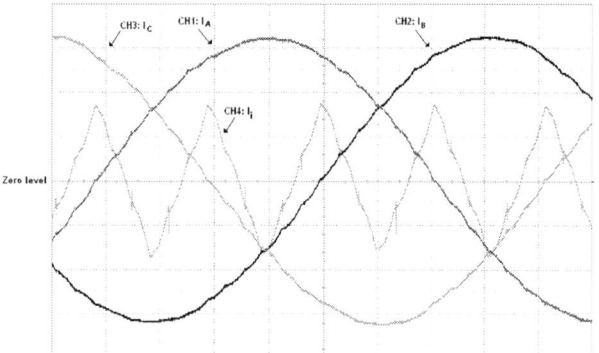

Figure 13 Input line currents and injection current I_i, V_{AsN}=115 V rms, F_S=400 Hz, P_O= 3.9 kW, Current scale CH1 to CH3: 5 A/div; CH4: 1 A/div, time scale: 200 μs/div

The current harmonics in each line, expressed as a percentage of the fundamental component, are shown in Figure 14, along with the theoretical harmonics for the turns-ratio N_V=3.055 and γ=10°, obtained using expressions (5-7). The harmonic limits of typical aerospace specifications [2] are also shown.

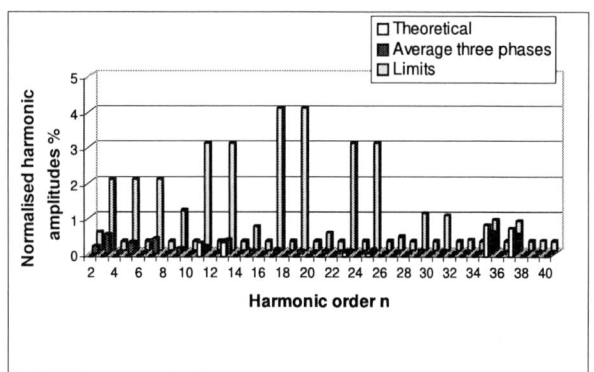

Figure 14 Normalised input line current harmonics, V_{AsN}=115 V rms, F_S=400 Hz, P_O= 3.95 kW.

Figure 15 shows some of the most relevant waveforms:

the voltage of the terminal A1, with respect to the mid-point of the output capacitors V_{A1G}, the voltage at the right-hand side of the input inductor of the phase A V_{AN}, and the line current of the phase A I_A. The measured waveforms are very similar in shape to the idealised waveforms presented in Figure 9.

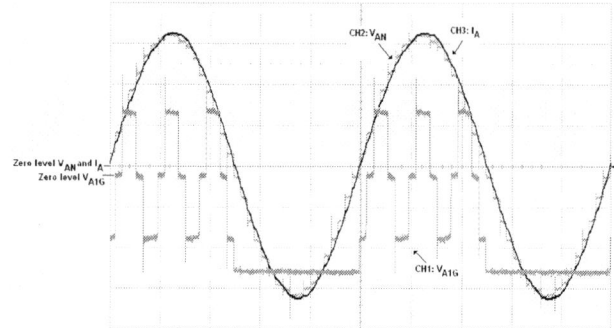

Figure 15 V_{AN}, I_A and V_{A1G}, V_{AsN}=115 V rms, F_S=400 Hz, P_O= 3.95 kW, Voltage scale: 50 V/div; Current scale: 5 A/div; time scale: 500 μs

The VA-rating of the voltage injection transformer was calculated to be 1.9% of the power throughput, based on the rms value of the winding currents and the rms value of the equivalent sinusoidal voltage that produces the same peak flux density, and is of the same frequency as the applied voltage to the transformer. The current that flows through the switch has an rms value of 1.5% of the output current, and an average value of 0.8% of the output current.

V. CONCLUSIONS

The converter presented in this paper combines a passive harmonic voltage injection technique with a low rating bi-directional injection switch that operates at six-times the supply frequency to obtain input currents that are characteristic of a 36-pulse converter. The VA-rating of the voltage injection transformer is less than 2% of the power throughput and the current rating of the injection switch has rms and average values of 1.5 % and 0.8 % of the output current. The current THD_I obtained in the experiments was below 1.2%. Although this converter requires control circuitry with associated sensing devices to synchronise the operation of the switch, there are no components connected in the series path of the main current. This converter has been analysed, and its operation and performance have been validated using a 3.95 kW, 400 Hz experimental prototype

VI. ACKNOWLEDGMENT

The authors gratefully acknowledge the help and support from Marc Holme and the team at Goodrich ESTC, and Giovanni Raimondi and Peter Crouchley from Goodrich Power Systems. They are also grateful to Dr David Trainer for his valued contributions at the beginning of this work.

VII. REFERENCES

[1] A. Emadi and M. Ehsani, "Aircraft power systems:

Technology, state of the art, and future trends," *IEEE AES Systems magazine*, vol. 15, no. 1, pp. 28-32, Jan 2000.

[2] E. Matheson and K. Karimi, "Power quality specification development for More Electric Airplane architectures", *Society of Automotive Engineers*, 2002-01-326.

[3] K. J. Karimi and A. C. Mong, "Modeling non-linear loads for aerospace power systems," *IEEE EnergyConversion Engineering Conference*, pp. 33-38, July 2002.

[4] D.A. Paice, *Power electronic converter harmonics: Multipulse methods for clean power*, IEEE Press, New York, 1995.

[5] F. J. Chivite, A.J. Forsyth and D. R. Trainer, "Analysis and practical evaluation of an 18-pulse rectifier for aerospace applications," *IEE International conference on Power Electronics Machines and Drives*, no. 498, pp. 338-343, April 2004

[6] B.M. Bird, J.F. Marsh, P.R. McLellan, "Harmonic reduction in multiple converters by tripple-frequency current injection.", *IEE Proceedings*, Volume 116, no. 10, 1969, pp. 1730-1734.

[7] S. Choi, P. N. Enjeti and I. J. Pitel, "Polyphase transformer arrangements with reduced kVA capacities for harmonic current reduction in rectifier-type utility interface," *IEEE Transactions on Power Electronics*, vol. 11, no. 5, pp. 680-690, Sept. 1996.

[8] F. J. Chivite, A. J. Forsyth, D. R. Trainer, "A simple, Passive 24-Pulse AC-DC Converter With Inherent Load Balancing Using Harmonic Voltage Injection", *IEEE Conference on Power Electronics Specialists*, Jun. 2005, pp: 76-82.

[9] K. Mino, Y. Nishida, and J.W. Kolar, "Novel Hybrid 12-Pulse Line Interphase Transformer Boost-Type with Controlled Output Voltage and Sinusoidal Utility Currents", *Proceedings of the IEEJ International Power Electronics Conference, IPEC'05*, Niigata, Japan April. 2005.

[10] Y.H. Liu, J. Arrillaga and N.R. Watson, "A new high-pulse voltage-sourced converter for HVdc transmission," *IEEE Transactions on Power Delivery*, vol. 18 no.4, pp. 1388-1393, October 2003.

[11] P. Horowitz and W. Hill, *The art of electronics*, Cambridge University Press, Cambridge, UK, 1989.

A 48-pulse converter using dc-ripple injection

F. Javier Chivite-Zabalza

Javier.Chivite-Zabalza@manchester.ac.uk

Andrew J. Forsyth

The University of Manchester
School of Electrical and Electronic Engineering
Sackville Street, PO BOX 88 Manchester M60 1QD UK

Abstract – **This paper presents a 3-phase, voltage-sourced, 48-pulse converter. The converter results from the combination of a series connected 12-pulse voltage-sourced rectifier with a passive voltage injection circuit, a passive current injection circuit and a bi-directional switch. The voltage and current injection circuits each use a single-phase rectifier bridge and a single-phase transformer circuit. Both the injection circuits and the bi-directional switch operate at six times the supply frequency and have a low rating. In this paper the converter operation is explained, analysed and simulated.**

I. INTRODUCTION

There has been a significant effort in recent times to reduce the harmonics drawn by rectifier equipment. Excessive harmonics in the power system increase the losses and thermal stress on the equipment, reduce the utilisation of the electric energy, disrupt control systems, interfere with communication systems, affect the electromagnetic torque of motors and generators, may cause unwanted resonant conditions and may cause the malfunctioning of system or plant equipment [1]. For these reasons, legislation and recommendations have been imposed upon manufacturers and users of equipment [2]. Many types of clean rectifiers with a low impact on the supply have been reported in the literature [3-7], which can be achieved by active, passive and hybrid means. There has been additional interest in these issues due to the increased amount of electrically powered equipment that is used in modern aircraft in the move to the future more electric aircraft (MEA). This has lead air-frame manufacturers to impose stringent power quality limits [8].

Multipulse rectifiers offer a relatively simple and reliable solution [7]. They are widely used in industrial applications [7, 9] and are being increasingly used in aerospace systems [10]. Typical aerospace systems, where weight, size and reliability are important considerations, operate from a variable frequency supply (VF) and require a performance characteristic of 12 or 18-pulse systems [8]. An interesting way to increase the number of pulses of a converter without adding more rectifier channels or complex winding arrangements is the use of ripple injection techniques. These typically consist of a low rated injection circuit that modifies the rectifier dc waveforms to eliminate certain harmonics from the line currents. They can for instance, improve the performance of a 12-pulse circuit to achieve input currents that are characteristic of a 24 or a 36-pulse converter [11-16].

This paper presents a 48-pulse converter based on a 12-pulse, voltage sourced rectifier, that is enhanced by the addition of active and passive harmonic injection circuits.

The magnetic components have a total VA rating of 50% of the power throughput. This rating is comparable with or higher than that in other reported solutions [7, 9, 12, 17]. However, due to its simple and compact configuration, the overall weight and size of the proposed circuit is comparable, or potentially smaller. The injection circuit combines previous work on a 36-pulse passive rectifier [14] with a bi-directional switch that operates at six-times the supply frequency. The operation of the circuit is explained and analysed, and is validated by simulation results.

II. OVERVIEW

Figure 1 presents an entirely passive 36-pulse converter which was presented in [14]. The converter is based on a 12-pulse voltage-sourced, series-connected converter that is enhanced by the addition of a passive ripple injection circuit to obtain a performance typical of a 36-pulse converter. The injection circuit combines two previously discussed techniques: current injection [15] and voltage injection [16], as explained in [14].

The converter in Figure 1 is a voltage-sourced topology in which the input currents are defined by the voltage that is impressed across the input line inductors. It is based on a 12-pulse converter that is obtained by combining in series the outputs two bridge rectifiers that have their ac supplies out of phase by 30° due to the action of a Y/Δ transformer. By sacrificing galvanic isolation, the Y/Δ transformer in the circuit of Figure 1 is rated to 50% of the power throughput since only one of the rectifiers is fed by its Δ secondary winding. As the other rectifier is connected in series with the primary Y winding, the ac inputs of the rectifiers are connected in series. The converter is enhanced by the addition of two injection circuits, namely current and voltage injection. They are each composed of a single-phase transformer and a single-phase rectifier. The rectifier of the current injection transformer has its output terminals connected in series with the main bridge rectifiers, whereas the output of voltage injection rectifier is connected directly to the converter output. The sensing windings of the injection transformers are connected to the ac inputs of their respective rectifiers, whereas the injection windings are connected in series, between the mid-point of the split dc-link capacitors and one of the ac inputs of the current injection bridge.

The effect of the current injection circuit is to impose a limit on the injection current I_i that may flow between the common point of the two rectifier bridges and the mid-point of the split dc-link capacitors. The result is to introduce two additional levels into the voltage waveforms at the output of the bridge rectifiers. Conversely, as [14] explains, the effect of the voltage injection circuit is to inject a $\pm V_O/N_V$

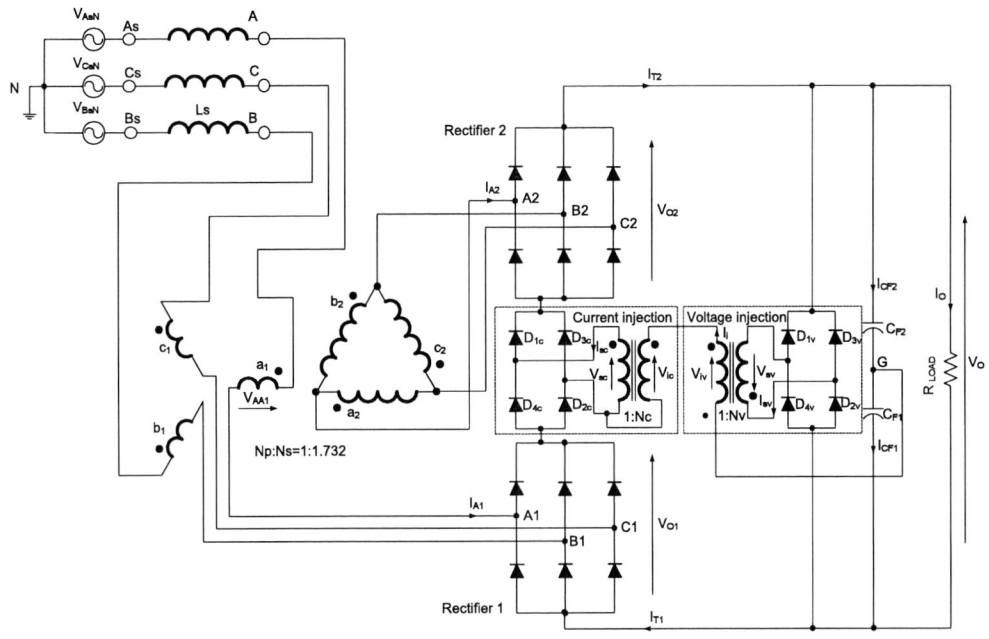

Figure 1   36-pulse converter using combined harmonic current and voltage injection

squarewave voltage V_i at six times the supply frequency at the outputs of the bridge rectifiers, the sign of the injected voltage being defined by the direction of the injection current I_i. When the voltage and current injection techniques are combined together, the output voltages of the main rectifiers take the shape of complementary 4-level voltage waveforms at six times the supply frequency as shown in Figure 2-(a).

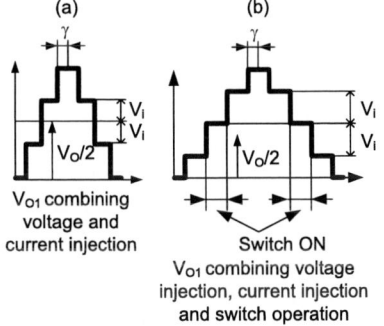

Figure 2   Formation of the multi-level voltage at the output of the main rectifiers in the 48-pulse converter

To confirm the converter operation, a 400 Hz, 15 kW, prototype was built and tested. The converter employed a Y/Δ 3-phase transformer with a Y to Δ turns-ratio of √3. The circuit relied on the leakage inductance of the transformer, which had a phase value of 111 µH when measured from the Y windings, and had a series resistance of 0.21 Ω. The current and voltage injection transformers where build using EPCOS N87 ETD 59 ferrite cores, where the turns-ratios where measured to be N_V=6.08 and N_C=10.7 respectively. The output dc-link capacitors had a value of 280 µF each, typical of aerospace applications. Figure 3 presents the three-phase currents and the output voltage, at an output power level of 15 kW, for a supply voltage of 115 V rms line-to-

neutral at 400 Hz. The total harmonic distortion of the input currents (THD$_I$) was measured using a high-bandwidth power analyser to have a value of 1.51 % and the power factor was 0.994.

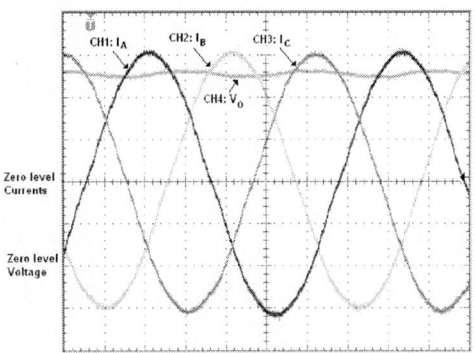

Figure 3   36-pulse converter using harmonic injection. Input line currents and output voltage. V_{AsN}=115 V rms, F_S=400 Hz, P$_O$= 15 kW, Current scale: 20 A/div; voltage scale :100 V/div., time scale: 400 µs/div

When a single bi-directional switch is connected in parallel with the injection winding of the voltage injection transformer, as shown in Figure 4, the performance of the converter is greatly improved. By choosing appropriate turns-ratio values for the injection transformers and an adequate conduction angle for the bi-directional switch, the input currents to the converter become like those of a 48-pulse rectifier. This is explained by an additional voltage level of $V_O/2$ that is added to the output voltages of the rectifiers as shown in Figure 2-b, which then take the shape of 5-level waveforms. Consequently, the voltages at the right-hand-side of the input inductors result in 48-pulse waveforms.

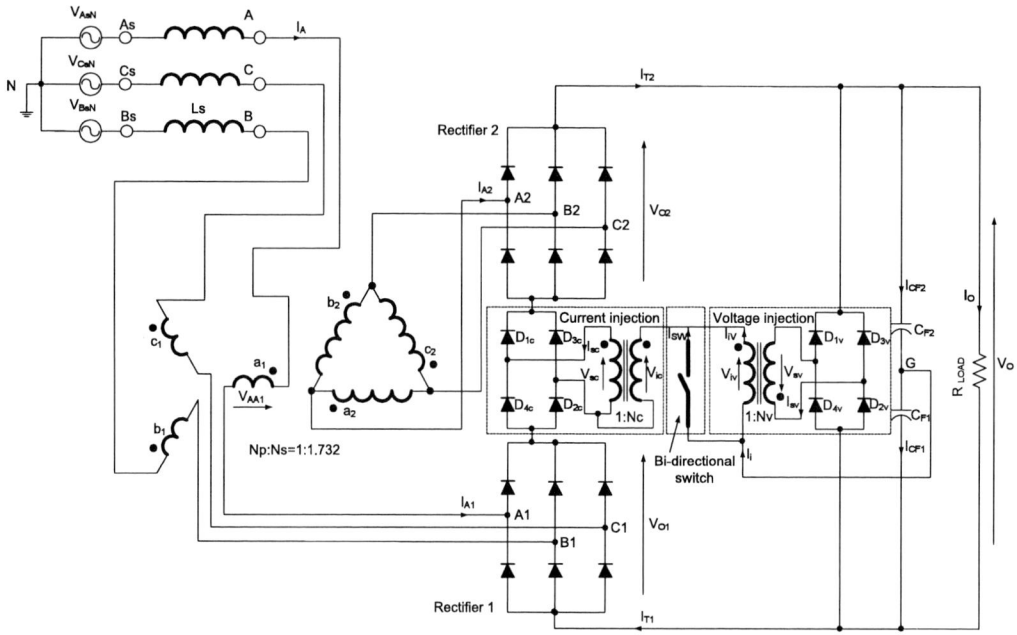

Figure 4   48-pulse converter using combined harmonic current and voltage injection and a single bi-directional switch

The next section explains the operation of the 48-pulse circuit of Figure 4. The operation of the 36-pulse circuit of Figure 1 is not described in this paper, but it can be considered as a particular case of the 48-pulse converter, resulting from the bi-directional switch being turned off.

III. OPERATION OF THE 48-PULSE CONVERTER WITH ACTIVE AND PASSIVE INJECTION

The operation of the 48-pulse converter shown in Figure 4 is explained based on Figure 5, which represents the topological configurations of the circuit, and on Figure 6, that shows the idealised converter waveforms. The explanation assumes that the input currents are sinusoidal and that the converter components are ideal.

The top waveforms in Figure 6 represent the full-wave rectified input currents to the two main bridge rectifiers. In this converter the bridge output currents will be slightly different to these waveforms as described below. The second waveform is the absolute value of the difference between I_1 and I_2. This waveform, which cannot be directly measured in the converter, is used for the purpose of the explanation and has approximately a triangular shape when sinusoidal input currents are assumed. The third waveform shows the output voltage of the bottom rectifier. The fourth waveform is the current I_i that flows through the injection winding of the current injection transformer. The currents seen as the fifth waveform are the actual output currents of the bridge rectifiers I_{T1} and I_{T2}. Finally, the last waveform represents the activation pulses to the bi-directional switch. The angle 2γ represents the angular periods when the voltage at the output of the rectifiers has collapsed, whilst δ is the angular conduction of the bi-directional switch.

The rectifiers in Figure 5 are represented by the two current sources, I_1 and I_2, and by bypass diodes. These represent the possibility of additional current flow through the output terminals of each rectifier due to the conduction of both diodes in a leg.

The explanation begins by describing configuration 1, Figure 5, which occurs when the current I_2 is substantially greater than the current I_1. Consequently, diodes D_{1C} and D_{2C} in the current injection circuit are in conduction, and a current I_2 flows through the sensing winding of the current injection transformer. By Ampere's law, a current $I_i=I_2/N_C$ flows through the injection windings of both transformers. Since the current I_i is smaller than the current difference I_2-I_1, a current $I_{by-pass}$ defined by (1) must flow through diode D_1 in Figure 5. That causes the output voltage V_{O1} across the bottom rectifier to be virtually zero.

$$I_{by-pass} = I_2 - \frac{I_2}{N_C} - I_1 \qquad (1)$$

The output current to the top rectifier, I_{T2}, equals I_2. However, the output current to the bottom rectifier, I_{T1}, is obtained by the summation of I_1 and $I_{by-pass}$ as:

$$I_{T1} = I_2 - \frac{I_2}{N_C} \qquad (2)$$

Since the injection current I_i is positive, a positive current I_{SV} will flow in the sensing winding of the voltage injection transformer and diodes D_{1V} and D_{2V} will be in conduction.

806

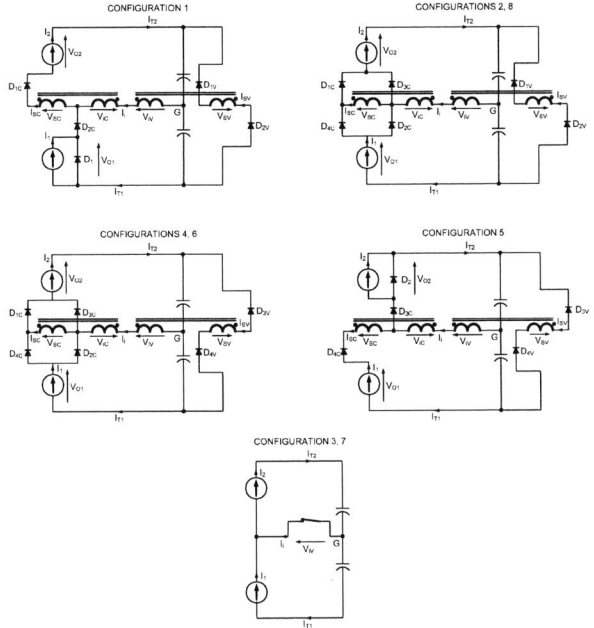

Figure 5   Equivalent circuits for the cyclic operation of the injection circuit

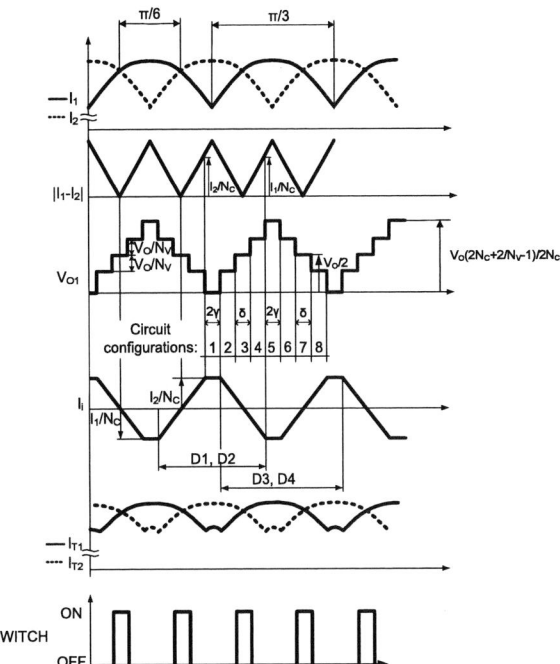

Figure 6   Idealised waveforms for the 48-pulse injection circuit

Consequently, the voltage on the injection winding of the voltage injection transformer will have a value of $-V_O/N_V$, and the voltage V_{iC} across the injection winding of the current transformer will have the value defined in (3). This results in the output voltage of top rectifier having the value defined by (4).

$$V_{iC} = (V_O/2 - V_O/N_V) \qquad (3)$$

$$V_{O2} = \frac{V_O}{2}\left[\frac{2N_C + \frac{2}{N_V} - 1}{N_C}\right] \approx V_O \qquad (4)$$

Configuration 1, Figure 5, ends when the injection current I_i equals the current difference I_2-I_1, that is, when $I_2/N_C=I_2-I_1$. Consequently, the angular duration 2γ can be determined by the choice of the current injection transformer turns-ratio N_C. The circuit then enters configuration 2, Figure 5, during which the instantaneous current difference is comparatively small, and can be provided by the injection current I_i. Therefore, the current through the injection windings of both transformers I_i and I_{iV} equals $I_i=I_{iV}=I_2-I_1$. The resulting currents flowing through the sensing winding of the current injection transformer is $I_{SC}=(I_2-I_1)/N_C$. To accommodate that current, diodes D_{1C} to D_{4C} need to be in conduction, causing the voltage across the current injection transformer windings to be virtually zero. Since the direction of the injection current I_i remains positive, the voltage V_{iV} still equals -V_O/N_V. Consequently, the output voltages to the bridge rectifiers are:

$$V_{O1} = \frac{V_O}{2}\left[1 - \frac{2}{N_V}\right] \qquad (5)$$

$$V_{O2} = \frac{V_O}{2}\left[1 + \frac{2}{N_V}\right] \qquad (6)$$

The circuit enters configuration 3, Figure 5, when the bi-directional switch is turned on. As seen in the equivalent circuit of Figure 5, the output voltages to both bridge rectifiers equal $V_O/2$. During that configuration, the current I_1 becomes greater than I_2 and the injection current I_i reverses direction. Configuration 3 has an angular duration of δ, defined by the switch control circuitry. When the bi-directional switch is turned off the circuit enters configuration 4, Figure 5. This configuration is similar to configuration 2, with the difference that the injection current I_i now flows in the opposite direction. Therefore, the voltage V_{iV} equals V_O/N_V, V_{O1} equals (6) and V_{O2} equals (5). During configurations 2, 3 and 4, the rectifier bridge output currents I_{T1} and I_{T2} equal I_1 and I_2. Configuration 4 terminates when the injection current I_i cannot provide the current difference I_1-I_2 any longer. At that point, $I_i = -I_1/N_C = I_2-I_1$. $I_{D3C}=I_2$ and $I_1 = I_{D4C}$. The circuit enters configuration 5, Figure 5, where any further increase of the injection current to support the current difference is impossible, and the upper by-pass diode D_2 begins conduction. Configuration 5 is a mirror image of configuration 1. The output voltage of the top rectifier, V_{O2} is virtually zero, and the output voltage of the bottom rectifier, V_{O1} equals:

807

$$V_{O1} = \frac{V_O}{2}\left[\frac{2N_C + \dfrac{2}{N_V} - 1}{N_C}\right] \approx V_O \qquad (7)$$

In this configuration the output current I_{T1} equals I_1, and the output current I_{T2} results from the summation of I_2 and $I_{by\text{-}pass}$, that is:

$$I_{T2} = I_1 - \frac{I_1}{N_C} \qquad (8)$$

This completes a half-cycle of the operation of the injection circuit. Configurations 6, 7 and 8 in the negative half-cycle, Figure 5, are equal to configurations 4, 3 and 2 respectively, with the same angular durations. Therefore, the output voltages V_{O1} and V_{O2} of the bridge rectifiers take the shape of 5-level waveforms. The angular duration of configurations 1 and 5, 2γ, can be calculated from the ideal waveforms of Figure 6 using the following expression [14]:

$$2\sin\left(\frac{\pi}{12} - \delta\right)\sin\frac{\pi}{12} = \frac{\cos(\delta)}{N_C} \qquad (9)$$
$$\text{for} \quad 0 < \delta \le \frac{\pi}{6}$$

Therefore, the angle γ is determined by the current injection turns-ratio N_C. The optimum values of N_V, N_C and δ are calculated in section V.

The following section describes how the 5-level rectifier bridge output waveforms V_{O1} and V_{O2} result in 48-pulse waveforms at the right-hand-side of the input line inductors.

IV. PRODUCTION OF THE 48-PULSE WAVEFORM

The main waveforms of the circuit of Figure 4 are presented in Figure 7. The first waveform is the voltage of the input terminal A1 with respect to the mid-point of the dc-link capacitors V_{A1G}. When the current I_{A1} is positive, the A1 terminal is connected to the output voltage of the rectifier and V_{A1G} is obtained by subtracting $V_O/2$ from V_{O1}. When the current I_{A1} becomes negative, the bottom diode in the leg conducts and V_{A1G} equals $-V_O/2$. The voltages V_{B1G} and V_{C1G}, seen as the third and fourth waveforms, are equal to V_{A1G} but are out phase by 120° and 240° respectively. The voltage of the A1 terminal with respect to the neutral point is V_{AN}, presented as the fifth waveform, and is obtained by subtracting the common mode voltage V_{NG} from V_{A1G}. The V_{NG} voltage, seen as the fourth waveform, is calculated in (10):

$$V_{NG} = \frac{V_{A1G} + V_{B1G} + V_{C1G}}{3} \qquad (10)$$

The voltage V_{A2G}, seen as the sixth waveform, can be obtained in a similar way to V_{A1G}. When the current I_{A2} is positive, the terminal A2 appears connected to the positive output terminal of the top rectifier and V_{A2G} equals $V_O/2$. When the current I_{A2} reverses direction, the terminal A2 appears connected to the bottom output terminal of the rectifier and V_{A2G} can is calculated as $V_{O2}-V_O/2$. The waveform V_{C2G}, shown as the sixth waveform, has the same shape as V_{A2G} but is out of phase by 240°. The waveform V_{C2G} is subtracted from V_{A2G} to calculate the voltage V_{A2C2},

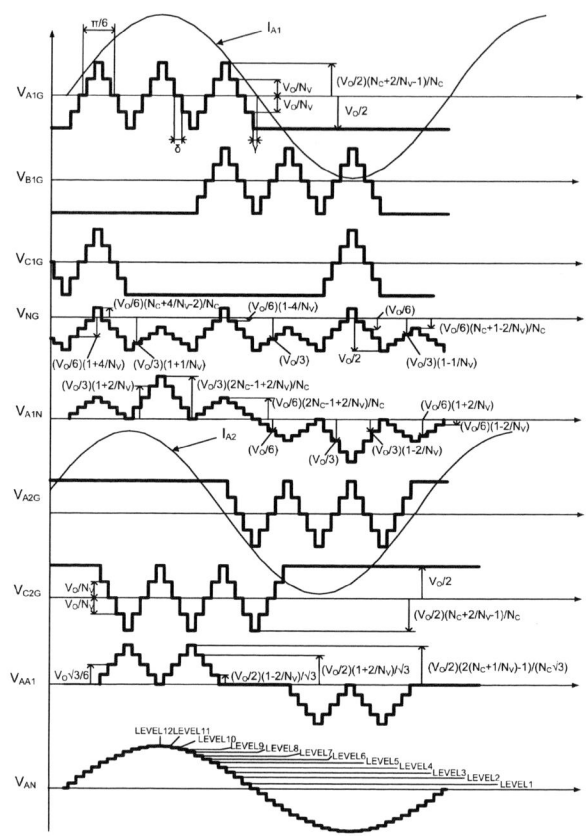

Figure 7   Production of the 48-pulse V_{AN} waveform

$V_{AN\text{-}LEVEL1}$	$V_{AN\text{-}LEVEL2}$	$V_{AN\text{-}LEVEL3}$	$V_{AN\text{-}LEVEL4}$
$\dfrac{V_O}{6}\left(1-\dfrac{2}{N_V}\right)$	$\dfrac{V_O}{6}$	$\dfrac{V_O}{6}\left(1+\dfrac{2}{N_V}\right)$	$\dfrac{V_O}{6}\left(\dfrac{2N_C+\dfrac{2}{N_V}-1}{N_C}\right)$
$V_{AN\text{-}LEVEL5}$	$V_{AN\text{-}LEVEL6}$	$V_{AN\text{-}LEVEL7}$	$V_{AN\text{-}LEVEL8}$
$\dfrac{V_O}{6}\left(\dfrac{2}{N_V}\left(1-\sqrt{3}\right)+1+\sqrt{3}\right)$	$\dfrac{V_O}{6}\left(1+\sqrt{3}\right)$	$\dfrac{V_O}{6}\left(\dfrac{2}{N_V}\left(\sqrt{3}-1\right)+1+\sqrt{3}\right)$	$\dfrac{V_O}{6}\dfrac{2\left(N_C+\dfrac{1}{N_V}\right)-1}{N_C}$
$V_{AN\text{-}LEVEL9}$	$V_{AN\text{-}LEVEL10}$	$V_{AN\text{-}LEVEL11}$	$V_{AN\text{-}LEVEL12}$
$\dfrac{V_O}{6}\left(\dfrac{2}{N_V}\left(\sqrt{3}-2\right)+2+\sqrt{3}\right)$	$\dfrac{V_O}{6}\left(2+\sqrt{3}\right)$	$\dfrac{V_O}{6}\left(\dfrac{2}{N_V}\left(2-\sqrt{3}\right)+2+\sqrt{3}\right)$	$\dfrac{V_O}{3}\dfrac{2N_C+\dfrac{2}{N_V}-1}{N_C}$

Table 1   Voltage levels for the V_{AN} waveform

depicted as the seventh waveform in Figure 7. Finally, the voltage V_{AN}, presented as the last waveform, is obtained as $V_{AN}=V_{A1N}+V_{A2C2}/\sqrt{3}$. The voltages V_{BN} and V_{CN} have the same shape as V_{AN}, but will be out of phase by 120° and 240° respectively. The voltage levels of the V_{AN} waveform are separately presented in Table 1 for clarity.

V. OPTIMUM TURNS-RATIO FOR THE INJECTION TRANSFORMERS AND SWITCH CONDUCTION ANGLE

To calculate the optimum turns-ratio values of the current and voltage injection transformers, N_C and N_V, and the conduction angle of the bi-directional switch, δ, the Fourier expression of the V_{AN} voltage at the right hand side of the input line inductors is calculated first. The Total Harmonic Distortion (THD$_V$) of that expression is then calculated for a wide number of N_V, N_C and δ values using the first 1000 harmonic components.

The line currents are determined by the voltages impressed across the line input inductors, that is, the difference between V_{AsN} and V_{AN} for the case of I_A. Assuming a sinusoidal supply and ideal line input inductors, the harmonic content of the current I_A will be proportional to $V_{AN}(n)/n$, where n is the harmonic order. However, as the harmonic order is increased the expression produces small values that are difficult to analyse. Therefore, the purity of the V_{AN} voltage, or its THD$_V$ is used instead. It is expected that, for a minimum THD$_V$, the most significant harmonics will be greatly minimised. The optimum point of the $V_{AN}(n)/n$ expression is also shown to occur at approximately the same point.

The complex Fourier coefficient expression of the V_{AN} voltage, An_{AN}, is calculated using (11) following the graphical procedure used to obtain the waveforms in Figure 7.

$$An_{AN} = \frac{1}{2\pi} \int_0^{2\pi} V_{AN}(\alpha) e^{-jn\alpha} d\alpha \qquad (11)$$

resulting in:

$$An_{AN} = \frac{-2}{3n\pi} j \left[K_A - K_B \right]$$

$$\left\{ \frac{3}{2} sin\left(\frac{n\pi}{2} \right) + \sqrt{3} \; sin\left(\frac{n\pi}{3} \right) \right\} \qquad (12)$$

$$for \; (n \neq 2k) \, \& \, (n \neq 3k)$$

$$An_{AIN} = 0 \; for \; (n \neq 2k) \; or \; (n \neq 3k) \; k = 1,2,3 \ldots$$

where:

$$K_A = \frac{1}{2} sin\left(\frac{n\pi}{2} \right) \qquad (13)$$

$$K_B = 2 \begin{bmatrix} sin\left(\frac{n\gamma}{2} \right) \\ \left\{ cos\left[\frac{n}{2}\left(\frac{\pi}{3} - \gamma \right) \right] - K_L \, cos\left(\frac{n\gamma}{2} \right) \right\} - \\ -\frac{4}{N_V} sin\left(\frac{n\pi}{12} \right) sin\left[\frac{n}{24}(\pi - 12\gamma - 6\delta) \right] \\ sin\left[\frac{n}{24}(\pi - 12\gamma + 6\delta) \right] \end{bmatrix} \qquad (14)$$

$$for \; (n \neq 2k) \, \& \, (n \neq 3k)$$

$$K_L = \frac{N_C + \frac{2}{N_V} - 1}{N_C} \qquad (15)$$

Figure 8 presents the variation of the V_{AN} THD$_V$ with respect to the activation angle of the switch δ, by keeping the current injection transformer turns-ratio N_C, and the voltage injection transformer turns-ratio N_V at their optimum values. Conversely, Figure 9 presents the V_{AN} THD$_V$ against N_V, with δ and N_C kept at their optimum values. Equally, Figure

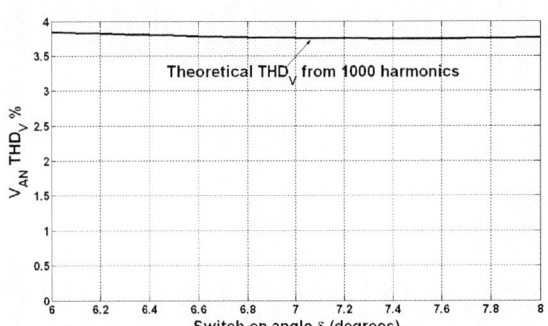

Figure 8   THD$_V$ (%) of V_{AN} versus the Switch ON time angle δ for N_V=4.1 and N_C=9.9

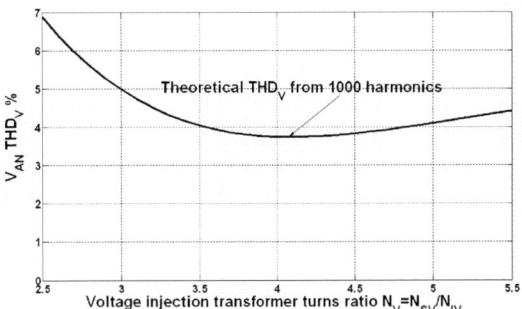

Figure 9   THD$_V$ (%) of V_{AN} versus the voltage injection transformer turns-ratio N_V for δ=7.5° and N_C=9.9

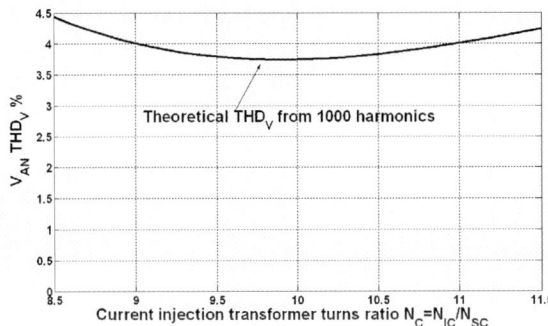

Figure 10   THD$_V$ (%) of V_{AN} versus the current injection transformer turns-ratio N_C for δ=7.5° and N_V=4.1

10 depicts the V_{AN} THD$_V$ against N_C, with δ and N_V kept at their optimum values.

A minimum value of the total harmonic distortion, THD$_V$ of the voltage V_{AN} of 3.74% is obtained for δ=7.5° N_V= 4.1, and N_C = 9.9, which results in an angle 2γ of 7.52°. The minimum current harmonic distortion, that is, the minimum THD of the V_{AN}/n function, which is related to the THD$_I$ of the I_A current, where n is the harmonic number, is obtained at the slightly different values of δ=7.3°, N_V= 4.2, and N_C = 10, which results in an angle 2γ of 7.58°. Nevertheless, that small difference between the voltage THD$_V$ and the current THD$_I$ can be neglected for practical purposes.

For these optimum values, the converter input voltages V_{AN}, V_{BN} and V_{CN} exhibit 48-pulse characteristics, with harmonics of the order of 48k±1. The relatively flat-minima in Figure 8, Figure 9, and Figure 10, show that the optimum design point is not especially sensitive to the values of δ, N_V and N_C, indicating that the circuit performance will be fairly robust to small errors in the parameters.

Using these optimum values, the VA rating of the voltage and current injection transformers is calculated as 1.4 % and 0.5 % of the power throughput respectively. The VA rating is obtained by multiplying the rms value of the winding currents by the rms value of a sinusoidal voltage at the transformer operating frequency that produces the same peak flux density as the actual voltage waveform. The rms and average values of the currents through the diodes of the voltage injection circuit are 1.2% and 0.7% of the output current, and 63% and 52% of the output current for the current injection circuit. As for the bi-directional switch, the rms and average currents represent 1% and 0.4% of the output current respectively. The peak value of the fundamental component of V_{AN} has a value of 1.548 V_O. Consequently, the theoretical no-load voltage for a 115 V line-to-neutral rms supply has a value of 251.8 V. The output voltage, the line currents, the power factor and the THD_I can be calculated for different load values using the method shown in [15].

VI. CIRCUIT SIMULATION

A SABER simulation using idealised components was used to verify the performance of the circuit. The line-to-neutral supply voltage was 115 V rms at 400 Hz, the load resistance was 3.6 Ω, the input inductors were chosen to be 100 μH, the turns-ratio of the current injection transformer was N_C=10, the turns-ratio of the voltage injection transformer was N_V=4.21, the conduction angle of the switch was δ=7.5°, and the dc-link capacitors were 141 μF.

Figure 11 shows the I_A line current and the V_{AN} voltage at the right-hand side of the input inductors, and Figure 12 shows some of the most relevant converter waveforms. The voltage waveform V_{AN} and the current I_A in Figure 11 have 48-pulse characteristics as predicted by the analysis.

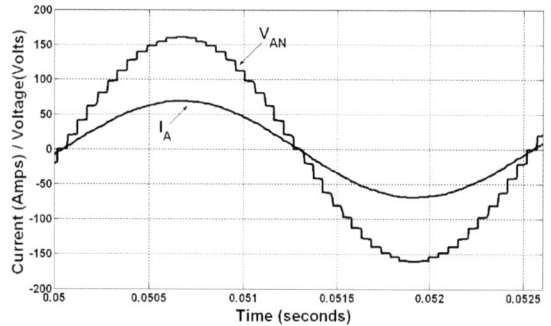

Figure 11 SABER Simulation waveforms V_{AsN}=115 V rms, F_S=400 Hz, P_O= 15 kW. Voltage V_{AN} and line current I_A

On the left-hand side of Figure 12, the first waveform (a) is the output voltage of the bottom rectifier bridge V_{O1}, the second waveform (b) shows the current I_{SC} flowing in the

sense winding of the current injection transformer, and the third waveform (c) shows the current I_{T1} that flows into the negative terminal of the bottom rectifier. The idealised sketches of V_{O1} and I_{T1} are depicted in Figure 6. The last waveform in the left column (d) shows the common mode voltage V_{NG} between the supply neutral and the mid-point of the output capacitors. On the right-hand side, the first two waveforms (e) and (f) show the voltage of the terminals A1 and A2 in Figure 4 with respect to the mid-point of the output capacitors V_{A1G} and V_{A2G} respectively. Also presented are the currents that flow through the rectifier output terminals, I_{A1} and I_{A2}. The third waveform (g) is the voltage V_{A1N} of terminal A1 with respect to the supply neutral N, and the last waveform in the right-hand column (h) shows the voltage V_{A2C2} between the terminals A2 and C2. The idealised sketches of waveforms (d), (e), (f), and (g) are presented in Figure 7. V_{AA1} in Figure 7 is the waveform V_{A2C2} presented in Figure 12-(h), divided by the turns-ratio of the main transformer, which is √3. The simulated waveforms are very similar in shape to the idealised waveforms presented in Figure 6 and Figure 7.

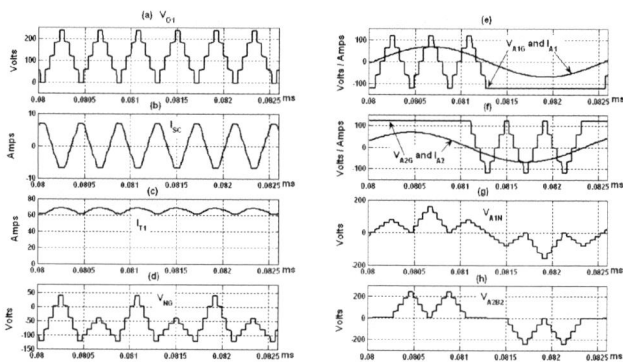

Figure 12 SABER Simulation waveforms 115V line to neutral supply,

VII. CONCLUSIONS

The converter presented in this paper is a 48-pulse converter that combines a passive 36-pulse converter with a bidirectional switch operating at six-times the supply frequency, to obtain input currents that are characteristic of a 48-pulse converter. The VA-ratings of the current and voltage injection transformers are only 0.5% and 1.4% of the power throughput respectively, and the rms and average currents of the injection switch are only 1% and 0.4% of the output current respectively. Control circuitry with associated sensing devices is required to synchronise the operation of the injection switch. The operation of this converter has been explained, analysed and validated by simulation.

VIII. ACKNOWLEDGMENT

The authors gratefully acknowledge the help and support from Marc Holme and the team at Goodrich ESTC, and Giovanni Raimondi and Peter Crouchley from Goodrich

Power Systems. They are also grateful to Dr David Trainer for his valued contributions at the beginning of this work.

IX. REFERENCES

[1] J. Arrillaga and N. R. Watson.; *Power system harmonics*, John Wiley and Sons inc., N.J.: 2003.

[2] IEEE recommended practices and requirements for harmonic control in electrical power systems," *IEEE Std 519-1992* , vol., no., pp.-, 12 Apr 1993.

[3] J. W. Kolar and H. Ertl, "Status of the techniques of three-phase rectifier systems with low effects on the mains," *IEEE Intelec conference*, no. 21, 1999.

[4] M. Rastogi, R. Naik and N. Mohan, "A comparative evaluation of harmonic reduction techniques in three-phase utility interface of power electronic loads," *IEEE Transactions on Industry Applications*, vol. 30, no. 5, pp. 1149 – 1155, Sep/Oct 1995.

[5] B. Singh, B. N. Singh, A. Chandra, K. Al-Haddad, A. Pandey and D. P. Kothari, "A review of three-phase improved power quality ac-dc converters," *IEEE Transactions on Industrial Electronics*, vol. 51, no. 3, pp. 641 – 659, June 2004.

[6] G. Gong, M. L. Heldwein, U. Drofenik, J. Minibock, K. Mino and J. W. Kolar, "Comparative evaluation of three-phase high-power-factor ac-dc converter concepts for application in future more electric aircraft," *IEEE Transactions on Industrial Electronics*, vol. 52, no. 3, pp. 727 – 736, June 2005.

[7] D. A. Paice, *Power electronic converter harmonics: Multipulse methods for clean power*, IEEE Press, New York, (1995).

[8] E. Matheson and K. Karimi, "Power quality specification development for More electric airplane architectures," *Society of Automotive Engineers*, no. 2002-01-326, 2002.

[9] G. L. Skibinski, N. Guskov and D. Zhou, "Cost effective multi-pulse transformer solutions for harmonic mitigation in ac drives," *IEEE Conference on Industry Applications*, vol. 3, pp. 1488-1497, Oct. 2003.

[10] K. J. Karimi and A. C. Mong, "Modeling non-linear loads for aerospace power systems," *IEEE EnergyConversion Engineering Conference*, pp. 33-38, July 2002.

[11] B.M. Bird, J.F. Marsh, P.R. McLellan, "Harmonic reduction in multiple converters by tripple-frequency current injection.", *IEE Proceedings*, Volume 116, no. 10, 1969, pp. 1730-1734.

[12] S. Choi, P. N. Enjeti and I. J. Pitel, "Polyphase transformer arrangements with reduced kVA capacities for harmonic current reduction in rectifier-type utility interface," *IEEE Transactions on Power Electronics*, vol. 11, no. 5, pp. 680-690, Sept. 1996.

[13] Y.H. Liu, J. Arrillaga and N.R. Watson, "A new high-pulse voltage-sourced converter for HVdc transmission," *IEEE Transactions on Power Delivery,* vol. 18 no.4, pp. 1388-1393, October 2003

[14] F.J. Chivite-Zabalza and A.J. Forsyth, "A passive 36-pulse ac-dc converter with inherent load balancing using combined harmonic voltage and current injection," *IEEE Transactions on Power Electronics*, vol. 22, no. 3, pp. 1027-1035, May 2007.

[15] F.J. Chivite-Zabalza, A.J. Forsyth and D.R. Trainer, "A simple, passive 24-pulse ac-dc converter with inherent load balancing," *IEEE Transactions on Power Electronics*, vol. 21, no. 2, pp. 430-439, March 2006.

[16] F.J. Chivite-Zabalza and A.J. Forsyth, "A simple, passive 24-pulse ac-dc converter with inherent load balancing using harmonic voltage injection," *IEEE Power Electronics Specialist Conference*, no. 36, pp 76-82, June 2005.

[17] M. Depenbrock, and C. Niermann "A new 18-pulse rectifier circuit with line-side interphase transformer and nearly sinusoidal line currents," *IPEC*, pp. 539-546, 1990

A Study of Different Possible Switched Mode Chopper Circuits for Multi-Magnet Based DC Electromagnetic Levitation System

Subrata Banerjee

Department of Electrical Engineering
National Institute of Technology
Durgapur-713209, INDIA
bansub2004@rediffmail.com

Dinkar Prasad

Emerson Network Power, India
Dinkar.Prasad@EmersonNetwork.co.in

Jayanta Pal

Department of Electrical Engineering
Indian Institute of Technology
Kharagpur-721302, INDIA
jpal@ee.iitkgp.ernet.in

Abstract– In this paper an overview of power amplifier for multi-magnet based DC electromagnetic levitation system has been presented. The general requirement of power amplifier for levitation system has been described. The proposed multi-magnet based single axis levitation scheme has been described in brief. Different possible switched mode power circuits for the multi-magnet based electromagnetic levitation system have been discussed and a comparative study has been made of the different topologies of power amplifiers based on their structure and practical results that derived from the proposed experimental set-up. In the actual work design, fabrication and testing of multi-magnet based electromagnetic levitation scheme has been made as a complete project. But in this paper the emphasis is given on the power amplifier portion due to limitation of space.

Index Terms– Electromagnetic levitation, switched mode chopper circuit, asymmetrical H-bridge converter, Electromagnetic interference, SISO control.

I. INTRODUCTION

The suspension of objects with no visible means of support due to magnetic force is termed as magnetic levitation or MAGLEV. Based on the basic principle, magnetic levitation may broadly be classified into two types, electrodynamic levitation and electromagnetic levitation [1-2]. The electrodynamic system actuates through repulsive forces. Most of such systems utilize superconducting magnets to generate the forces. In electromagnetic system, the levitation is produced due to the attractive force between electromagnets and ferromagnetic objects. In electromagnetic levitation (attraction system), the electromagnets are driven either by AC or DC source [1-2]. Although several experimental systems using AC sources have been built, these methods are considered to be suited for applications where mass of the suspended object is small. The severe constraints imposed by eddy-current losses in the magnet and the rather complex control circuitry for power modulation makes the AC method of stabilization inappropriate for heavy payloads [2]. In contrast, the explicit DC method, technically known as the DC electromagnetic levitation system (EMLS) [1-3], has a considerably simpler configuration with favourable power requirement. In DC EMLS, the current as well as

the attraction force of the electromagnet can be effectively controlled by utilizing a switched mode power amplifier.

In DC EMLS, the coil current for the magnets used in levitation needs to be precisely controlled to meet the attractive force demand. This calls for a fast DC to DC power amplifier that can be controlled in a closed loop fashion. For DC EMLS, both linear and switched mode type solid-state power amplifiers have been proposed [3-5] which have their own advantages and disadvantages. The linear power supplies produce less switching related noise and the overall electro-magnetic interference (EMI) is less when compared to the one produced by switched mode power supply (SMPS) circuits [7]. However, linear power supplies have considerably low energy efficiency in comparison to the SMPS. For high power magnets, use of linear amplifier will mean un-practically large switch ratings and heat sink ratings. A combination of switched supply and linear amplifier circuit may be thought of to further improve the efficiency of linear amplifier circuits. The switched mode DC-to-DC power supply (chopper) circuits are energy efficient but they generate switching related electromagnetic noise and pollute the position signals (as most position sensors are affected by electromagnetic noise). Once it has been decided to use the SMPS type circuit one must make a good layout of the power circuit to limit EMI [7] and some kind of electromagnetic shield may be provided between the chopper circuit and the position sensor. The other important consideration is the chopper switching frequency. The chopping frequency should be significantly higher than the frequency band of expected position signal to enable effective filtering of EMI generated by chopper from the low frequency position signal. The low pass filter cut-off frequency should be kept significantly higher than the position signal frequency but much lower than the switching noise frequency. Due to high chopper frequency the resulting current in the magnet coil is almost continuous resulting in a linear transfer function for the chopper being used. High chopper frequency eliminates low order harmonics from the coil current resulting in smooth current variation and less humming noise.

978-1-4244-0644-9/07/$25.00 ©2007 IEEE

II. DESCRIPTION OF PROPOSED MULTI-MAGNET BASED SINGLE AXIS LEVITATION SYSTEM

Though the single magnet, single axis levitation scheme may be useful for some industrial applications, majority of the applications require multi-axis levitation control where one may need to use a multiple-magnet based levitation system. This paper reports control of an electro magnetically levitated platform (vehicle structure) [6] that uses four attraction type magnets fixed at the four corners of the platform. The photograph of the experimental setup during stable levitation around 4 mm air-gap is shown in Fig.8. The overall structure of the electromagnetically levitated vehicle may consist of independent levitation, guidance and propulsion systems. However, this work considers only the singe axis magnetic levitation part. The four magnets are controlled independently through four identical controllers and stable levitation of the platform is achieved through single input and single output (SISO) control of each air-gap. Here the basic structure of each controller unit is same and the control methodology used for each controller is similar. The controller design part is not included here due to page limitation.

The prototype (total mass 14 kg) consists of four identical electromagnets placed at the four corners of a platform (Fig.8) and the structure is made to remain suspended at different air-gap positions under a ferromagnetic guide-way - the arrangement that is normally used for electromagnetic MAGLEV system. The minimum number of actuators required to levitate such a platform is four, but more than four actuators may be used for better reliability. In each case the current of the electromagnet is controlled through a single switch based DC to DC switched mode chopper circuit (Fig.6) utilizing an outer position control loop and an inner current feedback control loop as shown in the block diagram of Fig.1. A lead compensator [8] is used in cascade with the position control loop for maintaining overall closed loop stability. A linear inductive type position sensor is used to measure the actual gap between the magnet pole-face and the guide-way. Output of the position loop (current reference signal) is compared with the actual coil-current signal sensed by an LEM make (LA-55P) Hall-effect current sensor. For better dynamic response and steady state accuracy the current error is processed through a PI controller and its output is used to control the chopper output voltage through PWM control logic. The duty ratio of the MOSFET switches varies as the platform moves up and down within the electromagnetic field. When the platform moves upwards (beyond reference gap) the duty ratio of the MOSFET gate pulse is decreased, consequently the magnet current decreases and causes it to go down and vice-versa.

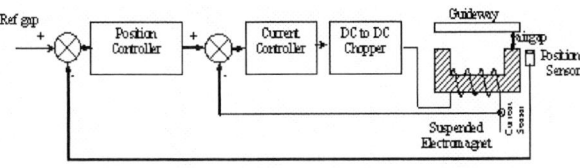

Fig.1 Block diagram of the proposed scheme

III. DESCRIPTION AND ANALYSIS OF DIFFERENT SMPS CIRCUITS USED IN MULTI-MAGNET BASED DC ELECTROMAGNETIC LEVITATION SYSTEM (EMLS).

The magnets used in EMLS are generally large and have large time constants (L/R ratio). The working air-gap between the magnet pole face and the fixed iron guide rails are small. The electro-magnetic forces are generally large and unless there is fast control of the magnet current the levitating object will either be falling on the ground structure or will be hitting the guide rails above. The magnet current needs to rise and fall in accordance with the control signal generated by the position controller. The expected variation in the magnet demand current (small signal component), over its nominal DC value, is expected to be band limited to around 10 Hz but it is better to have a current tracking capability in the range of up to 100 Hz [1]. The electrical time constant of the EMLS magnet being large, the amplifier needs to apply considerably large instantaneous voltages to the magnet coil (larger in comparison to the DC voltage required to maintain just the nominal current) for allowing quick control of the coil current.

A simple Buck type DC-to-DC chopper (class-D) circuit of Fig.2 (using a controlled switch and a freewheeling diode) will not be suitable to feed the levitation magnet as this chopper circuit can apply only unipolar voltage to the magnet coil. By adding suitable value of series resistance to the magnet coil one can make the coil voltage negative as the resistance drop applies negative voltage across the coil during freewheeling mode of the class-D chopper. However, the resistance will be dissipating significant amount of power and the energy efficiency will thus be low. But for multi-magnet based EMLS the total power circuit utilizing class-D chopper may be simpler and cost-effective.

A full bridge circuit having four controlled switches (Fig.3) can apply equal amount of positive and negative voltage to the load (magnet coil) while allowing coil current to be bi-directional. Electromagnetic attraction force is, however, independent of the coil-current direction and hence one may as well go for a cheaper asymmetrical bridge circuit that allows only one direction of load (coil) current. Moreover for the multi-magnet based levitation system the use of full bridge circuit requires many numbers of isolated power supply as well as gate drivers and the overall control circuit will be complicated.

The asymmetrical bridge circuit shown in Fig.4 requires only half the number of switches and diodes than the full bridge circuit and is capable of applying bi-directional load voltage similar to the full bridge circuit. Within each high frequency cycle when the switches are ON, positive voltage across the coil causes the coil current to rise and during OFF duration, coil current decays due to application of negative voltage. During current decay, through the diodes, part of the magnet energy is fed back to the supply and is not required to dissipate through any external resistance and thus the circuit is quite energy efficient. Because of its high

813

energy-efficiency this asymmetric converter is ideal for high power applications. Ohmic isolation is required between the gate drive signals of the two controlled switches and this call for a proper isolation and amplification stage for driving each of the two switches. Considering asymmetrical bridge circuit, one need to have four bridge circuits for four electromagnets, each having two controlled switches and two diodes (Fig.5). So the overall power circuit will require eight numbers of isolated gate drivers. Since many numbers of hard switching devices are to be used, a careful design for the layout is required to limit the EMI.

Instead of this power circuit topology, if the simplified circuit (Fig.6) with only one switch (and a diode with energy-dump circuit) per magnet is used, the overall saving in the power amplifier components may be significant. All the four switches of the simplified power circuit may be connected such that their gate drive signals may not need ohmic isolation. Moreover all four amplifiers may share a common energy dump capacitor-resistor circuit. The single switched based power circuit topology (Fig.6) has been utilized, for levitating the platform. When the switch for one magnet-coil is turned on, full supply voltage appears across that coil and the current in the coil increases. Thus energy flows from source to the load. During ON time the energy dump capacitor (10 μF, 400V) that holds some charge due to previous operation, discharges partially through the rheostat (185Ω). During OFF period of the switch, the coil current finds a path through fast-recovery power diode to the resistor capacitor combination. The capacitor voltage effectively appears across the coil with a negative polarity (reverse of the polarity during ON duration). With reversal of coil voltage the coil current starts decreasing at a slope decided by the coil inductance and the capacitor voltage.

With four coil circuit, using simplified chopper topology (Fig.6), it may make sense to invest on another simple circuit to control the dump capacitor voltage within desired limits or to feed back this energy to input supply. In spite of this additional modification there is expected to be overall saving in the power amplifier cost as compared to the four asymmetrical bridge circuit. Another advantage envisaged in favor of the simplified circuit is the overall reduction in the EMI as the number of switches is less which is all connected to the same common point. For the present work, for levitating the platform, the simplified chopper topology has been used. In this circuit each coil requires one switch and one diode, but the capacitor and the discharge resistor is common to all four circuits. MOSFETs are connected in common source configuration. The common source point can be directly connected to the control ground and gate drive isolation is not a must here. Fig.7 shows a modification over the circuit of Fig.6 to keep the capacitor voltage magnitude under control and nearly ripple free. This is achieved by replacing the simple energy dump resistor by a controlled resistor using a single switch chopper circuit.

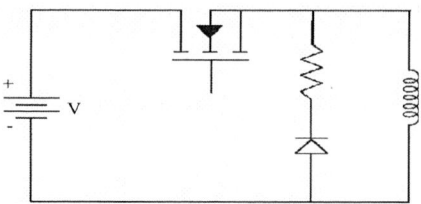

Fig.2 Simple class-D chopper circuit

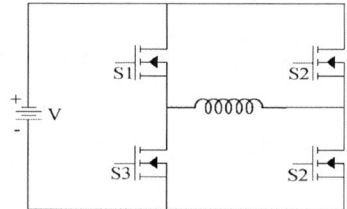

Fig.3 Full Bridge switched mode power amplifier

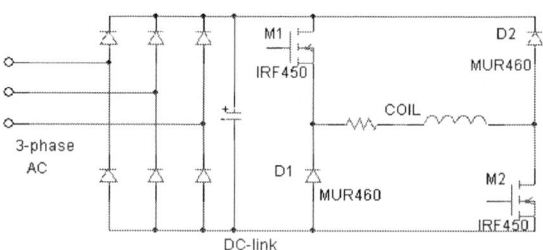

Fig.4 Asymmetrical bridge (H-bridge) converter circuit

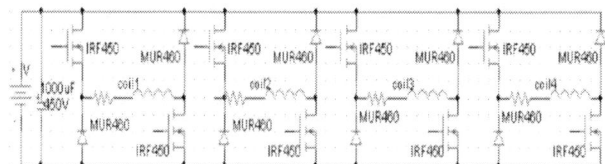

Fig.5 The asymmetrical H-bridge converter for four coils

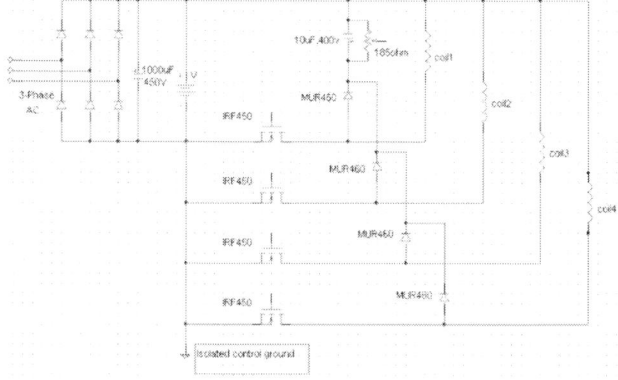

Fig.6 Proposed single switch based power converter for four coils

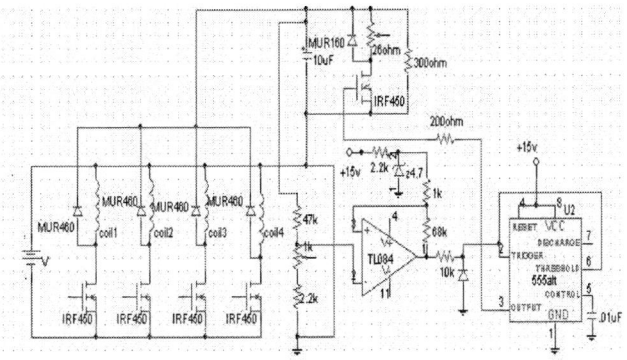

Fig. 7 Modified power circuit

IV. EXPERIMENTAL RESULTS

The proposed multi-magnet based levitated system (utilizing the single switch based power supply) has been successfully tested and its stable levitation was demonstrated around the designed operating gap of 4 mm. The photograph of the platform during stable levitation around 4 mm air-gap is shown in Fig.8. Fig.9 shows a typical oscillogram of one coil voltage and the capacitor voltage and Fig.10 shows rise and fall of one coil-current along with the coil voltage during stable levitation of platform when the DC input supply is 150 volts. The coil voltage is made equal positive and negative supply voltage (150 volts) with the predefined circuit parameters. The capacitor ripple is found to be 10% of the mean voltage. During OFF period of the switch, the capacitor starts charging with the flow of coil-energy into it. When the switch turns on, it discharges through the rheostat, thus, causing loss of voltage. During ON period of the switch, coil-voltage rises due to the energy transfer from source to the load. Similarly, during OFF period of the switch, coil-voltage falls (shown in oscillogram of Fig.10) due to energy transfer from the coil to the capacitor and a part of this energy is dissipated through the rheostat. The four position signals are given in Fig.11 during the dynamic operation of the platform.

Fig.8 Photograph of the setup during stable levitation

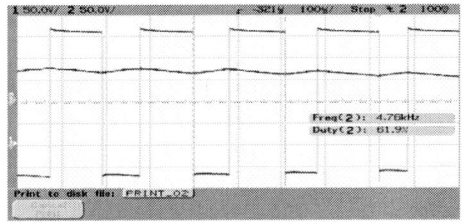

Fig.9 Coil voltage (CH2) and Capacitor voltage (CH1) during levitation

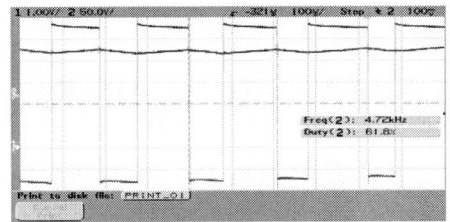

Fig.10 Coil voltage (CH2) and coil current (CH1) during stable levitation

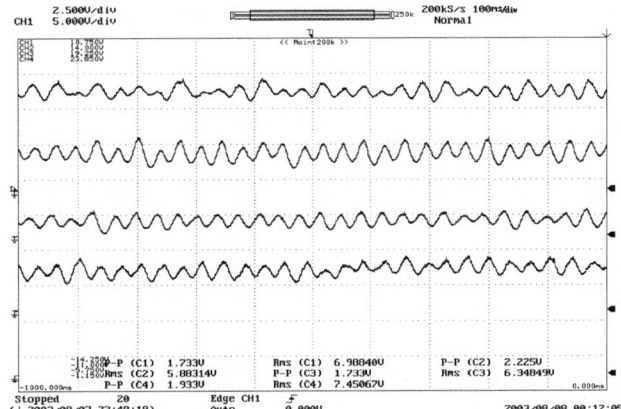

Fig.11 Four position signals during levitation

V. CONCLUSION

In this paper an overview of power amplifier for multi-magnet based DC EMLS has been presented. The general requirement of power amplifier for multi-coil system has been described. The proposed multi-magnet based levitation system has been described in brief. A comparison has been made between different possible configurations of switching power amplifiers based on their structure and experimental results with reference to multi-magnet based levitation system. The proposed single switch based switched mode power circuit is simpler in topology but it is not so energy efficient as the asymmetrical bridge circuit. However because of its simplicity and less cost it may find favor in applications involving small levitation power. There are some other advantages too associated with this new circuit, like, production of less electromagnetic noise and compactness of power circuit. There is a possibility of further modifying this circuit where the fixed resistor is replaced by a chopper controlled variable resistor. This will allow almost ripple free voltage across the energy dumping circuit capacitor. It is also possible to recover the energy stored in the energy dump capacitor and then the system will become energy efficient too.

815

REFERENCES

[1] B.V. Jayawant, 'Review lecture on electromagnetic suspension and levitation techniques', *Proc. R. Soc. Lond.* A416, 1988, pp.245-320.

[2] P.K. Sinha, *Electromagnetic Suspension, Dynamics and Control*, Peter Peregrinus Ltd., London, 1987.

[3] N. A. Shirazee and A. Basak, 'Electropermanent suspension system for acquiring large air-gaps to suspend loads', *IEEE Trans. on Magnetics*, Vol.31, No.6, Nov.1995, pp.4193-4195.

[4] S. Carabelli, F. Maddaleno and M. Muzzarelli, 'High-efficiency linear power amplifier for active magnetic bearings', *IEEE Trans. Industrial electronics'*, Vol.47, No.1, Feb.2000, pp.17-24.

[5] F. Maddaleno, S Carabelli and M Muzzarelli, 'A modified class G amplifier for active magnetic bearing', 13th annual applied power electronics conf, *Proc. APEC'97*, 15-19 Feb.1997, California, USA ,Vol.1, pp.534-538.

[6] A. Bittar and R. M. Sales, ' H_2 and H_∞ control for maglev vehicles', *IEEE Control Systems Magazine*, Vol.18, No. 4, Aug.1998, pp.18-25.

[7] N.Mohan, T.M.Undeland and W.P.Robbins, *Power Electronic Converters, Applications and Design*, Second Edition, IEEE Press and John Wiley & Sons, 1996.

[8] K. Ogata, Modern *Control Engineering*, Third Edition, Prentice Hall of India, New Delhi, 2000.

Power Supply with Potential Use in Magnetic Stimulation

Duarte M. Sousa[1] and António Ferraz[2]

[1] Instituto Superior Técnico, DEEC AC-Energia, TU Lisbon – Av. Rovisco Pais – 1049-001 Lisboa – Portugal
[2] Instituto Superior Técnico, DF/TagusPark, TU Lisbon, Av. Professor Cavaco Silva, 2780-990 Porto Salvo and Centro de Física da Matéria Condensada, Av. Prof. Gama Pinto 2, 1649-003 Lisboa, Portugal
[1]pcdsousa@mail.ist.utl.pt [2] antonio.ferraz@tagus.ist.utl.pt

Abstract-A power supply with potential use in the magnetic stimulation of biological systems is presented in this paper.

The effect of the magnetic stimulation depends on the geometry and orientation of the induced electric field, as well as on the current pulse waveform delivered by the stimulator (coil). These factors are very important to define the equipment requirements and characteristics, namely the topology of the power supply and the size and geometry of the coil.

The proposed solution is able to generate current pulses with variable amplitude and duration, according to a user-defined input. The main characteristics and the working principle, including the strategy of control, are described. In order to design, to analyse and to verify the performance of the developed power supply, both simulation and experimental results are presented.

The proposed solution is based on a topology that uses elements to store and transfer energy from the power source to the load. With the proposed topology, an adequate control strategy and right set of the power circuit parameters it is possible to obtain either unipolar waves or bipolar waves.

Index Terms—Power Supply, Magnetic Stimulation, Simulation.

I. INTRODUCTION

Magnetic Stimulation (MS) is a new tool that allows non-invasive stimulation of biological systems. This technique uses specific equipment, in which the most visible element is the coil that generates the magnetic pulses. The coils is driven by high-intensity current that is rapidly turned on and off through the discharge of capacitors. This current produces a time-varying magnetic field that lasts for about 100 to 200 microseconds. The typical magnetic field strength is about 2 Tesla [1-21].

Among the MS techniques, the Transcranial Magnetic Stimulation (TMS) is a tool for neuroscience which allows non-invasive stimulation of peripheral nerves and cerebral cortex [19]. Its purpose is to create a pulsed electric current induced by a time-varying magnetic filed and it has appeared in 1985 [21] as a brain mapping tool. It is nowadays considered as an established research tool in the cognitive neurosciences field, including the study of perception, attention, learning, plasticity, language and awareness [3, 16]. TMS has also being used in the study and treatment of several neuropsychiatric disorders, such

as movement disorders, epilepsy, both unipolar and bipolar disorders, anxiety disorders and schizophrenia [6, 8, 10, 13, 16-19]. Research aiming the use of TMS in other types of disorders and also directed to specific population sets has been reported for several times in the last few years [1, 4, 5, 11-12].

TMS is based on Faraday´s principle of electromagnetic induction [30]. According to it, an oscillating electric current passing through a coil produces an oscillating magnetic field which in turn induces an electric current in an adjacent coil. Using the TMS method, the first coil is placed over somebody's scalp and the brain tissue reacts like the second coil. The intensity and duration of the current and the geometry of the primary coil are responsible for the amplitude and shape of the oscillating magnetic field produced and thus determines the intensity and location of the electric current induced in the brain.

The use of this technique on biological tissues, requires a magnetic field as focal as possible to stimulate only well defined areas. The effect of the MS depends on the geometry and orientation of the induced electric field, as well on the current pulse waveform delivered by the stimulator. These factors are very important to define the equipment requirements and characteristics, namely the size and geometry of the coil and the topology of the power supply.

The main objective behind this work is to develop new and state of the art magnetic stimulation equipment. The main effort has been focused on the design of the coil and its cooling system [to be published] and in the development of the power supply. The design and implementation of the MS coil has been done using modern computational techniques to optimize their electrical and geometrical characteristics [28-29]. The tested topology of the power supply fulfils the requirements to be implemented in MS systems [22-27].

II. THE CIRCUIT

The main blocks of the power supply are a transformer, a rectifier, a DC/DC converter and the control circuit (Fig. 1). In order to command, control and synchronize the MS equipment, a computer and an I/O system are included in the MS equipment. In general, this application requires specific software.

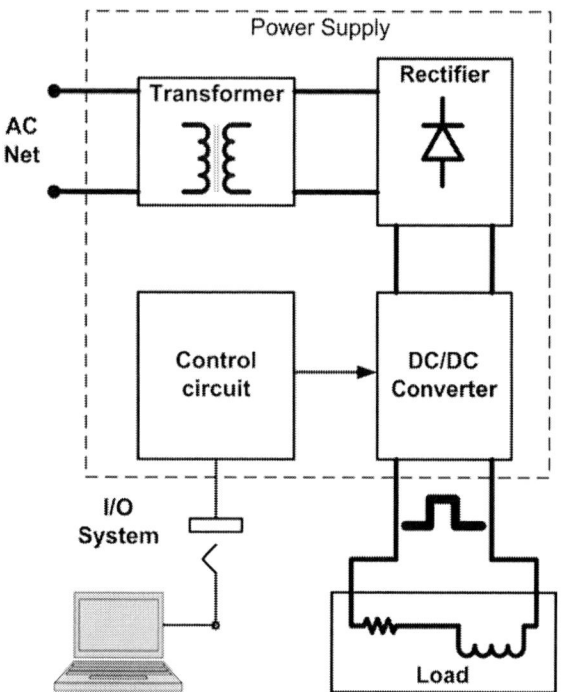

Fig. 1. Main blocks of a magnetic stimulation system.

Over viewing the power supplies of known magnetic stimulation systems [vg. 16, 23, 25, 31], the output current can have the following forms: unipolar waves, bipolar waves and damped waves, as exemplified in figure 2.

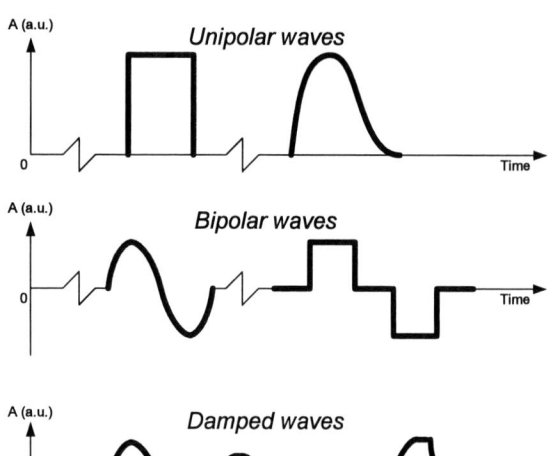

Fig. 2. Output current waveforms.

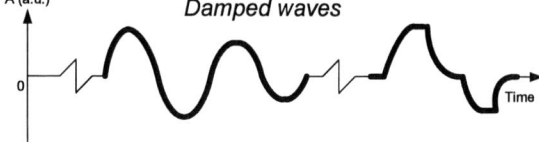

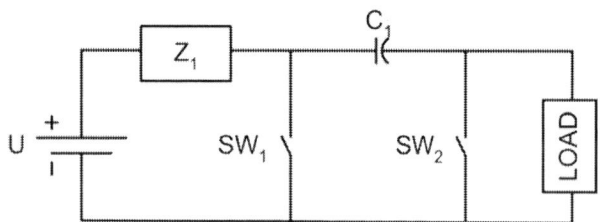

Fig. 3. Basic topology of the circuit.

This work started analysing the dynamic behaviour of a circuit, which basic topology is represented in figure 3.

Using a home-designed stimulator fulfilling the magnetic stimulation requirements, this basic circuit was implemented. Their parameters (some of them are adjustable) and the type of semiconductors used are shown in figure 4. The range of the parameters is shown in Table I.

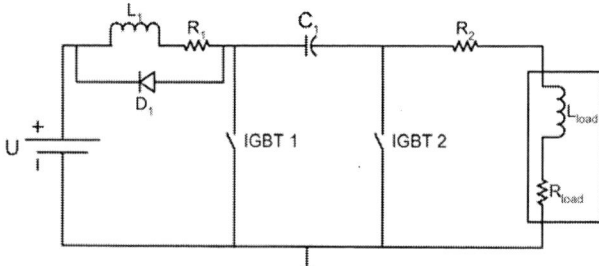

Fig. 4. Tested circuit.

III. SIMULATION AND EXPERIMENTAL RESULTS

The duty-cycles and the gate signals of the IGBTs were initially adjusted avoiding short-circuits and generating current pulses with the required square wave form in terms of amplitude and duration. To study and analyse the performance of such system, both experimental and simulation results were obtained. The implemented model (simulated using the ©Matlab/Simulink) includes dynamic models of the main power supply, the capacitors, the load (R_{load}) and the semiconductors.

The electrical equations representing the behaviour of the circuit are, in a first approach, the following:

- If $I_{Load} = 0$ (short-circuit of the load):

$$U_{power\_source} = R_1 I_1 + L_1 \frac{d\,I_1}{dt} + \frac{1}{C_1} \int I_1 dt \quad (1)$$

- If $I_{Load} \neq 0$:

$$U_{capacitor} = (R_{Load} + R_2) I_{Load} + L_{Load} \frac{d\,I_{Load}}{dt} \quad (2)$$

The adjustable parameters R_1, R_2, L_1 and C_1 depend on the circuit parameters (including the protection circuits of the semiconductors) and on the experimental conditions.

A. Working principle

In accordance to the state of the semiconductors (IGBT 1 and IGBT 2), the circuit represented in figure 4 can be unfolded into equivalent circuits. The situations corresponding to equations (1) and (2) occur when:

- Situation #1: IGBT 1 *ON/OFF* and IGBT 2 *ON*;
- Situation #2: IGBT 1 *ON* and IGBT 2 *ON/OFF*;

The equivalent circuit corresponding to situation #1 is represented in figure 5.

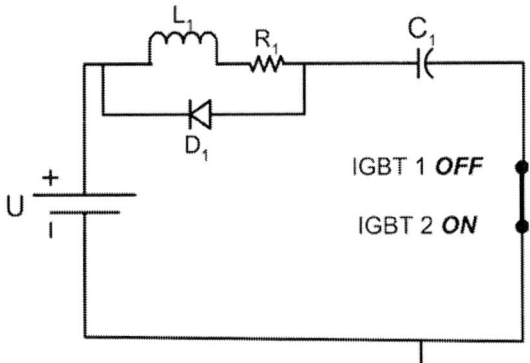

Fig. 5. Equivalent circuit with IGBT 1 OFF and IGBT 2 ON.

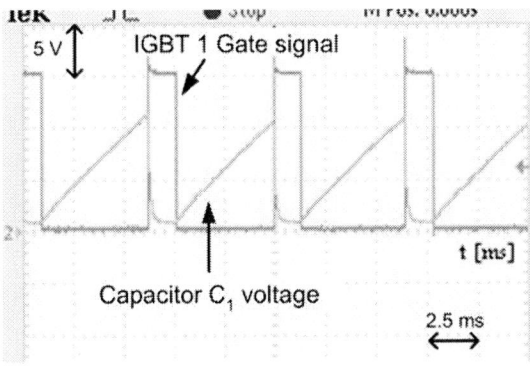

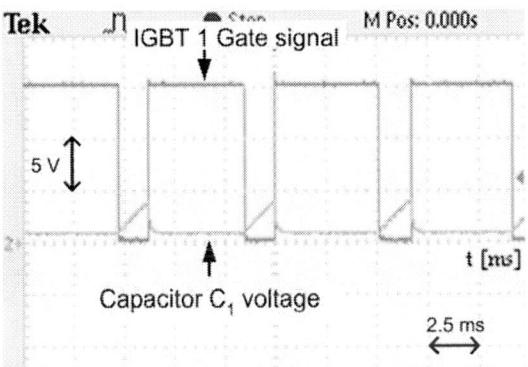

Fig. 6. Time evolution of the voltage of the capacitor C_1 for different duty-cycles (experimental results).

In this situation and according to the duty cycle of the IGBT 1 (for this analysis, IGBT 2 remains always ON), it is possible to vary the voltage of the capacitor C_1. In steady state, it can be assumed that this voltage can be controlled switching the IGBT 1 as represented in figures 6 (experimental results) and 7 (simulation results).

The equivalent circuit corresponding to situation #2 is represented in figure 8.

In this situation and according to the duty cycle of the IGBT 2 (for this analysis, IGBT 1 remains always ON), in steady state, it can be assumed that the time evolution of the capacitor C_1 voltage depends on the initial conditions, i.e., depends on the energy stored in the capacitor C_1 when

IGBT 1 is ON and the IGBT 2 is switched OFF. In this case, the dynamic behaviour of the circuit is described by equation (2) and is strongly dependent on the electrical parameters of the circuit (R_2, L_{load} and R_{load}).

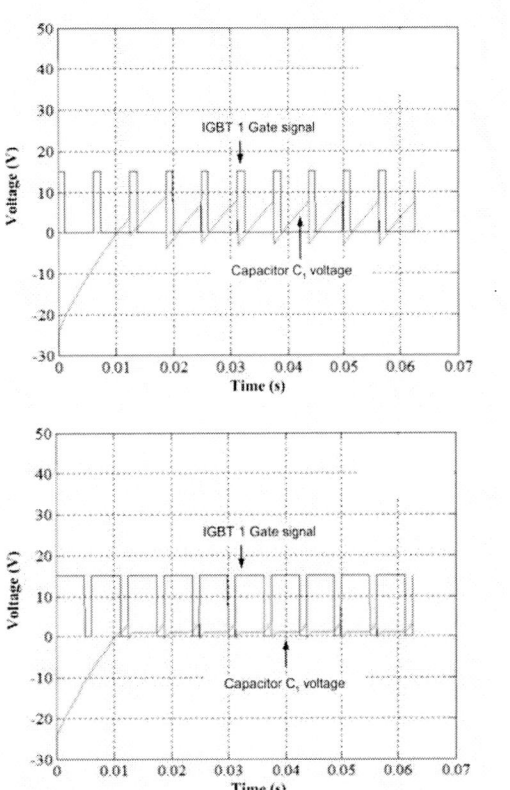

Fig. 7. Time evolution of the voltage of the capacitor C_1 for different duty-cycles (simulation results).

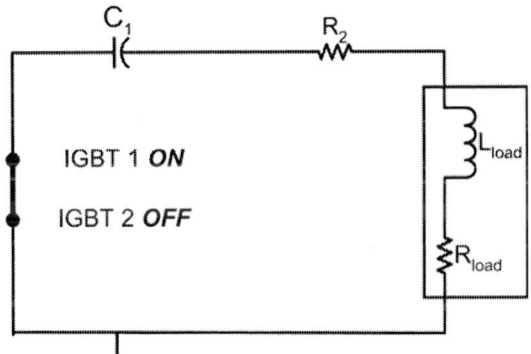

Fig. 8 Equivalent circuit with IGBT 1 ON and IGBT 2 OFF.

Taking into account the principles described above, the most important step to verify the performance of the circuit is to observe the load current.

B. *Load current*

To the proposed topology and taking into account the initial conditions, the electrical equation representing the circuit when IGBT 1 is ON and IGBT is OFF, is the following:

$$\frac{1}{C_1}\int_0^t I_{Load}(\tau)d\tau + I_{C_1}(0) =$$

$$\left(R_{Load} + R_2\right)I_{Load} + L_{Load}\frac{d\,I_{Load}}{dt} \qquad (3)$$

Corresponding to the expression:

$$\frac{1}{L_{Load}\,C_1}I_{Load}(t) =$$

$$\frac{R_{Load} + R_2}{L_{Load}}\frac{d\,I_{Load}(t)}{dt} + \frac{d\,I^2_{Load}(t)}{d\,t^2} \qquad (4)$$

To random load diagrams, some experimental and simulation results obtained to the load current are shown in figures 9, 11, 13 and 15 (simulation), and 10, 12 and 14 (experimental). From these results, it is important to point out that it is possible to obtain current pulses with square wave form and with adjustable duration. Furthermore, reasonable agreement between experimental and simulation results is observed, leading to the assumption that the developed models constitute a good approach and the circuit behaves as foreseen analytically.

However, the above working principle is valid only if an effective control of the semiconductors is reached.

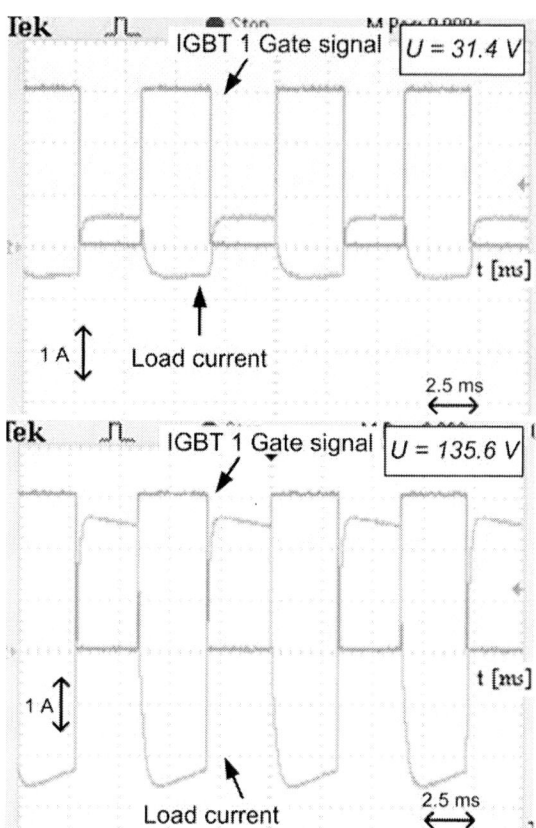

Fig. 10. Load current and IGBT 1 gate signal to the same duty cycle (0.52) and different power supply voltages (experimental results).

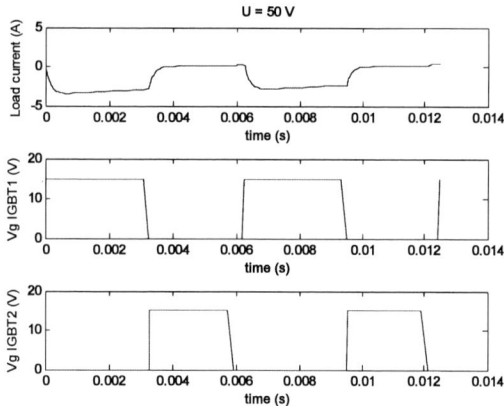

Fig. 9. Typical Load current and IGBTs gate signals (simulation results).

In order to verify the dynamic behaviour of the circuit and to point out the advantages and weaknesses of the proposed topology, different sets of experimental results and simulation results are shown.

TABLE I
PARAMETERS OF THE CIRCUIT

Parameter	U (V)	L_1 (mH)	R_1 (Ω)	C_1 (mF)	R_2 (Ω)	L_{loa} (μH)$_d$	R_{load} (mΩ)
Range	0 to 150	0 to 30	0 to 22	1	0 to 100	63.2	700

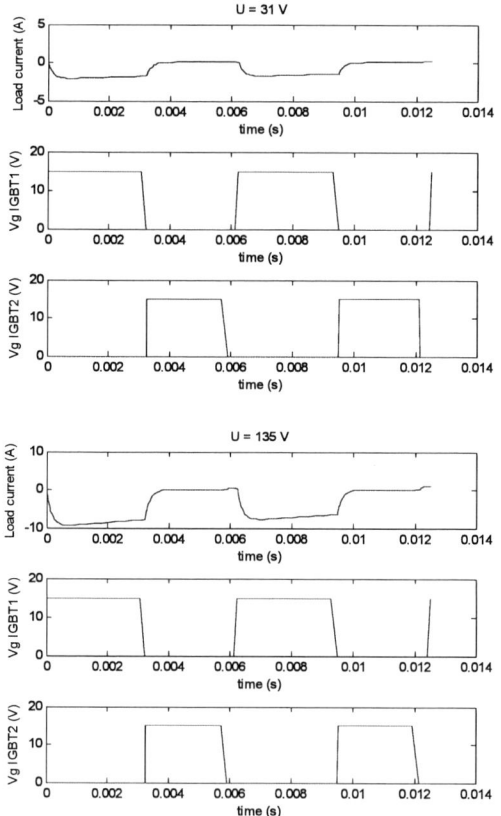

Fig. 11. Load current and IGBT 1 gate signal to the same duty cycle (0.52) and different power supply voltages (simulation results).

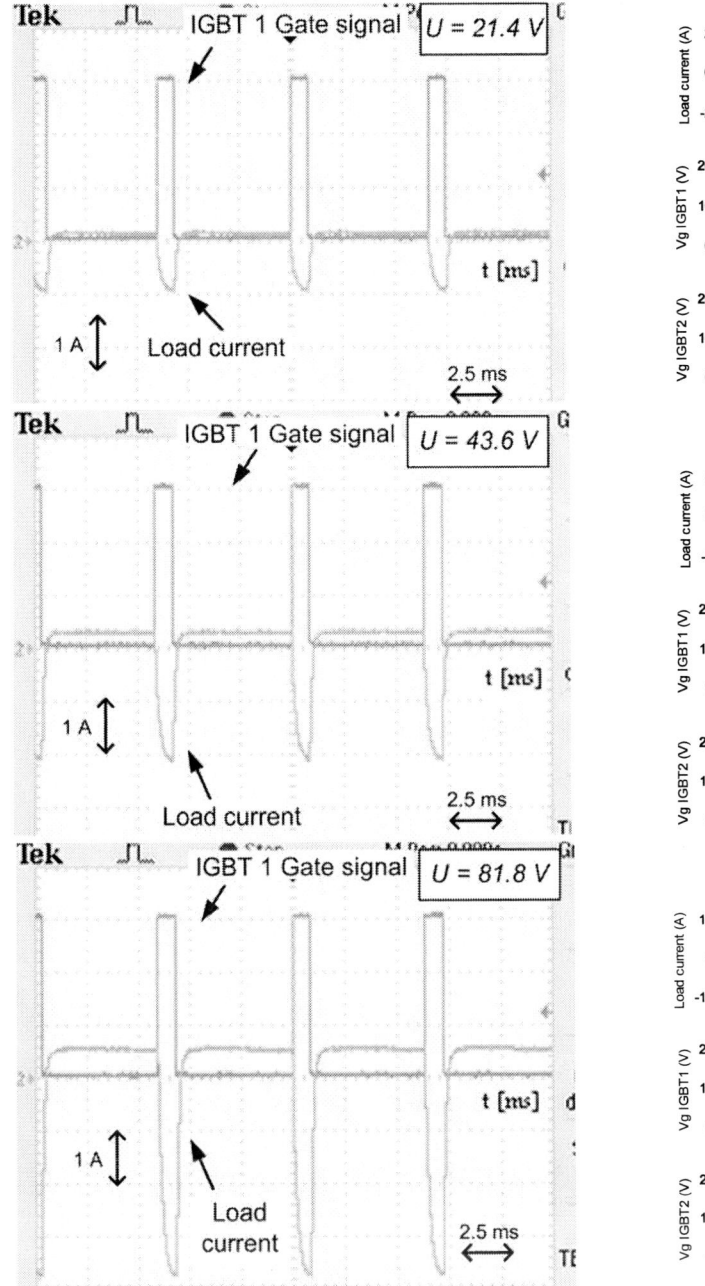

Fig. 12. Load current and IGBT 1 gate signal to the same duty cycle (0.12) and different power supply voltages (experimental results).

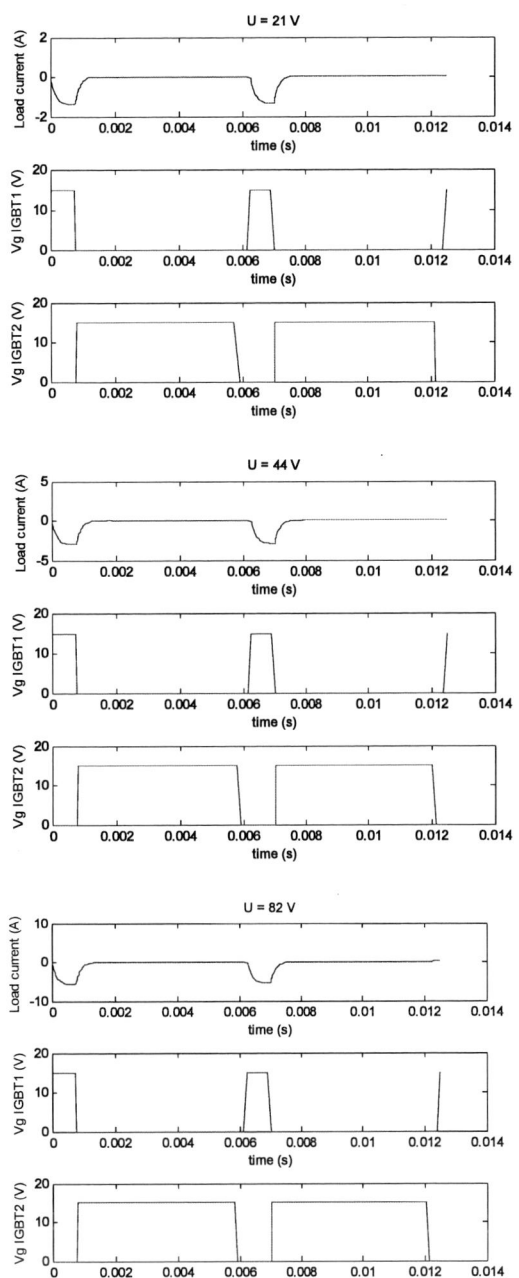

Fig. 13. Load current and IGBT 1 gate signal to the same duty cycle (0.12) and different power supply voltages (simulation results).

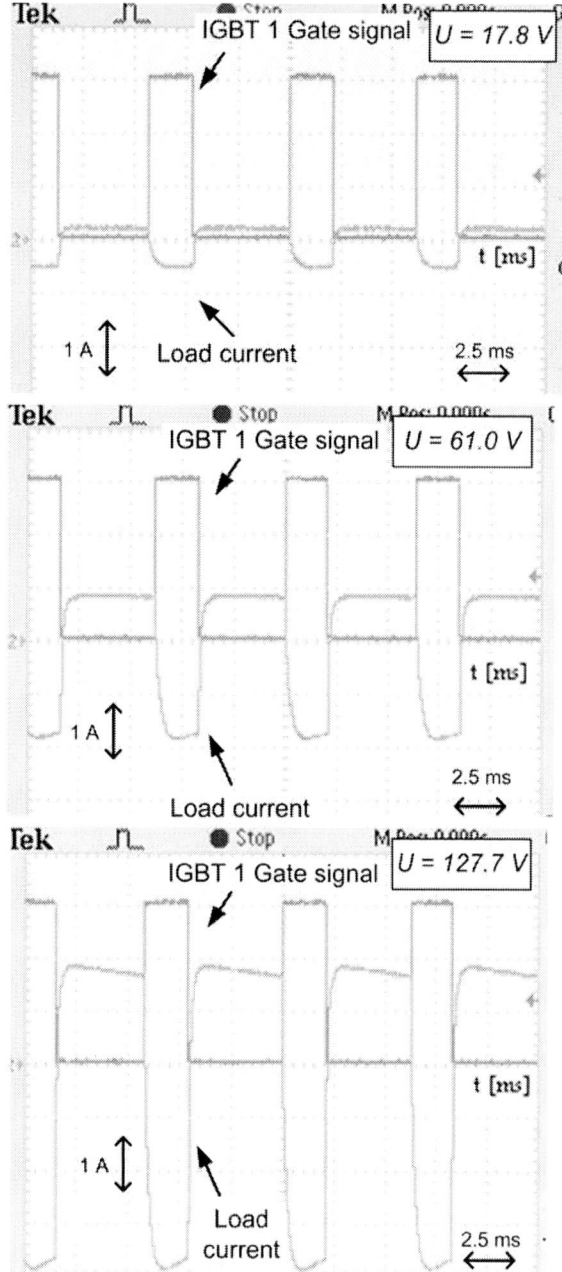

Fig. 14. Load current and IGBT 1 gate signal to the same duty cycle (0.32) and different power supply voltages (experimental results).

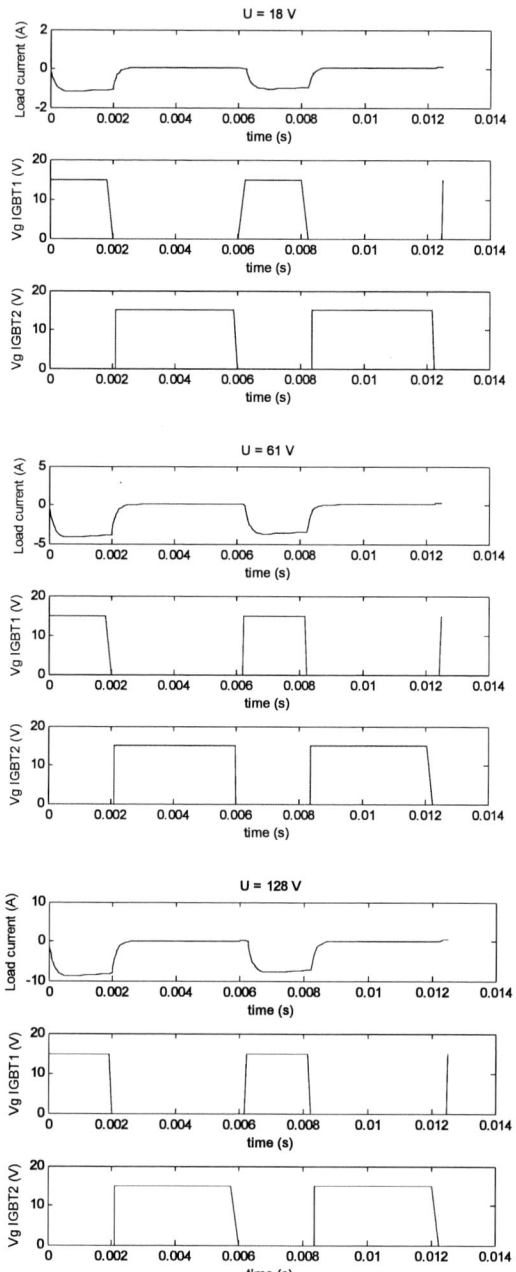

Fig. 15. Load current and IGBT 1 gate signal to the same duty cycle (0.32) and different power supply voltages (simulation results).

IV. CONCLUSIONS

With this work an untypical topology for magnetic stimulation is proposed. This topology used a DC power supply and switching semiconductors instead of the classical solutions, which use AC power supplies or resonant circuits. Analytic simulations together with the experimental results show that it is possible to achieve the technique requirements with the proposed circuit and that it is possible to adjust the current wave form (its duration and amplitude).

More accurate selection of the stimulation regions requires better localized magnetic fields. Therefore, a control strategy that allows an effective management of the magnetic field pulses (frequency, amplitude and duty-cycle) is mandatory.

In a detailed analysis of the proposed circuit, it is possible to identify aspects that can be improved, mainly the one related to the shape of the current pulses (i.e, unipolar vs. bipolar waveform). Furthermore, the nature of the stimulation coil used (air core) strongly influences the dynamic of the electric system. To test the system in order to not damage either the load or/and the semiconductors, the gate signals were set avoiding short-circuits. In addition, the semiconductors have adequate protection circuits.

Furthermore, it is important to point out that setting appropriately the parameters L_1, R_1 and R_2 in accordance to the parameters of the stimulation coil (L_{load} and R_{load}) and implementing an adequate control strategy, bipolar current pulses can be obtained from a power source based on the proposed topology.

As a final remark, it is important to refer that the proposed solution is based on a topology that uses elements to store and transfer energy from the power source to the load. With this topology, an adequate control strategy and a right set of the power circuit parameters, it is possible to obtain either unipolar waves or bipolar waves, which can be adjusted in frequency, duration and amplitude.

REFERENCES

[1] Bae, E. H. et al., Safety and tolerability of repetitive transcranial magnetic stimulation in patients with epilepsy: a review of the literature. *Epilepsy & Behavior* 10: 521-528, 2007.

[2] Moore, S. K. March, Psychiatry's Shocking New Tools. IEEE Spectrum. Volume 43, No. 3, pp. 18-25, 2006.

[3] Ueno, S., Sekino, M., Biomagnetics and bioimaging for medical applications. *Journal of Magneism and Magnetic Materials* 304: 122-127, 2006.

[4] Leo, R. J., Latif, T., 2007. Repetitive Trancranial Magnetic Stimulation (rTMS) in Experimentally Induced and Chronic Neuropathic Pain: A Review. *The Journal of Pain* 8: 453-459.

[5] Helmich, R. C. et al., Repetitive transcranial magnetic stimulation to improve mood and motor function in Parkinson's disease. *Journal of the Neurological Sciences* 248: 84-96, 2006.

[6] Rachid, F., Bertschy, G., Safety and efficacy of repetitive transcranial magnetic stimulation in the treatment of depression: a critical appraisal of the last 10 years. *Neurophysiologie Clinique* 36: 157-183, 2006.

[7] Hortobagyi, T. Bonato, P., Transcranial Magnetic Stimulation. *IEEE Engineering in Medicine and Biology Magazine*. Volume: 24; No. 1, pap. 20-21, 2005.

[8] Loo, C. L., Mitchell, P. B., A review of the efficacy of transcranial magnetic stimulation (TMS) treatment for depression, and current and future strategies to optimize efficacy. *Journal of Affective Disorders* 88: 255-267, 2005.

[9] Panescu, D., Vagus Nerve Stimulation for the Treatment of Depression. IEEE Engineering in Medicine and Biology Magazine. pp. 68-72, 2005.

[10] Haraldsson, H. M., et al., Transcranial Magnetic Stimulation in the investigation and treatment of schizophrenia: a review. *Schizophrenia Research* 71: 1-16, 2004.

[11] Garvey, M. A. et al., Transcranial magnetic stimulation in children. *European Journal of Paediatric Neurology* 8: 7-19 2004.

[12] Barclay, L., Tinnitus may respond to transcranial magnetic stimulation, *Ann. Neurol. online (53), 2002.*

[13] Fitzgerald, P. B., Brown, T. L. & Daskalakis, Z. J., The application of transcranial magnetic stimulation in psychiatry and neurosciences research. *Acta Psychiatrica Scandinavica* 105(5): 324-340, 2002.

[14] Janicak, P. G., et al., Repetitive Transcranial Magnetic Stimulation versus Electroconvulsive Therapy for Major Depression: Preliminary Results of a Randomized Trial. *Biol Psychiatry* 51:659-667, 2002.

[15] Bohning DE: Introduction and overview of TMS physics, in Transcranial Magnetic Stimulation in Neuropsychiatry, edited by George MS, Belmaker RH. Washington, DC, American Psychiatric Press, pp 13-44, 2000.

[16] George, M. S. & Belmaker, R. H. (eds.), *Transcranial Magnetic Stimulation in Neuropsychiatry*: 13-44. Washington DC: American Psychiatric Press, Inc, 2000.

[17] George, M. S. et al., Transcranial Magnetic Stimulation Applications in Neuropsychiatry. *Arc Gen Psychiatry* 56: 300-311, 1999.

[18] Topka, H. et al., Cerebellar-like terminal and postural tremor induced in normal man by transcranial magnetic stimulation. *Brain* 122: 1551-1562, 1999.

[19] George, M. S. et al., Transcranial magnetic stimulation: a neuropsychiatric tool for the 21st century. *J Neuropsychiatry Clin Neurosci* 8: 373-382, 1996.

[20] Ueno, S. et al., Localised stimulation of neural tissue in the brain by means of a paired configuration of time-varying magnetic fields. *J Appl. Phys.* 64: 5862-5864, 1988.

[21] Barker, A. T., et al., Non-invasive magnetic stimulation of human motor cortex. *Lancet* 1:1106-1107, 1985.

[22] Davey, K.R and Riehl, M.. "Suppressing the surface field during transcranial magnetic stimulation", *IEEE Transactions on Biomedical Engineering*, Volume: 53, Issue: 2, pp. 190- 194, 2006.

[23] Davey, K.R., Riehl, M. 2005. Designing transcranial magnetic stimulation systems, *IEEE Transactions on Magnetics*, Volume: 41, Issue: 3, pp. 1142- 1148.

[24] Ruohonen, J., Ollikainen, M., Nikouline, V., Virtanen, J., Ilmoniemi, R.J., Coil design for real and sham transcranial magnetic stimulation. *IEEE Transactions on Biomedical Engineering*. Volume: 47, Issue: 2, pp. 145-148, 2000.

[25] Ruohonen, J., 1998. *Transcranial Magnetic Stimulation: Modelling and New Techniques*, PhD Thesis, Espoo, Finland: University of Technology.

[26] Redondo, L.M. Margato, E. Silva, J.F., Low-voltage semiconductor topology for kV pulse generation using aleakage flux corrected step-up transformer, *Power Electronics Specialists Conference, 2000. PESC 00. 2000 IEEE 31st Annual*, vol. 1, pp. 326-331, 2000.

[27] A. Ferraz, L. V. Melo, D. M. Sousa, Design of Coils with Potential use in Transcranial Magnetic Stimulation, *9th International Conference on Enhancement and Promotion of Computational Methods in Engineering and Science*, Macau, 25-28 November, 2003.

[28] Wilson, M. N.. *Superconducting Magnets*. Oxford: Oxford Science Publications, 1997.

[29] Reddy, J. N. 1985. *An introduction to the Finite Element Method*. McGraw-Hill Book Co.

[30] Jackson, J. D., *Classical Electrodynamics* (2nd Edition). John Wiley & Sons, 1975.

[31] Nollet, H., et al., Transcranial magnetic stimulation: review of the technique, basic principles and applications, *The Veterinary Journal*,, vol. 166, pp. 28-42, 2003.

A Novel Maximum Power Point Tracking Method for the Photovoltaic System

Hurng-Liahng Jou*, *Member, IEEE*, Wen-Jung Chiang* and Jinn-Chang Wu**

* Department of Electrical Engineering
National Kaohsiung University of Applied Sciences
415 Jiangong Road, Kaohsiung 80778, Taiwan, Republic of China
** Department of Microelectronics Engineering
National Kaohsiung Marine University
142 Haijhuan Road, Nanzih District 81143, Taiwan, Republic of China

*Abstract--*A novel maximum power point tracking (MPPT) method for the photovoltaic system is proposed in this paper. The main feature of the proposed MPPT method is that only a current is required to be detected. Therefore, the proposed method has the advantage of simplifying the control circuit compared with the conventional perturbation and observation method where both current and voltage should be detected. A prototype based on the digital signal processor (DSP) controller is developed and tested to verify the performance of the proposed MPPT method. The experimental results show the proposed method has the expected performance.

*Index Terms--*photovoltaic, maximum power point tracking, digital signal processor

I. INTRODUCTION

The conventional energy sources for electrical power include hydroelectric, fossil fuels, and nuclear energy. The wide use of fossil fuels has resulted in the problem of greenhouse emissions worldwide. This also seriously damages the earth's environment. Besides, fossil fuels will be exhausted in the future, and their cost has obviously increased. The Kyoto agreement on global decrease of greenhouse emissions has prompted has prompted interest in and the importance of the issue of renewable energy sources to relieve the problem of greenhouse emissions. Photovoltaic is one of the important renewable energy sources [1-3]. The cost of the photovoltaic is on a falling trend and is expected to fall further as demand and production increases.

The power conversion interface is important for using the photovoltaic effectively. The key technologies

---

The work was financial support by the ABLEREX Electronics Corporation, Ltd.

for power conversion interface of the photovoltaic system include the technologies of DC/DC power converter, grid-connected DC/AC inverter, islanding detection and maximum power point tracking (MPPT).

The MPPT technology is addressed in this paper. Many MPPT technologies have been proposed, such as the voltage feedback method, power feedback method, perturbation and observation method, linear line approximation method, fuzzy logic control method, neural network method and practical measure method [4-13]. However, there are some disadvantages to these technologies. The conventional perturbation and observation method [11] requires at least two detected signals, a voltage and a current in tracing the maximum power point.

In order to simplifying the control circuit of conventional perturbation and observation MPPT method, a novel MPPT method for the solar photovoltaic system is proposed in this paper. The proposed MPPT method is that only a current is detected, therefore, it has the advantage of simplifying the control circuit compared with the conventional perturbation and observation methods where both current and voltage are detected. A prototype based on the DSP controller TMS320C2407 is developed and tested to verify the performance of the proposed MPPT method.

II. SYSTEM DESCRIPTION OF PHOTOVOLTAIC SYSTEM

Figure 1 shows the configuration of the DC/DC power converter for the proposed photovoltaic system. The proposed photovoltaic system is configured by a solar cell array and a DC/DC power converter. The DC/DC power converter is a boost type power converter, and it is applied to trace the maximum power point of the solar cell array. In general, the photovoltaic system can be divided into stand-alone and grid-connected types. The regulated DC power of the stand-alone photovoltaic system is used to charge a battery set and supply power to a DC load or an AC load by a DC/AC inverter. The output voltage of DC/DC power converter is clamped by the battery and varied slowly. The regulated DC power of the grid-connected photovoltaic system is sent to the grid-connected DC/AC inverter and injected into the utility. The output voltage of DC/DC power converter is regulated by the grid-connected DC/AC inverter. Therefore, the function of DC/DC power converter is addressed in the MPPT of solar cell array.

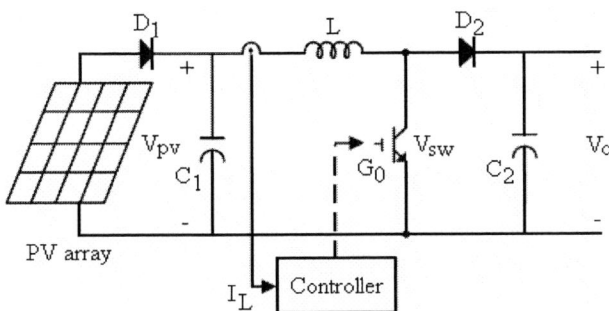

Fig. 1 The DC/DC power converter with the proposed MPPT method for a solar cell array.

III. OPERATION OF DC/DC POWER CONVERTER

Since the different operating voltage of the solar cell array will produce different output power, the ripple voltage of solar cell array will degrade the efficiency of the photovoltaic system. Thus, a DC capacitor is connected to the solar cell array in parallel to stabilize the operation voltage. The controller is used to produce a control signal to turn on or off the power electronic switch of the DC/DC power converter. If the power electronic switch is turned on, the inductor can be energized by the power of the solar cell array. Conversely, if the power electronic switch is turned off, the energy stored in the inductor is transferred to the output capacitor via the diode. So, the unregulated DC voltage of the solar cell array can be converted into a regulated higher DC voltage. If the power electronic switch is turned on, a voltage across the power electronic switch approximates zero. Conversely, when the power electronic switch is turned off, a voltage across the power electronic switch is almost equal to the output voltage of the DC/DC power converter because the diode is conducted. A square waveform between zero and the output voltage of the DC/DC power converter appears across the power electronic switch while turning the power electronic switch on and off. If the inductor current is continuous, the average voltage across the

power electronic switch can be derived as:

$$
\begin{aligned}
V_{sw} &= \frac{1}{T_s} \int_0^{T_s} V_{sw}(t)dt \\
&= \frac{t_{off}}{T_s} V_o = (1-D)V_o
\end{aligned}
\tag{1}
$$

where T_s, t_{off} and D are the switching period, turn-off time and duty ratio of the power electronic switch respectively, and Vo is the output voltage of DC/DC power converter. If the average voltage across the power electronic switch can be controlled to be proportional to the inductor current, the DC/DC power converter can be regarded as an active resistor (R). Thus, the average voltage across the power electronic switch can be represented as:

$$
V_{sw} = RI_L = \frac{t_{off}}{T_s} V_o
\tag{2}
$$

Because T_s and V_o are constant in (2), the DC/DC power converter can be operated as an active resistor while the turn-off time (t_{off}) is proportional to the inductor current. The active resistor of DC/DC power converter is controllable because turn-off time is a controllable parameter. This means the equivalent active resistance of the DC/DC power converter can be adjusted by adjusting the turn-off time.

Figure 2 shows the control block diagram of the DC/DC power converter. It includes a current detector, a multiplier, a PWM circuit and a driver circuit. Further, one of the multiplier input is connected to the output of the MPPT circuit. The output of the MPPT circuit is the active resistance control signal (K_m).

The output of the comparator is low and the power electronic switch is turned off while the modulation signal is lower than the high frequency carrier. At this time, the voltage across the power electronic switch is the output voltage of the DC/DC power converter. Conversely, the output of the comparator is high and the power electronic switch is turned on while the modulation signal is lower than the high frequency carrier. So, the voltage across the power electronic switch is zero. The turn-off time can be derived as:

$$
t_{off} = \frac{K_m I_L}{\hat{V}_{tri}}
\tag{3}
$$

where $\hat{V}_{tri}$ is the amplitude of the high frequency carrier. Substituting (3) into (1), then

$$
V_{sw} = \frac{K_m V_o}{T_s \hat{V}_{tri}} I_L
\tag{4}
$$

The average voltage across the power electronic switch

is proportional to the inductor current. From (4) and (2), the active resistor can be represented as:

$$R = \frac{K_m V_o}{T_s \hat{V}_{tri}} \qquad (5)$$

As seen in (5), the active resistor of DC/DC power converter is proportional to the active resistance control signal (K_m) get from the MPPT circuit. Therefore, the MPPT circuit can change the active resistor of DC/DC power converter so the real power supplied from the solar cell array can be adjusted. From Fig. 2, it can be found that only a feed forward control is used in the proposed DC/DC power converter, and it is simplified compared with the conventional DC/DC power converter.

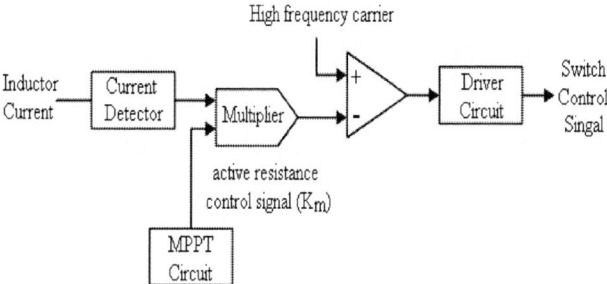

Fig. 2 The control block diagram of the proposed MPPT method.

VI. FLOWCHART OF THE PROPOSED MPPT METHOD

Figure 3 shows the flowchart of the proposed MPPT method. Referring to Fig. 3, first, an adjusting value $\Delta K_m(n)$ and the initial value $K_m(0)$ of the active resistance control signal are preset. The initial value $K_m(0)$ of the active resistance control signal is sent to the controller of the DC/DC power converter. Subsequently, after an interval, an average current (identified as "I_L") of the inductor is calculated. Since the active resistance of the DC/DC power converter is proportional to the active resistance control signal (K_m), the output power of the solar cell array is also proportional to the product of the square of the average inductor current I_L and the active resistance control signal. Thus, the initial output power $P(0)$ of the solar cell array can be gained by multiplying the square of the average inductor current I_L and the first value $K_m(0)$ of the active resistance control signal. For obtaining the practical output power of the solar cell array, the average inductor current must be calculated after an interval for stabilizing the inductor current of the DC/DC power converter when a new active resistance control signal is applied. The interval depends on the response of the DC/DC power converter.

Still referring to Fig. 3, the initial value $K_m(0)$ of the active resistance control signal is regarded as an old

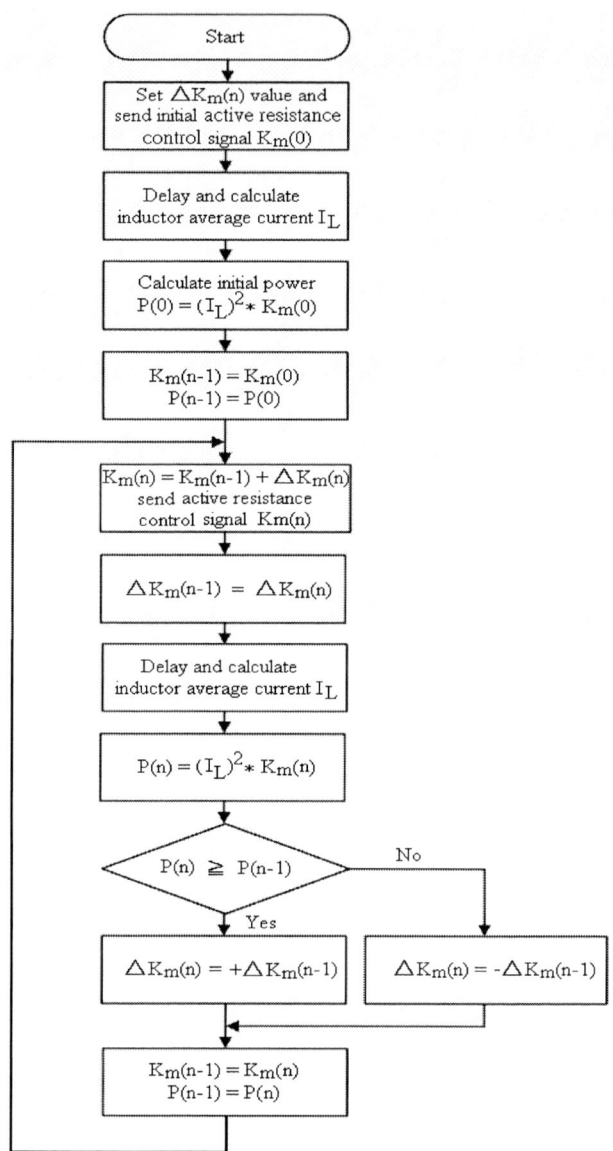

Fig.3 The flowchart of the proposed MPPT method.

value $K_m(n-1)$ while the initial output power $P(0)$ of the solar cell array is regarded as an old output power $P(n-1)$. In addition, a new value $K_m(n)$ of the active resistance control signal is obtained by adding the old value $K_m(n-1)$ of the active resistance control signal and a new interval value $\Delta K_m(n)$, then sent to the controller of the DC/DC power converter to act as an active resistance control signal. In this circumstance, the new interval value $\Delta K_m(n)$ has replaced the old interval value $\Delta K_m(n-1)$. Subsequently, after an interval, an average inductor current I_L is calculated, and the square of the average inductor current I_L and the new value $K_m(n)$ of the active resistance control signal are multiplied for obtaining a new output power $P(n)$ of the solar cell array.

To track the maximum power point, the new output power $P(n)$ of the solar cell array is compared with its old value P(n-1). If the output power P(n) of the solar cell array is greater than its old value P(n-1), $\Delta K_m(n)$ is equal to $+\Delta K_m(n-1)$. If the output power P(n) of the solar cell array is smaller than the old value P(n-1), $\Delta K_m(n)$ is equal to $-\Delta K_m(n-1)$.

Finally, the output power P(n) of the solar cell array has replaced the old value P(n-1), and the new value $K_m(n)$ of the active resistance control signal has also replaced the old value $K_m(n-1)$ at the same time. Subsequently, a new series of steps is repeated and circulated continuously by the previous steps until a maximum power point of output power is tracked. Once an operation point of the maximum power point is detected, the MPPT circuit controls the output power of the solar cell array continuous perturbation around the operation point of the maximum power point.

The proposed maximum power point tracking method detects the inductor current of the DC/DC power converter for tracking the maximum power point. Thus, it is useful that this method and circuit can simplify the entire structure.

V. EXPERIMENTAL RESULTS

To verify the performance of the proposed MPPT method, a prototype based on the DSP controller TMS320C2407 is developed and tested. The solar cell array is configured by four strings, and every string contains eight solar modules connected in series. Table 1 shows the main parameters of the prototype. A DC/AC inverter is connected to the output of DC/DC power converter to convert the output power of the photovoltaic system to utility, and it regulates the output voltage of DC/DC power converter at 200 V.

Table 1 Experiment parameters

Solar module	
Rate of maximum power	75W
Open voltage	21.3V
Short current	3.5A
DC/DC power converter	
Inductor	3mH
Input Capacitor	4700 μF
Switching Frequency	20KHz

Figure 4 shows the output voltage, output power of solar cell array and the inductor current of DC/DC power converter after applying the proposed MPPT method to the DC/DC power converter. As seen in Fig. 4, the output power of solar cell array is almost constant during the experiment period, and the maximum power shown in

Fig. 4 is around 1.73 KW. The output voltage of the solar cell array and the inductor current are about 115.4 V and 14.9 A at the maximum power point. Therefore, this verifies the proposed MPPT method can trace the practical maximum power of the solar cell array effectively.

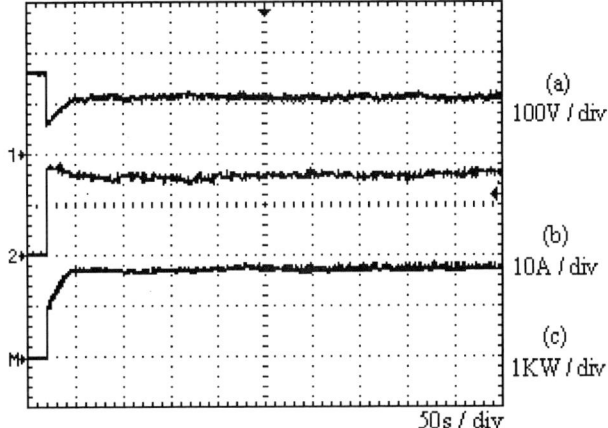

Fig. 4 Experimental result of the proposed MPPT method (a) voltage of solar cell array, (b) inductor current, (c) output power of solar cell array.

Figure 5 show the experimental result of the output power, output voltage of solar cell array and inductor current of DC/DC power converter under the condition that the solar cell array is shielded abruptly. As seen in Fig. 5, the output voltage of solar cell array and the inductor current of DC/DC power converter are changed to respond to the change of the power produced by the solar cell array. Then, the output power of DC/DC power converter is also changed immediately. This experimental result proves the proposed MPPT method can track the maximum power point fast and effectively.

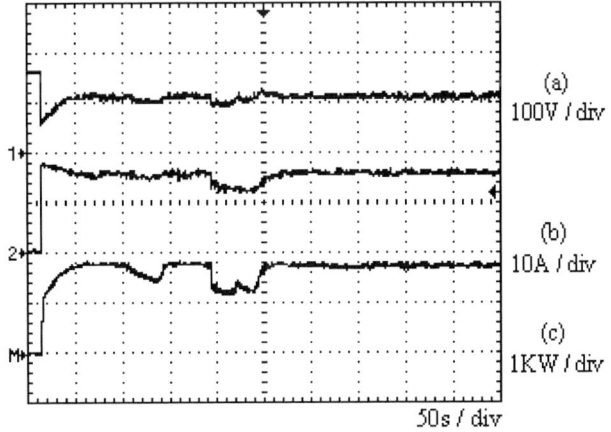

Fig. 5 Experimental result of the proposed MPPT method under part of the solar cell array shielding abruptly, (a) voltage of solar cell array, (b) inductor current, (c) output power of solar cell array.

VI. CONCLUSIONS

This paper proposed a novel MPPT method for the photovoltaic system. The proposed MPPT method has the feature that only the inductor current of the DC/DC power converter is required to be detected. Therefore, the proposed has can simplify the entire structure compared with the conventional perturbation and observation MPPT method requiring to detect both voltage and current to calculate the real power for detecting the maximum power point. The experimental results of the developed prototype verify that the performance of the proposed MPPT method is as expected.

REFERENCES

[1] B. Kroposki and R. DeBlasio, "Technologies for the new millennium: photovoltaics as a distributed resource," in *Proc. IEEE Power Engineering Society Summer Meeting*, Vol. 3, July 2000, pp. 1798-1801.

[2] M. P. Choi and A. Tan, "Photovoltaics Demonstration Projects," *Proc. of EMPD 98*, Vol. 2, 1998, pp.637-643.

[3] L. Castaner and S. Silvestre, "Modeling Photovoltaic System," *John Wiley & Sons Ltd*, 2002.

[4] Veerachary M., T. Senjyu, and K. Uezato, "Voltage-based maximum power point tracking control of PV system," *IEEE Trans. Aerosp. Electron. Syst.*, Vol. 38, No. 1, Jan. 2002, pp. 262–270.

[5] Z. Salameh, F. Dagher and W. A. Lynch, "Step-down maximum power point tracker for photovoltaic arrays," *Solar Energy*, Vol. 46, No. 5, 1991, pp. 278–282.

[6] F. Harashima and H. Inaba, "Micro processor controlled SIT inverter for solar energy system," *IEEE Trans. on Indus. Electron.*, Vol. 34, Feb. 1985, pp. 50-55.

[7] N. Femia, G. Petrone, G. Spagnuolo, and M. Vitelli, "Optimization of perturb and observe maximum power point tracking method," *IEEE Trans. on Power Electronics*, Vol. 20, No. 4, July 2005, pp. 963-973.

[8] K.H. Hussein, I. Muta, T. Hoshino and M. Osakada, "Maximum photovoltaic power tracking: an algorithm for rapidly changing atmospheric conditions," *IEE Proc. Generation, Transmission and Distribution*, Vol. 42, No. 1, Jan. 1995, pp. 59-64.

[9] T. Hiyama and K. Kitabayashi, "Neural network based estimation of maximum power generation from PV module using environment information," *IEEE Trans. on Energy Conv.*, Vol. 12, No. 3, September 1997, pp.241-247.

[10] N. Khaehintung, K. Pramotung, B. Tuvirat, and P. Sirisuk, "RISC-microcontroller built-in fuzzy logic controller of maximum power point tracking for solar-powered light-flasher applications," *IEEE IECON*, Vol. 3, Nov. 2004, pp. 2673-2678.

[11] M. E. Frederick and J. B. Jermakian, "Microprocessor control of multiple peak power tracking DC/DC converters for use with solar cell arrays," *U.S. Patent* 5327071, 1994.

[12] E. Koutroulis, K. Kalaitzakis and N. C. Voularis, "Development of a microcontroller-based, photovoltaic maximum power point tracking control system," *IEEE Trans. Power Electron.*, Vol. 16, No. 1, Jan. 1995, pp. 46-54.

[13] C. T. Pan, J. Y. Chen, C. P. Chu and Y. S. Huang, "A fast maximum power point tracker for photovoltaic power systems," *IEEE IECON*, Vol. 1, Nov. 1999, pp. 390-393.

Maximum Power Point Algorithm in PV Generation: An Overview

Hardik P. Desai*, and H. K. Patel**, Senior Member IEEE

* Research Scholar, SVNIT, Surat, India
** Professor, SVNIT, Surat, India

Abstract: Use of solar energy using photovoltaic (PV) arrays is emphasized increasingly and regarded as an important resource of power energy in the coming years. As the power supplied by PV arrays depends upon the insolation, temperature and array voltage, it is necessary to control the operating point to extract the maximum power from the PV arrays. Number of methods for Maximum Power Point Tracking (MPPT) has been reported in the literature. This paper aims to give a comprehensive comparison of different MPPT algorithms in terms of their tracking efficiencies, cost effectiveness, complexity of realization etc. Here the methods are classified according to the parameters they use in finding MPP, viz, PV voltage, PV current, output voltage, and output current.

Index Terms – Maximum Power Point Tracking algorithms, Photovoltaic power generation

I. INTRODUCTION

The increase of energy demands due to growth of industries and population, have led effort to find new sources of energy to permit reduction in the utilization of the natural resources of fuel. In this content, solar energy appears as an important alternative to meet the increase of energy demands. The PV generation is gaining increased importance as renewable energy source due to the merits of pollution free, low operating and maintenance cost, abundant and broadly available free energy source. Reduction in development cost of both PV cells and high power semiconductor devices provide further boost in this direction. PV generator has non linear I-V and P-V characteristics which depend on insolation intensity, temperature and ageing of the PV array. For each insolation level, there exists a unique operating point $V = V_{MPP}$ known as maximum power point (MPP), on the array characteristics that delivers the maximum power. As the cost of PV electricity high due to high installation cost, it is always of interest to operate the array at MPP for cost optimization. To extract maximum power, Maximum Power Point Tracker (MPPT) have been developed and reported in literature [2-5, 8]. This includes DC/DC converters (buck, boost, buck-boost etc.) and their control algorithms. In this paper various algorithms are studied and their simulation results are presented. Algorithms are classified based on the variable, they use for MPP tracking. Commonly used variables are: PV current, PV voltage, output current, output voltage. Comparative performance like tracking efficiency, cost effectiveness, complexity in realization is presented.

II. MPP METHODS

A. Look-up Table Method

In this case, the measured values of the PV generator's voltage and current are compared with those stored in the controlling system, which correspond to the operation at the maximum point, under predetermined climatologically conditions [1,2]. In one of the methods [5] I_{PV} is defined as a function of P_{PV} $I_{MPP} = f(P_{MAX})$. In this method, A PI type controller adjusts the duty cycle of the DC–DC converter. The zero error is reached when the current and power of the Photovoltaic generator are equal to the pre-determined values of I_{MPP} and P_{MAX}. Any change of the insolation or load results a disturbance of the tuned system and the PI controller again brings the system to its optimum operating point. Then, these algorithms have the disadvantage that a large capacity of memory is required for storage of the data. Moreover, the implementation must be adjusted for a panel PV specific. In addition, it is difficult to record and store all possible system conditions. But it has also some advantages. It is simple and system is able to perform fast tracking as all the data regarding maximum point are available.

B. Voltage based photovoltaic generator method

It is found that voltage of Photovoltaic generator at the MPP is approximately linearly proportional to its open-circuit voltage, Voc. The proportional constant depends on the fabrication technologies, solar cells technology, fill factor and the meteorological conditions, mainly.

$$V_{MPP} = k_1 * V_{OC} \qquad (1)$$

The PV generator's open-circuit voltage is measured by disconnecting the load from the system, with a certain frequency, measuring the value V_{OC}. Then the MPP is calculated, according to Eq. (1), and the operating voltage is adjusted to the maximum voltage point. This process will be repeated periodically. Although this method is apparently simple, it is difficult to choose an optimal value of the constant k_1. However, k_1 values ranging from 0.65 to 0.80, for polycrystalline PV modules. This method has as an advantage that it is simple and low-priced. It is required only to measure open circuit voltage. Nevertheless, its drawback is that the interrupted system operation yields power losses when scanning the entire control range.

978-1-4244-0644-9/07/$25.00 ©2007 IEEE

C. Current based photovoltaic generator method

A method similar to the above procedure is used in [4]. There is a linear relation between MPP and short circuit current of PV generator (4). So here output terminals should be short circuited to obtain I_{sc}, short circuit current. Like above method it required to measure only current. As does the previous method, the proportional constant depends on the fabrication technologies, solar cells technology, fill factor and the meteorological conditions, mainly. However, the constant, k2, can be considered to be around 0.85.

$$I_{MPP} = k_2 * I_{SC} \qquad (2)$$

This method offers the same advantages and disadvantages as the above method. If we do the circuit interruption is done within very short time, then this methods are very useful as cost is very low.

D. Voltage based photovoltaic generator method using test cell

In order to avoid possible drawbacks related to the frequent interruption of the system, an additional test cell (pilot cell) should be used [22]. Thus, the PV generator's open-circuit voltage (or short circuit current) is measured from the test cell, which is electrically isolated from the rest of the PV array. The resulting values of the K_3 (or K_4) will be applied to the main PV generator.

$$V_{MPP} = k_3 * V_{OC,cell} \qquad (3)$$
$$I_{MPP} = k_4 * I_{SC,cell} \qquad (4)$$

This method's advantage is that it is simple and economical; it uses only one feedback loop control. Moreover, it avoids the problems caused by the interruptions of the operation of the PV set out in the previous method. As a disadvantage, it assumes that the test cell has characteristics (I-V and P-V) identical to each cell of the PV generator main. Therefore, V_{OC} (or I_{SC}) of the test cell is considered proportional to V_{OC} (or I_{SC}) of the PV unit used in the selection of the MPP. If the assumption is incorrect, maximum power will not be extracted. In above last three methods, it is assumed that maximum power is independent of insolation intensity, temperature. But main advantage is, they are very simple and economical methods (as it requires only one sensor).

Now following methods include those methods that use PV voltage and/or current measurements. From those, and considering variations of the PV generator operating points, the optimum operating point is obtained. These algorithms have the advantage of being independent from the a priori knowledge of the PV generator characteristics.

E. A novel on-line MPP search algorithm

The algorithm [6] requires neither the measurement of temperature and solar irradiation level nor a PV array model as in the case of look up table method. In this algorithm, the main task is to determine the value of

reference maximum power and then the current power is compared with it. This difference is called maximum power error. In order to have the PV array be operated at its MPP the maximum power error should be zero or near zero. The operating power is the PV array output power to the load and is given as the multiplication of PV array output voltage by the current. Here, first RMP, Reference Maximum Power, is to be required. Since RMP is changed with variation in temperature and solar irradiation level, it is not a constant reference and has a non-linear uncertainty that makes the tracking of PV array reference maximum power difficult. To get reference maximum power, to find the maximum power error, flow chart, used in [6], given in Fig: 1 is used. If the reference MPP is changed due to changing temperature and solar irradiation level, the algorithm adjusts the array voltage and finds the new MPP. This algorithm will not be able to determine the PV array MPP if the load power or current is much smaller than the PV array MPP power and current. In this case, additional loads should be connected to increase the PV array current so that the PV array can be operated at the MPP. It is preferred that we can charge battery as an additional load.

F. Perturbation and Observe ("P&O") method

The "P&O" method is that which is most commonly used in practice by the majority of authors [7-13], among others. It is an iterative method of obtaining MPP. It

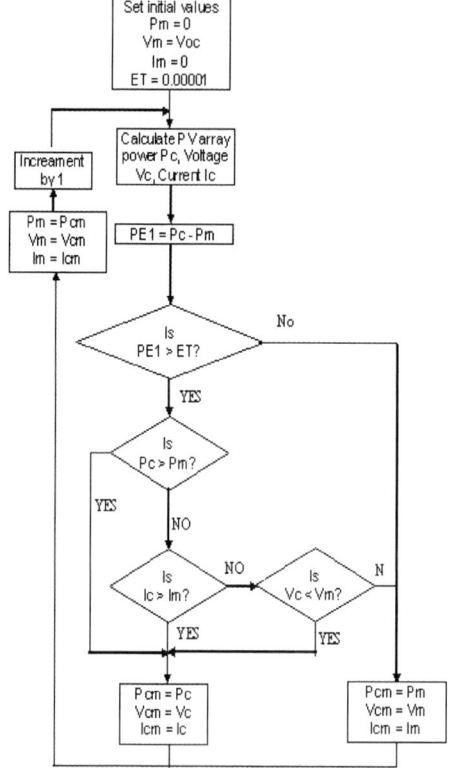

Fig : 1 Flow chart of the on line search algorithm

measures the PV array characteristics, and then perturbs the operating point of PV generator to encounter the change direction. The maximum point is reached when $dP_{PV}/dV_{PV} = 0$. There are many varieties, from simple to complex. A flowchart of the most basic form is shown in Fig: 2.

Doing this, the operating voltage of the PV generator is perturbed, by a small increment ΔV_{PV}, and the resulting change, ΔP_{PV}, in power, is measured. If ΔP_{PV} is positive, the perturbation of the operating voltage should be in the same direction of the increment. However, if it is negative, the system operating point obtained moves away from the MPPT and the operating voltage should be in the opposite direction of the increment. This means the array terminal voltage is perturbed every MPT cycle; therefore when the MPOP is reached, the P&O algorithm will oscillate around it, resulting in a loss of PV power, especially in cases of constant or slowly varying atmospheric conditions. This problem can be solved by improving the logic of the P&O algorithm to compare the parameters of two preceding cycles in order to check when the

MPOP is reached, and bypass the perturbation stage. Another way to reduce the power loss around the MPOP is to decrease the perturbation step, however, the algorithm will be slow in following the MPOP when the atmospheric conditions start to vary and more power will be lost. The advantages of this method can be summarized as: a previous knowledge is not required of PV generator characteristics; it is a relatively simple

method. Nevertheless, in their most simple form, at a steady state, the operating point oscillates around the MPP, giving rise to the wasting of some amount of available energy. In addition, it is an unsuitable method with rapidly changing atmospheric conditions. To eliminate the drawbacks listed above of the algorithm, a new algorithm for tracking maximum power point in photovoltaic systems can be used. This is a fast tracking algorithm, where an initial approximation of maximum power point (MPP) is quickly achieved using a variable step-size. Subsequently, the exact maximum power point can be targeted using any conventional method like the P & O or incremental conductance method. Thus, the drawback of a fixed small step-size over the entire tracking range is removed, results in reduction in number of iterations and much faster tracking compared to conventional methods. In this method [11,12] new component β (5) is used as main variable. We can extract maximum power from the PV array, with $dP_{PV}/dV_{PV} = 0$. Solving this equation and we can get,

$$\ln(i_{PV}/v_{PV}) - c * v_{PV} = \ln(I_0 * C) = \beta \qquad (5)$$

From above equation β is independent of insolation but depends on temperature. The flowchart for the proposed MPP tracking algorithm is shown in Fig. 3. If d_{new} is less than d_{min} or greater than d_{max}, then change to d_{min} to utilize the fact that MPP is close to the point. Steps to get maximum power are shown in fig. 3. The algorithm brings the operating point very close to the actual MPP with a few iterations. Then, perturbation and observation method or incremental conductance methods with fewer steps can be used to track the exact MPP. The proposed algorithm has the advantage of very fast convergence and

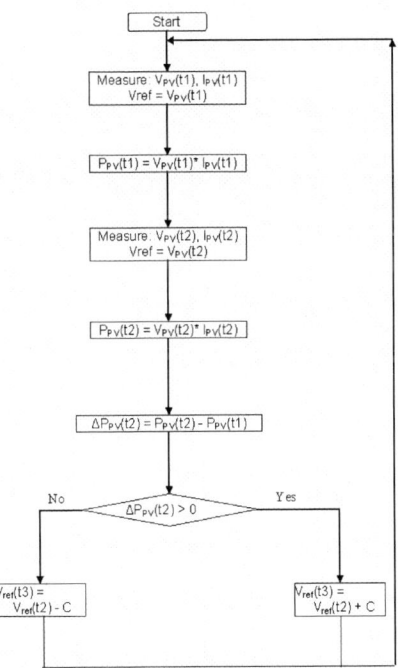

Fig. 2. Conventional Perturbation and Observe algorithm flowchart. C is the step of the perturbation.

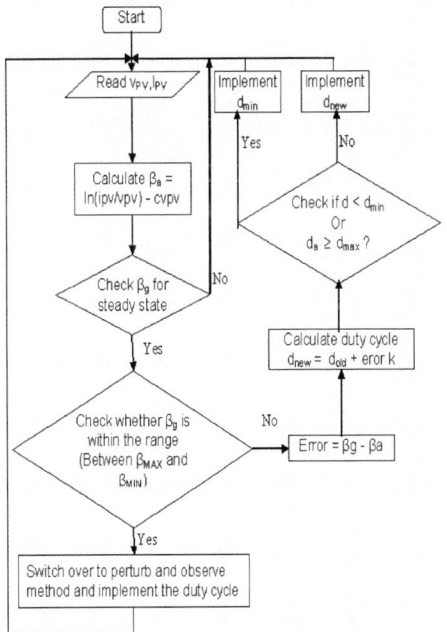

Fig. 3 Flow chart for the new tracking method using β

accurate tracking of MPP.

E. The MPP via capacitor identifier

This method is discussed in [14, 15]. In this method, estimating the capacitance based on the model reference adaptive system, and then we get the accurate capacitance in any time. The change in duty ratio is decided by identifying capacitance and increase or decrease in duty ratio is decided by P & o method or inc cond method. Fig : 4 shows the flowchart of proposed maximum power point tracking. First, the voltage and current of PV array and the capacitor current are detected. The capacitance C_S, is estimated and the estimated capacitance is used to correct the variation of duty ratio for MPPT as shown in the following equation. $\Delta D = \Delta D_n, * C_{sn}/C_s(k)$ where, ΔD_n, is the variation of duty ratio which is determined based on the nominal capacitance and C_{sn} is the nominal capacitance, The input power Ps(k) is calculated by the detected values and compared to that calculated at the previous sampling Ps(k-1). Here, a is the increased or decreased value of duty ratio and a(k - 1) is the same at the previous sampling. The relationship between a(k) and a(k-1) decides increase or decrease of the duty ratio.

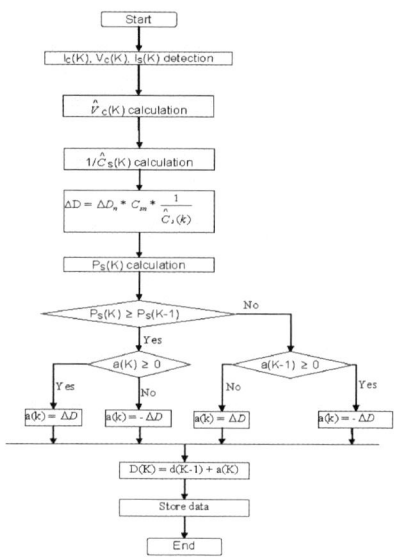

Fig. 4 Flow chart for MPP tracking via capacitor finder

F. The MPP via output parameter

In this method of maximum power point tracker, output voltage and/or current are used for control purposes, rather than for its input voltage and current. The output parameters simplify the maximum power point tracker controller. Moreover, using this approach [16-19], only one out of the two output parameters needs to be sensed. This observation is general and applies regardless of the power stage or the realization control algorithm. As controllable variable (vc) increases, the operation point changes along the curve in Fig. 5(a), from point 1 to the MPP (point 3). If (vc) keeps increasing, the current decreases as the operation point becomes point number 4, Fig. 5(b) and same is also shown in

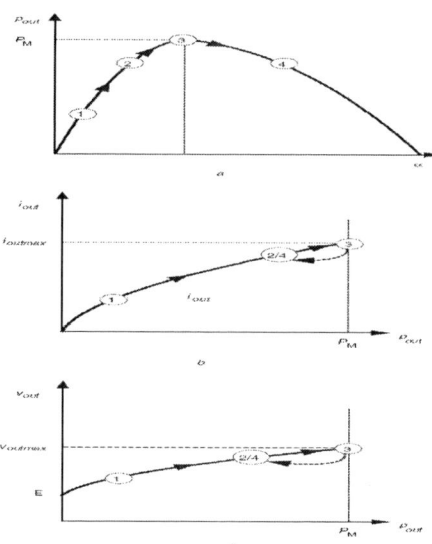

Fig. 5 Output power Vs controllable parameter (a), Output current Vs output power (b), Output voltage Vs output power(c)

voltage-power plane (Fig. 5(c)). Since the power changes, the output current also changes. Maximum current is reached at maximum power. As (vc) is further increased the operation point changes from point 3 to point 4, returning on the same i–p curve (Fig. 5(b)). For instance, operation points 2 and 4, though different, coincide on the i–p curve (Fig. 5(b)). Thus a control strategy, different from the conventional one, can be applied. It is possible to continuously change (vc) while observing i_out, output current. As long as i_out increases as a result of the (vc) variation, it should still be changing in the same direction. If i_out decreases as a result of changing (vc), the tendency of (vc) variation should be reversed. In this manner the operation point should converge to the MPP (i_{outmax}, P_M). Thus an MPPT algorithm may be based on a single output parameter. In this algorithm the output current or output voltage is maximized rather than the power. MPPT control via output parameters is possible and advantageous. Controlling MPPTs via its output parameters facilitates sensing of a single output parameter and removes the need for a multiplier in the controller. Single output parameter control is, in general, possible for nearly all practical load types, regardless of load nature. In small PV systems with an analogue controller, omission of the multiplier provides a major advantage as it simplifies the hardware. In PV systems with a digital controller, omission of the multiplier implies algorithm simplification. In both cases one of the usual employed sensors (voltage or current) is saved. Control algorithm, used in [19], is shown in fig : 6.

G. Incremental conductance Method

This algorithm is discussed in [13,20,21]. The basic idea is that at the MPOP the derivative of the power with respect to the voltage vanishes because the maximum power operating point (MPOP) is the maximum of the power curve. Also from Fig. 7 we note that to the left of

832

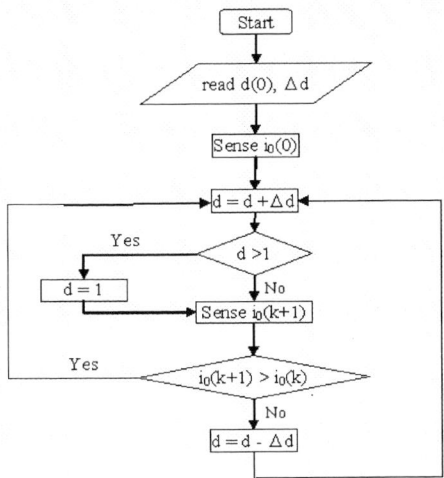

Fig. 6 Flow chart for MPP tracking using output current

the MPOP the power is increasing with the voltage, i.e. dP/dV > 0, and it is decreasing to the right of the MPOP, i.e. dP/dV < 0. This can be rewritten in the following simple equations:

dP/dV = 0 at the MPOP   (6)

dP/dV > 0 to the left of the MPOP   (7)

dP/dV < 0 to the right of the MPOP   (8)

These relations can further be written in terms of the array current and voltage using

dP/dV = d(IV)/dV = I + V dI/dV

(9)

Hence, the PV array terminal voltage can be adjusted relative to the MPOP voltage by measuring the incremental and instantaneous array conductance (dI/dV and I / V , respectively) and making use of eqns. 6-9. The detailed operation of the Inc Cond algorithm, used in [20], can be followed with reference to the flow chart of Fig: 8.The main advantage of this algorithm is that it offers a good result under rapidly changing atmospheric conditions. Also, it achieves lower oscillation around the MPP than the P&O method. It has as drawback that it requires complex control circuit. It is required two sensors for the measurement of voltage and current which might have resulted in a high cost system. However, now days there are many software/processors available for doing it much more cheaply.

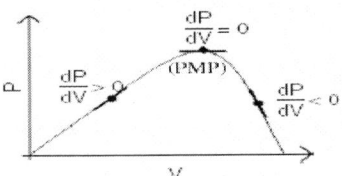

Fig. 7 Characteristics curve of the P–V photovoltaic generator with Variation of the dP/dV.

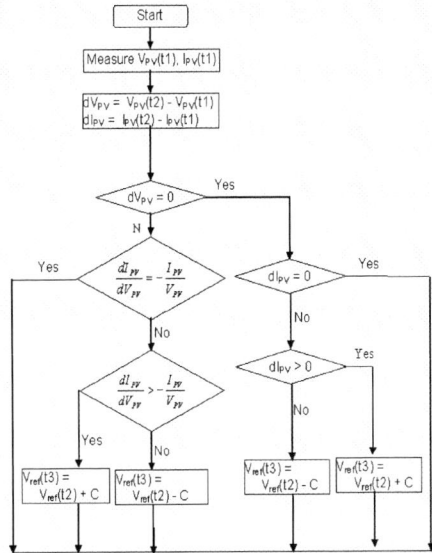

Fig. 8 Incremental conductance algorithm flowchart.

III. RESULTS

Voltage based maximum power point tracking method is simulated in the MATLAB Simulink software and the results are shown below. As a tracker DC/DC converter is used. Algorithm is checked for different conditions. For G=1, T=0 Fig: 8, 9 show results without and with MPPT respectively. Without MPPT, we get P_{mpp} = 38 W, and V_{MPP} = 10.68 V while with MPPT, P_{MPP} = 47.61 W, and V_{MPP} = 15.77 V which

is very nearer to MPOP. For G = 0.75, T = 25°C Fig 10 and 11 show results without and with MPPT respectively. Without MPPT, we get P_{MPP} = 27 W and V_{MPP} = 9.5 V while with MPPT, P_{MPP} = 34.41 W, and V_{MPP} = 14.16 V which is also nearer to the MPOP.

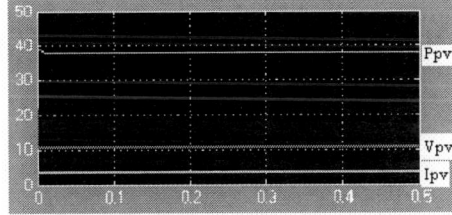

Fig. 8 Results without MPPT

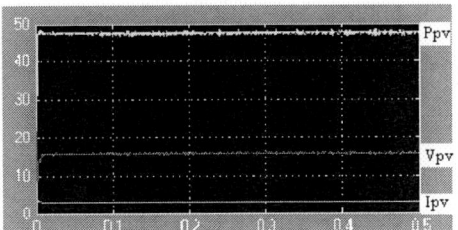

Fig. 9 Results with MPPT

TABLE I
COMPARISON OF DIFFERENT MPP ALGORITHMS

MPPT Algorithm	Cost	Measurement Parameter	No. of sensors required	Algorithm response	Complexity of realization	Final Point	Other Comments
Look-up table method	Higher	I_{pv}, V_{pv}, G, T	Min 4	Fast	Simple	At MPP	More storage required
Voltage based photovoltaic generator method	Economical	V_{OC}	1	Fast	Simple	Nearer to MPP	PV system must be interrupted
Current based photovoltaic generator method	Economical	I_{SC}	1	Fast	Simple	Nearer to MPP	PV system must be interrupted
A novel on-line MPP search algorithm	Medium	V_{pv}, I_{pv}	2	Fast	Increased	At MPP	Calculation of RMP is difficult
Perturbation and Observe ("P&O") method	Medium	V_{pv}, I_{pv}	2	Slow	Increased	Oscillate at MPP	More number of iterations, oscillation at MPP reduced by controlling β
The MPP via capacitor identifier	Higher	V_{pv}, I_{pv}, Capacitor current	2	Slow	Increased	Nearer to MPP	Apply where more numbers of modules are used
The MPP via output parameter	Economic	V_{OP} or I_{OP}	1	Fast	Simple	At MPP	Only one sensor required
Incremental conductance Method	Higher	V_{pv}, I_{pv}	2	Slow	Increased	At MPP	Good results under fast changing atmospheric condition, reduced oscillation at MPP

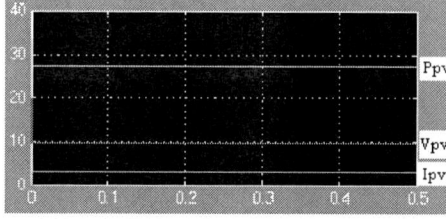

Fig:10 Results without MPPT

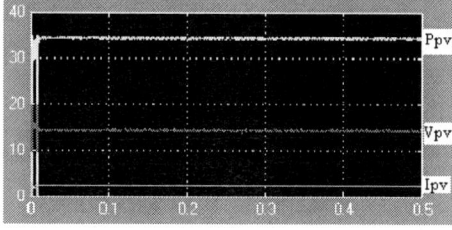

Fig. 11 Results with MPPT

IV. CONCLUSION

Here different methods of extracting maximum power from the photovoltaic source are discussed. As has been shown, there are many ways of distinguishing and grouping the methods for tracking the MPP to the PV generator. Comparison of all of them is shown in table I. Results show that by using maximum power point tracker, maximum power is extracted form PV source.

REFERENCES

[1] T. Hiyama nad K Kitabyashi, "Neural Network based estimation of maximum power generation", IEEE trans. On energy conversion, vol. 12, pp 241-247, sept 1997

[2] Eng. Hassan El-Sayed Ahmed Ibrahim, Assoc. Prof. Dr./ Faten F. Houssiny Microcomputer controlled buck regulator for maximum power point tracker for dc pumping system operates from photovoltaic system, 1999 IEEE International Fuzzy Systems Conference Proceedings August 22-25, 1999, Seoul, Korea

[3] Mohammad A S Masoum, Hooman Dehbonei, and Ewald F Fuchs, Theoretical and experimental analysis of Photovoltaic systems with voltage and current based maximum power point tracking, IEEE transaction on Energy conversion, vol 17 no 4 December 2002

[4] S M Alghuwainem, Matchhing of a DC motor to a photovoltaic generator using a step up converter with a current locked loop, 1993, IEEE

[5] H. Tarik Duru, A maximum power tracking algorithm based on Impp = f(Pmax) function for matching passive and active loads to a photovoltaic generator,solar energy 2005

[6] I H Atlas, A M Sharaf, A novel on line MPP search algorithm for PV arrays, IEEE transaction on Energy conversion vol 11, No 4 December 1996

[7] Nicola Femia, Giovanni Petrone, Giovanni Spagnuolo and Massimo Vitelli,Optimization of Perturb and Observe

Maximum Power Point Tracking Method, IEEE transaction on power Electronics Vol 20, No 4, July 2005

[8] Eftichios Koutroulis, koastas Kalaitzakis, Nicholas C Voulgaris,Development of a microcontroller based photovoltaic maximum power point tracking control system, IEEE Transaction on Power Electronics vol 16, no 1 January 2001

[9] V M Pacheco, L C Freitas, J B V ieira Jr, E AQ A Coelho and V J Farias,Stand alone Photo voltaic energy storage system with maximum power point tracking, 2003

[10] Jancarle L Sanntos, Fernando L M Antunes,Maximum Power Point Trackker for PV systems, World climate & energy event, 1-5 December 2003, Rio de Janeiro, Brazil

[11] Sachin Jain, Vivek Agarwal, New current control based MPPT technique for single stage grid connected PV systems, energy conversion and management, 2006

[12] Sachin Jain, Vivek Agarwal,A new algorithm for rapid tracking of Approximate maximum power point in photovoltaic systems IEEE power electronics letters, vol 2 no 1 march 2004

[13] Chihchiang Hua and Chihming Shen, Comparative study of peak power tracking techniques for solar storage system, IEEE 1998

[14] N Kasa, T Lida and H iwamoto, Maximum power point tracking with capacitor identifier for photovoltaic power system, IEEE proc. Electr power appl vol 147 no. 6 november 2000

[15] D P Hohm, M E ropp, Comparative study of maximum power point tracking algorithms using an experimental, programmable, maximum power point tracking test bed, IEEE 2000

[16] H valderrama-Blavi, C Alonso, L Martinez-Salamero, S Singer, B Estibals and J Maixe-Altes, AC-LFR concept applied to modular photovoltaic power conversion chains, IEE Proc-Electr. Power Appl. Vol. 149, No 6 November 2002

[17] Sigmond Singer, Member, IEEE, Shaul Ozeri, and Doron Shmilovitz, A Pure Realization of Loss-Free Resistor, ieee transactions on circuits and systems—i: regular papers, vol. 51, no. 8, august 2004

[18] S singer, R Giral, J Calvente, R Leyva, L Martinez-Salamero, D naunin, Maximum power point tracker based on a loss free resistor topology, Proceeding of the fifth European Space power conference, Taragana, spain, 21-25 september 1998

[19] D Shmilovitz, On the control of photovoltaic maximum power point tracker via output parameters, IEE Proc – Electr. Power appl. Vol 152 no 2, march 2005

[20] K H Hussein, I Muta, T Hoshino, M Osakada, Maximum photovoltaic power tracking: an algorithm for rapidly changing atmospheric conditions, IEE proc-Gener. Trans Distrib, vol 142, no 1 January 1995

[21] Yushaizad Yusof, Siti Hamizah Sayuti, Muhammad Abdul Latif and Mohd Zamri che wanik, Modelling and simulation of Maximum power point tracker for photovoltaic system, National power & Energy conf (PECon) 2004 Proceedings, Malaysia

[22] X. vallve and J. Serrasolses, design and operation of a 50 kwp PV rural electrification project for remote sites in Spain, solar energy vol. 59, nos. 1-3, pp. 111-119,1997

A DC-Module-Based Power Configuration for Residential Photovoltaic Power Application

Bangyin Liu, Shanxu Duan, and Yong Kang

College of Electrical and Electronic Engineering, Huazhong University of Science and Technology, Wuhan, P. R. China.

Abstract--The photovoltaic power system for residential and building integrated application is sensitive to the mismatch of the electrical parameters due to the partial shadows, different orientations, variations of current-voltage characteristics of the photovoltaic panels. As a result, the yield of the system decreases remarkably, in severe cases, the hot-spot problem can arise and the photovoltaic modules can be irreversibly damaged. In order to improved the system performance and maximize the energy converter efficiency under the mismatch condition, a novel dc-module-based photovoltaic power system is proposed. Firstly the system configuration and advantages are described in detail, and then the suitable topologies are presented and evaluated, and finally some design considerations and issues on the performance of the system are discussed.

Index Terms—Photovoltaic power system, power configuration, DC module, high step-up topology

I. INTRODUCTION

Growing concerns about environmental issues and the world energy crisis have attracted a great deal of interest in the development and application of the photovoltaic power system using the nonpolluting renewable solar energy. The output voltage and current of a photovoltaic panel are lower, in order to generate sufficiently high voltage to avoid further amplification, the photovoltaic panels are connected in series usually, and then these strings may be connected in parallel to reach expected power levels. Consequently, most of the photovoltaic power systems in the world are centralized, string and multi-string configurations [1][2], as is shown in Fig. 1.

The centralized system is shown in Fig. 1, where the photovoltaic array are connected to a single central inverter. The photovoltaic panels were connected in series (called a string) to generate a sufficient high voltage, then these strings, through blocking-diodes, were connected parallel in order to reach high power levels. The efficiency of the centralized system is high due to adopting single stage converter. Similar to the centralized system, the string system consists of a single string of photovoltaic panels and an inverter. The input voltage may be high enough to avoid further amplification. The multi-string system is an evolutional version of the string system. It connects several strings with separate dc-dc converter to a common inverter. These configurations have some disadvantages, such as centralized maximum power point tracking (MPPT) for a photovoltaic array or

string, hard to acquire the status information of each photovoltaic panel and mismatch losses between the photovoltaic panels. Furthermore, in the residential and building integrated application, the photovoltaic power system is easy to be affected by partial shadows created by surrounding buildings, trees and clouds, and the energy yield decreases. If the photovoltaic power system is not appropriately protected, hot-spot problem can arise, in severe cases, the system can be irreversibly damaged.

In order to improve the reliability and maximize the output capability, the optimum power configuration and balance of system should be developed. A novel system configuration based on photovoltaic dc modules is proposed in this paper. The novel dc-module-based system integrates the high step-up dc-dc converter with the photovoltaic panel into one electrical device, then assigned to a common inverter, and a compact and cost effective solution is achieved. The system configuration and advantages are described in detail in Section II. The suitable topologies are presented and evaluated in Section III. Some design considerations and issues on performance of the system are discussed in Section IV. Conclusions are drawn in Section V.

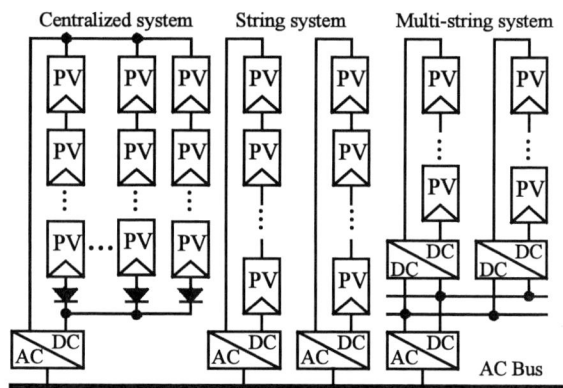

Fig. 1 Centralized, string and multi-string configuration

II. DC-MODULE-BASED PHOTOVOLTAIC POWER SYSTEM

A. System Configuration

Many dc modules and a common inverter constitutes the dc-module-based photovoltaic power system, as illustrated in Fig. 2. The dc module consists of a photovoltaic panel integrated with a dc-dc converter, controller and the unit of power line carrier communication (PLCC). these dc modules are connected to the common dc bus, then a single dc-ac inverter is

Project Supported by Delta Science & Technology Educational Development Program (DREK200501).

978-1-4244-0644-9/07/$25.00 ©2007 IEEE   836

required to connect to the local load and grid. The status information of the photovoltaic panel and its own converter is transmitted to the central inverter system by the component of PLCC. The human machine interface can provide control interface and the particular status information of the system.

The dc-dc converter boosts the lower output voltage (around 20 to 50V) of the photovoltaic panel to an appropriate voltage, such as 200V or 400V, and acts as the maximum power point tracker. Consequently, the conventional full-bridge dc-ac topology is suitable for the common inverter, which acts as the interface between the dc bus and the local load or grid (100V to 240V usually). The energy storage system or other distributed generation systems, such as wind and fuel cell system, can be extended easily by connecting to the dc bus according to the user's demand.

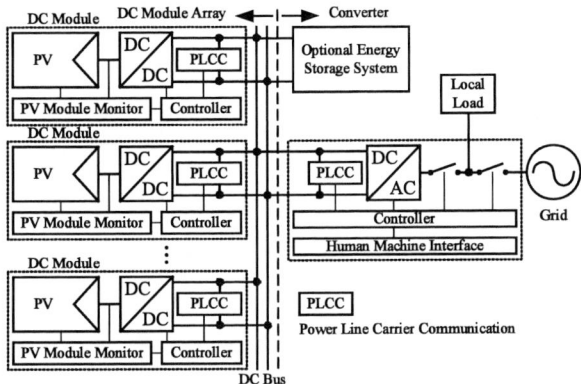

Fig. 2 DC-module-based photovoltaic power system

B. Advantages

The advantages of the proposed dc-module-based system include:

(1) Individual MPPT for every photovoltaic panel, which can increase the efficiency of solar utilization and allows photovoltaic panel to be installed on different positions and orientations;

(2) Lower sensitivity to perturbation of the environment and the mismatch of the photovoltaic panel's electrical parameters due to the partial shadow;

(3) Each dc module has an inherent data monitor. The status information of every photovoltaic panel and its own converter is easy to acquire, and it can be transmit to central monitor system by power line carrier communication. These advantages provide better performance of protection and maintenance for the photovoltaic power system;

(4) The dc module is easy to standardize and intended to be mass produced, thereby the manufacturing cost will be reduced. It can be designed as a plug-and-play device, which can be installed easily. Further enlarging system capacity are easily achieved.

(5) The dc module based power configuration support utilizing multiple sustainable energy sources, such as the energy storage system using lead-acid batteries or something else, fuel cell, wind power generator and so on, which can be connected to the common dc bus easily.

These advantages makes residential and building integrated photovoltaic application possible. The dc-module-based system are of interest in future hybrid generation application for residential users.

III. SUITABLE DC-DC TOPOLOGIES FOR DC MODULE

The maximum power point (MPP) voltage range of the general photovoltaic panel is 23V to 38V at a power generation of 100W to 300W, and their open-circuit voltage is below 45V usually. In order to achieve a high efficiency of energy conversion, the output voltage ripple should be below 8.5% of the MPP voltage[1][2]. According to these electrical characteristics of the photovoltaic panel and the grid voltage in different countries, the main specifications of the dc-dc converter for the dc module can be defined as:

(1) Input voltage: 20V to 50V

(2) Output voltage: 200V or 400V

(3) Nominal power: 100W to 300W

(4) Lower input voltage and current ripple

(5) High efficiency

The suitable high step-up dc-dc converter can be classified as: (1) Non-isolated topologies; (2) Isolated topologies. Some topologies are reviewed regarding the capabilities to meet the requirements of dc module

A. Non-isolated High Step-Up Topologies

The topology in Fig. 3 proposed in [3] integrates a multiphase voltage multiplier to achieve high static gain with low voltage stress. The interleaved technique is also used to minimize the current stress in all semiconductors. Therefore the efficiency of the topology is high, and the efficiency of 95% is achieved on a 24V input voltage, 200V or 400V output voltage and 400W prototype. The disadvantage of this topology is the complex structure and needs a number of semiconductors.

Fig. 4 shows a high step-up converter with couple-inductor. A passive regenerative snubber is utilized for absorbing the energy of stray inductance so that the switch duty cycle can be operated under a wide range. The reported efficiency is over 96.5% on a 25V to 400V and 100W prototype

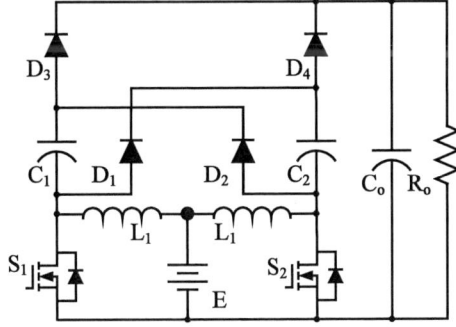

Fig. 3. The high step-up dc-dc converter proposed in [3]

As shown in Fig. 5, the topology is constructed on the basis of voltage-clamped and soft-switching techniques for alleviating the switching and conduction losses, to increase further the conversion efficiency. The reported

837

maximum efficiency is over 95% on a 26V to 39V input voltage, 200V output voltage, and 250W prototype.

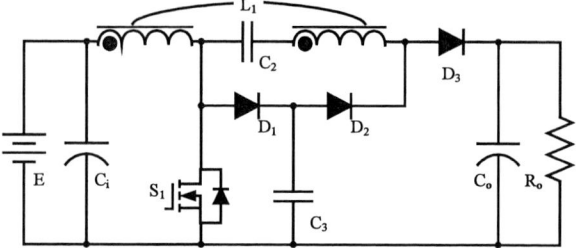

Fig. 4. The high step-up dc-dc converter proposed in [4]

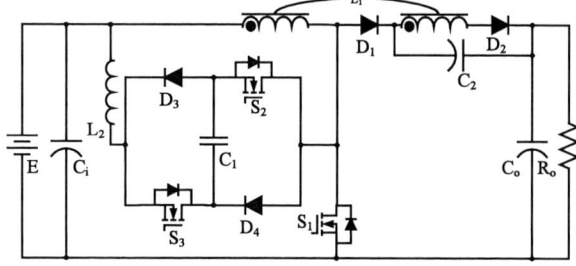

Fig. 5. The high step-up dc-dc converter proposed in [5]

B. Isolated High Step-Up Topologies

The isolated high step-up topologies include: (1) Voltage-fed topologies; (2) Current-fed topologies. The difference between the voltage-fed and current-fed topologies is the place of the filter inductor, which is on the secondary side of the voltage-fed converter and on the primary side of the current-fed converter. Comparing with the voltage-fed dc-dc converter, noticeable advantages of the current-fed converter for the high step–up application can be summarized as[6][7]:

(1) Immunity from transformer flux imbalance;

(2) Lower input current ripple and electrolyte capacitor size resulting in decrease of the overall converter size and manufacturing cost;

(3) A high voltage gain can be achieved with relatively low winding ratio, at the same time, the leakage inductance and size of the transformer decrease.

Consequently, the current-fed converter may be the better choice than the voltage-fed converter for the dc module.

The topology in Fig. 6 is current-fed push-pull converter. The topology has been widely applied by industry as battery chargers or power factor correction circuits. Some drawbacks of this topology include high output voltage ripple, requiring large filtering capacitors in the output, high voltage across the main switches, twice the output voltage referred to the primary, and high volt-ampere rating for the transformer[8].

The topology in Fig. 7 is current-fed full-bridge dc-dc converter[9]. The advantage of this topology is better utilization of the transformer due to bi-directional magnetization of the transformer. However, it increases the cost for the low power application, because the topology needs four active switches.

Fig. 8 shows the current-fed dual-bridge dc-dc

converter proposed in [7]. It characterize by small inductor, no dead-time operation, low output ripple current and no start-up problem. The main limitations of the new topology are that six power switches are used, and that input voltage range should remain within 2:1 in order to maintain the no dead-time property.

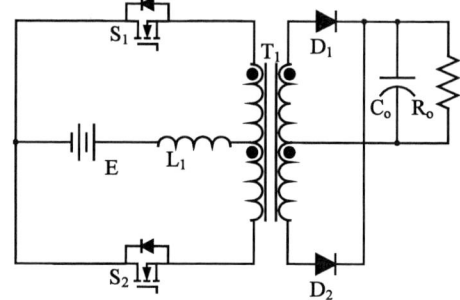

Fig. 6. Current-fed push-pull dc-dc converter

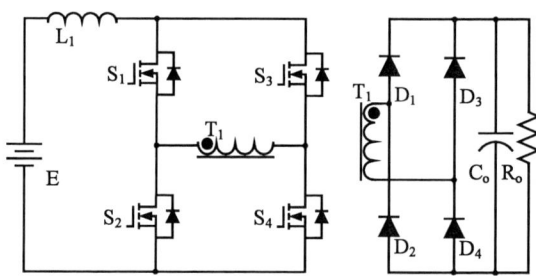

Fig. 7. Current-fed full-bridge dc-dc converter

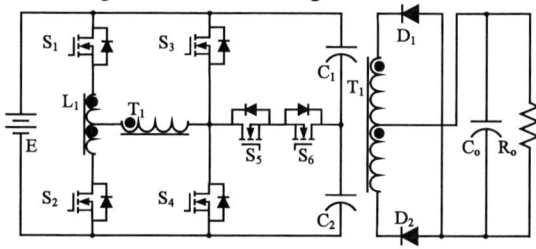

Fig. 8. Current-fed dual-bridge dc-dc converter proposed in [7]

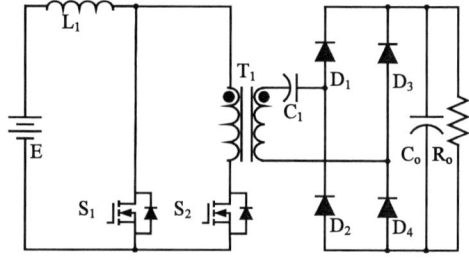

Fig. 9. Current-fed single-inductor dc-dc converter proposed in [10]

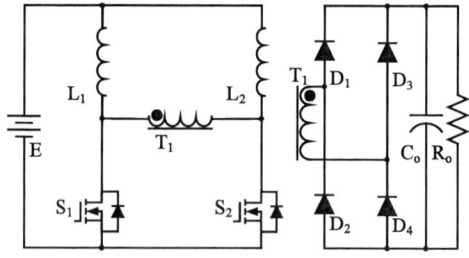

Fig. 10. Current-fed two-inductor dc-dc converter

838

The current-fed single-inductor dc-dc converter proposed in [10], as shown in Fig. 9, has only two active switches gated in a complementary and asymmetrical way, connected to the same ground. The main drawback of this topology is that the power is transferred by an electrolyte capacitor.

The topology in Fig. 10 is current-fed two-inductor converter[11]. This topology possesses significant advantages in the fuel cell and photovoltaic power application including (1) Only two active switches connected to the same ground; (2) Lower input current ripple, switch induction loss and switch voltage stress than other current-fed topologies; (3) High dc voltage gain and good transformer utilization.

According to the comparison of characteristics on these isolated topologies, the current-fed two-inductor converter seem to be the best candidate for dc module design. However, this topology has its own disadvantages, such as start problem and restricted load range. Some effective strategies have been reported in [6][12][13] to solve these problem.

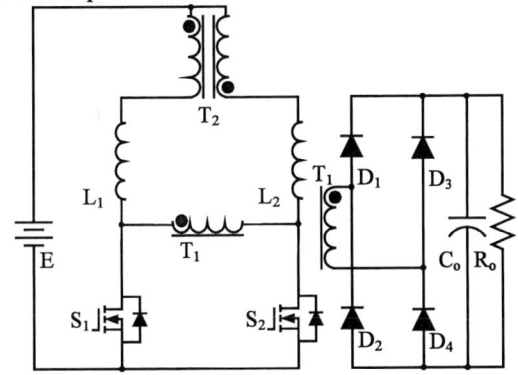

Fig.11 The improved current-fed two-inductor converter proposed in [12]

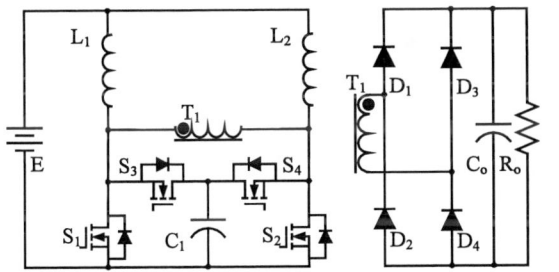

Fig.12. The improved current-fed two-inductor converter proposed in [6]

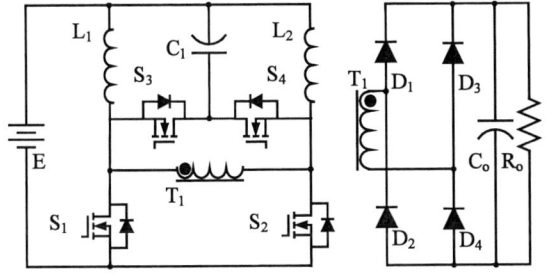

Fig. 13. The improved current-fed two-inductor converter proposed in [12]

Fig. 11 shows the improved current-fed two-inductor converter by adding a auxiliary transformer. The new two-inductor converter has no start problem and the full load range is achieved. The another two improved topologies proposed in [6] and [13] is shown in Fig. 12 and Fig. 13. A simply active clamp circuit is introduced, and the ZVS of all active switches is achieved. Therefore the converter efficiency is over 95% along the wide load range on a 200W, 24V to 200V prototype[13].

The phase-shifted parallel-input/series-output two-inductor converter, as shown in Fig. 14 proposed in [14], consists of two two-inductor converters and an auxiliary circuit producing an output voltage proportional to the amount of the phase-shift.

Whereby it has no start problem, and the efficiency is significantly improved.

In order to improve the power density of the converter, an integrated magnetic isolated two-inductor boost converter is proposed in [15] and shown in Fig. 15. In this topology, all magnetic components are integrated into one magnetic assembly, and it features wide power regulation range, reduction in the number of magnetic assemblies, and reduction in the number of windings on the primary side.

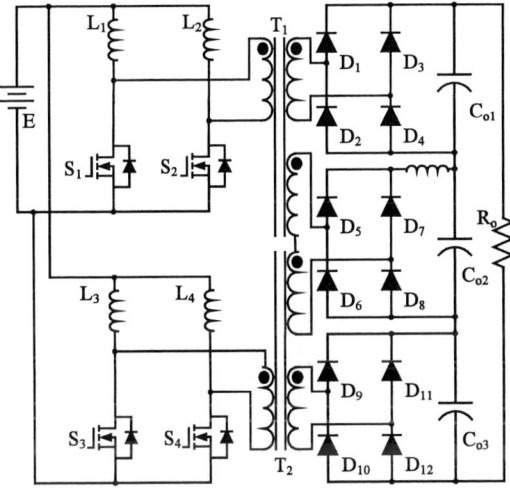

Fig. 14. The phase-shifted parallel-input/series-output two-inductor converter [14]

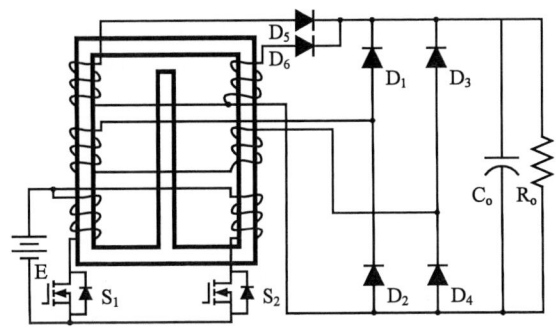

Fig. 15. The integrated magnetic isolated two-inductor boost converter [15]

IV. DISCUSSION

A. DC Module Versus AC Module

The ac module is an attractive technique for photovoltaic power application in recent years [1]. The ac module system consists of a number of parallel connected module integrated inverters, which is the integration of the inverter and photovoltaic panel into one electrical device. The ac module system has some common advantages with the dc-module-based system, such as individual MPPT and low mismatch losses. However, in order to acquire the status information of the ac module, the additional communication connections must be extend. It increases the complex of the system. If the PLCC is adopted to predigest the system, the output power quality of the ac module become a problem due to its direct connection to the local load and grid. Considering the reliability and performance of protection and maintenance for the photovoltaic power system, dc-module-based system is better than ac-module-based system.

B. Cascaded connection versus parallel connection

A cascaded dc-dc converter connection of photovoltaic modules is proposed in [15][16], as shown in Fig. 16(a). In the Fig. 16(a), the output voltage and power of the kth dc-dc converter are V_k and P_k, and the input voltage of the inverter is V ($V > V_k$). The V_k can be expressed as

$$V_k = \frac{P_k}{\sum_{j=1}^{n} P_j} V \qquad (1)$$

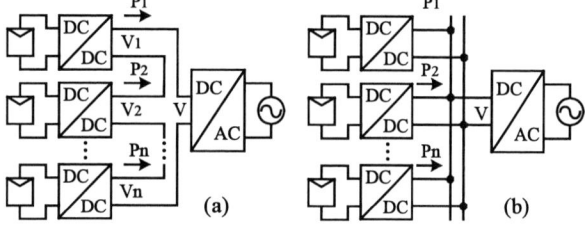

Fig. 16. Two type of photovoltaic power system. (a) Cascaded connection proposed in [15] and (b) Parallel connection proposed by this paper

The output voltage range of each dc-dc converter is $V_{\min} \le V_k \le V_{\max}$, therefore the normal operation condition of the dc-dc converter in the cascaded system is limited by

$$\frac{V_{\min}}{V} \le \frac{P_k}{\sum_{j=1}^{n} P_j} \le \frac{V_{\max}}{V} \qquad (2)$$

Inequation (2) shows that some dc-dc converter will cease transmitting energy, although the panel still can generate energy, under the mismatch condition of photovoltaic panels.

As an example, the simulation models of the two type photovoltaic power system according to the Fig. 16(a)

and Fig. 16(b) are established to evaluate their performance [17]. The parameters of the model are shown in Table. 1. The data of the photovoltaic panel is defined at standard test condition (25°C cell temperature, 1000W/m² sunlight intensity, and an air mass 1.5 solar spectral content).

Table. 1 The parameters of the simulation model

Parameters	Value
The number of the panels	3
DC-DC converter efficiency	100%
Output voltage range of cascaded dc-dc converter	20V-50V
Nominal maximum power	153W
MPP voltage	31.3V
MPP current	4.88A
Open-circuit voltage	42.5V
Short-circuit current	5.4A

The overall output power of the two type system under four different illumination condition , while the total illumination of the three photovoltaic panels is invariable (2100W/m2), is shown in Fig. 17. Fig. 17 shows that the output power of the cascaded system decrease gradually with the mismatch severity of the three photovoltaic panels. Under the fourth illumination condition, the output power of the cascaded system turn to zero. However, the output power of the paralleled system proposed by this paper is invariable almost. Consequently, the energy converter efficiency of the paralleled system is higher than cascaded system.

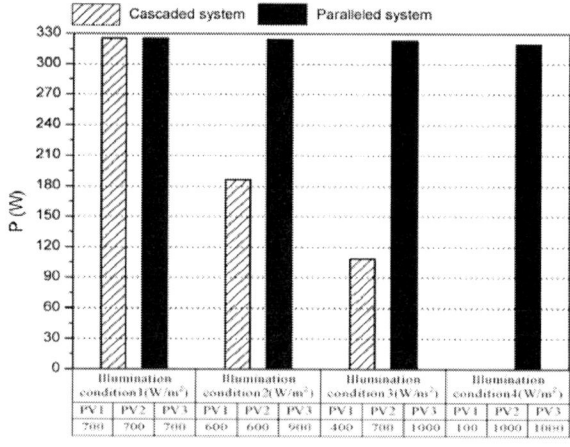

Fig. 17 The output power of the cascaded system and the paralleled system under different illumination condition

C. Isolation topologies versus non-isolation topologies

Grounding includes equipment and system grounding. Equipment ground is required in all countries, and system ground is required in some countries when the system voltage is over 50V. The topologies with electrical isolation between photovoltaic panel and grid is easy to satisfy the requirement of dual-grounding. Furthermore, The non-isolated topologies achieve high voltage gain by using interleaved and couple inductor technique generally[3][4][5]. Therefore, these topologies is more

840

complex than isolated topologies. According to these analysis, the isolated topologies is better than non-isolated topologies. On the other hand, in order to meet the requirement of ac utility voltage, which range is from 100V to 240V, in different countries, the output voltage of the dc module should be 200V and 400V optionally.

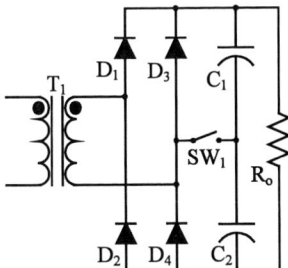

Fig. 18. The secondary circuit of dc module for different output voltage

The non-isolated topologies is difficult to realize it, however the isolated topologies can realize it easily using the circuit shown in Fig. 18. The SW1 is a output-voltage-select switch, which is always opened in the 100V to 120V line range and always closed in the 200V to 240V line range.

V. CONCLUSION

The dc-module-based photovoltaic power system, which integrates the high step-up dc-dc converter with a photovoltaic panel into one electrical device, then assigned to a common inverter, have some remarkable advantages for the residential and building integrated application, such as individual MPPT, lower sensitivity to mismatch, better system monitor and low cost etc. Based on the comparison and evaluation of some optional non-isolated and isolated topologies, the active clamp ZVS current-fed two-inductor dc-dc converters seem to be the better candidate for the dc module application due to its high performance and low cost.

REFERENCES

[1] Soeren baekhoej Kjaer, John K. Pedersen, and Frede Blaabjerg, A Review of Single-Phase Grid-Connected Inverters for Photovoltaic Modules, IEEE trans. on Industry Applications, vol. 41, pp. 1292-1306, October 2005.

[2] Soeren baekhoej Kjaer, Design and control of an inverter for photovoltaic applications, Ph.D. dissertation, Inst. Energy Technol., Aalborg University Aalborg East, Denmark, 2004/2005

[3] Luis Claudio Franco, Luciano Lopes Pfitscher, Roger Gules, A new high static gain nonisolated DC-DC converter, IEEE PESC'03, vol. 3, pp. 1367-1372, June 2003

[4] Rong-Jong Wai, Rou-Yong Duan, High Step-Up Converter With Couoled-Inductor, IEEE trans.on Power Electronics, vol. 20, no. 5, Sept. 2005

[5] R.-J. Wai and C.-Y. Lin, High-efficiency, high-step-up DC-DC converter for fuel-cell generation system, IEE proc.-Electr. Power Appl., vol. 152, no. 5, pp. 1371-1378, Sept. 2005.

[6] Jin-Tae Kim, Byoung-Kuk Lee, Tae-Won Lee, Su-Jin Jang, Soo-Seok Kim, and Chung-Yuen Won, An Active Clamping Current-Fed Half-Bridge Converter for Fuel-Cell Generation System, IEEE PESC'04, pp.4709-4714, 2004

[7] Wei Song and Brad Lehman, Current-Fed Dual-Bridge DC-DC Conver, IEEE trans. on Power Electronics, vol. 22, no. 2, pp. 461-469, March, 2007

[8] Wilson C.P. de Aragao Filho, Ivo Barbi, A comparison between two current-fed push-pull DC-DC converters-analysis, design and experimentation, INTELEC'96, pp. 313-320, Oct. 1996.

[9] R.Y. Chen, R.L. Lin, T.J. Liang, J.F. Chen, and K,C. Tseng,Current-fed Full-bridge Boost Converter with Zero Current Switching for High Voltage Applications, IEEE IAS 2005, pp. 2000-2006, 2005

[10] Peter mantobanelli and Ivo Barbi, A New Current-Fed, Isolated PWM DC-DC Converter, IEEE trans. on Power Electronics, vol. 11, no. 3, pp. 431-438,May 1996

[11] Quan Li, High Frequency Transformer Linked Converters for Photovoltaic Applications, Ph.D. dissertation, Central Queensland University, Australia, 2006

[12] Yungtaek Jang, and Milan M. Jovanovic´,New Two-Inductor Boost Converter With Auxiliary Transformer, IEEE trans. on Power Electronics, vol.19, no. 1, pp. 169-175, Jan. 2004

[13] Sang-Kyoo Han, Hyun-Ki Yoon,Gun-Woo Moon, Myung-Joong Youn, Yoon-Ho Kim, and Kang-Hee Lee, A New Active Clamping Zero-Voltage Switching PWM Current-Fed Half-Bridge Converter, IEEE trans. on Power Electronics, vol. 20, no. 6, pp. 1271-1279, Nov. 2005

[14] Jeong-il Kang, Chung-Wook Roh, Gun-Woo Moon, and Myung-Joong Youn,Phase-Shifted Parallel-Input/Series-Output Dual Converter for High-Power Step-Up Applications, IEEE trans. on Industrial Electronics, vol. 49, no. 3, pp. 649-652, June 2002

[15] Liang Yan, and Brad Lehman, An Integrated Magnetic Isolated Two-Inductor Boost Converter: Analysis, Design and Experimentation, IEEE trans. on Power Electronics, vol. 20, no. 2, March 2005

[16] Geoffrey R. Walker, Paul C. Sernia, Cascaded DC-DC Converter Connection of Photovoltaic Modules[J], IEEE trans. On Power Electronics, vol. 19, no. 4, July, 2004

[17] Eduardo Román, Ricardo Alonso, Pedro Ibañez, Sabino Elorduizapatarietxe, and Damián Goitia, Intelligent PV Module for Grid-Connected PV Systems, IEEE trans. on Industrial Electronics, vol. 53, NO.4, pp. 1066-1073, August 2006

[18] Volker Quaschning and Rolf Hanitsch, Numerical Simulation of Current-Voltage Characteristics of Photovoltaic Systems with Shaded Solar Cells, Solar Energy, vol. 56, no. 6, pp. 513-520, 1996

Analysis and Improvement of Maximum Power Point Tracking Algorithm Based on Incremental Conductance Method for Photovoltaic Array

Bangyin Liu, Shanxu Duan, Fei Liu, and Pengwei Xu

College of Electrical and Electronic Engineering, Huazhong University of Science and Technology, Wuhan, P. R. China.

Abstract—Photovoltaic (PV) array in the photovoltaic power system (PVPS) has the nonlinear current-voltage characteristic which is affected by the panels temperature and irradiance conditions. To improve the energy converter efficiency of the PVPS, the maximum power point tracking algorithm should be adopted to maximize the yield of the PV array. The incremental conductance (INC) method is widely used in the PVPS due to its succinctness and high tracking efficiency. The conventional INC algorithm using a fixed iteration step-size is impossible to achieve rapid dynamic response and good steady tracking accuracy simultaneously, because if the step-size is increased for rapid dynamic response, the tracking accuracy is decreased and vice versa. An improved algorithm with variable step-size is proposed to solve the problem in this paper. The contrastive experimental results between the improved and the conventional INC algorithm are presented to demonstrate its effectiveness.

Index Terms-- Maximum power point tracking (MPPT), photovoltaic power system (PVPS), photovoltaic array, incremental conductance method (INC), variable step-size.

I. INTRODUCTION

Photovoltaic power system (PVPS) is attracting more and more interest, due to using nonpolluting, renewable and free solar energy, with the growing concerns about environmental issues and the world energy crisis. The PV array in the PVPS present nonlinear I-V characteristic, and the output power changes with the illumination and ambient temperature. Many MPPT techniques have been proposed to extract maximum power from the PV array and transfer it to the load at all times [1].

The incremental conductance techniques are widely used due to its ease to implementation and high tracking efficiency [1-13]. Fig. 1 shows the flowchart of the INC algorithm which is based on the fact that the sum of instantaneous conductance (I/V) and the incremental conductance ($\Delta I/\Delta V$) is zero at the MPP, positive an the left of the MPP, and negative on the right. The condition $\Delta I/\Delta V = -I/V$ is in practice hard to exactly satisfied because of noise and errors [3], therefore the condition is

usually replaced by $|\Delta I/\Delta V + I/V| < \varepsilon$, where ε is a small positive numbers [11]. With this algorithm at steady state the operating point, as is shown in Fig. 2, may be located at a certain point in the interval BC or oscillated between the interval AB and the interval CD. Therefore the step-size ΔV_{ref} are determined with the consideration of the tradeoff between steady tracking accuracy and dynamic response. If the step-size ΔV_{ref} is increased for rapid dynamic response to extract more energy under rapidly changing atmospheric (e.g. solar radiation and temperature) conditions, the tracking accuracy is decreased due to the higher oscillation around the MPP and vice versa.

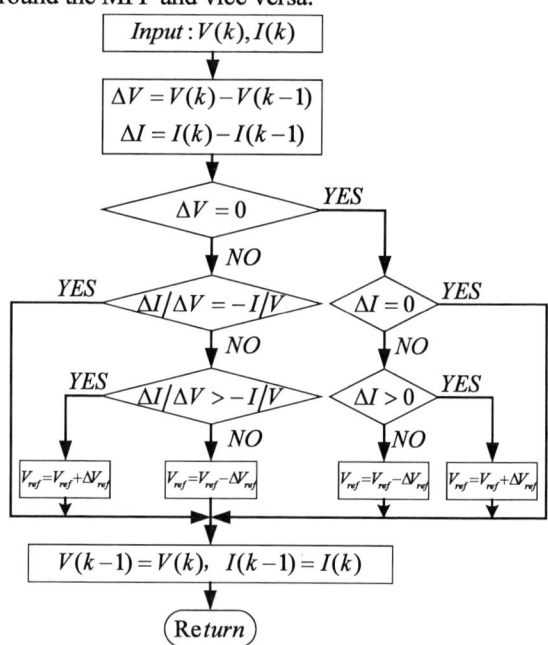

Fig. 1. Incremental conductance algorithm as shown in [1] and [2].

An improved INC algorithm is presented in this paper. The proposed algorithm uses variable perturbation step-size to eliminate the tradeoff between the dynamic response and steady state oscillation. The perturbation step-size is automatically tuned according to the inherent PV array characteristics. If the operating point is far from MPP, the step-size is increased to achieve fast dynamic

---

Project Supported by Delta Science & Technology Educational Development Program (DREK200501).

978-1-4244-0644-9/07/$25.00 ©2007 IEEE

response. Whereas the operating point is near to the MPP, the step-size becomes very small to reduce the steady state oscillation and improve the energy converter efficiency of the PVPS. The conventional fixed step-size and improved variable step-size INC algorithms are implemented in a DSP controlled MPPT test plant and the contrastive experimental results are presented to demonstrate its effectiveness.

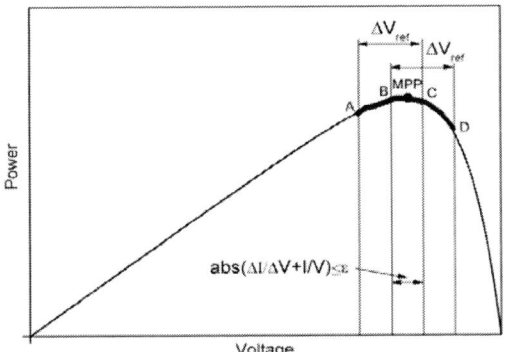

Fig. 2. The INC MPPT operating point trajectory.

II. VARIABLE STEP-SIZE INC ALGORITHM

A. Basic idea

According to the mathematical model of the PV array [14], the normalized current-voltage, power-voltage and absolute derivative of the power-voltage characteristics of a PV array are plotted in Fig. 3. From the curves shown in Fig. 3, the absolute derivative of the power-voltage is large when the operating point is away from MPP, and monotonically decrease as the MPP is approached, as given by

$$\begin{cases} dP/dV > 0 & ,left\ of\ MPP \\ dP/dV = 0 & ,at\ MPP \\ dP/dV < 0 & ,right\ of\ MPP \end{cases} \quad (1)$$

From Fig. 1, it can be seen that the perturbation step-size should be positive (negative) to locate the MPP when operating on the left (right) of MPP. Considering (1), it is evident that the dP/dV is essentially suited for using as step-size after proper scaling [4].

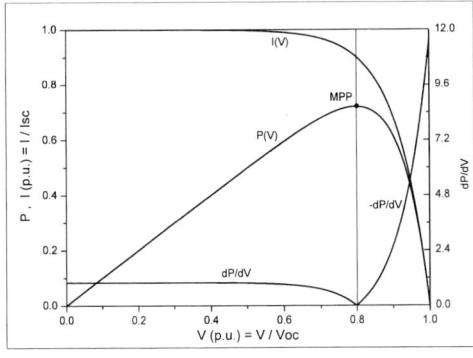

Fig. 3. The normalized current-voltage, power-voltage and absolute derivative of the power-voltage characteristics of a PV array.

B. The Variable Step-size INC Algorithm

The flowchart of the variable step-size INC algorithm is shown in Fig. 4, where ε_1, ε_2 and ε_3 are small positive numbers near to zero, α and β is the accelerate factors. Since

$$\frac{dP}{dV} = I + V\frac{dI}{dV} \approx V(\frac{I}{V} + \frac{\Delta I}{\Delta V}) \quad (2)$$

The update rule of the V_{ref} can be expressed as

$$V_{ref} = V_{ref} + V_{step} \quad (3)$$

where

$$V_{step} = \begin{cases} \alpha V\left(\frac{I}{V} + \frac{\Delta I}{\Delta V}\right) & ,\Delta V \neq 0 \\ \beta \cdot \Delta I & ,\Delta V = 0 \end{cases} \quad (4)$$

The direction of perturbation embedded in the variable step-size V_{step} itself, therefore the flowchart of the variable step-size INC algorithm is conciser than the fixed step-size one.

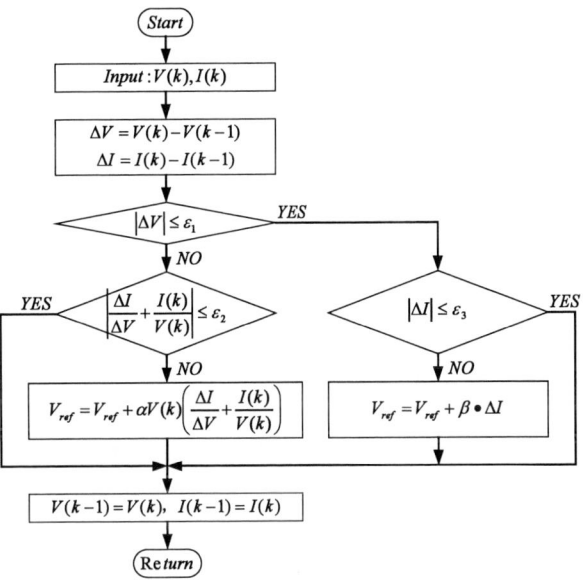

Fig. 4. The flowchart of the variable step-size INC algorithm

C. Parameters Design

The accelerate factor α in the update rule determines the performance of the variable step-size INC algorithm. Manual tuning of accelerate factor α is difficult to obtain optimal results for different MPPT systems. To ensure satisfactory tracking performance it is essential that α be automatically tuned during startup process. A simple method to determine α can be expressed as

$$\alpha \leq \frac{V_{step\_max}}{\left|dP/dV\right|_{max}} \quad (5)$$

843

where V_{step_max} is the allowable maximum step-size under the fixed step operating condition. $|dP/dV|_{max}$ can be evaluated by

$$\left|\frac{dP}{dV}\right|_{max} \approx \left|\frac{P|_{Vref=mV_{OC}} - P|_{Vref=V_{OC}}}{mV_{OC} - V_{OC}}\right| = \left|\frac{mI|_{Vref=mV_{OC}}}{m-1}\right| \quad (6)$$

where m is a positive number near to 1, e.g. 0.98, V_{OC} is the open circuit voltage of PV array. A smaller α will exhibits a slightly slower dynamic response than a larger one.

A change in ΔI indicates a change in atmospheric conditions. Under varying atmospheric conditions the relationship between V_{MPP} and V_{OC} of the PV array is near linear [1]. Therefore the accelerate factor β should be a small positive number near to zero.

The iteration period of the INC algorithm should be set higher than a proper threshold in order to avoid instability of the MPPT algorithm. It must be ensured that, after each iteration, the MPPT system reaches the steady-state before the next iteration is done [15].

III. EXPERIMENTAL RESULTS

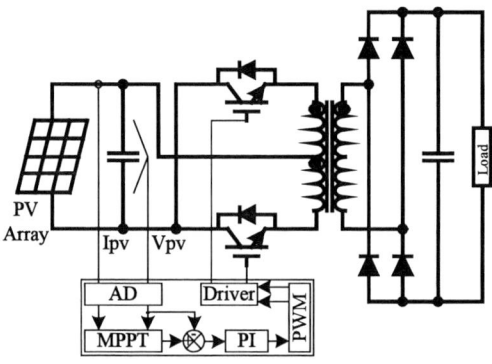

Fig. 5. The experimental system for MPPT algorithms test

The DSP (TMS320LF2407A) controlled high-frequency isolated push-pull dc-dc converter is built to act as the maximum power point tracker, as is shown in Fig. 5, and the PV array is composed of three series connected PV modules, which electrical parameters are shown in Tab. 1. The iteration period of the two INC algorithms are 0.25 second and the fixed step size of the conventional method is 0.3V. In order to simulate the step changed illumination level, the PV array is switched between two and three PV modules. The contrastive experimental results are presented in Fig. 6, and the measured data is shown in Tab. 2.

Fig. 6(a) and (b) show the start process of the tracker. The proposed algorithm tracks the MPP within 3 second, while the conventional one spends about 16.5 second. The proposed algorithm tracks the new MPP within 1.4 second, and the conventional one spends about 10 second, as is shown in Fig. 6(c) and (d), at step increased illumination level. From Fig. 6(e) and (f), it can be seen that the proposed algorithm tracks the new MPP within

1.5 second, and the conventional one spends about 8.5 second at step decreased illumination level. Therefore the proposed algorithm has the faster tracking speed than the conventional one. As is shown in Tab. 1, it is clear that both the conventional and the improved INC algorithms can track the accurate MPP, and the maximum tracking error is limited within ± 4 W.

Tab. 1 The Electrical parameters of the PV Module

Parameter	Value
Maximum power	120Wp
Open circuit voltage	21.6V
Short circuit current	8.9A
Voltage at MPP	17.3V
Current at MPP	6.9A

Tab. 2. Experimental MPPs of Photovoltaic array

Fig.2	Experimental MPP	Measured MPP
(a)	0 to 212.8W	0 to 216.5W
(b)	0 to 210.6W	0 to 209.9
(c)	122.1 to 177.6W	121.9 to 181.1W
(d)	160.6 to 240.9W	162.6 to 243.9W
(e)	177.6 to 122.1W	173.7 to 118.3W
(f)	204.8 to 134.6W	201.4 to 135.9W

IV. CONCLUSIONS

An improved incremental conductance algorithm with a variable step size is proposed for PV array applications in this paper. It automatically adjusts the step size according to the operating point of the PV array. When the operating point is far from the MPP of the PV array, the step size is great and the operating point of the PV array approaches quickly to the MPP, and when it is near the MPP, the step size is small and the operation point of the PV array is located near the MPP. The design issue of the variable step-size parameters is discussed and a simple design rule is proposed. The presented contrastive experimental results of the improved and conventional INC algorithm verified the validity of the proposed algorithm.

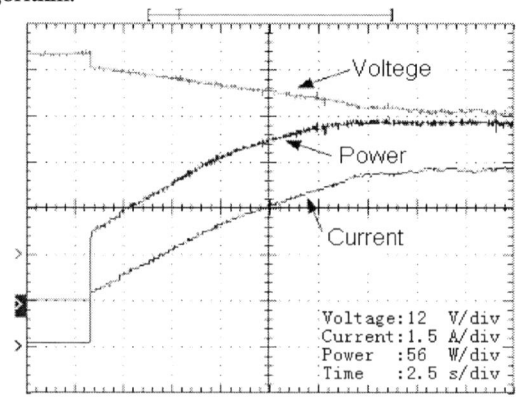

(a) Start process using the conventional INC algorithm

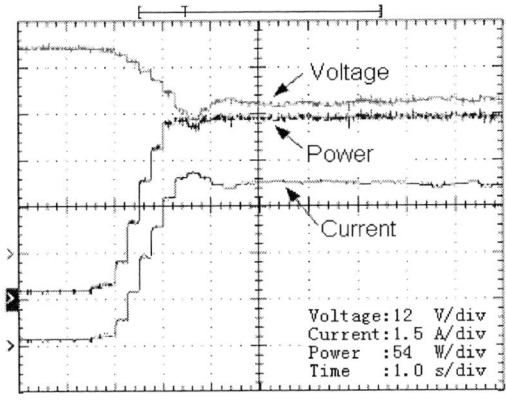

(b) Start process using the proposed INC algorithm

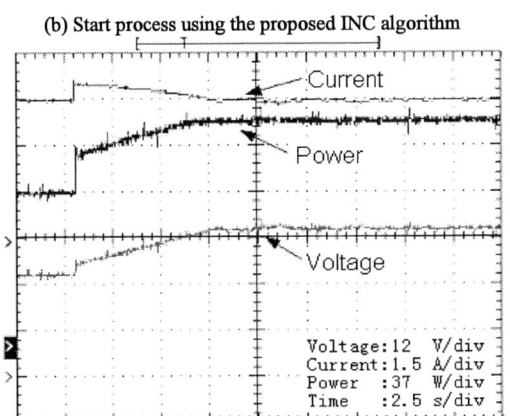

(c) Dynamic characteristic of the conventional INC algorithm at step increased illumination level

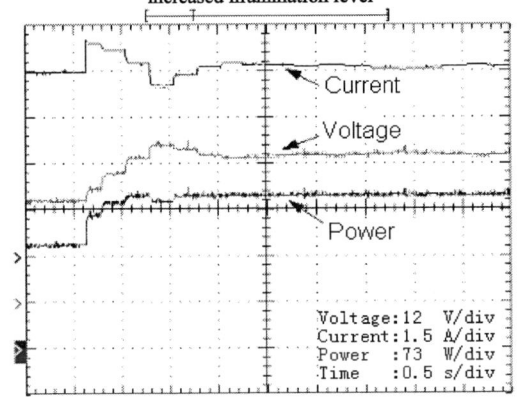

(d) Dynamic characteristic of the proposed INC algorithm at step increased illumination level

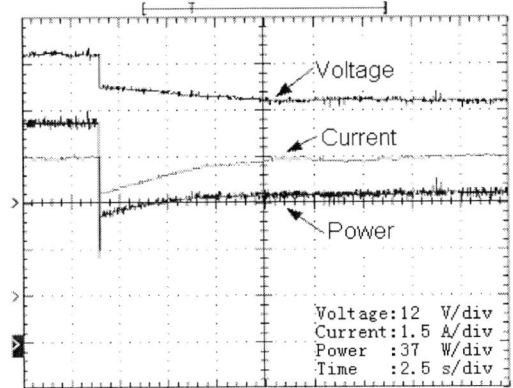

(e) Dynamic characteristic of the conventional INC algorithm at step decreased illumination level

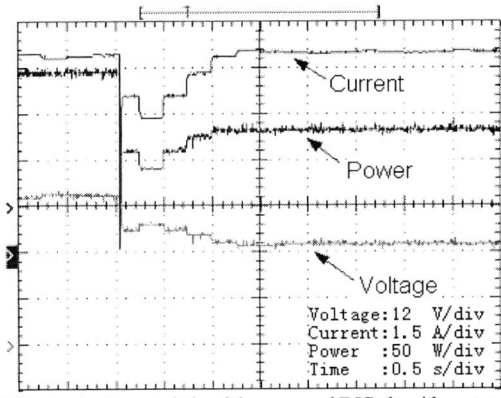

(f) Dynamic characteristic of the proposed INC algorithm at step decreased illumination level

Fig. 6 The output voltage, current and power of the PV array.

REFERENCES

[1] T. Esram and P. L. Chapman, "Comparison of Photovoltaic Array Maximum Power Point Tracking Techniques," IEEE TRANSACTIONS ON ENERGY CONVERSION, vol. 22, no. 2, pp. 439-449, Jun. 2007.

[2] K. H. Hussein and I. Mota, "Maximum photovoltaic power tracking: An algorithm for rapidly changing atmospheric conditions," in IEE Proc. Generation Transmiss. Distrib., 1995, pp. 59–64.

[3] D. P. Hohm and M. E. Ropp, "Comparative study of maximum power point tracking algorithm," in Proc. 28th IEEE Photovoltaic Specialists Conf., Sept. 2000, pp. 1699-1702.

[4] Ashish Pandey, Nivedita Dasgupta and Ashok K. Mukerjee, "Design Issues in Implementing MPPT for improved Tracking and Dynamic Performance", in Proc. IECON'06, 2006, pp. 4387-4391

[5] A. Brambilla, M. Gambarara, A. Garutti, and F. Ronchi, "New approach to photovoltaic arrays maximum power point tracking," in Proc. 30th Annu. IEEE Power Electron. Spec. Conf., 1999, pp. 632–637.

[6] K. Irisawa, T. Saito, I. Takano, and Y. Sawada, "Maximum power point tracking control of photovoltaic generation system under non-uniform insolation by means of monitoring cells," in Conf. Record Twenty-Eighth IEEE Photovoltaic Spec. Conf., 2000, pp. 1707–1710.

[7] T.-Y. Kim, H.-G. Ahn, S. K. Park, and Y.-K. Lee, "A novel maximum power point tracking control for photovoltaic power system under rapidly changing solar radiation," in IEEE Int. Symp. Ind. Electron., 2001, pp. 1011–1014.

[8] Y.-C. Kuo, T.-J. Liang, and J.-F. Chen, "Novel maximum-power-point tracking controller for photovoltaic energy conversion system," IEEE Trans. Ind. Electron., vol. 48, no. 3, pp. 594–601, Jun. 2001.

[9] G. J. Yu, Y. S. Jung, J. Y. Choi, I. Choy, J. H. Song, and G. S. Kim, "A novel two-mode MPPT control algorithm based on comparative study of existing

algorithms," in Conf. Record Twenty-Ninth IEEE Photovoltaic Spec. Conf., 2002, pp. 1531–1534.

[10] K. Kobayashi, I. Takano, and Y. Sawada, "A study on a two stage maximum power point tracking control of a photovoltaic system under partially shaded insolation conditions," in IEEE Power Eng. Soc. Gen. Meet., 2003, pp. 2612–2617.

[11] W. Wu, N. Pongratananukul, W. Qiu, K. Rustom, T. Kasparis, and I. Batarseh, "DSP-based multiple peak power tracking for expandable power system," in Eighteenth Annu. IEEE Appl. Power Electron. Conf. Expo., 2003, pp. 525–530.

[12] Jae Ho Lee , HyunSu Bae and Bo Hyung Cho, Advanced Incremental Conductance MPPT Algorithm with a Variable Step Size, EPE-PEMC 2006, pp. 603-607

[13] Nobuyoshi Mutoh, Masahiro Ohno, Takayoshi Inoue. A method for MPPT control while searching for parameters corresponding to weather conditions for PV generation systems. IEEE Transactions on Industrial Electronics, 2006, vol. 53, no. 4, pp. 1055-1065.

[14] J.A. Gow, and C.D. Manning, Development of a Photovoltaic Array Model for use in power-electronics simulation studies, IEE Proc.-Electr. Power Appl., Vol. 146, No. 2, pp. 193-200, March 1999

[15] Nicola Femia, Giovanni Petrone, Giovanni Spagnuolo and Massimo Vitelli, "Optimization of Perturb and Observe Maximum Power Point Tracking Method," IEEE Transactions on Power Electronics, 2005, vol. 20, No. 4, pp. 963-973.

Application of Maximum Power Point Tracker with Self-organizing Fuzzy Logic Controller for Solar-powered Traffic Lights

Noppadol Khaehintung[*] and Phaophak Sirisuk[**]

[*] Dept. of Control and Instrumentation Engineering, Faculty of Engineering,
[**]Dept. of Computer Engineering, Faculty of Engineering,
Mahanakorn University of Technology, Bangkok, Thailand 10530
E-mail: noppadol@mut.ac.th and phaophak@mut.ac.th

Abstract-- This paper presents the development of Maximum Power Point Tracking (MPPT) using an adjustable Self-Organizing Fuzzy Logic Controller (SOFLC) for a Solar-powered Traffic Light Equipment (SPTLE) with an integrated Maximum Power Point Tracking (MPPT) system on a low-cost microcontroller. The proposed system is integrated with a boost converter for realizing of high performance SPTLE, whose adaptability properties are very attractive for operation of a solar array power tracking in dynamic environments. The proposed MPPT scheme obtained by varying the duty ratio for DC-DC boost converter has been successfully implemented on a low-cost PIC16F876A RISC-microcontroller. Experimental results of the hardware prototypes for SPTLE, light flasher and light chevron, with commercial solar array show that our proposed MPPT using SOFLC as compared with Fuzzy Logic Controller (FLC) in terms of tracking speed with 92% of overall system efficiency.

Index Terms--- Maximum Power Point Tracking, Fuzzy Logic Controller, Self-Organizing Fuzzy Logic Controller, Microcontroller.

I. INTRODUCTION

Today, photovoltaic has become one of the strongest candidates as a secondary energy source. This is because the problem of fossil energy depletion becomes more severe. The term photovoltaic refers to the phenomenon involving the conversion of sunlight into electrical energy via a solar cell [1]. Under certain temperature and light intensity, there is only single maximum-power point (MPP) in a normal cell. Therefore, maximum power point tracking (MPPT) of the solar cell is essential as far as the system efficiency is concerned.

Recently, various MPPT techniques have been implemented on a microcontroller unit (MCU) in several solar-powered applications [2]. For example, a RISC microcontroller was employed to realize MPPT using a Perturbation and Observation Method (P&O) method for a battery charging application in [3]. For a transportation industry, one of sectors that gain benefits from such a system, a solar-powered light–flasher (SPLF) was developed in [4]. Besides, a hill-climbing algorithm, which is similar to P&O method, was also implemented on RISC microcontroller for an illumination application in [5]. The sophisticated Artificial Intelligent (AI)

methods, such as Artificial Neural Network (ANN) [6] and Fuzzy Logic Control (FLC) [3,7-9], have been developed for solar-powered applications. For FLC, an inference engine is time-consuming. Thus, the relation between input and output of FLC can be stored in a memory-limited lookup table (LUT). The implementations of FLC stored in LUT for MPPT have been successfully implemented in [3] and [8,9] for a solar-power battery charger (SPBC) and an SPLF, respectively. Comparatively, the conventional MPPT methods can give poorer performances, but implementation is always easier. AI methods, on the other hand, perform better, but their structure is generally more complicated and requires relatively high performance processor. Therefore, AI is not suitable for some applications where cost is a prime concern. Furthermore, they still lacks of the adaptability required for MPPT controller to efficiently deal with time-varying environments.

An alternative to overcome the problem of adaptability is a Self-Organizing Fuzzy Logic Controller (SOFLC) originally proposed by Procyk and Mamdani in [10]. By self-organizing, it is meant that the controller can recursively adjust its associated fuzzy rule in accordance with a desired response [11]. Besides, the technique is simple and can be efficiently realized by Look-Up Table (LUT), offering a cost-effective solution to hardware implementation. In [12], the authors introduced an application of SOFLC for MPPT in a solar-powered battery charging system Nonetheless, the applications a stand alone of solar-powered system has been not investigated.

In this paper, the implementation of the Self-Organizing Fuzzy Logic Controller for a Solar-powered Traffic Light Equipment (SOFLC-SPTLE) with built-in MPPT is presented. A low-cost PIC16F876A RISC MCU [13] is employed for the algorithm processing, and it is integrated to a boost converter to form a solar-powered battery charging system. There is no external sensory unit required for the system. Despite of its cost-effectiveness, the experimental results demonstrate that the system performance is outstanding.

The remainder of the paper is organized as follows. A typical solar power system and FLC for MPPT technique is reviewed in Section II. The proposed SOFLC-SPTLE is introduced in Section III. In Section IV, experimental

978-1-4244-0644-9/07/$25.00 ©2007 IEEE

results obtained from SOFLC-SPTLE prototypes are shown. Finally, conclusions are drawn in Section V.

II. MPPT FOR A SOLAR ARRAY

A. The Characteristic of a Solar Array

The characteristic of a solar array can be comprehensively described by its operating curve known as an I-V curve, supplied by manufacturers to represent the solar array behaviours. It shows the relationship between output voltage (V_s) and current (I_s) of the solar array at some light intensity and temperature as depicted in Fig. 1. It can be observed that under a certain light intensity and temperature there is a unique point located at the knee of the I-V curve, at which the maximum power can be generated from the solar array. Thus, a mechanism is required to track those underlying points so that an optimal operation of the overall system can be achieved.

B. A Fuzzy Logic Controller for MPPT

Generally speaking, MPPT attempts to move an operation point of a PV array as close to the MPP or "*knee*" of the I-V curve shown in Fig. 1 as possible. Mathematically, this is equivalent to finding the point, where the derivative dP/dV_s is equal to zero. To this end, let us first denote the error function by

$$e_c(n) = \frac{P(n) - P(n-1)}{V_s(n) - V_s(n-1)}, \qquad (1)$$

and the associated change of error as:

$$\Delta e_c(n) = e_c(n) - e_c(n-1), \qquad (2)$$

Therefore, by forcing (1) and (2) to zero, MPPT can be achieved. The operation may be performed via a fuzzy logic controller. In particular, instead of finding the underlying derivative the following fuzzy algorithm based upon the meta-rule can be employed [8]. That is, "*If the last change in the duty ratio ($D(n)$) has caused the power to rise, keep moving the duty ratio ($D(n)$) in the same direction; otherwise, if it has caused the power to drop move it in the opposite direction.*" This can be translated into the following fuzzy control rule as:

Rule (i) : IF $e_c(n)$ is A_i and $\Delta e_c(n)$ is B_l
THEN $\Delta D(n)$ is C^i, $\qquad (3)$

where, A_i, B_l, and C^i represent fuzzy sets including positive big (PB), positive small (PS), zero (ZE), negative big (NB), and negative small (NS). Also, Table I collects the fuzzy rule for $\Delta D(n)$.

Note that the number of fuzzy function should be optimally determined with considering the interrelation of control accuracy and calculation capacity. The selection of the membership function of each fuzzy set is based on trial-and-error basis so that the region of interest is covered appropriately as shown in Fig. 2 (a) and (b).

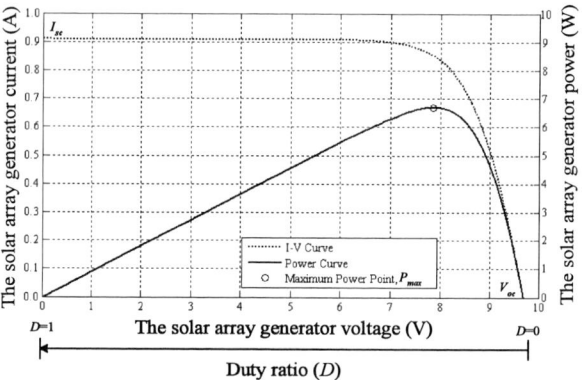

Fig. 1. I-V and Power curve under a certain light intensity and temperature.

TABLE I
A FUZZY RULE BASE TABLE FOR MPPT

$e_c(n)$ \ $\Delta e_c(n)$	NB	NS	ZE	PS	PB
NB	PB	PB	PS	PS	PS
NS	ZE	ZE	PS	PS	PS
ZE	NS	NS	ZE	PS	PS
PS	NS	NS	NS	ZE	ZE
PB	NB	NB	NS	NS	NS

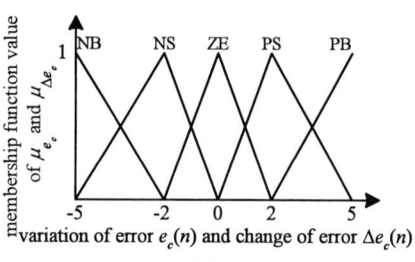

(a)

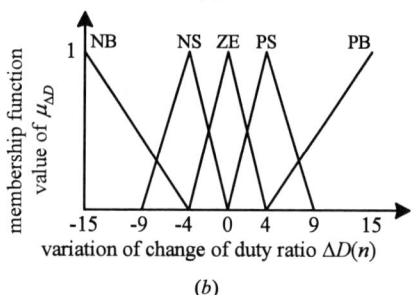

(b)

Fig. 2. Membership function of FLC for MPPT (a) membership function of $e_c(n)$ and $\Delta e_c(n)$ and (b) membership function of $\Delta D(n)$.

III. IMPLEMENTATION OF SOFLC-SPTLE

A. Hardware Circuitry

The circuit configuration of SOFLC-SPTLE and a hardware prototype used in this research work are shown in Fig. 3 and Fig. 4, respectively. A key of the proposed system is the cost-effectiveness PIC16F876A RISC MCU as the main control unit. The boost converter circuit used in the battery charging mechanism comprises one inductor (L=15mH), two capacitors (C_1=100μF, C_2=100 μF), and two Schottky diodes D_1 and D_2. Besides, a power MOSFET BUZ11 operating at 31.25 kHz is used

848

as a switching device S. Finally, a commercial-grade battery bank rated at 12V, 7.5Ahr is used for the power storage selected in accordance with the solar array specification [8]. Fig 4 (a) and (b) illustrate the SOFLC-SPTLE prototypes of the solar-powered light flasher (SPLF) and the solar-powered light Chevron (SPLC), respectively.

In our design, MCU performs two main functions. First, it determines an optimal duty ratio D for generation of PWM signal for controlling the switch S in such a way that the operating point of the solar array is moved towards the maximum power point. Moreover, the controller must consistently detect the battery voltage to avoid battery overcharging. When the battery is fully charged, the controller will switch off the switch S. The LED array is divided into n sub-networks. The PIC16F876A microcontroller generates three different on-off patterns (LSQ) for each LED network corresponding to the battery level as shown in Fig. 5.

B. An SOFLC for MPPT on MCU

Basically, the SOFLC is the FLC that allows its fuzzy rule to be recursively updated. As hardware implementation of FLC can be efficiently made using a Look-Up table (LUT), the idea behind the self-organization may be clearly described as an adjustment mechanism that updates the values in the table, based on the current performance of the controller. If the performance is poor, the responsible table value should be punished, such that next time that cell of the table is visited, the control signal will be better.

By using fuzzy inference techniques, the relation of $e_c(n)$, $\Delta e_c(\text{n})$ and $\Delta D(n)$ is calculated by [9]

$$\Delta D = \frac{\sum_{l=1}^{M} \Delta D^l w_l}{\sum_{l=1}^{M} w_l},\qquad(4)$$

where, $w_l = \min[\mu_{e_c}(e_c(n)), \mu_{\Delta e_c}(\Delta e_c(n))]$ is the compatibility (weighting factor) and ΔD^l is a value corresponding to the membership function of $\Delta D(n)$. Invoking (4) and some scaling, $\Delta D(n)$ corresponding to each pair of $e_c(n)$ and $\Delta e_c(n)$ is pre-calculated as shown in Table II. In FLC context, this look up table is often referred to as a control table ($\mathbf{F}$). In our proposed SPTLE, the table is stored in the MCU built-in E^{2}PROM. Since both input variables $e_c(n)$ and $\Delta e_c(n)$ are chosen in the interval [-5,5] which is divided into 11 entries, LUT ends up with 11×11 entries, resulting in memory usage of 121 bytes.

As discussed previously, self-organizing mechanism can be integrated into the FLC via another table, namely performance-measure table ($\mathbf{P}$). Each entry of $\mathbf{P}$ contains the amount required for refining each entry of the control table ($\mathbf{F}$). In our work, the table is constructed based on extensive experiments and is shown in Table III. Similarly, $\mathbf{P}$ is realized using the MCU built-in ROM. At each recursion, the control table is thus updated as:

$$\mathbf{F}(i,j)_{n-d} = \mathbf{F}(i,j)_{n-d} + \mathbf{P}(i,j)_n,\qquad(5)$$

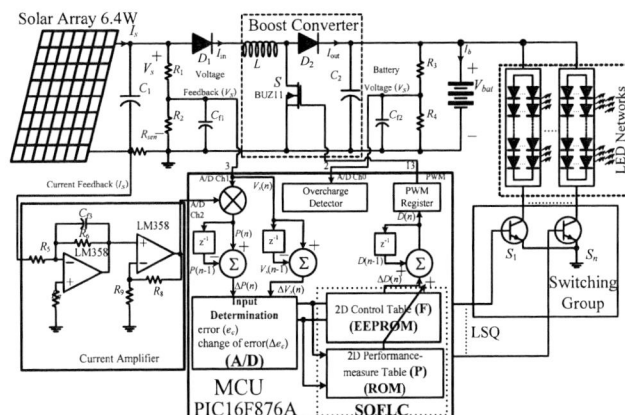

Fig. 3. The proposed SOFLC-SPTLE circuit configuration.

(a)

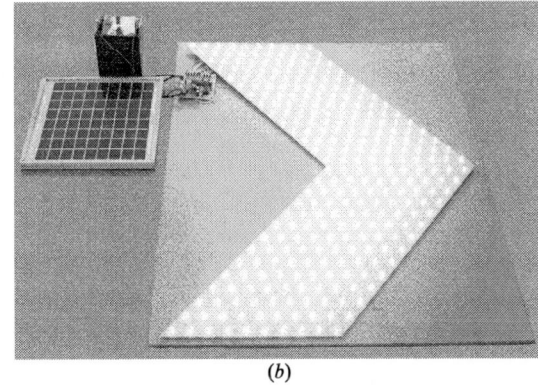

(b)

Fig. 4. The SOFLC-SPTLE prototypes (a) SOFLC-SPLF and (b) SOFLC-SPLC.

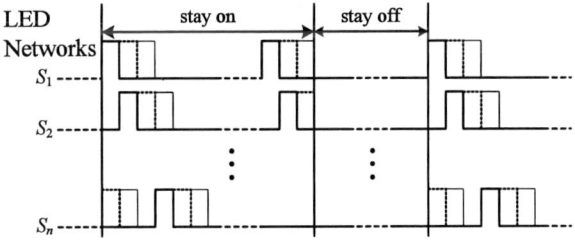

Fig. 5. LED on-off patterns.

where, d is called the delay-in-penalty. Finally, the MCU computes the change of duty ratio $\Delta D(n)$ to adapt the duty ratio ($D(n)$) for PWM register in PIC16F876A RISC MCU via

$$D(n) = D(n-1) + \Delta D(n)\qquad(6)$$

TABLE II
A LUT OF MPPT (**F**)

		$\Delta e_c(n)$										
		-5	-4	-3	-2	-1	0	1	2	3	4	5
$e_c(n)$	-5	11	11	11	11	8	5	5	5	5	5	5
	-4	8	8	8	8	6	5	5	5	5	5	5
	-3	5	6	6	5	5	5	5	5	5	5	5
	-2	0	0	0	0	2	5	5	5	5	5	5
	-1	-2	-2	-2	-2	0	2	4	5	5	5	5
	0	-5	-5	-5	-5	-2	0	2	5	5	5	5
	1	-5	-5	-5	-5	-4	-2	0	2	2	2	2
	2	-5	-5	-5	-5	-5	-5	-2	0	0	0	0
	3	-7	-8	-8	-7	-6	-5	-3	-2	-2	-2	-2
	4	-10	-9	-9	-10	-7	-5	-4	-3	-3	-3	-3
	5	-11	-11	-11	-11	-8	-5	-5	-5	-5	-5	-5

TABLE III
A LUT OF PERFORMANCE MEASUREMENT (**P**)

		$\Delta e_c(n)$										
		-5	-4	-3	-2	-1	0	1	2	3	4	5
$e_c(n)$	-5	3	3	3	3	2	1	1	1	1	1	1
	-4	2	2	2	2	1	1	1	1	1	1	1
	-3	1	1	1	1	1	1	1	1	1	1	1
	-2	0	0	0	0	1	1	1	1	1	1	1
	-1	-1	-1	-1	-1	0	0	0	1	1	1	1
	0	-1	-1	-1	-1	0	0	0	1	1	1	1
	1	-1	-1	-1	-1	0	0	0	1	1	1	1
	2	-1	-1	-1	-1	-1	-1	-1	0	0	0	0
	3	-1	-1	-1	-1	-1	-1	-1	-1	-1	-1	-1
	4	-2	-2	-2	-2	-1	-1	-1	-1	-1	-1	-1
	5	-3	-3	-3	-2	-2	-1	-1	-1	-1	-1	-1

IV. EXPERIMENTAL RESULTS

For performance evaluation, the MPPT software was developed using C-programming language and was programmed into a low-cost PIC16F876A RISC MCU [13]. Our experiments revealed that the control unit including boost converter consumes only 35mW. The power consumption of the hardware prototypes are given in Table IV. Using the solar array whose characteristics is shown in Table V, tracking performances of MPPT using FLC and the proposed SOFLC were examined. Comparative memory usage was found for realization of MPPT using the conventional FLC and proposed SOFLC as shown in Table VI.

In the experiments, a light source was turned on at t=0s and tracked powers from both controllers were observed using a digital storage oscilloscope. A sampling interval for both algorithms was selected to be 50ms. Fig. 6 (a) and (b) illustrate the tracked powers obtained from FLC and the proposed SOFLC, respectively. With reference to the maximum power point at 5.78W, it may be seen that while the rise time of tracked power of the conventional FLC is approximately 0.18s, the rise time achieved from our proposed SOFLC is only 0.15s, approximately. That is tracking speed is improved by 16%, approximately.

As regards the overall efficiency of MPPT using SOFLC, a practical scenario where light intensity varies with time was assumed. Specifically, light intensity was set at 45mW/cm^2, 85mW/cm^2 and 30mW/cm^2 from 0s to 2s, 2s to 4s, and 4s onwards, respectively at 35°C tem-

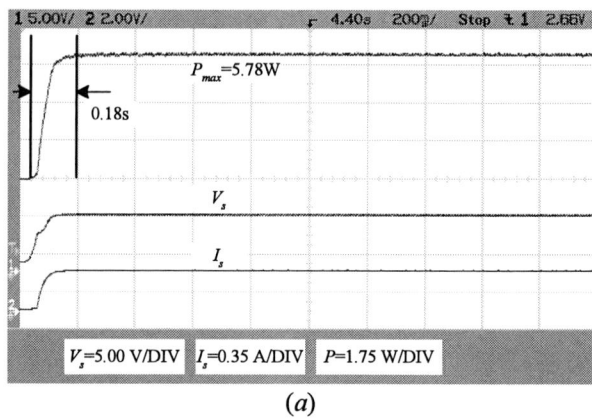

(a)

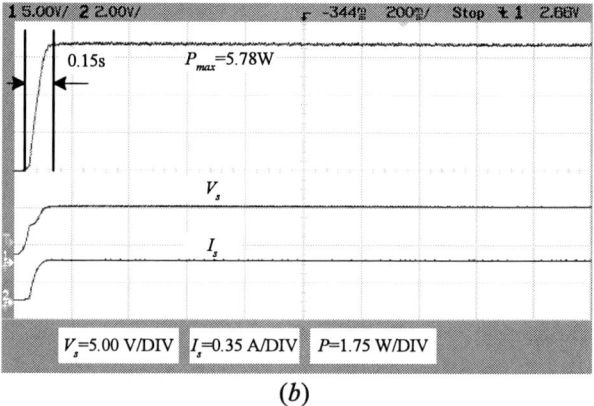

(b)

Fig. 6. Tracked power from MPPT using (a) FLC and (b) the proposed SOFLC under constant light intensity.

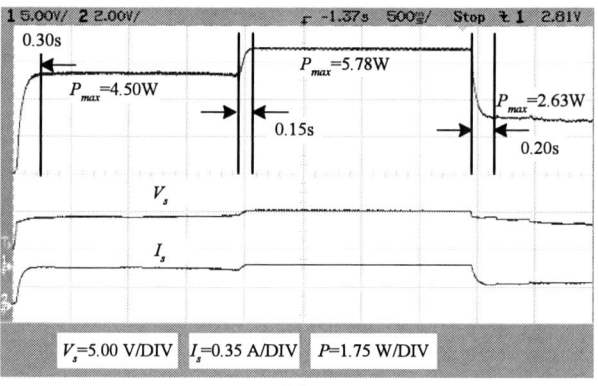

(a)

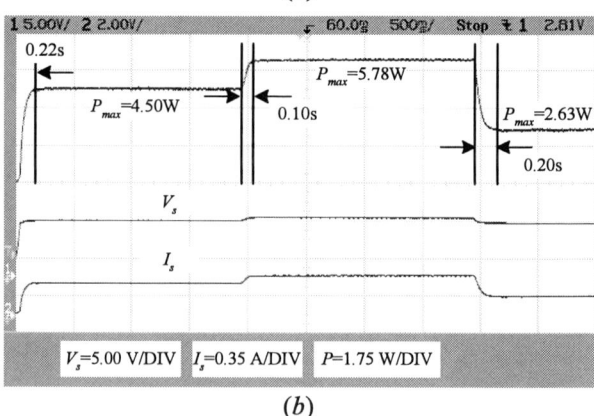

(b)

Fig. 7. Tracked power from MPPT using (a) FLC and (b) the proposed SOFLC under varied light intensity.

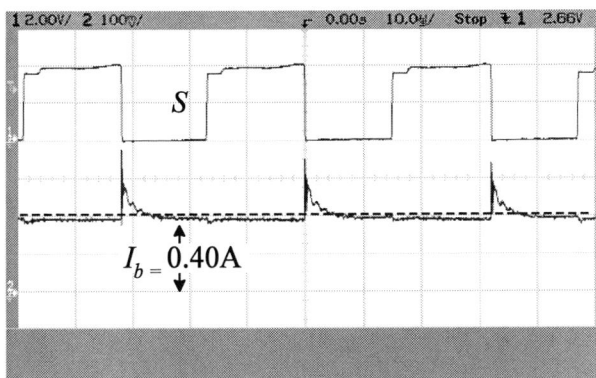

$I_b = 0.40A$

Fig. 8. Battery current and controlled switch signal.

TABLE IV
THE POWER CONSUMPTION OF THE HARDWARE PROTOTYPES

Equipment	Power Consumption	Light Density Generator
SOFLC-SPLF	2.00W	2,200LUX
SOFLC-SPLC	2.20W	2,500LUX

TABLE V
SOLAR ARRAY SPECIFICATION (25°C, 100mW/cm²)

Parameters	Definitions
Maximum power, P_{max}	6.70W
Voltage at Maximum power point, V_m	7.90V
Current at Maximum power point, I_m	0.81A
Short circuit current, I_{sc}	0.91A
Open circuit voltage, V_{oc}	9.70V
Model size	275×270×26mm

TABLE VI
COMPARISON OF PROGRAM MEMORY CONSUMPTION FOR MPPT ALGORITHM

MPPT algorithm	Percentage of Memory Consumption	
	RAM	ROM
FLC	13%	43%
The proposed SOFLC	16%	47%

perature. The dynamic environment resulted in three maximum power points of 4.50W, 5.78W, 2.63W, respectively. Fig 7 (a) and (b) depict the obtained power, which compare between FLC and SOFLC. From these results, it is seen that the transient behaviour of MPPT is improved with our proposed SOFLC. In particular, a settling time is 0.30s, 0.15s and 0.20s for FLC, while the settling time of the SOFLC reduces to 0.22s, 0.10s and 0.20s.

In order to verify the efficiency of the overall system, the steady state power received by the battery was measured. The DC-link current flowing through the battery at steady state is shown in Fig. 8. It is observed that the battery voltage, V_{bat}, measured after the boost converter is 12.50V, and the battery current is 0.40A, approximately. Hence, the output power fed into the battery becomes 5.0W. This means the overall efficiency of the proposed system is more than 92%, so that the output power solar array is 5.78W.

V. CONCLUSIONS

In this paper, the development of solar-powered traffic light equipment (SPTLE), the solar-powered light flasher (SPLF) and the solar-powered light Chevron (SPLC), using self-organizing fuzzy logic controller (SOFLC) with maximum power point tracking (MPPT) has been presented. The proposed SOFLC has been successfully implemented on a low-cost PIC16F876 RISC-microcontroller-based system. The experimental results clearly show that the proposed SOFLC offers significantly faster tracking speed than fuzzy logic controller (FLC) under comparable resources even operating in a dynamic environment. Furthermore, the efficiency of the whole system is well above 92%. Then, the SOFLC-SPLE and SOFLC-SPLC realize upon our proposed controller suitable for real life applications.

REFERENCES

[1] T. Markvart, *Solar Electricity*, John Wiley & Sons, 1994.
[2] E. Koutroulis, K. Klaitzakis and N.C. Voulgaris, "Development of Microcontroller-Based Photovoltaic Maximum Power Point Tracking Control System," *Trans. on IEEE Power Electronics*, Vol. 16, No. 1, pp. 46-54, 2001.
[3] N. Khaehintung, K. Pramotung and P. Sirisuk, "RISC microcontroller built-in fuzzy logic controller for maximum power point tracking in solar-powered battery charger," *Proc. of IEEE TENCON 2004 Conf.*, pp. 637-640, 2004.
[4] N. Khaehintung, B. Tuvirat, K. Pramotung and P. Sirisuk, "A Low-Cost Solar-Powered Light-Flasher with Built-in Maximum Power Point Tracking," *Tech. Digest of International PVSEC-14*, pp. 867-868, Thailand, 2004.
[5] M.G. Simoes and N.N. Franceschetti, "A RISC-MICROCONTROLLER based Photovoltaic System for Illumination Application," *Proc. of IEEE APEC Conf.*, pp. 1151–1156, 2000.
[6] P. Petchjatuporn, W. Ngamkham, N. Khaehintung, P. Sirisuk and W. Kiranon, "A Solar-powered Battery Charger with Neural Network Maximum Power Point Tracking Implemented on a Low-Cost PIC-microcontroller," *Proc. of IEEE PEDS 2005 Conf.*, pp. 507-510, 2005.
[7] T.F. Wu, C.H. Chang and Y.K. Chen, "A Fuzzy-Logic-Controller Single-Stage Converter for PV-Powered Lighting System Applications," *Trans. on IEEE Industrial Electronics*, Vol. 47, No. 2, pp. 287-296, 2000.
[8] N. Khaehintung, K. Pramotung, B. Tuvirat and P. Sirisuk, "RISC-microcontroller built-in fuzzy logic controller of maximum power point tracking for solar-powered light-flasher applications," *Proc. of IEEE IECON 2004 Conf.*, pp. 2673-2678, 2004.
[9] N. Khaehintung and P. Sirisuk, "Implementation of Maximum Power Point Tracking using Fuzzy Logic Controller for Solar-Powered Light-Flasher Applications," *Proc. of IEEE MWSCAS 2004 Conf.*, pp. III-171-174, 2004.
[10] T. J. Procyk and E. H. Mamdani, "A linguistic self-organizing process controller," *Automatica*, Vol.15, pp.15-30, 1979.
[11] N. Khaehintung, C. Kangsajian, P. Sirisuk and A. Kunakorn. "Grid-connected Photovoltaic System with Maximum Power Point Tracking using Self-Organizing Fuzzy Logic Controller," *Proc. of IEEE PEDS 2005 Conf.*, pp. 517-521, 2005.
[12] K. Pramotung N. Khaehintung, M. Aorpimai and P. Sirisuk "RISC Implementation of Self-Organizing Fuzzy Logic Controller for a Solar-powered Battery Charger with Maximum Power Point Tracking," *Proc. of the 3rd ECTI-2006 Conf.*, pp. 741-744 , 2006.
[13] Microchip, PIC16F87XA Datasheet, 2003.

Supply-side Current Harmonics Control of Three Phase PWM Boost Rectifiers Under Distorted and Unbalanced Supply Voltage Conditions

Xinhui Wu, *Student Member, IEEE,* *Sanjib K. Panda, *Senior Member, IEEE,*
Jianxin Xu, *Senior Member, IEEE*

Department of Electrical and Computer Engineering, National University of Singapore,
4 Engineering Drive 3, Singapore 117576
email: *skpanda@ieee.org, tel: (65)-65166484

Abstract—This paper presents a hybrid current control scheme to minimize the even order harmonics at the dc link voltage and odd order harmonics in the line currents under the distorted and unbalanced supply voltage conditions. The hybrid current control scheme consists of a conventional PI and a repetitive controller (RC). Based on the mathematical model of the three-phase PWM boost rectifier in the positive and negative synchronous rotating frames, the influence of the distorted supply voltages on the line side currents has been investigated from the analytical point of view. The control task is divided into: (a) *dc-link voltage harmonics control* and (b) *line side current harmonics control*. In *voltage harmonics control*, a reference current calculation algorithm has been derived accordingly to ensure that the dc link voltage is maintained constant and the supply side power factor is kept close to unity. In *current harmonics control*, a plug-in repetitive controller is designed to achieve low THD line currents of the three phase PWM boost rectifier. The proposed analysis and hybrid current control scheme have been validated by experimental test results on a 1.6 kVA laboratory based PWM rectifier. Test results obtained confirm that the proposed control scheme enhances the performance of the PWM rectifier over the conventional one.

I. INTRODUCTION

The boost PWM rectifier has been increasingly employed for high-performance applications in recent years, because it offers the possibility of low line current distortions with unity power factor operation and constant dc-link voltage with a small output filter capacitor. However, all the advantages of the PWM boost rectifier are valid only with the assumption of balanced input supply voltage conditions. The unbalanced input supply voltages lead to the appearance of even order harmonics at the dc output and odd order harmonics in the input line currents [2]. Nevertheless, the unbalanced and distorted input supply voltage conditions occur frequently, particularly in a weak ac system. Since EN 50160 [12] standard allows 6% low order supply voltage harmonics, and supply voltage sags/swells within ±10%, the normal performance of the PWM rectifier should be ensured under such varying and unbalanced supply voltage conditions.

In order to guarantee that the three phase PWM rectifier is able to keep the dc link voltage constant and not to inject more harmonic currents into the distribution system under the distorted and unbalanced supply voltage conditions, there are two approaches feasible. One approach is to use bulky filter circuits to remove the ripples in the output voltage and input currents [8]. However, it would slow down the dynamic response of the PWM boost rectifier [1]. The other alternative is to use active control schemes to minimize the harmonics so that small size input/output filters can be used, which would improve the dynamic response of the PWM rectifier.

Extensive studies have been carried out to investigate the influence of the unbalanced input supply voltages on the output dc link voltage. Various control schemes have been reported on the performance of PWM rectifier [1]-[6]. As the even order harmonics at the dc output voltage is linked with the odd order harmonics in the ac input current by the instantaneous power [7], most of the studies tried to regulate the dc link voltage [1], [2] or the instantaneous power flow of the system [3]-[6]. Consequently, the harmonics in the ac line currents can be eliminated as a side effect. Few control schemes have been considered to directly deal with the harmonics in the ac line currents under the unbalanced input supply conditions. Therefore, when the supply voltages are distorted, those control schemes can hardly handle the current harmonics on the supply side.

In this paper, a hybrid control scheme is proposed to effectively minimize the line current harmonics and the dc link voltage harmonics under the distorted and unbalanced operating conditions. Based on the mathematical model of the three-phase PWM boost rectifier in the positive and negative synchronous rotating frames [6], the influence of the distorted supply voltages on the line side currents has been investigated from the analytical point of view and discussed in Section III. Accordingly, the control objectives can be separated as *voltage harmonics control* and *current harmonics control* and are discussed in Section IV. The algorithm of the reference current calculation is based on the *voltage harmonics control* to make sure that the dc link voltage is maintained constant and the supply side power factor is kept close to unity under the unbalanced supply voltage operating conditions. The *current harmonics control* is used to eliminate the current harmonics in the line side. A plug-in repetitive controller (RC) is designed and employed to achieve the low THD line currents of the three phase PWM boost rectifier under the distorted supply voltage conditions. All the proposed analysis and control

978-1-4244-0644-9/07/$25.00 ©2007 IEEE

scheme have been validated by experimental test results on a 1.6 kVA laboratory based PWM rectifier.

II. MODELING OF THREE PHASE BOOST PWM CONVERTER UNDER GENERAL SUPPLY VOLTAGE CONDITIONS

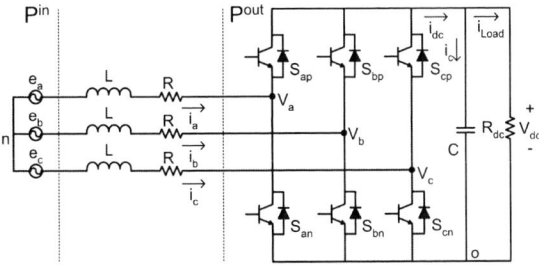

Fig. 1. Three phase PWM AC-DC boost rectifier circuit

Fig. 1 shows the power circuit configuration of a three phase PWM AC-DC boost rectifier. It is assumed that the converter is feeding a resistive load. If zero-sequence voltage is assumed to be absent, then the system can be resolved into the positive and negative rotating space vectors in the synchronous frame, respectively.

Therefore, the system can be expressed in the positive and negative rotating synchronous d-q frame as,

$$
\begin{bmatrix} i_d^p \\ i_q^p \end{bmatrix} = \begin{bmatrix} -\frac{R}{L} & \omega \\ -\omega & -\frac{R}{L} \end{bmatrix} \begin{bmatrix} i_d^p \\ i_q^p \end{bmatrix} +
$$
$$
\begin{bmatrix} \frac{1}{L} & 0 \\ 0 & \frac{1}{L} \end{bmatrix} \begin{bmatrix} e_d^p \\ e_q^p \end{bmatrix} + \begin{bmatrix} -\frac{v_{dc}}{L} & 0 \\ 0 & -\frac{v_{dc}}{L} \end{bmatrix} \begin{bmatrix} S_d^p \\ S_q^p \end{bmatrix}
$$
$$
\begin{bmatrix} i_d^n \\ i_q^n \end{bmatrix} = \begin{bmatrix} -\frac{R}{L} & -\omega \\ \omega & -\frac{R}{L} \end{bmatrix} \begin{bmatrix} i_d^n \\ i_q^n \end{bmatrix} + \quad (1)
$$
$$
\begin{bmatrix} \frac{1}{L} & 0 \\ 0 & \frac{1}{L} \end{bmatrix} \begin{bmatrix} e_d^n \\ e_q^n \end{bmatrix} + \begin{bmatrix} -\frac{v_{dc}}{L} & 0 \\ 0 & -\frac{v_{dc}}{L} \end{bmatrix} \begin{bmatrix} S_d^n \\ S_q^n \end{bmatrix}
$$

where the variables e, i and S represent the supply voltage, current and average switching function, respectively. Superscript p and n designate the positive-sequence and the negative-sequence, respectively, and subscripts α and β represent the variables in the stationary α-β frame, and d and q represent the same in the rotating d-q frame.

The dc side current expression can be derived by expressing i_α, i_β, S_α and S_β into the rotating synchronous d-q frame and defining ωt as θ:

$$
\begin{aligned}
i_{dc} &= i_\alpha S_\alpha + i_\beta S_\beta \\
&= (i_d^n \cos 2\theta + i_q^n \sin 2\theta + i_d^p) S_d^p \\
&+ (i_q^n \cos 2\theta - i_d^n \sin 2\theta + i_q^p) S_q^p \\
&+ (i_d^p \cos 2\theta - i_q^p \sin 2\theta + i_d^n) S_d^n \\
&+ (i_d^p \sin 2\theta + i_q^p \cos 2\theta + i_q^n) S_q^n
\end{aligned} \quad (2)
$$

Finally, the dynamics of the dc link voltage [6] can be presented as:

$$
\begin{aligned}
C\dot{v}_{dc} &= -\frac{1}{R_{dc}} v_{dc} + i_{dc} \\
&= -\frac{v_{dc}}{R_{dc}} + \begin{bmatrix} i_d^n \cos 2\theta + i_q^n \sin 2\theta + i_d^p \\ i_q^n \cos 2\theta - i_d^n \sin 2\theta + i_q^p \\ i_d^p \cos 2\theta - i_q^p \sin 2\theta + i_d^n \\ i_d^p \sin 2\theta + i_q^p \cos 2\theta + i_q^n \end{bmatrix}' \begin{bmatrix} S_d^p \\ S_q^p \\ S_d^n \\ S_q^n \end{bmatrix}
\end{aligned} \quad (3)
$$

Eqn.(3) gives the complete picture about the dc link voltage variation of the PWM rectifier. When it works under the ideal balanced operating condition, v_{dc} only contains the average dc quantity with all the negative sequence components being maintained at zero. While under the distorted and unbalanced operating conditions, it leads to the presence of the harmonic components and the negative sequence components both in currents as well as average switching function. Correspondingly, these components cause the even order harmonic components at the output of the rectifier as can be seen from the second term in (3).

III. INFLUENCE OF DISTORTED SUPPLY VOLTAGES

In the power distribution networks, voltage distortions due to current harmonics is becoming a major issue because of the large numbers of nonlinear loads. Electrical power converters such as variable speed drives, SCR drives, etc., are the largest contributors to harmonic distortions. It is not uncommon to have current THD levels as high as 25% within some industrial settings [13].

The three phase balanced supply voltages which include kth-order harmonics with an amplitude E_k and angle offset θ_k can be represented as,

$$
\begin{aligned}
e_a &= \sum_{\substack{n=1 \\ k=1,6n-1,6n+1}}^{\infty} E_k \sin(k\omega t + \theta_k) \\
e_b &= \sum_{\substack{n=1 \\ k=1,6n-1,6n+1}}^{\infty} E_k \sin(k\omega t - k \cdot \tfrac{2\pi}{3} + \theta_k) \quad (4) \\
e_c &= \sum_{\substack{n=1 \\ k=1,6n-1,6n+1}}^{\infty} E_k \sin(k\omega t + k \cdot \tfrac{2\pi}{3} + \theta_k)
\end{aligned}
$$

According to the traditional method in [9] to extract the symmetrical components, the balanced distorted three phase voltages e_a, e_b and e_c appear as positive sequence component. The negative sequence component will not be affected by the supply voltage harmonic components.

By using the *Park's transformation*, the supply voltages can be expressed as $e_{\alpha k}^p$ and $e_{\beta k}^p$ for the different kth-order harmonics in the stationary reference frame, α-β, when **k = 6n-1**,

$$
e_{\alpha k}^p = \frac{\sqrt{6} E_k}{2} \sin(k\omega t + \theta_k) \quad (5)
$$

$$
e_{\beta k}^p = \frac{\sqrt{6} E_k}{2} \cos(k\omega t + \theta_k) \quad (6)
$$

853

Similarly, when **k = 1** and **k = 6n+1**,

$$e^p_{\alpha k} = \frac{\sqrt{6}E_k}{2}\sin(k\omega t + \theta_k) \tag{7}$$

$$e^p_{\beta k} = -\frac{\sqrt{6}E_k}{2}\cos(k\omega t + \theta_k) \tag{8}$$

It can be seen from (5)-(8) that the different voltage harmonics in the stationary *a-b-c* frame will cause the corresponding harmonics in the stationary α-β frame. Moreover, for $(6n-1)$th order harmonics such as 5th and 11th order harmonics, $e^p_{\alpha k}$ is lagging with $e^p_{\beta k}$ by 90°, while for $(6n+1)$th order harmonics such as 7th and 13th, $e^p_{\alpha k}$ is leading with $e^p_{\beta k}$ by 90°.

By using *Clark's Transformation*, the supply voltages can be resolved as e^p_{dk} and e^p_{qk} in the rotating *d-q* frame, when **k = 1**,

$$e^p_{d1} = \frac{\sqrt{6}E_1}{2}\sin\theta_1 \tag{9}$$

$$e^p_{q1} = -\frac{\sqrt{6}E_1}{2}\cos\theta_1 \tag{10}$$

k = 6n-1,

$$e^p_{dk} = \frac{\sqrt{6}E_k}{2}\sin[(k+1)\omega t + \theta_k]$$

$$= \frac{\sqrt{6}E_k}{2}\sin(6n\omega t + \theta_k) \tag{11}$$

$$e^p_{qk} = \frac{\sqrt{6}E_k}{2}\cos[(k+1)\omega t + \theta_k]$$

$$= \frac{\sqrt{6}E_k}{2}\cos(6n\omega t + \theta_k) \tag{12}$$

k = 6n+1,

$$e^p_{dk} = \frac{\sqrt{6}E_k}{2}\sin[(k-1)\omega t + \theta_k]$$

$$= \frac{\sqrt{6}E_k}{2}\sin(6n\omega t + \theta_k) \tag{13}$$

$$e^p_{qk} = -\frac{\sqrt{6}E_k}{2}\cos[(k-1)\omega t + \theta_k]$$

$$= -\frac{\sqrt{6}E_k}{2}\cos(6n\omega t + \theta_k) \tag{14}$$

Eqns.(9)-(14) show that the fundamental voltage will appear as a constant signal in the *d-q* frame, while the other voltage harmonics will result in the corresponding ac signals in the rotatory *d-q* frame. For a specified *n*, the $(6n-1)$th and $(6n+1)$th order harmonics will produce the $6n$th order harmonics in the *d-q* frame. For example, 5th and 7th order voltage harmonics in *a-b-c* frame appear as 6th order harmonics in the *d-q* frame, and similarly the 11th and 13th order voltage harmonics appear as 12th order harmonics in the *d-q* frame.

Use **k = 6n-1** as a example, the space vector of the positive sequence supply voltages $\overrightarrow{e^p_k}$ in the rotating synchronous frame can be defined as,

$$\overrightarrow{e^p_k} = e^p_{qk} + je^p_{dk} = E_k e^{j(6n\omega t + \theta_k)} \tag{15}$$

Since the currents caused by the supply voltage harmonics are small in comparison with the fundamental line currents, and also the operating value of the dc link voltage is relatively high, so the variation of the dc link voltage can be neglected according to (3). Based on the assumption that v_{dc} is constant, then (1) becomes a set of linear equations. The equations (1) in the positive and negative rotating synchronous *d-q* frame can be expressed in the space vector format,

$$\dot{\overrightarrow{i^p}} = (-\frac{R}{L} - j\omega)\overrightarrow{i^p} + \frac{1}{L}\overrightarrow{e^p} - \frac{v_{dc}}{L}\overrightarrow{S^p}$$

$$\dot{\overrightarrow{i^n}} = (-\frac{R}{L} + j\omega)\overrightarrow{i^n} + \frac{1}{L}\overrightarrow{e^n} - \frac{v_{dc}}{L}\overrightarrow{S^n} \tag{16}$$

The current response in the frequency domain can be derived as,

$$\overrightarrow{i^p}(s) = \frac{1}{Ls - (-R - jL\omega)}(\overrightarrow{e^p}(s) - v_{dc}\overrightarrow{S^p}(s))$$

$$\overrightarrow{i^n}(s) = \frac{1}{Ls - (-R + jL\omega)}(\overrightarrow{e^n}(s) - v_{dc}\overrightarrow{S^n}(s)) \tag{17}$$

Because of the linear characteristics of (1), the effect of each frequency component on the system can be considered separately. For *k*th-order harmonics, the supply voltage in the frequency domain can be obtained from (15),

$$\overrightarrow{e^p_k}(s) = E_k e^{j\theta_k}\frac{1}{s - j6n\omega} \tag{18}$$

The control signal $\overrightarrow{S^p_k}$ can be designed to compensate the effect of supply voltage harmonic $\overrightarrow{e^p_k}$, which will be discussed in Section IV. When the system works in open loop, $\overrightarrow{S^p_k} = 0$, from (17) and (18), the corresponding positive sequence current response in the time domain is,

$$\overrightarrow{i^p_k}(t) = Ae^{j6n\omega t} - Ae^{-\frac{R}{L}t - j\omega t)} \tag{19}$$

where $A = \frac{E_k e^{j\theta_k}}{R + j6nL\omega + jL\omega}$

From (19), we can see that the first term of (19) will decay exponentially to zero and only the second term will remain at the steady state. Moreover, when the *k*th order harmonics appear in the supply voltage, the corresponding $6n$ order harmonics will generate in the currents of the rotating synchronous *d-q* frame. Consequently, this $6n$ order harmonics will reflect as the *k*th order harmonics in the *a-b-c* frame. Also, the amplitude of the positive sequence current is decided by the impedance of the line side. When the frequency of the supply voltage is higher, the amplitude of the corresponding current is smaller. Therefore, the high frequency harmonics effect can be neglected because of high impedance, while the low frequency harmonics should be paid more attention. When the system works under the distorted and unbalanced operating conditions, the negative sequence currents can be analyzed in the same way.

IV. CONTROL STRATEGY

The control objective to operate the three phase PWM rectifiers under the distorted and unbalanced supply voltage conditions can be concluded to eliminate the even-order harmonics at the dc link voltage and odd-order harmonics in the line currents. The control signals S_d^p, S_q^p, S_d^n and S_q^n can be divided into two parts, *voltage harmonics control signals*, S_{dv}^p, S_{qv}^p, S_{dv}^n and S_{qv}^n, those are used to take care of the even-order harmonics at the dc link voltage, and *current harmonics control signals*, S_{di}^p, S_{qi}^p, S_{di}^n and S_{qi}^n, those are supposed to eliminate the odd-order harmonics in the line currents.

Substituting the control signals, the PWM rectifier model in (1) and (3) under the distorted and unbalanced supply voltage can be rewritten as,

$$
\begin{bmatrix} \dot{i}_d^p \\ \dot{i}_q^p \end{bmatrix} = \begin{bmatrix} -\frac{R}{L} & \omega \\ -\omega & -\frac{R}{L} \end{bmatrix} \begin{bmatrix} i_d^p \\ i_q^p \end{bmatrix} +
$$
$$
\begin{bmatrix} \frac{1}{L} & 0 \\ 0 & \frac{1}{L} \end{bmatrix} \begin{bmatrix} e_{d1}^p \\ e_{q1}^p \end{bmatrix} + \begin{bmatrix} -\frac{v_{dc}}{L} & 0 \\ 0 & -\frac{v_{dc}}{L} \end{bmatrix} \begin{bmatrix} S_{dv}^p \\ S_{qv}^p \end{bmatrix} +
$$
$$
\begin{bmatrix} \frac{1}{L} & 0 \\ 0 & \frac{1}{L} \end{bmatrix} \begin{bmatrix} \sum_{k\neq1} e_{dk}^p \\ \sum_{k\neq1} e_{qk}^p \end{bmatrix} + \begin{bmatrix} -\frac{v_{dc}}{L} & 0 \\ 0 & -\frac{v_{dc}}{L} \end{bmatrix} \begin{bmatrix} S_{di}^p \\ S_{qi}^p \end{bmatrix}
$$
$$
\begin{bmatrix} \dot{i}_d^n \\ \dot{i}_q^n \end{bmatrix} = \begin{bmatrix} -\frac{R}{L} & -\omega \\ \omega & -\frac{R}{L} \end{bmatrix} \begin{bmatrix} i_d^n \\ i_q^n \end{bmatrix} +
$$
$$
\begin{bmatrix} \frac{1}{L} & 0 \\ 0 & \frac{1}{L} \end{bmatrix} \begin{bmatrix} e_{d1}^n \\ e_{q1}^n \end{bmatrix} + \begin{bmatrix} -\frac{v_{dc}}{L} & 0 \\ 0 & -\frac{v_{dc}}{L} \end{bmatrix} \begin{bmatrix} S_{dv}^n \\ S_{qv}^n \end{bmatrix} +
$$
$$
\begin{bmatrix} \frac{1}{L} & 0 \\ 0 & \frac{1}{L} \end{bmatrix} \begin{bmatrix} \sum_{k\neq1} e_{dk}^n \\ \sum_{k\neq1} e_{qk}^n \end{bmatrix} + \begin{bmatrix} -\frac{v_{dc}}{L} & 0 \\ 0 & -\frac{v_{dc}}{L} \end{bmatrix} \begin{bmatrix} S_{di}^n \\ S_{qi}^n \end{bmatrix} \tag{20}
$$

$$
C\dot{v}_{dc} + \frac{v_{dc}}{R_{dc}} =
$$
$$
\begin{bmatrix} i_d^n \cos 2\theta + i_q^n \sin 2\theta + i_d^p \\ i_q^n \cos 2\theta - i_d^n \sin 2\theta + i_q^p \\ i_d^p \cos 2\theta - i_q^p \sin 2\theta + i_d^n \\ i_d^p \sin 2\theta + i_q^p \cos 2\theta + i_q^n \end{bmatrix}' \begin{bmatrix} S_{dv}^p + S_{di}^p \\ S_{qv}^p + S_{qi}^p \\ S_{dv}^n + S_{di}^n \\ S_{qv}^n + S_{qi}^n \end{bmatrix} \tag{21}
$$

In order to get rid of the odd order harmonics and only keep the fundamental waveform in the ac line currents, the currents in the rotating synchronous frame, i_d^p, i_q^p, i_d^n, i_q^n should be regulated to a fixed certain value. Therefore, the effect from the supply voltage harmonics must be compensated by current harmonics control signals, S_{di}^p, S_{qi}^p, S_{di}^n and S_{qi}^n, in (20). The other part of control signals, S_{dv}^p, S_{qv}^p, S_{dv}^n and S_{qv}^n, is constant to ensure the required line currents. Since the dc link voltage v_{dc} is much higher than the supply voltage harmonics, current harmonics control signals S_{di}^p, S_{qi}^p, S_{di}^n and S_{qi}^n are so small that the effect of this part control signals can be neglected in (21). We can see from (21) that voltage harmonics control signals, S_{dv}^p, S_{qv}^p, S_{dv}^n and S_{qv}^n, are responsible to eliminate the even order harmonics at the dc link voltage.

A. DC Link Voltage Harmonics Control

In the steady-state condition, the variables at the dc side can be replaced by the sum of dc and ac signals as $v_{dc} = \overline{v_{dc}} + \Delta v_{dc}$ and $i_{dc} = \overline{i_{dc}} + \Delta i_{dc}$. The dc signals $\overline{v_{dc}}$

and $\overline{i_{dc}}$ are represented as the average value of the dc side voltage and current. Only considering the contribution from the voltage harmonics control signals, S_{dv}^p, S_{qv}^p, S_{dv}^n and S_{qv}^n in (21), because the currents i_d^p, i_q^p, i_d^n, i_q^n are expected to be constant, the differential equation (21) can be solved. Finally, by neglecting the exponentially decaying term, the ac component of the dc link voltage can be derived as (22) according to [6],

$$
\Delta V_{dc} = \frac{R_{dc}\sqrt{M^2 + N^2}}{\sqrt{1 + 4\omega^2 C^2 R_{dc}^2}} \sin(2\omega t + \phi_v) \tag{22}
$$

where
$$
M = -i_d^p S_{qv}^n + i_q^p S_{dv}^n + i_d^n S_{qv}^p - i_q^n S_{dv}^p
$$
$$
N = i_d^p S_{dv}^n + i_q^p S_{qv}^n + i_d^n S_{dv}^p + i_q^n S_{dv}^p
$$

The amplitude of the second order harmonic term ΔV_{dc} is decided by the value of M and N. Therefore, only when the conditions, $M = 0$ and $N = 0$, are satisfied, the ac component of the dc link voltage is eliminated and a constant dc output voltage can be maintained. In the mean time, the input average active power P_o^{in} from the fundamental supply voltage determines the dc link voltage level and the average input reactive power Q_o^{in} decides the input power factor.

Therefore, the control laws can be expressed by a set of four linear equations in a matrix form as the following [6],

$$
\begin{bmatrix} P_o^{in} \\ Q_o^{in} \\ M \\ N \end{bmatrix} = \begin{bmatrix} P_o^{out} + P_{loss} \\ k_{pf} P_o^{in} \\ 0 \\ 0 \end{bmatrix}
$$
$$
= \begin{bmatrix} E_{d1}^p & E_{q1}^p & E_{d1}^n & E_{q1}^n \\ E_{q1}^p & -E_{d1}^p & -E_{q1}^n & E_{d1}^n \\ S_{qv}^n & -S_{dv}^n & -S_{qv}^p & S_{dv}^p \\ S_{dv}^n & S_{qv}^n & S_{dv}^p & S_{qv}^p \end{bmatrix} \begin{bmatrix} I_d^p \\ I_q^p \\ I_d^n \\ I_q^n \end{bmatrix} \tag{23}
$$

where $k_{pf} = \frac{\sqrt{1-pf^2}}{pf}$, during the unity power factor operation, k_{pf} is equal to zero, and S_{dv}^p, S_{qv}^p, S_{dv}^n and S_{qv}^n are the switching signals obtained in the previous step.

The whole control block diagram of the cascaded dual frame current regulator with a voltage regulator is shown in Fig. 2. The reference input average active power P_o^{in} can be obtained from the outer voltage loop. Then the estimation of the input average active power can be used in the reference current calculation according to (23). The current control calculation algorithm is implemented in the current reference calculating block of the control scheme to provide the four reference commands (I_d^{p*}, I_q^{p*}, I_d^{n*}, I_q^{n*}) for the inner current control loop. The inner loop is made up of two parallel positive and negative sequence d-q synchronous frame current regulators.

B. Line Side Current Harmonics Control

In practice, 5th and 7th order voltage harmonics are dominant in the power distribution network. As discussed in Section III, 5th and 7th order voltage harmonics in a-b-c frame cause 6th order current harmonics in the d-q

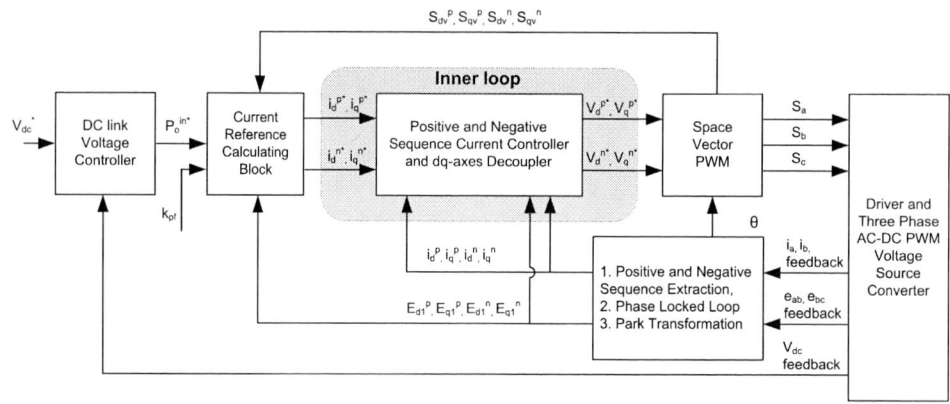

Fig. 2. Cascaded control block diagram of three phase PWM AC-DC boost rectifier

frame. The input current harmonics control is mainly to eliminate the 6th order harmonics in the d-q frame line currents. Therefore, the repetitive control (RC) method [10] is employed to regulate the currents I_d^p, I_q^p, I_d^n and I_q^n and remove the periodic errors.

The idea of repetitive control is to use the information of the preceding cycle to improve the control performance in the present cycle. The control law of the RC is given by,

$$u_{k+1}(z) = u_k(z) + K_{rc}e_{k+1}(z) \qquad (24)$$

where $u_{k+1}(z)$ and $u_k(z)$ are z-transforms of the control signals from the repetitive control during cycle $k+1$ and k, respectively. K_{rc} is the learning gain, and e_{k+1} is the tracking error during cycle $k+1$.

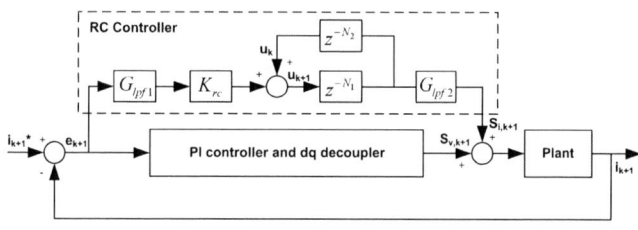

Fig. 3. Block diagram of PI controller with a plug-in type repetitive controller

The proposed current control scheme consists of the conventional PI controller together with a plug-in type repetitive controller for PWM rectifier as shown in Fig. 3. The transfer function of the plug-in repetitive controller $G_{rc}(z)$ is

$$G_{rc}(z) = G_{lpf1}(z)G_{lpf2}(z)\frac{K_{rc}z^{-N_1}}{1 - z^{-N}} \qquad (25)$$

where N is the number of the samples for each cycle, which is equal to $N_1 + N_2$. $G_{lpf1}(z)$ and $G_{lpf2}(z)$ are the low-pass filters to reject the noise. N_1 is used to compensate the phase lag caused by the two low-pass filters.

In this hybrid control scheme, the linear PI controller provides majority of the control efforts during the transient stage. Once the system enters the steady state, the repetitive controller takes over the control effort and tries to reduce the steady-state tracking error from cycle to cycle. Therefore, the

6th order harmonics in the currents I_d^p, I_q^p, I_d^n and I_q^n can be removed.

V. EXPERIMENT RESULTS

To verify the feasibility of the proposed control scheme, the experiments were conducted under the unbalanced operating conditions. The complete control scheme was implemented on a dSPACE platform (DS1104). The DSP processor TI TMS320F240 was employed with 10 kHz sampling rate in the control board. The IGBT module adopted the symmetric PWM modulation with 20 kHz switching frequency from the control signals. The system parameters used in the experiment platform are summarized in Table I.

TABLE I
PARAMETERS USED IN THE EXPERIMENT

Parameters	Value	Parameters	Value
R	0.3 Ω	R_{dc}	225 Ω
L	0.005 H	C	200 μF
E_{rms}	80 V	$V_{dc,ref}$	400 V
K_{pc}	4	K_{ic}	200
K_{pv}	0.03	K_{iv}	4.59

Fig. 4 and Fig. 5 show the performance of PWM rectifier before and after using the proposed hybrid control scheme when 10% of the 7th order harmonic voltage is injected in the supply voltage. Without using the repetitive control, Fig. 4 gives the picture of the dc link voltage and ac line currents of the system. The distorted supply voltages cause 5th order harmonics and 7th order harmonics in the ac line currents to be 5% and 21%, respectively. The total harmonics distortion (THD) of the phase a input current is 22.68%. When the plug-in repetitive controller is activated, the current harmonics are reduced significantly as can be seen in Fig. 5. From the frequency spectra of the harmonics in the ac line currents, we can see that the 5th order harmonic component decreases from 5% to 1.02% and 7th order harmonics is from 21% to 1.13%. And the THD of the phase input current is reduced to 5.53%.

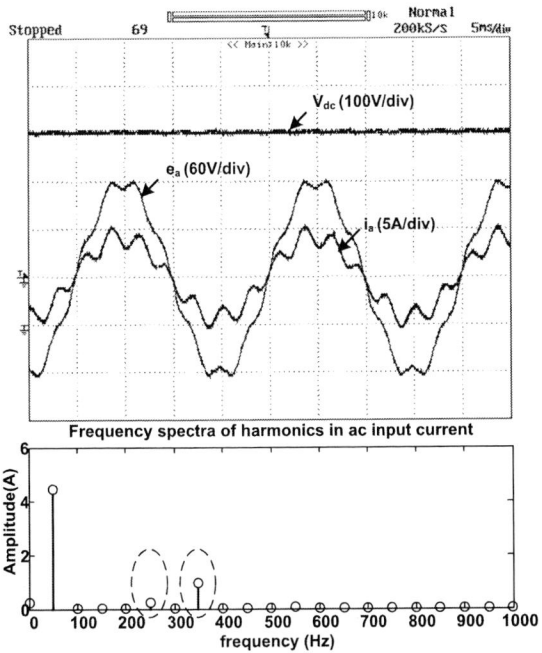

Fig. 4. Experiment results of dc link voltage and ac line current before using the repetitive controller

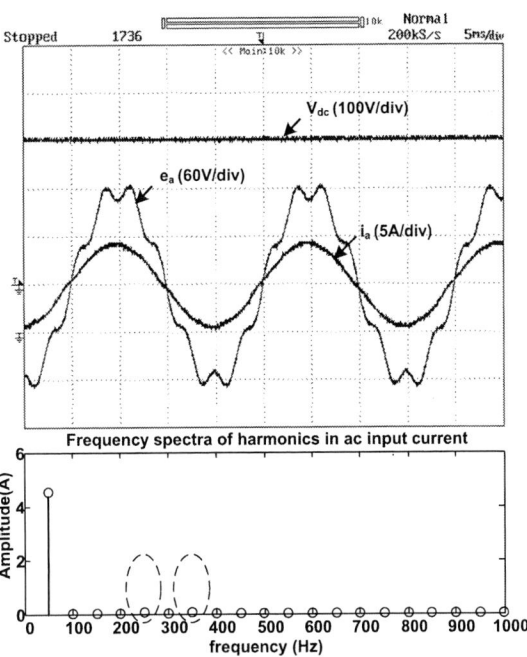

Fig. 5. Experiment results of dc link voltage and ac line current after using the repetitive controller

Therefore, the plug-in repetitive controller can minimize the harmonics in the supply current side and provide better performance under the distorted supply voltage operating conditions.

VI. CONCLUSIONS

This paper proposes a hybrid current control scheme, consisting of a conventional PI controller together with a plug-in repetitive controller, to effectively minimize the even order harmonics at the dc link voltage and the odd order harmonics in the line current under the distorted and unbalanced supply voltage operating conditions. Based on the analysis of the distorted supply voltage, the predominant voltage harmonics, 5th and 7th order harmonics result in the same frequency, 6th order harmonics on the line side currents in the rotating synchronous d-q frame. Therefore, under the distorted and unbalanced supply voltage operating conditions, the control task can be divided into *voltage harmonics control* and *current harmonics control*. The algorithm of the reference current calculation is based on *voltage harmonics control* to make sure that the dc link voltage is maintained constant and the supply side power factor is kept close to unity. A plug-in repetitive controller (RC), as *current harmonics control*, is designed and employed to improve the line current waveform of the three phase PWM boost rectifier under the distorted supply voltage conditions. The hybrid control scheme can eliminate the harmonics in the line current while maintaining the dc link voltage constant.

REFERENCES

[1] P.N.Enjeti and S.A.Choudhury, "A new control strategy to improve the performance of a PWM AC to DC Converter under unbalanced

operating conditions," *IEEE Trans. Power Electronics*, Vol. 8, pp. 493-500, 1993.

[2] P.Rioual, H.Pouliquen and J.P.Louis, "Regulation of a PWM rectifier in the unbalanced network state using a generalized model," *IEEE Trans. Power Electron.*, Vol. 11, pp. 495-502, May, 1996.

[3] H.Song and K.Nam, "Dual current control scheme for PWM converter under unbalanced input voltage conditions," *IEEE Trans. Ind. Electron.*, Vol. 46, pp. 953-959, Oct, 1999.

[4] Y.Suh, V.Tijeras and T.A.Lipo, "A nonlinear control of the instantaneous power in dq synchronous frame for PWM ac/dc converter under generalized unbalanced operating conditions," *Proc. IEEE-IAS Annual Meeting*, Vol. 2, pp. 1189 - 1196, Oct. 2002

[5] A.V.Stankovic and T.A.Lipo, "A generalized control method for input simultaneous unbalanced input voltages and input impedances," *Proc. IEEE Conf. Rec. PESC 2001*, Vol. 3, pp. 1309-1314, 2001.

[6] X.H.Wu, S.K.Panda and J.X.Xu, "Development of a new mathematical model of three phase PWM boost rectifier under unbalanced supply voltage operating conditions," *Proc. IEEE Conf. Rec. PESC 2006*, Vol. 2, pp. 1391-1398, April. 2006

[7] X.H.Wu, S.K.Panda and J.X.Xu, "Analysis and control of the output instantaneous power for three phase PWM boost rectifier under unbalanced supply voltage conditions," *Proc. IEEE Conf. Rec. IECON 2006*, Vol. 2, pp. Nov. 2006

[8] L.Mihalanche, "A high performance DSP controller for three-phase PWM rectifiers with ultra low input current THD under unbalanced and distorted input voltage," *Proc. IEEE IAS annual meeting 2005.*, vol.1, pp. 138-144, Jan, 2005.

[9] D.Vincenti and Hua Jin, "A three-phase regulated PWM rectifier with on-line feedforward input unbalance correction," *IEEE Trans. Industrial Electron.*, Vol. 41, pp. 526-532, Oct, 1994.

[10] K.Zhou and D.Wang, "Digital repetitive controlled three-phase PWM rectifier," *IEEE Trans. Power Electron.*, Vol. 18, pp. 309-316, Jan, 2003.

[11] IEEE Std 1459-2000 IEEE Trial-Use Standard Definitions for the Measurement of Electric Power Quantities Under Sinusoidal, Nonsinusoidal, Balanced, or Unbalanced Conditions, 2000

[12] European standard EN-50160 Voltage characteristics of electricity supplied by public distribution systems, CENELEC, Brussels, Belgium, 1994

[13] M.H.J.Bollen Understanding Power Quality Problem: Voltage Sags and Interruptions, IEEE Press, New York, 1999

A Two-stage Converter with a Coupled-Inductor

Hirotaka Nakanishi*, Yoshihiro Tomihisa*, Terukazu Sato*

Takashi Nabeshima*, Kimihiro Nishijima*, and Tadao Nakano*
* Oita University
700.Dannoharu.Oita.Japan 870-1192

Abstract—In this paper, a novel multi-stage converter with a coupled-inductor is presented in order to reduce the ripple currents in the switches and the output capacitor. The operating principle is described in detail and the steady state analysis is performed for a 12V to 1.5V prototype buck converter. From the experiments, it is confirmed that ripple currents in switches are greatly reduced and power efficiency is improved compared with the conventional multiphase converter. Furthermore, from the small signal analysis, it is found that the transfer function of the duty ratio of the first stage to the output voltage of the second stage becomes nearly quadratic similar to the single buck converter even though composed of two cascaded stages, so the converter with feedback circuit is inherently stable.

Index Terms-- coupled-inductor, two-stage converter, transfer function, ripple current

I. INTRODUCTION

In recent years, multiphase converters with coupled inductor have been widely used for the electronic devices that require fast response for large load current change such as CPUs. However, since the multiphase converter has more than two converters in parallel, the current unbalance among them has been one of the significant issues to be resolved. At the same time, the development of the DC-DC converter having a large conversion ratio is aspired to improved power efficiency [1-6].

In this paper, a novel multi-stage converter with a coupled-inductor is presented in order to reduce the ripple currents in the switches and the output capacitor. The operating principle is described in detail and the steady state analysis is performed for a 12V to 1.5V prototype buck converter. From the experiments, it is confirmed that ripple currents in switches are greatly reduced and power efficiency is improved compared with the conventional multiphase converter. Furthermore, from the small signal analysis, it is found that the transfer function of the duty ratio of the first stage to the output voltage of the second stage becomes nearly quadratic similar to the single buck converter even though composed of two cascaded stages, so the converter with feedback circuit is inherently stable.

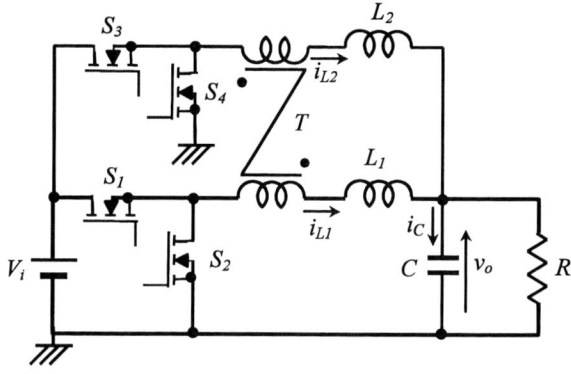

Fig.1. Conventional two-phase buck converter with coupled inductor.

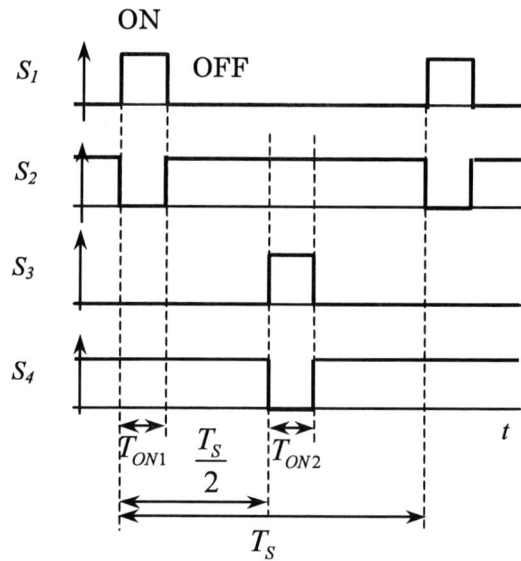

Fig.2. States of the switches.

II. CIRCUIT AND OPERATION

A. Conventional Multiphase Converter with a Coupled Inductor

For the sake of simplicity, consider the two-stage converter with a coupled inductor as shown Fig.1. The switches S_1, S_2, S_3 and S_4 are driven as shown in Fig.2. As seen in the figure it has two converters and the inductor of each converter is coupled together so that the ripple currents in the switches and inductors should be reduced. Thus, by means of coupling the inductors, ripple

978-1-4244-0644-9/07/$25.00 ©2007 IEEE   858

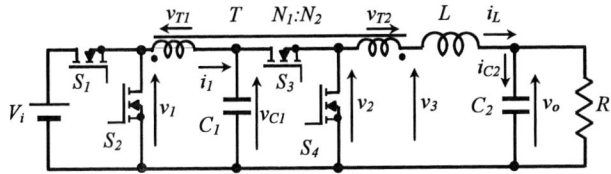

Fig.3. Proposed two-stage converter with coupled inductor.

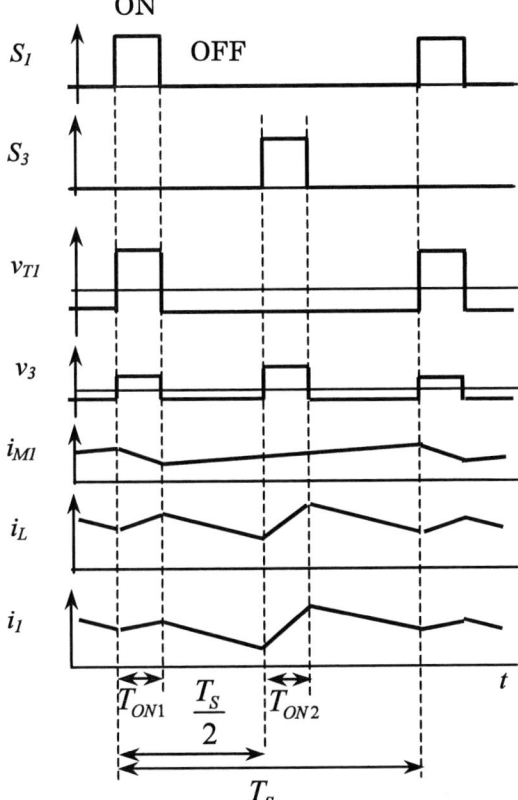

Fig.4. Voltage and current waveforms.

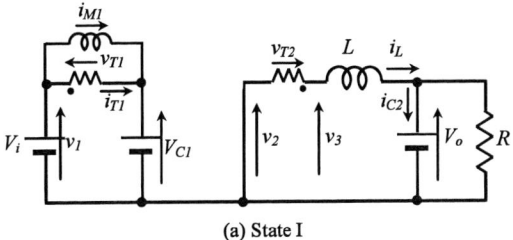

(a) State I

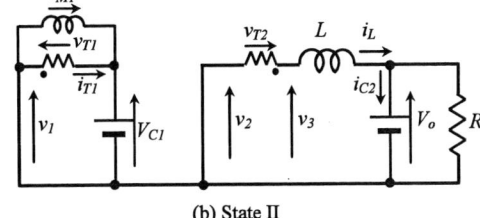

(b) State II

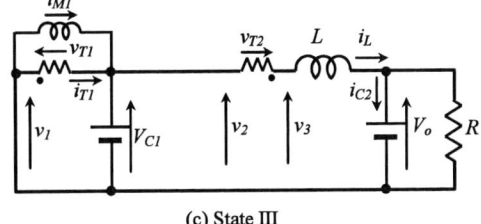

(c) State III

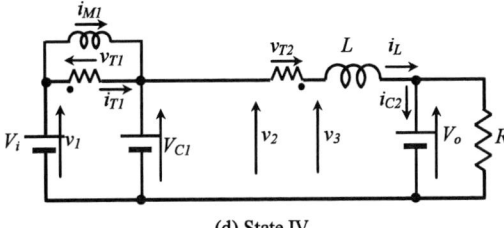

(d) State IV

Fig.5. Equivalent circuits for each stage.

currents become smaller and the efficiency in light load is improved. In addition, since smaller inductors can be employed under conditions of the same amount of ripple currents, the fast response is obtained for the load current change. One thing that must be taken into account is that some kind or another devices to overcome the current unbalance of each converter is implemented.

B. Two-stage Converter with a Coupled Inductor

In order to settle an above-mentioned problem, a two-stage converter with a coupled inductor is proposed. Fig.3 shows the circuit configuration of proposed two-stage converter in which the filter inductors of the first stage and the second stage are coupled with the turn's ratio of N_1 and N_2. On-off states of the switches are as same as those of the conventional two-stage converter as shown in Fig.2. By coupling the inductors, ripple currents in the switches and inductors in both stages are reduced dramatically. It is obvious that there is no need to consider current unbalance. The DC value of the current

in second stage is determined by duty ratio of the switch S_3. On the other hand, the AC component is determined by turn's ratio of transformer T.

III. STEADY STATE ANALYSIS

A. States of Operation

Before starting the analysis, the following assumptions are introduced [7]:

(1) Switches are ideal, i.e., they are represented as shot circuit in a state of ON and open circuits in a state of OFF. The active region of switches is omitted.

(2) Winding resistances and stray capacitance between to windings are negligible.

(3) Filter capacitors are large enough so large that the voltage across the them can be considered as a DC voltage source.

Definitions of symbols used in this paper are as follows:
V_i: dc input voltage,
v_1: voltage across S_2,

859

v_{T1}: voltage across the primary winding of transformer,

v_{T2}: voltage across the secondly winding of transformer,

i_1: current in the primary winding of transformer,

v_{C1}: voltage across C_1,

v_2: voltage across S_4,

v_3: voltage of the output filter input,

i_L: currents in the filter inductor L,

i_{C2}: currents in the output capacitor C_2,

v_o: output voltage,

i_{M1}: transfer magnetizing current,

D_1: duty circle of S_1,

D_2: duty circle of S_3,

T_S: switching period.

Fig.4 shows key waveforms of voltages and currents. We divide switching period into the following fore states: Fig.5. shows the equivalent circuit for each state. From the figure, the following equation are obtained:

1) state I: S_1 and S_4 are ON. S_2 and S_3 are OFF.

$$V_i = L_M \frac{di_{M1}}{dT} + V_{C1} \tag{1}$$

$$-\frac{N_2}{N_1} L_M \frac{di_{M1}}{dT} + L \frac{di_L}{dT} + V_o = 0 \tag{2}$$

2) state II: S_2 and S_4 are ON. S_1 and S_3 are OFF.

$$0 = L_M \frac{di_{M1}}{dT} + V_{C1} \tag{3}$$

$$-\frac{N_2}{N_1} L_M \frac{di_{M1}}{dT} + L \frac{di_L}{dT} + V_o = 0 \tag{4}$$

3) state III: S_2 and S_3 are ON. S_1 and S_4 are OFF.

$$0 = L_M \frac{di_{M1}}{dT} + V_{C1} \tag{5}$$

$$-\frac{N_2}{N_1} L_M \frac{di_{M1}}{dT} + L \frac{di_L}{dT} + V_o = V_{C1} \tag{6}$$

4) state IV: S_2 and S_4 are ON. S_1 and S_3 are OFF.

$$0 = L_M \frac{di_{M1}}{dT} + V_{C1} \tag{7}$$

$$-\frac{N_2}{N_1} L_M \frac{di_{M1}}{dT} + L \frac{di_L}{dT} + V_o = 0 \tag{8}$$

B. Voltage Conversion ratio

In the steady state, the initial values and the final values of the inductor currents are equal, so the following equations are obtained;

$$\frac{V_{C1}}{V_i} = D_1 \tag{9}$$

$$\frac{V_o}{V_i} = D_2 \frac{V_{C1}}{V_i} = D_1 D_2 \tag{10}$$

Equation (9) is the same as the expression of output

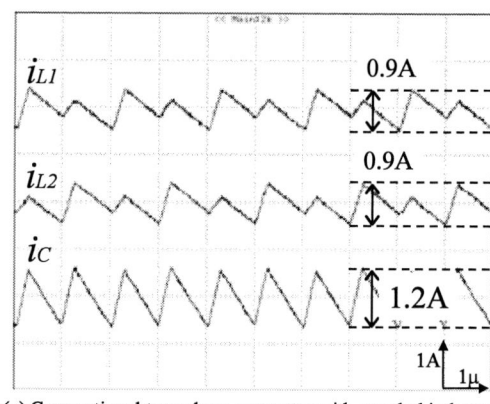

(a) Conventional two-phase converter with coupled inductor.

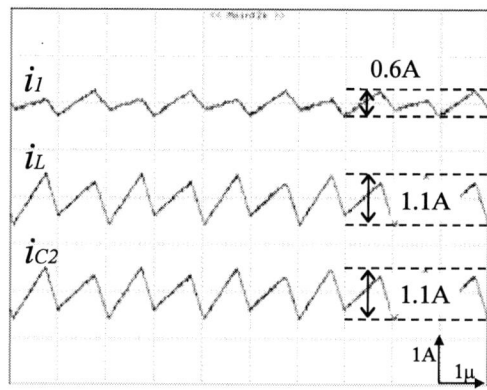

(b) Proposed two-stage converter with coupled inductor.

Fig.6. Waveforms of ripple currents.

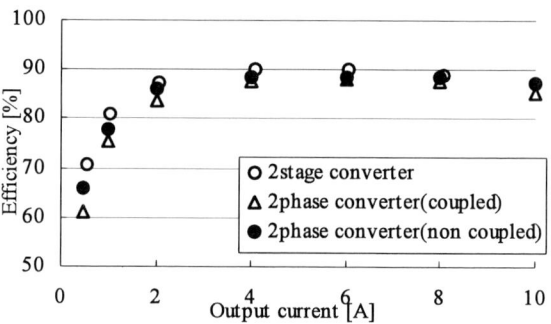

Fig.7. Efficiency characteristics.

voltage of the two-stage converter without coupled inductor.

C. Ripple Current

From Fig.4 and equation (1) through (10), the peak to peak value of the ripple current in inductor L is easily obtained as follow:

$$\Delta i_{Lp-p} = D_1 D_2 (1 - D_2) \frac{V_i}{L} T_s \tag{11}$$

D. Experimental Results

Table.1 shows the parameters used in the experiments. Fig.6 shows the waveforms of ripple currents of conventional two-phase buck converter and proposed

TABLE I
CIRCUIT PARAMETERS OF TWO-STAGE CONVERTER

SYMBOL	VALUE
V_I	12V
V_o	1.5V
$T\ (=N_1:N_2)$	3:1
L_M	10μH
L	0.75μH
C_1	400μF
C_2	400μF
R	1 Ω

converter. From the figure, it is seen that the ripple currents in proposed converter is as small as those of conventional one.

Efficiency characteristics of the propose converter are shown Fig.7. From the figure, the efficiency of the proposed converter in the light load region is higher than that of the conventional one because the ripple current in the first stage converter is small and the conducting losses of S_1 and S_2 are reduced.

IV. SMALL SIGNAL AC CHARACTERISTICS

A. Control to Output Transfer Function

Applying the state space averaging method, the output voltage of proposed two-stage converter in Fig.2 is as follows;

$$V_o = V_2 \frac{1}{1 + s\dfrac{L}{R} + s^2 L C_2} \qquad (12)$$

where,

$$V_2 = \left(D_2 - \frac{N_2}{N_1}\right)V_1 + \frac{N_2}{N_1}D_1 V_i \qquad (13)$$

So, from equation (12) and (13),

$$V_o = \frac{1}{1 + s\dfrac{L}{R} + s^2 L C_2}\left[\left(D_2 - \frac{N_2}{N_1}\right)V_1 + \frac{N_2}{N_1}D_1 V_i\right] \qquad (14)$$

The transfer function of the output voltage of the second stage from the duty cycle of the first stage is obtained as follows;

$$\frac{\Delta V_o(s)}{\Delta D_1(s)} = \frac{1}{1 + s\dfrac{L}{R} + s^2 L C_2}\left[\left(D_2 - \frac{N_2}{N_1}\right)\frac{\Delta V_1(s)}{\Delta D_1(s)} + \frac{N_2}{N_1}V_i\right] \qquad (15)$$

The order of equation (15) is higher than quadratic because it contains $V_i(s)/\Delta D_1(s)$, however, setting $D_2=N_2/N_1$, this term is eliminated and then it becomes almost quadratic.

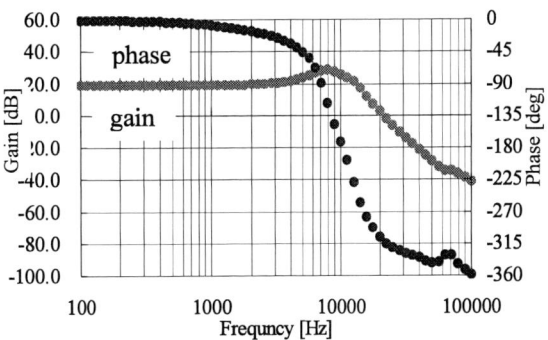

Fig.8. Bode plots of $\Delta V_o(s)/\Delta D_1(s)$ without inductor coupling.

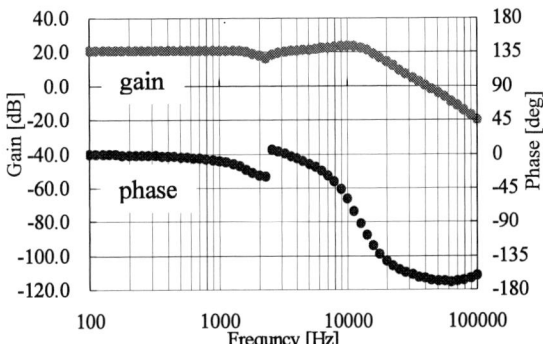

Fig.9. Bode plots of $\Delta V_o(s)/\Delta D_1(s)$ with inductor coupling.

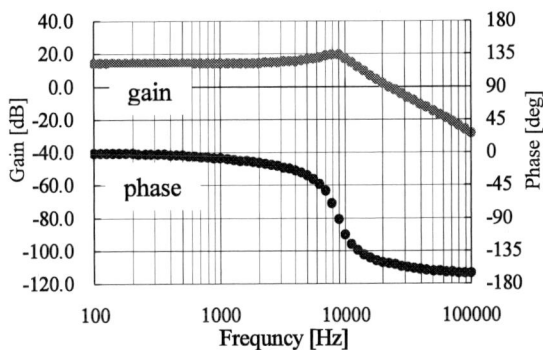

Fig.10. Bode plots of $\Delta V_o(s)/\Delta D_2(s)$ with inductor coupling.

B. Experimental Result

Fig.8 shows the bode plots of $\Delta V_o(s)/\Delta D_1(s)$ of the two-stage converter without inductor coupling and Fig.9 shows those with inductor coupling. As stated in the analysis, by coupling the inductor, the order of the transfer function becomes nearly quadratic similar to the buck converter even though composed of two cascaded stages. Fig.10 shows the bode plots of $\Delta V_o(s)/\Delta D_2(s)$ of the proposed converter with inductor coupling. These characteristics are exactly the same as those of the single stage buck converter.

V. CONCLUSION

A novel two-stage converter with a coupled inductor is presented in this paper. The steady state analysis is

861

performed and examined by experiment, and confirmed that ripple currents are reduced and the power efficiency in the light load region is improved. From the experimental results of small signal AC characteristics, transfer function of the duty cycle to the output voltage is nearly quadratic, so the converter with feedback circuit is inherently stable. Therefore, the proposed converter is well suited to fast response and high efficiency power supplies.

REFERENCES

[1] Jia Wei and Fred C.Lee, "Two-Stage Voltage Regulator for Laptop Computer CPUs and the Corresponding Advanced Control Schemes to Improved Light-Load", IEEE APEC.04, pp.1294-1300.

[2] Yuancheng Ren, Ming Xu, Kaiwei Yao, Yu Meng, Fred C.Lee and Jinghong Guo, "Two- Stage Approach for 12V VR", IEEE APEC'04, pp.1306-1312.

[3] P.Alou, J.A.Cobos, R.Prieto, O.Garcia and J.Uceda, "A Two-Stage Voltage Regulator Nobel with Fast Transient Respnse Capability", IEEE PESC'03, pp.138-143.

[4] S.Abe, J.Yamamoto, T.Zaitsu, T.Ninomiya, "Extension of Bandwidth of Two-Stage DC-DC Converter with Low-Voltage / High-Current Output", IEEE PESC'03, pp.1593-1598.

[5] S.Abe, J.Yamamoto, T.Zaitsu, T.Ninomiya, "Fast Transient Response of Two-Stage DC-DC Converter with Low-Voltage / High-Current Output", IEEE ISIE'03, pp.417-421.

[6] S.Abe, T.Ninomiya, J.Yamamoto, T.Uematu, "Transient Response Comparison of Conventional and Output-Inductor less Two-Stage DC-DC Converter with Low-Voltage / High-Current Output", IEEE CIEP'04, pp.67-70.

[7] Pit-Leong wong; Peng Xu; Yang, P,; Lee, F.C.; "Performance improvements of interleaving VRMs with coupling inductors," IEEE Transactions on Power Electronics, Issue 4, July 2001, vol.16, pp449-507.

Three-Phase AC to DC Converter with Minimized DC Bus Capacitor and Fast Dynamic Response

U. Kamnarn*, Y. Kanthaphayao** and V. Chunkag***

*Rajamangala University of Technology Lanna, 128 Huay Keew Road, Muang Distric, Chiang Mai 50300, Thailand
** Rajamangala University of Technology Suvarnabhumi, 7/1 Sonyai, Muang Distric, Nonthaburi 11000, Thailand
*** King Mongut's Institute of Technology North Bangkok, 1518 Pibulsongkram Road, Bangsue, Bangkok 10800,Thailand

Abstract--The analysis and design of a nearly unity power-factor and fast dynamic response of the modular three-phase AC to DC converter with minimized DC bus capacitor is discussed. The proposed system significantly improves the dynamic response of the converter to load steps with minimized DC bus capacitor. It chooses output capacitor according to the output voltage droop and overshoot. This is confirmed by 600 W prototype modular three-phase ac to dc converter comprising three 200 W single-phase SEPIC rectifier modules with dc bus output capacitor 470 µF and experimental implementation.

Index-- Fast Dynamic, minimized capacitor, module rectifier, power factor correction.

I. INTRODUCTION

In recent years, single-phase switch-mode AC to DC power converters have been increasingly used in industrial, commercial, residential, aerospace, electronic equipment and telecommunication systems [1]. Generally, a single high-power power supply has some disadvantage such as the large capacitance. Therefore, a single-phase parallel configuration using PFC boost, SEPIC and CUK topology for DC DPS have been developed [2-3]. However, a modular three-phase AC to DC power converter is becoming popular for low voltage or medium power applications [4]. It often requires isolation transformer, high power factor, low input current Total Harmonic Distortion (THDi), high efficiency and high power density. Such a modular development approach has the following advantages : well proven and reliable single-phase converter technology can be used immediately, no major change to existing production line is required, power expandability offers great flexibility in the development of power converter products for different power levels, less requirements for maintenance and repair of power converter modules because of the use of standard single-phase converter units, standard single-phase converter

units do not require the high-voltage devices that are normally needed in specially designed three-phase converters.

In this paper, a nearly unity power-factor and fast dynamic response of the modular three-phase AC to DC converter using three single-phase isolated SEPIC rectifier modules with minimized DC bus capacitor is proposed. The dynamic behavior of output voltage will be described with different values of output capacitance (Highest case: C_2= 13,600 µF and Lowest case: C_2=150 µF) by experimental results.

II. Description of power balance control applied to modular three-phase AC to DC converter.

The block diagram of modular three-phase AC to DC converter based on power balance control technique is shown in Fig. 1. The average small-signal analysis of the proposed system is based on a proposed control technique.

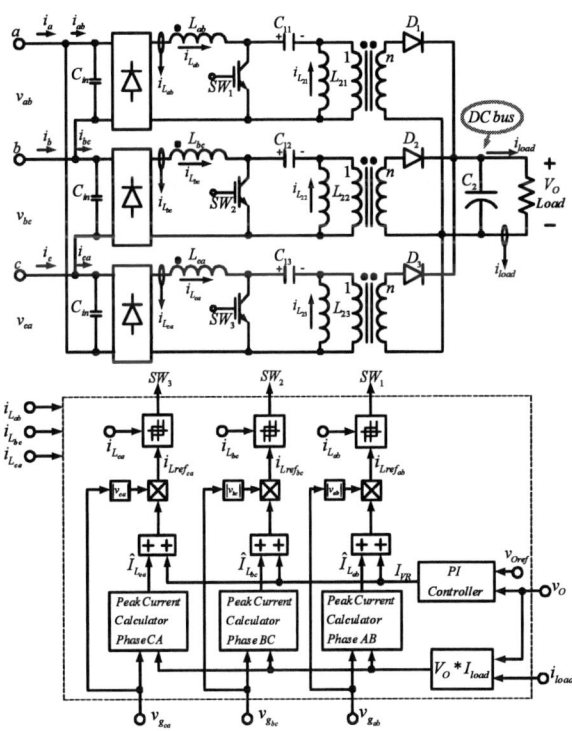

Fig. 1. System configuration block diagram of the modular three-phase AC to DC converter with minimized DC bus capacitor based on power balance control technique.

The authors wish to acknowledge the financial support received from the National Electronics and Computer Technology Center (NECTEC) under contract NT-B-22-E4-22-49-04 as well as Rajamangala University of Technology Lanna, Chiang Mai, THAILAND, Rajamangala University of Technology Suvarnabhumi, THAILAND and King Mongkut's Institute of Technology North Bangkok THAILAND.

The model of the proposed system is:

$$\sum_{ij=ab}^{ca} V_{gij} I_{L_{ij}} = V_o I_o \tag{1}$$

When, $ij = ab, bc$ and ca, V_{gij} is rectifier voltage, $I_{L_{ij}}$ is inductor current, V_o is DC output voltage, I_o is average output current over a half-line cycle. The peak value of the inductor current is

$$\hat{I}_{L_{ij}} = \frac{K_2 V_o I_{load}}{3V_{gij}} \tag{2}$$

$$\hat{I}_{Lref_{ij}} = \hat{I}_{L_{ij}} + I_{VR} \tag{3}$$

$\hat{I}_{Lref_{ij}}$ is peak value of the inductor reference current, $\hat{I}_{L_{ij}}$ is peak value of the inductor current, I_{VR} is correcting signal of PI controller, and K_2 is conversion gain. The dynamic equation of the output voltage is

$$\sum_{i=1}^{3} I_{oi} = C_2 \frac{dV_o}{dt} + I_{load} \tag{4}$$

Let the $\overline{V}$ represents steady-state value and $\tilde{v}$ is the introduced perturbation. Applying the perturbations in (1), (2), (3), and (4), and performing the small-signal approximation ($\tilde{v} \cdot \tilde{v} = 0$), results in :

$$\tilde{i}_o = \frac{3K_1 \overline{\hat{I}}_{Lref_{ij}}}{\overline{V}_o} \tilde{v}_{gij} + \frac{3K_1 \overline{V}_{gij}}{\overline{V}_o} \tilde{\hat{i}}_{Lref_{ij}} - \frac{\overline{I}_o}{\overline{V}_o} \tilde{v}_o \tag{5}$$

$$\tilde{i}_o = C_2 \frac{d\tilde{v}_o}{dt} + \tilde{i}_{load} \tag{6}$$

$$\tilde{i}_{Lref_{ij}} = \tilde{\hat{i}}_{L_{ij}} + \tilde{i}_{VR} \tag{7}$$

$$\tilde{\hat{i}}_{L_{ij}} = \frac{K_2 \overline{V}_o}{3\overline{V}_{gij}} \tilde{i}_{load} + \frac{K_2 \overline{I}_{load}}{3\overline{V}_{gij}} \tilde{v}_o - \frac{K_2 \overline{V}_o \overline{I}_{load}}{3\left(\overline{V}_{gij}\right)^2} \tilde{v}_{gij} \tag{8}$$

The transfer function of the proposed system is

$$\frac{\tilde{v}_o}{\tilde{v}_{oref}} = \frac{3G_{VR} \overline{V}_{gij} K_1}{\overline{V}_o C_2 S + 3G_{VR} \overline{V}_{gij} K_1 k_{fb}} \tag{9}$$

Here, a PI controller is chosen for voltage regulation.

$$G_{VR}(s) = \frac{k_p \left(S + \omega_Z\right)}{S} \tag{10}$$

Then, feedback transfer function is

$$k_{fb} = \frac{R_2}{R_1 + R_2} \tag{11}$$

Plant Transfer Function is

$$PTF(s) = \frac{3\overline{V}_{gij} K_1 k_{fb}}{\overline{V}_o C_2 S} \tag{12}$$

Open Loop Transfer Function is

$$OLTF(s) = \frac{3G_{VR}(s)\overline{V}_{gij} K_1 k_{fb}}{\overline{V}_o C_2 S} \tag{13}$$

III. The Selection of DC output capacitor

The value of the output capacitor is generally determined by several factors, the hold-up time, the output ripple voltage and dc output voltage. In this paper, the output dc capacitance is determined according to the hold-up time Δth, so the relationship between the value of energy-storage capacitor, C_2, it should be selected as follows.

$$C_2 \geq \frac{2P_o \Delta th}{V_o^2 - V_{o,min}^2} \tag{14}$$

As can be from (14) the parameter is determined by Laplace's equation and suppose Δth is a function of V_o. The partial is replaced by ordinary derivative equation as follows.

$$\frac{\partial^2 \Delta th}{\partial V_o^2} = 0$$

$$\frac{d^2 \Delta th}{dV_o^2} = 0 \tag{15}$$

From equation (15) integrate twice will get the result as follows.

$$\Delta th = A V_o + B \tag{16}$$

From equation (16) A and B are the constant value of a second order differential equations. The constant value are considered by the boundary conditions. The hold-up time (16) is depends on the output voltage, variable A and B, so we can find the constant value by the principle of the crossover frequency is considered as follows. The frequency at which the gain of open loop system is equal 1 or 0 dB, this crossover frequency must be below the switching frequency to respond quickly to the transients such as a sudden change of the load. In this paper, it chooses less than half the switching frequency, but the usual practice is to fix crossover frequency at one-fourth to one-fifth of the switching frequency [5]. Next, the hold-up time is considered to minimize DC bus capacitor as follows.

$$\Delta th = \frac{1}{f_{sc}} \qquad (17)$$

f_{sc} is the crossover frequency.

In case I, suppose $\Delta th = 0$ and $V_o = 0$ so, from equation (16) the constant value B is obtained in (18)

$$0 = A(0) + B$$
$$B = 0 \qquad (18)$$

For case II, consider dc bus capacitor is an optimum to the transient response. Suppose $V_o = V_o$ and chooses at one-fourth of the switching frequency, therefore

$$f_{sc} = \frac{f_s}{4}$$

then $\Delta th = \frac{4}{f_s}$

And then substituting (17) and (18) in (16) the constant value A is

$$A = \frac{4}{f_s V_o} \qquad (19)$$

And in case III, Consider the dc bus capacitor is affected the output voltage droop, ΔV_{droop} because of the step load change from minimum to maximum. Suppose $V_o = \Delta V_{droop}$ and Δth, we choose the crossover frequency less than half switching in (17) is

$$f_{sc} = \frac{f_s}{2}$$

then $\Delta th = \frac{2}{f_s}$

And then substituting (17) and (18) in (16) the constant value A is

$$A = \frac{2}{f_s \Delta V_{droop}} V_o \qquad (20)$$

Finally case IV, the dc bus capacitor is affected the output voltage overshoot, $\Delta V_{overshoot}$ because of the step load change from maximum to minimum. Suppose $V_o = \Delta V_{overshoot}$ and Δth have the same case III.

$$f_{sc} = \frac{f_s}{2}$$

then $\Delta th = \frac{2}{f_s}$

As the same case III the constant value A is

$$A = \frac{2}{f_s \Delta V_{overshoot}} V_o \qquad (21)$$

Then substituting (18),(19),(20) and (21) in (16) we can found Δth are

In case II: $\Delta th = \frac{4}{f_s} \qquad (22)$

And case III: $\Delta th = \frac{2V_o}{f_s \Delta V_{droop}} \qquad (23)$

Finally case IV: $\Delta th = \frac{2V_o}{f_s \Delta V_{overshoot}} \qquad (24)$

Equation (22) to (24) are implies that the hold-up time for determined the output capacitor in (14) to fast dynamic response.

IV. Experimental Results of the Modular three-phase AC to DC converter and its behavior.

This section presents experimental results of dynamic behavior of the proposed system. The objective is to find the limits of the dynamic characteristics of these proposed system when priority is to improve the output voltage regulation, minimized DC output capacitor with low total harmonic distortion and high input power-factor. A 600 W prototype modular three-phase AC to DC converter comprising three 200 W single-phase isolated SEPIC rectifier modules has been designed and built. Each converter operates in continuous conduction mode (CCM) together with hysteresis current control. The circuit parameters and normal operating conditions for the proposed system are : $V_{ab}=V_{bc}=V_{ca}=220$ V, 50 Hz, $C_{21}=C_{22}=C_{23}=1\,\mu F$, turn ratio(n)=0.5, L_{ab}=5.08 mH, L_{21}=2.14 mH, L_{bc}=5.04mH, L_{22}=2.51mH, L_{ca}=5.00 mH, and L_{23}=2.16 mH, at switching frequency about 30 kHz, the output DC voltage, $V_o = 48$ V , $\Delta V_{overshoot} = +5\%$ and $\Delta V_{droop} = -5\%$. So, we can find the dc bus capacitor from (16) which Δth equals (22),(23) and (24) respectively. The value capacitance C_2 are found as follows:

$$C_2 = \frac{2 \times 600 \times 4}{30 \times 10^3 (48^2 - 45.6^2)} = 712.25\,\mu F$$

$$C_2 = \frac{2 \times 600 \times 2 \times 48}{30 \times 10^3 \times 45.6 \times (48^2 - 45.6^2)} = 374.86\,\mu F$$

$$C_2 = \frac{2 \times 600 \times 2 \times 48}{30 \times 10^3 \times 50.4 \times (48^2 - 45.6^2)} = 339.17\,\mu F$$

Pick C_2 difference values of output capacitor (highest value: C_2=13,600 µF and lowest value: C_2=150 µF).

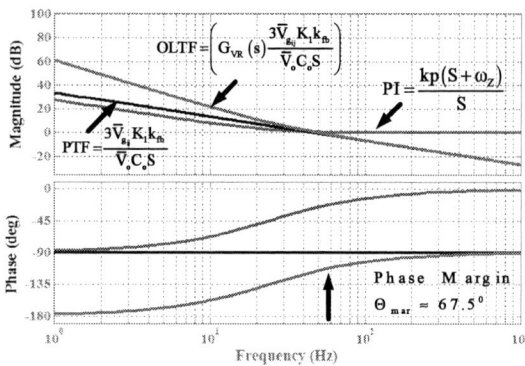

Fig. 2. Bode plot of the modular three-phase ac to dc converter.

The frequency response analysis is used to describe the stability of the proposed system. The voltage loop is designed to have a crossover frequency of 60 Hz. It also gives a 67.5° phase margin (k_p =1 and ω_z =150) and shown in Fig. 2

A. Performance evaluation

The main performance features of the proposed system for input power-factor and input current total harmonic distortion (THDi) are plotted in Fig.3 as a function of the different DC capacitor size (highest value : C_2=13,600 μF, and lowest value : C_2=150 μF) at rated input and output voltage and output power (220 V, 48 V and 600 W). They are measured with the Digital power meter YOKOGAWA model 2531A. The following can be concluded

1. Power factor at rated load is nearly unity, greater than 0.99 for all difference values of output capacitors
2. Input current THD$_i$ remains low, lower than 4% at rated load (600 W), for twenty-one values of output capacitor.

The transient operation of the modular three-phase ac to dc converter depends to a large extent on the dc bus capacitor. The influence of its value on ΔV_{droop}, $\Delta V_{overshoot}$ and settling time is shown in Fig. 4. The voltage droop or overshoot is typically unsymmetrical. As expected the settling time decreases as the capacitance increases, down to 100 μs for $C_2 \geq 3,000$ μF, and the ΔV_{droop}, $\Delta V_{overshoot}$ decreases as well. The voltage ΔV_{droop} is higher than $\Delta V_{overshoot}$ for $C_2 <$ 1,000 μF. At low values of dc bus capacitor with $C_2 < 470$μF, the total ΔV_{droop} and $\Delta V_{overshoot}$ is larger than 5% and Δt of the settling time is very large. However, for high values of dc bus capacitor with 3,000μF ≤ C_2 ≤ 13,600μF, the voltage droop and overshoot is nearly zero and Δt of the settling time is nearly zero as well.

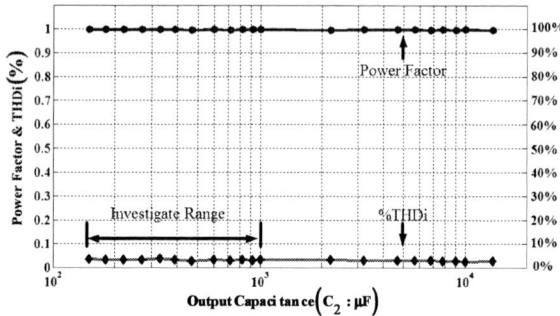

Fig. 3. Experimental results of performance as a function of DC capacitor size (V_s = 220 V, V_o = 48 V and P_o = 600 W).

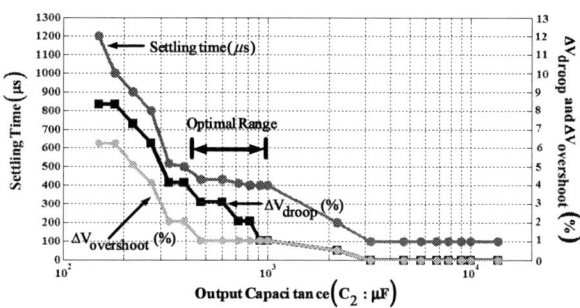

Fig. 4. Performance as a function of dc capacitor size under condition step load change from 60W(10%) to 600W(100%) and vice versa. ($V_s = 220\,V$ and $V_o = 48$ V).

B. Investigated range based on step transient response

This section deals with this tradeoff between output capacitance and output voltage regulation for the dynamic limits of the proposed system. The experimental results of the proposed system operating at step load change from 60 W to 600 W and 600 W to 60W under six different output capacitors (150 μF, 270 μF, 330 μF, 470 μF, 820 μF and 1,000 μF) are shown in Fig. 5(a) to (f), and Fig. 6 (a) to (f) respectively. The comparison between the results indicates that the small voltage dips and less variant in response trajectories are yielded by applying power balance control technique.

V. Discussions

Table I shows the transient performance indexes at investigated range of DC bus capacitors. It can be observed that the settling time, percentage output undershoot and overshoot decrease, as the DC bus capacitor increases. Steady state and transient response conditions of the proposed system with minimized DC bus output capacitor are shown in Fig. 7(a) to (f), respectively. Figure 7(a) shows input voltage and three-input current waveforms of the proposed system. Three input currents are in phase with their relative input voltages and nearly sinusoidal. Thus, the input power factor approaches unity. They have been illustrated that the proposed system achieved a high power factor. Figure 7(b) and (c) shows three-phase input currents and three-inductor currents at full load. Due to the effects of difference inductance values and parasitic in input

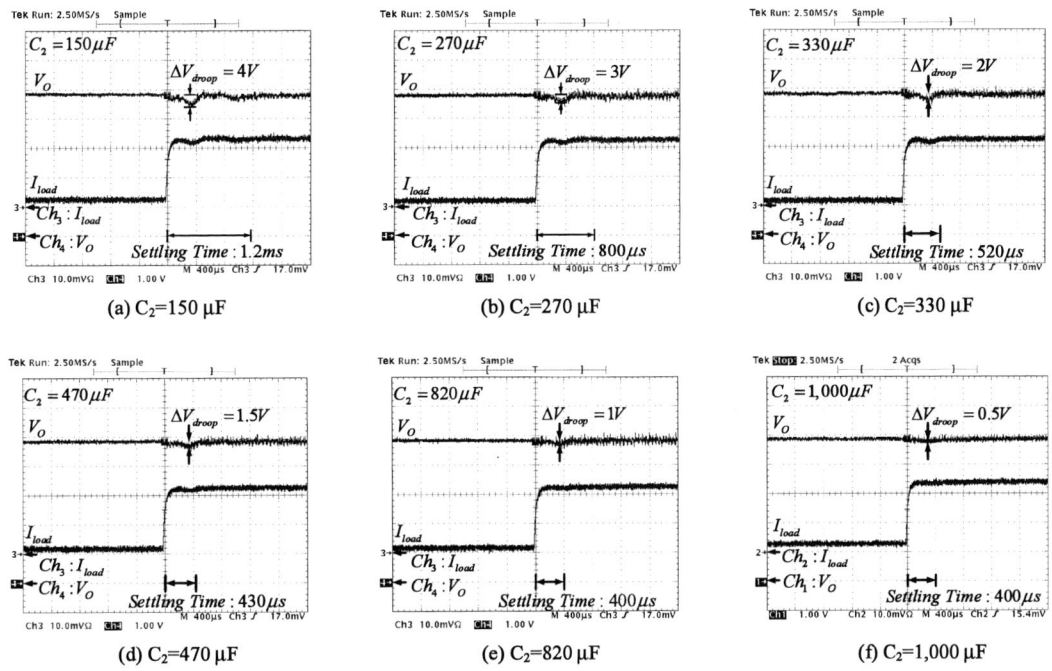

Fig.5. Experimental results of DC bus voltage (V_o : $10V/div$) and load current (I_{load} : $5A/div$, $Time$: $400\mu s/div$) at step load change from 10%(60 W) to 100%(600 W) (parameter of PI controller k_p =1 and ω_Z =150).

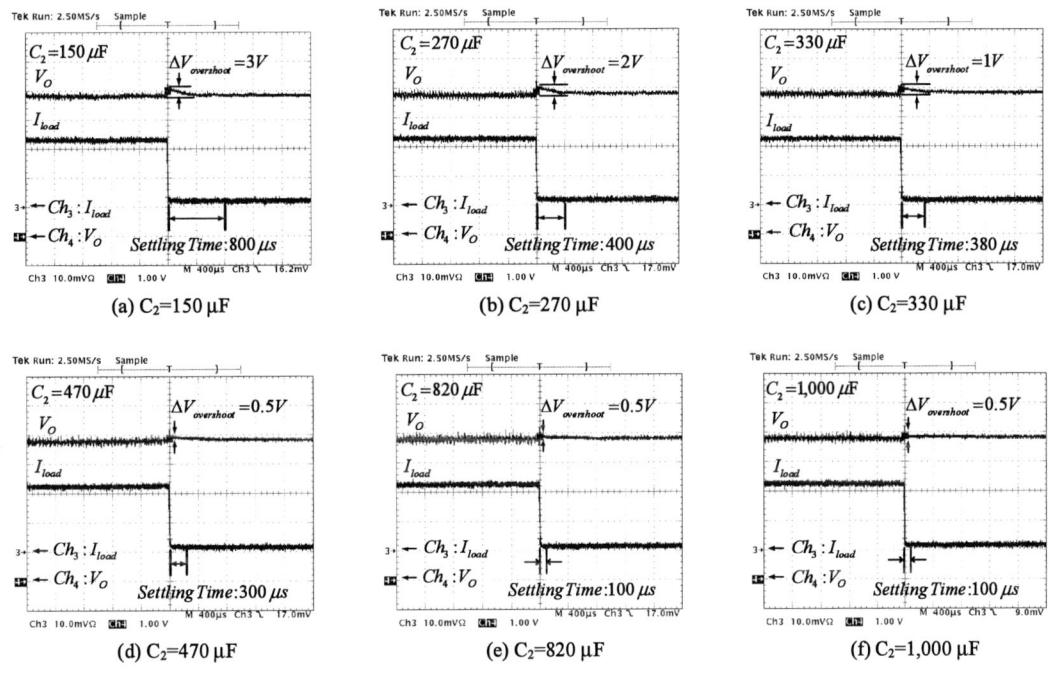

Fig.6. Experimental results of DC bus voltage (V_o : $10\,V/div$) and load current (I_{load} : $5A/div$, $Time$: $400\mu s/div$) at step load change from 100%(600 W) to 10%(60 W) (parameter of PI controller k_p =1 and ω_Z =150).

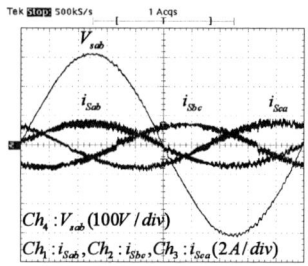 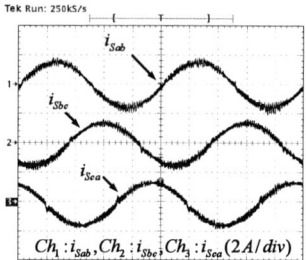 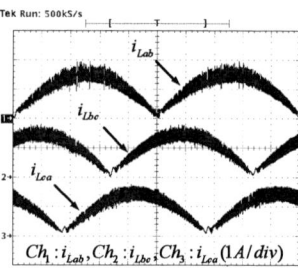

(a) Input voltage and current, Time:2 ms/div.   (b) Three input currents, Time:4 ms/div.   (c) Three inductor currents, Time:2 ms/div.

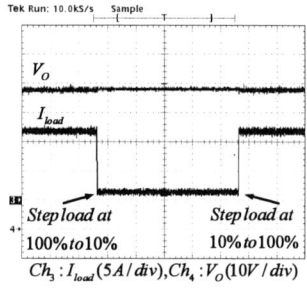 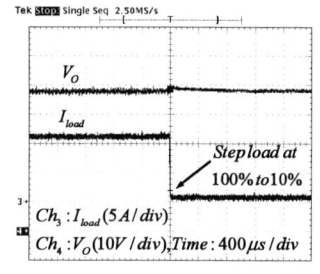 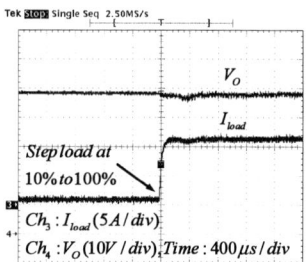

(d) Output voltage and load current, Time:100 ms/div.   (e) Step load change from 600 W to 60 W   (f) Step load change from 60 W to 600 W

Fig. 7. Experimental results of modular three-phase AC to DC converter with minimized DC output capacitor (C_2=470 µF).

inductors, these three input currents and three-individual inductor currents are not exactly equal.However, all input and inductor currents have approximately same amplitude. Figure 7(d) shows the dynamic response of the output voltage and load current of the proposed system due to step load changes at the load current between 100% to 10% and vice versa. Figure 7(e) and Fig. 7(f) shows closed up of transient response condition due to step load changes at 100% to 10% and 10% to 100%, respectively. It can be seen that such scheme is effective and fast transient response characteristic. The voltage droop or overshoot is typically unsymmetrical. The voltage droop and overshoot are 3.12% and 0.52%, respectively. The settling time is 430 µs. The main performance features of the proposed system for C_2=470 µF, the input power factor at rated load is high, greater than 0.99, the overall efficiency remains high, approximate 90%, and THD_i less than 4% at maximum load. The harmonic spectrum for the input current is measured, and it meets the regulation of IEC 61000-3-2 Class A limits.

VI. Conclusions

The paper has demonstrated that, with a minimized DC bus capacitor and power balance control technique, excellent power factor correction,

module load sharing under load transitions and high overall efficiency can be achieved for a modular three-phase AC to DC converter using three single-phase isolated SEPIC rectifier modules. In this case, DC bus capacitor C_2=470 µF according to ΔV_{droop} and $\Delta V_{overshoot}$, the DC output voltage is regulated to have good transient responses by the designed PI controller with power balance control technique. It should be noted that this dynamic response is fast enough for many conventional industrial applications and hence, no second stage is needed. Thus, the proposed regulator is relatively feasible in medium-power applications, which has been verified by the measured results.

VII. References

[1] J. Sebastian, M. Jaureguizar, and J. Uceda, "An overview of power factor correction in single-phase off-line power supply systems", IEEE IECON '94, pp. 1688-1693, September. 1994.

[2] A. Newton, T.C. Green, and D. Andrew, "AC/DC Power Factor Correction Using Interleaved Boost & CUK Converters", IEE PEVSD '00, pp. 293-298.

[3] U. Kamnarn, and V. Chunkag, "Analysis and Design of a Parallel CUK Power Factor Correction Circuit Based on Power Balance Control Technique", Industry Application, IEEJ Transaction on, Vol. 126-D, Number 5, May 2006, pp. 533-540.

[4] Y.K. Eric Ho, S.Y.R. Hui, and Yim-Shu Lee, "Characterization of Single-Stage Three-Phase Power-Factor-Correction Circuit Using Modular Single-Phase PWM DC-to-DC Converters", IEEE Trans. On Power Electronics, pp. 62-71, January 2000.

[5] Abraham I.Pressman, Switching power Supply Design, McGraw-Hill,1998,pp.437-440.

Table I.
COMPARISIONS OF TRANSIENT PERFORMANCE INDEXES.

C_2 (µF)	Settling time (µs)	% output undershoot	% output overshoot	Max. load current (A)
150µF	1,200	8.33	6.25	12.5
270µF	800	6.25	4.16	12.5
330µF	520	4.16	2.08	12.5
470µF	430	3.12	1.04	12.5
820µF	400	2.08	1.04	12.5
1,000µF	400	1.04	1.04	12.5

A Simple Effective Duty Cycle Controller for High Power Factor Boost Rectifier

Hussain S. Athab, *IEEE Member*, P. K. Shadhu Khan, *senior IEEE Member*
Faculty of Engineering, Multimedia University, 63100 cyberjaya, Malaysia
hussain@mmu.edu.my; poritosh@ieee.org

Abstract – **This paper proposes a simple low-cost modulating duty cycle analog controller to reduce line harmonics for high power factor boost rectifier. The proposed method eliminates the need for current sensing, and simultaneously offers the performance results comparable to those of continuous conduction mode (CCM) scheme. This scheme also maintains the simplicity comparable to that of discontinuous conduction (DCM). Only the output voltage and the rectified input voltage are monitored to vary the duty cycle of the boost switch within a line cycle so that the third-order harmonic, which is the lowest order harmonic (LOH) of the input current, is reduced. As a result, the total harmonic distortion (THD) of the line current as well as the input power factor is improved. The proposed method is developed for constant switching frequency boost rectifier. Simulation and experimental results are presented to verify the effectiveness of the proposed control method.**

Index Terms— **AC/DC Converter, Boost Rectifier, Power Factor Correction (PFC), Switch-Mode Power Supply, Total harmonic Distortion (THD).**

I. INTRODUCTION

The AC-DC rectifier consisting of a diode bridge rectifier followed by a filter capacitor is cheap and robust, but demands a harmonic-rich ac line current and poor power factor. To overcome the problem, passive and active power factor correction (PFC) circuits have been proposed. In general, active methods are more efficient, lighter in weight, and less expensive than passive circuit methods [1]. Recent international regulations governing the harmonic content of the input current drawn by electrical equipment have inspired the development of many new active PFC circuits employing switched mode DC-DC converters. There are different topologies for implementing active PFC techniques including the boost converter [2] and the buck converter. For reasons of simplicity and its popularity, the boost converter is used to improve the input power factor. In boost circuit Fig 1, the switching device handles only a portion of the output power and this property can be used to increase the efficiency of the converter. To have unity power factor, the boost circuit must maintain the input current proportional to

the input voltage. This proportionality is maintained in a closed loop manner using feedback, such as in the common continuous conduction mode (CCM). Various techniques are available for CCM active current shaping, most of which are now supported by dedicated integrated circuits. For Example, using the average current-mode control Fig.1 [3], [4], very low current harmonic distortion in a wide range of input line voltages and output load currents can be readily obtained. However, the average current-mode has relatively complex implementation and mainly requires sensing the inductor current.

The simplest technique for power factor correction is to operate the converter in discontinuous conduction mode (DCM) at constant switching frequency and constant duty cycle [5], [6]. For converters operated in DCM, the input current is related to the input voltage by a factor which depends on duty ratio. Rectifiers employing converters of this type are called automatic current shapers. The DCM boost rectifier falls into the class of automatic current shapers. It is not, however, an ideal automatic current shaper in that the input current is not directly proportional to the input voltage for constant duty ratio. The input current exhibits a low frequency harmonics ($3'_{rd}$, $5'_{th}$, etc) due to the slow discharge of the inductor current after the switch is turn off. Another drawback of DCM operation is high current stress and higher noise caused by the pulsating inductor current.

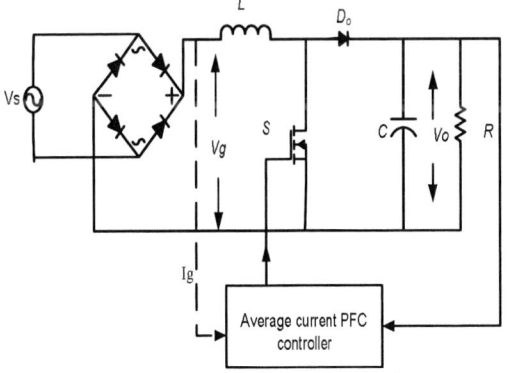

Fig. 1 Power factor correction circuit employing boost converter

A number of control techniques are reported in the literature to improve the performance of DCM boost

978-1-4244-0644-9/07/$25.00 ©2007 IEEE   869

rectifiers. Among them second harmonic injection [7], varying duty cycle (divider approach) [8], multiplier approach [9] and operation at CCM-DCM boundary (variable switching frequency) [10] using input voltage sensing. These can give high quality input current at the cost of complicating the control circuitry.

The purpose of this paper is to propose a simple low cost controller to reduce the line harmonics for single-phase single-switch boost rectifier. The proposed method eliminates the need for current sensing, and simultaneously offers the performance results comparable to CCM scheme. Its simplicity is comparable to that of DCM operation. The results showed that the proposed technique has similar performances to those of non-linear carrier control (NLC) technique presented in [11]. The duty cycle is varied within a fixed switching period such that the low-order harmonics of the input current is attenuated. As a result, the total harmonic distortion (THD) as well as input power factor is improved.

II. BRIEF REVIEW OF BOOST CONVERTER OPERATION AND ANALYSIS OF THE NEW CONTROL STRATEGY

The topology of boost converter is shown in Fig.1. The diode bridge rectifies the ac line voltage, and the Power switch S, the inductor L, and the output diode D_o operate as a boost chopper. The ripple in the output voltage is reduced by using the capacitor C. The load is assumed to be purely resistive (R). The input line-filter reduces the high-frequency components in the input current. The power switch is operated at a constant switching frequency of 20 kHz, and the output voltage is varied by varying the duty cycle. It is assumed that the boost converter operates at continuous conduction mode and the switching frequency is much higher than the line frequency. Hence, the input voltage can be assumed as a constant during one switching cycle. Based on these assumptions, when the switch S, is on or off, the circuit of Fig. 2 (a) or (b) are obtained and the inductor current in Fig. 2 (c) can be described as in Eq. (1) and (2), respectively.

$$L\frac{di_L}{dt}=V_g \qquad 0 \le t \prec dT_s \qquad (1)$$

$$L\frac{di_L}{dt}=V_g-V_o \qquad dT_s \le t \prec T_s \qquad (2)$$

Where V_g is the rectified input voltage, V_o is the output voltage, i_L inductor current, d duty cycle, and T_s is the switching period.

Fig. 3 shows the average equivalent circuit model of boost converter. Therefore, based on this model, the boost converter can be modeled using two differential equations for the inductor current and output dc capacitor, as shown in Eq. (3) and (4).

$$L\frac{di_L}{dt}=V_g-V_o(1-d) \qquad (3)$$

$$C\frac{dV_o}{dt}=i_L(1-d)-\frac{V_o}{R} \qquad (4)$$

Where $(1-d)$ is the off time of the switch S.

The main objectives of the converter control system are to produce an input AC current with low harmonics, with high power factor, and to control the average load voltage. To achieve the first objective, the inductor current (i_L) is required to follow a reference as in Eq. (5)

$$i_L^* = \left| i_m \sin(\omega t) \right| \qquad (5)$$

The proposed method is to generate the switching function $(1-d(t))$ or $d(t)$, which can be derived from Eq. 3, as follows:

$$[1-d(t)]=\left(\frac{1}{V_o}\right)\left(V_g-L\frac{di_L^*}{dt}\right) \qquad (6)$$

Or

$$d(t)=1-\left(\frac{1}{V_o}\right)\left(V_g-L\frac{di_L^*}{dt}\right) \qquad (7)$$

The PWM switching signal for the converter is generated by comparing $d(t)$ with a triangular carrier. The amplitude $i_m(t)$ of Eq. (5) is generated by a proportional Integral (PI) regulator for the output voltage control as follows:

$$i_m(t)=k_p \Delta v_o + k_i \int \Delta v_o dt \qquad (8)$$

Where, $\Delta v_0 = v_o - V_{ref}$.

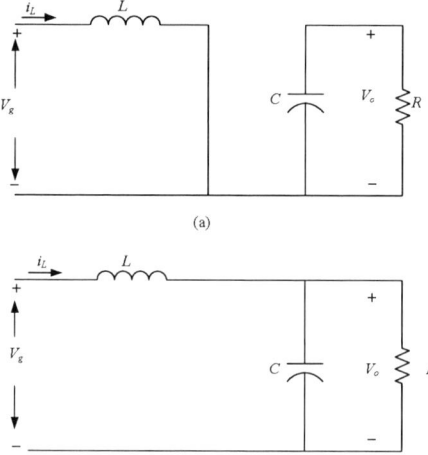

(a)

(b)

870

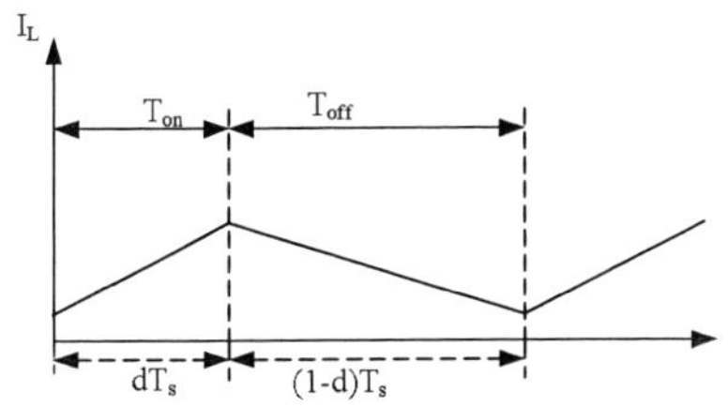

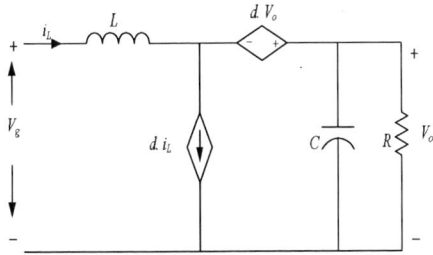

(c)

Fig. 2 Boost converter circuit: (a) switch is on, (b) switch is off, and (c) Inductor current

Fig. 3 The average equivalent circuit model of boost converter

III. IMPLEMENTATION OF THE CONTROL STRATEGY

Fig. 4 shows the boost converter with the new control method. The PWM controller detects the bridge rectifier output voltage and the dc-side voltage across the load (R) and calculates the operations in the right hand side of Eq. (6), which is actually the off-time (1-d(t)) of the switch There is no need to monitor the inductor current. The division by V_o in Eq. (6) can be easily accomplished by modulating the amplitude of the carrier triangular waveform with V_o. Therefore, it offers an inexpensive method to perform the division operation. A conventional PI-controller in cascade with a low pass filter regulator is used to regulate the output voltage. The purpose of the filter is to suppress the high frequency ripple in the output voltage (V_o) due to switching.

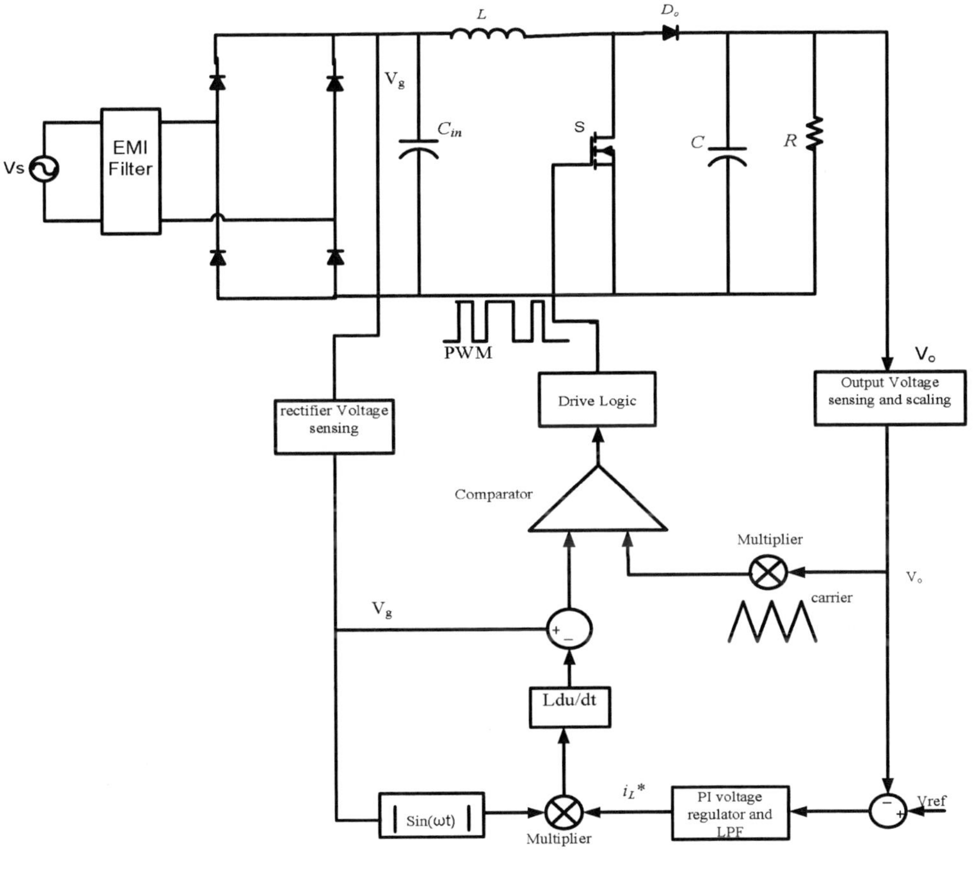

Fig. 4 Boost converter with the proposed control method

IV SIMULATION AND EXPERIMENTAL RESULTS

871

Simulation and experimental results are provided to verify the effectiveness of the proposed control algorithm. Both simulation model and experimental prototype are designed for an output power of a 500W and output dc voltage of a 220V. The input inductor, L = 500 μH, and the output capacitor voltage is 440μF. The ac source voltage is 110 Vrms with 50Hz. The input filter section is L-C type with L=1 mH, and C = 5μF and it mainly reduces the harmonics which are artifacts of the switching. A filter capacitor, C_{in} =8 μF is placed between the bridge rectifier and the boost inductor L, in order to reduce the high frequency ripple of the rectifier voltage V_g. C_{in} can also reduce the peak inductor current as well as the power switch peak current. In the new control scheme, the PWM signal is obtained by monitoring the input and output voltages only. Since the power factor is a steady-state quantity, the dc output voltage and the rectified input voltage are used. There is no need to monitor the current. The converter is simulated using Simulink toolbox of Matlab. The simulation results of the input voltage, input current and the inductor current for the proposed modulated duty cycle are shown in Fig. 5. Fig. 6 shows the simulated harmonic contents of the input current. The corresponding experimental results are shown in Fig 7 and Fig.8. The input power factor is 0.99 and the total harmonic distortion THD is 2.58%. Based on the simulation and experimental results, the line current is sinusoidal with nearly unity power factor.

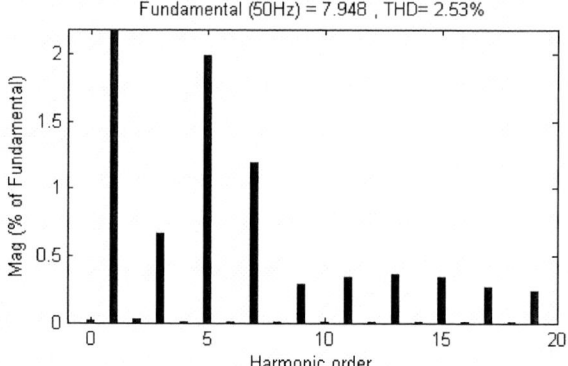

Fig. 6, harmonics content of the ac source current with proposed modulated duty cycle

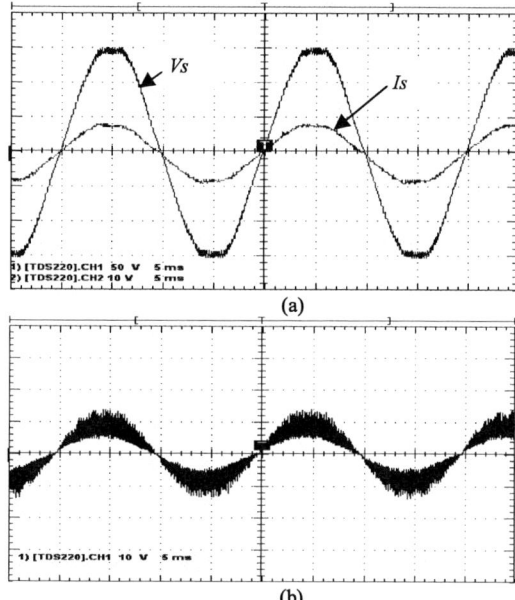

Fig. 7, Experimental Results for proposed modulated duty cycle
(a)   Ac source voltage and current (b) Inductor current

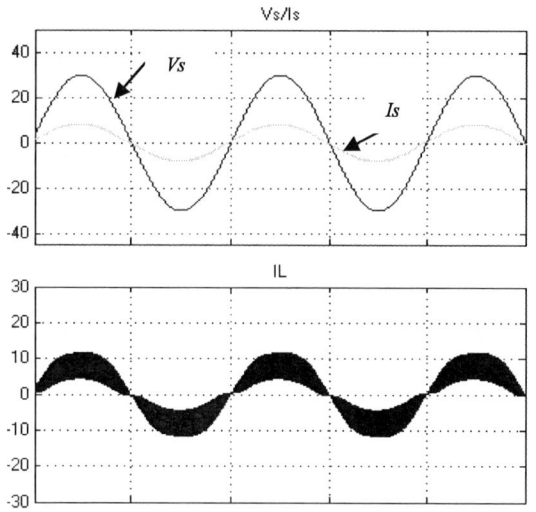

Fig. 5, Simulation results of AC source Voltage (scale 1:5) and current (top trace), and the Inductor current with proposed modulated duty cycle (bottom trace)

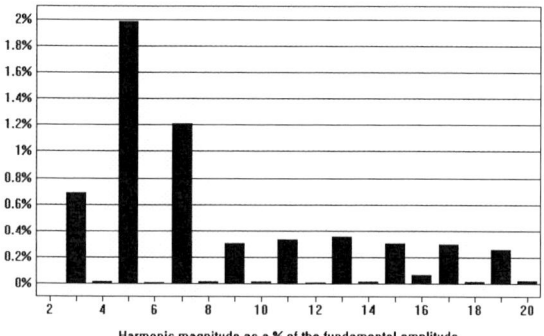

Fig. 8, harmonics contents of the ac source current with proposed modulated duty cycle. THD 2.58%

V. CONCLUSION

The paper proposes a simple, low cost harmonic analog controller to modulate the duty cycle of the boost switch such that the third-order harmonic of the input current is reduced and the overall THD is improved. The proposed method eliminates the need for current sensing and simultaneously offers the performance results comparable to those of CCM scheme. Its simplicity is comparable to that of DCM operation. The PWM control signal is obtained by monitoring the input and output voltages only. The results showed that the proposed technique has similar performances to those of non-linear carrier control (NLC) technique. The duty cycle is varied within a fixed switching period such that the low-order harmonics of the input current is attenuated. Moreover, the duty cycle variations are naturally synchronized with the input voltage without using additional expensive phase-detecting, phase-locking circuits. It is found that simulated and experimental results match closely. Based on the simulation and experimental results, the line current is almost sinusoidal with nearly unity power factor.

REFERENCES

[1] O.Gracia, J. A. Cobos, R. Prieto, and J. Uceda, "Single-phase power factor correction: a survey," *IEEE Trans. Power Electron.,* Vol. 18, no. 3, pp. 749-755, May 2003.

[2] Yang, Z. H. and P. C. Sen,, "Recent developments in high power factor switch -mode converters," *IEEE Canadian Conf. Electrical Computer Engineering,* Vol. 2, 477–488, May 1998.

[3] R. Mammano and R. Neidroff, "Improving input power factor- A new active controller simplifies the task," *in Proc. Power conversion,* Oct. 1989, pp. 100-109.

[4] J. Bazinet and J. A. O'Conner, "Analysis and design of a zero voltage transition power factor correction circuit," *in Proc. IEEE,* APEC 1994, pp. 591-597.

[5] D. Chambers and D. Wang, "Dynamic P.F. correction in capacitor input off-line converters," *in Proc. Pwercon,* Miami, FL, May1979.

[6]] Qiao, C.; Smedley, K.M., *"*A Topology Survey of Single-Stage Power Factor Corrector with **a** Boost Type Input-Current Shaper," *IEEE Transactions on Power Electronics.* Volume: 16 Issue: 3, May 2001, Page(s): 360-368.

[7] DeFeng weng, and S. Yuvarajian, "Constant switching frequency AC-DC converter using second harmonic injected PWM " *IEEE trans. In Power Electronics, Vol.* 11, no. 1 Jan. 1996.

[8] Yu-Kang Lo, Sheng-Yuan Ou and T. Song, "Varying duty cycle control for discontinuous conduction mode boost rectifier" *in Proc. IEEE, PEDS,* Vol. 1, pp, 149-151, Oct. 2001.

[9] M. Ferdowasi, and A. Emadi, "Estimative current mode control technique for DC-DC converters operating in discontinuous conduction mode," *in IEEE power electronics letters,* Vol. 2, No.1, pp. 20-23, March 2004.

[10] Chen, D- S. and Lai, J-S, "A study power correction boost converter operating at CCM-DCM mode", *inProc. IEEE Southeastcon'93,* April 1993.

[11] D. Maksimoviv, Y. Jang., and R. Erickson, "Non-linear carrier control for high power factor boost rectifiers," *in Proc IEEE,* APEC' 95 Conf.pp. 635-641

A Cost Effective Method of Reducing Total Harmonic Distortion (THD) in Single-Phase Boost Rectifier

Hussain S. Athab, *IEEE Member*, P. K. Shadhu Khan, *senior IEEE Member*
Faculty of Engineering, Multimedia University, 63100 cyberjaya, Malaysia
hussain@mmu.edu.my; poritosh@ieee.org

Abstract-- **Due to its simplicity, the discontinuous conduction mode boost rectifier is potentially the least expensive active line-harmonics reducing circuit. The line current however, shows considerable distortion when the peak input voltage is close to the output voltage. This paper proposes a simple, low-cost method to reduce the line harmonics. A periodic voltage signal is injected in the control circuit to vary the duty cycle of the boost switch within a line cycle so that the third-order harmonic of the input current is reduced and the THD is improved. The proposed technique eliminates the additional harmonic generator, phase detecting and phase-locking circuits, which is proposed in the literature. Instead we can utilize the output voltage of the rectifier in the boost converter in order to modulate the duty cycle of the boost switch. As a result, the injected signal is naturally synchronized with line current. Simulation and experimental results are presented to confirm the validity of the method.**

Index Terms—**AC/DC converter, Boost Rectifier, Power Factor Correction (PFC), Switched-mode Power Supply (SMPS), Total harmonic Distortion (THD).**

I. INTRODUCTION

An ac to dc converter consisting of a line frequency diode bridge rectifier with a large output filter capacitor is cheap and robust, but demands a harmonic rich ac line current. As a result, the input power factor is poor. Due to problems associated with low power factor and harmonics, harmonic standards and guidelines, which will limit the amount of current distortion allowed into the utility, is introduced. Thus the simple diode rectifiers may not in use. To correct the poor power factor and reduce high harmonic current contents, passive and active circuits can be used. In general, active methods are more efficient, lighter in weight, and less expensive than passive circuit methods [1].

In active power factor correction techniques approach, switched mode power supply (SMPS) technique is used to shape the input current in phase with the input voltage. Basically in this technique power factor correcting cell makes the load behave like a resistor leading to near unity power factor. Fig. 1 shows the circuit diagram of basic active power correction technique [2]. There are different topologies for implementing active power factor correction techniques including the boost converter [2] and the buck converter. For reasons of simplicity and its popularity, the boost converter is used to improve the power factor. In boost circuit, the switching device handles only a portion of the output power and this property can be used to increase the efficiency of the converter. The boost converter may be designed to operate either in the continuous conduction mode (CCM) or in the discontinuous conduction mode (DCM). Compared with the CCM approach, a converter operating in DCM provides a simpler control scheme, which requires only one (voltage) control loop to modulate the on-time, Fig 2. Furthermore, operating a boost converter in discontinuous mode avoids the output diode reverse recovery problem and alleviates the high switching loss in continuous mode operation. One drawback of DCM PFC approach is that its input current waveform is not always purely sinusoidal. The input current will contain certain distortion due to the modulation of inductor current discharging time. This waveform distortion is found to be a function of the ratio of the peak line voltage to the output voltage of the PFC circuit.

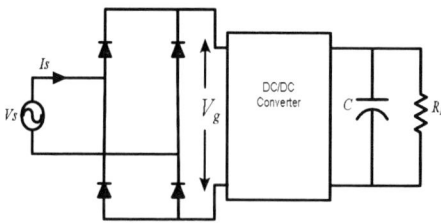

Fig. 1, Active PFC Technique

Generally, to reduce the harmonic current contents of DCM boost converter, the duty cycle of the rectifier switch needs to be properly modulated during a rectified line period instead of being kept constant. Recently, a number of duty cycle modulation techniques for the DCM boost rectifier have been introduced to reduce the total

978-1-4244-0644-9/07/$25.00 ©2007 IEEE

harmonic distortion (THD) of the input current for single-phase and three-phase systems [3]-[6]. Specifically, the approach based on variable switching frequency control was presented and analyzed in [6]. However, since the switching frequency directly depends on the input voltage and output power variations, the variable switching frequency method suffers from very wide frequency range which decreases the efficiency and makes the rectifier design and control circuit more complex. To improve the performance of the DCM boost converter at constant switching frequency, harmonic injection methods have been introduced [3]-[5], which gives high quality input current at the cost of complicating the control circuitry.

The purpose of this paper is to propose a new simple, low cost harmonic reduction method, which has the simplicity of voltage follower technique. In the proposed method a periodic signal proportional to the rectified ac line voltage is injected into the control circuit to modulate the duty cycle of the power switch, S, such that the amplitude of third-order harmonic of the line current is reduced and THD is improved. The generation of the injected signal with unity amplitude is simplified by sensing the output voltage of the bridge rectifier V_g (Fig.1). As a result, the additional circuit required to generate and synchronize the second-order harmonic signal, Fig. 2, proposed in [3] is eliminated. Experimental and simulated results show the effectiveness of the proposed method.

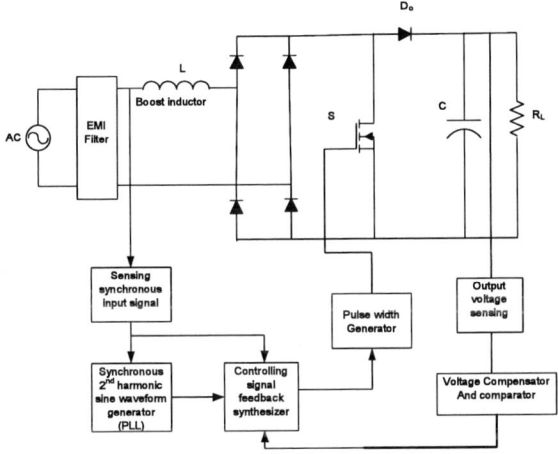

Fig. 2, single-phase boost converter proposed in [3]

II. ANALYSIS OF THE CURRENT WAVEFORM DISTORTION BOOST RECTIFIER

The voltage follower (DCM) boost converter is shown in Fig. 3, assuming that the converter operates in DCM with a constant switching frequency, f_s. The diode bridge rectifier is used to rectify the ac line voltage, and the semiconductor power switch S, the inductor L, and the output diode D_o operate as a boost chopper. The Power switch, S, is operated at high switching frequency and the output voltage is regulated by varying the duty cycle of

the switch, S. A capacitor C_o is used to reduce the ripple in the output voltage. The EMI input filter is used to filter out the high frequency components in the input current.

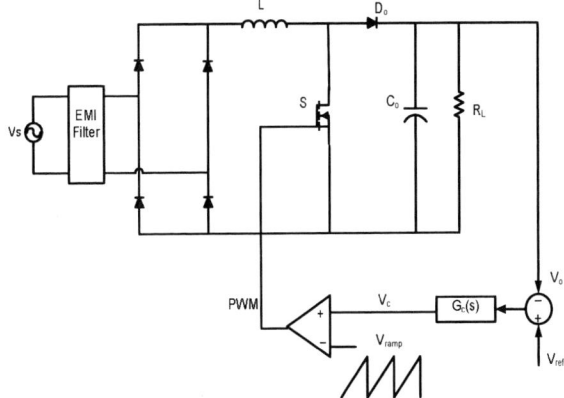

Fig. 3, Single-phase single switch DCM boost converter

In voltage-follower PFC circuit, the on-time, T_{on}, is designed to change slowly and is almost constant over an ac line cycle. In general, the input current in constant switching frequency boost converter is composed of a charging component and discharging component, as shown in Fig. 4. Assuming sinusoidal input voltage ($V_s = V_p \sin\omega t$), the peak inductor current, I_{pk}, is determined as

$$I_{pk} = \frac{V_s \cdot T_{on}}{L} = \frac{V_p \cdot D \cdot T}{L} \sin \omega t \qquad (1)$$

Where T is the period of a switching cycle, D is the duty cycle. The peak inductor current follows an envelope of the input voltage. At the end of on-time, the inductor current is discharged to the output and is reset by a voltage of $V_o - V_s$, where V_o is the output voltage of the boost converter. The discharging time Td, is:

$$T_d = \frac{I_{pk}}{(V_o - V_s)/L} = \frac{V_s T_{on}}{V_o - V_s} \qquad (2)$$

The ac line, in effect, sees an average inductor current waveform due to the presence of the input filter capacitor and the stray line inductance.

$$I_{in}(\text{avg}) = I_{on}(\text{avg}) + I_d(\text{avg}) \qquad (3)$$

Where I_{in}(avg) is the line current, I_{on} (avg) is the average of inductor current during the on-time, and I_d(avg) is the average of inductor current during discharge time. From Eq. (1),

$$I_{on}(\text{avg}) = \frac{I_{pk}}{2} \frac{T_{on}}{T} = \frac{V_s D^2 T}{2L} \qquad (4)$$

From Eq. (2)

$$I_d(\text{avg}) = \frac{I_{pk}}{2} \frac{T_d}{T} = \frac{V_s^2 D^2 T}{2L(V_o - V_s)} \qquad (5)$$

Therefore, $I_{in}(avg) = \dfrac{D^2T}{2L}\left(\dfrac{V_o V_s}{V_o - V_s}\right)$

$I_{in}(avg) = \dfrac{D^2TV_p}{2L}\left(\dfrac{M\sin\omega t}{M - |\sin\omega t|}\right)$

$I_{in}(avg) = k.\left(\dfrac{M\sin\omega t}{M - |\sin\omega t|}\right)$ (6)

Where $k = \dfrac{D^2TV_p}{2L}$ and $M = \dfrac{V_o}{V_p}$ is the voltage gain.

From Eq. (6), it can be seen that when the ratio M is large the current waveform is almost sinusoidal, Fig. 5.

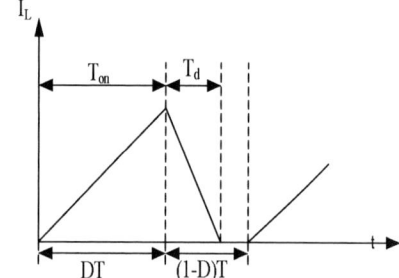

Fig. 4, Inductor current in one switching period

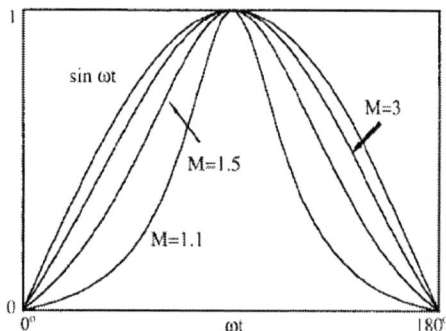

Fig. 5, Normalized line current waveform for half cycle

Using Fourier analysis, I_{in} can be shown to contain several low-order harmonics. The significant lower-order harmonics are the third, fifth, and the seventh. Among the three lower-order harmonics, the third harmonic component has the largest amplitude and has to be attenuated using PWM.

III. NEW LOW COST INJECTION SIGNAL IMPLEMENTAION

In the proposed circuit as shown in Fig. 6, a signal $d(t)$ proportional to the rectified ac line voltage is injected into the control circuit to modulate the duty cycle of the power switch, S, such that the amplitude of third-order harmonic of the line current is reduced and THD is improved. In this paper, the generation of the injected signal with unity

amplitude is simplified by eliminating the additional circuit required to generate the second-order harmonic signal proposed in [3], Fig 2.

It also eliminates the need to employ phase detecting and phase-locking circuits to properly synchronize the injected signal with rectifier input current. The proposed harmonic injection technique uses a voltage signal which is proportional to the rectified ac input voltage. As a result, the injected signal is naturally synchronized with the input voltage. Therefore, to modulate the duty cycle of the power switch, we can employ the output of the diode bridge rectifier of the power stage, which contains a second-order harmonic and higher-order components such as 4th, 6th, 8th,…etc.

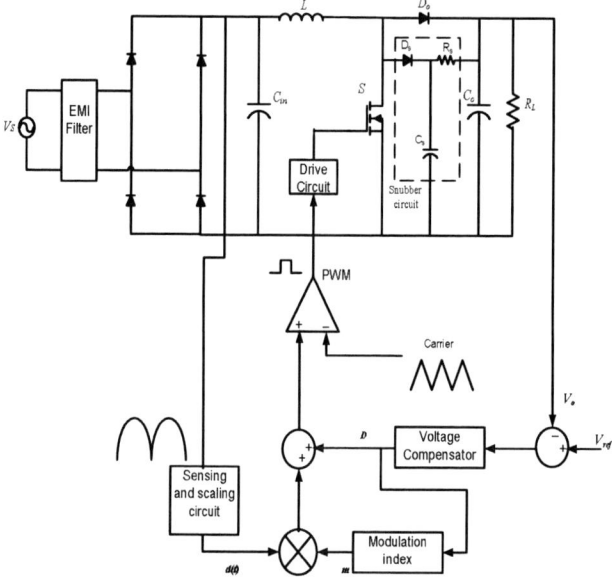

Fig. 6, Single-phase single switch boost converter with proposed modulated duty cycle

Therefore the injected signal can be expressed as:

$d(t) = m\displaystyle\sum_{n=2,4,\dots}^{\infty}\dfrac{-1}{\pi(n-1)(n+1)}\cos(n\omega t)$

or   (7)

$d(t) = m\displaystyle\sum_{n=2,4,\dots}^{\infty}\dfrac{1}{\pi(n-1)(n+1)}\sin(n\omega t + \dfrac{3\pi}{2})$

Where $0 < m < 1$ is the modulation index, and ω is the input frequency. A filter capacitor, C_{in}, is placed between the bridge rectifier and the boost inductor L, in order to reduce the high frequency ripple of the injected signal. C_{in} can also reduce the peak inductor current as well as the power switch peak current due the DCM operation. Therefore, using the modulated duty cycle, which optimizes the time-on of the power switch, and the capacitor C_{in}, the switching loss of the power switch, can be reduced. It is Possible to get the optimal values of modulation index m and the capacitor C_{in}, which make the total harmonic distortion THD as small as possible for a given values of input and output voltages. The optimal values of m and C_{in} that result in a THD of less than 4% are found out through the simulation and later implemented in the experimental converter. The modulation of the duty ratio during a line cycle can be expressed as:

$$D_{\mathrm{mod}}(t) = D[1 + d(t)] \tag{8}$$

Where D_{mod} is the modulated duty cycle, and D is duty cycle in the absence of the modulation. By substituting D in Eq. (6) with the modified duty cycle D_{mod} defined in Eq. (8), the average input current in the presence of the signal injection can be described as

$$I_{in}(\mathrm{inj}) = I_{in}(\mathrm{avg})[1 + d(t)]^2 \tag{9}$$
$$\therefore I_{in}(\mathrm{inj}) \approx I_{in}(\mathrm{avg})[1 + 2d(t)]$$

Where $[d(t)]^2$ term is neglected, since it is much smaller than the unity. Since the third-order harmonic is the dominant harmonic with constant switching frequency PWM control, the input current can be approximately expresses as

$$I_{in} = D^2 I_1 \sin(\omega t) + D^2 I_3 \sin(3\omega t) \tag{10}$$

Where I_1 and I_3 are constant values. Substituting for D with the modified duty cycle D_{mod}, and by setting n=2 in Eq. (7) for the second-order harmonic, since it has the highest magnitude, the input current can be rewritten as

$$I_{in} = I_1[1 + 2m\sin(2\omega t + (3\pi/3))]\sin(\omega t) \tag{11}$$
$$+ I_3[1 + 2m\sin(2\omega t + (3\pi/3))]\sin(3\omega t)$$

Simplifying Eq. 11 yields

$$I_{in} = (I_1 + mI_1 + mI_3)\sin(\omega t)$$
$$+ (I_3 - mI_1)\sin(3\omega t) - mI_3\sin(5\omega t) \tag{12}$$

or

$$I_{in} = I_1'\sin(\omega t) + I_3'\sin(3\omega t) - mI_3\sin(5\omega t)$$

Furthermore, if we consider the fifth-order harmonic in input current Eq. (10), the term $[-mI_3\sin(5\omega t)]$ can reduce the amplitude of this harmonic component by small amount of $(-mI_3)$. Therefore, the total harmonic distortion (THD) of the input current with the modulated duty cycle is given by

$$THD' \approx \sqrt{(I_3 - mI_1)^2 + (mI_3)^2}/I_1' \tag{13}$$

Obviously THD' with injection is much smaller than $THD \approx \dfrac{I_3}{I_1}$ without injection.

IV. SIMULATION AND EXPERIMENTAL RESULTS

The performance of the harmonic reduction with harmonic injected PWM of a boost converter Fig. 6 was verified on both a simulation model and an experimental prototype with the following parameters:

- Input voltage: 106V/50Hz
- Output voltage:215V
- Output power: 500W
- Switching frequency: 20kHz
- Input inductor: 130μH
- Output capacitor: 440μF

In the experimental prototype, a MOSFET (IRFB22N50A) is used as the main switch. The input bridge rectifier is constructed using MUR1540 ultra fast recovery and DSEI60-10A is the output diode. The control circuit for harmonic injection is very simple. As shown in Fig. 6, based on the voltage compensator designed for constant switching frequency PWM, a synchronized harmonic signal with unity amplitude, which contains a second-order harmonic and higher-order components, is used. A multiplier is used to modify the amplitude of the signal with the modulation index, and an adder is used to combine the injected signal with the voltage feedback. In the new control scheme, the PWM signal is obtained by monitoring the input and output voltages only. Since the power factor is a steady-state quantity, the dc output voltage and the rectified input voltage are used. There is no need to monitor the current as in the case of continuous current mode of operation. Thus the present method eliminates the use of a current sensor. Therefore, the monitor circuit is very simple. The converter was simulated using Matlab/Simulink for both constant and modulated duty cycle. The simulation results of the input voltage and current with constant duty cycle are presented in Fig. 7. The corresponding results with the proposed modulated duty cycle are shown in Fig 8. The corresponding experimental results are recorded using a digital storage oscilloscope and presented in Fig. 9 and Fig. 10.

The peak inductor current for discontinuous current mode of operation is usually high. However, using the capacitor C_{in} between the bridge rectifier and the boost inductor the peak current can be reduced. With modulated duty cycle and the capacitor C_{in} the inductor current can be continuous and this is one of the significant advantages of the proposed method. The frequency spectrums of the input current for both constant and modulated duty cycle are presented in Fig. 11- Fig. 14 respectively. As we can see that the dominate third order-harmonic is attenuated using the modulated duty cycle. The value of THD is reduced below 3% and as a result the input power factor is improved.

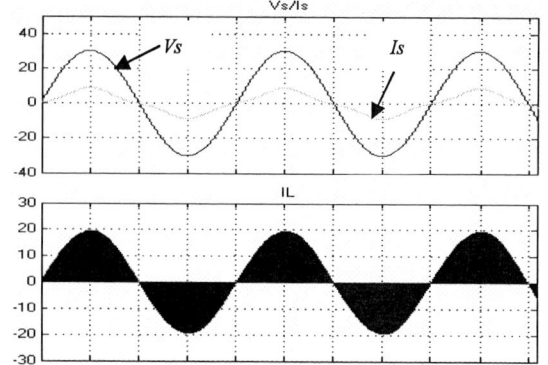

Fig. 7 Simulated input Voltage (scale 1:5), filtered input current and the inductor current with constant duty cycle

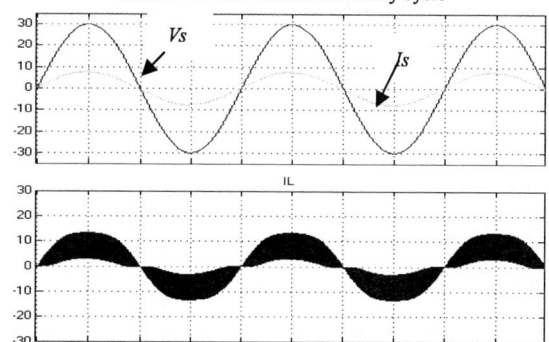

Fig. 8 Simulated input Voltage (scale 1:5), filtered input current and the inductor current with proposed duty cycle

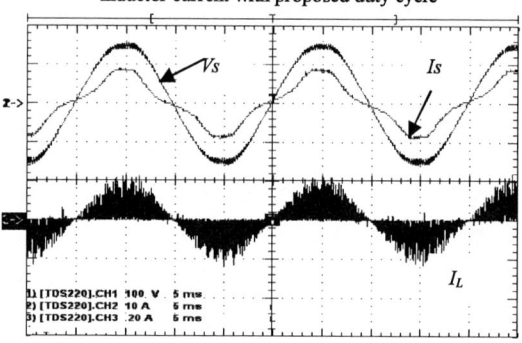

Fig. 9 measured input Voltage, filtered input current and the inductor current with constant duty cycle

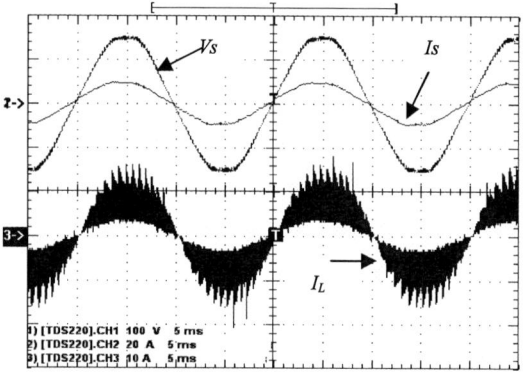

Fig. 10 measured input Voltage, filtered input current and the inductor current with the proposed duty cycle

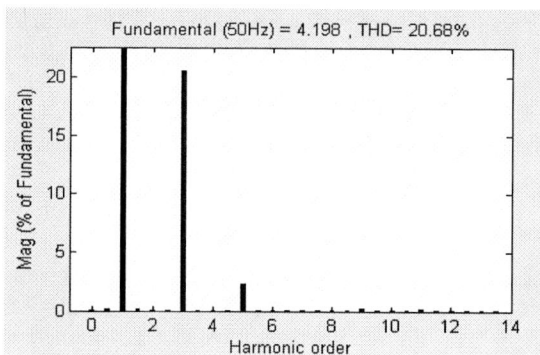

Fig. 11 Frequency spectrum of the simulated input current with constant duty cycle

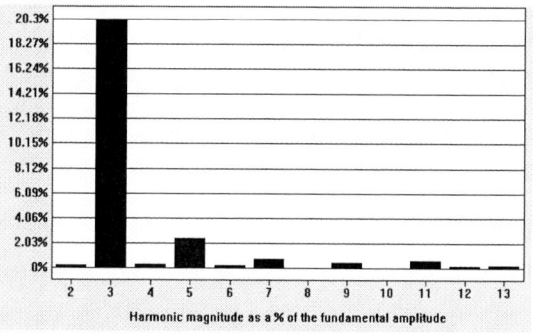

Fig. 12 Frequency spectrum of the measured input current with constant duty cycle, THD=20.59%

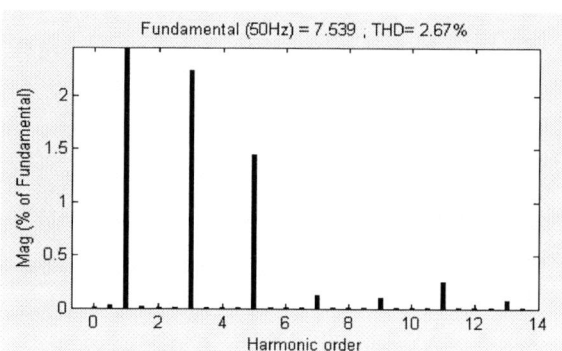

Fig. 13 Frequency spectrum of the simulated input current with the proposed duty cycle

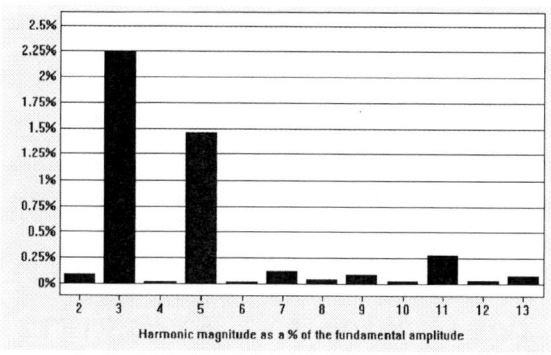

Fig. 14 Frequency spectrum of the measured input current with the proposed duty cycle, THD=2.7%

V. CONCLUSION

The paper reviews the topology, operation, and analysis of the single-phase single-switch DCM boost rectifier using constant switching frequency control, and determines the current distortion caused by the slow discharge of the inductor. The paper proposes a simple, low cost harmonic injection analog controller to modulate the duty cycle of the boost switch such that the third-order harmonic of the input current is reduced and the overall THD is improved. It is found that third-order harmonic, which is the lowest order harmonic (LOH), can be attenuated by adjusting the modulation index (m) shown in Eq.7. The high ripple inductor current of the DCM operation is reduced by using a suitable value of the capacitor C_{in}, which is optimized through the simulation model. It is found that C_{in} can reduce the THD by some extend, and it makes the inductor current to be continuous. Moreover, the injected duty cycle variations are naturally synchronized with the input voltage without using an additional harmonic generator and expensive phase-detecting, phase-locking circuits. It is found that simulated and experimental results match closely. It is also found that the THD is less than 3%.

REFERENCES

[1] O.Gracia, J. A. Cobos, R. Prieto, and J. Uceda, "Single-phase power factor correction: a survey," *IEEE Trans. Power Electron.*, Vol. 18, no. 3, pp. 749-755, May 2003.

[2] Yang, Z. H. and P. C. Sen,, "Recent developments in high power factor switch -mode converters," *IEEE Canadian Conf. Electrical Computer Engineering*, Vol. 2, 477–488, May 1998.

[3] DeFeng weng, and S. Yuvarajian, "Constant switching frequency AC-DC converter using second harmonic injected PWM." *IEEE trans. In Power Electronics*, Vol. 11, no. 1 Jan. 1996.

[4] J. Sun, N. Frohleke, and H. Grotstollen," Harmonic reduction techniques for single switch three-phase boost rectifiers," *in Conf. rec. 1996 IEEE Industrial Applications society* . pp. 1225-1232.

[5] Q. Huang and F.C. Lee, "Harmonic reduction in a single-switch three-phase boost rectifier with high order harmonic injected PWM," *in IEEE power electronic specialist conf. (PESC) Rec,* 1996, pp 1266-1271.

[6] D. S. L Simonetti, , J. Sebastian, and J. Uceda, " single switch three-phase power factor under variable switching frequency and discontinuous input current," *in IEEE PESC Rec,* 1993, pp 657-66.

Comparison of Different Methods to Detect Static Air Gap Asymmetry in Inverter Fed Induction Machines

T.M. Wolbank, P. Macheiner

Vienna University of Technology, Gusshausstrasse 25/372, A-1040, Vienna, AUSTRIA

Abstract-- **Monitoring of air gap asymmetry in induction machines is very challenging even considering steady state and ideal line fed operation. Considering also inverter fed and dynamic operation situation gets even worse. In this investigation two different attempts to detect static eccentricity are compared with respect to their sensitivity at different machines. The first method is based on Fourier transform of the phase currents also denoted motor current signature analysis (MCSA) in literature. The second is based on the exploitation of the transient current response of the machine to an excitation with voltage pulses caused by the inverter switching. It is shown that the performance of the MCSA depends on the number of slots per pole whereas the indicator obtained from the transient current response is not affected by this parameter.**

Index Terms—**induction machine, drive, monitoring, eccentricity.**

I. INTRODUCTION

In modern industrial drives the combination of voltage source inverter and ac machine is nowadays commonly applied. With the gain in performance when using fast switching power electronic devices and high dynamic control algorithms, there is however, also a drawback in terms of system reliability compared to the classical mains-fed applications. Due to additional stress placed on all machine components the occurrences of fault conditions are reported to clearly increased.

In addition also the demands of the customers on overall drive system reliability has generally risen and it is extremely high when looking at applications like x-by-wire and security critical industrial processes. In an increasing number of applications it is thus indispensable to apply effective on-line monitoring systems.

Disregard security aspects, the worst case scenario from industrial point of view is a sudden interruption of industrial processes resulting in a loss of production. Preventive maintenance routines are one means to avoid such unscheduled production stops. They are however, linked also with considerable costs. On-line monitoring is generally unable to prevent the failure of a single component from happening, but it can detect a deterioration of the operating state of a component and thus predict a fault condition in an early stage. As a result not only the repair costs can be reduced by avoiding a total breakdown of a system as well as damage to otherwise unaffected components. Also unscheduled down times can be avoided and the costs of preventive maintenance can be reduced by realizing predictive maintenance.

In order to usefully apply a monitoring method some conditions have to be met. The method has to be able to detect a deterioration of the system state on-line during normal operation of the drive, which implies various ranges of torque and speed as well as steady state and transient operation.

When considering induction machines and their fault conditions three different scenarios can be distinguished. According to studies [1] a breakdown of the bearings account for almost 50% of all motor failures, followed by insulation faults with 35% and rotor cage defects with 10%. Unfortunately the most frequent fault - the bearing defect - is also the most difficult one to detect because of the interaction of the fault condition with possible torque oscillations on the stator current, as shown in [2]. As a result the methods practically applicable to detect bearing degradation or air gap asymmetry are only a few.

The majority of previous investigations and proposed methods to detect air gap eccentricity [2] - [14] are based on spectral analysis of the electrical quantities. They have the advantage that they work well with standard current sensors already available in most drive applications. However, as Fourier transform technique is not sufficient to represent non stationary signals their performance degrade during non stationary operation and they are thus mainly applicable to line-fed operation of the machine. The application of Wavelet transform or other more sophisticated algorithms of spectral analysis is able to slightly improve the dynamic properties of the result (for example [15]), however, the problem of the interaction of fault induced harmonics with harmonics introduced by the inverter, the load, or the control algorithm still exists. In addition it was reported that the applicability of spectral analysis based methods strongly depends on the design (slot number) of the machine.

Another way to monitor eccentricity is to use additional sensors mounted to the machine. The signals obtained are shaft voltages, axial magnetic flux [16], or vibration that are usually also processed by Fourier transform. The only method to detect bearing defect accepted in industrial environment so far is the vibration spectrum. However, for that method extensive measurements have to be made upon commissioning of

978-1-4244-0644-9/07/$25.00 ©2007 IEEE

each drive/load combination to have the reference spectra of the healthy bearings for comparison. This commissioning as well as the mounting of additional sensors is very cost intensive, it is thus not commonly applied.

An attractive alternative especially for the application to inverter fed drives is the exploitation of the transient electrical properties of the machine. By comparing the transient current slope of the three phases resulting from the excitation with voltage pulses - caused by the switching of the inverter – a signal can be obtained that contains the information on all asymmetries present in the machine that influence the transient machine reactances which includes air gap asymmetry.

The goal of this investigation is to compare the results obtained with the Fourier analysis of the fundamental wave currents with the signal resulting from the modulation of the transient current slope on the same machine and at the same points of operation as well as the same level of rotor eccentricity.

In the first part a short review of the two monitoring methods considered is given. Then measurement results made on different induction machines with 5,5kW rated power that are equipped with special bearing housings to adjust and verify the rotational axis of the rotor are presented to show the sensitivity of the detection signals to asymmetries in the air gap.

II. DETECTION OF AIR GAP ASYMMETRY – HARMONIC ANALYSIS

As soon as the air gap of an induction machine is not uniform along the circumference the condition of air gap asymmetry exists. This asymmetry may be static if it is spatially fixed and/or dynamic if the spatial position is changing with time or rotor position. In practical operation usually both types of asymmetries are present simultaneously. The reason for air gap asymmetry may be found in the manufacturing process, a shaft misalignment between machine and load, or also bearing currents induced by the inverter operation that may also lead to a rapid degradation of the bearings. The problem of this type of fault conditions is the resulting unbalanced magnetic pull leading to acoustic noise and vibration at the bearings which adds to the main reason for the asymmetry and then results in a subsequent bearing failure.

During the manufacturing of the machine usually a relative eccentricity of less than 10% is achieved by the production process as well as the quality control management. In general this value is accepted as normal. If the level is above 20% this would be considered as unacceptable and if a level of 50% is detected the corresponding motor has a serious problem and should be replaced or serviced immediately [3].

The eccentricity leads to anomalies in the air gap flux density and thus influences the inductances of the machine. As a result, harmonics in the stator current are generated, which can be detected for example by spectral analysis. The frequencies of the harmonics due to the asymmetries caused by the slotting and eccentricity can

be calculated according to the following eq. (1), where f_{stator} denotes the machine supply frequency and R gives the number of rotor slots [5].

$$f_{asym} = f_{stator}\left((kR \pm n_d)\left(\frac{1-s}{p}\right) \pm n_w \right) \qquad (1)$$

The value of k gives the number of the slot harmonic and n_d is the order of the eccentricity with 0 for static and 1,2,3... for dynamic eccentricity. The pole pair number and the per unit slip are given by p and s respectively. The order of the stator magneto-motive force (mmf) is represented by n_w. With eq. (1) the magnitude of the side bands of the slot frequencies caused by an eccentricity can be calculated.

If the side bands of the stator frequency caused by the existence of static and dynamic eccentricity are to be detected [3], the following eq. (2) can be used with f_{mech} as the rotational frequency of the rotor and the value of k being 1,2,3....

$$f_{asym} = f_{stator} \pm k f_{mech} \qquad (2)$$

In practical operation usually the application of (2) is preferred due to the higher magnitudes of the corresponding side bands when compared to the slotting related side bands (1).

When working in the frequency domain the above mentioned side bands can be detected and changes in their magnitude used as eccentricity indicator. However, caution has to be taken with some side effects coming from the operation of the machine. A changing load torque may also result in current harmonics similar to that calculated with the above equations, especially (2). In [4] a method is proposed to reduce this influence on the eccentricity indicator via decoupling. In addition changes in the modulation index of the PWM may also introduce similar harmonics during transient states. As a consequence steady state operation or only slow changing operating conditions are thus preferred for the application of the frequency-domain-based methods.

For the evaluation of the current spectrum it is possible to use either the phase values or the current space phasor as input to the spectral analysis. Usually the usage of phase values is preferred in literature. Thus for the comparison carried out in this investigation only the phase current values have been considered.

Changing the input signal from samples taken from the fundamental wave to measurements of the transient current slope leads to a different type of detection signal that also has different properties and sensitivity to air gap eccentricity.

III. DETECTION OF AIR GAP ASYMMETRY – TRANSIENT CURRENT REACTION

If a machine is fed by a voltage source inverter it is persistently excited with voltage pulses caused by the PWM and the current control scheme. The reaction of the machine, a transient current change can be sampled and exploited to extract the values of the three transient phase reactances. Thus a signal is obtained that reacts very sensitive to asymmetries in the air gap length and/or saturation level.

Assuming a symmetrical machine the reaction of the machine to voltage pulses is determined by the stator equation.

$$\underline{v}_S = r_S \cdot \underline{i}_S + l_l \cdot \frac{d\underline{i}_S}{d\tau} + \frac{d\underline{\lambda}_R}{d\tau} \qquad (3)$$

The transient change in the machine current $d\underline{i}_S/d\tau$ is thus determined by the voltage $\underline{v}_S$ applied, whose magnitude corresponds to the dc link voltage during each active inverter switching state, and whose direction is determined by the switching state. In addition also the value of the leakage inductance l_l, the stator resistance voltage drop $r_S \cdot \underline{i}_S$, as well as the voltage induced by the time derivative (back emf) of the rotor flux $\underline{\lambda}_R$. influence the transient current slope.

As the dc link voltage, back emf, as well as the stator resistance can be assumed constant during two subsequent voltage pulses of only a few ten µs duration, their influence can be eliminated leaving dominantly only the influence of the leakage inductance l_l . This elimination can be done each time that two subsequent pulses pointing in different spatial directions are sampled.

The signal after that elimination is influenced by the mean value of the transient phase reactances as well as by their modulations. Comparing the actual values of the three phases an asymmetry information can be obtained. In a faultless machine there are already some asymmetries detectable caused by inherent saliencies as well as the measurement setup.

These inherent saliencies are caused by spatial saturation of the lamination, the slotting of the machine, as well as anisotropy of the material or geometrical design. Each of these saliencies is directly connected in its spatial direction to either the electrical flux/current angle or the rotor angle and in addition their magnitudes also depend on the point of operation. This signal behavior is individual for each type of machine and deterministic. It can thus be identified to remove all asymmetric components in the signal of a faultless machine.

If, in addition to the inherent saliencies, an air gap asymmetry is present, there is a distinct change in the different signal components. When looking at machines with one pole pair the offset component will dominantly be affected offering the possibility not only to detect air gap asymmetry but also to clearly identify its direction. For machines with higher pole pair numbers the influence of eccentricity on the offset components is clearly reduced due to geometrical reasons. At the same time however, the influence on the other signal components (saturation and slotting) is increased [12] leading to a change in these signal components as preferred fault indicator.

IV. COMPARISON OF METHODS

Both methods described in the previous chapters where tested on induction machines with 5,5kW rated

power with 1, 2 and 3 pole pairs respectively equipped with special bearing housings to adjust different levels and orientations of rotor eccentricity. The laminations of the machines have 36 stator slots and 28 rotor slots. The results are given in the following figures. The eccentricity was realized by two eccentric rings on each side of the machine. The actual air gap length was then verified on four different positions along the air gap using a digital camera.

In Fig. 1 and Fig. 2 the two methods are compared on a machine with two pole pairs. The machine was operated at steady state and no load to obtain ideal conditions for the method using the Fourier transform. When looking at the signals obtained from the transient current response (Fig. 1) it can be seen that the signal components related to both saturation ($+4^{th}$) as well as that of the inter-modulation of saturation and slotting (-32nd) show a distinct change when eccentricity is present.

It has however to be stressed that for the application of this method it is not necessary to work in the frequency domain. This representation has only been chosen to obtain an easier comparison of the results.

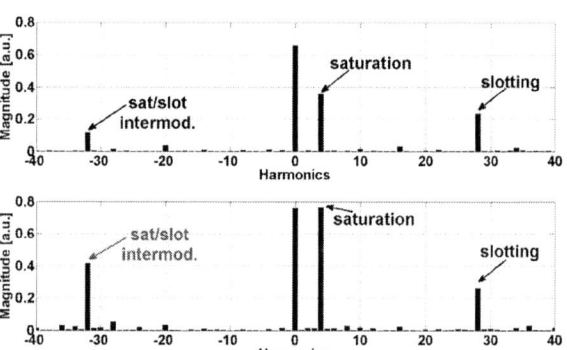

Fig. 1: Fault indicator obtained from transient current response. Machine with two pole pairs; without (upper) and with (lower) 50% air gap eccentricity.

Looking at Fig. 1 (upper diagram) it can be seen that the inherent asymmetries of saturation and slotting are prominent together with the signal offset caused by the mean value of the transient reactances. The intermodulation of the saturation and the slotting effect is relatively small but still clearly detectable.

Introducing a static eccentricity of 50% leads to a distinct change (lower diagram). As can be seen the offset is increased by about 10%. In addition the magnitude of the saturation component is strongly increased by almost 100%. The biggest impact of eccentricity can be observed in the intermodulation component that is increased by almost 400%. The slotting component is the only one whose magnitude is almost independent from the eccentricity.

The same operating condition has been adjusted in Fig. 2. To ensure constant frequency for the spectral transformation the machine was operated with constant voltage, frequency as well as zero load. To avoid distortions due to over-modulation, the fundamental frequency was set to 25Hz corresponding to half rated frequency. Then the harmonic spectrum of one phase current was calculated.

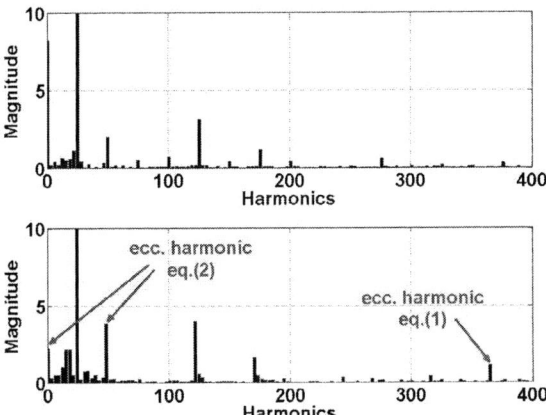

Fig. 2: Fault indicator obtained from current spectrum. Machine with two pole pairs; without (upper) and with (lower) 50% air gap eccentricity.

Again the upper and lower diagram of the figure gives the result for the symmetrical case and an eccentricity of 50%. For the spectrum of the stator current (Fig. 2) this pole pair/ slot number ratio is unfavorable making a detection of the eccentricity difficult even during steady state operation. Comparing the upper and the lower diagram it can be seen that the 50th harmonic ($f_{el}+2f_{mech}$ according to (2)) is increased by around 90% by the eccentricity. The 365th harmonic (according to (1)) is changed by about 300%. The main problem with the line current harmonics is the dominating influence of the fundamental wave which in that case had a magnitude of 240, what is in the range of 100 times bigger than the fault indicating harmonics. Comparing the ratio of the magnitudes for Fig. 1 it can be seen that there all signal components are in the same range of magnitude.

Basically also the signal offset is changed by eccentricity as indicated in Fig. 2. It has to be stressed that the signal offset also includes non-ideal sensor behavior as well as inverter non-linearity. Thus for application to eccentricity detection the usage of the signal offset is usually avoided.

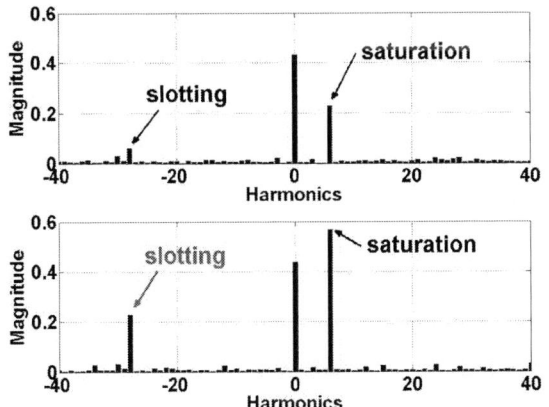

Fig. 3: Fault indicator obtained from transient current response. Machine with three pole pairs; without (upper) and with (lower) 50% air gap eccentricity.

Changing the pole number from 4 to 6 leads to different results as can be seen in the following Fig. 3 and Fig. 4.
Again the same point of operation (no load, rated flux) was used and the same eccentricity (50%) adjusted (lower diagrams).
Looking at the results obtained from the transient current slope again the three main signal components are visible. For this machine the intermodulation component is not detectable leaving only the offset, the saturation and the slotting. When introducing the eccentricity the slotting component is increased by about 400%.

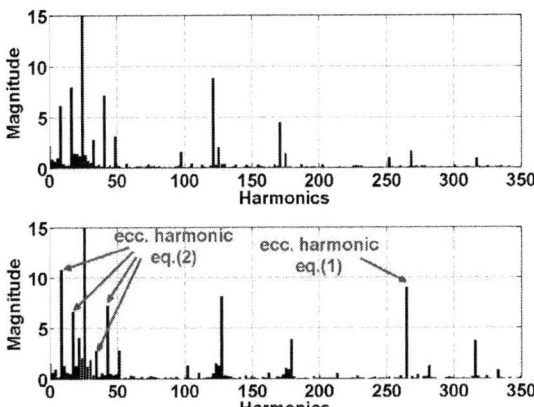

Fig. 4: Fault indicator obtained from current spectrum. Machine with three pole pairs; without (upper) and with (lower) 50% air gap eccentricity.

The offset now stays independent from the eccentricity and the saturation component is increased to approximately 150% . The fundamental frequency again was set to about 25Hz.

The fundamental wave again is dominating with a magnitude of 270. The side bands resulting from (2) are almost unchanged by eccentricity except for the 8Hz (25 - 2*(25/3)) that is increased by 75%. The slotting related side band (315Hz) according to (1) shows a very distinct change of about 900% compared to the symmetrical case.

As the reference value for the speed control of the load machine was set by a potentiometer the two fundamental frequencies for the symmetrical and the eccentricity measurement are not identical. This results in a shift of the slot harmonic side band from 300Hz to 315Hz in the two diagrams of Fig. 4.

For that machine (6 pole) as well as operating condition the two methods can be considered equal in their performance. It has however, again to be stressed that the steady state operation is a requirement for the sensitivity of the method based on the harmonic analysis. Whereas the transient current change method does not rely on the frequency domain but can also be realized in the time domain leading to almost constant performance even during transient operation.

Finally also a machine with two poles was investigated. As can be seen in the two diagrams of Fig. 5 the modulations of the slotting as well as that of the saturation stay almost independent from the eccentricity.

The main fault indicator now is the signal offset that is raised by 240% when the eccentricity of 50% is introduced. It has also to be stressed that for the two pole machine also the direction of the signal offset delivers information on the eccentricity and has thus also to be considered. For the measurement depicted in Fig. 5 the angle changes from ~0° to -39°.

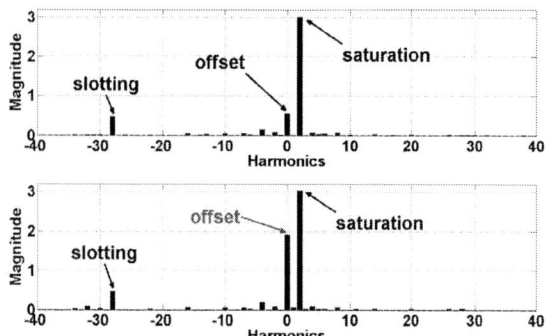

Fig. 5: Fault indicator obtained from transient current response. Machine with one pole pair; without (upper) and with (lower) 50% air gap eccentricity.

Finally in Fig. 6 the phase current spectrum is given for the two pole machine. As already mentioned in Fig. 2, also here the signal offset delivers information on the eccentricity. For the reasons given above its application is usually avoided. The dominating fundamental wave again has a magnitude of around 290. The significant 3rd harmonic is probably caused by a winding asymmetry. As the machine is additionally equipped with tapped windings. The harmonic related to (1) shows almost no influence on the introduced eccentricity.

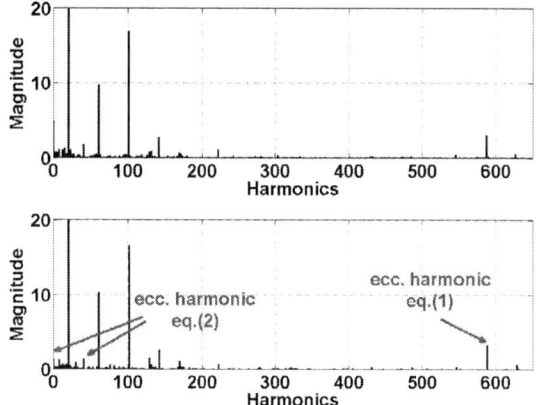

Fig. 6: Fault indicator obtained from current spectrum. Machine with one pole pair; without (upper) and with (lower) 50% air gap eccentricity.

V. CONCLUSIONS

Two different attempts to detect static eccentricity were compared with respect to their sensitivity at different machines. The measurements were performed only at steady state operation to give optimum performance for the Fourier transform. The method based on Fourier transform of the phase currents is practically applicable only for specific number of slots / pole pairs. The method based on the transient current response to

voltage pulses is not affected by this design parameter.

Looking at the magnitude of the different signal components the fundamental wave in the current spectrum is about 100 times bigger than the signal component correlated to eccentricity making a detection difficult in practical operation. For the transient current response all components are in the same range of magnitude what is a clear advantage with respect to noise immunity.

In addition its performance is not limited to steady state operation. In steady state under ideal operating conditions both methods show more or less the same sensitivity. When changing the operating conditions like the machine design, leaving steady state operation, or when considering asymmetrical supply voltage of the machine by the inverter due to dead time or the dynamics introduced by the control loops of current and/or torque, the transient current response shows clear advantages.

ACKNOWLEDGEMENT

The authors gratefully acknowledge the financial support of the Austrian Science Foundation - "Fonds zur Förderung der wissenschaftlichen Forschung" (FWF) - under grant no. P17595.

REFERENCES

[1] G.B. Kliman, W.J. Premerlani, B. Yazici, R.A. Koegl, J. Mazereeuw, "Sensorless, Online Motor Diagnostics", *IEEE Computer Applications in Power*, Vol.10, No.2, pp.39-43, (1997)

[2] R.R. Schoen, Th.G. Habetler, F. Kamran, R.G. Bartheld, "Motor Bearing Damage Detection Using Stator Current Monitoring", *IEEE Transactions on Industry Applications*, Vol.31 No.6, pp.1274-1279, (1995)

[3] W.T. Thomson, A. Barbour, "On-Line Current Monitoring and Application of a Finite Element Method to Predict the Level of Static Airgap Eccentricity in Three-Phase Induction Motors", *IEEE Transactions on Energy Conversion*, Vol.13 No.4, pp.347-357, (1998)

[4] A. Stavrou, J. Penman, "Modelling Dynamic Eccentricity in Smooth Air-Gap Induction Machines", *Proceedings of International Electric Machines and Drives Conference IEMDC*, Vol.1, pp.864-871, (2001)

[5] R.R. Schoen, Th.G. Habetler, "Evaluation and Implementation of a System to Eliminate Arbitrary Load Effects in Current-Based Monitoring of Induction Machines", *IEEE Transactions on Industry Applications*, Vol.33 No.6, pp.1571-1577, (1997)

[6] A.B. Yazici, G.B. Kliman, "An Adaptive Statistical Time-Frequency Method for Detection of Broken Bars and Bearing Faults in Motors Using Stator Current", *IEEE Transactions on Industry Applications*, Vol.35 No.2, pp.442-452, (1999)

[7] S. Nandi, S. Ahmed, H.A. Toliyat, "Detection of Rotor Slot and Other Eccentricity Related Harmonics in a Three Phase Induction Motor with Different Rotor Cages", *IEEE Transactions on Energy Conversion*, Vol.16 No.3, pp.253-260, (2001)

[8] A. Ferrah, P.J. Hogben-Laing, K.J. Bradley, G.M. Asher, M.S. Woolfson, "The Effect of Rotor Design on Sensorless Speed Estimation Using Rotor Slot Harmonics Identified by Adaptive Digital Filtering Using The Maximum Likelihood Approach", *Proceedings of IEEE Industry Applications Annual Meeting*, New Orleans, pp.128-135, (1997)

[9] A.J. Cardoso, E.S. Saraiva, "Computer-Aided Detection of Airgap Eccentricity in Operating Three-Phase Induction Motors by Park's Vector Approach", *IEEE Transactions on Industry Applications*, Vol.29 No.5, pp.897-901, (1993)

[10] Schroedl M.; "Sensorless Control of AC Machines at Low Speed and Standstill based on the Inform Method", *IEEE IAS Annual Meeting*, San Diego, Vol.1, pp.270-277, (1996)

[11] Th.M. Wolbank, R. Woehrnschimmel, "Investigating the dependence of induction machines transient reactances on stator teeth saturation reference to sensorless control'; *IEEE Power Electronics Specialists Conference*, Vol.2, pp.828-833, (2001)

[12] Th.M. Wolbank, P. Macheiner, "Detection of airgap asymmetry in induction machines with different pole pair number using the current step response" *Proceedings of International Electric Machines and Drives Conference, IEMDC*, (2007)

[13] D.G. Dorell, W.T. Thomson, S. Roach, "Combined Effects of Static and Dynamic Eccentricity on Airgap Flux Waves and the Application of Current Monitoring to Detect Dynamic Eccentricity in 3-Phase Induction Motors", *Electrical Machines and Drives Conference*, pp.151-155, (1995)

[14] H.A. Toliyat, M.S. Arefeen, A.G. Parlos, "A Method for Dynamic Simulation and Detection of Air-Gap Eccentricity in Induction Machines", Proceedings of Industry Applications Annual Meeting IAS, Vol.1, pp.629-636, (1995)

[15] J.A. Antonino-Daviu, M. Riera-Guasp, J.R. Folch, M.P. MolinaPalomares, "Validation of a New Method for the Diagnosis of Rotor Bar Failures via Wavelet Transform in Industrial Induction Machines", *IEEE Transactions on Industry Applications*, Vol.42, No.4, pp. 990-996 (2006)

[16] H. Henao, G.A. Capolino, C. Martis, "On the stray flux analysis for the detection of the three-phase induction machine faults", *Proceedings of IEEE Industry Applications Conference*, IAS, Vol. 2, pp.1368-1373, (2003)

Analysis of the Synchronous Torques in a Split Phase Induction Motor

P. Scavenius Andersen*, D. G. Dorrell**, N. C. Weihrauch* and P. E. Hansen*

*Danfoss Compressors GmbH, Flensburg, Germany
**University of Glasgow, Glasgow, UK

Abstract—This paper puts forward a method for calculating the synchronous torque dips in a split-phase induction machine. First it derives the equivalent circuits so that the torque speed/curve can be obtained over a full speed range (including asynchronous torque oscillations). When the currents are resolved these are used to calculate the synchronous torques from a set of interactions between the machine MMFs and the slot permeances. This gives the synchronous torques (speed and magnitude) which can be superimposed onto the torque/speed curve. The method is tested experimentally and found to give reasonable results.

Index Terms—Split phase induction motors, asynchronous torques, synchronous torques.

I. Introduction

The split phase motor, where there is a single phase main winding and an orthogonal auxiliary winding which is connected in parallel with the main winding (often with a series resistor or capacitor) is the preferred drive for a vast array of water and chemical pumping application as well as refrigeration applications, in a range possibly up to a few kW. They are cheap and relatively efficient and do not require a 3-phase supply.

Many analysis techniques still assume a sinusoidal winding so that only the fundamental forwards and backwards rotating MMFs are modelled. The usual ways to analyze these machines use the cross-field or revolving field techniques. These two methods are well documented by Veinott [1]. For a balanced two-phase machine the 5th MMF harmonic asynchronous torque rotates forwards and the 7th backwards, etc, (whereas for a three-phase machine the 5th rotates backwards and the 7th rotates forwards) For a split-phase machine, there may also be a substantial a 3rd harmonic MMF asynchronous torque dip as well as other odd harmonics that rotate both forwards and backwards [2].

Recently, with the drive for energy efficiency, and also design improvement, the operation and simulation of the split phase motor has attracted more detailed interest [3]-[6]. These address the issues of asynchronous torque dips as mentioned above and also inter-bar rotor currents. Here, we will investigate the issue of synchronous torque spikes.

This work is forms part of the PhD studies of Mr Scavenius Andersen and he is grateful for the help and support of Danfoss Compressors GmbH.

II. Analysis

The analysis builds on the theory put forward in [4] and [6]. The permeance waves due to the slotting are described in the Appendix. First, it is illustrated how the permeance harmonics due to slotting can interact with the MMF to produce flux waves of the correct pole number to interact with other MMF waves (or iven the souce MMF) which rotate at different rotational velocities. At certain rotor speeds the MMF and permeance-sourced flux waves will rotate at the same speeds to generate constant torque. Therefore a synchronous locking torque is a pulsating torque where at certain speeds the frequency of the oscillation is zero. Then the paper goes on to examine explicit cases.

A. Interaction of MMF and permeance waves

The machine MMF will interact with the slot permeance coefficient:

$$B = F_m \times P \tag{1}$$

where the MMF F_m of order m is rotating. Consider a general permeance coefficient and a pth harmonic forward-rotating component of MMF so that

$$B(t, \theta) = F_{fp} \cos(\omega t - p\theta) \times P_x \cos(x\theta - y\omega_r t)$$

$$= \frac{F_{fp} P_x}{2} \begin{bmatrix} \cos((\omega + y\omega_r)t - (p+x)\theta) \\ + \cos((\omega - y\omega_r)t - (p-x)\theta) \end{bmatrix} \tag{2}$$

Where, for different MMF and permeance harmonics:

Term 1: $y = 0$ $x = 0$
Term 2: $y = 0$ $x = m N_s$
Term 3: $y = n N_r$ $x = n N_r$
Term 4: $y = -n N_r$ $x = m N_s - n N_r$
Term 5: $y = n N_r$ $x = m N_s + n N_r$

and the stator slot number is N_s and the rotor slot number is N_r. It can be seen that the air-gap has two resulting rotating fields in (2) per Term, whose number of pole-pairs and rotational speeds are given by

$$n_{f1} = p + x \qquad n_{f2} = p - x$$

$$\omega_{f1} = \frac{\omega + y\omega_r}{p + x} \qquad \omega_{f2} = \frac{\omega - y\omega_r}{p - x}$$

If the pole-pair number is negative this corresponds to the reverse direction. ω_{f1} and ω_{f2} are the rotational speeds of the fields. They create net torque with rotor MMFs of same pole number when their speeds coincide, i.e., when

$$\omega_{r1} = \frac{\omega_{f1}(p+x) - \omega}{y} \quad \text{with } \omega_{f1} = \pm \frac{\omega}{n_{f1}} \tag{3}$$

In the same way, the interaction of the pth harmonic backwards-rotating MMF with the various permeance coefficient terms will result in a field set given by

$$B(t,\theta) = F_{bp}\cos(\omega t + p\theta) \times P_x \cos(x\theta - y\omega_r t)$$

$$= \frac{F_{bp}P_x}{2}\left[\begin{array}{c}\cos((\omega + y\omega_r)t + (p-x)\theta) \\ + \cos((\omega - y\omega_r)t + (p+x)\theta)\end{array}\right] \quad (4)$$

which is also composed of two rotating fields per Term, whose number of pole-pairs and rotational speeds are given by

$$n_{b1} = p - x \qquad \omega_{b1} = -\frac{\omega + y\,\omega_r}{p-x} = \frac{\omega + y\,\omega_r}{x-p}$$

$$n_{b2} = p + x \qquad \omega_{b2} = \frac{\omega - y\omega_r}{-x-p}$$

with p = MMF harmonic under consideration. In summary:

- The forward and backward magnetizing MMF with p pole-pairs each interact with the general permeance term with x pole pairs. (In total, 2 MMFs per harmonic)
- For each interaction, two counter rotating flux densities with

$$n_b = (p+x) \text{ or } (p-x) \quad (5)$$

  pole pairs are created. (In total, 4 flux densities per harmonic $[p,m,n]$)

- The rotational speed of each flux density is determined by the rotor speed. Each flux density may produce average torque with either the forward or the backward rotor MMF of same pole number when their speeds coincide, i.e. at either positive or negative synchronous speed of the rotor MMF. (8 speeds per harmonic $[p,m,n]$)

This means, that for each harmonic component of winding MMF, eight components of synchronous locking torque need to be considered (i.e., Terms 4 and 5 for (2) and (4) – Term 2 is the main torque-producing field that is un-modulated by the permeance and Terms 2 and 3 produce field harmonics that are too high to interact and produce synchronous torques).

B. MMF, permeance and torque

The synchronous torque components arise as a combination of MMFs interacting with flux densities of various harmonic orders. In order to calculate the magnitude of the torque components a detailed knowledge of harmonic magnetization as well as rotor MMFs and permeances is required.

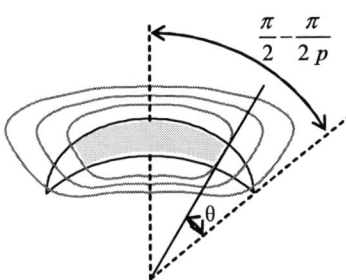

Fig. 1. Harmonic pitch definition.

C. Magnetizing MMF of the pth harmonic winding

The magnitude of the forward revolving, p^{th} harmonic magnetizing (or resulting) MMF can be calculated from[6]:

$$F_{mag\,p} = \int_0^{\pi/p} \frac{pN_{p\,main}}{4}\sqrt{2}\,\overline{I}_{mag\,p\,main}\sin(p\theta)d\theta$$

$$+ \int_0^{\pi/p} \frac{pN_{p\,main}}{4}\frac{-j\beta_p}{\sin\left(\frac{p\pi}{2}\right)}\sqrt{2}\,\overline{I}_{mag\,p\,aux}\sin(p\theta)d\theta \quad (6)$$

$$= \left|\frac{N_{p\,main}}{\sqrt{2}}\left(\overline{I}_{mag\,p\,main} + \frac{-j\beta_p}{\sin\left(\frac{p\pi}{2}\right)}\overline{I}_{mag\,p\,aux}\right)\right|$$

where $pN_{p\,main}/4$ is the winding amplitude of half of the pth harmonic winding. (the total winding number is N_p). In addition there is the harmonic winding ratio β_p, which is defined by

$$\beta_p = \frac{N_{p\,aux}}{N_{p\,main}} \quad (6)$$

And the current is solved for the winding harmonics as given in [4] (the focus of this paper is the solution of the synchronous torques rather than the main torque and current components).

Having established an expression for resulting MMFs, the flux density component (with $p \pm x$ pole-pairs) can be calculated from the permeance. Therefore an expression for the actual permeances is needed

D. Permeance and permeance coefficient

The air-gap flux of one pole pitch from the pth harmonic winding is considered. Firstly, this can be expressed by magnetizing MMF and air-gap permeance $Perm$ as

$$\Phi_p = \frac{2}{\pi}F_p\,Perm$$

$$= \frac{2I}{\pi}\int_0^{\pi/p} N_p\sin(p\theta)d\theta\,Perm \quad (8)$$

$$= \frac{4IN_p}{\pi p}\,Perm$$

The factor $2/\pi$ takes into account the sinusoidal distribution of the flux, since it represents the ratio between the areas of a sine function and a rectangular function with equal amplitude.

Secondly, a relationship between the flux linkage and the total flux of one pole pitch of the pth harmonic winding must be found. Fig. 1 shows a pole pitch of a pth harmonic, sinusoidally distributed winding. At angle θ, the shaded area of the coil is linked by a number of flux lines, as shown in red. At angle θ, the linked winding number is

$$N_{linked}(\theta) = 2\int_\theta^{\frac{\pi}{2}} N_p\sin(p\theta)d\theta = \frac{2N_p}{p}\cos(p\theta) \quad (9)$$

and the total flux is

$$\Phi(\theta) = \Phi_p \sin(p\theta) \tag{10}$$

Observing a small section with width $d\theta$ at angle θ, the flux linkage is

$$d\Psi(\theta) = \frac{d}{d\theta} N_{linked}(\theta) \cdot \frac{d}{d\theta} \Phi(\theta) d\theta$$
$$= -2\hat{N}_p \sin(p\theta) \cdot p\Phi_p \cos(p\theta) \tag{11}$$

From this, the total flux linkage of one pole pitch can be found by integrating over the entire pole pitch, i.e.:

$$\Psi_p = \int_{\frac{\pi}{2}-\frac{\pi}{p2}}^{\frac{\pi}{2}} d\Psi(\theta) = -2pN_p\Phi_p \int_{\frac{\pi}{2}-\frac{\pi}{p2}}^{\frac{\pi}{2}} \sin(p\theta)\cos(p\theta) d\theta$$

$$= -pN_p\Phi_p \int_{\frac{\pi}{2}-\frac{\pi}{p2}}^{\frac{\pi}{2}} \sin(2p\theta) d\theta$$

$$= \frac{N_p\Phi_p}{2} \left[\cos(p\pi) - \cos(p\pi - \pi) \right]$$

$$= \frac{N_p\Phi_p}{2} \left[2\sin\left(\pi\left(\frac{2p-1}{2}\right)\right)\sin\left(-\frac{\pi}{2}\right) \right]$$

$$= N_p\Phi_p \tag{12}$$

Solving for the pitch flux, and introducing the relationship between flux linkage, current and inductance gives

$$\Phi_p = \frac{\Psi_p}{N_p} = \frac{X_{p,mag}}{N_p\omega} I = \frac{r L_{stk} \pi \mu_0 \left(\frac{N_p}{p}\right)^2}{N_p l_{gap}} I \tag{13}$$

Solving for *Perm* yields

$$\frac{4I\hat{N}_p}{p\pi} Perm = I \frac{rL_{stk}\pi\mu_0\hat{N}_p}{p^2 l_{gap}} \quad \Rightarrow$$

$$Perm = \frac{rL_{stk}\pi^2\mu_0}{4pl_{gap}} \tag{14}$$

This expression relates to the flux rather than the flux density, which is sought here. Therefore, the relationship between permeance and permeance coefficient is the same as that between flux and flux density *for one pole pitch*, i.e.:

$$\frac{\Phi}{p} = \int B dA = rL_{stk} \int_0^{\pi/p} B\sin(p\theta)d\theta = \frac{rL_{stk}2B}{p} = \alpha B \tag{15}$$

In order to obtain an expression for the permeance coefficient, the following comparisons are made. The flux is given by

$$\Phi_p = F \times Perm = \alpha B \tag{16}$$

and the flux density is given by

$$B_p = F \times P \tag{17}$$

Dividing (16) by (17) yields

$$\frac{\Phi_p}{B_p} = \frac{Perm}{P} = \alpha \tag{18}$$

from which

$$P = \frac{Perm}{\alpha} = \frac{\pi^2\mu_0}{8l_{gap}} \tag{19}$$

where l_{gap} is the amplitude of the harmonic gap lengths.

E. Rotor MMF calculation

The rotor MMF is expressed in terms of its referred value, since only referred currents are present in the equivalent circuit (Fig. 2). Hence

$$F_{Rn_b} = \int_0^{\pi/n_b} \frac{n_b N_{n_b\,main}}{4} \sqrt{2}\, \overline{I}'_{Rn_b\,main} \sin(n_b\theta)d\theta$$

$$+ \int_0^{\pi/n_b} \frac{pN_{n_b\,main}}{4} \frac{-j\beta n_b}{\sin\left(\frac{n_b\pi}{2}\right)} \sqrt{2}\, \overline{I}'_{Rn_b\,aux} \sin(n_b\theta)d\theta \tag{20}$$

$$= \left| \frac{N_{n_b\,main}}{\sqrt{2}} \left(\overline{I}'_{Rn_b\,main} + \frac{-j\beta_{n_b}}{\sin\left(\frac{n_b\pi}{2}\right)} \overline{I}'_{Rn_b\,aux} \right) \right|$$

Where n_b is given by (5). The relationship between MMF and current distribution amplitude is given by

$$\hat{I}_{Rn_b} = \frac{n_b F_{Rn_b}}{2}$$

$$= \left| \frac{n_b N_{n_b\,main}}{2\sqrt{2}} \left(\overline{I}_{Rn_b\,main} + \frac{-j\beta_{n_b}}{\sin\left(\frac{n_b\pi}{2}\right)} \overline{I}_{Rn_b\,aux} \right) \right| \tag{22}$$

Where $N_{n_b\,main}$ is the total winding number of one pole pitch of the n_bth harmonic winding.

F. Torque calculation

From (22) and (1) the magnitudes of the locking torques can be determined. The maximum value of locking torque occurs when the flux density and the current distribution are in phase, and the minimum value occurs when they are in anti-phase. The magnitude is given by

$$T_{sync\,p} = \pm rL \int_0^{2\pi} \hat{I}_{R\,p} \sin(p\theta)\hat{B}_p \sin(p\theta)d\theta$$

$$= \pm rL\hat{I}_{R\,p}\hat{B}_p\pi \tag{23}$$

G. Calculation synchronous locking torques

As mentioned in the previous sections, several independent synchronous locking torques can be calculated from the pth magnetizing MMF, which interacts with each permeance term to create flux density of harmonic order n_b. The torque arises from the rotor MMF of the same harmonic order n_b. The magnitude of the torque presupposes that the magnitudes of the various flux densities and MMFs are known. Hence it is necessary to obtain realistic values for these.

This method can be applied to several methods of calculation as a post-processing method of assessing the synchronous torques.

III. SIMULATIONS

The equivalent circuit including winding harmonics is used to simulate the machine. This includes four separate winding harmonics of both forward and backwards rotation. This gives a system of twenty equations when considering both main and auxiliary windings. The fundamental and third winding harmonic variables are always considered, since these, or at least the fundamental, will dictate the line current drawn from the supply. In addition, the winding harmonics of order p (which is the one that interacts with the permeance harmonic to produce flux) and n_b (which is the winding harmonic of the same order as the resulting flux) are considered in order to obtain the values needed for torque calculation. This is illustrated in Fig. 2, where the sub-circuits representing four forward harmonic windings of the main winding are shown. In the full circuit used for calculation, four backwards windings are also included and the auxiliary winding circuit.

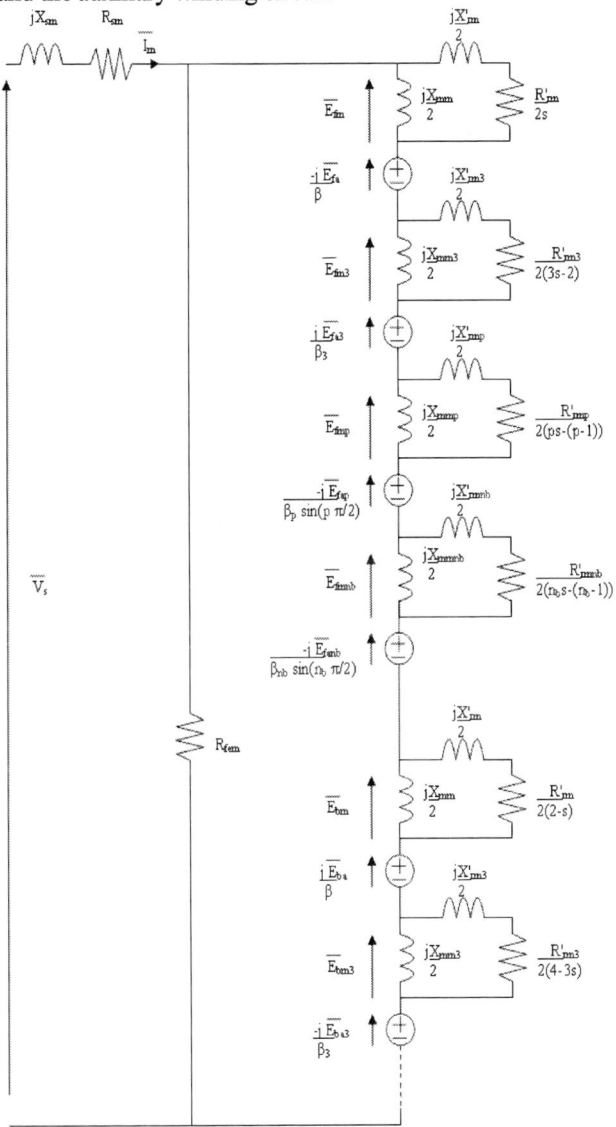

Fig. 2. Equivalent circuit for the main winding and forwards-rotating flux waves (and two backwards-rotating waves together with linkages with the auxiliary windings).

A. Simulation models

Several machines with different types of rotor were simulated and verified experimentally. In the simulations, four machines were investigated. These were a 220 V 50 Hz machine with 20 and 24 bar rotors (unskewed) and a 115 V 60 Hz machine with 18 and 24 bar rotors (unskewed).

For experimental verification a set of rotors were constructed which had 18 rotor bars (unskewed), 24 rotor bars (unskewed) and 28 bar rotors with 1 slot skew. These are put forward in Section IV.

The synchronous locking torques were calculated after the solution of the torque-speed curve and solution of the main and auxiliary winding currents. These are used to obtain the different components of the synchronous locking torques which are summed (since several occur at the same speed) and superimposed on the torque-speed curves.

B. 220 V machine 20 and 24 bar rotor simulations

This motor is designed for run capacitor operation. A resistor is used in series with the auxiliary during the Start mode. The number of stator slots is 24 and the mechanical air gap is 0.28 mm. Fig. 3 shows the synchronous locking torques superimposed on the steady state torque/speed curves for 20 rotor slots, which results in limited synchronous locking torque, as well as for 24 rotor slots (equal to stator slot number) which results in severe locking at standstill.

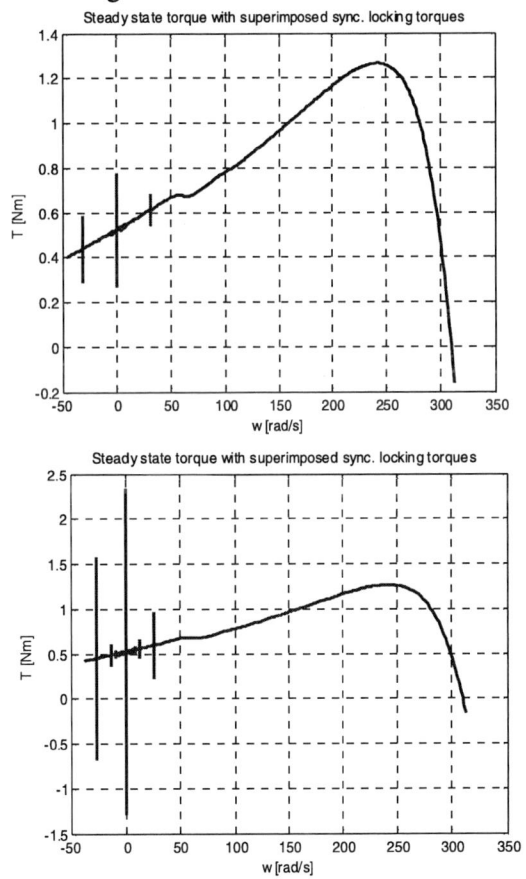

Fig. 3. Synchronous torque simulations with 20 rotor bars (top) and 24 rotor bars (bottom) – 220 V machine (start mode).

C. 115 V machine with 18 and 24 bar rotor simulations

This motor is a 115 V 60 Hz motor designed for single phase run operation. A series resistor is used with the auxiliary winding during starting . The number of stator slots is 24 and the mechanical air gap is 0.36 mm. Fig. 4 shows the synchronous locking torques superimposed on the steady state torque/speed curves for 18 rotor slots, which results in synchronous locking torque, as well as for 24 rotor slots (equal to stator slot number) which results in severe locking at standstill. The steady-state torques themselves show considerable asynchronous torque dips.

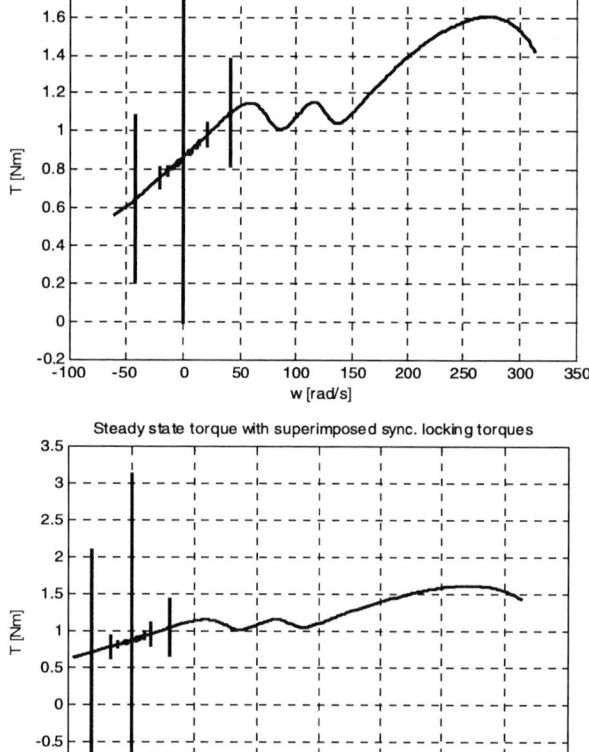

Fig. 4. Synchronous torque simulations with 18 rotor bars (top) and 24 rotor bars (bottom) – 115 V machine (start mode).

IV. EXPERIMENTAL RESULTS – 18, 24 AND 28 BAR ROTOR MACHINES

The simulation technique was verified experimentally as shown in Fig. 5. This is for a 230 V 50 Hz machine with an 18 rotor bar machine. Five nominally identical 18 bar unskewed rotors were constructed for this machine as well as five 24 bar unskewed rotors. Five similar production machines with 28 bar rotors and skew were also tested. Space constraints prevent a full description of the methods of testing for synchronous torques and this will be reported at a later date. The averages across the rotors are shown in the results here.

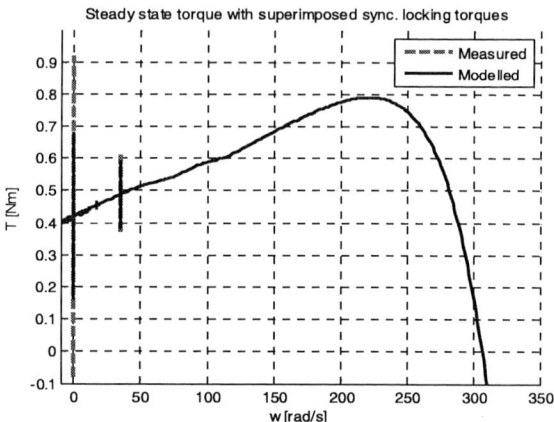

Fig. 5. Synchronous torque simulations for 18 slot experimental machine.

The results of the simulations when compared to the measurements are shown for the 18 and 24 bar rotors in Fig. 6. It can be seen that there are locking torques at zero speed and 27 rad/sec (258 rpm) for the 24 bar rotor and 35 rad/sec (335 rpm) for the 18 bar rotor. There is experimental error in these and the experimental results are higher than the simulation predictions.

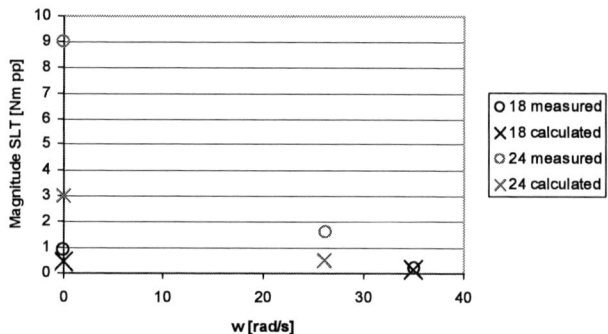

Fig. 6. Synchronous torque simulations and measurements for 18 and 24 bar rotor machines.

Deceleration torque/speed tests were carried out on the different rotors. Fig. 7 shows the deceleration tests for the 18 bar rotors from 500 rpm (right) down to 0 rpm. It can be seen that there are synchronous torques around 335 rpm as well as the locking torques at zero speed. Fig. 8 shows the deceleration tests for the 24 bar rotors. This shows the synchronous torques at about 258 rpm (the speed axis is 0 to 500 rpm). However the torque is from - 4 Nm to 2 Nm which illustrates the large synchronous locking torque at zero speed due to slot cogging.

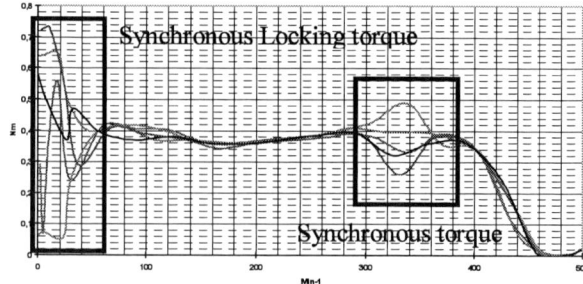

Fig. 7. Deceleration tests for 18 bar rotors – the y axis is torque 0 to 0.8 Nm and max speed (right) is 500 rpm.

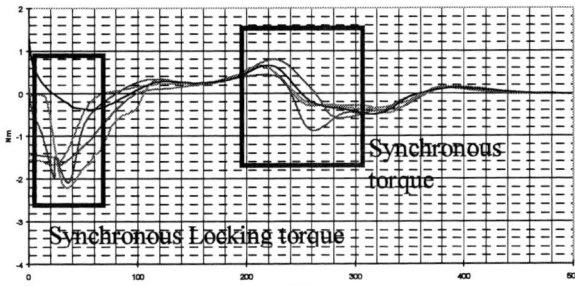

Fig. 8. Deceleration tests for 24 bar rotors – the y axis is torque -4 to 2 Nm and max speed (right) is 500 rpm.

Fig. 9 shows the deceleration test for the production machine with 28 skewed bars. It can be seen here that both the synchronous locking torque, and the characteristic synchronous torque at about 220 rpm are much reduced compared to the unskewed rotors, hence illustrating the importance of rotor skew.

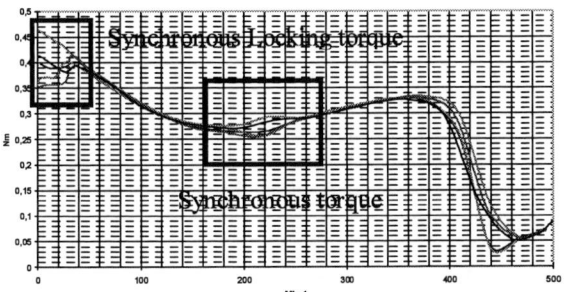

Fig. 9. Deceleration tests for 28 skewed bar rotors – the y axis is torque -0 to 0.5 Nm and max speed (right) is 500 rpm.

V. CONCLUSIONS

The paper puts forward an analytical method for modelling split phase induction motors that includes the calculation of synchronous torque spikes and dips from permeance harmonic considerations. This will be an aid to motor designers when assessing a design, and in particular the stator/rotor slot number combination. The methods are verified experimentally using rotors with varying bar numbers.

REFERENCES

[1] C. G. Veinott, *Theory and Design of Small Induction Motors*, McGraw-Hill Book Company, 1959.

[2] P. L. Alger, *Induction Machines, Their Behavior and Uses*, Gordon and Breach Publishers, Third Edition, 1995, ISBN 2-88449-199-6.

[3] D. G. Dorrell "Analysis of split-phase induction motors using an impedance matrix", *IEE Power Electronics Machines and Drives Conference PEMD*, Edinburgh, March 2004 (on CD).

[4] P. Scavenius Andersen and D. G. Dorrell, "Modelling of Split-Phase Induction Machine using Rotating Field Theory", ICEM 2006, Crete, September 2007

[5] S. Williamson and C. Y. Poh, "The effect of interbar currents in a permanent split capacitor motor", IEEE Transactions on Industry Applications, Volume 42, Issue 2, March-April 2006 pp 423 – 428.

[6] P. Scavenius Andersen, *Modelling and Analysis of Asynchronous and Synchronous Torques in Split-Phase Induction Machines*, PhD thesis, University of Glasgow, 2007.

[7] F. W. Carter, "A Note on Airgap and Interpolar Induction", JIEE, No. 29, pp 925, 1900.

[8] F. W. Carter, "The Magnetic Field of the Dynamo-Electric Machine", JIEE, No. 64, pp 100, 1926.

[9] B. Heller and V. Hamata, *Harmonic Field Effects in Induction Machines*, Elsevier Scientific Publishing Company, 1977.

APPENDIX

F.W. Carter described a relationship between the mechanical air-gap and the equivalent air gap which takes stator and rotor slots into account [7][8]. Here, the air-gap as seen by the electromagnetic field is investigated in a little more detail.

A. Air-gap length harmonics

In order to asses the variation of the air-gap magnetic field, a 2D finite element model is created which has an actual stator geometry but assumes a non-slotted rotor. This is shown in Fig. A.1. The red slots carry a uniform current density of opposite magnitude, whereas the grey slots carry no current. Hence, a quasi-square wave MMF is created.

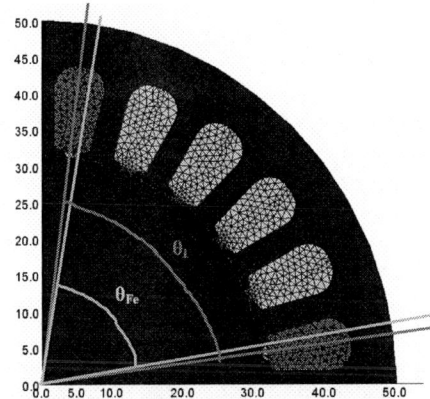

Fig. A.1. 2D finite element analysis of machine with smooth rotor-definition of θ_I and θ_{Fe}

A static analysis of this model results in an air-gap flux density versus angle for a single tooth pitch as given in Fig. A.2.

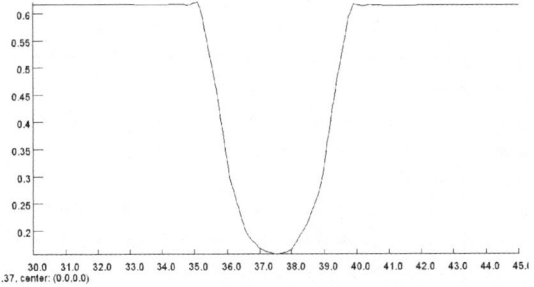

Fig. A.2. Air-gap flux density from FEA (y-axis is in Tesla)

From this flux density variation, the radial air gap length can be extracted from Ámpere's law using

$$l_g(\theta) = \frac{\theta_I}{\theta_{Fe}} \frac{\mu_0}{2} \frac{I_{tot}}{B_g(\theta)} \qquad (A.1)$$

The angles θ_I and θ_{Fe} are defined in Fig. A.1 and the ratio takes into account the concentration of flux due to the slotting, since the angular span of the MMF is larger than the span of the flux due to the slot openings.

Extracting the corresponding air-gap length from Fig. A.2 and performing a harmonic analysis of the resulting function results in a spectrum of harmonic air-gap lengths as given by Fig. A.3.

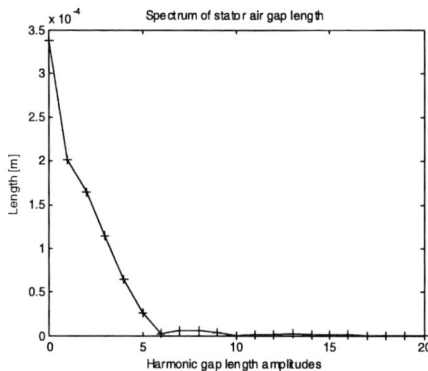

Fig. A.3. Harmonic air-gap lengths.

For the analysis, it is necessary to express the air-gap length variation mathematically. From Fig. A.2, a suitable approximation is indicated by the dotted blue line in Fig. A.4.

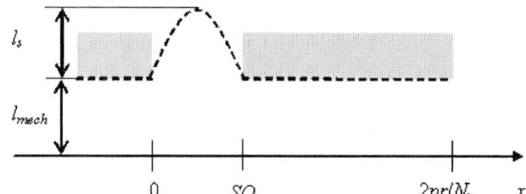

Fig. A.4. Air gap length approximation.

A suitable mathematic expression over the span $0 < x < SO$ is

$$f(x) = l_{mech} + l_s \sin\left(\frac{\pi}{SO} x\right) \qquad (A.2)$$

And for $SO \le x \le 2\pi r/N_s$:

$$f(x) = l_{mech} \qquad (A.3)$$

where l_{mech} is the mechanical air-gap and l_s is the effective air-gap as determined using the Carter factor. This actually corresponds to a constant term, i.e., the zero order harmonic of Fig. A.3.

Performing a Fourier analysis of the periodic function described by (A.2) and (A.3) will result in a spectrum which can be directly compared with the one in Fig. A.3 (as obtained from the finite element analysis). Any general magnitude deviation can be accounted for by multiplying the harmonic lengths with a correction factor. In Fig. A.5, blue graph shows the spectrum of the series defined by (A.2) and (A.3) directly, whereas the red graph shows the same spectrum, but where the harmonic coefficients have been multiplied with by a factor of 1.3.

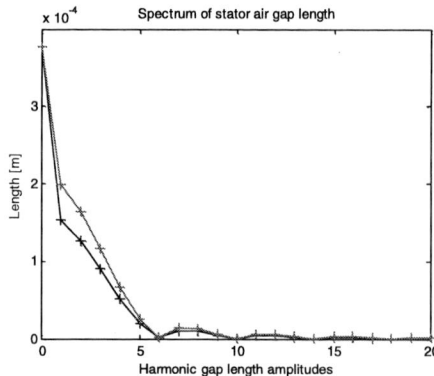

Fig. A.5. Comparison of harmonic air-gap lengths with adjustments (blue is the original harmonic magnitude and red multiplies the harmonic lengths by a factor of 1.3)

It is seen that the corrected spectrum (red graph) is in good agreement with the reference spectrum in Fig. A.3. Therefore, the mathematical functions in (A.2) and (A.3) for describing the air-gap length variation will be used in the following analysis. Generally, it can be observed from the spectra that both even and odd harmonics are present. For the lower harmonics the magnitudes are significant.

Since the spectrum shows the length variation over one tooth pitch, the fundamental value corresponds to the N_s'th harmonic when observing the entire air-gap where N_s is the number of stator slots. Hence, for the entire air-gap, the order of the harmonic is N_s times higher than shown in Fig A.5.

B. A.2. Air-gap permeance

The reluctance variation of the air-gap is proportional to the air-gap length. The permeance is the inverse of the reluctance and it will have a similar spectrum as shown in Fig. A.5. This spectrum represents an air-gap section corresponding to one tooth pitch. For an air-gap around the whole motor air-gap circumference, the harmonic series has a fundamental of order N_s (the number of stator slots) rather that unity. This is shown in Fig. A.6. This shows the air-gap permeance and the corresponding spectrum for an air-gap with 24 slots.

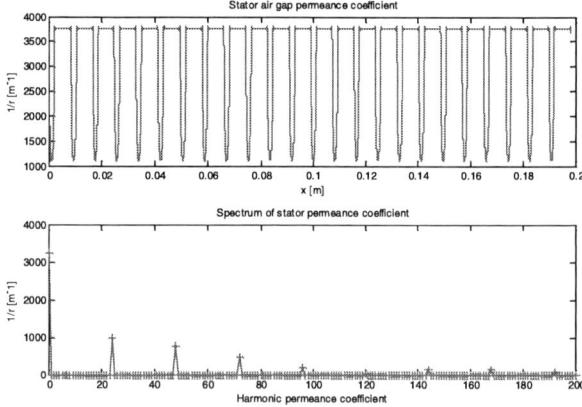

Fig. A.6. Air-gap permeance with harmonic decomposition

A similar spectrum will exist for the rotor, with a fundamental corresponding to the number of bar slots, and the fundamental and harmonics will rotate with the rotor itself.

C. Representation of permeance harmonics

In the following derivation, the combined effects of stator and rotor permeance are described analytically. The approach developed by [2] is used and extended. The total air-gap reluctance is a series connection of independent reluctance terms. However, for the total air-gap permeance, the inversion process means that the stator slot, rotor slot and average permeance terms may be described by a parallel connection of individual permeances, i.e.:

$$P = \left(\frac{1}{P_r} + \frac{1}{P_s} + \frac{1}{P_g} \right)^{-1} = \frac{P_r \, P_s \, P_g}{P_s \, P_g + P_r \, P_g + P_r \, P_s} \quad (A.4)$$

This type of air-gap harmonic inversion was also studied by Heller and Hamata [9]. By substitution, and for the moment ignoring higher slot harmonics:

$$P_s = P_{0s} + P_{ms} \cos\left(mN_s\theta\right) \quad (A.5)$$

and

$$P_r = P_{0r} + P_{mr} \cos\left(mN_r\theta - \omega_r t\right) \quad (A.6)$$

Therefore the resulting air-gap permeance will be

$$P = \frac{\left(P_{0r} + P_{mr}\cos\left(mN_r\theta - \omega_r t\right)\right)\left(P_{0s} + P_{ms}\cos\left(mN_s\theta\right)\right)P_g}{\left[\begin{array}{l}\left(\begin{array}{l}P_{0s} + \\ P_{ms}\cos\left(mN_s\theta\right)\end{array}\right)P_g + \left(\begin{array}{l}P_{0r} + \\ P_{mr}\cos\left(mN_r\theta - \omega_r t\right)\end{array}\right)P_g + \\ \left(P_{0r} + P_{mr}\cos\left(mN_r\theta - \omega_r t\right)\right)\left(P_{0s} + P_{ms}\cos\left(mN_s\theta\right)\right)\end{array}\right]}$$

$$= \frac{\left\{\begin{array}{l}P_{0r}P_{0s}P_g + P_g P_{0s} P_{mr}\cos\left(mN_r\theta - \omega_r t\right) \\ P_g P_{0r} P_{ms}\cos\left(mN_s\theta\right) + \\ P_{mr}\cos\left(mN_r\theta - \omega_r t\right)P_{ms}\cos\left(mN_s\theta\right)\end{array}\right\}}{\left\{\begin{array}{l}P_{0s}P_g + P_{0r}P_g + P_{0r}P_{0s} + \left(P_{0r} + P_g\right)P_{ms}\cos\left(mN_s\theta\right) + \\ + \left(P_g + P_{0s}\right)P_{mr}\cos\left(mN_r\theta - \omega_r t\right) + \\ + P_{mr}\cos\left(mN_r\theta - \omega_r t\right)P_{ms}\cos\left(mN_s\theta\right)\end{array}\right\}}$$

$$(A.7)$$

Both the numerator and denominator of (A.7) contain contributions of constant and time-space-varying permeance terms. For the denominator, however, the time-space-varying permeance terms are small compared to the constant terms and are therefore ignored in the following analysis. Hence, the combined-effect air-gap permeance can be written as

$$P = P_0\left[1 + \sum_{m=1}^{M} \frac{P_{Sm}}{P_0}\cos\left(mN_S\theta\right)\right] \times$$

$$\times \left[1 + \sum_{m=1}^{M} \frac{P_{Rm}}{P_0}\cos\left(mN_R\left(\theta - \omega_r t\right)\right)\right] \quad (A.8)$$

The terms belonging to the stator are fixed in space, whereas the rotor terms will rotate with rotor speed. This gives rise to several terms of permeance where

$$P = P_0 + P_{S1}\cos\left(N_S\theta\right) + P_{S2}\cos\left(2N_S\theta\right) +$$

$$+ P_{S3}\cos\left(3N_S\theta\right) + \cdots$$

$$+ P_{R1}\cos\left(N_R\left(\theta - \omega_r t\right)\right) +$$

$$+ P_{R2}\cos\left(2N_R\left(\theta - \omega_r t\right)\right) + \cdots$$

$$+ \frac{P_{S1}P_{R1}}{P_0}\cos\left(N_S\theta\right)\cos\left(N_R\left(\theta - \omega_r t\right)\right) +$$

$$+ \frac{P_{S2}P_{R2}}{P_0}\cos\left(2N_S\theta\right)\cos\left(2N_R\left(\theta - \omega_r t\right)\right) + \cdots$$

$$+ \frac{P_{S1}P_{R2}}{P_0}\cos\left(N_S\theta\right)\cos\left(2N_R\left(\theta - \omega_r t\right)\right) +$$

$$+ \frac{P_{S1}P_{R3}}{P_0}\cos\left(N_S\theta\right)\cos\left(3N_R\left(\theta - \omega_r t\right)\right) + \cdots$$

$$+ \frac{P_{S2}P_{R1}}{P_0}\cos\left(2N_S\theta\right)\cos\left(N_R\left(\theta - \omega_r t\right)\right) +$$

$$+ \frac{P_{S2}P_{R3}}{P_0}\cos\left(2N_S\theta\right)\cos\left(3N_R\left(\theta - \omega_r t\right)\right) + \cdots$$

$$+ \frac{P_{S3}P_{R1}}{P_0}\cos\left(3N_S\theta\right)\cos\left(N_R\left(\theta - \omega_r t\right)\right) +$$

$$+ \frac{P_{S3}P_{R2}}{P_0}\cos\left(3N_S\theta\right)\cos\left(2N_R\left(\theta - \omega_r t\right)\right) + \cdots$$

$$+ \frac{P_{S4}P_{R1}}{P_0} \cdots$$

$$(A.9)$$

so that

$$P_{m,n} = \frac{P_{Sm}P_{Rn}}{2P_0}\cos\left(\left[mN_S - nN_R\right]\theta + nN_R\omega_r t\right)$$

$$+ \frac{P_{Sm}P_{Rn}}{2P_0}\cos\left(\left[mN_S + nN_R\right]\theta - nN_R\omega_r t\right)$$

$$(A.10)$$

where m and n are stator and rotor slot harmonics. Altogether, the total air-gap permeance can be approximated using the following five terms:

$$P = P_0 + \sum_{m=1}^{M} P_{Sm}\cos\left(mN_S\theta\right) + \sum_{n=1}^{N} P_{Rn}\cos\left(nN_R\left(\theta - \omega_r t\right)\right)$$

$$+ \sum_{m=1}^{M}\sum_{n=1}^{N} \frac{P_{Sm}P_{Rn}}{2P_0}\left[\begin{array}{l}\cos\left(\left[mN_S - nN_R\right]\theta + nN_R\,\omega_r t\right) \\ + \cos\left(\left[mN_S + nN_R\right]\theta - nN_R\,\omega_r t\right)\end{array}\right]$$

$$(A.11)$$

This includes the higher slot harmonics. The first term corresponds to the average air-gap permeance, calculated using the Carter factor. This is the only term which has a non-zero mean value. The second and third terms correspond to stator and rotor slot harmonics. These will have mN_S and nN_R pole-pairs. The fourth and fifth terms contain combinations of the stator and rotor harmonic permeances. These are permeance waves consisting of waves of $|mN_S + nN_R|$ pole pairs for the fourth term and $|mN_S - nN_R|$ pole pairs for the fifth term.

On-Line Diagnosis of Three-Phase Closed Loop Induction Motor Drives Using an Eigenvalue αβ–Vector Approach

J. F. Martins*, V. Fernão Pires* and A. J. Pires*

* LabSEI-Escola Superior de Tecnologia de Setúbal / Instituto Politécnico de Setúbal,
Campus do IPS, Estefanilha, 2914-508 Setúbal, PORTUGAL

Abstract– Non-invasive diagnosis of induction machine stator winding turn faults is a well established technique, allowing accurate results and cost-savings. This diagnosis is based on the analysis of the machine input line currents. However inverter-fed machines are becoming a common solution for industrial purposes. Power electronics technological advances allow good performance with competitive costs. Closed-loop drives represent a problem when current based fault detection techniques are used because in this kind of drives the controller regulates the machine input current. In order to keep the detection procedure non-invasive one should monitor the machine input voltage. In this paper the αβ-vector approach fault detection method is extended to inverter-fed induction machines. Results are presented in order to fully understand the advantages of the proposed method.

Index Terms — Closed loop drives, Eigenvalue αβ-vector approach fault detection, Induction motor drives, On-line diagnosis.

I. INTRODUCTION

Preventive maintenance of electrical motors plays a very important role within industrial life. This requires the monitoring of their operation for detection of abnormal electrical and mechanical conditions that indicate, or may lead to, a failure of the system. The goal of predictive maintenance is the early identification of unsatisfactory working conditions, before it results in total system failure.

Induction machine drives can be divided into open-loop and closed-loop drives. Open-loop drives are essentially based on the well-known V/Hz operation. High performance closed-loop drives (FOC for example) usually consider an outer control loop, that generates an input line current reference, and an inner current control loop that makes the drive supplied by a controlled current source.

In the last years monitoring induction motors issues become very important in order to reduce maintenance costs and prevent unscheduled downtimes. Therefore, there has been a substantial amount of research to provide new condition monitoring techniques for three phase induction motors [1,2,3,4]. Fault diagnosis in a closed-loop drive presents difficulties that are not presented when a constant 50/60Hz power supply or an open-loop

drive (the machine is also supplied by a voltage source) are considered. Some authors have addressed this problem using soft computing techniques or extensions of the MCSA method (requiring the analysis of the input electrical quantities spectrum) [5,6].

The αβ-vector approach was already successfully used to detect induction motor faults, considering constant speed faults [7] and open-loop variable speed drives [8]. In this paper the αβ-vector approach fault detection method is extended to inverter-fed induction machines. Results are presented in order to fully understand the advantages of the proposed method.

II. EIGENVALUE METHODOLOGY

In three-phase induction motors without neutral connection the analysis of this system can be simplified transforming the three-phase stator currents or voltages to an equivalent two-phase system, using the Concordia transformation in order to maintain invariant power [9]. Therefore, the αβ stator quantities are given by (1), where x denote currents or voltages.

$$\begin{cases} x_\alpha = \sqrt{\dfrac{2}{3}}\left(x_a - \dfrac{1}{2}x_b - \dfrac{1}{2}x_c \right) \\ x_\beta = \dfrac{1}{\sqrt{2}}x_b - \dfrac{1}{\sqrt{2}}x_c \end{cases} \tag{1}$$

$$x_c = -x_a - x_b$$

Considering a constant frequency and balanced power supply, the corresponding current Concordia vector pattern is a circle centered at the origin of the αβ coordinates. However, under abnormal working conditions the previous settings are no longer valid and the circle pattern no longer appears. Still considering a constant frequency power supply, for an induction motor with a stator winding fault the current αβ–vector pattern assumes an elliptic pattern [10].

Principal Component Analysis (PCA) is a common statistical method usually considered for data analysis [11]. This method was introduced by Pearson in 1901 [12] and its use is nowadays widespread within a diversity of areas, from engineering to economics. By defining the eigenvectors, this technique is able to obtain the main directions of the data sample on the space-vector. It also allows the definition of significant values (eigenvalues) that weight the spread of the data sample through the main directions.

The first step is to obtain a data sample matrix (2), where the number of significant samples n corresponds to the number of rows of matrix S. The input currents – i_α and i_β – form the columns of matrix S. The first sample will be $[i_\alpha(t_0)\ i_\beta(t_0)]$, where t_0 denotes the initial time instant and Δt subsequently denotes the sample interval.

$$S = \begin{bmatrix} i_\alpha(t_0) & i_\beta(t_0) \\ i_\alpha(t_0+\Delta t) & i_\beta(t_0+\Delta t) \\ \vdots & \vdots \\ i_\alpha(t_0+(n-1)\Delta t) & i_\beta(t_0+(n-1)\Delta t) \end{bmatrix} \quad (2)$$

After establishing the correlation matrix of S, denoted as E (3), the eigenvectors and the respective eigenvalues of E are calculated.

$$E = S^T.S \quad (3)$$

The principal component is the one where the data has more energy and the second principal direction is the one with the less energy. One should note that can be as many principal directions as one chooses. Obviously only the first ones are of interest, because they carry the most of the data energy.

The diagnosis of the three-phase induction motor with stator turn faults will be performed using the obtained data eigenvalues. If the motor is healthy the current $\alpha\beta$-vector pattern is a circle, reflecting the constant amplitude of the $\alpha\beta$ current components, and the respective eigenvalues are equal. Their equality reflects the fact that both principal directions carry the same energy. Otherwise, whenever a stator winding fault condition occurs the input current $\alpha\beta$–vector pattern becomes an ellipse because the amplitude of the $\alpha\beta$ current components will no longer be equal to each other. In this condition the energy associated with the first two principal directions is not the same, assuming the respective eigenvalues distinct values.

The PCA obtained eigenvalues can be used to infer the fault severity. This will be given by a severity index (4), where λ_{high} and λ_{low} denote respectively the highest and lowest of the two eigenvalues. The severity index varies between zero and one, being the absence of any fault reported by $s_i=0$.

$$s_i = 1 - \frac{\lambda_{low}}{\lambda_{high}} \quad (4)$$

III. EIGENVALUE METHODOLOGY FOR CLOSED-LOOP DRIVES

For closed-loop drives the induction machine electrical source can be seen as a controlled current source. Assuming that the line input current is controlled, even in a stator winding fault situation, the current $\alpha\beta$-vector pattern is always a circle making it impossible to detect the fault. However, being the current controlled, in a faulty situation the electrical quantity that appears

distorted is the input voltage. In an ideal current source drive this distortion is easily detected, however in an IGBT PWM inverter-fed drive this is an extremely difficult task. One should recall that in these drives only eight voltage vectors are allowed. The $\alpha\beta$-vector approach proposed methodology is able to accurately detect stator winding faults within current controlled inverter-fed drives.

Let one first assume that the induction machine is supplied by an ideal current source (Figure 1). Figure 2 presents the $\alpha\beta$-vector supplied current and voltage pattern for a healthy motor. The obtained current and voltage eigenvalues are identical and both severity indexes report zero.

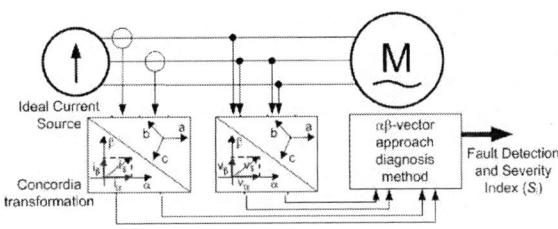

Fig. 1. Eigenvalue methodology applied to an induction motor drive with an ideal current source.

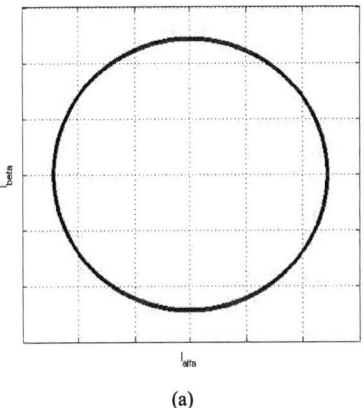

(a)

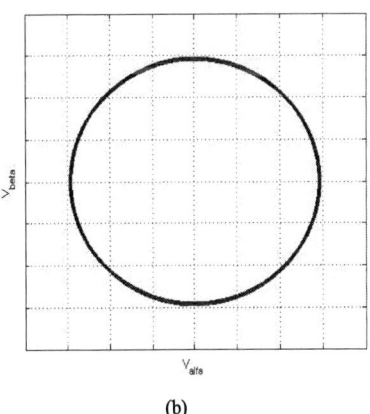

(b)

Fig. 2. $\alpha\beta$-vector patterns for a healthy motor with an ideal current source. (a) Current; (b) Voltage.

895

In the presence of a winding stator fault the αβ-vector current and voltage pattern are presented in Figure 3. In this case the current eigenvalues remain equal (implying a zero severity index) but the voltage eigenvalues present distinct values (and a consequent non-zero severity index) denoting a stator fault. The αβ-vector current pattern remains unchanged but the αβ-vector voltage pattern changes depending on the faulty phase and on the fault extend.

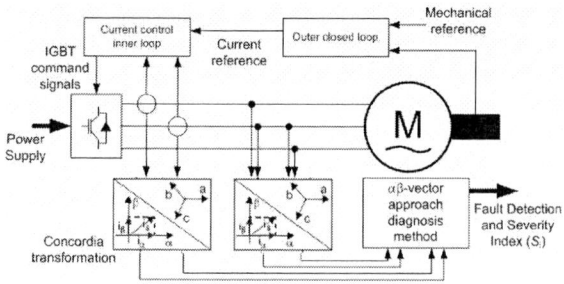

Fig. 4. Eigenvalue methodology applied to an induction motor drive with closed loop current control.

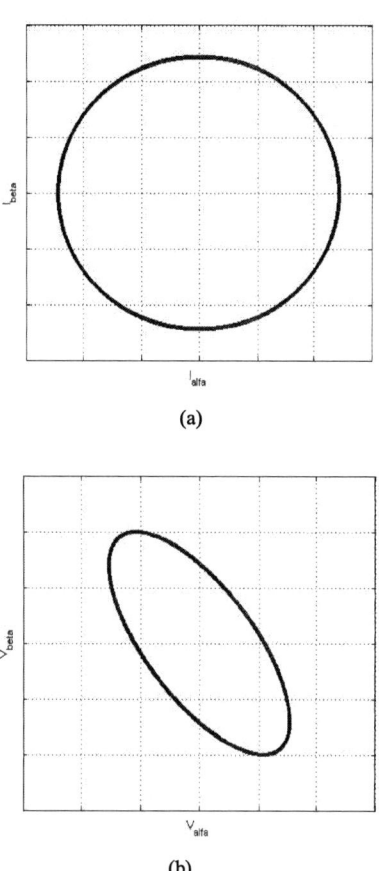

Fig. 3. αβ-vector patterns for a stator winding fault (motor with an ideal current source). (a) Current; (b) Voltage

However in an IGBT PWM inverter-fed drive (Figure 4) the input line voltage is not sinusoidal but a square wave presenting only eight vectors. Due to this fact the input voltage αβ–vector pattern will not become an ellipse for a stator winding fault. This can be seen by several tests where the motor will have stator winding fault. So, as expected from these tests obtained patterns of the current αβ-vector will not change. Figure 5 presents these current patterns for the increasing severity stator winding fault. The obtained correspondent current eigenvalues are also similar.

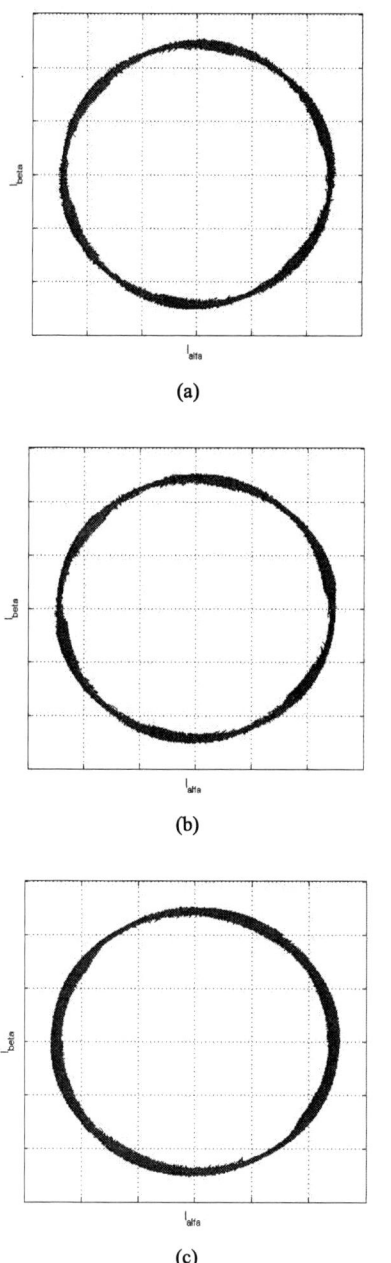

Fig. 5. αβ-vector current patterns for increasing severity stator winding fault (induction motor drive with closed loop current control). (a) Healthy motor; (b) Minor fault; (c) Severe fault.

However, although the αβ-vector voltage pattern remains the same, there will be some changes on the voltage variable. From Figure 6 it is possible to verify that the αβ-vector voltage pattern does not have a change for the same increasing severity stator winding fault. This is due to the fact that the eight vectors PWM inverter forces the voltage to assume the same values for each situation. This invalidates any expertise analysis or pattern recognition method.

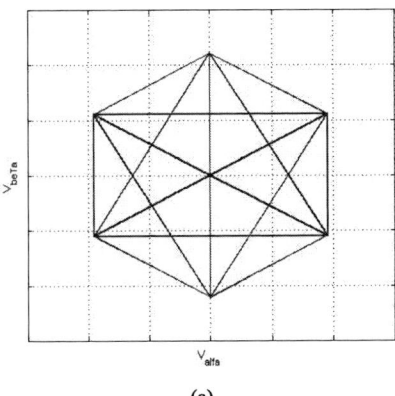

(a)

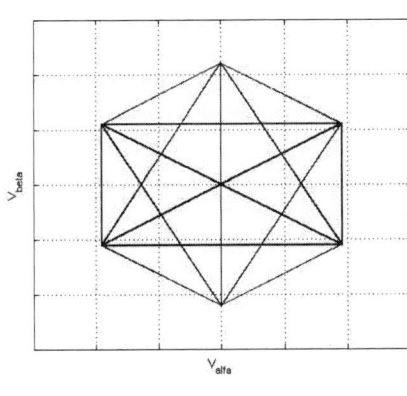

(b)

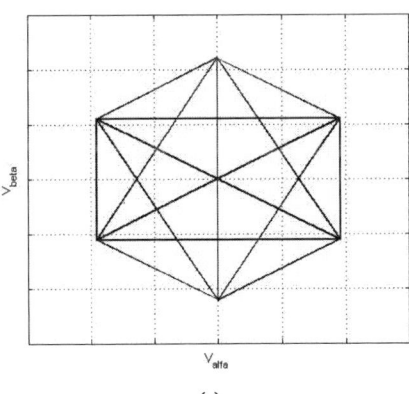

(c)

Fig. 6. αβ-vector voltage patterns for increasing severity stator winding fault (induction motor drive with closed loop current control). (a) Healthy motor; (b) Minor fault; (c) Severe fault.

However, in the three considered cases the voltage time evolution is not the same, meaning that the eight available voltage vectors are not equally applied to the machine. So, in the αβ-vector voltage space there are some directions that carry more energy than others. The proposed αβ–vector approach fault detection method is suitable to be used in this kind of drives.

For the increasing severity stator winding fault presented in Figure 6 the following results were obtained.

Table I. Obtained eigenvalues for increasing severity stator winding fault

Voltage					
Healthy motor		Minor fault		Severe fault	
416,3	418,2	370,1	403,5	317,8	446,4
Current					
Healthy motor		Minor fault		Severe fault	
5,6	5,6	5,6	5,6	5,6	5,6

These values imply a voltage severity index of 0.0046 for a healthy motor, 0.0827 for a minor stator winding fault and 0.2881 for a severe stator winding fault. For the same essay the correspondent current severity index results are 0.0000, 0.0020 and 0.0082, for a healthy motor, a minor stator winding fault and severe fault, respectively. The results show the suitability of the voltage severity index for the analysis of the described situation.

IV. CONCLUSIONS

For high performance closed-loop drives, with an inner current control loop, the observation of the input line current is not a good tool for inferring stator winding faults. The closed loop control hides the fault influence over the current. On the other hand this fault influence appears on the supplied voltage. However, as the IGBT supplied voltage presents only eight voltage vectors the previous influence is masked. The proposed αβ-vector approach fault detection method is able to easily detect and infer the severity of stator winding faults in closed-loop induction motor drives.

REFERENCES

[1] G. B. Kliman, R. A. Koegl, J. Stein, R. D. Endicott, M.W. Madden,, "Noninvasive detection of broken rotor bars in operating induction motors," *IEEE Trans. Energy Conv.*, vol. EC-3, no. 4, pp. 873–879, Dec. 1988.

[2] G. B. Kliman, J. Stein, "Methods of motor current signature analysis," *Electric Machines and Power Systems*, vol. 20, nº5, pp 463-474, September 1992.

[3] S. Nandi, H. A. Toliyat, "Condition monitoring and fault diagnosis of electrical machines - A review," *IEEE Industry Applications Conference 1999*, vol. 1, 1999, pp. 197–204.

[4] B. M. El Hadremi, "A review of induction motor signature analysis as a medium for fault detection", *IEEE Trans. on Ind. Electron.*, vol. 47, pp 984-993, Oct. 2000.

[5] Rangaranjan M. Tallam, Thomas G. Habetler, Ronald G. Harley, "stator winding turn-fault detection for closed-loop induction motor drives", *IEEE Trans. on Ind. App.*, vol. 39, May 2003.

[6] Alberto Bellini, Fiorenze Filippetti, Giovanni franceschini, Carla Tassoni, "Classification of diagnostic indexes for field orientated induction motor drives", Symposium on Diagnostics for Electric Machine, Power Electgronics and Drives SDEMPED2003, Atlanta, USA, August 2003.

[7] V. Fernão Pires, J. F. Martins and A. J. Pires, "On-Line Diagnosis of Three-Phase Induction Motor Using an Eigenvalue αβ–Vector Approach" *Proc. of IEEE ISIE 2005*, Dubrovnik, Croatia, June 2005.

[8] J. F. Martins, V. Fernão Pires, A. J. Pires, "PCA-Based On-Line Diagnosis of Induction Motor Stator Fault Feed by PWM Inverter" *Proc. of IEEE ISIE 2006*, Montreal, Canada, July 2006.

[9] R. H. Park, "Two reaction Theory of Synchronous Machines – Generalized Method of Analysis – Part I", AIEE Trans., vol. 48, pp. 716-727, July 1990.

[10] A. J. M. Cardoso, S. M. A. Cruz, J. F. S. Carvalho, E. S. Saraiva, "Rotor Cage Fault Diagnosis in Three-Phase Induction Motors by Park's Vector Approach," IEEE Industry Applications Conference 1995, vol. 1, 1995, pp. 642–646.

[11] K. Pearson, "On Lines and Planes of Closest Fit to Systems of Points in Space", Philosophical Magazine, vol. 2, pp. 559-572, 1901.

[12] I. T. Jolliffe, *Principal Component Analysis*, New York, Springer-Verlag, 1986.

Design and Development of a 36-Pulse AC-DC Converter for Vector Controlled Induction Motor Drive

Bhim Singh, *Senior Member*, IEEE, Sanjay Gairola

Abstract— **In this paper, an autotransformer based 36-pulse AC-DC converter fed vector controlled induction motor drives (VCIMD's) is presented for reduction of harmonic-currents. A 36-pulse converter is realized using a specially designed autotransformer and two 18-pulse diode bridge rectifiers. The effect of load variation on VCIMD is also studied to demonstrate the effectiveness of the proposed AC-DC converter. A set of power quality indices on input AC mains and on DC bus for a VCIMD fed from 6-pulse and 18-pulse AC-DC converters are also given to compare their performance. It is observed that input current total harmonic distortion (THD) less than 5% is possible with the proposed topology at varying-loads. A laboratory prototype of this 36-pulse AC-DC converter is developed and test results are presented to validate the design and simulation model. It improves power quality at AC mains and meets IEEE-519 standard requirements at varying loads.**

Index Terms—**18-pulse, 36-pulse, AC-DC converter, power quality, autotransformer, VCIMD.**

I. INTRODUCTION

THE vector control of an induction motor drive has been widely used [1] in the applications such as in heating, ventilation and air-conditioning (HVAC) systems, blowers, pumps for waste water treatment plants etc. A vector controlled induction motor drive (VCIMD) employs insulated gate bipolar transistor (IGBT) based voltage source inverter (VSI) and it is fed from six-pulse diode bridge rectifier as shown in Fig. 1. The six-pulse diode-bridge rectifier suffers from problems of poor power factor and injection of harmonic currents into AC mains. To reduce harmonics in to AC mains various standards such as IEEE standard 519 [2] and IEC 61000-3-2 [3] have appeared for limiting voltage and current distortions.

The rapid development of power electronic controllers for this purpose has attracted the attention of researchers towards the power quality improvement at the AC-mains to meet IEEE standard requirements. Several topologies in 12-pulse and 18-pulse (and some in 24-pulse and 30-pulse) have been reported

Bhim Singh is with Department of Electrical Engineering, Indian Institute of Technology, Delhi, New-Delhi-110016, India (e-mail: bhimsinghr@gmail.com).
Sanjay Gairola is with Department of Electrical and Electronics Engineering, Krishna Institute of Engineering and Technology, Ghaziabad (U.P.)-201106, India (e-mail: sanjaygairola@gmail.com).

in the literature [4-14]. These topologies are based either on phase-multiplication or phase-shifting or pulse-doubling or their combination. However it is observed that the total harmonic distortion (THD) of input current in these converters up to 18-pulse AC-DC converter configuration is more than 5% when operating at light load or the source impedance is small. An 18-pulse AC-DC converter has also been presented in [11], however, the THD of the supply current with this topology is reported to vary from 6.9% to 13.1% which is not within IEEE-519 standard limits. Another 18-pulse autotransformer based AC-DC converter is reported in [14] which has THD variation of 4.15% to 7.06% from full-load to light-load (20% of full-load).

However, some applications have very stringent power quality specifications [7] such as in defence equipments. To meet these stringent power quality specifications it is inevitable to go beyond 24-pulse AC-DC converter configuration. A 38-pulse AC-DC converter is discussed [7] but that employs unsymmetrical polygon transformer. Another 18-pulse AC-DC converter configuration is reported in the literature [6] for aerospace applications that employs autotransformer rating of 56% of output power. Therefore, it is suggested that higher pulse converter configuration must be used to meet IEEE-519 standard requirements. A single autotransformer capable of producing 36-pulse rectified output is not seen in the literature and is presented here.

An 18-pulse AC-DC converter for VCIMD is shown in Fig. 2. This topology uses a polygon auto-transformer that feeds three 6-pulse diode bridge converters, which are connected on DC side by three interphase transformers to absorb the instantaneous voltage difference of 6-pulse DC ripples. This topology is also studied and simulated to compare the performance with proposed topology described later.

In this paper, a 36-pulse AC-DC converter is proposed employing a novel polygon autotransformer as shown in Fig. 3. The proposed transformer is capable of feeding two sets of diode bridges (each having nine legs) that can be connected in parallel. These two 18-pulse AC-DC converter bridges are connected on DC side by two interphase transformers. This parallel connection produces a 36-pulse AC-DC conversion. As the power to the load is transferred at low voltage levels, the parallel bridge configuration is used. The designed converter system is modeled and simulated in MATLAB to demonstrate its power quality improvement at AC mains. A laboratory prototype of 36-pulse AC-DC converter is developed and test results are presented to validate the design and simulation model.

978-1-4244-0644-9/07/$25.00 ©2007 IEEE

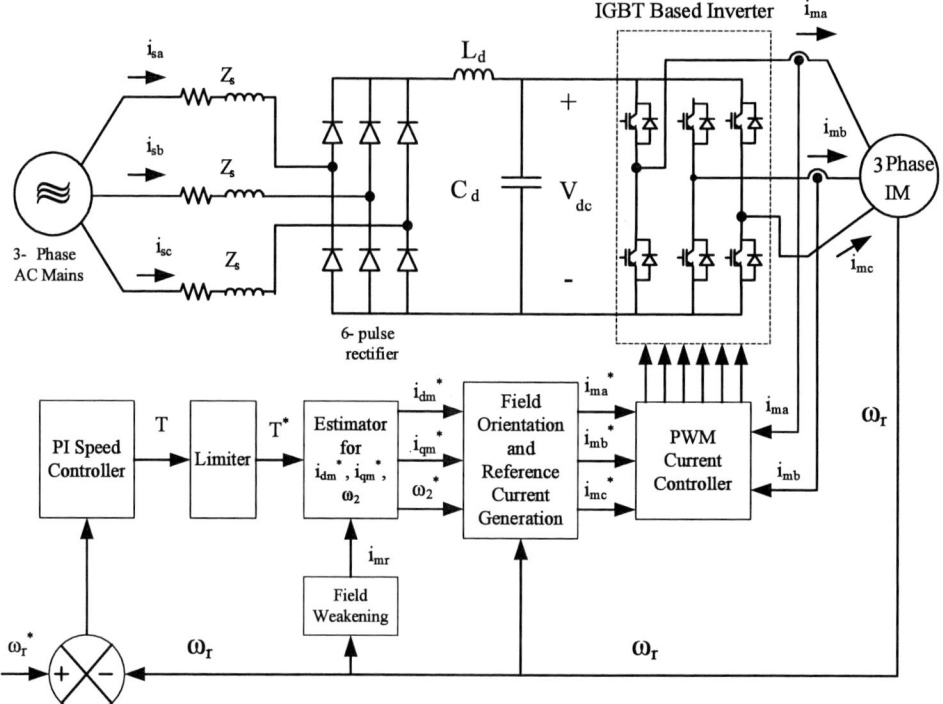

Fig. 1 A 6-pulse diode bridge rectifier fed vector controlled induction motor drive.

II. PROPOSED 36-PULSE AC-DC CONVERSION APPROACH

Fig. 2 shows an 18-pulse AC-DC converter system, which uses a polygon autotransformer, three parallel connected 6-pulse diode bridges and an IGBT based voltage source inverter fed VCIMD. The proposed 36-pulse AC-DC converter for an IGBT based VSI fed VCIMD is shown in Fig. 3. The autotransformer arrangement and its phasor diagram are shown in Fig.4.

A. Design of Proposed Autotransformer for 36-Pulse AC-DC Converter

Fig. 4 shows the schematic of proposed 36-pulse polygon autotransformer scheme and its graphical representation depicting angular position of various phasors. Two nine-leg diode bridge converters I and II are connected to two sets of nine-phase autotransformer outputs at (B11, B12, B13, B14, B15, B16, B17, B18, B19) and (B21, B22, B23, B24, B25,

B26, B27, B28, B29). B1x and B2x (where 1 and 2 corresponds to bridge I and II respectively and 'x' is integer 1 to 9) are terminals connected to the 18-pulse AC-DC converters I and II respectively. These two sets are displaced by 10° from each other and at -5° and +5° respectively from input voltage of phase A, phasor V_p. The number of turns for every winding is determined as a function of the phase voltage, V_p. These winding voltages, as marked in Fig. 4, are expressed by following relationships.

Consider that the input phase voltage is V_p ($=V_{LL}/\sqrt{3}$, where V_{LL} is line rms supply voltage) and two set of nine voltages fed to each bridge be V_{BXY} (V_{B11},V_{B12}, V_{B13}, V_{B14}, V_{B15}, V_{B16},V_{B17}, V_{B18}, V_{B19} to the nine leg bridge converter I and V_{B21},V_{B22}, V_{B23}, V_{B24}, V_{B25} , V_{B26}, V_{B27}, V_{B28}, V_{B29} to the converter II). The magnitude of phasors V_{BXY} is same and lies on the circle having radius $|V_R|$. The magnitude of V_{BXY} can be found in terms of V_p from the phasor diagram as:

$$|V_{BXY}| = |V_R| = (\sin 30°/\sin 145°) |V_p| = 0.8717 |V_p| \quad (1)$$

Now, considering the following set of three phase supply voltages as:

$$V_A = V_p\angle 0°, \ V_B = V_p\angle -120°, \ V_C = V_p\angle 120° \quad (2)$$

Two set of required voltages for converters I and II are:

$$V_{B11} = V_R \angle 5°, \ V_{B12} = V_R \angle -35°, \ V_{B13} = V_{B13}\angle -75°,$$

$$V_{B14} = V_R \angle -115°, \ V_{B15} = V_R \angle -155°, \ V_{B16} = V_R \angle 165°, \quad (3)$$

$$V_{B17} = V_R \angle 125°, \ V_{B18} = V_R \angle 85°, \ V_{B19} = V_R \angle 45°$$

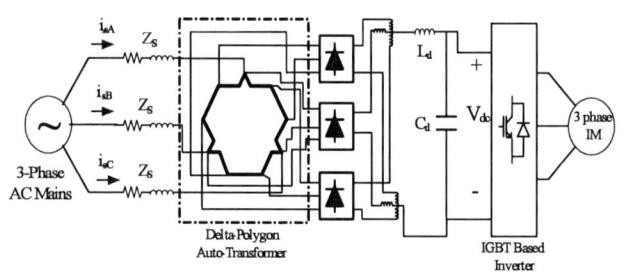

Fig. 2 A polygon autotransformer based 18-pulse AC-DC converter fed VCIMD.

900

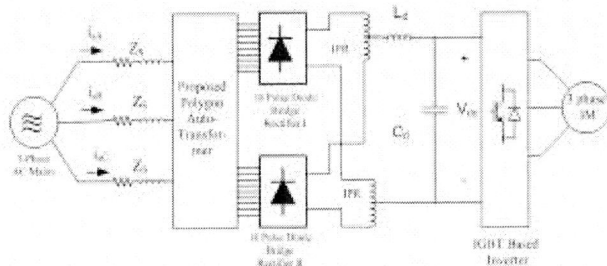

Fig. 3 Proposed polygon auto-transformer based 36-pulse AC-DC converter fed VCIMD.

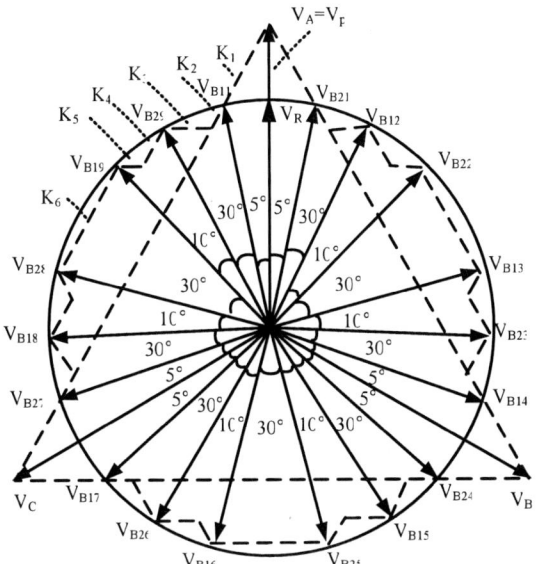

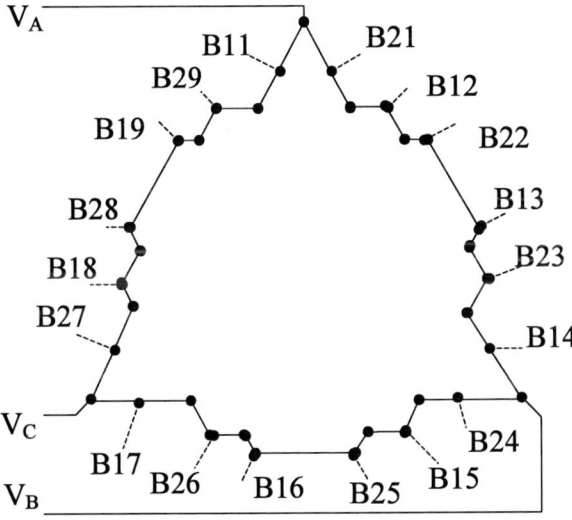

Fig. 4 Phasor representation and winding arrangement for the auto-transformer that produces two sets of nine phases for proposed 36-pulse AC-DC converter topology.

$$V_{B21} = V_R \angle -5°, V_{B22} = V_R \angle -45°, \; V_{B23} = V_R \angle -85°,$$

$$V_{B24} = V_R \angle -125°, \; V_{B25} = V_R \angle -165°, \; V_{B26} = V_R \angle 155°, \quad (4)$$

$$V_{B27} = V_R \angle 115°, \; V_{B28} = V_R \angle 75°, V_{B29} = V_R \angle 35°$$

These voltages for various phasors shown in Fig. 4 are related by following relations:

$$V_{B11} = K_1 V_p \angle 5° \quad (5)$$

$$V_{B21} = V_{B11} \angle -10° \quad (6)$$

$$V_{B12} = V_{B21} - K_2 V_{AB} + K_3 V_{BC} \quad (7)$$

$$V_{B22} = V_{B12} - K_4 V_{AB} + K_5 V_{BC} \quad (8)$$

$$V_{B13} = V_{B22} - K_6 V_{AB} \quad (9)$$

$$V_{B23} = V_{B21} \angle -50° \quad (10)$$

$$V_{B14} = V_{B21} \angle -110° \quad (11)$$

where,

$$V_{AB} = \sqrt{3} V_A \angle 30°, \; V_{BC} = \sqrt{3} V_B \angle 30°, \; V_{CA} = \sqrt{3} V_C \angle 30°$$

Eqns. (5-11) give the values of constants K_1 to K_6 for desired phase shift as:

$$K_1 = 0.1519, K_2 = 0.1782, K_3 = 0.3349, K_4 = 0.1127, \quad (12)$$
$$K_5 = 0.060, K_6 = 0.4512$$

The values of these constants, K_1 to K_6, determine the winding turns as a fraction of supply phase voltage.

III. VECTOR CONTROLLED INDUCTION MOTOR DRIVE

Fig.1 shows the schematic diagram of an indirect vector controlled induction motor drive (VCIMD). In the rotor flux-oriented reference frame the d-component of the stator current reference vector $i_{dm}{}^*$ is obtained as:

$$I_{dm}{}^* = i_{mr} + \tau_r (\Delta i_{mr} / \Delta T) \quad (13)$$

The closed loop PI speed controller compares the reference speed ($\omega_r{}^*$) with motor speed (ω_r) and generates reference torque T* (after limiting it to the suitable value) as:

$$T^*{}_{(n)} = T^*{}_{(n-1)} + K_p\{\omega_{e(n)} - \omega_{e(n-1)}\} + K_I \omega_{(n)} \quad (14)$$

where $T^*{}_{(n)}$ and $T^*{}_{(n-1)}$ are the output of the PI controller (after limiting it to a suitable value) and $\omega_{e(n)}$ and $-\omega_{e(n-1)}$ refer to speed error ($\omega_r{}^* - \omega_r$) at the nth and $(n-1)^{th}$ instants. K_P and K_I are the proportional and integral gain constants.

The q-component of the stator current reference vector $i_{qm}{}^*$ is obtained from the output of the PI controller as:

$$i_{qm}{}^* = T^* / (K \, i_{mr}) \quad (15)$$

where, $K = (3/2)(P/2)\{M/(1+\sigma_r)\}$ \quad (16)

The rotor slip frequency and rotor flux angle are calculated as:

$$\omega_2{}^* = i_{qm}{}^* / (\tau_r \, i_{mr}) \quad (17)$$

$$\psi_{(n)} = \psi_{(n-1)} + (\omega_2 + \omega_r) \Delta T \quad (18)$$

where i_{mr} is the magnetizing current, $\omega_2{}^*$ is the slip speed of rotor, ω_r is the angular velocity of rotor P, M and σ_r are the number of poles, mutual inductance and rotor leakage factor respectively, $\psi_{(n)}$ and $\psi_{(n-1)}$ are the value of rotor flux angles at n^{th} and $(n-1)^{th}$ instants respectively and ΔT is the sampling time.

These d and q axes currents ($i_{dm}{}^*$, $i_{qm}{}^*$) are in synchronously rotating reference frame and these are converted into stationary reference frame three phase currents (i_{ma}^*, i_{mb}^*, i_{mc}^*) as given below:

$$i_{ma}{}^* = -i_{qm}{}^* \sin\psi + i_{dm}{}^* \cos\psi \quad (19)$$

$$i_{mb}^* = \{(-\cos\psi \, \sqrt{3} \, \sin\psi) i_{dm}{}^* + (\sin\psi + \sqrt{3} \cos\psi) i_{qm}{}^*\}/2 \quad (20)$$

$$i_{mc}^* = -(i_{ma}{}^* + i_{mb}{}^*) \quad (21)$$

These three phase reference currents generated by the vector controller are compared with the sensed motor currents (i_{ma}, i_{mb} and i_{mc}). The calculated current errors are:

$$i_{mke} = i_{mk}^* - i_{mk}, \text{ where k = a, b, c.} \quad (22)$$

These current errors are amplified and fed to the PWM current controller, which controls the on and off periods of different switches in VSI. The VSI generates the PWM voltages being fed to the motor to develop the desired torque for running the motor at a reference speed under various loading conditions.

IV. MATLAB BASED SIMULATION

The proposed AC-DC converter with the VCIMD is simulated in MATLAB environment along with SIMULINK and power system block set (PSB) toolboxes, as shown in Fig.5. The AC-DC converter system is fed from 460V, 60Hz AC supply. A three-phase induction motor of a rating of 50HP (37.3kW) is controlled using a vector controller. The VCIMD details are given in the Appendix. This VCIMD is fed from 6-pulse, 18-pulse and proposed 36-pulse AC-DC converters. The source impedance has been kept at a practical value of 3% in all simulations. The waveforms depicting dynamic response of VCIMD fed from six-pulse, eighteen–pulse and thirty-six-pulse AC-DC converters are shown in Figs. 6-8. The current waveforms along with its harmonic spectrum and THD at light (20% of full-load) and full-load for the three converters are given in Figs. 9-14. The power quality indices and rating of

magnetics are obtained for these converters and tabulated in (Tables I-III) to study the effect of load variation.

V. EXPERIMENTAL VALIDATION

To validate the proposed 36-pulse AC-DC converter configuration, a prototype has been developed in the laboratory for a 7.5kW rating in three-phase 230V, 50Hz system. The developed thirty six-pulse AC-DC converter is tested with equivalent resistive load and extensive tests are conducted to study its performance. The results are recorded using Fluke 43B Power Analyzer. The recorded waveforms showing power quality parameters at light load and full-load are shown in Figs. 15-16. It can be seen that the current waveform is nearly sinusoidal at light load as well as full load and show similar trends as simulated results. The test results showing effect of load variation on various power quality parameters are given in Table IV. These experimental results show that the THD of AC mains current with the 36-pulse AC-DC converter varies in the range of 3.3% to 3.9% on load variation while displacement factor (DPF) and power factor (PF) are almost unity. The details of developed autotransformer for 36-pulse AC-DC converter are given Appendix.

VI. RESULTS AND DISCUSSION

The power quality indices obtained from simulations of the

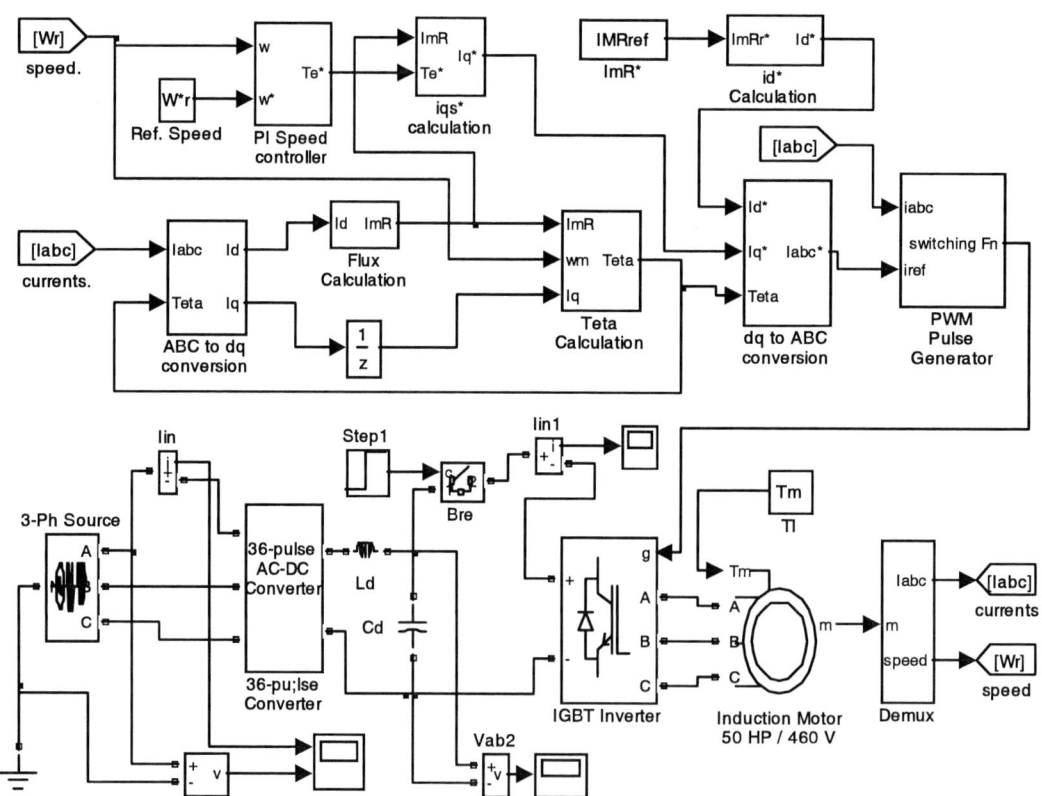

Fig. 5 Matlab model of proposed 36-pulse AC-DC converter system feeding VCIMD.

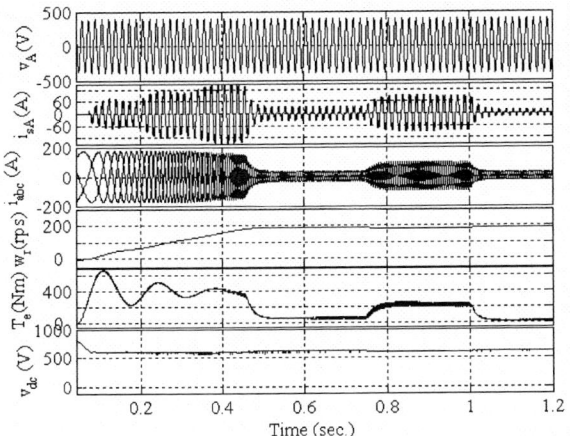

Fig. 6 The waveforms depicting dynamic response of 6-pulse diode rectifier fed VCIMD with load perturbation--supply phase voltage v_A, source current i_{sA}, motor currents i_{abc}, speed ω_r, developed electromagnetic torque T_e and DC link voltage v_{dc}.

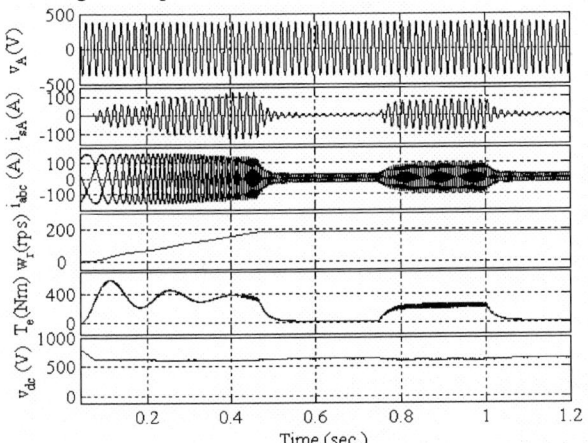

Fig. 7 The waveforms depicting dynamic response of 18-pulse diode rectifier fed VCIMD with load perturbation--supply phase voltage V_A, source current i_{sA}, motor currents i_{abc}, speed w_r, developed electromagnetic torque T_e and DC link voltage v_{dc}.

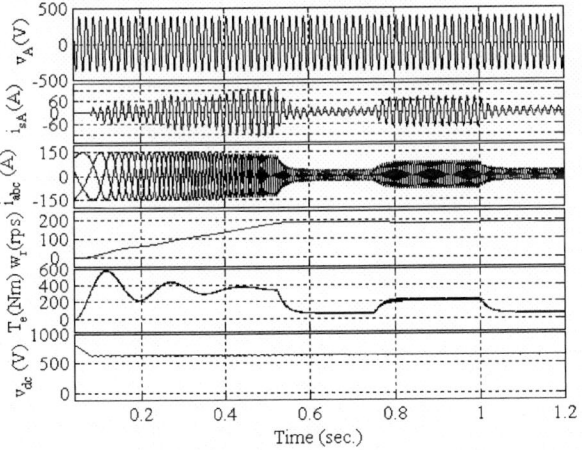

Fig. 8 The waveforms depicting dynamic response of 36-pulse diode rectifier fed VCIMD with load perturbation--supply phase voltage V_A, source current i_{sA}, motor currents i_{abc}, speed w_r, developed electromagnetic torque T_e and DC link voltage v_{dc}.

proposed 36-pulse AC-DC converter are given in Table I. The various waveforms depicting dynamic response of the 6-pulse, 18-pulse and 36-pulse AC-DC converter systems feeding

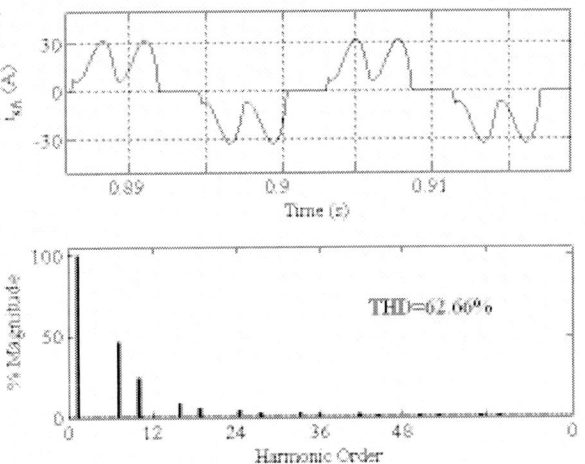

Fig. 9 Input current waveform of 6-pulse AC-DC converter at light load and its harmonic spectrum.

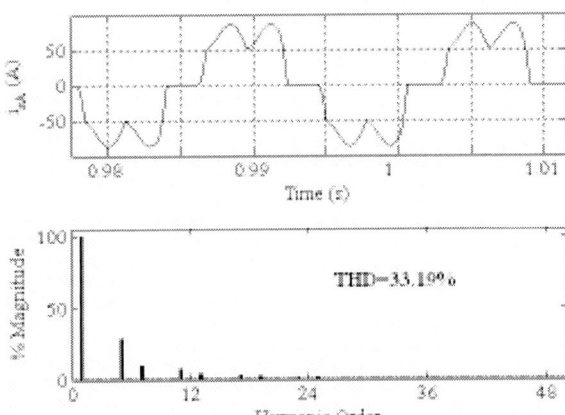

Fig. 10 Input current waveform of 6-pulse AC-DC converter at full load and its harmonic spectrum.

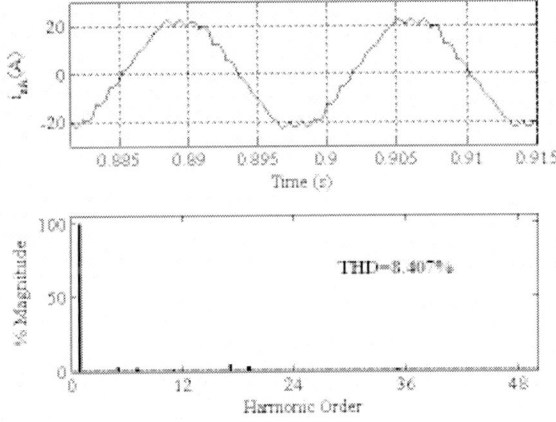

Fig. 11 Input current waveform of 18-pulse AC-DC converter at light load and its harmonic spectrum.

VCIMD with load perturbation are shown in Figs.6-8 respectively. Figs. 9 and 10 show the input current waveform i_{sA} of 6-pulse AC-DC converter and its harmonic spectrum. It clearly shows the dominant 5th and 7th harmonics. The THD of AC mains current at full-load is 33.19% which deteriorates to 62.66% at light load (20% of full-load). Figs. 11 and 12 show

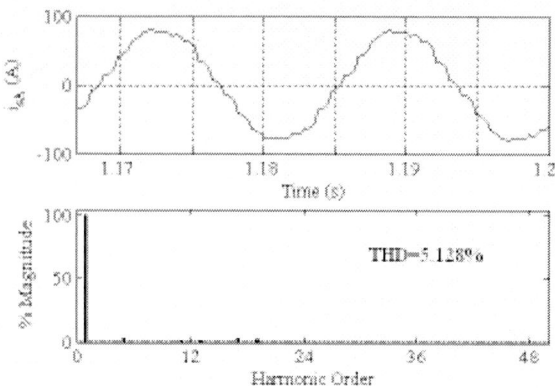

Fig. 12 Input current waveform of 18-pulse AC-DC converter at full load and its harmonic spectrum.

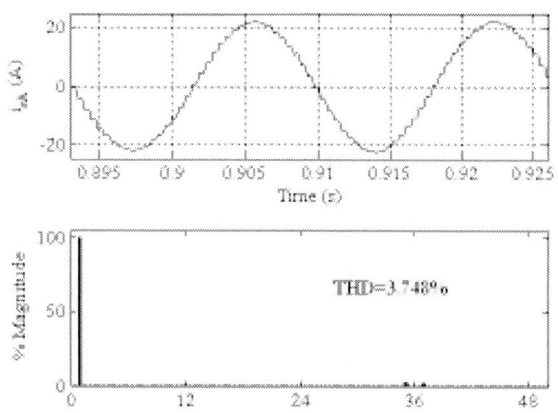

Fig. 13 Input current waveform of 36 pulse AC-DC converter at light load and its harmonic spectrum.

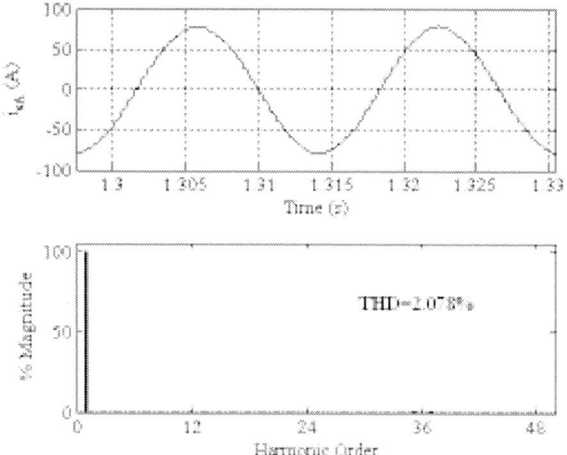

Fig. 14 Input current waveform of 36 pulse AC-DC converter at full load and its harmonic spectrum.

the current waveforms and harmonic spectra of an 18-pulse AC-DC converter at light-load and full load resulting in an improved performance. The THD variation with load in this topology is observed to be 5.128% to 8.407%.

Figs. 13-14 show the AC input supply current waveform i_{sA} of the proposed 36-pulse converter at light load and full-load respectively. At light load the 36 steps in one cycle of AC

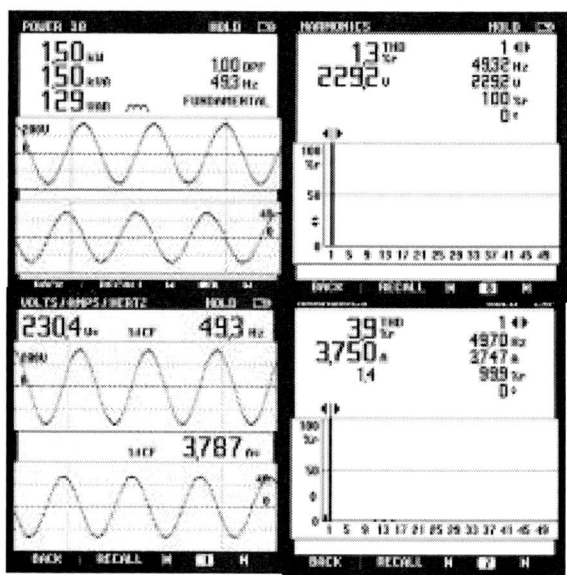

Fig. 15 Input waveforms of 36 pulse AC-DC converter at light load and harmonic spectrum of voltage and current.

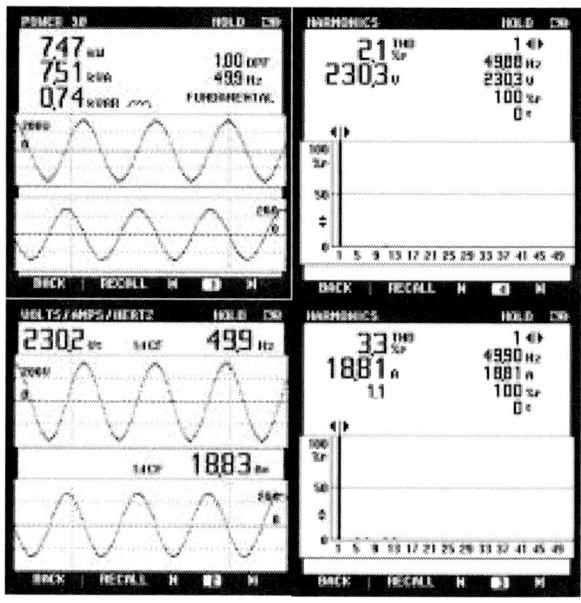

Fig. 16 Input waveforms of 36 pulse AC-DC converter at full load and harmonic spectrum of voltage and current.

mains current are clearly visible and THD is only 3.748%. The dominant 35th and 37th harmonics of negligible magnitude can be seen in Fig 13. The THD of input current reduces to 2.038% at full load as observed from Fig. 14.

The power quality indices, viz., total harmonic distortion of supply current (THD$_i$), total harmonic distortion of supply voltage (THD$_v$), distortion factor (DF), displacement factor (DPF) and power factor (PF) at different loads can be seen in Table I. It can be seen that the THD of input AC current of proposed 36-pulse AC-DC converter system is significantly reduced. THD$_i$ of AC mains current drawn by proposed 36-pulse AC-DC converter system at all loads is less than 4%. The power factor is observed to be order of 0.997 at varying loads in proposed 36-pulse AC-DC converter as given in Table I.

TABLE I

COMPARISON OF POWER QUALITY PARAMETERS OF THE LOAD FED FROM PROPOSED AC-DC CONVERTER.

Sr. No.	Topology	Load	THD V_{ac} (%)	AC Mains Current I_{sA} (A)	THD of I_{sA}(%)	Distortion Factor, DF	Displacement Factor, DPF	Power Factor, PF	DC Voltage (V_{dc})	Ripple Factor, RF (%)
1	36-pulse	20	1.417	15.64	3.748	.9992	.9996	.9988	639.1	.050
		40	1.895	25.32	3.142	.9993	.9992	.9985	638.5	.050
		60	2.262	35.26	2.906	.9993	.9989	.9982	637.8	.052
		80	2.526	45.09	2.468	.9994	.9986	.9980	637.1	.059
		100	2.704	55.34	2.038	.9994	.9985	.9979	636.4	.065

TABLE II

COMPARISON OF POWER QUALITY PARAMETERS OF THE LOAD FED FROM DIFFERENT AC-DC CONVERTERS.

Sr. No.	Topology	THD V_{ac} %	%THD of I_{sA}, at		DF		DPF		PF		DC Voltage (V)	
			Full Load	Light Load	Full Load	Light Load	Full Load	Light Load	Full Load	Light Load	Full Load	Light Load
1	6-pulse	6.441	33.19	62.66	.9490	.8469	.9321	.9776	.9821	.828	609.8	617.1
2	18-pulse	3.199	5.128	8.407	.9962	.9978	.9829	.9870	.9808	.9835	615.1	635.9
3	36-pulse	2.704	2.038	3.748	.9994	.9992	.9985	.9996	.9979	.9988	636.5	639.1

TABLE III

COMPARISON OF ACTIVE POWER MAGNETICS RATING IN DIFFERENT AC-DC CONVERTERS.

Topology	Main Transformer rating, % of load	Interphase transformer rating, % of load	Total
6-pulse	-	-	-
18-pulse	38.31	10.7	49.01
36-pulse	42.36	0.86	43.22

The comparison of power quality indices of proposed converter is made with 6-pulse and 18-pulse AC-DC converters and shown in Table II. It can be observed that the 36-pulse AC-DC converter has definitely improved performance. The size of active magnetics involved in different converters is also given in Table III. It can be seen that a total of 43.22% of magnetics is needed to achieve THD$_i$ <4% (at full-load) in the 36-pulse AC-DC converter. This

rating is less than many other topologies of non-isolated AC-DC converters for inverter fed AC motor drives.

Test results obtained from extensive experimentation carried out on the prototype validate the simulation results. Table IV shows that the THD variation from no load to full-load is from 3.79% to 3.3% and power factor is also of order of 0.996. It can be seen in Figs. 15 and 16 that the voltage as well as current waveforms are nearly sinusoidal. These test results show similar trends as simulated results and thus validate the developed design and model of the proposed 36-pulse AC-DC converter.

VII. CONCLUSIONS

The polygon autotransformer used for proposed 36-pulse AC-DC converter system have small rating and needs only two interphase reactors unlike the 18-pulse pulse AC-DC converter where three such reactors are required. The resulting system has exhibited a high level performance with clean power characteristics to be used in uncontrolled front-end rectifiers. The total harmonic distortion of input current from test results

TABLE IV

TEST RESULTS OF THE PROPOSED AC-DC CONVERTERS WITH VARYING LOAD AT 230V AC MAINS VOLTAGE.

Topology	Load, kW	THD V_{ac} (%)	AC Mains Current I_{sA} (A)	THD of I_{sA}(%)	Crest Factor, CF	Displacement Power Factor, DPF	Power Factor, PF	DC Voltage (V_{dc})	Load Current A	DC Power kW
36-pulse AC-DC Converter	1.50	1.3	3.787	3.9	1.4	1.00	0.9967	306.7	4.486	1.38
	2.04	1.3	5.21	3.8	1.4	1.00	0.9979	311.1	6.345	1.98
	3.10	1.5	7.84	3.5	1.4	1.00	0.9962	307.7	9.60	2.96
	4.01	1.7	10.12	3.4	1.4	1.00	0.9962	304.7	12.46	3.79
	5.13	1.7	12.91	3.3	1.4	1.00	0.9963	300.1	15.80	4.76
	6.16	1.9	15.48	3.3	1.4	1.00	0.9963	299.4	19.07	5.72
	7.13	2.0	18.02	3.3	1.4	1.00	0.9963	295.6	22.19	6.54
	7.47	2.1	18.83	3.3	1.4	1.00	0.9963	294.7	23.14	6.82

905

is observed to be less than 4% at varying loads which is well within IEEE-519 standard requirements. The input power factor is improved to value higher than 0.99 at varying loads. Test results have shown a close match with the simulation results thus validating its power quality improvement capability.

APPENDIX

A. Motor and controller specifications:
Three-phase squirrel cage induction motor −50 hp (37.5 kW), 3-phase, 4-pole, Y-connected, 460 V, 60 Hz, R_s = 0.087 ohms, R_r = 0.228 ohms, X_{ls} = 0.3016 ohms, X_{lr} =0.3016 ohms, X_m = 13.08 ohms, J = 0.8 kg-m^2.
PI controller: K_p =45 , K_i = 0.01
DC link parameters: L_d =0.6mH, C_d =3200μF.

B. Design of Autotransformers:
Autotransformer rating: 3.2 kVA
Transformer design details:
Flux Density: 0.8Tesla, Current Density: 2.3A/mm^2,
Core size:
E-Laminations: Length= 23.5cm, Width= 16cm
I-Laminations: Length= 23.5cm, Width= 4cm
Area of cross-section of core= 58 cm^2(7.6 cm X 8.6 cm)
Autotransformer winding details-

Winding Voltage	Number of turns	Gauge of wire (SWG)
K_1* V_A	36	16
K_2 * V_A	42	16
K_3* V_A	79	16
K_4 * V_A	27	18
K_5* V_A	14	18
K_6 * V_A	107	20

REFERENCES

[1] P. Vas, *Sensorless vector and direct torque control*, Oxford University Press, 1998.

[2] IEEE Standard 519-1992, *IEEE Recommended Practices and Requirements for Harmonic Control in Electrical Power Systems, IEEE Inc.*, New York, 1993.

[3] IEC 61000-3-2:2004, *Limits for harmonic current emissions, International Electromechanical Commission*, Geneva, 2004.

[4] D. A. Paice, Power Electronic Converter Harmonics: Multipulse Methods for Clean Power, IEEE Press, New York, 1996.

[5] R. P. Burgos, A.Uan-Zo-li, F. Lacaux, A Roshan, F.Wang and D. Boroyevich, "Analysis of New Step-Up and Step-Down 18-pulse Direct Asymmetric Autotransformer-Rectifiers", *in Proc. of IEEE conf. IAS- 2005*, 2-6 Oct., 2005, vol. 1, pp. 145-152.

[6] F. J. Chivite-Zabalza, A. J. Forsyth, D. R. Trainer, "Analysis and Practical Evaluation of an 18-Pulse Rectifier for Aerospace Applications", *in Proc. PEMD* (Conf. Publ. no. 498), 31 March- 2 April 2004, vol. 1, pp. 338-343.

[7] R. Hammond L. Johnson, A. Shimp and D. Harder, "Magnetic solutions to line current Harmonic reduction", in *Proc. of Conf. Power Conversion-1994,* Sept., 1994, pp. 354-364

[8] Sewan Choi, Bang sup Lee and Prasad N. Enjeti, "New 24-pulse Diode Rectifier Systems for Utility Interface of High Power AC Motor Drives", *IEEE Trans. on Ind. Appl.*, vol.33, no.2, pp. 531-541, March/April 1997.

[9] P.W. Hammond, "Autotransformer", U.S. Patent No. 5619407, April 8, 1997.

[10] D.A. Paice, "Transformers for multipulse AC/DC converters", U. S. Patent No. 6101113, 8 August, 2000.

[11] G. R. Kamath, B. Runyan and Richard Wood, "A compact autotransformer based 12-pulse rectifier circuit", in *Proc. 2001 of IEEE IECON, Conf.*, 2001, pp. 1344-1349.

[12] G. R. Kamath, D. Benson and Richard Wood, "A Novel Autotransformer based 18-pulse Rectifier Circuit", in *Proc. 2001 of IEEE IECON, Conf.*, 2002, pp. 795-801.

[13] E. P. Wiechmann and P. E. Aqueveque, "Filterless high current rectifier for electrolytic applications", in *Proc. of IEEE conf. IAS 2005*, vol.1, 2-6 Oct. 2005, pp.198-203.

[14] B. Singh, G. Bhuwaneswari and V. Garg, "Multipulse improved-power-quality AC-DC converters for vector controlled induction motor drives" , *IEE Proc. –Electr Power Appl.*, vol. 153, no. 1, pp. 88-96, Jan. 2006.

BIOGRAPHIES

Bhim Singh (SM'99) was born in Rahamapur, U. P., India in 1956. He received B. E. degree in Electrical engineering from University of Roorkee, India in 1977 and M. Tech. and Ph. D. degrees from Indian Institute of Technology (IIT), New Delhi, in 1979 and 1983, respectively. In 1983, he joined as a Lecturer and in 1988 became a Reader in the Department of Electrical Engineering, University of Roorkee. In December 1990, he joined as an Assistant Professor, became an Associate Professor in 1994 and Professor in 1997 at the Department of Electrical Engineering, IIT Delhi. His field of interest includes power electronics, electrical machines and drives, active filters, static VAR compensator, analysis and digital control of electrical machines.

Prof. Singh is a Fellow of Indian National Academy of Engineering (INAE), Institution of Engineers (India) (IE (I)) and Institution of Electronics and Telecommunication Engineers (IETE), a Life Member of Indian Society for Technical Education (ISTE), System Society of India (SSI) and National Institution of Quality and Reliability (NIQR) and Senior Member of IEEE (Institute of Electrical and Electronics Engineers).

Sanjay Gairola was born in Chandigarh, India in 1968. He received B.E. degree in Electrical engineering from M.N. Regional Engineering College, Allahabad in 1991 and M.Tech. degree from Indian Institute of Technology (IIT), New Delhi, in 2001. In 1997, he joined as a Lecturer in the Department of Electrical Engineering, Krishna Institute of Engineering and Technology (KIET), Ghaziabad, U.P., India. In January 2004, he became Assistant Professor. He is a Life Member of Indian Society for Technical Education (ISTE). Presently he is also a research scholar in the Department of Electrical Engineering, IIT Delhi, pursuing for his Ph.D. degree. His field of interest includes power electronics, electric machines and drives.

Comparison of Outer- and Inner-Rotor Switched Reluctance Machines

Martin D. Hennen, Rik W. De Doncker
Institute of Power Electronics and Electrical Drives
RWTH Aachen University
Jaegerstrasse 17-19, 52066 Aachen, Germany
Phone: +49 241 8097153 Fax: +49 241 8092203
Email: he@isea.rwth-aachen.de

Topic Area

Analysis and design of electrical machines

Abstract—**In applications with high torque requirements an outer-rotor design has several advantages compared to an inner-rotor one. A feasibility study of both designs for an in-wheel direct drive application is presented. The torque density depending on slot area, fill factor and air gap radius is analyzed. The influence of cooling on the torque capability of the inner- and outer-rotor machine is shown. A comparison of the thermal behavior will show if the cooling of the outer-rotor machine is adequate for high torque applications. Finite element simulations, linked with analytical software are used to calculate and compare the torque of the machines.**

I. INTRODUCTION

Two different designs of switched reluctance machines (SRMs) are investigated. The most common SRM is the inner-rotor switched reluctance machine. The simple and robust structure, the lack of magnets or brushes make the SRM competitive, especially in high speed applications. However, the SRM can also be used at low speeds with very high torque densities. The given application is an in-wheel direct drive for railway traction. This means operation at low speeds when high torque is required. With outer-rotor machines it is possible to increase the torque per ampere ratio and hence reduce the volt-ampere requirements of the converter. Previous publications have analyzed the outer-rotor design for other machine types and applications like in-wheel direct drives with permanent magnet synchronous machines [1], wind turbines as permanent-magnet generators [2] or starter-alternator with an outer-rotor switched reluctance machine [3].

To design a machine for high-torque traction applications, the diameter of the machine has to be as large as possible. A rough relationship between torque, diameter and length of an SRM is described in [4]. It is shown that torque increases with the square of the diameter. This relation can be used if the machine is scaled completely and hence includes the change of size of the teeth and the slot area. In this paper, two designs for a given maximum outer radius are compared. Consequently, torque depends on the air gap radius which builds the lever arm for the rotational force. To compare the machines, torque for a set operating point and different air gap radii is calculated. Regarding the inner-rotor design, the

air gap radius is limited by the space needed for the coils inside the stator and furthermore by the cooling inside the housing surrounding the laminations. The outer-rotor design has the benefit that coils and cooling can be placed near the shaft, increasing the possible air gap radius. The influence on slot area, fill factor, cooling and the resulting torque of the machines is analyzed in the following paragraphs.

II. COMPARISON OF GEOMETRY

To derive the coherences between torque and type of machine, the next sections will illustrate the differences of the two machine types. Torque is proportional to the rotational force and the radius of the air gap. The rotational force can be calculated with the magnetomotive force (m.m.f.) and the shape of the stator and rotor teeth. Hence, if the shape of the teeth is constant, torque is proportional to the m.m.f. and the air gap radius. The m.m.f. is thermally limited by the maximum current inside the windings and this current is dependent on the resulting copper and iron losses and the quality of the cooling.

Figure 1 shows a part of the cross section to define the radii and the slot area of both machine types. In the following cases the radii r_1 and r_2 are constant. The only radius that changes in the analysis is the radius of the air gap r_{gap}. Depending on this radius, the height of the stator teeth can be calculated. To compare both designs, the stator tooth height can be expressed in percent of the gap between the radius r_1 and r_2 as described in (1).

$$s_{\text{height}} = \frac{h_{\text{st}}}{r_2 - r_1} \cdot 100\,\%$$ (1)

The resulting slot area can be calculated depending on s_{height} as shown in Fig. 2. The air gap radius of the outer-rotor machine increases from $0\,\%$ to $100\,\%$, opposite to the inner-rotor machine. It can be noticed, that at any stator teeth heigth, the inner-rotor machine has a larger slot area. This is due to the increasing circumference of the stator teeth for the inner-rotor machine.

The maximum difference of slot area at $50\,\%$ stator tooth height is approx. $30\,\%$. This would lead to a $30\,\%$ higher current density for the outer-rotor machine at a constant m.m.f.. The optimum percentage of stator tooth height for

978-1-4244-0644-9/07/$25.00 ©2007 IEEE

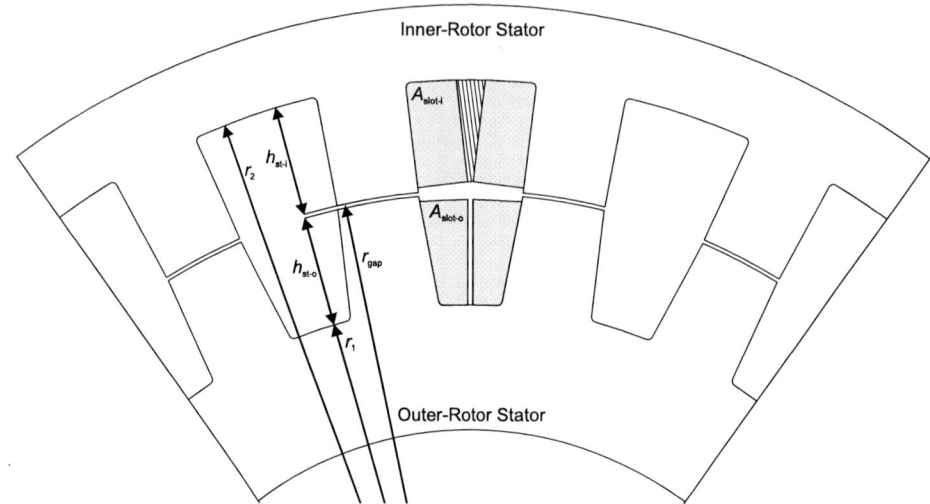

Fig. 1.   SRM Geometry

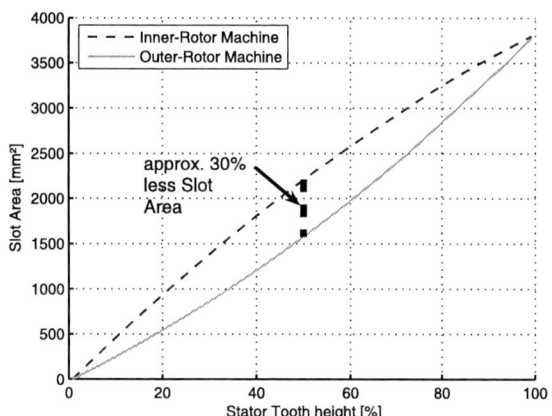

Fig. 2.   Slot Area over percent of Stator Tooth height

each design is a trade off between the ratio of the inductance at the aligned and the unaligned position and the slot area. The inductance ratio will decrease varying s_{height} between 50-100 % in contrast to the slot area. A typical design has a percentage between 60-70 %. In the following, inner- and outer-rotor machines with an optimized stator tooth height are analyzed. Since torque from an SRM depends on saturation effects, the machines were simulated with the analytical software PC-SRD from Glasgow University. To compare two machines with different slot areas and air gap radii, the torque density of the machines for a constant current density are computed. Case 1 in section IV-A shows the results for two machines with a current density of 6.3 A/mm^2 and a copper fill factor of 50 %.

An advantage of an outer-rotor SRM is the geometry of the slot. It is known that most SRMs have concentrated coils. The coils can be pre-wound and afterwards mounted in the stator. For an inner-rotor SRM the slot area cannot be completely utilized. As Fig. 1 shows, the triangular area between the coils

cannot be filled, hence the fill factor is reduced. Figure 3 shows the resulting reduction of fill factor for the inner-rotor machine depending on the stator tooth height s_{height}.

In Fig. 1 it can be seen that the slot area of the outer-rotor machine can be completely filled with pre-wound coils. Calculating the slot area including the reduction of fill factor (Fig. 4) shows that the advantage of a higher slot area for the inner-rotor machine is reduced. Assuming a stator tooth height percentage of more than 67 %, the outer-rotor machine even has a larger slot area and air gap radius than the inner-rotor machine. The influence of this fill factor reduction on the torque density of the inner-rotor machine from case 1 will be shown in case 2 in section IV-B.

The last two cases assume the same outer radius for both machines. In most applications, the maximum outer radius of the housing is given. If the machine is liquid cooled, the maximum radius of the stator and hence the air gap is limited by the space needed for the cooling. As shown in Fig. 5, the outer-rotor machine can fit the cooling between inner radius of the stator and shaft, hence, does not limit the air gap radius. The necessary height for the cooling reduces the maximum outer radius of the stator by approx. 6 % compared to a housing without cooling. The resulting air gap radius of the machine is approx. 16 % smaller including the previous existent difference of inner- and outer-rotor machine. The influence on the produced torque of the smaller machine is analyzed in case 3 in section IV-C.

So far, the space for cooling is considered, but not the cooling itself. Regarding the influence of the cooling, the assumption of an equal current density in both machines is not reasonable. The different location of the cooling leads to a different thermal resistance for both machines. As a result, the losses inside the machines have to be adapted to the capability of the cooling. If the thermal resistance is higher, the losses inside the machine have to be reduced. Section III shows the calculation of the cooling for both types of machines. Case 4

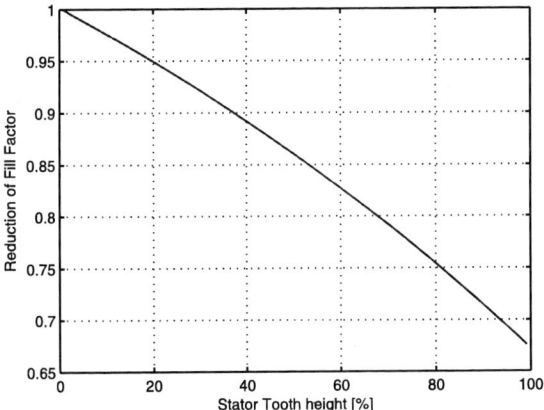

Fig. 3.   Reduction of Fill Factor for the Inner-Rotor SRM

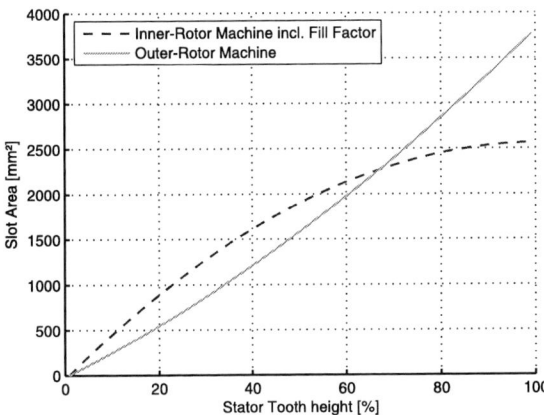

Fig. 4.   Slot Area over percent of Stator Tooth height including Fill Factor

in section IV-D shows the results for the machines including the capability of the cooling.

III. Comparison of Heat Transfer

To compare the inner- and outer-rotor machine it is not necessary to build a complete thermal model of the two machines. The influences of the position of the cooling and the heat sources are derived analytically. The difference of the resulting thermal resistance is important, which is dependent on the geometries of the stator and the cooling. To estimate the resistance, the geometry can be reduced to a simple tubular model. Figure 5 shows the simplified radii of the cooling r_c and the heat source r_h and the direction of the heat flux q.

Two types of heat transfer are analyzed. The first type is the conduction of heat through the stator joke. The second type is the resistance of the contact between the lamination and the cooling. Starting with Fourier's Law in cylindrical coordinates (2) and a convective boundary condition from [5], one can see that the heat flux falls off inversely with the radius. This is because the flux passes through an increasingly large surface.

For the machine model with inner radius r_c, outer radius r_h

and length l, the heat transfer rate can be calculated (3). The thermal resistance consists of two elements. The resistances for the conduction through the joke R_{tcond} (4) and the resistance at the contact area between the joke and the cooling R_{tconv} (5). Both resistances depend on the radii of the stator joke and the radius of the boundary to the cooling. R_{tcond} is constant as long as the ratio of the two radii is constant. Comparing both machines for a constant stator tooth height, the outer-rotor machine has a smaller ratio and hence a higher conduction resistance. The radius of the boundary to the cooling is also significantly reduced. The resistances can be calculated for the inner- and outer-rotor machine from the previous cases inserting the correct radii. The proportional factor v in (6) will show the difference between the cooling capability of the two types of machines. The results including this factor will be shown in case 4 in section IV-D.

$$q = -k\frac{\partial T}{\partial r} = k\frac{\Delta T}{\frac{k}{hr_c} + ln(r_c/r_h)}\frac{1}{r} \qquad (2)$$

$$Q = 2\pi r l q = \frac{\Delta T}{\frac{1}{h2\pi r_c l} + \frac{ln(r_c/r_h)}{2\pi k l}} \qquad (3)$$

$$R_{tcond} = \frac{ln(r_c/r_h)}{2\pi k l} \qquad (4)$$

$$R_{tconv} = \frac{1}{\bar{h}2\pi r_c l} \qquad (5)$$

$$v = \frac{R_{tcondIR} + R_{tconvIR}}{R_{tcondOR} + R_{tconvOR}} \qquad (6)$$

IV. Comparison of Electromagnetic Behavior

To calculate the influence of the air gap radius and the slot area, PC-SRD was used in combination with the finite element software Flux2D from Cedrat. Figure (6) and (7) show the two simulated machines. To improve the results from PC-SRD, the FEM simulated flux-linkage depending on current and rotor position was imported in PC-SRD. The next sections will illustrate the influence on torque and current densities for the previous mentioned cases 1-4. It can be noticed, that the comparison of the torque per ampere ratio for the same outer and inner radii is only depending on the air gap radius. Hence, the torque per ampere will always be better for the outer-rotor machine. This decreases the volt-ampere requirement of the converter and hence size and cost. The values of torque per ampere for the example machines are shown in Table I. The calculated torque densities in the next cases are always related to the volume of the machine including the housing and the bearings.

A. Case 1:

For a constant outer and inner radius r_1 and r_2 and a constant tooth width, the air gap radius is optimized for each machine. For the outer-rotor machine, the radius is 260 mm and 230 mm for the inner-rotor machine (12 % less, 67 % percent stator tooth height). To compare the two machines,

909

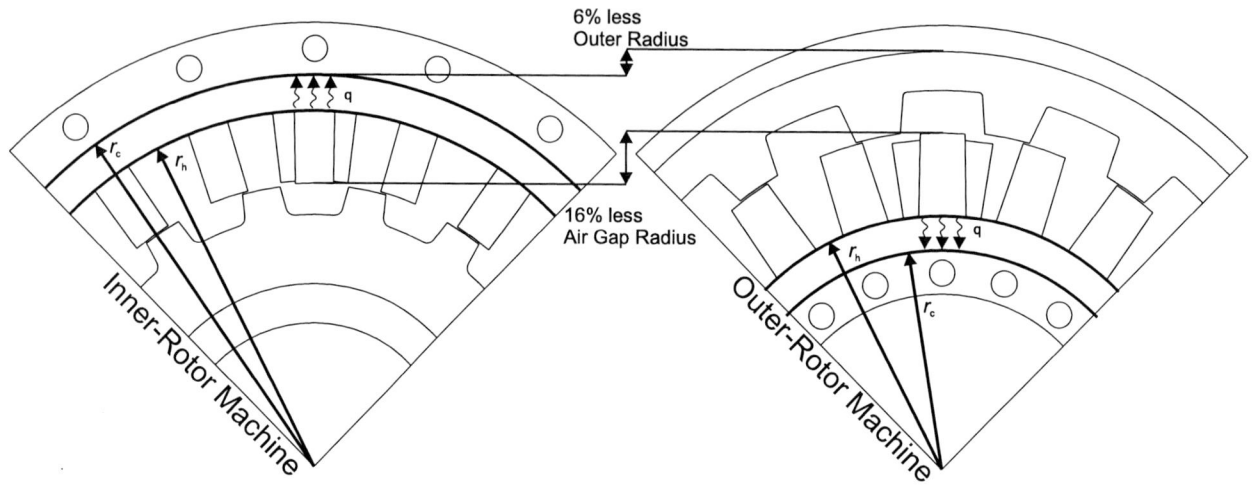

Fig. 5. Comparison of Geometries including Housing

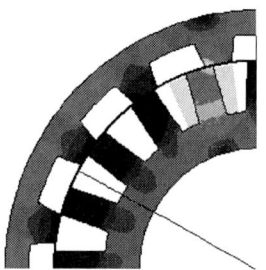

Fig. 6. FEM-Simulated Outer-Rotor Machine

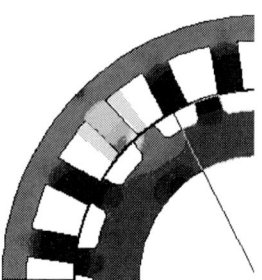

Fig. 7. FEM-Simulated Inner-Rotor Machine

the performance at nominal point with the same number of turns, the same turn-on and turn-off angles, the same dc link voltage and a constant fill factor of 50 % is simulated. Table I gives the results at 350 rpm and a constant current density of 6.3 A/mm^2. The slot area of the inner-rotor machine, as can be seen from Fig. 2, is approx. 25 % higher compared to the outer-rotor machine. Hence, the current applied to the inner-rotor machine to obtain the same current density inside the windings can be set 25 % higher. The resulting torque is approx. 13 % higher. The inner-rotor machine produces more torque despite the smaller air gap radius. Hence, neglecting fill factor, housing diameter and cooling the inner-rotor machine has a higher torque density than the outer-rotor machine.

B. Case 2:

Now, the reduced fill factor, i.e. the effective slot area due to pre-wound coils (Fig. 4) is considered. The assumed fill factor of 50 % for the inner-rotor machine in case 1 is reduced by 20 %. The resulting fill factor of 40 % leads to less maximum ampere-turns and hence, less torque produced by the machine. Both machines have a similar effective slot area. Table I shows that for approx. 97 - 98 A, the outer-rotor machine now produces approx. 6 % more torque. Hence, including the reduction of fill factor, the outer-rotor machine has higher torque density than the inner-rotor machine.

C. Case 3:

Considering the height of the cooling located at the circumference of the stator in this example, the required space is assumed to decrease the maximum outer stator radius by approx. 6 %. Figure 5 shows the cross section of the inner- and outer-rotor machine including the cooling and housing. The slot area is decreased by approx. 17 %. Compared to the outer-rotor machine from case 1, the produced torque for 6.3 A/mm^2, including the reduction of fill factor is reduced by 26 %. Thus, if the machines outer radius is set by the application, the possibility to design the cooling between stator and inner-radius has the effect that the torque density of the outer-rotor machine can be higher than that of the inner-rotor machine.

D. Case 4:

This advantage of a higher torque density in case 3 only exists if the cooling itself is neglected. In section III it was calculated, that the outer-rotor machine has a higher thermal resistance than the inner-rotor machine. This effect has to be included in the simulations in PC-SRD. To do a fair comparison, the losses inside the inner-rotor machine can be higher, assuming the same cooling temperature for both machines. Calculating the ratio of the thermal resistances v for these example machines will show, that the outer-rotor machine has

TABLE I
SPECIFICATIONS

	Case 1-4	Case 1	Case 2	Case 3	Case 4
Design	Outer-Rotor Machine	Inner-Rotor Machine	Inner-Rotor Machine incl. FF	Inner-Rotor Machine incl. FF and Housing	Inner-Rotor Machine incl. FF, Housing, Cooling
R_{gap}	260 mm	230 mm	230 mm	216 mm	216 mm
A_{slot}	2170 mm^2	2730 mm^2	2730 mm^2	2270 mm^2	2270 mm^2
Fill Factor	50 %	50 %	40 %	40 %	40 %
Torque	1245 Nm	1403 Nm	1175 Nm	922 Nm	1245 Nm
Current RMS	97.3 A	123 A	98 A	82 A	116 A
Torque Density	14.3 Nm/Liter	16.1 Nm/Liter	13.5 Nm/Liter	10.6 Nm/Liter	14.3 Nm/Liter
Torque/Ampere Ratio	12.8 Nm/A	11.4 Nm/A	12 Nm/Liter	11.2 Nm/A	10.7 Nm/A
Current Density	6.3 A/mm^2	6.3 A/mm^2	6.3 A/mm^2	6.3 A/mm^2	9 A/mm^2
Efficiency	92 %	90.6 %	90.7 %	90.1 %	87 %

a 56 % higher thermal resistance. This is directly correlated with the losses. If the current of the inner-rotor machine is increased until both machines produce the same torque, the resulting losses of the inner-rotor machine are approx. 70 % higher. All in all, the two machines have the same torque density, but the outer-rotor machine has a significantly higher torque/ampere ratio and hence efficiency, as can be seen in Table I.

The last cases have shown that if the cooling and the reduced fill factor is considered the outer-rotor machine is superior to the inner-rotor machine. These results prove, that in the application at hand, the outer-rotor machine has to be preferred.

V. MECHANICAL CONSTRUCTION

Depending on the application, the construction of an outer-rotor machine is more complex compared to a standard inner-rotor machine. For applications which do not allow a direct integration of machine and drive train, the connection to a load has to be realized with a special fixture which connects the rotating rotor with a shaft. The axle which holds the stator also has to be used to channel the terminals and the cooling inside the machine. Additionally, the outer-rotor can be a safety risk unless a case is built around it, which costs additional space.

For applications like in-wheel direct drives, wind turbines or hybrid electric vehicles, the complexity of the mechanical construction is not a draw back. The outer-rotor can be integrated with the drive train and hence no complex mechanical construction is necessary. The previous outer-rotor machine was designed for an in-wheel direct drive railway traction application. The analysis of both designs has shown, that for this application the outer-rotor machine is particularly suitable.

VI. CONCLUSION

For the same outer radius, fill factor and current density the inner-rotor machine can have a higher torque density.

However, if pre-wound coils are used, the slot area of the inner-rotor machine is not fully available for the windings. The reduced fill factor decreases the torque density. The air gap radius of the inner-rotor machine is limited by the space needed for the coils and the cooling inside the housing. Regarding the outer-rotor machine, both can be placed near the shaft. On the one hand, this allows a higher air gap radius and hence a higher torque per ampere ratio. On the other hand, the cooling capability of the outer-rotor machine is reduced compared to the inner-rotor machine. Different simulations have shown that, including the reduced fill factor, the air gap radius and the cooling, both machines have a similar torque density. An advantage of the outer-rotor machine is the increased torque per ampere ratio and hence increased efficiency. This also reduces the size and costs of the converter. Consequently, the outer-rotor machine is well suitable for applications which require high torque densities and efficiencies. It has to be decided, depending on the application, if the mechanical complexity of the outer-rotor machine is a drawback or not.

REFERENCES

[1] M. Terashima, T. Ashikaga, T. Mizuno, K. Natori, N. Fujiwara, and M. Yada, "Novel motors and controllers for high-performance electric vehicle with four in-wheel motors," *IEEE Transactions on Industrial Electronics*, vol. 44, no. 1, pp. 28–38, 1997.
[2] J. Chen, C. V. Nayar, and L. Xu, "Design and finite-element analysis of an outer-rotor permanent-magnet generator for directly coupled wind turbines," *IEEE Transactions on Magnetics*, vol. 36, pp. 3802–3809, 2000.
[3] R. B. Inderka and R. W. De Doncker, "Outside-rotor switched reluctance machine for minimal hybrid vehicle application," *17th International Electric-Vehicle Symposium (EVS17), Montréal/Canada*, 2000.
[4] T. J. E. Miller, *Switched Reluctance Motors and their Control*. Magna Physics Publications, 1994.
[5] J. H. Lienhard, *A Heat Transfer Textbook*. Cambridge Massachusetts: Phlogiston Press, 2006.

Optimization of Predesign of Switched Reluctance Machines Cross Section Using Genetic Algorithms

Satit Owatchaiphong
Christian Carstensen
and Rik W. De Doncker
Institute of Power Electronics and Electrical Drives
RWTH Aachen University
Jaegerstrasse 17-19, 52066 Aachen, Germany
Phone: +49 241 8096982
Email: ow@isea.rwth-aachen.de

Abstract—Genetic algorithms (GA) have been applied in optimization of machine designs since the first publication in 1975 [1]. In this paper, a practical implementation of this search technique in predesign of switched reluctance machines is presented. An optimized design was found by means of GA based on an objective function for maximizing an average torque of the machine, where dimensions and a thermal loading are specified. Moreover, an auxiliary objective for the most preferable geometries is utilized, well supporting a vector format of the model in GA. The simulation results verified and demonstrated the efficacy of the proposed strategy.

Index Terms—switched reluctance machine, genetic algorithm, global optimal design

I. INTRODUCTION

Recently, switched reluctance machines (SRMs) are used more and more in industrial applications and appliances. SRMs are durable and have a simple construction, as well as low manufacturing costs. The rotor of a SRM is normally a stack of soft magnetic steel sheets without any coils or magnets. However, one disadvantage of SRM against induction machines is its strongly non-linear behavior, since the machine is normally operated in the saturation region of the magnetic material. Hence, it is very difficult to precisely calculate machine performances, which are applied for sizing the dimensions of machines. [2] [3] [4] propose methods to determine the machine performances and to design the machine. Even so, a systematical solution to determine the optimal design is still a controversial topic. Since every parameter of SRMs has more or less influence on the machine performance, designing the machine by determining one parameter sequentially is a work-around method, which requires great experience of the designers to find the well optimized model. By considering an idea of a global optimization, varying all design parameters over a reasonable range is only possible in case few parameters are considered. If many parameters are considered, varying all parameters requires so much computing time that this approach is not practical. In this paper, an application of a search technique of the global optimization based on genetic algorithms is proposed.

After Holland released the first genetic algorithm to the public, it has been used widely in optimization problems. It

has been a successful solution in machine design, e.g. in [5] [6]. In the next section, an introduction and functionality of GA are presented. The advantage of the global optimization by this technique is applied for the optimization of SRMs. The optimization process based on GA was implemented. The functionality and the efficacy of the proposed process were verified by the simulation results.

II. INTRODUCTION AND FUNCTIONALITY OF GA

Genetic algorithms are stochastic search processes, based on fundamental principles of natural selection and genetic evolution. A method that GA uses for creating an individual vector, biologically called chromosome, makes it different from deterministic search processes. The use of this search algorithm was applied first for optimizing designs. In particular, it has presented high-efficiency in electromagnetic applications [7].

For an optimization process, GA performs its search technique from generation to generation. Each generation is composed by a group of individual vectors. A chain of design parameters which are referred as genes forms the individual vector or chromosome. Every chromosome represents a model in a universe, while its dimension is equal to the number of design parameters.

Similar to other optimization processes, the model has to be evaluated and scored. In terms of GA, the score is called fitness, which is an important value for being selected as a survival. The more the fitness is, the better the possibility for the chromosome to survive to form new chromosomes for the next generation. The reforming process, called reproduction, is a powerful search technique of GA [8]. The reproduction basically includes three processes, which are parent selection, crossover and mutation.

A. Parent Selection

As mentioned, some chromosomes will be chosen as survivals for generating new chromosomes in the reproduction. This selection process called parent selection. Unlike selection processes with deterministic search methods that select the most fits, a selection decision of the parent selection in GA is made by probability. A higher reproductive chance is naturally given to the most fit individuals but the chance still remains for the least fits. It can be explained by a roulette game. A number of slots, which are belonging to each individual, are proportional to its fitness. The most fit individual has the highest number of slots in the roulette wheel. It means that the fittest has the highest probability to be selected as the parent for producing the next generation. However, the least fit individual still has a chance for being selected by this method. It prevents this search technique from being trapped by local optimal points, since the genes of each parent are copied to children in the next generation.

B. Crossover

Crossover imitates a recombination process of natural evolution. Two corresponding parents from the parent selection exchange their information or genes to make two new individuals or chromosomes, called children. Some say that the crossover is the most important process in GA. Unless the crossover operation does exist, the results are no longer a genetic algorithm.

C. Mutation

Creation of new children by only one corresponding chromosome is the mutation process. There are also many methods of mutation purposed. The most obvious one is an alternation of characteristic of concerning genes. Although a mutation rate is naturally rare and random, but favorable mutations that confer some advantages to the chromosome in which they occur are rare, being sufficient to provide the variation necessary for natural selection and thus evolution [8].

III. CHROMOSOME REPRESENTATION OF MACHINE PARAMETERS

This research is being applied in optimizing of predesign in switched reluctance machines. The purpose is to approximate and optimize the machine cross section for the required torque. A 3-phase 6/4 SRM is used as an examined object. Other requirements of this model will be described in the next section. As mentioned before, time expense and designer's experiences for determining a well optimized design are strongly required. Even so it is still a difficult problem for experienced SRM designers, because of machine's non-linear behaviors. This research presents simulation results of the search technique using GA. In the first stage, machine parameters are selected and encoded as genes, and then formed as the individual chromosome.

The representation of a chromosome is a coding scheme of design variables, which are the machine parameters. The coding scheme may be varied according to the nature of the target. In general, a bit string encoding or binary encoding is the most classic method used by GA. The conventional GA operations are also developed on the basis of this fundamental structure. Moreover, because of its simplicity and traceability, a trivial binary encoding yields acceptable results. Therefore,

the bit string encoding is adopted in many applications. However, studies have shown that Gray code encoding performs better. The difference between Gray code and normal binary encoding is the meaning of individual changes in the universe, when one bit is altered [8]. Since each position in a binary number has different multiple factors, one bit change of most significant bit (MSB) means the largest step change in the gene. In contrast to binary code, one bit change in Gray code has no difference for each position. This characteristic prohibits the problem of discarding the best chromosome, which can occur by the reproduction process, especially by the mutation. Hence, in this paper, Gray code is chosen as the chromosome representation. The chromosome is composed of 9 machine parameters, as shown in Fig. 1. This chromosome is defined as

$$X = [R_{sh}, y_r, h_r, h_s, y_s, airgap, \beta_r, \beta_s, L_{stk}] \qquad (1)$$

Where $R_{sh}, y_r, h_r, h_s, y_s, airgap, \beta_r, \beta_s$ and L_{stk} are rotor shaft radius, rotor yoke thickness, rotor pole depth, stator pole depth, stator yoke thickness, airgap, rotor and stator pole angle and stack length of machine, respectively. (1) presents gene defined parameters, utilized for the most advantage of generating the models in GA. Because of this additive method, it assures validation of every model from randomly generating chromosomes. The feasible ranges of the machine parameters are roughly discussed in [2].

IV. OBJECTIVE FUNCTIONS

To examine the proposed process, simulation boundaries and machine performances are assigned. The process has to maximize machine capacity represented as an average torque of a 3-phase 6/4 SRM for the mentioned conditions. In this

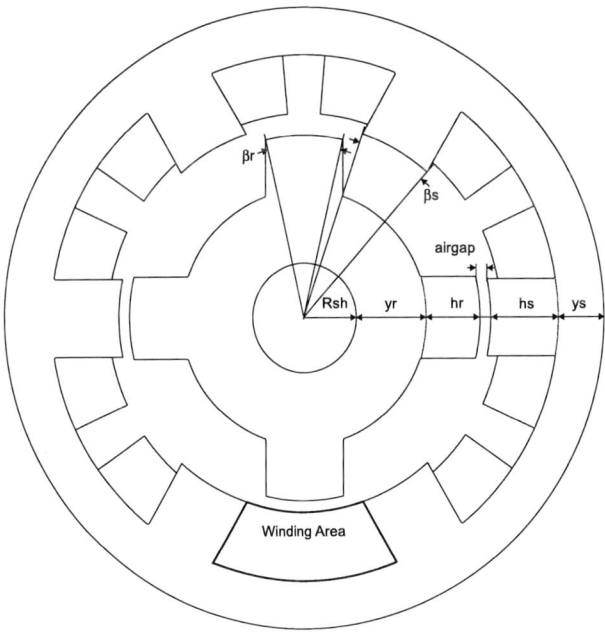

Fig. 1.   Cross section and design parameters of a 3-phase 6/4 SRM

study, an outer dimension of the machine is given. The current density of a winding is specified by a cooling type of the machine. In [2], the conventional torque output equation is used for determining a preliminary machine dimension.

$$T = k \cdot D^2 \cdot L_{\text{stk}} \qquad (2)$$

where T is average torque. k, D and L_{stk} are output coefficient, rotor diameter and machine stack length, respectively. While k depends on several design parameters, described in [2], $D^2 \cdot L_{\text{stk}}$ is the rotor volume of the machine. It can be implied that the torque is proportional to the rotor volume. However, by increasing the rotor volume, the winding area is decreased as the outer dimension of the machine is limited. Thus, it affects the current density. Fig. 2 presents the average torque at one specific current density by varying the rotor diameter, while stator and rotor yoke thickness, as well as rotor pole depth are kept constant. It presents the optimal rotor pole angle, in which the model produces the highest average torque.

The average torque of machine is scored by an objective function defined in (3). A torque reference determined by an average torque at the specific current density is applied for making the per-unit for this objective function. The per-unit is very useful for multi-objective problems [9]. The torque objective function is defined as

$$f_1 = f(\frac{T}{T_{\text{ref}}}) \qquad (3)$$

Where T_{ref} is a torque reference. Due to the non-linear behavior of SRMs, machine performance cannot be determined by a few analytical equations. The torque of SRMs is calculated by a flux linkage diagram, proposed in [2] [3] [10]. The current density is set at 8 A/mm^2 for fan cooled machines. The varied ranges of the current density are presented based on the cooling method in [2]. Besides torque and current density, a kind of auxiliary objective function is utilized for

the assigned dimensions. Geometric parameter limitations and preferred dimensions are defined as other objectives. While the additive method removes any condition for generating the model, some dimensions are most preferred or they are even limited by several applications. The machine parameters are examined and scored by these objectives. The unsatisfactory models will be thrown away by the means of the survival rules in GA. This objective is used to format the mentioned chromosome. In this study, the outer dimension of the machine is limited. Thus, the auxiliary objective function is defined as

$$f_2 = f(\frac{X}{X_{\text{ref}}}) \qquad (4)$$

Where X is the examined parameter. In this case, The specified outer stator radius and stack length of the machine are defined as the reference values, X_{ref}. The stator radius,R_{s} is defined as

$$R_{\text{s}} = R_{\text{sh}} + y_{\text{r}} + h_{\text{r}} + h_{\text{s}} + y_{\text{s}} + airgap \qquad (5)$$

Hence, the fitness is calculated by

$$fitness = w_1 \cdot f_1 + w_2 \cdot f_2 \qquad (6)$$

Where w_1 and w_2 are weighted average of each objective fitness due to difference priority of each objective. Fig. 3 and Fig. 4 present scoring functions of both objectives. The torque is proportionally scored by a linear increasing slope. It simply means that the model producing more torque receives the higher score. In contrast to torque, the machine dimensions are not aimed for obtaining the smallest possible size, but it is for regulating the machine dimensions. Because of this, every model in the specified ranges gets the same positive score. Otherwise, a negative point will be applied to the model. To prevent the objectives from compensating each others, a condition is applied to the scoring function of the torque objective. In the case that the model produces more than 120 percent of the torque reference, the score becomes constant and still remains at the same score as the model producing 120 percents of the torque reference, as presented in Fig. 3. This prohibits the occurrence of large machines receiving very high fitness scores, even if their outer dimensions extremely exceed the limitations since the score of the torque objective compensates for the negative score of the dimension objective.

V. Simulation Results

The proposed process was implemented in MATLAB, known as a standard program for calculation. The chromosome is first generated and then exported into PCSRD, a machine design commercial program. In PCSRD, machine characteristics are calculated. The calculation results are imported into MATLAB. The fitness value of the model is evaluated by the proposed functions. The implemented process has been used to optimize the 3-phase 6/4 SRM by considering the objective functions of the average torque and the outer dimension of the machine. The process has shown the promising success in the design optimization for the assignment, as shown in

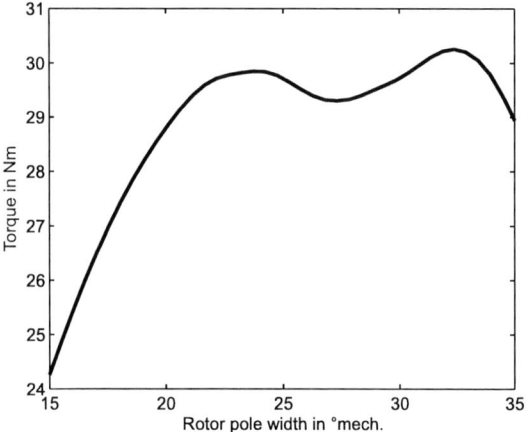

Fig. 2. Average torque of the 3-phase 6/4 SRM at 8 A/mm^2, varying the rotor pole width

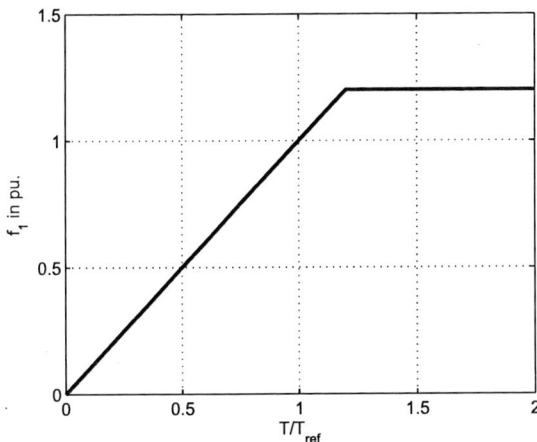

Fig. 3. Objective function

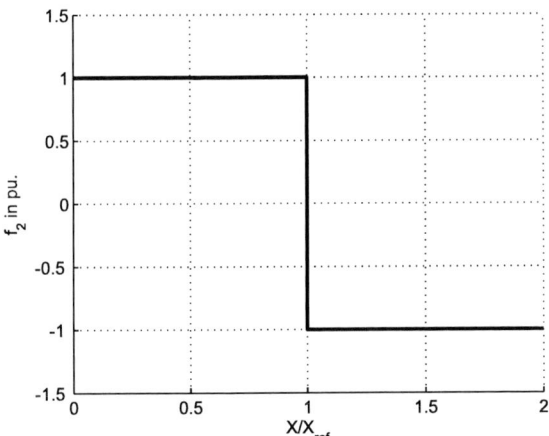

Fig. 4. Auxiliary objective function

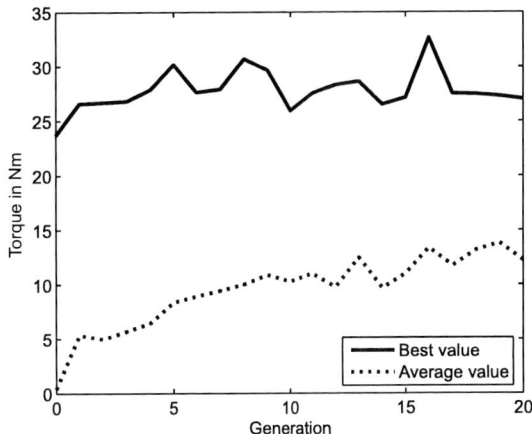

Fig. 5. Change of average torque of best and average fitness values

Fig. 5. The tendency of the average torque was increasing for the past generations. It represents the improvement of the optimization by this search technique, while the best model was found at the 17th generation. Afterwards, the best fitness model was constructed by PCSRD for observing the machine performances.

To present the improvement of this process, another 3-phase 6/4 SRM was constructed as a reference model by the conventional method proposed in [2] [3] [4] [10]. The model

Comparison lists	Conventionally optimized model	GA optimized model
Outer diameter in mm.	170	168.8
Stack length in mm.	120	119.5
Average torque in Nm	30.92	32.12
Current density in A/mm^2	8	8

TABLE I
COMPARISON BETWEEN OPTIMIZED MODEL BY GA AND CONVENTIONAL METHOD

was first initiated by a combination of the rotor radius and the rotor pole arc angle, which produces the highest average torque at a current density of 8 A/mm^2. Then, the other parameters were determined related with the rotor radius and the pole arc angle. Finally, the model was finely adjusted. The two machines were examined under the same simulation conditions. The simulation results present that the optimized model by GA improves the average torque of the machine by 3.75 percents from the conventionally optimized model. It is important to notice that the torque was increased, while the outer dimension of the optimized model still not reaches at its limits, as shown in Table I. Although the outer diameter and stack length are smaller than the limitation, but the purpose was not to optimize the machine size. Hence, it can be considered as incompleteness of the proposed process. According to the presented auxiliary objective function, the outer dimensions of machine are limited but it does not lead to fully use of all available volume. Although GA does not have learning ability, but modification of fitness function may guide GA to reach its objective. Fig. 6 presents a modified objective function for scoring the outer dimension of model. Oversize models still receive a negative point, but a higher score is applied for the model that its outer dimension is closed to the objective. However, because of time limit, the modified objective was not implemented and tested in this process.

Furthermore, the models were observing for the average torque over a wide range of the current density. The relationship between the torque and current density is determined as shown in Fig. 7 and Fig. 8. It can be seen that the model optimized by the proposed process generates the higher average torque for current densities up to 34 A/mm^2. After that, the conventionally optimized model gained the higher torque.

VI. CONCLUSION

In this paper, the use of genetic algorithms was utilized in the machine designing process. The optimization process was implemented by this search algorithm. Machine geometry

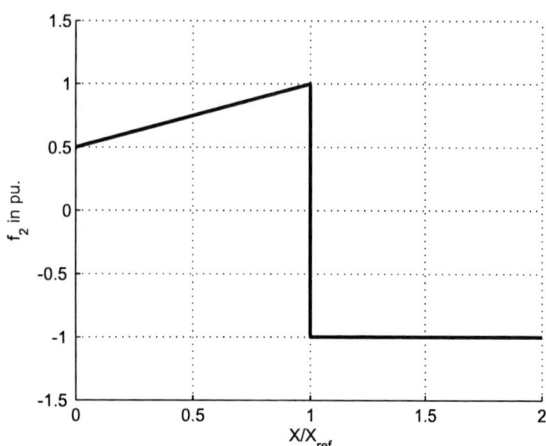

Fig. 6. Modified auxiliary objective function

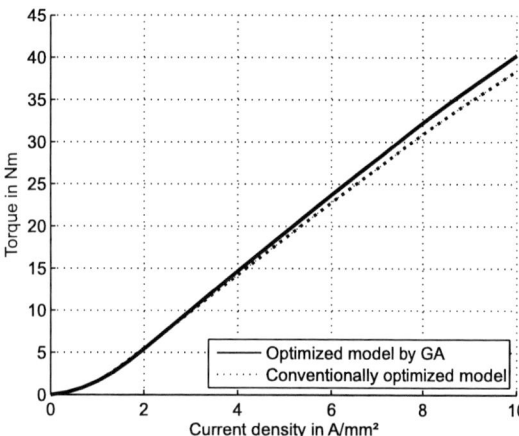

Fig. 7. Average torque as a function of current density from 0-10 A/mm^2

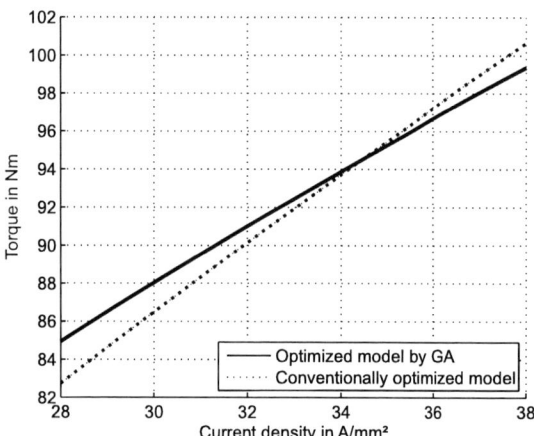

Fig. 8. Average torque as a function of current density from 28-38 A/mm^2

parameters are considered as the design variables and formed in the proposed chromosome format. The implemented application successfully optimized the examined machine for the specified assignments. The machine performance, especially the average torque, was improved by this global optimization for the switched reluctance machine, which consists of many design parameters. The comparison results between the GA optimized model and the conventionally optimized model was discussed. It verified the functionality of the proposed method. Moreover, this method overcomes some difficulties of the optimization for the multi-objective problem, and can compensate the lack of designer's experience, effectively.

REFERENCES

[1] J. H. Holland, *Adaptation in natural and artificial systems.* Cambridge, MA, USA: MIT Press, 1992.
[2] T. J. E. Miller, *Switched Reluctance Motors and Their Control.* London: Oxford University Press, 1993.
[3] ——, "Optimal design of switched reluctance motors," *IEEE Transaction on Industrial Electronics*, vol. 49, no. 1, pp. 15–27, February 2002.
[4] M. N. Anwar, I. Husain, and A. V. Radun, "A comprehensive design methodology for switched reluctance machines," *IEEE Transaction on Industrial Applications*, vol. 37, no. 6, pp. 1684–1692, November 2001.
[5] B. Mirzaeian, M. Moallem, V. Tahani, and C. Lucas, "Multiobjective optimization method based on a genetic algorithm for switched reluctance motor design," *IEEE Transactions on magnetics*, vol. 38, no. 3, pp. 1524–1527, May 2002.
[6] O. Mohammed, "Ga optimization in electric machines," *Electric Machines and Drives Conference Record*, May 1997.
[7] G. F. Uler, "Genetic algorithms in the design optimization of electromagnetic devices," Ph.D. dissertation, Miami, FL, USA, 1994.
[8] K. F. Man, K. S. Tang, and S. Kwong, *Genetic Algorithms, Concepts and Designs.* Springer, 1999.
[9] D. E. Goldberg, *Genetic Algorithms in Search, Optimization and Machine Learning.* Boston, MA, USA: Addison Wesley Publishing Company, Inc., 1989.
[10] N. H. Fuengwarodsakul, J. O. Fiedler, S. E. Bauer, and R. W. D. Doncker, "New methodology in sizing and predesign of switched reluctance machines using normalized flux-linkage diagram," *Industry Applications Conference*, vol. 4, pp. 2704–2711, October 2005.

Shaft Position for an 8/6 Switched Reluctance Machine: Theoretical concept, FEM analysis and Experimental results.

Silviano Rafael*, P.J. Costa Branco**, A.J. Pires*

* LabSEI, Escola Superior de Tecnologia de Setúbal, Instituto Politécnico de Setúbal,
**Instituto Superior Técnico, Portugal

Abstract– The Switched Reluctance Machine (SRM) was, during this last decade, the target of various researchers' attention. Several authors have published mainly research results analysing torque ripple reduction, speed control, noise reduction, magnetic characteristics, and the SR machine operation without position sensor. However, very few works have studied the SR machine in terms of its shaft position. This paper presents this thematic. The static torque, experimentally obtained, is analysed followed by the determination of which phases must be excited in order to immobilize the shaft. Finite Element Method (FEM) is used to calculate the current values, function of the various shaft positions. Experimental tests were done in order to validate the theoretical assumptions and FEM results. Finally, the results are discussed.

Index Terms—Switched Reluctance Machine, Shaft Position, Concept of position control for SRM, Torque equilibrium.

I. INTRODUCTION

The SRM has desirable features such as simple construction, high reliability, and low cost production and maintenance. Some inherent advantages are: robust rotor structure, simple winding configuration, short coil ends and high torque inertia. Recent studies suggest that the SRM presents better performance and efficiency when compared with classical electrical machines for the same dimensions. However it presents some disadvantages such as noise and torque ripple. Some researchers have presented diverse solutions for these problems [1, 2 and 3]. Some studies demonstrate the good speed and torque behaviour [4]. In terms of its application one can verify an increment in the domain of the electrical appliances and in the automobile industry [5]. Overall this means that its mechanical characteristic is attractive.

However, in the servomotor functioning service, the SRM don't yet found its place because still few works exist to demonstrate its potentiality. This work aims to be a contribution on this field and consists mainly on the torque study and FEM analysis in order to place the machine shaft in an intended angular position.

II. SRM CHARACTERIZATION

The functioning principle of this machine is based on the reluctance variations of its magnetic circuit that depends on the rotor position [6]. In fact, this magnetic reluctance circuit is dependent on its geometry and also on some constructive parameters, like the dimension of the magnetic circuit, the type of ferromagnetic material that is used on its structure, its thickness and lamination factor that form the magnetic circuit, etc.

The SRM in study is composed by 8 stator poles, 6 rotor poles and 4 phases.

III. MATHEMATICAL MODEL

The model is considered complex due to its magnetic circuit non linearity. The SRM electrical equations of one phase can be expressed as:

$$V_k = Ri_k + \frac{\partial \lambda_k(\theta, i_k)}{dt} \qquad (1)$$

where Vk is the phase voltage, ik is the phase current, R is the resistance per phase, k is the active phase and λk (,θ, ik) is the phase linkage flux. Using the chain rule, (1) can be rewritten as:

$$V_k = Ri_k + \frac{\partial \lambda_k(\theta, i_k)}{di} \frac{di_k}{dt} + \frac{\partial \lambda_k(\theta, i_k)}{di} \frac{d\theta}{dt} \qquad (2)$$

The torque developed by one excited phase is determined by the variation of the magnetic coenergy produced in its magnetic circuit in order to the variation of the position and it is expressed in (3).

$$T_{(\theta, i_k)} = \left. \frac{\partial W'_{(\theta, i_k)}}{\partial \theta} \right|_{i=const} \qquad (3)$$

The magnetic coenergy is characterized by the following expression (4).

$$W'_{(i, \theta)} = \left. \int_0^i \lambda_k(\theta, i_k) di \right|_{\theta=const} \qquad (4)$$

IV. TORQUE CHARACTERISTICS

To analyse the SRM behaviour and to define the necessary conditions for the rotor position operation, it is important to know its torque characteristic (3).

Torque curves characterizing the SRM can be obtained with different techniques [6, 7]. In this study, the SRM static torque was measured. A metallic arm system carefully balanced and linked with the shaft structure was used. The displacement of the balanced arm was very limited. The torque engine was calculated from the knowledge of the arm distance to a known weight that was placed in order to equilibrate the developed torque by the machine when the phase was excited with a specific current value. This procedure was repeated for some positions (-30° +30°) and, in each position, some values of current were applied (1 to 18A), allowing to obtain the curves in Fig. 1. It was a meticulous and much delayed work.

In Fig. 1, the 0° position corresponds to the aligned poles position of phase 1. Now, one knows the static torque curves developed by one phase. Considering that the next three phases have identical static torque curves, the 4 phases static torque curves can be obtained only taking into account the angular shift position of each phase, 15° each one. This shift position is imposed by the geometry of the SRM. In Fig. 2 the phases overlapping angles are observed in two near pole phases and the torque values are positive for the phase sequence 1, 4, 3 and 2. For a reverse direction, with negative torque values, the phase sequence is 1, 2, 3 and 4.

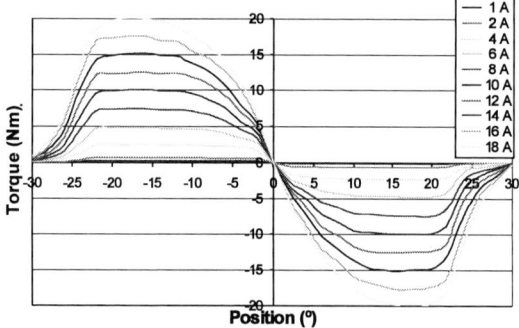

Fig. 1. Torque curves family of one phase.

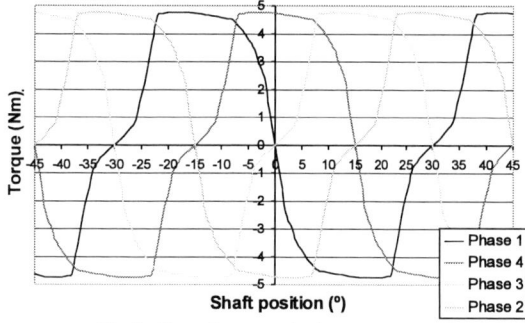

Fig. 2. Four phase nominal torque curve.

V. ROTOR POSITION WITHOUT LOAD - THEORETICAL CONCEPT

The difficulty increases when one tries to position the rotor without load, because this machine, of concentrated polar flux, when excited goes to rotor positions correspondent to the alignment of their poles. These main alignment positions are multiple of 15°. That means there are up to 24 alignment pole positions in 360°. The intermediate positions are only obtained due to the interaction of two phases. For example, it could be done exciting two phases, one at each time in a high frequency, in a way that the average resultant torque is null in the intended position.

Another technique would be to excite the two phases

simultaneously in order that $T_{avg} = T_j + T_k$, being j the phase number such that $T_j > 0$ and k the opposite phase such that $T_k < 0$, producing $T_{avg} = 0$, that keeps the shaft in the desired position. For our initial study, only the static torque curve developed for the rated current of the SRM (6A) is used.

To determine which combinations of phases are needed, let's analyze Fig. 2. The principle is that phase 1 of Fig. 2 will be the torque reference for comparison with other phases in angular position intervals. Diverse angular intervals are detected that can contribute for a steady-state form of the rotor positioning. For a more easy understanding, it is presented in Fig. 3, phases 1 and 2, which were extracted from Fig. 2. One can analyze in Fig. 4 the case related with the excitation of phases 1 and 4 in any angular position intervals delimited by two consecutive null torque points.

Each phase when excited tends to align poles. Between -15° and 0° one verifies that the produced torques, by both phases, have opposite signals. This is represented by the blue zone in Fig. 3 and it is stable because when phase 1 tends to the poles alignment, phase 2 increases its opposite torque and vice versa. This tends to equilibrium. In the remaining position of Fig. 3 there are no more zones where these two phases can be in static equilibrium. For example, between 0° and 15° phase 2 facilitates the poles alignment of phase 1 because the developed torques signals are the same. Another unstable torque zone is defined for example, between 15° and 30°, when one of the phases tends to the poles alignment and its opponent torque tends to diminish, never accomplish equilibrium in any one middle term shaft position.

The same procedure is put in practice for the other pairs of phases. For example in Fig. 4 the torque curves of phases 1 and 4 are presented. It can be verified that

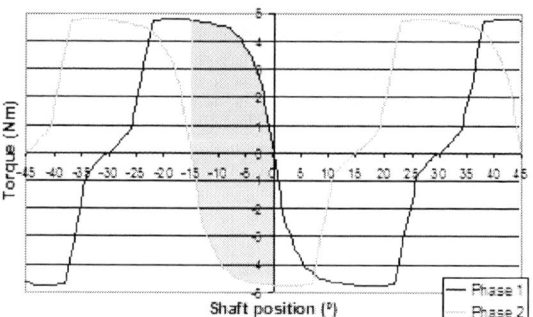

Fig. 3. Intersection of nominal torque curves of phase 1 and phase 2.

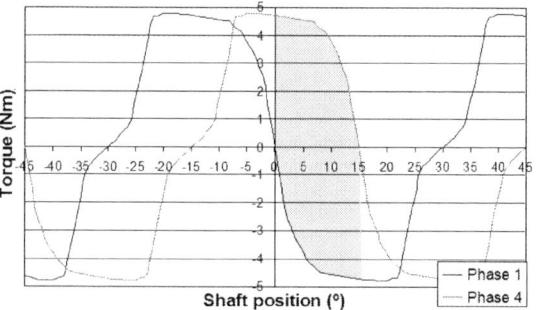

Fig. 4. . Intersection of nominal torque curves of phase 1 and phase 4.

between 0° and 15° it is possible to have the static equilibrium in any position. Finally, observing the developed torques of phases 1 and 3 it is verified that they are antagonist because when one phase tends to poles alignment, another one tends to the unalignment, decreasing the torque opposition. So, it is unstable in all angular position.

Table 1 presents the torque signals and phases, separated in angular position intervals, to have position equilibrium during one polar step.

TABLE I
TORQUE SIGNAL DEVELOPED IN ANGULAR POSITION INTERVALS

Angle (°)	0° to 15°	15° to 30°	30° to 45°	45° to 60°
Positive Torque	Phase 4	Phase 3	Phase 2	Phase 1
Negative Torque	Phase 1	Phase 4	Phase 3	Phase 2

VI. APPLICATION OF FINITE ELEMENTS METHOD

The interest of the Finite Elements Method (FEM) application is already well known [8]. We use this method for obtain the total torque in various rotor positions with different phase current values. These currents are reproduced in the experimental test. Validation and verification of the FEM model are made comparing the experimental torque curve with that from the FEM simulation. Fig. 5 shows that the approach is acceptable.

The error of the torque obtained with FEM is more raised between -30° and -22°. This is not important because this area is not used. The main zone is situated between -22° and 0° where the error is in average lesser or equal to 2 % for each current value. In the same way it is valid for the values between 0° and +22°. The work development with FEM was important to test, in a way to determine and to produce the procedures to apply in the experimental assay. Simultaneously the behaviour of the machine for some positions, as shown in Fig. 6 for 6° and 13°, was foreseen. The procedure consisted, for one given shaft position, to keep constant the phase current (called phase 4 in diagonal alignment) and to excite another one (called phase 1 in vertical alignment) with a constant current value that iteratively diminishes until getting a value of torque that it is lesser than friction torque.

The application results of the theoretical concepts with FEM are presented in section VIII.

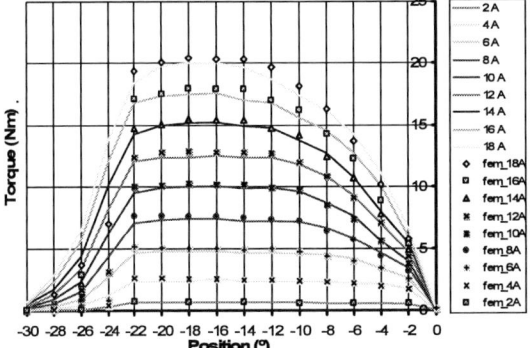

Fig. 5. Comparison of experimental torque curves and FEM torque curves results.

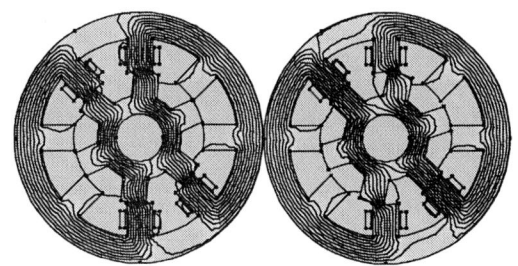

Fig. 6. FEM applied to SRM 8/6 geometry; left 6° and right 13°.

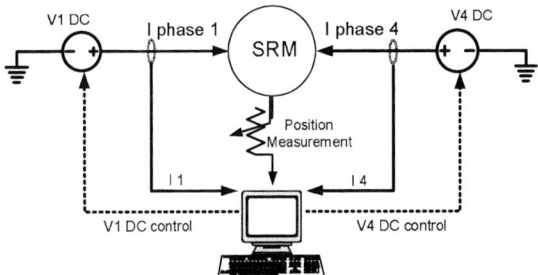

Fig. 7. Diagram set up.

VII. EXPERIMENTAL PROCEDURES

For testing and validating the presented study, an experimental test was developed based on the previous FEM model. It consisted on applying a direct current in two contiguous phases, phase 1 and phase 4, of the machine producing the flux lines representation like in Fig. 6, and using the principle diagram set up showed in Fig. 7. A software application in Visual Basic was developed and through a NI DAQ card the current applied to phase 1 was controlled, keeping constant the current applied to phase 4. Phase 1 current value was changed in steps of 0,2 A approximately. The phase current was controlled in order to remain constant, while the torque equilibrium position developed by each phase was settled. Simultaneously it was registered the rotor position by one precision potentiometer of 10 turns. This potentiometer was linked with the motor shaft through two gears forming a multiplier relation of 1: 4.

VIII. FEM AND EXPERIMENTAL RESULTS

In this section, the experimental and correspondent FEM results are presented and analyzed.

Table 2 presents the current values that will have to be controlled in the two phases in order to produce the equilibrium for determined position. For example, when phase 1 was excited with a current of 3,5 A (2nd column) and phase 4 (3rd column) with a constant current value of 3 A, the total torque was 0,0212 Nm. However the total torque produced in the shaft is lesser than friction torque. The friction torque was experimentally determined through the accumulation of some know masses until the threshold of the very soft shaft movement. This friction torque, thus determined, was 0,043 Nm. It is observed in table 2 that, despite the torque ones being positive or negative, they are not bigger than the friction torque. Therefore, in the angular position the shaft is in static equilibrium.

Table 3 presents the obtained values of the total torque when the next current value is applied. For example when the shaft is in the 0º position, which means that the poles of phase 1 are aligned, and if one applies a current value lesser than the one for equilibrium , for example 3,3 A, it generates a total torque of 0,126 Nm in the direction to the poles alignment of phase 4. According to table 3 the resultant torque values are bigger than torque friction in all shaft positions. This produces the shaft movement until reaching a new equilibrium torque point in the next position.

TABLE II
TOTAL TORQUE AND CURRENT VALUES FOR EQUILIBRIM POSITION OF FEM RESULTS

Position (º)	Phase 1 (A)	Phase 4 (A)	Torque (Nm)
0	3,5	3	0,02121
2	3,3	3	-0,01157
6	3,1	3	-0,03315
10	2,9	3	0,03713
13	2,7	3	0,04122
15	2,5	3	0,01631

TABLE III
TOTAL TORQUE VALUE WHEN NEXT CURRENT VALUE IS APPLIED

Position (º)	Phase 1 (A)	Phase 4 (A)	Torque (Nm)
0	3,3	3	0,1263
2	3,1	3	0,12105
6	2,9	3	0,4599
10	2,7	3	0,10355
13	2,5	3	0,08802

The experimental results of the currents in phase 1 and phase 4 are presented in Fig. 8. The existing current noise signal in phase 1 made more difficult the accurate adjustments. Notice that in the first current step the rotor pole was aligned up with stator phase 1, in agreement with Fig. 9, due to the higher current value in this phase. In the same way in the last current step the rotor pole was aligned up with the stator pole of phase 4. For these extreme positions it would be enough to excite only one phase. However, for the intermediate positions, two phases, in minimum, are necessary for verifying the equilibrium of the acting torque. These torque values depend on the individual current values in each position

In table 4 it is presented the used current values in each equilibrium shaft position. It is verified that the average values of experimental current are slightly less than the values of the current utilized through the FEM analysis. The maximum error is close to 2% what it is considered acceptable for the current forecast.

TABLE IV
PHASE CURRENT COMPARISON TABLE

Position (º)	Experimental Phase 1 Current (A)	FEM Phase 1 Current (A)
0	3,43	3,5
2	3,26	3,3
6	3,08	3,1
10	2,89	2,9
13	2,7	2,7
15	2,51	2,5

IX. CONCLUSIONS

This work presents the study and a method to position the rotor shaft of a 8/6 Switched Reluctance Machine for any angle. The SRM used in this work was characterized in terms of torque/current/position. The study was developed in order to place the no load motor shaft in any angular position, suggesting that it becomes possible when two phases are correctly excited. The analysis of the motor torque curves shows the possibilities of equilibrium when two well defined phases are excited simultaneously in order to position the shaft in definitive angular intervals. FEM analysis is used for defining the experimental procedures and determines iteratively the value of the current for each shaft position. This study is strengthened by the presentation of experimental results that demonstrate what previously was argued and presented.

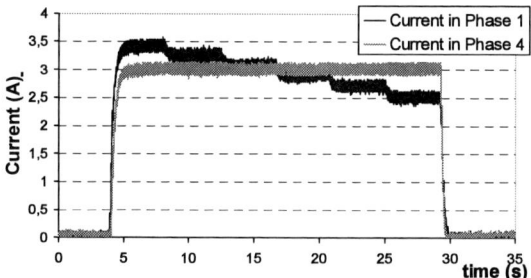

Fig. 8. Experimental current values applied to phase 1 and 4.

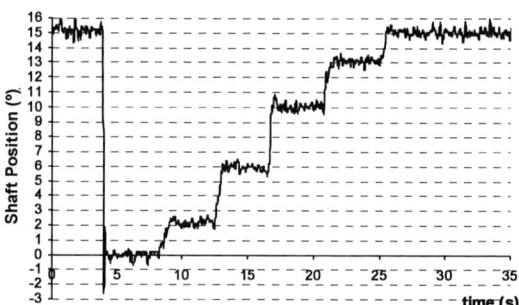

Fig. 9. Experimental results of shaft position.

REFERENCES

[1] Sanada R., Morimoto S., Takeda Y., Matsui N., "Novel rotor pole design of switched reluctance motors to reduce the acoustic noise", *IEEE conference Industry Applications*, Vol 1, pp 107-113, Oct. 2000.

[2] Chai J., Lin Y., Liaw C., "Comparative study of switching controls in vibration and acoustic noise reduction for switched reluctance motor", IEE Proceedings – Electric Power Applications, Vol 153, issue 3, pp 348-360, May 2006.

[3] Sahin F., Ertan H., Leblebicioglu K., "Optimum geometry for torque ripple minimization of switched reluctance motors", IEEE Transactions On Energy Conversion, Vol 15, issue 1, pp 30-39, March 2000.

[4] Rahman K., Fahimi B., Suresh G., Rajarathnam A., Ehsani M.,"Advantages of switched reluctance motor applications to EV and HEV; design and control issues." *IEEE Transactions on Industry Applications*, vol 36, issue 1, pp 111-121, Jan 2000.

[5] Nisai H., Marcus M., Robert B., Rik W., "High Dynamic four quadrant switched reluctance drive based on DITC", *IEEE transactions On Industry applications*, Vol 41, issue 5, pp 1232-1242, Sept 2005.

[6] Chancharoensook P.,Rahman M., "Magnetisation and static torque characterization of a four phase switched reluctance motor: experimental investigations", *Proc. of the 4th International Conf. Power Electronics and Drives*, vol 2, pp 456-460, Oct 2001.

[7] Srinivas K., Arumugam R., "Analysis and characterization of switched reluctance motors: partI – Dynamic, static and frequency spectrum analyses2, IEEE Transactions on Magnetics, vol 41, issue 4, pp 1306-1320, April 2005. Vol 15, issue 1, pp 30-39, March 2000.

[8] Baltazar Perreira, Silviano Rafael, Armando Pires e Paulo Branco, "Obtaining the magnetic characteristics of an 8/6-switched reluctance machine: from FEM analysis and experimental tests", *IEEE Transactions on Industrial Electronics*, Vol 52, nº 6, pp 1635-1643, December 2005.

Sensorless Control of Brushless Doubly-Fed Reluctance Machines using an Angular Velocity Observer

Milutin G Jovanović
Northumbria University
School of Computing, Engineering and Inf. Sciences
Newcastle upon Tyne NE1 8ST, UK

David G Dorrell
The University of Glasgow
Dept of Electronics and Electrical Engineering
Glasgow G12 8LT, UK

Abstract—The brushless doubly-fed reluctance machine (BDFRM) has been considered by academic and industrial communities as a potential alternative to conventional doubly-excited wound rotor induction machines (DEWRIM) in variable speed applications with limited speed ranges such as large pumps and wind turbines. While offering similar cost benefits, afforded by the use of partially-rated power electronics, the BDFRM has the following important advantages over DEWRIM – brushless design and consequent maintenance-free operation. The main purpose of this paper is to further improve the reliability and cost-effectiveness of the BDFRM drive by proposing a new observer based algorithm for speed and direct torque (and flux) control (DTC) of this machine without a shaft position sensor using a maximum power factor control strategy as a case study. The developed sensorless control scheme has been experimentally verified on a small BDFRM prototype, and obtained test results have shown that it can perform very well down to zero applied frequency to the inverter-fed winding this being difficult or impossible to achieve with traditional DTC concepts.

Index Terms—Brushless Doubly Fed; Reluctance Machines; Sensorless Control; Velocity Observer.

I. INTRODUCTION

A classical wound rotor induction machine (WRIM) with a power electronic converter in the rotor circuit (either for external resistance variations or grid-connection) has been traditionally used in applications with restricted variable speed requirements like pumps and wind energy conversion systems. The fact that the inverter only needs to handle a speed-range dependent portion of the machine total power output (this is known as slip power recovery property) means that its rating, and therefore size and cost, can be minimised in these cases. However, the presence of brushes and slip rings represents the main reliability problem of this machine, and may be an obstacle of its wider use, for example, in off-shore wind turbine installations where operation and maintenance costs can be significant [1].

Brushless doubly-fed machines (BDFM), on the other hand, do not suffer from this limitation and may be an attractive solution for the above or similar applications [2]–[4]. While the BDFM stator windings are standard, sinusoidally distributed as with WRIM or any other synchronous machine, the rotor design is quite distinct

and appears in two main forms as illustrated in Fig. 1: reluctance type or cage type with a special 'nested' structure. The BDFM with a former rotor configuration has been referred to as Brushless Doubly-Fed Reluctance Machine (BDFRM) [5], [6], and the other as Brushless Doubly-Fed Induction Machine (BDFIM) [7]–[10].

It can be seen from Fig. 1 that the BDFM has two stator windings of different pole numbers and generally different applied frequencies: the primary or power winding, the power flow through which dictates the operating mode of the machine (i.e. motoring or generation), and the secondary or control winding with power electronics for bi-directional power flow in either mode. Unlike a conventional machine, the rotor pole number must be equal to half the total number of stator poles in order to provide rotor position dependent magnetic coupling between the stator windings, a pre-requisite for the machine to produce torque.

If one compares the two BDFM types, the BDFRM is superior in many respects: (1) It can use the existing production lines of modern commercially available synchronous reluctance machine (SyncREL) rotors (although this design option may not be the most optimal for the BDFRM [6]) which may bring down its manufacturing costs relative to the BDFIM; (2) The cage-less 'cold' rotor is more mechanically robust, allows simpler modeling and control[1] as well as higher efficiency of the BDFRM [13]. This paper will therefore limit its scope to the BDFRM as a more promising machine technology than its counterpart.

The relative modeling simplicity and clear performance advantages over the BDFIM identified above, have stimulated research on, and resulted in the successful development of, almost all well-established control methods for the BDFRM that have been traditionally applied to other AC machines: scalar control [3], vector controllers [4], [11] for implementation of various optimal control strategies [3], [5], [14] using a shaft encoder, and more recently, direct torque control (DTC) [15], [16]. To the best of the authors' knowledge, the only sensorless speed control algorithm for the BDFRM has been reported in

---

[1]Field-oriented control of the primary winding reactive power and electromagnetic torque is inherently decoupled in the BDFRM [11], [12] in contrast to the BDFIM.

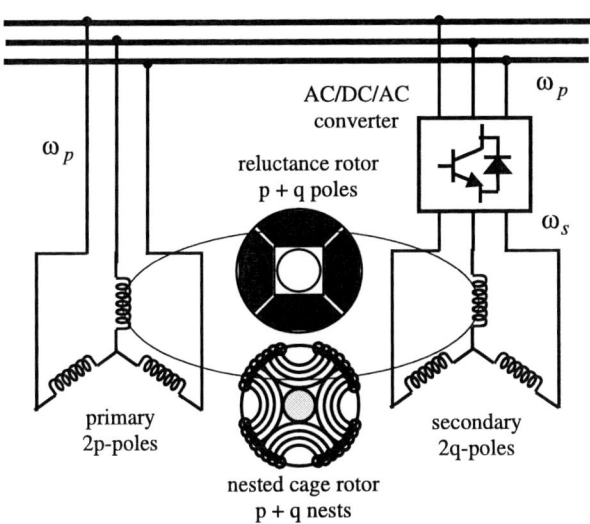

Fig. 1. A structural diagram of the BDFM with reluctance and cage rotors

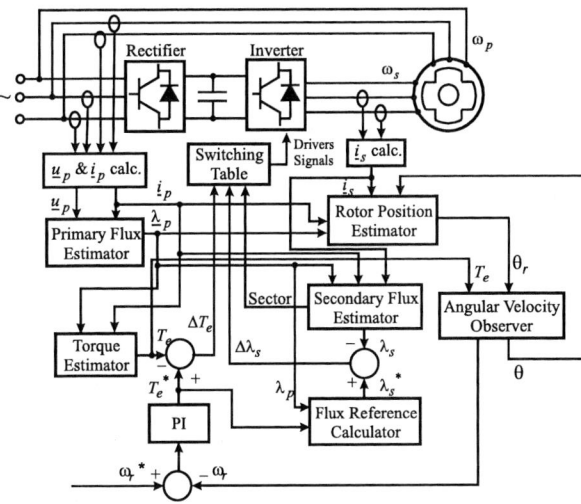

Fig. 2. Sensorless speed and direct torque control of BDFRM

[17], but without supporting test results to demonstrate its practical realisation. A DTC scheme proposed and simulated in [15], and experimentally verified in [16], has been shown to overcome the usual deficiencies of the traditional DTC approaches and has allowed stable machine operation down to zero applied frequency of the inverter-fed (secondary) winding (Fig.2). However, while sensorless control of torque and flux has been achieved in [15], [16], the speed feedback information required for speed control has been derived from rotor position measurements and not estimates.

This paper is complementary in nature to [15], [16], and can be treated as a comprehensive extension to the author's previous theoretical work on sensorless DTC [18]. Unlike [18], where the maximum torque per inverter ampere property has been considered, the conditions for maximum primary power factor control will be developed in the following and their successful practical implementation evidenced by presented experimental results. The latter will clearly show how a conventional load model based observer [19] can be effectively used for the machine speed identification from the estimated rotor position to achieve true encoder-less speed control in real-time.

II. DYNAMIC MODEL

The space-vector equations in a stationary reference frame, and the fundamental angular velocity relationship for the BDFRM torque production, are [20]–[22]:

$$\underline{u}_p = R_p \underline{i}_p + \frac{d\underline{\lambda}_p}{dt} = R_p \underline{i}_p + \left.\frac{d\underline{\lambda}_p}{dt}\right|_{\theta_p \text{ const}} + j\omega_p \underline{\lambda}_p \quad (1)$$

$$\underline{u}_s = R_s \underline{i}_s + \frac{d\underline{\lambda}_s}{dt} = R_s \underline{i}_s + \left.\frac{d\underline{\lambda}_s}{dt}\right|_{\theta_s \text{ const}} + j\omega_s \underline{\lambda}_s \quad (2)$$

$$\underline{\lambda}_p = L_p \underline{i}_p + L_{ps} \underline{i}_s^* e^{j\theta_r} = \lambda_p e^{j\theta_p} \quad (3)$$

$$\underline{\lambda}_s = L_s \underline{i}_s + L_{ps} \underline{i}_p^* e^{j\theta_r} = \lambda_s e^{j\theta_s} \quad (4)$$

$$\omega_r = d\theta_r/dt = p_r \omega_{rm} = \omega_p + \omega_s \quad (5)$$

where $L_{p,s,ps}$ represent the 3-phase inductances of the grid-connected (primary or power) and inverter-fed (secondary or control) windings [5], [22], ω_{rm} is the rotor angular velocity (rad/s) at which the machine develops useful torque, p_r is the number of rotor poles[2], and $\omega_{p,s}$ are the applied frequencies to the windings. Notice that $\omega_s > 0$ for super-synchronous operation and $\omega_s < 0$ if the machine is operated below the synchronous speed. At synchronous speed $\omega_s = 0$ i.e. the secondary side is DC supplied as with a classical $2p_r$-pole synchronous machine. The 'negative' secondary frequency in the sub-synchronous mode simply means the opposite phase sequence of the secondary to the primary winding.

III. DTC PRINCIPLES

A detailed description and performance evaluation of the DTC scheme for the BDFRM developed by the author and his colleagues can be found in [15], [16], [18]. The key points of this work and expressions relevant for DTC are reproduced below for convenience of analysis in this paper.

A. Fundamentals

By omitting the differential and $e^{j\theta_r}$ terms in (1)-(4), one obtains BDFRM equations in an arbitrary reference frame rotating at ω_p:

$$\underline{\lambda}_p = L_p \underline{i}_p + L_{ps} \underline{i}_{sp}^* \quad (6)$$

$$\underline{\lambda}_s = L_s \underline{i}_s + L_{ps} \underline{i}_{ps}^* = \sigma L_s \underline{i}_s + \underbrace{\frac{L_{ps}}{L_p} \underline{\lambda}_p^*}_{\underline{\lambda}_{ps}} \quad (7)$$

where $\sigma = 1 - L_{ps}^2/(L_p L_s)$ is the leakage factor, λ_{ps} is the primary flux linking the secondary winding (approximately constant for the grid-connected primary winding), $\underline{i}_{sp}^*$ and $\underline{i}_{ps}^*$ are the complex conjugates of the coupled current vectors from the secondary to the primary winding

[2]This is equal to the sum of the windings pole-pairs and, unlike a conventional machine, can be odd (for example, $p_r = 3$ in case of a 4/2-pole stator) due to an unusual operating principle [20]–[22].

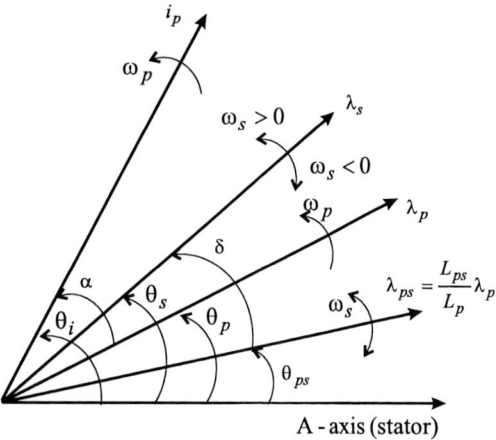

Fig. 3. Characteristic phasors in a stationary reference frame

side and vice-versa respectively such that $\underline{i}_{sp} = \underline{i}_s$ and $\underline{i}_{ps} = \underline{i}_p$ in the corresponding reference frames. Note from Fig. 3 that $\underline{\lambda}_{ps}$ rotates at ω_s and not at ω_p as the source current phasor, $\underline{i}_p$. This frequency transformation results from the modulation process of the stator mmf waveforms (of different both temporal and spatial pole numbers) via the reluctance rotor and represents the basic mechanism of magnetic coupling and torque production in the BDFRM [20]–[22].

The fundamental DTC concept can be understood from the following torque expression(s) for the BDFRM:

$$T_e = \frac{3}{2} p_r \left| \underline{\lambda}_{ps} \times \underline{i}_s \right| = \frac{3p_r}{2\sigma L_s} \left| \underline{\lambda}_{ps} \times \underline{\lambda}_s \right| \qquad (8)$$

$$= \frac{3p_r}{2\sigma L_s} \frac{L_{ps}}{L_p} \lambda_p \lambda_s \sin\delta = \frac{3p_r}{2\sigma L_s} \lambda_{ps} \lambda_s \sin\delta \qquad (9)$$

where λ_{ps} and λ_s phasors are both rotating at ω_s (Fig. 3) so that useful torque can be produced. Instantaneous torque variations are achieved through δ i.e. $\theta_s = \delta + \theta_{ps}$ as $\theta_{ps} \approx$ const. over a short DTC interval, and particularly at low ω_s values in limited speed range applications of the BDFRM.

B. Secondary Flux for Maximum Power Factor

One of the BDFRM's main attributes is its power factor control capability [12]. The power factor in the secondary winding is directly related to the inverter size, but is irrelevant to the outside utility network (since the inverter effectively isolates the secondary from the mains supply). However the power factor of the primary winding is of great importance to the utility grid (especially in weak networks) in the light of reactive power requirements. To minimise the total current loading (and thus losses) for a given real power demand, it is therefore desirable to keep the primary power factor at or, as close as possible to, unity.

Using the primary flux oriented forms of (6)-(9) (i.e. by setting $\lambda_{pq} = 0$), one can derive the secondary flux expression for optimisation of the primary power factor or any other performance indicator of the machine. It can be shown [11], [12] that the maximum primary power

factor (MPPF) i.e. no reactive power flow through the primary winding, $Q_p = \frac{3}{2} \frac{\omega_p \lambda_p}{L_p} (\lambda_p - L_{ps} i_{sd}) = 0$, is achieved if $i_{sd} = \lambda_p / L_{ps}$. Under this condition, the MPPF secondary flux reference for a desired torque (Fig.2) can be expressed as:

$$\lambda_s^* = \sqrt{\underbrace{\left(\frac{L_s}{L_{ps}} \cdot \lambda_p \right)^2}_{\lambda_{sd}} + \underbrace{\left(\frac{\sigma L_{ps}}{1-\sigma} \cdot \frac{2T_e^*}{3p_r \lambda_p} \right)^2}_{\lambda_{sq}}} \qquad (10)$$

One can see from (10) that $\lambda_{sd} \approx$ const. irrespective of the machine loading due to the primary winding grid connection i.e. $\lambda_p \approx$ const. This fact is important as it means that the torque producing λ_{sq} component can be controlled indirectly via λ_s but in a stationary (and not rotating) frame. In other words, the DTC can be optimised since in this case vector control problem is reduced to a single variable effectively becoming scalar in nature.

C. Estimation Techniques

The use of a partially-rated inverter imposes the BD-FRM operation in a narrow speed range (typically 2:1 or so) around the synchronous speed ($\omega_{syn} = \omega_p/p_r$), and therefore at small ω_s according to (5). In order to avoid the well-known voltage integration problems and flux estimation inaccuracies of traditional DTC methods at low frequencies, the secondary flux magnitude, λ_s, and its stationary frame angle, θ_s, have been estimated using (1), (3) and (4). The resultant stationary frame expressions of importance for the control (Fig. 2) are:

$$\underline{\lambda}_s = \lambda_s e^{j\theta_s} = L_s \underline{i}_s + \underline{i}_p^* \cdot \frac{\underline{\lambda}_p - L_p \underline{i}_p}{\underline{i}_s^*} \qquad (11)$$

$$\underline{\lambda}_p = \lambda_p e^{j\theta_p} = \int \left(\underline{u}_p - R_p \underline{i}_p \right) dt \qquad (12)$$

where the primary voltage and winding current phasors can be easily determined from measurements [15], [16], [18]. The penalty to pay for by-passing (2) in the proposed estimation approach is the dependence of (11) on the knowledge of the winding inductances $L_{p,s}$ [16]. Note that $\underline{\lambda}_p$ can be accurately estimated using (12) over the entire speed range of the machine (i.e. irrespective of ω_s) due to the primary winding connection to a fixed voltage and frequency grid. For smaller machines having inherently higher resistances, R_p should be taken into account in (12) to minimise $\underline{\lambda}_p$ estimation errors and consequent sensitivity effects which may be associated with (11) under the conditions of some control strategies (other than that considered here) as reported in [16].

For the scope of this paper, another significant benefit of greater control freedom, afforded by the accessibility of both BDFRM windings, is the possibility of sensorless speed control [18]. The rotor angle, θ_r, can be retrieved from (3) as follows:

$$\left. \begin{array}{l} \theta_{r_1} = \tan^{-1} \dfrac{\mathrm{Im}\left[(\underline{\lambda}_p - L_p \underline{i}_p) \underline{i}_s \right]}{\mathrm{Re}\left[(\underline{\lambda}_p - L_p \underline{i}_p) \underline{i}_s \right]} \\ \theta_{r_2} = \theta_{r_1} + \pi \end{array} \right\} \qquad (13)$$

The raw position estimates are then input to a Luenberger type PI observer [19] to predict the rotor angular velocity

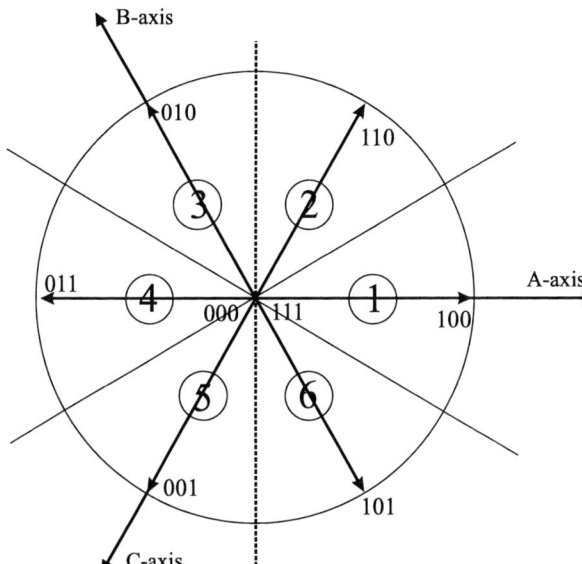

Fig. 4. Sectorial division of the stationary frame plane and binary coding of active voltage vectors attributed to each of the six $\pi/3$ sectors

TABLE I
OPTIMUM SWITCHING LOOK-UP TABLE

Comparator		Sector location of $\underline{\lambda}_s$ determined by (11)					
$\Delta\lambda_s$	ΔT_e	1	2	3	4	5	6
1	1	110	010	011	001	101	100
1	0	101	100	110	010	011	001
0	1	010	011	001	101	100	110
0	0	001	101	100	110	010	011

at the minimum switching rate are given in Table I and Fig. 4.

It is important to identify the meanings of binary outputs in (16) and the actions of the respective switching vectors from Table I. The function of row '1' voltage vectors is to increase and '0' vectors to decrease the machine instantaneous torque in its actual (not absolute) sense where the torque acting counter-clockwise is assumed positive. The reasons for applying active (non-zero) vectors only can be summarised in the following (refer to [15], [16], [18] for further details):

- A higher rate of change of secondary current/torque and superior controller performance down to synchronous speed (i.e. $\omega_s = 0$) relative to the zero vector conditions.
- A 2-level structure of the torque comparator (as opposed to a 3-level configuration with zero vectors [15], [18]) and easier implementation of (16) across Simulink/dSPACE platforms.
- A simpler DTC algorithm as knowledge of the machine speed is not required for torque control (Fig.2). This is not the case while using zero vectors due to their ambiguous effect on torque variations in super- and sub-synchronous modes.

The only limitation of using solely active vectors is the higher switching rate and therefore increased inverter losses for the same torque ripple [15], [16], [18].

IV. EXPERIMENTAL RESULTS

The sensorless algorithm in Fig. 2 has been implemented and executed in dSPACE® for the MPPF control strategy (see Section III-B) on a small 6/2-pole BDFRM prototype (refer to Appendix for details). It should be mentioned that the starting circuitry is not shown in Fig. 2. If a partially rated inverter is used then auxiliary contactors are usually needed to short the secondary terminals directly or through external resistors and allow the BDFRM start as an unloaded wound rotor induction machine. After reaching steady-state (determined by the respective no-load slip) near synchronous speed (750 rpm), the contactors are opened and the inverter connected with the control enabled. This self-starting procedure is required to prevent the inverter overloading during start-up. Alternatively, one could start the machine with the shorted primary windings using the inverter, and then self-synchronize it to the grid for doubly-fed operation in a manner similar to commercial DEWRIM drives [23].

$\omega_r = d\theta/dt$ used for the speed control as shown in Fig.2. An excellent dynamic response and low pass filtering abilities of this observer, anticipated by simulations in [18], have been experimentally verified by the results presented in this paper. It should be mentioned here that the rotor position information is only required for speed estimation and not for torque control being stator frame based as usual for DTC.

The torque comparator input (Fig. 2) can be generated using an appropriate control form of (9):

$$T_e = \frac{3}{2}p_r\left|\underline{\lambda}_p \times \underline{i}_p\right| = \frac{3}{2}p_r\left(\lambda_{pd}i_{pq} - \lambda_{pq}i_{pd}\right) \quad (14)$$

where the subscripts 'pd' and 'pq' indicate the respective components in a stationary frame aligned with the A-phase axes of the windings[3]. This expression is virtually parameter independent, it is unaffected by the switching ripples from the secondary winding side, and most importantly, it relies solely on the primary quantities of fixed line frequency. As such, it can provide quality torque estimates at either machine speed.

D. Switching Strategy

In the DTC algorithm of Fig.2, the comparators outputs can be defined as:

$$\Delta\lambda_s = \begin{cases} 1, & \lambda_s^* - \lambda_s \geq \Delta\lambda \\ 0, & \lambda_s^* - \lambda_s \leq -\Delta\lambda \end{cases} \quad (15)$$

$$\Delta T_e = \begin{cases} 1, & T_e^* - T_e \geq \Delta T \\ 0, & T_e^* - T_e \leq -\Delta T \end{cases} \quad (16)$$

where $\Delta\lambda$ and ΔT indicate the flux and torque hysteresis bands respectively. The binary codes and angular positions of the voltage vectors to be applied to the secondary winding for a particular combination of (15) and (16)

[3]These are normally accommodated in the same slots to avoid space displacement.

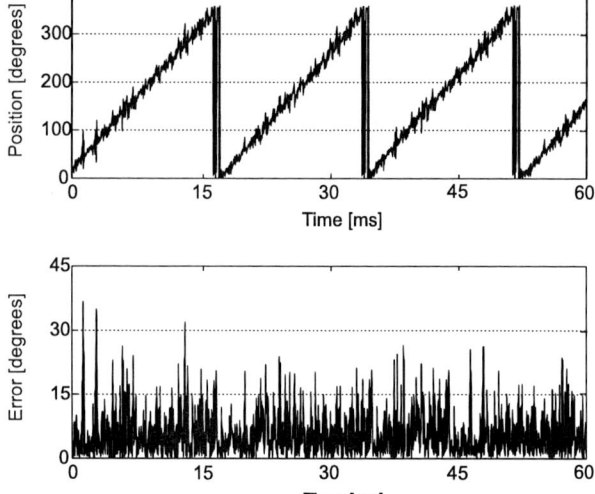

Fig. 5. Estimated position and estimator absolute errors at 850 rpm (f_s=6.7 Hz)

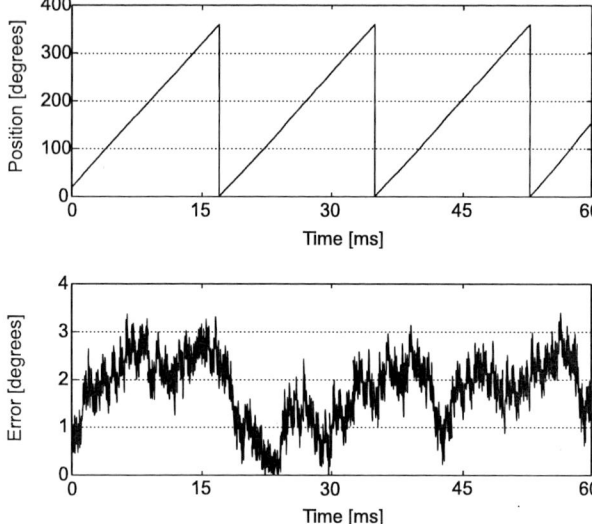

Fig. 6. Observed position and observer errors corresponding to Fig. 5

The experimental results were generated at 10 kHz sampling rate (i.e. it took no more than 100 μs for the processor to calculate the control in Fig. 2) but the inverter variable switching frequency, typical for DTC, was much below this figure. The speed estimates from the observer (and the speed controller output i.e. desired torque values) were updated at 2 kHz which was quite sufficient to satisfy modest dynamic requirements of the drive system inertia. The lower speed control rate has also added benefits in terms of improving the quality of rotor position and speed estimation as discussed below. The preliminary tests were conducted for the unloaded machine as the primary intention was to assess the sensorless control applicability.

The plots in Fig. 5 represent the rotor angles (θ_r) obtained from (13), and their absolute variations from encoder[4] measurements. The raw estimates, θ_r, are notably noisy, the error spikes being occasionally larger than 30°. Despite these ripples, which have been found to be mainly due to the practical effects such as measurement noise and quantization as well as sensitivity to parameter knowledge inaccuracies, the average estimation error is still reasonably low ($\approx 7°$).

The effectiveness of the observer as a low-pass filter is evident from Fig. 6, and a significant improvement in accuracy is achieved by processing θ_r through it. The average error is reduced to approximately 1.5° with the maximum values being about 3.4° or less. The main reason for such a high accuracy are the high quality estimates being fed into the observer by the position estimator which, similarly to the latter, works in a closed-loop fashion as illustrated in Fig. 2. The observer last prediction, θ, has served as a reference while selecting the best out of 10 raw estimates available per 500 μs speed control interval (as there are two possible solutions

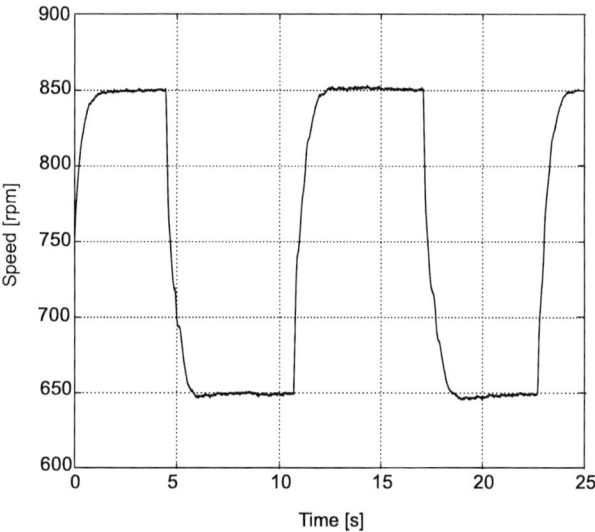

Fig. 7. BDFRM response to a varying speed reference between 850 rpm and 650 rpm

for θ_r according to (13) calculated each 100 μs) i.e. the one having the least absolute deviation from θ. Therefore, the estimator block itself carries out the first filtering of noisy θ_r before inputting the best estimate to the observer for further processing. The filtered θ_r values are actually plotted out in Fig. 5.

In order to demonstrate the validity and high accuracy of the sensorless algorithm in a limited speed range around synchronous speed (750 rpm) at low secondary frequencies(f_s), the machine was operated in super- and sub-synchronous modes at $f_s = 6.7$ Hz. The respective speed waveform in Fig. 7 clearly illustrates the good controller performance with very little overshoot under transient conditions.

Fig. 8 shows similar results to Fig. 7 but for changing desired speed values between 950 rpm, 750 rpm and 550 rpm. In this case, the speed limits correspond to $f_s \approx 13.3$ Hz in either mode. It can be seen that the machine can

---

[4]A shaft position sensor was used simply for monitoring purposes and not for control. For this reason, it is not shown in Fig. 2.

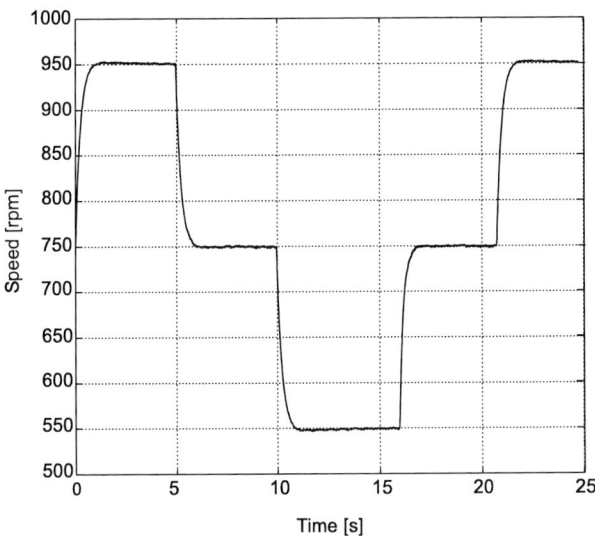

Fig. 8. Sensorless control performance down to synchronous speed

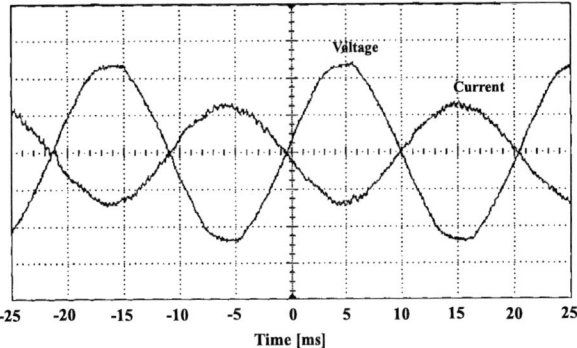

Fig. 9. Unity primary power factor voltage and current waveforms

be effectively controlled over the considered speed range, including synchronous speed (750 rpm) when $f_s = 0$. A reliable low frequency operation of the BDFRM is an important merit of the proposed sensorless scheme, and certainly represents a significant advantage over traditional DTC and many other back-emf based control methods having difficulties (or simply not working) in this frequency region even in sensor speed mode. It should be emphasised that the gains of both the speed PI regulator and the observer must be lowered and appropriately tuned as instability and divergency of the control algorithm may otherwise occur due to noisy input estimates. This trade-off results in low bandwidth control and relatively modest dynamic response of the machine which, fortunately, is quite acceptable for the target applications where steady-state performance is of more interest.

Finally, the oscilloscope traces for primary winding voltage and current in Fig. 9 clearly demonstrate that the intended maximum primary factor operation has been successfully achieved. Note that the respective waveforms are smooth, virtually switching ripple-free (due to the relatively week magnetic coupling between the windings being inherent with this particular machine), at line frequency (50-Hz). Similar results could be obtained for unity (or even leading) line power factor control in which case the secondary side would be entirely responsible for the machine magnetisation by providing the necessary reactive power to the primary (or to the grid) at the expense of increased inverter loading.

V. COMPUTER SIMULATIONS

The BDFRM load tests are currently in progress but the results are still not available in a publishable form. For this reason, the plots generated using Simulink are presented below to illustrate potentially good performance of the sensorless algorithm under loading conditions as well. Although the load torque is speed dependent in the target applications (e.g. pumps and wind turbines), it will be assumed fixed (10-Nm) for convenience of

analysis. This approximation is justified by the fast controller response to relatively slow speed and load torque variations encountered in reality. In other words, from a torque control viewpoint the operating conditions would be quasi-stationary.

Fig. 10 shows the machine response to varying speed reference of about 6 rad/s (57-rpm) above and below synchronous speed (750-rpm = 78.54-rad/s) with the control enabled at 2.5-s. Prior to the inverter connection and control activation, by analogy to the real-time situation considered in the previous section, the BDFRM was operated with the shorted secondary terminals as a wound rotor induction machine at synchronous speed (i.e. at zero slip unlike the real machine case) due to the use of an ideal model (i.e. with iron losses ignored) for the simulation studies. One can notice that the machine speed accurately attains the desired values with no or very little overshoot during transients. Such a high control quality is expected given the very similar controller behavior achieved experimentally (Fig. 7), foremost in steady-state. It is notable, however, that despite the load existence the simulated dynamic performance is still much superior to that in Fig. 7 for the unloaded machine allowing almost instant change of the operating modes (from super- to sub-synchronous and vice-versa). This can be explained by the fact that control trade-offs associated with measurement noise and other practical effects (such as the observer de-tuning which deemed necessary to preserve the algorithm convergence in the presence of noisy rotor position estimates as discussed previously) have not been taken into account in the simulations.

The torque and secondary flux control is equally effective as illustrated in Figs. 11 and 12 with the control variables being maintained within the user specified hysteresis bands, characteristic for DTC. An excellent tracking of the desired trajectories is more than evident from the same figures.

VI. CONCLUSIONS

The importance and main contribution of the paper is the development and experimental verification of a sensorless speed algorithm with direct torque control (DTC) suitable for low frequency operation of the BDFRM in applications with limited variable speed capability.

927

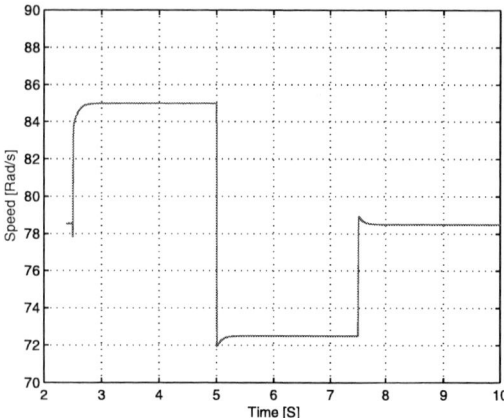

Fig. 10. Simulated sensorless control performance at 10-Nm load

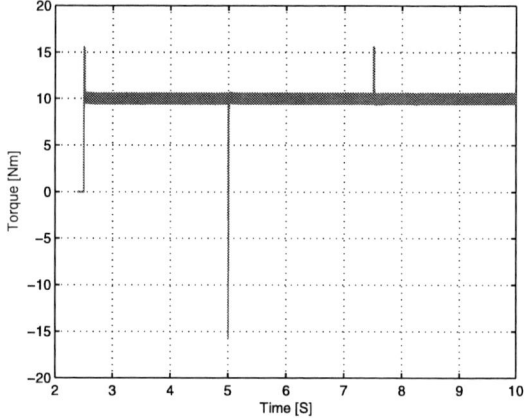

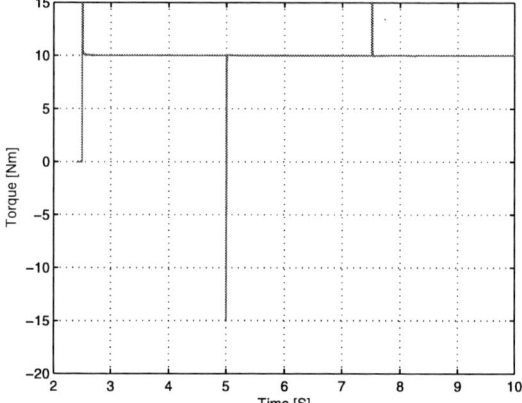

Fig. 11. Simulated torque control performance at fixed load torque: actual (top) and reference (bottom)

The preliminary test results for an unloaded BDFRM prototype have clearly demonstrated the good controller performance (especially in steady state), and additional simulation studies have undoubtedly suggested its promising potential under loading conditions of the machine as well.

The advantages and key properties of the rotor speed (and position) estimation technique and dedicated sensorless control scheme can be summarised in the following:

- Applicability over the entire speed range of the machine down to synchronous speed.
- The rotor position is estimated on-line at the control rate allowing the controller to effectively replace the encoder measurements.
- The algorithm does not require the injection of any special signals or special inverter switching techniques used for many other sensorless methods.
- The high instantaneous accuracy of both the position and angular velocity estimates at either speed is achieved by employing a standard load model based observer.
- The proposed estimation method is versatile and can serve as a basis for sensorless vector control where an accurate rotor position knowledge is required.

Further tests are currently in progress to evaluate the controller performance for different loading conditions of the machine and validate the simulation results presented here. The outcomes of this work will be published in our future papers.

APPENDIX
BDFRM TEST RIG

The laboratory test system for a 'proof-of-concept' BDFRM driving an 'off-the-shelf' DC load machine is presented in Fig.13. The 6-pole primary and 2-pole secondary windings are both rated at 1.5 kW, 2.5 A, 415 V, 50 Hz. The 4-pole axially-laminated reluctance rotor [15] and the stator have been custom designed and built. The machine design is not optimal as the main focus of the project being undertaken has been on control aspects. A standard IGBT voltage source inverter supplying the

secondary winding is controlled by a high performance DS1103 PPC controller board from dSPACE®. A high precision incremental encoder, mounted on the drive load side, has been used for shaft position sensing/speed detection and served only for monitoring purposes.

The BDFRM parameters relevant for the control [16] have been identified by applying off-line testing methods for conventional slip ring induction machines [13].

REFERENCES

[1] P.Bauer, S. de Haan, C.R.Meyl, and J.T.G.Pierik, "Evaluation of electrical systems for off-shore windfarms," *IAS Annual Meeting*, Rome,Italy,2000.

[2] B.Gorti, D.Zhou, R.Spée, G.Alexander, and A.Wallace, "Development of a brushless doubly-fed machine for a limited speed pump drive in a waste water treatment plant," *Proc. of the IEEE-IAS Annual Meeting*, pp. 523–529, Denver,Colorado,October 1994.

[3] M.G.Jovanović, R.E.Betz, and J.Yu, "The use of doubly fed reluctance machines for large pumps and wind turbines," *IEEE Transactions on Industry Applications*, vol. 38, pp. 1508–1516, Nov/Dec 2002.

[4] L. Xu and Y. Tang, "A novel wind-power generating system using field orientation controlled doubly-excited brushless reluctance machine," *Proc. of the IEEE IAS Annual Meeting*, Houston, Texas, October 1992.

[5] R.E.Betz and M.G.Jovanović, "The brushless doubly fed reluctance machine and the synchronous reluctance machine - a comparison," *IEEE Transactions on Industry Applications*, vol. 36, pp. 1103–1110, July/August 2000.

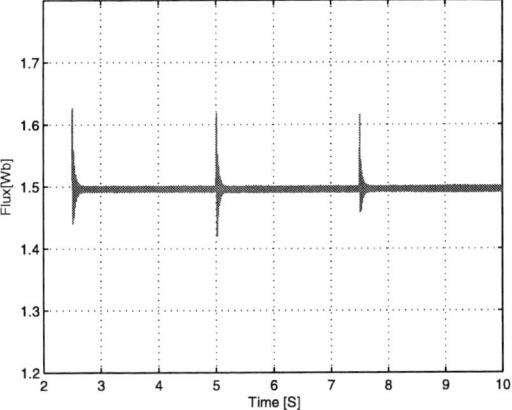

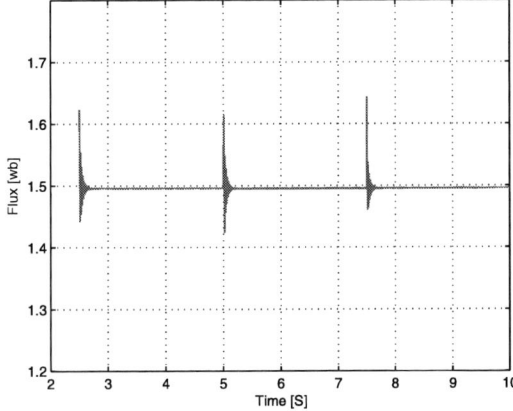

Fig. 12. Simulated flux waveforms at fixed load torque: actual (top) and reference (bottom)

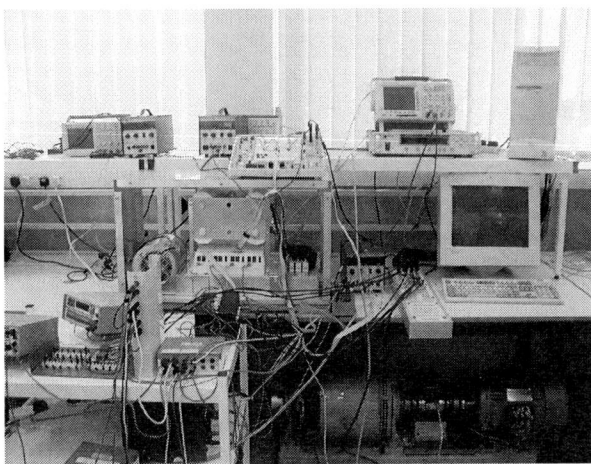

Fig. 13. BDFRM Test System

[6] E.M.Schulz and R.E.Betz, "Optimal torque per amp for brushless doubly fed reluctance machines," *CD-ROM Proc. of IEEE-IAS Annual Meeting*, Hong Kong, October 2005.

[7] S. Williamson, A. Ferreira, and A. Wallace, "Generalised theory of the brushless doubly-fed machine. part 1: Analysis," *IEE Proc.-Electric Power Applications*, vol. 144, pp. 111–122, March 1997.

[8] P.C.Roberts, R.A.McMahon, P.J.Tavner, J.M.Maciejowski, and T.J.Flack, "Equivalent circuit for the brushless doubly fed machine (BDFM) including parameter estimation and experimental verification," *IEE Proc.-Electr. Power Appl.*, vol. 152, pp. 933–942, July 2005.

[9] R.A.McMahon, P.C.Roberts, X.Wang, and P.J.Tavner, "Performance of BDFM as generator and motor," *IEE Proc.-Electr. Power Appl.*, vol. 153, pp. 289–299, March 2006.

[10] J. Poza, E. Oyarbide, D. Roye, and M. Rodriguez, "Unified reference frame dq model of the brushless doubly fed machine," *IEE Proc.-Electr. Power Appl.*, vol. 153, pp. 726–734, Sept 2006.

[11] L. Xu, L. Zhen, and E. Kim, "Field-orientation control of a doubly excited brushless reluctance machine," *IEEE Transactions on Industry Applications*, vol. 34, pp. 148–155, Jan/Feb 1998.

[12] M.G.Jovanović and R.E.Betz, "Power factor control using brushless doubly fed reluctance machines," *Proc. of the IEEE-IAS Annual Meeting*, Rome, Italy, October 2000.

[13] F.Wang, F.Zhang, and L.Xu, "Parameter and performance comparison of doubly-fed brushless machine with cage and reluctance rotors," *IEEE Transactions on Industry Applications*, vol. 38, pp. 1237–1243, Sept/Oct 2002.

[14] R.E.Betz and M.G.Jovanović, "Theoretical analysis of control properties for the brushless doubly fed reluctance machine," *IEEE Transactions on Energy Conversion*, vol. 17, pp. 332–339, Sept 2002.

[15] M.G.Jovanović, J.Yu, and E.Levi, "Direct torque control of brushless doubly fed reluctance machines," *Electric Power Components and Systems*, vol. 32, pp. 941–958, October 2004.

[16] M.G.Jovanović, J.Yu, and E.Levi, "Encoderless direct torque controller for limited speed range applications of brushless doubly fed reluctance motors," *IEEE Transactions on Industry Applications*, vol. 42, pp. 712–722, May/June 2006.

[17] Y. Liao and C. Sun, "A novel position sensorless control scheme for doubly fed reluctance motor drives," *IEEE Transactions on Industry Applications*, vol. 30, pp. 1210–1218, Sept/Oct 1994.

[18] M.G.Jovanović, J.Yu, and E.Levi, "A doubly-fed reluctance motor drive with sensorless direct torque control," *IEEE International Electric Machines and Drives Conference (IEMDC)*, Madison, Wisconsin, June 2003.

[19] R. Lorenz and K. Patten, "High-resolution velocity estimation for all-digital, ac servo drives," *IEEE Trans. on Industry Applications*, vol. IA-27, pp. 701–705, July/August 1991.

[20] F. Liang, L. Xu, and T. Lipo, "D-q analysis of a variable speed doubly AC excited reluctance motor," *Electric Machines and Power Systems*, vol. 19, pp. 125–138, March 1991.

[21] Y. Liao, L. Xu, and L. Zhen, "Design of a doubly-fed reluctance motor for adjustable speed drives," *IEEE Transactions on Industry Applications*, vol. 32, pp. 1195–1203, Sept/Oct 1996.

[22] R.E.Betz and M.G.Jovanović, "Introduction to the space vector modelling of the brushless doubly-fed reluctance machine," *Electric Power Components and Systems*, vol. 31, pp. 729–755, August 2003.

[23] L.Morel, H.Godfroid, A.Mirzaian, and J.M.Kauffmann, "Double-fed induction machine: Converter optimisation and field oriented control without position sensor," *IEE Proc. - Electr. Power Appl.*, vol. 145, pp. 360–368, July 1998.

A Half-Bridge PV System with Bi-direction Power Flow Controlling and Power Quality Improvement

C.-L. Shen and S.-T. Peng

Department of Electrical Engineering

Nan-Jeon Institute of Technology

Yen-Shui, Tainan, Taiwan, R.O.C.

E-mail: ea001@mail.njtc.edu.tw

Tel: 886-6-6523111 ext. 531

Abstract--In this paper a half-bridge photovoltaic (PV) system is proposed, which can not only process power bi-directionally but improve power quality. According to varying insolation, the system conditions real power for dc and ac loads to accommodate different amount of PV power. Furthermore, the system eliminates current harmonics and improves power factor simultaneously. As compared with conventional PV inverter, the total number of active switches and current sensors can be reduced so that its cost is lowered significantly. For current command determination, a linear-approximation method (LAM) is applied to avoid complicated calculation and achieve maximum-power-point tracking (MPPT) feature. For controlling, a direct-source-current-shaping (DSCS) algorithm is presented to shape the waveform of line current. Simulation results and practical measurements demonstrate the feasibility of the proposed half-bridge PV system.

Index Terms—Half-bridge PV system, current harmonics, power factor.

I. INTRODUCTION

Solar energy is clean, pollution-free and inexhaustible so that developing solar energy power system can solve the energy crisis of exhausting in fossil fuel. Recently, photovoltaic arrays are widely used for power supply [1]-[14]. PV systems can be briefly classified into stand-alone and grid-connection types. Owing to more flexibility in power conditioning, the study on the grid-connection type stimulates many interests. Fig. 1 shows the configuration of a conventional grid-connection PV system, which consists of multiple stages, leading to low efficiency, large volume and high cost. To improve part of the disadvantages, some researchers have designed two-stage configurations, as shown in Fig. 2. For further efficiency improvement and cost reduction, single-stage PV system has been developed [15]-[18], of which block diagram is shown in Fig. 3. Even though the structure of a single-stage PV system is simpler than that of a two-stage one, a couple of active switches, current sensors and corresponding drivers are still needed in the power stage.

In this paper a half-bridge single-stage PV system is proposed to reduce the total number of active switches

and current sensors. As a result, the proposed PV inverter system is compact and cost can be reduced significantly. Furthermore, the proposed system can not only process real power bi-directionally but improves power factor and eliminates harmonic currents. To draw maximum power from PV arrays, a linear-approximation method (LAM) is developed to complete maximum-power-point tracking (MPPT). Based on the LAM, a reference dc-link voltage is chosen. With an outer-voltage controller, source current commands are determined, which avoids optimal current determination from complicated calculations. A direct-source-current-shaping (DSCS) algorithm is applied to perform wave-shaping for bi-direction power flow controlling and power factor improvement. A prototype is established, simulated, tested and measured. The simulation results and experimental measurements have verified the feasibility of the proposed PV system.

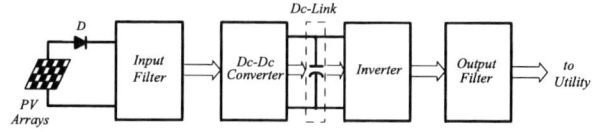

Fig. 1. A block diagram of a conventional grid-connection PV system.

Fig. 2. Illustration for a two-stage grid-connection PV system.

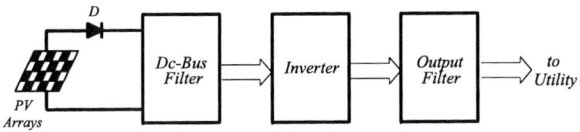

Fig. 3. A block diagram to represent a single-stage grid-connection PV system.

II. CONFIGURATION OF THE PROPOSED PV SYSTEM

Fig. 4 illustrates the configuration of the proposed PV power system, which consists of a dc-bus filter, a half-bridge inverter, an output filter, and a system controller. The half-bridge inverter, which contains two active switches and two dc-voltage-divided capacitors, can process real power bi-directionally. That is, the inverter either transfers PV power to ac side or draws power from utility for dc loads. In addition, the inverter performs current harmonics eliminating and power factor correcting to improve power quality. The dc-bus filter suppresses dc-link voltage fluctuations and filters out ac components on the dc side for accurate MPPT, while the output filter serves as an interface between the inverter and the utility to prevent inrush current from occurring. According to dc-link, reference and line voltages, the system controller implemented in a DSP chip calculates current commands and then, determines appropriate switch signals to perform wave-shaping. A conceptual block diagram of the system controller is shown in Fig. 5.

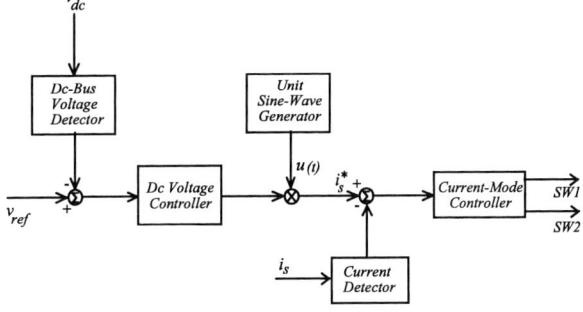

Fig. 5. A block diagram of the system controller.

III. OPERATION PRINCIPLE OF THE PV SYSTEM

To understand the operation principle of the proposed PV system, power flow controlling is discussed. The PV system processes real power, reactive power and distortion power simultaneously. Fig. 6 is the power tetrahedron diagram, which shows the relationship among these types of power. In Fig. 6, the S stands for apparent power and is expressed as

$$S = \sqrt{(\overline{p})^2 + (\overline{q})^2 + (h_{pw})^2} \qquad (1)$$

where $\overline{p}$, $\overline{q}$ and h_{pw} denote real power, reactive power and distortion power, respectively. According to different insolation, the PV system can deal with power bi-directionally. Based on $\overline{p}$ - $\overline{q}$ -h_{pw} coordinate frame, Fig. 7 shows a trajectory to indicates operation points varying with insolation. From point a to b, during the interval of high insolation the PV system generates solar power to supply dc loads and ac loads, and inject real power into utility. In addition, the half-bridge inverter processes reactive power and distortion power for ac loads so as to improve power factor. A corresponding power flow is illustrated in Fig. 8. From point b to c, during the interval of medium insolation the system supplies power for dc loads and part of real power for ac loads and the insufficient draws from utility. Fig. 9 is the related power flow. From point c to d, during the interval of low insolation the PV arrays cannot feed total amount of dc demanded power so that the inverter transforms ac power to dc one for dc loads and deals with reactive power and distortion power for ac loads simultaneously. The corresponding power flow is shown in Fig. 10. At point d, during the interval of no insolation the inverter processes real power for dc loads and deals with reactive power and distortion power for ac loads. Fig. 11 shows the power flow direction.

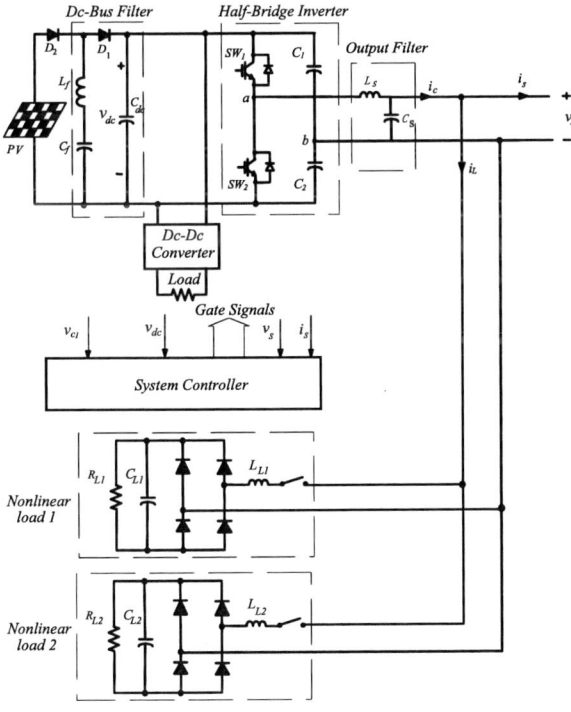

Fig. 4. Configuration of the proposed PV inverter system.

931

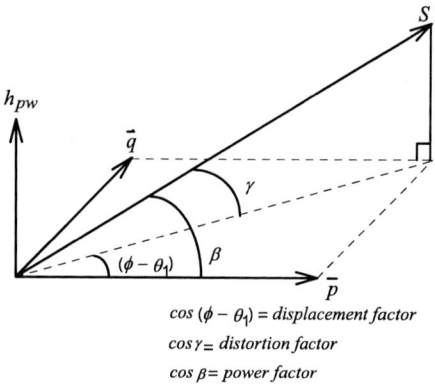

$cos\,(\phi - \theta_1) = displacement\ factor$

$cos\,\gamma = distortion\ factor$

$cos\,\beta = power\ factor$

Fig. 6. Power tetrahedron diagram.

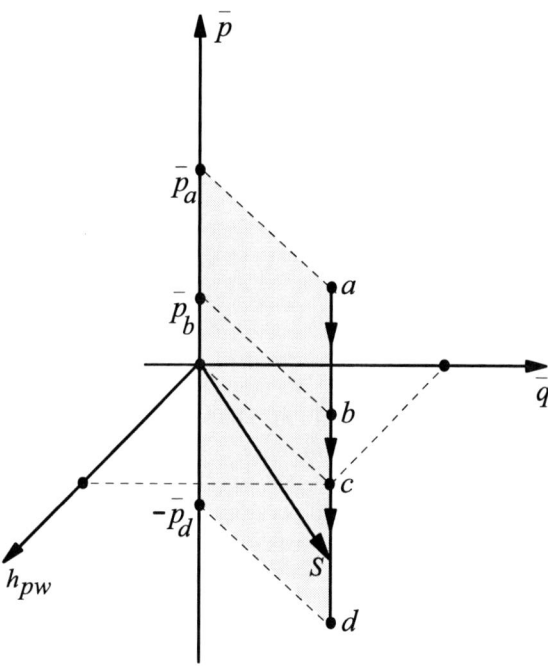

Fig. 7. A trajectory to indicate operation points varying with insolation.

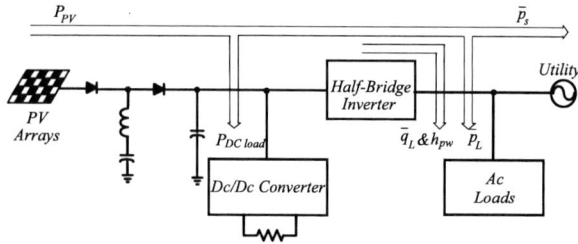

Fig. 8. Illustration of power flow during the interval of high insolation.

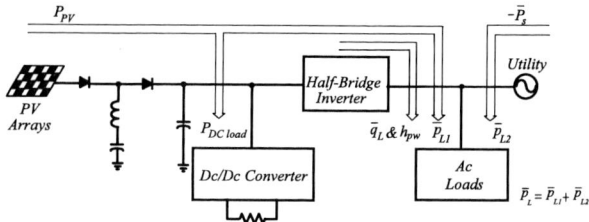

Fig. 9. Illustration of power flow during the interval of medium insolation.

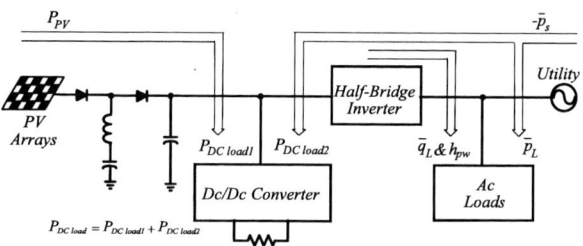

Fig. 10. Illustration of power flow during the interval of low insolation.

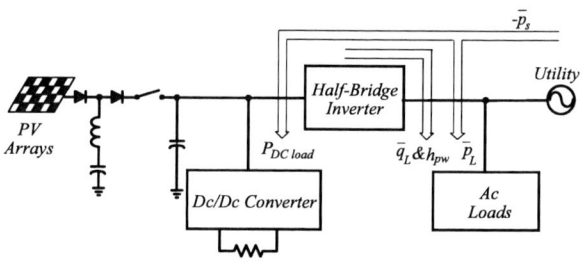

Fig.11. Illustration of power flow during the interval of no insolation.

IV. DERIVATION OF CURRENT COMMANDS

In the PV system, once a current command is determined, the output current of the half-bridge inverter will trace the waveform of the reference current to perform power flow controlling and power quality improvement. In the followings, an optimal current command is derived.

According to the current and voltage definitions shown in Fig. 4, the line voltage $v_s(t)$ and non-linear load current $i_L(t)$ are expressed as

$$v_s(t) = \sqrt{2}V_{rms}\sin(\omega t - \phi) \qquad (2)$$

and

$$i_L(t) = \sum_{n=1}^{\infty}\sqrt{2}I_n\sin(n\omega t - \theta_n), \qquad (3)$$

respectively. Then, the load instantaneous real power

$(p_L(t))$ and instantaneous reactive power $(q_L(t))$ can be calculated as follows:

$$p_L(t) = v_s(t)i_L(t)$$

$$= V_{rms}I_1 \cos(\phi - \theta_1) - V_{rms}I_1 \cos(2\omega t + \phi + \theta_1)$$

$$+ \sum_{n=2}^{\infty} 2V_{rms}I_n \sin(n\omega t + \theta_n)\sin(\omega t + \phi)$$

$$= \overline{p}_L + \tilde{p}_L, \tag{4}$$

where

$$\overline{p}_L = V_{rms}I_1 \cos(\phi - \theta_1), \tag{5}$$

and

$$\tilde{p}_L = -V_{rms}I_1 \cos(2\omega t + \phi + \theta_1)$$

$$+ \sum_{n=2}^{\infty} 2V_{rms}I_n \sin(n\omega t + \theta_n)\sin(\omega t + \phi). \tag{6}$$

Notation $\overline{p}_L$ represents the constant part and $\tilde{p}_L$ denotes the variant component. The instantaneous reactive power can be obtained by multiplying the nonlinear load current with a 90°-shifted voltage as follows:

$$q_L(t) = v_s'(t)i_L(t)$$

$$= V_{rms}I_1 \sin(\phi - \theta_1) - V_{rms}I_1 \sin(2\omega t + \phi + \theta_1)$$

$$- \sum_{n=2}^{\infty} 2V_{rms}I_n \sin(n\omega t + \theta_n)\cos(\omega t + \phi)$$

$$= \overline{q}_L + \tilde{q}_L, \tag{7}$$

where $v_s'(t)$ is the line voltage shifted by $90°$, $\tilde{q}_L$ is the constant part and $\tilde{q}_L$ is the variant component of instantaneous reactive power. Apparent power is determined by

$$S = V_{rms}\sqrt{\sum_{n=1}^{\infty} I_n^2}$$

$$= \sqrt{[V_{rms}I_1 \cos(\phi - \theta_1)]^2 + [V_{rms}I_1 \sin(\phi - \theta_1)]^2 + \sum_{n=2}^{\infty} V_{rms}^2 I_n^2}, \tag{8}$$

in which the first, second and third terms are the square of real, reactive and distortion power, respectively. The reactive and distortion power of a nonlinear load will be supplied by the PV system. As a result, a compensated line current, of which amplitude depends on PV power is purely sinusoidal and in phase with line voltage. It can be determined by

$$i_s^* = \frac{\sqrt{2}\left(p_{MPPT} - \overline{p}_L(t)\right)}{V_{rms}} \sin(\omega t - \phi). \tag{9}$$

In addition, a corresponding inverter output current is expressed as

$$i_c^* = \frac{\sqrt{2}\left(p_{MPPT} - \overline{p}_L(t)\right)}{V_{rms}} \sin(\omega t - \phi) + i_L, \tag{10}$$

where p_{MPPT} is the maximum power drawn from the PV arrays and can be represented as

$$p_{MPPT} = \left(v_{PV}(t) \cdot i_{PV}(t)\right)_{\max} \tag{11}$$

In (9) and (10), the difference between p_{MPPT} and $\overline{p}_L(t)$ decides the amplitudes of current commands, which can be also obtained from the comparison of the dc-link voltage with a reference voltage. A linear-approximation method (LAM) to achieve maximum power point tracking (MPPT) is illustrated in Fig. 12, determining the reference voltage corresponding to a maximum power point. Output power of the PV arrays is proportional to insolation. In this paper, insolation is detected by a photodiode to converter luminance into current, which can be detected by a resistor. According to Fig. 12, a MPPT voltage v_{ref}' is determined as

$$v_{ref}' = \frac{k}{m}i_p + \alpha, \tag{12}$$

where m represents the slope of the approximation line, α stands for the crossover point with output-voltage axes, i_p is the output current of a photodiode, and k is a coefficient. A maximum power point also varies with temperature. Hence, the v_{ref}' should be modified according to the relationship between the voltage obtained from (12) and temperature, as shown in Fig. 13. Therefore, a reference voltage can be found by

$$v_{ref} = \zeta v_{ref}' + \beta, \tag{13}$$

where ζ is temperature compensation coefficient and notation β is a constant value.

Output power (W)

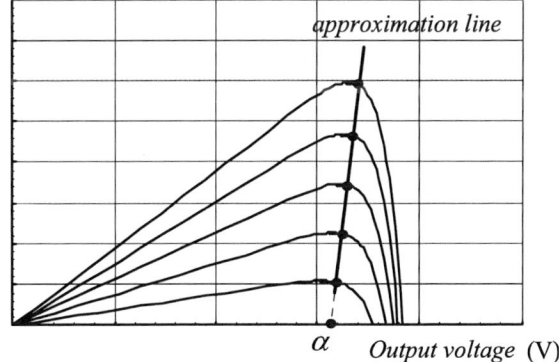

Fig. 12. Illustration of the LAM to achieve MPPT feature.

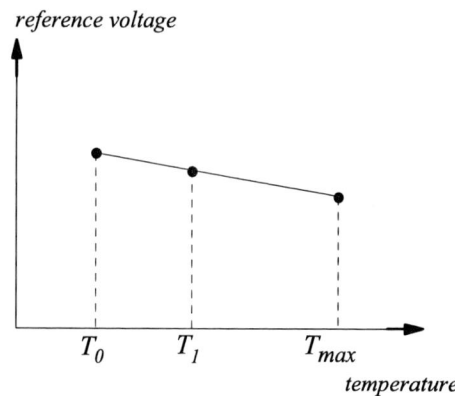

Fig. 13. Illustration of the relationship between reference voltage and temperature.

V. SIMULATED AND EXPERIMENTAL RESULTS

An example of 110V 60Hz half-bridge PV system is designed, simulated and implemented. Component values and important parameters are determined as:

power switches : IGBT, TOSHIBA GT25Q101, 1200V/25A,
PV arrays: SHARP NT-KR5EX (12 pieces in series),
f_s = 20 kHz,
$C_1 = C_2 = C_{dc}$ = 940 μF, C_f = 880 μF,
L_s = 4 mH, L_f = 2 mH,
v_{ref}: from 395 V to 420 V,
PV power: from 200 W to 1.8 kW.

Nonlinear loads are connected to utility, of which power dissipation is 650 W. Fig. 14 shows the waveform of the load current. During the interval of high insolation, PV arrays generate 1.8 kW. In addition, a reference dc-link voltage v_{ref} is 420 V based on the LAM for MPPT. The simulated line current and the corresponding inverter current are shown in Figs. 15 and 16, respectively. In this period, PV system supplies total amount of demanded power for dc and ac loads, and inject real power into utility. Simultaneously, PV system compensates reactive power and distortion power for non-linear loads to improve power factor. From Fig. 16, it can be observed that line current is sinusoidal and in phase with line voltage. That is, high power factor is achieved and PV power can be injected into utility. During the interval of medium insolation, output power of the PV arrays is 800 W and reference voltage v_{ref} is 404 V. PV system provides total amount of power for dc load and part of real power for nonlinear loads. The simulated line current is shown in Fig. 17, while Fig. 18 is the corresponding inverter current. From Fig. 17, it can be found that the line current is purely sinusoidal and 180° out of phase to line voltage. That is, insufficient power for ac load is fed from utility and power factor correction is performed by the half-bridge inverter simultaneously. During the interval of low insolation, PV power is 200 W and a reference dc-link voltage is 395 V. Fig. 19 shows

the simulated line current and Fig. 20 is the corresponding inverter current. Once there is no insolation, the reference voltage is 395 V and the half-bridge inverter provides reactive power and distortion power for non-linear loads and draws real power from utility for dc loads. The line current and inverter current are shown in Figs. 21 and 22, in turn. Figs. 23 and 24 present the practical measurements of line currents during the intervals of high insolation and medium insolation, respectively.

CONCLUSIONS

A half-bridge PV inverter is presented in this paper. As compared with full-bridge one, the total number of active switches is reduced by half so that the system configuration is simplified and its cost is lowered significantly. The LAM is applied to obtain an optimal reference voltage for the determination of a current command and to achieve MPPT feature, which avoids sophisticated calculation. The DSCS algorithm performs wave-shaping to achieve power quality improvement. Simulation results and practical measurements have demonstrated the feasibility of the proposed PV system.

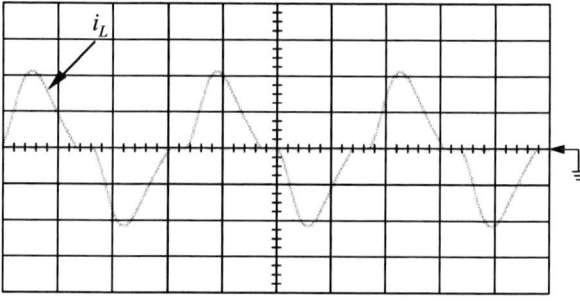

(i_L: 5 A/div, time: 5 ms/div)

Fig. 14. Load current while nonlinear loads are connected to utility.

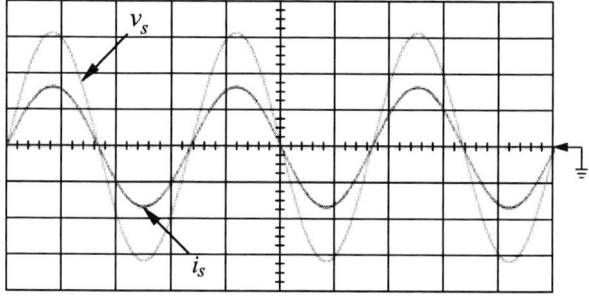

(v_s: 50 V/div, i_s: 5 A/div, time: 5 ms/div)

Fig. 15. Filtered line current and line voltage during the interval of high insolation.

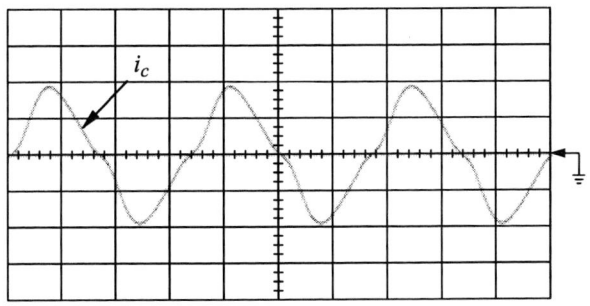

(*i_c*: 10 A/div, time: 5 ms/div)

Fig. 16. The corresponding inverter current during the interval of high insolation.

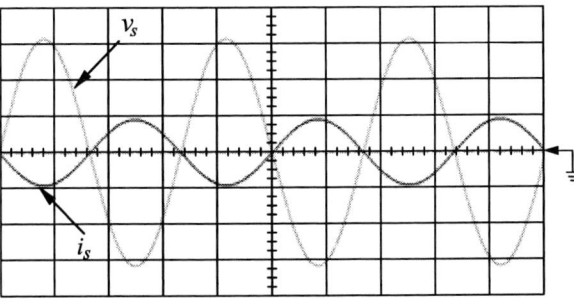

(*v_s*: 50 V/div, *i_s*: 5 A/div, time: 5 ms/div)

Fig. 17. Filtered line current and line voltage during the interval of medium insolation.

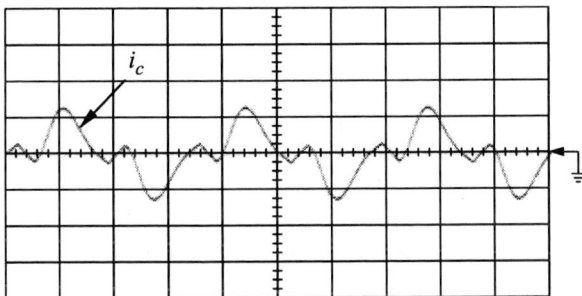

(*i_c*: 5 A/div, time: 5 ms/div)

Fig. 18. The corresponding inverter current during the interval of medium insolation.

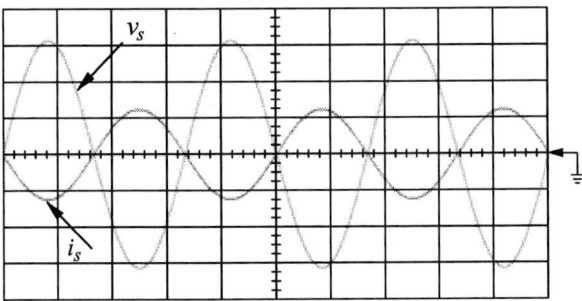

(*v_s*: 50 V/div, *i_s*: 10 A/div, time: 5 ms/div)

Fig. 19. Filtered line current and line voltage during the interval of low insolation.

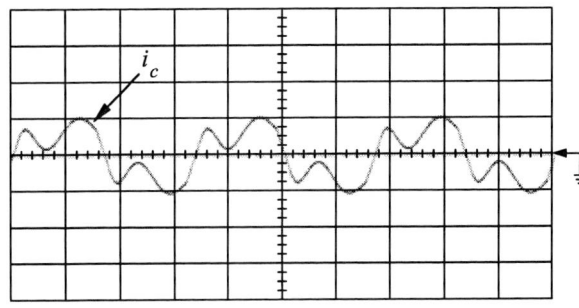

(*i_c*: 5 A/div, time: 5 ms/div)

Fig. 20. The corresponding inverter current during the interval of low insolation

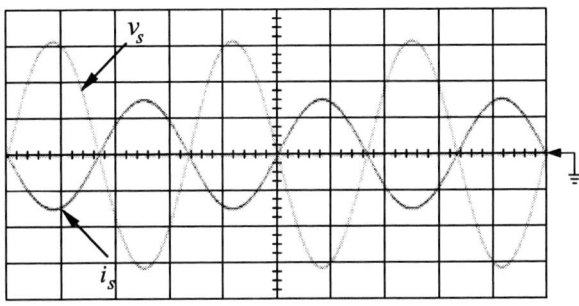

(*v_s*: 50 V/div, *i_s*: 10 A/div, time: 5 ms/div)

Fig. 21. Filtered line current and line voltage during the interval of no insolation.

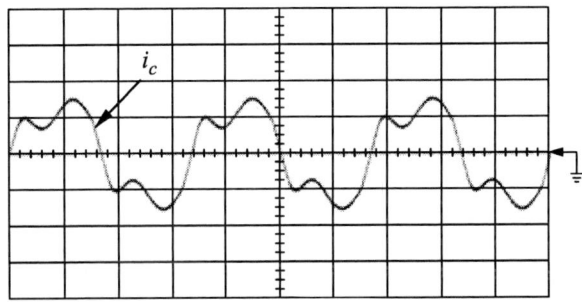

(*i_c*: 5 A/div, time: 5 ms/div)

Fig. 22. The corresponding inverter current during the interval of no insolation

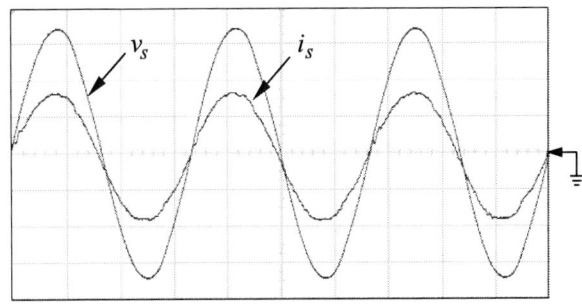

(*v_s*: 50 V/div, *i_s*: 5 A/div, time: 5 ms/div)

Fig. 23. Experimental result: the filtered line current and line voltage during the interval of high insolation.

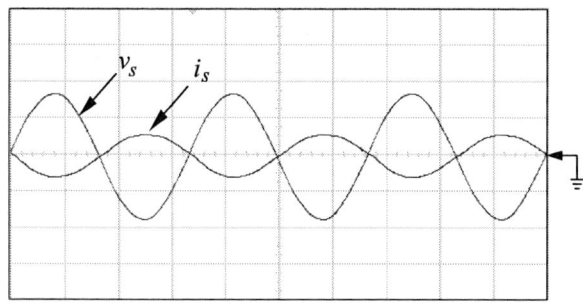

(v_s: 100 V/div, i_s: 10 A/div, time: 5 ms/div)

Fig. 24. Experimental result: the filtered line current and line voltage during the interval of medium insolation.

REFERENCES

[1] L. Asiminoaei, R. Teodorescu, F. Blaabjerg, and U. Borup, " A digital controlled PV-inverter with grid impedance estimation for ENS detection," *IEEE Transactions on Power Electronics*, vol. 20, nov. 2005, pp. 1480-1490.

[2] T.-F. Wu, C.-L. Shen, Chien-Hsuan Chang and Jei-Yang Chiu, " A 1ϕ3W Grid-Connection PV Power Inverter with Partial Active Power Filter," *IEEE Transactions on Aerospace and Electronic Systems*, vol. 39, Apr. 2003, pp. 635-646.

[3] S. A. Danie and N. AmmasaiGounden, " A novel hybrid isolated generating system based on PV fed inverter-assisted wind-driven induction Generators," *IEEE Transactions on Energy Conversion*, vol. 19, June 2004, pp. 416-422.

[4] H. Koizumi, T. Mizuno, T. Kaito, Y. Noda, N. Goshima, M. Kawasaki, K. Nagasaka and K. Kurokawa, " A Novel Microcontroller for Grid-Connected Photovoltaic Systems," *IEEE Transactions on Industrial Electronics*, vol. 53, Dec. 2006, pp. 1889-1897.

[5] Tsai-Fu Wu, Hung-Shou Nien, Chih-Lung Shen and Tsung-Ming Chen, " A single-phase inverter system for PV power injection and active power filtering with nonlinear inductor consideration," *IEEE Transactions on Industry Applications*, vol. 41, July-Aug. 2005, pp. 1075-1083.

[6] P. P. Barker and J. M. Bing, " Advances in solar photovoltaic technology: an applications perspective," *IEEE Proceedings of Power Engineering Society General Meeting*, vol. 2, June 2005, pp. 1955-1960.

[7] Ho, B.M.T. and Henry Shu-Hung Chung, " An integrated inverter with maximum power tracking for grid-connected PV systems," *IEEE Transactions on Power Electronics*, vol. 20, July 2005, pp. 953-962.

[8] P. G. Barbosa, H. A. C. Braga, Rodrigues, Md. C. B. and E. C. Teixeira, " Boost current multilevel inverter and its application on single-phase grid-connected photovoltaic systems," *IEEE Transactions on Power Electronics*, vol. 21, July 2006, pp. 1116-1124.

[9] J. J. Negroni, C. Meza, D. Biel and F. Guinjoan, " Control of a buck inverter for grid-connected PV systems: a digital and sliding mode control approach," *Proceedings of the IEEE International Symposium on Industrial Electronics*, vol. 2, June 2005, pp. 736-744.

[10] C. Rodriguez, Amaratunga and G.A.J., " Dynamic maximum power injection control of AC photovoltaic modules using current-mode control," *IEE Proceedings of Electric Power Applications*, vol. 153, Jan. 2006, pp. 83-87.

[11] A. Kotsopoulos, Heskes, P. J. M. and M. J. Jansen, " Zero-crossing distortion in grid-connected PV inverters," *IEEE Transactions on Industrial Electronics*, vol. 52, April 2005, pp. 558-565.

[12] N. Kasa, T. Iida and Liang Chen, " Flyback Inverter Controlled by Sensorless Current MPPT for Photovoltaic Power System," *IEEE Transactions on Industrial Electronics*, vol. 52, Aug. 2005, pp. 1145-1152.

[13] N. Femia, D. Granozio, G. Petrone, G. Spagnuolo and M. Vitelli, " Optimized one-cycle control in photovoltaic grid connected applications," *IEEE Transactions on Aerospace and Electronic Systems*, vol. 42, July 2006, pp. 954-972.

[14] A. O. Zue and A. Chandra, " Simulation and stability analysis of a 100 kW grid connected LCL photovoltaic inverter for industry." *IEEE Proceedings of Power Engineering Society General Meeting*, June 2006, pp. 1-6.

[15] Tsai-Fu Wu, C.-L. Shen, Hung-Shou Nei and Guang-Feng Li, "A 1ϕ3W Inverter with Grid Connection and Active Power Filtering Based on Nonlinear Programming and FZPD Algorithm," *IEEE Transactions on Power Electronics*, vol. 20, Jan. 2005, pp. 218-226.

[16] Yang Chen and K. M. Smedley, " A cost-effective single-stage inverter with maximum power point tracking," *IEEE Transactions on Power Electronics*, vol. 19, Sept. 2004, pp. 1289-1294.

[17] Kiranmai, K. S. Phani and M. Veerachary, " A Single-Stage Power Conversion System for the PV MPPT Application," *Proceedings of the IEEE International Conference on Industrial Technology*, Dec. 2006, pp. 2125-2130.

[18] R. Gonzalez, J. Lopez, P. Sanchis and L. Marroyo, " Transformerless Inverter for Single-Phase Photovoltaic Systems," *IEEE Transactions on Power Electronics*, vol. 22, March 2007, pp. 693-697.

Response of DSTATCOM
under Voltage Flicker In Farm Wind

K. Aodsup[1], P. N. Boonchiam[1], A. Sode-Yome[2], P. Kongsuk[3] and N. Mithulananthan[4]

[1]Department of Electrical Engineering, Rajamangala University of Technology Thanyaburi, Pathunthani, Thailand
[2]Department of Electrical Engineering, Siam University, Bangkok, Thailand
[3]Industrial Technology Field of Study, Rajamangala University of Technology Tawan-ok, Thailand
[4]Energy Field of Study, Asian Institute of Technology, Thailand

Abstract—This paper presents the response of distribution static synchronous compensator (DSTATCOM). The connection of wind turbines to distribution systems may affect the voltage quality offered to the consumers. One of the factors contributing to this effect is the rapid variations of the wind turbine output power, which cause respective fluctuations in the supply voltage referred to as flicker. This paper presents the design, control and analysis of a DSTATCOM enhanced with an energy storage device when combined with a wind farm comprising fixed speed induction generators, In this paper it is shown that the DSTATCOM, controlled via a decoupled vector control technique, is an effective way of reducing voltage flicker emissions at the point of common coupling, removing the wind speed fluctuations and improving the transient stability of wind farm. Energy storage rating requirements are discussed in relation to the level of power quality obtained.

Index terms- DSTATCOM, power quality, wind farm

I. INTRODUCTION

Wind turbines connected to electrical grids may affect the power quality of the supply, as a result of the fluctuating character of their output power. This contains both periodic component and random variations, resulting in corresponding fluctuations of the voltage magnitude along the feeder where the wind turbines are connected. Fluctuations in the frequency range between 0.5 and 35 Hz contribute to the light flickering effect, referred to as "flicker". Flicker is evaluated according to IEC 60860 standard [1][2]. A common measure of its severity is the short-term flicker index, P_{st} measured over 10 min periods, whereas in certain cases the long-term (120 mm) index, P_{st} is also applicable. The drawback characteristic of wind turbine are the higher mechanical stress, and the enhanced effect of power fluctuations caused by the blades passing the tower, power variations caused by variable with speeds. Real and reactive power injection is thus necessary to compensate for these effects.

Static compensators (STATCOMs) are promising technology that can control grid voltage more rapidly than synchronous generators and thyristor based static var compensators [3] and can improve power quality [4]. STATCOMs are used at a transmission level [5] and distribution level, where they are known as DSTATCOM [6]. Without energy storage, a DSTATCOM can control the voltage by exchanging reactive power with the grid, but their capacity to exchange active power is very limited due to the small energy that can be stored in a conventional DC-link capacitor. DSTATCOM enhanced with energy storage can give additional benefits, such as increased capacity to damp electro-mechanical oscillation, increased power quality and reliability of system.

Wind farm combined with a DSTATCOM comprising an energy storage device placed onshore, could balance the reactive and real power needed to boost the voltage at the point of common coupling (PCC), mitigate flicker emissions and eliminate wind speed fluctuations. This paper describes analysis and control design of DSTATCOM to control the voltage at the PCC, keep the power constant at the load and mitigate flicker emissions from intermittent wind speeds. The MATLAB/simulink software package is used to simulate the proposed compensation strategies. A vector control technique based on the decoupling of real and reactive power.

II. THE GRID

The study case power system considered in this paper is shown in Fig. 1. The medium voltage distribution grid is represented by its Thevenin equivalent, consisting of a voltage source E_{th} and the series impedance Z_{th}. A concentrated local load is connected at the PCC, corresponding to the consumer loads in the nearby area. Although the local load could have been included in the Thevenin equivalent of the grid, it is modeled here independently in order to investigate its effect on the PCC.

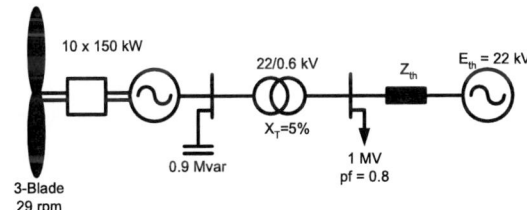

Fig. 1. Study case power system.

A 1.5 MW wind farm consisting of ten 150 kW wind turbines connected to a 22 kV distribution system exports power to a 230 kV grid through a 20 km, 22 kV feeder. A 0.9 Mvar (Q=50) filter are connected at the 575 V generation bus. The turbine parameters specifying ratings of power components of the wind turbine are saved in a Matlab programme. This file is automatically executed at simulation start so that parameters for the 10x150 kW turbines are loaded

978-1-4244-0644-9/07/$25.00 ©2007 IEEE

in our Matlab workspace. The wind turbine is equipped with a 3-blade fixed-pitch rotor, rotating at 29 rpm. The rotor aerodynamic characteristics are simulated using its static aerodynamic power coefficient, $C_P(\lambda)$, curve, resulting in the power curve shown in Fig. 2.

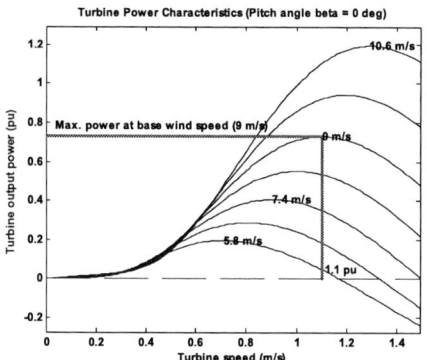

Fig. 2. Wind turbine power curve.

III. FLICKER CALCULATION

The assessment of the short-term flicker index, Pst, is performed as specified in IEC 60868 and briefly described in this section, for the sake of completeness. In Fig.3, the UIE/IEC flickermeter block diagram is shown. Input is a 10-min time series of the voltage at the evaluation node, which may be expressed as

$$u(t) = \left[U_0 + \Delta U(t) \right] \sin \left(\omega t + \psi \right) \quad (1)$$

where U_0 is the average node voltage magnitude, $\Delta U(t)$ the superimposed amplitude variations, $\omega = 2\pi f$ the system frequency and ψ the initial phase angle. Using equation (1), the output of the first block of the flickermeter can be expressed as:

$$\frac{u^2(t)}{U_0} = \frac{U_0}{2} + \Delta U(t) - \left[\frac{U_0}{2} + \Delta U(t) \right] \cos \left[2 \left(\omega t + \psi \right) \right] \quad (2)$$

The constant $U_0/2$ in the above expression is eliminated by the high pass filter (1st order, 0.5 Hz cut-off frequency), whereas the double power frequency component is filtered out by the subsequent low-pass filter block (6th order Butterworth, 35 Hz cut-off frequency). Hence, output of the third block are the voltage magnitude variations, $\Delta U(t)$, in the frequency range 0.5 to 35 Hz.

The next block in diagram of Fig. 3 is a weighting function, simulating the perception ability of the human eye vs. the frequency of the disturbing signal. The peak of this curve is located at 8.8 Hz. Since the irritation caused is proportional to the square of the voltage magnitude fluctuations, the output of the weighting function block is squared and led to a 1st order lag, representing the memory tendency of the human brain. Its output is the time series of the instantaneous flicker sensation, $P_{st}(t)$.

The calculation of the short term flicker index, P_{st}, required then a simple statistical processing of the $P_{st}(t)$ time series. First the cumulative duration curve of the $P_{st}(t)$ values id found. The P_{st} index is then given by

$$P_{st} = \sqrt{0.0314 P_{0.1s} + 0.0525 P_{1s} + 0.0657 P_{3s} + 0.28 P_{10s} + 0.08 P_{50s}} \quad (3)$$

where P_x is the x % percentile (i.e. the flicker level which is exceeded for x % of the time), calculated from the $P_{st}(t)$ duration curve. The subscript "s" denoted smoothed values, obtained by averaging neighboring values of the duration curve, as described in IEC 60868-0.

Fig. 3. UIE/IEC flickermeter functional diagram.

IV. MATHEMATICAL MODEL OF DSTATCOM

DSTATCOM shown in Fig.4 is modeled using the *dq* transformation as defined in [8]. The inverter block is treated as an ideal lossless voltage source converter. The inverter switching losses are represented by a resistance R_f in parallel with a DC-link energy store supercapacitor. The transformer leakage reactance and the inverter series resistance are represented by L_f and R_f, respectively

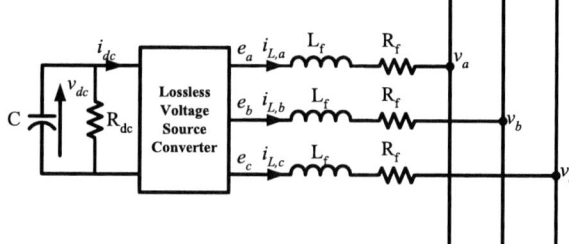

Fig. 4. Equivalent circuit of DSTATCOM.

The AC-side equations can be written in terms of the instantaneous variables shown in Fig. 1.

$$\frac{d}{dt} \begin{bmatrix} i_a \\ i_b \\ i_c \end{bmatrix} = \frac{R_f}{L} \begin{bmatrix} i_a \\ i_b \\ i_c \end{bmatrix} + \frac{1}{L} \begin{bmatrix} e_a - v_a \\ e_b - v_b \\ e_c - v_c \end{bmatrix} \quad (4)$$

Equation 4 is converted into per-unit values and transform to rotating frame as shown in (5)

$$\frac{d}{dt} \begin{bmatrix} i_d \\ i_q \end{bmatrix} = \begin{bmatrix} -\dfrac{R_f \omega_b}{L} & \omega \\ -\omega & -\dfrac{R_f \omega_b}{L} \end{bmatrix} \begin{bmatrix} i_d \\ i_q \end{bmatrix} + \frac{\omega_b}{L} \begin{bmatrix} e_d - v \\ e_q \end{bmatrix} \quad (5)$$

where v is the *rms*-value of the grid voltage.

The instantaneous power at the DC and AC side of the DSTATCOM inverter must be equal and therefore the following power balance equation must be hold:

$$P_{dc} = P_{ac} \Rightarrow v_{dc} i_{dc} = \frac{3}{2}\left(e_d i_d + e_q i_q\right) \tag{6}$$

From Fig. 2 and using p.u. values, the DC side circuit equation is derived as follows:

$$\frac{dv_{dc}}{dt} = -\frac{1}{C} \cdot \frac{v_{dc}}{R_f} - \frac{1}{C} i_{dc} \tag{7}$$

$$\frac{dv_{dc}}{dt} = -\omega_b C \left(i_{dc} + \frac{v_{dc}}{R_f} \right) \tag{8}$$

The state space equation for DSTATCOM is obtained:

$$\frac{d}{dt}\begin{bmatrix} i_d \\ i_q \\ v_{dc} \end{bmatrix} = [A]\begin{bmatrix} i_d \\ i_q \\ v_{dc} \end{bmatrix} - \frac{\omega_b}{L}\begin{bmatrix} v \\ 0 \\ 0 \end{bmatrix} \tag{9}$$

Linearising around an operation point and using the fact that no active power is exchanged at the equilibrium point it is possible to get a decoupled expression for id and i_q [2].

$$\frac{d}{dt}\begin{bmatrix} \Delta i_d \\ \Delta i_q \end{bmatrix} = \begin{bmatrix} \dfrac{R_f \omega}{L} & 0 \\ 0 & \dfrac{R_f \omega}{L} \end{bmatrix}\begin{bmatrix} \Delta i_d \\ \Delta i_q \end{bmatrix} + \begin{bmatrix} c_1 \\ c_2 \end{bmatrix} \tag{10}$$

It can be seen from equation (10) that no cross-coupling between id and i_q occurs. Also, if c_1 and c_2 are expressed as in equation (11), a real and reactive power decoupled vector control strategy can be achieved since i_d can be used to regulate the active power and i_q to control the reactive power flow.

$$\begin{bmatrix} c_1 \\ c_2 \end{bmatrix} = \begin{bmatrix} \left(k_p + \dfrac{k_i}{s}\right)\left(i_d^* - i_d\right) \\ \left(k_p + \dfrac{k_i}{s}\right)\left(i_q^* - i_q\right) \end{bmatrix} \tag{11}$$

The relation between the DC-side voltage and the line to neutral AC side voltage of the inverter terminal can be expressed as follows:

$$e_d' = m v_{dc} \cos\alpha \tag{12}$$

$$e_q' = m v_{dc} \sin\alpha \tag{13}$$

The modulation index m is the magnitude ratio of the DC and AC side voltage and α is the angle by which the inverter voltages lead or lag the bus voltages. A vector control scheme controls id and i_q by varying the modulation index and angle, as shown in Fig. 5. A phase-locked-Loop (PLL) is used to provide the reference voltage angle θ, thereby synchronizing the current vectors to the fundamental of the PCC voltages

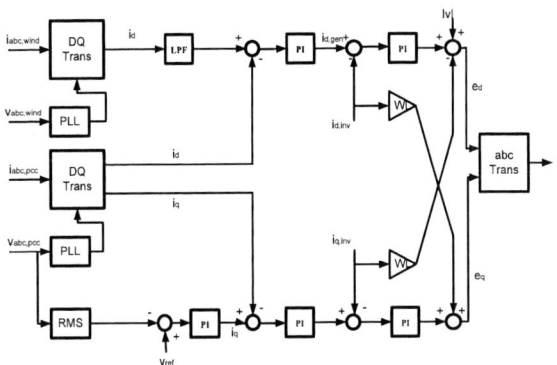

Fig. 5 DSTATCOM vector control scheme-LPF: Low Pass Filter, PI: Proportional Integral regulator.

The wind turbines current are subjected to a low pass filter, which has a selectable time constant to achieve different cut-off frequency. The filter i_{wt} signal provides the real power reference of the control scheme. The functionality of this LPF can be explained as follows. Selecting a 10-4 Hz cut-off makes he input current reference virtually constant representing effectively the mean of i_{wt}. In this case, the control system will try to perfectly smooth the power at the PCC by maximum variation of the real power component of the DSTATCOM current i_{dinv} to compensate for real power fluctuation in i_{wt}. On the other hand, a LPF with 1 Hz cut-off relaxes significantly the bandwidth of control, in other words allowing some fluctuation at PCC, resulting in less power compensation and requiring less energy storage.

V. Energy Storage Sizing

For a practical application it is important to minimize the energy storage rating to reduce the cost and size. The required energy storage can be derived from the modulation in energy exchange by the STATCOM which is necessary to reduce flicker to an acceptable level for a certain site. The DC-link energy is used in order to assess the energy requirement of constant and LPF references, over a specific period.

$$E_{dc} = \int P_{dc} dt = \int V_{dc} I_{dc} dt \tag{14}$$

A minimum value for DC voltage must chosen to avoid an excessive current rating of the power electronic converter, and to be able to control P and Q independently. A minimum value of 90% of the nominal DC voltage has been chosen here. The value of capacitance can then be calculated from the measured peak to peak DC energy as follows.

$$E_{dc} = \frac{1}{2} C \left(V_{dc}^2 - \left(0.9 V_{dc}^2 \right) \right) \tag{15}$$

VI. SIMULATION RESULT

Simulation used power distribution of Thailand sizing 22 kV 50 Hz .Connect to transformer 100 kVA . DSTATCOM can compensation voltage flicker more 10 sec with used capacitor sizing 40 mF and control dc voltage at dc link. Fig. 6 shows during voltage flicker at 40 ms . Fig. 7 show current inject to system when during voltage flicker at 40 ms and Fig. 8 shows voltage curve of phase A when DSTATCOM compensation . From simulation DSTATCOM can compensation voltage flicker.

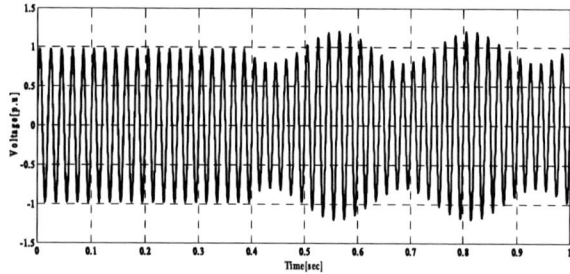

Fig. 6. Voltage Flicker at Phase A

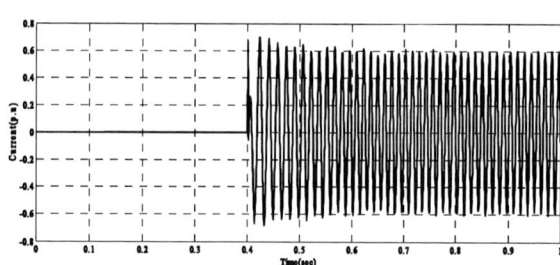

Fig. 7. Output current of the D-STATCOM.

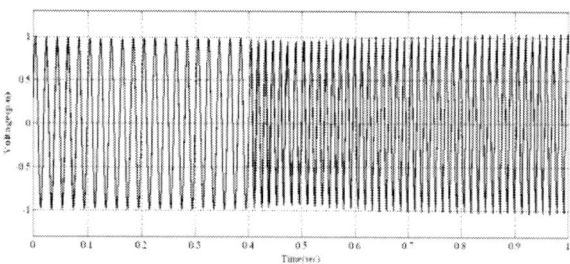

Fig. 8. Voltage waveform before and during the compensation

VII. CONCLUSIONS

The use of an energy storage device linked to a DSTATCOM to provide real power as well as reactive power flow was explored in this paper. The results have shown that the DSTATCOM device has the ability to significantly reduce the power fluctuation generated by wind farm comprising fixed speed induction generators, while also keeping the voltage at the PCC at 1 p.u. In addition, the wind farm's fault ride through capability can be greatly improved by the DSTATCOM. For commercial application a more detailed design and technical can economic analysis needs to be carried out. Taking into consideration maintenance and operation costs, this would allow a comparison between a wind farm using fixed speed induction generators combined with DSTATCOM onshore, and wind farms which inherently generate less flicker due to the smoothing effect of the rotor with variable speed.

REFERENCES

[1] CIGRE Study Committee, "Development of Dispersed Generation and Consequence for Power System," Final Report of WG C6.01 (ex 37.33), July, 2003.
[2] J.G. Slootweg, W.L. Kling; "Aggregated Modeling of Wind Parks in Power system Dynamic Simulations", Delft University of Technology, the Netherlands
[3] Rui M.G. Castro, J.M. Ferreira de Jesus; "An Aggregated wind Park Model", Technical University of Lisbon, Portugal
[4] C. Abbey, B. Khodabakhchian, F. Zhou; "Transient modelling and comparison of wind generator topologies", International conference on power system transients (IPST'05), Canada
[5] J.G. Slootweg, W.L. Kling; "Modelling of Large Wind Farms in Power System Simulation", Delft University of Technology, the Netherlands
[6] N. Mithulananthan, Claudio A. Canizares; "Comparison of PSS, SVC, and STATCOM controllers for damping power system oscillation", IEEE Trans. Power Systems, October 2002
[7] P. Garcia-Gonzalez, A. Garcia C.; "Control system for a PWM-based STATCOM", Universidad Pontificia Comillas de Mardrid, Spain.
[8] P.S. Sensarma, K.R. Padiyar; "Analysis and erformance Evaluation of a Distribution STATCOM for ompensating Voltage Fluctuations", The Department of Electrical Engineering, Indian Institute of Science, India

A Comparative Study of Fixed Speed and Variable Speed Wind Energy Conversion Systems Feeding the Grid

S.S. Murthy* *Senior Member, IEEE*, Bhim Singh** *Senior Member, IEEE* , P.K. Goel[$], S.K. Tiwari[$$]

* Professor, Department of Electrical Engineering, IIT New Delhi-110016, India, email: ssmurthy @iitd.ac.in
** Professor, Department of Electrical Engineering, IIT New Delhi-110016, India, email: bsingh@iitd.ac.in
[$] Director, Ministry of Power, Government of India, email: puneet3866@yahoo.com
[$$] Deputy. Manager, NTPC Ltd., Noida, UP email: shailendraket@rediffmail.com

Abstract-- **In the early development of the wind energy, the majority of the wind turbines have been operated at constant speed. Subsequently, the number of variable-speed wind turbines installed in wind farms has increased. In this paper, a comparative study of fixed and variable speed wind turbines incorporating squirrel cage induction generator take into consideration all realistic constraints has been presented. These two systems are modeled and simulated using Matlab. The variable speed induction generator system has been modeled using power electronic converters and vector control technology combined with peak power extraction technique (PPET) enabling the system to run at the most optimal speed. The results are presented for 55kW induction generator, and variable speed operation is highlighted as preferred mode of operation.**

Index Terms-- **Wind energy conversion system, Doubly fed induction generator, Fixed speed induction generator, Variable speed induction generator.**

I. NOMENCLATURE

R_s, X_{ls}	Stator resistance and leakage reactance
R_r', X_{lr}'	Rotor resistance and leakage reactance referred to stator
H	Inertia constant in sec.
Xm	Mutual inductance
ω_{base}, P_{base}	Base value of speed and power
V_{base}, V_{base}	Base value of voltage and current

II. INTRODUCTION

Use of wind energy as electrical energy started in 1925, when wind farms coupled to DC generators have been used as battery charger in the US farmlands. With the formation of a national a.c. grid in US, the use of DC generators declined and fixed speed synchronous generators came in operation by 1939 [1]. Large scale use of wind energy for electric power generation started in the late seventies, after sudden spur in oil prices. Initial period was dominated by fixed speed induction generators. Mid nineties saw advent of variable speed induction generators as well as synchronous generators with the advancement of power electronics and reactive power compensator technology [2].

The main components of wind energy conversion process are turbine and generator. The characteristic of turbine is such that it has the best conversion efficiency at certain rotational speed corresponding to certain wind speed and the efficiency drops either way. Till today, the most of the wind energy conversion systems use fixed speed induction generators (FSIG) which do not run efficiently during the most of the time and draw large reactive power. There have been grid failures attributed to large reactive power drawn by wind farms. Now a days variable speed induction generators using power electronic (PE) converter can be made to run at desired speed, by which turbine can be made to run efficiently at all wind speeds. Power electronic converter incurs extra losses and injects harmonics into the grid. There are many configurations using power electronics [3]. PE converter is a frequency changer scheme by AC-DC-AC configuration, which decouples machine from the grid. The variable speed WECS could either use doubly fed induction generator (DFIG) or stator controlled variable speed induction generator (SCVSIG). DFIG utilizes wound rotor machine with PE converter on the rotor, whereas SCVSIG has PE converter on the stator. Of the two variable speed induction generators, SCVIG is popular in sub MW range, the range in which FSIG is also operational.

SCVSIG offers following advantages vis-à-vis DFIG.

- Rotor bars have more thermal and electrical withstand limit, as a result more power can be generated with same rating of wound rotor machine.
- It is cheap, rugged and easily available.
- Rotor bars are less prone to failure.
- No brush loss.
- Lower weight and inertia.
- More controllable speed range 2-3 against 1.5-2 of DFIG.
- Being larger capacity converter they can provide more reactive power and can help in grid stability.

Here the performance of stator controlled variable speed induction generator (SCVIG) is analyzed and compared with that of FSIG.

978-1-4244-0644-9/07/$25.00 ©2007 IEEE

III. System Details

The generator specifications used for the study are given in Table I [4].

TABLE I
GENERATOR SPECIFICATION

Power	55kW	Voltage	415V
Pole	6	Class	A
Connection	Delta	frequency	50 Hz.
Insulation	F	H	1.5 sec
R_s	0.019	R_r'	0.0164
X_{ls}	0.069	X_{lr}'	0.087
X_m	3.0	V_{base}	415V
ω_{base}	104.72	I_{base}	93A
VA_{base}	38600	P_{base}	55000

The schematic diagram of grid connected WECS is shown in Fig. 1

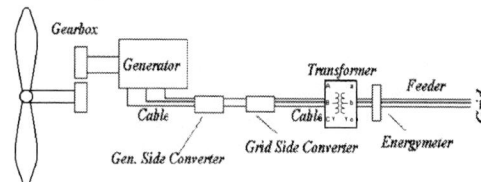

Fig. 1. Torque-speed characteristic of an induction motor.

For fixed speed systems, the converters are absent. The specifications of each component of the WECS are given in the following paragraphs.

A. Turbine and Gear

Turbine rotor is 15 m diameter. The coefficient of performance, which is the measure of aerodynamic efficiency of turbine, is given by following formula [5].

$$cp(\lambda, \beta) = 0.73(\frac{151}{\lambda i} - 0.002 * \beta - 13.2)e^{-(18.4/\lambda i)}$$

where

$$\frac{1}{\lambda i} = \frac{1}{(\lambda + 0.08\beta)} - \frac{0.035}{\beta^3 + 1} \qquad \lambda = \frac{\omega r}{\eta V_W}$$

λ (the tip speed ratio) is the ratio of tip speed of turbine to speed of wind, and β is the blade pitch angle. Based on above formula, the turbine is modeled. It is found that the turbine has optimum tip speed ratio of 7.059 and the maximum coefficient of performance of 0.444.

The frictional losses in the gear box consist of mess and churning losses and are 2% of the full load power. Mesh loss of gear is constant and churning loss depends on speed. The turbine losses are given by the following formula [6]:

Turbine losses $= A + B * \omega$ A=0.0066 B=0.0133

For the same power rating, gear ratio may not be the same for fixed speed and variable speed turbines for the best mechanical conversion of wind power.

For FSIG, if high gear ratio is selected, the turbine speed will be slower and it will capture energy from lower wind speeds more efficiently, and its co-efficient of performance will fall in high wind speed regions. On the other hand, if low gear ratio is selected, the turbine will rotate faster and not be able to capture energy efficiently for low wind speeds, which is at maximum period of time. There are schemes where an induction generator with two windings is used for this purpose to extract substantial power in full range. An optimization has to be done based on available wind data.

For SCVIG, the generator should operate at line voltage either less or equal to supply voltage. Hence gear ratio should be selected such that maximum conversion efficiency corresponds to maximum wind speed and occurs near synchronous speed. The turbine specifications are given in Table II.

TABLE II
TURBINE SPECIFICATIONS

	FSIG	SCVIG
Speed Range	5-12 m/sec	5-12 m/sec
Shut down speed	22.4 m/sec	22.4 m/sec
Cut out Speed	4.9 m/sec.	4.9 m/sec.
Diameter	15 m	15 m
Gear Ratio	17:1	12:1
Turbine Speed	59-60 RPM	59-60 RPM

B. Cable

The turbine is typically located at 30 m height. The ength of LT cable is taken as 50 meter. The cable is 3*35 sq*mm Aluminum cable. The cable impedance is 0.0317 Ohms (0.0071 p.u.) at 60^{0}C temperatures, taking @ of 0.634 ohms/Km.

C. Transformer

Delta/star transformer of 0.415/11kV 63 kVA. The series parameters are (0.01157+j0.03) of 38.6KVA.

D. Feeder

Machine is connected to 11 KV utility grids. The impedance of line is (1.021+j0.382) ohms/km [4] and taking 2 km feeder length:

Z_F= 2.042+j0.764 ohms
=0.00195+j0.0007311 p.u. on 11kV /38.6 kVA base.

E. Grid

Short circuit MVA-250 and Impedance angle for 33 KV line is 50 deg:

Impedance=11000/(1.7321*13.121x10^3)
=0.484 $\angle$50
= (0.314+j0.37) Ohms.
=0.00010967+j0.000118 p.u. on 11kV/38.6 kVA base.

IV. Model Implementation

Model implementation of both fixed speed and variable speed is done in matlab simulink. Modeling of the fixed speed system is done taking all the above components. The parameters of interest are efficiency, reactive power and starting current. Simulation studies are done without any reactive power compensator.

For the variable speed system incorporating SCVIG, the heart of the system is the two converters, namely machine side converter and grid side converter. These converters facilitate machine to operate in all 4 quadrants.

Both converters are vector controlled. Machine side converter provides reactive power to the machine in addition to speed control. Grid side converter helps for power evacuation in addition to maintaining DC bus voltage. Both the converters use IGBT switching devices.

A. Machine Side Converter

This converter uses indirect field oriented vector control based on Voltage Source converter (VSC) using PWM technology. After abc-dqo transformation, torque equation becomes.

$$T_{el} = \frac{3}{2}\left(\frac{p}{2}\right)\left(\frac{L_m}{L_r}\right)\left(\lambda_{dr} i_{qs} - \lambda_{qr} i_{ds}\right) \tag{1}$$

For transforming stator quantities, we require rotor field position which can be calculated as [7]

$$\delta = \int_0^t (\omega_m + \omega_r)\,dt \tag{2}$$

$\omega_m = rotor\ speed\ in\ rad/\sec$

$\omega_r = synchronous\ speed\ of\ rotor\ current\ in\ rad/\sec$

$$= \frac{L_m i_q}{T_r \lambda_{dr}} \tag{3}$$

If all the electrical parameters of stator are transformed to synchronously rotating frame with stator direct axis aligned to rotor field, then

$$\lambda_{qr} = 0$$

$$T_{el} = \frac{3}{2}\left(\frac{p}{2}\right)\left(\frac{L_m}{L_r}\right)\left(\lambda_{dr} i_{qs}\right) \tag{4}$$

Above equation is analogous to dc machine with i_{ds} is the flux producing component and i_{qs} is the armature current component producing torque.

$$i_{qs} = \frac{T_{el}}{(3/2)(P/2)(L_m/L_r)\lambda_{dr}} \tag{5}$$

Speed set point is generated from wind speed.

Since $\quad \omega_{sp} = \frac{\lambda \eta V_w}{r} \qquad \eta = $ Gear Ratio.

The difference of instant and desired speed set point after processing though PI controller gives signal proportional to torque.

To obtain uniform acceleration and deceleration, i_{qs} equivalent of instant torque is subtracted is subtracted from the $i_{qs}*$ to get the torque component net accelerating or decelerating torque. i_{qs} equivalent of instant torque is found by control diagram shown in fig.2

The working flux linkage depends on applied voltage/

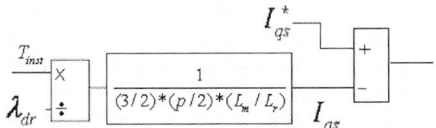

Fig. 2 Generation of i_qs*

phase.

$$Vph = 4.44 f \lambda_{dr} K_w$$

Taking V_{ph}=415 K_w=0.9 one may get $\lambda_{dr} = 2.07$

We take $\lambda_{dr} = 2.1$ web-turns.

Instant value of flux linkage is given by

$$\lambda_{dr}{}^* = \frac{L_m I_{sd}}{(1 + T_r s)} \tag{6}$$

$$T_r = \frac{(L_m + L_r')}{R_r} = \frac{L_{rr}}{R_r} \quad =\text{Open Circuit rotor Time constant}$$

The reference current ids* is obtained by processing the difference of set value of λ_{dr} (i.e. 2.1 weber-turns) and actual λ_{dr} through PI controller.

The reference line current signal of machine attained by inverse transformation of iqs* and ids* in stator frame. The transformation angle δ is the same which was used in abc-dqo transformation.

The PWM signal is generated though HB controller of actual line current and reference line current signals.

B. Grid Side Converter

The grid side converter uses direct field oriented vector control. Here no position sensors are required. The converter modulates DC bus voltage which is tuned for 800 V in addition to maintain unity power. The orientation of the reference frame is done along the supply voltage vector to obtain decoupled control over active and reactive power. This is achieved by direct field oriented control. [8]. By doing so direct and quadrature component of current is used to modulate active power and reactive power respectively.

The motoring convention, line voltage is sum of converter voltage and impedance drop.

$$V_{abc} = R_f i_{abc} + L_f \frac{di}{dt} + V_{abc\_con} \tag{7}$$

After transformation to dqo frame along supply voltage vector

$$V_{dq} = r_f i_{dq} + L_f \begin{pmatrix} 0 & -\omega \\ \omega & 0 \end{pmatrix} i_{dq} + V_{dq\_con} \tag{8}$$

$$V_{d\_con} = V_d - i_d r_f + i_q \omega L_f \tag{9}$$

$$V_{q\_con} = V_q - i_q r_f - i_d \omega L_f \tag{10}$$

L_f and r_f is the inductance and resistance of all the elements between converter and transformer output. It includes filter and transformer series impedance. Presence of inductive element also enables boost operation of AC to DC. This essential for bidirectional power flow

$$V_{dc} I_{dc} = \frac{3}{2} V_d I_d \tag{11}$$

Since orientation of the reference current is along supply voltage vector, I_d refers to active power flow. Here V_d is determined by depth of voltage sag and DC bus voltage is controlled by modulating the converter direct axis current component. I_q is proportional to reactive power [7].

DC bus is maintained at 800V. To enable bidirectional power flow, the controller must work as AC-DC boost converter. For this Vdc>1.635 V_{Line}[9]. The amount of voltage sag (800-V_{dc_actual}) though PI controller gives Id*. The setpoint of reactive power component, Iq* is zero(corresponding to zero reactive power flow).

Reference voltage before converter, V_{d_con}*is computed by processing difference of I_d* and Id through PI controller. Similarly V_{q_con} is also computed. Subsequently V_d* are computed by difference of V_{d_con} and V_{d_con}* and V_q* by difference of V_{q_con} and V_{q_con}*.

V_d* and V_q* line voltage in dq frame to maintain desired power flow at unity power factor. The reference phase voltage forms modulating signals on unity voltage scale to IGBT bridge. It is obtained by inverse transformation of V_d* and V_q*. To avoid over modulation, modulating signal is limited to unity without affecting the phase($\phi = \tan^{-1}(V_q*/V_d*)$). This is achieved by changing V_d* and V_q* to polar form and limiting it value and again changing to Cartesian form [10].

V. PERFORMANCE ANALYSIS

Measurement of power is done at outlet of transformer. The comparison is done under following ambient conditions:

Temperature-30 ° Celsius.
Height above Sea level-100 +22 Meter.
Atmospheric Pressure-99.93891 kPa
Density of Air-1.1520 Kg/m3

A. System With FSIG

Performance analysis is carried transient and steady state conditions. In transient analysis like starting, efficiency not so important. During steady state analysis efficiency and power quality is more important. Power measurement/ calculation is done at following points.

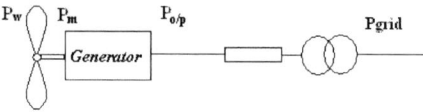

Fig. 3 Points of measurement/calculation of power.

P_w=wind power = $0.5\pi r^2 \rho V w^3$
P_m=mechanical input power
P_o/p=machine output power
P_w-P_m=Turbine conversion loss

1) Starting Transients

FSIG is started as motor, hence like induction motor it takes heavy current. The point of interest is the starting current (which also signifies the reactive power) and time taken to reach. machine in steady state. Fig. 4 shows the amount of drawn starting current and time taken to reach steady state speed, when the wind speed is 8 m/sec.

2) Steady State Performance At Various Wind Speeds

Steady state performance of the machine is done by subjecting the machine to wind speed at regular time

interval and measuring each parameters. The results of the simulation have been tabulated in Table III and IV.

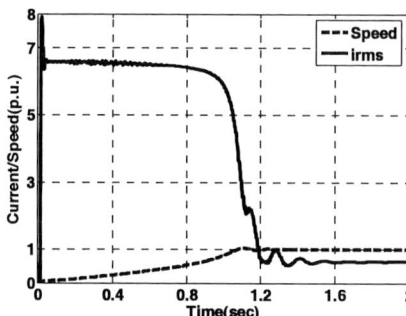

Fig. 4 Starting transients in FSIG

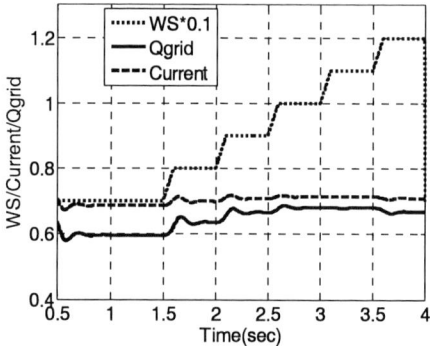

Fig. 5. Effect on current and reactive power

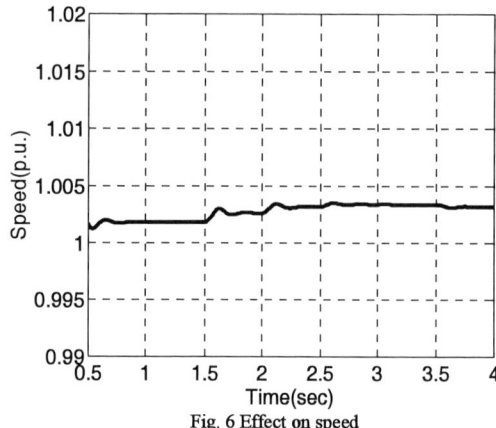

Fig. 6 Effect on speed

B. System With SCVSIG

Like FSIG, simulation study is repeated with SCVSIG. Measurement of Power is done as shown in Fig. 8.

1) Starting Transients

Vector controlled machine produces high starting torque. As a result of higher electromagnetic torque steady state speed is attained early as shown in Fig.8.

2) Steady State Performance at Various Wind Speed

Just like FSIG steady state performance of variable speed system by subjecting machine to different wind speed at regular interval. Wind speed is changed form 6 to 12 m/sec as shown in Fig 9. The speed corresponding to wind speed is given by Fig. 10. The line current and

944

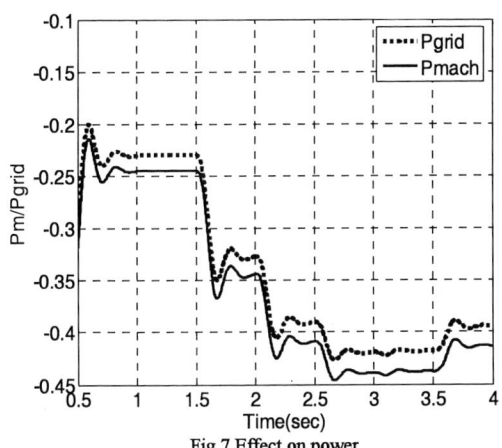

Fig 7 Effect on power

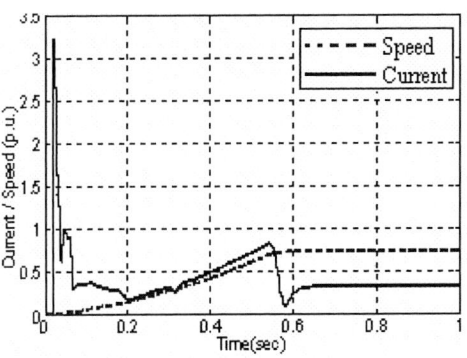

Fig. 8 Effect on Speed and Starting Current

TABLE III
PERFORMANCE OF FSIG

Wind Speed	Current	P_{OUTPUT}	Reactive Power	P_{EXP}	p.f.
6	0.560	-0.0313	0.680	-0.018	0.043
7	0.585	-0.2088	0.695	-0.194	0.37
8	0.652	-0.3850	0.702	-0.368	0.58
9	0.727	-0.5375	0.717	-0.516	0.66
10	0.793	-0.6519	0.732	-0.626	0.67
11	0.837	-0.7234	0.742	-0.694	0.67
12	0.857	-0.7533	0.747	-0.724	0.67

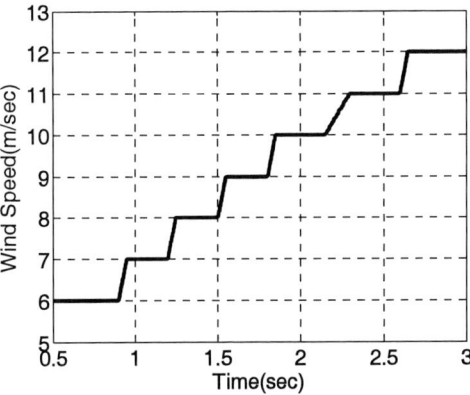

Fig. 9 Wind profile testing.

TABLE IV
EFFICIENCY CALCULATION

Wind Speed	Pmech	P_{OUTPUT}	Machine Losses (%)	P_{EXP}**	Other Losses (%)	Gross Efficiency(%)
6	0.057	0.031	45.08	0.018	23.49	5.01
7	0.233	0.209	10.38	0.194	7.08	30.6
8	0.410	0.385	6.09	0.368	4.41	38.6
9	0.567	0.537	5.20	0.516	4.00	38.2
10	0.689	0.652	5.38	0.626	3.97	33.8
11	0.770	0.723	6.05	0.694	4.06	28.2
12	0.808	0.753	6.76	0.724	3.88	22.6

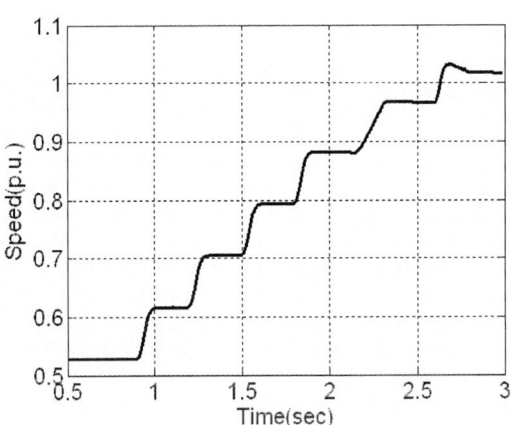

Fig. 10. Generator speed at various wind speed

reactive power drawn in motoring convention is given by Fig. 11. From the curve of Figs 11 and 12, one can see that whenever there is a change in wind speed, the machine enters motoring mode and there is spike observed in current and power. But actually wind has substantial inertia, and so above high rate is impossible. When have slow rate of change of speed as shown in Fig 13 when wind speed is changing at rate of 0.5 m/sec² no spike is observed. The current and the power at the grid

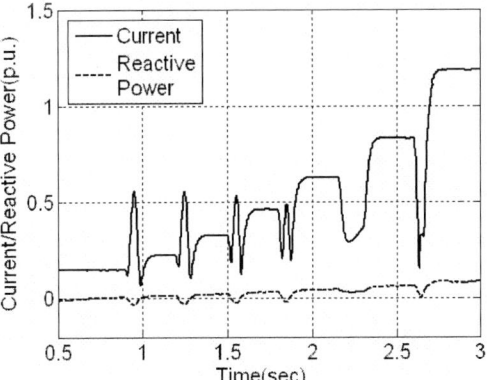

Fig. 11 Effect on reactive power and current

945

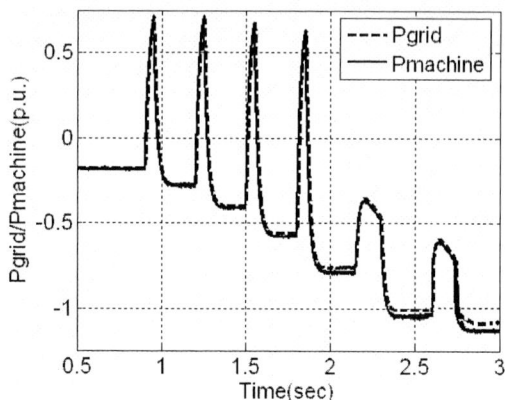

Fig. 12 Effect on power

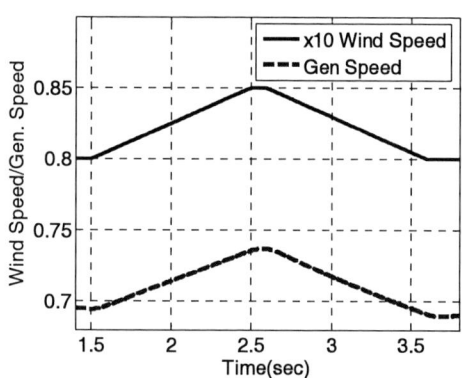

Fig. 13 Effect on speed

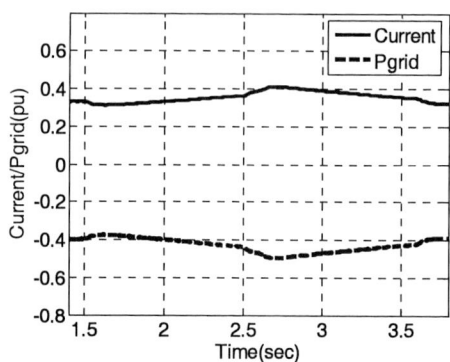

Fig. 14 Effect on current and output power

TABLE VI
EFFICIENCY CALCULATION

Wind Speed m/sec	Pmech(p.u.)	P_{OUTPUT}	Machine Loss(%)	P_{EXP}*	Other Losses(%)	Gross Efficiency (%)
6	-0.175	-0.171	2.30	0.1581	7.43	39.61
7	-0.278	-0.271	2.30	0.2535	6.47	40.00
8	-0.413	-0.395	4.33	0.3734	5.23	39.47
9	-0.590	-0.557	5.52	0.5230	5.83	38.83
10	-0.808	-0.756	6.43	0.7174	4.77	38.83
11	-1.07	-0.972	9.14	0.8955	7.15	36.41
12	-1.39	-1.273	8.42	1.1583	8.25	36.28

are shown in Fig. 14. The results of the simulation have been given in Tables V, VI and VII.

Losses in Cable and Transformer are as $3(0.0071*I_{mac}2+0.011*I_{Line}^2)$

Converter Loss=Other Electrical Loss-Losses in transformer and cable

3) Performance of Converter on
a) Analysis of Harmonics

Switching by PE devices inject harmonics in the utility grid. Harmonics content is measured by Total Harmonics Distortion (THD). The FFT of line current harmonics for wind speed 6 and 11 m/sec is shown in Fig. 15. One may find that THD decreases with increase of current. The

switching frequency is constant and effect of switching and filter becomes less in proportion to line current as line current and wind speed increases. Machine current can also have harmonics as a result of switching by machine side converter. The harmonics of content machine current is well below IEEE 519 limit like line current. Fig. 16 shows harmonics content of machine current when Wind speed is 6 m/sec.

VI. RESULT AND CONCLUSION

From Figure 4 and 8, one may see that during starting FSIG takes 6 p.u. current for 1.1 sec. On the other hand SCVSIG takes less than 1 .u. (Except during first quarter

TABLE V
PERFORMANCE OF SCVSIG

Wind Speed m/sec	Speed (p.u.)	Current	P_{OUTPUT} (p.u.)	Reactive Power	P_{EXP} (p.u.)	p.f.
6	0.528	0.150	-0.171	0.011	-0.168	0.994
7	0.612	0.210	-0.272	0.021	-0.253	0.997
8	0.708	0.315	-0.395	0.028	-0.373	0.998
9	0.799	0.440	-0.557	0.045	-0.523	0.995
10	0.890	0.605	-0.756	0.060	-0.717	0.996
11	0.981	0.778	-0.972	0.079	-0.896	0.997
12	1.02	0.988	-1.273	0.101	-1.158	0.997

TABLE VII
MACHINE AND CONVERTER LOSS

Wind Speed	$W_{ELECT\ LOSS}$ FSIG(%)	$W_{ELECT\ LOSS}$ SCVSIG(%)	Converter Loss (p.u.)*	Converter Loss(%)
6	68.3	7.73	0.0067	3.84
7	17.4	8.77	0.0099	3.64
8	10.5	9.56	0.0137	3.32
9	10.2	11.3	0.0207	3.51
10	9.4	11.2	0.0317	3.92
11	10.1	16.3	0.0462	4.31
12	10.6	16.6	0.0676	4.86

946

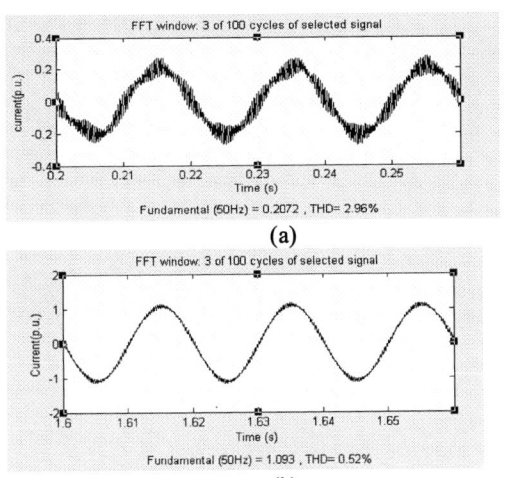

(a)

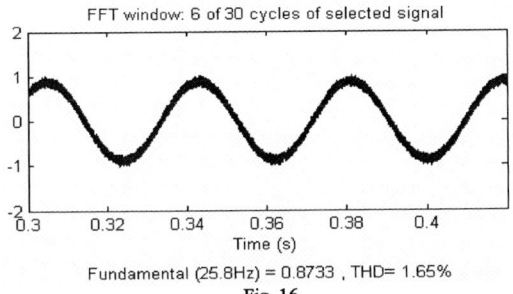

(b)

Fig. 15 Line current at (a) Wind 6m/sec. (b) Wind speed 11 m/sec

TABLE VIII
THD AT DIFFERENT LOADS

Wind Speed	Current(p.u)	THD (%)
6	0.171	2.96
7	0.272	3.56
8	0.395	2.43
9	0.557	1.60
10	0.756	1.50
11	0.972	0.52
12	1.273	0.50

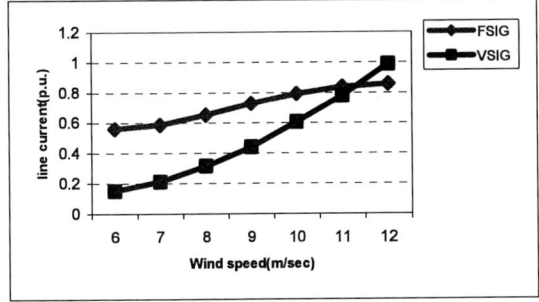

Fig. 16

cycle). The starting time is also lower. Hence FSIG is under more electrical and thermal stress. The line current is considerably higher in FSIG than VSIG, since SCVSIG is incorporated with reactive power compensator on the line side converter as evident from the Figs. 16 and 17

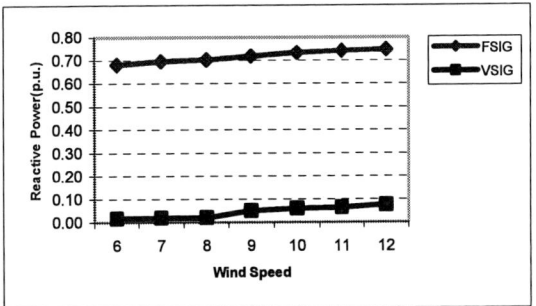

Fig. 17 Comparison of Reactive power

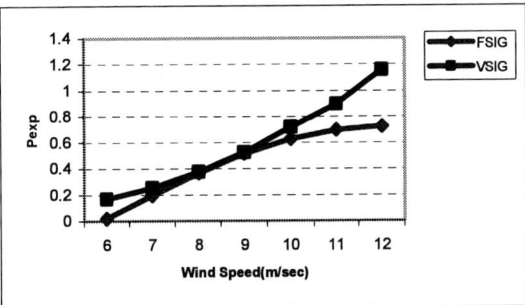

Fig. 18 Comparison of exported power

For FSIG the mechanical conversion efficiency varies considerably with wind speed. The system is not able to extract energy efficiently in higher speed range as a result, output exported power vs wind speed curve saturates in higher speed range. Further, FSIG is not able to utilize its capacity as shown in Fig. 18.

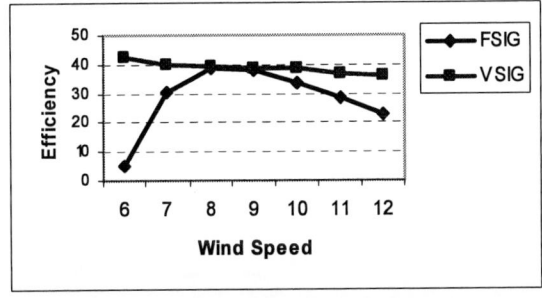

Fig. 19 Comparison of gross efficiency.

For SCVSIG the mechanical conversion efficiency can be made high limited only in higher wind speed limit. In high speed limit (>=11.54 m/sec) with the help of control of turbine combined with controller action, generator speed is not allowed to increase beyond 1.02 p.u. corresponding to 50 Hz. stator frequency.

For fixed speed machine the frequency or speed dependant loss (Hysteresis, Frictional loss of Gear, and Frictional and windage loss of Generator) is almost constant But in SCVSIG, above losses reduce with fall of frequency/speed hence net losses are lower as evident from the curve of Fig. 19

The converter loss occurs only in case of SCVSIG. The machine side converter utilizes hysterisis band controller where switching frequency is not constant. The grid side converter utilizes constant switching frequency. So converter loss is 4-5% and result agrees with Anders [6]. Moreover it is found that converter loss decreases with increase of load. Net loss is observed in range of 10-15%.

Fig. 16 Comparison of Line current

APPENDIX

Converter Data

Line Filter Inductance-1.8 mH

DC capacitor-4000 μH

Grid Side Converter Switching Frequency-2550 Hz.

REFERENCES

[1] "The History and State of the Art of Variable-Speed Wind Turbine Technology" Technical report of RNEL, Department of energy US.

[2] Helge Wittholz and David Pan, "A Study of supply-chain capabilities in Canadian Wind Power Industry", SYNOVA International Business Development November 2004.

[3] F. Blaabjerg, Z. Chen, S.B. Kjer "Power electronics as efficient interface of renewable source", IPEMC conference 2004.

[4] SS Murthy, CS Jha, P.S. Nagendra Rao, "Analysis of Grid connected Induction Generators Driven by Hydro/Wind Turbine under realistic system constraints", IEEE/PES 1989.

[5] J. G. Slootweg S. W Haan, H. Polinder, W. L Kling, "General Model for Representing Variable Speed Wind Turbines in Power System Dynamics Simulations.", IEEE Trans. on Power Systems 2003, vol. 18, No. 1, February.

[6] Anders Grauers, "Efficiency of three wind energy generator systems" IEEE Transactions on Energy, 1995.

[7] B.K.Bose, "Power Electronics Converter and A.C. Drives" Pearson publication, 2005.

[8] Marta Molinas, Bjarne Naess, William Gullvik, Tore Undeland "Cage Induction Generators for Wind Turbines with Power Electronics Converters in the Light of the New Grid Codes", Norwegian University Of Science And Technology Conversion, Vol. 11, No. 3, September 1996.

[9] Ned Mohan, Tore M. Underland, and William P. Robbins "Power Electronics-Converters, Applications, and Design", John Wiley & Sons, Inc.2004.

[10] Math works, Inc. Simulink version 7.3(R-2006).

BIOGRAPHY

S.S. Murthy (SM' 87) was born in Karnataka, India, in 1946. He received his B.E, M.Tech. and Ph.D.degrees respectively from Bangalore University, Indian Institute of Technology (IIT)Bombay, and IIT Delhi.

He has been with IIT Delhi since 1970 and was the Chairman of the Department of Electrical Engineering from 1998-2001. He has held assignment in University of New castle, UK, University of Calgary Canada, ERDA Barodra and Kirloskar Electric, Banglore. He holds 4 patents on the SEIG, Micro Hydel Applications and a novel-braking scheme. He has also transferred technology of self-excited and grid connected induction generators to industry for low-and medium-power generation under standalone or grid connected mode. He has completed several industry sponsored research and constancy projects dealing with electrical machines, drives, and energy systems. Recently, he was instrumental in establishing start-of-the-art energy audit and energy conservation facilities at IIT under World Bank funding. His areas of interest include electric machines, drives, special machines, power electronic applications, renewable energy systems, energy efficiency and conservation.

Dr. Murthy has received many awards notable being-ISTE/Maharashtra Govt. Award for outstanding research and IETE/Bimal Bose Award for contribution in Power Electronics. He has made significant contributions to professional societies, including being General Chair of the 1st IEEE International Conference on Power Electronics, Drive and Energy Systems (PEDES' 96) held in January 1996 in New Delhi. He is a Senior Member of IEEE, Fellow of IEE Life Fellow of the Institution of Engineers (India), and Life Member of ISTE.

Bhim Singh (SM'99) was born in Rahamapur, India, in 1956. He received the B.E (Electrical) degree from the University of Roorkee, Roorkee, India, in 1977 and the M.Tech and Ph.D. degrees from the Indian Institute of Technology (IIT) Delhi, New Delhi, India, in 1979 and 1983, respectively. In 1983, he joined the Department of Electrical Engineering, University of Roorkee, as a Lecturer, and in 1988 became a Reader. In December 1990, he joined the Department of Electrical Engineering, IIT Delhi, as an Assistant Professor. He became an Associate Professor in 1994 and Professor in 1997. His area of interest includes power electronics, electrical machines and drives, active filters, FACTS, HVDC and power quality.

Dr. Singh is a fellow of Indian National Academy of Engineering (INAE), the Institution of Engineers (India) (IE(1)), and the Institution of Electronics and Telecommunication Engineers (IETE), a life member of the Indian Society for Technical Education(ISTE), the System Society of India (SSI), and the National Institution of Quality and Reliability (NIQR) and Senior Member of Institute of Electrical and Electronics Engineers (IEEE).

Puneet K. Goel was born in New Delhi, India in 1966. He received his M. Tech degree in Power Apparatus and System from IIT Delhi, MBA (Finance) from IGNOU and MS in Electrical Engineering from USC, Los Angeles. He joined Indian Administrative Service in the year 1991. Presently he is working as Director (Thermal) in Ministry of Power. Government of India. His main research interests are distributed generation, small hydro, and wind energy and power generation from biomass.

S.K. Tiwari was born in Siwan, Bihar, India in the year 1975. He completed his B.E. in 1996 from Govt. Engineering College, Bihar, India and joined Central Power Utility (NTPC) in the year 1997. He is presently working as Dy. Manager in Renewable and Distributed Generation Department, NTPC. His main research interests are Distributed Generation and Power Electronics

Author Index

A

Abdi, Ehsan 1096
Abe, Seiya 1388
Abjadi, N. R. 1442
Achara, P. 394
Adélaide, L. 569
Adya, A. 731
Afjei, E. 722
Agarwal, Pramod 1810
Agarwal, Vineeta 1891
Ahmadian, H. Molla 1147
Ahn, Jin-Woo 1857
Almardy, M. 677
Alonge, F. 959
Amaral, Acácio M. R. 587, 643
Amirudin, Dessy 534
Amrane, F. 569
An, Young-Joo 1527
Andersen, P. Scavenius 886
Ang, Y. 382
Ang, Yong-Ann 376
Aodsup, K. 937
Arab, G.R. 1449
Aree, P. 703
Arvindan, A. N. 480
Ashaibi, Ahmed Ali 1242
Ataei, S. 722
Athab, Hussain S. 869, 874
Attaviriyanupap, Pathom 1102
Auger, Francois 1368
Ayob, S. M. 1274, 1363, 1682
Azli, N. A. 475, 1041, 1274, 1363, 1682

B

Bac, Nguyen Xuan 1501
Baharom, R. 1626
Baiju, M.R. 1047
Banerjee, Subrata 812
Bartholet, M.T. 257
Batzies, Ekkehard 1316
Bhat, A.K.S. 677
Bhuvaneswari, G. 310
Bi, C. 1082
Bina, M. Tavakoli 465, 1060, 1065, 1799
Binder, A. 249
Bingham, C. M. 382
Bingham, Chris 376
Binh, Tran Cong 1195
Biswas, S. K. 1352, 1605
Blaabjerg, Frede 226, 541, 1247, 1376
Boonchiam, P. N. 937, 1851
Boonyaroonate, Itsda 1078, 1383
Bosing, M. 1160
Branco, P.J. Costa 917
Brauer, Helge J. 716
Buatti, Gustavo M. 643

Bunlaksananusorn, C. 977, 1575

C

Cangemi, T. 959
Cardoso, A. J. Marques 587, 643
Carstensen, Christian 912
Chalermyanont, Kusumal 295
Champa, P. 703
Chan, K. W. 1394, 1727, 1804
Chan, Shun-Yu 305
Chang, David 305
Chang, Tsin-Yuan 581
Chang, Y. D. 1321
Chao, Ma Xian 665
Chatratana, S. 1495
Chaudhari, M. A. 1708
Chen, Jiaxin 510
Chen, L. 1538
Chen, Sufen 1262
Chen, W. C. 440
Chen, Wei 1636
Chen, Y. H. 456, 1278
Chen, Y. M. 1321
Chenfeng, Yang 745
Cheng, Chien-Lung 749, 1703
Cheng, K. W. Eric 1691, 1697
Cheng, K.W.E. 1727
Cheng, Qiang 1330
Chengfeng, Yang 270, 427
Chereau, Vinciane 1368
Chern, Shyi-Ching 749, 1703
Cheung, N.C. 1727
Cheung, Norbert C. 1691, 1697
Chi, Chien-An 388
Chiang, Wen-Jung 824
Chien, F.T. 660
Chin, Li-Yuan 305
Chivite-Zabalza, F. Javier 788, 796, 804
Cho, B. H. 401
Cho, Kyu Min 665, 1142
Choi, S. J. 401
Choudhary, Sonika 1757
Chrin, P. 1575
Chudamani, R. 1827
Chudoung, Nakharet 1213
Chun, Tae-Won 1857
Chunkag, V. 863
Ciobotaru, M. 226
Colak, Baris 1507
Corradini, L. 600
Cosic, A. 1301
Cruden, A. 1182

D

Dahlan, N.Y. 527
Dahono, P. A. 1267
Dahono, Pekik Argo 534

Author Index

Dai, Z. 1885
Dananjayan, P. 626
Davat, Bernard 1
Deb, N. K. 1352, 1605
Dehbonei, H. 1657
Deleroi, W. 1495
Deng-Em, S. 1851
Deni, ... 534, 1267
Densei-Lambda, K.K. 280
Desai, Hardik P. 829
Dhomane, G. A. 1590
Dick, Christian P. 448
D'ippolito, F. 959
Ditmanson, C. 556
Doki, Shinji 999
Doncker, R. W. De 907, 912, 1160
Doncker, Rik W. De ... 213, 327, 333, 448, 710, 716
Dong, Lei 1691
Dong, Ming-Chui 607, 614
Dong, Yang 1340
Dorkmai, Pramoch 697
Dorrell, David G 886, 922, 1167, 1174
Duan, S.X. 551
Duan, Shanxu 836, 842
Dwivedi, Avneesh 1757
Dzung, Phan Quoc 1195, 1202, 1501

E

Ekkaravarodome, Chainarin 1383
Ertan, H. Bulent 1507
Eskandari, B. 1060, 1065

F

Fang, D. Z. 1394, 1804
Fang, Kuo-Lun 1610, 1712
Fang, Tzu-Hsuan 1717
Fei, Wanmin 350, 354, 1672
Ferraz, Antonio 817
Ferreira, O.C. 1017
Fidler, Peter 327
Fingerhuth, S. 1160
Finney, S.J. 299, 1242
Finney, Steve J. 1255
Foroosh, S. Chini 1465
Forsyth, Andrew J. 788, 796, 804
Foster, M. P. 382
Foster, Martin 376
Fuengwarodsakul, Nisai H. 710
Fukuda, Shoji 1070
Fukushima, K. 1885

G

Gairola, Sanjay 738, 899
Gao, F. 1247, 1376
Garg, Vipin 310
Geethalakshmi, B. 626

Goel, P.K. 941
Gonthier, L. 322
Gopinath, Anish 1047
Goyal, Devendra 1520
Grant, D M 368
Grantham, Colin 1284
Gruber , W. 574
Guan, Xiaohan 994, 1752
Gueldner, H. 1006
Guldner, H. 556
Guo, Youguang 275, 510, 1662
Gupta, H.O. 1810
Gupta, J.R.P 731

H

Hai, Quach Thanh 1033
Hajian, M. 1449
Hamzah, M.K. 527, 1626
Hamzah, N.R. 527, 1626
Han, Ying-Duo 607, 614
Hansen, P. E. 886
Haque, M. Tarafdar 620
Harada, Y. 1885
Hasegawa, Masaru 1543
Hellinger, R. 1006
Hennen, Martin D. 716, 907
Hew, W.P. 1514
Heyun, Lin 270, 427, 745
Higuchi, Kohji 280
Hinkkanen, Marko 406
Hirokawa, Masahiko 1388
Hirota, Atsushi 1740
Ho, Shine-Tzong 498, 1788
Hoang, Nguyen Minh 1195, 1501
Hofmann, W. 781, 1538
Hotait, Hadi A 299, 1255
Hothongkham, Prasopchok 1236
Hsieh, C. T. 1567
Hsu, Chih-Jen 286
Huang, P. L. 1321
Hung, Tsung-You 1762, 1767
Hwu, K. I. 338, 456, 692, 1278

I

Idris, Z. 527
Iov, F. 226
Ishitobi, Manabu 504
Islam, S. 1834
Iso, Osamu 1102

J

Jang, B.H. 1657
Jangjaempradit, Saksit 1641
Jangwanitlert, A. 989, 1412
Janjornmanit, Suchart 1327
Jayashree, E. 1555

Author Index

Jeevananthan, S. 1221
Jegathesan, V. 1677
Jerome, Jovitha 1677
Jeung, Giwoo 1092
Jian, Guo 270, 427, 745
Jou, Hurng-Liahng 493, 824
Jovanovic, Milutin G 922
Junge, Christian 1533
Jwo, W. S. 1560

K

Kadir, M. N. Abdul 1514
Kaewsingha, Aswin 1327
Kamnarn, U. 863
Kamper, M.J. 420, 1017, 1295
Kando, M. 488
Kang, Yong 551, 836
Kano, Masaru 414
Kanthaphayao, Y. 863
Kanzi, K. 465
Karunakar, K 1620
Karutz, P. 574
Kasal, Gaurav Kumar 357
Kasper, K. A. 1160
Kavitha, A. 595
Kazimierczuk, Marian K 1136
Kennel, R.M. 1017
Kerz, O. 363
Khaehintung, Noppadol 847, 1429
Khajeh, A. 1455
Khalil, Ahmed G. Abo- 1471
Khan, P. K. Shadhu 869, 874
Khan-Ngern, Werachet 460, 1335
Khomfoi, Surin 1055, 1228
Khun, C. 488
Kim, Dong-Hun 1092
Kim, H. S. 1846
Kim, Hee Jun 665, 1142
Kim, Heung-Geun 1092
Kim, I. C. 1846
Kim, In Dong 1092
Kinnaraes, V. 1483
Kinnares, V. 1356, 1489
Kinnares, Vijit 1236
Kittiratsatcha, S. 977
Ko, S. H. 1657, 1846, 1846
Ko, T.K. 1657
Ko, Yi-Pin 388
Kobayashi, Takayuki 265
Kock, H.W. De 1017
Koenig, Andreas 327
Kohama, Teruhiko 1417
Kok, W. Sae- 368
Kolar, J.W. 257, 574
Kongsuk, P. 937
Kongthawornwattana, P. 977
Konig, Andreas 448

Krein, Philip T. 221
Krismadinata, ...1290
Kubota, Hisao 265
Kulvitit, Youthana 342, 697
Kumar, S. Ganesh 1632
Kumar, S. Krishna 1632
Kumchaiyo, Ruthapong 1078
Kunakorn, Anantawat 1429
Kuo, J. S. 440
Kuo, Jian-Long 1717, 1722
Kurokawa, F. 968, 1398
Kusuhara, Yoshito 954
Kwok, K. W. 1727
Kwok, Y. L. 1727
Kwon, Soon Kurl 504

L

Laczynski, T. 1645
Lafzi, A. 620
Lai, Y. M. 1262
Lai, Yen-Shin 1586
Lakhdari, Z. 569
Lan, Yi-Hung 749
Lee, Chien-Min 1586
Lee, Dong-Choon 1471
Lee, Dong-Hee 1527, 1857
Lee, Hong Hee 1027, 1033
Lee, S. R. 1657, 1846
Lee, S. W. 1657, 1846
Lee, Yuang-Shung 286, 388
Lei, Dong 949, 1340, 1697
Lei, Yuzhou 291, 1777
Leibfried, T. 363, 726
Lenke, Robert U. 213
Lenwari, W. 470
Leou, Rong Ceng 546
Lerdudomsak, Smith 999
Li, X. 551
Li, Y. J. 440
Liang, C. 1376
Liao, C.N. 660
Liao, Xiaozhong 1691
Lijie, Wang 949
Lim, P. Y. 475, 1041
Lim, S. H. 1846
Lim, T.C. 299
Lin, Chang-Hua 1610, 1712, 1762, 1767
Lin, Chih-Hong 1549
Lin, H. C. 1567
Lin, Hung-Chih 581
Lin, Min 1423
Lipo, Thomas A. 1308
Liu, B.Y. 551
Liu, Bangyin 836, 842, 1636
Liu, Dikai 275
Liu, Fei 842
Liu, Maw-Yang 1610, 1712

Author Index

Liu, Xian-Lin 1722
Liu, Yi-Hwa 546
Liu, Yuanchao 291, 1615, 1752, 1777
Liu, Z. 551
Loh, P. C. 1247, 1376, 1620
Loh, Poh Chiang 541
Loron, Luc 1368
Lu, Haiyan 275
Lu, Y. 1727
Lu, Zhengyu 354, 1088
Luomi, Jorma 406

M

Ma, Yu 632, 1088, 1601
Macheiner, P. 880
Madawala, U. K. 648, 654
Makany, Ph. 569
Makino, Tomoaki 1740
Manmek, Thip 773
Mao, Peng 1615
Markadeh, Gh. R. Arab 1442
Marques, Gil D. 636
Martin, F. 363
Martins, J. F. 894, 1875
Masoum, Amir S. 767
Masoum, M.A.S. 767, 1834
Massoud, A.M. 299
Massoud, Ahmed M. 1255
Massoud, Ahmed 1242
Matsui, Keiju 1543
Matsui, N. 1398
Matsui, Y. 1109
Matsui, Yasuaki 1102
Matsuo, K. 1686
Matsuse, K. 394, 685
Matsuse, Kouki 521, 1460
Mattavelli, P. 600
Mattavelli, Paolo 760
Mcmahon, Richard 1096
Medagam, Peda V 1477
Mekhilef, S. 1514
Meng, Peipei 632
Mertens, A. 1645
Meyer, Christoph 213
Milani, A. Roshan 620
Miri, A. M. 726
Mirmousa, H. 1404
Mishima, Tomokazu 563
Mithulananthan, N. 937
Mittal, A.P. 731
Mittal, Raghu K. 1757
Miura, T. 1686
Miyamoto, Hiroyuki 1773
Moallem, Ali 983, 1147
Modak, J. P. 1708
Moghani, J. S. 1455
Mondal, N. 1352, 1605

Moon, Y.H. 1657
Morimoto, Masayuki 1641, 1773
Morita, Katsuaki 1773
Moses, Paul S. 767
Mossner, K. 363
Mudannayake, Chathura P 773
Mun, Sang Pil 504, 563
Mura, Florian 213
Muraoka, Hidekazu 563
Murthy, S.S. 941, 1123, 1757

N

Nabeshima, Takashi 858, 1423, 1734
Naetiladdanon, Sumate 755
Nagai, Satoshi 1740
Nakagawa, Shin 954
Nakanishi, Hirotaka 858, 1734
Nakano, Kazushi 280
Nakano, Tadao 858, 1734
Nakaoka, Mutsuo 504, 563
Nakayama, Asahi 954
Nandhakumar, R. 1221
Nathakaranakule, Adisak 1383
Navi, K. 722
Nazarzadeh, Jalal 1782
Neammanee, B. 1129
Neuhaus, Christoph R. 710
Ngern, W. Khan- 488, 515, 1667, 1794
Ngoc, Ha Pham 1102
Nguyen, Binhminh 1434
Nho, Eui-Chel 1527
Nho, Nguyen Van 1027, 1033
Nia, S.Hosein 1449
Ninomiya, T. 1885
Ninomiya, Tamotsu 954, 1388, 1417
Nishijima, Kimihiro 858, 1423, 1734
Nishimura, Jun 1460
Noguchi, Toshihiko 414, 1595, 1651
Noor, S.Z. Mohammad 1626
Norigoe, I. 1885
Nussbaumer, T. 257, 574

O

Obata, S. 671
Ogura, K. 1109
Oh, Won Seok 1142
Oka, Kazuo 521, 1460
Okuma, Shigeru 999
Omori, Hideki 563
Opanuruk, Puckapon 342
Oranpiroj, Kosol 318
Owatchaiphong, Satit 912
Ozdemir, Engin 1055
Ozdemir, Sule 1055

Author Index

P

Pai, Kai-Jun 1762, 1767
Pal, Jayanta 812
Palandurkar, M.V 1708
Panda, Sanjib K. 852
Park, Hong-Geuk 1471
Park, J. H. 401
Pashajavid, E. 465
Passal, A. 322
Patel, H. K. 829
Pavitra, G. 1757
Peng, S.T. 930
Phuong, Le Minh 1195, 1202, 1501
Piboonwattanakit, K. 1667
Piippo, Antti 406
Pinto, A.J.P. 1123
Pires, A. J. 894, 917
Pires, V. Fernao 894, 1875
Plum, Thomas 327, 333
Pothana, Aravind 1152
Pothi, N. 1208
Pourboghrat, Farzad 1477
Prasad, Dinkar 812
Prasertsit, Anuwat 295
Premrudeepreechacharn, Suttichai 318, 1208
Pusorn, W. 1851

Q

Qian, Zhaoming 632, 1088, 1601
Qu, Yilong 1822, 1880

R

Rafael, Silviano 917
Rahim, Nasrudin Abd 1290
Rahimzadeh, S. 1799
Rahman, Muhammed Fazlur 1284
Rakpenthai, C. 1208
Ramalingam, C.S. 1827
Ramli, M. Z. 1274
Randewijk, P.J. 420, 1744
Rentzsch, M. 556
Ribeiro, Antonio C 636
Ribeiro, Hugo 643
Ritchie, E. 1167
Rizqiawan, Arwindra 534
Rockhill, Andrew A. 1308
Rong, Runjie 541
Rossouw, F.G. 1295
Rost, J. 1006

S

Sadarangani, Chandur 1012, 1301
Saggini, S. 600
Saha, Bishwajit 504, 563
Saito, Y. 671

Sakulhirirak, D. 515
Salam, Z. 1274, 1363, 1682
Sanajit, N. 1412
Sangampai, Pairote 295
Sangwongwanich, Somboon 1213
Sankar, S. Siva. 1632
Sano, Kohji 1595
Saparon, A. 527, 1626
Saritsiri, Kritsada 460
Sato, Akira 1651
Sato, S. 1109
Sato, Terukazu 858, 1423, 1734
Sawatpipat, P. 1840
Sawetsakulanond, B. 1356, 1483, 1489
Schmidt, I. 1115
Schneider, T. 249
Scholler, Tobias 1316
Schroder, D. 1187
Schuster, H. 1187
Sebastiao, Pedro J 636
Sekine, T. 671
Sekiya, Hiroo 1136
Selvaraj, Jeyraj 1290
Senicar, Florian 1533
Sera, D. 226
Sezgin, Volkan 1507
Shah, Laxman 1182
Shahbazi, M. 1455
Shao, Shiyi 1096
Shariatmadar, S. Mohammad 1782
Sharma, Deepen 973
Sharma, V. K. 480
Shen, C.L. 930
Shi, Hu 949
Shiang, J. Z. 433, 1560
Shibano, Yusuke 265
Shisha, Samer 1012
Shuang, Gao 949, 1697
Shuhua, Fang 270, 427, 745
Silber, S. 257
Silva, J. Fernando 1875
Sim, J. M. 401
Sing, Bhim 941
Singer, A. 781
Singh, Bhim 58, 310, 357, 731, 738, 899, 1520, 1816
Singhal, Varun 1816
Sinha, S. 1352, 1605
Sirisuk, Phaophak 847, 1429
Sirisumrannukul, S. 1495
Skorokhod, Y.Y. 1868
Sode-Yome, A. 937
Soh, C.S. 1082
Soltani, J. 1442, 1449
Somsiri, P. 703
Son, Kwang-Myoung 1471
Songboonkaew, J. 989
Soter, Stefan 1533
Soulard, J. 1301

A-5

Author Index

Sousa, Duarte M. 636, 817
Srisongkram, W. 1851
Stone, D. A. 382
Stone, David 376
Stumberger, R.H. 1346
Su, Ching-Hung 1610, 1712
Su, Y.-H. 433
Subsingha, W. 1851
Sudhakar, S. Bala 1581
Sudmee, W. 1129
Suetsugu, Tadashi 1136
Sugawara, A. 1109
Sugimura, Hisayuki 504, 563
Sukita, S. 968
Sumner, M. 470
Sun, J. Q. 1394
Sun, Yu-Hua 493
Supriatna, E. G. 1267
Suryawanshi, H. M. 1590
Svechkarenko, D. 1301

T

Ta, Minh C. 1434
Tahami, F. 1147, 1465
Tai, Sio-Un 607, 614
Takeda, T. 1109
Takegami, Eiji 280
Tan, K. 1834
Tan, Siew-Chong 1262
Tan, Weipu 1822, 1880
Taniguchi, T. 1686
Tansatit, Tanvaa 342, 697
Tarateeraseth, V. 515
Tarnekar, S. G. 1708
Tayjasanant, T. 1840, 1862
Tedeschi, Elisabetta 760
Tenca, Pierluigi 1308
Teng, Jen-Hao 305
Teng, L. Y. 1041
Tenti, P. 600
Tenti, Paolo 760
Teo, K.K. 1082
Teodorescu, R. 226, 1247
Teshnizi, Hesameddin Mirzaee 983
Theinmontri, Surapon 295
Thrimawithana, D. J. 648, 654
Tiwari, S.K. 941
To, Huu-Phuc 1284
Tolbert, Leon M. 1055, 1228
Tomihisa, Yoshihiro 858, 1734
Tomioka, Satoshi 280
Tomita, H. 671
Trevisan, D. 600
Tsai, Y.T. 660
Tse, Chi K. 1262
Tseng, S. Y. 440, 433, 1321, 1560, 1567
Tsukakoshi, K. 1885

Tsunesada, Ryota 1417
Tungpimonrut, K. 703

U

Ueda, Shigeta 1070
Ulinuha, A. 1834
Uma, G. 595, 1555, 1632
Uyaisom, C. 1794

V

Vadirajacharya, K. 1810
Vaigundamoorthi, M. 1555
Vargas, Ismael Araujo- 788, 796
Vasudevan, Krishna 1152, 1827
Veerachary, M. 973, 1581
Veszpremi, K. 1115
Vilathgamuwa, D M 1247, 1620
Vinh, Pham Quang 1195, 1202, 1501
Viriya, P. 394, 685
Vishwakarma, Alok 1891
Vogelsberger, M.A. 1346
Volskiy, S.I. 1868
Vorlander, M. 1160

W

Walker, J. A. 1167
Wang, Chengzhi 1636
Wang, Chien-Ming 1610, 1712, 1762, 1767
Wang, Hua 1662
Wang, Peng 541
Wang, Qi 350
Wang, R-J. 420
Wang, Shoufang 1672
Wang, Shuhong 275
Wang, Shun-Chung 546
Wang, Xixi. 1691
Wangsathitwong, S. 1495
Watanabe, Kazushi 280
Watanabe, Takayuki 1136
Wegener, Ralf..1533
Weihrauch, N. C. 886
Welker, Volkmar 1316
Weller, A. 1006
Westermaier, C. 1187
Williams, Barry W 299, 1182, 1242, 1255
Wipasuramonton, P. 703
Wolbank, T.M. 880, 1346
Wong, Man-Chung 607, 614
Wu, Jinn-Chang 493, 824
Wu, Ming-Yi 1703
Wu, T. F 1321
Wu, Xinhui 852

X

Xiaozhong, Liao 949, 1340, 1697

Author Index

Xie, Xiaogao 1601
Xiping, Liu 270, 427, 745
Xu, Hai 1142
Xu, Jianxin 852
Xu, Pengwei 842
Xu, Yun 1636

Y

Yachiangkam, Samart 1327
Yang, C. M. 1560
Yang, Yihan 1822, 1880
Yau, Y. T. 338, 692
Yeh, Jim-Chwen 749, 1703
Yeon, Jae Eul 665
Yingkayun, Krisda 318
Yongyuth, N. 685
Yoothanom, N. 515
Yoshida, Takatsugu 1070
Yoshimura, S. 671
Yoshioka, Satoshi 1543
Yossombut, K. 1840
Yun, S. T. 401

Z

Zanchetta, P. 470
Zhan, Yuedong 1662
Zhang, Dongyan 994
Zhang, H.B. 299
Zhang, Junming 632
Zhang, Weiping 291, 994, 1330, 1615, 1752, 1777
Zhang, Xiaofeng 1088
Zhang, Xiaoqiang 291, 1777
Zhang, Yanli 350, 354, 1672
Zhao, Xusen 1752
Zhijun, E. 1804
Zhu, G.R. 551
Zhu, Jianguo 275, 510, 1662
Zirn, Oliver 1316
Zolghadri, M.R. 1404
Zolghadri, Mohammadreza 983
Zoller, T. 726
Zou, Yunping 1636